AC PAOLONE

EXERCISE
PHYSIOLOGY

Energy, Nutrition, and Human Performance

WILLIAM D. McARDLE

Professor, Department of Health and Physical Education
Queens College of the City University of New York
Flushing, New York

FRANK I. KATCH

Professor, Department of Exercise Science
University of Massachusetts
Amherst, Massachusetts

VICTOR L. KATCH

Professor, Department of Movement Science
Division of Kinesiology
Associate Professor, Pediatrics
School of Medicine
Director, Weight Control Clinic
University of Michigan
Ann Arbor, Michigan

Lea & Febiger
200 Chester Field Parkway
Malvern, PA 19355-9725
(215) 251-2230
1-800-444-1785

Library of Congress Cataloging-in-Publication Data

McArdle, William D.
    Exercise physiology: energy, nutrition, and human performance /
    William D. McArdle, Frank I. Katch, Victor L. Katch.—3rd ed.
        p.   cm.
    Includes bibliographical references.
    Includes index.
    ISBN 0–8121–1351–9
    1. Exercise—Physiological aspects.   I. Katch, Frank I.
II. Katch, Victor L.   III. Title.
    [DNLM: 1. Exercise—physiology.   2. Nutrition.   3. Sports
Medicine.   QT 260 M478e]
QP301.M375   1991
612'.044—dc20
DNLM/DLC
for Library of Congress   90–13355
                                                               CIP

Print No.   4   3   2   1

EXERCISE PHYSIOLOGY

Energy, Nutrition, and Human Performance

third edition

1991

LEA & FEBIGER PHILADELPHIA/LONDON

To the professors and scientists who made a difference: J. Ball, A.R. Behnke, D. Benson, J.A. Faulkner, D. Fleming, G.F. Foglia, F.M. Henry, E.D. Michael, Jr., H.J. Montoye, G.Q. Rich III, R.H. Salmons, and E. Wallis. Also to our wives and families: Kathy (Theresa, Amy, Kevin, and Jennifer), Kerry (David, Kevin, and Ellen), and Heather (Erika and Leslie)

PREFACE

As in previous editions, our primary goal for the 3rd edition of *Exercise Physiology: Energy, Nutrition, and Human Performance* is to integrate basic concepts and relevant scientific information to provide the foundation for understanding nutrition, energy transfer, and exercise and training. To be relevant to both the beginning and the more advanced student, we have tried to balance our discussions of theoretical foundations (with ample documentation) and practical applications. What initially began as a fine tuning of the material to bring it in line with expanding scientific knowledge has resulted in significant modifications as we tracked the emerging areas related to the broad scope of exercise physiology and sports medicine. In the area of sports nutrition and exercise, for example, we present the latest findings on the performance effects of specific feedings prior to, during, and in recovery from exercise. The composition of sports beverages has also been evaluated in terms of the effect on fluid replacement and temperature regulation. New information is also presented in the controversial area of protein requirements in exercise and training. From the health perspective, we continue to focus on the interaction between nutrition, physical activity, and the all-important blood lipid profile.

Within the framework of exercise performance and bioenergetics, we have expanded our presentation of the assessment of anaerobic power, anaerobic capacity, and the testing of aerobic fitness. We have also attempted to integrate research on the potential influence of genetics, gender, and age on the various aspects of physiologic function and exercise capacity. The chapter on the hormonal response to both exercise and training has been significantly reworked and expanded to provide a more comprehensive introduction to this complex area. The sections on muscular strength and conditioning now include plyometrics and resistance training for children; a noteworthy expansion is the presentation of specific factors related to strength improvements with training in men and women and the responsiveness of the elderly to such training.

In the section on ergogenic aids, we have upgraded the discussion of the performance effects of caffeine and bicarbonate drinks, as well as the latest findings on steroid use among adolescent and adult males and females. A section also has been added on newer approaches to enhance performance that utilize biosynthetic growth hormone and erythropoietin hormone.

In the area of body composition, sections are included on bioelectrical impedence analysis and other less costly techniques. Body density assessment of body composition has been revisited with an expanded discussion of potential sources of error in the assumptions that underlie this common laboratory procedure. The sections on the physique of champion athletes now include the latest findings on triathletes and football players; for gender comparisons, we contrast body builders, ballet dancers, and untrained adolescents.

The sections on obesity, weight control, and exercise have been rewritten to include current information on the variations in human obesity, as well as the potential health risks of excess body weight, body fat, and regional fat distribution.

The final chapter, which deals with exercise, aging, and health, now incorporates a discussion of the important role of physical activity epidemiology in the study of the health-related aspects of exercise. New sections also include risk factor assessment, interaction of risk factors, perceived exertion, responses of the elderly to various forms of training, and the recent research that relates regular physical activity to health and longevity.

The appendices have been significantly upgraded, and new sections have been added. For the nutritive value of common foods, the most significant addition (not found in any prior textbooks) is the grouping of foods into nine major categories (breads; cakes, cookies, candy bars, and desserts; cereals; cheese; fish; fruits; meats and eggs; milk and dairy products; vegetables) based on 1-ounce portions. Thus, all foods are now comparable for the important nutrients on a per ounce basis. There is also a new food category that consists of soups, sandwiches, salad dressings, oils, and condiments. We also present new caloric information on alcoholic and nonalcoholic beverages including, from the last edition, foods from a number of popular fast-food restaurants.

There are three new appendices: conversions to metric and SI units, a listing of frequently cited journals in the exercise sciences (91 journal entries and abbreviations), and a step-by-step procedure for calculation of the new ponderal somatogram that is discussed in the chapter on body composition. New figures and tables have been added where applicable, and all of the figures from the 2nd edition have been redrawn. We wish to acknowledge Robert Starnes, Rick Soldin, and Sammy Doyle, electronic artists par excellence at Arcata Graphics, Kingsport, Tennessee, for expert technical assistance in helping us to create the figures and artwork. They used a 256 grayscale scanner and an Apple Macintosh II X with 8 megabytes of RAM running Aldus Freehand enhanced with 32-bit color processing. As devoted Macintosh users from the early days of the first Mac, working with them was a dream come true; we hope that the quality of the artwork has enhanced the educational relevance of the textbook.

Finally, our ever increasing gratitude and respect for Lea & Febiger is once again acknowledged, with special thanks to Tom Colaiezzi and his staff for continued support and production excellence; to Jack Spahr, Sr. for patience and leadership, to John Spahr, Jr. for all of his efforts on our behalf, and to the many people behind the scenes at Lea & Febiger for their assistance in so many ways. To our current and former students and our many colleagues in physical education and the exercise and sport sciences, we say thank you for the privilege of your close association and friendship over the years. We appreciate your support and the many constructive ideas that have been incorporated in the 3rd edition. To the new generation of scientists with an interest in exercise physiology, we hope you will share your enthusiasm for the new frontiers of knowledge with your students and colleagues. Together, we can make a difference in unraveling the many secrets that are waiting to be discovered.

Flushing, New York
Amherst, Massachusetts
Ann Arbor, Michigan

William D. McArdle
Frank I. Katch
Victor L. Katch

CONTENTS

P A R T

1 EXERCISE PHYSIOLOGY

S E C T I O N   I   NUTRITION: THE BASE FOR HUMAN PERFORMANCE

1	CARBOHYDRATES, FATS, AND PROTEINS	**5**
	CARBOHYDRATES	**6**
	FATS	**19**
	PROTEINS	**29**
2	VITAMINS, MINERALS, AND WATER	**44**
	VITAMINS	**44**
	MINERALS	**50**
	WATER	**60**
3	OPTIMAL NUTRITION FOR EXERCISE	**68**

S E C T I O N   I I   ENERGY FOR PHYSICAL ACTIVITY

4	ENERGY VALUE OF FOOD	**85**
5	INTRODUCTION TO ENERGY TRANSFER	**92**
6	ENERGY TRANSFER IN THE BODY	**101**
	PHOSPHATE BOND ENERGY	**101**
	ENERGY RELEASE FROM FOOD	**107**
7	ENERGY TRANSFER IN EXERCISE	**123**
8	MEASUREMENT OF HUMAN ENERGY EXPENDITURE	**145**
9	HUMAN ENERGY EXPENDITURE DURING REST AND PHYSICAL ACTIVITY	**158**
	ENERGY EXPENDITURE AT REST	**158**
	ENERGY EXPENDITURE IN PHYSICAL ACTIVITY	**165**

10 ENERGY EXPENDITURE DURING WALKING, JOGGING, RUNNING, AND SWIMMING **174**

11 INDIVIDUAL DIFFERENCES AND MEASUREMENT OF ENERGY CAPACITIES **199**

S E C T I O N   I I I   SYSTEMS OF ENERGY DELIVERY AND UTILIZATION

12 PULMONARY STRUCTURE AND FUNCTION **235**

13 GAS EXCHANGE AND TRANSPORT **254**
GASEOUS EXCHANGE IN THE LUNGS AND TISSUES **254**
TRANSPORT OF OXYGEN **260**
TRANSPORT OF CARBON DIOXIDE **266**

14 DYNAMICS OF PULMONARY VENTILATION **270**
REGULATION OF PULMONARY VENTILATION **270**
PULMONARY VENTILATION DURING EXERCISE **278**
ACID-BASE REGULATION **285**

15 THE CARDIOVASCULAR SYSTEM **292**

16 CARDIOVASCULAR REGULATION AND INTEGRATION **313**

17 FUNCTIONAL CAPACITY OF THE CARDIOVASCULAR SYSTEM **326**

18 SKELETAL MUSCLE: STRUCTURE AND FUNCTION **348**

19 NEURAL CONTROL OF HUMAN MOVEMENT **367**

20 THE ENDOCRINE SYSTEM AND EXERCISE **384**

P A R T
2 APPLIED EXERCISE PHYSIOLOGY

S E C T I O N   I V   ENHANCEMENT OF ENERGY CAPACITY

21 TRAINING FOR ANAEROBIC AND AEROBIC POWER **423**

22 MUSCULAR STRENGTH: TRAINING MUSCLES TO BECOME STRONGER **452**
MEASURING AND IMPROVING MUSCULAR STRENGTH **452**
ADAPTATIONS WITH RESISTANCE TRAINING **475**

23 SPECIAL AIDS TO PERFORMANCE AND CONDITIIONING **497**

S E C T I O N   V   WORK PERFORMANCE AND ENVIRONMENTAL STRESS

24   EXERCISE AT MEDIUM AND HIGH ALTITUDE   **529**
25   EXERCISE AND THERMAL STRESS   **547**
         MECHANISMS OF THERMOREGULATION   **547**
         THERMOREGULATION AND ENVIRONMENTAL STRESS
         DURING EXERCISE   **556**
26   SPORT DIVING   **580**

S E C T I O N   V I   BODY COMPOSITION, ENERGY BALANCE, AND WEIGHT CONTROL

27   BODY COMPOSITION ASSESSMENT   **599**
28   PHYSIQUE, PERFORMANCE, AND PHYSICAL ACTIVITY   **634**
29   OBESITY AND WEIGHT CONTROL   **656**
         OBESITY   **656**
         WEIGHT CONTROL   **674**
30   PHYSICAL ACTIVITY, HEALTH, AND AGING   **698**

APPENDICES

A   THE METRIC SYSTEM AND SI UNITS   743
        THE METRIC SYSTEM   743
        SI UNITS   757

B   NUTRITIVE VALUE FOR COMMON FOODS, ALCOHOLIC AND NONALCO-
        HOLIC BEVERAGES, AND SPECIALTY AND FAST-FOOD ITEMS   761
        NUTRITIVE VALUES FOR COMMON FOODS   761
        NUTRITIVE VALUES FOR ALCOHOLIC AND NONALCOHOLIC BEVERAGES   779
        NUTRITIVE VALUES FOR SPECIALTY AND FAST-FOOD ITEMS   782

C   METABOLIC COMPUTATIONS IN OPEN-CIRCUIT SPIROMETRY   796

D   ENERGY EXPENDITURE IN HOUSEHOLD, RECREATIONAL, AND SPORTS
        ACTIVITIES (in kcal $\cdot$ min^{-1})   804

E   CORRECTION FOR WATER DENSITY AT DIFFERENT TEMPERATURES   812

F   BODY COMPOSITION   813

G COMPUTERIZED MEAL AND EXERCISE PLAN 820

H FREQUENTLY CITED JOURNALS IN EXERCISE PHYSIOLOGY 828

I THE BODY PROFILE 830
ANTHROPOMETRIC MEASUREMENTS 830
COMPUTATIONS 832
GRAPHIC ANALYSIS 834

INDEX 837

PART 1

EXERCISE PHYSIOLOGY

NUTRITION: THE BASE FOR HUMAN PERFORMANCE

In a textbook dealing with the physiology of human performance, it is becoming more common to find a section devoted to the basics of human nutrition. We feel strongly that this topic is of considerable importance and should serve as the starting point for this book.

Proper nutrition forms the foundation for physical performance; it provides both the fuel for biologic work and the chemicals for extracting and using the potential energy contained within this fuel. Food also provides the essential elements for the synthesis of new tissue and the repair of existing cells.

Some may argue that adequate nutrition for exercise can easily be achieved through the intake of a well-balanced diet and that it therefore is of little consequence in the study of exercise performance. We maintain, however, that the study of exercise, when viewed within the framework of energy capacities, must be based on an understanding of the sources of food energy and the role of nutrients in the process of energy release. With this perspective, it becomes possible for the exercise specialist to appreciate the importance of "adequate" nutrition and to evaluate critically the validity of claims concerning nutrient supplements and special dietary modifications for enhancing physical performance. Because various food nutrients provide energy and regulate physiologic processes associated with exercise, it is tempting to link dietary modification to improvement in athletic performance. Too often individuals spend considerable time and "energy" striving for the optimum in exercise performance, only to fall short due to inadequate, counterproductive, and sometimes harmful nutritional practices. Finally, it is now apparent that aspects of sound nutritional practice impact on a variety of disease conditions to which regular exercise may make an important, positive contribution.

In the chapters that follow, we look at the six broad classifications of nutrients: carbohydrates, fats, proteins, vitamins, minerals, and water. We attempt to answer the following questions: What are they? Where are they found? What are their functions? What specific role do they play in physical activity?

1

CARBOHYDRATES, FATS, AND PROTEINS

The carbohydrate, fat, and protein nutrients consumed daily provide the necessary energy to maintain body functions both at rest and during various forms of physical activity. Aside from their role as biologic fuel, these nutrients (called **macronutrients** by nutritionists) also play an important part in maintaining the structural and functional integrity of the organism. In this chapter, each of these nutrients is discussed in terms of general structure, function, and source in specific foods in the diet. Emphasis is placed on their importance in sustaining physiologic function during physical activity.

BASIC STRUCTURE OF NUTRIENTS

All biologic systems are composed of cells that engage in similar activities required to maintain the integrity and life of the cell. Although different cells possess specialized functions that necessitate special structures, the basic life-sustaining processes among all cells are similar. Cells with diverse functions also have similar chemical compositions. All cells are composed of essentially the same chemicals, differing only in the proportion and arrangement of these chemicals.

ATOMS: NATURE'S BUILDING BLOCKS

Of the 103 different types of atoms or elements, nitrogen, hydrogen, carbon, and oxygen make up 3, 10, 18, and 65% of the body mass, respectively. These atoms play the *major* role in the chemical composition of nutrients and comprise the structural units for most biologically active substances in the body.

Molecules are formed from the union of two or more atoms. The specific atoms as well as the arrangement of these atoms give the molecule its properties. Glucose is glucose because of the arrangement of 24 atoms of three different kinds within its molecule. **Chemical bonding,** e.g., between atoms of hydrogen and oxygen in the water molecule, involves a common sharing of electrons between atoms. **It is this force of attraction created between the positive and negative charges of atoms that forms the basis for bonding and provides the "cement" that keeps the atoms and molecules within a substance from readily coming apart.** When these forces are altered due to the removal, transfer, or exchange of certain electrons, energy is provided to power cellular functions. When two or more molecules are chemically bound, a larger aggregate of matter or a **substance** is formed. This substance may take the form of a gas, a liquid, or a solid, depending on the force of interaction between molecules.

CARBON: THE VERSATILE ELEMENT

All of the nutrients except water and minerals contain carbon. In fact, almost all of the biologic substances within the body are composed of compounds containing carbon. Carbon atoms have an almost unlimited ability to share chemical bonds with other carbon atoms and with atoms of other elements to form large carbon-chain molecules.

Fats and carbohydrates are formed from specific linkages of carbon atoms with atoms of hydrogen and oxygen. With the addition of nitrogen and certain mineral substances, a protein molecule is formed. These atoms of carbon, hydrogen, oxygen, and nitrogen are the **organic** building blocks from which the nutrients are made.

P A R T

1

Carbohydrates

THE NATURE OF CARBOHYDRATES

Carbohydrates, as the name implies, are composed of carbon and water. Atoms of carbon, hydrogen, and oxygen combine to form carbohydrate compounds. The basic chemical structure of a simple sugar molecule consists of a chain of from 3 to 7 carbon atoms with the hydrogen and oxygen atoms attached singly. The most typical sugar, **glucose,** is illustrated in Figure 1–1. The glucose molecule consists of 6 carbon, 12 hydrogen, and 6 oxygen atoms ($C_6H_{12}O_6$). Each of the carbon atoms has four bonding sites that can link to other atoms, including carbon atoms. Carbon bonds not linked to other carbons are "free" to hold hydrogen (which has only one bond site itself), oxygen (which has two bond sites), or an oxygen-hydrogen combination, termed a hydroxyl (OH).

Fructose and **galactose** are two other simple sugars that have the same chemical formula as glucose, with a slightly different carbon-to-hydrogen-to-oxygen linkage. This alteration in atomic arrangement makes fructose, galactose, and glucose different substances.

KINDS AND SOURCES OF CARBOHYDRATES

There are three kinds of carbohydrates: **monosaccharides, oligosaccharides,** and **polysaccharides.** Each form of carbohydrate is distinguished by the number of simple sugars in combination within the molecule.

MONOSACCHARIDES

More than 200 monosaccharides have been found in nature.[119] The most common of the monosaccharides, glucose, fructose, and galactose, were de-

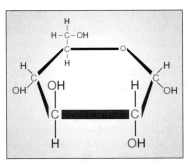

FIG. 1–1.
Three-dimensional structure of the simple sugar molecule glucose. The molecule resembles a hexagonal plate to which H and O atoms are attached.

scribed previously. Glucose, also called dextrose or blood sugar, is formed as a natural sugar in food or is produced in the body as a result of digestion of more complex carbohydrates or via the process of **gluconeogenesis,** whereby it is synthesized from the carbon skeletons of other compounds. After absorption by the small intestines, glucose can be used directly by the cell for energy, stored as glycogen in the muscles and liver, or converted to fats for energy storage.

Fructose, or fruit sugar, is present in large amounts in natural form in fruits and honey and is the sweetest of the simple sugars. Although some fructose is absorbed directly into the blood from the digestive tract, it is all slowly converted to glucose in the liver. Galactose is not found freely in nature; rather, it must be produced from milk sugar in the mammary glands of lactating animals. In the body, galactose is converted to glucose for energy metabolism.

OLIGOSACCHARIDES

The major types of oligosaccharides are the **disaccharides** or double sugars, which are formed from the combination of two monosaccharide molecules.

The mono- and disaccharides collectively make up what is commonly referred to as the **simple sugars.** These sugars are packaged under a variety of guises—brown sugar, corn syrup, invert sugar, honey, and "natural sweeteners."

In the structure of each of the disaccharides, glucose is one of the simple sugars. The three principal disaccharides are:

> **Sucrose**—glucose + fructose
> **Lactose**—glucose + galactose
> **Maltose**—glucose + glucose

Sucrose is the most common dietary disaccharide and contributes up to 25% of the total quantity of ingested calories in the United States. It occurs naturally in most foods containing carbohydrates, especially in beet and cane sugar, brown sugar, sorghum, maple syrup, and honey. Honey, while sweeter than table sugar owing to its greater fructose content, is no more superior nutritionally or as an energy source.

Lactose is found in natural form only in milk and is often called milk sugar. It is the least sweet of the disaccharides. Lactose can be artificially processed and is often found in carbohydrate-rich, high-calorie liquid meals.

Maltose occurs in malt products and in germinating cereals. It is considered a negligible carbohydrate in terms of its contribution to the carbohydrate content of the average person's diet.

POLYSACCHARIDES

Three or more simple sugar molecules form a polysaccharide. In fact, from 300 to 26,000 monosaccharide molecules can be linked together or polymerized to form a polysaccharide. There are generally two classifications of polysaccharides, plant and animal.

Plant Polysaccharides

Two common forms of plant polysaccharides are **starch** and **fiber.**

Starch. Starch is the most familiar form of plant polysaccharide. It is found in seeds, corn, and in the various grains from which bread, cereal, spaghetti, and pastries are made. Large amounts are also present in peas, beans, potatoes, and roots, where it serves as an energy store for future use by plants. Plant starch is still the most important dietary source of carbohydrate in the American diet, accounting for approximately 50% of the total carbohydrate intake. Starch intake, however, has decreased about 30% since the turn of the century, whereas the consumption of more simple sugars such as sucrose has correspondingly increased from 31% to about 50%.

Cellulose. Cellulose and most other nonstarch fibrous materials that are generally resistant to human digestive enzymes are another form of polysaccharide. They are found exclusively in plants and make up the structural part of leaves, stems, roots, seeds, and fruit coverings. The various fibers differ widely in physical and chemical characteristics and in physiologic action; they are found mostly within the cellular walls as cellulose, hemicellulose, pectin, and the noncarbohydrate lignin, whereas other fibers such as mucilage and gums are found within the plant cell itself.

Health Implications. Although technically not a nutrient, **dietary fiber** has received considerable attention by researchers and the lay press. Much of this interest originated from studies that linked high levels of fiber intake with a lower occurrence of obesity, diabetes, intestinal disorders, and heart disease.[71,79] Because the Western diet is high in fiber-free animal foods and loses much of its natural fiber through processing, it is speculated that this accounts for the prevalence of intestinal disorders in this country compared to countries that consume a more primitive-type diet high in unrefined, complex carbohydrates.[103] For example, the typical American diet contains a daily fiber intake of about 12 grams (g)* whereas diets from Africa and India range between 40 and 150 g per day.[71,82] Fibers hold considerable water and thus give "bulk" to the food residues in the small intestines, often increasing stool weight and volume by 40 to 100%. This bulking action may aid gastrointestinal functioning by exerting a scraping action on the gut wall, binding or diluting harmful chemicals or inhibiting their activity, or by shortening the transit time for the passage of food residues (and possibly carcinogenic materials) through the digestive tract. This may reduce the chances of contracting colon cancer and various other gastrointestinal diseases later in life.[16,39]

Fiber intake may also lower serum cholesterol in humans,[9,25,75] especially the **water-soluble** mucilagenous fibers such as pectin and guar gum present in oats, beans, brown rice, peas, carrots, and a variety of fruits. For men with elevated blood lipids, for example, adding 100 g of oat bran to the daily diet caused a 13% reduction in serum cholesterol and favorably affected the ratio of the blood's lipoprotein components.[75] More recently, daily intake of quar gum fiber has been shown to reduce total cholesterol by 10%, specifically by lowering the low-density lipoprotein component of the cholesterol profile.[127] In contrast, the **water-insoluble** fibers, such as cellulose, hemicellulose, and lignin and cellulose-rich products like wheat bran showed no cholesterol-lowering effect.[71,129] The precise mechanism is unclear by which dietary fibers favorably affect serum cholesterol. It may be that the addition of fiber simply replaces other cholesterol-ridden items in the diet. On the

---

* Scientific measurement is generally presented in terms of the metric system. Appendix A shows the relationship between metric units and English units that are relevant to the material presented in this book. Also presented are some common expressions of work, energy, and power.

other hand, some fibers may actually hinder cholesterol absorption while others may reduce cholesterol metabolism in the gut. These actions would depress the synthesis of cholesterol while at the same time facilitating the excretion of existing cholesterol bound to the fiber in the feces.[43] Research also indicates that dietary fiber decreases the total energy consumed in subsequent meals. Consuming a fiber-rich breakfast, for example, decreases the calories consumed both during breakfast and during a buffet lunch 3.5 hours later.[85a]

Present nutritional wisdom maintains that a dietary fiber intake of about 30 g per day is an important part of a well-structured diet. Table 1–1 gives the fiber content of some common foods. Perhaps as research progresses and the analysis of the fiber content of various foods is refined, a recommended daily requirement for specific fibers will be established. Excessive fiber intake, however, is not prudent, especially for individuals with marginal levels of nutrition. Increased fiber intake decreases the absorption of calcium, iron, magnesium, phosphorous, and certain trace minerals.[72]

Animal Polysaccharides

Glycogen is the polysaccharide synthesized from glucose in the process of **glucogenesis** and stored in the tissues of animals. Glycogen molecules are usually large and range in size from a few hundred to thousands of glucose molecules linked together, much like the links in a chain of sausages, with some branch points for additional glucose linkage. In well-nourished humans, approximately 375 to 475 g of carbohydrate are stored in the body. Of this, approximately 325 g are muscle glycogen, 90 to 110 g are liver glycogen, and only 15 to 20 g are present as blood glucose.[46] As each gram of glycogen contains 4 calories of energy, the average person stores between 1,500 and 2,000 calories of energy within the bonds of the carbohydrate molecule. This is about enough energy to power a 20-mile run.

Several factors determine the rate and quantity of either glycogen synthesis or breakdown. During exercise, the carbohydrate stored as muscle glycogen is used as a source of energy for the specific muscle in which it is stored. In the liver, in contrast, glycogen is reconverted to glucose (under the control

TABLE 1–1.
FIBER CONTENT OF SOME COMMON FOODS LISTED IN ORDER OF OVERALL FIBER CONTENT

	Serving Size	Total Fiber (g)	Soluble Fiber (g)	Insoluble Fiber (g)
100% bran cereal	½ cup	10.0	0.3	9.7
Peas	½ cup	5.2	2.0	3.2
Kidney beans	½ cup	4.5	0.5	4.0
Apple	1 small	3.9	2.3	1.6
Potato	1 small	3.8	2.2	1.6
Broccoli	½ cup	2.5	1.1	1.4
Strawberries	¾ cup	2.4	0.9	1.5
Oats, whole	½ cup	1.6	0.5	1.1
Banana	1 small	1.3	0.6	0.7
Spaghetti	½ cup	1.0	0.2	0.8
Lettuce	½ cup	0.5	0.2	0.3
White rice	½ cup	0.5	0	0.5

of a specific **phosphatase** enzyme) and transported in the blood for eventual use by the working muscles. The term **glycogenolysis** is used to describe this reconversion process, which provides a rapid supply of glucose for muscular contraction during all forms of work. When liver and muscle glycogen is depleted through dietary restriction or exercise, glucose synthesis from the structural components of the other nutrients, especially proteins, tends to increase through gluconeogenesis. Hormones, especially insulin, play an important part in the regulation of liver and muscle glycogen stores by controlling the level of circulating blood sugar.

Because comparatively little glycogen is stored in the body, its quantity can be modified considerably through the diet. For example, a 24-hour fast or low-carbohydrate, normal-calorie diet results in a large reduction in glycogen reserves.[65] On the other hand, maintaining a carbohydrate-rich diet for several days enhances the body's carbohydrate stores to a level almost twice that obtained with a normal, well-balanced diet.[13] In fact, the upper limit for glycogen storage in the body is about 15 g per kilogram (kg) of body weight. This is a capacity of 1100 g for an average-size man. The effect of enhanced carbohydrate storage on exercise performance is discussed in a later section of this chapter.

RECOMMENDED INTAKE OF CARBOHYDRATES

Figure 1–2 illustrates the carbohydrate content of selected foods. Cereals, cookies, candies, breads, and cakes are rich carbohydrate sources. Because the values are based on carbohydrate percentage in relation to the total weight, including water content, fruits and vegetables appear to be less valuable sources of carbohydrates. The dried portion of these foods, however, is almost pure carbohydrate.

The typical American diet consists of approximately 40 to 50% of the total calories as carbohydrate. For a sedentary 70-kg person, for example, this amounts to approximately 300 g of carbohydrate per day. For active people and those involved in exercise training, about 60% of the daily caloric

FIG. 1–2.
Percentage of carbohydrates in commonly served foods.

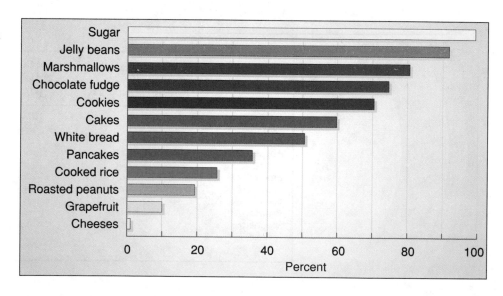

intake (400 to 600 g) should be in the form of carbohydrates, predominantly of the complex variety.

Important dietary carbohydrate sources are generally fruits and vegetables, although this in no way represents the "state of affairs" for all individuals. In fact, it has been estimated that the average American consumes about 50% of carbohydrate as simple sugars in the predominant form of sucrose and high-fructose corn syrup. This amount represents more than 16 teaspoons a day or 60 pounds (lb) of table sugar and 46 lb of corn syrup each year as contrasted to a 4 lb per person intake of table sugar 100 years ago!

Excessive fermentable carbohydrate, especially sucrose in the diet is a main **cause** of tooth decay.[38] In addition, excessive dietary sugar is believed to be involved in a variety of other disease processes, most notably diabetes, obesity, and coronary heart disease.[113] One way to reduce surcrose intake is to substitute fructose, a monosaccharide that is about twice as sweet as table sugar. This provides equal sweetness with fewer calories. In addition, fructose does not stimulate pancreatic insulin secretion and may be taken up by the muscle without assistance of insulin. As a result, blood-glucose level remains fairly stable after fructose ingestion.[92]

ROLE OF CARBOHYDRATES IN THE BODY

Carbohydrates serve several important functions related to exercise performance.

ENERGY SOURCE

The main function of carbohydrate is to serve as an energy fuel for the body. The energy derived from the breakdown of glucose and glycogen is ultimately used to power muscular contraction as well as all other forms of biologic work.[116]

It is important that adequate amounts of carbohydrates are ingested routinely to maintain the body's relatively limited glycogen stores. If too few carbohydrates are ingested, glucose is then obtained from glycogen breakdown and the carbohydrate reserves become depleted. In contrast, following a meal, excess carbohydrates may be readily converted to muscle and liver glycogen. Once the capacity of the cell for glycogen storage is reached, the excess sugars are converted and stored as fat. This action helps explain how body fat increases when excess calories in the form of carbohydrates are consumed. This process occurs even if the diet is low in fat.

PROTEIN SPARING

Carbohydrate also provides a "protein sparing" effect. Under normal conditions, protein serves a vital role in the maintenance, repair, and growth of body tissues, and to a considerably lesser degree, as a nutrient source of energy. When carbohydrate reserves are reduced, however, metabolic pathways exist for the synthesis of glucose from protein and the glycerol portion of the fat molecule. This process of gluconeogenesis provides a metabolic option for augmenting carbohydrate availability in the face of depleted glycogen stores. This becomes increasingly important in prolonged endurance exercise. The price that is paid, however, is a temporary reduction in the body's protein "stores," especially muscle protein. In extreme conditions this can cause a significant reduction in lean tissue and an accompanying load on the kidneys as they excrete the nitrogen-containing byproducts of protein breakdown. **Adequate intake and use of carbohydrates aid in the maintenance of tissue protein.**

METABOLIC PRIMER

Another function of carbohydrate is to serve as a "primer" for fat metabolism. Certain products from the breakdown of carbohydrate must be available to facilitate the metabolism of fat. If insufficient carbohydrate metabolism exists, either through limitation in the transport of glucose into the cell, which occurs in diabetes, or depletion of glycogen through improper diet or prolonged exercise, the body begins to mobilize fat to a greater extent than it can use. The result is incomplete fat metabolism and the accumulation of acid by-products called **ketone bodies.**[76] This situation may lead to a harmful increase in the acidity of body fluids, a condition called acidosis— or more specifically with regard to fat breakdown, **ketosis.** More is said of the role of carbohydrate as a primer for fat metabolism in Chapter 6.

FUEL FOR THE CENTRAL NERVOUS SYSTEM

Carbohydrate is essential for the proper functioning of the central nervous system. Under normal conditions and in short-term starvation, the brain uses blood glucose almost exclusively as a fuel and essentially has no stored supply of this nutrient. As starvation or low carbohydrate intake progresses, adaptations in metabolism occur, and after about 8 days the brain is able to use a relatively large amount of fat for its fuel requirement.[21,105] There is even some indication that adaptations take place in skeletal muscle that increase its ability to burn fat during exercise and concurrently spare muscle glycogen.[94]

The symptoms of a modest reduction in blood glucose (**hypoglycemia**) include feelings of weakness, hunger, and dizziness. This condition impairs exercise performance and may partially explain the fatigue associated with prolonged exercise. Sustained and profound low blood sugar can cause irreversible brain damage. Because of the specific role played by glucose in generating energy for use by nerve tissue, blood sugar is regulated within narrow limits.

CARBOHYDRATE BALANCE IN EXERCISE

The fuel mixture during exercise depends on the intensity and duration of effort, as well as the fitness and nutritional status of the exerciser.[2,68] With the use of biochemical and biopsy techniques, it has been possible to study the contributions of various nutrients to the energy demands of physical activity. The biopsy technique permits the sampling of specific muscles with little interruption in exercise. Thus, through serial sampling of the same muscle in the same person, the role of intramuscular nutrients can be carefully evaluated during exercise. Under most conditions, exercise brings about a marked increase in the production and release of glucose by the liver and its subsequent use by muscle.[131]

INTENSE EXERCISE

Stored muscle glycogen and blood-borne glucose are the prime contributors of energy in the early minutes of exercise, in which oxygen supply does not meet the demands for aerobic metabolism, and during high-intensity exercise. Blood glucose, for example, may supply 30 to 40% if the total energy of exercising muscles.[46] As illustrated in Figure 1.3, during the initial stage of exercise, the uptake of circulating blood glucose by the muscles increases sharply and continues to increase as exercise progresses. By the fortieth minute of exercise, the glucose uptake has risen to between 7 and 20 times the uptake at rest, depending on the intensity of the exercise. The increase in the percentage contribution of carbohydrate during intense exercise is explained largely by the fact that it is the **only** nutrient that provides energy when the oxygen supplied to the muscles is insufficient in relation to the oxygen needs. The specifics of this role in energy release are explained in Chapter 6.

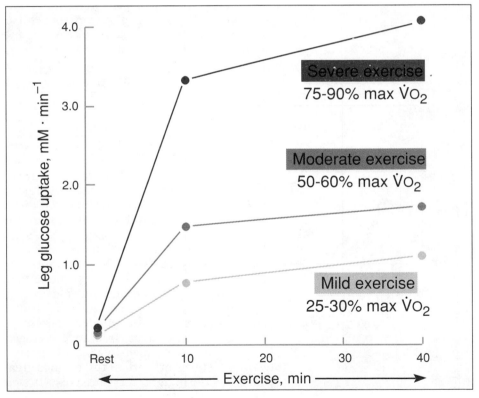

FIG. 1–3.
Relationship between blood glucose uptake by the leg muscles and exercise duration and intensity. Here intensity is expressed as a percent of one's maximum capacity to consume oxygen or max $\dot{V}O_2$. (From Felig, P., and Wahren, J.: Fuel homeostasis in exercise. Seminar in Medicine of the Beth Israel Hospital, Boston. N. Engl. J. Med., *293:*1078, 1975.)

MODERATE AND PROLONGED EXERCISE

In the transition from rest to submaximal exercise almost all of the energy is supplied from glycogen stored in the active muscles. During the subsequent 20 minutes or so, liver and muscle glycogen provide about 40 to 50% of the energy requirement with the remaining requirement provided by fat breakdown. As exercise continues and glycogen stores become reduced, however, an increasingly greater percentage of energy is supplied through fat metabolism. Eventually, glucose output by the liver fails to keep pace with its use by the muscles and blood glucose concentration slowly begins to fall.[4] In fact, the level of circulating blood glucose may actually fall to hypoglycemic levels (<45 mg glucose per 100 ml blood) after 90 minutes of continuous exercise.[44]

Fatigue occurs if exercise is performed to the point at which the glycogen in the liver and specific muscles becomes severely lowered, even though sufficient oxygen is available to the muscles and the potential energy from stored fat remains almost unlimited.[120] Endurance athletes commonly refer to this sensation of fatigue as "bonking" or "hitting the wall." Because the phosphatase enzyme is not present to allow for glycogen transfer between muscles, the relatively inactive muscles maintain their glycogen content. It is unclear why the depletion of muscle glycogen coincides with the point of fatigue in prolonged submaximal exercise. The functions of blood glucose as energy for the central nervous system and muscle glycogen as a "primer" in fat metabolism may provide part of the answer.

Effect of Diet on Muscle Glycogen Stores

Ingested carbohydrates represent an energy nutrient that is readily available to exercising muscles.[67] Figure 1–4 shows the results of one experiment in which the initial muscle glycogen stores were varied in 6 subjects through dietary manipulation.[13] In one condition, the normal caloric intake was maintained for 3 days but the major quantity of calories was supplied in the form of fat with less than 5% as carbohydrate. In the second condition, the 3-day diet was normal and contained the recommended daily percentages of carbohydrate, fat, and protein. In the third diet, 82% of the calories were provided in the form of carbohydrates. The glycogen content of the **quadriceps femoris** muscle of the leg, determined by needle biopsy, averaged 0.63, 1.75, and 3.75 g of glycogen per 100 g wet muscle as a result of the high-fat, normal, and high-carbohydrate diets, respectively.

Endurance capacity on the bicycle ergometer varied considerably depending upon the diet each person consumed during the 3 days prior to the exercise test. With the normal diet, moderate exercise could be tolerated for an average of 114 minutes, whereas endurance averaged only 57 minutes with the high-fat diet. The endurance capacity of subjects who were fed the high-carbohydrate diet was more than three times greater than when the same subjects consumed the high-fat diet. Such results clearly demonstrate the importance of muscle glycogen for prolonged and intense exercise lasting more than an hour, and emphasize the important role of nutrition in establishing the appropriate energy reserves for both long-term exercise and heavy training.[31]

A diet deficient in carbohydrates rapidly depletes muscle and liver glycogen and subsequently affects performance in intense short-term exercise as well as in prolonged submaximal endurance activities.[64] These observations are important not only for athletes but also for individuals who have modified

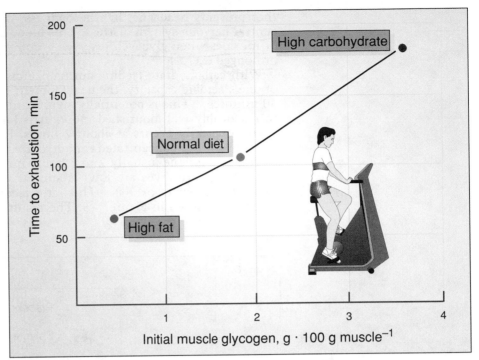

FIG. 1–4.
Effects of a low-carbohydrate diet, a mixed diet, and a high-carbohydrate diet on the glycogen content of the quadriceps femoris and the duration of exercise on a bicycle ergometer. (From Bergstrom, J., et al.: Diet, muscle glycogen and physical performance. Acta Physiol. Scand., *71*:140, 1967.)

their diet so the normal recommended percentage of carbohydrates becomes reduced. Reliance on starvation diets or on other potentially harmful diets such as high-fat, low-carbohydrate diets, "liquid-protein" diets, or water diets is counterproductive for weight control, exercise performance, optimal nutrition, and good health.[70] Such low-carbohydrate diets make it extremely difficult from the standpoint of energy supply to participate in vigorous physical activity or training.[32]

Administration of Oral Glucose Before and During Exercise

Some controversy exists as to the contribution to exercise performance of ingesting sugar-rich drinks.[44,47,81,130]

During Exercise. It appears that such drinks consumed during exercise benefit relatively high intensity, long-term aerobic performance.[36,95,115,117] In low intensity exercise the beneficial effect is negligible for carbohydrate feeding. This is because such exercise is fueled mainly by the oxidation of fat with relatively small demand on carbohydrate breakdown. In high intensity exercise, when the demand on glycogen is much greater, the glucose feedings provide supplementary carbohydrate. This may either spare muscle glycogen because the ingested glucose is used as fuel to power exercise[3,35,81,109] or help to maintain a more optimal level of blood glucose that prevents headache, lightheadedness, nausea, and other symptoms of central nervous system distress. This blood glucose also supplies the needs of muscles when glycogen reserves become depleted later in the period of prolonged exercise.[26,62,106]

With carbohydrate feeding during exercise at an intensity of 60 to 80% of one's aerobic capacity, the development of fatigue is postponed by 15 to 30 minutes.[37] This is potentially significant to endurance athletes, because in reasonably well-nourished individuals fatigue from moderately intense exercise usually occurs at about 2 hours. Recent research shows that the effect of a single concentrated carbohydrate feeding about 30 minutes before anticipated fatigue (usually 2 to 3 hours into exercise, when blood glucose and glycogen reserves are low) is similar to that of regular carbohydrate ingestion earlier in exercise. This later feeding restores the level of blood glucose as shown in Figure 1–5. The result is an increase in carbohydrate

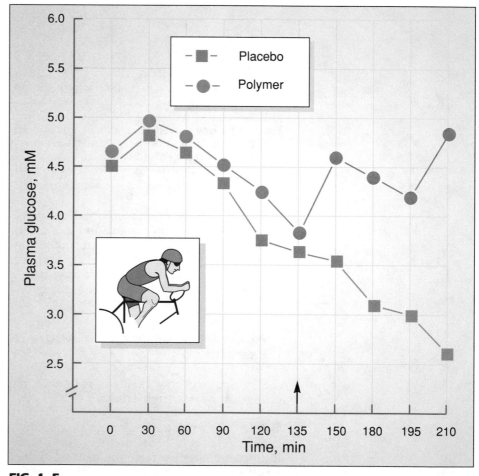

FIG. 1–5.
Average plasma glucose concentration during prolonged high-intensity aerobic exercise when subjects were fed (↑) a placebo (■) or glucose polymers (●) (3 g per kg body weight in a 50% solution). (From Coggan, A.R., and Coyle, E.F.: Metabolism and performance following carbohydrate ingestion late in exercise. Med. Sci. Sports Exerc., *21*:59, 1989.)

availability and a delay in fatigue as blood glucose serves the energy needs of the working muscles.[26] There appears to be a limit, however, to the intensity of exercise at which carbohydrate feedings are beneficial. The endurance benefits are most effective during exercise at about 75% of aerobic capacity. If a higher intensity of effort is maintained in the earlier stage of exercise, subjects were forced to lower exercise intensity to about the 75% level during the final phases so they could maintain the benefits from the carbohydrate feedings.[27] Research also shows that repeated feedings of carbohydrate in solid form (43 g sucrose with 400 ml water) at the beginning and at 1, 2, and 3 hours of exercise maintains blood glucose and reduces glycogen depletion during 4 hours of cycling; sprint performance to exhaustion at the end of the activity was also enhanced.[47,59] These findings illustrate clearly that **carbohydrate feedings during exercise contribute to metabolism in a way that either conserves the muscle's glycogen content for later use or maintains blood glucose to be used later as exercise progresses and muscle glycogen becomes depleted.** This is reflected by improved endurance at a relatively high steady pace or a greater sprint capacity toward the completion of effort. In a marathon, the biochemical ability to sustain effort and sprint to the finish could determine the winner.

What to Drink. A variety of carbohydrate supplements have been selected, yet none has been more effective than glucose.[37] The "sugar drink" usually recommended ranges between an isotonic 5% solution that can be made by adding 50 g of either glucose, fructose, or sucrose (table sugar) to one liter of water to a 25 to 50% concentrated hypertonic solution.[14,108] A practical recommendation is to ingest a strong sugar solution (70 g sugar in 140 ml water) 20 to 30 minutes after the start of exercise followed by less-concentrated solutions containing about 24 g of carbohydrate (8 oz of a 5% solution every 15 minutes) over 30-minute intervals during exercise.[35,120]

When sugars are consumed during exercise, there is no overreaction in insulin response and resulting hypoglycemia normally observed with pre-exercise feedings. More than likely, this occurs because the hormones of the sympathetic nervous system released during exercise tend to inhibit the release of insulin. Concurrently, exercise appears to augment a muscle's ability to take up glucose so that the added glucose provided by feeding moves into the cells with a lower insulin requirement.

Prior to Exercise. The benefits of ingesting glucose- or sucrose-rich drinks are observed only when they are consumed at intervals during the activity. In fact, drinking a strong sugary solution in the 30 minutes prior to exercise often hinders one's endurance capability.[31,48,73] Paradoxically, the muscle glycogen reserves become prematurely depleted at a more rapid rate when using pre-event sugar drinks in contrast to drinking plain water. Actually, the best performances were turned in with water as the pre-exercise drink!

A concentrated challenge of simple sugars prior to exercise causes a dramatic rise in blood sugar 5 to 10 minutes after ingestion. This rise leads to an overshoot in the release of insulin from the pancreas and actually produces a decline in blood sugar (hypoglycemia) as glucose moves rapidly into the muscle cells.[60,138] At the same time, insulin inhibits the mobilization and utilization of fat for energy.[48] Consequently, when exercise begins, intramuscular carbohydrate is used to a much greater degree than under normal conditions. This utilization of intramuscular carbohydrate causes glycogen depletion and fatigue to occur earlier than it normally would. If more than

30 minutes is provided after initiation of pre-exercise glucose feedings, endurance performance might not be impaired.[48,52]

Because fructose is absorbed more slowly from the gut than either glucose or sucrose, and causes only a minimal insulin response with essentially no decline in blood glucose, there is debate as to whether fructose would be beneficial as an immediate pre-exercise carbohydrate feeding.[61,78,98,104] Although the theoretical rationale for the use of fructose appears strong, the ergogenic benefits are inconclusive concerning such feedings. What is important, however, is that the consumption of high-fructose beverages is often accompanied by significant gastrointestinal distress that in itself can negatively affect exercise performance.

Glucose Feedings and Water Uptake. Of considerable importance with regard to sugar drinks is their potential negative effect on water absorption from the digestive tract. Because the emptying of fluid from the stomach into the small intestines, where it is absorbed, is inhibited by the concentration of particles in solution (**osmolality**), even a small amount of glucose retards the movement of water out of the stomach[34,100] This could be deleterious during prolonged exercise in the heat, when adequate intake **and** absorption of fluid are of prime importance to the health and safety of the athlete. In this regard, research is encouraging in that the negative effects of sugar molecules on gastric emptying can be reduced[49,118,124] and plasma volume is maintained during prolonged exercise in the heat[115] if a glucose polymer solution is used in formulating the drink. The number of particles in solution with polymerized glucose is greatly reduced, which facilitates the movement of water from the stomach to the intestines for absorption. There is considerable debate concerning the optimal beverage for rehydration during and after warm-weather exercise,[22,24,95,97] because it has yet to be shown that the slower rate of gastric emptying observed when ingesting certain electrolyte-glucose solutions (and to a lesser degree glucose polymer solutions) compromises fluid uptake in the intestines to the extent that thermoregulation and exercise performance are impaired. More is said concerning fluid replacement in Chapter 25.

SUMMARY

1. Atoms are the basic building blocks of all matter, and they combine to form molecules. Most cells are composed of the same chemicals, differing only in proportion and arrangement.
2. Carbon, hydrogen, oxygen, and nitrogen are the primary structural units of most of the biologically active substances in the body. Specific combinations of carbon with oxygen and hydrogen form carbohydrates and fats, whereas other combinations with the addition of nitrogen and minerals make proteins.
3. Simple sugars consist of a chain of from 3 to 7 carbon atoms with hydrogen and oxygen in the ratio of 2 to 1. Glucose, the most common simple sugar, contains a 6-carbon chain as $C_6H_{12}O_6$.
4. There are three kinds of carbohydrates: monosaccharides (sugars like glucose and fructose); oligosaccharides, (disaccharides like sucrose, lactose, and maltose); and polysaccharides that contain three or more simple sugars to form starch, fiber, and glycogen.
5. Glycogenolysis refers to the process of reconverting glycogen to glucose,

whereas gluconeogenesis refers to the process of glucose synthesis, especially from protein sources.

6. Americans typically consume 40 to 50% of their total calories as carbohydrates. This is generally in the form of fruits, grains, and vegetables, although greater sugar intake in the form of sweets (simple sugars) is common and possibly harmful.

7. Carbohydrates, which are stored in limited quantity in liver and muscle, serve (1) as a major source of energy, (2) to spare the breakdown of proteins, (3) as a metabolic primer for fat metabolism, and (4) as the fuel for the central nervous system.

8. Muscle glycogen and blood glucose are the primary fuels during intense exercise. The body's glycogen stores also serve an important role in energy balance in sustained, high levels of aerobic exercise such as marathon running, distance cycling, and swimming.

9. Individuals involved in heavy training should consume about 60% of their daily calories (400 to 600 g) as carbohydrates, predominantly in complex form.

10. Sugar drinks consumed during exercise may enhance performance by maintaining blood sugar levels which can then be used by active muscles as glycogen becomes depleted. These drinks, however, have been shown to retard the movement of liquid out of the stomach, which ultimately may upset the body's fluid balance. This issue, however, has not been resolved.

P A R T

2 Fats

THE NATURE OF FATS

A molecule of fat possesses the same structural elements as the carbohydrate molecule except that the linking of the specific atoms is markedly different. Specifically, the ratio of hydrogen to oxygen is considerably higher in the fat compound. For example, the common fat **stearin** has the formula $C_{57}H_{110}O_6$

KINDS AND SOURCES OF FATS

According to common classification, fats can be placed into one of three main groups: **simple fats, compound fats,** and **derived fats.** Fats are found in both plants and animals, are generally greasy to the touch, and are insoluble in water.

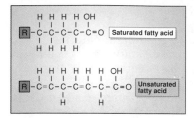

FIG. 1–6.
The major structural difference between saturated and unsaturated fatty acids is the presence or absence of double bonds between the carbon atoms. R represents the glycerol portion of the fat molecule. Although the fatty acids in the body contain between 14 to 22 carbons, the most prevalent contain 16 to 18.

SIMPLE FATS

The simple fats are often called "neutral fats" and consist primarily of **triglycerides.** Triglycerides, the most plentiful fat in the body, constitute the major storage form of fat (more than 95% of the body fat is in the form of triglycerides). A triglyceride molecule consists of two different clusters of atoms. One cluster is **glycerol,** a 3-carbon molecule. Glycerol in itself is not a fat because it is readily soluble in water. Attached to the glycerol molecule are three clusters of carbon-chained atoms termed **fatty acids.**

When glycerol and fatty acids are joined in the synthesis of the triglyceride molecule, three molecules of water are formed. Conversely, during **hydrolysis,** when the fat molecule is cleaved into its constituents, three molecules of water are added at the point where the fat molecule is split. The basic structures of the two kinds of fatty acid molecules, **saturated** and **unsaturated,** are shown in Figure 1–6. All foods with fat contain a mixture of saturated and unsaturated fats.

Saturated Fatty Acids

A **saturated** fatty acid contains only single bonds between carbon atoms; the remaining bonds attach to hydrogen. The fatty acid molecule is said to be saturated because it holds as many hydrogen atoms as is chemically possible.

Saturated fats are found primarily in animal products including beef, lamb, pork, and chicken. Saturated fats are also present in egg yolk and in the dairy fats of cream, milk, butter, and cheese. Coconut and palm oil, vegetable shortening, and hydrogenated margarine are sources of saturated fat from the plant kingdom and are present to a relatively high degree in commercially prepared cakes, pies, and cookies.

Unsaturated Fatty Acids

Fatty acids containing one or more double bonds along the main carbon chain are classified as **unsaturated.** In this case, each double bond in the carbon chain reduces the number of potential hydrogen-binding sites; therefore the molecule is said to be unsaturated with respect to hydrogen. If only one double bond is present along the main carbon chain, as with canola, olive, and peanut oil, the fatty acid is said to be **monounsaturated.** If there are two or more double bonds along the main carbon chain, as with safflower, sunflower, soybean, and corn oil, the fatty acid is said to be **polyunsaturated.**

Fats from plant sources are generally unsaturated and tend to liquify at room temperature. In general, the less firm the fat the greater the degree of unsaturation. Unsaturated fats that are present as liquids are called oils. Unsaturated oils can be changed to semisolid compounds by the chemical process of **hydrogenation.** This process reduces a double bond in the unsaturated fat to a single bond to allow more hydrogen atoms to attach to the carbon atoms in the chain. Consequently, the fat behaves as a saturated fat. The most common hydrogenated fats include lard substitutes and margarine.

Fats in The Diet

The amount of saturated fat consumed in the typical American diet has steadily increased to the point that the average person now consumes about 15% of total calories or over 50 lb of saturated fat per year, most of which is animal in origin. This is in contrast to groups like the Tarahumara Indians of Mexico whose high complex, unrefined carbohydrate diet contains only 2% of the total calories as saturated fat.[28] Coinciding with this increased consumption of saturated fats has been an increase in coronary heart disease. This relationship has led many nutritionists and medical personnel to suggest replacing at least a portion of the saturated fat in one's diet with fats that are unsaturated.[51a]

Attention has focused recently on the health benefits of ingesting two 20-carbon polyunsaturated fatty acids, eicosapentaenoic acid and docosahexenoic acid. They belong to an **omega-3** family of fatty acids found primarily in the oils of cold water fish such as tuna, herring, sardines, and mackerel. It is now apparent that regular intake of fish and fish oil has a beneficial effect on the various aspects of one's lipid profile[63a,107,132a] and overall heart disease risk.[14a,80,93] One proposed protective mechanism is that fish oil helps prevent clots from forming on artery walls.

It has become common practice to cook with and ingest fats derived primarily from vegetable sources, such as corn oil. This approach, however, may be too simplistic, because increasing evidence suggests that total fat intake, both saturated and unsaturated, may be a diabetes and heart disease "risk" and that all fat should be reduced. Concern has been expressed concerning the association of high-fat diets (both saturated and unsaturated fats) with breast and colon cancer as well as the possibility that such diets promote the growth of other cancers as well.

Reducing the fat content of the diet may also provide benefits for weight control. Due to the energy requirements of various metabolic pathways, the body is especially efficient in converting excess calories of dietary fat to stored fat.[122] Consequently, greater increases in body fat occur when the diet is high in fat content compared to an equivalent caloric excess in the form of carbohydrate. As a general recommendation for both adults and children, it is probably prudent to consume no more than 30% of the total calories in the form of fat with a concomitant increase in fruits and complex carbohydrates. Of this fat intake, about 30% or less should be saturated fat.

The saturated and unsaturated fatty acid content of various sources of dietary fat are listed in Table 1–2. Of all the dietary fats, several polyunsaturated fatty acids, most prominently **linoleic acid,** a fatty acid present in cooking and salad oils, must be consumed in the diet, because it cannot be synthesized by the body. This is an **essential fatty acid** required for ensuring the integrity of cell membranes and for growth, reproduction, skin maintenance, and general body functioning.

COMPOUND FATS

Compound fats are composed of a neutral fat in combination with other chemicals. One such group, the **phospholipids,** consists of a combination of one or more fatty acid molecules with phosphoric acid and a nitrogenous base. These fats are formed in all cells, though most are synthesized in the liver. In addition to helping maintain the structural integrity of the cell, phospholipids are important in blood clotting and in the structure of the insulating sheath around nerve fibers.

Other compound fats are the **glucolipids,** which are fatty acids bound with carbohydrate and nitrogen, and the **lipoproteins,** formed primarily in the liver from the union of either triglycerides, phospholipids, or cholesterol with protein. **The lipoproteins are important because they constitute the main form of transport for fat in the blood.** If blood lipids (Greek: lipos

TABLE 1–2.
COMMON DIETARY SOURCES OF FAT

Food	Percent Fat	Percent Saturated	Percent Unsaturated
Animal Sources			
Beef heart	6	50	50
Veal cutlet	10	50	50
Chicken	10–17	30	70
Beef	16–42	52	48
Lamb	19–29	60	40
Ham, sliced	23	45	55
Pork	32	45	55
Butter	81	55	36
Plant Sources			
Carrots	0	0	0
Potato chips	35	25	75
Cashew nuts	48	18	82
Peanut butter	50	25	75
Margarine	81	26	66
Corn oil	100	7	78
Cottonseed oil	100	21.5	71.5
Olive oil	100	14	86
Soybean oil	100	14	71.5

meaning fat) were not bound to protein or some other substance they would float to the top like cream in nonhomogenized milk.

Specifically, the **high density lipoproteins** (HDL) contain the least amount of cholesterol. **Low** and **very low density lipoproteins** (LDL and VLDL, respectively) contain the greatest fat and least protein components. The LDLs, which normally carry 60 to 80% of the total cholesterol, have the greatest affinity for the arterial wall. They help to carry cholesterol into the arterial tissue to become chemically modified and ultimately cause a proliferation of underlying smooth muscle cells and further changes that damage and narrow the artery in the process of coronary heart disease. The HDLs may operate to **protect against heart disease** in two ways: (1) to carry cholesterol away from the arterial wall for degradation to bile in the liver and subsequently excreted by the intestines, and (2) to compete with the LDL fragment for entrance into the cells of the arterial wall.

The quantity of LDL and HDL, as well as the specific ratio of these plasma lipoproteins to each other and to total cholesterol, may provide a more meaningful signal than cholesterol per se in predicting the probability of contracting coronary heart disease.[56,89] This ratio is improved with a low-calorie, low-saturated fat diet. Regular and moderate levels of aerobic exercise may also increase the HDL level and favorably affect the LDL/HDL ratio. This topic is discussed more fully in Chapter 30.

DERIVED FATS

This group of fats includes substances derived from the simple and compound fats. The most widely known of the derived fats is **cholesterol,** a sterol found only in animal tissue that contains no fatty acids but exhibits some of the physical and chemical characteristics of fat. Thus, from a dietary viewpoint, it is considered a fat. Cholesterol is present in all cells and is either consumed in foods (exogenous cholesterol) or is synthesized within the cell (endogenous cholesterol). Even when an individual maintains a "cholesterol-free diet," the rate of endogenous cholesterol synthesis may vary from 0.5 to 2.0 g per day. More can be produced, especially if the diet is high in saturated fat, which facilitates cholesterol synthesis by the liver. While the liver is the major organ for cholesterol synthesis (about 70%), other tissues—including the walls of arteries and intestines—can synthesize this derived fat. It has been determined that this rate of synthesis is sufficient for body needs; hence, a severe reduction in dietary intake of cholesterol, except in infants, probably is not harmful.

Functions of Cholesterol

Cholesterol is normally required in many complex bodily functions, including the building of cell membranes, the synthesis of vitamin D and the adrenal gland hormones as well as estrogen, androgen, and progesterone, the hormones responsible for male and female secondary sex characteristics. Cholesterol also plays an intimate role in the formation of the bile secretions that emulsify fat during digestion.

The richest source of cholesterol in foods is egg yolk. Cholesterol is also plentiful in red meats and organ meats such as liver, kidney, and brains and in shellfish, especially shrimp, as well as in dairy products such as ice cream, cream cheese, butter, and whole milk. **Cholesterol is not present in any foods of plant origin.**

Cholesterol–Heart Disease Controversy

A high level of serum cholesterol and the cholesterol-rich LDL molecule are powerful predictors of coronary artery disease.[74,90,125] In fact, the relation between serum cholesterol and death from coronary artery disease is not related to some threshold level but, instead is continuous and graded so that any lowering of this blood lipid may offer heart disease protection.[86a,126] Numerous animal studies indicate that diets high in cholesterol and saturated fat raise serum cholesterol in "susceptible" animals and eventually produce a degenerative process characterized by the formation of cholesterol-rich deposits called **plaque** on the inner lining of the medium and larger arteries. This process is termed **atherosclerosis,** and it leads to a narrowing and eventual closure of these vessels. In humans, a reduced saturated fat and cholesterol intake generally has a lowering effect on serum cholesterol,[6,58] although for most people the effect is modest.[20] Similarly, dietary mono- and polyunsaturated fatty acids may also exert a cholesterol-lowering effect.

A cause-and-effect relationship between serum cholesterol and heart disease has been shown in a controlled 7- to 10-year investigation of nearly 4,000 apparently healthy middle-aged men with elevated serum cholesterol.[87,88] In this study, lowering blood cholesterol by 25% significantly reduced the chances of a heart attack, and if an attack occurred the chances for survival were improved. With the aid of diet and the cholesterol-lowering drug cholestyramine, the reduction in blood cholesterol caused the rate of heart disease to fall by 50%. In fact, the improvement in one's coronary heart disease risk was closely related to the decrease in cholesterol by the factor of 1:2—a 1% reduction in cholesterol caused a 2% reduction in risk! These findings, which have been corroborated from other clinical trials,[23,50] are encouraging because they provide an important "missing link" in the diet-heart disease theory and support the effort of health professionals to cause people to attain and maintain reduced serum lipids through diet, exercise, and weight control.[11] Controlling blood cholesterol may also be important for children because cholesterol level in childhood is a strong predictor of that level in adulthood.[57] Although the cholesterol value considered by some to be within the acceptable range for the adult American male is about 230 to 240 mg per 100 ml of serum, many experts feel that this value is much too high, with a value below 200 mg being more desirable.[6,114]

RECOMMENDED DAILY CHOLESTEROL INTAKE

The American Heart Association has recommended the consumption of no more than 300 mg ($\frac{1}{100}$ oz) of cholesterol each day, limiting intake to no more than 100 mg per 1000 calories of food ingested. This is almost the amount of cholesterol contained in the yolk of one large egg and just about one half the cholesterol ingested by the average American male. A reduction in daily cholesterol intake towards 150 to 200 mg may be even more desirable.[134]

While it is definitely prudent to apply appropriate methods of nutrition, exercise, and weight control to achieve recommended levels of both plasma lipids and lipoproteins, the cholesterol-heart disease controversy is far from resolved. For example, there are no controlled studies of cholesterol lowering and heart disease risk in women and it is not known whether cholesterol lowering is necessary or effective for the elderly. Recommendations to the general population are generally based on inferences generated from studies of middle-aged men with high cholesterol levels. It is disturbing to note that in clinical trials of middle-aged men who were administered powerful cholesterol-lowering drugs, the significant reduction in heart disease was not accompanied by an increased longevity; this was because, for reasons unknown, the group receiving treatment had an increase in the incidence of violent and accidental deaths as well as an excess of gastrointestinal disorders.

TABLE 1–3.
CHOLESTEROL AND SATURATED FAT CONTENT OF SOME COMMON FOODS

Foods	Serving Size	Cholesterol (mg)	Saturated Fat (g)
Beef kidney	3 oz	680	3.8
Beef liver	3 oz	370	2.5
Egg	medium	275	1.7
Shrimp	3 oz	128	0.2
Beef hot dog	3 oz	75	9.9
Lean beef	3 oz	73	3.7
Ice cream	1 cup	59	8.9
Lean fish	3 oz	43	0.8
Whole milk	1 cup	33	5.1
Butter	1 tbsp	31	7.1
Chocolate bar	3 oz	18	16.3
Yogurt	1 cup	14	2.3
Skim milk	1 cup	4	0.3
Peanut butter	1 tbsp	0	1.5
Margarine	1 tbsp	0	2.1

The cholesterol and saturated fat content of some common foods is presented in Table 1–3.

FATS IN FOOD

Figure 1–7 shows the approximate percentage contribution of some of the common food groups to the total fat content of the typical American diet. The "visible" fat-containing substances (butter, lard, cooking oil, and mayonnaise) contribute 30% or more to the normal dietary intake of fat, whereas "invisible" fat in meat, eggs, milk, cheese, nuts, vegetables, and cereals contributes the remaining 70%. Vegetable fat generally contributes about 34% of the daily fat intake, whereas the remaining 66% is from animal fat. In the United States, the fat consumed in the diet represents approximately 40 to 50% of the total caloric intake or about 115 lb of fat consumed per person each year.

ROLE OF FAT IN THE BODY

The most noteworthy functions of body fat include (1) providing the body's largest store of potential energy, (2) serving as a cushion for the protection of vital organs, and (3) providing insulation from the thermal stress of cold environments.

ENERGY SOURCE AND RESERVE

Fat constitutes the ideal cellular fuel because each molecule carries large quantities of energy per unit weight, is easily transported and stored, and is readily converted into energy. At rest in well-nourished individuals, fat may provide as much as 80 to 90% of the body's energy requirement. One gram of fat contains about 9 calories of energy, more than **twice** the energy storage capability of an equal quantity of carbohydrate or protein. This is largely due to the greater quantity of hydrogen in the fat molecule compared to the carbohydrate or protein molecule. As discussed later in Chapters 6 and 7,

FIG. 1–7.
Contribution from the major food sources to the fat content of the American diet.

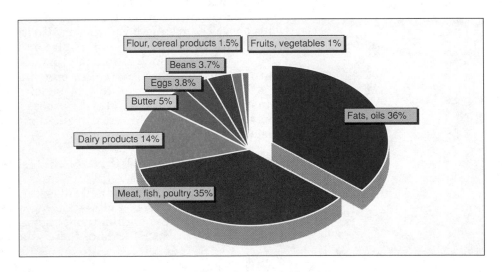

it is the oxidation of these hydrogen atoms that provides the energy required for bodily functions at rest and during exercise. It should be recalled that three molecules of water are produced and liberated when a fat molecule is synthesized from the union of glycerol and three fatty acid molecules. In contrast, when glycogen is formed in the cell from glucose, 2.7 g of water are stored with each gram of glycogen. Thus, fat is a relatively water-free, concentrated fuel, whereas glycogen is hydrated and very heavy relative to its energy content.

Fat content of the body constitutes approximately 15% of the body mass of males and 25% of females. Consequently, the potential energy stored in the fat molecules of an average college-aged male is about 100,000 calories. Most of this fat is available for energy, especially during prolonged exercise. It would be sufficient fuel to power a run from New York City to Madison, Wisconsin. Contrast this to the limited 2,000-calorie energy reserve of stored carbohydrate that could provide energy for a 20-mile run! Viewed from a different perspective, the energy reserves from the body's carbohydrate and fat could power running for about 1.6 and 119 hours, respectively.[101] Excess nutrients, other than fats, are readily converted to fat for storage. In this way, fat serves as the major storehouse of excess nutrient energy. As was the case with carbohydrates, the use of fat as a fuel "spares" protein for its important functions of tissue synthesis and repair.

PROTECTION AND INSULATION

Fat serves as a protective shield against trauma to the vital organs such as the heart, liver, kidneys, spleen, brain, and spinal cord. Up to 4% of the total body fat serves in this capacity. Body fat located in storage depots just below the skin serves an important insulating function, determining one's ability to tolerate extremes of cold exposure.[91] Swimmers who specialized in swimming the English Channel showed only a slight drop in body temperature while resting in cold water and essentially no drop in body temperature while swimming.[111] In contrast, the temperature of leaner, non-Channel swimmers dropped considerably under both conditions. The insulatory layer of fat, which in many individuals assumes a greater proportion of total body weight as age increases, is probably of little value except for those engaged in cold-related activities, such as deep-sea divers, ocean or channel swimmers, or arctic inhabitants. In fact, in most instances, excess body fat is a liability in terms of temperature regulation. This is especially apparent during sustained exercise in air when the body's heat production can be increased 20 times above the resting level. In this situation, heat flow from the body is greatly retarded by the shield of insulation from subcutaneous fat.

For some athletes such as football linemen, excess fat storage provides an additional cushion that may aid in protection from the normal hazards of the sport. This possible protective benefit, however, must be evaluated against the liability imposed by the excess "dead weight" in terms of both energy expenditure and thermal regulation and their possible detrimental effects on exercise performance.

VITAMIN CARRIER AND HUNGER DEPRESSOR

Dietary fat serves as a carrier and transport medium for the fat-soluble vitamins—vitamins A, D, E, and K—and ingesting about 20 g a day can serve this purpose. Thus, the elimination or significant reduction of fat from the diet can lead to a reduced level of these vitamins that may ultimately lead to vitamin deficiency. Dietary fat is also believed necessary for the absorption of vitamin A precursors from nonfat sources such as carrots.

Because fat emptying from the stomach takes about 3.5 hours after ingestion, some fat in the diet helps delay the onset of "hunger pangs" and contributes to the feeling of satiety after a meal. This is one reason why reducing diets containing moderate amounts of fat are considered more successful than low-fat diets.

FAT BALANCE IN EXERCISE

Fatty acids released from triglycerides in the fat storage sites and delivered in the circulation to muscle tissue as free fatty acids (FFA) bound to blood albumin, as well as the triglycerides stored in the muscle cell itself, contribute considerably to the energy requirements of exercise. During brief periods of relatively moderate exercise, energy is derived in approximately equal amounts from carbohydrate and fat. As exercise continues for an hour or more and carbohydrates become depleted, there is a gradual increase in the quantity of fat utilized for energy. In prolonged exercise, fat (mainly as FFA) may supply nearly 80% of the total energy required. This is probably brought about by a small drop in blood sugar and subsequent decrease in insulin and increase in glucagon output by the pancreas. This ultimately reduces glucose metabolism and stimulates the liberation and subsequent breakdown of fats for energy.[117]

The data in Figure 1–8 show that the uptake of fatty acids by working muscles rises during 1 to 4 hours of moderate exercise. In the first hour of exercise, about 50% of the energy was supplied by fat. As exercise continued into the third hour, fat contributed up to 70% of the total energy requirement.

SUMMARY

1. Fats, like carbohydrates, contain carbon, hydrogen, and oxygen atoms, but the ratio of hydrogen to oxygen is much higher. For example, the fat stearin has the formula $C_{57}H_{110}O_6$. Fat molecules are composed of one glycerol molecule and three fatty acid molecules.
2. Fats are synthesized by plants and animals. They can be classified into three groups: simple fats (glycerol + 3 fatty acids), compound fats composed of simple fats in combination with other chemicals (phospholipids, glucolipids, and lipoproteins), and derived fats like cholesterol, which is made from simple and compound fats.
3. Saturated fatty acids contain as many hydrogen atoms as is chemically possible; thus, the molecule is said to be saturated with respect to hydrogen. Saturated fats are present primarily in animal meat, egg yolk, dairy fats, and cheese. High intakes of saturated fats have been linked to elevated blood cholesterol and the development of coronary heart disease.
4. Unsaturated fatty acids contain fewer hydrogen atoms attached to the carbon chain. Instead, the carbon atoms are joined by double bonds, and they are said to be either mono- or polyunsaturated with respect to hydrogen. Increasing the proportion of these fats in the diet may offer protection against heart disease risk.
5. It is now apparent that lowering blood cholesterol, especially that carried by the LDL, provides significant protection against coronary heart disease.

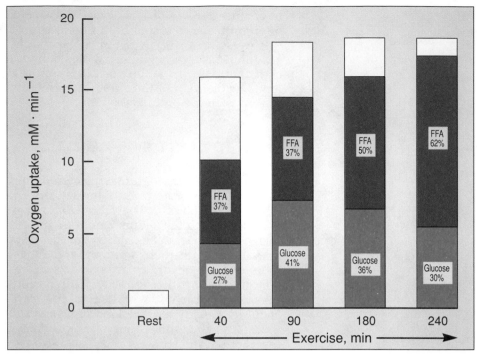

FIG. 1–8.
Uptake of oxygen and nutrients by the legs during prolonged exercise. Shaded areas represent the proportion of total oxygen uptake contributed by the oxidation of free fatty acids (FFA) and blood glucose. Open portions indicate the oxidation of non-blood-borne fuels (muscle glycogen and intramuscular fats and proteins). (From Ahlborg, G., et al.: Substrate turnover during prolonged exercise in man. J. Clin. Invest, 53:1080, 1974.)

6. Fats provide the largest nutrient store of potential energy to power biologic work. They protect vital organs and provide insulation from the cold. Fat also acts as the carrier of the fat-soluble vitamins, A, D, E, and K.
7. During light and moderate exercise, fat contributes about 50% of the energy requirement. As exercise continues, the role of stored fat becomes more important, and, during prolonged work, the fatty acid molecules may provide more than 80% of the energy needs of the body.

P A R T

3

Proteins

THE NATURE OF PROTEINS

Proteins, from the Greek word meaning "of prime importance," are like carbohydrates and fats in that they contain atoms of carbon, oxygen, and hydrogen. In addition, proteins also contain nitrogen, which makes up approximately 16% of the molecule, along with sulfur, phosphorous, and iron. Just as the carbohydrate glycogen is formed by the linkage of many simpler glucose subunits, so also is the protein molecule polymerized from its "building blocks," the **amino acids.** These amino acids are linked by peptide bonds in long chains in various forms and chemical combinations to make up the numerous protein structures.

Of the 20 different amino acids required by the body, each contains an **amino radical** and a radical called an **organic acid.** The amino radical is composed of two hydrogen atoms attached to nitrogen (NH_2), whereas the organic acid radical (technically termed a carboxyl group) is made up of one carbon atom, two oxygen atoms, and one hydrogen atom (COOH). The remainder of the amino acid molecule may take on several different forms and is often referred to as the **side chain** of the amino acid molecule.

Two different amino acids, **alanine** and **leucine,** are shown in Figure 1–9. Each protein contains the basic amino and organic acid group with different structural side chains. **It is the specific structure of this side chain that gives an amino acid its particular characteristics.**

Because there are so many ways the 20 amino acids can combine to form a particular protein, there is almost an infinite number of possible proteins depending on the combination of amino acids. For example, if we consider only proteins formed from the linkage of three different amino acids, there could be 20^3 or 8000 different proteins! With few exceptions, the proteins in the body are composed of numerous linkages of amino acids.

KINDS OF PROTEIN

Eight amino acids (9 in children and stressed older adults) cannot be synthesized in the body and therefore must be provided preformed in foods. These are called **essential amino acids.** One may see them listed on food supplement labels; they are isoleucine, leucine, lysine, methionine, phenylalanine, threonine, tryptophan, and valine; in addition, cystine and tyrosine are synthesized in the body from methionine and phenylalanine, respectively. Furthermore, infants cannot synthesize histidine. The remaining nine amino acids that can be manufactured within the body are termed **nonessential.** This does not mean that they are unimportant, but simply that they can be synthesized in the body from compounds ordinarily available and at a rate that meets the demands for normal growth.

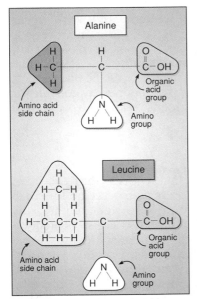

FIG. 1–9.
Two different amino acids, alanine and leucine. Each protein contains the basic amino and organic acid group with different structural side chains.

Proteins that contain the essential amino acids can be found in the cells of both animals and plants. There is nothing "better" about a specific amino acid from an animal compared to the same amino acid of vegetable origin. Plants make their protein by incorporating nitrogen contained in the soil to synthesize amino acids. Carbon, oxygen, and hydrogen are available from the air and water. In contrast, animals do not have a broad capability for protein synthesis and thus must derive their protein from ingested sources.

The appropriate amino acids must be available at the time of protein synthesis for a specific protein to be constructed. Protein nutrients that contain all of the essential amino acids in terms of quantity and in the correct ratio to maintain nitrogen balance and allow for tissue growth and repair are known as **complete proteins** (higher quality). An **incomplete protein** (lower quality) lacks one or more of the essential amino acids. Diets containing predominantly incomplete protein may eventually result in protein malnutrition, even though they are adequate in caloric value and protein quantity.

Protein Sources

Sources of complete protein are eggs, milk, meat, fish, and poultry. The mixture of essential amino acids present in eggs has been judged to be the best among food sources; hence, eggs are given the highest quality rating of 100 for comparison with other foods. Some common sources of dietary protein in the diet are rated in Table 1–4. Presently, almost two thirds of dietary protein comes from animal sources, whereas 80 years ago protein came equally from both the plant and animal kingdom. This present-day reliance on animal protein is largely responsible for the relatively high intake of cholesterol and saturated fat.

The "biologic value" of food refers to the completeness with which the food supplies essential amino acids. Foods of high-quality protein are largely of animal origin whereas vegetable proteins (lentils, dried beans and peas, nuts, and cereals) are incomplete in terms of one or more essential amino acids and thus have a relatively lower biologic value. It should be understood, however, that **all** of the essential amino acids can be obtained by consuming a **variety** (grains, fruits, and vegetables) of plant foods, each with a different quality and quantity of amino acids.

The Vegetarian Approach

True vegetarians or **vegans** obtain all of their nutrients from the plant kingdom and in the form of dietary supplements. Vegans make up less than 1% of the population of the United States. There are, however, an increasing number of competitive and champion athletes whose diet consists predominately of nutrients from varied plant sources as well as some dairy products.[102,123] In fact, two thirds of the people in the world are adequately nourished on essentially vegetarian diets using only small amounts of animal protein. Well-balanced vegetarian and vegetarian-type diets provide relatively large amounts of carbohydrate often crucial to the endurance athlete and others involved in heavy training. Except for the nutrients calcium, phosphorus, iron and vitamin B_{12} (produced by bacteria in the digestive tract of animals), a strict vegetarian's nutritional problem is one of getting ample high quality proteins. This is easily resolved with a **lactovegetarian** diet that allows the addition of milk and related products such as ice cream, cheese, and yogurt. The lactovegetarian approach minimizes the problem of getting

**TABLE 1–4.
RATING OF COMMON
SOURCES OF DIETARY
PROTEIN**

Food	Protein Rating
Eggs	100
Fish	70
Lean beef	69
Cow's milk	60
Brown rice	57
White rice	56
Soybeans	47
Brewer's hash	45
Whole-grain wheat	44
Peanuts	43
Dry beans	34
White potato	34

FIG. 1–10.
Contribution from the major food sources to the protein content of the American diet.

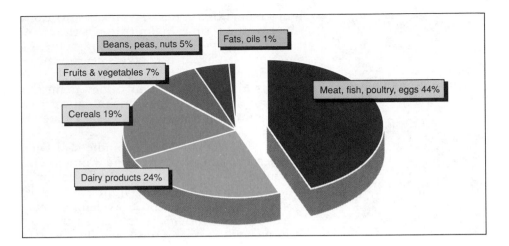

Beans, peas, nuts 5%

Fats, oils 1%

Fruits & vegetables 7%

Cereals 19%

Meat, fish, poultry, eggs 44%

Dairy products 24%

sufficient high quality protein and increases the intake of calcium, phosphorus, and vitamin B_{12}. By adding an egg to the diet (**ovolactovegetarian diet**), an intake of high-quality protein is assured.

The contribution of various food groups to the protein content of the American diet is shown in Figure 1–10. By far, the greatest intake of protein comes from animal sources, whereas only about 30% comes from vegetable sources.

RECOMMENDED INTAKE OF PROTEINS

Despite the beliefs of many coaches and trainers,[12] there is probably no benefit from eating excessive amounts of protein. In fact, increasing the protein intake by more than three times the recommended level had no effect in enhancing work capacity during intensive training.[29] For athletes, muscle mass is **not** increased simply by eating high-protein foods. Additional calories in the form of protein are, after deamination, used directly for energy or converted to fat and stored in the subcutaneous depots. In fact, excessive protein may be harmful because the metabolism of large quantities of this nutrient may place a strain on liver and kidney function.[7]

Many Americans consume more than twice the protein requirement. For both endurance and resistance trained athletes, many of whom consume considerable quantities of food, the diet may contain more than three times the required protein.[99,121]

The RDA: A Liberal Standard

The RDA for protein as well as for the various vitamins and minerals required by the body are standards for nutrient intake expressed as a daily average developed by the Food and Nutrition Board of the National Research Council/National Academy of Science and revised ten times since 1943. RDA levels are believed to represent a liberal yet safe level of excess to meet the nutritional needs of practically all **healthy** people. It is important to understand that the RDA reflects nutritional needs of a population over a long period of time. Malnutrition is the cumulative result of weeks, months, and even years of reduced nutrient intake. Also, an individual who regularly consumes a diet that contains certain nutrients below the RDA standards may not necessarily be malnourished. The RDA is best viewed as a probability statement for adequate nutrition: as nutrient intake falls below the RDA the probability for malnourishment for that person is increased, and this probability becomes progressively greater as the nutrient intake becomes lower.

Table 1–5 shows the Recommended Dietary Allowances for protein for adolescent and adult men and women. On the average, a daily intake of about 0.8 g of protein per kg body weight is recommended. (To determine your protein requirement, multiply your body weight in pounds by 0.37) This recommendation holds even for people who are overweight. Generally, the protein requirement as well as the quantity of the required essential amino acids decreases with age. On the other hand, for infants and growing children, the daily recommended intake of protein amounts to 2.0 to 4.0 g per kg body weight, whereas pregnant women and nursing mothers should increase their protein intake by 20 and 10 g, respectively. Likewise, stress, disease, and injury increase the protein requirement.

Although little experimental evidence supports the practice of protein supplementation, it is possible that growing athletes, athletes involved in strength development programs that enhance muscle tissue growth and endurance programs that increase protein breakdown, and those subjected to recurring trauma may need a slightly larger protein intake.[92,96,128] This requirement, however, is more than likely met by the generally increased food intake of these athletes to compensate for the increased energy expenditure associated with training.

Preparations of Simple Amino Acids

The common practice among some weight lifters, body builders, and other power athletes of consuming liquids, powders, or pills of predigested protein is a waste of money and may actually be counterproductive in terms of the desired outcome. For example, many of these preparations contain proteins

TABLE 1–5.
RECOMMENDED DIETARY ALLOWANCES OF PROTEIN FOR ADOLESCENT AND ADULT MEN AND WOMEN

Recommended Amount	Men		Women	
	Adolescent	Adult	Adolescent	Adult
Grams of protein per kg body weight	0.9	0.8	0.9	0.8
Grams per day based on average weight[a]	59	56	50	44

[a] Average weight is based on a "reference" man and woman.[70] For adolescents (age 14–18), average weight is approximately 65.8 kg (145 lb) for males and 55.7 kg (123 lb) for females. For adult men, average weight is 70 kg (154 lb). For adult women, average weight is 55 kg (120 lb).

that are predigested to simple amino acids through chemical action in the laboratory. The belief is that the simple amino-acid molecule is absorbed more easily and rapidly by the body and in some magical way is available rapidly to facilitate the expected muscle growth brought on by training. This, however, is not the case. Dietary proteins are absorbed rapidly in the intestines by the healthy body when they are part of the more complex di- and tripeptide molecules compared to the simple amino-acid molecule. The intestinal tract is better able to handle protein in its more complex form, whereas a concentrated amino-acid solution draws water into the intestines. This process can cause irritation, cramping, and diarrhea. Simply stated, **amino acid supplementation in any form above the RDA has not been shown experimentally to improve strength, power, muscle mass, or endurance.**

ROLE OF PROTEIN IN THE BODY

Protein makes up about 12 to 15% of the body mass. There is, however, considerable variability in the protein content of different cells. A brain cell is only about 10% protein. Red blood cells and muscle cells, on the other hand, may contain as much as 20% of their total weight as protein. The protein content of skeletal muscle, which represents about 65% of the body's total protein, can be increased dramatically with the systematic application of resistance training exercises.[53]

Amino acids provide the major substance for the synthesis of cellular components as well as new tissue. The process of building tissue is termed **anabolism,** and the amino acid requirement for anabolic processes can vary considerably. During periods of rapid growth, as occurs in infancy and childhood, over one third of the protein intake is retained for tissue anabolism. As the growth rate declines, so does the percentage of protein retained for growth-related processes. Once an optimal body size is attained and growth stabilizes, there is still a continuous turnover of tissue protein.

Proteins are present in all cells and are the primary constituents that make up cell membranes as well as internal cellular material. The proteins found in the nuclei of the cell (nucleoproteins) transmit hereditary characteristics and are responsible for continued protein synthesis within the cell. The hair, skin, nails, tendons, and ligaments are special forms of **structural** proteins. The **globular** proteins make up the nearly 2000 different enzymes that catalyze or speed up chemical reactions. These compounds are critical in regulating the breakdown of fats, carbohydrates, and proteins to release energy. Blood plasma also contains specialized proteins. For example, the plasma proteins thrombin, fibrin, and fibrinogen are intimately involved in blood clotting. Within the red blood cell, the oxygen-carrying compound hemoglobin contains the large protein molecule globin.

Proteins also play an important role in regulating the acid-base quality of the body fluids. This buffering function is important during vigorous exercise when large quantities of acid metabolites are formed. Proteins are essential for muscle contraction; actin and myosin are the structural proteins that "slide" past each other as the muscle shortens during movement. Amino acids are essential building blocks of certain hormones and are needed for the activation of selected vitamins that play a key role in metabolic and physiologic regulation.

DYNAMICS OF PROTEIN METABOLISM

Although the main function of dietary protein resides in its contribution of amino acids to various anabolic processes, protein is also broken down or catabolized for energy. In reasonably well-nourished individuals at rest, the breakdown of protein contributes between 2 to 5% of the body's total energy requirement. During this **catabolism,** protein must first be degraded into its amino acid components. Nitrogen is then stripped from the amino acid molecule in the process of deamination in the liver and excreted from the body as **urea.** In muscle, enzymes are available that facilitate the removal of nitrogen from certain amino acids and then subsequently pass the nitrogen to other compounds in the process of **transamination.** In both deamination and transamination, the resulting carbon skeleton of the non-nitrogenous amino acid residue can then be further degraded during energy metabolism. When the intake of nitrogen (protein) equals nitrogen excretion, a **nitrogen balance** exists. If the body is in **positive nitrogen balance,** where nitrogen intake is greater than nitrogen excretion, then new tissue is being synthesized and protein is retained. This circumstance is often observed in children, during pregnancy, in recovery from illness, and as a result of intensive resistance training when protein components of the muscle cell are being synthesized.

It is unlikely whether the body can develop a protein reserve, as is the case with fat in adipose tissue and to some extent carbohydrate. Nevertheless, individuals fed a diet with adequate protein have a higher content of muscle and liver protein than individuals fed a low-protein diet. Also, by use of radioactive protein (injecting protein that has one or several of its carbon atoms "tagged"), it has been shown that certain proteins are more easily recruited for energy metabolism, whereas others are relatively "fixed" as cellular constituents and cannot be used without tissue damage.[45]

The protein in nervous and connective tissue is essentially fixed whereas muscle and liver protein can be altered and used for energy. This fact helps explain the rapid muscle shrinkage or **atrophy** during periods of inactivity, and the loss in lean tissue by persons on reducing diets, especially if the diet is extremely low in carbohydrate or protein or the person remains sedentary during the weight loss period.[77]

A greater output of nitrogen relative to its intake indicates the use of protein for energy and a possible encroachment on the body's available amino acids, primarily those in skeletal muscle. Such a **negative nitrogen balance** can exist at levels of protein intake above the standards established as the minimum requirement. This could occur if the body catabolizes protein because of a lack of other energy nutrients. For example, an individual may consume adequate protein but too little carbohydrate and fat. Consequently, protein is used as a primary energy fuel, the result being a negative protein or nitrogen balance. The protein sparing role of dietary fat and carbohydrate discussed previously is especially important during periods of growth and high-energy output, such as occurs in intensive training. Also, in starvation, the greatest negative nitrogen balance is observed. As a result, **starvation diets or diets with reduced carbohydrate result not only in the depletion of glycogen reserves, but also in a possible protein deficiency and accompanying loss of muscle tissue.**

PROTEIN BALANCE IN EXERCISE AND TRAINING: IS THE RDA REALLY ENOUGH?

Nutritionists and exercise physiologists have long maintained that the RDA for protein represents a standard liberal enough to provide for any amino acid molecules catabolized for energy during exercise or required for the augmented protein synthesis following exercise.[10,54,84] The contention that protein is used only to a limited extent as an energy fuel has generally been based on two observations: (1) protein's primary role is to provide the amino-acid building blocks for tissue synthesis, and (2) the findings of early studies that concluded only minimal protein breakdown during endurance exercise as reflected by the quantity of urinary nitrogen in the immediate 24-hour recovery period. In addition, theoretical computations of the protein required to sustain tissue synthesis brought about by resistance training,[42] as well as some experimental evidence,[17,19] support the position that the RDA for protein provides a "margin of safety" for both anabolic and catabolic requirements of exercise and training. **This of course is predicated on the basis that energy intake is adequate for the added needs of exercise.** If energy intake is inadequate during heavy training, even an augmented protein intake may be insufficient to maintain nitrogen balance. To this extent, dieting could negatively affect training regimens geared to increase muscle mass or maintain a high level of strength and power.[132]

A review on the topic of protein balance in exercise,[83,85] however, presents a compelling argument that protein is used as a significant energy substrate, particularly during exercise of long-duration and into recovery. These conclusions were based on studies that expand the classic method of determining protein breakdown through urea excretion as well as using radioactive tracers that "label" the CO_2 portion of the amino-acid molecule. For example, the output of labeled CO_2 from amino acids injected or ingested increased during exercise in a manner proportional to the metabolic rate.[133] It is also now apparent that as exercise progresses the concentration of plasma urea increases. This increase is coupled with a dramatic rise in nitrogen excretion in sweat without any change in urinary-nitrogen excretion.[63,84,112] These observations account for earlier conclusions of minimal protein breakdown during endurance exercise because the early studies only measured nitrogen in the urine. As shown in Figure 1–11, the sweat mechanism clearly is an important means for excreting the nitrogen from protein breakdown during exercise. Furthermore, urea production may not accurately reflect all aspects of protein breakdown. The rate of oxidation of both plasma and intracellular leucine, an essential amino acid, increased significantly during moderate exercise in humans, regardless of a change in urea production.[136]

Figure 1–11 also shows that the use of protein for energy in exercise was greatest when subjects exercised in the glycogen-depleted state. This emphasizes the important role of carbohydrate as a protein sparer[39] and suggests that the demand on protein "reserves" in exercise is linked to carbohydrate availability. Certainly this would become an important factor in endurance exercise or in frequent heavy training in which glycogen reserves become greatly reduced.[29,128,135] In fact, when the process for gluconeogenesis is chemically blocked, hypoglycemia results and endurance performance is greatly impaired.[69]

The pattern of protein catabolism during exercise and the fact that the greatest protein breakdown occurs when the glycogen reserves are low in-

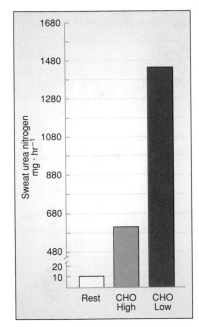

FIG. 1–11.
Excretion of urea in sweat at rest, and during exercise after carbohydrate loading (CHO High) and carbohydrate depletion (CHO Low). (From Lemon, P.W.R., and Nagle, F.J.: Effects of exercise on protein and amino acid metabolism. Med. Sci. Sports Exerc., *13*:141, 1981.)

dicates that the metabolic mixture in endurance exercise eventually reflects that observed during short-term starvation. As in endurance exercise, glucose output from the liver is maintained largely by the process of gluconeogenesis. More than likely this reflects the body's attempt to provide glucose for central nervous system functioning. **These observations would certainly support the importance of a high-carbohydrate diet as a means to conserve muscle protein for athletes who engage in protracted and hard training.** The use of protein for energy and the depression of protein synthesis in heavy exercise in such cases may help to explain why individuals involved in resistance training to augment muscle size generally refrain from significant training involving endurance-type exercise.

The beginning phase of an exercise training program also places an increased demand on body protein due perhaps to both muscle injury and metabolic requirements.[83,110] Whether this effect is transient or reflects a true long-term increase in the protein requirement above that provided by the RDA is an area of current controversy.[18,51,128]

The Alanine-Glucose Cycle.

Certain proteins in the body are not readily available for energy. Proteins in muscle, however, are more labile, and when the demand arises, can enter the process of energy metabolism.[25a,46,55] Figure 1–12 shows the influence of exercise on the release of the amino acid alanine from the leg muscles. Clearly, an increased alanine output is related to the severity of the exercise; as exercise increases in intensity, alanine output correspondingly increases.

A model has been proposed that alanine **indirectly** serves the energy requirements of exercise.[45] Amino acids within the muscle are converted to glutamate and then to alanine. The alanine released from the exercising muscle is transported to the liver where it is deaminated. The remaining

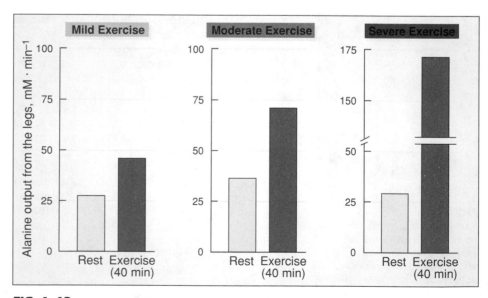

FIG. 1–12.
Influence of 40 minutes of exercise at various work intensities on estimated alanine release from the leg muscles. As the severity of exercise increases, there is a corresponding increase in alanine release. (From Felig, P., and Wahren, J.: Amino acid metabolism in exercising man. J. Clin. Invest. *50:*2703, 1971.)

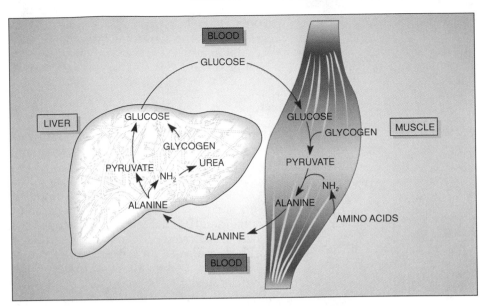

FIG. 1–13.
The alanine-glucose cycle in exercise. Alanine, which is synthesized in muscle from glucose-derived pyruvic acid, is released to the blood and converted to glucose and urea in the liver. This glucose is then released into the blood and delivered to the muscle to serve the energy needs of the cell. During exercise, the increased production and output of alanine from muscle helps supply glucose fuel molecules. (From Felig, P., and Wahren, J.: Amino acid metabolism in exercising man. J. Clin. Invest., *50*:2703, 1971.)

carbon skeleton is converted to glucose (gluconeogenesis) and then released to the blood and delivered to the working muscles. The carbon fragments from the amino acids that form alanine can then be oxidized for energy within the specific muscle cell. The sequence of the **alanine-glucose cycle** is summarized in Figure 1–13. After 4 hours of continuous light exercise, the liver's output of alanine-derived glucose can account for 45% of the total glucose released from the liver; with more intense exercise gluconeogenic precursors could account for 60% of the glucose output.[2] **Energy derived from the alanine-glucose cycle may supply as much as 10 to 15% of the total exercise requirement.** Furthermore, nonprotein precursors, namely lactate, glycerol, and pyruvate, also serve the process of gluconeogenesis.

In summary, protein breakdown above the resting level occurs during both endurance and resistance training exercise to a degree greater than previously thought.[18,41,83,92] This is especially true when carbohydrate and energy stores are low. Such observations support the importance of a high intake of carbohydrate and adequate calories for individuals involved in heavy training, as adequate carbohydrate aids in conserving muscle protein. Little is known concerning the actual protein requirements of individuals who train 4 to 6 hours a day by resistive-type exercise to develop muscular size, strength, and power. Recent evidence suggests that their requirement may be only slightly greater than for sedentary individuals.[128] It is also possible that despite an increased utilization of protein for energy during heavy training adaptations occur that augment the efficiency with which the body utilizes dietary protein to enhance amino acid balance.[5,15] Whether there will be a modification of the current protein RDA for specific athletic groups awaits further research. Until such data become available, it seems reasonable to acknowledge the heretofore unrecognized significant role of protein as a potential energy fuel.

SUMMARY

1. Proteins differ chemically from fats and carbohydrates in that they contain nitrogen in addition to other elements such as sulfur, phosphorous, and iron.
2. Proteins are formed from subunits called amino acids. The body requires 20 different amino acids, each containing an amino radical (NH_2) and an organic acid radical called a carboxyl group (COOH). In addition to NH_2 and COOH, amino acids contain a side-chain molecule that gives the amino acid its particular chemical characteristics.
3. There is almost an infinite number of combinations for the 20 different amino acids; thus, there is almost an infinite number of possible protein structures.
4. Eight of the 20 amino acids cannot be synthesized in the body. These are the essential amino acids and must be consumed in the diet.
5. Proteins are found in the cells of all animals and plants. Proteins containing all the essential amino acids are called complete (high-quality) proteins; the others are called incomplete (low-quality) proteins. Animal proteins such as those found in eggs, milk, cheese, meat, fish, and poultry are examples of high-quality, complete proteins.
6. All of the essential amino acids can be obtained by consuming a variety of plant foods, each with a different quality and quantity of amino acids.
7. Proteins provide the building blocks for the synthesis of essentially all cellular material (anabolism). Under certain conditions, the amino acids also contribute their "carbon skeletons" for energy metabolism (catabolism).
8. Certain proteins, especially those in nervous and connective tissue, are not generally sacrificed in energy metabolism. The amino acids alanine and glutamic acid, however, play a key role in providing carbohydrate fuel for exercise, especially prolonged, submaximal exercise. This is achieved through the process of gluconeogenesis. In prolonged exercise, the alanine-glucose cycle may account for as much as 60% of the total glucose released by the liver.
9. Protein breakdown during exercise becomes apparent when the body's carbohydrate reserves are low. Such findings further support the wisdom of maintaining optimal glycogen stores during training.
10. Future research must determine whether the increased demand on body protein with training is sufficient to increase the protein requirement above that provided by the RDA for individuals who maintain adequate energy intake.

REFERENCES

1. Acheson, K.J., et al.: Glycogen storage capacity and de novo lipogenesis during massive carbohydrate overfeeding in man. *Am. J. Clin. Nutr.*, 48:240, 1988.
2. Ahlborg, G., et al.: Substrate turnover during prolonged exercise in man. *J. Clin. Invest.*, 53:1080, 1974.

3. Ahlborg, G., and Felig, P.: Influence of glucose ingestion on the fuel-hormone response during prolonged exercise. *J. Appl. Physiol.*, 41:683, 1976.

4. Ahlborg, G., and Felig, P.: Lactate and glucose exchange across the forearms, legs, and splanchnic bed during and after prolonged leg exercise. *J. Clin. Invest.*, 69:45, 1982.

5. Albert, J.K., et al.: Exercise-mediated tissue and whole body amino acid metabolism during intravenous feedings in normal men. *Clin. Sci.*, 77:113, 1989.

6. American Heart Association Steering Committee for Medical and Community Programs: Risk Factors and Coronary Disease. *Circulation*, 62:449A, 1980.

7. Anderson, C.F., et al.: Nutritional therapy for adults with renal disease. *JAMA*, 223:68, 1973.

8. Anderson, J.W., et al.: Hypolipidemic effects of high-carbohydrate, high-fiber diets. *Metabolism*, 29:551, 1980.

9. Anderson, J.W., and Chen, W.L.: Plant fiber: carbohydrate and lipid metabolism. *Am. J. Clin. Nutr.*, 32:346, 1979.

10. Åstrand, P.O., and Rodahl, K.: *Textbook of Work Physiology*. New York, McGraw-Hill, 1986.

11. Barnard, R.J., et al.: Effects of an intensive exercise and nutrition program on patients with coronary artery disease: five-year follow up. *J. Cardiac Rehab.*, 3:183, 1983.

12. Bentivegna, A., et al.: Diet, fitness, and athletic performance. *Phys. Sportsmed.*, 7:99, 1979.

13. Bergstrom, J., et al.: Diet, muscle glycogen and physical performance. *Acta Physiol. Scand.*, 71:140, 1967.

14. Bergstrom, J., and Hultman, E.: Nutrition for maximal sports performance. *JAMA*, 221:999, 1972.

14a. Bønaa, K.H., et al.: Effect of eicosapentaenoic and docosahexaenoic acids on blood pressure in hypertension. *N. Engl. J. Med.*, 322:795, 1990.

15. Brooks, G.A.: Amino acid and protein metabolism during exercise and recovery. *Med. Sci. Sports Exerc.*, 19:S150, 1987.

16. Burkitt, D.: Dietary Fiber. In *Medical Applications of Clinical Nutrition*. New Canaan, CT, Keats, 1983.

17. Butterfield, G.E.: Whole body protein utilization in humans. *Med. Sci. Sports Exerc.*, 19:S157, 1987.

18. Butterfield, G., and Calloway, D.H.: Physical activity improves protein utilization in young men. *Br. J. Nutr.*, 51:171, 1984.

19. Butterfield-Hodgden, G., and Calloway, D.H.: Protein utilization in men under two conditions of energy balance and work. *Fed. Proc.*, 39:377, 1977.

20. Caggiula, A.W., et al.: The multiple risk factor intervention trial (Mr. Fit):IV. Intervention on blood lipids. *Prev. Med.*, 10:443, 1981.

21. Cahill, G.E., Jr., and Aoki, T.T.: Partial and total starvation. In *Assessment of Energy Metabolism in Health and Disease*. Columbus, OH, Ross Laboratories, 1980.

22. Candas, V., et al.: Hydration during exercise: effects on thermal and cardiovascular adjustments. *Eur. J. Appl. Physiol.*, 55:113, 1986.

23. Canner, P.L., et al.: Fifteen year mortality in Coronary Drug Project patients: long-term benefit with niacin. *J. Am. Coll. Cardiol.*, 8:1245, 1986.

24. Carter, J.E., and Gisolfi, C.V.: Fluid replacement during and after exercise in the heat. *Med. Sci. Sports Exerc.*, 21:532, 1989.

25. Cassidy, M., et al.: Morphological aspects of dietary fibers in the intestines. In *Advances in Lipid Research*. Edited by R. Paoletti and D. Kaitchevsky. New York, Academic Press, 1982.

25a. Christensen, H.N.: Role of amino acid transport and counter transport in nutrition and metabolism. *Physiol. Rev.*, 70:43, 1990.

26. Coggan, A.R., and Coyle, E.F.: Effect of carbohydrate feedings during high-intensity exercise. *J. Appl. Physiol.*, 65:1703, 1988.

27. Coggan, A.R., and Coyle, E.F.: Metabolism and performance following carbohydrate ingestion late in exercise. *Med. Sci. Sports Exerc.*, 21:59, 1989.

28. Conner, W.E., et al.: The plasma lipids, lipoproteins, and diet of the Tarahumara Indians of Mexico. *Am. J. Clin. Nutr.*, 31:1131, 1978.

29. Consolazio, C.F., et al.: Protein metabolism during intensive physical training in the young adult. *Am. J. Clin. Nutr.*, 28:29, 1975.

30. Costill, D.L., et al.: Muscle glycogen utilization during prolonged exercise on successive days. *J. Appl. Physiol.*, 31:834, 1971.
31. Costill, D.L., et al.: Effects of elevated plasma FFA and insulin on muscle glycogen usage during exercise. *J. Appl. Physiol.*, 43:695, 1977.
32. Costill, D.L., et al.: Effects of repeated days of intensified training on muscle glycogen and swimming performance. *Med. Sci. Sports Exerc.*, 20:249, 1988.
33. Costill, D.L., and Miller, J.M.: Nutrition for endurance sport: carbohydrate and fluid balance. *Int. J. Sports Med.*, 306:895, 1982.
34. Costill, D.L., and Saltin, B.: Factors limiting gastric emptying during rest and exercise. *J. Appl. Physiol.*, 36:679, 1974.
35. Coyle, E.F., et al.: Carbohydrate feeding during prolonged strenuous exercise can delay fatigue. *J. Appl., Physiol.*, 55:230, 1983.
36. Coyle, E.F., et al.: Muscle glycogen utilization during prolonged strenuous exercise when fed carbohydrate. *J. Appl. Physiol.*, 61:165, 1986.
37. Coyle, E.F., and Coggan, A.R.: Effectiveness of carbohydrate feeding in delaying fatigue during prolonged exercise. *Sports Med.*, 1:446, 1984.
38. Crapo, P.A.: Sugar and sugar alcohols. *Contemp. Nutr.*, 7 (12):1981.
39. Davies, C.T.M., et al.: Glucose inhibits CO_2 production from leucine during whole-body exercise in man. *J. Physiol.*, 332:40, 1982.
40. Deutsch, R.M.: *Realities of Nutrition.* Palo Alto, CA, Bull Publishing, 1976.
41. Dohm, G.L., et al.: Time course of changes in gluconeogenic enzyme activities during exercise and recovery. *Am. J. Physiol.*, 249:E6, 1985.
42. Durnin, J.V.G.A.: Protein requirements and physical activity. In *Nutrition, Physical Fitness and Health.* Edited by J. Parizkova and V.A. Rogozkin. Baltimore, MD, University Park Press, 1978.
43. Eastwood, M.A., and Kay, R.M.: An hypothesis for the action of dietary fiber along the gastrointestinal tract. *Am. J. Clin. Nutr.*, 32:364, 1979.
44. Felig, P., et al.: Hypoglycemia during prolonged exercise in normal men. *N. Engl. J. Med.*, 306:895, 1982.
45. Felig, P., and Wahren, J.: Amino acid metabolism in exercising man. *J. Clin. Invest.*, 50:2703, 1971.
46. Felig, P., and Wahren, J.: Fuel homeostasis in exercise. *N. Engl. J. Med.*, 293:1078, 1975.
47. Fielding, R.A., et al.: Effect of carbohydrate feeding frequencies and dosage on muscle glycogen use during exercise. *Med. Sci. Sports Exerc.*, 17:472, 1985.
48. Foster, C., et al.: Effects of pre-exercise feedings on endurance performance. *Med. Sci. Sports*, 11:1, 1979.
49. Foster, C., et al.: Gastric emptying characteristics of glucose polymer solutions. *Res. Quart.*, 51:299, 1980.
50. Frick, M.H., et al.: Helsinki Heart Study: primary-prevention trial with gemfibrozil in middle-aged men with dyslipidemia. Safety of treatment, changes in risk factors, and incidence of coronary heart disease. *N. Engl. J. Med.*, 317:3217, 1987.
51. Friedman, J.E., and Lemon, P.W.R.: Effect of chronic endurance exercise on retention of dietary protein. *Int. J. Sports Med.*, 10:118, 1989.
51a. Ginsberg, H.N.: Reduction of plasma cholesterol levels in normal men on an American Heart Association Step 1 diet or a Step 2 diet with added monounsaturated fat. *N. Engl. J. Med.*, 322:574, 1990.
52. Gleeson, M., et al.: Comparision of the effects of pre-exercise feedings of glucose, glycerol, and placebo on endurance and fuel homeostasis in man. *Eur. J. Appl. Physiol.*, 55:645, 1986.
53. Goldberg, A.L., et al.: Mechanism of work-induced hypertrophy of skeletal muscle. *Med. Sci. Sports*, 7:185, 1975.
54. Goldspink, D.F.: The influence of activity on muscle size and protein turnover. *J. Physiol.* (London), 264:283, 1977.
55. Goodman, M.N., and Ruderman, N.B.: Influence of muscle use on amino acid metabolism. In *Exercise and Sport Science Review*, Vol. 10. Edited by R.L. Terjung. Philadelphia, Franklin Institute, 1982.
56. Gordon, D.J., et al.: High-density lipoprotein cholesterol and coronary heart disease in hypercholesterolemic men: The Lipid Research Clinics Coronary Primary Prevention Trial. *Circulation*, 74:12176, 1986.

57. Grundy, S.M., et al.: Comparison of monounsaturated fatty acids and carbohydrates for reducing raised levels of plasma cholesterol in man. *Am. J. Clin. Nutr.*, 47:965, 1988.
58. Hall, J., et al.: Effects of diet and exercise on peripheral vascular disease. *Phys. Sportsmed.*, 10:90, 1982.
59. Hargreaves, M., et al.: Effect of carbohydrate feedings on muscle glycogen utilization and exercise performance. *Med. Sci. Sports Exerc.*, 16:219, 1984.
60. Hargreaves, M., et al.: Effect of fructose ingestion on muscle glycogen usage during exercise. *Med. Sci. Sports Exerc.*, 17:360, 1985.
61. Hargreaves, M., et al. Effect of pre-exercise carbohydrate feedings of endurance cycling performance. *Med. Sci. Sports Exerc.*, 19:33, 1987.
62. Hargreaves, M., and Briggs, C.A.: Effect of carbohydrate ingestion on exercise metabolism. *J. Appl. Physiol.*, 65:1553, 1988.
63. Harlambie, G., and Sensor, L.: Metabolic Changes in man during long-distance swimming. *Eur. J. Appl. Physiol.*, 43:115, 1980.
63a. Harris, W.S., et al.: Fish oils in hypertriglyceridemia: a dose response study. *Am. J. Clin. Nutr.*, 51:399, 1990.
64. Heigenhauser, G.J.F., et al.: Effect of glycogen depletion on the ventilatory response to exercise. *J. Appl. Physiol.*, 54:470, 1983.
65. Hultman, E.: Liver as a glucose supplying source during rest and exercise, with special reference to diet. In *Nutrition, Physical Fitness, and Health*. Edited by J. Parizkova and V.A. Rogozkin. Baltimore, MD, University Park Press, 1978.
66. Ivy, J.L.: Endurance improved by ingestion of glucose polymer supplement. *Med. Sci. Sports Exerc.*, 11:466, 1983.
67. Jandrain, B., et al.: Metabolic availability of glucose ingested three hours before prolonged exercise in humans. *J. Appl. Physiol.*, 56:1314, 1984.
68. Jansson, E., and Kaijser, L.: Effect of diet on the utilization of blood-borne and intramuscular substrates during exercise in man. *Acta Physiol. Scand.*, 115:19, 1982.
69. John-Adler, H.B., et al.: Reduced running endurance in gluconeogenesis-inhibited rats. *Am. J. Physiol.*, 251, 1986.
70. Katch, F.I., and McArdle, W.D.: *Nutrition, Weight Control, and Exercise*. 3rd ed., Philadelphia, Lea & Febiger, 1988.
71. Kay, R.M.: Dietary fiber. *J. Lipid Res.*, 23:221, 1982.
72. Keis, C.: Edible fiber: practical problems. *Contemp. Nutr.*, 7(2), 1982.
73. Keller, K., and Schwarzkopf, R.: Pre-exercise snacks may decrease exercise performance. *Phys. Sportsmed.*, 12:89, 1984.
74. Keys, A.: Coronary heart disease—the global picture. *Atherosclerosis*, 22:149, 1975.
75. Kirby, R.W., et al.: Oat-bran intake selectively lowers serum low-density liprotein cholesterol concentrations of hypercholesterolemic men. *Am. J. Clin. Nutr.*, 34:824, 1981.
76. Koeslag, J.H.: Post-exercise ketosis and the hormone response to exercise: a review. *Med. Sci. Sports Exerc.*, 14:327, 1982.
77. Konstantin, N., et al.: Effects of dieting and exercise on lean body mass, oxygen uptake, and strength. *Med. Sci. Sports Exer.*, 17:466, 1985.
78. Koivisto, V.A., et al.: Glycogen depletion during prolonged exercise: influence of glucose, fructose, or placebo. *J. Appl. Physiol.*, 58:731, 1985.
79. Kromhout, D., et al.: Dietary fiber and 10 year mortality from coronary heart disease, cancer, and all causes. *Lancet*, 2:518, 1982.
80. Kromhout, D., et al.: The inverse relationship between fish consumption and 20-year mortality from coronary heart disease. *N. Engl. J. Med.*, 312:1205, 1985.
81. Krzentowski, B., et al.: Availability of glucose given orally during exercise. *J. Appl. Physiol.*, 56:315, 1984.
82. Lanza, E., et al.: Dietary fiber intake in the U.S. population. *Am. J. Clin. Nutr.*, 46:790, 1987.
83. Lemon, P.W.R.: Protein and exercise: Update 1987. *Med. Sci. Sports Exerc.*, 19:S179, 1987.
84. Lemon, P.W.R., and Mullin, J.P.: The effect of initial muscle glycogen levels on protein catabolism during exercise. *J. Appl. Physiol.*, 48:624, 1980.
85. Lemon, P.W.R., and Nagle, F.J.: Effects of exercise on protein and amino acid metabolism. *Med. Sci. Sports Exerc.*, 13:141, 1981.
85a. Levine, A.S., et al.: Effect of breakfast cereals on short-term food intake. *Am. J. Clin. Nutr.*, 50:1303, 1989.

86. Levine, L., et al.: Fructose and glucose ingestion and muscle glycogen use during submaximal exercise. *J. Appl. Physiol.*, 55:1767, 1983.
86a. Levy, D., et al.: Stratifying the patient at risk for coronary disease: new insights from the Framingham Heart Study. *Am. Heart J.*, 119:712, 1990.
87. Lipid Research Clinics Program: The Lipid Research Clinics coronary primary prevention trial results. I. Reduction in incidence of coronary heart disease. *JAMA*, 251:351, 1984.
88. Lipid Research Clinics Program: The Lipid Research Clinics coronary primary prevention trial results. II. The relationship of reduction in incidence of coronary heart disease to cholesterol lowering. *JAMA*, 251:365, 1984.
89. Manninen, V., et al.: Lipid alternations and decline in the incidence of coronary heart disease in the Helsinki Heart Study. *JAMA*, 260:641, 1988.
90. Martin, M.J., et al.: Serum cholesterol, blood pressure, and mortality: implications from a cohort of 361, 662 men. Lancet, 2:933, 1986.
91. McArdle, W.D., et al.: Thermal adjustment to cold-water exposure in resting men and women. *J. Appl. Physiol.*, 56:1565, 1984.
92. Meridith, C.N., et al.: Dietary protein requirements and body protein metabolism in endurance-trained men. *J. Appl. Physiol.*, 66:2850, 1989.
93. Metha, J., et al.: Eicosapentanoic acid: its relevance in atherosclerosis and coronary artery disease. *Am. J. Cardiol.*, 59:155, 1987.
94. Miller, V.C., et al.: Adaptations to high-fat diet that increase exercise endurance in male rats. *J. Appl. Physiol.*, 56:78, 1984.
95. Mitchell, J.B., et al.: Effects of carbohydrate ingestion on gastric emptying and exercise performance. *Med. Sci. Sports Exerc.*, 20:110, 1988.
96. Molé, P., and Johnson, R.: Disclosure of dietary modification of an exercise-induced protein catabolism in man. *J. Appl. Physiol.*, 31:185, 1971.
97. Murray, R.: The effects of consuming carbohydrate-electrolyte beverages on gastric emptying and fluid absorption during and following exercise. *Sports Med.*, 4:322, 1987.
98. Murray, R., et al.: The effects of glucose, fructose, and sucrose ingestion during exercise. *Med. Sci. Sports Exerc.*, 21:275, 1989.
99. Nationwide Food Consumption Survey, 1977–78: Preliminary Report No. 2, Food and Nutrient Intakes of Individuals in One Day in the United States, Spring, 1977, U.S. Department of Agriculture, Science and Education Administration, Washington, DC, September 1980.
100. Neufer, P.D., et al.: Effects of exercise and carbohydrate composition on gastric emptying. *Med. Sci. Sports Exerc.*, 18:658, 1986.
101. Newsholme, E.: The regulation of intracellular and extracellular fuel supply during sustained exercise. *Ann. N.Y. Acad. Sci.*, 301:81, 1977.
102. Nieman, D.: Vegetarian dietary practices and endurance performance. *Am. J. Clin. Nutr.*, 48:754, 1988.
103. Ohi, G., et al.: Changes in dietary fiber intake among Japanese in the 20th century: a relationship to prevalence of diverticular disease. *Am. J. Clin. Nutr.*, 32:115, 1983.
104. Okano, G., et al.: Effect of pre-exercise fructose ingestion on endurance performance in fed man. *Med. Sci. Sports Exerc.*, 20:105, 1988.
105. Owen, O.E., et al.: Brain metabolism during fasting. *J. Clin. Invest.*, 46:1589, 1967.
106. Pallikarikas, N., et al.: Remarkable metabolic activity of oral glucose during long-duration exercise in humans. *J. Appl. Physiol.*, 60:1035, 1986.
107. Phillipson, B.E., et al.: Reduction of plasma lipids, lipoproteins, and apoproteins by dietary fish oils in patients with hypertriglyceridema. *N. Engl. J. Med.*, 312:1210, 1985.
108. Pirnay, F., et al.: Oxidation of orally administered "naturally labeled" ^{13}C-glucose during prolonged muscular exercise. In *Biochemistry of Exercise*. Edited by J. Poortmans and G. Niset. Baltimore, MD, University Park Press, 1981.
109. Pirnay, F., et al.: Fate of exogenous glucose during exercise of different intensities in man. *J. Appl. Physiol.*, 53:1620, 1982.
110. Pivarnik, J.M., et al.: Urinary 3-methylhistidine excretion increases with repeated weight training exercise. *Med. Sci. Sports Exerc.*, 21:283, 1989.
111. Pugh, L.G.C.E., and Edholm, O.G.: The physiology of channel swimmers. *Lancet*, 2:761, 1955.
112. Refsum, H.E., et al.: Changes in plasma amino acid distribution and urinary amino acid excretion during prolonged heavy exercise. *Scand. J. Clin. Invest.*, 39:407, 1979.

113. Reiser, S., et al.: Isocaloric exchange of dietary starch and sucrose in humans. *Am. J. Clin. Nutr.*, 32:1659, 1979.
114. Rifkind, B.M., and Segal, P.: Lipid Research Clinics program reference values for hyperlipidemia and hypolipidemia. *JAMA*, 250:1869, 1983.
115. Ryan, A.J., et al.: Gastric emptying during prolonged exercise in the heat. *Med. Sci. Sports Exerc.*, 21:51, 1989.
116. Saltin, B., and Gollnick, P.D.: Fuel for muscular exercise: role of carbohydrate. In *Exercise, Nutrition, and Energy Metabolism*. Edited by E.S. Horton, and R.L. Terjung. New York, Macmillan, 1988.
117. Saudek, C.D., and Felig, P.: The metabolic events of starvation. *Am. J. Med.*, 60:117, 1976.
118. Seiple, R.S., et al.: Gastric-emptying characteristics of two glucose polymer-electrolyte solutions. *Med. Sci. Sports Exerc.*, 15:366, 1983.
119. Sharon, N.: Carbohydrates. *Sci. Am.*, 243:90, 1980.
120. Sherman, W.M., and Costill, D.L.: The marathon: dietary manipulation to optimize performance. *Am. J. Sports Med.*, 12:44, 1984.
121. Short, S.H., and Short, W.R.: Four-year study of university athletes' dietary intake. *J. Am. Diet. Assoc.*, 82:632, 1983.
122. Sims, E.A.H., and Danforth, Jr., E.: Expenditure and storage of energy in man (perspective). *J. Clin. Invest.*, 79:1019, 1987.
123. Slavin, J.L. et al.: Nutritional practices of women cyclists including recreational riders and elite racers. In *Sport, Health, and Nutrition*. Edited by F.I. Katch. Champaign, IL, Human Kinetic Publishers, 1986.
124. Sole, C.C., and Noakes, T.D.: Faster emptying for glucose-polymer and fructose solutions than for glucose in humans. *Eur. J. Appl. Physiol.*, 58:605, 1989.
125. Stamler, J.: Population studies. In *Nutrition, Lipids, and Coronary Heart Disease*. Edited by R.I. Levy et al. New York, Raven Press, 1979.
126. Stamler, J., et al.: Is relationship between serum cholesterol and risk of premature death from coronary heart disease continuous or graded? Findings in 356,222 primary screenees of the Multiple Risk Factor Intervention Trial (Mr Fit). *JAMA*, 256:2823, 1986.
127. Superko, H.R., et al.: The effect of solid and liquid gum on the reduction of plasma cholesterol in patients with moderate hypercholesterolemia. *Am. J. Cardiol.*, 62:51, 1988.
128. Tarnopolosky, M.A., et al.: Influence of protein intake and training status on nitrogen balance and lean body mass. *J. Appl. Physiol.*, 64:187, 1988.
129. Vahouny, G.V.: Dietary fiber, lipid metabolism, and atherosclerosis. *Fed. Proc.*, 41:2801, 1982.
130. Van Handel, P.J., et al.: Fate of ^{14}C-glucose ingested during prolonged exercise. *Int. J. Sports. Med.*, 1:127, 1980.
131. Vranic, M., and Berger, M.: Exercise and diabetes mellitus. *Diabetes*, 28:147, 1979.
132. Walberg, J.L., et al.: Macronutrient content of a hypoenergy diet affects nitrogen retention and muscle function in weight lifters. *Int. J. Sports Med.*, 9:261, 1988.
132a. Warner, J.G., Jr., et al.: Combined effect of aerobic exercise and omega-3 fatty acids in hyperlipidemic persons. *Med. Sci. Sports Exerc.*, 21:498, 1989.
133. White, T.P., and Brooks, G.A.:[u-^{14}C] glucose-alanine and leucine oxidation in rats at rest and two intensities of running. *Am. J. Physiol.*, 240:E155, 1981.
134. Williams, O.D., et al.: Common methods, different populations: The Lipid Research Clinics program prevalence study. *Circulation*, 62(Suppl. 4):18, 1980.
135. Wolfe, R.R., et al.: Glucose metabolism in man: response to intravenous glucose infusion. *Metabolism*, 28:210, 1979.
136. Wolfe, R.R., et al.: Isotopic analysis of leucine and urea metabolism in exercising humans. *J. Appl. Physiol.*, 52:458, 1982.
137. Wolfe, R.R., et al.: Role of changes in insulin and glucagon in glucose homeostasis in exercise. *J. Clin. Invest.*, 77:900, 1986.
138. Yannick, C., et al.: Oxidation of corn starch, glucose, and fructose ingested before exercise. *Med. Sci. Sports Exerc.*, 21:45, 1989.

2 VITAMINS, MINERALS, AND WATER

The effective regulation of all metabolic processes requires a delicate blending of food nutrients in the watery medium of the cell. Of special significance in the metabolic mixture are the **micronutrients**—the small quantities of vitamins and minerals that play highly specific roles in facilitating energy transfer. These substances are readily obtained in the foods consumed in well-balanced meals. With proper nutrition from a variety of food sources, there is little need to consume vitamin and mineral supplements; such practices are both physiologically and economically wasteful.

1 Vitamins

THE NATURE OF VITAMINS

The importance of vitamins was known long before scientists isolated and classified them. For example, Hippocrates advocated the ingestion of liver to cure night blindness, and the consumption of various citrus fruits and fresh vegetables. In 1897, scientists observed that a regular diet of polished rice caused the disease **beriberi,** whereas the addition of the thiamine-rich rice polishings cured it. In the early 19th century, the disease **scurvy** was eliminated by adding lemons—then called limes—to the diet of British sailors. It was not until 1932, however, that the protective ascorbic acid or vitamin C was isolated from lemon juice.

The formal discovery of vitamins revealed they were organic substances needed by the body in minute amounts. The vitamins have no particular chemical structure in common and are often considered accessory nutrients because they neither supply energy nor contribute substantially to the body's mass. With the exception of vitamin D, the body cannot manufacture vitamins; hence, they must be supplied in the diet or through supplementation.

Much of the food we consume contains an abundant quantity of vitamins. For example, green leaves and roots of plants manufacture vitamins during the process of photosynthesis. Animals obtain their vitamins from the plants, seeds, grains, and fruits they eat, or from the meat of other animals that have previously consumed these foods. Several vitamins, most notably A, D, niacin, and folacin, are activated in the presence of **provitamins.** The best known of the provitamins are the **carotenes,** the yellow and yellow-orange pigment-precursors of vitamin A that give color to vegetables and fruits such as carrots, squash, corn, pumpkins, apricots, and peaches.

KINDS OF VITAMINS

Thirteen different vitamins have been isolated, analyzed, classified, and synthesized, and recommended dietary intakes established. These vitamins are classified as **fat-soluble** and **water-soluble.** The fat-soluble vitamins are vitamins A, D, E, and K; the water-soluble vitamins are vitamin B_6 (pyridoxine), thiamine (B_1), riboflavin (B_2), niacin (nicotinic acid), pantothenic acid, biotin, folacin (folic acid), cobalamin (B_{12}), and vitamin C (ascorbic acid).

FAT-SOLUBLE VITAMINS

The daily ingestion of fat-soluble vitamins is not absolutely necessary because these substances are dissolved and stored in the fatty tissues of the body. In fact, it may take years for symptoms of a fat-soluble vitamin insufficiency to become evident. Fat-soluble vitamins are often obtained in dietary fat, and consuming a so-called "fat-free diet" could certainly speed up the development of an insufficiency.

At the other extreme, excessive intake of fat-soluble vitamins can be harmful. For example, daily ingestion of moderate to large doses of vitamins A and D can eventually have serious toxic effects. In young children, excessive vitamin A intake (called hypervitaminosis A) causes irritability, swelling of the bones, weight loss, and dry itchy skin. In adults, symptoms can include nausea, headache, drowsiness, loss of hair, diarrhea, and loss of calcium from bones, causing brittleness. Discontinuance of such high intakes of vitamin A reverses these symptoms.

The fat-soluble vitamins should **not** be consumed in excess without proper medical supervision. Although an "overdose" from vitamins E and K is rare, it is generally believed that intakes above the recommended level are of no benefit.

WATER-SOLUBLE VITAMINS

The other group of vitamins (grouped together as vitamin B-complex and vitamin C) is classified as water-soluble. These vitamins act as **coenzymes** which are small molecules that can combine with a larger protein molecule to make an active enzyme. They are similar to their fat-soluble counterparts because they are composed of atoms of carbon, hydrogen, and oxygen. They also contain nitrogen and other elements such as sulfur and cobalt. Because of their water solubility they are transported in the body fluids and not stored to an appreciable extent. Consequently, these vitamins should generally be consumed on a daily basis, and an excessive intake is eventually voided in the urine.

ROLE OF VITAMINS IN THE BODY

Whereas vitamins contain no useful energy for the body, they generally serve as essential links to help regulate the chain of metabolic reactions that facilitate the release of energy bound in the food molecule and control the process of tissue synthesis. **Because vitamins can be used repeatedly in metabolic reactions, the vitamin needs of athletes are generally no greater than the requirements of sedentary people.**

The vitamins, bodily functions, dietary requirements, and their major food sources are listed in Table 2–1. It is important to emphasize that an adequate

TABLE 2–1.
WATER- AND FAT-SOLUBLE VITAMINS, THEIR RECOMMENDED DAILY INTAKE, DIETARY SOURCES, MAJOR BODILY FUNCTIONS, AND EFFECTS OF DEFICIENCIES AND EXCESSES[a]

Vitamin	RDA for Healthy Adult (age 19–50) Male and Female[b] (mg)	Dietary Sources	Major Body Functions	Deficiency	Excess
Water-Soluble					
Vitamin B-1 (Thiamine)	1.5 1.1	Pork, organ meats, whole grains, legumes	Coenzyme (thiamine pyrophosphate) in reactions involving the removal of carbon dioxide	Beriberi (peripheral nerve changes, edema, heart failure)	None reported
Vitamin B-2 (Riboflavin)	1.7 1.3	Widely distributed in foods	Constituent of two flavin nucleotide coenzymes involved in energy metabolism (FAD and FMN)	Reddened lips, cracks at corner of mouth (cheilosis), lesions of eye	None reported
Niacin	19 15	Liver, lean meats, grains, legumes (can be formed from tryptophan)	Constituent of two coenzymes involved in oxidation-reduction reactions (NAD and NADP)	Pellagra (skin and gastrointestinal lesions, nervous, mental disorders)	Flushing, burning and tingling around neck, face, and hands
Vitamin B-6 (Pyridoxine)	2.0 1.6	Meats, vegetables, whole-grain cereals	Coenzyme (pyridoxal phosphate) involved in amino acid and glycogen metabolism	Irritability, convulsions, muscular twitching, dermatitis near eyes, kidney stones	None reported
Pantothenic Acid	4–7[c] 4–7	Widely distributed in foods	Constituent of coenzyme A, which plays a central role in energy metabolism	Fatigue, sleep disturbances, impaired coordination, nausea (rare in man)	None reported
Folacin	0.2 0.2	Legumes, green vegetables, whole-wheat products	Coenzyme (reduced form) involved in transfer of single-carbon units in nucleic acid and amino acid metabolism	Anemia, gastrointestinal disturbances, diarrhea, red tongue	None reported
Vitamin B-12	0.002 0.002	Muscle meats, eggs, dairy products, (not present in plant foods)	Coenzyme involved in transfer of single-carbon units in nucleic acid metabolism	Pernicious anemia, neurologic disorders	None reported

Vitamin	RDA for Healthy Adult (age 19–50) Male and Female[b] (mg)	Dietary Sources	Major Body Functions	Deficiency	Excess
Water-Soluble					
Biotin	0.03–0.10[c]	Legumes, vegetables, meats	Coenzymes required for fat synthesis, amino acid metabolism, and glycogen (animal starch) formation	Fatigue, depression, nausea, dermatitis, muscular pains	None reported
Vitamin C (Ascorbic acid)	60[d] 60	Citrus fruits, tomatoes, green peppers, salad greens	Maintains intercellular matrix of cartilage, bone, and dentine. Important in collagen synthesis.	Scurvy (degeneration of skin, teeth, blood vessels, epithelial hemorrhages)	Relatively non-toxic. Possibility of kidney stones
Fat Soluble					
Vitamin A (Retinol)	1.0 0.8	Provitamin A (beta-carotene) widely distributed in green vegetables. Retinol present in milk, butter, cheese, fortified margarine	Constituent of rhodopsin (visual pigment). Maintenance of epithelial tissues. Role in mucopolysaccharide synthesis	Xerophthalmia (keratinization of ocular tissue), night blindness, permanent blindness	Headache, vomiting, peeling of skin, anorexia, swelling of long bones
Vitamin D	0.01[e] 0.01	Cod-liver oil, eggs, dairy products, fortified milk, and margarine.	Promotes growth and mineralization of bones. Increases absorption of calcium	Rickets (bone deformities) in children. Osteomalacia in adults.	Vomiting, diarrhea, loss of weight, kidney damage
Vitamin E (Tocopherol)	10 8	Seeds, green leafy vegetables, margarines, shortenings	Functions as an antitoxidant to prevent cell-membrane damage.	Possibly anemia	Relatively non-toxic
Vitamin K (Phylloquinone)	0.08 0.06	Green leafy vegetables. Small amount in cereals, fruits, and meats	Important in blood clotting (involved in formation of active prothrombin)	Conditioned deficiencies associated with severe bleeding; internal hemorrhages	Relatively non-toxic. Synthetic forms at high doses may cause jaundice

[a] Recommended Dietary Allowances. Revised 1989. Food and Nutrition Board, National Academy of Sciences—National Research Council, Washington, D.C.

[b] First values are for males.

[c] Because there is less information on which to base allowances these figures are given in the form of ranges.

[d] 100 for adults who smoke.

[e] 0.005 mg for adults 25 and older.

quantity of all vitamins is available for those individuals who consume normal, well-balanced meals. This is true regardless of age and level of physical activity. Indeed, there appears to be no need for individuals who expend considerable energy in vigorous exercise to consume extra vitamins in the form of special foods or supplements. Also, at high levels of physical activity, food intake is generally increased to sustain the added energy requirements of exercise. If this added food is obtained through well-balanced meals, a proportionate increase in vitamin and mineral intake is assured.[70]

Several possible exceptions to this general rule should be noted. Vitamin C and the B-vitamin folacin (folic acid) are in foods that usually make only a minimal contribution to the caloric content of the American diet and whose availability varies by season. Adequate intake of these vitamins can be assured if the daily diet contains fresh fruit and uncooked or steamed vegetables. For individuals on meatless diets, a small amount of milk, milk products, or eggs should be included because vitamin B_{12} is only available in foods of animal origin.

VITAMIN SUPPLEMENTS: THE COMPETITIVE EDGE?

It is well established that vitamin supplements can reverse the symptoms of vitamin deficiency. Also, vitamins synthesized in the laboratory are no less effective for bodily functioning than vitamins from natural sources. Once a nutritional deficiency is cured, supplements do not further improve a normal status. In addition, over 40 years of research has not supported the wisdom of using vitamin supplements to improve exercise performance or the ability to train in nutritionally adequate healthy people.[5,53,60,84,91] When vitamin intake is at recommended levels, supplements neither improve exercise performance nor necessarily increase the blood levels of these nutrients.[93] The facts have become clouded by the "testimonials" of coaches and elite athletes to the effect that their success was due to a particular dietary modification that usually included specific vitamin supplements.

VITAMINS AND EXERCISE PERFORMANCE

Because of the key role of the B-complex vitamins in co-enzyme form in important energy-yielding reactions during fat, carbohydrate, and protein breakdown as well as hemoglobin synthesis and red blood cell production, it has been tempting to speculate that an increase in the intake of these vitamins would enhance energy release and lead to improved physical performance.[92] The belief that "if a little is good, more must be better," has led many coaches, athletes, fitness enthusiasts, and even some scientists to advocate the use of vitamin supplements. Although 35% of "Big Ten" coaches recommend supplements for their players,[96] and about 45% of the competitors actually take them, this approach is simply not supported by research findings or by the overwhelming majority of professional nutritionists.[71,78]

Supplementing with vitamin B_6, an essential cofactor in glycogen and amino acid metabolism, at levels above the RDA was of no benefit to the metabolic mixture metabolized by women during high intensity aerobic exercise.[53] No exercise benefit has been shown for vitamins other than the B-complex group such as vitamins C and E. Vitamin C, for example, is a factor in the synthesis of collagen and the stress-related adrenal hormone, norepinephrine. Studies have shown that supplements of vitamin C had negligible effects on endurance performance and on the rate, severity, and duration of injuries compared to treatment with a placebo.[31] It has never been firmly established with careful research that a deficiency state for vitamin E exists for normal individuals, let alone that vitamin E supplements are beneficial to stamina, circulatory function, energy metabolism, aging, or sexual potency.[50]

MEGAVITAMINS

Although individuals who eat a well-balanced diet do not need to take additional vitamins, most nutritionists feel that taking a multivitamin capsule of the recommended quantity of each vitamin does no harm. For some people, the psychologic effects may even be beneficial. It is of great concern, however, that some athletes resort to taking **megavitamins,** or doses of at least **tenfold** and up to 1,000 times the RDA in the hope that "supercharging" with vitamins will improve exercise performance.

This practice can be harmful except in cases of specific serious medical illness. Once the enzyme systems that are catalyzed by specific vitamins are saturated, the excess vitamins in the megadose function as chemicals in the body.[39] For example, a megadose of water-soluble vitamin C can raise serum uric acid levels and precipitate gout in people predisposed to this disease. Also, some American blacks, Asians, and Sephardic Jews have a genetic metabolic deficiency that can be activated to hemolytic anemia by excesses of vitamin C.[17] In addition, vitamin C supplementation appears to have little effect on the frequency and duration of the common cold, although vitamin C appears to act as an antihistamine to reduce cold symptoms. In individuals who are iron-deficient, megadoses of vitamin C destroy significant amounts of vitamin B_{12}.[40,62] In healthy people, vitamin C supplements frequently irritate the bowel and cause diarrhea.

It is now believed that an excessive intake of vitamin B_6 may produce liver disease and nerve damage;[81] excessive riboflavin (B_2) can impair vision, whereas a megadose of nicotinic acid inhibits the uptake of fatty acids by cardiac muscle during exercise.[39] Possible side effects of vitamin E megadose include headache, fatigue, blurred vision, gastrointestinal disturbances, muscular weakness, and low blood sugar.[38] Because vitamin E is usually found associated with unsaturated fats, it is difficult even to "construct" a vitamin E-deficient diet. The toxicity to the nervous system of megadoses of vitamin A and the damaging effects to the kidneys of excess vitamin D have been well demonstrated.[83]

It is indeed troublesome that recent data from the U.S. National Health Interview Survey indicate that more than 30% of American adults use vitamin and/or mineral supplements, often at potentially toxic dosages.[91a] If vitamin supplementation has a role for physically active individuals, it may only be for those who have marginal vitamin stores. Well-controlled research, however, is needed to fully determine if and under what circumstances such supplementation would be beneficial.

Perhaps the misuse and abuse of vitamins by individuals hoping to improve athletic performance can be put in proper perspective by the following quotation.[70] "The sale of vitamins is probably the biggest rip-off in our society today. Their only effect would appear to be a highly enriched sewage around athletic training or competition sites."

SUMMARY

1. Vitamins are organic substances that neither supply energy nor contribute to the body's mass, but that serve crucial functions in almost all body processes. Vitamins must be obtained from food or from dietary supplementation.
2. Vitamins are synthesized by plants, and are also found in animals that produce them from precursor substances known as provitamins.

3. There are 13 known vitamins classified as either water or fat soluble. The fat-soluble vitamins are vitamins A, D, E, and K; vitamin C and the B-complex vitamins are water soluble.
4. Fat-soluble vitamins taken in excess accumulate in the tissues and eventually can be toxic. Except in relatively rare and specific instances, excesses of water-soluble vitamins are generally nontoxic and are eventually excreted in the urine.
5. Vitamins regulate metabolism, facilitate energy release, and are important in the process of bone and tissue synthesis.
6. Research generally shows that vitamin supplementation (above that obtained in the well-balanced diet) is not related to improved exercise performance or potential for training. In fact, excessive dosage of fat-soluble, and in some instances water-soluble vitamins, can result in serious illness.

P A R T

2 **Minerals**

THE NATURE OF MINERALS

In addition to the organic elements oxygen, carbon, hydrogen, and nitrogen, approximately 4% of the body's weight, or about 2 kg for a 50 kg woman, is composed of a group of 22 mostly metallic elements collectively called **minerals.** Most of the minerals are found in living cells, although not all are necessarily essential for life. The most important minerals are those found in enzymes, hormones, and vitamins. Minerals appear in combination with organic compounds, for example, calcium phosphate in bone, or singularly such as free calcium in intracellular fluids.

In the body, minerals are classified as **major minerals**—those present in large quantities and that have known biologic functions, and as **trace minerals**—those present in minute quantities (less than 0.05% body mass). Any excess of minerals is useless to the body and may even be toxic. RDAs for many minerals have been established. It is believed that if the requirements for these and other nutrients are met, the remaining minerals can be obtained in adequate amounts.

Most minerals, major or trace, occur freely in nature, mainly in the waters of rivers, lakes, and oceans, in topsoil, and beneath the earth's surface. Minerals can be found in the root systems of plants and trees and in the body structure of animals who consume plants and water containing the minerals. In humans, absorption of minerals takes place in the small intestines.

KINDS AND SOURCES OF MINERALS

The important minerals and their functions, food sources, and daily requirements are listed in Table 2–2. As with vitamins, mineral supplements are generally not needed because most minerals are readily available in the

TABLE 2–2.
THE IMPORTANT MINERALS IN THE BODY, THEIR RECOMMENDED DAILY INTAKE, DIETARY SOURCES, MAJOR BODILY FUNCTIONS, AND THE EFFECTS OF DEFICIENCIES AND EXCESSES[a]

Mineral	Amount in Adult Body (g)	RDA for Healthy Adult (age 19–50) Male and Female[b] (mg)	Dietary Sources	Major Body Functions	Deficiency	Excess
Calcium	1,500	1200[c] 1200	Milk, cheese, dark-green vegetables, dried legumes	Bone and tooth formation Blood clotting Nerve transmission	Stunted growth Rickets, osteoporosis Convulsions	Not reported in man
Phosphorus	860	1200[c] 1200	Milk, cheese, meat, poultry, grains	Bone and tooth formation Acid-base balance	Weakness, demineralization of bone Loss of calcium	Erosion of jaw (fossy jaw)
Sodium	64	1100–3300	Common salt	Acid-base balance Body water balance Nerve function	Muscle cramps Mental apathy Reduced appetite	High blood pressure
Magnesium	25	350 280	Whole grains, green leafy vegetables	Activates enzymes. Involved in protein synthesis	Growth failure Behavioral disturbances Weakness, spasms	Diarrhea
Iron	4.5	10 15	Eggs, lean meats, legumes, whole grains, green leafy vegetables	Constituent of hemoglobin and enzymes involved in energy metabolism	Iron-deficiency anemia (weakness, reduced resistance to infection)	Siderosis Cirrhosis of liver
Fluorine	2.6	1.5–4.0[d]	Drinking water, tea, seafood	May be important in maintenance of bone structure	Higher frequency of tooth decay	Mottling of teeth increased bone density Neurologic disturbances
Zinc	2	15 12	Widely distributed in foods	Constituent of enzymes involved in digestion	Growth failure Small sex glands	Fever, nausea, vomiting, diarrhea
Copper	0.1	1.5–3.0[d] 1.5–3.0[d]	Meats, drinking water	Constituent of enzymes associated with iron metabolism	Anemia, bone changes (rare in man)	Rare meabolic condition (Wilson's disease)
Selenium	0.013	0.070 0.055	Seafood, meat, grains	Functions in close association with Vitamin E	Anemia (rare)	Gastrointestinal disorders, lung irritation
Iodine	0.011	0.15	Marine fish and shellfish, dairy products, many vegetables	Constituent of thyroid hormones	Goiter (enlarged thyroid)	Very high intakes depress thyroid activity
Molybdenum	0.009	0.075–0.25[d]	Legumes, cereals, organ meats	Constituent of some enzymes	Not reported in man	Inhibition of enzymes
Chromium	0.006	0.05–0.25[d]	Fats, vegetable oils, meats	Involved in glucose and energy metabolism	Impaired ability to metabolize glucose	Occupational exposures: skin and kidney damage
Water	40,000 (60% of body weight)	1.5 liters per day	Solid foods, liquids, drinking water	Transport of nutrients Temperature regulation Participates in metabolic reactions	Thirst, dehydration	Headaches, nausea Edema High blood pressure

[a] Recommended Dietary Allowances, Revised 1989. Food and Nutrition Board, National Academy of Sciences—National Research Council, Washington, D.C.
[b] First values are for males.
[c] 800 mg for adults 25 and older.
[d] Because there is less information on which to base allowances these figures are given in the form of ranges.

TABLE 2–3. RECOMMENDED DIETARY ALLOWANCES FOR IRON

	Age	Iron (mg)
Children	1–10	10
Males	11–18	12
	19+	10
Females	11–50	15
	51+	10
	Pregnant	30[a]
	Lactating	15[a]

[a] Generally, this increased requirement cannot be met by ordinary diets; therefore, the use of 30 to 60 mg of supplemental iron is recommended. (From Food and Nutrition Board; Recommended Dietary Allowances. Washington, D.C., National Academy of Sciences, revised, 1989.)

foods we eat and the water we drink. Some supplementation may be necessary, however, in geographic regions where the soil or water supply is poor in a particular mineral. For example, in certain regions of the United States, particularly the basin of the Great Lakes and Pacific Northwest, sources of the mineral **iodine** are relatively poor. Iodine is taken up by the thyroid gland to become part of **thyroxine,** a hormone that exerts an accelerating influence on the cells' resting metabolic level. A deficiency in this mineral can easily be prevented by adding iodine to the water supply or to table salt (iodized salt).

A common mineral deficiency in this country results from a lack of iron in the diet (Table 2–3). It is estimated that 30 to 50% of American women of child-bearing age suffer some form of iron insufficiency. In most instances, appropriate iron supplementation can be achieved with a diet rich in iron-containing foods such as beans, peas, dried uncooked fruits, leafy green vegetables, egg yolk, and meats, especially liver, kidney, and heart.

ROLE OF MINERALS IN THE BODY

Whereas vitamins activate chemical processes without becoming part of the products of the reactions they catalyze, minerals tend to become incorporated within the structures and working chemicals of the body. Minerals serve three broad roles in the body. They provide **structure** in the formation of bones and teeth. In a **functional** sense they are intimately involved in maintaining normal heart rhythm, muscular contractility, neural conductivity, and the acid-base balance of the body. Mineral nutrients also play a **regulatory** role in cellular metabolism. They serve as important parts of enzymes and hormones that modify and regulate cellular activity. Figure 2–1 shows various minerals that participate in catabolic and anabolic cellular processes.

Minerals are important in activating numerous reactions that release energy during the breakdown of carbohydrates, fats, and proteins. In addition to the processes of catabolism, minerals are essential for the synthesis of biologic nutrients; glycogen from glucose, fats from fatty acids and glycerol, and proteins from amino acids. Without the essential minerals, the fine balance would be disrupted between catabolism and anabolism.

Minerals also form important constituents of hormones. An inadequate thyroxine production due to iodine deficiency could significantly slow the body's resting metabolism. In extreme cases, this reduced level of energy output could predispose a person to develop obesity. The synthesis of insulin, the hormone that facilitates glucose uptake by the cells, requires the mineral zinc whereas the digestive acid, hydrochloric acid, is formed from the mineral chlorine. In the subsequent sections, specific functions are described for several of the more important minerals.

CALCIUM

Calcium is the most abundant mineral in the body. Calcium combines with phosphorus to form the bones and teeth. These two minerals represent about 75% of the body's total mineral content or about 2.5% of body mass. In its ionized form, calcium plays an important role in muscular contraction and the transmission of nerve impulses. It also activates several enzymes, and is essential for blood clotting and for the transport of fluids across cell membranes.

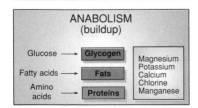

FIG. 2–1.
Minerals involved in the catabolism (breakdown) and anabolism (build-up) of nutrients.

Osteoporosis: Calcium, Estrogen, and Exercise

Although it is true that growing children need more calcium per unit body weight on a daily basis than adults, it is also clear that many adult men and women are deficient in their intake of this mineral. As a general guideline, adolescents and young adults need 1200 mg of calcium daily (800 mg for adults past age 24) or about as much calcium as in four 8-oz glasses of milk. In reality, however, calcium is one of the most frequently lacking nutrients. For example, more than 75% of adults consume less than the daily requirement, and about 25% of all females in the United States consume less than 300 mg of calcium on any given day.[37] As a result, the body draws upon its calcium reserve in bone to restore the deficit. If an imbalance is prolonged, the condition of **osteoporosis** (literally meaning porous bones) eventually sets in as the bone loses its mineral mass and progressively becomes porous and brittle; it eventually breaks under the stresses of normal living. Among older individuals, especially women past the age of 60, osteoporosis has reached almost epidemic proportions. For example, osteoporosis accounts for more than 1.2 million fractures yearly including 500,000 spinal fractures and 230,000 hip fractures. The increased susceptibility to osteoporosis among older women is closely associated with the decrease in estrogen production that accompanies the menopause.[58,66,73] The precise manner by which estrogen exerts its protective effect on bone is not known, although it is believed to enhance calcium absorption and limit its withdrawal from bone.

A Progressive Disease. Osteoporosis, which affects about 24 million Americans, begins early in life because the average teenager consumes suboptimal calcium to support the growing bones. This imbalance worsens into adulthood and by middle age, adult women consume only about one-third of the calcium they require for optimal bone maintenance. Starting at about age 50, the average man experiences about a 0.4% bone loss each year, while the female begins to lose twice this amount starting at age 35. For men, the normal rate of bone mineral loss does not usually pose a problem until the eighth decade of life. Women, however, become extremely susceptible to the ravages of osteoporosis at the menopause when bone loss accelerates to 1.0 to 3.0% per year. At this rate, an average woman loses about 15% of her bone mass in the first decade after menopause; for some women a loss of as much as 30% of the bone mineral mass is observed by age 70.[85]

A prime defense against bone loss with age is adequate calcium intake throughout life.[1,51,74] The previous 800-mg RDA has recently been upgraded to 1200 mg for ages 11 to 24. Many experts also recommend a further increase to between 1200 and 1500 mg for estrogen-deprived women after menopause to assure a positive calcium balance later in life,[36,88] although it is not clear whether calcium supplements are of benefit in the absence of adequate estrogen.[34] Milk and milk products, sardines and canned salmon, kidney beans, and dark green leafy vegetables are good calcium sources. Calcium supplements can also help to correct dietary deficiencies regardless of whether the extra calcium comes from supplements (calcium carbonate recommended) or from food products.[59] Adequate availability of vitamin D facilitates calcium metabolism, while excessive consumption of meat, salt, coffee, and alcohol inhibits calcium absorption. In postmenopausal women, estrogen supplements serve as a treatment for severe osteoporosis. Estrogen therapy must be continued for prolonged periods, thus increasing the risk for cancers of the uterus, breast, and other organs. Thus, hormone treatment for osteoporosis is often viewed as a more dramatic approach.

Exercise Is Helpful. It is important to note that **regular exercise helps to slow the rate of aging of the skeleton.** Regardless of age, individuals who maintain an active lifestyle have significantly greater bone mass compared to sedentary counterparts,[6,9,20,34,48,67,72a,94] and this benefit is maintained into the seventh and even eighth decade of life![41,86] In fact, the decline in vigorous exercise with the sedentary lifestyle associated with advancing age closely parallels the age-related loss of bone mass. Exercise can provide a safe and potent stimulus to maintain and even increase bone mass in adults. Especially beneficial is exercise of a weight-bearing nature, which includes walking, running, dancing, rope skipping, or activities such as circuit resistance training in which significant muscular force can be generated against the long bone of the body.[54]

Muscular forces acting on specific bones appear to modify their metabolism at the point of stress. For example, the leg bones of older cross-country runners show greater bone mineral content compared to the bones of less active counterparts. Likewise, the playing arm of tennis players and the throwing arm of baseball players show greater bone thickness compared to the less used, non-dominant arm. Prevailing theory considers bone to behave as a piezoelectric crystal that converts mechanical stress into electrical energy. The electrical changes created when bone is mechanically stressed stimulate the activity of bone-forming cells which leads to a buildup of calcium. The quantity of bone buildup depends on both the magnitude of the force as well as its frequency of application. Chemical factors produced in bone itself may also contribute to bone formation. Certain characteristics or risk factors that predispose one to osteoporosis include: white or Asian female, slight build, sedentary, early menopause, cigarette and alcohol abuse, and family history.

Is Too Much Training Harmful? An apparent paradox between exercise and bone dynamics has been noted for premenopausal women who train intensely and reduce body mass and body fat to a point at which the menstrual cycle actually ceases, a condition termed **secondary amenorrhea.**[20a] It has been suggested that the hormonal imbalances associated with the cessation of menstruation remove estrogen's protective effect on bone and makes these women especially vulnerable to calcium loss and a possible decrease in bone mass.[14,32] Concurrently, nutritional factors (e.g., low protein and fat intake) magnify the problem.[25] If amenorrhea persisted, then the benefits of exercise on bone mass would be negated, and the risk of musculoskeletal injuries would increase during exercise and osteoporosis would set in at an early age.

PHOSPHORUS

Besides its important function in combining with calcium to give rigidity to bones and teeth, phosphorus is an essential component of high-energy compounds **adenosine triphosphate** (ATP) and **creatine phosphate** (CP). ATP and CP are crucial in supplying the energy for all forms of biologic work. Phosphorus also participates in the buffering of the acid end products of energy metabolism. For this reason some coaches and trainers recommend the consumption of special "phosphate drinks" in the hope of improving performance by reducing the effects of acid production. In Chapter 23, more is said about the applicability of buffering drinks to the improvement of exercise performance.

MAGNESIUM

Magnesium plays a vital role in glucose metabolism by facilitating the formation of muscle and liver glycogen from blood-borne glucose. Magnesium also participates as a cofactor in the breakdown of glucose, fatty acids, and amino acids during energy metabolism. Furthermore, magnesium is important in the synthesis of fats and proteins and in stabilizing the neuromuscular system in terms of nerve conduction and muscle contraction.

IRON

From 3 to 5 g or about $\frac{1}{6}$ of an ounce of iron is normally contained in the body. About 80% of this iron is in functionally active compounds predominantly combined with **hemoglobin** in the red blood cells.[8] This iron-protein compound increases the oxygen-carrying capacity of blood about 65 times. Iron serves other important exercise-related functions aside from its role in oxygen transport in red blood cells. Iron is a structural component of **myoglobin** (about 5% of total iron), a compound similar to hemoglobin, which aids in the storage and transport of oxygen within the muscle cell. Iron is present in small amounts in specialized substances called **cytochromes** that function as catalysts in the energy transfer systems operating within the cell. About 20% of the body's iron is not combined with functionally active compounds; this constitutes the iron stores located in the liver, spleen, and bone marrow as **hemosiderin** and **ferritin.** It is these stores that replenish iron lost from the functional compounds and that provide the iron reserve in periods of insufficient dietary intake.

The athlete should be sure to include iron-rich foods in the daily diet. People who do not take in enough iron or who have limited rates of iron absorption or high rates of iron loss can develop a condition in which the concentration of hemoglobin in red blood cells is reduced. This extreme condition of iron insufficiency, commonly called **iron deficiency anemia,** is characterized by general sluggishness, loss of appetite, and a reduced capacity for sustaining even mild exercise.[30,56,82] With "iron therapy," both the hemoglobin content of the blood and the exercise response can be brought back to normal levels. Table 2–3 gives the recommendations for daily iron intake for children and adults.

Females: A Population at Risk

Inadequate iron intake frequently occurs among young children, teenagers, and females of child-bearing age, including groups of physically active women.[15] A moderate iron deficiency anemia is common during pregnancy when there is an increased demand for iron for both mother and fetus. In addition, females usually lose between 5 and 45 mg of iron during the menstrual cycle.

The additional 5 mg of iron per day intake requirements for females compared to males would increase the average monthly intake by about 150 mg. Because between 10 to 15% of the iron ingested is absorbed (depending on one's iron status, form of iron, and composition of the meal), the female would then have available to her each month an additional 20 to 25 mg of iron for the synthesis of red blood cells lost during menstruation. When this added iron requirement for the female is combined with the fact that the normal American diet contains only about 6 mg of iron in each 1000 calories of food ingested, it is not surprising that the average adult female consumes only about 12 mg of iron daily. This accounts for the 30 to 50% of American women who have significant dietary iron insufficiencies.

Source of Iron Is Important

While iron absorption from the intestines varies with iron need, a considerable difference in iron absorption occurs in relation to the composition of the diet. For example, only between 2 to 10% of the iron obtained from the plant kingdom (ferric or non-heme iron) is absorbed, whereas 10 to 35% of animal or heme (ferrous) iron is absorbed.[52] (The presence of heme iron also increases the absorption or bioavailability of dietary iron from non-heme sources.) This places women on vegetarian-type diets at a higher risk of developing iron insufficiency. In fact, recent research indicates that vegetarian athletes have a poorer iron status than counterparts who consumed the same quantity of iron from predominantly animal sources.[87] This problem can be alleviated somewhat by including vitamin C-rich foods in the diet because ascorbic acid increases the solubility of non-heme iron and makes it available for absorption at the alkaline pH of the intestines. The ascorbic acid in one glass of orange juice, for example, stimulates a 3-fold increase in non-heme iron absorption from a breakfast meal.[80]

Exercise-Induced Anemia: Fact or Fiction?

Because of the great interest in endurance activities combined with the increased participation of women in such sports, research has focused on the influence of hard training on the body's iron status.[63,68,72] The term "sports anemia" is frequently used to describe reductions in hemoglobin to levels approaching **clinical anemia** (12 g and 13 g of hemoglobin per 100 ml blood for women and men, respectively) that are believed due to intense training. Some researchers maintain that exercise training creates an added demand for iron that outstrips its intake.[15] As a result, iron reserves are taxed which can eventually lead to a fall in hemoglobin levels[7] and/or a reduction in iron containing compounds within the cell's energy transfer system.[21] Of concern is the possibility that those individuals most susceptible to an "iron drain" may ultimately experience a reduced capacity for exercise due to the crucial role of iron in both oxygen transport and utilization.[13]

It is postulated that heavy training creates an augmented iron demand due to a loss of iron in sweat[26,47] (which is probably minimal[11]), or the loss of hemoglobin in the urine due to the actual destruction of red blood cells with increased temperature, spleen activity, and circulation rates as well as from mechanical trauma caused by pounding of the feet on the running surface.[27,55,65] In addition, reports indicate that there may be some gastrointestinal bleeding following long distance running[12,89] which appears to be unrelated to age, gender, or performance time.[76] Any such increase in iron loss would certainly stress the body's iron reserves for the synthesis of new red blood cells. This iron loss would be particularly significant to women who, as a group, have the greatest requirement and lowest intake of this important mineral.

To support the possibility of an exercise-induced anemia, some data indicate that suboptimal hemoglobin concentration and hematocrit are more prevalent among endurance athletes.[42,45] On closer scrutiny, however, it appears that the observed reductions in hemoglobin concentration are tran-

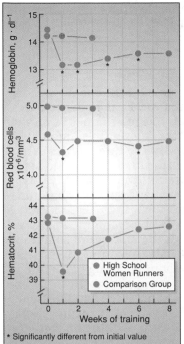

FIG. 2–2.
Hemoglobin, RBC count, and hematocrit during the competitive season. (From Puhl, J.L., et al.: Erythrocyte changes during training in high school women cross-country runners. Res. Q. Exerc. Sport, 54:484, 1981.)

sient and occur in the early phase of training and then return to pretraining values.[29,72] This general response for hematologic characteristics is illustrated in Figure 2–2 for a group of high-school female cross-country runners during a competitive season. It is interesting to note that the decrease in hemoglobin concentration generally parallels the disproportionately large expansion in plasma volume in relation to total hemoglobin with training.[64] For example, just 4 days of submaximal exercise training increases plasma volume by 20%, whereas the volume of red blood cells remains unchanged.[33] Consequently, whereas **total** hemoglobin (an important factor in endurance performance[44]) may actually increase with training, its concentration in the plasma decreases.[10] Despite this apparent dilution of hemoglobin, aerobic capacity and exercise performance consistently increase during training.[23,29]

Although there may be some mechanical destruction of red blood cells with vigorous exercise and some loss of iron in sweat, it has yet to be verified whether these factors are of sufficient magnitude to strain an athlete's iron reserves and precipitate anemia if iron intake is normal.[4a,51a,69,95] A recent study of a relatively large number of female athletes indicated that the prevalence of iron deficiency did not differ in comparisons among specific athletic groups or with a nonathletic control group.[75] Because adolescent and premenopausal women have a relatively high iron requirement, and because many consume inadequate dietary iron, any increase in iron loss with training could strain an already limited iron reserve. Whereas hematologic characteristics of athletes often range within normal limits, measures of iron reserve show that some of these athletes, especially the females, may be at risk for iron deficiency.[15,29] This finding does not mean that **all** men and women in training should take supplementary iron or that all indications of "sports anemia" are the result of iron intake deficiencies or iron loss. It does suggest, however, that the iron status of athletes should at least be monitored. This is best achieved by periodic evaluation of both hemotoligical characteristics as well as iron reserves which are indicated by measuring serum ferritin levels.[16,37]

Iron supplementation in a person whose diet is sufficient in this mineral does not necessarily lead to an increase in hemoglobin or hematocrit,[69] or other measures of iron status.[35] Even in instances of mild iron deficiency without anemia, iron supplementation may not enhance exercise capacity or aerobic performance.[61] When supplements are administered they should not be used indiscriminately because excessive iron can accumulate to toxic levels in the body and cause serious side effects.

SODIUM, POTASSIUM, AND CHLORINE

The minerals sodium, potassium, and chlorine are collectively termed **electrolytes** because they are dissolved in the body as electrically charged particles called ions. Sodium and chlorine are the chief minerals in blood plasma and extracellular fluid. A major function of these electrolytes is to modulate fluid exchange within the body's various fluid compartments. This allows for a constant, well-regulated exchange of nutrients and waste products between the cell and its external fluid environment. Potassium is the chief intracellular mineral.

Perhaps the most important function of the mineral electrolytes sodium and potassium is their role in establishing the proper electrical gradients across cell membranes. This electrical difference between the interior and exterior of the cell is required for the transmission of nerve impulses, for the stimulation and contraction of muscle, and for the proper functioning of glands. The electrolytes are also important in maintaining the permeability of the cell membranes and controlling the balance between the acid and base qualities of the body fluids, especially the blood.

Sodium: How Much is Enough?

In general, if sodium intake is low the hormone **aldosterone** acts on the kidney to conserve sodium. Conversely if sodium intake is high the excess is excreted in the urine. Salt balance, consequently, is maintained at normal levels throughout a wide range of intakes. For certain susceptible individuals, however, this is not always the case and excessive sodium intake is not adequately regulated. This intake, combined with other factors, tends to increase fluid volume and elevate blood pressure to levels that may pose a health risk.[90] This sodium-induced hypertension that occurs in about one-third of people suffering from hypertension, is prevalent in the United States and Japan. Here, sodium intake regularly exceeds the recommended level for adults of between 1100 to 3300 mg per day, or the amount of sodium in 0.5 to 1.5 teaspoons of salt (about 40% of table salt is sodium). In fact, for the person consuming the typical Western diet about 4500 mg of sodium or about 8 to 12 g of salt are ingested each day,[28] a value that is 10 times the 500 mg of sodium that the body actually needs. This large sodium intake is primarily due to the heavy reliance placed on table salt in processing, curing, cooking, seasoning, and storing foods. Aside from table salt, common sodium-rich dietary sources are monosodium glutamate (MSG), soy sauce, condiments, canned foods, baking soda, and baking powder.

For decades, one low-risk first line of defense in the treatment of high blood pressure was to eliminate all excess sodium in the diet. Sodium is so widely distributed naturally in foods that it is easy to obtain the daily requirement without relying on salt. By reducing sodium intake it is believed that sodium and fluid levels in the body are concomitantly reduced which may favorably lower blood pressure. If dietary constraints are ineffective, drugs that induce a water loss (diuretics) are the next line of defense. In addition to reducing sodium and fluid in the body, however, diuretics also cause a loss in other minerals, particularly potassium. For a patient using diuretics, a potassium-rich diet is a must. Although the effectiveness of sodium restriction for controlling hypertension in the general population is presently debated among medical specialists, it does appear that certain individuals are "salt sensitive." They respond favorably in terms of blood pressure when dietary sodium is reduced.

MINERALS AND EXERCISE PERFORMANCE

For normal individuals receiving the RDA of minerals, there is no evidence that supplementation benefits exercise performance. An important consequence of prolonged exercise, especially in hot weather, is the loss of water and mineral salts, primarily sodium and some potassium chloride in sweat. Excessive water and electrolyte losses impair heat tolerance and exercise performance and can lead to severe dysfunction in the form of heat cramps, heat exhaustion, or heat stroke. The yearly toll of heat-related deaths during spring and summer football practice provides a tragic illustration of the importance of fluid and electrolyte replacement. It is not uncommon for an athlete to lose anywhere from 1 to 5 kg of water each practice session or during a game as a result of sweating. This fluid loss corresponds to a depletion of 1.5 to 8.0 g of salt, because each kilogram (1 liter) of sweat generally contains about 1.5 g of salt. **The crucial and immediate need in these situations is to replace the water lost through sweating.**

Vigorous exercise triggers a rapid and coordinated release of the hormones vasopressin, renin, and aldosterone, that reduce sodium and water loss through the kidneys.[22] Even under extreme conditions such as marathon running in warm weather in which sweat output may be as high as 2 l per hour, sodium conservation by the kidneys is increased and electrolytes that are lost can usually be replenished by adding a slight amount of salt to the fluid ingested or to the normal daily food intake. In one study of runners during a 20-day road race in Hawaii, plasma minerals were maintained at normal levels when the athletes consumed an unrestricted diet without mineral supplements.[24] This finding, as well as those of others,[43] indicates that ingesting so-called athletic drinks is of no special benefit in replacing most of the minerals lost through sweating when individuals maintain a well-balanced diet. In fact, research shows that most individuals unconsciously consume more salt when the need exists. For fluid losses in excess of 4 or 5 kg and for prolonged periods of work in the heat, salt supplements may be necessary and can be achieved with a 0.1 to 0.2% salt solution by adding about 0.3 teaspoon of table salt per liter of water.[2] Although a potassium deficiency may occur with intense exercise,[49,79] the appropriate potassium level with most exercise and training is generally assured by consuming a diet containing normal amounts of this mineral[18,19] A glass of orange or tomato juice replaces almost all of the calcium, potassium, and magnesium lost in about 3 liters (7 lb) of sweat.

For chromium and zinc, strenuous exercise may place a drain on the body's content of these essential trace minerals required for normal carbohydrate and lipid metabolism. In one study,[3] urinary losses of zinc and chromium were 1.5- and 2–fold higher, respectively, on the day of a strenuous 6-mile run compared to a rest day. Although this loss does not necessarily mean that athletes should suppplement these micronutrients, it is possible that for most people who have marginal zinc and chromium intakes any further loss with strenuous exercise would need to be replaced to prevent an overt deficiency. Undoubtedly, we may see an increase in research in the fascinating area of trace mineral metabolism in general and the effects of exercise and training on mineral requirements in particular.[4,51a]

■■■■ SUMMARY

1. About 4% of the body's weight is composed of 22 elements called minerals. Minerals are a part of enzymes, hormones, and vitamins; they are found in muscles, connective tissues, and all body fluids.
2. Minerals occur freely in nature, in the waters of rivers, lakes, and oceans, and in soil. They are absorbed into the root system of plants and eventually incorporated into the tissues of animals that consume these plants.
3. A primary function of minerals is in metabolism where they serve as important parts of regulatory enzymes. Minerals provide structure in the formation of bones and teeth and are also important for the synthesis of the biologic nutrients, glycogen, fat, and protein.
4. With a balanced diet, there is generally adequate mineral intake, except perhaps in geographic locations where there is an absence of certain minerals such as iodine.

5. There is some evidence that about 40% of the women of child-bearing age in this country suffer from iron insufficiency. This could lead to iron deficiency anemia, a condition that would significantly affect exercise performance.

6. It is not clear whether regular physical activity creates a significant drain on the body's iron reserves. If this proves to be the case, females who, as a group, have the greatest iron requirements and lowest intake could be at risk of developing anemia. Assessment of the body's iron status should include an evaluation of both hematological characteristics and iron reserves.

7. Among older individuals, particularly women, the disease of osteoporosis has reached almost epidemic proportions. Adequate calcium intake and regular exercise provide an effective defense against bone loss at any age.

8. Paradoxically, women who train intensely and reduce body weight to the point where menstruation is adversely affected often show advanced bone loss at an early age.

9. As a result of excessive sweating caused by exercise, there are large losses of body water and related minerals. These should be replaced during and following exercise. In most instances specific mineral supplementation is not required.

P   A   R   T

3   Water

WATER IN THE BODY

Although water does not contribute to the nutrient value of food per se, it is still important in describing food composition and energy balance. The energy content of a particular food tends to be inversely related to its water content. As a general rule, foods that are high in water are low in calories. For this reason, the energy content of any food is sometimes expressed per "dry weight" of the food.

From 40 to 60% of an individual's body weight is water. Water constitutes 65 to 75% of the weight of muscle and less than 25% of the weight of fat. Consequently, differences in total body water between individuals are largely due to variations in body composition (that is, differences in lean versus fat tissue).

There are two main water "compartments" in the body; **intracellular,** referring to inside the cell and **extracellular,** referring to outside the cell. The extracellular fluid includes the blood plasma and lymph, saliva, fluids in the eyes, fluids secreted by glands and the intestines, fluids that bathe the nerves of the spinal cord, and fluids excreted from the skin and kidneys. Of the total body of water, an average of 62% is located intracellularly and 38% extracellularly.

FUNCTIONS OF BODY WATER

Water is truly a remarkable nutrient that is essential to life. Without water, death occurs within days. It serves as the body's transport and reactive medium. Diffusion of gases always takes place across surfaces moistened by water. Nutrients and gases are transported in aqueous solution; waste products leave the body through the water in urine and feces. Water has tremendous heat-stabilizing qualities. It can absorb a considerable quantity of heat with only a small change in temperature. Water lubricates our joints. Because it is essentially noncompressible, it helps give structure and form to the body through the turgor it provides for body tissues.

WATER BALANCE: INTAKE VERSUS OUTPUT

The water content of the body remains relatively stable within an individual over time. Although water output may frequently exceed water intake, this imbalance is quickly adjusted with appropriate fluid intake to restore the body's fluid level balance. The sources of water intake and output are shown in Figure 2–3.

WATER INTAKE

Normally, about 2.5 liters of water are required each day for a fairly sedentary adult in a normal environment. This water is supplied from three sources: (1) from liquids, (2) in foods, and (3) during metabolism.

Water from Liquids

The average individual normally consumes 1200 milliliters (ml) or 41 oz of water each day. Of course, during exercise and thermal stress, fluid intake can increase five or six times above normal. There are reports of an individual losing 13.6 kg (30 lb) of water weight during a 2-day, 17-hour, 55-mile run across Death Valley, California.[77] With proper fluid ingestion, however, including salt supplements, the actual body weight loss was only 1.4 kg. In this example, fluid loss and replenishment amounted to between 3.5 and 4 gallons of liquid!

Water in Foods

Most foods, especially fruits and vegetables, contain large quantities of water. Such foods as lettuce, pickles, green beans, and broccoli are examples of foods that have a high water content, whereas the water contained in butter, oils, dried meats, and chocolate, cookies, and cakes is relatively low.

Metabolic Water

When food molecules are degraded for energy, carbon dioxide and water are formed. This water is termed metabolic water and accounts for about 25% of the daily water requirement of a sedentary person. The complete breakdown of 100 g of carbohydrate, protein, and fat produces 55, 100, and 107 g of metabolic water, respectively. As mentioned in Chapter 1, each gram of glycogen is hydrated with 2.7 g of water. Consequently, this water also becomes available when glycogen is used for energy.

WATER OUTPUT

Water is lost from the body in urine, through the skin, as water vapor in the expired air, and in feces.

Water Loss in Urine

Under normal conditions the kidneys reabsorb about 99% of the 140 to 160 liters of filtrate formed each day; thus, the volume of urine excreted by the kidneys ranges from 1000 to 1500 ml or about 1.5 quarts per day.

It has been estimated that 15 ml of water are required to eliminate 1 g of solute. Thus, a portion of water in urine is "obligated" in order to rid the body of metabolic solutes such as urea, an end product of protein breakdown. In this way, the use of large quantities of protein for energy (as would occur with a high-protein diet) can actually speed up the body's dehydration during exercise.

FIG. 2–3.
Water balance in humans.

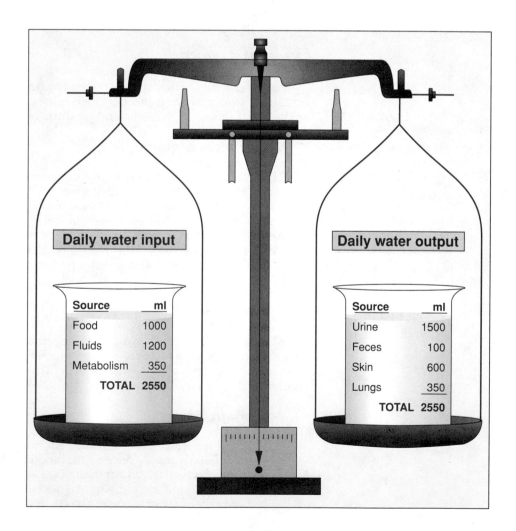

Water Loss Through the Skin

A small quantity of water, perhaps 350 ml, seeps from the deeper tissues through the skin to the body's surface. This continuous loss of water has been termed **insensible perspiration.**

Water is also lost through the skin in the form of sweat produced by specialized sweat glands located beneath the skin. Evaporation of sweat provides the refrigeration mechanism to cool the body. Under normal conditions, 500 to 700 ml of sweat are secreted each day. This by no means reflects sweating capacity because 8 to 12 liters of sweat (about 10 kg at a rate of 1 liter per hour) can be produced during prolonged exercise in a hot environment.

Water Loss as Water Vapor

The amount of insensible water loss through small water droplets in exhaled air is 250 to 350 ml per day. Exercise also affects this source of water loss. For physically active persons, 2 to 5 ml of water are lost from the respiratory passages each minute during strenuous exercise.[57] This amount varies considerably with climate, being less in hot, humid weather and greatest in cold temperatures where the inspired air contains little moisture or at altitudes where ventilatory volumes are significantly elevated at rest and during exercise.

Water Loss in Feces

Between 100 and 200 ml of water are lost through intestinal elimination because approximately 70% of fecal matter is water. The remainder is composed of nondigestible material; this includes bacteria from the digestive process and the residues of digestive juices from the intestines, stomach, and pancreas. With diarrhea or vomiting, the water loss can increase to 1,500 to 5,000 ml.

WATER REQUIREMENT IN EXERCISE

The most serious consequence of profuse sweating is the loss of body water. The amount of water lost through sweating depends on the severity of physical activity as well as on the environmental temperature. The relative humidity of the surrounding air is also an important factor affecting the efficiency of the sweating mechanism in temperature regulation. The term **relative humidity** refers to the water content of the air. During conditions of 100% relative humidity, the air is completely saturated with water vapor. Thus, evaporation of fluid from the skin to the air is impossible, and this important avenue is closed for body cooling. Under such conditions, sweat beads on the skin and eventually rolls off. On a dry day, the air can hold a considerable amount of moisture, and there is rapid evaporation of fluid from the skin. Thus, the sweat mechanism functions at optimal efficiency and body temperature is more easily controlled. A more detailed discussion about sweating, temperature regulation, and fluid replacement with exercise is presented in Chapter 25. It is important to note at this point, however, that blood volume is reduced when sweating causes a fluid loss of 2 to 3% of body mass. This loss of fluid places a significant strain on circulatory function that ultimately impairs the capacity for both exercise and thermoregulation.

SUMMARY

1. Water makes up 40 to 60% of the total body mass. Muscle is 72% water by weight, whereas water represents only about 20 to 25% of the weight of fat.
2. Of the total body water, roughly 62% is located intracellularly (inside the cells) and 38% extracellularly in the plasma, lymph, and other fluids outside the cell.
3. Normal daily water intake of about 2.5 liters is supplied from (1) liquid intake (1.2 liters), (2) food (1.0 liters), and (3) metabolic water produced during energy-yielding reactions (0.3 liters).
4. Water is lost from the body each day (1) in the urine (1 to 1.5 liters), (2) through the skin as insensible perspiration (0.50 to 0.70 liters), (3) as water vapor in expired air (0.25 to 0.30 liters), and (4) in feces as about 70% of fecal matter is water (0.10 liters).
5. Food and oxygen are always supplied in aqueous solution and waste products always leave via a watery medium. Water also helps give structure and form to the body and plays a vital role in temperature regulation.
6. Exercise in hot weather greatly increases the body's water requirement. In extreme conditions the fluid needs can increase five or six times above normal.

REFERENCES

1. Aloai, J.F.: Exercise and skeletal health. *J. Am. Geriatrics Soc.*, 29:104, 1981.
2. American College of Sports Medicine: Position statement on prevention of heat injuries during distance running. *Med. Sci. Sports Exerc.*, 16:ix, 1984.
3. Anderson, R.A., et al.: Strenuous running: acute effects on chromium, copper, zinc, and selected variables in urine and serum of male runners. *Biol. Trace Element Res.*, 6:327, 1984.
4. Anderson, R.A., and Guttman, H.N.: Trace minerals and exercise. In *Exercise, Nutrition, and Energy Metabolism.* Edited by E.S. Horton and R.L. Terjung. New York, Macmillan, 1988.
4a. Balban, E.P., et al.: The frequency of anemia and iron deficiency in the runner. *Med. Sci. Sports Exerec.*, 21:643, 1989.
5. Belko, A.Z.:Vitamins and exercise: an update. *Med. Sci. Sports Exerc.*, 19:S191, 1987.
6. Block, J.E., et al.: Determinants of bone density among athletes engaged in weight-bearing and non-weight-bearing activity. *J. Appl. Physiol.*, 67:1100, 1989.
7. Blum, S.M., et al.: The effects of fitness-type exercise on iron status in adult women. *Am. J. Clin. Nutr.*, 43:456, 1986.
8. Bothwell, T.H., et al.: *Iron Metabolism in Man.* Boston, Blackwell, 1979.
9. Brewer, V., et al.: Role of exercise in prevention of involutional bone loss. *Med. Sci. Sports Exerc.*, 15:455, 1983.
10. Brotherhood, J., et al.: Hematological status of middle- and long-distance runners. *Clin. Sci. Mol. Med.*, 48:139, 1975.
11. Brune, M., et al.: Iron loss in sweat. *Am. J. Clin. Nutri.*, 43:438, 1986.
12. Buckman, M.T.: Gastrointestinal bleeding in long distance runners. *Ann. Intern. Med.*, 101:127, 1984.
13. Buick, F.J., et al.: Effects of induced erythrocythemia on aerobic work capacity. *J. App. Physiol.*, 48:636, 1980.

14. Cann, E.C.: Decreased spinal mineral content in amenorrheic women. *JAMA*, 251:626, 1984.
15. Clement, D.B., and Asmundson, R.C.: Nutritional intake and hematological parameters in endurance runners. *Phys. Sportsmed.*, 10:37, 1982.
16. Clement, D.B., and Sawchuk, L.L.: Iron status and sports performance. *Sports Med.*, 1:65, 1984.
17. Clinical Nutrition: Vitamin C. Toxicity. *Nutr. Rev.*, 34:236, 1977.
18. Costill, D.L.: Nutritional requirements for endurance athletes. In *Toward an Understanding of Human Performance.* Edited by E.J. Burke. Ithaca, NY, Movement Publications, 1977.
19. Costill, D.L., et al.: Dietary potassium and heavy exercise: effects on muscle water and electrolytes. *Am. J. Clin. Nutr.*, 36:266, 1982.
20. Dalsky, G.P., et al.: Weight-bearing exercise training and lumbar bone mineral content in postmenopausal women. *Ann. Intern. Med.*, 108:824, 1988.
20a. Dalsky, G.P., et al.: Effect of exercise on bone: permissive influence of estrogen and calcium. *Med. Sci. Sports Exerc.*, 22:281, 1990.
21. Davies, K.J.A., et al.: Muscle mitochondrial bioenergetics, oxygen supply, and work capacity during dietary iron deficiency and repletion. *Am. J. Physiol.*, 242:E418, 1982.
22. De Souza, M.J., et al.: Menstrual status and plasma vasopressin, renin activity, and aldosterone exercise responses. *J. Appl. Physiol.*, 67:736, 1989.
23. Dressendorfer, R.H., et al.: Development of pseudoanemia in marathon runners during a 20-day road race. *JAMA*, 246:1215, 1981.
24. Dressendorfer, R.H., et al.: Plasma mineral levels in marathon runners during a 20-day road race. *Phys. Sportsmed.*, 10:113, 1982.
25. Drinkwater, B.L., et al.: Bone mineral content of amenorrheic and eumennorrheic athletes. *N. Engl. J. Med.*, 311:277, 1984.
26. Ehn, L., et al.: Iron status of athletes involved in intense physical activity. *Med. Sci. Sports Exerc.*, 12:61, 1980.
27. Falsetti, H.L., et al.: Hematological variations after endurance running with hard- and soft-soled running shoes. *Phys. Sportsmed.*, 11:118, 1983.
28. Federation of American Societies for Experimental Biology: Evaluation of the health aspects of sodium chloride and potassium chloride as food ingredients. Prepared for Bureau of Foods, Food and Drug Administration, US Department of Health, Education and Welfare, Washington, DC, 1979.
29. Frederickson, L.A., et al.: Effects of training on indices of iron status of young female cross-country runners. *Med. Sci. Sports Exerc.*, 15:271, 1983.
30. Gardner, G.W., et al.: Cardiorespiratory, hematological and physical performance responses of anemic subjects to iron treatment. *Am. J. Clin. Nutr.*, 28:982, 1975.
31. Gey, G.O., et al.: Effects of ascorbic acid on endurance performance and athletic injury. *JAMA*, 211:105, 1970.
32. Gonzalez, E.R.: Premature bone loss found in some nonmenstruating sports-women. *JAMA*, 248:513, 1982.
33. Green, H.J., et al.: Training induced hypervolemia: lack of an effect on oxygen utilization during exercise. *Med. Sci. Sports Exerc.*, 19:202, 1987.
34. Haliova, L., and Anderson, J.J.B.: Lifetime calcium intake and physical activity habits: independent and combined effects on the radial bone of healthy premenopausal Caucasian women. *Am. J. Clin. Nutr.*, 49:534, 1989.
35. Haymes, E.M., et al.: Training for cross-country skiing and iron status. *Med. Sci. Sports Exerc.*, 18:162, 1986.
36. Heaney, R.P., et al.: Menopausal changes in calcium balance performance. *J. Lab. Clin. Med.*, 92:953, 1978.
37. Heaney, R. P., et al.: Calcium nutrition and bone health in the elderly. *Am. J. Clin. Nutr.*, 36:986, 1982.
38. Herbert, V.: Toxicity of vitamin E. *Nutr. Rev.*, 35:158, 1977.
39. Herbert V.: Megavitamin therapy. *Contemp. Nutr.*, 2(10):1977.
40. Herbert, V., et al.: Destruction of vitamin B by vitamin C. *Am. J. Clin. Nutr.*, 30:297, 1977.
41. Huddleston, A.L., et al.: Bone mass in lifetime tennis athletes. *JAMA*, 244:1107, 1980.
42. Hundig, A., et al.: Runner's anemia and iron deficiency. *Acta Med. Scand.*, 209:315, 1981.
43. Johnson, H.L., et al.: Effects of electrolyte and nutrient solutions on performance and metabolic balance. *Med. Sci. Sports Exerc.*, 20:26, 1988.

44. Kanstrup, L.-L., and Ekblom, B.: Blood volume and hemoglobin concentration as determinants of maximal aerobic power. *Med. Sci. Sports Exerc.*, 16:256, 1984.

45. Krebs, P., et al.: The acute and prolonged effects of marathon running on 20 blood parameters. *Phys. Sportsmed.*, 10:58, 1982.

46. Krolner, B., et al.: Physical exercise as a prophylaxis against involutional vertebral bone loss: a controlled trial. *Clin. Sci.*, 64:541, 1983.

47. Lamanca, J.J., et al.: Sweat iron loss of male and female runners during exercise. *Int. J. Sports Med.*, 9:52, 1988.

48. Lane, N.E., et al.: Long-distance running, bone density, and osteoarthritis. *JAMA*, 255:1147, 1986.

49. Lane, H.W., et al.: Effect of physical activity on human potassium metabolism in a hot and humid environment. *Am. J. Clin. Nutr.*, 31:838, 1978.

50. Lawrence, J.D., et al.: Effects of α-tocopherol acetate on the swimming endurance of trained swimmers. *Am. J. Clin. Nutr.*, 28:205, 1975.

51. Lee, C.J., et al.: Effects of supplementation of the diets with calcium and calcium-rich foods on bone density of elderly females with osteoporosis. *Am. J. Clin. Nutr.*, 34:819, 1981.

51a. Lukaski, H.C., et al.: Physical training and copper, iron, and zinc status of swimmers. *Am. J. Clin. Nutr.*, 51:1093, 1990.

52. Lynch, S.R., et al.: Iron status of elderly Americans. *Am. J. Clin. Nutr.*, 36:1032, 1982.

53. Manore, M.M., and Leklem, J.E.: Effect of carbohydrate and vitamin B_6 on fuel substrates during exercise in women. *Med. Sci. Sports Exerc.*, 20:233, 1988.

54. Martin, R.K., et al.: Load-carrying effects on the adult beagle tibia. *Med. Sci. Sports Exerc.*, 13:343, 1981.

55. McCabe, M.E., et al.: Gastrointestinal blood loss associated with running a marathon. *Dig. Dis. Sci.*, 31:1229, 1986.

56. McDonald, R., et al.: Effects of iron deficiency and exercise on myoglobin in rats. *Eur. J. Appl. Physiol.*, 52:414, 1984.

57. Mitchel, J., et al.: Respiratory weight loss during exercise. *J. Appl. Physiol.*, 32:474, 1972.

58. Nachtigall, L.E., et al.: Estrogen replacement therapy: I. A ten-year prospective study in the relationship to osteoporosis. *Obstet. Gynecol.*, 53:277, 1979.

59. National Dairy Council: Food vs pills vs fortified foods. *Dairy Council Digest*, 58:7, 1987.

60. Nelson, R.A.: What athletes should eat. Unmixing folly and facts. *Phys. Sportsmed.*, 3:67, 1975.

61. Newhouse, I.J., et al.: The effects of prelatent/latent iron deficiency on physical work capacity. *Med. Sci. Sports Exerc.*, 21:263, 1987.

62. Newmark, H.L., et al.: Stability of vitamin B_{12} in the presence of ascorbic acid. *Am. J. Clin. Nutr.*, 29:645, 1976.

63. Nickerson, H.J., and Tripp, H.J.: Iron deficiency in adolescent cross-country runners. *Phys. Sportsmed.*, 11:60, 1983.

64. Oscai, L.B., et al.: Effect of exercise on blood volume. *J. Appl. Physiol.*, 24:622, 1968.

65. O'Toole, M.L., et al.: Hemolysis during triathlon races: its relation to race distance. *Med. Sci. Sports Exerc.*, 20:172, 1988.

66. Owen, R.A., et al.: The national cost of acute care of hip fractures associated with osteoporosis. *Clin. Orthop.*, 150:172, 1980.

67. Oyster, N., et al.: Physical activity and osteoporosis in post-menopausal women. *Med. Sci. Sports. Exerc.*, 16:44, 1984.

68. Pate, R.: Sports anemia: A review of the current literature. *Phys. Sportsmed.*, 11:115, 1983.

69. Pate, R., et al.: Dietary iron supplementation in women athletes. *Phys. Sportsmed.*, 7:16, 1979.

70. Percey, E.C.: Ergogenic aids in athletics. *Med. Sci. Sports*, 10:298, 1978.

71. Position of the American Dietetic Association: Nutrition for physical fitness and athletic performance for adults. *ADA Reports*, 87:933, 1987.

72. Puhl, J.L., et al.: Erythrocyte changes during training in high school women cross-country runners. *Res. Q. Exer. Sport*, 52:484, 1981.

72a. Raab, D.M., et al.: Bone mechanical properties after exercise training in young and old rats. *J. Appl. Physiol.*, 68:130, 1990.

73. Regelson, W., and Sines, F.: *Intervention in the Aging Process*. New York, Alan R. Liss, 1983.

74. Riis, B., et al.: Does calcium supplementation prevent postmenopausal bone loss? A double-blind controlled clinical study. *N. Engl. J. Med.*, 36:173, 1987.
75. Risser, W.L., et al.: Iron deficiency in female athletes: Its prevalence and impact on performance. *Med. Sci. Sports. Exerc.*, 20:116, 1988.
76. Robertson, J.O., et al: Fecal blood loss in response to exercise. *Br. Med. J.*, 295:303, 1987.
77. Robinson, S.: Cardiovascular and respiratory reactions to heat. In *Physiological Adaptations*. Edited by M.K. Yousef et al. New York, Academic Press, 1972.
78. Roe, D.A.: Vitamin requirements for increased physical activity. *Exercise, Nutrition, and Energy Metabolism.* Edited by E. Horton and R.L. Terjung. New York, Macmillan, 1988.
79. Rose, K.: Warning for millions: intense exercise can deplete potassium. *Phys. Sportsmed.*, 3:67, 1975.
80. Rossander, L., et al.: Absorption of iron from breakfast meals. *Am. J. Clin. Nutr.*, 32:2484, 1979.
81. Schawmberg, H., et al.: Sensory neuropathy from pyridoxine abuse: a new megavitamin syndrome. *N. Engl., J. Med.*, 309:445, 1983.
82. Schoene, R.B., et al.: Iron repletion decreases maximal exercise lactate concentrations in female athletes with minimal iron-deficiency anemia. *J. Lab. Clin. Med.*, 102:306, 1983.
83. Schrimshaw, N.S., and Young, V.R.: The requirements of human nutrition. *Sci. Am.*, 235:50, 1976.
84. Serfass, W.C.: Nutrition for the athlete. *Contemp. Nutr.*, 2:1977.
85. Smith, E.L.: Exercise for prevention of osteoporosis: a review. *Phys. Sportsmed.*, 10:72, 1982.
86. Smith, E.L., et al.: Physical activity and calcium modalities for bone mineral increase in aged women. *Med. Sci. Sports Exerc.*, 13:60, 1981.
87. Snyder, A.C., et al.: Importance of dietary iron source on measures of iron status among female runners. *Med. Sci. Sports Exerc.*, 21:7, 1989.
88. Spencer, H., and Kramer, L.: NIH Consensus Conference: Osteoporosis. *J. Nutr.*, 116:316, 1986.
89. Stewart, J.G., et al.: Gastrointestinal blood and anemia in runners. *Ann Intern. Med.*, 100:843, 1984.
90. Tobian, L.: Dietary salt and hypertension. *Am. J. Clin. Nutr.*, 32:2659, 1979.
91. Trembly, A., et al.: The effects of riboflavin supplementation on the nutritional status and performance of elite swimmers. *Nutr. Res.*, 4:201, 1984.
91a. Use of vitamin and mineral supplements in the United States. *Nutr. Rev.*, 70:43, 1990.
92. Van Dam, B.: Vitamins and sports. *Br. J. Sports Med.*, 12:74, 1978.
93. Weight, L.M., et al.: Vitamin and mineral supplementation: effect on running performance of trained athletes. *Am. J. Clin. Nutr.*, 47:192, 1988.
94. Williams, J.A.: The effect of long-distance running upon appendicular bone mineral content. *Med. Sci. Sports. Exerc.*, 16:223, 1984.
95. Wirth, J.C., et al.: The effects of physical training on the serum iron levels of college-age women. *Med. Sci. Sports*, 10:223, 1978.
96. Wolf, W., and Lohman, T.G.: Nutritional practices of coaches in the Big Ten. *Phys. Sportsmed.*, 7:112, 1979.

OPTIMAL NUTRITION FOR EXERCISE

An optimal diet is defined as one in which the supply of required nutrients is adequate for tissue maintenance, repair, and growth without an excess energy intake. Only in the last few years has it been possible to obtain a reasonable estimate of the specific nutrient needs for individuals of different ages and body sizes, with considerations for individual differences in digestion, storage capacity, nutrient metabolism, and daily levels of energy expenditure. Dietary recommendations for athletes may be further complicated by the specific energy requirements of a particular sport as well as by the athlete's dietary preferences. Truly, there is no one diet for optimal exercise performance. Sound nutritional guidelines must be followed, however, in planning and evaluating food intake.

NUTRIENT REQUIREMENT

Many coaches make dietary recommendations based on their own "feelings" and past experiences rather than rely on available evidence. This problem is compounded because athletes often have either inadequate or incorrect information concerning prudent dietary practices as well as the role of specific nutrients in the diet.[23] Although research in this area is far from complete, the general consensus is that active people and athletes do not require additional nutrients beyond those obtained in a balanced diet.[16] This is important because a large number of adult Americans exercise regularly to keep fit. In fact, research indicates that active Americans, including those involved in exceptional endurance activities, consume typical diets that are remarkably similar in composition to those consumed by their more sedentary counterparts.[2,13,14,21] As shown in Table 3–1, the main difference is that they eat more of the same foods which results in a larger total **quantity** of food consumed to support the extra energy required by training. In essence, sound nutrition for athletes is sound human nutrition. For the endurance athlete, special consideration must be given to maintaining adequate and regular carbohydrate intake.

TABLE 3–1.
COMPARISON OF CARBOHYDRATE, FAT, PROTEIN, AND CALORIC INTAKE OF MIDDLE-AGED MALE AND FEMALE RUNNERS AND SEDENTARY CONTROLS[a]

	Runners	Sedentary Controls
Males		
Calories (kcal · day^{-1})	2959*	2361
Protein (g · day^{-1})	102.1	93.6
Protein (%)	13.8*	15.8
Fat (g · day^{-1})	134.4*	109.0
Fat (%)	40.8	41.5
Carbohydrate (g · day^{-1})	294.6*	225.7
Carbohydrate (%)	39.8	38.6
Cholesterol (mg · 1000 kcal^{-1})	175.0	190.0
Saturated fat (g · 1000 kcal^{-1})	16.2	16.0
Polyunsaturated fat (g · 1000 kcal^{-1})	9.0	9.3
Females		
Calories (kcal · day^{-1})	2386*	1871
Protein (g · day^{-1})	82.2	76.7
Protein (%)	14.2*	17.4
Fat (g · day^{-1})	110.7*	83.0
Fat (%)	41.1	40.3
Carbohydrate (g · day^{-1})	234.3*	174.7
Carbohydrate (%)	39.5	39.1
Cholesterol (mg · 1000 kcal^{-1})	190.0	205.0
Saturated fat (g · 1000 kcal^{-1})	16.8	16.5
Polyunsaturated fat (g · 1000 kcal^{-1})	8.5	7.9

[a] % calories do not total 100% because alcohol calories constitute the difference.
* Values for runners are significantly different from controls.
(From: Blair, S.N., et al.: Comparisons of nutrient intake in middle-aged men and women runners and controls. *Med. Sci. Sports. Exerc.*, 13:310, 1981.)

RECOMMENDED NUTRIENT INTAKE

Figure 3–1 shows the recommended basic nutrient intake for protein, fat, and carbohydrate, as well as the general category of food sources for these nutrients. These guidelines provide for the necessary vitamin, mineral, and protein requirements even though the energy content of this food intake amounts to only about 1200 calories per day. (A calorie is a unit of heat used to express the energy value of food.) In terms of average values for young adult Americans, the total daily energy requirement is about 2100 for women and 2700 calories for men. **Once the basic nutrient requirements are met (as recommended in Fig. 3–1), the extra energy needs of the person can be supplied for a variety of food sources based on individual preference.**

Protein

As we discussed in Chapter 1, the standard recommendation for protein intake is 0.8 g protein per kilogram body weight. A person who weighs 77 kg (170 lbs) would therefore require about 57 g or 2.0 ounces of protein daily. Assuming that even during strenuous exercise there is relatively little protein

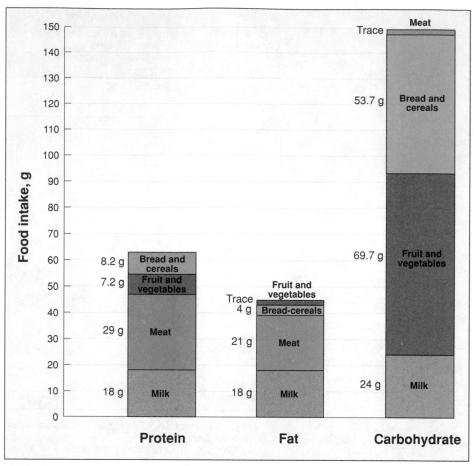

FIG. 3–1.
Recommendations for basic carbohydrate, fat, and protein intake as well as food sources for these nutrients. (From Howe, P.: Basic Nutrition in Health and Disease. Philadelphia, W.B. Saunders Co., 1971.)

loss through energy metabolism (an assumption that is probably not entirely correct), the preceding protein recommendation is probably adequate for both active and sedentary persons. Also, the protein intake in the average American diet significantly exceeds the recommended protein requirement, and the athlete's diet is usually two to five times in excess of the protein intake considered optimal!

Fat

Standards for optimal fat intake have not been firmly established because relatively little is known concerning the human requirement for this nutrient. The amount of dietary fat varies widely according to personal taste, money spent on food, and the availability of fat-rich foods. For example, only about 10% of the energy in the average diet of people living in Asia is furnished by fat, whereas in the United States, Canada, Scandinavia, Germany, and France, fat accounts for 40 to 45% of the caloric intake. **Many nutritionists** believe that to promote optimal health, fat intake should not exceed 30% of the energy content of the diet. Of this, at least 70% should be in the form of unsaturated fats. To attempt to eliminate "all" fat from the diet, however, may be unwise and detrimental in terms of exercise performance. With low-fat diets, it may be difficult to increase one's intake of carbohydrate and protein to furnish sufficient energy to maintain a stable body weight and muscle mass during strenuous training. Also, because the essential fatty acids, most notably linoleic acid, and many vitamins gain entrance to the body through dietary fat, a low-fat diet could eventually result in a relative state of malnutrition.

Carbohydrate

The prominence of carbohydrates in the diet also varies widely throughout the world, depending upon factors such as the availability and relative cost of fat and protein-rich foods. Carbohydrate-rich foods such as grains, starchy roots, and dried peas and beans are usually the cheapest foods in relation to their energy value. In the Far East, carbohydrates (rice) contribute 80% of the total caloric intake, whereas in the United States only about 40 to 50% of the energy requirement comes from carbohydrates. Most evidence suggests that there is no health hazard in subsisting chiefly on carbohydrates (starches), provided that the essential amino acids, minerals, and vitamins are also present in the diet. In fact, the diet of the relatively primitive Tarahumara Indians of Mexico is very high in complex carbohydrates (75% of calories) and crude fiber (19 g/day) and correspondingly low in cholesterol (71 mg/day), fat (12% of calories) and saturated fat (2% of calories).[6] These people are noted for their remarkable physical endurance: They reportedly run distances of up to 200 miles in competitive soccer-type sports events that often last several days! This type of diet may offer health benefits to those who partake of it: Particularly notable among the Tarahumaras is the virtual absence of hypertension, obesity, and death from cardiac and circulatory complications. On the other hand, in terms of energy requirements, there is no evidence that more than a small amount of carbohydrate need be present in the diet if daily physical activity level is low. If, however, **the individual is physically active, the "prudent" diet should contain at least 50 to 60% of its calories in the form of carbohydrates, predominantly starches.** In strenuous training for specific sports and prior to competition, the carbohydrate intake may even be increased above this recommended level to ensure adequate glycogen stores. The specific dietary-exercise techniques for facilitating glycogen storage are presented in Chapter 23.

CARBOHYDRATE NEEDS IN PROLONGED, SEVERE TRAINING

Athletes training for endurance activities such as distance running, swimming, cross-country skiing, or cycling frequently experience a state of chronic fatigue in which successive days of hard training become exceedingly more difficult. This "staleness" may be related to a gradual depletion of the body's carbohydrate reserves with repeated strenuous training, even though the athlete's diet contains the recommended percentage of carbohydrate.[22] As shown in Figure 3–2, after three successive days of running 16.1 km (10 miles) a day, the glycogen in the thigh muscle was nearly depleted. This occurred although the daily food intake of the runners contained 40 to 60% carbohydrates. In addition, by the third day, the quantity of glycogen used

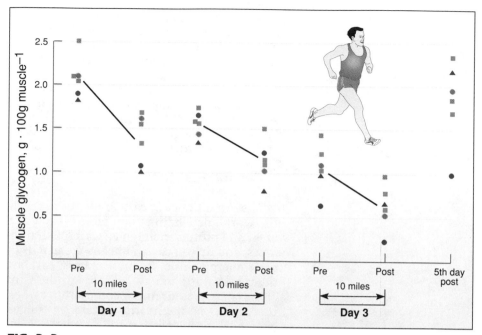

FIG. 3–2.

Changes in muscle glycogen concentration for 6 male subjects before and after each 16.1-km run performed on 3 successive days. Muscle glycogen was measured 5 days after the last run and is referred to as "fifth day post." (From Costill, D.L., et al.: Muscle glycogen utilization during prolonged exercise on successive days. J. Appl. Physiol., *31*:834, 1971.)

during the run was much less than on the first day. Presumably, the energy for work was supplied predominantly by the body's fat reserves. Because glycogen synthesis is related to dietary carbohydrate intake, some researchers have recommended increasing the daily intake of carbohydrates to 70% of total calories (612 g for 3500 calories) or higher to prevent the gradual depletion of the body's glycogen stores and induce protein sparing with successive days of hard training.[3,4,8,17] With more moderate yet regular training at least 400 to 500 g of carbohydrate should be ingested. This generally will be equivalent to between 50 and 60% of caloric intake.

Even if the diet is high in carbohydrates, muscle glycogen is not rapidly restored to the pre-exercise level. Although liver glycogen is restored rapidly,[7] at least 48 hours are required to restore muscle glycogen levels after prolonged, exhaustive exercise.[17] From the values in Figure 3–2, some individuals may require more than 5 days to re-establish muscle glycogen levels if the diet contains only moderate amounts of carbohydrates. Unmistakably, if a person performs heavy exercise on successive days, daily allowances must be adjusted to permit optimal glycogen resynthesis and the maintenance of high quality training.[10] **In addition, at least 2 days of rest or lighter exercise and high carbohydrate intake must be provided to establish the pre-exercise muscle glycogen levels.** This certainly provides a nutritional justification for the recommendation of many coaches and trainers to gradually reduce or taper the intensity of workouts several days prior to competition.

THE FOUR-FOOD-GROUP PLAN: THE ESSENTIALS OF GOOD NUTRITION

A practical approach to sound nutrition is to categorize foods that make similar nutrient contributions and then provide servings from each category in the daily diet. Key principles of good eating are variety and moderation.

This can be readily achieved by use of the Four-Food-Group Plan illustrated in Table 3–2. As long as the recommended number of servings from the variety provided in each group is supplied, and cooking and handling are proper, adequate nutrition is assured. More of these and other foods can be used as needed for growth, for activity, and for desirable weight.

Table 3–3 presents examples of three daily low calorie menus formulated from the guidelines of the basic diet plan shown in Table 3–2. Provided the foods are stored and prepared in a proper manner, these menus provide **all** essential nutrients, even though the energy value of each is well below the average adult requirement. In fact, these menus serve as excellent nutritional models for reducing diets. For active individuals whose daily energy requirement may be as large as 5000 calories, all that need be done once the essentials are provided is to increase the quantity of food consumed with greater emphasis given to carbohydrate-rich food sources. This is achieved either by increasing the size of portions, the frequency of meals or snacks, or the variety of nutritious foods consumed at each meal.

TABLE 3–2.
THE FOUR-FOOD-GROUP PLAN—THE FOUNDATION FOR A GOOD DIET

Food Category	Examples	Recommended Daily Servings[c]
I. Milk and milk products[a]	Milk, cheese, ice cream, sour cream, yogurt	2[e]
II. Meat and high-protein[b]	Meat, fish, poultry, eggs—with dried beans, peas, nuts, or peanut butter as alternatives	2
III. Vegetables and fruits[d]	Dark green or yellow vegetables; citrus fruits or tomatoes	4
IV. Cereal and grain food	Enriched breads, cereals, flour, baked goods, or whole-grain products	4

[a] If large quantities of milk are normally consumed, *fortified* skimmed milk should be substituted to reduce the quantity of saturated fats.
[b] Fish, chicken, and high-protein vegetables contain significantly less saturated fats than other protein sources.
[c] A basic serving of meat or fish is usually 100 g or 3.5 oz of edible food; 1 cup (8 oz) milk; 1 oz cheese; 1/2 cup fruit, vegetables, juice; 1 slice bread; 1/2 cup cooked cereal or 1 cup ready-to-eat cereal.
[d] One should be rich in vitamin C; at least one every other day rich in vitamin A.
[e] Children, teenagers, and pregnant and nursing women—4 servings.

TABLE 3–3.
THREE DAILY MENUS FORMULATED FROM GUIDELINES ESTABLISHED BY THE FOUR-FOOD-GROUP PLAN[a]

3 Meals a Day	5 Meals a Day	6 Small Meals a Day
Breakfast	**Breakfast**	**Breakfast**
½ cup unsweetened grapefruit juice 1 poached egg 1 slice toast 1 teaspoon butter or margarine ½ cup skim milk tea or coffee, black	½ grapefruit ⅔ cup bran flakes 1 cup skim or low-fat milk or other beverage	½ cup orange juice ¾ cup ready-to-eat cereal ½ cup skim milk tea or coffee, black
Lunch	**Snack**	**Mid-Morning Snack**
2 ounces lean roast beef[b] ½ cup cooked summer squash 1 slice rye bread 1 teaspoon butter or margarine 1 cup skim milk 10 grapes	1 small package raisins ½ bologna sandwich	½ cup low fat cottage cheese
	Lunch	**Lunch**
Dinner	1 slice pizza carrot sticks 1 apple 1 cup skim or low-fat milk	2 ounces sliced turkey on 1 slice white toast 1 teaspoon butter or margarine 2 canned drained peach halves ½ cup skim milk
3 ounces poached haddock[b] ½ cup cooked spinach tomato and lettuce salad 1 teaspoon oil + vinegar or lemon 1 small biscuit 1 teaspoon butter or margarine ½ cup canned drained fruit cocktail ½ cup skim milk	**Snack**	**Mid-Afternoon Snack**
	1 banana	1 cup fresh spinach and lettuce salad 2 teaspoons oil + vinegar or lemon 3 saltines
	Dinner	**Dinner**

	baked fish with
	    mushrooms (3 oz)[b]
	baked potato
	2 teaspoons margarine
	½ cup broccoli
	1 cup tomato juice or skim or
	    low-fat milk
		1 cup clear broth
		3 ounces broiled chicken breast[b]
		½ cup cooked rice with
		    1 teaspoon butter or margarine
		¼ cup cooked mushrooms
		½ cup cooked broccoli
		½ cup skim milk

		Evening Snack
		1 medium apple ½ cup skim milk
Total Calories: about 1200	**Total Calories: about 1400**	**Total Calories: about 1200**

[a] Each menu provides *all* essential nutrients; the energy or caloric value of the diet can be easily increased by increasing the size of the portions, the frequency of meals, or the variety of foods consumed at each sitting.
[b] Cooked weight

EXERCISE AND FOOD INTAKE

For individuals who engage regularly in moderate to intense physical activity, it is relatively easy to match food intake with the daily level of energy expenditure. Lumbermen, for example, who expend about 4500 calories daily, unconsciously adjust their caloric intake to balance closely their energy output. Consequently, body mass remains stable despite an extremely large food consumption. The balancing of food intake to meet a new level of energy output takes about a day or so, during which time a new energy equilibrium is attained. Apparently, this fine balance between energy expenditure and food intake is not maintained in sedentary people.[15] Here, the caloric intake generally exceeds the daily energy expenditure. This lack of precision in regulating food intake at the low end of the physical activity spectrum probably accounts for the "creeping obesity" commonly observed in highly mechanized and technically advanced societies.

It was reported that the daily food intake of **athletes** in the 1936 Olympics averaged more than 7000 calories, or roughly three times the average daily intake.[1] For these competitors, the protein intake amounted to 19% of the total caloric intake (320 g), whereas fat contributed 35% (270 g) and carbohydrate contributed 46% (800 g). These caloric values are often quoted and used to justify what appears to be an enormous food requirement of athletes in training. However, these figures are only estimates because objective dietary data were not presented in the original report. In all likelihood, they are inflated estimates of the energy expended (and required) by the athletes.[13] For example, distance runners who train upward to 100 miles per week (6 min per mile pace at about 15 calories per min) probably do not expend more than 800 to 1300 calories each day in excess of their normal energy requirement. For these endurance athletes, the daily food intake should supply about 4000 calories. Figure 3–3 presents data on energy intake from a large sample of elite male and female endurance, strength, and team sport athletes in the Netherlands. For male athletes energy intake ranged between 2900 and 5900 kcal per day, while the daily intake of female competitors ranged between 1600 and 3200 kcal. With the exception of the relatively high daily energy intake of athletes at extremes of performance and training such as triathletes and Tour de France cyclists, daily caloric intake generally did not exceed 4000 kcal for the men and 3000 kcal for the women.[11]

Eat More, Weigh Less. As was shown in Table 3–1, the caloric intake of 61 middle-aged men and women who ran an average of 60 kilometers per week amounted to about 40 to 60% more calories per kg of body weight compared to sedentary controls. This larger daily caloric intake for the runners was accounted for by the extra energy required to run between 8 and 10 kilometers daily. Paradoxically, the active men and women who ate considerably more on a daily basis weighed considerably less than subjects who were less active. Such data are generally consistent with other studies of active people and add further evidence to the strong argument that regular exercise provides an effective means by which a person can actually "eat more yet weigh less" and maintain a lower percentage of body fat.[5,21,24] It appears that active people maintain a lighter and leaner body and a healthier heart disease risk profile despite an increased intake of the typical American diet. The important role of exercise for weight control is discussed more fully in Chapter 29.

A more objective evaluation of the nutrient intake of highly trained athletes as estimated from dietary recall is presented in Table 3–4. These relatively large athletes were in the peak of training and were in good health as judged by physiologic, medical, and performance measures.[22] The average of 4663 calories consumed each day exceeded the average American adult's intake by about 67%. Protein intake exceeded by 137% the recommended requirement, whereas the above-average fat and carbohydrate intake supplied the energy requirements of training. The intake of all vitamins and minerals

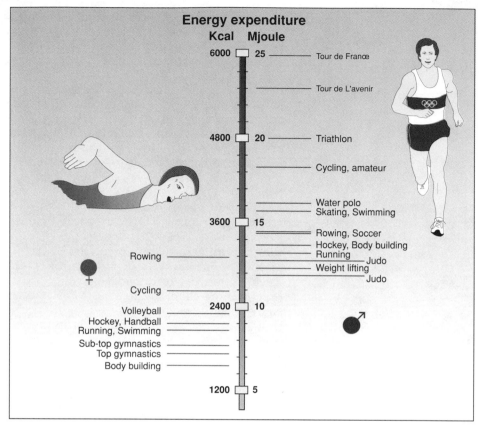

FIG. 3–3.
Daily energy intake in MJoule (kcal) per day in elite male and female endurance, strength, and team sport athletes. (From: van Erp-Baart, A.M.J., et al.: Nationwide survey on nutritional habits in elite athletes. Int. J. Sports Med., *10*:53, 1989.)

significantly exceeded recommendations of the Food and Nutrition Board. This "excess" of nutrients is due largely to the increased food intake and in no way suggests a "requirement" above the recommended guide-lines.

THE PREGAME MEAL

The main purpose of the pregame meal is to provide the athlete with adequate carbohydrate energy and assure optimal hydration. Within this framework, fasting prior to competition or training makes no sense physiologically as it rapidly leads to a depletion of liver and muscle glycogen. The food preferences of the athlete, the "psychologic set" of the competition, and the digestibility of the foods should be considered. As a general rule, foods that are high in fat and protein content should be eliminated from the diet on the day of competition because these foods are digested slowly and remain in the digestive tract for a longer time than foods containing similar amounts of energy in the form of carbohydrates.

A consideration must be made concerning the timing of the pre-event meal. With the increased stress and tension that usually accompany com-

TABLE 3–4.
COMPARISON OF THE NUTRIENT INTAKES OF UNITED STATES OLYMPIC-CALIBER ATHLETES WITH RDA VALUES[a,b]

	Total Calories	Protein (g)	Carbo-hydrates (g)	Fat (g)	Minerals (mg)		Vitamins (mg)				
					Ca	Fe	Thiamin	Ribo-flavin	Niacin	C	A$_{IU}$
Athletes (N = 16) Mean	4663	237	446	203	2158	29	3.3	4.9	43	165	8850
RDA	2700 (4250)[c]	89.6[d]	100[e]	130[f]	1200	10	1.5	1.7	19	60	5000
Absolute amount greater than RDA	+1963 (+413)[c]	+147.4	+346	+73	+958	+19	+1.8	+3.2	+24	+105	+3840
Percent greater than RDA	+73 (+9.7)[c]	+62	+346	+56	+80	+190	+120	+188	+126	+175	+77

[a] Modified from Ward, P., et al.: U.S.A. Discus Camp; Preliminary Report. Track Field Q. Rev. 76:29, 1976.

[b] Recommended allowance proposed by the Food and Nutrition Board. Recommended Dietary Allowances. Washington, D.C. National Academy of Sciences, 1989.

[c] Recommended daily caloric intake for persons engaged in heavy manual work or athletic training. Canadian Council on Nutrition. Dietary Standard of Canada. Can. Bull. Nutr. 6:1, 1964.

[d] Based on National Research Council standards of 0.8 g/kg body mass per day. The average mass of the athletes was 110.7 kg.

[e] Minimal requirement proposed by the Food and Nutrition Board of the National Research Council to prevent undesirable metabolic responses.

[f] Based on 25% of total caloric intake.

petition, there may be a significant decrease in blood flow to the stomach and small intestines and an accompanying decrease in absorption from the digestive tract. **A 3–hour period should be sufficient to provide for adequate absorption of the pre-event meal.**

PROTEIN OR CARBOHYDRATE?

Many athletes are psychologically accustomed and even dependent on the "classic" pregame meal that usually consists of steak and eggs. Although this meal may be satisfying to the athlete, coach, and restauranteur, its benefits in terms of exercise performance have yet to be demonstrated. In fact, such a meal, which is actually low in carbohydrates may be detrimental to optimal performance. On the other hand, high carbohydrate feedings in the pre-exercise meal may contribute to improved performance.[18]

There are several reasons for modifying or even abolishing the high-protein, pregame meal in favor of one high in carbohydrates. For one thing, the normal overnight fast results in a significant depletion of carbohydrate stores in liver and muscle which must be replenished by exogeneous carbohydrate sources via food intake. Carbohydrates are also digested and absorbed more rapidly than either proteins or fats. Thus, this food is available for energy faster and may also reduce the feeling of fullness usually experienced following a meal. Furthermore, the digestion, absorption, and assimilation of a high-protein meal elevates the resting metabolic rate considerably more than similar intake of a high-fat, high-carbohydrate meal. This metabolic heat adds additional strain to the body's heat-dissipating mechanisms that could be detrimental to exercise performance in hot weather. Concurrently, the breakdown of protein for energy facilitates dehydration during exercise. This is because the by-products of amino acid breakdown demand water for urinary excretion.

Equally important for favoring carbohydrate intake is because this is the main nutrient energy source for intense exercise and is also important in prolonged exercise. The pregame meal should provide adequate quantities of carbohydrate to assure a normal level of blood glucose and sufficient glycogen "energy reserves" for most activities, **provided that the athlete has maintained a nutritionally sound diet throughout training.** The pre-event meal cannot correct nutritional deficiencies and inadequate macronutrient composition during the weeks prior to competition. For the endurance athlete, the pre-event meal should be high in carbohydrates to assure peak storage of liver and muscle glycogen. This can be achieved in conjunction with other specific exercise-dietary modifications for "carbohydrate loading" discussed more fully in Chapter 23.

As was discussed in Chapter 1, a moderately large glucose challenge just before an endurance activity may actually impair subsequent performance. For example, the riding time of young men and women on a bicycle ergometer was reduced 19% when they consumed a 300-ml solution containing 75 g of glucose 30 minutes before exercise, compared to similar trials preceded by the same volume of water or a liquid meal of protein, fat, and carbohydrate.[12] This negative effect of glucose feeding is probably mediated by an increased insulin output following the glucose challenge and the inhibitory effects of insulin on fat mobilization. Thus, if large quantities of sugars are consumed prior to competition, several hours must be provided to permit assimilation of this nutrient and the re-establishment of metabolic and hormonal balance.

LIQUID MEALS

Commercially prepared liquid meals offer an alternate and seemingly effective approach to the pre-event meal. These foods are generally well balanced in nutritive value. They are high in carbohydrate yet contain enough fat and protein to contribute to a feeling of satiety. Because they are in liquid form, they also contribute to the athlete's fluid needs. The liquid meal is also advantageous because it is digested rapidly and completely, leaving essentially no residue in the intestinal tract. This approach to proper nutrition on the day of competition is especially effective during day-long meets such as in swimming and track, or in some tennis and basketball tournaments. In these situations, an athlete may have relatively little time (or interest) for food. Liquid meals may also provide a practical approach to the supplementation of the daily caloric intake of athletes who have difficulty maintaining body weight or who are desirous of increasing body mass.[20]

COMPUTERIZED MEAL AND EXERCISE PLAN

Nutritionists and exercise specialists have applied computer technology in the formulation of well-balanced meals and exercise programs for weight control. An example of such a plan is illustrated in Appendix G. The formulation of daily menus is based on the Four Food Group Plan and dietary exchange method developed by the American Dietetic Association. Rather than prescribing a particular food plan, the computerized dietary plan allows the person to select the **specific** foods he will eat from a basic list of the most common foods. Combined with age, weight, height, weight loss desired, and current level of physical activity, the computer prepares three nutritious meals for breakfast, lunch, and dinner for a 14-day period. The menu varies from day to day. The meals are balanced for nutrient intake of carbohydrate, fat, protein, vitamins, and minerals and are designed so that the individual can reduce excess weight (fat) at a safe but steady level. The 12 to 18-page printout includes a weight-loss curve that is not a straight line, but rather is curvilinear to account for changes that occur in metabolism and body composition as one progresses through the program of exercise and reduced food consumption. Also included are beginner, intermediate, or advanced programs of exercise for walking, jogging, running, cycling, or swimming, and other aerobic activities that include racquetball, circuit training, squash, badminton, basketball, downhill skiing, tennis, golf, and aerobic dancing.

Advantages of Computerized Planning

In addition to speed and reasonable accuracy, there are three major advantages of utilizing a computer in the dietary and exercise prescription.

1. Because the meal plans are based on the dietary exchange method, foods of one's preference can easily be substituted within a given meal to offer tremendous variety in planning the daily menus. Such flexibility still maintains the integrity of the nutritional adequacy of the meals and assures a constancy to the recommended level of caloric intake.
2. The major difference between the computer meal plan for dieting and the myriad of typical "diets" is that with the former, there is direct participation in the creation of well-balanced meals without having to follow a preset selection of foods chosen by someone else, often some self-appointed "expert." People who become actively involved in planning their daily meals are much more likely to remain with the program than if nutritional choices are unavailable. If a person selects to eat a particular food, then he probably will; having to eat disliked foods as part of a particular diet regimen is negative reinforcement and usually leads to failure (reject the diet) before achieving a particular goal (15-lb weight loss). The same is true for the exercise. If the plan calls for jogging, and jogging is despised with a passion, what are the chances of sticking with it? Slim of course. We all know men and women who refuse to exercise regularly because they have been led to believe that long-distance running is the only beneficial exercise. With a choice of activities, as well as an appropriate starting level, however, the chances are excellent for rapid progress. To this end the computer stands ready with its variety of **choices.**
3. The computerized exercise prescription allows a person to interchange between activities and still expend about the same number of calories. For example, if it rains one day and jogging is not preferred, then the computer allows alternative activities like swimming, cycling, or racquetball to be performed at about the same caloric expenditure. The example in Appendix G shows the equivalency in gross caloric output between walking, cycling, and nine sport activities. Without a computer, such calculations would literally require hundreds of hours, particularly because the exercise output is matched to the dietary plan, and both are individualized to the needs of the particular person desiring a particular weight loss.

■ SUMMARY

1. Many dietary options are available for obtaining the required nutrients for tissue maintenance, repair, and growth. Within rather broad limits, the nutrient requirements of athletes and other individuals engaged in training programs can be achieved with a balanced diet. With well-planned menus, the necessary vitamin, mineral, and protein requirements can be met with a food intake of about 1200 calories a day. Additional food can then be consumed to meet the energy needs that fluctuate, depending on the daily level of physical activity.
2. The recommended protein intake is about 0.8 g of protein per kg body weight. For the average man and woman, this is a liberal requirement and represents about 12% of the daily total caloric intake.
3. Athletes generally consume 2 to 5 times the recommended protein intake because their proportionately greater caloric intake usually provides proportionately more protein.
4. Precise recommendations for fat and carbohydrate intake have not been established. A prudent recommendation is that no more than 30% of the daily calories be obtained from fats; of this, the majority should be in the form of unsaturated fatty acids. For people who are physically active, 60% or more of the calories should come from carbohydrates, particularly polysaccharides. This will generally represent between 400 and 600 g on a daily basis.
5. Successive days of prolonged, hard training may gradually deplete the body's carbohydrate reserves, even if the recommended carbohydrate intake is maintained. This could lead to a training "staleness" because muscle glycogen may take several days to return to normal levels following a single session of prolonged exercise.
6. Because sustained vigorous exercise can increase the metabolic rate 20 to 25 times, the most important factor determining the daily caloric requirement is one's level of physical activity. In all likelihood, however, the caloric requirements of athletes in the most strenuous sports do not exceed 4000 calories per day unless body mass is considerable. Such a high caloric intake usually supplies well above the RDA requirements for protein, vitamins, and minerals.
7. The pre-event meal should include foods that are readily digested and contribute to the energy and fluid requirements of exercise. For this reason, the meal should be high in carbohydrate and relatively low in fat and protein. Clearly, the typical low-carbohydrate "steak-and-eggs diet" does not meet the requirements for optimal pre-event nutrition.
8. Commercially prepared liquid meals offer a practical approach to pregame nutrition and caloric supplementation. These "meals" are well balanced in nutritive value, contribute to fluid needs, and are absorbed rapidly, leaving practically no residue in the digestive tract.
9. Two or three hours should be sufficient time to permit digestion and absorption of the pre-event meal.

■ REFERENCES

1. Abrahams, A.: The nutrition of athletes. *Br. J. Nutr.*, 2:266, 1948.
2. Blair, S.N., et al.: Comparison of nutrient intake in middle-aged men and women runners and controls. *Med. Sci. Sports Exerc.*, 13:310, 1981.
3. Brouns, F., et al.: Eating, drinking, and cycling. A controlled Tour de France simulation study, Part I. *Int. J. Sports Med.*, 10:532, 1989.

4. Brouns, F., et al.: Eating, drinking, and cycling. A controlled Tour de France simulation study, Part II. Effect of diet manipulation. *Int. J. Sports Med.*, 10:541, 1989.
5. Brownell, K.D., and Stunkard, A.J.: Physical activity in the development and control of obesity. In *Obesity*. Edited by A.J. Stunkard. Philadelphia, W.B. Saunders, 1980.
6. Connor, W.E., et al.: The plasma lipids, lipoproteins, and diet of the Tarahumara Indians of Mexico. *Am. J. Clin. Nutr.*, 31:1131, 1978.
7. Costill, D.L.: Nutritional requirements for endurance athletes. In *Toward an Understanding of Human Performance*. Edited by E.J. Burke. Ithaca, N.Y., Mouvement Publications, 1977.
8. Costill, D.L., and Miller, J.: Nutrition for endurance sports: carbohydrate and fluid balance. *Int. J. Sports Med.*, 1:2, 1980.
9. Costill, D.L., et al.: Muscle glycogen utilization during prolonged exercise on successive days. *J. Appl. Physiol.*, 31:834, 1971.
10. Costill, D.L., et al.: The role of dietary carbohydrate in muscle glycogen resynthesis after strenuous running. *Am. J. Clin. Nutr.*, 34:1831, 1982.
11. Erp-Baart, A.M.J. van, et al.: Nationwide survey on nutritional habits in elite athletes. Part I. Energy, carbohydrate, protein, and fat intake. *Int. J. Sports Med.*, 10:53, 1989.
12. Foster, C., et al.: Effects of pre-exercise feedings on endurance performance. *Med. Sci. Sports*, 11:1, 1979.
13. Grandjean, A.C.: Macronutrient intakes of U.S. athletes compared with the general population and recommendations made for athletes. *Am. J. Clin. Nutr.*, 49:1070, 1989.
14. Hartung, G.H., et al.: Effects of marathon running, jogging, and diet on coronary risk factors in middle-aged men. *Preventive Med.*, 10:316, 1981.
15. Mayer, J., et al.: Relation between caloric intake, body weight and physical work in an industrial male population in West Bengal. *Am. J. Clin. Nutr.*, 4:169, 1956.
16. Percy, E.C.: Ergogenic aids in athletics. *Med. Sci. Sports*, 10:298, 1978.
17. Piehl, K.: Glycogen storage and depletion in human skeletal muscle fibers. *Acta Physiol. Scand.*, Suppl. 402, 1974.
18. Sherman, W.M., et al.: Effect of carbohydrate in four hour pre-exercise meals. *Med. Sci. Sports Exerc.*, 20:5157, 1988.
19. Sherman, W.M., and Costill, D.L.: The marathon: dietary manipulation to optimize performance. *Am. J. Sports Med.*, 12:44, 1984.
20. Smith, N.J.: *Food for Sport*. Palo Alto, CA, Bull Publishing, 1976.
21. Vodak, P.A., et al.: HDL-cholesterol and other plasma lipid and lipoprotein concentration in middle-aged male and female tennis players. *Metabolism*, 29:745, 1980.
22. Ward, P., et al.: U.S.A. discus camp: preliminary report. *Track Field Q. Rev.*, 76:29, 1976.
23. Werblow, J.A.: Nutritional knowledge, attitudes, and food patterns of women athletes. *J. Am. Diet. Assoc.*, 73:242, 1978.
24. Wood, P.D., et al.: Exercise and plasma lipoproteins: a one-year randomized, controlled trial. *Council on Epidemiology Newsletter*, 30:20, 1981.

GENERAL REFERENCES

Haskell, W., et al. (Eds.): *Nutrition and Athletic Performance*. Palo Alto, CA, Bull Publishing, 1982.

Horton, E.S., and Terjung, R.L. (Eds.): *Exercise, Nutrition, and Energy Metabolism*. New York, Macmillan, 1988.

Linder, M.C. (Ed.): *Nutritional Biochemistry and Metabolism*. New York, Elsvier, 1985.

Parizkova, J., and Rogozkin, V.A.: *Nutrition, Physical Fitness and Health*. Baltimore, MD, University Park Press, 1978.

Shils, M.E., and Young, V.R. (Eds.): *Modern Nutrition in Health and Disease*, 6th ed. Philadelphia, Lea & Febiger, 1988.

Tuckerman, M.M., and Turco, S.J.: *Human Nutrition*. Philadelphia, Lea & Febiger, 1983.

Williams, M.H.: *Nutritional Aspects of Human Physical and Athletic Performance*. Springfield, IL, Charles C Thomas, 1985.

ENERGY FOR PHYSICAL ACTIVITY

Considerable energy is generated rapidly for short periods of time by biochemical reactions that do not consume oxygen. This form of rapid anaerobic energy is crucial in maintaining a high standard of performance in sprint activities and in other all-out bursts of exercise. In comparison, in longer duration aerobic exercise, energy must be extracted from food through reactions that require oxygen. To be effective, the physiologic conditioning process necessitates an understanding of how energy is generated to sustain exercise, the sources of that energy, and the energy requirements of a particular activity.

This section presents a broad overview of how cells extract chemical energy bound within the food nutrients and use it to power biologic work. Emphasis is placed on the importance of the food nutrients and the processes of energy transfer in sustaining physiologic function during light, moderate, and strenuous exercise.

4   ENERGY VALUE OF FOOD

All biologic functions require energy. Because the carbohydrate, fat, and protein food nutrients contain the energy that ultimately powers biologic work, it is possible to classify both food and physical activity in terms of **energy.** This chapter deals with the quantification of food energy.

MEASUREMENT OF FOOD ENERGY

UNIT OF MEASUREMENT, THE CALORIE

A calorie is a measure used to express the heat or energy value of food and physical activity. It is defined as the amount of heat necessary to raise the temperature of 1 kg (1 liter) of water 1° C, from 14.5 to 15.5° C. Thus, a calorie is more accurately termed a **kilogram calorie** or kilocalorie (abbreviated **kcal**). For example, if 300 kcal is the caloric value of a particular food, then the energy trapped within the chemical bonds of this food, if released, would change the temperature of 300 liters of water by 1° C. The accepted international standard for expressing energy is the **joule.** To convert kcal to kilojoules (kJ) multiply the kcal value by 4.2. Different foods contain different amounts of energy. One-half cup of apricot nectar, for example, has a caloric value of 70 kcal (294 kJ) and therefore contains the equivalent heat energy to increase the temperature of 70 liters of water 1° C.

GROSS ENERGY VALUE OF FOODS

Many laboratories throughout the world have used a **bomb calorimeter,** similar to that shown in Figure 4–1, to measure the gross or total energy value of the various food nutrients. The principle behind this method of **direct calorimetry** is simple—the food is burned and the heat liberated measured.

As depicted in the figure, a piece of food is placed inside the chamber, which is then charged with oxygen. A fuse and an electric current are used to ignite the food-oxygen mixture. As the food burns, the heat liberated is

FIG. 4—1
A bomb calorimeter is used to determine the energy value of foods.

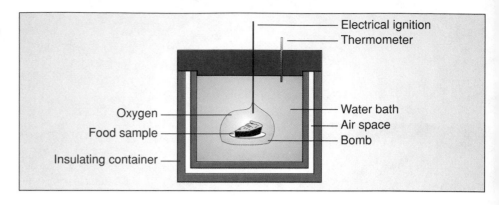

absorbed by a water jacket surrounding the bomb. Because the calorimeter is fully insulated from the outside environment, any increase in water temperature directly reflects the heat liberated during the oxidation of the specific food nutrient.

The heat liberated by the burning or **oxidation** of food is referred to as its **heat of combustion** and represents the total energy value of the food. For example, when one teaspoon of margarine is completely burned in the calorimeter, 100 kcal of heat energy is released. This is enough energy to raise 1.0 kg or 2.2 lb of ice water to the boiling point. It is important to point out that although the pathways for oxidation are quite different between the intact organism and the bomb calorimeter, the quantity of energy liberated in the breakdown of food is the same and is independent of the pathways by which combustion occurs.

Heat of Combustion: Fat

The heat of combustion for fat varies somewhat, depending upon the structural composition of the particular fatty acids that make up the triglycerides. For example, 1 g of either beef or pork fat yields 9.50 kcal, whereas the oxidation of 1 g of butterfat liberates 9.27 kcal. The average caloric value for one gram of fat in meat, fish, and eggs is 9.50 kcal per gram, in dairy products, 9.25 kcal per gram, and in vegetables and fruits, 9.30 kcal per gram. **The average heat of combustion per gram of fat oxidized in the calorimeter is generally considered to be 9.4 kcal.**

Heat of Combustion: Carbohydrate

The heat of combustion for a carbohydrate also varies depending upon the arrangement of atoms in the particular carbohydrate molecule. The simple carbohydrate glucose has a heat of combustion of 3.74 kcal per gram, whereas a gram of glycogen and starch liberates 4.19 and 4.20 kcal, respectively. **A value of 4.2 kcal is generally used to represent the heat of combustion for a gram of carbohydrate.**

Heat of Combustion: Protein

The energy released during the combustion of the protein portion of food also varies depending on two factors: (1) the kind of protein in the food, and (2) the relative proportions of the protein and nonprotein nitrogenous substances present. Many common proteins contain approximately 16% nitrogen with a corresponding heat of combustion that averages 5.75 kcal per gram. These proteins are found in foods such as eggs, meat, corn (maize), and beans (jack, lima, navy, soy). Proteins in the other food items have a somewhat higher nitrogen content such as most nuts and seeds (18.9%), whole-kernel wheat, rye, millets, and barley (17.2%), whereas other foods contain a slightly lower percentage of nitrogen such as whole milk (15.7%) or bran (15.8%). **A general value for the heat of combustion of protein is 5.65 kcal per gram.**

Comparing the Energy Value of Nutrients

Based on the average heats of combustion for the three nutrients (carbohydrate, 4.2 kcal $\cdot$ g^{-1}; fat, 9.4 kcal $\cdot$ g^{-1}; protein, 5.65 kcal $\cdot$ g^{-1}), it is clear that the complete oxidation of fat in the bomb calorimeter liberates about 65% more energy per gram than protein, and 120% more energy than an equal weight of carbohydrate. Recall from Chapter 1 that the fat molecule contains considerably more hydrogen atoms than either the carbohydrate or protein molecule. The fatty acid palmitic acid, for example, has the structural formula $C_{16}H_{32}O_2$. For this molecule as well as for other fat molecules, the ratio of hydrogen atoms to oxygen atoms is always considerably greater than 2 to 1. Compared with carbohydrate and protein molecules, there are more hydrogen atoms in fat molecules that can be cleaved away and oxidized to generate energy for the body's needs.

From the previous discussion, it is evident that the energy content of food that contains a considerable amount of fat is greater than that of relatively fat-free foods. For example, one cup of whole milk contains 160 kcal, whereas the same quantity of skimmed milk contains only 90 kcal. If someone who normally consumes one quart of whole milk each day switches to skimmed milk, the total calories ingested each year would be reduced by an amount equal to about 25 pounds of body fat!

NET ENERGY VALUE OF FOODS

The energy value of a particular food is not exactly the same when its heat of combustion (gross energy value) as determined by direct calorimetry is compared with the **net** energy actually available to the body. This is especially true for proteins because the body cannot utilize this nutrient's nitrogen component. Thus, nitrogen combines with hydrogen to form urea, which is then excreted in the urine. The elimination of hydrogen represents a loss of about 17% potential energy that reduces the heat of combustion of protein in the body to approximately 4.6 kcal per gram instead of the 5.65 kcal released during its complete oxidation in the bomb calorimeter. Because carbohydrates and fats contain no nitrogen, their physiologic fuel value is **identical** to the heat of combustion determined by bomb calorimetry.

Coefficient of Digestibility

Another consideration in determining the ultimate caloric yield from the food nutrients is the efficiency of the digestive process. The **coefficient of digestibility** refers to the proportion of ingested food that is actually digested and absorbed to serve the metabolic needs of the body. The remainder is not absorbed completely from the intestinal tract and is voided in the feces.

Furthermore, the energy available from the macronutrients is reduced when the meal has a high fiber content. This may be due to fiber's effect on speeding the transit of food through the intestines thus reducing the time for absorption. Fiber may also cause mechanical erosion of intestinal mucosa which must be replaced by energy-requiring processes.

Table 4–1 presents the digestibility coefficients as well as the heats of combustion and net energy values of nutrients in various food groups. As can be seen, there are different coefficients of digestibility for the different food nutrients. The relative percentages of each food nutrient that are completely digested and absorbed are 97% for carbohydrate, 95% for fat, and 92% for protein with little difference noted in comparisons between obese and lean individuals. It should be kept in mind that these values are only

TABLE 4–1.
FACTORS FOR DIGESTIBILITY, HEATS OF COMBUSTION, AND NET PHYSIOLOGIC ENERGY VALUES[b] OF PROTEIN, FAT, AND CARBOHYDRATE

Food Group	Digestibility, (%)	Heat of Combustion. (kcal · g^{-1})	Net Energy (kcal · g^{-1})
Protein			
Meats, fish	97	5.65	4.27
Eggs	97	5.75	4.37
Dairy products	97	5.65	4.27
Animal food	97	5.65	4.27
Cereals	85	5.80	3.87
Legumes	78	5.70	3.47
Vegetables	83	5.00	3.11
Fruits	85	5.20	3.36
Vegetable food	85	5.65	3.74
Total Food	92	5.65	4.05
Fat			
Meat and eggs	95	9.50	9.03
Dairy products	95	9.25	8.79
Animal food	95	9.40	8.93
Vegetable food	90	9.30	8.37
Total Food	95	9.40	8.93
Carbohydrate			
Animal food	98	3.90	3.82
Cereals	98	4.20	4.11
Legumes	97	4.20	4.07
Vegetables	95	4.20	3.99
Fruits	90	4.00	3.60
Sugars	98	3.95	3.87
Vegetable food	97	4.15	4.03
Total Food	97	4.15	4.03

[a] Net physiologic energy values are computed as the coefficient of digestibility × heat of combustion adjusted for energy loss in urine.
[b] From Merrill, A. L., and Watt, B. K., Energy value of foods . . . basis and derivation. Agricultural Handbook No. 74, Washington, D.C., U.S. Department of Agriculture, 1973.

averages, and some variability can be expected in efficiency percentages for any food within a particular category. For proteins in particular, the digestive efficiency ranges from a low of about 78% for legumes to a high of 97% for protein from animal sources. This relatively low coefficient of digestibility for plant protein has been used by some to argue for the use of vegetables in a diet for weight loss. It also emphasizes the importance of maintaining adequate quantity and quality of protein for those individuals on a vegetarian-type diet.

From the data in Table 4–1, **the average net energy value for carbohydrates, fats, and proteins ingested in the diet can be rounded to the simple whole numbers 4, 9, and 4 kcal per gram, respectively.** These values are referred to as the **Atwater general factors** and have been traditionally used by nutritionists for almost 60 years to represent the energy available to the body from food nutrients. Except when exact energy values are desired, as in preparing experimental or therapeutic diets, the Atwater general factors can be used to estimate the net energy value of typical foods consumed in the American diet. When alcohol is consumed, a value of 7 kcal is given for each g (ml) of pure (200 proof) alcohol ingested.

Caloric Value of a Meal

The caloric content of any portion of food can be determined using the Atwater factors if the composition and weight of the food are known. Table 4–2 illustrates the method for calculating the kcal value of 100 grams (3.5 oz) of vanilla ice cream. Based on laboratory analysis, vanilla ice cream contains approximately 4% protein, 13% fat, and 21% carbohydrate, with the remaining 62% essentially water. Thus, each gram of ice cream contains 0.04 g protein, 0.13 g fat, and 0.21 g carbohydrate. Using these compositional values and the Atwater factors, the kcal value per gram of ice cream is determined as follows: The net kcal values indicate that 0.04 g of protein contains 0.16 kcal (0.04×4.0 kcal $\cdot$ g^{-1}), 0.13 g fat equals 1.17 kcal (0.13×9 kcal $\cdot$ g^{-1}), and 0.21 g carbohydrate yields 0.84 kcal (0.21×4.0 kcal $\cdot$

TABLE 4–2.
METHOD OF CALCULATING THE CALORIC VALUE OF A FOOD FROM ITS COMPOSITION OF NUTRIENTS

Food: ice cream (vanilla)
Weight: three-fourths cup = 100 grams

	Composition		
	Protein	**Fat**	**Carbohydrate**
Percentage	4%	13%	21%
Total grams	4	13	21
In one gram	.04 g	.13 g	.21 g
Calories per gram	.16	1.17	.84

$$(.04 \times 4.0 \text{ kcal}) + (.13 \times 9.0 \text{ kcal}) + (.21 \times 4.0 \text{ kcal})$$

Total calories per gram: .16 + 1.17 + .84 = 2.17 kcal
Total calories per 100 grams: 2.17 × 100 = 217 kcal

g^{-1}). Consequently, the total energy value for each gram of vanilla ice cream equals 2.17 kcal (0.16 + 1.17 + 0.84). For a 100-g serving, the caloric value is 100 times as large, or 217 kcal. This method of computation can be used to estimate the kcal value for any food serving. Of course, increasing or decreasing portion sizes, or adding extra-rich sauces, candies, or fruits, or calorie-free substitutes, would affect the caloric content accordingly.

Use of Appendix B

This procedure for computing the kcal of foods is time-consuming and laborious. Because of this, the United States Department of Agriculture has evaluated and compiled the nutritive value of over 2,500 food items. Appendix B presents a sample listing of some of the more common foods consumed in the American diet. One should keep in mind that caloric and nutritive values for specific food dishes such as chicken or beef tacos are computed from standard recipes and may vary considerably, depending on taste preference and method of preparation.

If Appendix B is examined carefully, it is observed that there are large differences in the energy values of various foods. To consume an equal number of calories from different foods often requires a tremendous intake of a particular food. For example, in order to consume 100 kcal from each of four common foods, carrots, green peppers, medium-sized eggs, and mayonnaise, one must eat 5 carrots, 4 cups of green peppers, 1¼ eggs, and only 1 tablespoon of mayonnaise.

To meet the daily energy requirements of an average sedentary adult, one would have to consume about 400 celery stalks, 120 carrots, 96 cups of green pepper, 30 eggs, but only 1½ cup of mayonnaise or 8 ounces of salad oil. This illustrates dramatically that foods high in fat contain considerable calories compared to food low in fat and correspondingly high in water content.

Another important consideration is that a calorie is a measure of food energy, regardless of its source. Thus, 100 calories from mayonnaise is no more fattening than the 100 calories contained in 5 cups of raw cabbage. The more one eats of a given food, the more calories one consumes. It is just that only a small quantity of fatty foods represents a considerable quantity of calories—thus, these foods are considered fattening. An individual's caloric intake is simply equal to the sum of **all** calories consumed, be they from a small or large quantity of food.

■ SUMMARY

1. A calorie or kilocalorie (kcal) is a measure of heat used to express the energy value of food. This food energy is directly measured in a bomb calorimeter.
2. The heat of combustion represents the heat liberated by the complete oxidation of food. For fats, carbohydrates, and proteins, these gross energy values are 9.4, 4.2 and 5.65 kcal per gram, respectively.
3. The coefficient of digestibility is the proportion of ingested food actually digested and absorbed to be used by the body. This represents about 97% for carbohydrates, 95% for fats, and 92% for proteins. Thus, the net energy values are 4, 9, and 4 kcal per gram for carbohydrate, fat, and protein, respectively. These values are referred to as the Atwater general factors and are used to estimate the net energy value of typical foods in the diet.

4. With the Atwater calorific values, it is possible to compute the caloric content of any meal—as long as the carbohydrate, fat, and protein composition is known.
5. From an energy standpoint, a calorie is a unit of heat energy regardless of the food source. Thus, it is incorrect to consider 300 kcal of chocolate ice cream any more fattening than 300 kcal of watermelon, 300 kcal of pepperoni pizza, or 300 kcal of bagels and sour cream.

GENERAL REFERENCES

Consolazio, C.F., Johnson, R., and Pecora, L.: *Physiological Measurements of Metabolic Functions in Man.* New York, McGraw-Hill, 1963.

Guthrie, H.A. (Ed.): *Introductory Nutrition.* 7th ed., St. Louis, C.V. Mosby, 1988.

Guyton, A.C.: *Textbook of Medical Physiology.* 7th ed., Philadelphia, W.B. Saunders, 1986.

Hamilton, E.M., Whitney, E.N., and Sizer, F.S.: *Nutrition: Concepts and Controversies,* 3rd ed. St. Paul, Minn., West, 1985.

Hickson, J.F., Jr., and Wolinsky, I. (Eds.). *Nutrition in Exercise and Sport.* Boca Raton, FL, CRC Press, 1989.

Krause, M.V., and Mahan, K.I.: *Food, Nutrition and Diet Therapy: A Textbook of Nutritional Care.* Philadelphia, W.B. Saunders, 1984.

Miles, C.W., et al.: Effect of dietary fiber on the metabolizable energy of human diets. *J. Nutr.,* 118:1075, 1988.

Reed, P.B.: *Nutrition: An Applied Science.* St. Paul, MN, West, 1980.

Shils, M.E., and Young, V.R.: *Modern Nutrition in Health and Disease,* 7th ed. Philadelphia, Lea & Febiger, 1988.

5 INTRODUCTION TO ENERGY TRANSFER

The ability to swim, run, or ski long distances is framed largely by one's capacity to extract energy from the food nutrients and transfer it to the contractile elements of skeletal muscle. Likewise, specific energy-transferring capabilities determine success in weight lifting, sprinting, jumping, and football line play. Although our main frame of reference in this text is muscular activity, the direct transfer of chemical energy is required to power **all** forms of biologic work.

The sections that follow introduce general concepts dealing with bioenergetics. They provide the basis for understanding energy metabolism during physical activities.

ENERGY, THE CAPACITY FOR WORK

Unlike the physical properties of matter, it is difficult to define **energy** in concrete terms of size, shape, or mass. Rather, the term energy suggests a dynamic state related to a condition of **change,** because the presence of energy is revealed only when a change has taken place. Within this context, **energy relates to the ability to perform work.** As work increases, the transfer of energy also increases so that a change occurs.

The **first law of thermodynamics** states that energy is neither created nor destroyed, but is transformed from one form to another. In essence, this is the immutable principle of the **conservation of energy** that applies to both living and nonliving systems. The large amount of chemical energy in fuel oil, for example, is readily converted to heat energy in the home oil burner. In the body, however, all of the chemical energy trapped within the bonds of the food nutrients is not immediately lost as heat; rather, a large portion is conserved as chemical energy and then changed into mechanical energy (and then ultimately heat energy) by the action of the musculoskeletal system. The underlying principle is that energy is **not** produced, consumed, or used up; it is merely transformed from one form into another.

POTENTIAL AND KINETIC ENERGY

The total energy of any system consists of two components, **potential energy** and **kinetic energy.** Potential energy can be energy of position, such as that possessed by a stone at the top of a hill; it can also be light energy, electric energy, or bound energy within the internal structure of a substance. **When potential energy is released, it is transformed into kinetic energy or energy of motion.** The conversion between potential and kinetic energy can take many forms. For example, the rearrangement of the chemical structure of

FIG. 5–1.
High-grade potential energy capable of performing work is degraded to kinetic energy.

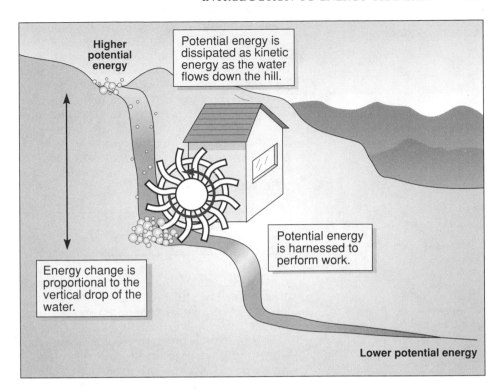

> Higher potential energy
>
> Potential energy is dissipated as kinetic energy as the water flows down the hill.
>
> Potential energy is harnessed to perform work.
>
> Energy change is proportional to the vertical drop of the water.
>
> Lower potential energy

a substance occurs when the bonds between atoms are broken. This results in the release of potential energy with a concomitant increase in kinetic energy. In some instances, the bound energy in one substance can be directly **transferred** to other substances to increase the potential energy of these molecules. Energy transfers of this type are necessary for the body's chemical work of **biosynthesis.** In this process, specific building-block molecules are activated and join other molecules in the synthesis of important biologic compounds. Some of these compounds serve structural needs, while others can then serve the energy needs of the cell.

Figure 5–1 depicts the relationship between potential and kinetic energy as related to position. Relatively calm water at sea level has little potential energy and could not drive the waterwheel. As water from a mountain stream flows downhill and over the fall, however, the water's potential energy is transformed into kinetic energy. Some of this energy is harnessed to turn the waterwheel. When the distance of the falling water increases, more kinetic energy is generated. Water at the top of the fall has more potential and less kinetic energy than water at the bottom, and vice versa. Thus, the potential energy stored in the water atop the fall changes into kinetic energy as it moves from a higher to a lower level.

ENERGY-RELEASING AND ENERGY-CONSERVING PROCESSES

Any physical or chemical process that results in a release of energy to its surroundings is termed **exergonic.** Exergonic reactions can be viewed as "downhill" processes; they result in a decline in **free energy,** that is, "useful" energy for work. Processes that store or absorb energy are termed **endergonic;**

these reactions represent "uphill" processes and proceed with an increase in free energy. In some instances, exergonic processes can be **coupled** or linked with endergonic reactions so that some of the energy is transferred to the endergonic process. **These coupled reactions are of considerable importance to the body because they serve as the means of conserving a large portion of the chemical energy in food nutrients in a usable form.** This subject is discussed more fully in the next chapter.

In exergonic chemical reactions, potential energy stored within specific atoms is released; the liberated energy is equal to the difference in the potential energy of the reactant and product substances. For example, the union of hydrogen and oxygen to form water releases 68 kcal of free energy in the following reaction:

$$H_2 + O \longrightarrow H_2O \qquad -68 \text{ kcal per mole}$$

(The negative sign means that heat is lost from the system as the reaction proceeds. One mole or gram molecular weight represents the molecular weight of a substance in grams. For example, a mole of glucose weighs 180 g.)

This process is reversible; in the reverse endergonic reaction, the chemical bonds of the water molecule are broken and the original hydrogen and oxygen are set free. To achieve this, 68 kcal of energy must be supplied to each mole of water. This "uphill" process of energy transfer provides the hydrogen and oxygen atoms with their original energy content to satisfy the principle of the conservation of energy.

$$H_2 + O \longleftarrow H_2O \qquad +68 \text{ kcal per mole}$$

The process of energy transfer in humans follows the same principles outlined in the waterfall-waterwheel example. The carbohydrate, fat, and protein nutrients possess considerable potential energy. Through the action of specific enzymes, there is a progressive loss of potential energy from the nutrient molecule and a corresponding increase in kinetic energy as product substances are formed. With the aid of appropriate transfer systems, a portion of this chemical energy is harnessed or **conserved** in new compounds that are then used for biologic work.

The transfer of potential energy is unidirectional; it always proceeds so that the capacity of the total energy to perform work decreases. The tendency of potential energy to degrade to kinetic energy with a lower capacity for work represents a statement of the **second law of thermodynamics.** A good example is the car battery. The electrochemical energy stored within its cells is slowly released, even if the battery is not used. The energy from sunlight is also continually degraded to heat energy when light strikes and is absorbed by a surface. Food and other chemicals are excellent stores of potential energy. This energy, however, is eventually released as the compounds decompose through normal oxidative processes. Energy, like water, always runs downhill, whereby its potential energy is decreased. **Ultimately, all of the potential energy in a system is degraded to the nonusable form of kinetic energy or heat.**

INTERCONVERSIONS OF ENERGY

Because the total energy in an isolated system remains constant, a decrease in one form of energy is matched by an equivalent increase in another form. In the process of energy conversion, however, a loss of potential energy from one source can result in a temporary increase in the potential energy of another source. Clearly, vast quantities of potential energy can be harnessed in nature for useful purposes. Even under such conditions, the tendency for **entropy** is dominant and the net flow of energy in the biologic world results in the degradation of potential energy.

Entropy is a measure of the continual process of energy change; all chemical and physical processes proceed in a direction where total randomness or disorder increases and the energy available to do work decreases. In coupled reactions that occur during biosynthesis, part of a system may show a decrease in entropy whereas another part shows an increase. There is, however, no circumventing the second law: The entire system always shows a net **increase** in entropy.

FORMS OF ENERGY

Figure 5–2 shows that energy can be categorized into one of six forms: chemical, mechanical, heat, light, electric, and nuclear.

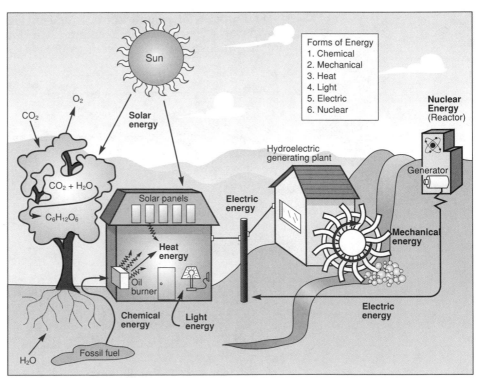

FIG. 5–2.
Interconversions of energy forms.

EXAMPLES OF ENERGY CONVERSIONS

The conversion of energy from one form to another occurs readily in both the inanimate and animate worlds. The most fundamental examples of energy conversion in living cells are the processes of **photosynthesis** and **respiration.**

Photosynthesis

In the sun, with a temperature of several million degrees Farenheit, part of the potential energy stored in the nucleus of the hydrogen atom is released by the process of nuclear fusion. This energy in the form of gamma radiation is then converted to radiant energy.

As shown in Figure 5–3, the pigment chlorophyll in green plants absorbs solar (radiant) energy and transforms it into the chemical potential energy of carbohydrates through the reactions of photosynthesis. This endergonic process that involves the synthesis of glucose from carbon dioxide and water and the release of oxygen (subsequently used by animals in the process of energy metabolism), requires an input of 686 kcal of energy per mole of glucose synthesized. Carbohydrates can be converted to fats and proteins for storage in the plant as a reserve of potential energy for future use. Animals then ingest the plant nutrients to serve their own energy needs. In essence, solar energy, coupled with the process of photosynthesis, powers the animal world with food and oxygen.

Respiration

The process of respiration outlined in Figure 5–4 is the reverse of photosynthesis. During these exergonic reactions, the chemical energy stored in the glucose, fat, and protein molecules is extracted in the presence of

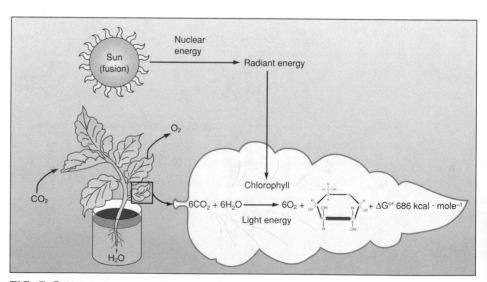

FIG. 5–3.
The endergonic process of photosynthesis. Endergonic reactions that require energy, as in the synthesis of glucose in green plants, have a positive standard free energy change (useful energy) indicated as $+\Delta G^{\circ\prime}$.

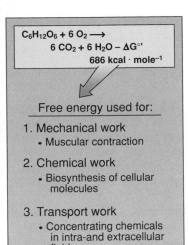

FIG. 5–4.
The exergonic process of respiration. Exergonic reactions that release potential energy, such as the burning of gasoline or the combustion of glucose, have a negative standard free energy change (i.e., reduction in total energy available to do work) indicated as $-\Delta G^{\circ\prime}$.

oxygen. For glucose, this results in a release of 686 kcal per mole oxidized. **A portion of the energy released during respiration can be conserved in other chemical compounds and then converted to mechanical work in the body by the action of muscles; the remaining energy flows to the environment as heat.**

BIOLOGIC WORK IN HUMANS

The energy released during cellular respiration in humans is used to sustain biologic work. This work can take one of three familiar forms: (1) mechanical work of muscle contraction, (2) chemical work that involves the synthesis of cellular molecules, or (3) transport work that concentrates various substances in the intra- and extracellular fluids.

Mechanical Work

The most obvious example of energy transformation in the body is the mechanical work generated by muscle contraction. The protein filaments of the muscle fibers directly convert chemical energy into mechanical energy. However, this is not the body's only form of mechanical work. In the cell nucleus, for example, contractile elements similar to those found in muscle literally tug at the chromosomes to facilitate the process of cell division. Mechanical work is also performed by specialized structures such as cilia that are part of many cells.

Chemical Work

Chemical work is performed by all cells for growth and maintenance. Cellular components are continually synthesized as other components are destroyed. This biosynthesis opposes the tendency toward entropy and requires an input of energy for the bonding together of cellular elements.

Transport Work

Much less conspicuous than mechanical or chemical work is the work of transporting or concentrating substances in the body. Cellular materials normally flow from an area of high concentration to one of low concentration. In this passive process, called **diffusion,** no energy is required. For proper physiologic functioning, however, certain chemicals must also be transported "uphill" against their concentration gradients; that is, from an area of lower to one of higher concentration. The term **active transport** is usually associated with this process. Secretion and reabsorption in the kidney tubules require active transport, as does the establishment of proper electrochemical gradients about cell membranes. A continual expenditure of stored chemical energy is required to accomplish these "quiet" forms of biologic work.

FACTORS THAT AFFECT THE RATE OF BIOENERGETICS

The rate at which chemical energy in the food nutrients is extracted, conserved, and transferred to the contractile filaments of skeletal muscle determines the intensity at which exercise can progress. **The sustained pace of the marathon runner, or the rapid speed of the sprinter, is a direct expression** of the body's capacity to transfer chemical energy into mechanical work. Some factors that affect the rate of energy release during chemical reactions are now considered.

THE MASS ACTION EFFECT

The effect of the concentration of chemicals on the frequency of a particular chemical reaction reflects the **law of mass action** and is called the **mass action effect.** It is not uncommon to find certain substances in the body that are tied to several reactions; thus, the products of one reaction become reactant substances for other reactions. Thus, changing only the concentration of one substance can profoundly affect a number of reactions. Also, certain chemicals play key roles in a whole chain of chemical events. Oxygen, for example, exerts a significant mass action effect on reactions required for energy transfer. If the oxygen supply to the tissues is diminished, several chemical processes cease and the net energy available for biologic work is reduced dramatically.

ENZYMES: THE BIOLOGIC CATALYSTS

Enzymes are highly specific protein **catalysts** that accelerate the speed of a chemical reaction. The compound changed in an enzyme-regulated reaction is called the **substrate** for that enzyme. Enzymes do not make reactions occur that could not otherwise take place under proper conditions; rather, they facilitate the interaction of substrates that normally would occur at a much slower rate. In a way, enzymes reduce the required activation energy so that the reaction rate is changed, even though the total energy released per reaction remains unaltered.

Enzymes possess the unique property of not being readily altered by the reactions in which they participate. Consequently, the turnover of enzymes in the body is relatively slow and specific enzymes are continually reused. Because enzymes are proteins, they are adversely affected by acidity and temperature, especially heat. Another important characteristic of enzymes is their **specificity** in interacting with other substances. For example, the breakdown of glucose to carbon dioxide and water requires 19 different chemical reactions, each catalyzed by a specific enzyme. In humans, about 900 different enzymes have been identified with each performing its specific role in a different chemical reaction. The activation of molecules is the important characteristic of enzyme-regulated processes and ensures that a reaction occurs more readily. It is only through the action of enzymes that cells function as chemical engines.

ROLE OF COENZYMES

Some enzymes are totally inactive in the absence of additional substances termed **coenzymes.** These are complex, nonprotein, organic substances that facilitate enzyme action by helping to bind the substrate with its specific enzyme.

The action of a coenzyme is less specific than an enzyme because the coenzyme can act in a number of different reactions. It can act as a "co-binder," or it can serve as a temporary carrier of intermediary products of the reaction. For example, the hydrogen atoms and electrons split from the nutrient substrates during energy metabolism are temporarily carried by the coenzyme **nicotinamide adeninedinucleotide,** or NAD^+, to form NADH.

The electrons are then passed to special carrier molecules that ultimately deliver them to molecular oxygen.

Vitamins serve a major role as coenzymes in the transfer of chemical energy. As coenzymes, vitamins **do not** directly provide chemical energy, although in a sense, they "make the reactions go."

MEASURING ENERGY RELEASE IN HUMANS

The gain or loss in the heat of a system provides a simple means for determining the energy change of any process. In the breakdown of foods, for example, the energy change can be determined **directly** in a calorimeter. This heat energy is usually expressed in terms of calories or kilocalories.

Because the complete combustion of food is achieved at the expense of molecular oxygen, the heat generated in these exergonic reactions can be conveniently and accurately estimated by measuring oxygen consumption. This forms the basis of **indirect calorimetry** and enables one to infer the energy metabolism of humans at rest and during many forms of physical activity. The use of direct and indirect calorimetry for determining heat production in humans is discussed in Chapter 8.

SUMMARY

1. Energy is defined as the ability to perform work and therefore is revealed only when change takes place.
2. Energy exists in either a potential or kinetic form. Potential energy is the energy associated with substances because of structure or position, whereas kinetic energy is the energy of motion. Potential energy is measured when it is transformed into kinetic energy.
3. There are six forms of energy: chemical, mechanical, heat, light, electric, and nuclear. Each form of energy can be converted or transformed into another form.
4. Exergonic energy reactions result in a transfer of energy to the surroundings. Endergonic energy reactions result in the storage, conservation, or increase in free energy. All potential energy is ultimately degraded into kinetic heat energy. In living organisms, however, a portion of this chemical energy is conserved within the structure of new compounds that are then used for biologic work.
5. Energy transfer in humans generally takes one of three forms: chemical (biosynthesis of cellular molecules), mechanical (muscle contraction), and transport (transfer of substances between cells).
6. The main factors regulating the rate at which chemical energy is extracted from the food nutrients are: (1) mass action effect, which depends on the concentration gradients of substances, (2) biologic catalysts (enzymes) that accelerate the speed of chemical reactions, and (3) coenzymes that facilitate specific enzyme action.

GENERAL REFERENCES

Åstrand, P.O., and Rodahl, K.: *Textbook of Work Physiology*. 3rd ed. New York, McGraw-Hill, 1986.

Brooks, G.A., and Fahey, T.D.: *Exercise Physiology: Human Bioenergetics and its Applications*. New York, Wiley, 1984.

Lehninger, A.L.: *Bioenergetics: The Molecular Bases of Biological Energy Transformations.* Menlo Park, Calif., W.A. Benjamin, 1971.
Lehninger, A.L.: *Principles of Biochemistry.* New York, Worth Publishers, 1982.
Stryer, L.: *Biochemistry.* 2nd ed., San Francisco, Freeman, 1988.
Vander, A.J., et al.: *Human Physiology.* 2nd ed., New York, McGraw-Hill, 1985.

6

ENERGY TRANSFER IN THE BODY

The human body must be continually supplied with chemical energy to perform its many complex functions. Energy derived from the oxidation of food is not released suddenly at some kindling temperature because the body, unlike a mechanical engine, **cannot** use heat energy. If this were the case, the body fluids would actually boil and our tissues would burst into flames. Rather, the chemical energy trapped within the bonds of carbohydrates, fats, and proteins is extracted in small amounts during complex, enzymatically controlled reactions that occur in the relatively cool, watery medium of the cell. This process reduces the loss of energy as heat and provides for much greater efficiency in energy transformations, thereby enabling the body to make direct use of chemical energy. In a sense, energy can be supplied to the cells as it is needed. The story of how the body maintains its continuous energy supply begins with the special carrier for free energy, ATP.

1 Phosphate Bond Energy

THE ENERGY CURRENCY, ADENOSINE TRIPHOSPHATE

The energy in food is not transferred directly to the cells for biologic work. Rather, this "nutrient energy" is harvested and funneled through the energy-rich compound **adenosine triphosphate** or ATP. The potential energy within the ATP molecule is then utilized for **all** the energy-requiring processes of the cell. This energy receiver-energy doner cycle, in essence, represents the two major energy-transforming activities in the cell: (1) to form and conserve ATP from the potential energy in food, and (2) to use the chemical energy in ATP for biologic work.

Figure 6–1 shows the ATP molecule formed from a molecule of adenine and ribose, called adenosine, linked to three phosphate molecules. The bonds that link the two outermost phosphates are termed high-energy bonds because they represent a considerable quantity of potential energy within the ATP molecule.

101

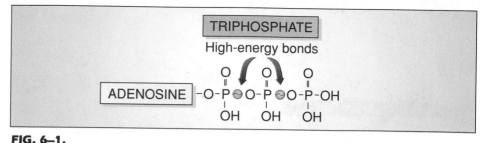

FIG. 6–1.
Simplified structure of ATP, the energy currency of the cell. The symbol ~ represents the high-energy bonds.

When ATP joins with water in a process called **hydrolysis,** catalyzed by the enzyme **adenosine triphosphatase,** the outermost phosphate bond is broken and a new compound is formed **adenosine diphosphate** or *ADP.* In this reaction, approximately 7.3 kcal of free energy are liberated per mole of ATP degraded to ADP.

$$\text{ATP} + \text{H}_2\text{O} \xrightarrow[\text{ase}]{\text{ATP}} \text{ADP} + \text{P} \quad - 7.3 \text{ kcal per mole}$$

The free energy liberated in the hydrolysis of ATP is a measure of the energy difference between the reactants and the end products. Because considerable energy is generated in this reaction, ATP is often referred to as a **high-energy phosphate.** Infrequently, additional energy is released when another phosphate is split from ADP. In some reactions of biosynthesis, the two terminal phosphates from ATP are simultaneously donated in the construction of new cellular material. The remaining molecule with a single phosphate group is **adenosine monophosphate** or *AMP.*

The energy liberated during ATP breakdown is directly transferred to other energy-requiring molecules. In muscle, for example, this energy activates specific sites on the contractile elements causing the muscle fiber to shorten. Because energy from ATP is harnessed to power **all** forms of biologic work, ATP is considered the cell's "energy currency." The general role of ATP as energy currency is illustrated in Figure 6–2.

The splitting of the ATP molecule takes place whether oxygen is available or not. This reaction is immediate, **nonaerobic,**[*] and energy-liberating. The cell's capacity for ATP breakdown enables it to generate energy for immediate use; this would not occur if oxygen were required at all times for energy metabolism. For this reason, all types of exercise can be performed immediately without consuming oxygen, such as sprinting for a bus or lifting a heavy barbell.

THE ENERGY RESERVOIR, CREATINE PHOSPHATE

Only a small quantity of ATP is stored within the cell; hence, it must be resynthesized at the rate it is used. This situation provides a sensitive mechanism for regulating energy metabolism in the cell. By maintaining only a

---

[*] The term nonaerobic more precisely describes the breakdown of the phosphagens. The term *anaerobic* is properly used to describe reactions of glycolysis in which glucose is degraded to lactic acid.

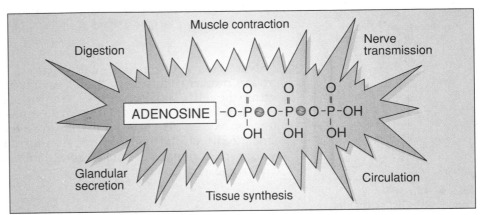

FIG. 6–2.
ATP is the energy currency for all forms of biologic work.

small amount of ATP, its relative concentration (and corresponding concentration of ADP) is altered rapidly with any increase in a cell's energy metabolism. This change, in turn, immediately stimulates the breakdown of stored nutrients to provide energy for ATP resynthesis. In this way, energy metabolism increases rapidly in the early stages of exercise.

The total quantity of ATP within the body at any one time is about 85 g (3 oz.) This amount provides only enough energy to perform maximum exercise for several seconds. Because ATP cannot be supplied by way of the blood or from other tissues, it must be recycled continuously within each cell. Some of this energy for ATP resynthesis is supplied rapidly and without oxygen by the transfer of chemical energy from another high-energy phosphate compound called **creatine phosphate,** or CP (Fig. 6–3). The cell's concentration of CP is about three to five times greater than that of ATP. For this reason, CP is considered the high-energy phosphate "reservoir."

The CP molecule is similar to the ATP molecule in that a large amount of free energy is released when the bond between the creatine and phosphate molecules splits. Because CP has a higher free energy of hydrolysis than ATP, its phosphate is donated directly to ADP to re-form ATP. This reaction is catalyzed by the enzyme **creatine kinase.** The arrows pointing in opposite directions show that the reactions are reversible. If sufficient energy is available, creatine (C) and phosphate (P) can be joined to re-form CP. The same is true for ATP; the top reaction illustrates the union of ADP and P to re-form ATP (Fig. 6–3).

It should now be apparent that human energy dynamics involves the transfer of energy by means of chemical bonds. Potential energy is released by the splitting of bonds and conserved by the formation of new bonds. Some energy lost by one molecule can be transferred to the chemical structure of another and does not appear as heat. Compounds relatively low in potential energy can be "juiced-up" by energy transfer from the high-energy phosphates to accomplish biologic work. Adenosine triphosphate serves as the ideal energy-transfer agent; in one respect it "traps" in its phosphate bonds a large portion of potential energy in the original food molecule, yet readily transfers this energy to other compounds to raise them to a higher level of activation. This transfer of energy in the form of phosphate bonds is termed **phosphorylation.** The energy for phosphorylation is ultimately generated by the oxidation of carbohydrates, fats, and proteins consumed in the diet.

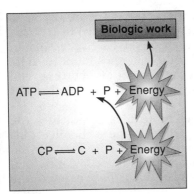

FIG. 6–3.
ATP and CP are nonaerobic sources of phosphate bond energy. The energy from the breakdown of CP is used to rebond ADP and P to reform ATP.

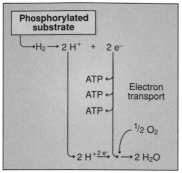

FIG. 6–4.
A general scheme for the oxidation of hydrogen and accompanying electron transport. In this process, oxygen is reduced and water is formed.

CELLULAR OXIDATION

Hydrogen atoms are continually stripped from the carbohydrate, fat, and protein nutrient substrates during energy metabolism. Carrier molecules within the cell's "energy factories," the **mitochondria,** remove electrons from hydrogen and eventually pass them to molecular oxygen. To complete the process, oxygen also accepts hydrogen to form water. Much of the energy generated in cellular oxidation (that is, the transfer of electrons from hydrogen to oxygen) is trapped or conserved as chemical energy in the form of high-energy phosphates.

ELECTRON TRANSPORT

The general scheme for the oxidation of hydrogen and the accompanying **electron transport** to oxygen is illustrated in Figure 6–4. During cellular oxidation, hydrogen atoms are not merely turned loose in the cell fluid. Rather, the release of hydrogen from the nutrient substrate is catalyzed by highly specific **dehydrogenase** enzymes. The electrons (energy) from hydrogen are picked up in pairs by the coenzyme part of the dehydrogenase, which is usually the vitamin B (niacin) containing coenzyme, **nicotinamide adenine-dinucleotide** or NAD^+. While the substrate is being oxidized and losing hydrogen (electrons), NAD^+ is gaining a hydrogen and two electrons and being reduced to NADH; the other hydrogen appears in the cell fluid as H^+.

The other important electron acceptor in the oxidation of the food fragments is **flavin adenine dinucleotide,** or *FAD.* Flavin adenine dinucleotide is derived from the B-vitamin riboflavin. This coenzyme also catalyzes dehydrogenations and accepts pairs of electrons. Unlike NAD^+, however, FAD accepts both hydrogens to become $FADH_2$.

The NADH and $FADH_2$ formed in the breakdown of the food nutrients are energy-rich molecules because they carry electrons that have a high-energy-transfer potential. On the inner membranes of the mitochondria, the electrons carried by NADH and $FADH_2$ are then passed in "bucket brigade" fashion by a series of iron-protein electron carriers, the **cytochromes.** The iron portion of each cytochrome can exist in either its oxidized or reduced state as ferric (Fe^{+++}) or ferrous (Fe^{++}) ion, respectively. By accepting an electron, the ferric portion of a specific cytochrome becomes reduced to its ferrous form. In turn, this donates its electron to the next cytochrome, and so on down the line. By shuttling between these iron forms, the cytochromes transfer electrons.

The transport of electrons by specific carrier molecules constitutes the **respiratory chain.** This is the final common pathway where the electrons extracted from hydrogen are passed to oxygen. For each pair of hydrogen atoms, two electrons flow down the chain and reduce one atom of oxygen to form water. Of the five specific cytochromes, only the last, **cytochrome oxidase** (cytochrome A_3), can discharge its electron directly to molecular oxygen. Figure 6–5A shows the route for the oxidation of hydrogen and the accompanying electron transport and energy transfer in the respiratory chain. **Free energy is released in the respiratory chain in relatively small amounts, and in several of the electron transfers, energy is conserved by the formation of high-energy phosphate bonds.**

FIG. 6–5.
How energy is harnessed. (A) In the respiratory chain, much of the chemical energy stored within the hydrogen atom is conserved in the formation of high-energy phosphate bonds. (B) Energy from falling water is used to turn the waterwheel.

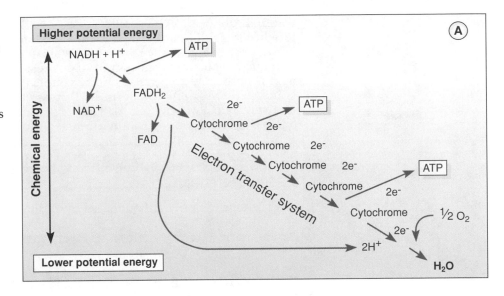

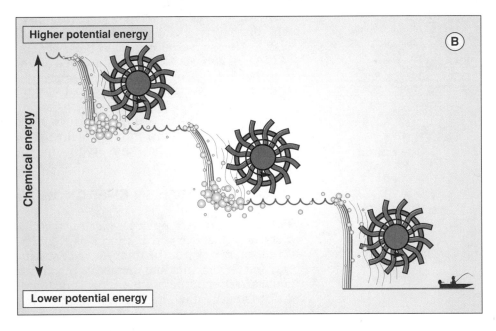

OXIDATIVE PHOSPHORYLATION

Oxidative phosphorylation is the process by which ATP is synthesized during the transfer of electrons from NADH and FADH$_2$ to molecular oxygen. This important process represents the cell's primary means for extracting and trapping chemical energy in the form of high-energy phosphates. In fact, over 90% of ATP synthesis is accomplished in the respiratory chain by oxidative reactions coupled with phosphorylation.

In a way, the process of oxidative phosphorylation can be likened to a waterfall divided into several separate cascades by the intervention of waterwheels at different heights. As depicted in Figure 6–5*B*, the waterwheels harness the energy of the falling water; similarly, the electrochemical energy generated in electron transport from one respiratory chain component to the next is harnessed and transferred or coupled to ADP. The energy in NADH is transferred to ADP to reform ATP at three distinct coupling sites during electron transport (Fig. 6–5*A*). This oxidation of hydrogen and subsequent phosphorylation can be summarized as follows:

$$NADH + H^+ + 3 ADP + 3 P + \tfrac{1}{2} O_2 \longrightarrow NAD^+ + H_2O + 3 ATP$$

If hydrogen is originally donated by $FADH_2$, only two molecules of ATP are formed for each hydrogen pair oxidized. This occurs because $FADH_2$ enters the respiratory chain at a lower energy level at a point beyond the site of the first ATP synthesis.

EFFICIENCY OF ELECTRON TRANSPORT AND OXIDATIVE PHOSPHORYLATION

Approximately 7 kcal of energy are required for the synthesis of each mole of ATP. Because 3 moles of ATP are generated in the oxidation of 1 mole of NADH, about 21 kcal are conserved as chemical energy. In total, 52 kcal are liberated during the oxidation of a mole of NADH. Thus, the relative **efficiency** of electron transport-oxidative phosphorylation for harnessing chemical energy is approximately 40% (21 kcal ÷ 52 kcal × 100). The remaining 60% of the energy is lost to the body as heat. Considering that the steam engine transforms its fuel into useful energy with an efficiency of only about 30%, the value of 40% for the human body represents high efficiency.

ROLE OF OXYGEN IN ENERGY METABOLISM

Three prerequisites must be met for the continual resynthesis of ATP: (1) a donor of electrons in the form of NADH (or $FADH_2$) must be available, (2) adequate oxygen, the final electron and hydrogen acceptor, must be present, and (3) enzymes and metabolic machinery must be present in sufficient concentration to make the energy transfer reactions "go." When these conditions are satisfied, hydrogen is continually shuttled down the respiratory chain to molecular oxygen during the breakdown of the food substrate.

In strenuous exercise, when the rate of oxygen delivery (prerequisite #2 above) or utilization (prerequisite #3 above) is often inadequate, a relative imbalance is created between the release of hydrogen and its final acceptance by oxygen. In a sense, electron flow down the respiratory chain begins to "back up" with an accumulation of hydrogens bound to NAD^+. As is discussed in a subsequent section, pyruvic acid temporarily binds these excess hydrogens (electrons) to form lactic acid; this permits oxidative phosphorylation by electron transport to continue.

The function of oxygen in energy metabolism is clearly to serve as the final electron acceptor in the respiratory chain and combine with hydrogen to form water. This process which involves a considerable transfer of usable energy through oxidative phosphorylation in the respiratory chain, is referred to as **aerobic metabolism.** In one sense, this term is misleading because oxygen does not participate directly in the synthesis of ATP. The presence of oxygen at the "end of the line," however, largely determines one's capability for sustained aerobic energy release during exercise.

SUMMARY

1. The energy contained within the chemical structure of carbohydrates, fats, and proteins is not released in the body suddenly at some kindling temperature; rather, it is released slowly in small amounts during complex enzymatically controlled reactions. This allows for greater efficiency in energy transfer.
2. About 40% of the potential energy in the food nutrients is transferred to the compound ATP. When the terminal phosphate bond of the ATP is broken, the free energy liberated is harnessed to power all forms of biologic work. Thus, ATP is considered the body's energy currency, although its quantity is limited.
3. Creatine phosphate interacts with ADP to form ATP and thus serves as an energy reservoir to replenish ATP rapidly.
4. Phosphorylation is the process by which energy is transferred in the form of phosphate bonds. In this process, ADP and creatine are continually recycled into ATP and CP.
5. Cellular oxidation occurs in the mitochondria and involves the transfer of electrons from hydrogen to molecular oxygen. This results in the release and transfer of chemical energy to form the high-energy phosphates.
6. In the aerobic resynthesis of ATP, oxygen's primary role is to serve as the final electron acceptor in the respiratory chain and to combine with hydrogen to form water.

P A R T

2 Energy Release from Food

The energy generated in the breakdown of the food nutrients serves one purpose—to phosphorylate ADP to re-form the energy-rich compound ATP (Fig. 6–6). Although the breakdown of various food nutrients during energy metabolism is geared toward generating phosphate-bond energy, the specific pathways of degradation differ depending on the nutrients metabolized. In the sections that follow, we show how the potential energy in the food nutrients is extracted and utilized to synthesize ATP.

ENERGY RELEASE FROM CARBOHYDRATES

The primary function of carbohydrate is to supply energy for cellular work. Our discussion of nutrient energy metabolism begins with carbohydrates for several reasons: (1) Carbohydrate is the only nutrient whose stored energy can be used to generate ATP anaerobically. This is important in vigorous

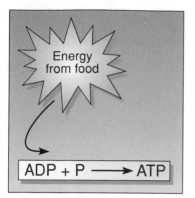

FIG. 6–6.
Food energy is used for the resynthesis of ATP.

exercise that requires rapid energy release above levels supplied by aerobic metabolic reactions. In this case, stored glycogen and blood glucose must supply the main portion of energy for ATP resynthesis. (2) During light and moderate exercise, carbohydrates supply about one-half of the body's energy requirements. (3) A continual breakdown of some carbohydrate is required so that fat nutrients can be processed through the metabolic mill and used for energy. It is not uncommon in prolonged exercise such as marathon running for a participant to experience fatigue—a state associated with a depletion of glycogen in the muscles and liver.

The complete breakdown of one mole of glucose to carbon dioxide and water yields a maximum of 686 kcal of chemical free energy, or energy available to do work.

$$C_6H_{12}O_6 + 6\,O_2 \longrightarrow 6\,CO_2 + 6\,H_2O \quad -\Delta G^{\circ\prime}\ \textbf{686 kcal per mole}$$

In the cell, however, the complete breakdown of glucose to its simple, stable end products is accompanied by the conservation of energy in the form of ATP. Because 7.3 kcal are required to synthesize each mole of ATP from ADP and inorganic phosphate, it would be theoretically possible to form 94 moles of ATP per mole of glucose by coupling all of the energy in glucose to phosphorylation (686 kcal ÷ 7.3 = 94). In the muscle, however, only about 38%, or 263 kcal of energy, is actually conserved in phosphate bonds with the remainder dissipated as heat. Consequently, 36 moles of ATP are regenerated in glucose breakdown (263 kcal ÷ 7.3 = 36) with an accompanying gain in free energy of 263 kcal.

Anaerobic vs Aerobic

There are two stages for glucose degradation in the body. The first stage involves the breakdown of a glucose molecule to two molecules of pyruvic acid. **These reactions involve energy transfers that do not require oxygen; they are termed anaerobic.** In the second phase of glucose catabolism, the pyruvic acid molecules are further degraded to carbon dioxide and water. **Energy transfers from these reactions involve electron transport and the accompanying oxidative phosphorylation; they are thus referred to as aerobic.**

ANAEROBIC ENERGY FROM GLUCOSE: GLYCOLYSIS

When a glucose molecule enters a cell to be used for energy, it undergoes a series of chemical reactions collectively termed **glycolysis.** When this series of reactions starts with stored glycogen the process is called **glycogenolysis.** These reactions, summarized in Figure 6–7, occur in the watery medium of the cell outside of the mitochondrion. In a sense, this process represents a more primitive form of energy transfer that is well developed in amphibia, reptiles, fish, and diving mammals. In skeletal muscle the breakdown of stored glycogen for energy is regulated and limited by the enzyme **phosphorylase.**[2] The activity of this enzyme is greatly influenced by the action of **epinephrine,** a hormone of the sympathetic nervous system.

In the first reaction, ATP acts as a phosphate donor to phosphorylate glucose to **glucose 6-phosphate.** At this point, in most tissues of the body, the glucose molecule is "trapped" in the cell. (For example, liver and, to a

FIG. 6—7.
Glycolysis: The breakdown of the 6-carbon glucose molecule to two, 3-carbon molecules of pyruvic acid.

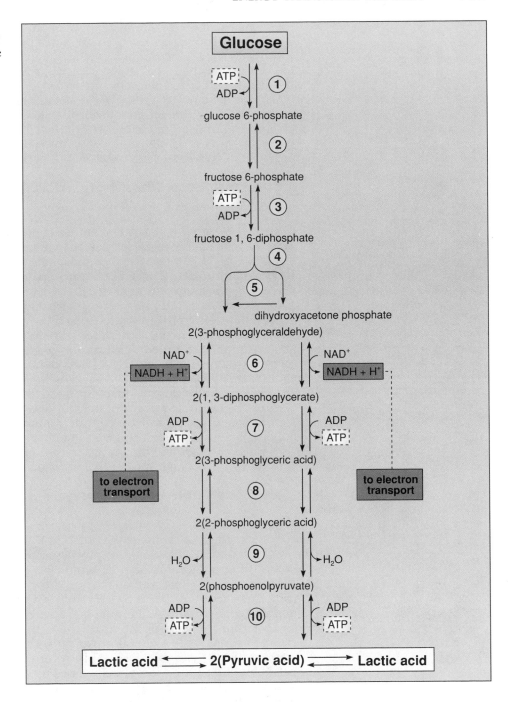

small extent, kidney cells contain the enzyme phosphatase that can split the phosphate from the glucose 6-phosphate and thus make glucose available to leave the cell and to be transported throughout the body.) The glucose molecule can now be linked together, or polymerized, with other glucose molecules to form glycogen, the storage form of glucose. This process is dependent on the enzyme **glycogen synthetase.** In energy metabolism, however, glucose 6-phosphate is changed to an isomere, **fructose 6-phosphate.** At this stage, no energy has been extracted but energy has been incorporated into the original glucose molecule at the expense of one molecule of ATP. In a sense, phosphorylation has "primed the pump" so that energy metabolism can proceed. Then, controlled by the enzyme **phosphofructokinase** (PFK), the fructose 6-phosphate molecule is changed to fructose 1, 6-diphosphate. It is the activity level of PFK that probably places a limit on glycolysis during all-out exercise. Fructose 1, 6-diphosphate then splits into **two** phosphorylated molecules with three carbon chains; these are further degraded in five successive reactions to pyruvic acid.

Substrate-Level Phosphorylation in Glycolysis

Most of the energy generated in the reactions of glycolysis is insufficient to resynthesize ATP and is lost to the body as heat. In reactions 7 and 10, however, the energy released from the glucose intermediates is sufficient to stimulate the direct transfer of a phosphate group to ADP. This generates a total of four molecules of ATP. **Because two molecules of ATP were lost in the initial phosphorylation of the glucose molecule, the net energy transfer from glycolysis results in a gain of two molecules of ATP.** These specific energy transfers from substrate to ADP by phosphorylation do not require oxygen. Rather, energy is directly transferred to phosphate bonds in the anaerobic process called **substrate-level phosphorylation.** This process of energy conservation during glycolysis operates at an efficiency of about 30%.

Only about 5% of the total ATP generated in the complete breakdown of the glucose molecule is formed during glycolysis. Due to the high concentration of glycolytic enzymes and the speed of these reactions, however, significant energy for muscle contraction can be provided **rapidly** from glycolysis. The athlete sprinting at the end of the mile run relies heavily on this form of anaerobic energy transfer—which is possible only from the breakdown of the body's carbohydrate stores by glycolytic reactions.

Hydrogen Release in Glycolysis

During glycolysis, two pairs of hydrogen atoms are stripped from the substrate and their electrons passed to NAD^+ to form NADH (Fig. 6–7). Normally, if these electrons were processed directly through the respiratory chain, three molecules of ATP would be generated for each molecule of NADH oxidized. The mitochondrion, however, is impermeable to NADH formed in the cytoplasm during glycolysis. Consequently, the electrons from extramitochondrial NADH must be **shuttled** indirectly into the mitochondria. Within the cells of the heart the extramitochondrial hydrogen is delivered to NAD^+. In skeletal muscle, this route ends with electrons being passed to FAD to form $FADH_2$. This in turn transfers its electrons at a point below the first formation of ATP (see Fig. 6–5A). Thus **in skeletal muscle two, rather than three, ATP molecules are formed when cytoplasmic NADH is oxidized by the respiratory chain. Because two molecules of NADH are formed in glycolysis, four molecules of ATP are generated aerobically by subsequent electron transport-oxidative phosphorylation.**

Formation of Lactic Acid

During moderate levels of energy metabolism, sufficient oxygen is available to the cells. Consequently, most of the hydrogens (electrons) stripped from the substrate and carried by NADH are oxidized within the mitochondria and passed to oxygen to form water. In a biochemical sense a "steady state," or more precisely a "steady rate," exists because hydrogen is oxidized at about the same rate as it is made available. Biochemists frequently refer to this condition as **aerobic glycolysis** with pyruvic acid being the predominant end product. Even at rest or in mild exercise some lactic acid is continually formed by the energy metabolism of red blood cells that contain no mitochondria and limitations posed by enzyme activity and the equilibrium constants for chemical reactions. **In this case, lactic acid does not build up because its removal rate equals its rate of production.**

In strenuous exercise, when the energy demands exceed either the oxygen supply or its rate of utilization, all of the hydrogen joined to NADH cannot be processed through the respiratory chain. Continued release of anaerobic energy in glycolysis depends on the availability of NAD^+ for the oxidation of **3-phosphoglyceraldehyde** (see reaction 6, Fig. 6–7); otherwise, the rapid rate of glycolysis would "grind to a halt." Under conditions of **anaerobic glycolysis,** NAD^+ is "freed" as pairs of "excess" hydrogens combine with pyruvic acid in an additional step, catalyzed by the enzyme **lactic dehydrogenase,** to form lactic acid in the reversible reaction:

$$\overset{\displaystyle O}{\overset{\displaystyle \|}{CH_3-C-COOH}} + NADH + H^+ \xrightarrow[\text{dehydrogenase}]{\text{lactic}}$$

(Pyruvic acid)

$$\overset{\displaystyle OH}{\overset{\displaystyle |}{\underset{\displaystyle \underset{\displaystyle H}{|}}{CH_3-C-COOH}}} + NAD^+$$

(Lactic acid)

The temporary storage of hydrogen with pyruvic acid is a unique aspect of energy metabolism because it provides a ready "sump" for the disappearance of the end products of anaerobic glycolysis. Also, once lactic acid is formed in the muscle, it diffuses rapidly to the blood where it is buffered and carried away from the site of energy metabolism. In this way, glycolysis can proceed to supply additional anaerobic energy for the resynthesis of ATP. This avenue for extra energy is only temporary, however, because as the level of lactic acid (more precisely lactate) in the blood and muscles increases, the regeneration of ATP cannot keep pace with its utilization, fatigue sets in, and exercise must stop. Fatigue is probably mediated by **increased acidity** that inactivates various enzymes involved in energy transfer as well as the muscle's contractile properties.[1,3]

Lactic acid should not be viewed as a metabolic "waste product." On the contrary, it is a valuable source of chemical energy that accumulates and is retained in the body during heavy physical exercise. When sufficient oxygen is once again available, as in recovery or when the pace of exercise is slowed, hydrogens attached to lactic acid are picked up by NAD^+ and eventually oxidized. Consequently, lactic acid is readily reconverted to pyruvic acid and used as an energy source. In addition, the potential energy in the lactate

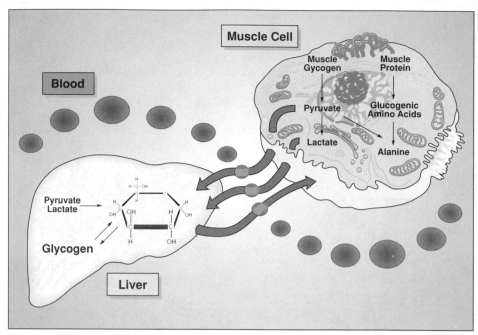

FIG. 6–8.
The Cori cycle provides for the removal of lactic acid formed in muscle with a resulting increase in available glucose.

and pyruvate molecules formed in muscle during exercise can be conserved and the carbon skeletons of these molecules used for the synthesis of glucose in the gluconeogenic process of the **Cori cycle** illustrated in Figure 6–8. This cycle not only provides a means for lactic acid removal but also for augmenting blood glucose and muscle glyogen.

THE KREBS CYCLE

Because the anaerobic reactions of glycolysis release only about 10% of the energy within the glucose molecule, an additional means for extracting the remaining energy is available. This is provided when the pyruvic acid molecules are **irreversibly** converted to a form of acetic acid, **acetyl-CoA.** This intermediate compound enters the second stage of carbohydrate breakdown known as the **Krebs cycle** (or more descriptively, the citric acid or tricarboxylic acid cycle). As shown schematically in Figure 6–9, the main function of the Krebs cycle is to degrade the acetyl-CoA substrate to carbon dioxide and hydrogen atoms within the mitochondria. The hydrogen atoms are then oxidized in processes involving electron transport-oxidative phosphorylation with the subsequent regeneration of ATP.

As shown in Figure 6–10, pyruvic acid is prepared for entrance into the Krebs cycle by joining with the vitamin B-derivative **coenzyme A** (A for acetic acid) to form the 2-carbon compound acetyl-CoA. In the process, two hydrogens are released and their electrons transferred to NAD^+ and one molecule of carbon dioxide is formed as follows:

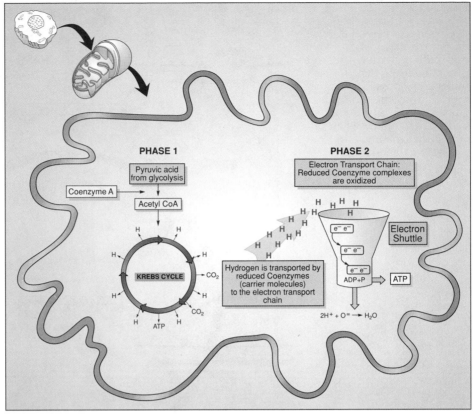

FIG. 6–9.
In the mitochondria the Krebs cycle generates hydrogen atoms in the breakdown of acetyl-CoA (Phase 1). These hydrogens are then oxidized via the aerobic process of electron transport-oxidative phosphorylation, and significant quantities of ATP are regenerated (Phase 2).

$$\text{Pyruvic acid} + \text{NAD}^+ + \text{CoA} \longrightarrow \text{Acetyl-CoA} + \text{CO}_2 + \text{NADH} + \text{H}^+$$

When the acetyl portion of acetyl-CoA joins with **oxaloacetic acid,** it forms **citric acid** (the same citric acid found in citrus fruits), a 6-carbon compound that then proceeds through the Krebs cycle. The cycle is continued because the original oxaloacetic molecule is retained and joins with a new acetyl fragment.

For each molecule of acetyl-CoA entering the Krebs cycle, two carbon dioxide molecules and four pairs of hydrogen atoms are cleaved from the substrate. One molecule of ATP is also regenerated directly by substrate level phosphorylation from Krebs cycle reactions (see reaction 7, Fig. 6–10). As summarized at the bottom of Figure 6–10, for the two pyruvic acid molecules formed in glycolysis, a total of four hydrogens are released in the formation of acetyl-CoA, and 16 hydrogens are released in the Krebs cycle. **In essence, the most important function of the Krebs cycle is to generate**

FIG. 6–10.
"Flowsheet" for release of hydrogen and carbon dioxide during the degradation of one molecule of pyruvic acid in the mitochondrion. Because two molecules of pyruvic acid are formed from one molecule of glucose during glycolysis, all values are doubled when computing the net gain from pyruvate catabolism.

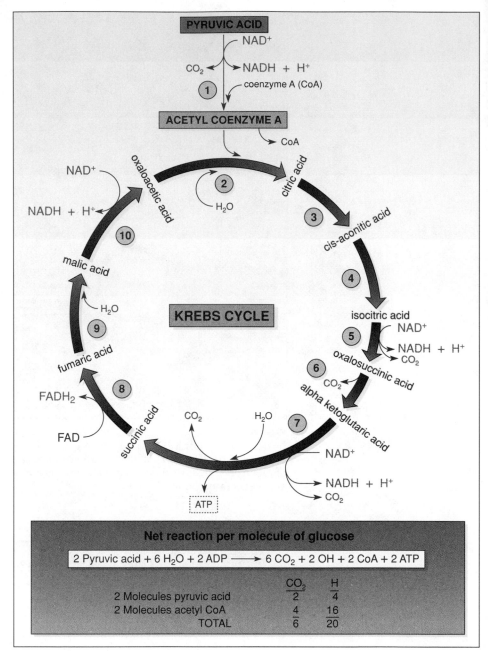

Net reaction per molecule of glucose

$$2 \text{ Pyruvic acid} + 6 \text{ H}_2\text{O} + 2 \text{ ADP} \longrightarrow 6 \text{ CO}_2 + 2 \text{ OH} + 2 \text{ CoA} + 2 \text{ ATP}$$

	$\frac{CO_2}{2}$	$\frac{H}{4}$
2 Molecules pyruvic acid	2	4
2 Molecules acetyl CoA	4	16
TOTAL	6	20

electrons (hydrogens) for their passage to the respiratory chain by means of NAD$^+$ and FAD.

Molecular oxygen does not participate directly in the reactions of the Krebs cycle. The major portion of the chemical energy in pyruvic acid is transferred to ADP through the aerobic process of electron transport-oxidative phosphorylation. As long as there is an adequate oxygen supply with enzymes and substrate available, NAD$^+$ and FAD are regenerated and Krebs cycle aerobic metabolism proceeds.

TOTAL ENERGY TRANSFER FROM GLUCOSE CATABOLISM

The pathways for energy transfer during the breakdown of a glucose molecule in skeletal muscle are summarized in Figure 6–11. A net of 2 ATP is formed from substrate phosphorylation in glycolysis and, similarly, 2 ATP are generated during acetyl-CoA degradation in the Krebs cycle. The 24 released hydrogen atoms can be accounted for as follows: (1) the 4 hydrogens (2 NADH)

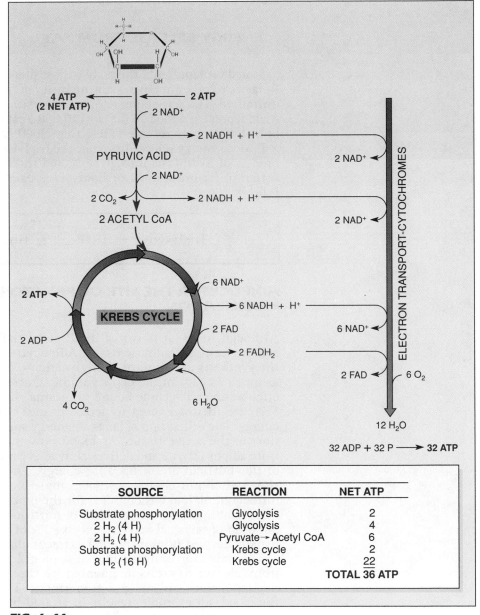

SOURCE	REACTION	NET ATP
Substrate phosphorylation	Glycolysis	2
2 H$_2$ (4 H)	Glycolysis	4
2 H$_2$ (4 H)	Pyruvate → Acetyl CoA	6
Substrate phosphorylation	Krebs cycle	2
8 H$_2$ (16 H)	Krebs cycle	<u>22</u>
		TOTAL 36 ATP

FIG. 6–11.
Adenosine triphosphate (ATP) yield from energy transfer during the complete oxidation of glucose.

generated outside the mitochondria during glycolysis yield 4 ATP (6 ATP in heart muscle) during oxidative phosphorylation, (2) the 4 hydrogens (2 NADH) released in the mitochondria as pyruvic acid are changed to acetyl-CoA to yield 6 ATP, (3) 12 of the 16 hydrogens (6 NADH) released in the Krebs cycle yield 18 ATP, and (4) the remaining 4 hydrogens joined to FAD (2 FADH$_2$) in the Krebs cycle yield 4 ATP. **The net ATP yield from the complete breakdown of the glucose molecule in skeletal muscle is 36 molecules of ATP; 4 ATP molecules are formed directly from substrate phosphorylation (via glycolysis and the Krebs cycle) whereas 32 ATPs are generated during oxidative phosphorylation.**

ENERGY RELEASE FROM FAT

Stored fat represents the body's most plentiful source of potential energy. Relative to other nutrients, the quantity of fat available for energy is almost unlimited. The actual fuel reserves from stored fat in an average young adult male represent about 90,000 to 110,000 kcal of energy. In contrast, the carbohydrate energy reserve is less than 2000 kcal.

Prior to energy release from fat, the triglyceride molecule is cleaved in the process of hydrolysis into glycerol and three fatty acid molecules. This reaction of fat breakdown or **lipolysis** is catalyzed by the enzyme **lipase** as follows:

$$\text{Triglyceride} + 3H_2O \xrightarrow{\text{lipase}} \text{Glycerol} + 3 \text{ fatty acids}$$

ADIPOCYTES: THE SITE OF FAT STORAGE AND MOBILIZATION

Although some fat is stored in all cells, the most active supplier of fatty acid molecules is adipose tissue. **Adipocytes,** or fat cells, are specialized for the synthesis and storage of triglycerides. Triglyceride fat droplets occupy as much as 95% of the cell's volume. Once the fatty acids diffuse into the circulation and become bound to plasma albumin, these **free fatty acids,** or *FFA*, are then delivered to active tissues where they are metabolized for energy. The utilization of fat as an energy substrate varies closely with blood flow in the active tissue. As blood flow increases more FFA are removed from adipose tissue and delivered to active muscle, hence greater quantities of this nutrient are utilized for energy. This is especially the case for the slow-twitch muscle fibers whose ample blood supply and large, numerous mitochondria make them ideal for the process of fat breakdown.

Depending on a person's state of nutrition and fitness, and the level and duration of physical activity, 30 to 80% of the energy for biologic work is usually derived from intra- and extracellular fat molecules (see Chap. 1).

The activation of lipase and subsequent lipolysis and mobilization of FFA from adipose tissue is augmented by the hormones epinephrine, norepinephrine, glucagon, and growth hormone. The injection of epinephrine into the blood, for example, results in a rapid increase in plasma FFA. Because plasma concentrations of these hormones are increased during exercise, this mechanism for lipase activation provides the muscle with a supply of a potent energy substrate. It appears that lipase activation (and thus the regulation of fat breakdown) is catalyzed by an intracellular mediator, **adenosine 3′,5′-cyclic monophosphate** or **cyclic AMP.** Cylic AMP is activated by the various fat mobilizing hormones which themselves do not enter the cell.[4] Alterations in lipase activity may partly explain the enhanced utilization of fat observed with aerobic exercise training. A more detailed evaluation of hormone regulation in exercise is presented in Chapter 20.

CATABOLISM OF GLYCEROL AND FATTY ACIDS

Figure 6–12 summarizes the pathways for the degradation of the glycerol and fatty acid fragments of the triglyceride molecule.

Glycerol can be accepted into the anaerobic reactions of glycolysis as 3-phosphoglyceraldehyde and degraded to pyruvic acid. In this process, ATP is formed via substrate phosphorylation and hydrogen atoms are released to NAD^+; pyruvic acid is then oxidized in the Krebs cycle. In total, 19 ATP molecules are synthesized in the complete breakdown of the glycerol mol-

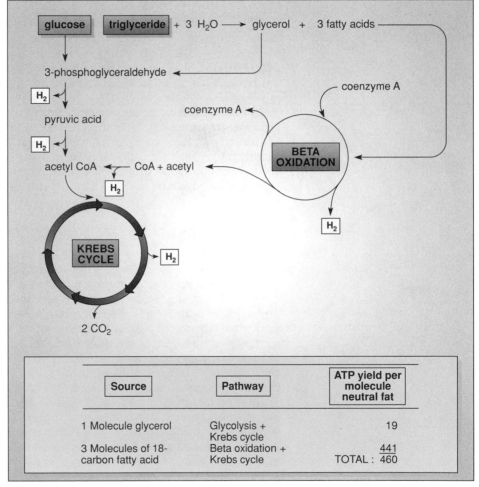

Source	Pathway	ATP yield per molecule neutral fat
1 Molecule glycerol	Glycolysis + Krebs cycle	19
3 Molecules of 18-carbon fatty acid	Beta oxidation + Krebs cycle	441
		TOTAL : 460

FIG. 6–12.

General scheme for the degradation of the glycerol and fatty acid fragments of neutral fat. Note that the released hydrogens go to electron transport.

ecule. Glycerol also serves an important function in providing carbon skeletons for the synthesis of glucose. The gluconeogenic role of glycerol is important when carbohydrate is restricted in the diet or in long-term exercise which places a significant drain on glycogen reserves.

The **fatty acid** molecule undergoes transformation to acetyl-CoA in the mitochondrion in a process called **beta oxidation.** This process involves the successive release of 2-carbon acetyl fragments split from the long chain of the fatty acid. Adenosine triphosphate is used to phosphorylate the reactions, water is added, hydrogens are passed to NAD^+ and FAD, and the acetyl fragment joins with coenzyme A to form acetyl-CoA, the same acetyl unit generated from glucose breakdown. This process is repeated over and over until the entire fatty acid molecule is degraded to acetyl-CoA, which then enters the Krebs cycle directly to be metabolized. The hydrogens released during fatty acid catabolism are oxidized through the respiratory chain.

It is important to note that the breakdown of fatty acids is directly associated with oxygen uptake. Oxygen must be available to accept hydrogen so beta oxidation can proceed. Under anaerobic conditions, hydrogen remains with NAD^+ and FAD, and fat catabolism is halted.

TOTAL ENERGY TRANSFER FROM FAT CATABOLISM

For each 18-carbon fatty acid molecule, 147 molecules of ADP are phosphorylated to ATP during beta oxidation and Krebs cycle metabolism. Because there are three fatty acid molecules in each triglyceride molecule, 441 ATP molecules are formed from the fatty acid component of neutral fat (3 × 147 ATP). Also because 19 molecules of ATP are formed during glycerol catabolism, a total of 460 molecules of ATP are generated for each neutral fat molecule catabolized for energy. This is a considerable energy yield considering that only 36 ATP are formed during the catabolism of a glucose molecule in skeletal muscle. The efficiency of energy conservation for fatty acid oxidation is about 40%, a value similar to that of glucose.

ENERGY RELEASE FROM PROTEIN

As mentioned in Chapter 1, protein can serve an important role as an energy substrate during sustained exercise and heavy training. To provide energy, the amino acids (primarily leucine, isoleucine, valine, glutamine, and aspartate) must first be converted to a form that can readily enter the pathways for energy release. Whereas the main site for this deamination (removal of nitrogen) is the liver, skeletal muscles also contain the enzymes for removing nitrogen in amino acids and passing it to other compounds in the process of transamination. In this way certain amino acids can be used directly in muscle for energy. In fact, the levels of these enzymes for **transamination** adapt to training which may further facilitate the use of protein as an energy substrate. Once the amino or nitrogen-containing group is removed from the amino acid, the remaining "carbon skeleton" is usually one of the reactive compounds that can contribute to the formation of high-energy phosphate bonds. Alanine, for example, loses its amine group and gains a double-bond oxygen to form pyruvic acid; glutamine forms alpha ketoglutaric acid; and aspartate forms oxaloacetic acid—all of these end products are Krebs cycle intermediates.

THE METABOLIC MILL—INTERRELATIONSHIPS AMONG CARBOHYDRATE, FAT, AND PROTEIN METABOLISM

The Krebs cycle plays a much more important role than simply the degradation of pyruvic acid produced during glucose catabolism. The Krebs cycle provides the means by which fragments of other organic compounds formed from the breakdown of fats and proteins can be effectively metabolized for energy. As illustrated in Figure 6–13, the deaminated residues of excess amino acids enter the Krebs cycle at various intermediate stages, whereas the glycerol fragment of fat catabolism gains entrance via the glycolytic pathway. Fatty acids are oxidized by beta oxidation to acetyl-CoA which then directly enters the Krebs cycle.

This sketch of the "metabolic mill" also shows the interconversions between fragments of the various nutrients and the possible routes for substrate synthesis. Excess carbohydrates, for example, provide the glycerol and acetyl

FIG. 6–13.
The metabolic mill: interconversions between carbohydrates, fats, and proteins.

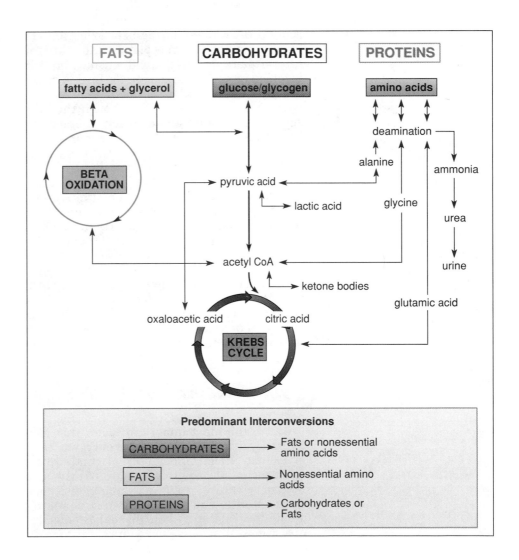

fragments for the synthesis of neutral fat. Acetyl-CoA can also function as the starting point for the synthesis of cholesterol and many hormones. Because the conversion of pyruvic acid to acetyl-CoA is not reversible, however, (notice the one-way arrow), fatty acids **cannot** be used to synthesize glucose. Amino acids with carbon skeletons resembling Krebs cycle intermediates are deaminated and synthesized to glucose. This finding is especially true for the amino acid alanine (see Chap. 1).

FATS BURN IN A CARBOHYDRATE FLAME

One interesting aspect of the metabolic mill is that the breakdown of fatty acids depends somewhat on a continual background level of carbohydrate catabolism. It should be recalled that acetyl-CoA enters the Krebs cycle by combining with oxaloacetic acid (generated mainly by carbohydrate catabolism) to form citric acid. The degradation of fatty acids via the Krebs cycle continues only if sufficient oxaloacetic acid is available to combine with acetyl-CoA formed during beta oxidation. The pyruvic acid formed during glucose metabolism may play an important role in furnishing this oxaloacetic intermediate (Fig. 6–13), and when carbohydrate levels fall oxaloacetic acid levels may become inadequate: In this sense, "fats burn in a carbohydrate flame." It is also likely that there may be a rate limit to fatty-acid utilization by the exercising muscle. Although this limit can be greatly enhanced by aerobic-type training, the power generated only by fat breakdown never appears to equal that generated by the combination of both fat and carbohydrate breakdown together. Thus, when muscle glycogen is depleted the maximum power output of muscle must fall.

An appreciable reduction in carbohydrate breakdown, which could occur in prolonged exercise such as marathon running or consecutive days of heavy training, starvation, dietary elimination of carbohydrates (as advocated with high-fat, low-carbohydrate "ketogenic diets"), or diabetes, also seriously limits the capacity for energy transfer. This occurs despite the fact that large amounts of fatty acid substrate are available in the circulation. In instances of extreme carbohydrate restriction or depletion, the acetate fragments produced in beta oxidation begin to build up in the extracellular fluids because they cannot be accommodated in the Krebs cycle. These fragments are readily converted to ketone bodies, some of which are excreted in the urine. If this condition of ketosis persists, the acid quality of the body fluids can increase to potentially toxic levels.

WHAT REGULATES ENERGY METABOLISM?

By far, the most important factor that controls the breakdown of carbohydrate, fat, and amino acids for energy release in the cell is the concentration of cellular ADP. Under normal conditions, the transfer of electrons and subsequent release of energy are tightly coupled to ADP phosphorylation. In general, unless ADP is available and phosphorylated to ATP, electrons do not shuttle down the respiratory chain to oxygen. This particular mechanism for respiratory control makes considerable "sense," because any increase in ADP signals a need for energy to restore the ATP levels. Conversely, a low level of cellular ADP indicates a relatively low energy requirement.

SUMMARY

1. The food nutrients provide a source of potential energy to rejoin ADP and free phosphate to form ATP.
2. The complete breakdown of 1 mole of carbohydrate liberates 686 kcal of energy. Of this, about 263 kcal (38%) is conserved in the bonds of ATP; the remainder is dissipated as heat.
3. In the reactions of glycolysis in the cell cytoplasm, 2 ATP are formed in the anaerobic process of substrate-level phosphorylation.
4. In the second stage of carbohydrate breakdown that occurs within the mitochondrial complex, pyruvic acid is converted to acetyl-CoA, which is then processed through the Krebs cycle. The hydrogens released during glucose breakdown are oxidized via the respiratory chain and the energy generated is coupled to phosphorylation.
5. In the complete breakdown of carbohydrate in skeletal muscle a total of 36 ATP molecules are formed.
6. When hydrogen atoms are oxidized at the rate they are formed a biochemical "steady state" is said to exist. In heavy exercise when the oxidation of hydrogen does not keep pace with its rate of production, pyruvic acid temporarily combines with hydrogen to form lactic acid.
7. The complete breakdown of a fat molecule yields about 460 ATP molecules. Fatty acid metabolism is directly associated with oxygen uptake; that is, the reactions are aerobic.
8. Protein can also serve as an important energy substrate. After nitrogen is removed from the amino acid molecule in the process of deamination, the remaining carbon skeleton can enter the Krebs cycle for the aerobic production of ATP.
9. Numerous interconversions are possible among the various food nutrients. The exception is fatty acids that cannot be used for the synthesis of glucose.
10. A certain level of carbohydrate breakdown is required for fats to be metabolized continually for energy in the metabolic mill. To this extent, "fats burn in a carbohydrate flame."

REFERENCES

1. Bertocci, L.A., and Gollnick, P.D.: pH effect on mitochondria and individual enzyme function. *Med. Sci. Sports Exerc.*, 17:244, 1985.
2. Chasiotis, D.: Role of cyclic AMP and inorganic phosphate in the regulation of muscle glycogenolysis during exercise. *Med. Sci. Sports Exerc.*, 20:545, 1988.
3. Mainwood, G.W., and Renaud, J.M.: The effect of acid-base on fatigue of skeletal muscle. *Can. J. Physiol. Pharmacol.*, 63:403, 1985.
4. Shepherd, R.E., and Bah, M.D.: Cyclic AMP regulation of fuel metabolism during exercise: regulation of adipose tissue lipolysis during exercise. *Med. Sci. Sports Exerc.*, 20:531, 1988.

GENERAL REFERENCES

Brooks, G.A., and Fahey, T.D.: *Exercise Physiology: Human Bioenergetics and its Applications.* New York, Wiley, 1984.
Giese, A.C.: *Cell Physiology.* Philadelphia, W.B. Saunders, 1979.
Horton, E.S., and Terjung, R.L. (Eds.): *Exercise, Nutrition, and Energy Metabolism.* New York, Macmillan, 1988.

Lehninger, A.L.: *Bioenergetics: The Molecular Basis of Biological Energy Transformations,* Menlo Park, Calif., W.A. Benjamin, 1971.

Lehnigner, A.L.: *Principles of Biochemistry.* New York, Worth Publishers, 1982.

Marieb, E.N.: *Human Anatomy and Physiology.* Redwood City, Benjamin Cummings, 1989.

Mott-Smith, M.: *The Concept of Energy Simply Explained.* New York, Dover Press, 1964.

Shils, M.E., and Young, V.R.: *Modern Nutrition in Health and Disease,* 7th ed. Philadelphia, Lea & Febiger, 1988.

Stryer, L.: *Biochemistry.* 2nd ed. San Francisco, W.H. Freeman, 1988.

Vander, A.J., et al.: *Human Physiology.* 2nd ed. New York, McGraw-Hill, 1985.

7

ENERGY TRANSFER IN EXERCISE

Physical activity by far provides the greatest demand for energy. In sprint running and swimming, for example, the energy output from the working muscles may be as much as 120 times higher than at rest. During less intense but sustained exercise such as marathon running, the energy requirement increases some 20 to 30 times above rest. Depending on the intensity and duration of exercise, and the fitness of the participant, the relative contributions of the body's various means for energy transfer differ markedly.

IMMEDIATE ENERGY: THE ATP-CP SYSTEM

Performances of short duration and high intensity such as the 100-yard dash, 25-yard swim, or weight lifting require an immediate and rapid supply of energy. This energy is provided almost exclusively from the high-energy phosphates ATP and CP stored within the specific muscles activated during exercise.

Approximately 5 millimoles (mmol) of ATP and 15 mmol of CP are stored within each kilogram of muscle.[40] For a 70-kg person who has a muscle mass of 30 kg, this is between 570 and 690 mmol of high-energy phosphates. If we assume that 20 kg of muscle are activated during exercise, then there is sufficient stored phosphate energy to walk briskly for 1 minute, run a cross-country race for 20 to 30 seconds, or perform all-out exercises such as sprint running and swimming for about 6 seconds.[5] In the 100-yard dash, for example, the body cannot maintain maximum speed for longer than this time, and the runners may actually be slowing down in the last portion of the race. In this situation, the quantity of intramuscular phosphate may significantly influence one's ability to generate intense energy for a short duration.

All sports require utilization of the high-energy phosphates, but many activities rely almost exclusively on this means for energy transfer. For example, success in football, weight lifting, field events, baseball, and volleyball all require a brief maximal effort during the performance. It is difficult to imagine an end run in football or a pole vault without the capability for generating energy rapidly from the stored phosphagens. For sustained exercise and for recovery from an all-out effort, additional energy must be generated for ATP replenishment. To this end, the stored carbohyrates, fats, and proteins stand ready to continually recharge the phosphate pool.

Magnetic Resonance Spectroscopy to Study Exercise Muscle Metabolism. Magnetic resonance spectroscopy is an exciting new non-invasive method to continuously measure the relative concentrations in

FIG. 7–1.
A. The wrist flexor muscles are placed on a surface coil in a superconducting magnet. The subject grasps a handle that is attached to an isokinetic dynamometer (constant velocity, variable force output). The subject watches a strip chart recorder that provides feedback on the level of force production.
B. Example of magnetic resonance spectroscopy spectra for ATP, PCr, and inorganic phosphate during rest and two levels of exercise. (From McCully K.K. et al., Application of ^{31}P magnetic resonance spectroscopy to the study of athletic performance. *Sports Med.*, 5:312, 1988.)

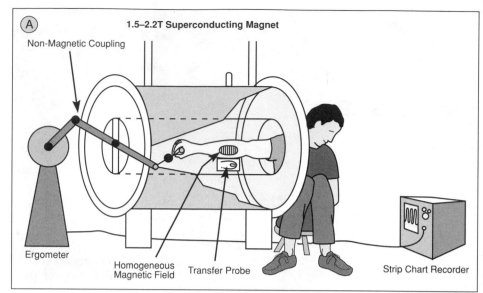

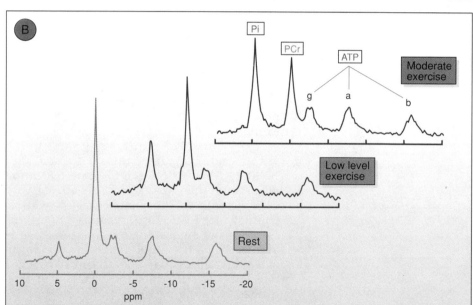

muscle of the high-energy phosphate compounds as well as other metabolic events during exercise.[52,53] Figure 7–1A illustrates the method during wrist flexion exercise. The muscles used primarily during the exercise are placed over a superconducting magnet, and the subject performs exercise under conditions that control for tension output, contraction speed, and exercise duration. By applying specific radiofrequency pulses within the strong magnetic field one can determine the concentrations of various biochemically active compounds. Figure 7–1B shows the results for ATP, CP(PCr), and inorganic phosphate during rest and low and moderate intensity exercise. With magnetic resonance spectroscopy, the areas under the peaks correspond to the relative concentrations of the free phosphorus compounds, including the three phosphorus atoms of ATP. Such studies of the ratio of inorganic phosphate to phosphocreatine can evaluate the rate of mitochondrial respiration. Currently this technique is being applied to study muscle injury, glycolytic metabolism, and the effects of training on the intricacies of muscle metabolic function.[54,55a]

SHORT-TERM ENERGY: THE LACTIC ACID SYSTEM

The high-energy phosphates must continually be resynthesized at a rapid rate for strenuous exercise to continue beyond a brief time period. In such intense exercise, the energy to phosphorylate ADP comes mainly from glucose and stored glycogen during the anaerobic process of glycolysis with the resulting formation of lactic acid. In a way, this mechanism of lactic acid formation "buys time." It allows for the rapid formation of ATP by substrate-level phosphorylation, even though the oxygen supply is inadequate or the energy demands outstrip the capacity for ATP resynthesis aerobically. This anaerobic energy for ATP resynthesis can be thought of as reserve fuel that is brought into use by the athlete "kicking" the last portion of a mile run. It is also of critical importance in supplying the rapid energy above that available from the stored phosphagens during a 440-yard run or 100-yard swim.[1] **The most rapidly accumulated and highest lactic acid levels are reached during exercise that can be sustained for 60 to 180 seconds.** As the intensity of "all-out" exercise decreases, thereby extending the work period, there is a corresponding decrease in both the rate of buildup and the final level of lactic acid.[46]

Lactate Accumulation. Lactic acid does not necessarily accumulate at all levels of exercise. Figure 7–2 illustrates the general relationship between oxygen consumption, expressed as a percentage of maximum, and blood lactic acid during light, moderate, and heavy exercise in endurance athletes and untrained subjects. During light and moderate exercise, the energy demands of both groups are adequately met by reactions that use oxygen. In biochemical terms, the ATP for muscular contraction is made available predominantly through energy generated by the oxidation of hydrogen. Any lactic acid formed in light exercise is rapidly oxidized. As such, the blood lactic acid level remains fairly stable even though oxygen consumption increases.

Lactic acid begins to accumulate and rise in an exponential fashion at about 55% of the healthy, untrained subject's maximal capacity for aerobic metabolism.[15,18] The usual explanation for the increase in lactic acid is based on the assumption of a relative tissue hypoxia in heavy exercise. It is argued that under these conditions of oxygen deficiency the energy requirement is partially met by a predominance of anaerobic glycolysis as the release of hydrogen begins to exceed its oxidation down the respiratory chain. Consequently, excess hydrogens are passed to pyruvic acid and lactic acid accumulates.[49,49a]* This increase in lactic acid becomes greater as exercise

* An alternate explanation for lactic acid buildup is based on studies that use radioactive tracers to label the carbon in the carbohydrate molecule.[7,20] These studies show that lactic acid is formed continuously at rest and in light exercise. Under aerobic conditions, however, lactic acid formation is matched by its rate of removal so the concentration of lactic acid remains fairly stable. With aerobic training, lactic acid removal keeps pace with its production; a buildup or accumulation occurs only at higher levels of exercise.[62a]

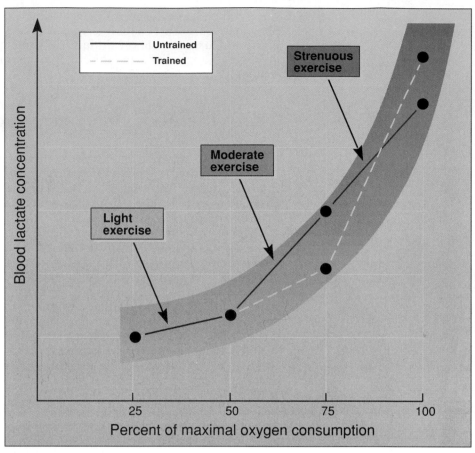

FIG. 7–2.
Increases in blood lactate concentration at different levels of exercise expressed as a percentage of maximal oxygen consumption for trained and untrained subjects.

becomes more intense and the muscle cells cannot meet the additional energy demands aerobically. This pattern is essentially similar for the trained subjects except the threshold for lactate buildup, termed the anaerobic threshold, or more precisely, **blood lactate threshold,** occurs at a higher percentage of the athlete's aerobic capacity.[13,16,22,26] This favorable response could be due to the endurance athlete's genetic endowment (type of muscle fiber) or specific local adaptations with training that would favor the production of less lactic acid,[17,65] as well as a more rapid rate of removal at any particular exercise level.[20] For example, it is well documented that capillary density as well as the size and number of mitochondria increase with endurance training, as does the concentration of various enzymes and transfer agents involved in aerobic metabolism,[28,34,37,38] and this training response may be unimpaired by the aging process.[57,67] Such alterations certainly enhance the cell's capacity to generate ATP aerobically, especially through the breakdown of fatty acids, and may extend the percentage of one's maximum that can be sustained before the **onset of blood lactate accumulation** (OBLA).[41,60] Trained endurance athletes, for example, perform at exercise intensities that represent about 80 to 90% of their maximum capability for aerobic metabolism.[12,15,64]

It is also suggested that lactate formed in one part of a working muscle can be oxidized by other fibers in the same muscle or by less active neighboring muscle tissue.[39,43,47] These adjustments and training adaptations would certainly help to keep lactate levels low during exercise and would also provide an important means for glucose conservation in prolonged work.[11a] The concept of the blood lactate threshold and its relation to endurance performance is developed more fully in Chapter 14.

Lactate-Producing Capacity. The ability to generate a high lactic acid level in all-out exercise is increased with specific "anaerobic training" and subsequently reduced with detraining. Well-trained athletes have shown that after they perform maximal short-term exercise, the blood lactate level is 20 to 30% higher than in untrained subjects under similar circumstances. The mechanism for this response is unknown, but it may be due to large differences in motivation level accompanying the trained state as well as about a 20% increase in enzymes involved in glycolysis, specifically phosphofructokinase, observed as a result of anaerobic-type training. Because lactic acid is continuously removed during and after exercise at a varied rate among individuals, it is unlikely that blood lactate measured at a particular time in recovery gives the full picture of an individual's capacity for anaerobic metabolism.[26] It is also likely that the increased intramuscular glycogen stores that accompany the trained state allow for a greater contribution of energy via anaerobic glycolysis.[63] Although increases in enzymes of the anaerobic pathway have been reported with sprint-type training,[23,43] these changes do not appear as impressive as the changes in aerobic enzymes with endurance training.[38]

LONG-TERM ENERGY: THE AEROBIC SYSTEM

Although the energy released in glycolysis is rapid and does not require oxygen, relatively little ATP is resynthesized in this manner. Consequently, aerobic reactions provide the important final stage for energy transfer, especially if vigorous exercise proceeds beyond several minutes.

OXYGEN CONSUMPTION DURING EXERCISE

The curve in Figure 7–3 illustrates the oxygen consumption during each minute of a relatively slow jog continued at a steady pace for 10 minutes. Oxygen consumption rises rapidly during the first minutes of exercise. Between the third and fourth minute a plateau is reached, and the oxygen consumption remains relatively stable for the rest of the exercise period. The flat portion or plateau of the oxygen consumption curve is generally considered the **steady state** or more precisely, the **steady rate.** This steady rate reflects a balance between the energy required by the working muscles and the rate of ATP production via aerobic metabolism. In this region, oxygen-consuming reactions supply the energy for exercise, and any lactic acid produced is either oxidized or reconverted to glucose, presumably in the liver and possibly kidneys. **Under steady-rate metabolic conditions, lactic acid accumulation is minimal.**

Many believe that once a steady rate is attained, exercise could continue indefinitely if the individual had the willpower to continue. This, of course, is based on the premise that a steady rate of aerobic metabolism is the only

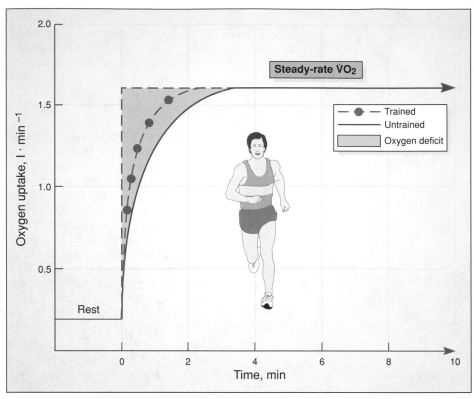

FIG. 7–3.
Time course of oxygen uptake during a continuous jog at a relatively slow pace for 10 minutes. The shaded area indicates the "oxygen deficit" or the extra quantity of oxygen that would have been consumed had the oxygen uptake reached a steady rate immediately.

factor determining one's capacity for sustained submaximal exercise. Other factors, however, must be considered. Fluid loss and electrolyte depletion often become significant limiting factors, especially during work in the heat. Also of considerable importance to prolonged exercise is maintaining adequate fuel reserves, particularly blood glucose for central nervous system function, and glycogen in the liver and in specific muscles utilized in the activity. Once a muscle's glycogen reserves are depleted, its work capability is dramatically reduced.

There are many steady rate levels. For some, the spectrum of steady rates might range from lying in bed to pushing a power lawn mower. On the other hand, at the upper limit the elite endurance runner can maintain a steady rate of aerobic metabolism throughout a 26-mile run averaging less than 5 minutes per mile or cover a 658 mile ultramarathon averaging 118 miles a day for $5\frac{1}{2}$ days! These magnificent accomplishments are determined largely by the athlete's ability to deliver and utilize oxygen.

Oxygen Deficit

The curve of oxygen consumption shown in Figure 7–3 does not increase instantaneously to a steady rate at the start of exercise. In fact, in the beginning stages of work the oxygen uptake is considerably below the steady-rate level, even though the energy required to perform the exercise presumably remains unchanged throughout the exercise period. This lag in oxygen uptake should not be surprising, however, because the immediate energy for muscular work is **always** provided directly by the immediate and non-oxygen-consuming breakdown of ATP in the muscle. Oxygen becomes important only in subsequent reactions of energy transfer when it serves as an electron acceptor and combines with the hydrogens generated during glycolysis, beta oxidation of fatty acids, or the reactions of the Krebs cycle.

The oxygen deficit can be viewed quantitatively as the difference between the total oxygen actually consumed during exercise and the total that would have been consumed had a steady rate of aerobic metabolism been reached immediately at the start. The energy provided during the deficit phase of exercise represents nonaerobic energy (that is, immediate energy from the stored phosphates[56] plus anaerobic energy from glycolysis) that is utilized until a steady rate is reached between oxygen consumption and the energy demands of exercise.

Figure 7–4 depicts the relationship between the size of the oxygen deficit and the contribution of energy from both the ATP-CP and lactic acid energy systems.[58] As shown, the high-energy phosphates are substantially depleted

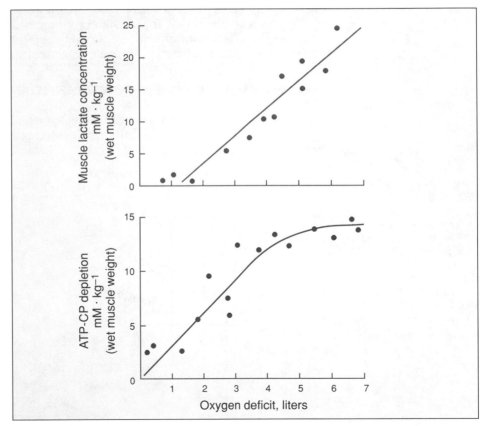

FIG. 7–4.
ATP and CP depletion and lactic acid formation in relation to the calculated oxygen deficit. (From Pernow, B., and Karlsson, J.: Muscle ATP, CP and lactate in submaximal and maximal exercise. In Muscle Metabolism During Exercise. Edited by B. Pernow, and B. Saltin. New York, Plenum Press, 1971.)

by exercise that generates about a three- to four-liter oxygen deficit. Consequently, this exercise can continue only on a "pay-as-you-go" basis with ATP being continually replenished through the breakdown of the food nutrients by oxidative phosphorylation or glycolysis. Interestingly, lactic acid begins to increase in exercising muscle well before the phosphates reach their lowest levels. This finding indicates that glycolysis contributes anaerobic energy in the early stages of vigorous exercise, even before the full utilization of the high-energy phosphates. **These observations show that energy for exercise is not merely the result of a series of energy systems that "switch on" and "switch off," but rather, the smooth blending with considerable overlap from one mode of energy transfer to another.**

Oxygen Deficit in Trained and Untrained Individuals

It is generally observed that oxygen consumption during light and moderate exercise is similar in the trained and untrained once the steady rate is reached. The trained person, however, reaches the steady rate more rapidly[30,35] and has a **smaller** oxygen deficit for the same exercise compared to the untrained (Fig. 7–3). Consequently, the total oxygen consumed during exercise would be greater for the trained person, and, presumably, the anaerobic component of energy transfer would be proportionately smaller.[25] This facilitated level of aerobic metabolism in the early stages of exercise may be the result of cellular adaptations with endurance training, many of which are known to increase the capacity of muscle to generate ATP aerobically.[37,38]

MAXIMAL OXYGEN CONSUMPTION (max $\dot{V}_{O_2}$)

Figure 7–5 depicts the oxygen consumption response during a series of constant-speed runs up six hills, each progressively steeper than the next. (In the laboratory, these "hills" can be simulated by increasing the elevation

FIG. 7–5.
Oxygen consumption during exercise of increasing intensity up to the maximal oxygen consumption. This occurs in the region where a further increase in work is not accompanied by an additional increase in oxygen consumption.

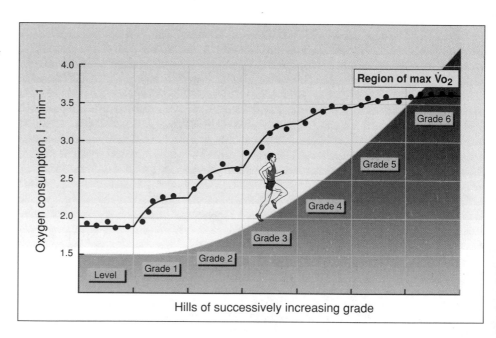

of a treadmill or step bench or increasing the resistance to pedaling a bicycle ergometer.) Each successive hill requires a greater energy output and thus places an additional load on the runner's capacity for aerobic metabolism. During the first several hills, the increases in oxygen consumption are linear and in direct proportion to the severity of exercise. Although the runner is able to maintain running speed up the last two hills, the oxygen consumption does not increase to the same extent observed for the previous hills. In fact, no increase is noted for the run up the last hill. **The point at which the oxygen consumption plateaus and shows no further increase (or increases only slightly) with an additional workload is called the maximal oxygen consumption, maximal oxygen uptake, maximal aerobic power, or, simply, max $\dot{V}O_2$.** It is generally assumed that this represents the person's capacity for the aerobic resynthesis of ATP. Additional work is accomplished only by the energy transfer reactions of glycolysis with the resulting formation of lactic acid. Under these conditions, the runner soon becomes exhausted and unable to continue.

The **max $\dot{V}O_2$** provides a quantitative statement of an individual's capacity for aerobic energy transfer. As such, it is one of the more important factors that determines one's ability to sustain high-intensity exercise for longer than 4 or 5 minutes. In subsequent chapters, we discuss various aspects of aerobic power, including its measurement and its role in exercise performance.

FAST- AND SLOW-TWITCH MUSCLE FIBERS

By means of surgical biopsy, which extracts about 20 to 40 mg of tissue (the size of a grain of rice) biochemists and exercise physiologists have studied the functional and structural characteristics of human skeletal muscle. This study has led to the identification of two distinct types of muscle fibers,[21,47] the proportion of which probably remains fairly constant throughout life. One type is a **fast-twitch** (FT) fiber. This fast-contracting fiber has two basic subdivisions and possesses a high capability for the anaerobic production of ATP during glycolysis. These fibers are activated during change-of-pace and stop-and-go activities such as basketball and ice hockey as well as during all-out exercise that requires rapid, powerful movements that depend almost exclusively on the energy generated from anaerobic metabolism.

The second major classification is the **slow-twitch** (ST) muscle fiber. This type is a predominantly aerobic fiber with a relatively slow speed of contraction compared to its fast-twitch counterpart. The capacity of these fibers to generate ATP aerobically is intimately related to their numerous and large mitochondria and to the high levels of enzymes required to sustain aerobic metabolism. This is particularly true for the capacity of these fibers to metabolize fatty acids. The primary role of the slow-twitch fiber is to sustain continuous endurance-type activities that require a steady rate of aerobic energy transfer. In fatigue associated with distance running, glycogen depletion occurs primarily in the slow-twitch fibers.[14,29] This pattern of glycogen depletion has also been observed in the arms of wheelchair-dependent athletes during prolonged exercise.[61] More than likely, it is the concentration of slow-twitch muscle fibers that greatly contributes to the high levels of exercise before the OBLA commonly observed among successful endurance athletes.[47,59] It should be kept in mind that most sport activities require a blend of powerful and sustained muscular contractions and that both types of muscle fibers are used.

From the preceding discussion, it would seem that the predominant fiber type in specific muscles is an important factor that determines success in a particular sport or activity. This idea is discussed more fully in Chapter 18 as well as other considerations concerning each type of muscle fiber and their various subdivisions.

THE ENERGY SPECTRUM OF EXERCISE

Figure 7–6 illustrates the relative contribution of anaerobic and aerobic energy sources during various durations of maximal exercise. In addition, the data in Table 7–1 show the approximate energy yield from these systems of energy transfer. Although these data were originally obtained from laboratory experiments that involved running and bicycling, they can easily be related to other activities by drawing the appropriate relationships in terms of time. For example, a 2- to 7-second period of all-out exercise represents most plays that occur in football or a "solo dash" in soccer, ice hockey, or basketball. All-out exercise for about 1 minute incorporates the 440-yard dash in track, the 100-yard swim and, possibly, a full-court press at the end of a basketball game.

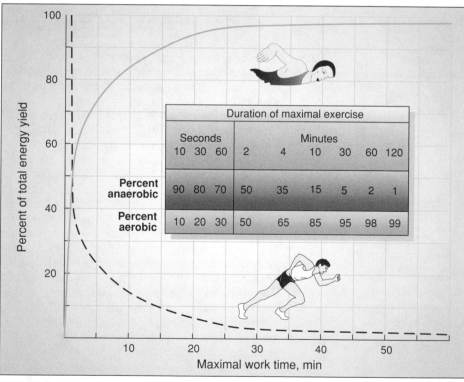

| Duration of maximal exercise | | | | | | | | |
| Seconds | | | Minutes | | | | | |
10	30	60	2	4	10	30	60	120
Percent anaerobic 90	80	70	50	35	15	5	2	1
Percent aerobic 10	20	30	50	65	85	95	98	99

FIG. 7–6.

Relative contribution of aerobic and anaerobic energy during maximal physical activity of various durations. It should be noted that 2 minutes of maximal effort requires about 50% of the energy from aerobic and anaerobic processes. (Adapted from Åstrand, P.O., and Rodahl, K.: Textbook of Work Physiology. New York, McGraw-Hill Book Company, © 1977.)

At one extreme, the total energy for exercise is supplied almost entirely by the intramuscular phosphagens. In intense exercise that lasts 2 minutes, about half of the energy is supplied by the ATP-CP and lactic acid systems, whereas aerobic reactions supply the remainder. Under these conditions, it is desirable to possess a high capacity for both aerobic and anaerobic metabolism. Intense exercise of an intermediate duration performed for 5 to 10 minutes, as in middle distance running and swimming, basketball, or soccer, results in a greater demand for aerobic energy. Performances of long duration such as marathon running, distance swimming, cycling, recreational jogging, or hiking require a fairly constant supply of aerobic energy with little reliance on the mechanism of lactic acid formation.

An understanding of the energy demands of various activities provides some explanation as to why a world-record-holder in the one-mile run is not necessarily a noted distance runner. Conversely, premier marathon runners are generally unable to run a mile in less than 4 minutes yet can complete 26 miles at a 5-minute per mile pace. The appropriate approach to exercise training is to analyze an activity in terms of its specific energy components and then improve those systems to assure optimal physiologic and metabolic adaptations. **An improved capacity for energy transfer usually translates into improved exercise performance.**

OXYGEN CONSUMPTION DURING RECOVERY: THE "OXYGEN DEBT"

After exercise, bodily processes do not immediately return to resting levels. In submaximal exercise, recovery is rapid and often proceeds unnoticed. If the activity is particularly stressful, such as running a half-mile race or trying to swim 200 yards as fast as possible, the body requires considerable time to return to rest. Recovery from both moderate and strenuous exercise is associated largely with the specific metabolic and physiologic processes that result from each form of exercise.

The oxygen consumption during exercise and recovery from moderate and strenuous work is shown in Figure 7–7. During light exercise, when the oxygen deficit is small, the quantity of oxygen consumed in recovery is also small, and the pre-exercise metabolic rate is achieved rapidly. During the exhaustive exercise illustrated in the bottom curve, a steady rate of aerobic metabolism cannot be attained. (In this instance, it is necessary to estimate the steady-rate oxygen consumption as being the max V_{O_2} in order to calculate the oxygen deficit.) In this situation, anaerobic energy transfer is large and lactic acid accumulates. Complete recovery requires considerable time from this type of exercise. In recovery from either light, moderate, or strenuous exercise, the oxygen consumed in excess of the resting value has been termed the **oxygen debt.** The oxygen debt is indicated by the shaded area under the recovery curve and is calculated as the total oxygen consumed in recovery minus the total oxygen theoretically consumed at rest during the recovery period.

For example, if a total of 5.5 liters of oxygen were consumed in recovery until the resting value of .310 liters per minute was reached, and the recovery required 10 minutes, the oxygen debt would be 5.5 liters minus (.310ℓ × 10 min), or 2.4 liters. This result means that the preceding exercise plus the metabolic activities that occurred during recovery caused the consumption of an additional 2.4 liters of oxygen before the pre-exercise state was reached.

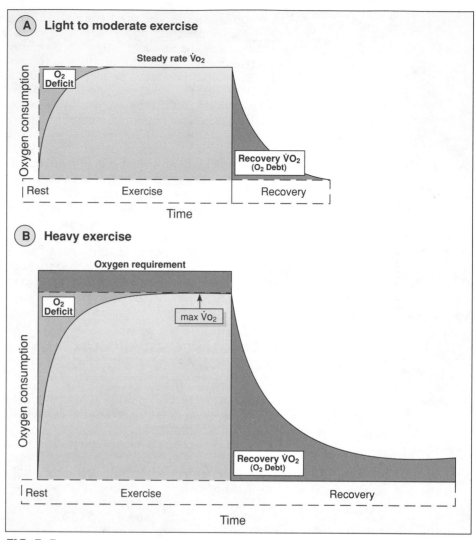

FIG. 7–7.
Oxygen consumption during and in recovery from light to moderate steady-rate exercise (A), and heavy exercise with the resulting buildup of lactic acid (B). The dark purple area represents that portion of the exercise oxygen requirement that cannot be supplied by aerobic metabolism.

An important assumption that underlies the calculation of the oxygen debt is that resting oxygen uptake remained essentially unchanged during exercise and recovery. As we shall see, this assumption is not entirely correct, especially in recovery from strenuous exercise.

The recovery curves in Figure 7–7 illustrate two important characteristics of oxygen uptake during recovery: (1) If the previous exercise was primarily aerobic (with little disruption in body temperature), about one-half of the total recovery oxygen consumption is repaid within 30 seconds; within several minutes, the recovery is complete (fast component). (2) Recovery from strenuous exercise, where lactic acid and body temperature have increased considerably, presents a somewhat different picture. In addition to the fast

TABLE 7–1.
RELATIVE CONTRIBUTION OF ANAEROBIC AND AEROBIC PROCESSES TO TOTAL ENERGY OUTPUT DURING MAXIMAL EXERCISE OF DIFFERENT DURATIONS*

Work Time Maximal Exercise	Energy Output (kcal)		
	Anaerobic Processes	Aerobic Processes	Total
10 s	20	4	24
1 min	30	20	50
2 min	30	45	75
5 min	30	120	150
10 min	25	245	270
30 min	20	675	695
60 min	15	1200	1215

* From Gollnick, P.D., and Hermansen, L.: Biochemical adaptions to exercise: Anaerobic metabolism. In Exercise and Sport Science Reviews, Vol. 1. Edited by J.H. Wilmore. New York, Academic Press, 1973.

component of recovery oxygen consumption, a second, slower phase is termed the "slow" component. Depending on the intensity and duration of exercise, this phase of recovery may take up to several hours, or even a day, before the pre-exercise oxygen consumption level is once again established.

METABOLIC DYNAMICS OF OXYGEN DEBT

A precise biochemical explanation of the recovery oxygen consumption, especially the role of lactic acid, is not possible because the specific chemical dynamics of oxygen debt are still unclear.[24]

Traditional Concepts

The term oxygen debt was first coined by the Nobel Prize scientist Archibald Vivian Hill in 1922. A.V. Hill, as well as others, discussed energy metabolism during exercise and recovery in financial-accounting terms.[36] Within this framework, the body's carbohydrate stores were likened to energy "credits." If these stored credits were expended during exercise, then a "debt" was incurred. The greater the energy "deficit," or use of available stored energy credits, the larger the energy debt incurred. The recovery oxygen uptake was thought to represent the metabolic cost of repaying this debt— hence the term oxygen debt.

In more concrete terms, the accumulation of lactic acid during the anaerobic component of exercise represented the utilization of the stored energy credit, glycogen. The ensuing oxygen debt was believed to serve two purposes: (1) to re-establish the original carbohydrate stores (credits) by resynthesizing approximately 80% of the lactic acid back to glycogen in the liver, and (2) to catabolize the remaining lactic acid through the pyruvic acid-Krebs cycle pathway. The ATP generated in this process presumably was used to power the resynthesis of lactic acid to glycogen in the Cori cycle. This early explanation of the dynamics of recovery oxygen consumption has often been termed the "lactic acid theory of oxygen debt."

In 1933, subsequent to the work of Hill, researchers at the Harvard Fatigue Laboratory attempted to explain their observations that the initial portion of the recovery oxygen was consumed before blood lactic acid began to decrease.[51] In fact, they showed that it was possible to incur an oxygen debt of almost 3 liters without any appreciable elevation in lactic acid. To resolve these findings, two phases of oxygen debt were proposed: (1) **alactic or alactacid oxygen debt** (without lactic acid buildup), and (2) lactic acid or **lactacid oxygen debt.** It is noteworthy that these two explanations were based on speculation, because these researchers were unable to measure ATP and CP replenishment or the relationship between lactic acid and glucose and glycogen levels. Essentially, the following model has served to explain the energetics of oxygen debt for nearly 60 years:

1. **Alactacid debt**—The alactacid portion of the oxygen debt depicted for steady-rate exercise in the top graph in Figure 7–7, or for the rapid phase of recovery from strenuous exercise (bottom graph), was attributed to the restoration of the high-energy phosphates ATP and CP depleted during exercise. The energy for this restoration comes from the aerobic breakdown of the food nutrients during recovery. A small portion of the recovery oxygen is also used to reload the muscle myoglobin as well as the hemoglobin in the blood returning from previously active tissues.
2. **Lactacid debt**—In keeping with the explanation by A.V. Hill, the major portion of the lactic acid oxygen debt was thought to represent the reconversion of lactic acid to glycogen in the liver.

Controversy with Traditional Explanation of Oxygen Debt

Several relationships must be established to support the contention that an aerobic energy deficit in exercise is temporarily compensated for by energy from anaerobic sources that are then resynthesized in recovery. For example, only a moderate relationship exists between the degree of anaerobiosis in exercise (oxygen deficit) and the excess oxygen uptake in recovery (oxygen debt.)[4,32]

To accept the traditional explanation for the lactacid phase of the oxygen debt, it must also be clearly established that the major portion of lactic acid produced in exercise is **actually** resynthesized to glycogen in recovery, as Hill and others had speculated. This has never been shown.[8] In fact, when radioactive lactic acid is infused into rat muscle, more than 75% of this substrate appears as radioactive carbon dioxide, whereas only 25% is synthesized to glycogen.[10] In experiments with humans, no substantial replenishment of glycogen was observed 10 minutes after strenuous exercise, even though blood lactate levels were significantly reduced. **Apparently a major portion of lactic acid is oxidized for energy.** It is well established that heart, liver, kidney, and skeletal muscle tissues use lactate as an energy substrate during both exercise and recovery.[62a]

Contemporary Concepts of Oxygen Debt

There is no doubt that the elevated aerobic metabolism in recovery is necessary to restore the body to its pre-exercise condition and is largely the result of the preceding metabolic and physiologic events during exercise.[62] In moderate exercise, the recovery oxygen consumption serves to replenish the high-energy phosphates depleted by exercise,[31,50] whereas in strenuous exercise, some oxygen is used to resynthesize a portion of lactic acid to glycogen. This mechanism for gluconeogenesis takes place during exercise and is noteworthy among trained groups.[20] The main source for re-establishing pre-exercise glycogen levels, however, is the carbohydrate in the diet, **not** resynthesized lactic acid.

A significant portion of the recovery oxygen consumption is also attributed to physiologic processes that actually take place **during** recovery. The considerably larger oxygen debt in relation to oxygen deficit in exhaustive exercise is probably the result of such factors. Body temperature, for example, is elevated about 3° C (5.4° F) during vigorous, longer duration aerobic exercise and can remain elevated for several hours in recovery. This can have a direct stimulating effect on metabolism and can cause a significant increase in recovery oxygen consumption.[3,9,30]

Other factors also affect recovery oxygen consumption. Perhaps as much as 10% of the recovery oxygen goes to reload the blood as it returns from the exercised muscles. An additional 2 to 5% restores the oxygen dissolved in body fluids and the oxygen bound to myoglobin in the muscle itself. In very intense exercise, the volume of air breathed increases 8 to 10 times above rest and remains elevated for some time in recovery. Thus, the respiratory muscles also require more oxygen for the work of breathing during recovery than they normally require at rest.[48] The heart also works harder and requires a greater oxygen supply during recovery. Tissue repair and the redistribution of the ions calcium, potassium, and sodium within the muscle and other body compartments require energy, whereas the residual effects of the hormones epinephrine, norepinephrine, and thyroxine released in exercise may continue to affect metabolism for a considerable time in recovery.[27] In essence, all of the physiologic systems that are activated to meet the demands of exercise also increase their own particular need for oxygen during recovery. **The oxygen debt, or more accurately the recovery oxygen consumption, reflects both the anaerobic metabolism of exercise and the respiratory, circulatory, hormonal, ionic, and thermal adjustments that occur in recovery.**

IMPLICATIONS OF OXYGEN DEBT FOR EXERCISE AND RECOVERY

An understanding of the dynamics of the recovery oxygen consumption provides a basis for structuring work intervals and optimizing recovery. No appreciable lactic acid accumulates with either steady-rate aerobic exercise or brief 5 to 10-second bouts of all-out work. Consequently, recovery is rapid (fast component), and work can begin again without the hindering effects of fatigue. In contrast, longer periods of anaerobic exercise are performed at the expense of lactate buildup in the blood and exercising muscles, as well as a significant disruption in other physiologic processes. In this situation, recovery oxygen consumption consists of both fast and slow components, and considerably more time is required for complete recovery. This can pose a problem in sports such as basketball, hockey, soccer, tennis, and badminton, because a performer pushed to a high level of anaerobic metabolism may not fully recover during brief rest periods such as times out, between points, or even half-time breaks.

Procedures for speeding recovery from exercise can generally be categorized as either **active** or **passive**. In active recovery (often called "cooling-down" or "tapering-off"), submaximal aerobic exercise is performed in the belief that this continued movement in some way prevents muscle cramps and stiffness and facilitates the recovery process. With passive recovery, the person usually lies down with the hope that complete inactivity may reduce the resting energy requirements and thus "free" oxygen for the recovery process. Modifications of active and passive recovery have included the use of cold showers, massages, specific body positions, and the ingestion of cold liquids.

Optimal Recovery from Steady-Rate Exercise

For most people, exercise performed at an oxygen consumption below 55 to 60% of max $\dot{V}O_2$, can generally be performed at a steady rate with little lactic acid buildup. Recovery from this exercise entails the resynthesis of high-energy phosphates, replenishment of oxygen in the blood, body fluids, and muscle myoglobin, and a small energy cost to sustain circulation and ventilation. In this situation, recovery is more rapid with passive procedures[45] because exercise would only serve to elevate total metabolism and delay recovery to the resting level.

Optimal Recovery from Non-Steady-Rate Exercise

When exercise intensity exceeds about 55 to 60% of max $\dot{V}O_2$, a steady rate of aerobic metabolism is no longer maintained, lactic acid formation exceeds its rate of removal, and lactic acid accumulates. As work intensity increases, the level of lactic acid rises sharply and the exerciser soon becomes exhausted. Although the mechanisms of fatigue during intense anaerobic exercise are poorly understood, the level of blood lactate does provide a fairly objective indication of the relative strenuousness of exercise and may also reflect the adequacy of the recovery process.[42]

Lactic acid removal is accelerated by active aerobic recovery exercise.[19,26,44] Apparently, the optimal level of recovery exercise is between 29 and 45% of the max $\dot{V}O_2$ for bicycle exercise,[4,55] and 55 to 60% of max $\dot{V}O_2$ when the recovery involves treadmill running. The variation between these two forms of exercise is probably due to the more localized nature of bicycling which is generally reflected by a lower lactate accumulation threshold in this activity.[33] Figure 7–8 illustrates blood lactate recovery patterns for trained males who performed 6 minutes of supermaximal work on a bicycle ergometer.

Active recovery involved 40 minutes of continuous exercise at either 35 or 65% of max $\dot{V}O_2$. A combination of 65% of max (7 minutes) followed by 35% of max (33 minutes) was also used to evaluate whether a higher-intensity exercise interval early in recovery would expedite the removal of lactic acid from tissue to the blood. Clearly, **moderate aerobic exercise in recovery facilitates lactate removal compared to passive recovery.** The combination of higher intensity followed by lower intensity exercise was of no more benefit than a single moderate intensity. It is important to note that if exercise in recovery is too intense and above the lactate threshold, it is of no added benefit and may even prolong recovery by increasing lactic acid formation.[19,26] In a practical sense, if left to their own choice, people voluntarily select their own optimal intensity of recovery exercise.[7] In fact, the most effective approach may be to individualize recovery exercise to an intensity just slightly below a person's lactate-accumulation threshold rather than basing the exercise level on some preselected percentage of max $\dot{V}O_2$ or maximum heart rate.

The reasons are not clear for the benefits of active recovery compared to

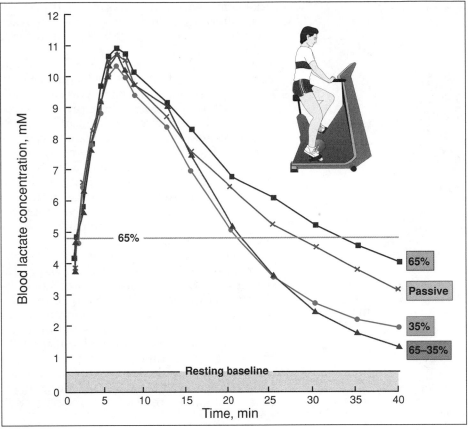

FIG. 7–8.
Blood lactate concentration following maximal exercise during passive and active exercise recoveries at 35 and 65, and a combination of 35% and 65% of max $\dot{V}O_2$. (Modified from Dodd, S., et al.: Blood lactate disappearance at various intensities of recovery exercise. J. Appl. Physiol.: Respirat. Environ. Exercise Physiol. 57:1462, 1984)

passive recovery. The facilitated removal of lactate may be the result of an increased perfusion of blood through "lactate-using" organs like the liver and heart. In addition, increased flow of blood through the muscles in active recovery would certainly enhance lactic acid removal because muscle tissue can use this substrate and oxidize it via Krebs cycle metabolism.[2,26]

INTERMITTENT EXERCISE

Several approaches can be taken to enable a person to perform significant amounts of normally exhaustive exercise, while at the same time reducing the contribution from anaerobic energy transfer through glycolysis and subsequent lactic acid buildup. One means is to train the aerobic systems to increase their capacity to sustain exercise at a high rate of aerobic energy transfer. A dramatic example of this high steady-rate capability is the performance of elite marathon runners, distance swimmers, and cross-country skiers. Another approach to performing work that would normally cause exhaustion within 3 to 5 minutes if performed continuously is to exercise in an intermittent manner using a pre-established spacing of work and rest intervals.[11] This technique is popular in conditioning programs and is known as **interval training.** Here, various work-to-rest intervals using "supermaximal" exercise are applied to overload the various systems of energy transfer. For example, with all-out work bouts of up to 8 seconds, the intramuscular phosphates provide the major portion of energy, and reliance is minimal on the glycolytic pathway. Therefore, recovery is rapid (fast component), and another bout of heavy exercise can begin after only a brief recovery period.

The results of a series of laboratory experiments using various combinations of work and rest intervals during intermittent exercise are summarized in Table 7–2. On one day, the subject ran at a speed that would normally exhaust him within 5 minutes. About 0.8 mile was covered during this continuous run, and the runner attained a maximal oxygen consumption of 5.6 liters per minute. A relative state of exhaustion plus a high level of anaerobic metabolism were verified by the high lactic acid level shown in the last column of the table.

On another day, the same fast speed was maintained, but the exercise was performed intermittently with periods of 10 seconds of exercise and 5 seconds of recovery. With a 30-minute protocol of intermittent exercise, the actual duration of running amounted to 20 minutes and the distance covered was 4 miles compared to less than 5 minutes and 0.8 miles when the run was performed continuously! This work output capability is even more impressive if one considers that the blood lactate level remained low even though the oxygen consumption was quite high, averaging 5.1 liters per minute (91% of max V_{O_2}) throughout the 30-minute period. Thus, a relative balance had been achieved between the energy requirements of work and the level of aerobic energy transfer within the muscle during the work and rest intervals.

Clearly, by manipulating the duration of work and rest intervals, a specific energy transfer system can be emphasized and overloaded. When the rest interval was extended from 5 to 10 seconds, the oxygen consumption averaged 4.4 liters per minute; with 15 second work and 30-second recovery intervals,

TABLE 7–2.
RESULTS OF AN EXPERIMENT DEALING WITH INTERMITTENT EXERCISE*

Exercise: Rest Periods	Total Distance Run (Yards)	Average Oxygen Consumption ($l \cdot min^{-1}$)	Blood Lactate Level (mg · 100 ml blood^{-1})
4 min continuous	1422	5.6	150
10 s exercise 5 s rest	7294	5.1	44
10 s exercise 10 s rest	5468	4.4	20
15 s exercise 30 s rest	3642	3.6	16

* From data of Christenson, E.H., et al.: Intermittent and continuous running. Acta Physiol. Scand., 50:269, 1960, as reported in Åstrand, P.O., and Rodahl, K.: Textbook of Work Physiology. New York, McGraw-Hill, 1970, p. 384.

only a 3.6 liter oxygen consumption was noted. In each case of 30 minutes of intermittent exercise, however, the runner achieved a longer distance and a much lower lactic acid level compared to the same work performed continuously. The specific application of the principles of intermittent exercise to both aerobic and anaerobic training and sports performance is discussed in Chapter 21.

SUMMARY

1. The major energy pathway for ATP production differs depending on the intensity and duration of exercise. In intense exercise of short duration (100-yard dash, weight lifting), the energy is derived from the already present stores of intramuscular ATP and CP (immediate energy system). For intense exercise of longer duration (1 to 2 min), energy is generated mainly from the anaerobic reactions of glycolysis (short-term energy system). As exercise progresses beyond several minutes, the aerobic system predominates and oxygen consumption becomes an important factor (long-term energy system).
2. Humans possess different types of muscle fibers, each with unique metabolic and contractile properties. The two major fiber types are (1) low glycolytic–high oxidative, slow-twitch fibers, and (2) low oxidative–high glycolytic, fast-twitch fibers. Intermediate fibers with overlapping metabolic characteristics of the fast-twitch type are also present.
3. By understanding the energy spectrum of exercise, it is possible to train for specific improvement of the appropriate energy system.
4. A steady rate of oxygen uptake represents a balance between the energy requirements of the working muscles and the aerobic resynthesis of ATP. The difference between the oxygen requirement and the actual oxygen consumed is called the oxygen deficit.
5. The maximum capacity for the aerobic resynthesis of ATP is quantitatively measured as the maximum oxygen consumption or max $\dot{V}O_2$. This is one of the important indicators of one's ability for sustained exercise.
6. Following exercise, the oxygen uptake remains elevated above the resting level. This recovery oxygen consumption reflects the metabolic characteristics of the preceding exercise as well as the physiologic alterations caused by that exercise.
7. Moderate exercise performed during recovery (active recovery) facilitates the recovery process compared to passive procedures. In most situations, this recovery is reflected in a faster removal of lactic acid.

REFERENCES

1. Alpert, N.R.: Lactate production and removal and the regulation of metabolism. *Ann. NY Acad. Sci.*, 119:995, 1965.
2. Baldwin, K.M., et al.: Glycolytic enzymes in different types of skeletal muscle: adaptation to exercise. *Am. J. Physiol.*, 225:962, 1973.
3. Belcastro, A.N., and Bonen, A.: Lactic acid removal rates during controlled and uncontrolled recovery exercise. *J. Appl. Physiol.*, 39:932, 1975.
4. Berg, W.E.: Individual differences in respiratory gas exchange during recovery from moderate exercise. *Am. J. Physiol.*, 149:597, 1947.
5. Bergstrom, J., et al.: Energy rich phosphagens in dynamic and static work. In *Muscle Metabolism During Exercise*. Edited by B. Pernow and B. Saltin. New York, Plenum Press, 1971.

6. Bonen, A., and Belcastro, A.N.: Comparison of self-selected recovery methods on lactic acid removal rates. *Med. Sci. Sports*, 8:176, 1976.
7. Brooks, G.A.: Anaerobic threshold: review of the concept and directions for future research. *Med. Sci. Sports Exerc.*, 17:22, 1985.
8. Brooks, G.A., and Gaesser, G.A.: Endpoints of lactate and glucose metabolism after exhausting exercise. *J. Appl. Physiol.*, 49:1057, 1980.
9. Brooks, G.A., et al.: Temperature, skeletal muscle mitochondrial functions and oxygen debt. *Am. J. Physiol.*, 220:1053, 1971.
10. Brooks, G.A., et al.: Glycogen synthesis and metabolism of lactic acid after exercise. *Am. J. Physiol.*, 224:1162, 1973.
11. Christenson, E.H., et al.: Intermittent and continuous running. *Acta Physiol. Scand.*, 50:269, 1960.
11a. Coggan, A.R., et al.: Endurance training decreases plasma glucose turnover and oxidation during moderate-intensity exercise in men. *J. Appl. Physiol.*, 68:990, 1990.
12. Conley, D.L., et al.: Physiological correlates of female road racing performance. *Res. Q. Exerc. Sport*, 52:441, 1981.
13. Costill, D.L.: Metabolic responses during distance running. *J. Appl. Physiol.*, 28:251, 1970.
14. Costill, D.L., et al.: Glycogen depletion patterns in human muscle fibers during distance running. *Acta Physiol. Scand.*, 89:374, 1973.
15. Costill, D.L., et al.: Fractional utilization of the aerobic capacity during distance running. *Med. Sci. Sports*, 5:248, 1973.
16. Coyle, E.F.: Blood lactate threshold in some well trained ischemic heart disease patients. *J. Appl. Physiol.*, 54:18, 1983.
17. Davies, K.J.A., et al.: Biochemical adaptation of mitochondria, muscle and whole-animal respiration to endurance training. *Arch. Biochem. Biophys.*, 209:539, 1981.
18. Davis, J.A., et al.: Anaerobic threshold alterations caused by endurance training in middle-aged men. *J. Appl. Physiol.*, 46:1039, 1979.
19. Dodd, S., et al.: Blood lactate disappearance at various intensities of recovery exercise. *J. Appl. Physiol.*, 57:1462, 1984.
20. Donovan, C.M., and Brooks, G.A.: Endurance training affects lactate clearance, not lactate production. *Am. J. Physiol.*, 244 (Endocrinol. Metab. 7): E83, 1983.
21. Edgerton, V.R.: Exercise and growth and development of muscle tissue. In *Physical Activity: Human Growth and Development.* Edited by G.L. Rarick. New York, Academic Press, 1973.
22. Farrell, P.A., et al.: Plasma lactate accumulation and distance running performance. *Med. Sci. Sports*, 11:338, 1979.
23. Fournier, M.: Skeletal muscle adaptation in adolescent boys: sprint and endurance training and detraining. *Med. Sci. Sports Exerc.*, 14:453, 1982.
24. Gaesser, G.A., and Brooks, G.A.: Metabolic basis of excess post-exercise oxygen consumption: a review. *Med. Sci. Sports Exerc.*, 16:29, 1984.
25. Girandola, R.N., and Katch, F.I.: Effects of physical conditioning on changes in exercise and recovery O_2 uptake and efficiency during constant-load ergometer exercise. *Med. Sci. Sports*, 5:242, 1973.
26. Gladden, L.B.: Lactate uptake by skeletal muscle. In *Exercise and Sport Sciences Reviews*, Vol. 17. Edited by K.B. Pandolf. New York, Macmillan, 1989.
27. Gladden, L.B., et al.: Norepinephrine increases and canine skeletal muscle $\dot{V}_{O_2}$ during recovery. *Med. Sci. Sports Exerc.*, 14:371, 1982.
28. Gollnick, P.D., and Saltin, B.: Significance of skeletal muscle oxidative enzyme enhancement with endurance training. *Clin. Physiol.*, 2:1, 1983.
29. Gollnick, P.D., et al.: Glycogen depletion patterns in human skeletal muscle fibers after varying types and intensities of exercise. In *Metabolic Adaptation to Prolonged Exercise.* Edited by H. Howard and J. Poortmans. Basel, Birkhausen Verlag, 1975.
30. Hagberg, J.M., et al.: Faster adjustment to and recovery from submaximal exercise in the trained state. *J. Appl. Physiol.*, 48:218, 1980.
31. Harris, R.R., et al.: The time course of phosphorylcreatine resynthesis during recovery of the quadriceps muscle in man. *Pflüger's Arch.*, 367:137, 1976.
32. Henry, F.M.: Aerobic oxygen consumption and alactic debt in muscular work. *J. Appl. Physiol.*, 3:427, 1951.

33. Hermansen, L., and Stensvold, I.: Production and removal of lactate during exercise in man. *Acta Physiol. Scand.*, 86:191, 1972.

34. Hickson, R.C.: Skeletal muscle cytochrome c and myoglobin, endurance, and frequency of training. *J. Appl. Physiol.*, 51:746, 1981.

35. Hickson, R.C., et al.: Faster adjustment of O_2 uptake to the energy requirement of exercise in the trained state. *J. Appl. Physiol.*, 44:877, 1978.

36. Hill, A.V., et al.: Muscular exercise, lactic acid and the supply and utilization of oxygen. *Proc. R. Soc. Lond. (Biol.)* 96:438, 1924.

37. Holloszy, J.O.: Biochemical adaptation to exercise: Aerobic metabolism. In *Exercise and Sport Science Reviews*. Vol. 1. Edited by J. Wilmore. New York, Academic Press, 1973.

38. Holloszy, J.O., and Coyle, E. F.: Adaptations of skeletal muscle to endurance training and their metabolic consequences. *J. Appl. Physiol.*, 56:831, 1984.

39. Houston, M.E., et al.: Physiological and muscle enzyme adaptations to two different intensities of swim training. *Eur. J. Appl. Physiol.*, 46:283, 1981.

40. Hultman, E.: Studies on muscle metabolism of glycogen and active phosphate in man with special reference to exercise and diet. *Scand. J. Clin. Lab. Invest.*, Suppl. 94, 1967.

41. Ivy, J.L., et al.: Muscle respiratory capacity and fiber type as determinants of lactate threshold. *J. Appl. Physiol.*, 48:523, 1980.

42. Jacobs, I.: Blood lactate: implications for training and sports performance. *Sports Med.*, 3:10, 1986.

43. Jacobs, I., et al.: Sprint training effects on muscle myoglobin, enzymes, fiber types, and blood lactate. *Med. Sci. Sports Exerc.*, 19:368, 1987.

44. Jervell, O.: Investigation of the concentration of lactic acid in blood and urine under physiologic and pathologic conditions. *Acta Med. Scand.*, 24:1, 1928.

45. Jorfeldt, L.: Metabolism of L(+)-lactate in human skeletal muscle during exercise. *Acta Physiol. Scand.*, Suppl. 338, 1970.

46. Karlsson, J.: Lactate and phosphagen concentrations in working muscle of man. *Acta Physiol. Scand.*, Suppl. 358, 1971.

47. Karlsson, J., and Jacobs, I.: Onset of blood lactate accumulation during muscular exercise as a threshold concept. I. Theoretical considerations. *Int. J. Sports Med.*, 3:190, 1982.

48. Katch, F.I., et al.: The influence of the estimated oxygen cost of ventilation on oxygen deficit and recovery oxygen intake for moderately heavy bicycle ergometer exercise. *Med. Sci. Sports*, 4:71, 1972.

49. Katz, A., and Sahlin, K.: Regulation of lactic acid production during exercise. *J. Appl. Physiol.*, 65:509, 1988.

49a. Katz, A., and Sahlin, K.: Role of oxygen in regulation of glycolysis and lactate production in human skeletal muscle. In *Exercise and Sport Sciences Reviews*. Vol. 18. Edited by K.B. Pandolf. Baltimore, Williams & Wilkins, 1990.

50. Mahler, M., and Homsher, E.: Metabolic rate changes in parallel with creatine level during non-steady states in frog skeletal muscle. *Fed. Proc.*, 41:979, 1979.

51. Margaria, R., et al.: The possible mechanism of contracting and paying the oxygen debt and the role of lactic acid in muscular contraction. *Am. J. Physiol.*, 106:687, 1933.

52. McCully, K.K., and Boden, B.P.: Forearm muscle metabolism of rowers versus control subjects. *Magn. Reson. Med.*, 2:593, 1987.

53. McCully, K.K., et al.: Detection of muscle injury in humans with 31-P magnetic resonance spectroscopy. *Muscle Nerve*, 11:212, 1988.

54. McCully, K.K., et al.: Wrist flexor muscles of elite rowers measured with magnetic resonance spectroscopy. *J. Appl. Physiol.*, 67:926, 1989.

55. McLellan, T.M., and Skinner, J.S.: Blood lactate removal during active recovery related to aerobic threshold. *Int. J. Sports Med.*, 3:224, 1982.

55a. Minotti, J.R., et al.: Training-induced skeletal muscle adaptations are independent of systemic adaptations. *J. Appl. Physiol.*, 68:289, 1990.

56. Molé, P.A., et al.: In vivo[31] P-NMR in human muscle: transient patterns with exercise. *J. Appl. Physiol.*, 59:101, 1985.

57. Örlander, J., and Aniansson, A.: Effects of physical training on skeletal muscle metabolism and ultrastructure in 70- to 75-year-old men. *Acta Physiol. Scand.*, 109:149, 1980.

58. Pernow, B., and Karlsson, J.: Muscle ATP, CP and lactate in submaximal and maximal exercise. In *Muscle Metabolism During Exercise*. Edited by B. Pernow and B. Saltin. New York, Plenum Press, 1971.

59. Sjödin, B., and Jacobs, I.: Onset of blood lactate accumulation and marathon running performance. *Int. J. Sports Med.*, 2:23, 1981.
60. Sjödin, B., et al.: Muscle enzymes, onset of blood lactate accumulation (OBLA) after training at OBLA. *Eur. J. Appl. Physiol.*, 49:45, 1982.
61. Skirnar, G.S., et al.: Glycogen utilization in wheelchair-dependent athletes. *Int. J. Sports Med.*, 3:215, 1982.
62. Stainsby, W.N., and Barclay, J.K.: Exercise metabolism: O_2 deficit, steady level O_2 uptake and O_2 uptake in recovery. *Med. Sci. Sports*, 2:177, 1970.
62a. Stainsby, W.N., and Brooks, G.A.: Control of lactic acid metabolism in contracting muscles and during exercise. In *Exercise and Sport Sciences Reviews*. Vol. 18. Edited by K.B. Pandolf. Baltimore, Williams & Wilkins, 1990.
63. Tan, M.H., et al.: Muscle glycogen repletion after exercise in trained normal and diabetic rats. *J. Appl. Physiol.*, 57:1404, 1984.
64. Wasserman, K., et al.: Respiratory physiology of exercise: metabolism, gas exchange and ventilatory control. *Int. Rev. Resp. Physiol.*, 111:149, 1981.
65. Weltman, A.: The lactate threshold and endurance performance. *Adv. Sports Med. Fitness*, 2:91, 1989.
66. Weltman, A., et al.: Recovery from maximal effort exercise: lactate disappearance and subsequent performance. *J. Appl. Physiol.*, 47:677, 1979.
67. Young, J.C., et al.: Maintenance of the adaptation of skeletal muscle mitochondria to exercise in old rats. *Med. Sci. Sports Exerc.*, 15:243, 1983.

8 MEASUREMENT OF HUMAN ENERGY EXPENDITURE

METHODS OF MEASURING THE BODY'S HEAT PRODUCTION

The quantity of energy generated by the body during rest and muscular effort can be accurately determined by several different methods. These methods are broadly classified as **direct and indirect calorimetry.**

DIRECT CALORIMETRY

All of the processes that occur within the body result ultimately in the production of heat. Heat production and metabolism can be viewed consequently in a similar context. Human heat production can be measured directly in a calorimeter similar to the bomb calorimeter described in Chapter 4 to determine the energy content of food. The calorimeter illustrated in Figure 8–1 consists of an airtight, thermally insulated living chamber. The heat produced and radiated by the person is removed by a stream of cold water that flows at a constant rate through tubes coiled near the ceiling of the chamber. The difference in the temperature of water entering and leaving the chamber reflects the person's heat production. Humidified air is continually supplied and circulated while the expired carbon dioxide is removed by chemical absorbents. Oxygen is added to the air before it re-enters the calorimeter to maintain a normal oxygen supply.

The techniques of direct calorimetry, although highly accurate and of great theoretical importance, are impractical for studies of human energy expenditure during various sport, recreational, and occupational activities. In these situations, indirect methods are almost always used.

INDIRECT CALORIMETRY

All energy metabolism in the body ultimately depends on the utilization of oxygen. Thus, by measuring a person's oxygen consumption at rest and under steady-rate exercise conditions, it is possible to obtain an indirect estimate of energy metabolism because the anaerobic energy yield is very small under such conditions.[1]

Studies with the bomb calorimeter have shown that approximately 4.82 kcal of heat are liberated when a blend of carbohydrate, fat, and protein is burned in one liter of oxygen. This calorific value for oxygen varies only slightly even with large variations in the metabolic mixture. If we assume the combustion of a mixed diet, then for convenience in calculations, a value of 5 kcal per liter of oxygen consumed can be used as an appropriate con-

FIG. 8–1.
Human calorimeter used to
measure heat production.

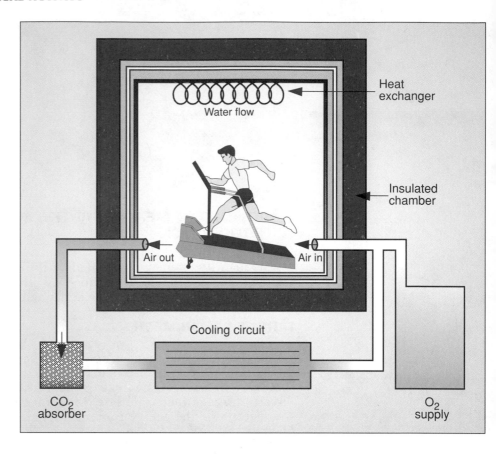

version factor for estimating the body's energy expenditure. This energy-oxygen equivalent, 5 kcal per liter of oxygen, is the convenient yardstick for transposing any aerobic exercise to a caloric frame of reference. In fact, indirect calorimetry through oxygen consumption measurement is the means by which the caloric stress of most activities has been evaluated.

Although the techniques for indirect calorimetry are relatively simple and inexpensive compared to direct measurement in the human calorimeter, both measures give comparable results.[6] **Closed-circuit and open-circuit spirometry** represent the two applications of indirect calorimetry.

Closed-Circuit Spirometry

The method of closed-circuit spirometry illustrated in Figure 8–2 is routinely used in hospitals and other laboratory settings where resting estimates are made of energy expenditure. The subject breathes and rebreathes from a prefilled container or spirometer of oxygen. This method is considered a "closed system" because the person rebreathes only the gas in the spirometer. Carbon dioxide in the exhaled air is absorbed by a cannister of soda lime (potassium hydroxide) placed in the breathing circuit: a drum that revolves at a known speed is attached to the spirometer to record changes in the volume of the system as oxygen is consumed.

During exercise, it is exceedingly difficult to measure oxygen consumption with closed-circuit spirometry. The spirometer is bulky, the subject must

FIG. 8–2.
Spirometer used to measure oxygen consumption by the closed-circuit method.

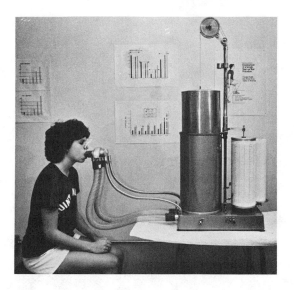

remain close to the equipment, resistance offered by the circuit to the large breathing volumes required by exercise is considerable, and the rate of carbon dioxide removal may be inadequate during moderate and heavy exercise. For these reasons, the method of open-circuit spirometry is the most widely used to measure exercise oxygen consumption.

Open-Circuit Spirometry

With this method the subject does not rebreathe from a container of oxygen as in the closed-circuit method but instead, inhales ambient air that has a constant composition of 20.93% oxygen, 0.03% carbon dioxide, and 79.04% nitrogen; this nitrogen fraction also includes the small quantity of inert gases. Because oxygen is utilized during energy-yielding reactions and carbon dioxide is produced, the exhaled air contains less oxygen and more carbon dioxide than the inhaled air. Thus, an analysis of the difference in composition between the exhaled air and the ambient air brought into the lungs reflects the body's constant release of energy. The open-circuit method provides a relatively simple means to measure oxygen consumption and indirectly determine energy metabolism. Two common techniques for open-circuit spirometry in exercise make use of either (1) a light-weight, portable spirometer that is actually worn during an activity, or (2) the "Douglas bag" or balloon method, that is used routinely to collect expired air under laboratory conditions. This also includes computerized methods for sampling and analyzing small aliquots of expired air.

Portable Spirometer. The box-shaped portable spirometer shown in Figure 8–3 was originally used to estimate the energy requirements of people working in different industrial jobs and thus provide an equitable basis for food rationing in Germany during the 1940s. The unit weighs about 3.6 kg and is usually worn on the back. By means of a two-way breathing valve, ambient air is inspired while the expired air passes through the gas meter that measures the volume and also collects a small gas sample. This sample is later analyzed for oxygen and carbon dioxide content, and oxygen consumption and energy expenditure are computed for the measurement period.

FIG. 8–3.
Portable spirometer used to measure oxygen consumption by the open-circuit
method during (A) golf, (B) leisure cycling, (C) sit-up, and (D) calisthenic exercise.
(Courtesy of Exercise Physiology Laboratory, Department of Exercise Science,
University of Massachusetts, MA).

The attractive aspect of the portable spirometer is that the subject has
considerable freedom of movement in a variety of diverse activities such as
mountain climbing, downhill skiing, golf, and gardening. The equipment
does become cumbersome during vigorous activity, and there is some question as to the accuracy of the measurement of air flow through the meter
during rapid breathing rates in heavy exercise.

Douglas Bag.   The Douglas bag method is shown in Figure 8–4A-D. The
subject shown in Figure 8–4C is running on a motor-driven treadmill. A

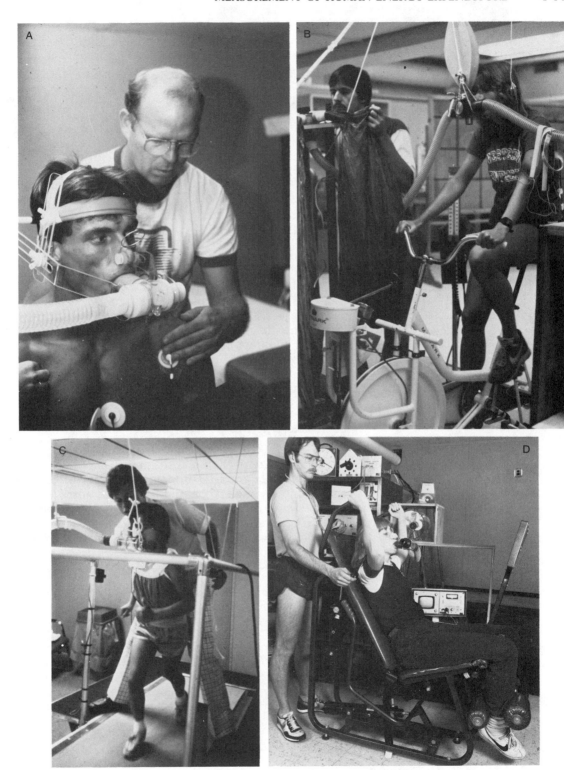

FIG. 8—4.
Measurement of oxygen consumption by open-circuit spirometry during (*A*)
recovery from maximal exercise in a champion body builder, (*B*) cycle ergometry,
(*C*) treadmill exercise in children, and (*D*) hydraulic resistive exercise. [(*A*) is
courtesy of L.E. Smith and (*B–D*) are courtesy of P. Freedson].

special headgear is worn to which a two-way, high-velocity, low-resistance breathing valve is attached(8–4A). Ambient air is breathed through one side of the valve while expired air moves out the other side and passes into either large canvas or plastic Douglas bags or rubber meterologic balloons, or directly through a gas meter that measures the volume of expired air. A small sample of expired air is collected and analyzed for its oxygen and carbon dioxide composition. As with all indirect calorimetric techniques, energy expenditure is computed from oxygen consumption using the appropriate calorific transformation. Figure 8–4 shows expired air collection by the open-circuit method in recovery and during exercise on a bicycle ergometer, treadmill, and hydraulic resistive exercise machine.

DIRECT VERSUS INDIRECT CALORIMETRY

Studies of energy metabolism using both direct and indirect calorimetry provide convincing evidence for the validity of the indirect method as a means to estimate energy metabolism during rest and exercise. At the turn of the 20th century, the two methods of calorimetry were compared in experiments conducted for 40 days on three individuals who lived in a calorimeter similar to the one shown in Figure 8–1. The subjects' daily caloric output averaged 2,723 kcal when measured directly by heat exchange and 2,717 kcal when computed indirectly by closed-circuit measures of oxygen consumption, a difference of only 0.22%! Other experiments with animals and humans utilizing moderate exercise also demonstrated a close agreement between direct and indirect methods, and in most instances, the difference was less than 1%.

THE RESPIRATORY QUOTIENT

Because of inherent chemical differences in the composition of carbohydrates, fats, and proteins, different amounts of oxygen are required to oxidize completely the carbon and hydrogen atoms in the molecule to the end products, carbon dioxide and water. Thus, the quantity of carbon dioxide produced in relation to oxygen consumed varies somewhat depending on the substrate metabolized. This ratio of metabolic gas exchange is termed the **respiratory quotient** or RQ.

$$RQ = \frac{CO_2 \text{ produced}}{O_2 \text{ consumed}}$$

The usefulness of the RQ during rest and submaximal aerobic exercise is that it serves as a convenient although perhaps general guide to the nutrient mixture being catabolized for energy.[2,4] Also, because the caloric equivalent for oxygen differs somewhat depending on the nutrient oxidized, one must know the RQ **and** the amount of oxygen consumed to estimate precisely the body's heat production.

RQ FOR CARBOHYDRATE

Because the ratio of hydrogen to oxygen atoms in carbohydrates is always the same as in water, that is 2 to 1, all of the oxygen consumed by the cells is used to oxidize the carbon in the carbohydrate molecule to carbon dioxide.

Consequently, during the complete oxidation of a glucose molecule, six molecules of carbon dioxide are produced and six molecules of oxygen are consumed. The overall equation for this reaction is:

$$C_6H_{12}O_6 + 6\,O_2 \rightarrow 6\,CO_2 + 6\,H_2O$$

Because the gas exchange in this reaction is equal, the RQ for carbohydrate is unity or 1.00:

$$RQ = \frac{6\,CO_2}{6\,O_2} = 1.00$$

RQ FOR FAT

The chemical composition of fats differs from carbohydrates because fats contain considerably fewer oxygen atoms in proportion to atoms of carbon and hydrogen. Consequently, when fat is degraded or oxidized, relatively more oxygen is required to oxidize fat to carbon dioxide and water. When palmitic acid, a typical fatty acid, is oxidized to carbon dioxide and water, 16 carbon dioxide molecules are produced for every 23 oxygen molecules consumed. This exchange is summarized by the equation:

$$C_{16}H_{32}O_2 + 23\,O_2 \rightarrow 16\,CO_2 + 16\,H_2O$$

$$RQ = \frac{16\,CO_2}{23\,O_2} = 0.696$$

Generally, the RQ value for fat is considered to be 0.70.

RQ FOR PROTEIN

In the body, proteins are not simply oxidized to carbon dioxide and water during energy metabolism. Rather, the protein is first deaminated in the liver and the nitrogen and sulfur fragments are excreted in the urine and feces. The resulting "keto acid" fragments are then oxidized to carbon dioxide and water to provide energy to sustain metabolism. As was the case with fat metabolism, these short-chain keto acids require more oxygen for complete combustion in relation to carbon dioxide produced. The protein albumin oxidizes as follows:

$$C_{72}H_{112}N_2O_{22}S + 77\,O_2 \rightarrow$$
Albumin

$$63\,CO_2 + 38\,H_2O + SO_3 + 9\,CO(NH_2)_2$$
Urea

$$RQ = \frac{63\,CO_2}{77\,O_2} = 0.818$$

The general value for the RQ of protein is 0.82.

NONPROTEIN RQ

The RQ computed from the compositional analysis of expired air usually reflects some blend of carbohydrates, fats, and proteins. The precise contribution can be determined for each of these nutrients to energy metabolism. For example, approximately 1 g of urinary nitrogen is excreted for every 6.25 g of protein metabolized for energy. **Each g of excreted nitrogen represents**

a carbon dioxide production of approximately 4.8 liters and an oxygen consumption of about 6.0 liters. Within this framework, the following example illustrates the stepwise procedure used in calculating the **nonprotein** elements in the RQ: that is, that portion of the respiratory exchange attributed to the combustion of **only** carbohydrate and fat.

The following calculation is based on data from a subject who consumes 4.0 liters of oxygen and produces 3.4 liters of carbon dioxide during a 15-minute rest period. During this time, 0.13 g of nitrogen are excreted in the urine.

Step 1. $4.8 \; \ell \; CO_2 \cdot g^{-1}$ protein metabolized $\times$ 0.13 g = 0.62 $\ell \; CO_2$ produced in the catabolism of protein

Step 2. $6.0 \; \ell \; O_2 \cdot g^{-1}$ protein metabolized $\times$ 0.13 g = 0.78 $\ell \; O_2$ consumed in the catabolism of protein

Step 3. Nonprotein CO_2 produced = 3.4 $\ell \; CO_2$ − 0.62 $\ell \; CO_2$ = 2.78 $\ell \; CO_2$

Step 4. Nonprotein O_2 consumed = 4.0 $\ell \; O_2$ − 0.78 $\ell \; O_2$ = 3.22 $\ell \; O_2$

Step 5. Nonprotein RQ = 2.78 ÷ 3.22 = 0.863

Table 8–1 presents the thermal (energy) equivalents for oxygen consumption for different nonprotein RQ values as well as the actual percentage of fat and carbohydrate utilized for energy. For the nonprotein RQ of 0.863 computed in the previous example, 4.875 kcal are liberated per liter of oxygen consumed. Also, for this RQ, 54.1% of the "nonprotein" calories are derived from carbohydrate and 45.9% from fat. The total 15-minute heat production at rest attributed to the metabolism of fat and carbohydrate is 15.70 kcal (4.875 kcal $\cdot \; \ell^{-1} \times$ 3.22 $\ell \; O_2$); the energy from the breakdown of protein is equal to 3.51 kcal (4.5 kcal $\cdot \; \ell^{-1} \times$ 0.78 $\ell \; O_2$). Consequently, the total energy from both protein and nonprotein nutrients during the 15-minute period is 19.21 kcal (15.70 kcal nonprotein + 3.51 kcal protein).

Interestingly, had the thermal equivalent for a mixed diet with an RQ of 0.82 been used in the caloric transformation, or if the RQ had been obtained simply from the total respiratory gas exchange and applied to Table 8–1 without considering the protein component, the estimated energy expenditure during this period would have been about 19.3 kcal (4.825 kcal $\cdot \; \ell^{-1} \times$ 4.0 $\ell \; O_2$; assuming a mixed diet)—a difference of only 0.5% from the value obtained with the more elaborate and time-consuming method requiring urinary nitrogen analysis.

Although the use of Table 8–1 assumes a nonprotein RQ, in most cases the gross metabolic RQ calculated without measures of urinary and other nitrogen introduces only minimal error because the contribution of protein to energy metabolism is usually small.

HOW MUCH FOOD WAS METABOLIZED FOR ENERGY?

The last two columns of Table 8–1 present the conversions for the nonprotein RQ to grams of carbohydrate and fat metabolized per liter of oxygen consumed. For the subject with an RQ of 0.86, this represents approximately 0.62 and 0.25 g of carbohydrate and fat, respectively. For the 3.22 liters of oxygen consumed during the 15-minute period of rest, 2.00 g of carbohydrate (3.22 $\ell \; O_2 \times$ 0.62) and 0.80 g of fat (3.22 $\ell \; O_2 \times$ 0.25) were metabolized for energy.

TABLE 8-1.
THERMAL EQUIVALENT OF OXYGEN FOR NONPROTEIN RESPIRATOR QUOTIENT, INCLUDING PERCENT KCAL AND GRAMS DERIVED FROM CARBOHYDRATE AND FAT[a]

Nonprotein RQ	KCAL per Liter Oxygen Consumed	Percentage KCAL Derived from		Grams per Liter O_2 Consumed	
		Carbohydrate	Fat	Carbohydrate	Fat
0.707	4.686	0	100	0.000	.496
.71	4.690	1.10	98.9	.012	.491
.72	4.702	4.76	95.2	.051	.476
.73	4.714	8.40	91.6	0.90	.460
.74	4.727	12.0	88.0	.130	.444
.75	4.739	15.6	84.4	.170	.428
.76	4.751	19.2	80.8	.211	.412
.77	4.764	22.8	77.2	.250	.396
.78	4.776	26.3	73.7	.290	.380
.79	4.788	29.9	70.1	.330	.363
.80	4.801	33.4	66.6	.371	.347
.81	4.813	36.9	63.1	.413	.330
.82	4.825	40.3	59.7	.454	.313
.83	4.838	43.8	56.2	.496	.297
.84	4.850	47.2	52.8	.537	.280
.85	4.862	50.7	49.3	.579	.263
.86	4.875	54.1	45.9	.621	.247
.87	4.887	57.5	42.5	.663	.230
.87	4.887	57.5	42.5	.663	.230
.88	4.899	60.8	39.2	.705	.213
.89	4.911	64.2	35.8	.749	.195
.90	4.924	67.5	32.5	.791	.178
.91	4.936	70.8	29.2	.834	.160
.92	4.948	74.1	25.9	.877	.143
.93	4.961	77.4	22.6	.921	.125
.94	4.973	80.7	19.3	.964	.108
.95	4.985	84.0	16.0	1.008	.090
.96	4.998	87.2	12.8	1.052	.072
.97	5.010	90.4	9.58	1.097	.054
.98	5.022	93.6	6.37	1.142	.036
.99	5.035	96.8	3.18	1.186	.018
1.00	5.047	100.0	0	1.231	.000

[a] From Zuntz, N.: Pflugers Arch. Physiol., 83:557, 1901.

RQ FOR A MIXED DIET

During activities that range from complete bed rest to mild, aerobic exercise such as walking or slow jogging, the RQ seldom reflects the oxidation of pure carbohydrate or pure fat. Instead, a mixture of these nutrients is usually used, and the RQ is intermediate in value between 0.70 and 1.00 **For most purposes, an RQ of 0.82 from the metabolism of a mixture of 40% carbohydrate and 60% fat can be assumed, and the caloric equivalent of 4.825 kcal per liter of oxygen can be applied in energy transformations.** By use of this midpoint value, the maximum error possible in estimating energy metabolism from oxygen consumption would be only about 4%. Of course, if greater precision is required, the actual RQ can be calculated and Table 8–1 consulted to obtain the exact caloric transformation, as well as the percentage contribution of carbohydrate and fat to the metabolic mixture.

RESPIRATORY EXCHANGE RATIO (R)

The calculation of RQ is based on the assumption that the exchange of O_2 and CO_2 measured at the lungs reflects the actual gas exchange from nutrient metabolism in the cell. Under steady rate exercise conditions this assumption is reasonably valid.[2] Factors that disturb the normal metabolic relationship between these gases, however, may spuriously alter this exchange ratio. Respiratory physiologists have termed the ratio of carbon dioxide produced to oxygen consumed under such conditions, when the exchange of oxygen and carbon dioxide at the lungs no longer reflects the oxidation of specific foods in the cells, the **Respiratory Exchange Ratio *or* R—even though this ratio is calculated in exactly the same manner as the RQ.**

For example, an increase in carbon dioxide elimination occurs during hyperventilation (see Chap. 14) since the response of breathing is disproportionate to the metabolic demands of a particular situation. As a result of this overbreathing, the normal level of carbon dioxide in the blood is reduced because the gas is "blown off" in the expired air. This increase in carbon dioxide elimination is not accompanied by a corresponding increase in oxygen consumption; thus, there is a disproportionate increase in the respiratory exchange ratio that cannot be attributed to the oxidation of foodstuffs. In such cases, the R usually increases above 1.00.

Exhaustive exercise presents another situation in which R can rise significantly above 1.00. The lactic acid generated during anaerobic exercise is buffered or "neutralized" by sodium bicarbonate in the blood to maintain the proper acid-base balance (see Chap. 14) in the reaction:

$$HLA + NaHCO_3 \rightleftharpoons NaLa + H_2CO_3$$
$$\updownarrow$$
$$H_2O + CO_2$$
$$\downarrow$$
$$\text{Lungs}$$

During this process, carbonic acid, a weaker acid, is formed. In the pulmonary capillaries, carbonic acid breaks down to its components, carbon dioxide and water, and the carbon dioxide exits through the lungs. This buffering process adds "extra" carbon dioxide to that quantity normally released during energy metabolism, and the R moves toward and above 1.00.

In rare instances, the exchange ratio of a person gaining body fat while maintaining a high carbohydrate diet also exceeds 1.00. In this situation, oxygen is liberated when carbohydrates are converted to fat as the excess calories become stored. This extra oxygen then can be used in energy metabolism; consequently, less atmospheric oxygen is consumed even though the normal metabolic complement of carbon dioxide is released during energy metabolism.

It is also possible to obtain relatively low R values. For example, carbon dioxide tends to be retained in the cells and body fluids following very strenuous anaerobic exercise to replenish the bicarbonate used to buffer lactic acid. This action reduces the expired carbon dioxide and may cause the respiratory exchange ratio to dip below 0.70.

FIG. 8–5.
Computerized systems approach for the collection, analysis, and output of respiratory and metabolic data. A. Subject performing 30% of left arm maximal voluntary contraction to evaluate the pressor response during left leg dynamic exercise. Metabolic measurements, including blood pressure, were recorded continuously during exercise and recovery. B. Metabolic measurements during recovery from dynamic leg exercise performed at different percentages of quadriceps maximum voluntary contraction. C. Oxygen uptake and blood pressure measured during cycle ergometry. The computerized approach can be used in most recreational, household, industrial, and sport activities of healthy subjects throughout the day as well as in hospital patients. Photos courtesy of Dr. Dan Mistry, M.D. and Dawn Gillis, Amherst, MA, Sensormedics Corporation, Anaheim, CA, and Hydra Fitness Industries, Belton, TX.

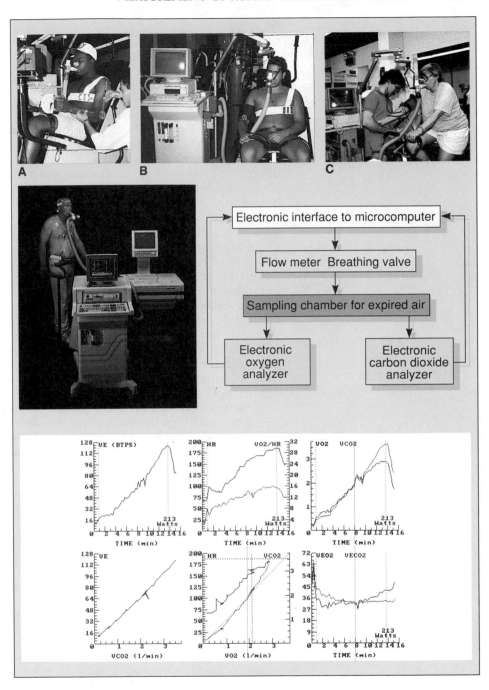

COMPUTERIZED SYSTEMS APPROACH IN METABOLIC DETERMINATIONS

The newest approach to indirect calorimetry makes use of computer technology and microelectronic instrumentation for the collection, measurement, and computation of respiratory and metabolic data.[5,7] With the system illustrated in Figure 8–5, a computer is interfaced with four measuring devices: (1) an automated system that continuously samples expired air, (2) a flow meter for measuring the volume of expired air, and (3) rapid electronic oxygen and (4) carbon dioxide gas analyzers for measuring the fractional concentration of gases in the expired air sample. The output data from the measuring devices are either fed directly into the computer or can be "punched in" by an operator who observes and records the output from the various analyzers. The computer is "preprogrammed" to perform all of the necessary computations for oxygen consumption, carbon dioxide production, and caloric expenditure. A printed output of the subject's data can occur simultaneously during exercise to provide a record of all necessary computations by the time the data collection is completed. Several research studies have verified that the automated-computerized approach gives reproducible and accurate measurements when compared against the more time-consuming techniques of the Douglas bag or balloon methods previously described.[3,7] More advanced systems include, in addition to respiratory gas analyzers, automated blood pressure, temperature, and heart rate monitors, and automatic ergometers that are preprogrammed as to workload and duration of exercise.

Although there are tremendous advantages to computerized systems in terms of ease of operation and speed of data analysis, there are also distinct disadvantages that include the high cost of equipment and delays due to system breakdowns. **Regardless of the sophistication of a particular "automated" system, the output data are only as good as the accuracy of the measuring devices. In large part, this depends on careful and frequent calibration of the electronic equipment using previously established standards.**

METABOLIC CALCULATIONS

Much of the study of exercise physiology involves the measurement of **energy metabolism.** Measurement of the oxygen and carbon dioxide content of expired air and either the inspired or expired breathing volume provides the basic data for determining the respiratory gas exchange and oxygen consumption and for inferring the body's rate of energy expenditure. Appendix C presents the step-by-step method and rationale for metabolic calculations based on experimental data utilizing methods of open-circuit spirometry.

SUMMARY

1. Direct and indirect calorimetry are the two methods for determining the body's rate of energy expenditure. With direct calorimetry, the actual heat production is measured in an appropriately insulated calorimeter. Indirect calorimetry infers energy expenditure from measurements of oxygen up-

take and carbon dioxide production, using either closed or open-circuit spirometry.

2. Because of their chemical composition, each nutrient requires different amounts of oxygen in relation to carbon dioxide produced during oxidation. The ratio of CO_2 produced to O_2 consumed is called the respiratory quotient and provides an important clue to the nutrient mixture catabolized for energy. The RQ for carbohydrate is 1.00, for fat 0.70, and for protein 0.82.

3. For each RQ value, there is a corresponding calorific value for each liter of oxygen consumed. This value provides for a high degree of accuracy in determining energy expenditure during exercise.

4. In strenuous exercise, the RQ may not be representative of specific substrate utilization because of nonmetabolic production of carbon dioxide, as occurs during the buffering of lactic acid.

REFERENCES

1. Brooks, G.A., et al.: Estimation of anaerobic energy production and efficiency in rats during exercise. *J. Appl. Physiol.*, 56:520, 1984.
2. Jansson, E.: On the significance of the respiratory exchange ratio after different diets during exercise in man. *Acta Physiol. Scand.*, 114:103, 1982.
3. Kannagi, T., et al.: An evaluation of the Beckman Metabolic Cart for measuring ventilation and aerobic requirements during exercise. *J. Cardiac Rehab.*, 3:38, 1983.
4. Livesey, G., and Elia, M.: Estimation of energy expenditure, net carbohydrate utilization and net fat oxidization and synthesis by indirect calorimetry: evaluation of errors with special reference to detailed composition of fuels. *Am. J. Clin. Nutr.*, 47:608, 1988.
5. Norton, A.C.: Portable equipment for gas exchange. In *Assessment of Energy Metabolism in Health and Disease.* Columbus, OH, Ross Laboratories, 1980.
6. Snellen, J.W.: Studies in human calorimetry. In *Assessment of Energy in Health and Disease.* Columbus, OH, Ross Laboratories, 1980.
7. Wilmore, J.H., et al.: An automated system for assessing metabolic and respiratory function during exercise. *J. Appl. Physiol.*, 40:619, 1976.

HUMAN ENERGY EXPENDITURE DURING REST AND PHYSICAL ACTIVITY

Figure 9–1 illustrates that the total energy expended during a day is determined by the influence of 3 factors: (1) the resting metabolic rate (that includes basal and sleeping conditions plus the added cost of arousal), (2) the thermogenic influence of food consumed, and (3) the energy expended during and in recovery from physical activity above the resting state.

Energy Expenditure at Rest

BASAL METABOLIC RATE

There is a minimum level of energy required to sustain the body's vital functions in the waking state. This energy requirement is called the **basal metabolic rate,** or simply **BMR.** The BMR reflects the body's heat production and is determined indirectly by measuring oxygen consumption under fairly stringent conditions. For example, the person is measured in the postabsorptive state. That is, food is not eaten for at least 12 hours prior to the measurement so there is no increase in metabolism due to the energy required for the digestion, absorption, and assimilation of the ingested nutrients. To reduce other calorigenic influences, the person should remain relatively inactive prior to the BMR test. The actual test is conducted with the subject resting supine in a comfortable environment; after about 30 minutes, oxygen consumption is measured for a 10-minute period. Values for oxygen consumption during the BMR test usually range between 160 and 290 ml per minute (0.8 to 1.43 kcal $\cdot$ min^{-1}), depending upon a variety of factors, especially the size of the subject.

Use of the BMR establishes the important energy baseline for constructing a sound program of weight control by use of diet, exercise, or the effective combination of both. In most instances, so-called basal values measured under controlled conditions are only slightly lower than resting values measured 3 to 4 hours following a light meal. For our purposes, the terms basal and resting metabolic rate are used interchangeably.

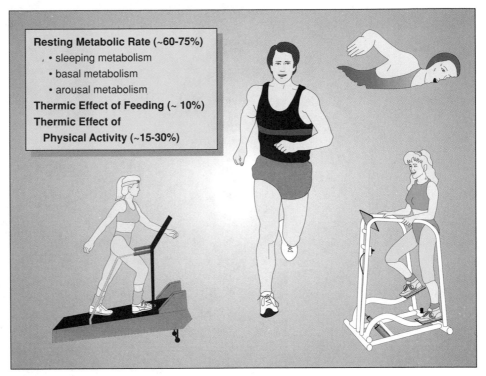

Resting Metabolic Rate (~60-75%)
- sleeping metabolism
- basal metabolism
- arousal metabolism

Thermic Effect of Feeding (~ 10%)

Thermic Effect of Physical Activity (~15-30%)

FIG. 9–1.
Approximation of components of daily energy expenditure for an average 70 kg man.

METABOLISM AT REST

Influence of Body Size. In the late 1800s, it was observed that the energy metabolism at rest was proportional to the surface area of the body. This "surface area law" was illustrated in a series of experiments in which the energy metabolism of a dog and a man was determined during a 24-hour period. As expected, the total amount of heat generated by the larger man was about 200% greater than that generated by the dog. When heat production was expressed in relation to surface area, however, the metabolic difference between the man and the dog was reduced to only 10%. Similar results have been obtained for other species of animals that differ considerably in size. This provided the basis for the common practice of expressing resting metabolic rate (energy expenditure) in terms of body surface area.

The results of numerous experiments have provided data with respect to average values of basal metabolism for men and women of different ages. These data are presented in Figure 9–2 and are expressed as hourly values of heat production per square meter of surface area (kcal per m^2 per h). Whereas they do represent averages established from measurements of large numbers of men and women, a person's BMR estimated from these curves is generally

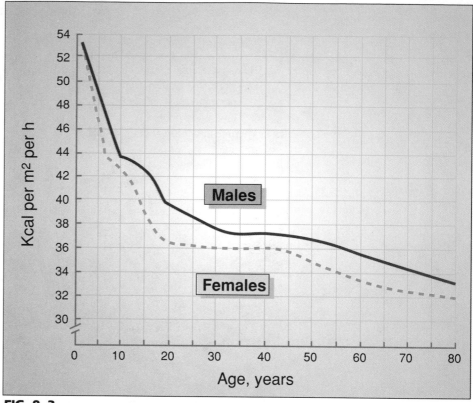

FIG. 9–2.
Basal metabolic rate as a function of age and gender. (Data from Altman, P.L., and Dittmer, D.S.: Metabolism. Bethesda, Md., Federation of American Societies for Experimental Biology, 1968.)

within 10% of the actual value obtained from measurements under strict laboratory conditions.

Figure 9–2 reveals that resting metabolism is about 5 to 10% lower in women than in men. This does not reflect a true gender difference in the metabolic rate of specific tissues. Rather, it is largely because women generally possess more body fat than men of similar size, and fat is metabolically less active than muscle. In fact, the BMR differences between genders are essentially eliminated when the metabolic rate is expressed per unit of "fat-free" or lean body mass. Differences in body composition also largely explain the 2% decrease in BMR per decade usually observed through adulthood.[21] In comparisons between young and middle-aged endurance-trained men where there was no difference in lean body mass, however, measures of BMR were essentially similar.[29]

Although body composition differences largely explain the BMR differences between men and women, the curves in Figure 9–2 can be used to estimate adequately a person's resting metabolic rate. For example, between the ages of 20 to 40, the BMR of men averages about 38 kcal per m^2 per hour, whereas for women the corresponding value is 35 kcal. For greater precision, the value for a specific age can be read directly from the appropriate curve. The estimated metabolic rate per hour is obtained when this value for BMR is multiplied by the person's surface area. This provides important

FIG. 9–3.
Nomogram to estimate body surface area from stature and body mass. [Reproduced from "Clinical Spirometry" (as prepared by Boothby and Sandiford of the Mayo Clinic), through the courtesy of Warren E. Collins, Inc., Braintree, Mass.]

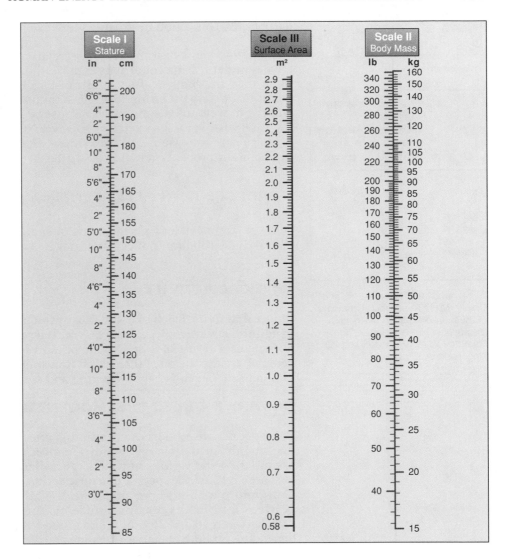

baseline information for determining daily rates of energy expenditure as well as appropriate requirements for caloric intake.

Figure 9–3 illustrates a simple method for determining surface area from stature and body mass. Surface area is determined by locating stature on Scale I and mass on Scale II. These two points are then connected with a straight edge and the intersection at Scale III gives the surface area expressed in square meters (m^2). For example, if stature is 185 cm and mass is 75 kg, surface area according to scale III on the nomogram would be 1.98 m^2.

ESTIMATE OF DAILY RESTING ENERGY EXPENDITURE

To estimate a person's resting energy expenditure in terms of total kcal, the appropriate BMR value in Figure 9–2 should be multiplied by the surface area computed from stature and mass. For a 55-year-old woman, the estimated BMR is 32 kcal per m^2 per hour. If her surface area was 1.40 m^2, the hourly energy expenditure would be 44.8 kcal per hour (32 kcal × 1.40 m^2). On a

**TABLE 9–1.
OXYGEN CONSUMPTION
OF VARIOUS TISSUES AT
REST FOR A 64.9
KG MAN***

Organ	Oxygen Used (ml · min⁻¹)	Percent of Resting Metabolism
Liver	67	27
Brain	47	19
Heart	17	7
Kidneys	26	10
Skeletal muscle	45	18
Remainder	48	19
	250	100

* From data of Passmore, R., and Draper, M.H.: The chemical anatomy of the human body. *In* Biochemical Disorders in Human Disease, 2nd ed., Edited by R.H.S. Thompson and E.J. King. London, Churchill, 1964.

daily basis, this amounts to an energy expenditure of 1075 kcal (44.8 kcal × 24).

Table 9–1 shows estimates of the relative energy needs of various body tissues under resting conditions expressed in terms of oxygen consumption. Note that the brain and skeletal muscles consume about the same total quantity of oxygen during rest, even though the brain weighs only 1.6 kg and the muscle mass constitutes almost 50% of the body mass. This is not the case with vigorous exercise, however, because the energy generated by muscles can increase nearly 120 times whereas the total energy expended by the brain probably remains unchanged.

FACTORS THAT AFFECT ENERGY EXPENDITURE

Many factors affect a person's rate of energy metabolism, including physical activity, dietary-induced thermogenesis, climate, and pregnancy and lactation.

PHYSICAL ACTIVITY

Physical activity has by far the most profound effect on human energy expenditure. For example, world-class athletes nearly double their daily caloric outputs as a result of 3 or 4 hours of hard training. In fact, most of us can generate metabolic rates that are 10 times the resting value during sustained "big muscle" exercise as in running and swimming.

DIETARY-INDUCED THERMOGENESIS

For most people food has a stimulating effect on energy metabolism that is due mainly to the energy-requiring processes of digesting, absorbing, and assimilating the various nutrients. This **dietary induced thermogenesis** has also been termed the **specific dynamic action** effect or **SDA** effect and reaches a maximum within 1 hour after a meal. While considerable variability exists between individuals, the magnitude of dietary induced thermogenesis can vary between 10 and 35% of the ingested food energy in normal individuals depending on both the quantity and type of food eaten.[2,22,41] A meal of pure protein, for example, elicits a thermic effect that is nearly 25% of the total calories in the protein itself.[14] This large thermogenesis, that may persist for a considerable time following a protein meal, is due largely to the digestive processes as well as the extra energy required by the liver to assimilate and synthesize protein or deaminate amino acids.

The calorigenic effect of protein ingestion has been used by some to argue for a high-protein diet for weight reduction. They maintain that because protein has a relatively high thermic effect, fewer calories are ultimately available to the body compared to a meal of similar caloric value consisting predominantly of fat or carbohydrate. Although this point has some validity, many other factors must be considered in formulating a sound program for weight loss—not to mention the potentially harmful strain on kidney and liver function that could result from excessive protein intake. For one thing, well-balanced nutrition requires a blend of carbohydrate, fat, and protein as well as appropriate quantities of vitamins and minerals. In addition, if physical activity is used in conjunction with dietary modification, it is important to maintain carbohydrate intake to power both rapid and sustained forms of exercise.

Research indicates that individuals who have poor control of body mass often have a blunted thermic response to eating that may, over a period of years, contribute to the accumulation of body fat.[15,34-37] Also, the magnitude of dietary induced thermogenesis may be significantly lower in exercise-trained individuals compared to untrained counterparts.[25,33,40] Whereas such a "training effect" may seem counter-productive to the potential of exercise for weight control, it may reflect a calorie sparing adaptation to conserve energy and glycogen during periods of increased physical activity. The important point is that for a physically active person the thermic effect of any food represents only a small portion of the total daily energy expenditure.

CALORIGENIC EFFECT OF FOOD ON EXERCISE METABOLISM

Dietary-induced thermogenesis has been compared in exercising and resting subjects after consuming meals of similar size and composition. In one study,[3] six men performed submaximal, moderate exercise on a bicycle ergometer before breakfast on one day; then on separate days, exercise was performed 30 minutes after a breakfast containing either 350, 1000, or 3000 kcal. The following results were obtained: (1) breakfast increased the resting metabolism by 10%, (2) variations in the caloric value of the meal had no influence on the thermic effect, and (3) when exercise was performed following a meal of 1000 or 3000 kcal, energy expenditure was larger compared to exercise without prior food ingestion. This calorigenic effect of food amounted to nearly two times the thermogenic value of the food at rest. Apparently, exercise augments dietary-induced thermogenesis. This finding is in agreement with previous findings in which the thermic response to a 1,000 kcal meal averaged 28% of the basal requirement at rest and increased to 56% of the basal requirement after exercise was performed.[30] It appears that the potentiation of the thermic effect of eating with exercise may not be as great in the obese.[44] An important area for research is to determine the extent to which an exercise augmented dietary-induced thermogenesis contributes to weight control in active people despite their relatively large caloric intakes. Within the framework of the available though limited research, it seems reasonable to encourage exercise after eating to stimulate caloric expenditure for weight control.

CLIMATE

Environmental factors also influence resting metabolic rate. For example, the resting metabolism of people in a tropical climate is generally 5 to 20% higher than for counterparts living in a more temperate area. Exercise in the heat also imposes a small additional metabolic load, causing the oxygen consumption to be about 5% higher compared to the same work performed in a thermoneutral environment. This finding is probably due to the effects of an elevated core temperature per se, as well as to the additional energy required for sweat-gland activity and altered circulatory dynamics during work in the heat.

Cold environments exert a significant influence on metabolic rate both at rest and during exercise, the extent of which depends largely on a person's body fat content and amount and type of clothing worn. During extreme cold stress at rest, metabolic rate can double or triple with shivering because the body attempts to maintain a stable core temperature. The effects of cold during exercise are most evident in cold water because it is often difficult to maintain core temperature in this environment.[28]

PREGNANCY

With an increased number of women involved in physically demanding occupations as well as exercise and sports, it is important to establish prudent guidelines concerning exercise during pregnancy. Researchers are only now beginning to understand the impact of exercise and training on both mother and fetus. One area of interest is the degree to which pregnancy affects the metabolic cost and physiologic strain imposed by exercise. In one investigation,[23] 13 women were studied from the sixth month of pregnancy to 6 weeks after the birth of the baby. Physiologic measures taken every 4 weeks included heart rate and oxygen consumption during bicycle and treadmill exercise. The increased heart rate and oxygen consumption during walking coincided with body weight changes in pregnancy, whereas no differences were noted in weight-supported bicycle exercise during the period of investigation. These findings and those of others suggest that maternal cardiovascular dynamics follow normal response patterns[1,42] and that pregnancy offers no greater physiologic stress to the mother during moderate exercise other than that provided by the additional weight gain and possible encumbrance of fetal tissue.[12,19] Certainly, when body mass is supported, the exercise response for heart rate and oxygen consumption for the mother during pregnancy is essentially identical to that observed prior to and following birth. As pregnancy progresses, on the other hand, the increase in maternal body weight adds significantly to exercise effort with weight-bearing exercise such as walking, jogging, and stair climbing.[16]

Exercise Effects on Fetus. Another area of concern among exercise specialists and physicians is the impact of vigorous exercise of the mother on the developing fetus, particularly placental blood flow. Certainly any factor that could temporarily compromise fetal blood supply would be of concern in counseling pregnant women concerning exercise. Although research in this area with humans is obviously sparse, other species of mammals have been studied. In one investigation,[5] treadmill exercise to exhaustion caused a fall in both uterine blood flow and oxygen pressure in near-term pregnant ewes. Despite this response, however, oxygen uptake by the uteroplacental tissues and fetus was maintained by a facilitated unloading of oxygen from the available blood supply. At intense exercise levels fetal responses generally followed those of the mother and there was no change in the fetal oxygen supply. In animals, however, with one umbilical artery tied off to restrict placental circulation, a significant reduction was noted in fetal oxygen supply.[13] The researchers concluded that vigorous maternal exercise was well tolerated by the fetus, but could be potentially harmful to a fetus with some limitation of the umbilical circulation.

It appears that for a previously active, healthy, low-risk woman during an uncomplicated pregnancy, moderate aerobic exercise does not produce circulatory alterations that compromise fetal oxygen supply. Performed on a regular basis, such aerobic exercise not only maintains fitness but may actually generate a training effect.[8,24] Because vigorous exercise probably diverts some blood from the uterus that could pose a hazard for a fetus with restricted placental blood flow, it would be prudent for a pregnant woman to exercise in moderation, especially if the pregnancy is compromised to any degree.[38] In addition, exercise late in pregnancy may magnify the normal maternal hypoglycemic response by increasing glucose uptake by maternal skeletal muscle; in the extreme, this response could adversely affect the fetus.[6,39] It is also possible that a decrease in uterine blood flow or an elevation in maternal core temperature with extended exercise could compromise the dissipation of heat from the fetus through the placenta.[4,20] Because such hyperthermia can negatively affect fetal development, especially in the first trimester of pregnancy, it is recommended that during warm weather, pregnant women should exercise in the cool part of the day, for shorter intervals, while maintaining regular fluid intake. Within this framework, exercise in water would serve as an ideal form of maternal exercise.

Regular aerobic exercise can play an important role in maintaining the functional capacity, optimal body weight, and general well being of a pregnant woman. It remains unclear, however, whether extremes of maternal exercise are beneficial to the developing fetus, or if exercise during pregnancy enhances the course of pregnancy including labor, delivery, and outcome.[7,10,17,20,24]

SUMMARY

1. One's total daily energy expenditure is the sum total of the energy required in basal and resting metabolism, thermogenic influences (especially the thermic effect of food), and the energy generated in physical activity.
2. The BMR is the minimum energy required to maintain vital functions in the waking state. The BMR is only slightly lower than the resting metabolism and is proportionate to the surface area of the body. It is also related to age and generally is higher for men than for women; these influences are largely due to variation in lean body mass.
3. Different organs use different amounts of oxygen during rest and exercise. At rest, muscles require about 20% of the total oxygen uptake. During exercise, however, the oxygen uptake of skeletal muscle increases 100 times above rest; this represents close to 85% of the oxygen uptake.
4. Four major factors that affect a person's metabolic rate are physical activity, dietary-induced thermogenesis, climate, and pregnancy. Physical activity has, by far, the greatest influence.
5. For the previously active, healthy woman, moderate aerobic exercise does not appear to compromise fetal oxygen supply. It remains unclear, however, whether extremes of maternal exercise are beneficial to the course of pregnancy or to the child in the early period after birth.

P A R T
2

Energy Expenditure in Physical Activity

CLASSIFICATION OF PHYSICAL ACTIVITIES BY ENERGY EXPENDITURE

All of us at one time or another have done some type of physical work that we would classify as exceedingly "difficult." This physical work might be walking up a flight of stairs, shoveling snow to clear the driveway, running to catch a bus, loading and unloading a truck, digging a trench to fix an underground pipe, skiing through a blizzard, or climbing a steep mountain. Intensity and duration are two important factors in rating the difficulty or strenuousness of a particular task. For example, the same number of calories may be required to complete a 26-mile marathon at various running speeds. One runner, however, might exert considerable energy running at maximum pace and complete the race in a little more than 2 hours. Another runner of equal fitness might select a slower more comfortable pace and complete the run in 3 hours. In this example, the **intensity** of the exercise is the factor distinguishing the manner in which a specific work task is completed. In another situation, two people may run at the same speed, but one person may run twice as long as the other. In this situation, exercise **duration** becomes the important consideration.

Several classification systems have been proposed for rating the difficulty of sustained physical activity in terms of its strenuousness. One recommendation is that work tasks be rated by the ratio of energy required for work to the resting or basal energy requirement. With this system, **light work** for men is defined as that eliciting an oxygen consumption (or energy expenditure) up to three times the resting requirement. **Heavy work** is categorized as that requiring six to eight times the resting metabolism, whereas maximal work is considered as any task requiring an increase in metabolism nine times or more above the resting level. As a frame of reference, most industrial jobs and household tasks require less than three times the energy expenditure at rest. For women, the work classifications in terms of units of resting metabolism are slightly lower owing to their generally lower level of aerobic capacity.

The MET. The five-level classification system presented in Table 9–2 is based on the energy required by untrained men and women performing different tasks.[11] Because 5 kcal is approximately equal to one liter of oxygen

TABLE 9–2.
FIVE-LEVEL CLASSIFICATION OF PHYSICAL ACTIVITY IN TERMS OF EXERCISE INTENSITY*

Level	Energy Expenditure			
	Men			
	kcal · min⁻¹	l · min⁻¹	ml · kg⁻¹ · min⁻¹	METS
Light	2.0–4.9	0.40–0.99	6.1–15.2	1.6–3.9
Moderate	5.0–7.4	1.00–1.49	15.3–22.9	4.0–5.9
Heavy	7.5–9.9	1.50–1.99	23.0–30.6	6.0–7.9
Very heavy	10.0–12.4	2.00–2.49	30.7–38.3	8.0–9.9
Unduly heavy	12.5–	2.50–	38.4–	10.0–
	Women			
	kcal · min⁻¹	l · min⁻¹	ml · kg⁻¹ · min⁻¹	METS
Light	1.5–3.4	0.30–0.69	5.4–12.5	1.2–2.7
Moderate	3.5–5.4	0.70–1.09	12.6–19.8	2.8–4.3
Heavy	5.5–7.4	1.10–1.49	19.9–27.1	4.4–5.9
Very heavy	7.5–9.4	1.50–1.89	27.2–34.4	6.0–7.5
Unduly heavy	9.5–	1.90–	34.5–	7.6–

*l · min⁻¹ based on 5 kcal per liter of oxygen; ml · kg⁻¹ · min⁻¹ based on 65-kg man and 55-kg woman; one MET is equivalent to the average resting oxygen consumption.

consumed, it is also possible to present this five-stage classification in terms of liters of oxygen consumed per minute, or milliliters of oxygen consumed per kilogram of body mass per minute $(ml \cdot kg^{-1} \cdot min^{-1})$, or METS, **a MET being defined as a multiple of the resting metabolic rate.** Thus, 1 MET is equivalent to the resting oxygen consumption that, for an average man and woman, is approximately 250 and 200 ml per minute, respectively. Work at 2 METS requires twice the resting metabolism or about 500 ml of oxygen per minute for a man, and 3 METS is three times the resting energy expenditure, and so on. For slightly more accurate classifications, the MET can be expressed in terms of oxygen consumption per unit of body mass with 1 MET equal to approximately $3.6 \ ml \cdot kg^{-1} \cdot min^{-1}$.

DAILY RATES OF AVERAGE ENERGY EXPENDITURE

Table 9–3 shows averages for daily rates of energy expenditure for men and women living in the United States. Between the ages of 23 and 50 the "average" man expends between 2700 and 3000 kcal per day, whereas his female counterpart expends about 2000 to 2100 kcal. As can be seen in the bottom portion of this table, nearly 75% of the average person's day, regardless of gender, is spent in activities requiring only a light expenditure of energy. For most Americans, energy expenditure rarely climbs significantly above the resting level, with walking being the most prevalent form of exercise. Indeed, our citizens have all too appropriately been termed *Homo sedentarius.*

TABLE 9–3.
AVERAGE RATES OF ENERGY EXPENDITURE FOR MEN AND WOMEN LIVING IN THE UNITED STATES*

	Age (years)	Mass (kg)	Mass (lb)	Stature (cm)	Stature (in)	Energy Expenditure (kcal)
Men	15–18	66	145	176	69	2800
	19–22	70	154	177	70	2900
	23–50	70	154	178	70	2700
	51+	70	154	178	70	2400
Women	15–18	55	120	163	64	2100
	19–22	55	120	163	64	2100
	23–50	55	120	163	64	2000
	51+	55	120	163	64	1800

Average Time Spent During the Day for Men and Women

Activity	Time (h)
Sleeping and lying	8
Sitting	6
Standing	6
Walking	2
Recreational: sports or exercises	2

* Data from Food and Nutrition Board, National Research Council, Recommended Dietary Allowances, 8th rev. ed., National Academy of Sciences, Washington, D.C., 1980.

TABLE 9–4.
DAILY RATES OF ENERGY EXPENDITURE FOR VARIOUS OCCUPATIONS*

| | Occupation | Energy Expenditure, kcal per day | | |
		Average	Minimum	Maximum
Men	Elderly retired	2330	1750	2810
	Office workers	2520	1820	3270
	Coal mine clerks	2800	2330	3290
	Laboratory technicians	2840	2240	3820
	Older industrial workers	2840	2180	3710
	University students	2930	2270	4410
	Building workers	3000	2440	3730
	Steel workers	3280	2600	3960
	Army cadets	3490	2990	4100
	Older peasants (Swiss)	3530	2210	5000
	Farmers	3550	2450	4670
	Coal miners	3660	2970	4560
	Forestry workers	3670	2860	4600
Women	Older housewives	1990	1490	2410
	Middle-aged housewives	2090	1760	2320
	Laboratory assistants	2130	1340	2540
	Assistants in department store	2250	1820	2850
	University students	2290	2090	2500
	Factory workers	2320	1970	2980
	Bakery workers	2510	1980	3390
	Older peasants (Swiss)	2890	2200	3860

* Data from Durnin, J.V.G.A., and Passmore, R.: Energy, Work and Leisure. London, Heinemann Educational Books, 1967.

Energy Expenditure Grouped by Occupation. Dietary surveys of nutrient intake and corresponding values of energy expenditure for various occupational groups of different ages have been reported by various teams of researchers. Such surveys provide a comprehensive evaluation of energy expenditure, as the data in Table 9–4 illustrate. Included are data based on 7-day observations of Swiss peasants and British army cadets. Daily energy expenditure was estimated by determining the time spent in each activity during the day, and the average energy expended for the activity.

Even though the average energy expenditure increases in ascending order in the job classifications presented in Table 9–4, there is considerable variability for men and women in any one classification. With university students, for example, some individuals expend less energy per day than the average for elderly retired persons, whereas other students expend energy that exceeds the average for farmers, coal miners, and forestry workers. These variations are largely explained by the intensity and duration of activities performed outside of work, especially those related to recreational pursuits.

ENERGY COST OF HOUSEHOLD, INDUSTRIAL, AND RECREATIONAL ACTIVITIES

A list of energy expenditures, expressed in terms of body mass for common household activities, selected industrial tasks, and popular recreational and sports activities is presented in Appendix D. These activities illustrate the large variation in energy expenditure that occurs with participation in various forms of physical activity. The caloric values represent averages that can vary considerably depending on factors such as skill, pace, and fitness level.

The value listed in the column that corresponds to a particular body mass is the caloric cost of the activity for 1 minute. These are gross energy values because also included is the cost of rest for the 1-minute period. The total cost of participating in an activity is estimated by multiplying the value listed in the table by the number of minutes of participation. For example, if a 70 kg man spends 30 minutes vacuuming (carpet sweeping), his total caloric expenditure of 102 kcal for this household activity would be determined by multiplying the value of 3.4 kcal per minute by 30. The same individual would spend approximately 690 kcal during a 50-minute judo workout, but only 90 kcal while sitting quietly and watching television for 2 hours. Golf requires about 6.0 kcal per minute, or 360 kcal per hour, for a person who weighs 70 kg. The same person expends almost twice this energy, or 708 kcal per hour, while swimming the backstroke. Viewed somewhat differently, 25 minutes of swimming the backstroke requires about the same number of calories as playing golf for one hour. If the pace of either the swim or the golf game is increased, the energy expenditure also increases proportionally.

EFFECT OF BODY MASS

Body mass is an important factor that affects the energy expended in many forms of exercise. This factor can be seen in Appendix D, where the energy cost of a particular exercise is generally greater for heavier people, especially in weight-bearing exercise like walking and running where the person must transport his or her body mass during the activity. This finding is clearly illustrated in Figure 9–4, which shows that the energy cost of walking increases directly with body mass. In fact, for people of the same body mass, the variation in oxygen consumption is so small that the energy expenditure during walking can be predicted from body mass with high accuracy. This effect of added mass on energy metabolism and exercise performance occurs whether a person gains weight "naturally" as body fat,[18] or as an acute added load as sports equipment or as a weighted vest on the torso.[9,31] With stationary cycling, however, weight is supported, and the influence of body mass on energy cost is less extreme (about 5% higher due to extra energy required to lift heavier limbs). Certainly, for heavy people desiring to use exercise in a program for weight loss, weight-bearing forms of exercise can provide a considerable caloric expenditure.

It can be seen from Appendix D that the energy cost for cross-country running ranges between 8.2 kcal per minute for a 50-kg person to almost twice as much at 16.0 kcal for a person weighing 98 kg. If, however, the energy requirement is expressed in relation to body mass as $kcal \cdot kg^{-1} \cdot min^{-1}$, this difference is essentially eliminated, and the energy cost averages about $0.164 \ kcal \cdot kg^{-1} \cdot min^{-1}$. By expressing energy cost in this manner (that is, per unit of body mass), the differences between individuals are greatly reduced regardless of age, race, gender, and body mass. The **total** number of calories expended by the heavier person, however, is still considerably larger than that expended by a lighter counterpart, simply because the body mass must be transported in the activity—and this requires proportionately more total energy.

USE OF HEART RATE TO ESTIMATE ENERGY EXPENDITURE

For each person, heart rate and oxygen consumption tend to be linearly related throughout a large portion of the aerobic work range. If this precise rela-

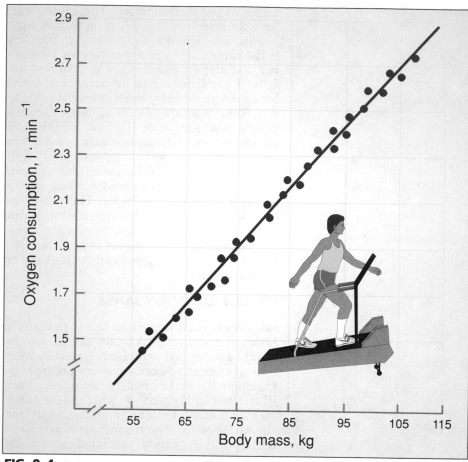

FIG. 9–4.
Relationship between body mass and oxygen consumption measured during treadmill walking. (From Laboratory of Applied Physiology, Queens College, N.Y.)

tionship is known, the exercise heart rate can be used to estimate oxygen consumption (and then compute energy expenditure) during other forms of similar activity such as exercise performed while running, walking, cycling, and swimming. This approach can be used when the oxygen consumption cannot be measured during the activity.

The data for two members of a nationally ranked women's basketball team during a laboratory treadmill test are presented in Figure 9–5.[27] For each woman, the heart rate increased linearly, with each increase in oxygen consumption being accompanied by a proportionate increase in heart rate. Even though both heart rate-oxygen consumption lines are essentially linear, however, the same heart rate does not correspond to the same level of oxygen consumption. This is because the slope or rate of change of the line differs considerably among people. For each increase in heart rate, the oxygen consumption of subject A increases to a much greater extent than for subject B. The significance of this difference and its relation to cardiovascular fitness are discussed in Chapters 11, 17, and 21. The important point is that if heart rate is known, oxygen consumption can often be estimated with reasonable accuracy. For player A, an exercise heart rate of 140 beats per minute cor-

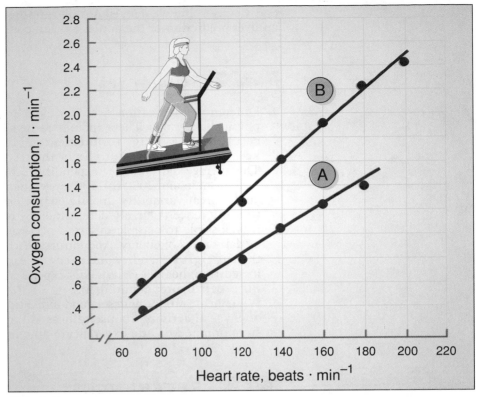

FIG. 9–5.
Linear relationship between heart rate and oxygen consumption for two subjects. Measurements were made while running on a treadmill. The elevation of the treadmill was increased by 2.5% every 3 minutes. (From Laboratory of Applied Physiology, Queens College, N.Y.)

responds to an oxygen consumption of 1.08 liters per minute, whereas the same heart rate for player B would be related to a 1.6-liter oxygen consumption.

Although the technique for using heart rate to estimate energy expenditure is practical, it may be of limited use for research purposes because its validity has yet to be adequately established for more than a few general activities. One of the major problems is determining the degree of similarity between the laboratory test to establish the heart rate-oxygen consumption line and the specific activity to which it is applied. Factors other than oxygen consumption can influence the heart rate. These factors include temperature, emotions, food intake, body position, the muscle groups exercised, whether the exercise is continuous or stop-and-go, or whether the muscles are contracting isometrically or in a more rhythmic manner. For an activity such as aerobic dance, for example, heart rates during dancing at a particular oxygen consumption are significantly higher than heart rates at a comparable oxygen consumption on the treadmill.[32] In arm exercise or when muscles are contracting statically as in a straining-type exercise, heart rates are **consistently higher** when compared to rhythmic leg exercise at any particular submaximal oxygen consumption. Consequently, when a heart rate during a task requiring upper body or static exercise is applied to a heart rate-oxygen consumption line developed during running or cycling, the result would be an **over prediction** of the actual oxygen consumption.[26]

SUMMARY

1. Different classification systems exist for rating the strenuousness of physical activities. These include ratings based on (1) the ratio of energy cost to resting energy requirement, (2) the oxygen requirement in ml · kg^{-1} · min^{-1}, or (3) multiples or the resting metabolic rate or METS.
2. The average daily energy expenditure is estimated to be 2700 to 2900 kcal for men and 2000 to 2100 for women between the ages of 15 and 50 years. Great variability in daily energy expenditure exists, however, and this difference is largely determined by one's physical activity level.
3. It is possible to classify different occupations as well as athletic groups by daily rates of energy expenditure. Within any classification, however, there is great variability due to energy expended in recreational pursuits. In addition, heavier individuals expend more energy in physical activity than their lighter counterparts.
4. While heart rate provides a good indication of the relative strenuousness of physical activity, its usefullness is limited in predicting oxygen consumption and caloric expenditure among diverse forms of exercise.

REFERENCES

1. Artal, R., et al.: Exercise in pregnancy: maternal cardiovascular and metabolic responses in normal pregnancy. *Am. J. Obst. Gynecol.*, 140:123, 1981.
2. Belko, A., et al.: Effect of energy and protein intake and exercise intensity on the thermic effect of food. *Am. J. Clin. Nutr.*, 43:863, 1986.
3. Bray, G.: The acute effects of food intake on energy expenditure during cycle ergometry. *Am. J. Clin. Nutr.*, 27:254, 1974.
4. Bullard, J.A.: Exercise and pregnancy. *Canadian Family Physician*, 27:977, 1981.
5. Clapp, J.F., III.: Acute exercise stress in the pregnant ewe. *Am. J. Obst. Gynecol.*, 136:489, 1980.
6. Clapp, J.F. III: Thermoregulatory and metabolic responses to jogging prior to and during pregnancy. *Med. Sci. Sports Exerc.*, 19:124, 1987.
7. Clapp, J.F., III, and Dickstein, S.: Endurance exercise and pregnancy outcome. *Med. Sci. Sports Exerc.*, 16:556, 1984.
8. Colling, C.A., et al.: Maternal and fetal responses to maternal aerobic exercise program. *Am. J. Obstet. Gynecol.*, 145:702, 1983.
9. Cureton, K.J., and Sparling, P.B.: Distance running performance and metabolic responses to running in men and women with excess weight experimentally equated. *Med. Sci. Sports*, 12:288, 1980.
10. Dressendorfer, R.H.: Physical training during pregnancy and lactation. *Phys. Sportsmed.*, 6:74, 1978.
11. Durnin, J.V.G.A., and Passmore, R.: *Energy, Work and Leisure*. London, Heinmann, 1967.
12. Emerson, K., et al.: Caloric cost of normal pregnancy. *Obst. Gynecol.*, 40:786, 1972.
13. Emmanouilides, G.C., et al.: Fetal responses to maternal exercise in sheep. *Am. J. Obst. Gynecol.*, 112:130, 1982.
14. Flatt, J.B.: Energetics of intermediary metabolism. In *Assessment of Energy Metabolism in Health and Disease*. Columbus, OH, Ross Laboratories, 1980.
15. Garrow, J.S.: *Energy Balance and Obesity in Man*. New York, Elsevier, 1978.
16. Gorski, J.: Exercise during pregnancy: maternal and fetal responses. A brief review. *Med. Sci. Sports Exerc.*, 17:407, 1985.

17. Hall, D.C., and Kaufmann, D.A.: Effects of aerobic and strength conditioning on pregnancy outcomes. *Am. J. Obstet. Gynecol.*, 157:1199, 1987.

18. Hanson, J.S.: Exercise responses following production of experimental obesity. *J. Appl. Physiol.*, 35:587, 1973.

19. Hutchinson, P.L., et al.: Metabolic and circulatory responses to running during pregnancy. *Phys. Sportsmed.*, 9:55, 1981.

20. Jarrett, J.C., and Spellacy, W.N.: Jogging during pregnancy: an improved outcome? *Obstet. Gynecol.*, 61:705, 1983.

21. Keys, A., et al.: Basal metabolism and age of adult men. *Metabolism*, 22:579, 1973.

22. Kleiber, M.: *The Fire of Life: An Introduction to Animal Energetics.* Huntington, NY, Krieger, 1975.

23. Knuttgen, H.G., and Emerson, K., Jr.: Physiological response to pregnancy at rest and during exercise. *J. Appl. Physiol.*, 36:549, 1974.

24. Kulpa, P.J., et al.: Aerobic exercise in pregnancy. *Am. J. Obstet. Gynecol.*, 156:1395, 1987.

25. LeBlanc, J., et al.: Hormonal factors in reduced post prandial heat production of exercise trained subjects. *J. Appl. Physiol.*, 56:772, 1984.

26. Maas, S., et al.: The validity of the use of heart rate in estimating oxygen consumption in static and in combined static/dynamic exercise. *Ergonomics*, 32:141, 1989.

27. McArdle, W.D., et al.: Aerobic capacity, heart rate, and estimated energy cost during woman's competitive basketball. *Res. Q.*, 42:178, 1971.

28. McArdle, W.D., et al.: Metabolic and cardiovascular adjustment to work in air and water at 18, 25 and 33°C. *J. Appl. Physiol.*, 40:85, 1976.

29. Meredith, C.N., et al.: Body composition and aerobic capacity in young and middle-aged endurance-trained men. *Med. Sci. Sports Exerc.*, 19:557, 1987.

30. Miller, D.S., et al.: Gluttony 2: Thermogenesis in overeating man. *Am. J. Clin. Nutr.*, 20:1223, 1967.

31. Montgomery, D.L., et al.: The effect of added weight on ice hockey performance. *Phys. Sportsmed.*, 10:91, 1982.

32. Parker, S.B., et al.: Failure of target heart rate to accurately monitor intensity during aerobic dance. *Med. Sci. Sports Exerc.*, 21:230, 1989.

33. Poehlman, E.T., et al.: Resting metabolic rate and post prandial thermogenesis in highly trained and untrained males. *Am. J. Clin. Nutr.*, 47:793, 1988.

34. Rothwell, N.J., and Stock, M.J.: Luxuskonsumption, diet-induced thermogenesis and brown fat: the case in favor. *Clin. Sci.*, 64:64, 1983.

35. Schutz, Y., et al.: Diet-induced thermogenesis measured over a whole day in obese and non-obese women. *Am. J. Clin. Nutr.*, 40:542, 1984.

36. Segal, K.R., et al.: Thermic effects of food and exercise on lean and obese men of similar lean body mass. *Am. J. Physiol.*, 252:E110, 1987.

37. Shetty, P.S., et al.: Post prandial thermogenesis in obesity. *Clin. Sci.*, 60:519, 1981.

38. Speroff, L.: Can exercise cause problems in pregnancy and menstruation? *Contemp. Obstet. Gynecol.*, 16:57, 1980.

39. Treadway, J.L., and Young, J.C.: Decreased glucose uptake in fetus after maternal exercise. *Med. Sci. Sports Exerc.*, 21:140, 1989.

40. Trembly, A., et al.: Diminished dietary thermogenesis in exercise-trained human subjects. *Eur. J. Appl. Physiol.*, 52:1, 1983.

41. Tuckerman, M.M., and Turco, S.J.: *Human Nutrition.* Philadelphia, Lea & Febiger, 1983.

42. Vetland, M.D., et al.: Maternal cardiovascular dynamics. *Am. J. Obstet. Gynecol.*, 81:856, 1974.

43. Wolfe, L.A., et al.: Physiological interactions between pregnancy and anaerobic exercise. In *Exercise and Sport Sciences Reviews.* Edited by K.B. Pandolf. Vol. 17. New York, Macmillan, 1989.

44. Zahorska-Markiewicz, B.: Thermic effect of food and exercise in obesity. *Eur. J. Appl. Physiol.*, 44:231, 1980.

10

ENERGY EXPENDITURE DURING WALKING, JOGGING, RUNNING, AND SWIMMING

The total amount of energy expended each day depends largely on the type and duration of one's physical activity. The following sections detail the energy expenditure of the popular activities: walking, running, and swimming. Aside from being competitive sports, these exercises take on special significance for they are commonly prescribed for weight control, conditioning, and cardiac rehabilitation.

GROSS AND NET ENERGY EXPENDITURE

The following example illustrates the use of oxygen-consumption measures to estimate the energy requirements of swimming. A 25-year-old man swimming at a moderate and steady pace that requires an average oxygen consumption of 2 liters per minute would, with 40 minutes of swimming, consume a total of 80 liters of oxygen. This oxygen value can easily be transposed to an energy value by using the approximate calorific transformation of 5 kcal of energy generated per liter of oxygen consumed. Thus, the swimmer would expend about 400 kcal (80 liters × 5 kcal) during the swim. This, however, is not the cost of the swim per se because this total or **gross energy expenditure** also includes energy that would have been expended if the person had rested and not swum. To obtain a clearer picture of the cost of exercise or the **net energy expenditure,** one must obtain an estimate of the resting metabolic rate and subtract this from the gross energy cost of the exercise.

$$\begin{array}{c} \text{Net energy} \\ \text{expenditure} \end{array} = \begin{array}{c} \text{Gross energy} \\ \text{expenditure} \end{array} - \begin{array}{c} \text{Resting energy} \\ \text{expenditure} \\ \text{(for equivalent} \\ \text{time period)} \end{array}$$

By knowing the swimmer's size (mass, 65 kg; stature, 174 cm), his surface area of 1.78 m^2 is computed from the nomogram in Figure 9–3. When this value is multiplied by the average BMR for young men of 38 kcal per m^2 per hour (Fig. 9–2), an estimated resting energy expenditure of 67.6 kcal per hour (1.78 m^2 × 38 kcal) is obtained or about 45 kcal per 40-minute swimming period. The energy expended solely for the swim is then computed as gross energy expenditure (400 kcal) minus the 40-minute resting value (45 kcal). This results in a **net** energy expenditure for the swimming period of 355 kcal.

When a constant-load exercise is performed at light to moderate intensity, the oxygen consumption rises rapidly at the start and then levels off and remains at a relatively steady rate throughout the activity period. When exercise stops, oxygen consumption decreases rapidly with the recovery oxygen consumption being approximately equal to the quantity of oxygen not consumed in the early adjustment to exercise (oxygen deficit). Energy expenditure can therefore be estimated from only one or two measures of oxygen consumption during the steady-rate phase of relatively moderate exercise. In activities with considerable variation in pace such as tennis, soccer, or basketball, more frequent measures of oxygen consumption must be made to estimate accurately the total energy expenditure. In vigorous exercise, when the energy requirements exceed the capacity for aerobic energy transfer, considerable anaerobic energy is generated and lactic acid accumulates. In this situation, energy cost estimates require measurements of oxygen consumption in both exercise and recovery. The precision of these estimates is limited, however, because near-exhaustive exercise of long duration causes physiologic, metabolic, and thermal adjustments that can cause the oxygen consumption to remain elevated for up to 24 hours. This can increase the total energy cost by about 10% compared to the same exercise performed aerobically.

ECONOMY OF MOVEMENT AND MECHANICAL EFFICIENCY IN EXERCISE

The concept of efficiency can generally be viewed as the relationship between input and resulting output. In an economic sense, efficiency of operation relates to the cost required to produce goods in relation to the money generated from the sale of such goods. In another context, the auto industry is always striving to optimize the aerodynamic design of its vehicles to improve efficiency of operation which is reflected in the important miles-per-gallon rating. In terms of efficiency of human movement, the basic concern is the quantity of energy required to perform a particular task in relation to the actual work accomplished. In a sense this is illustrated in assessing the ease of movement of highly trained athletes. It does not require a trained eye to qualitatively discriminate the ease of effort in comparisons of elite swimmers, skiers, cyclists, and dancers with less skilled counterparts who seem to expend considerable "wasted energy" to perform the same task.

Economy

A fairly simple and commonly used means to establish differences between individuals in the **economy** of physical effort is to evaluate the oxygen consumed while performing a particular exercise.[59] This approach is useful during steady-rate exercise in which the oxygen consumed during the activity closely mirrors the energy expended. For example, at a given submaximal speed of running, cycling, or swimming, an individual with greater economy for movement consumes less oxygen to perform the task. This is important in longer-duration exercise where success largely depends on the aerobic capability of the individual and the oxygen requirements of the task. All else being equal, any adjustment in training that improves the economy of effort directly translates into improved performance.

No single biomechanical factor can account for variation in running economy.[58] Even among trained runners, significant variation is observed among individuals in economy at a particular speed.[12,44] In addition, economy for running can be improved with training.[13,17] An ideal training program enhances both the economy of movement and the aerobic capacity to perform the exercise task.

Efficiency

Another approach to evaluate the relationship between input (metabolic energy expenditure) and resulting mechanical output in exercise is to compute the actual **mechanical efficiency** of human movement. This provides an indication of the percentage of total energy expended that can produce external work. Within this context:

$$\text{Mechanical efficiency (\%)} = \frac{\text{Actual mechanical work accomplished}}{\text{Input of energy}} \times 100$$

The external work or energy output accomplished is expressed as force acting through a vertical distance $(F \times D)$, usually recorded as foot pounds or kilogram meters (kg m). This is fairly easy to determine during cycle ergometry or in lifting the body mass as in stair climbing or bench stepping. In horizontal walking or running the computation of external work output is not possible. External work is not being accomplished because the reciprocal movements of the legs and arms negate each other and the body does not achieve a net gain in terms of vertical distance. If a person walks or runs up a grade the work component can be estimated from the body mass and the vertical distance or lift that was achieved during the duration of the test.

The energy input portion of the efficiency equation is commonly inferred from the steady-rate oxygen consumption during the exercise. To obtain common units for expressing either work or mechanical power (which is quite difficult to quantify for many physical activities), the oxygen consumption is converted into energy units (roughly 1 ℓ O_2 = 5 kcal; see Table 8–1 for precise calorific transformations based on RQ) which are then converted to units of work (1 kcal = 3087 ft $\cdot$ lb^{-1} or 426.4 kg m in a perfect machine).

For example, suppose 13,300 kg m of work were generated during a 15-minute ride on a stationary bicycle and the net oxygen consumed to achieve this work totaled 25 liters (RQ = 0.88). To create common units of measurement the oxygen consumed is converted to a corresponding work output. From Table 8-1, note that at an RQ of 0.88, each liter of oxygen consumption generates an energy equivalent of 4.9 kcal. Therefore, 25 liters of oxygen consumption generates 122.5 kcal of energy (25 × 4.9 kcal) during the 15-minute ride. As the work equivalent of 1 kcal is 426.4 kg m in a perfectly efficient machine, the work input is computed as 52,234 kg m (122.5 × 426.4 kg m), and mechanical efficiency is computed as follows:

$$\text{Mechanical efficiency} = \frac{13,300 \text{ kg m}}{52,234 \text{ kg m}} \times 100$$

$$= 25.5\%$$

As with all machines, the efficiency of the human body for producing mechanical work is significantly less than 100%. The biggest factor that affects efficiency is the energy required to overcome internal and external

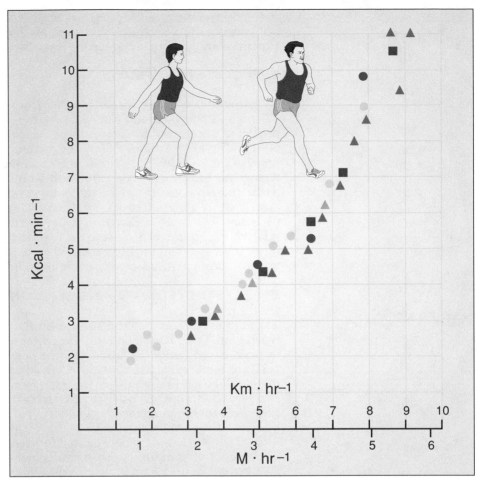

FIG. 10–1.
Energy expenditure walking on the level at different speeds. Different symbols represent the mean values from various studies reported in the literature.

friction. This is essentially wasted energy because it does not produce work output; as a result, work output is always less than work input. In general, the efficiency of human locomotion in walking, running, and cycling ranges between 20 to 30%.

ENERGY EXPENDITURE DURING WALKING

Walking is the most common form of exercise. For most individuals, it represents the major type of physical activity that falls outside the realm of sedentary living. Figure 10–1 displays the research from five countries on the energy expenditure of men who walked at speeds ranging from 1.0 to 10 kilometers per hour (0.62 to 6.2 mph). The relationship between walking speed and oxygen consumption is approximately linear between speeds of 3.0 and 5.0 kilometers per hour (1.86 to 3.10 mph); at faster speeds, walking becomes less efficient and the relationship curves in an upward direction that indicates a greater caloric cost per unit of distance traveled. This finding accounts for the observation that, per unit distance traveled, the total calories expended are greater at the faster, less efficient walking speeds.[5,20]

EFFECT OF BODY MASS

At horizontal walking speeds ranging from 3.2 to 6.4 km per hour (2.0 to 4.0 mph), energy expenditure for people who differ in body mass can be predicted accurately by an equation based on the combined data in Fig. 10–1 and on other studies.[19,29,47] The predicted values for energy expenditure are listed in Table 10–1; these are accurate to within 15% of the actual energy expenditure for both men and women of different sizes. On a daily basis, therefore, estimates of the energy expended in walking could be in error by only about 50 to 100 kcal, assuming that the person walks 2 hours each day. The table is easy to use, is formulated on sound research, and is relatively accurate for assessing the caloric cost of walking within the speed range indicated and for body masses up to 91 kg (200 lb). For heavier individuals, extrapolations can be made but with some loss in accuracy.

EFFECTS OF TERRAIN AND WALKING SURFACE

The influence of terrain and surface on the energy cost of walking is summarized in Table 10–2. Economy is similar for level walking on a grass track or on a paved surface. Walking in the sand, however, is almost **twice** as costly as walking on a hard surface, while walking in soft snow elevates the metabolic cost 3-fold compared to similar walking on a treadmill.[54] Certainly, a brisk walk along a beach or in freshly fallen snow would provide an excellent exercise stress in programs designed to "burn up" calories or improve physiologic fitness.

In another series of experiments,[50] it was clearly demonstrated that the energy cost of walking on a treadmill at 5.86 km per hour (3.6 mph) and 2.93 km per hour (1.8 mph) was no different from normal walking on a hard surface at the same speeds. This indicates that people can generate essentially the same exercise stress either by walking on the level or walking at the same speed and distance on an exercise treadmill. Such results also lend further support to the use of laboratory data to quantify human energy expenditure in "real life" situations.

TABLE 10–1.
PREDICTION OF ENERGY EXPENDITURE (kcal · min⁻¹) BASED ON SPEED OF LEVEL WALKING AND BODY MASS*

Speed		Body Mass						
mph	km · hr⁻¹	kg 36 / lbs 80	45 / 100	54 / 120	64 / 140	73 / 160	82 / 180	91 / 200
2.0	3.22	1.9	2.2	2.6	2.9	3.2	3.5	3.8
2.5	4.02	2.3	2.7	3.1	3.5	3.8	4.2	4.5
3.0	4.83	2.7	3.1	3.6	4.0	4.4	4.8	5.3
3.5	5.63	3.1	3.6	4.2	4.6	5.0	5.4	6.1
4.0	6.44	3.5	4.1	4.7	5.2	5.8	6.4	7.0

* Data from Passmore, R., and Durnin, J.V.G.A.: Human energy expenditure. Physiol. Rev. 35:801, 1955.

TABLE 10–2. EFFECT OF DIFFERENT TERRAIN ON THE ENERGY EXPENDITURE OF WALKING BETWEEN 5.2 AND 5.6 km · h⁻¹

Terrain	Correction Factor[a,b]
Paved road (similar to grass track)	0.0
Ploughed field	1.5
Hard snow	1.6
Sand dune	1.8

[a] First entry from Passmore, R., and Durnin, J.V.G.A.: Human energy expenditure. Physiol. Rev., 35:801, 1955. Last three entries from Givoni, B., and Goldman, R.F.: Predicting metabolic energy cost. J. Appl. Physiol., 30:429, 1971.
[b] The correction factor is a multiple of the energy expenditure for walking on a paved road or grass track. For example, the energy cost of walking in a ploughed field is 1.5 times that of walking on the paved road.

EFFECTS OF FOOTWEAR

It is considerably more costly to carry weight on the feet or ankles than to carry similar weight attached to the torso.[55] For example, for an increment in weight equal to 1.4% of body mass placed on the ankles, the energy cost of walking increased an average of 8% or nearly 6 times greater than if the same weight were carried on the torso.[30] In a practical sense, the energy cost of locomotion during walking and running is significantly increased by wearing boots compared to running shoes. Simply adding an additional 100 g to each shoe causes a 1% increase in oxygen uptake during moderate running. The implications of these findings for the design of running shoes, hiking and climbing boots, and work boots traditionally required in professions such as mining, forestry, and the military is clear—small changes in shoe weight produce large changes in the economy of locomotion. Furthermore, shoes with differing cushioning properties also affect the economy of movement with a softer-soled running shoe reducing the oxygen cost of running at moderate speed by about 2.4% compared to a similar shoe with a firmer cushioning system—even though the softer-soled shoe was 31 g heavier per pair.[22]

USE OF HAND-HELD AND ANKLE WEIGHTS

The impact force on the leg while running is equal to about 3 times body mass whereas the level of leg shock with walking is only about 30% of this value.

Walking

For many men and women the use of ankle weights increases the energy cost of walking to values usually observed for running.[43] This is beneficial to people who desire to use only walking as a relatively low-impact training modality, yet require intensities of effort higher than can be provided at normal walking speeds. Hand-held weights also increase the metabolic and physiologic cost of walking. There is some indication, however, that this procedure disproportionately elevates systolic blood pressure, perhaps due to the elevated intramuscular tension with gripping the weight. This could restrict the use of hand-held weights for individuals with existing hypertension or coronary heart disease.[23]

Running

Considering the relatively small increase in energy expenditure during hand- or ankle-weighted running, it would seem that simply increasing the unweighted running speed or distance would be a more desirable alternative if running is the preferred mode of exercise. This would certainly reduce the injury potential from increased impact force in the weighted condition and eliminate the added discomfort of carrying weights.[11]

COMPETITION WALKERS

The energy expenditure of five Olympic-caliber walkers has been studied at various speeds while walking and running on a treadmill.[41] In actual competition, the walking speed of these athletes averaged 13.0 kilometers per hour (11.5 to 14.8 km · h⁻¹ or 7.1 to 9.2 mph) over distances ranging from

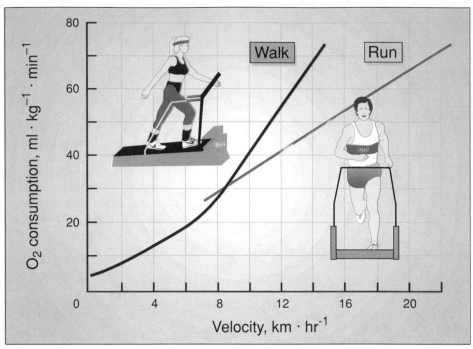

FIG. 10–2.
Relationship between oxygen consumption and velocity for walking and running on a treadmill in competition walkers. (From Menier, D.R., and Pugh, L.G.C.E.: The relation of oxygen intake and velocity of walking and running in competition walkers. J. Physiol., *197:*717, 1968.)

1.6 to 50 km. This was a relatively fast speed, because the winner of the 20-km walk at the 1988 Korea Olympics averaged a record 14.4 km per hour (8.9 mph) during this 12.4-mile walk. As illustrated in Figure 10–2, the break point in efficiency of locomotion between walking and running for these competitive race-walkers is between 8.0 and 9.0 km · hr.[-1] These data, plus biomechanical evidence,[7] support the contention that the crossover speed at which running becomes more economical than walking remains about the same for both conventional walking and competitive styles of walking (Fig. 10–3). In addition, for racewalkers the oxygen consumption during tread-mill walking at competition speeds was only slightly lower than the highest oxygen uptake measured for these athletes during treadmill running. Also, the relationship between oxygen consumption and walking at speeds above 8 km per hour (4.97 mph) was approximately linear, but the slope of the line was twice as steep compared to running at the same speeds. Although these athletes were able to walk at velocities up to 16 km per hour (9.94 mph) and attain oxygen uptakes as high as those achieved while running, **the economy of walking faster than 8 km per hour was one-half of that for running at similar speeds.** The attainment of similar values for max $\dot{V}O_2$ in both racewalking and running in the elite competitive walkers further supports the model for aerobic training **specificity** because in nontrained subjects aerobic capacity measured during a graded walking test is generally signif-icantly lower compared to that achieved with running.[24,39]

Competition walkers are able to achieve such high yet uneconomical rates of movement, unattainable with conventional walking, by use of a special

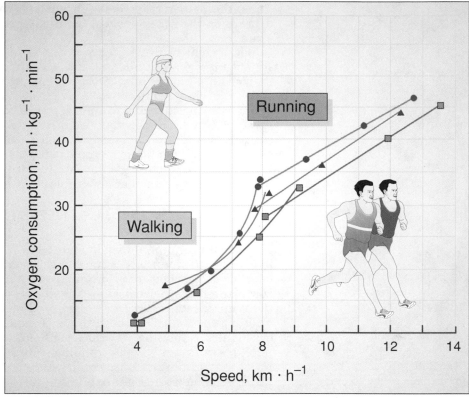

FIG. 10–3.

Relationship between oxygen uptake and speed of horizontal walking and running in men and women. Different symbols represent values from various studies reported in the literature. (From Falls, H.B., and Humphrey, L.D.: Energy cost of running and walking in young women. Med. Sci. Sports, 8:9, 1976.)

gait that involves a "rolling" of the hips. Among elite race walkers, variations in economy for walking contribute more to performance in this sport than previously observed among competitive runners.[24]

ENERGY EXPENDITURE DURING RUNNING

Running can be performed at various levels of intensity depending on terrain, weather, training goals, and the performer's fitness level. The energy expenditure for running has been quantified in two ways: (1) during performance of the actual activity, and (2) on a treadmill in the laboratory where running speed and grade can be precisely controlled. Jogging and running are essentially qualitative terms that relate to the speed at which movement is performed. This difference is determined largely by the relative aerobic energy demands required in raising and lowering the body's center of gravity and accelerating and decelerating the limbs during the run. At identical running speeds, a highly conditioned distance runner runs at a lower percentage of maximal aerobic capacity than an untrained runner, even though the oxygen consumption during the run may be similar for both people.

Thus, the demarcation between what is considered jogging and running depends on the fitness level of the participant; a jog for one person could be a run for another.

Independent of fitness, however, it is more economical from an energy standpoint to discontinue walking and to begin to jog or run at speeds greater than about 8 km per hour (5 mph). This is illustrated in Figure 10–3, which shows the relationship between oxygen consumption and horizontal walking and running for men and women at speeds ranging from 4 to 14 km per hour (2.5 to 8.7 mph). In the data depicted by the triangles,[35] the lines relating oxygen consumption and speed of walking and running intersect at a running speed of 8.5 km per hour (5.3 mph), whereas for the values from competition walkers shown by circles,[41] the "break point" between the economy of walking and running occurs at about 9 km per hour (5.6 mph).

THE ECONOMY OF RUNNING FAST OR SLOW

The data for running shown in Figure 10–3 illustrate an important principle in relation to running speed and energy expenditure. **Because the relationship between oxygen consumption and speed of running is linear, the total caloric cost of running a given distance at a steady-rate oxygen consumption is about the same whether the pace is fast or slow.** In simple terms, if one runs a mile at a speed of 10 miles per hour, it requires about twice as much energy per minute as running at 5 miles per hour; however, the runner finishes the mile in 6 minutes whereas running at the slower speed requires twice the time, or 12 minutes. Consequently, the energy cost of the mile is about the same. This holds true not only for horizontal running but also for running at inclines that range from −45 to +15%.[18,35] **For horizontal running, the net energy cost (that is, excluding the resting requirement) per kilogram of body mass per kilometer traveled is approximately 1 kcal, or 1 kcal · kg^{-1} · km^{-1}.**[35] Thus, for an individual who weighs 78 kg, the net energy requirement for running 1 km would be about 78 kcal, regardless of the running speed. Expressed in terms of oxygen consumption, this would amount to 15.6 liters of oxygen consumed per kilometer (1 liter O_2 = 5 kcal). In comparing the energy cost of locomotion per unit distance traveled, it is well documented that for both men and women it is more costly to **run** than to **walk** a given distance.[4,28]

ENERGY COST VALUES

Table 10–3 presents values for the **net** energy expended during running for 1 hour at various speeds. Running speeds are expressed as kilometers per hour, miles per hour, as well as the number of minutes required to complete one mile at a particular running speed. The boldface values are the net calories expended to run one mile for a given body mass; as mentioned before, this energy requirement is fairly constant and independent of running speed. **Thus, for a person who weighs 62 kg, running a 26-mile marathon requires about 2600 kcal whether the run is completed in just over 2 hours or 4 hours!**

For a heavier person, the energy cost per mile increases proportionately. This certainly supports the role of exercise as a caloric stress for the relatively unfit, overfat individual who may want to increase energy expenditure for purposes of weight control. For example, if a 102-kg person jogs 5 miles each day at any comfortable pace, 163 kcal are expended for each mile completed, or a total of 815 kcal for the 5-mile run. Increasing or decreasing the speed

TABLE 10–3.
NET ENERGY EXPENDITURE PER HOUR FOR HORIZONTAL RUNNING[a]

Body Mass kg	lb	km · hr⁻¹[b] mph min per mile kcal per mile	8 4.97 12:00	9 5.60 10:43	10 6.20 9:41	11 6.84 8:46	12 7.46 8:02	13 8.08 7:26	14 8.70 6:54	15 9.32 6:26	16 9.94 6:02
50	110	**80**	400	450	500	550	600	650	700	750	800
54	119	**86**	432	486	540	594	648	702	756	810	864
58	128	**93**	464	522	580	638	696	754	812	870	928
62	137	**99**	496	558	620	682	744	806	868	930	992
66	146	**106**	528	594	660	726	792	858	924	990	1056
70	154	**112**	560	630	700	770	840	910	980	1050	1120
74	163	**118**	592	666	740	814	888	962	1036	1110	1184
78	172	**125**	624	702	780	858	936	1014	1092	1170	1248
82	181	**131**	656	738	820	902	984	1066	1148	1230	1312
86	190	**138**	688	774	860	946	1032	1118	1204	1290	1376
90	199	**144**	720	810	900	990	1080	1170	1260	1350	1440
94	207	**150**	752	846	940	1034	1128	1222	1316	1410	1504
98	216	**157**	784	882	980	1078	1176	1274	1372	1470	1568
102	225	**163**	816	918	1020	1122	1224	1326	1428	1530	1632
106	234	**170**	848	954	1060	1166	1272	1378	1484	1590	1696

[a] The table is interpreted as follows: For a 50-kg person, the net energy expenditure for running for 1 hour at 8 km · hr⁻¹ or 4.97 mph is 400 kcal; this speed represents a 12-minute per mile pace. Thus, in 1 hour 5 miles would be run and 400 kcal would be expended. If the pace was increased to 12 km · hr⁻¹, 600 kcal would be expended during the hour of running.

[b] Running speeds are expressed as kilometers per hour (km · hr⁻¹), miles per hour (mph), and minutes required to complete each mile (min per mile). The values in boldface type are the net calories expended to run 1 mile for a given body mass, independent of running speed.

(within the broad range of steady-rate paces) simply alters the length of the exercise period; it has little effect on the **total energy expended.**

STRIDE LENGTH, STRIDE FREQUENCY, AND SPEED

Running speed can be increased in one of three ways: (1) by increasing the number of steps taken each minute (**stride frequency**), (2) by increasing the distance between steps (**stride length**), or (3) by increasing **both** the length and frequency of strides. Although the third option may seem the obvious means to increase running speed, several carefully conducted experiments have provided objective data concerning this question.

In 1944, the stride pattern was evaluated for the Danish champion in the 5- and 10-km running events.[5] At a running speed of 9.3 km per hour (5.8 mph), stride frequency for this athlete was 160 per minute with a corresponding stride length of 97 centimeters (cm) (38.2 in). When running speed was increased 91% to 17.8 km per hour (11.1 mph), stride frequency increased only 10% to 176 per minute, whereas an 83% increase was observed in stride length to 168 cm. Similarly, for another well-trained runner who performed at the same speeds, there was a 12 and 81% increase in stride frequency and stride length, respectively, as the faster speed was attained.

Figure 10–4A displays graphically the interaction between stride frequency and stride length as running speed increases. Doubling the running speed from 10 to 20 km per hour increases stride length by 85% whereas stride frequency only increases by about 9%. Increases in speed above 23 km per hour (14.3 mph) were achieved mainly by augmenting stride frequency. **Except at rapid speeds, running speed is increased mainly by lengthening the stride.**

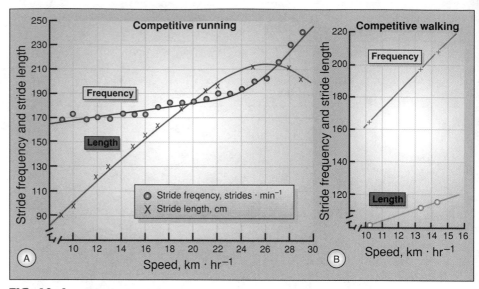

FIG. 10–4.
A, Stride frequency and stride length plotted as a function of speed. B, Data for an Olympic walker. (From Hogberg, P.: Length of stride, stride frequency, flight period and maximum distance between the feet during running with different speeds. Int. Z. Angew. Physiol., *14*:431, 1952.)

The competitive walker, on the other hand, does not increase speed in the same way as a runner. Figure 10–4*B* shows the stride length-stride frequency relationship for walking at speeds from 10 to 14.4 km per hour. The subject was an Olympic-medal-winner in the 10-km walk. When walking speed increased from 10 to 14.4 km per hour, the corresponding increase in frequency and length of stride was 27% and 13%, respectively. At faster speeds, there was an even greater increase in stride frequency because, unlike running, where the body glides through the air, competition walking requires that the back foot remain on the ground until the front foot makes contact. Thus, lengthening the stride becomes an ineffective means to increase speed in competitive walking. Owing to this standardization of style in competitive walking, additional energy must be expended to move the leg rapidly forward; this requires a corresponding involvement of the trunk and arm musculature and explains why it is more economical from an energy standpoint to run than walk at speeds greater than 8 or 9 km per hour (Fig. 10–3).

OPTIMUM STRIDE LENGTH

For running at a constant speed, there seems to be an optimum combination of stride length and frequency that is largely dependent on the person's mechanics or "style" of running, and this optimum cannot be determined accurately from an individual's body measurements.[9] Generally, however, it is more costly to overstride than to understride. Figure 10–5 shows values for oxygen consumption plotted in relation to different stride lengths that the runner altered during submaximal running at a relatively fast, constant speed of 14 km per hour (8.7 mph).

For this runner, a stride length of 135 cm was associated with an oxygen consumption of 3.35 liters per minute. Oxygen consumption increased 8%

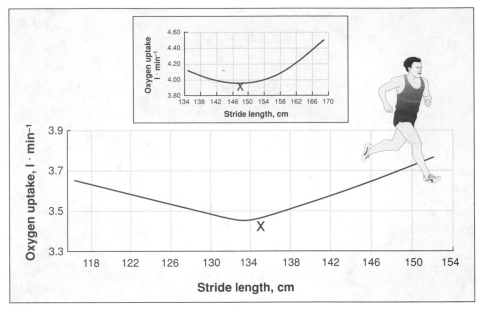

FIG. 10–5.
Oxygen consumption while running at 14 km per hour plotted as a function of different stride lengths. The insert curve is a plot of oxygen consumption at a speed of 16 km per hour at different stride lengths. (From Hogberg, P.: How do stride length and stride frequency influence the energy output during running? Int. Z. Angew. Physiol., *14*:437, 1952.)

when stride length was **shortened** to 118 cm; a 12% increase in oxygen consumption was noted when the distance between steps **lengthened** from 135 to 153 cm. The curve in the insert graph shows a similar pattern for oxygen consumption when running speed was increased to 16 km per hour at stride lengths that varied from 135 to 169 cm. Decreasing the length of the stride from this runner's optimum of 149 cm to 135 cm increased oxygen consumption by 4.1%. Aerobic requirements increased almost 13% by lengthening the stride to 169 cm. As might be expected, the most economical stride length for a particular running speed was the one selected by the runner (marked in the figure by an X). Lengthening the stride above that normally used by the runner produced a larger increase in oxygen consumption than when the stride was shortened below the optimum length. Thus, to urge a runner who shows signs of fatigue to "Lengthen your stride!" in order to maintain speed is actually counterproductive in terms of oxygen cost.

At least for well-trained runners, it would seem that the best procedure is to let them run at the stride length they have selected through years of practice;[25] this generally produces the most economical running performance blended to individual variations in body mass, inertia of limb segments, and anatomic development.[8] Consequently, **there is no "best" style characteristic of elite runners!** Biomechanical analysis may help the athlete correct minor irregularities in movement patterns while running;[46] this would certainly be of considerable practical importance to the competitive runner.

In two recent studies, human locomotion was evaluated in terms of how closely it followed the principles of a force-driven harmonic oscillator.[27a,b] A major tenet of this principle is that the minimum energy cost of walking

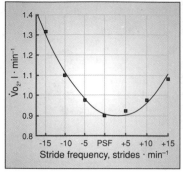

FIG. 10–6.
Oxygen uptake as a function of the preferred stride frequency (PSF). The oxygen uptake increases as the individual deviates from −15% to +15% of SC/F selected stride frequency. (Data from Holt K.G., et al.[27a,b].)

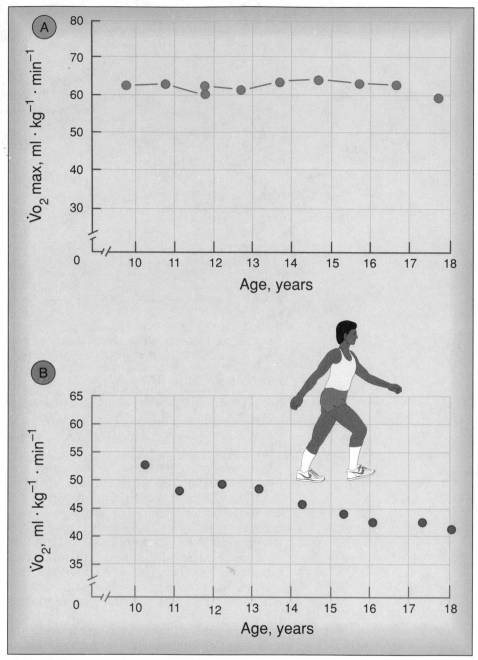

FIG. 10–7.
Effects of growth on max $\dot{V}O_2$ (A) and submaximal oxygen consumption in running at 202 m · min⁻¹ (B). (From Daniels, J., et al.: Differences and changes in $\dot{V}O_2$ among runners 10 to 18 years of age. Med Sci Sports, *10*:200, 1978.

occurs at the individual's natural stride frequency, and therefore, at the natural stride length. Figure 10–6 shows that the preferred stride frequency (PSF) and stride length are indeed associated with the minimum oxygen uptake; note that the oxygen uptake becomes progressively higher as the person deviates from −15% to +15% from the self selected PSF.

RUNNING ECONOMY: CHILDREN AND ADULTS, TRAINED AND UNTRAINED

Comparisons of running economy are usually made by determining the steady-rate oxygen consumption during running at various speeds on a motorized treadmill. In general, boys and girls are less economical in running compared to adults because they require between 20 to 30% more oxygen per unit of body mass to run at a particular speed.[1,17,31] These differences have been attributed to greater stride frequency among children, as well as differences in mechanics that could contribute to their inferior movement economy.[21,52] As shown in Figure 10–7B, running economy improves steadily throughout years 10 to 18. This, in part, explains the relatively poor performance of young children in distance running as well as the progressive improvement in endurance performance of children throughout adolescence, despite the fact that the aerobic capacity in relation to body mass (ml $O_2 \cdot kg^{-1} \cdot min^{-1}$; Fig. 10–7A) remains relatively constant during this time.[16,17]

At a particular speed, elite endurance runners generally run at an oxygen consumption that is lower than less trained or less successful counterparts of the same age. This has been shown for 8 to 11-year-old cross-country runners as well as adult marathoners.[14,34] In fact, distance athletes as a group

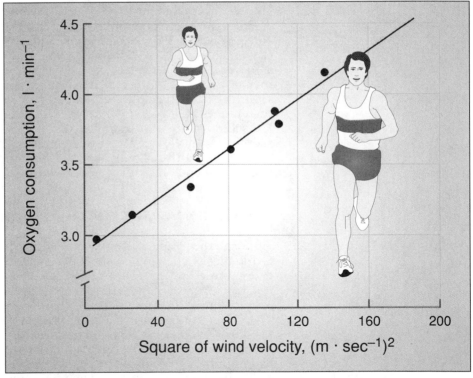

FIG. 10–8.
Oxygen consumption as a function of the square of wind velocity while running at 15.9 km per hr against various headwinds. (From Pugh, L.G.C.E.: Oxygen intake and treadmill running with observations of the effect of air resistance. J. Physiol., *207*:823, 1970.)

tend to run with between 5 to 10% more economy than well-trained middle-distance runners.[12]

AIR RESISTANCE

Anyone who has run into a head wind intuitively knows that more energy is expended trying to maintain a given pace compared with running in calm weather or with the wind at one's back. The magnitude of the effect of air resistance on the energy cost of running varies with three factors: (1) air density, (2) the runner's projected surface area, and (3) the square of the wind velocity. Depending on running speed, overcoming air resistance accounts for 3 to 9% of the total energy requirement of running in calm weather.[26,48] Running into a head wind creates an additional energy expense. As shown in Figure 10–8, the mean oxygen consumption while running at 15.9 km per hour in calm conditions was 2.92 liters per minute.[49] This increased by 5.5% to 3.09 liters per minute against a 16-km per hour "head wind" (9.9 mph) and 4.1 liters per minute while running against the strongest wind (41 mph); this represented a 41% additional expenditure of energy to maintain running velocity! This influence of headwind on the oxygen cost of running verifies the wisdom of some athletes who select to run in a more aerodynamically desirable position directly behind a competitor.

Some may argue that the negative effects of running into a headwind are counterbalanced on one's return with the tailwind. This is not the case, however, as the energy cost of cutting through a headwind is significantly greater than the reduction in exercise oxygen consumption observed with an equivalent wind velocity at one's back. Wind tunnel tests have shown that modification of clothing or even trimming one's hair can improve aerodynamics and reduce wind resistance effects by as much as 6%. This may result in a significant increase in running performance.[32]

At higher altitudes, wind velocity has less effect on energy expenditure than it does at sea level due to the reduced air density at higher elevations. For speed skaters, for example, the oxygen cost of skating at a particular speed is always lower at altitude compared to sea level.[2,10] Overcoming air resistance at altitude only becomes important at the faster skating speeds. In all likelihood, this altitude-effect would also be the case for running, cross-country skiing, and cycling.

TREADMILL VERSUS TRACK RUNNING

Although the treadmill is used almost exclusively to evaluate the physiology of running, a question exists concerning the validity of this procedure for determining the energetics of running and for relating this to performance on a track. For example, is the energy required to run a given speed or distance on a treadmill the same as that required to run on a track in calm weather? To answer this question, eight distance runners were studied on both a treadmill and track at three submaximal speeds of 180, 210, and 260 meters per minute (6.7, 7.8, and 9.7 mph), and during a graded exercise test to determine possible differences between treadmill and track running on maximal oxygen consumption.[40] The track runs were performed under environmental conditions similar to those in the laboratory. The results for one submaximal running speed and for maximal exercise are summarized in Table 10–4.

From a practical as well as a statistical standpoint, there were no measurable differences in the aerobic requirements of submaximal running (up

TABLE 10—4
COMPARISION OF AVERAGE METABOLIC RESPONSES DURING TREADMILL AND TRACK RUNNING*

Measurement	Treadmill	Track	Difference
Submaximal Exercise			
Oxygen consumption, $ml \cdot kg^{-1} \cdot min^{-1}$	42.2	42.7	+0.5
Respiratory exchange ratio	0.89	0.87	−0.02
Running speed, $m \cdot min^{-1}$	213.7	216.8	+3.1
Maximal Exercise			
Oxygen consumption, $l \cdot min^{-1}$	4.40	4.44	+0.04
$ml \cdot kg^{-1} \cdot min^{-1}$	66.9	66.3	−0.3
Ventilation, $l \cdot min^{-1}$, BTPS	142.5	146.5	+4.0
Respiratory exchange ratio	1.15	1.11	−0.04

* Adapted from McMiken, D.F., and Daniels, J.T.: Aerobic requirements and maximum aerobic power in treadmill and track running. Med. Sci. Sports, 8:14, 1976.

to 286 m per min) on the treadmill or track (either on level or up a grade), or between the maximal oxygen consumption measured in both forms of exercise under similar environmental conditions.[3,40] It is still possible, however, that at the faster running speeds achieved during endurance competition, the influence of air resistance becomes considerable and the oxygen cost of track running may become greater compared to "stationary" running on a treadmill at the same speed.

MARATHON RUNNING

The first place finisher in the men's division of the 1989 New York City Marathon set a course record of 2:08:1. The average speed of 4:53 minutes per mile over the 26.2 mile course is an outstanding achievement in terms of human performance. Not only does this pace require a steady-rate aerobic metabolism that exceeds the aerobic capacity of the average male college student, but it also represents about 75 to 90% of the marathoner's aerobic power that must be maintained for just over 2 hours! These athletes have an average aerobic capacity of approximately 4.4 liters per minute or 70 to 84 $ml \cdot kg^{-1} \cdot min^{-1}$.[15]

Two long-distance runners were measured during a marathon to determine the energy expenditure per minute and the total caloric cost of the run.[37] For these racers, oxygen consumption was determined every 3 miles by use of the balloon technique of open-circuit spirometry illustrated in Figure 8—4. Their marathon times were 2 h: 36 min: 34 s and 2 h: 39 min: 28 s; their maximal oxygen uptakes measured during treadmill running were 4.43 liters per minute (70.5 $ml \cdot kg^{-1} \cdot min^{-1}$) and 4.66 liters per minute (73.9 $ml \cdot kg^{-1} \cdot min^{-1}$), respectively. During the marathon, the first runner maintained an average speed of 270.5 meters (m) per minute (10.0 mph), which required an oxygen consumption equal to 80% of his maximal aerobic power. For the second runner, whose average speed was slower at 266.1 m per minute (9.92 mph), the aerobic energy requirement per minute averaged 78.3% of maximum. For both men, the energy requirement for running the marathon was between 2300 to 2400 kcal.

For distance runners who train about 100 miles per week, or slightly less than the distance of four marathons at close to competitive speeds, the weekly caloric expenditure from exercise is about 10,000 kcal. For the serious marathon runner who trains year-round, the total energy expended in training for 4 years prior to an Olympic competition would be close to two million calories! It is not surprising then that these superior athletes have such a low quantity of body fat (3 to 5% of body mass for men). As illustrated in Table 10–3, **the total expenditure of energy for a marathon run remains fairly constant for individuals of similar body size, regardless of running speed.** Obviously, persons with low aerobic capacities must maintain a slower running speed; thus, they will need more time to finish the marathon.

SWIMMING

Swimming exercise differs in several important respects from walking or running.[27] One obvious difference in swimming is that energy must be expended to maintain buoyancy and at the same time to generate horizontal movement by the use of the arms and legs, either in combination or separately. Other differences include the requirements for overcoming the **drag forces** that impede the movement of an object through a fluid. The amount of drag depends on the fluid medium and on the size, shape, and velocity of the object. All of these differences contribute to a total mechanical efficiency in front crawl swimming that ranges between only 5 and 9.5%.[56,57a] Within this framework, **the energy cost of swimming a given distance is about four times greater than running the same distance.**

METHODS OF MEASUREMENT

For short swims, such as 25 yards swum at different velocities, subjects need not breathe during the swim, and energy expenditure has been roughly estimated from oxygen consumption during 20 to 40 minutes of recovery. For swims of longer duration, including 12- to 14-hour endurance swims, energy expenditure has been computed from oxygen consumption measured by open-circuit spirometry (Fig. 8–4) during portions of the swimming performance. In studies conducted in the pool, the researcher walks alongside the swimmer and carries the portable gas collection equipment.[38] In another form of swimming exercise, illustrated in Figure 10–9A, the subject remains stationary, attached or "tethered" to a cable and pulley system by means of a belt worn around the waist.[33] The amount of weight attached to the cable can be increased periodically, thereby forcing the swimmer to exert more effort to keep from being pulled back. With the ergometer illustrated in Figure 10–9B, the swimmer is paced from a moving platform. A system of pulleys is attached to the platform, which in turn connects with the subject. The platform can be positioned either in front of or behind the swimmer; the force exerted by the swimmer must be sufficient to maintain the pulley at a given height.

Figure 10–10 shows a subject swimming in a flume or "swim-mill." Water is circulated and its velocity can vary from a slow swimming speed to a near-record pace for a free-style sprint. Also, water temperature in the 38,000-liter swim-mill can be varied from 10 to 40°C, and photographic analysis of

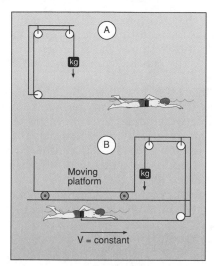

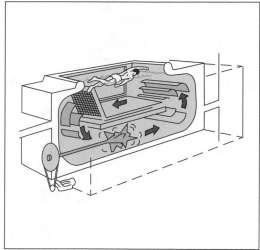

FIG. 10–9 (above left).
A, Apparatus used to measure energy expenditure during tethered swimming. B, Platform procedure to study swimming dynamics. The platform moves at a constant speed about an angular pool 60 meters in circumference. (From DiPrampero, P.E., et al.: Energetics of swimming in man. J. Physiol., 237:1, 1974.)

FIG. 10–10 (above right).
The swimming flume. (From Åstrand, P.O., and Englesson, S.: A swimming flume. J. Appl. Physiol., 33:514, 1972.)

stroke mechanics is possible through windows on the side of the flume beneath the water's surface. Values for aerobic capacity measured by either tethered, free, or flume swimming are essentially identical.[6] This means that either of these can be used to evaluate the functional capacity of the aerobic system during swimming.

ENERGY COST AND DRAG

Both the flume and platform apparatus have been particularly useful for quantifying the effects of total body drag during both tethered and "free" swimming.

The total drag force encountered by the swimmer consists of three components: (1) **Wave drag** is caused by waves that build up in front of and form hollows behind the swimmer as he or she moves through the water. This component of drag is not a significant factor when swimming at slow velocities, but its influence becomes greater at faster swimming speeds. (2) **Skin friction drag** is produced as the water slides over the surface of the skin. Even at relatively fast swimming velocities, the quantitative contribution of skin friction drag to the total drag is probably small. However, recent research supports the common practice of swimmers "shaving down" to reduce skin friction drag. Removal of body hair reduced drag, thus decreasing the energy cost and physiologic demands during swimming.[53a] (3) **Viscous pressure drag** contributes substantially to counter the propulsive efforts of the swimmer at slow velocities. It is caused by the separation of the thin sheet of water, or boundary layer, adjacent to the swimmer.[42] The pressure differential created in front of and behind the swimmer represents the viscous

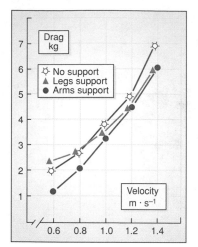

FIG. 10–11.
Water drag in three different
prone positions in relation to
towing velocity. (From
Holmér, I.: Energy cost of arm
stroke, leg kick, and the whole
stroke in competitive
swimming styles. Eur. J. Appl.
Physiol., 33.105, 1974.)

pressure drag. Its effect is probably reduced in highly skilled swimmers who have learned to "streamline" their stroke. Such streamlining with improved stroke mechanics reduces the separation region by moving the separation point closer to the trailing edge of the water. This is similar to what occurs when an oar slices through the water with the blade parallel rather than perpendicular to the flow of water. It has recently been shown that the wet suits worn by triathletes during the swim portion of the competition resulted in a 14% reduction in drag. This effect could largely explain the significantly faster swim times of athletes when using a wet suit.[58]

As shown in Figure 10–11, a curvilinear relationship exists between body drag and the velocity at which a swimmer is towed through the water. As velocity increased above 0.8 m per second, drag was reduced when the legs and arms were supported by flotation devices that placed the entire body in a horizontal position. In general, the drag force is about 2 to 2.5 times higher in swimming than in passive towing. These observations have been recently confirmed in studies of Olympic-caliber performers.[57]

ENERGY COST, SWIMMING VELOCITY, AND SKILL

Elite swimmers are able to swim a particular stroke at a given velocity at a lower oxygen consumption than are relatively untrained or recreational swimmers. This is illustrated in Figure 10–12A for the breaststroke, front crawl, and back crawl with subjects representing three levels of ability. One subject was a recreational swimmer who did not participate in swim training; the trained subject swam on a daily basis and was a top Swedish swimmer; the elite swimmer was a European champion. Except for the breaststroke, the elite swimmer was able to swim at a given speed with a lower oxygen uptake than his trained and untrained counterpart. Figure 10–12B shows that for the two trained athletes swimming at any particular speed, the breaststroke was the most costly; this was followed by the backstroke, with the front crawl being the least "expensive" of the three strokes studied.

EFFECTS OF WATER TEMPERATURE

Swimming in relatively cold water places the swimmer under thermal stress and brings about metabolic and cardiovascular adjustments that are different from those observed in warmer water.[27] These responses are geared primarily toward maintaining a relative consistency in core temperature because heat flow from the body is considerable, especially at water temperatures below 25° C (77° F). Heat loss is marked in lean subjects who benefit less from the insulation of subcutaneous fat.[45]

Figure 10–13 shows the oxygen consumption during breaststroke swimming at water temperatures of 18, 26, and 33° C. Regardless of swimming speed, the highest values for oxygen consumption occurred in cold water. This "extra" oxygen cost of swimming in cold water is due almost entirely to the energy expended in shivering as the body attempts to regulate core temperature. It appears that the optimal water temperature for competitive swimming for individuals of average body composition is 28 to 30° C (82 to 86° F). Within this temperature range, the metabolic heat generated in exercise is easily transferred to the water, yet the gradient for heat flow does not cause a significant increase in energy cost or change in core temperature due to cold stress.

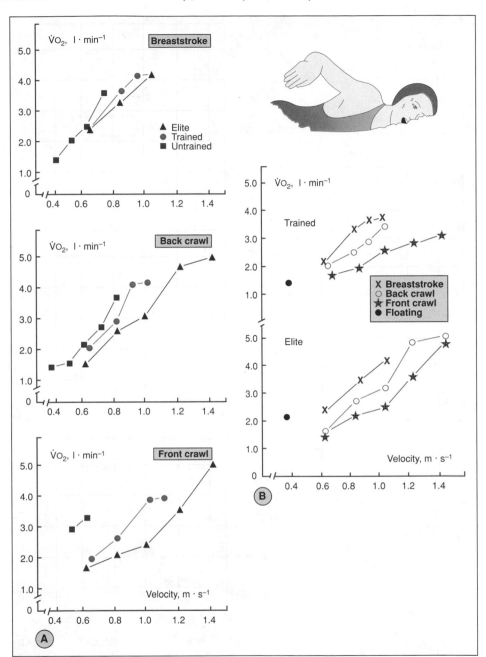

FIG. 10–12.

A, Oxygen consumption as a function of speed for the breaststroke, front crawl, and back crawl in subjects who represented three levels of skill ability. B, Oxygen consumption for two trained swimmers during three competitive strokes. (From Holmér, I.: Oxygen uptake during swimming in man. J. Appl. Physiol., 33:502, 1972.)

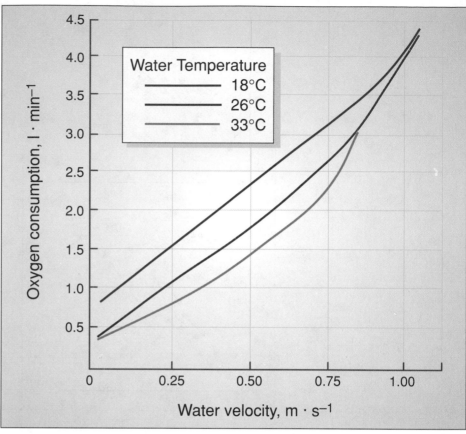

FIG. 10–13.
Energy expenditure for the breast-stroke at three water temperatures in relation to velocity. (From Nadel, E.R., et al.: J. Appl. Physiol., *36:*465, 1974.)

EFFECTS OF BUOYANCY: MEN VERSUS WOMEN

Women of all ages possess on the average significantly more total body fat than men. Because fat floats and muscle and bone sink in the water, the average woman therefore gains a hydrodynamic lift and floats more easily than her male counterpart. It is likely that this difference in body fat and thus in buoyancy is partly responsible for the greater economy for swimming observed in women.[51] For example, women can swim a given distance at about 30% lower energy cost than men; expressed another way, women can achieve a higher swimming velocity than men for the same level of energy expenditure.

It is also possible that the distribution of body fat in women[6a] is such that their legs float high in the water, making them more horizontal or "streamlined," whereas the leaner men's legs tend to swing down and float lower in the water. This lowering of the legs to a deeper position would increase body drag and thus reduce swimming economy.

Such differences in flotation may also help explain the "gender difference" in swimming economy and the fact that present swimming performances of women are more closely approaching those of men. The potential hydrodynamic benefits enjoyed by women become especially noteworthy in longer distances in which swimming economy and body insulation are important. In fact, the record for swimming the English Channel of 7 hours:40 minutes is presently held by a female compared to the men's record of 8 hours:12 minutes!

ENDURANCE SWIMMERS

Distance swimming in ocean water poses a severe metabolic, physiologic, and thermal challenge to the swimmer. In one investigation of nine distance swimmers,[49] data were secured both under race conditions and in a salt-water pool at swimming speeds ranging from 2.6 to 4.9 km per hour. During the race, the competitors maintained a constant stroke rate and pace until the last few hours when fatigue set in. From detailed observations on one male subject, the average speed of 2.85 km per hour during a 12-hour swim required an average oxygen consumption of approximately 1.7 liters per minute, or an equivalent energy expenditure of 8.5 kcal per minute. Consequently, the gross caloric requirement for the 12-hour swim was about 6120 kcal (8.5 kcal $\times$ 60 min $\times$ 12 h). The net caloric cost of swimming the English Channel, assuming a resting energy expenditure of 1.2 kcal per minute (0.260 l $O_2 \cdot min^{-1}$), was slightly in excess of 5200 kcal. This is about two times the calories consumed while running a marathon.

SUMMARY

1. Energy expenditure can be expressed in gross as well as net terms. Total or gross values include the resting energy requirement, whereas net energy expenditure is the energy cost of the activity per se excluding the resting value.
2. The relationship between walking speed and oxygen consumption is essentially linear. Walking surface also has an influence, because walking on sand requires about twice the energy expenditure as walking on hard surfaces. The energy cost is proportionally larger for heavier people.
3. It is more economical, from an energy standpoint, to jog-run rather than to walk at speeds greater than 8 km $\cdot$ h^{-1}. The difference between jogging and running depends on the fitness level of the participant; a jog for one person may be a run for another.
4. For many people, hand-held and ankle weights can increase the energy cost of walking to values usually observed for running. This would benefit those desiring to use only walking as a low-impact form of exercise training.
5. The total caloric cost for running a given distance is about the same whether the pace is fast or slow. For horizontal running, the net energy expenditure is about 1 kcal $\cdot$ kg$^{-1} \cdot$ km^{-1}.
6. It generally costs less energy to shorten running stride and to increase the number of steps in order to maintain a constant running speed rather than to lengthen the stride and reduce stride frequency. An individual will usually "select" the combination of length and frequency that produces optimal economy.
7. Overcoming air resistance accounts for 3 to 9% of the cost of running in calm weather. This percentage increases considerably if a runner attempts to maintain pace while running into a head wind.
8. Children generally require significantly more oxygen to transport their

body mass while running compared to adults. This relatively lower running economy accounts for the poor endurance performance of children compared to adults with similar aerobic capacity.

9. The energy required to run a given distance or speed on a treadmill is about the same as that required to run on a track under identical weather conditions.

10. The energy expended swimming a given distance is about four times greater than that expended running the same distance. This is because the swimmer must expend considerable energy to maintain buoyancy and overcome the drag forces that impede movement.

11. There are significant differences between men and women for active body drag, mechanical efficiency, and net oxygen consumption during swimming. Women swim a given distance at about 30% lower energy cost than men.

12. Elite swimmers expend fewer calories to swim a given stroke at any velocity. The optimal water temperature for most competitive swimming is 28 to 30° C (82 to 86° F).

REFERENCES

1. Åstrand, P.O.: *Experimental Studies of Physical Working Capacity in Relation to Sex and Age.* Copenhagen, Munksgaard, 1952.
2. Åstrand, P.O., and Rodahl, K.: *Textbook of Work Physiology.* 3rd ed. New York, McGraw-Hill, 1986.
3. Bassett, D.R., et al.: Aerobic requirements of overground versus treadmill running. *Med. Sci. Sports Exerc.,* 17:477, 1985.
4. Bhambhani, Y., and Singh, M.: Metabolic and cinematographic analysis of walking and running in men and women. *Med. Sci. Sports Exerc.,* 17:131, 1985.
5. Bøje, O.: Energy production, pulmonary ventilation, and length of steps in well-trained runners working on a treadmill. *Acta Physiol. Scand.,* 7:362, 1944.
6. Bonen, A., et al.: Maximal oxygen uptake during free, tethered, and flume swimming. *J. Appl. Physiol.,* 48:232, 1980.
6a. Campaigne, B.N.: Body fat distribution in females: metabolic consequences and implications for weight loss. *Med. Sci. Sports Exerc.,* 22:291, 1990.
7. Cavagna, G.A., and Franzetti, P.: Mechanics of competition walking. *J. Physiol.* (London), 315:243, 1981.
8. Cavanagh, P.R., and Kram, R.: Mechanical and muscular factors affecting the efficiency of human movement. *Med. Sci. Sports Exerc.,* 17:326, 1985.
9. Cavanagh, P.R., and Kram, R.: Stride length in distance running: velocity, body dimensions, and added mass effects. *Med. Sci. Sports Exerc.,* 21:467, 1989.
10. Cerretelli, P.: Limiting factors to oxygen transport on Mount Everest. *J. Appl. Physiol.,* 40:658, 1976.
11. Claremont, A.P., and Hall, S.J.: Effects of extremity loading on energy expenditure and running mechanics. *Med. Sci. Sports Exerc.,* 20:161, 1988.
12. Conley, D.L., and Krahenbuhl, G.S.: Running economy and distance running performance of highly trained athletes. *Med. Sci. Sports Exerc.,* 12:357, 1980.
13. Conley, D. L., et al.: Training for aerobic capacity and running economy. *Phys. Sportsmed.,* 9:107, 1981.
14. Costill, D.L.: Physiology of marathon running. *JAMA,* 221:1024, 1972.
15. Costill, D.L., and Fox, E.L.: Energetics of marathon running. *Med. Sci. Sports,* 1:81, 1969.
16. Cunningham, D.A.: Development of cardiorespiratory function in circumpubertal boys: a longitudinal study. *J. Appl. Physiol.,* 56:302, 1984.
17. Daniels, J., et al.: Differences and changes in $\dot{V}O_2$ among runners 10 to 18 years of age. *Med. Sci. Sports,* 10:200, 1978.
18. Davies, C.T.M., et al.: The physiological response to running downhill. *Eur. J. Appl. Physiol.,* 32:187, 1974.

19. Falls, H.B., and Humphrey, L.D.: Energy cost of running and walking in young women. *Med. Sci. Sports*, 8:9, 1976.
20. Fellingham, G.W., et al.: Calorie cost of walking and running. *Med. Sci. Sports*, 10:132, 1978.
21. Fortney, V.L.: The kinematics of the running pattern of two, four, and six year old children. *Res. Q. Exerc. Sport*, 54:126, 1983.
22. Frederick, E.C., et al.: Lower oxygen demands of running in soft-soled shoes. *Res. Q. Exerc. Sport*, 57:174, 1986.
23. Graves, J.E., et al.: The effect of hand-held weights on the physiological responses to walking exercise. *Med. Sci. Sports Exerc.*, 19:260, 1987.
24. Hagberg, J.M., and Coyle, E.F.: Physiological determinants of endurance performance as studied in competitive race walkers. *Med. Sci. Sports Exerc.*, 15:287, 1983.
25. Heinert, L.D., et al.: Effect of stride length variation on oxygen uptake during level and positive grade treadmill running. *Res. Q. Exerc. Sport*, 59:127, 1988.
26. Hill, A.V.: The air resistance to a runner. *Proc. R. Soc. Lond. (Biol.)*, 102:380, 1927.
27. Holmér, I.: Physiology of swimming man. *Exerc. Sport Sci. Rev.* Vol. 7. Edited by R.S. Hutton and D.I. Miller. Philadelphia, Franklin Institute Press, 1980.
27a. Holt, K.G., et al.: The force driven harmonic oscillator as a model for human location. *Hum. Move. Sciences*, in press.
27b. Holt, K.G., et al.: Predicting the minimal energy costs of human walking. *Med. Sci. Sports Exerc.*, in press.
28. Howley, E.T., and Glover M.E.: The caloric cost of running and walking one mile for men and women. *Med. Sci. Sports*, 6:235, 1974.
29. Jankowski, L., et al.: Accuracy of methods for estimating O_2 cost of walking in coronary patients. *J. Appl. Physiol.*, 33:672, 1972.
30. Jones, B.H., et al.: Energy cost of walking and running in boots and shoes. *Ergonomics*, 27:895, 1984.
31. Krahenbuhl, G.S., and Pangrasi, R.: Characteristics associated with running performance in young boys. *Med. Sci. Sports Exerc.*, 15:488, 1983.
32. Kyle, C.R., and Caiozzo, V.J.: The effect of athletic clothing aerodynamics upon running speed. *Med. Sci. Sports Exerc.*, 18:509, 1986.
33. Magel, J.R., et al.: The specificity of swim training on maximum oxygen uptake. *J. Appl. Physiol.*, 36:753, 1974.
34. Mayers, N., and Gutin, B.: Physiological characteristics of elite prepubertal cross-country runners. *Med. Sci. Sports*, 11:172, 1979.
35. Margaria, R., et al.: Energy cost of running. *J. Appl. Physiol.*, 18:367, 1963.
36. Margaria, R.: *Biomechanics and Energetics of Muscular Exercise.* Oxford, Clarendon Press, 1976.
37. Maron, M., et al.: Oxygen uptake measurements during competitive marathon running. *J. Appl. Physiol.*, 40:836, 1976.
38. McArdle, W.D., et al.: Metabolic and cardiorespiratory response during free swimming and treadmill walking. J. Appl. Physiol., 30:733, 1971.
39. McArdle, W.D., et al.: Comparison of continuous and discontinuous treadmill and bicycle tests for $\dot{V}O_2$ max. *Med. Sci. Sports*, 5:156, 1973.
40. McMiken, D.F., and Daniels, J.T.: Aerobic requirements and maximum aerobic power in treadmill and track running. *Med. Sci. Sports*, 8:14, 1976.
41. Menier, D.R., and Pugh, L.G.C.E.: The relation of oxygen intake and velocity of walking and running in competition walkers. *J. Physiol.* (London), 197:717, 1968.
42. Miller, D.L.: Biomechanics of swimming. In *Exercise and Sport Sciences Reviews.* Vol. 3. Edited by J.H. Wilmore and J. Keogh. New York, Academic Press, 1975.
43. Miller, J.E., and Stamford, B.A.: Intensity and energy cost of weighted walking vs. running for men and women. *J. Appl. Physiol.*, 62:1947, 1987.
44. Morgan, D.W., et al.: Ten kilometer performance and predicted velocity at $\dot{V}O_2$ max among well-trained male runners. *Med. Sci. Sports Exerc.*, 21:78, 1989.
45. Nadel, E., et al.: Energy exchanges of swimming man. J. Appl. Physiol., 36:465, 1974.
46. Nelson, R.C., and Gregor, R.J.: Biomechanics of distance running: a longitudinal study. *Res. Q.*, 47:471, 1976.
47. Pandolf, K.B., et al.: Predicting energy expenditure with loads while standing or walking very slowly. *J. Appl. Physiol.*, 43:577, 1977.

48. Pugh, L.G.C.E.: Oxygen uptake in track and treadmill running with observations on the effect of air resistance. *J. Physiol.* (London), 207:823, 1970.
49. Pugh, L.G.C.E., and Edholm, O.G.: The physiology of channel swimmers. *Lancet*, 2:761, 1955.
50. Ralston, H.J.: Comparison of energy expenditure during treadmill walking and floor walking. *J. Appl. Physiol.*, 15:1156, 1960.
51. Rennie, D.W., et al.: Energetics of swimming in man. In *Swimming II.* Edited by L. Lewille and J. Clarys. Baltimore, University Park Press, 1975.
52. Rowland, T.W., et al.: Physiologic responses to treadmill running in adult and prepubertal males. *Int. J. Sports Med.*, 8:292, 1987.
53. Rowland, T.W., and Green, G.M.: Physiological responses to treadmill exercise in females: adult-child differences. *Med. Sci. Sports Exerc.*, 20:474, 1988.
53a. Sharp, R.L., and Costill, D.L.: Influence of body hair removal on physiological responses during breaststroke swimming. *Med. Sci. Sports Exerc.*, 21:576, 1989.
54. Smolander, J., et al.: Cardiorespiratory strain during walking in snow with boots of differing weights. *Ergonomics*, 32:319, 1989.
55. Soule, R., and Goldman, R.: Energy cost of loads carried on the head, hands, or feet. *J. Appl. Physiol.*, 27:5, 1969.
56. Toussaint, H.M., et al.: Propelling efficiency of front crawl swimming. *J. Appl. Physiol.*, 65:2056, 1988.
57. Toussaint, H.M., et al.: Active drag related to velocity in male and female swimmers. *J. Biomechanics*, 21:435, 1988.
57a. Toussaint, H.M. et al.: The mechanical efficiency of front crawl swimming. *Med. Sci. Sports Exerc.*, 22:402, 1990.
58. Toussaint, H.M., et al.: Effect of triathlon wet suit on drag during swimming. *Med. Sci. Sports Exerc.*, 21:325, 1989.
59. Williams, K.R., and Cavanagh, P.R.: Relationship between distance running mechanics, running economy, and performance. *J. Appl. Physiol.*, 63:1236, 1987.

INDIVIDUAL DIFFERENCES AND MEASUREMENT OF ENERGY CAPACITIES

We all possess the capability for anaerobic and aerobic energy metabolism, although the capacity for each form of energy transfer varies considerably among individuals. This between-individual variability underlies the concept of **individual differences** in metabolic capacity for exercise. It also appears that a person's capacity for energy transfer (and for many physiologic functions, for that matter) is not simply a general factor, but is highly dependent on the form of exercise with which it is trained and evaluated.[63,64,71,78] A high maximum oxygen uptake in running, for example, does not necessarily assure a similar metabolic power when different muscle groups are activated, as in swimming and rowing. This is an example of **specificity** of metabolic capacity. However, individuals with a high aerobic power in one activity often possess an above average aerobic power in other activities in relation to other individuals. This is an illustration of the **generality** of physiologic function and performance. Figure 11–1 illustrates the **specificity-generality** concept with respect to energy capacities. The nonoverlapped areas represent specificity of physiologic function, and the overlapped areas represent generality of function. For each of the energy systems there is more specificity than generality; it is rare to find individuals who possess tremendous capabilities in markedly different types of activities (e.g., sprinting and long-distance running).

Because of specificity, training for high aerobic power probably contributes little to one's capacity to generate anaerobic energy, and vice versa. The effects of systematic training are highly specific in terms of neurologic, physiologic, and metabolic demands. Terms such as "speed," "power," and "endurance" must therefore be defined carefully within the context of the specific movement patterns and the specific metabolic and physiologic requirements of an activity.

In this chapter, the capacity of the various energy transfer systems discussed in Chapter 6 and 7 are evaluated, with special reference to **measurement, specificity,** and **individual differences.**

OVERVIEW OF ENERGY TRANSFER CAPACITY DURING EXERCISE

All-out exercise for up to 2 minutes duration is powered mainly by the **immediate** and **short-term** energy systems. Both systems operate anaerobically because their transfer of chemical energy does not require molecular oxygen. Generally, there is greater reliance on anaerobic energy for fast movements or when there is resistance to movement at a given speed. This principle is shown in Figure 11–2 which illustrates the relative involvement

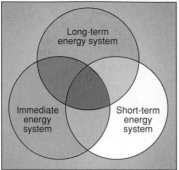

FIG. 11–1.
Illustration of the Specificity-
Generality concept with
respect to the three energy
systems. If only two systems
are considered, their overlap
represents generality and the
remainder denotes specificity.

of the anaerobic and aerobic energy transfer systems for different durations of **all-out exercise.** At the initiation of movement performed at high or low speed, the stored phosphates, ATP and CP, provide immediate and nonaerobic energy for muscle contraction. After the first few seconds of movement, an increasingly greater proportion of energy is generated by the glycolytic energy system. For exercise to continue, a greater demand is then placed on the aerobic metabolic pathways for purposes of ATP resynthesis. All activities and sports performances can be classified on the anaerobic-to-aerobic continuum. Some activities require the capacity of more than one energy system, whereas others rely predominately on a single system of energy transfer. All activities use some percentage from each of the energy systems, however, depending on intensity and duration. Of course the higher the intensity and shorter the time, the greater the demand on nonaerobic energy transfer.

ANAEROBIC ENERGY: THE IMMEDIATE AND SHORT-TERM ENERGY SYSTEMS

PERFORMANCE EVALUATION OF THE IMMEDIATE (ATP-CP) ENERGY SYSTEM

Performance tests that cause maximal activation of the ATP-CP energy system have been developed to provide practical "field tests" to evaluate the capacity of this immediate means for energy transfer.[45,46,83] These tests

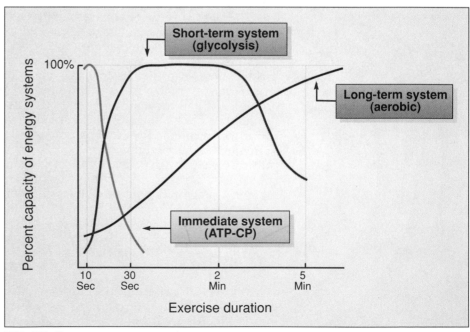

FIG. 11–2.
The various energy systems and their involvement during all-out exercise of different durations.

are generally referred to as **power tests;** power in this context is defined as the time-rate of doing work, or work accomplished per unit time. The formula for power output is:

$$P = \frac{F \times D}{T}$$

where F is the force generated, D is the distance through which the force is moved, and T is the duration of the work period. Power can be expressed in **watts,** which is equal to 0.73756 ft-lb $\cdot$ sec^{-1}, 0.01433 kcal $\cdot$ min^{-1}, 1.341×10^{-3} hp (or 0.0013 hp), and 6.12 kg-m $\cdot$ min^{-1}. Assumptions underlying performance estimates of the ATP-CP system of energy delivery include: (1) at maximal power output all ATP is regenerated by the ATP-CP system and accordingly, all power generated during this time can be attributed to this system, and (2) there is enough ATP and CP to support maximal performance for approximately 6 to 8 seconds.

Stair-Sprinting Power Tests

Researchers have proposed that muscular power can be measured by sprinting up a flight of stairs.[44,67] As illustrated in Figure 11–3, the subject runs up a staircase as fast as possible taking three steps at a time. The external work done in this test is the total vertical distance the body is lifted up the stairs; this distance for six stairs is usually about 1.05 m.

The power output of a 65-kg woman who traverses six steps in 0.52 seconds is computed as follows:

$$F = 65 \text{ kg}$$
$$D = 1.05 \text{ m}$$
$$T = 0.52 \text{ s}$$
$$\text{Power} = \frac{65 \text{ kg} \times 1.05 \text{ m}}{0.52 \text{ s}}$$
$$\text{Power} = 131.3 \text{ kgm per s (1287 watts)}$$

Because the power score in the stair-sprinting test is influenced significantly by the person's body mass, the heavier person will necessarily have a higher power score if several individuals achieve the same speed. This implies that the heavier person has a more highly developed immediate energy system than a lighter person who may cover the same vertical distance in the same time. Because there is no direct evidence to support this contention, caution is urged in interpreting differences in stair-sprinting power scores and making inferences about individual differences in ATP-CP energy capacity. **The test may be better suited for evaluating individuals of similar body mass, or the same people before and after a specific training regimen designed to develop rapid nonaerobic leg power.**

Jumping-Power Tests

For years, jump tests such as the **Sargent jump-and-reach test** or a **standing broad jump** have been common elements in many physical fitness test batteries.[84] The Sargent jump is scored as the difference between a person's standing reach and the maximum jump-and-touch height. In the case of the

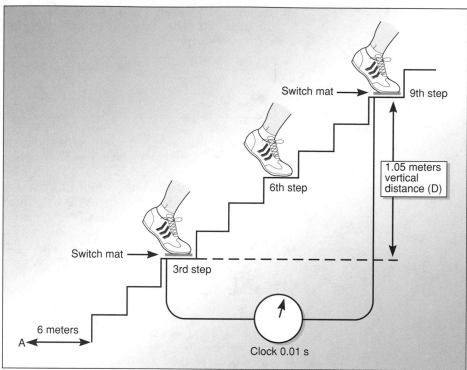

FIG. 11–3.
Stair-stepping power test.[44] The subject begins at point A and runs as rapidly as possible up a flight of stairs, taking three steps at a time. The time to cover the distance between stair 3 and stair 9 is recorded to the nearest 0.01 second by the use of switch mats placed on the steps. Power output is the product of the subject's mass (F) and vertical distance (D), divided by the time (T). (From Mathews, D.K., and Fox, E.L.: The Physiological Basis of Physical Education and Athletics. Philadelphia, W.B. Saunders, 2nd Edition, 1976.)

broad jump, the score is the horizontal distance covered in a leap from a semicrouched position. Although both tests purport to measure leg power, they probably fail to achieve this goal. For one thing, with the jump tests, power generated in propelling the body from the crouched position occurs only in the time the feet are in contact with the surface. **It is doubtful whether this brief period is sufficient to evaluate a person's non-aerobic ATP and CP capacity.** Also, no relationship has been established between jump-test scores and ATP-CP levels or depletion patterns.

Recent experiments in one of our laboratories have evaluated the vertical jump maneuver to assess different patterns of muscular force production in the eccentric and concentric phases of vertical jumping, including the relationship of the force curve "signature" to selected measures of anaerobic performance. Figure 11–4 displays the force curve over time for the standing vertical jump from measurements made on a force platform. The subject, with hands on hips, crouches to a knee angle of 110 to 130 degrees and then forcibly extends upward, attempting to attain the highest vertical distance possible. The curve in the top panel is a typical tracing of the force pattern as the subject goes from a standing position to a crouch until the upward thrust that is generated by concentric contractions of the quadriceps. The

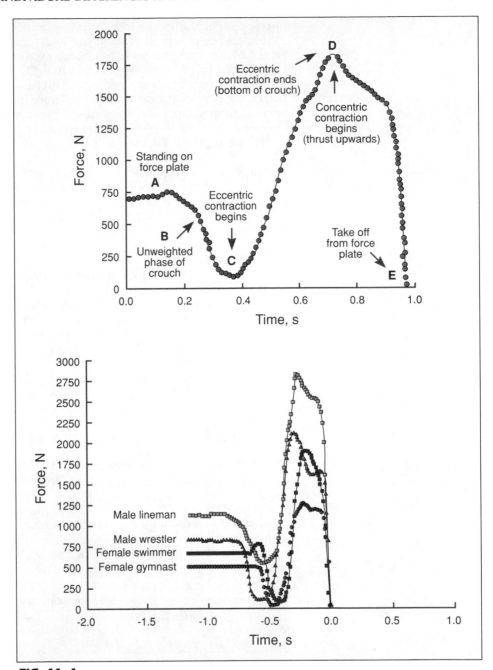

FIG. 11–4.

Force-time curve signature during the standing vertical jump. Top panel: typical force curve. The subject stands on the force platform (A) and then begins to crouch (B). During the first phase of the crouch, the force decreases because the initial downward movement is unweighted. When the movement begins to slow down (C), the muscular action becomes eccentric. This continues to point D, where the subject is at the bottom of the crouch (no velocity of movement) just prior to the upward thrust. From this point on, the contractions are concentric until take-off from the force plate (E). Bottom panel: force-time curve for two male and two female athletes who differed markedly in overall body size. (Unpublished data courtesy of R. Hintermeister, Muscle Dynamics Laboratory, Department of Exercise Science, University of Massachusetts, Amherst.)

bottom panel shows the force curve signature for two male and two female athletes, each of whom differed markedly in prior training history and body size. Current experiments are underway to analyze the components of the force curve to better explain individual differences in the signature of the jumping pattern that may relate to athletic capability. The rate and specific pattern of developing force might be more important factors in performance success than simply the maximum height attained in the jump.

Other Power Tests

As indicated in Figure 11–2, any performance involving all-out exercise of 6 to 8 seconds duration can probably be considered indicative of the person's capacity for immediate power from the high-energy phosphates in the specific muscles activated.[22] Other examples of such tests are sprint running or cycling, shuttle runs, or even certain more localized movements such as arm cranking.

Inter-Relationships Among Power Tests

If the various power tests measure the same metabolic capacity, then individuals ranking high on one test should rank correspondingly high on a second or third test. Although information on this topic is incomplete, the available data indicate that those who do well on one power performance test tend to do well on another, but the correlation is generally not strong. Table 11–1 shows the interrelationship (expressed statistically as a correlation coefficient) between several tests that supposedly measure immediate power output. The relationship ranges from poor to good, indicating some commonality between tests and suggesting that each may be measuring a similar metabolic quality. Of practical significance is the fairly strong relationship between scores on the stair-sprinting power test and the 40-yard dash. **Clearly, almost the same information can be obtained by sprint running on a track compared with the more elaborate set-up in the stair sprint.**

Several factors may explain why the interrelationship between the other test scores is not high. For one thing, **human performance is highly task-specific.** From a metabolic and performance standpoint, this means the best sprint runner is not necessarily the best sprint swimmer, sprint cyclist, "stair sprinter," or "arm cranker." Even though the energy to power each performance is generated by the same metabolic reactions, these reactions are

TABLE 11–1.
CORRELATIONS AMONG TESTS THAT ARE SUPPOSED TO MEASURE IMMEDIATE NONAEROBIC POWER OUTPUT[a]

Variables (n = 31 males)	40-Yard Dash	Sargent Jump Test	Power Bicycle Test
1. Stair-Sprinting power test	−.88[b]	.56	.69
2. 40-yard dash	—	−.48[b]	−.62[b]
3. Sargent jump test	—	—	.31

[a] From the Applied Physiology Laboratory, University of Michigan.
[b] A negative correlation coefficient means that for the group of individuals, a high score earned on one test is associated with a low score on the other test. For the correlations with the 40-yard dash, a negative correlation means a good performance on one test is associated with a low 40-yard run time, and a low score in running (time) is a good performance.,

isolated within the specific muscles activated by the exercise. It is also important to emphasize that each specific test requires different neurologic or skill components that tend to cause the scores to be more variable.

We have suggested that power tests might be used to show changes in an athlete's performance resulting from specific training. Such tests also offer an excellent means for self-testing and motivation and often provide the actual exercise for training the immediate energy system. With many football teams, for example, the 40-yard dash is often used as a criterion to evaluate a player's **speed.** Although there are many types of "speed" that need to be evaluated, these test scores may provide some information for the evaluation of a player, even though it has yet to be established that 40-yard speed in a straight line is related to overall football ability for players at similar positions! A run test of shorter duration (up to 20 yards) or one with frequent changes in direction, may turn out to be an equal or more suitable performance measure of overall ability.

Physiologic Evaluation of the ATP-CP Energy System

Besides performance estimates of the capacity of the immediate energy system, several physiologic and biochemical measures have been used. These include (1) estimating the size of the ATP and CP pools, (2) their depletion rates in response to all-out exercise of short duration, (3) the oxygen deficit calculated from the oxygen uptake curve (4), and (5) the alactic portion of the recovery oxygen uptake curve.[53,106] Of these measures, the depletion rates of ATP and CP appear to be the most direct and correlate the highest with performance estimates of the immediate energy system. It should be realized that it is very difficult to obtain precise quantitative physiologic data during all-out, short-duration exercise. Consequently, we must rely heavily on the apparent validity of the performance measures as true physiologic markers of ATP-CP energy production.

SHORT-TERM ENERGY SYSTEM OF GLYCOLYSIS

As depicted in Figure 11–2, when all-out exercise continues longer than a few seconds, increasingly more energy for ATP resynthesis is generated from the **short-term energy system** through the anaerobic reactions of glycolysis. This is not to say that aerobic metabolism is unimportant at this stage of exercise or that the oxygen-consuming reactions have not been "switched-on." To the contrary, Figure 11–2 shows an increase in the contribution of aerobic energy very early in exercise. But it is in all-out exercise that the energy requirement significantly exceeds the energy generated by the oxidation of hydrogen in the respiratory chain. As a result, the anaerobic reactions of glycolysis predominate and large quantities of lactic acid accumulate within the active muscle and ultimately in the blood.

Unlike tests for maximal oxygen consumption, no specific criteria exist to indicate that a person has reached a maximal anaerobic effort. In fact, it is highly likely that one's score on such a test is influenced by the level of self motivation as well as the test environment.[101] **The level of blood lactate is the most common indicator of the activation of the short-term energy system.** If the test is given in a controlled environment under standardized conditions, scores for anaerobic power appear to reach acceptable levels for test reliability.[24]

Performance Tests for Glycolytic Power

Performances requiring substantial activation of the short-term energy system are those demanding maximal work for up to 3 minutes.[5,46] All-out runs and cycling exercise have usually been used to test this energy capacity, although weight lifting (repetitive lifting of a certain percentage of maximum) and shuttle-runs have also been used. Because of the effects that factors such as age, skill, motivation, and body size have on performance, it is difficult

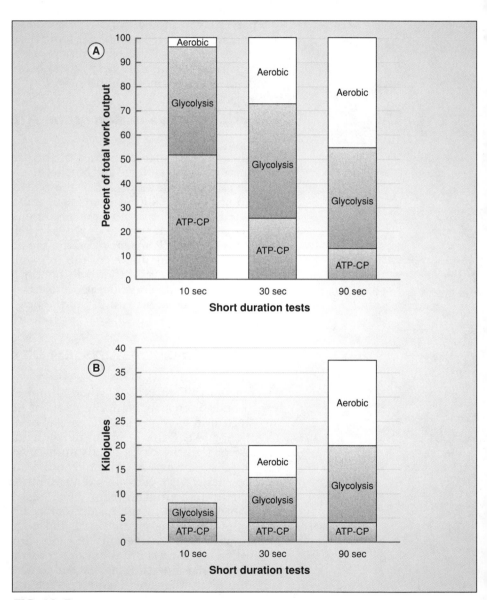

FIG. 11–5.
Relative contribution of each energy system to work output in three short tests as (A) percent of total work output and (B) kilojoules of energy. Test results based on protocol of Katch Test (see text, p. 207). (Data from the Applied Physiology Laboratory, University of Michigan.)

to select a suitable criterion test and to develop appropriate norms to evaluate the glycolytic energy system. Also, within the framework of exercise specificity, short-term anaerobic capacity for an arm and upper body activity like rowing or swimming cannot be adequately assessed with a test that makes maximum use of the leg muscles. The performance test must be similar to the activity for which the energy capacity is being evaluated. In most cases the actual activity can serve as the test.

An all-out cycling test of short duration to estimate the power and capacity of the anaerobic energy systems was first described in 1973 as the **Katch test.**[53] This work was extended in subsequent years[54] and resulted in a test in which the frictional resistance against the flywheel was preset at a high load (6 kp for men and 5 kp for women), and subjects attempted to turn as many revolutions as possible in 40 seconds. Revolution rate was recorded continuously, with the peak power achieved representing **anaerobic power** and the total work done representing **anaerobic capacity.** A later modification of this procedure, the **Wingate test,** involves 30 seconds of all-out supermaximal exercise performed on either an arm-crank or leg-cycle ergometer.[5,43] The resistance to exercising is based on body mass (0.075 kg per kg body mass) and is applied after initial inertia and unloaded frictional resistance are overcome. The assessment of **peak power** output represents the highest mechanical power generated during any 3- to 5-second period of the test; **average power** output is the arithmetic average of the total power generated during the 30-second test period. Rate of **fatigue,** or the rate of decline in power relative to the peak value, can also be computed.[5] As in the Katch test, the underlying assumption for this testing is that the value for peak power output represents the energy-generating capacity of the nonaerobic high-energy phosphates whereas the average power score is determined by one's glycolytic capacity.[42] Some of the highest reported all-out cycle ergometer power scores have been produced by elite volleyball players and ice hockey players.

Figure 11–5 presents the relative contribution of each metabolic pathway during three different duration all-out cycle ergometer tests. The lower portion shows the results in estimated kilojoules of energy, and the upper portion shows the results in percent of the total work output. Note the progressive change in the percentage contribution of each of the energy systems to the total work output as the duration of effort changes.

Performance scores on the Wingate test are reproducible, and validity is moderate using a variety of other measures of "anaerobic capacity."[77] Normative standards have recently been published for both peak power and average power for young adult men and women.[68] The mechanism is unknown to explain the relatively poorer performance of children on this test compared to adolescents and young adults. It is possible that both lower concentrations and rate of utilization of glycogen in children provide part of the answer.[27,42]

Blood Lactic Acid Levels

As was shown previously in Figure 7–3, the blood lactate level remains relatively low during steady-rate exercise up to about 55% of the maximum oxygen uptake. As the max $\dot{V}O_2$ is approached, there is a precipitous increase in the amount of lactic acid in the blood.

The data in Figure 11–6 were obtained from 10 college men who performed nine all-out bicycle ergometer rides of different durations (Katch Test) on different days. The subjects were highly motivated; many were involved in

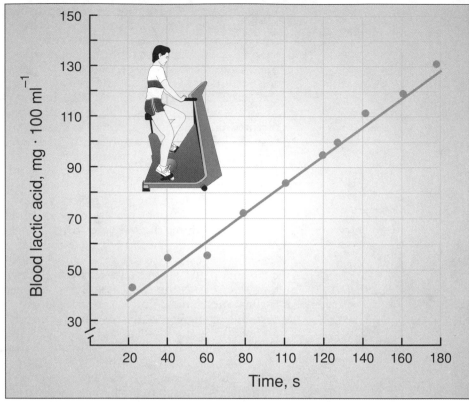

FIG. 11–6.
Blood lactate levels in all-out bicycle exercise of different durations. Each value represents the average of 10 subjects. (From the Applied Physiology Laboratory, University of Michigan.)

conditioning programs and some were varsity athletes. The men were unaware of the duration of each test but were instructed and urged to turn as many revolutions as possible. Venous blood-lactate was measured before and immediately after each test and throughout recovery. The plotted points are the average of the values for lactic acid obtained for each of the tests. Blood lactate levels increased in direct proportion with the duration (and total work output) of the all-out exercise. At the end of 3 minutes of cycling, the amount of lactic acid in the blood was highest and averaged about 140 mg in each 100 ml of blood.

Glycogen Depletion

Because the short-term energy system largely depends on glycogen stored in the specific muscles activated by exercise, the pattern of glycogen depletion in these muscles also provides an indication of the contribution of glycolysis to exercise. Figure 11-7 illustrates that the rate of glycogen depletion in the quadriceps femoris muscle during bicycle exercise is closely related to exercise intensity. With steady-rate exercise at about 30% of max $\dot{V}O_2$, a considerable reserve of muscle glycogen remains, even after 180 minutes of

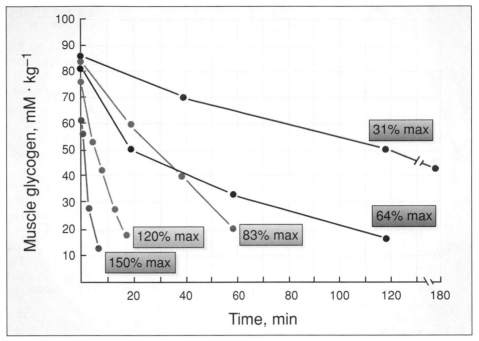

FIG. 11–7.
Glycogen depletion from the lateral portion of the quadriceps femoris muscle in bicycle exercise of different intensities and durations. The lightest workload results in some depletion of muscle glycogen. However, the most rapid and greatest glycogen depletion occurs with short-term, intense exercise. (Adapted from Gollnick, P.D.: Selective glycogen depletion pattern in human muscle fibers after exercise of varying intensity and at varying pedalling rates. J. Physiol., 241:45, 1974.)

exercise. Because metabolism in this relatively light exercise is essentially aerobic, large quantities of fatty acids are used for energy and the drain is only moderate on stored glycogen. At the two heaviest workloads, however, the most rapid and pronounced glycogen depletion is observed. This makes sense from a metabolic standpoint because glycogen is the only stored nutrient that provides anaerobic energy for the resynthesis of ATP; clearly, this substrate has high priority in the "metabolic mill" during strenuous exercise.

Changes in total muscle glycogen such as those illustrated in Figure 11–7, may not give a precise indication of the degree of glycogen breakdown in **specific fibers** within the muscle. Depending on the intensity of exercise, glycogen depletion occurs selectively in fast- and slow-twitch fibers.[79] For example, during all-out exercise (1-min sprints on a bicycle ergometer at a very heavy load), the fast-twitch fibers are activated to provide the predominant power for the exercise. Because of the anaerobic nature of this work, glycogen content in these fibers is almost totally depleted. In contrast, during moderate to heavy prolonged aerobic exercise, the slow-twitch fibers are always the first to become glycogen-depleted.[36] This **specificity** in glycogen utilization (and depletion) makes it difficult to evaluate glycolytic involvement from changes in a muscle's total glycogen content before and after exercise.

INDIVIDUAL DIFFERENCES IN THE CAPACITY FOR ANAEROBIC ENERGY TRANSFER

Several factors contribute to differences among individuals in their capacity to generate short-term anaerobic energy. These include differences in previous training, motivation, and the capacity to buffer acid metabolites.

Effects of Training

A comparison of the anaerobic capabilities of trained and untrained subjects is presented in Figure 11–8.[45] After short-term maximal exercise on the bicycle ergometer, the trained subjects always exhibited higher levels of muscle and blood lactic acid, as well as greater depletion of muscle glycogen. In short-term, intense forms of exercise, better performances are usually associated with higher levels of blood lactate.[47] These results support the belief that training for short-term, all-out exercise enhances one's capacity to generate energy from the glycolytic system.[18,31] This is important because in sprint-and middle-distance activities, individual differences in anaerobic capacity can account for large performance differences.

Buffering of Acid Metabolites

When anaerobic energy transfer predominates, lactic acid accumulates, and the acidity of muscle and blood increases. This has a dramatic negative influence on the intracellular environment and thus, on the contractile capability of exercising muscles.[35] This had led to the speculation that anaerobic training may enhance short-term energy capacity by the mechanism of increasing the body's alkaline reserve. This would theoretically enable greater lactic acid production because it could be buffered more effectively. Although this reasoning seems appealing, only a small increase in alkaline reserve has been noted in athletes as compared to sedentary counterparts.[32] Furthermore, there is no appreciable change in alkaline reserve following hard physical training. **The general consensus is that trained people have a buffering capability within the range expected for healthy untrained individuals.**

It is interesting that relatively short-term, high-intensity exercise performance can be enhanced significantly by temporarily altering acid-base balance in the direction of alkalosis.[100] This was achieved by ingesting a buffering solution of sodium bicarbonate prior to an 800-meter race . The significantly faster run times were accompanied by higher levels of blood lactate and extracellular H^+ which suggested an increased anaerobic energy contribution. More is said in Chapter 23 concerning the possible ergogenic effects of this procedure.

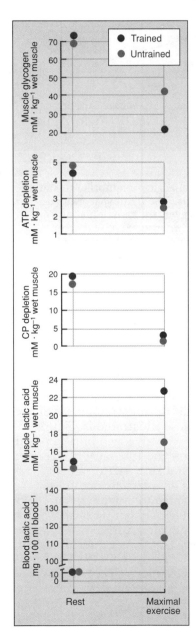

FIG. 11–8.
Depletion of various anaerobic substrates and increases in muscle and blood lactic acid during maximal exercise for trained and nontrained subjects. In each case, the trained subjects exhibited a greater increase in anaerobic metabolism or a depletion in anaerobic substrates while the depletion of the high-energy phosphates was essentially the same. (From Karlsson, J., et al.: Muscle metabolites during submaximal and maximal exercise in man. Scand. J. Clin. Lab. Invest., 26:382, 1971.)

Motivation

Individuals with greater "pain tolerance," "toughness," or ability to "push" beyond the discomforts of fatiguing exercise definitely accomplish more anaerobic work. These people usually generate greater levels of blood lactate and glycogen depletion; they also score higher on tests of short-term energy capacity. Motivational factors, which are difficult to categorize or quantify, play a key role in superior performance at all levels of competition.

AEROBIC ENERGY: THE LONG-TERM ENERGY SYSTEM

As shown in Figure 11–9, persons who engage in sports that require sustained, high-intensity exercise generally have a large capacity for aerobic energy transfer. The highest maximal oxygen uptakes are generally recorded for men and women who compete in distance running, swimming, bicycling, and cross-country skiing.[3] These athletes have almost double the aerobic capacity of a sedentary group! This is not to say that the max $\dot{V}_{O_2}$ is the only determinant of aerobic work capacity. Other factors, especially those at the muscular level, such as the number of capillaries, enzymes, and fiber type, exert a strong influence on the capacity to **sustain** high levels of aerobic exercise.[41] The max $\dot{V}_{O_2}$, however, provides important information on the capacity of the long-term energy system. In addition, the attainment of max $\dot{V}_{O_2}$ requires integration of the ventilatory, cardiovascular, and neuromuscular systems; this gives maximal oxygen uptake significant physiologic as well as metabolic meaning.[75] **In many ways, it has become one of the fundamental measures in exercise physiology.**

MEASUREMENT OF MAXIMAL AEROBIC POWER

The maximal oxygen uptake can be determined by a variety of work tasks that activate large muscle groups as long as the exercise is of sufficient intensity and duration to engage maximal aerobic energy transfer. The usual forms of exercise include treadmill walking or running, bench stepping, or cycling. The max $\dot{V}_{O_2}$, however, has also been measured during free, tethered, and flume swimming.[9,62] and swim-bench ergometry,[33] as well as during rowing,[15] ice skating,[29] and arm-crank exercise.[45,85,93] Considerable research effort has been directed toward the development and standardization of tests for maximal aerobic power and toward the establishment of norms for this measure in relation to age, gender, state of training, and body composition.

Criteria for max $\dot{V}_{O_2}$

To be reasonably sure that a person has reached the maximum capacity for aerobic metabolism during specific exercise (that is, achieved a "true" max $\dot{V}_{O_2}$), a levelling-off or peaking-over in oxygen uptake should be achieved (Fig. 11–10). When the generally accepted criteria for the attainment of max $\dot{V}_{O_2}$ are not met, or the test performance appears limited by local factors rather than central circulatory dynamics, the term **peak $\dot{V}_{O_2}$** is usually used. Peak $\dot{V}_{O_2}$ refers to the highest value of oxygen consumption measured during the test.

The max $\dot{V}_{O_2}$ test shown in Figure 11-10 involved progressive increases in treadmill exercise, and the test was terminated when the subject would

FIG. 11–9.
Maximal oxygen consumption of male and female olympic-caliber athletes and healthy sedentary subjects. (Adapted from Saltin, B., and Åstrand, P.O.: Maximal oxygen uptake in athletes. J. Appl. Physiol., 23:353, 1967.)

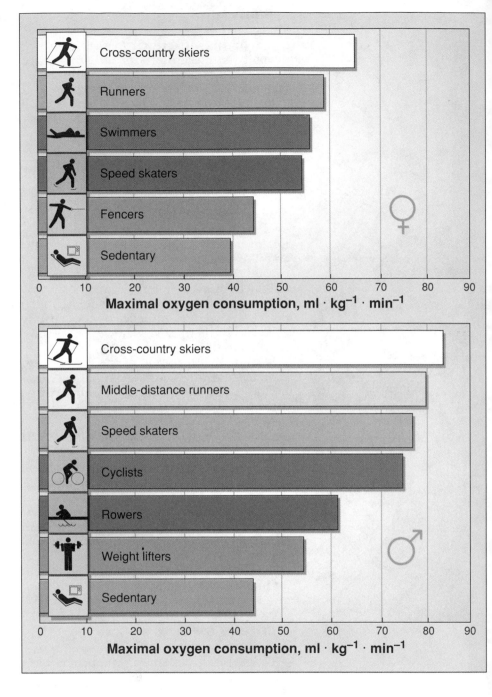

not complete the full duration of a particular work block. For the average oxygen consumption values of 18 subjects plotted in Figure 11–10, the highest oxygen uptake was reached before the subjects attained their maximum exercise level. This peaking-over criterion generally substantiates that the max $\dot{V}O_2$ has been reached.

In many instances, however, a peaking-over or slight decrease in oxygen

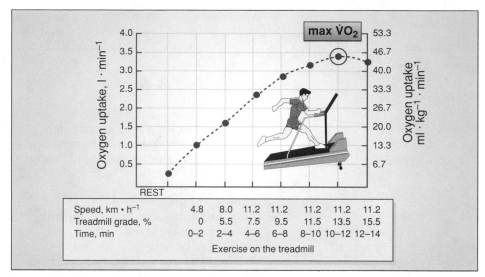

FIG. 11–10.
Peaking-over in oxygen uptake with increasing work output on the treadmill.
Each point represents the average oxygen uptake of 18 sedentary male subjects.
(From the Applied Physiology Laboratory, University of Michigan.)

consumption is not observed at the highest work level. Often, the highest oxygen consumption is recorded in the last minute of exercise. Consequently, additional criteria have been suggested based on changes in oxygen consumption with increasing work. In one regard, max $\dot{V}_{O_2}$ is considered to have been reached when oxygen consumption fails to increase by some value usually expected from previous observations with the particular test.[75,91] It is also argued that to accept an oxygen uptake value as being maximum, blood lactic-acid levels should reach 70 or 80 mg per 100 ml of blood, or higher. Less precise but more easily measured criteria are usually applied that are reasonable indicators that the oxygen consumption is near-maximum. These include the attainment of the age-predicted maximum heart rate (Fig. 21–6), or a respiratory exchange ratio (R) in excess of 1.00.

TESTS OF AEROBIC POWER

Numerous tests have been devised and standardized for the measurement of max $\dot{V}_{O_2}$. Performance on these tests should be independent of strength, speed, body size, and skill, with the exception of specialized tests such as swimming, rowing, and ice skating.

The max $\dot{V}_{O_2}$ test may require a continuous 3- to 5-minute "supermaximal" effort, but it usually consists of increments in effort (**graded exercise**) to the point where the subject will no longer continue to exercise. Some researchers have perhaps imprecisely termed this end point "exhaustion." It should be kept in mind, however, that it is the performer who, for whatever reason, terminates the test. This decision is often influenced by a variety of psychologic or motivational factors that may not necessarily reflect physiologic strain. We have found that it takes considerable urging and prodding to get subjects to the point at which acceptable criteria can be demonstrated for the attainment of max $\dot{V}_{O_2}$. Practical experience has shown that high motivation and a relatively large anaerobic output are generally required to demonstrate a plateau in oxygen consumption during the max $\dot{V}O_2$ test. This applies to untrained people in particular who do not normally perform strenuous exercise with its associated discomforts.

Comparison of Tests

Maximal oxygen uptake tests are usually performed (1) continuously—with no rest between work increments, or (2) discontinuously—with the subject resting several minutes between work periods. The data in Table 11–2 show the results of a systematic comparison of max $\dot{V}O_2$ scores measured by six common continuous and discontinuous treadmill and bicycle procedures.[71]

Although there was only a small 8-ml difference in maximal oxygen uptake between the continuous and discontinuous bicycle tests, the max $\dot{V}O_2$ during bicycle exercise averaged 6.4 to 11.2% below values on the treadmill. The largest difference between any of the three treadmill running tests was only 1.2%. The walking test, on the other hand, elicited max $\dot{V}O_2$ scores about 7% **above** values on the bicycle but 5% **below** the average for the three running tests.

A common complaint of subjects on both the continuous and discontinuous bicycle tests was a feeling of intense local discomfort in the thigh muscles during heavy work. Many subjects stated that this was the major factor limiting their ability to perform further work on the ergometer. In the walking test, the common complaint was severe local discomfort in the lower back and calf muscles, especially walking at the higher treadmill elevations. Local discomfort was not common in the running tests, and subjects complained more of a general fatigue that was usually categorized as feeling "winded." From a standpoint of ease of administration, the continuous treadmill run appears to be the test of preference for testing the aerobic capacity of large numbers of healthy subjects. The total time to administer the test averaged a little over 12 minutes, whereas discontinuous running tests averaged about 65 minutes. Subjects seemed to "tolerate" the continuous test well and preferred the shorter time period for testing. In fact, research indicates that max $\dot{V}O_2$ can be reached with a continuous exercise protocol where exercise intensity is increased progressively in 15-second intervals.[28] With this approach, the total test time for either bicycle or treadmill exercise averages about 5 minutes!

TABLE 11–2.
AVERAGE MAXIMAL OXYGEN UPTAKES FOR 15 COLLEGE STUDENTS DURING CONTINUOUS AND DISCONTINUOUS TESTS ON THE BICYCLE AND TREADMILL*

Variable	Bike, Discontinuous	Bike, Continuous	Treadmill, Discontinuous Walk-Run	Treadmill, Continuous Walk	Treadmill, Discontinuous Run	Treadmill, Continuous Run
max $\dot{V}O_2$, ml·min⁻¹	3691 ± 453	3683 ± 448	4145 ± 401	3944 ± 395	4157 ± 445	4109 ± 424
max $\dot{V}O_2$, ml·kg⁻¹·min⁻¹	50.0 ± 6.9	49.9 ± 7.0	56.6 ± 7.3	53.7 ± 7.5	56.6 ± 7.6	55.5 ± 6.8

Values are means ± standard deviation
* Adapted from McArdle, W.D., et al.: Comparison of continuous and discontinuous treadmill and bicycle tests for max $\dot{V}O_2$. Med. Sci. Sports, 5:156, 1973.

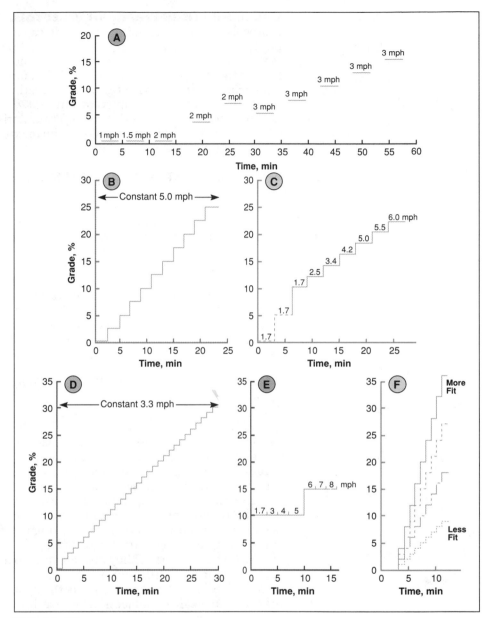

FIG. 11–11.

Six commonly used treadmill procedures. A, Naughton protocol.[76] Three minute exercise periods of increasing work rate alternate with 3 minute rest periods. The exercise periods vary in grade and speed. B, Åstrand protocol.[2] The speed is constant at 5 mph. After 3 minutes at 0% grade, the grade is increased 2½% every 2 minutes. C, Bruce protocol.[13] Grade and/or speed are changed every 3 minutes. The 0% and 5% grades are omitted in healthier subjects. D, Balke protocol.[4] After one minute at 0% grade and one minute at 2% grade, the grade is increased 1% per minute, all at a speed of 3.3 mph. E, Ellestad protocol.[26] The initial grade is 10% and the later grade is 15% while the speed is increased every 2 or 3 minutes. F, Harbor protocol.[11] After 3 minutes of walking at a comfortable speed, the grade is increased at a constant preselected amount each minute: 1%, 2%, 3%, or 4%, so that the subject reaches his maximum $\dot{V}O_2$ in approximately 10 min. (From Wasserman, K., et al.: Principles of Exercise Testing and Interpretation. Philadelphia, Lea & Febiger, 1987.)

Commonly Used Treadmill Protocols. Six commonly used treadmill protocols for the assessment of aerobic capacity in both normals and cardiac patients are summarized in Figure 11–11.[98] A feature common to these tests is the manipulation of exercise duration and treadmill speed and grade. The Harbor treadmill protocol (example F) is a unique application, and is referred to as a ramp test.[14,21] With this procedure, the grade is increased each minute up to 10 minutes by a constant amount that ranges from 1 to 4% grade depending on the exerciser's fitness. This relatively quick procedure produces a linear increase in oxygen uptake to the maximum level and is well tolerated by both normal persons and cardiac patients.[14]

FACTORS THAT AFFECT MAXIMAL AEROBIC POWER

Many factors influence the maximal oxygen uptake score. Of these, the most important are the mode of exercise and the person's heredity, state of training, body composition, gender, and age.

Mode of Exercise

It is generally accepted that the variations in max $\dot{V}_{O_2}$ during different forms of exercise reflect the quantity of muscle mass activated.[8,60] In experiments in which the max $\dot{V}_{O_2}$ was determined on the same subjects during different forms of exercise, the highest max $\dot{V}_{O_2}$ was obtained with treadmill exercise. Bench-stepping, however, generated max $\dot{V}_{O_2}$ scores identical to values obtained on the treadmill and significantly higher than those obtained on the bicycle ergometer.[48] With arm-crank exercise, the aerobic capacity value reaches only about 70% of one's treadmill performance.[93] For skilled but untrained swimmers, the maximal oxygen uptake during swimming was generally about 20% below treadmill values.[62,72] There is a definite **test specificity** in this form of exercise because trained collegiate swimmers achieved max $\dot{V}_{O_2}$s swimming that were only 11% below their treadmill values,[69] and some elite competitive swimmers can equal or even exceed their treadmill max $\dot{V}_{O_2}$ scores during a swimming test.[62] Similarly, a distinct exercise and training specificity is noted among competitive racewalkers who are capable of achieving an oxygen consumption during walking that is similar to their max $\dot{V}_{O_2}$ values during treadmill running.[73] Furthermore, if competitive cyclists are permitted to cycle at the fast pedal frequencies at which they compete, they also are able to achieve max $\dot{V}_{O_2}$ values equivalent to their treadmill scores.[37,90]

In the laboratory, the treadmill is the apparatus of choice for determining max $\dot{V}_{O_2}$ in healthy subjects. Exercise intensity is easily determined and regulated. Compared to other forms of exercise, the treadmill makes it easier for subjects to achieve one or more of the criteria for establishing that the max $\dot{V}_{O_2}$ has in fact been attained. In field experiments bench stepping or bicycle exercise is a suitable alternative.

Heredity

The question is frequently raised concerning the relative contribution of natural endowment to physiologic function and exercise performance.[2] For example, to what extent did heredity determine the extremely high aerobic capacities of the endurance athletes in Figure 11–9? Certainly these exceptionally high levels of functional capacity do not only reflect the effects of training. Although the answer is far from complete, some researchers have focused on the question of how genetic variability accounts for differences between individuals in physiologic and metabolic capacity.

Studies were made of 15 pairs of identical twins (who presumably had the same heredity since they came from the same fertilized egg) and 15 pairs of fraternal twins (who do not differ from ordinary siblings because they result from the separate fertilization of two eggs) raised in the same city and whose parents were of similar socioeconomic backgrounds. It was concluded that heredity alone accounted for up to 93% of the observed differences in aerobic capacity as measured by the max $\dot{V}_{O_2}$! In addition, the capacity of the short-term energy system of glycolysis and the maximum heart rate were shown to be genetically determined by about 81 and 86%, respectively.[57,58] Subsequent investigations of larger groups of brothers, fraternal twins, and identical twins have shown a significant but smaller effect of inherited factors on aerobic capacity and endurance performance. The genetic effect is estimated at about 40% for max $\dot{V}_{O_2}$, 50% for maximum heart rate, and 70% for physical working capacity.[11] Muscle fiber composition of identical twins is similar, whereas wide variation exists in fiber type among fraternal twins and brothers.[61] Future research might determine the upper limit of genetic determination, but available data currently indicate a significant contribution from inherited factors to both functional capacity and exercise performance.[10] It now appears that a large portion of the sensitivity of maximal aerobic and anaerobic power, as well as the adaptations of most muscle enzymes to training, is genotype-dependent.[12,38,80,87,92] In other words, members of the same twin-pair generally show the same response to training. Genetic makeup plays such a predominant role in determining the training response that it is almost impossible to predict a particular individual's response to a given training stimulus.

State of Training

The max $\dot{V}_{O_2}$ score must be evaluated relative to the person's state of training at the time of measurement. Improvements in aerobic capacity with training generally range between 6 and 20%, although increases have been reported as large as 50% above pretraining levels. This subject is discussed further in Chapter 21.

Gender

Max $\dot{V}_{O_2}$ values for women are typically 15 to 30% below scores for men.[39,96] Even among trained athletes, this difference ranges between 15 and 20%.[6,39] These differences, however, are considerably larger if the max $\dot{V}_{O_2}$ is expressed as an absolute value $(l \cdot min^{-1})$ rather than relative to body mass $(ml \cdot kg^{-1} \cdot min^{-1})$.[25,99] Among world class male and female cross-country skiers, for example, a 43% lower max $\dot{V}_{O_2}$ value for women (6.54 vs 3.75 $l \cdot min^{-1}$) is reduced to 15% (83.8 vs 71.2 $ml \cdot kg \cdot min^{-1}$) when the body mass of the athletes is used in the ratio expression of max $\dot{V}_{O_2}$.

The apparent gender difference in max $\dot{V}_{O_2}$ has generally been ascribed to differences in body composition and hemoglobin content. Untrained young adult women, for example, generally possess about 26% body fat whereas the corresponding value for men averages 15%.[50] Although trained athletes have a lower percentage of fat, women still possess significantly more body fat than their male counterparts. Thus, the male is generally able to generate more total aerobic energy simply because he possesses a relatively large muscle mass and less fat than the female.

Men also have a 10 to 14% greater concentration of hemoglobin than women.[1] This difference in the oxygen-carrying capacity of the blood potentially enables the male to circulate more oxygen during exercise and thus gives him a slight edge in aerobic capacity.

Although lower body fat and higher hemoglobin provide the male with some advantage in aerobic power, we must look for other factors to explain fully the difference between the genders. One possible explanation is the difference in the normal physical activity level between the "average" male and the "average" female. It can be convincingly argued that due to social constraints, the opportunities for women to exercise have been considerably less than for men. In fact, even among prepubertal children, boys are significantly more active in daily life than female counterparts.[34] Despite these possible limitations, however, the aerobic capacity of active females is generally higher than that of sedentary males. In fact, female cross-country skiers have max $\dot{V}O_2$ scores 40% higher than the untrained male![6] Even among the so-called normal population, considerable variability among the genders is observed, and the max $\dot{V}O_2$ scores of many women exceed the values for the less-fit men.

Body Composition

It is estimated that 69% of the differences in max $\dot{V}O_2$ scores among individuals can be explained simply by differences in body mass, 4% by differences in stature, and 1% by variations in lean body mass.[105] Thus, it is usually not meaningful to compare exercise performance[20] or the absolute value for oxygen consumption among individuals who differ in body size or composition. This has led to the common practice of expressing oxygen uptake in terms of body size—either in relation to surface area, body mass, lean body mass, or limb volume. As can be seen from Table 11–3, for an untrained man and woman who differ considerably in body mass, the percent difference in the max $\dot{V}O_2$, expressed in liters per minute, is 43%. When the max $\dot{V}O_2$ is expressed in relation to body mass ($ml \cdot kg^{-1} \cdot min^{-1}$), the woman is still about 20% lower than the man. If aerobic capacity is expressed in relation to lean body mass, however, the difference between the two subjects is reduced still more. This is especially true when men and women of equal training status are compared. These findings have also been noted for men and women during arm-cranking exercise.[97] When oxygen consumption values during maximal exercise were corrected for variations in arm and shoulder size, no difference in peak $\dot{V}O_2$ was observed between the genders. **These findings suggest that the differences in aerobic capacity between men and women are largely a function of the size of the contracting muscle mass.**

TABLE 11–3.
TRADITIONAL WAYS OF EXPRESSING OXYGEN UPTAKE

	Female	Male	Female v Male % Difference
Max $\dot{V}O_2$, $l \cdot min^{-1}$	2.00	3.50	−43
Body mass, kg	50	70	−29
Percent fat	25	15	+67
Lean body mass, kg	37.5	59.5	−37
Max $\dot{V}O_2$, $ml \cdot kg^{-1} \cdot min^{-1}$	40.0	50.0	−20
Max $\dot{V}O_2$, $ml \cdot kg\ LBM^{-1} \cdot min^{-1}$	53.3	58.8	−93

On the other hand, one must not be misled into believing that simply expressing the aerobic or endurance performance capacity by some measure of body composition will automatically "adjust" such criterion measures for the observed gender differences. The crucial test is to ascertain whether the gender differences are real (i.e., biological in origin) or are due to factors other than true inherited characteristics.

Are Gender Differences in Aerobic Capacity Biological?

The traditional ways of expressing oxygen uptake presented in Table 11–3 do not necessarily answer the question of whether gender differences in oxygen uptake are biologically inherent or simply attributable to differences in muscle size and body composition. This is because statistical "adjustments" do not truly eliminate the gender differences. They only express the criterion trait such as aerobic capacity or muscular strength relative to whatever divisor is used (e.g., body mass, lean body mass, or cross-sectional muscle area).

An experimental approach to evaluate this important topic would be to compare the physiologic responses of males and females who did not differ in body size, body composition, prior physical activity, or training history. In this way, there would be no need to express oxygen consumption as a ratio score relative to body size or composition. If dividing the aerobic capacity by body mass or lean body mass truly "adjusts" for gender differences in aerobic capacity, then matching males and females on these measures prior to testing for gender differences would eliminate the body size effects on aerobic capacity. Consequently, if body size were no longer a factor because such traits were matched, then it follows that there should be no gender differences in aerobic capacity.

In a recent experiment,[55] 10 pairs of sedentary and endurance-trained males and females were compared for aerobic capacity after matching for age, stature, body mass, lean body mass, and prior training history. In addition, aerobic capacity was adjusted for the observed gender difference in hemoglobin (Hb) concentration. Because part of the gender difference in aerobic capacity can be explained by differences in Hb concentration,[102] these differences must be considered when evaluating subsequent gender differences in aerobic capacity.

Figure 11–12 illustrates the effect of matching males and females for either body mass or lean body mass on the percentage difference in aerobic capacity measured during an incremental treadmill test. These differences in aerobic capacity between the genders matched for body mass were 25.3% (sedentary) and 22.1% (trained). After adjustment for differences in Hb concentration, the gender differences persisted; they were reduced slightly to 18.4% for the sedentary group and 12.8% for the trained group. When the males and females were matched on lean body mass, the gender differences were still substantial; they were 18.4% (sedentary) and 20.5% (trained). When the differences in aerobic capacity were adjusted further for differences in Hb concentration (Adj max $\dot{V}_{O_2}$), the gender differences were reduced somewhat, but were still 10.6% (sedentary) and 11.1% (trained).

These results demonstrate that gender differences persist in aerobic capacity even after matching for body mass, lean body mass, and Hb concentration. These findings raise the possibility that gender differences in aerobic capacity reflect traits that are biologically inherent and unalterable. This is not to say that such measures cannot be altered with training, because of course they can; rather, the results suggest that it may not be appropriate

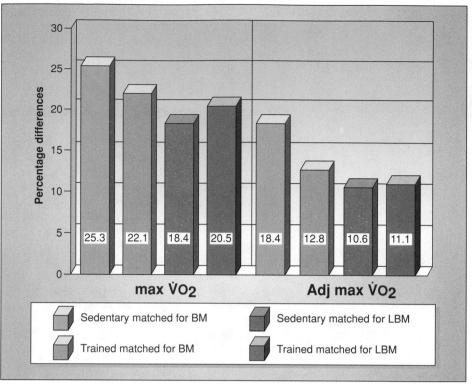

FIG. 11–12.
Percentage differences in max $\dot{V}O_2$, including the adjustment for hemoglobin (Adj max $\dot{V}O_2$), in sedentary and trained males and females matched for body mass (BM) and lean body mass (LBM). (From Keller, B.A.: The influence of body size variables on gender differences in strength and maximum aerobic capacity. Unpublished doctoral dissertation. University of Massachusetts, Amherst, 1989.)

to expect "gender-free" differences in aerobic capacity. In Chapter 22, data are presented on gender differences in muscular strength based on experiments of matching males and females on measures of body size and composition.

Age

The maximal oxygen uptake is not spared the effects of aging.[49,86,89] Although inferences from cross-sectional studies of people of different ages are somewhat limited, the available data provide insight into the possible effects of aging on physiologic function. As shown in Figure 11–13, the maximal oxygen uptake in liters per minute rapidly increases during the growth years. Longitudinal studies of children and aerobic power have been few.[7] The available data indicate that the absolute max $\dot{V}O_2$ ranges from about $1.0 \, l \cdot min^{-1}$ at age 6 years to about $3.2 \, l \cdot min^{-1}$ at 16 years. Average max $\dot{V}O_2$ in girls peaks at about age 14 and begins to decline thereafter. At age 14 the differences in max $\dot{V}O_2$ ($l \cdot min^{-1}$) between males and females approximates 25%, and will reach 50% by age 16 years. When expressed relative to body mass the max $\dot{V}O_2$ remains constant at about $53 \, ml \cdot kg^{-1} \cdot min^{-1}$ between the ages of 6 and 16 years for boys. On the other hand, relative max $\dot{V}O_2$

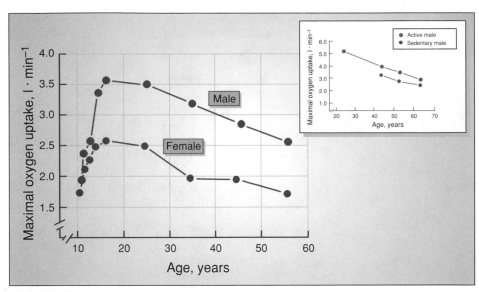

FIG. 11–13.
Maximal oxygen uptake as a function of age and level of activity in male and female subjects. (From Hermansen, L.: Individual differences. In Fitness, Health, and Work Capacity. International Standards for Assessment. Edited by L.A. Larson. New York, Macmillan Publishing Co., 1974. The inset graph was redrawn from tabled data of Åstrand, P.O., and Rodahl, K.R.: Textbook of Work Physiology. New York, McGraw-Hill Book Co.)

gradually decreases with age in girls from 52.0 ml $\cdot$ kg^{-1} $\cdot$ min^{-1} at 6 years to 40.5 ml $\cdot$ kg^{-1} $\cdot$ min^{-1} at age 16 years. The most commonly offered explanation for the discrepancy in relative max $\dot{V}_{O_2}$ between boys and girls as they advance in age is the relatively greater accumulation of fat in females.[59]

After age 25, the max $\dot{V}_{O_2}$ declines steadily at about 1% per year so that by age 55 it is about 27% below values reported for 20-year-olds. The data in the insert graph indicate that while active adults retain a relatively high max $\dot{V}_{O_2}$ at all ages, a progressive decline in this physiologic capacity occurs with advancing years. In one study of 8 women who averaged about 80 years of age, max $\dot{V}_{O_2}$ averaged 13.4 ml $\cdot$ kg^{-1} $\cdot$ min^{-1},[30] or about 3.7 METS! Research is accumulating, however, to indicate that one's habitual level of physical activity is more a determinant of aerobic capacity than chronological age per se.[74,94,95] The age-related effects on physiologic function are discussed more fully in Chapter 30.

TESTS TO PREDICT MAX $\dot{V}_{O_2}$

The direct measurement of max $\dot{V}_{O_2}$ requires an extensive laboratory and specialized equipment, as well as considerable motivation on the part of the subject. Consequently, these tests are not suitable for measuring large groups of untrained subjects in a field situation. In addition, such heavy exercise could pose a potential hazard to adults who are not medically cleared or who are exercised without appropriate safeguards or supervision. In view of these considerations, tests have been devised to **predict** the max $\dot{V}_{O_2}$ from performance measures such as walking and running endurance, or from easily obtained heart rates during or immediately after exercise.[16,40,103,104] The tests are easy to administer, can be used with large groups of men or women, and usually require submaximal exercise.

Walking Tests

With interest in "fitness walking" reaching a zenith in the 1980s, walking tests to predict max $\dot{V}_{O_2}$ were developed.[56] In a study on a large sample of males and females the following equation emerged to predict max $\dot{V}_{O_2}$ from walking speed and other variables:

$$\text{max } \dot{V}_{O_2} = 6.9652 + (0.0091 \times WT) - (0.0257 \times AGE) + (0.5955 \times GENDER) - (0.224 \times T1) - (0.0115 \times HR1\text{-}4)$$

where max $\dot{V}_{O_2}$ is in $l \cdot min^{-1}$; WT is body weight in pounds; Age in years; Gender: 0 = female, 1 = male; T1 is time for the 1-mile track walk expressed as minutes and hundredths of a minute; HR1-4 is the heart rate in beats $\cdot$ min^{-1} at the end of the last quarter-mile.

The equation for max $\dot{V}_{O_2}$ expressed in $ml \cdot kg^{-1} \cdot min^1$ is:

$$\text{max } \dot{V}_{O_2} = 132.853 - (0.0769 \times WT) - (0.3877 \times AGE) + (6.315 \times GENDER) - (3.2649 \times T1) - (0.1565 \times HR1\text{-}4)$$

For these equations the multiple correlation was R = 0.92 and the standard error of predicting an individual's max $\dot{V}_{O_2}$ was $\pm 0.335\ l \cdot min^{-1}$ (4.4 ml $\cdot$ $kg^{-1} \cdot min^{-1}$). Since the study population ranged in age from 30 to 69 years, this prediction method is applicable to a large segment of the population.

To illustrate this walking prediction test, suppose the following data were collected on a 30 year old female:

> Body mass = 155.5 lb
> T1 = 13.56 min
> HR1-4 = 145 beats $\cdot min^{-1}$

Substituting in the equation to predict max $\dot{V}_{O_2}$ in $ml \cdot kg^{-1} \cdot min^{-1}$ results in a value of 42.3 as follows:

$$\text{max } \dot{V}_{O_2} = 132.853 - (0.0769 \times 155.5) - (0.3877 \times 30.0) + (6.315 \times 0) - (3.2649 \times 13.56) - (0.1565 \times 145)$$

$$\text{max } \dot{V}_{O_2} = 132.853 - (11.96) - (11.63) + (0) - (44.27) - (22.69)$$

$$\text{max } \dot{V}_{O_2} = 42.3\ ml \cdot kg^{-1} \cdot min^{-1}$$

Endurance Runs

Like walking tests, runs of various durations or distances can be used to evaluate aerobic fitness.[9,51,52] Such tests are based on the reasonable notion that the distance one is able to run in a specified time (in excess of 5 or 6 minutes) is determined by the ability to maintain a high, steady-rate level of oxygen uptake. This in turn is largely based on one's maximum capacity to generate energy aerobically. Using this rationale, a field performance test was devised to evaluate aerobic fitness in military personnel.[4] The object of this test was to run as far as possible in 15 minutes. In 1968, Cooper shortened the run time to 12 minutes.[16]

In his original validation studies, Cooper observed an apparently strong correlation between the max $\dot{V}_{O_2}$ of Air Force personnel and the distance they could run-walk in 12 minutes. A correlation coefficient of r = 0.90 was reported between the 12-minute run-walk distance and max $\dot{V}_{O_2}$ in 47 men who varied considerably in age (17 to 54 years), body mass (52 to 123 kg), and max $\dot{V}_{O_2}$ (31 to 59 ml · kg^{-1} · min^{-1}). This same correlation was also observed in 9 ninth-grade boys.[23] Other investigators, however, have been unable to demonstrate such a strong relationship between "Cooper test" scores and aerobic capacity. One study, for example, measured 11- to 14-year-old boys and reported a correlation of r = 0.65.[65] For a group of 26 female athletes, the correlation between the run-walk scores and max $\dot{V}_{O_2}$ was r = 0.70,[66] whereas for 36 untrained college women a similar correlation was observed of r = 0.67.[52]

We would like to point out that a simple correlation of run-walk scores with max $\dot{V}_{O_2}$ does not take into account age and body-weight factors. These latter variables are in themselves related to the run-walk and max $\dot{V}_{O_2}$ scores. To avoid spurious correlations, the proper statistical evaluation must be performed. In fact, when the original data of Cooper were restricted to the same age range as those in the preceding study of 36 women,[16,52] the computed correlation was reduced significantly from r = 0.90 to r = 0.59!

The prediction of aerobic capacity should be approached with caution when using running performance. The need to establish a consistent level of motivation and effective pacing is critical for inexperienced subjects. Some individuals may achieve an optimal pace so they do not run too fast in the early part of a run and are therefore forced to slow down or even stop due to lactic-acid buildup as the test progresses. Other individuals may begin too slowly and continue that way so that their final performance score reflects inappropriate pacing or motivation rather than physiologic and metabolic capacity. In addition, max $\dot{V}_{O_2}$ is not the only variable that determines endurance running performance. Factors such as body mass and body fatness,[19] running efficiency, and the all-important percentage of one's aerobic capacity that can be sustained without lactic-acid buildup, all contribute significantly to successful running.[17,52]

Predictions Based on Heart Rate

Common tests to predict max $\dot{V}_{O_2}$ use the exercise or postexercise heart rate with a standardized regimen of submaximal exercise performed either on a bicycle, treadmill, or step test. These tests make use of the essentially linear relationship between heart rate and oxygen consumption for various intensities of light to moderately heavy exercise. The slope of this line (rate of heart rate increase) reflects the individual's aerobic fitness. The max $\dot{V}_{O_2}$ can then be estimated by drawing a straight line through several submaximum points that relate heart rate and oxygen consumption (or exercise intensity) and then extending this line to some assumed maximum heart rate for the particular age group.

Figure 11–14 illustrates the application of this "extrapolation" procedure for trained and untrained subjects. The heart rate-oxygen consumption line was drawn from four submaximal measures during bicycle exercise. Although each person's heart rate-oxygen consumption line tends to be linear, the **slope** of the individual lines can differ considerably. Consequently, a person with relatively high aerobic fitness can do more work and achieve

FIG. 11–14.

Application of the linear relationship between submaximal heart rate and oxygen consumption to predict max $\dot{V}O_2$.

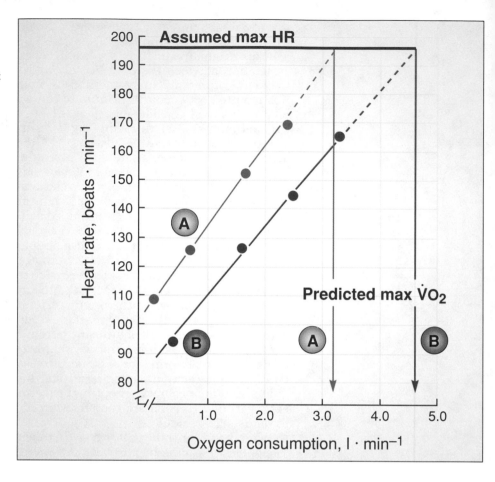

higher oxygen consumption before reaching a heart rate of 140 or 160 beats per minute than a less "fit" person. Also, because the heart rate increases linearly with the intensity of exercise, the person with the smallest increase in heart rate tends to have the largest work capacity and hence the highest max $\dot{V}O_2$. For the two subjects illustrated in Figure 11–14, max $\dot{V}O_2$ was predicted by extrapolating the line to a heart rate of 195 beats per minute—the assumed maximum heart rate for subjects of college age.

The accuracy of predicting max $\dot{V}O_2$ from submaximal exercise heart rate is limited by the following assumptions:

1. **Linearity of the heart rate-oxygen consumption (exercise intensity) relationship.** This assumption is met to a large degree, especially for various intensities of light to moderate exercise. In some subjects, the heart rate-oxygen consumption line curves or asymptotes at the heavier work loads in a direction that indicates a larger than expected increase in oxygen consumption per unit increase in heart rate. The oxygen consumption actually increases more than would be predicted through linear extrapolation of the heart rate-oxygen consumption line. Thus, the predicted max $\dot{V}O_2$ in these subjects would be underestimated.
2. **Similar maximum heart rates for all subjects.** The standard deviation is approximately ± 10 beats per minute about the average maximum heart

rate of individuals of the same age. Therefore, the max $\dot{V}_{O_2}$ of a person with an actual maximum heart rate of 185 beats per minute would be overestimated if the heart rate-oxygen consumption line were extrapolated to 195 or 200 beats per minute. The opposite would be true for a subject with an actual maximum heart rate of 210 beats per minute. Maximum heart rate also decreases with age. Unless this age effect is considered, older subjects will be consistently overestimated by assuming a maximum heart rate of 195 beats per minute, which is the appropriate heart rate maximum for 25-year-olds. More is said in Chapter 30 concerning the effect of age on maximum heart rate.

3. **Assumed constant economy or mechanical efficiency.** In cases in which submaximal oxygen consumption is not measured, but is instead estimated from work load, the predicted max $\dot{V}_{O_2}$ may be in error by the magnitude of variability in mechanical efficiency. A subject with poor mechanical efficiency (oxygen consumption at submaximal work higher than assumed) will be underestimated in terms of max $\dot{V}_{O_2}$, because heart rate will be elevated due to the added oxygen cost of the inefficient exercise. The variation among individuals in oxygen consumption during walking, stepping, or cycling does not usually exceed ±6%. In addition, seemingly small modifications in test procedures can have profound effects on the metabolic cost of exercise. Allowing individuals to support themselves with the treadmill handrails, for example, can reduce the oxygen cost of exercise by as much as 30%.[107]

4. **Day-to-day variation in heart rate.** Even under highly standardized conditions, the variation in submaximal heart rate is about ±5 beats per minute with day-to-day testing at the same exercise load.

Within the framework of these limitations, **the max $\dot{V}_{O_2}$ predicted from submaximal heart rate is generally within 10 to 20% of the person's actual value.** Clearly, this is not acceptable accuracy for research purposes. These tests, however, are well-suited for purposes of screening and classification in terms of aerobic fitness. Recently, this technique has been successfully applied in estimating aerobic capacity during pregnancy.[82]

The Step Test

The heart rate in recovery from a standardized bout of stepping is a practical and effective way to classify people in terms of aerobic fitness. With the use of "prediction equations" applied to step-test results, the max $\dot{V}_{O_2}$ can also be estimated with a reasonable degree of accuracy.

We have used a simple 3-minute step test (The Queens College Step Test) to evaluate the heart-rate response of thousands of college men and women.[70] The test was developed using the gymnasium bleachers (16¼ inches high) so that large numbers of students could be tested at the same time. Each stepping cycle was performed to a four-step cadence, "up-up-down-down." The women performed 22 complete step-ups per minute, which were regulated by a metronome at 88 beats per minute. Because the average man tended to be more fit for step-up exercise, the cadence was set at 24 step-ups per minute or 96 beats per minute on the metronome. The step test was begun after a brief demonstration and practice period. At the completion of stepping, the students remained standing, and pulse rate was measured for a 15-second period from 5 to 20 seconds into recovery. Recovery heart rate was converted to beats per minute (15 sec HR × 4) and compared to the established percentile rankings presented in Table 11–4.

TABLE 11–4.
PERCENTILE RANKINGS FOR RECOVERY HEART RATE (HR) AND PREDICTED MAXIMAL OXYGEN CONSUMPTION FOR MALE AND FEMALE COLLEGE STUDENTS*

Percentile Ranking	Recovery HR, Female	Predicted max $\dot{V}O_2$ $(ml \cdot kg^{-1} \cdot min^{-1})$	Recovery HR, Male	Predicted max $\dot{V}O_2$ $(ml \cdot kg^{-1} \cdot min^{-1})$
100	128	42.2	120	60.9
95	140	40.0	124	59.3
90	148	38.5	128	57.6
85	152	37.7	136	54.2
80	156	37.0	140	52.5
75	158	36.6	144	50.9
70	160	36.3	148	49.2
65	162	35.9	149	48.8
60	163	35.7	152	47.5
55	164	35.5	154	46.7
50	166	35.1	156	45.8
45	168	34.8	160	44.1
40	170	34.4	162	43.3
35	171	34.2	164	42.5
30	172	34.0	166	41.6
25	176	33.3	168	40.8
20	180	32.6	172	39.1
15	182	32.2	176	37.4
10	184	31.8	178	36.6
5	196	29.6	184	34.1

* From McArdle, W.D., et al.: Percentile norms for a valid step test in college women. Res Q, 44:498, 1973.

Based on the essentially linear relationship between heart rate and oxygen consumption, one would expect a person with a low exercise heart rate and recovery step test score to be further away from max $\dot{V}O_2$ than someone with a relatively high heart rate response. In other words, the lower the heart rate response to a standard work task, the higher is the max $\dot{V}O_2$. To evaluate the validity of the step test as a reflection of a person's aerobic capacity, we measured the max $\dot{V}O_2$ of a group of men and women who had also been evaluated by the step test. For the women in this group, the relationship between each subject's max $\dot{V}O_2$ and step-test score is shown in Figure 11–15. The results clearly indicated that knowledge of a person's max $\dot{V}O_2$ could be obtained from the score on the step test. Subjects with high recovery heart rates tended to have lower maximal oxygen uptakes, whereas a faster recovery (lower heart rate) tended to be associated with a relatively high max $\dot{V}O_2$. To enable us to predict the max $\dot{V}O_2$ $(ml \cdot kg^{-1} \cdot min^{-1})$ from step-test results for similar groups of men and women, the following equations were written:

Men: max $\dot{V}O_2$ = 111.33 − (0.42 × step-test pulse rate, beats · min^{-1})

Women: max $\dot{V}O_2$ = 65.81 − (0.1847 × step-test pulse rate, beats · min^{-1})

To simplify these conversions, the screened columns of Table 11–4 present the predicted maximal oxygen uptake values for men and women determined from recovery heart rate scores. In terms of accuracy of prediction, one can be 95% confident that the predicted max $\dot{V}O_2$ will be within about ±16% of the person's true max $\dot{V}O_2$.

When a high degree of accuracy is required, the maximal oxygen uptake

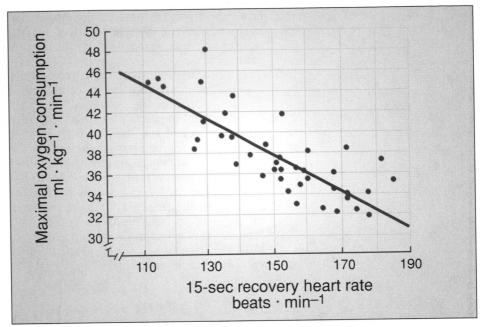

FIG. 11–15.
Scattergram and line of "best fit" relating step-test heart rate score and maximal oxygen consumption in college women. (From McArdle, W.D., et al.: Reliability and interrelationships between maximal oxygen uptake, physical work capacity, and step test scores in college women. Med. Sci. Sports, 4:182, 1972.)

should be measured directly in the laboratory with an appropriate, graded exercise test. When this is impractical, or when accuracy can be sacrificed somewhat, prediction tests are valuable for classification purposes.

SUMMARY

1. The concepts of **individual differences** and **specificity** are important for understanding capacities for anaerobic and aerobic power. Individual differences refer to real differences among individuals in contrast to the instability of a measure for any one individual. Specificity refers to metabolic and physiologic function that is not a general factor but one dependent on a host of factors.
2. The contribution of anaerobic and aerobic energy transfer depends largely on the intensity and duration of exercise. During strength and power-sprint activities, the primary energy transfer involves the immediate and short-term energy systems. The long-term aerobic energy system becomes progressively more important in activities that last longer than 2 minutes.
3. The capacity of each energy system can be estimated using appropriate physiologic measurements and performance tests; these can be used to evaluate a capacity at a particular time or to show changes consequent to specific training programs.
4. The stair-sprinting power test is commonly applied to measure power output generated by stored intramuscular high-energy phosphates. Peak

power and average power capacity from the glycolytic pathway can be evaluated by the 30-second, all-out Wingate test. Interpretation of test results must be made within the framework of the exercise specificity principle.

5. Training status, motivation, and acid-base regulation are factors that may explain differences among individuals in the capacities of the immediate and short-term energy systems.

6. The maximal oxygen uptake provides reliable and important information on the power of the long-term energy system as well as the functional capacity of various physiologic support systems. Consideration for type and amount of exercise performed and related physiologic functioning is required to assure that a "true" score has been attained.

7. The maximal aerobic power is influenced by heredity, state and type of training, age, gender and body composition. Each factor contributes uniquely to an individual's max $\dot{V}_{O_2}$.

8. Field methods for predicting max $\dot{V}_{O_2}$ should be viewed with caution; however, they may provide useful information in the absence of more valid laboratory methods.

REFERENCES

1. Åstrand, P.O.: Human physical fitness with special reference to age and sex. *Physiol. Rev.,* 36:307, 1956.
2. Åstrand, P.O., and Rodhal, K.: *Textbook of Work Physiology.* New York, McGraw-Hill, 1986.
3. Åstrand, P.O., and Saltin, B.: Maximal oxygen uptake and heart rate in various types of muscular activity. *J. Appl. Physiol.,* 16:977, 1961.
4. Balke, B., and Ware, R.W.: An experimental study of fitness of Air Force personnel. *U.S. Armed Forces Med. J.,* 10:675, 1959.
5. Bar-Or, O.: The Wingate anaerobic test: An update on methodology, reliability, and validity. *Sports Med.,* 4:381, 1987.
6. Bergh, V.: The influence of body mass in cross-country skiing. *Med. Sci. Sports Exerc.,* 19:324, 1987.
7. Beunen, G., and Malina, R.: Growth and physical performance relative to timing of the adolescent spurt. In *Exercise and Sport Sciences Reviews.* Vol 16. Edited by K.B. Pandolf. New York, Macmillan, 1988.
8. Blomquist, C.G., et al.: Similarity of the hemodynamic responses to static, and dynamic exercise of small muscle groups. *Circ. Res.,* 48 (Suppl. I):87, 1982.
9. Bonen, A., et al.: Maximal oxygen uptake during free, tethered, and flume swimming. *J. Appl. Physiol.,* 48:232, 1980.
10. Bouchard, C., and Lortie, J.: Heredity and human performance. *Sports Med.,* 1:38, 1984.
11. Bouchard, C., et al.: Aerobic performance in brothers, dizygotic and monozygotic twins. *Med. Sci. Sports Exerc.,* 18:639, 1986.
12. Bouchard, C., et al.: Genetic effects in human skeletal muscle fiber type distribution and enzyme activities. *Can. J. Physiol. Pharmacol.,* 64:125, 1986.
13. Bruce, R.A.: Exercise testing of patients with coronary artery disease. *Ann. Clin. Res.,* 3:323, 1971.
14. Buchfuhrer, M.J., et al.: Optimizing the exercise protocol for cardiopulmonary assessment. *J. Appl. Physiol.,* 55:1558, 1983.
15. Carey, P., et al.: Comparison of oxygen uptake during maximal work on the rowing ergometer. *Med. Sci. Sports,* 6:101, 1974.
16. Cooper, K.: Correlation between field and treadmill testing as a means for assessing maximal oxygen intake. *JAMA,* 203:201, 1968.
17. Costill, D.L.: Physiology of marathon running. *JAMA,* 221:1024, 1972.
18. Cunningham, D., and Faulkner, J.A.: The effect of training on aerobic and anaerobic metabolism during a short exhaustive run. *Med. Sci. Sports,* 1:65, 1969.

19. Cureton, K.J., et al.: Effect of experimental alterations in excess weight on aerobic capacity and distance running performance. *Med. Sci. Sports,* 10:194, 1978.
20. Cureton, K.J., et al.: Body fatness and performance differences between men and women. *Res. Q.,* 50:333, 1979.
21. Davis, J.A., et al.: Effect of ramp slope on measurement of aerobic parameters from the ramp exercise test. *Med. Sci. Sports Exerc.,* 14:339, 1982.
22. Di Prampero, P.E., et al.: Maximal muscular power, aerobic and anaerobic, in 116 athletes performing in the XIXth Olympic games in Mexico. *Ergonomics,* 13:665, 1970.
23. Doolittle, T.L., and Bigbee, R.: The twelve-minute run-walk: a test of cardiorespiratory fitness of adolescent boys. *Res. Q.,* 39:41, 1968.
24. Dotan, R., and Bar-Or, O.: Climatic heat stress and performance in the Wingate anaerobic test. *Eur. J. Appl. Physiol.,* 44:237, 1980.
25. Drinkwater, B.: Physiological responses of women to exercise. In *Exercise and Sport Science Reviews.* Vol. 1. Edited by J. Wilmore. New York, Academic Press, 1973.
26. Ellestad, M.H.: *Stress Testing.* 2nd Ed. Philadelphia, F.A. Davis, 1980.
27. Eriksson, B.O.: Muscle metabolism in children: a review. *Acta paediatr. Scand.* (Suppl.), 283:20, 1980.
28. Fairshter, R.D., et al.: A comparison of incremental exercise tests during cycle and treadmill ergometry. *Med. Sci. Sports Exerc.,* 15:549, 1983.
29. Ferguson, H.J., et al.: A maximal oxygen uptake test during ice skating. *Med. Sci. Sports,* 1:207, 1969.
30. Foster, V.L., et al.: The reproducibility of $\dot{V}O_2$ max, ventilatory, and lactate threshold in elderly women. *Med. Sci. Sports Exerc.,* 18:425, 1986.
31. Fournier, M.: Skeletal muscle adaptation in adolescent boys: endurance training and detraining. *Med. Sci. Sports Exerc.,* 14:453, 1982.
32. Full, F., and Herxheimer, H.: Ueber die Alkalireserve. *Klin. Wochenschr.,* 5:228, 1926.
33. Gergley, T., et al.: Specificity of arm training on aerobic power during swimming and running. *Med. Sci. Sports Exerc.,* 16:349, 1984.
34. Gilliam, T.B., et al.: Physical activity patterns as determined by heart rate monitoring in 6–7 year-old children. *Med. Sci. Sports Exerc.,* 13:65, 1981.
35. Gollnick, P.D., and Saltin, B.: Significance of skeletal muscle oxidative enzyme enhancement with endurance training. *Clin. Physiol.,* 2:1, 1982.
36. Gollnick, P.D., et al.: Glycogen depletion pattern in human skeletal muscle fiber after heavy exercise. *J. Appl. Physiol.,* 34:615, 1973.
37. Hagberg, J.M., et al.: Comparison of three procedures for measuring $\dot{V}O_2$ max in competitive cyclists. *Eur. J. Appl. Physiol.,* 39:47, 1978.
38. Hammel, P., et al.: Heredity and muscle adaptation to endurance training. *Med. Sci. Sports Exerc.,* 18:690, 1986.
39. Hermansen, L.H., and Anderson, L.: Aerobic work capacity in young Norwegian men and women. *J. Appl. Physiol.,* 20:425, 1965.
40. Hermiston, R., and Faulkner, J.A.: Prediction of maximal oxygen uptake by stepwise regression technique. *J. Appl. Physiol.,* 30:833, 1971.
41. Holloszy, J.O., and Coyle, E.F.: Adaptations of skeletal muscle to endurance exercise and their metabolic consequences. *J. Appl. Physiol.,* 56:831, 1984.
42. Inbar, O., and Bar-Or, O.: Anaerobic characteristics in male children and adolescents. *Med. Sci. Sports Exerc.,* 18:264, 1986.
43. Jacobs, I., et al.: Lactate in human skeletal muscle after 10 and 30s of supramaximal exercise. *J. Appl. Physiol.,* 55:365, 1983.
44. Kalamen, J.L.: Measurement of maximum muscular power in man. Unpublished doctoral dissertation. Ohio State University, Columbus, 1968.
45. Karlsson, J., and Saltin, B.: Lactate, ATP, and CP in working muscles during exhaustive exercise in man. *J. Appl. Physiol.,* 29:598, 1970.
46. Karlsson, J., et al.: Muscle metabolites during submaximal and maximal exercise in man. *Scand. J. Clin. Lab. Invest.* 26:385, 1971.
47. Karlsson, J., et al.: Relevance of muscle fibre type to fatigue in short intense and prolonged exercise in man. In *Human Muscle Fatigue: Physiological Mechanisms.* London, Pitman Medical, 1981.
48. Kasch, F.W., et al.: A comparison of maximal oxygen uptake by treadmill and step test procedures. *J. Appl. Physiol.,* 21:1387, 1966.

49. Kasch, F.W., et al.: A longitudinal study of cardiovascular stability in active men aged 45 to 65 years. *Phys. Sportsmed.*, 16:117, 1988.

50. Katch, F.I., and McArdle, W.D.: *Nutrition Weight Control and Exercise.* 3rd Ed. Philadelphia, Lea & Febiger, 1988.

51. Katch, F.I., et al.: Relationship between individual differences in steady pace endurance running performance and maximal oxygen intake. *Res. Q.*, 44:206, 1973.

52. Katch, F.I., et al.: Maximal oxygen intake, endurance running performance, and body composition in college women. *Res. Q.*, 44:301, 1973.

53. Katch, V.L.: Kinetics of oxygen uptake and recovery for supramaximal work of short duration. *Int. Z. Angew. Physiol.*, 31:197, 1973.

54. Katch, V. L., et al.: Optimal test characteristics for maximal anaerobic work on the bicycle ergometer. *Res. Q.*, 48:319, 1977.

55. Keller, B.A.: The influence of body size variables on gender differences in strength and maximum aerobic capacity. Unpublished doctoral dissertation. University of Massachusetts, Amherst, 1989.

56. Kline, G., et al.: Estimation of $\dot{V}O_{2max}$ from a one-mile track walk, gender, age, and body weight. *Med. Sci. Sports Exerc.*, 19:253, 1987.

57. Klissouras, V.: Heritability of adaptive variation. *J. Appl. Physiol.*, 31:338, 1971.

58. Klissouras, V., et al.: Adaptation to maximal effort: genetics and age. *J. Appl. Physiol.*, 35:288, 1973.

59. Krahenbuhl, G.S., et al.: Developmental aspects of maximal aerobic power in children. In *Exercise and Sport Sciences Reviews.* Edited by R.L. Terjung. New York, Macmillan, 1985.

60. Lewis, S.F., et al.: Cardiovascular responses to exercise as functions of absolute and relative work load. *J. Appl. Physiol.*, 54:1314, 1983.

61. Lortie, G., et al.: Muscle fiber type composition and enzyme activities in brothers and monozygotic twins. In *Proceedings of the 1984 Olympic Scientific Congress*, Vol. 4. Edited by R.M. Malina and C. Bouchard. *Sport and Human Genetics*, Champaign, IL, Human Kinetics Publishers, 1986.

62. Magel, J.R., and Faulkner, J.A.: Maximum oxygen uptake of college swimmers. *J. Appl. Physiol.*, 22:929, 1967.

63. Magel, J.R., et al.: Specificity of swim training on maximum oxygen uptake. *J. Appl. Physiol.*, 38:151, 1975.

64. Magel, J.R., et al.: Metabolic and cardiovascular adjustment to arm training. *J. Appl. Physiol.*, 45:75, 1978.

65. Maksud, M.G., and Coutts, K.D.: Application of the Cooper twelve-minute run-walk to young males. *Res. Q.*, 42:54, 1971.

66. Maksud, M.G., et al.: Energy expenditure and $\dot{V}O_2$ max of female athletes during treadmill exercise. *Res. Q.*, 47:692, 1976.

67. Margaria, R., et al.: Measurement of muscular power (anaerobic) in man. *J. Appl. Physiol.*, 21:1662, 1966.

68. Maud, P.J., and Schultz, B.B.: Norms for the Wingate anaerobic test with comparisons in another similar test. *Res. Q. Exerc. Sport*, 60:144, 1989.

69. McArdle, W.D., et al.: Metabolic and cardiorespiratory response during free swimming and treadmill walking. *J. Appl. Physiol.*, 30:733, 1971.

70. McArdle, W.D., et al.: Reliability and inter-relationships between maximal oxygen intake, physical work capacity, and step-test scores in college women. *Med. Sci. Sports*, 4:182, 1972.

71. McArdle, W.D., et al.: Comparison of continuous and discontinuous treadmill and bicycle tests for max $\dot{V}O_2$. *Med. Sci. Sports*, 5:156, 1973.

72. McArdle, W.D., et al.: Specificity of run training on $\dot{V}O_2$ max and heart rate changes during running and swimming. *Med. Sci. Sports*, 10:16, 1978.

73. Menier, D.R., and Pugh, L.G.C.E.: The relation of oxygen intake and velocity of walking in competition walkers. *J. Physiol. (London)*, 197:717, 1968.

74. Meredith, C.N., et al.: Body composition and aerobic capacity in young and middle-aged endurance-trained men. *Med. Sci. Sports Exerc.*, 19:557, 1987.

75. Mitchell, J., et al.: The physiological meaning of the maximal oxygen intake test. *J. Clin. Invest.*, 37:538, 1958.

76. Naughton, J., et al.: Treadmill exercise in assessment of patients with cardiac disease. *Am. J. Cardiol.*, 30:757, 1972.
77. Nebelsick-Gullett, L.J., et al.: A comparison between methods of measuring anaerobic work capacity. *Ergonomics*, 31:1413, 1988.
78. Pechar, G.S., et al.: Specificity of cardiorespiratory adaptation to bicycle and treadmill training. *J. Appl. Physiol.*, 36:753, 1974.
79. Piehl, K.: Glycogen storage and depletion in human skeletal muscle fibers. *Acta Physiol. Scand.*, Suppl., 402:1, 1974.
80. Prud'homme, D., et al.: Sensitivity of maximal aerobic power to training is genotype-dependent. *Med. Sci. Sports Exerc.*, 16:489, 1984.
81. Rowland, T.W.: Aerobic response to endurance training in prepubescent children: a critical analysis. *Med. Sci. Sports Exerc.*, 17:493, 1985.
82. Sady, S.P., et al.: Prediction of $\dot{V}O_2$ max during cycle exercise in pregnant women. *J. Appl. Physiol.*, 65:657, 1988.
83. Saltin, B.: Metabolic fundamentals in exercise. *Med. Sci. Sports*, 5:137, 1973.
84. Sargent, D.A.: Physical test of man. *Am. Phys. Ed. Rev.*, 26:188, 1921.
85. Sawka, M.N.: Physiology of upper body exercise. *Exerc. Sport. Sci. Rev.*, 14:175, 1986.
86. Schulman, S.P., and Gerstenblith, G.: Cardiovascular changes with aging: The response to exercise. *J. Cardiopulmonary Rehabil.*, 19:12, 1989.
87. Simoneau, J.A., et al.: Inheritance of human skeletal muscle and anaerobic capacity adaptation to high-intensity intermittent training. *Int. J. Sports Med.*, 7:167, 1986.
88. Sjördin, B.: Lactate dehydrogenase in human muscle. *Acta Physiol. Scand.*, Suppl., 436, 1976.
89. Stamford, B.A.: Exercise in the elderly. In *Exercise and Sport Sciences Reviews.* Vol. 16. New York, Macmillan, 1988.
90. Strømme, S.B., et al.: Assessment of maximal aerobic power in specially trained athletes. *J. Appl. Physiol.*, 42:833, 1977.
91. Taylor, H.L., et al.: Maximal oxygen intake as an objective measure of cardiorespiratory performance. *J. Appl. Physiol.*, 8:73, 1955.
92. Thiabault, M.C., et al.: Inheritance of human skeletal muscle adaptation to isokinetic strength training. *Hum. Hered.*, 36:341, 1986.
93. Toner, M.N. et al.: Cardiorespiratory responses to exercise distributed between the upper and lower body. *J. Appl. Physiol.*, 54:1403, 1983.
94. Upton, S.J., et al.: Comparative physiological profiles among young and middle-aged female distance runners. *Med. Sci. Sports Exerc.*, 16:67, 1984.
95. Vaccaro, P., et al.: Physiological characteristics of masters female distance runners. *Phys. Sportsmed.*, 9:105, 1981.
96. Vogel, J.A., et al.: An analysis of aerobic capacity in a large United States population. *J. Appl. Physiol.*, 60:494, 1986.
97. Washburn, R.A., and Seals, D.R.: Peak oxygen uptake during arm cranking in men and women. *J. Appl. Physiol.*, 56:954, 1984.
98. Wasserman, K., et al.: *Principles of Exercise Testing and Interpretation.* Philadelphia, Lea & Febiger, 1987.
99. Wells, C.L., and Plowman, S.A.: Sexual differences in athletic performance: biological or behavioral? *Phys. Sportsmed.*, 11:52, 1983.
100. Wilkes, D., et al.: Effect of acute induced metabolic alkalosis on 800-m racing time. *Med. Sci. Sports Exerc.*, 4:277, 1983.
101. Wilmore, J.H.: Influence of motivation on physical work capacity and performance. *J. Appl. Physiol.*, 24:459, 1968.
102. Woodson, R.D.: Hemoglobin concentration and exercise capacity. *Am. Rev. Resp. Dis.* (Suppl. 129):72, 1984.
103. Wyndham, C.H.: Submaximal tests for estimating maximum oxygen uptake. *Can. Med. Assoc. J.*, 96:736, 1967.
104. Wyndham, C.H., et al.: Studies of the maximum capacity of men for physical effort. Part I. A comparison of methods of assessing the maximum oxygen intake. *Int. Z. Angew. Physiol.*, 22:285, 1966.
105. Wyndham, C.H., and Hegns, A.J.A.: Determinants of oxygen consumption and maximum oxygen intake of caucasians and Bantu males. *Int. Z. Angew. Physiol.*, 27:51, 1969.

106. Yamamoto, S.H., et al.: Quantitative estimation of anaerobic and oxidative energy metabolism and contraction characteristics in intact human skeletal muscle in response to electrical stimulation. *Clin. Physiol.*, 3:227, 1983.
107. Zeimetz, G.A., et al.: Quantifiable changes in oxygen uptake, heart rate, and time to target heart rate when hand support is allowed during treadmill exercise. *J. Cardiac Rehab.*, 11:525, 1985.

SYSTEMS OF ENERGY DELIVERY AND UTILIZATION

Most sport, recreational, and occupational activities require a moderately intense yet sustained energy release. This energy for the phosphorylation of ADP to ATP is provided by the **aerobic** breakdown of carbohydrates, fats, and proteins. Unless a steady rate can be achieved between oxidative phosphorylation and the energy requirements of the activity, an anaerobic-aerobic energy imbalance develops, lactic acid accumulates, tissue acidity increases, and fatigue quickly ensues. The ability to sustain a high level of physical activity without undue fatigue depends on two factors: (1) the capacity and integration of the various physiologic systems for oxygen delivery, and (2) the capacity of the specific muscle cells to generate ATP aerobically.

Understanding the role of the ventilatory, circulatory, muscular, and endocrine systems during exercise enables us to appreciate individual differences in exercise capacity in essentially aerobic activities. Knowing the energy requirements of exercise and the corresponding physiologic adjustments also provides a sound basis for formulating a fitness program as well as for evaluating one's status before and during such a program.

12

PULMONARY STRUCTURE AND FUNCTION

Chapters 12, 13, and 14 deal with the process of pulmonary ventilation and gas transport, and elaborate the mechanisms whereby oxygen is supplied and extracted from the external environment and exchanged for almost equal quantities of carbon dioxide. This capability for pulmonary ventilation contributes significantly to the regulation of the internal environment at rest and during physical activity.

SURFACE AREA AND GAS EXCHANGE

If the oxygen supply of humans depended only on diffusion through the skin, it would be impossible to sustain the basal energy requirement, let alone the 3- to 4-liter gas exchange each minute necessary to run at a 5-minute per mile pace in a 26-mile marathon. Within the relatively compact human body, the needs for gas exchange are met by the remarkably effective **ventilatory system.** This system, depicted in Figure 12–1, regulates the gaseous state of the body's "external" environment to provide aeration of body fluids during rest and exercise.

ANATOMY OF VENTILATION

The process by which ambient air is brought into and exchanged with the air in the lungs is termed **pulmonary ventilation.** Air entering through the nose and mouth flows into the conductive portion of the ventilatory system where it is adjusted to body temperature, filtered, and almost completely humidified as it passes through the **trachea.** This air-conditioning process continues as the inspired air passes into two **bronchi,** the large tubes that serve as primary conduits in each of the two lungs. The bronchi further subdivide into numerous **brochioles** that conduct the inspired air through the tortuous and narrow route until it eventually mixes with the existing air in the **alveoli,** the terminal branches of the respiratory tract.

THE LUNGS

The lungs provide the surface between the blood and the external environment. Although the lung volume varies between 4 and 6 liters (about the amount of air contained in a basketball), its moist surface area is considerable. The lungs of an average-sized person weigh about 1 kg yet if spread out as in Figure 12–2, this tissue would cover a surface of 60 to 80 m^2. This is

235

FIG. 12–1.
The ventilatory system and gas exchange.

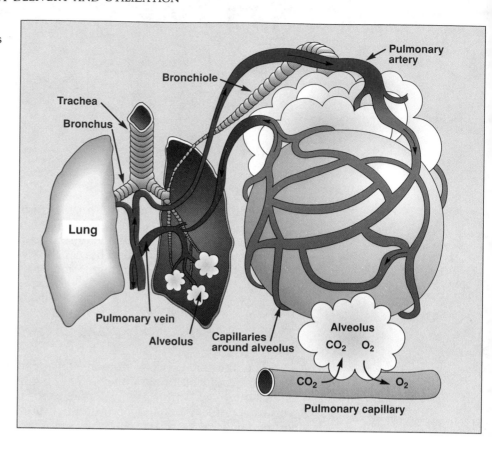

FIG. 12–2.
The lungs provide a large surface for gas exchange.

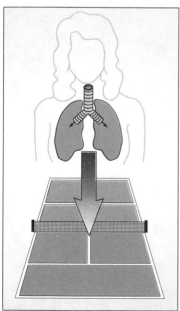

about 35 times greater than the surface of the person and would be sufficient to cover almost half a tennis court or an entire badminton court!

This highly vascularized, moist surface fits within the relatively small confines of the chest cavity by means of numerous infoldings so that lung membranes actually fold over onto themselves. The interface for the aeration of blood is considerable because during any second of maximal exercise, there is probably no more than one pint of blood in the fine network of blood vessels that surround the lung tissue.

THE ALVEOLI

There are more than 300 million alveoli. These elastic, thin-walled, membranous sacs provide the vital surface for gas exchange between the lungs and the blood. Alveolar tissue has the largest blood supply of any organ in the body. Millions of short, thin-walled capillaries and alveoli lie side by side with air moving on one side and blood on the other. Diffusion occurs through the extremely thin barrier of these alveolar and capillary cells, and this diffusion distance is essentially maintained throughout varying levels of exercise. Also, small pores within each alveolus enable the interchange of gas between adjacent alveoli. This provides for the indirect ventilation of some alveoli that may have been damaged or blocked as a result of disease.

During each minute at rest approximately 250 ml of oxygen leave the alveoli and enter the blood and about 200 ml of carbon dioxide diffuse in the reverse direction into the alveoli. During heavy exercise in trained endurance athletes, almost 25 times this quantity of oxygen is transferred across the alveolar membrane. The primary function of ventilation during rest and exercise is to maintain a fairly constant and favorable concentration of oxygen and carbon dioxide in the alevolar chambers. This assures effective gaseous exchange before the blood leaves the lungs to be transported throughout the body.

MECHANICS OF VENTILATION

Figure 12–3 illustrates the physical principle that underlies the act of breathing. Two lung-shaped balloons are suspended in a jar whose glass bottom has been replaced by a thin rubber membrane. When the membrane is pulled down the jar's volume increases, the air pressure within the jar becomes less than the air outside the jar, and air rushes in causing the balloons to inflate. Conversely, if the elastic membrane is allowed to recoil, the pressure in the jar temporarily increases and air rushes out. A considerable

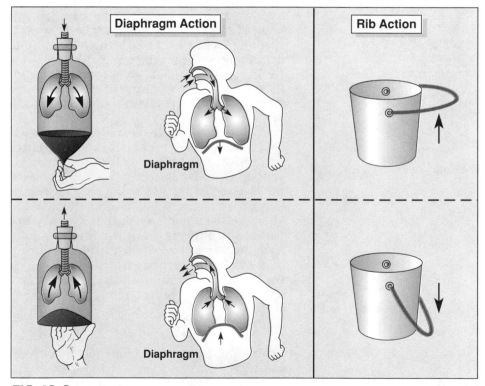

FIG. 12–3.
Mechanics of breathing. During inspiration, the chest cavity increases in size due to the raising of the ribs and the lowering of the muscular diaphragm. During exhalation, the ribs swing down and the diaphragm returns to the relaxed position. This reduces the volume of the thoracic cavity and air rushes out. The movement of the rubber bottom of the jar simulates the action of the diaphragm and causes air to enter and leave the two balloons; the movement of the bucket handle simulates the action of the ribs.

volume of air can be exchanged within the balloons in a given time period if the depth and rate of the descent and ascent of the rubber membrane are increased. This is essentially how ambient air and alveolar air are exchanged in the lungs.

The lungs are not merely suspended in the chest cavity as in the example with the balloons. Rather, the pressure differential between the air within the lungs and the lung-chest wall interface causes the lungs to adhere to the interior of the chest wall and literally follow its every movement. Thus, any change in the volume of the thoracic cavity causes a corresponding change in lung volume. The lungs depend on accessory means for altering their volume because they contain no muscles. The volume of the lungs is altered during **inspiration** and **expiration** by the action of voluntary muscles.

INSPIRATION

A large dome-shaped sheet of muscle called the **diaphragm** serves the same purpose as the rubber membrane of the jar. This muscle makes an airtight separation between the abdominal and thoracic cavities. During inspiration, the diaphragm muscle contracts, flattens out, and moves downward toward the abdominal cavity by as much as 10 cm. This movement causes the chest cavity to enlarge and become elongated. Consequently, the air in the lungs expands and its pressure, referred to as **intrapulmonic pressure,** becomes reduced slightly below atmospheric pressure. The lungs inflate as air is literally sucked in through the nose and mouth, with the degree of filling depending on the magnitude of the inspiratory movements. For healthy men, the maximum pressure generated by the inspiratory muscles ranges between 80 and 140 mm Hg.[35] Inspiration is completed when thoracic cavity expansion ceases and the intrapulmonic pressure increases to equal atmospheric pressure.

During exercise, the ribs and sternum also assist in the action of inspiration. The contraction of the **scaleni** and **external intercostal** muscles between the ribs causes the ribs to rotate and lift up and away from the body.[21] This action is similar to the movement of the handle lifted up-and-away from the side of the bucket at the right in Figure 12–3. The descent of the diaphragm, the upward swing of the ribs, and the outward thrust of the sternum all cause the volume of the chest cavity to increase with a subsequent inhalation of ambient air. It is not uncommon to see an athlete bend forward from the waist to facilitate breathing following an exhausting exercise. More than likely, this serves two purposes: (1) it facilitates the flow of blood to the heart, and (2) it minimizes the antagonistic effects of gravity on the usual upward direction of inspiratory movements.

EXPIRATION

Expiration, the process of air movement from the lungs, is predominantly a passive process during rest and light exercise. It results from the recoil of the stretched lung tissue and the relaxation of the inspiratory muscles. This causes the sternum and ribs to swing down and the diaphragm to move back toward the thoracic cavity. These movements decrease the size of the chest cavity and compress alveolar gas so that air moves out through the respiratory tract into the atmosphere. Expiration is completed when the compressive forces of the expiratory musculature are no longer acting, and intrapulmonic pressure decreases to atmospheric pressure. During ventilation in heavy exercise, the **internal intercostals** and **abdominal** muscles act powerfully on the ribs and abdominal cavity, respectively, to cause a reduction in thoracic dimensions. Thus, exhalation occurs more rapidly and to a more pronounced depth.

No major differences are observed in ventilatory mechanics between men and women or among people of different ages. At rest in the supine position, most people are "abdominal" or diaphragmatic breathers, whereas in the upright position, the action of the ribs and sternum becomes more apparent.[60] The rapid alterations in thoracic volume required during heavy exercise are accomplished mainly through the movement of the rib cage. This suggests that the muscles of the ribs are capable of more rapid action than the diaphragm and the abdominal muscles. The position of the head and back naturally adapted by long-distance runners (forward lean from the waist, neck flexed, and head extended forward with mandible parallel to ground) favors pulmonary ventilation during heavy exercise.[30]

Valsalva Maneuver. The expiratory muscles also play an important role in the ventilatory maneuvers of coughing and sneezing, as well as in the stabilization of the abdominal and chest cavities during the lifting of a heavy weight. During quiet breathing, the intrapulmonic pressure may fall only about 2 or 3 mm Hg during the inspiratory cycle and increase a similar amount above atmospheric pressure during exhalation. If, however, the **glottis** is closed following a full inspiration and the expiratory muscles are maximally activated, the compressive forces of exhalation can increase the **intrathoracic** pressure by more than 150 mm Hg above atmospheric pressure with somewhat higher pressures generally noted within the abdominal cavity.[31a] (The glottis is the narrowest part of the larynx through which air passes into and out of the trachea.) This forced exhalation against a closed glottis, termed the **Valsalva maneuver,** commonly occurs in weight lifting and in other activities that require a rapid and maximum application of force for a short duration. The fixation of the abdominal and chest cavities with this maneuver probably enhances the action of muscles that are attached to the chest.

Physiologic Consequences of the Valsalva. As illustrated in Figure 12-4, the increase in intrathoracic pressure during a Valsalva manuever is transmitted through the thin walls of the veins that pass through the thoracic region. Because venous blood is under relatively low pressure, these veins are compressed and blood flow is significantly reduced into the heart. This reduction in venous return and subsequent fall in arterial blood pressure can diminish the blood supply to the brain and frequently produces dizziness, "spots before the eyes," and even fainting with straining-type exercises. Normal blood flow is re-established (with perhaps even an "overshoot" in flow) once the glottis is opened and intrathoracic pressure is released.[61,66]

With the onset of the Valsalva maneuver at the start of the lift (Figure 12–4C), blood pressure rises abruptly as the elevated intrathoracic pressure forces blood from the heart into the arterial system; stroke volume and blood pressure then fall sharply due to the reduced venous return from the thoracic veins.[66] This temporary increase in blood pressure within the heart and arteries of the chest at the immediate onset of the Valsalva is probably compensated for by a proportionate pressure increase on their outside walls caused by the elevated intrathoracic pressure. The Valsalva maneuver usually accompanies straining-type muscular exertion similar to that performed in isometric and heavy resistance exercise. Such exercise greatly increases resistance to blood flow in the active muscles during the sustained contrac-

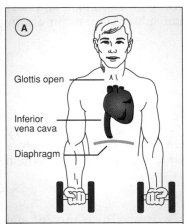

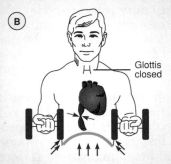

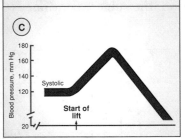

FIG. 12–4.
The Valsalva maneuver significantly reduces the return of blood to the heart because the increase in intrathoracic pressure collapses the vein that passes through the chest cavity. A, normal breathing, B, straining exercise with accompanying Valsalva, C, blood pressure response prior to and during straining-type exercise.

tion.[23a] In fact, intramuscular fluid pressure increases linearly with all levels of isometric contraction force up to maximum.[59] This causes a significant **rise** both in arterial blood pressure and in the work load of the heart throughout the exercise, even if the Valsalva is not performed.[38] This is one reason why **individuals with heart and vascular disease should refrain from all-out straining exercises such as isometrics or heavy lifting, and should seek more rhythmic muscular activity that results in a steady flow of blood and only moderate increases in arterial blood pressure and subsequent strain on the heart.**

LUNG VOLUMES AND CAPACITIES

The various lung volume measures that reflect one's ability to increase the depth of breathing are illustrated in Figure 12–5. The subject is breathing from a calibrated recording spirometer, similar to that described in Chapter 8 (Fig. 8–2), for measuring oxygen consumption by the closed-circuit method.

STATIC LUNG VOLUMES

The bell of the spirometer falls and subsequently rises as air is inhaled and exhaled from it. This provides a record of the ventilatory volume and breathing rate. The volume of air moved during either the inspiratory or expiratory phase of each breath is termed **tidal volume** (TV). It is indicated by the rise and fall of the spirometer in the first portion of the record. Under resting conditions, tidal volumes usually range between 0.4 and 1.0 liters of air per breath.

After several tracings are recorded for tidal volume, the subject is asked to inspire as deeply as possible following a normal inspiration. This additional volume of about 2.5 to 3.5 liters above the inspired tidal air represents one's reserve ability for inhalation and is termed the **inspiratory reserve volume** (IRV). After the measurement of IRV, the normal breathing pattern is once again established. Following a normal exhalation, the subject continues to

FIG. 12–5.
Static measures of lung volume.

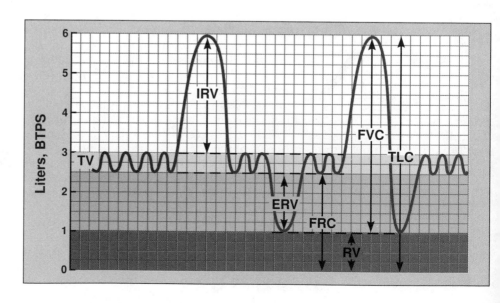

exhale and forces as much air as possible from the lungs. This is the **expiratory reserve volume** (ERV) that ranges between 1.0 and 1.5 liters for an average-sized man. **During exercise, encroachment on both inspiratory and expiratory reserve volumes, particularly the inspiratory volume, provides for a considerable increase in tidal volume.**

The total volume of air that can be voluntarily moved in one breath, from full inspiration to maximum expiration, or vice versa, is termed the **forced vital capacity** (FVC). This consists of the tidal volume plus the inspiratory and expiratory reserve volumes. Although values for vital capacity vary considerably with body size as well as with the position of the body during the measurement, average values are usually 4 to 5 liters in healthy young men and 3 to 4 liters in young women. Vital capacities of 6 to 7 liters are not uncommon for tall individuals and values of 7.6 and 8.1 liters have been reported for a professional football player and an Olympic gold medalist in cross-country skiing, respectively.[4,69] The large lung volumes of such athletes probably reflect genetic influences, because it is questionable whether static lung volumes can be changed significantly with training.[1]

Residual Lung Volume

When one exhales as deeply as possible, there is still a volume of air that remains in the lungs. This volume that cannot be exhaled is the **residual lung volume** (RV). It averages between 1.0 and 1.2 liters for women and 1.2 and 1.4 liters for men, although values of 0.96 to 2.46 liters have been reported for apparently healthy professional football players.[70] Residual lung volume tends to increase with age, whereas the inspiratory and expiratory reserve volumes become proportionally smaller. The loss in breathing reserve and the concomitant increase in residual volume are generally attributed to a decrease in the elastic components of the lung tissue with aging.[67] This, however, is probably not entirely an aging phenomenon per se as recent evidence indicates that endurance training in older athletes may alter the decline in static and dynamic lung functions associated with aging.[31]

The residual lung volume serves an important physiologic function because it allows for an uninterrupted exchange of gas between the blood and alveoli; this prevents fluctuations in blood gases during phases of the breathing cycle, including deep breathing. The residual lung volume plus the vital capacity constitute the **total lung capacity** (TLC).

The residual lung volume cannot be measured directly from spirographic tracings, but can be determined by several indirect techniques that involve rebreathing a known volume of gas containing either helium or pure oxygen.

With the **helium method,** the subject expires normally. The air remaining in the lungs at this end-normal expiration position is called the **functional residual capacity** (FRC) and includes the known expiratory reserve volume and the unknown residual volume. The subject then rebreathes a known helium mixture for about 5 minutes. Expired carbon dioxide is absorbed and the exact amount of oxygen consumed is replaced continuously to maintain a constant rebreathing volume. The functional residual lung volume can be computed easily from the dilution of the original helium mixture. This volume minus the expiratory reserve volume is equal to the residual lung volume.

The **oxygen dilution method** is more rapid than the helium method although it is similar in principle.[69] With this technique, the residual volume is determined from the dilution of the lung's original nitrogen concentration.

Dilution is achieved by the rapid rebreathing of a volume of about 5 liters of pure oxygen.

The residual lung volume is temporarily increased during and in recovery from an acute bout of both short- and long-term exercise.[22,42,65] In one study, a 21, 17, and 12% increase in residual lung volume was noted at 5, 15, and 30 minutes, respectively, following a maximal treadmill test.[14] Although the precise reason for an increase in residual lung volume with exercise is unknown, the closure of the small peripheral airways and an accumulation of pulmonary extravascular fluid with exercise (preventing a person from achieving a complete exhalation), may be operating factors. It is important to note that any temporary increase in residual lung volume (which is reversed within 24 hours) could significantly impact on subsequent computations of body volume for body composition studies (see Chapter 27).

Lung Volume Averages for Men and Women

As is the case with many anatomic and physiologic measures, lung volumes vary with age, gender, and body size, especially stature. For this reason, **lung volumes should only be evaluated in relation to norms based on age, gender, and size.** Average values for various lung volumes for men and women are presented in Table 12–1.

DYNAMIC LUNG VOLUMES

In appraising the adequacy of ventilation, the important consideration is the individual's ability to sustain high levels of airflow, rather than the quantity of air brought into the lungs in one breath. Dynamic ventilation depends on two factors: (1) the maximum "stroke volume" of the lungs, or the vital capacity, and (2) the speed with which this volume can be moved. The velocity of airflow, in turn, depends on the resistance offered by the respiratory passages to the smooth flow of air and the resistance offered by both the chest and lung tissue to a change in shape during breathing. When lung disease is present, pulmonary reserve is so great that patients rarely show symptoms of distress until a large part of their ventilatory capacity is lost.

TABLE 12–1.
AVERAGE LUNG VOLUMES AND CAPACITIES IN HEALTHY RECUMBENT SUBJECTS[a,b]

Measure	Males (20–30 Years)	Females (20–30 Years)	Males (50–60 Years)
Tidal volume	600	500	500
Inspiratory capacity	3600	2400	2600
Inspiratory reserve volume	3000	1900	2100
Expiratory reserve volume	1200	800	1000
Vital capacity	4800	3200	3600
Residual volume	1200	1000	2400
Functional residual capacity	2400	1800	3400
Total lung capacity	6000	4200	6000
RV/TLC × 100	20%	24%	40%

[a] Volumes are averages in milliliters for male and female subjects with surface areas of 1.7 m^2 and 1.6 m^2, respectively.
[b] Modified and produced with permission from Comroe, J.H., Jr., et al.: The Lung, 2nd edition. Copyright © 1962 by Year Book Medical Publishers, Inc., Chicago.

In fact, distance running can be engaged in regularly and successfully even in the presence of mild airway obstruction.[39]

FEV-to-FVC Ratio (FEV$_{1.0}$/FVC)

Normal values for vital capacity can be achieved by individuals with severe lung disease if no time limit is placed on the ventilatory maneuver. For this reason, physicians usually obtain a more "dynamic" measure of lung function such as the percentage of the forced vital capacity that can be expired in 1 second. This measure, termed **forced expiratory volume** (FEV$_{1.0}$/FVC), provides an indication of expiratory power and overall resistance to air movement in the lungs. Normally, about 85% of the vital capacity can be expelled in 1 second. With severe obstructive lung disease such as emphysema or bronchial asthma, the FEV$_{1.0}$ is considerably reduced and may often represent less than 40% of the vital capacity.[41,62] **Usually the demarcation point for airway obstruction is the point at which less than 70% of the vital capacity can be expired in 1 second.**[20]

MVV

Another dynamic test of ventilatory capacity requires rapid and deep breathing for 15 seconds. This 15-second volume is then extrapolated to the volume that would have been breathed had the subject continued for 1 minute and represents the **maximum voluntary ventilation** (MVV). The MVV is usually about 25% higher than the ventilation volume observed during maximal exercise. This is because the ventilatory system is **not** stressed maximally in exercise. The MVV of healthy, college-aged men is usually 140 to 180 liters of air per minute, whereas values for women are 80 to 120 liters per minute. Male members of the United States Nordic Ski Team averaged 192 liters per minute with an individual high MVV of 239 liters per minute.[28] Patients with obstructive lung disease, on the other hand, can only achieve about 40% of the MVV predicted normal for their age and size.[36] Specific exercise therapy is beneficial for such patients because the strength and endurance of the respiratory muscles and the MVV can be increased by exercises that train the breathing musculature in both normals and pulmonary patients.[2,34,62]

For patients, this training adaptation translates into an enhanced capacity for exercise with less physiologic strain.[54,63] For example, in one study of patients with emphysema, simple training of the inspiratory muscles caused significant improvement in exercise tolerance in both bicycle and treadmill exercise.[9] An important benefit of exercise training or specific training of the breathing musculature for patients who have obstructive lung disease is the progressive desensitization to the fear of breathlessness and a greater self-control of respiratory symptoms.[51]

LUNG FUNCTION, TRAINING, AND EXERCISE PERFORMANCE

Although some measures of lung function are sensitive indices of the severity of obstructive and restrictive lung disease, and do increase somewhat in patients and normal people with training,[9,57] they are of little use in predicting fitness or performance provided that the values fall within the normal range. For example, no difference was noted in the average vital capacities of prepubescent and Olympic wrestlers and trained middle-distance athletes, and in those of untrained, healthy subjects.[53,55,58] Furthermore, values for both static and dynamic lung function of accomplished marathoners and other endurance-trained athletes were no different compared to untrained controls of comparable body size.[31,40] Surprisingly, players from a professional football team averaged only 94% of their predicted vital capacity, with the defensive backs achieving only 83% of values predicted "normal" for their body size.[69]

Swimming and diving may be more conducive to the development of larger than normal static lung volumes.[18a] In these sports, the inspiratory muscles are probably strengthened as they work against additional resistance caused by the weight of water that compresses the thoracic cage. Relatively large vital capacities have been reported for skindivers and competitive swimmers.[5,12,17,53]

The inability to predict exercise performance of healthy individuals from lung function measures was demonstrated with a large group of teenage boys and girls following summer camp.[19] When lung volumes and capacities were adjusted for body size, there was essentially no relationship between lung function and various track performances, including a distance run. Similarly, in studies of marathon runners,[33,40] essentially no difference existed between the athletes' actual values for lung function measures and the values of sedentary subjects of similar body size. These observations, which are summarized in Table 12–2, are further supported by the fact that several of the top finishers in the Boston Marathon had below "normal" vital capacities.[24] Also, when variations are considered in body size, there is no relationship between maximal oxygen uptake and either vital capacity or maximum ventilation volume for healthy, untrained subjects.[27]

Although the feelings of fatigue in strenuous exercise are frequently related to feeling "out of breath" or "winded," it appears that the normal capacity for pulmonary ventilation does not limit exercise performance. The larger-than-normal lung volumes and breathing capacities of some athletes have generally been attributed to differences in genetic endowment[26] and may reflect strengthened respiratory muscles due to specific exercise training.

TABLE 12–2.
ANTHROPOMETRIC DATA, PULMONARY FUNCTION, AND RESTING VENTILATION IN 20 MARATHONERS AND CONTROLS[a]

Measure	Runners	Controls	Difference[b]
Anthropometric			
Age, y	27.8	27.4	0.4
Stature, cm	175.8	176.7	0.9
Surface area, m^2	1.82	1.89	0.07
Pulmonary Function			
FVC, l	5.13	5.34	0.21
TLC, l	6.91	7.13	0.22
FEV$_{1.0}$, l	4.32	4.47	0.15
FEV$_{1.0}$/FVC, %	84.3	83.8	0.5
MVV, l · min^{-1}	179.8	176.0	3.8
Resting Ventilation			
Ve, l · min^{-1}	11.9	11.9	0.9
Breathing rate, breaths · min^{-1}	10.9	11.1	0.2
Tidal volume, l	1.16	1.06	0.10

[a] From Mahler, D.A., et al.: Ventilatory responses at rest and during exercise in marathon runners. J. Appl. Physiol., 52:388, 1982.
[b] Differences not statistically significant.

Training May Benefit Ventilatory Endurance. Exercise training may be beneficial for improving the ability to sustain high levels of **submaximal** ventilation.[10,43,57] Twenty weeks of regular run training improved the ventilatory muscle endurance in healthy adult men and women approximately 16%. This may be partly due to the documented increase in the aerobic enzyme levels of ventilatory muscle.[50] It appears that the reduction in submaximal exercise ventilation combined with an enhanced endurance of the respiratory musculature with training delays the ventilatory fatigue observed in both short- and long-term exercise.[16,37,44] Ventilatory fatigue in the untrained may relate to the feelings of breathlessness and local pulmonary discomfort because blood lactate is significantly elevated for sedentary subjects who ventilate to levels observed in heavy exercise.[45]

EXERCISE AND THE ASTHMATIC

Asthma is a disease that plagues 10 million Americans, most of whom are children. A high level of fitness does not confer immunity from this ailment as about 11% of the U.S. athletes in the 1984 Olympic Games suffered from asthma or exercise-induced bronchospasm.[68] The disease is characterized by a hyperirritability of the pulmonary airways that is usually manifested by coughing, wheezing, or shortness of breath.

Unfortunately for many asthmatics, exercise is a potent stimulus for bronchoconstriction. With exercise, catecholamines from the sympathetic nervous system are released to produce a relaxation effect on the smooth muscle of the pulmonary airways. The initial bronchodilitation with exercise occurs in both the normal person and the asthmatic. With the asthmatic, however, bronchodilitation is then followed by bronchospasm, excessive mucous secretion, and subsequent constriction. An acute episode of airway obstruction is often apparent in the 10 minutes following exercise; recovery usually occurs between 30 to 90 minutes.[48] One technique for diagnosing an exercise-induced asthmatic response is simply to provide progressive increments of exercise. After each exercise bout a spirometric evaluation of FVC and $FEV_{1.0}$ is made during a 10- to 20-minute recovery period. Generally, a 15% reduction in pre-exercise values confirms the diagnosis of **exercise-induced asthma.**[15]

Sensitivity to Thermal Gradients

Many mechanisms are postulated for the bronchospastic response to exercise. One of the more attractive theories relates to the rate and magnitude of alterations in heat exchange brought about by the increased ventilation with exercise. As the incoming breath of air moves down the pulmonary pathway, heat and water are transferred from the respiratory tract as the air is warmed and humidified. The net effect of this air conditioning process is a cooling of the respiratory mucosa. In heavy exercise, large volumes of incompletely conditioned inspired air place a tremendous burden on the smaller airways and muscosal temperature decreases. In fact, there is a quantitative association between heat loss from the airways during exercise and the subsequent constriction of the bronchioles in the susceptible asthmatic patient.[15,48] At the end of exercise this airway cooling is then followed by abrupt rewarming; it appears that the thermal gradient provided by the combination of cooling and subsequent rewarming stimulates the release of chemical mediators to cause bronchospasm in susceptible individuals.[11,47]

The Environment Makes a Difference

The magnitude of the exercise-induced bronchospasm is considerably blunted when exercise is performed in a humid compared to a dry environment, regardless of air temperature.[7,8,13] In fact, the bronchospastic response in one study was totally abolished when the patients inhaled ambient air that was fully saturated with water vapor.[64] This is certainly ironic because it is commonly believed that a dry climate is most suitable for the asthmatic. This also explains why walking or jogging on a warm humid day or swimming in an indoor pool is usually well tolerated by asthmatics, whereas outdoor winter sports usually trigger an asthmatic attack. A prolonged warmup may also prove beneficial to the asthmatic athlete.[32]

A clearer understanding of the factors related to exercise-induced asthma will ultimately enable the physician to "prescribe" the optimum environment and exercise intensity so the asthmatic can obtain both physical and psychological benefits of regular exercise.[29] In addition, **appropriate medications** are available to limit the degree of bronchoconstriction in susceptible individuals who want to exercise on a regular basis without affecting their performance.[49]

Although exercise training cannot eliminate or "cure" an asthmatic condition, the training process can increase pulmonary airflow reserve and reduce ventilatory work by potentiating bronchodilation during exercise. This may enable the asthmatic to maintain high airflows throughout exercise so as to endure in submaximal activities despite an impaired pulmonary function.[56]

POST-EXERCISE COUGHING

Exercise is frequently associated with a dryness in the throat and coughing during the recovery period. This is common following exercise in cold weather. The phenomenon of post-exercise coughing is directly related to the overall rate of water loss from the respiratory tract (rather than respiratory heat loss) associated with the large ventilatory volumes during exercise.[6]

PULMONARY VENTILATION

MINUTE VENTILATION

During quiet breathing at rest, the breathing rate can average 12 breaths per minute, whereas the tidal volume averages about 0.5 liters of air per breath. Under these conditions, the volume of air breathed each minute, or **minute ventilation** ($\dot{V}E$), is 6 liters.

$$\begin{array}{ccc} \text{Minute} \\ \text{ventilation } (\dot{V}E) \\ \text{6 liters} \cdot \text{min}^{-1} \end{array} = \begin{array}{c} \text{Breathing} \\ \text{rate} \\ 12 \end{array} \times \begin{array}{c} \text{Tidal} \\ \text{volume} \\ 0.5 \end{array}$$

Significant increases in minute ventilation result from an increase in either the depth or rate of breathing, or both. During strenuous exercise, the breathing rate of healthy young adults usually increases to 35 to 45 breaths per minute, although rates as high as 60 to 70 breaths per minute are noted for elite athletes during maximal exercise. Tidal volumes of 2.0 liters and larger are common during exercise. Consequently, with increases in breathing rate and tidal volume, the minute ventilation can easily reach 100 liters or about 17 times the resting value. In well-conditioned male endurance athletes, ventilation may increase to 160 liters per minute in response to maximal exercise. In fact, ventilation volumes of 200 liters per minute have been reported in several research studies, and a high of 208 liters was observed for a professional football player during maximal bicycle exercise.[69] Even with these large minute ventilations, **the tidal volume rarely exceeds 55 to 65% of the vital capacity for both trained and untrained.**[18,23]

ALVEOLAR VENTILATION

A portion of the air in each breath does not enter the alveoli and thus is not involved in gaseous exchange with the blood. This air that fills the nose, mouth, trachea, and other nondiffusable conducting portions of the respiratory tract is contained within the **anatomic dead space.** In healthy subjects, this volume averages 150 to 200 ml or about 30% of the resting tidal volume. The composition of dead-space air is almost identical to that of ambient air except that it is fully saturated with water vapor.

Because of the dead-space volume, approximately 350 ml of the 500 ml of ambient air inspired in the tidal volume at rest enters into and mixes with the existing alveolar air. This does not mean that only 350 ml of air enters and leaves the alveoli with each breath. On the contrary, if the tidal volume is 500 ml, 500 ml of air enters the alveoli but only 350 ml is fresh air. This represents about one seventh of the total air in the alveoli. Such a relatively small and seemingly inefficient **alveolar ventilation** prevents drastic changes in the composition of alveolar air and assures a consistency in arterial blood gases throughout the entire breathing cycle.

The minute ventilation does not always reflect the actual alveolar ventilation. This is shown in Table 12–3. In the first example of shallow breathing, the tidal volume is reduced to 150 ml, yet it is still possible to achieve a 6-liter minute ventilation if the breathing rate is increased to 40 breaths per minute. The same 6-liter minute volume can also be achieved by decreasing the breathing rate to 12 breaths per minute and increasing the tidal

TABLE 12–3.
INTER-RELATIONSHIP BETWEEN TIDAL VOLUME, BREATHING RATE, AND PULMONARY VENTILATION

Condition	Tidal Volume (ml)	×	Breathing Rate (breaths · min^{-1})	=	Minute Ventilation (ml · min^{-1})	−	Dead Space Ventilation (ml · min^{-1})	=	Alveolar Ventilation (ml · min^{-1})
Shallow breathing	150		40		6000		(150 ml × 40)		0
Normal breathing	500		12		6000		(150 ml × 12)		4200
Deep breathing	1000		6		6000		(150 ml × 6)		5100

volume to 500 ml. On the other hand, by doubling tidal volume and halving the ventilatory rate, as in the example of deep breathing, the 6-liter minute ventilation is again achieved. Each of these ventilatory adjustments, however, drastically affects alveolar ventilation. In the example of shallow breathing, all that has been moved is the dead-space air: no alveolar ventilation has taken place. In the other examples, the breathing is deeper and a larger portion of each breath enters into and mixes with the existing alveolar air. **It is this alveolar ventilation that determines the gaseous concentrations at the alveolar-capillary membrane.**

Dead Space versus Tidal Volume

The preceding examples for alveolar ventilation were overly simplified because a constant dead space was assumed despite changes in tidal volume. Actually, the anatomic dead space increases as tidal volume becomes larger; it actually can double during deep breathing due to some stretching of the respiratory passages with a fuller inspiration.[3] This increase in dead space is still proportionately less than the increase in tidal volume. Consequently, deeper breathing provides for more effective alveolar ventilation than does similar minute ventilation achieved only through an increase in breathing rate.

Physiologic Dead Space

Adequate gas exchange between the alveoli and the blood requires ventilation that is well matched to the quantity of blood perfusing the pulmonary capillaries. For example, at rest, approximately 4.2 liters of air ventilate the alveoli each minute, whereas an average of 5.0 liters of blood flow through the pulmonary capillaries. In this instance, the ratio of alveolar ventilation to pulmonary blood flow, termed the **ventilation-perfusion ratio,** is approximately 0.8 (4.2 ÷ 5.0). This ratio means that each liter of pulmonary blood is matched by an alveolar ventilation of 0.8 liters. In light exercise, the ventilation-perfusion ratio is maintained at about 0.8, whereas in heavy exercise there is a disproportionate increase in alveolar ventilation. For healthy subjects, the ventilation-perfusion ratio may increase above 5.0 with fairly uniform distribution of blood flow to assure adequate aeration of the blood returning in the venous circulation.

In certain instances, a portion of the alveoli may not function adequately in gas exchange due to either (1) an underperfusion of blood, or (2) an inadequate ventilation relative to the size of the alveoli. This portion of the alveolar volume with a poor ventilation-perfusion ratio is termed the **physiologic dead space.** As illustrated in Figure 12–6, the physiologic dead space in the healthy lung is small and can be considered negligible. Physiologic dead space, however, can increase to as much as 50% of the tidal volume. This occurs with **inadequate perfusion** during hemorrhage or blockage of the pulmonary circulation with an embolism, or with **inadequate ventilation** that occurs in emphysema, asthma, and pulmonary fibrosis. When a relatively large physiologic dead space is due to a decreased functional alveolar surface, as occurs in emphysema, excessive ventilation is noted even at low exercise levels. In fact, many of these patients may never stress their circulatory systems maximally owing to ventilatory muscle fatigue from hyperventilation. Adequate gas exchange becomes imposssible when the total dead space of the lung exceeds 60% of the lung volume.

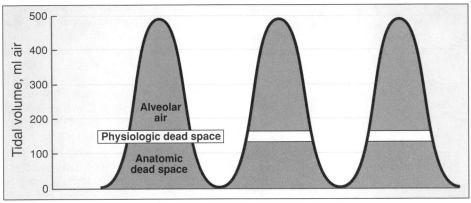

FIG. 12–6.
Distribution of tidal volume in healthy subject at rest. This tidal volume includes about 350 ml of ambient air that mixes with alveolar air, about 150 ml of air in the larger air passages (anatomic dead space), and a small portion of air distributed to either poorly ventilated or perfused alveoli (physiologic dead space).

Depth versus Rate

Alveolar ventilation during exercise is maintained through an increase in both the rate and depth of breathing. In moderate exercise, well-trained athletes achieve adequate alveolar ventilation by increasing tidal volume with only a small increase in breathing rate.[25] With deeper breathing, alveolar ventilation may increase from 70% of the minute ventilation at rest to over 85% of the total exercise ventilation. Figure 12–7 shows that the increase

FIG. 12–7.
Tidal volume and subdivisions of pulmonary air at rest and during exercise. (Modified from Lambertsen, C.J.: Physical and mechanical aspects of respiration. In Medical Physiology. Edited by V.B. Mountcastle. St. Louis, C.V. Mosby Co., 1968.)

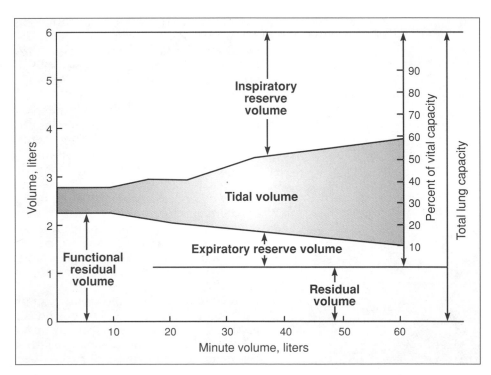

in tidal volume in exercise is due largely to encroachment on the inspiratory reserve volume with an accompanying but smaller decrease in the end-expiratory level. With more intense exercise, increases in tidal volume begin to plateau at about 60% of vital capacity and minute ventilation is further increased through an increase in breathing frequency. These adjustments occur unconsciously; each individual develops a "style" of breathing in which the respiratory frequency and tidal volume are blended to provide effective alveolar ventilation. Conscious attempts to modify breathing during general physical activities such as running are usually doomed to failure and probably are of no benefit in terms of performance. In fact, conscious manipulation of breathing would be detrimental to the exquisitely regulated physiologic adjustments to exercise. **At rest and in exercise, each individual should breathe in the manner that seems most natural.**

SUMMARY

1. The lungs provide a large interface between the body's internal fluid environment and the gaseous external environment. At any one second, there is probably no more than a pint of blood in the pulmonary capillaries.
2. Pulmonary ventilation is geared to maintain favorable concentrations of alveolar oxygen and carbon dioxide to assure adequate aeration of the blood flowing through the lungs.
3. Pulmonary airflow depends upon small pressure differences between ambient air and air within the lungs. These differences are brought about by the action of various muscles that act to alter the dimensions of the chest cavity.
4. A forced exhalation against a closed glottis is called a Valsalva maneuver. It can cause a large increase in pressure within the chest and abdominal cavities that compresses the thoracic veins, thereby significantly reducing venous return to the heart. The straining-type muscular effort usually accompanying the Valsalva maneuver temporarily elevates blood pressure and places an added work load on the heart. Thus, individuals with heart and vascular disease should refrain from straining exercises such as isometrics.
5. Lung volumes vary with age, gender, and body size, especially stature, and should only be evaluated in relation to norms based on these factors.
6. Tidal volume is increased during exercise by encroachment on both the inspiratory and expiratory reserve volumes. Even when a person breathes to vital capacity, air still remains in the lungs at maximal exhalation. This residual lung volume allows for an uninterrupted exchange of gas during all phases of the breathing cycle.
7. Forced expiratory volume and maximum voluntary ventilation give a dynamic picture of one's ability to sustain high levels of airflow. They serve as excellent screening tests to detect possible lung disease.
8. Tests of static and dynamic lung function are of little use in predicting fitness and exercise performance, provided that the values fall within a normal range.
9. Exercise-induced bronchospasm is an obstructive phenomenon that depends on both the rate and magnitude of airway cooling and subsequent rewarming. This response is essentially eliminated when humidified air is breathed during the exercise task. In the asthmatic, exercise training can increase airflow reserve and reduce respiratory work.

10. Minute ventilation is a function of breathing rate and tidal volume. It averages 6 to 10 liters at rest, whereas in maximum exercise, increases in breathing rate and depth may produce ventilations as high as 200 liters per minute.
11. Alveolar ventilation is the portion of the minute ventilation that enters the alveoli and is involved in gaseous exchange with the blood.
12. The ratio of alveolar ventilation to pulmonary blood flow is termed the ventilation-perfusion ratio. At rest and in light exercise, this ratio is maintained at about 0.8. This indicates that each liter of pulmonary blood is matched by an alveolar ventilation of 0.8 liters. In heavy exercise, alveolar ventilation in healthy people increases disproportionately, and the ratio may reach 5.0.
13. At rest and in exercise, a healthy person should breathe in a manner that seems most natural.

REFERENCES

1. Adams, W.C.: Effect of a season of varsity track and field on selected anthropometric, circulatory and pulmonary function parameters. *Res. Q.*, 39:5, 1968.
2. Akabas, S.R., et al.: Metabolic and functional adaptation of the diaphragm to training with resistive loads. *J. Appl. Physiol.*, 66:529, 1989.
3. Asmussen, E., and Nielsen, M.: Physiological dead-space and alveolar gas pressures at rest and during muscular exercise. *Acta Physiol. Scand.*, 38:1, 1956.
4. Åstrand, P-O., and Rodahl, K.: *Textbook of Work Physiology.* New York, McGraw-Hill, 1986.
5. Bachman, J.C., and Horvath, S.M.: Pulmonary function changes which accompany athletic training programs. *Res. Q.*, 39:235, 1968.
6. Banner, A.S., et al.: Relation of respiratory water loss to coughing after exercise. *N. Engl. J. Med.*, 311:883, 1984.
7. Bar-Or, O., et al.: Effects of dry and humid climates on exercise induced asthma in children and adolescents. *J. Allergy Clin. Immunol.*, 6:163, 1977.
8. Bar-Yishay, E., et al.: Differences between swimming and running as stimuli for exercise-induced asthma. *Eur. J. Appl. Physiol.*, 48:387, 1982.
9. Belman, M. J., and Mittman, C.: Ventilatory muscle training improves exercise capacity in COPD patients. *Am. Rev. Respir. Dis.*, 121:273, 1980.
10. Bender, P.R., and Martin, B.J.: Maximal ventilation for exhausting exercise. *Med. Sci. Sports Exerc.*, 17:164, 1985.
11. Berk, J.L., et al.: Cold-induced bronchoconstriction: role of cutaneous reflexes vs. direct airway effects. *J. Appl. Physiol.*, 63:659, 1987.
12. Bjurström, R.L., and Schoene, R.B.: Control of ventilation in elite synchronized swimmers. *J. Appl. Physiol.*, 63:1019, 1987.
13. Bundgaard, A., et al.: Influence of temperature and relative humidity of inhaled gas on exercise-induced asthma. *Eur. J. Respir. Dis.*, 63:239, 1982.
14. Buno, M.J., et al.: The effect of an acute bout of exercise on selected pulmonary function measurements. *Med. Sci. Sports Exerc.*, 13:290, 1981.
15. Busse, W.: Exercise induced asthma. *Am. J. Med.*, 68:471, 1980.
16. Bye, R.T.P., et al.: Ventilatory muscles during exercise in air and oxygen in normal men. *J. Appl. Physiol.*, 56:464, 1984.
17. Carey, C.R., et al.: Effects of skin diving on lung volumes. *J. Appl. Physiol.*, 8:519, 1955.
18. Coates, J., et al.: Breathing frequency and tidal volume: relationship to breathlessness. In *Breathing: Hering-Breuer Centenary Symposium.* Edited by R. Porter. London, Churchill, 1970.
18a. Cordain, L., et al.: Lung volumes and maximal respiratory pressures in collegiate swimmers and runners. Res. Q. Exerc. Sport, 61:70, 1990.
19. Cummings, G.R.: Correlation of athletic performance with pulmonary function in 13 to 17 year old boys and girls. *Med. Sci. Sports*, 1:140, 1969.

20. Dosman, J., et al.: The use of helium-oxygen mixture during maximum expiratory flow to demonstrate obstruction of small airways in smokers. *J. Clin. Invest.*, 55:1090, 1975.
21. Farkas, G.A., et al.: Contractile properties of intercostal muscles and their functional significance. *J. Appl. Physiol.*, 59:528, 1985.
22. Farrell, P.A., et al.: The course of lung volume changes during prolonged treadmill exercise. *Med. Sci. Sports Exerc.*, 15:319, 1983.
23. Folinsbee, L.J., et al.: Exercise respiratory pattern in elite cyclists and sedentary subjects. *Med. Sci. Sports Exerc.*, 15:503, 1983.
23a. Gaffney, F.A., et al.: Cardiovascular and metabolic responses to static contraction in man. Acta Physiol. Scand., 138:249, 1990.
24. Gordon, B., et al.: Observations on a group of marathon runners with special reference to the circulation. *Arch. Intern. Med.*, 33:425, 1924.
25. Grimby, G.: Respiration in exercise. *Med. Sci. Sports*, 1:9, 1969.
26. Grimby, G., and Saltin, B.: Physiological effects of physical training in different ages. *Scand. J. Rehabil. Med.*, 3:6, 1971.
27. Grimby, G., and Soderholm, B.: Spirometric studies in normal subjects. 111. *Acta Med. Scand.*, 173:199, 1963.
28. Hanson, J.S.: Maximal exercise performance in members of the U.S. Nordic Ski Team. *J. Appl. Physiol.*, 33:592, 1973.
29. Haas, F., et al.: Effect of aerobic training on forced expiratory airflow in exercising asthmatic humans. *J. Appl. Physiol.*, 63:1230, 1987.
30. Hass, F., et al.: Effect of upper body posture on forced inspiration and expiration. *J. Appl. Physiol.*, 52:879, 1982.
31. Hagberg, J.M., et al.: Pulmonary function in young and older athletes and untrained men. *J. Appl. Physiol.*, 65:101, 1988.
31a. Harman, E.A., et al.: Intra-abdominal and intra-thoracic pressures during lifting and jumping. *Med. Sci. Sports Exerc.*, 20:195, 1988.
32. Katz, R.M.: Prevention with and without the use of medications for exercise-induced asthma. *Med. Sci. Sports Exerc.*, 18:331, 1986.
33. Kaufmann, D.A., et al.: Pulmonary function of marathon runners. *Med. Sci. Sports*, 6:114, 1974.
34. Kuns, T.G., et al.: Ventilatory muscle endurance training in normal subjects and patients with cystic fibrosis. *Am. Rev. Respir. Dis.*, 113:853, 1977.
35. Leech, J.A., et al.: Respiratory pressures and function in young adults. *Am. Rev. Respir. Dis.*, 128:17, 1983.
36. Levison, H., and Cherniack, R.: Ventilatory cost of exercise in chronic obstructive pulmonary disease. *J. Appl. Physiol.*, 25:21, 1968.
37. Loke, J., et al.: Respiratory muscle fatigue after marathon running. *J. Appl. Physiol.*, 52:821, 1982.
38. MacDougall, J.D., et al.: Arterial blood pressure response to resistance exercise. *J. Appl. Physiol.*, 58:785, 1985.
39. Mahler, D.A., et al.: Exercise performance in marathon runners with airway obstruction. *Med. Sci. Sports Exerc.*, 13:284, 1981.
40. Mahler, D.A., et al.: Ventilatory responses at rest and during exercise in marathon runners. *J. Appl. Physiol.*, 52:388, 1982.
41. Mahler, D.A., and Harper, A.: Prediction of peak oxygen consumption in obstructive airway disease. *Med. Sci. Sports Exerc.*, 20:574, 1988.
42. Maron, M.B., et al.: Alterations in pulmonary function consequent to competitive marathon running. *Med. Sci. Sports*, 11:244, 1979.
43. Martin, B.J., and Stager, J.M.: Ventilation endurance in athletes and non-athletes. *Med. Sci. Sports Exerc.*, 13:21, 1981.
44. Martin, B.J., et al.: Low exercise ventilation in endurance athletes. *Med. Sci. Sports*, 11:181, 1979.
45. Martin, B.J., et al.: Anaerobic metabolism in the respiratory muscles during exercise. *Med. Sci. Sports Exerc.*, 16:82, 1984.
46. McFadden, E. R.: Respiratory heat and water exchange: physiological and clinical implications. *J. Appl. Physiol.*, 54:331, 1983.
47. McFadden, E.R.: Exercise-induced asthma; recent approaches. *Chest*, 93:1282, 1988.

48. McFadden, J., and Ingram, R.H.: Exercise induced asthma: observation on the ventilatory stimulus. *N. Engl. J. Med.*, 301:763, 1978.

49. McKenzie, D.C., et al.: Salbutamol and treadmill performance in non-atopic athletes. *Med. Sci. Sports Exerc.*, 15:520, 1983.

50. Moore, R.L., and Gollnick, P.D.: Response of ventilatory muscles of the rat to endurance training. *Pflügers Arch.*, 392:268, 1982.

51. Moser, K., et al.: Results of a comprehensive rehabilitation program; physiological and functional effects on patients with chronic obstructive pulmonary disease. *Arch. Int. Med.*, 140:1596, 1980.

52. Ness, G.E., et al.: Cardiopulmonary function in prospective competitive swimmers and their parents. *J. Appl. Physiol.*, 37:27, 1974.

53. Newman, F., et al.: A comparison between body size and lung function of swimmers and normal school children. *J. Physiol. (London)*, 156:9, 1961.

54. Pardy, R.L., et al.: Respiratory muscle training compared with physiotherapy in chronic airflow limitation. *Am. Rev. Respir. Dis.*, 123:421, 1981.

55. Rasch, P.J., and Brandt, J.W.A.: Measurement of pulmonary function in United States Olympic free style wrestlers. *Res. Q.*, 28:279, 1957.

56. Reiff, D.B., et al.: The effect of prolonged submaximal warm-up on exercise-induced asthma. *Am. Rev. Respir. Dis.*, 139:479, 1989.

57. Robinson, E.P., and Kjeldgaard, J.M.: Improvement in ventilatory muscle function with running. *J. Appl. Physiol.*, 52:1400, 1982.

58. Sady, S., et al.: Physiological characteristics of high-ability prepubescent wrestlers. *Med. Sci. Sports Exerc.*, 16:72, 1984.

59. Sejersted, O.N., et al.: Intramuscular fluid pressure during isometric contraction of human skeletal muscle. *J. Appl. Physiol.*, 56:287, 1984.

60. Sharp, J.T., et al.: Relative contributions of rib cage and abdomen to breathing in normal subjects. *J. Appl. Physiol.*, 39:608, 1975.

61. Smith, M.A., et al.: Assessment of beat to beat changes in cardiac output during the Valsalva manoeuvre using bioimpedence cardiology. *Clin. Sci.*, 72:423, 1987.

62. Sonne, L.J., and Davis, J.A.: Increased exercise performance in patients with severe COPD following inspiratory resistive training. *Chest*, 81:436, 1982.

63. Spiro, S.G., et al.: An analysis of the physiological strain of submaximal exercise in patients with chronic obstructive lung disease. *Thorax*, 30:415, 1975.

64. Strauss, R.H., et al.: Influence of heat and humidity on the airway obstruction induced by exercise in asthma. *J. Clin. Invest.*, 61:433, 1978.

65. Stubbing, D.G., et al.: Pulmonary mechanics during exercise in normal males. *J. Appl. Physiol.*, 49:506, 1980.

66. Ten Harkel, A.D.J., et al.: Assessment of cardiovascular reflexes: influence of posture and period of preceding rest. *J. Appl. Physiol.*, 68:147, 1990.

67. Turner, J.M., et al.: Elasticity of human lungs in relation to age. *J. Appl. Physiol.*, 25:664, 1968.

68. Voy, R.O.: The U.S. Olympic committee experience with exercise-induced bronchospasm, 1984. *Med. Sci. Sports Exerc.*, 18:328, 1986.

69. Wilmore, J.H.: A simplified method for the determination of residual lung volume. *J. Appl. Physiol.*, 27:96, 1969.

70. Wilmore, J.H., and Haskell, W.L.: Body composition and endurance capacity of professional football players. *J. Appl. Physiol.*, 33:564, 1972.

13

Our supply of oxygen depends on the oxygen **concentration** in ambient air and its **pressure.** Ambient or atmospheric air remains relatively constant in terms of composition. It is composed of approximately 20.93% oxygen, 79.04% nitrogen (this includes small quantities of other inert gases that behave physiologically like nitrogen), 0.03% carbon dioxide, and usually small quantities of water vapor. The gas molecules move at relatively great speeds and exert a pressure against any surface with which they come in contact. At sea level, the pressure of the gas molecules in air is sufficient to raise a column of mercury to a height of 760 mm, or 29.9 inches. These barometric readings vary somewhat with changing weather conditions and are considerably lower at altitude (see Chapter 24).

Gaseous Exchange in the Lungs and Tissues

CONCENTRATIONS AND PARTIAL PRESSURES OF RESPIRED GASES

The molecules of a specific gas in a mixture of gases exert their own **partial pressure.** The total pressure of the mixture is the sum of the partial pressures of the individual gases. Partial pressure is computed as:

$$\text{Partial pressure} = \text{Percent concentration} \times \text{Total pressure of gas mixture}$$

AMBIENT AIR

The volume, percentage, and partial pressure of the gases in dry, ambient air at sea level are presented in Table 13–1. The partial pressure of oxygen is 20.93% of the total pressure of 760 mm Hg exerted by air, or 159 mm Hg (.2093 × 760 mm Hg); the random movement of the minute quantity of carbon dioxide exerts a pressure of only 0.2 mm Hg (.0003 × 760 mm Hg),

TABLE 13–1.
PARTIAL PRESSURE AND VOLUME OF THE GASES IN DRY AMBIENT AIR AT SEA LEVEL

Gas	Percentage	Partial Pressure (at 760 mm Hg)	Volume of Gas (ml · l^{-1})
Oxygen	20.93	159 mm Hg	209.3
Carbon dioxide	0.03	0.2 mm Hg	0.4
Nitrogen	79.04[a]	600 mm Hg	790.3

[a] Includes 0.93% argon and other trace rare gases.

whereas the molecules of nitrogen exert a pressure that would raise the mercury in a manometer about 600 mm (.7094 × 760 mm Hg). Partial pressure is usually denoted by a P in front of the gas symbol; the P_{O_2}, P_{CO_2}, and P_{N_2} in ambient air at sea level are 159, 0.2, and 600 mm Hg, respectively.

TRACHEAL AIR

Air becomes completely saturated with water vapor as it enters the nose and mouth and passes down the respiratory tract. This vapor dilutes the inspired air mixture somewhat. At a body temperature of 37°C, for example, the pressure of water molecules in humidified air is 47 mm Hg; this leaves 713 mm Hg (760 − 47) as the total pressure exerted by the inspired dry air molecules. Consequently, the effective P_{O_2} in **tracheal air** is lowered by about 10 mm Hg from its ambient value of 159 mm Hg to 149 mm Hg [.2093(760 − 47 mm Hg)]. Because carbon dioxide is almost negligible in inspired air, the humidification process has little effect on the inspired P_{CO_2}.

ALVEOLAR AIR

The composition of alveolar air differs considerably from the incoming breath of moist ambient air because carbon dioxide is continually entering the alveoli from the blood, whereas oxygen is leaving the lungs to be carried throughout the body. As shown in Table 13–2, alveolar air contains approximately 14.5% oxygen, 5.5% carbon dioxide, and about 80.0% nitrogen. After subtracting the vapor pressure of moist alveolar gas, the average alveolar P_{O_2} is 103 mm Hg [.145 (760 − 47 mm Hg)] and 39 mm Hg [.055 (760 − 47 mm Hg)] for P_{CO_2}. **These values represent the average pressures exerted by oxygen and carbon dioxide molecules against the alveolar side of the alveolar-capillary membrane.** They are not physiologic constants but vary somewhat with the phase of the ventilatory cycle, as well as with the adequacy of ventilation in various portions of the lung. It should be recalled

TABLE 13–2.
PARTIAL PRESSURE AND VOLUME OF DRY ALVEOLAR GASES AT SEA LEVEL (37°C)

Gas	Percentage	Partial Pressure (at 760–47 mm Hg)	Volume of Gas (ml · l^{-1})
Oxygen	14.5	103 mm Hg	145
Carbon dioxide	5.5	39 mm Hg	55
Nitrogen	80.0	571 mm Hg	800
Water vapor		47 mm Hg	

that a relatively large volume of air remains in the lungs after each normal exhalation. This functional residual capacity serves as a damper so that each incoming breath of air has only a small effect on the composition of alveolar air. Thus, **the partial pressure of gases remains relatively stable in the alveoli.**

MOVEMENT OF GAS IN AIR AND FLUIDS

In accordance with **Henry's Law,** the amount of gas that dissolves in a fluid is a function of two factors: (1) the **pressure** of the gas above the fluid, and (2) the **solubility** of the gas.

PRESSURE

Oxygen molecules continually strike the surface of the water in the three chambers illustrated in Figure 13–1. Because the pure water in container A contains no oxygen, a large number of oxygen molecules enter the water and become dissolved. Because dissolved gas molecules are also in random motion, some oxygen molecules leave the water. In chamber B, the net movement of oxygen is still into the fluid from the gaseous state. Eventually, the number of molecules entering and leaving the fluid becomes equal as in chamber C. When this occurs, the gas pressures are in **equilibrium,** and no net diffusion of oxygen occurs. Conversely, if the pressure of dissolved oxygen molecules exceeds the pressure of the free gas in the air, oxygen leaves the fluid until a new pressure equilibrium is reached.

SOLUBILITY

For two different gases at identical pressures, the number of molecules moving into or out of a fluid is determined by the solubility of each gas. For each unit of pressure favoring diffusion, approximately 25 times more carbon dioxide than oxygen will move into (or from) a fluid. Viewed another way, equal quantities of oxygen and carbon dioxide will enter or leave a fluid under significantly different pressure gradients for each gas. This is precisely what takes place in the body.

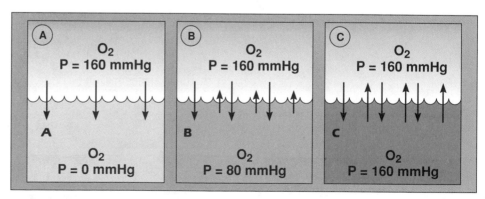

FIG. 13–1.
Solution of oxygen in water: A, when oxygen first comes in contact with pure water, B, after the dissolved oxygen is halfway to equilibrium with gaseous oxygen, and C, after equilibrium has been established.

GAS EXCHANGE IN THE LUNGS AND TISSUES

The exchange of gases between the lungs and the blood as well as their movement at the tissue level is due entirely to the passive process of diffusion. Figure 13–2 illustrates the pressure gradients favoring gas transfer in the body.

GAS EXCHANGE IN THE LUNGS

At rest, the pressure of oxygen molecules in the alveoli is about 60 mm Hg greater than in the venous blood that enters the pulmonary capillaries. Consequently, oxygen dissolves and diffuses through the alveolar membrane **into the blood.** Carbon dioxide, on the other hand, exists under a slightly greater pressure in returning venous blood than it does in the alveoli. The net diffusion of carbon dioxide is, therefore, from the blood **into the lungs.** Although the pressure gradient of 6 mm Hg for carbon dioxide diffusion is small compared to that for oxygen, adequate transfer of this gas is achieved rapidly due to its high solubility. Nitrogen, which is neither utilized nor produced in metabolic reactions, remains essentially unchanged in alveolar-capillary gas.

The process of gas exchange is so rapid in the healthy lung that an equilibrium between blood and alveolar gas occurs in less than 1 second, or at about the midpoint of the blood's transit through the lungs. Even during intense exercise, the speed at which red blood cells pass through the pulmonary capillaries is maintained within one-half the speed at rest. One important reason for this is that with increasing exercise intensity the pulmonary capillaries can increase the volume of blood contained within them to about 3 times the resting value.[2] Thus, the blood leaving the lungs to be delivered throughout the body contains oxygen at a pressure of approximately 100 mm Hg and carbon dioxide at 40 mm Hg, and these values vary little during vigorous exercise.*

GAS TRANSFER IN THE TISSUES

In the tissues, where oxygen is consumed in energy metabolism and an almost equal amount of carbon dioxide is produced, gas pressures can differ considerably from those in arterial blood. At rest, the average P_{O_2} in the fluid immediately outside a muscle cell rarely drops below 40 mm Hg, and the cellular P_{CO_2} averages about 46 mm Hg. In heavy exercise, however, the pressure of oxygen molecules in the muscle tissue may fall to about 3 mm Hg,[18] whereas the pressure of carbon dioxide approaches 90 mm Hg. **It is the pressure differences between gases in the plasma and tissues that establishes the gradients for diffusion.** Oxygen leaves the blood and diffuses toward the metabolizing cell while carbon dioxide flows from the cell to the blood. The blood then passes into the veins and is returned to the heart to be pumped subsequently to the lungs. As the blood enters the dense capillary network of the lungs, diffusion rapidly begins once again.

---

* The P_{O_2} of arterial blood is usually slightly lower than the alveolar P_{O_2} because some blood in the alveolar capillaries may pass through poorly ventilated alveoli. Also, the blood leaving the lungs is joined by venous blood from the bronchial and cardiac circulations. This small amount of poorly oxygenated blood has been termed **venous admixture.** Although its effect is small in healthy individuals, it does reduce the arterial P_{O_2} slightly below that in pulmonary end-capillary blood.

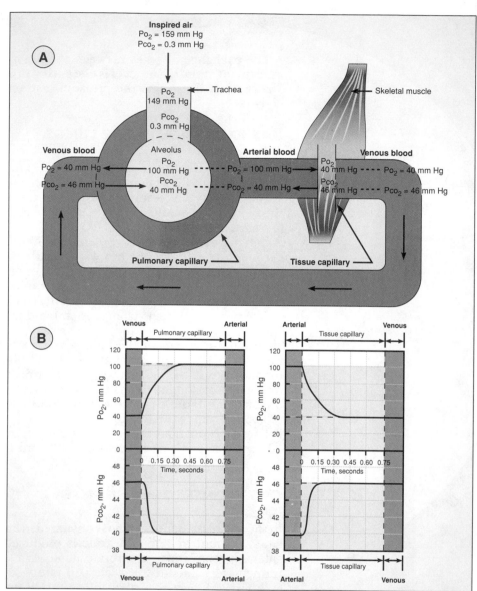

FIG. 13–2.

Pressure gradients for gas transfer in the body at rest. In A, the P_{O_2} and P_{CO_2} of ambient, tracheal, and alveolar air are shown as well as the gas pressures in venous and arterial blood and muscle tissue. Movement of gases at the alveolar-capillary and tissue-capillary membranes is always from an area of higher partial pressure to one of lower partial pressure. The time required for gas exchange is shown in B. At rest, blood remains in the pulmonary and tissue capillaries for about 0.75 s. During maximal exercise, the transit time is reduced to about 0.4 s, but this is still adequate for complete aeration of the blood. (From Mathews, D.K., and Fox, E.L.: The Physiological Basis of Physical Education and Athletics. Philadelphia, W.B. Saunders Co., 1976.)

The body does not attempt to rid itself completely of carbon dioxide. On the contrary, as the blood leaves the lungs with a P_{CO_2} of 40 mm Hg, it still contains about 50 ml of carbon dioxide in each 100 ml of blood. As can be seen in the next chapter, this "background level" of carbon dioxide is vital because it provides chemical input for the control of breathing through its effect on the respiratory center in the brain.

If it were not for our capacity to breathe, some average pressure would be reached between alveolar and blood gases and diffusion would cease. By bringing in another breath of air, however, the oxygen content of the alveoli is increased whereas the carbon dioxide is diluted. **By adjusting alveolar ventilation to metabolic demands, the composition of alveolar gas remains remarkably constant, even during strenuous exercise that increases oxygen uptake and carbon dioxide output by as much as 25 times.**

SUMMARY

1. In the lungs and tissues, gas molecules diffuse down their concentration gradients from an area of higher concentration (higher pressure) to one of lower concentration (lower pressure).
2. The partial pressure of a specific gas in a mixture of gases is proportional to the concentration of the gas and the total pressure exerted by the mixture.
3. The quantity of gas that dissolves in a fluid is determined by pressure and solubility. Because carbon dioxide is about 25 times more soluble in plasma than oxygen, large amounts of this gas move into and out of body fluids down a relatively small diffusion (pressure) gradient.
4. At rest and during exercise, adjustments in alveolar ventilation occur so the composition of alveolar gas remains constant. Oxygen and carbon dioxide pressures are maintained at about 100 mm Hg and 40 mm Hg, respectively. Because venous blood contains oxygen at lower and carbon dioxide at higher pressure than alveolar gases, oxygen diffuses into the blood and carbon dioxide diffuses into the lungs.
5. Gas exchange is so rapid in the healthy lung that equilibrium occurs at about the midpoint of the blood's transit through the pulmonary capillaries. Even with vigorous exercise the velocity of blood flow through the lungs generally does not restrict the full loading of oxygen and unloading of carbon dioxide.
6. At the tissues, the diffusion gradient favors the movement of oxygen from the capillary to the tissues and carbon dioxide from the cells into the blood. In exercise, these gradients are expanded and oxygen and carbon dioxide diffuse rapidly.

2 Transport of Oxygen

OXYGEN TRANSPORT IN THE BLOOD

Oxygen is carried in the blood in two ways: (1) in physical solution dissolved in the fluid portion of the blood, and (2) in loose combination with hemoglobin, the iron-protein molecule in the red blood cell.

OXYGEN IN SOLUTION

Oxygen is not particularly soluble in fluids. In fact, at an alveolar P_{O_2} of 100 mm Hg, only about 0.3 ml of gaseous oxygen dissolves in each 100 ml of plasma; this is equivalent to 3 ml of oxygen per liter of plasma. Because the average blood volume is about 5 liters, 15 ml of oxygen are carried dissolved in the fluid portion of the blood (3 ml per liter × 5). This is enough oxygen to sustain life for about 4 seconds! Viewed from a somewhat different perspective, if oxygen alone in physical solution was available to the body, about 80 liters of blood would have to be circulated each minute simply to supply the resting oxygen requirements. This rate is about 2 times higher than the maximum blood flow ever recorded for an exercising human!

The small quantity of oxygen transported in physical solution, however, does serve several important physiologic functions. The random movement of dissolved oxygen molecules establishes the P_{O_2} of the blood and tissue fluids. This pressure of dissolved oxygen plays a role in the regulation of breathing; it also determines the loading and subsequent release of oxygen from hemoglobin in the lungs and tissues, respectively.

OXYGEN COMBINED WITH HEMOGLOBIN

Metallic compounds are present in the blood of many species of animals and serve to augment the blood's oxygen-carrying capacity. In humans, this compound is **hemoglobin**, an iron-containing protein pigment. Hemoglobin, a main component of the body's 25 trillion red blood cells, increases the blood's oxygen-carrying capacity 65 to 70 times above that normally dissolved in plasma. Thus, for each liter of blood, about 197 ml of oxygen are temporarily "captured" by hemoglobin. Each of the four iron atoms in the hemoglobin molecule can loosely bind one molecule of oxygen in the reversible reaction:

$$Hb_4 + 4O_2 \rightleftarrows Hb_4O_8$$

This reaction requires no enzymes, and it occurs without a change in the valance of Fe^{++}, which would occur in the more permanent process of oxidation. **The oxygenation of hemoglobin to oxyhemoglobin depends entirely on the partial pressure of oxygen in solution.**

Oxygen-Carrying Capacity of Hemoglobin

In men, there are approximately 15 to 16 g of hemoglobin in each 100 ml of blood. The value is between 5 and 10% less for women and averages about 14 g per 100 ml of blood. This apparent "sex-difference" may account to some degree for the lower maximal aerobic capacity of women, even after considering differences in body weight and body fat.

Each gram of hemoglobin can combine loosely with 1.34 ml of oxygen. Thus, if the hemoglobin content of the blood is known, its oxygen-carrying capacity can easily be calculated as follows:

$$\begin{array}{ccc} \text{Blood's oxygen} & & \\ \text{capacity} & = & \text{Hemoglobin} \\ (\text{ml} \cdot 100\ \text{ml}^{-1}\ \text{blood}) & & (\text{g} \cdot 100\ \text{ml}^{-1}\ \text{blood}) \end{array} \times \begin{array}{c} \text{Oxygen capacity} \\ \text{of hemoglobin} \end{array}$$

$$20\ \text{ml O}_2 \quad = \quad 15 \quad \times\ 1.34\ \text{ml O}_2 \cdot \text{g}^{-1}$$

On the average, approximately 20 ml of oxygen would be carried with the hemoglobin in each 100 ml of blood when the hemoglobin is fully saturated with oxygen; that is, when all of the hemoglobin is converted to HbO_2.

The blood's oxygen transport capacity changes only slightly with normal variations in hemoglobin content. On the other hand, a significant decrease in the iron content of the red blood cell, as occurs in **iron deficiency anemia,** causes a decrease in the blood's oxygen-carrying capacity and corresponding reduced capacity for sustaining even mild aerobic exercise.[5] This finding is shown in Table 13–3. Twenty-nine iron-deficient-anemic men and women with low hemoglobin levels were placed in one of two groups; one group

TABLE 13–3.
HEMATOLOGIC AND EXERCISE HEART RATE RESPONSES OF ANEMIC SUBJECTS TO IRON TREATMENT*

Subjects	Hb (g per 100 ml blood) (average)	Peak Exercise Heart Rate (average)
Normal		
Men	14.3	119
Women	13.9	142
Iron-Deficient Men		
Pretreatment	7.1	155
Post-treatment	14.0	113
Iron-Deficient Women		
Pretreatment	7.7	152
Post-treatment	12.4	123
Iron-Deficient Men		
Preplacebo	7.7	146
Post-placebo	7.4	137
Iron-Deficient Women		
Preplacebo	8.1	154
Post-placebo	8.4	144

* From Gardner, G.W., et al.: Cardiorespiratory, hematological, and physical performance responses of anemic subjects to iron treatment. Am. J. Clin. Nutr., 28:982, 1975.

received intramuscular injections of iron over an 80-day period, whereas the placebo group received similar intramuscular injections of colored salt solution. A third group with normal levels of hemoglobin served as controls. All groups were tested during exercise prior to the experiment and after 80 days of either iron therapy or placebo treatment. The results show clearly that the anemic group given the iron supplement improved significantly in exercise response compared to their non-supplemented counterparts. Peak heart rate as measured during a 5-minute stepping performance decreased from 155 to 113 beats per minute for men and from 152 to 123 beats per minute for women. This translates into an average of 15% more oxygen delivered per heart beat.

Po₂ and Hemoglobin Saturation

Thus far, we have assumed that the hemoglobin becomes fully saturated with oxygen when exposed to alveolar gas. Figure 13–3 illustrates the **oxy-hemoglobin dissociation** curve that shows the saturation of hemoglobin with oxygen at various P_{O_2}s including that of normal alveolar-capillary gas

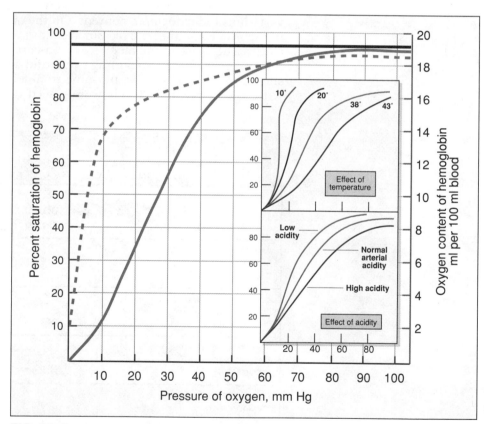

FIG. 13–3.

Percent saturation of hemoglobin (solid line) and myoglobin (dashed line) in relation to oxygen pressure. The quantity of oxygen carried in each 100 ml of blood under normal conditions is shown on the right ordinate. The insert curves indicate the effects of temperature and acidity in altering the affinity of hemoglobin for oxygen. Bold horizontal line at top indicates % saturation at alveolar P_{O_2} at sea level.

(P_{O_2} = 100 mm Hg). Shown on the right ordinate of this dissociation curve is the quantity of oxygen carried in each 100 ml of normal blood at a particular plasma P_{O_2}. Dissociation curves are usually established by exposing a small amount of blood in a sealed glass vessel to various pressures of oxygen. Once the blood-gas mixture has equilibrated in a water bath at a known temperature and acidity, the oxygen content and percent saturation of the blood are determined. Percent saturation is calculated as:

$$\text{Percent saturation} = \frac{O_2 \text{ combined with hemoglobin}}{O_2 \text{ capacity of hemoglobin}} \times 100$$

P_{O_2} in the Lung

Hemoglobin is about 98% saturated with oxygen at the normal alveolar P_{O_2} of 100 mm Hg. By applying this partial pressure value to the right ordinate of Figure 13–3, it is seen that for each 100 ml of blood leaving the lungs, hemoglobin carries about 19.7 ml of oxygen. Clearly, any additional increase in alveolar P_{O_2} contributes little to the quantity of oxygen already combined with hemoglobin. In addition to the oxygen bound to hemoglobin, the plasma of each 100 ml of arterial blood contains about 0.3 ml of oxygen in solution. This dissolved gas exerts a pressure of about 100 mm Hg when one breathes ambient air at sea level; this plasma P_{O_2} regulates the loading and unloading of hemoglobin. Thus, for healthy individuals who breathe ambient air at sea level, approximately 20.0 ml of oxygen are carried in each 100 ml of blood leaving the lungs; 19.7 ml are bound to hemoglobin, and 0.3 ml are dissolved in plasma. The percentage composition of centrifuged whole blood for plasma and red blood cells (**hematocrit**) as well as representative values for the quantity of oxygen carried in each component is shown in Figure 13–4.

Figure 13–3 also shows that the saturation of hemoglobin changes little until the pressure of oxygen falls to about 60 mm Hg. This flat, upper portion of the oxyhemoglobin curve provides a margin of safety to assure that the blood is adequately loaded with oxygen. Even if the alveolar P_{O_2} is reduced to 75 mm Hg, as could occur with certain lung diseases or when one travels to a higher altitude, the saturation of hemoglobin is only lowered about 6%. At an alveolar P_{O_2} of 60 mm Hg, hemoglobin is still 90% saturated with oxygen! Below this pressure, however, the quantity of oxygen that will combine with hemoglobin drops sharply.

P_{O_2} in the Tissues

At rest, the P_{O_2} in the cell fluids is approximately 40 mm Hg. Dissolved oxygen from the plasma diffuses across the capillary membrane through the tissue fluids into the cells. This reduces the plasma P_{O_2} below the P_{O_2} in the red blood cell, and hemoglobin is unable to maintain its high oxygen saturation. The released oxygen ($HbO_2 \rightarrow Hb + O_2$) moves out of the blood cells through the capillary membrane and into the tissues.

At the tissue-capillary P_{O_2} at rest (P_{O_2} = 40 mm Hg), hemoglobin holds about 70% of its total oxygen (Fig. 13–3). Blood leaving the tissues, therefore, carries about 15 ml of oxygen in each 100 ml of blood; nearly 5 ml of oxygen have been released to the tissues. This difference in the oxygen content of

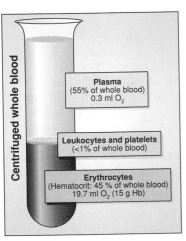

FIG. 13–4.
Major components and oxygen carried in 100 ml of whole blood.

arterial and mixed venous blood is termed the **arteriovenous oxygen difference,** or **a-$\bar{v}O_2$ difference.**

The a-$\bar{v}O_2$ difference at rest normally averages 4 to 5 ml of oxygen per 100 ml of blood. The large quantity of oxygen still remaining with hemoglobin provides an "automatic" reserve by which cells can immediately obtain oxygen should the metabolic demands suddenly increase. As the cell's need for oxygen increases in exercise, the tissue P_{O_2} becomes reduced and a larger quantity of oxygen is rapidly released. During vigorous exercise, for example, when the extracellular P_{O_2} decreases to about 15 mm Hg, only about 5 ml of oxygen remain bound to hemoglobin. As a result, the a-$\bar{v}O_2$ difference increases to 15 ml of oxygen per 100 ml of blood. When tissue P_{O_2} falls to 3 mm Hg during exhaustive exercise, virtually all of the oxygen is released from the blood that perfuses the active tissues. Clearly, without any increase in local blood flow, the amount of oxygen released to the muscles can increase almost three times above that normally supplied at rest—just by a more complete unloading of hemoglobin.

The Bohr Effect

The solid line in Figure 13–3 shows the oxyhemoglobin dissociation curve under resting physiologic conditions at an arterial pH of 7.4 and tissue temperature of 37°C. The insert curves depict other important characteristics of hemoglobin. Any increase in acidity, temperature, or concentration of carbon dioxide causes the dissociation curve to shift significantly downward and to the right. This phenomenon is called the **Bohr effect** after its discoverer and is a consequence of an alteration in the molecular structure of hemoglobin. The Bohr effect describes the reduced effectiveness of hemoglobin to hold oxygen, especially in the P_{O_2} range of 20 to 50 mm Hg. This is particularly important in vigorous exercise, because even more oxygen is released to the tissues with the accompanying increase in metabolic heat, carbon dioxide, and lactic acid. At the P_{O_2} in the alveoli the Bohr effect in pulmonary capillary blood is negligible. This is important because it allows hemoglobin to load completely with oxygen as blood passes through the lungs, even during maximal exercise.

Red-Blood-Cell 2,3-DPG

The substance **2,3-diphosphoglycerate,** or 2,3-DPG, is produced within the red blood cell during the anaerobic reactions of glycolysis. (Because the red blood cell contains no mitochondria, its energy is supplied via the anaerobic reactions of glycolysis; this contributes to a continual level of plasma lactate at rest.) This compound appears to bind loosely with subunits of the hemoglobin molecule, reducing its affinity for oxygen. This would theoretically increase the availability of oxygen to the tissues.[3] For a given decrease in P_{O_2}, therefore, more oxygen is released to the tissues. Individuals with cardiopulmonary disorders and those who live at high altitudes have an increased level of red blood cell 2,3-DPG.[11] This apparently provides a compensatory adjustment to facilitate oxygen release to the cells.

The presence of 2,3-DPG would also aid in oxygen transfer to the muscles during strenuous exercise which is limited by the capability to sustain aerobic metabolism. Conflicting results have been reported in comparing the 2,3-DPG level of trained and untrained subjects.[4,12,16,19] Significantly higher resting levels of this metabolic intermediate were observed in two groups of athletes than in untrained subjects.[19] The researchers also observed that the level of 2,3-DPG increased by 15% for the middle-distance runners following maximal exercise of short duration. Prolonged steady-rate exercise, on the other hand, resulted in a small decrease in 2,3-DPG in endurance athletes. Research also confirms that high-intensity exercise can cause a significant increase in 2,3-DPG,[9] thus supporting the idea that increases in this compound are adaptive responses to augment adequate oxygen for the active tissues. More than likely, this difference in the effects of exercise on hemoglobin's affinity for oxygen is due to the specific metabolic demands of exercise. Furthermore, females have been shown to have significantly higher levels of red blood cell 2,3-DPG compared to male counterparts of similar aerobic fitness. This possible gender difference might compensate for the lower hemoglobin levels routinely noted among females.[14]

MYOGLOBIN, THE MUSCLE'S OXYGEN STORE

Myoglobin is an iron-protein compound found in skeletal and cardiac muscle. Reddish muscle fibers have a high concentration of this respiratory pigment, whereas fibers deficient in myoglobin appear pale or white.[13] Myoglobin is similar to hemoglobin because it also combines reversibly with oxygen; however, each myoglobin molecule contains only one iron atom in contrast to hemoglobin that contains four atoms. Myoglobin adds additional oxygen to the muscle in the reaction:

$$Mb + O_2 \rightarrow MbO_2$$

Oxygen Released at Low Pressures

Aside from its function as an "extra" source of oxygen in muscle, myoglobin probably acts to facilitate the transfer of oxygen to the mitochondria, especially in the beginning of exercise and during very intense exercise when there is a considerable drop in cellular P_{O_2}.[21] It can be noted from the dissociation curve for myoglobin shown in Figure 13-3 (dashed line) that the line is not s-shaped, as was the case with hemoglobin, but instead forms a rectangular hyperbola. This shows that myoglobin binds and retains oxygen at low pressures much more readily than hemoglobin. During rest and moderate levels of exercise, myoglobin retains a high saturation with oxygen. For example, at a P_{O_2} of 40 mm Hg, myoglobin retains 95% of its oxygen. The greatest quantity of oxygen is released from MbO_2 when the tissue P_{O_2} drops to 5 mm Hg or less. Unlike hemoglobin, myoglobin does not demonstrate a "Bohr effect."

Effects of Training

As might be expected, slow-twitch fibers that have a high capacity to generate ATP aerobically contain relatively large quantities of myoglobin. With animals, the myoglobin level of muscle also appears to be related to the animal's level of physical activity.[6,10,15] The leg muscles of active hunting dogs contain more myoglobin than the muscles of sedentary house pets[20]; this is also the case for grazing cattle compared to cattle that are penned.[17] Whether myoglobin levels can be enhanced in humans as part of the adaptive response to training remains unclear.[1,7,8]

▬▬ SUMMARY

1. Hemoglobin, the iron-protein pigment in the red blood cell, increases the oxygen-carrying capacity of whole blood about 65 times that carried in physical solution dissolved in the plasma.

2. The small amount of oxygen dissolved in plasma exerts molecular movement and establishes the partial pressure of oxygen in the blood. This determines the loading at the lungs (oxygenation) and unloading at the tissues (deoxygenation) of hemoglobin.

3. The blood's oxygen transport capacity varies only slightly with normal variations in hemoglobin content. Iron deficiency anemia, however, significantly decreases the blood's oxygen-carrying capacity and consequently reduces aerobic exercise performance.

4. The s-shaped nature of the oxyhemoglobin dissociation curve shows that hemoglobin saturation changes very little until the P_{O_2} falls below 60 mm Hg. Because this low pressure occurs in the tissues, the quantity of oxygen bound to hemoglobin falls sharply. Thus, oxygen is released rapidly from capillary blood and flows into the tissues in response to the cells' metabolic demands.

5. At rest, only about 25% of the blood's total oxygen is released to the tissues; the remaining 75% returns to the heart in the venous blood. This is the arteriovenous oxygen difference and indicates that an "automatic" reserve of oxygen exists so cells can rapidly obtain oxygen should the metabolic demands increase suddenly.

6. Increases in acidity, temperature, carbon dioxide concentration, and red-blood-cell 2,3-DPG cause an alteration in the molecular structure of hemoglobin, thereby reducing its effectiveness to hold oxygen. Because these factors are accentuated in exercise, the release of oxygen to the tissues is further facilitated.

7. In skeletal and cardiac muscle, the iron-protein pigment myoglobin acts as an "extra" oxygen store. It releases its oxygen at low oxygen pressures and during strenuous exercise probably facilitates oxygen transfer to the mitochondria, when there is a considerable decrease in cellular P_{O_2}.

P A R T

3 **Transport of Carbon Dioxide**

CARBON DIOXIDE TRANSPORT IN THE BLOOD

Once carbon dioxide is formed in the cell, its only means for "escape" is through the process of diffusion and subsequent transport to the lungs in the venous blood. As with oxygen, a small amount of carbon dioxide is carried in physical solution in the blood plasma. Carbon dioxide is also transported combined with hemoglobin, and a large fraction combines with water and is delivered to the lung in the form of **bicarbonate.** Figure 13–5 illustrates the various means for transport of carbon dioxide from the tissues to the lungs.

FIG. 13–5.
Transport of carbon dioxide in the plasma and red blood cells as dissolved CO_2, bicarbonate, and carbamino compounds.

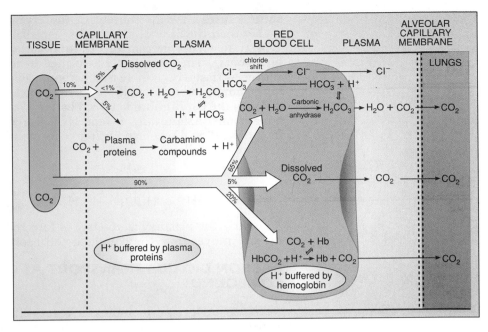

CARBON DIOXIDE TRANSPORT IN SOLUTION

Approximately 5% of the carbon dioxide produced during energy metabolism is carried as free carbon dioxide in solution in the plasma. Although this quantity is relatively small, it is the random movement of dissolved carbon dioxide molecules that establishes the P_{CO_2} of the blood.

CARBON DIOXIDE TRANSPORT AS BICARBONATE

Carbon dioxide in solution combines with water in a reversible reaction to form **carbonic acid.**

$$CO_2 + H_2O \rightleftharpoons H_2CO_3$$

This reaction is slow, and little carbon dioxide would be carried in this form if it were not for the action of **carbonic anhydrase,** a zinc-containing enzyme in the red blood cell. This catalyst accelerates the interaction of CO_2 and water about 5,000 times. In fact, the reaction reaches equilibrium while the blood cell is still moving through the tissue's capillary.

Once carbonic acid is formed in the tissues, most of it ionizes to hydrogen ions (H^+) and bicarbonate ions (HCO_3^-) as follows:

In Tissues

$$CO_2 + H_2O \xrightarrow{\text{Carbonic anhydrase}} H_2CO_3 \longrightarrow H^+ + HCO_3^-$$

The H^+ is then buffered by the protein portion of hemoglobin to maintain the pH of the blood within relatively narrow limits (see "Acid-Base Regulation," Chap. 14). Because HCO_3^- is quite soluble in blood, it diffuses from the red blood cell into the plasma in exchange for the chloride ion (Cl^-) that moves into the blood cell to maintain ionic equilibrium. This "chloride shift" causes the Cl^- content of the erythrocytes in venous blood to be higher than the arterial blood cells, especially during exercise.

Sixty to eighty percent of the total carbon dioxide is carried as plasma bicarbonate. The bicarbonate is formed in accordance with the law of mass action; as tissue P_{CO_2} increases, carbonic acid is formed rapidly. Conversely, in the lungs, carbon dioxide leaves the blood lowering the plasma P_{CO_2}. This disturbs the equilibrium between carbonic acid and the formation of bicarbonate ions. As a result, H^+ and HCO_3^- recombine to form carbonic acid. In turn, carbon dioxide and water re-form and carbon dioxide exits through the lungs as follows:

In Lungs

$$H^+ + HCO_3^- \longrightarrow H_2CO_3 \xrightarrow[\text{anhydrase}]{\text{Carbonic}} CO_2 + H_2O$$

Because the plasma bicarbonate is lowered in the pulmonary capillaries, the Cl^- moves from the red blood cell back into the plasma.

CARBON DIOXIDE TRANSPORT AS CARBAMINO COMPOUNDS

At the tissue level, carbon dioxide reacts directly with the amino acid molecules of blood proteins to form **carbamino compounds.** This is particularly true for the **globin** portion of hemoglobin that carries about 20% of the body's carbon dioxide as follows:

$$CO_2 + \underset{\text{(Hemoglobin)}}{HbNH} \longrightarrow \underset{\text{(Carbaminohemoglobin)}}{HbNHCOOH}$$

The formation of carbamino compounds is reversed as the plasma P_{CO_2} is lowered in the lungs. This causes carbon dioxide to move into solution and enter the alveoli. Concurrently, the oxygenation of hemoglobin reduces its binding ability for carbon dioxide. The interaction between oxygen loading and carbon dioxide release is termed the **Haldane effect.** This phenomenon facilitates the removal of carbon dioxide in the lung.

SUMMARY

1. A small amount of carbon dioxide is carried as free carbon dioxide in solution in the plasma. This dissolved carbon dioxide establishes the P_{CO_2} of the blood.
2. The major quantity of carbon dioxide is transported in chemical combination with water and is carried as bicarbonate as follows:

$$CO_2 + H_2O \rightarrow H_2CO_3 \rightarrow H^+ + HCO_3^-$$

In the lungs, this reaction is reversed and carbon dioxide leaves the blood and moves into the alveoli.
3. About 20% of the body's carbon dioxide combines with blood proteins, including hemoglobin, to form carbamino compounds.

REFERENCES

1. Coyle, E.F., et al.: Time course of loss of adaptations after stopping prolonged intense endurance training. *J. Appl. Physiol.*, 57:1857, 1984.
2. Dempsey, J.A.: Is the lung built for exercise? *Med. Sci. Sports Exerc.*, 18:143, 1986.

3. Dempsey, J.A., et al.: Muscular exercise, 2,3-DPG and oxy-hemoglobin affinity. *Int. J. Physiol.*, 30:34, 1971.
4. Fornaini, G., et al.: Glucose utilization in human erythrocytes during physical exercise. *Med. Sci. Sports Exerc.*, 13:323, 1981.
5. Gardner, G.W., et al.: Cardiorespiratory, hematological and physical performance responses of anemic subjects to iron treatment. *Am. J. Clin. Nutr.*, 28:982, 1975.
6. Hickson, R.C.: Skeletal muscle cytochrome c and myoglobin, endurance, and frequency of training. *J. Appl. Physiol.*, 51:746, 1981.
7. Jacobs, I.: Sprint training effects on muscle myoglobin, enzymes, fiber types, and blood lactate. *Med. Sci. Sports Exerc.*, 19:368, 1987.
8. Jansson, E., et al.: Myoglobin in the quadriceps femoris muscle of competitive cyclists and untrained men. *Acta Physiol. Scand.*, 114:627, 1982.
9. Klein, J.P., et al.: Hemoglobin affinity for oxygen during short-term exhaustive exercise. *J. Appl. Physiol.*, 48:236, 1980.
10. Lawrie, R.A.: Effect of enforced exercise on myoglobin in muscle. *Nature*, 171:1069, 1953.
11. Lenfant, C., et al.: Effect of altitude on oxygen binding by hemoglobin and on organic phosphate levels. *J. Clin. Invest.*, 47:2652, 1968.
12. Lijnen, P., et al.: Erythrocyte 2,3-diphosphoglycerate and serum enzyme concentrations in trained and sedentary men. *Med. Sci. Sports Exerc.*, 18:174, 1986.
13. Nemeth, P.M., and Lowry, O.H.: Myoglobin in individual human skeletal muscle fibers of different types. *J. Histochem. Cytochem.*, 32:1211, 1984.
14. Pate, R.R., et al.: A physiological comparison of performance—matched female and male distance runners. *Res. Q. Exerc. Sport*, 56:245, 1985.
15. Pattengale, P.K., and Holloszy, J.O.: Augmentation of skeletal muscle myoglobin by a program of treadmill running. *Am. J. Physiol.*, 213:783, 1967.
16. Rand, P.W., et al.: Influence of athletic training on hemoglobin-oxygen affinity. *Am. J. Physiol.*, 224:1334, 1973.
17. Shenk, J.H., et al.: Spectrophotometric characteristics of hemoglobins. *J. Biol. Chem.*, 105;741, 1934.
18. Stainsby, W.N., and Otis, A.B.: Blood flow, blood oxygen tension, oxygen uptake and oxygen transport in skeletal muscle. *Am. J. Physiol.*, 206:858, 1964.
19. Taunton, J.E., et al.: Alterations in 2,3-dpg and P_{50} with maximal and submaximal exercise. *Med. Sci. Sports*, 6:238, 1974.
20. Whipple, G.H.: The hemoglobin of striated muscle. 1. Variations due to age and exercise. *Am. J. Physiol.*, 76:693, 1926.
21. Wittenberg, B.A., et al.: Role of myoglobin in the oxygen supply to red skeletal muscle. *J. Biol. Chem.*, 250:9038, 1975.

Regulation of Pulmonary Ventilation

CONTROL OF VENTILATION

The rate and depth of breathing are exquisitely adjusted in response to the body's metabolic needs. In healthy individuals, the arterial pressures of oxygen, carbon dioxide, and pH are essentially regulated at the resting value regardless of exercise intensity. The mechanisms for this regulation, however, are complex and not fully understood. Intricate neural circuits relay information from higher centers in the brain, from the lungs themselves, and from other sensors throughout the body to contribute to the control of ventilation. In addition, the gaseous and chemical state of the blood that bathes the medulla and chemoreceptors in the aorta and carotid arteries act to mediate alveolar ventilation. As a result, relatively constant alveolar gas pressures are maintained even during exhaustive exercise. A schematic representation of the input for ventilatory control is shown in Figure 14–1.

NEURAL FACTORS

The normal respiratory cycle results from inherent automatic activity of inspiratory neurons whose cell bodies are located in the medial portion of the medulla. These neurons activate the diaphragm and intercostal muscles and the lungs inflate. The inspiratory neurons cease firing due to their own self-limitation as well as to the inhibitory influence from expiratory neurons also located in the **medulla.** As the lungs inflate, stretch receptors in lung tissue are also stimulated, especially in the bronchioles. These receptors act through afferent fibers to inhibit inspiration and stimulate expiration.

As the inspiratory muscles relax, exhalation occurs by the passive recoil of the stretched lung tissue and raised ribs. The activation of expiratory neurons and associated muscles that further facilitate expiration are synchronized with this passive phase. As expiration proceeds, the inspiratory center is progressively released from inhibition and once again becomes active.

FIG. 14–1.
Schematic representation of the mechanisms for ventilatory control.

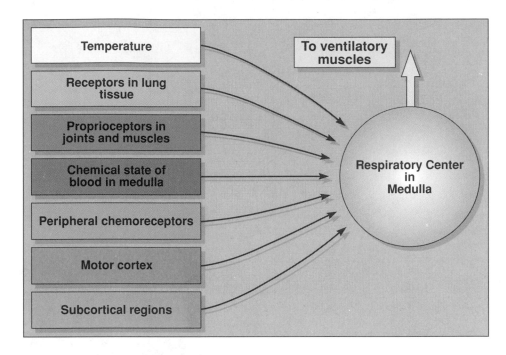

The inherent activity of the respiratory center cannot itself account for the smooth pattern of breathing in response to metabolic demands. The controlling influence of neurons in the cerebral hemispheres, the pons, and other regions of the brain also plays an important role in establishing the duration and intensity of the inspiratory cycle. For example, activation of the inspiratory center stimulates the pons region of the hindbrain, especially during labored breathing. The pons, in turn, relays excitatory impulses to the expiratory center, which hastens the exhalation phase of the breathing cycle.

HUMORAL FACTORS

Pulmonary ventilation at rest is largely regulated by the chemical state of the blood. Variations in arterial P_{O_2}, P_{CO_2}, acidity, and temperature activate sensitive neural units in the medulla and arterial system that adjust ventilation to maintain arterial blood chemistry within narrow limits.

Plasma P_{O_2} and Chemoreceptors

Inhalation of a gas mixture that contains 80% oxygen greatly increases alveolar P_{O_2} and causes about a 20% reduction in minute ventilation. Conversely, if the inspired oxygen concentration is reduced, ventilation increases, especially if the alveolar P_{O_2} falls below 60 mm Hg. It should be recalled that it is at this P_{O_2} that hemoglobin saturation begins to fall considerably.

Sensitivity to reduced oxygen pressures does not appear to reside in the respiratory center. Rather, it is more a result of the stimulation of peripheral **chemoreceptors.** As illustrated in Figure 14–2, these specialized neurons are located in the arch of the aorta and at the branching of the carotid arteries in the neck.

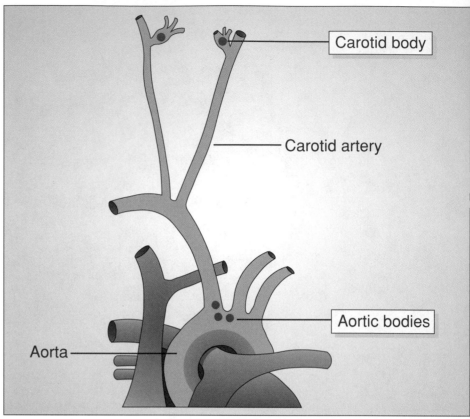

FIG. 14—2.
Cell bodies sensitive to a reduced plasma P_{O_2} are located in the aortic arch and bifurcation of the carotid arteries, and provide the body's first line of defense against arterial hypoxia.

A decrease in arterial P_{O_2}, as would occur when one ascends to high altitudes, activates the aortic and carotid receptors to increase ventilation. These chemoreceptors alone protect the organism against a reduced oxygen pressure in inspired air.

Aside from providing the early warning system against reduced oxygen pressure, the peripheral chemoreceptors also act to stimulate ventilation in response to increases in carbon dioxide, temperature, metabolic acidosis, and a fall in blood pressure.[69]

Plasma P_{CO_2} and Hydrogen Ion Concentration

At rest, the most important respiratory stimulus is the carbon dioxide pressure in arterial plasma. Small increases in P_{CO_2} in the inspired air cause large increases in minute ventilation. The resting ventilation is almost doubled, for example, by increasing the inspired P_{CO_2} to just 1.7 mm Hg (0.22% CO_2 in inspired air).

The regulation of ventilation by arterial P_{CO_2} is probably not mediated by the action of molecular carbon dioxide. Rather, ventilation appears to be controlled by plasma acidity that varies directly with the blood's CO_2 content. Recall that carbonic acid formed from carbon dioxide and water rapidly dissociates to bicarbonate ions and hydrogen ions. The increase in hydrogen ions, especially in the cerebrospinal fluid that bathes the respiratory areas, stimulates inspiratory activity. Then, as ventilation increases, CO_2 is eliminated, which in turn lowers the arterial hydrogen ion concentration.

HYPERVENTILATION AND BREATH-HOLDING

If a person breathholds after a normal exhalation, it takes about 40 seconds before the urge to breathe becomes so strong that the subject is forced to inspire. The desire to breathe is due mainly to the stimulating effects of increased arterial P_{CO_2} and H^+ concentration and **not** to the decreased P_{O_2} in the breathhold condition.[11] The "break point" for breathhold corresponds to an increase in arterial P_{CO_2} to about 50 mm Hg.

If this same person prior to breathhold consciously increases ventilation above the normal level, the composition of alveolar air changes and becomes more like that of ambient air. As a result of this overbreathing, or **hyperventilation,** alveolar P_{CO_2} may decrease to 15 mm Hg. This decrease creates a considerable diffusion gradient for the run-off of CO_2 from the venous blood entering the pulmonary capillaries. Consequently, a larger than normal quantity of CO_2 leaves the blood, and arterial P_{CO_2} becomes reduced significantly below normal levels. This extends the breathhold until the arterial P_{CO_2} and/or the H^+ concentration rise to the level to stimulate ventilation.

Hyperventilation and subsequent breathhold have been used by swimmers and divers in an attempt to improve performance. In sprint swimming, for example, it is undesirable from a mechanical viewpoint to roll the body and turn the head during the breathing phase of the stroke. Consequently, many swimmers hyperventilate on the starting blocks to prolong breathhold time during the swim. In sport diving, the intention of hyperventilation is the same as in competitive swimming—to extend breathhold time. In this sport, however, the results can be tragic.[12] As the length and depth of the dive increase, the oxygen content of the blood may be reduced to critically low values before arterial P_{CO_2} reaches the level to stimulate breathing and signal ascent to the surface. This reduction can cause the diver to lose consciousness before reaching the surface. Hyperventilation and other factors related to diving are discussed in Chapter 26.

REGULATION OF VENTILATION IN EXERCISE

CHEMICAL CONTROL

Chemical stimuli probably cannot entirely explain the increased ventilation, or **hyperpnea,** during physical activity. For example, even when artificial changes are made in P_{O_2}, P_{CO_2}, and acidity, increases in minute ventilation are not nearly as large as those observed in vigorous exercise.

During exercise, arterial P_{O_2} is **not** reduced to an extent that would increase ventilation due to chemoreceptor stimulation. In fact, in vigorous exercise, the large breathing volumes may cause the alveolar P_{O_2} to rise above the average resting value of 100 mm Hg. This rise is illustrated in Figure 14–3 where venous and alveolar P_{CO_2} and alveolar P_{O_2} are plotted in relation to oxygen consumption during a progressive exercise test. The slight **increase** in alveolar P_{O_2} in heavy exercise may hasten the equilibrium of alveolar

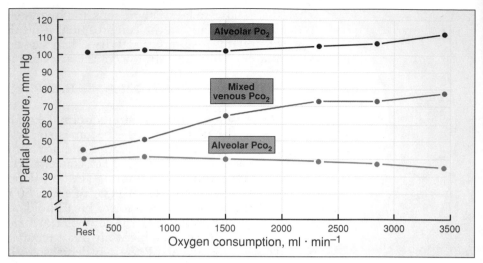

FIG. 14–3.

Values for the P_{CO_2} of mixed venous blood entering the lungs and the alveolar P_{O_2} and P_{CO_2} in relation to oxygen consumption during exercise. (Courtesy of Applied Physiology Laboratory, Queens College.)

blood gases; this facilitates the oxygenation of blood in the alveolar capillaries.

During light and moderate exercise, pulmonary ventilation is closely coupled to the metabolic rate in a manner that is proportional to the carbon dioxide production.[64] Under these conditions, the alveolar (and arterial) P_{CO_2} is generally maintained at about 40 mm Hg. In strenuous exercise, acidity and subsequent hydrogen ion concentration increases, which provides an additional ventilatory stimulus that usually **reduces** alveolar P_{CO_2} below this value, sometimes to as low as 25 mm Hg.[31] This would actually result in a decrease in arterial P_{CO_2} and a corresponding reduction in the ventilatory drive from carbon dioxide during exercise.

Thus, one might ask, how do the peripheral chemoreceptors exert their influence on ventilation during exercise? A possible explanation lies in the pattern of ventilation that causes the alveolar and capillary P_{CO_2} to be slightly lower at the end of inhalation and higher at the end of exhalation. Consequently, even though the **average** levels of arterial oxygen, carbon dioxide, and pH are well regulated in moderate exercise, the amplitude of cyclic changes or oscillations in these chemical factors during the breathing cycle may be detected by the chemoreceptors to influence breathing during exercise. This detection also may be facilitated by an increase in the sensitivity of the chemoreceptors as well as by other changes that affect the equilibrium of blood chemistry in exercise.[70]

NONCHEMICAL CONTROL

The changes in minute ventilation at the onset, during, and at the end of moderate exercise are shown in Figure 14–4. As exercise begins, ventilation increases so rapidly that it occurs almost within a single ventilatory cycle.*

---

* Some variation has been demonstrated in the rapidity of this initial ventilatory adjustment to exercise. Thus, the response should probably not be considered general, but rather a characteristic pattern of many people.[5,71]

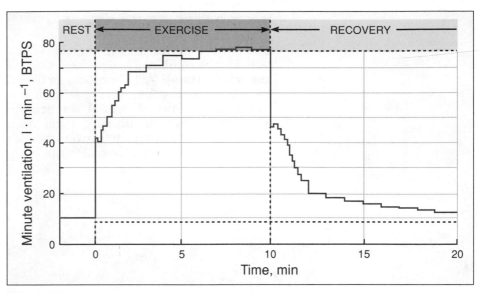

FIG. 14—4.
Rapid and slower components of ventilatory response during and in recovery from steady-rate exercise. (From Dejours, P.: Neurogenic factors in the control of ventilation during exercise. Circ. Res., XX, XXI (Suppl. 1): 146, 1967.)

This immediate change is followed by a plateau that lasts about 30 seconds and then by a gradual increase as the ventilation approaches a steady state. When exercise is terminated, ventilation decreases exponentially to a point about 40% of the steady-state value and then gradually returns to the resting level. The rapidity of the ventilatory response at the onset and cessation of exercise strongly suggests that this portion of exercise hyperpnea is mediated by input other than changes in arterial P_{CO_2} and H^+ content.

Neurogenic Factors

These factors include both cortical and peripheral influences.
Cortical influence—Neural outflow from regions of the motor cortex as well as cortical activation in anticipation of exercise stimulate the respiratory neurons in the medulla. This cortical outflow may act in concert with the demands of work to contribute to the abrupt increase in ventilation at the start of exercise.
Peripheral influence—Sensory input from joints, tendons, or muscles may influence the ventilatory adjustments to exercise. Although such peripheral receptors have not been identified, experiments involving passive limb movements, electrical stimulation of muscles, and voluntary exercise with the muscle's blood flow occluded, support the existence of such **mechano-receptors** in producing a reflex hyperpnea.[3]

Influence of Temperature

An increase in body temperature has a direct stimulating effect on the neurons of the respiratory center, and probably exerts some control over ventilation in prolonged exercise. The changes in ventilation at the beginning and end of exercise, however, are much more rapid than can be accounted for by changes in core temperature.

INTEGRATED REGULATION

The control of breathing in exercise is not the result of a single factor but rather is the combined and perhaps simultaneous result of several chemical and neural stimuli.[41,59] The model for respiratory control illustrated in Figure 14–5 suggests that neurogenic stimuli from the cerebral cortex or the exercising limbs cause the initial, abrupt increase in breathing at the beginning of exercise. After this initial change, minute ventilation gradually rises to a steady level that adequately meets the demands for metabolic gas exchange. Then, the regulation of alveolar gas pressures is probably maintained by

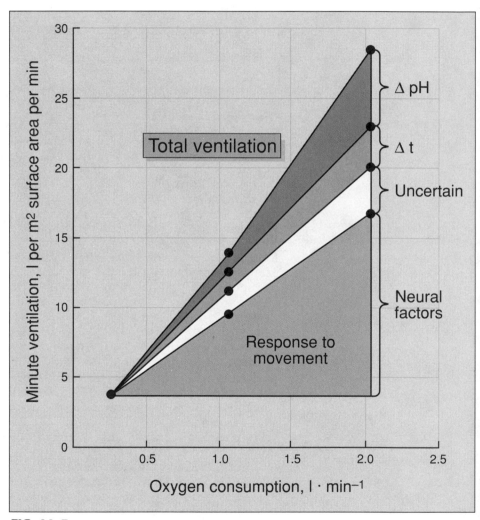

FIG. 14–5.
Composite of the ventilatory response to exercise. The contribution of changes in acidity (ΔpH) and temperature (Δt), as well as the effects of neurogenic stimuli from the cerebral regions and/or joints and muscles is estimated. The yellow-shaded wedge represents the ventilatory change not quantitatively accounted for by the other three factors. (From Lambertsen, C.J.: Interactions of physical, chemical, and nervous factors in respiratory control. In Medical Physiology. Edited by V.B. Mountcastle. St. Louis, C.V. Mosby Co., 1974.)

central and reflex chemical stimuli, especially those provided by temperature, carbon dioxide, and hydrogen ions.[59]

Some challenge has been provided to this theory of ventilatory control in exercise. It has been observed in dogs that ventilation increases immediately following an artificially induced, abrupt increase in blood flow through the heart, similar to that which occurs at the onset of exercise.[64,66] The rapidity of this ventilatory increase due to a surge in cardiac output suggests the presence of previously undetected chemoreceptors. The existence of such receptors would help explain the effective linkage between the cardiovascular and respiratory systems, as well as the rapid changes in breathing at the start and cessation of exercise.

SUMMARY

1. The normal respiratory cycle results from the inherent activity of neurons in the medulla. Superimposed on this neural output are intricate neural circuits that relay information from higher brain centers, from the lungs themselves, and from other sensors throughout the body.
2. At rest, several chemical factors act directly on the respiratory center or modify its activity reflexly through chemoreceptors to control alveolar ventilation. The most important factors are the level of arterial P_{CO_2} and acidity. A drop in arterial oxygen pressure as would occur during ascent to high altitude or in severe pulmonary disease will also provide a breathing stimulus.
3. Hyperventilation significantly lowers arterial P_{CO_2} and H^+ concentration. This prolongs breathhold time until normal levels of carbon dioxide and acidity are reached to stimulate breathing. Extended breathhold by hyperventilation should not be practiced during underwater swimming because the consequences can be deadly.
4. Ventilatory adjustments to exercise are augmented by nonchemical regulatory factors. These factors include (a) cortical activation in anticipation of exercise as well as outflow from the motor cortex when exercise begins, (b) peripheral sensory input from mechanoreceptors in joints and muscles, and (c) increases in body temperature.
5. Effective alveolar ventilation in exercise is the result of many neural and chemical factors that operate singularly and in combination. Each factor probably takes on greater importance at a particular phase of the adjustment to exercise.

Pulmonary Ventilation During Exercise

VENTILATION AND ENERGY DEMANDS

Physical activity affects oxygen consumption and carbon dioxide production more than any other form of physiologic stress. With exercise, large amounts of oxygen diffuse from the alveoli into the venous blood returning to the lungs. Conversely, considerable quantities of carbon dioxide move from the blood into the alveoli. Concurrently, ventilation increases to maintain the proper alveolar gas concentrations to allow for this increased exchange of oxygen and carbon dioxide.

VENTILATION IN STEADY-RATE EXERCISE

Figure 14–6 illustrates the relationship between oxygen consumption and minute ventilation during various levels of exercise up to the maximal oxygen uptake. During light and moderate steady-rate exercise, ventilation increases **linearly** with oxygen consumption and carbon dioxide production and averages between 20 and 25 liters of air for each liter of oxygen consumed. Under these conditions, ventilation is mainly increased by increasing tidal volume whereas at higher exercise levels breathing frequency takes on a more important role. With this adjustment in ventilation, there is complete aeration of blood because the alveolar P_{O_2} and P_{CO_2} remain at near resting values.[21]

The ratio of minute ventilation to oxygen consumption is termed the **ventilatory equivalent** and is symbolized $\dot{V}_E/\dot{V}_{O_2}$. In healthy young adults, this ratio is usually maintained at about 25 to 1 (that is, 25 liters of air breathed per liter of oxygen consumed) during submaximal exercise up to about 55% of the maximal oxygen uptake.[46,65] The ventilatory equivalent is progressively higher in younger children and averages about 32 liters in children 6 years old.[4,57]

During prone **swimming,** the ventilatory equivalents are significantly **lower** at all levels of energy expenditure.[27,45] This is due to the restrictive nature of swimming on breathing. This restriction may pose a problem in providing for adequate gas exchange at maximal swimming and may partly contribute to the generally lower maximal oxygen uptake during swimming compared to running.[39]

VENTILATION IN NON-STEADY-RATE EXERCISE

In more intense submaximal exercise, the minute ventilation takes a sharp upswing and increases disproportionately with increases in oxygen consumption. As a result, the ventilatory equivalent is greater than during steady-rate exercise and may increase to 35 or 40 liters of air per liter of oxygen consumed.

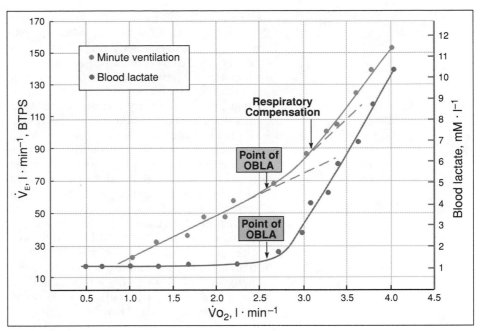

FIG. 14—6.
Pulmonary ventilation, blood lactate, and oxygen consumption during incremental exercise to the maximal oxygen uptake. The dashed line represents the extrapolation of the linear relationship between $\dot{V}_E$ and $\dot{V}_{O_2}$ observed during submaximal exercise. The "OBLA" is that point at which blood lactate begins to increase above the resting value and is detected by the point at which the relationship between ventilation and oxygen consumption deviates from linearity. "Respiratory Compensation" is a further increase in ventilation to counter the falling pH in anaerobic exercise.

Onset of Blood Lactate Accumulation (OBLA)

During steady-rate exercise, sufficient oxygen is supplied to and used by the working muscles. Under these conditions, lactic acid production does not exceed lactic acid uptake and no increase is noted in blood lactate. The exercise level or level of oxygen consumption that blood lactate begins to show a systematic increase above a resting or some slightly higher base-line level is termed the lactate threshold, or point of **onset of blood lactate accumulation,** or simply OBLA.[73] This normally occurs between 55 and 65% of the maximal oxygen uptake in healthy, untrained subjects and is often over 80% in more highly trained endurance athletes.[16,47,65]

Almost all of the excess lactic acid generated during anaerobic metabolism is buffered in the blood by sodium bicarbonate in the following reaction:

$$\text{Lactic acid} + NaHCO_3 \longrightarrow Na\ \text{lactate} + H_2CO_3$$
$$\Updownarrow$$
$$H_2O + CO_2$$

The excess, non-metabolic carbon dioxide released in this buffering reaction stimulates an increase in pulmonary ventilation and CO_2 is exhaled into the atmosphere.

The exact cause of the OBLA is controversial. It is often assumed that OBLA represents muscle anaerobiosis. Muscle lactic acid accumulation, however, is not necessarily linked to muscle anaerobiosis and the presence of an oxygen deficit. For example, lactic acid accumulation in the presence of adequate oxygenation of muscle implies an imbalance between lactic acid appearance in the blood and its subsequent rate of disappearance. This imbalance may be not only a result of muscle anaerobiosis, but rather simply the result of decreased lactic acid clearance in total, or increased lactic acid production in specific muscle fibers. Thus, caution is urged in interpreting too broadly the specific metabolic significance of the OBLA.

During exercise the OBLA is directly assessed by measuring the lactate level in the blood.[69] Because blood lactate accumulation is associated with changes in carbon dioxide production, blood pH, bicarbonate, hydrogen ion concentration, and respiratory exchange ratio, it has been suggested that these variables can also be used to indirectly assess OBLA.[32,67] Although these measures are related to OBLA, it is doubtful that each can be used to precisely denote the onset of anaerobic metabolism. Nevertheless, it is common practice to use "bloodless" techniques such as changes in the respiratory exchange ratio or ventilatory equivalent during incremental exercise to signal the onset of metabolic acidosis within exercising muscle. Even if the association between ventilatory dynamics and metabolic events is fortuitous, much useful information about exercise performance has resulted with application of these indirect prediction procedures.

Specificity of the Point of OBLA

As with many measures of physiologic function, OBLA is specific to the exercise task. Differences in OBLA have been observed in bicycle and treadmill exercise at all levels of oxygen consumption.[35] The influence of the specific form of exercise on the OBLA has also been shown for arm-cranking exercise when this type of work is compared to bicycle and treadmill exercise.[16] More than likely, these differences are the result of variations in the muscle mass activated in specific exercises. At a particular exercise level or rate of submaximal oxygen consumption, for example, the metabolic rate per unit of muscle mass would be higher in arm cranking and bicycle exercise than in treadmill walking or running. Thus, the OBLA would be reached at a lower oxygen uptake. **This indicates that different forms of exercise should not be used interchangeably to determine and quantify the point of OBLA.**

OBLA and Endurance Performance

As was discussed in Chapter 7 in comparisons of trained and untrained individuals, lactate begins to accumulate in the blood of a trained individual at both a higher level of oxygen consumption as well as a higher percentage of the maximal aerobic capacity.[58] In fact, the point of OBLA can be increased in training **without** a concomitant increase in max $\dot{V}O_2$ in both adults[14,15,23,34a] and children.[42] It appears that the point of OBLA and max $\dot{V}O_2$ are determined by different factors and that muscle fiber type, capillary density, and the alterations in skeletal muscle's oxidative capabilities with training play a major role in establishing the percentage of one's aerobic capacity that can be sustained in endurance exercise.[26,34,68]

This lack of association between aerobic capacity and OBLA has been shown in a study of trained cardiac patients compared to trained normal

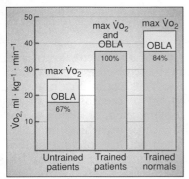

FIG. 14—7.
OBLA in relation to max $\dot{V}_{O_2}$ in trained and untrained patients who have heart disease, and in healthy trained subjects. (From Coyle, E.F., et al.: Blood lactate threshold in some well-trained ischemic heart disease patients. J. Appl. Physiol., *54*:18, 1983.)

counterparts.[10] Although the trained patients who had impaired cardiac function had a significantly lower max $\dot{V}_{O_2}$ than the normal men, the patients were able to run at the same speed (and achieve essentially the same endurance performance) as the normals without any accumulation of lactate in the blood. In fact, as shown in Figure 14—7, these patients were able to maintain close to a metabolic steady rate at a running speed that elicited their max $\dot{V}_{O_2}$. The point of OBLA represented 100% of max $\dot{V}_{O_2}$!

Two important factors influence endurance performance; (1) the maximal capacity to consume oxygen as reflected by the max $\dot{V}_{O_2}$, and (2) the maximal level for steady rate exercise which is the exercise intensity for OBLA. Traditionally, exercise physiologists have used the max $\dot{V}_{O_2}$ as the yardstick to gauge one's capacity for endurance exercise.[8] Although this measure generally relates to long-term exercise performance, it does not fully explain success in such performances. Experienced distance athletes generally compete at an exercise intensity just slightly above the point of OBLA.[10,37] This exercise level, rather than the point of max $\dot{V}_{O_2}$, is also an important factor for predicting success in endurance performance. **For men and women the exercise intensity at the point of OBLA is a consistent and powerful predictor of performance in aerobic exercise.**[18,19,31,36,60] This is clearly illustrated in a study of competitive racewalkers.[22] Here, the racewalking velocity and oxygen consumption at which blood lactate began to increase was highly correlated to 20-km performance. In fact, the race-walking velocity at OBLA predicted performance in the race to within 0.6% of the actual time! On the other hand, a subject's max $\dot{V}_{O_2}$ was a poor predictor of actual performance. Furthermore, changes in endurance performance with training are often more closely related to the training-induced changes in the exercise level for OBLA than to the changes in max $\dot{V}_{O_2}$.[1,61,68]

DOES VENTILATION LIMIT AEROBIC POWER?

If one's ability to breathe during exercise is inadequate, then the line relating pulmonary ventilation and oxygen consumption would curve in a direction opposite to that indicated in Figure 14—6, and the ventilation equivalent would decrease. Such a response would indicate a **failure** for ventilation to keep pace with oxygen consumption; in this instance, we would truly "run out of wind!" Actually, a healthy individual tends to overbreathe in relation to oxygen consumption during heavy exercise. This overbreathing was clearly illustrated in Figure 14—3, which demonstrated that the ventilatory adjustment to strenuous exercise generally results in a **decrease** in alveolar P_{CO_2} with a concomitant small **increase** in alveolar P_{O_2}. Even during maximal exercise, a considerable breathing reserve exists because pulmonary minute ventilation represents only about 60 to 85% of a healthy person's maximum capacity for breathing.[7,33,40]

Within this framework, it does not appear that pulmonary function is the "weak link" in the oxygen transport system of healthy individuals with an average to moderately high aerobic capacity. In the elite endurance athlete, however, in whom the cardiovascular and muscular adaptations to training have reached exceptional levels, the pulmonary system may be taxed maximally or even lag behind the functional capacity of the other "aerobic systems."[17,28,52,53,72] If such a condition occurs, then complete aeration of blood will not take place and pulmonary ventilation and/or pulmonary gas exchange can limit maximal exercise performance.

ENERGY COST OF BREATHING

Figure 14–8 shows the relationship between pulmonary ventilation and oxygen consumption during rest and submaximal exercise and its division into ventilatory and nonventilatory components.[38] At rest and in light exercise in healthy subjects, the oxygen requirement of breathing is small, averaging 1.9 to 3.1 ml of oxygen per liter of air breathed, or about 4% of the total energy expenditure. As the rate and depth of breathing increase, the cost of breathing rises to about 4 ml of oxygen per liter of ventilation, and may rise to as high as 9 ml of oxygen when ventilation exceeds 100 $1 \cdot min^{-1}$.[43]

RESPIRATORY DISEASE

The healthy person rarely senses the effort of breathing, even during moderate exercise. In respiratory disease, however, the work of breathing in itself may become an exhaustive exercise. In patients who have obstructive lung disease the cost of ventilation at rest may be three times that of normal patients and in light exercise it may increase to as much as 10 ml of oxygen for each liter of air breathed.[38] In severe pulmonary disease, the cost of breathing may easily reach 40% of the total exercise oxygen consumption. This would encroach on the oxygen available to the exercising, nonrespiratory muscles and seriously limit the exercise capabilities of these patients.

CIGARETTE SMOKING

The research relating smoking habits to exercise performance is meager, although most endurance athletes avoid cigarettes for fear of hindering performance due to a "loss of wind." The chronic cigarette smoker tends to show a decrease in dynamic lung function that, in severe instances, is manifested in obstructive lung disorders. Such pathologic processes, however, usually take some time to develop. Thus, with young smokers, chronic alterations in lung function may be minimal and insignificant in terms of their effect on physical performance. Other more **acute effects** of cigarette smoking may adversely affect exercise capacity.

Airway resistance at rest is increased as much as threefold in both chronic smokers and nonsmokers following 15 puffs on a cigarette during a 5-minute period.[48] This added resistance to breathing lasts an average of 35 minutes and probably has only a minor effect in light exercise where the oxygen cost of breathing is small. In vigorous exercise, however, this residual effect of smoking could be detrimental because the additional cost of breathing might become prohibitive. The increase in peripheral airway resistance with smoking is mainly due to the vagal reflex (possibly triggered from sensory stimulation of minute particles in cigarette smoke) and partially via the stimulation of parasympathetic ganglia by nicotine.[49]

The oxygen cost of breathing was studied in six habitual smokers immediately after they smoked two cigarettes and one day after abstinence from tobacco.[55] The subjects ran on a treadmill at a speed and grade that required approximately 80% of the individual's max $\dot{V}O_2$. Ventilation during the "smoking" and "non-smoking" runs was then increased in two ways: (1) subjects voluntarily hyperventilated during the run (Voluntary HV), and (2) hyperventilation was induced by increasing alveolar PCO_2 by having subjects breathe through a large-diameter tube that increased the anatomic dead

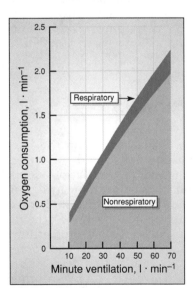

FIG. 14–8.
Relationship between minute ventilation and total oxygen consumption and its respiratory and nonrespiratory components during submaximal exercise in healthy subjects. (From Levison, H., and Cherniack, R.: Ventilatory cost of exercise in chronic obstructive pulmonary disease. J. Appl. Physiol., 25:21, 1968.)

space by about 1400 ml (Dead Space HV). The oxygen cost of the "extra" breathing was then determined as the difference between the normal oxygen consumption and the corresponding oxygen consumption in the hyperventilation experiments.

As shown in Table 14–1, the oxygen cost of breathing decreased between 13 and 79% as a result of abstinence. During exercise at 80% of maximal aerobic power, the energy requirement of breathing averaged 14% of the exercise oxygen consumption after smoking and only 9% in the "nonsmoking" trials for the heaviest smokers. Also, heart rates averaged 5 to 7% lower during exercise following one day of cigarette abstinence, and all subjects reported that they felt better exercising in the nonsmoking condition. Similar observations of acute impaired physiologic function during exercise after smoking have been noted by others.[56] It appears that a substantial reversibility of the increased oxygen cost of breathing with smoking can occur in chronic smokers with **only one day of abstinence.** Thus, if athletes are unable to eliminate smoking completely, they should at least stop on the day of the competition.

ADAPTATIONS IN BREATHING WITH TRAINING

Aerobic training brings about several changes in pulmonary ventilation during maximal and submaximal exercise.

1. **Maximal exercise.** As might be expected, maximal exercise ventilation increases with improvements in maximal oxygen uptake. This makes sense physiologically, because any increase in aerobic capacity results in a larger oxygen requirement and correspondingly larger production of carbon dioxide that must be eliminated through increased alveolar ventilation.
2. **Submaximal exercise.** Following only 4 weeks of training, a considerable **reduction** in the ventilatory equivalent is observed in submaximal exercise.[2] Consequently, a smaller amount of air is breathed at a particular

TABLE 14–1
THE OXYGEN COST OF HYPERVENTILATION (HV) IN "SMOKING" AND "NONSMOKING" EXERCISE THAT REPRESENTED APPROXIMATELY 80% OF EACH SUBJECT'S MAX $\dot{V}_{O_2}$[a]

	Smoking				Nonsmoking			
	Voluntary HV		Dead Space HV		Voluntary HV		Dead Space HV	
Subject	$\dot{V}_E$ $(l \cdot min^{-1})$	Cost $(ml \cdot l^{-1})$	$\dot{V}_E$ $(l \cdot min^{-1})$	Cost $(ml \cdot l^{-1})$	$\dot{V}_E$ $(l \cdot min^{-1})$	Cost $(ml \cdot l^{-1})$	$\dot{V}_E$ $(l \cdot min^{-1})$	Cost $(ml \cdot l^{-1})$
1	26.4	15.1	18.9	12.7	22.7	11.4	23.0	6.5
2	39.0	10.3	28.1	5.9	42.6	11.3	41.3	4.8
3	22.8	7.9	27.2	7.0	23.8	7.2	22.8	5.7
4	36.3	5.0	28.7	5.6	44.7	3.8	18.6	− 1.6[b]
5	52.7	13.5	26.7	12.4	75.2	6.1	22.8	5.7
6	22.4	8.5	27.3	1.1	23.2	3.4	30.1	3.0
Average	32.6	10.1	26.2	7.4	38.7	7.2	26.5	4.0

[a] From Rode, A., and Shephard, R.J.: The influence of cigarette smoking upon the oxygen cost of breathing in near-maximal exercise. Med. Sci. Sports, 3:51, 1971.
[b] The implication of the "negative" cost of $\dot{V}_E$ in this subject is that the added dead space reduced the cost of the normal exercise ventilation.

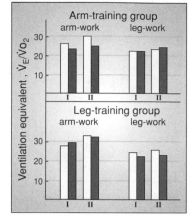

FIG. 14—9.
Ventilatory equivalents during light (I) and heavy (II) submaximal arm and leg exercise before and after arm training (top) and leg training (bottom). Red bars indicate post-training values. (From Rasmussen, B., et al.: Pulmonary ventilation, blood gases, and blood pH after training of the arms and the legs. J. Appl. Physiol., 38:250, 1975.)

rate of submaximal oxygen consumption; this reduces the percentage of the total oxygen cost of exercise attributable to breathing. Theoretically, this would be important in performing prolonged, vigorous exercise for two reasons: (1) it would reduce the fatiguing effects of exercise on the ventilatory musculature,[44] and (2) any oxygen freed from use by the respiratory muscles becomes available to the exercising muscles.

The mechanism is unknown for the training adaptations in ventilation in submaximal exercise. These changes, however, have been consistently observed in studies of adolescents and in both young and older men and women.[20,29,63] In general, the tidal volume becomes **larger** and breathing frequency is considerably **reduced.** Air remains in the lungs for a longer period of time between breaths. The result is an increase in the amount of oxygen extracted from the inspired air. The exhaled air of trained individuals often contains only 14 to 15% oxygen during submaximal exercise, whereas the expired air of untrained persons may contain 18% oxygen at the same work level. Obviously, the untrained person must ventilate proportionately more air to achieve the same submaximal oxygen uptake.

Ventilatory adaptations appear to be **specific** to the type of exercise used in training. When subjects performed either arm or leg exercise, the ventilation equivalent was always greater during arm exercise than during leg work (Fig. 14—9).[54] After training, the ventilatory equivalent was significantly reduced. This finding was noted, however, **only** during exercise that used the specifically trained muscle groups. For the group trained by arm ergometry, the ventilation equivalent was reduced only in arm exercise and vice-versa for the leg-trained group. This training adaptation was closely related to a less pronounced rise in blood lactic acid and heart rate in the specific training exercise. It is likely that the ventilatory adjustment to training results from local neural or chemical adaptations in the specific muscles trained through exercise.

▬▬▬ SUMMARY

1. In light to moderate exercise, ventilation increases linearly with oxygen consumption. The ventilatory equivalent is maintained at about 20 to 25 liters of air breathed per liter of oxygen consumed.
2. In non-steady-rate exercise, ventilation increases disproportionately with increases in oxygen consumption, and the ventilatory equivalent may reach 35 or 40 liters.
3. The eventual sharp upswing in ventilation during incremental exercise provides an effective, simple means for establishing a person's level for the onset of blood lactate accumulation (OBLA). This measure relates to the onset of anaerobiosis and the subsequent accumulation of lactic acid. Thus, one important aspect of aerobic fitness can be evaluated without a significant degree of metabolic acidosis or cardiovascular strain.
4. For healthy people, the oxygen cost of breathing is relatively small, even during the most severe exercise. In respiratory disease, however, the work of breathing becomes excessive and alveolar ventilation often becomes inadequate.
5. For individuals with average to moderate levels of aerobic fitness, pulmonary ventilation is not taxed to a degree that would limit optimal alveolar gas exchange in maximal exercise. For the elite endurance athlete, improvements in pulmonary dynamics may lag behind the exceptional

adaptations in cardiovascular and muscle function, and aeration may be compromised.

6. Airway resistance is greatly increased following cigarette smoking. This increases the oxygen cost of breathing, which could be detrimental during prolonged, vigorous exercise. Substantial reversibility of these smoking effects can occur with just one day of abstinence.

7. Training generally reduces the ventilatory equivalent in submaximal exercise. This "conserves" oxygen because the cost of breathing is lowered for a particular exercise task.

8. Ventilatory adjustments with training follow the principle of training specificity. A more efficient breathing pattern is generally observed only during the type of exercise used in training.

P A R T

3 **Acid-Base Regulation**

BUFFERING

Substances that dissociate and release H^+ are called **acids,** whereas a compound that can pick up or accept H^+ is a **base.** The term **buffering** is used to designate reactions that minimize changes in H^+ concentration, and the chemicals involved in preventing this change are termed **buffers.**

The acid-base quality of body fluids must be regulated within narrow limits because metabolism is highly sensitive to the H^+ concentration (pH) of the reacting medium. Blood pH is normally regulated on the slightly alkaline side of neutrality at a pH value of 7.4; this means that slightly more negative hydroxyl ions (OH^-) are present than H^+. An increase in pH above the normal average of 7.4 is the direct result of a decrease in H^+ concentration and is termed **alkalosis.** Conversely, an increase in H^+ concentration (decrease in pH) is referred to as **acidosis.**

A buffering system consists of a weak acid and the salt of that acid. The bicarbonate buffer, for example, is made up of the weak acid, **carbonic acid,** and the salt of that acid, **sodium bicarbonate.** Carbonic acid is formed when bicarbonate binds H^+. As long as the H^+ concentration remains elevated, the reaction produces the weak acid because the excess H^+ ions are bound in accordance with the general reaction:

$$H^+ + Buffer \rightarrow H\text{-}Buffer$$

If, however, the concentration of H^+ decreases (as occurs during hyperventilation when plasma carbonic acid decreases because carbon dioxide is eliminated from the blood), the buffering reaction moves in the opposite direction:

$$H^+ + Buffer \leftarrow H\text{-}Buffer$$

In the process, H^+ ions are released.

Much of the carbon dioxide generated in energy metabolism reacts with water to form the relatively weak carbonic acid. This then dissociates to H^+ and HCO_3^-. Likewise, lactic acid, a stronger acid, reacts with sodium bicarbonate to form **sodium lactate** and carbonic acid; in turn, carbonic acid dissociates and increases the H^+ concentration of the extracellular fluids. Other organic acids such as fatty acids dissociate and liberate H^+, as do the sulfuric and phosphoric acids produced during protein breakdown.

Three mechanisms control the acid-base quality of the internal environment: (1) chemical buffers, (2) pulmonary ventilation, and (3) kidney function.

CHEMICAL BUFFERS

Bicarbonate, phosphate, and protein chemical buffers provide the rapid first line of defense to maintain a consistency in the acid-base quality of the internal environment.

BICARBONATE BUFFER

The bicarbonate buffer system consists of carbonic acid and sodium bicarbonate in solution. In the buffering process, for example, hydrochloric acid (a strong acid) is changed into a much weaker acid by combining with sodium bicarbonate in the reaction:

$$HCl + Na\,HCO_3 \longrightarrow Na\,Cl + H_2CO_3$$
$$\updownarrow$$
$$H^+ + HCO_3^-$$

Buffering by sodium bicarbonate, therefore, produces only a slight reduction in pH. As mentioned previously, sodium bicarbonate in the plasma exerts a strong buffering action on lactic acid, the anaerobic metabolite. This causes the formation of sodium lactate and carbonic acid; consequently, a change in pH is minimized. Any additional increase in H^+ concentration (acidity) brought about by carbonic acid dissociation causes the dissociation reaction to move back in the opposite direction. In this situation, carbon dioxide is released into solution as follows:

Acidosis $H_2O + CO_2 \leftarrow H_2CO_3 \leftarrow H^+ + HCO_3^-$

An increase in plasma carbon dioxide or acidity immediately stimulates ventilation and the "excess" carbon dioxide is eliminated. Conversely, if the H^+ concentration is reduced and the body fluids become more alkaline, the ventilatory drive is inhibited, carbon dioxide is retained to combine with water, and the acidity is normalized.

Alkalosis $H_2O + CO_2 \rightarrow H_2CO_3 \rightarrow H^+ + HCO_3^-$

PHOSPHATE BUFFER

This buffer system consists of phosphoric acid and sodium phosphate. These chemicals act in a manner similar to that of the bicarbonate system. The phosphate buffer is particularly important in regulating the acid-base balance in the kidney tubules and intracellular fluids in which there is a relatively high concentration of phosphates.

PROTEIN BUFFER

Although carbonic acid produced from the union of water and carbon dioxide is a relatively weak acid, the H^+ released when it dissociates must be buffered in the venous blood. By far, the most important H^+ acceptor for this function is **hemoglobin.** Its potency for regulating acidity is almost six times greater than the other plasma proteins. In addition, when hemoglobin releases its oxygen to the cells, it becomes a weaker acid. This, in turn, increases its affinity for binding with H^+. The H^+ generated from the formation of carbonic acid in the erythrocyte combines readily with deoxygenated hemoglobin (Hb^-) in the reaction:

$$H^+ + Hb^-(Protein) \rightarrow HHb$$

Intracellular tissue proteins also contribute to the regulation of plasma pH. Some amino acids have free acidic radicals that, when dissociated, form OH^- that can react with H^+ to form water.

RELATIVE POWER OF CHEMICAL BUFFERS

Table 14–2 shows the relative power of the different chemical buffers in the blood, as well as blood plus interstitial fluids combined. As a frame of reference, the buffering power of the bicarbonate system is assumed to be 1.00.

PHYSIOLOGIC BUFFERS

The second line of defense in acid-base regulation is the ventilatory and renal systems. These systems provide buffering function only when a change in pH has already occurred.

VENTILATORY BUFFER

Any increase in the quantity of free H^+ in extracellular fluids and plasma directly stimulates the respiratory center and causes an immediate increase in alveolar ventilation. This adjustment rapidly reduces alveolar P_{CO_2} and causes carbon dioxide to be "blown off" from the blood. The reduction in plasma carbon dioxide facilitates the recombining of H^+ and HCO_3^-, thus lowering the free H^+ in the plasma. For example, if alveolar ventilation at rest is doubled by hyperventilation, the blood becomes more alkaline and pH increases by 0.23 units from 7.40 to 7.63. Conversely, reducing normal ventilation by one-half causes the blood to become more acidic by about 0.23 pH units. The potential magnitude of the ventilatory buffer has been estimated to be about twice that of the combined effect of all the chemical buffers.

TABLE 14–2. RELATIVE BUFFERING POWER OF CHEMICAL BUFFERS*

Chemical Buffer	Blood	Blood Plus Interstitial Fluids
Bicarbonate	1.0	1.0
Phosphate	0.3	0.3
Proteins (excluding hemoglobin)	1.4	0.8
Hemoglobin	5.3	1.5

* Modified from Guyton, A.C.: Medical Physiology. Philadelphia, W.B. Saunders Co., 1981.

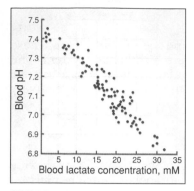

FIG. 14–10.
The relationship between blood pH and blood lactate concentration at rest and during various levels of intermittent exercise of short duration. (From Osnes, J.B., and Hermansen, L.: Acid-base balance after maximal exercise of short duration. J. Appl. Physiol., 32:59, 1972.)

RENAL BUFFER

The chemical buffers nullify the effects of excess acid only temporarily. The excretion of H^+ by the kidneys, although more time-consuming than the action of the chemical and ventilatory buffers, is important if the buffer reserve, or **alkaline reserve,** of the body is to be maintained. To this end, the kidneys stand as final sentinels. Acidity can be controlled by the renal buffer through complex chemical reactions in the renal tubules that involve alterations in the amounts of ammonia and H^+ secreted into the urine and in the amounts of alkali, chloride, and bicarbonate reabsorbed.

EFFECTS OF EXERCISE AND TRAINING

The regulation of pH becomes progressively more difficult in strenuous exercise where H^+ is increased from both carbon dioxide and lactic acid formation.[9] This occurs in the case of maximal, intermittent exercise of short duration, when blood lactate values can reach 30 mmol (about 270 mg of lactate per 100 ml of blood), or more.[24,25]

As seen in Figure 14–10, a negative linear relationship exists at rest and in various levels of intermittent exercise between blood lactate concentration and blood pH.[50] In these experiments, blood lactate varied between 0.8 mmol at rest (pH 7.43) and 32.1 mmol, or approximately 290 mg of lactate per 100 ml of blood during exhaustive exercise (pH 6.80). In the active muscle the pH is even lower, generally falling to a pH of about 6.4 or lower at the point of exhaustion.[24,62]

These results indicate that humans are able **temporarily** to tolerate pronounced disturbances in the acid-base balance, at least as low as a blood pH of about 6.80 (one of the lowest values reported for a human subject). The degree of acidosis at pH levels below 7.00 is not without consequence; many subjects experienced nausea, headache, and dizziness, as well as pain in the muscle groups involved in exercise.

Does Training Improve Buffering Capacity?

It is well known that vigorous anaerobic training enables a person to tolerate higher lactic acid levels and lower plasma pH values than were possible before training. It is tempting to speculate that such training may have some positive effect on the body's capability for acid-base regulation, perhaps through an enhancement of the chemical buffers or the alkaline reserve. But it has never been shown that buffering capacity becomes enhanced through specific modes of training. Improved anaerobic tolerance following a training program may be due to motivational influences; the repeated stress of vigorous training probably modifies one's mental attitude for tolerating the extreme discomfort of the acid condition.

SUMMARY

1. The acid-base quality of the body fluids is normally regulated within narrow limits by chemical and physiologic buffer systems.
2. The bicarbonate, phosphate, and protein chemical buffers provide the rapid first line of defense in acid-base regulation. These buffers consist

of a weak acid and the salt of that acid. In an acidic condition, their action converts strong acids to weaker acids and neutral salts.

3. The lungs and kidneys regulate pH when the chemical buffer system is stressed. Changes in alveolar ventilation can rapidly and significantly alter the free H^+ in extracellular fluids. In response to increased acidity, the renal tubules act as final sentinels and secrete H^+ into the urine and reabsorb bicarbonate.

4. Vigorous exercise creates a great demand for buffering; thus, the regulation of pH becomes progressively more difficult. From available evidence, however, it does not appear that physical training per se enhances the body's buffering capacity.

REFERENCES

1. Acevedo, E.O., and Goldfarb, A.H.: Increased training intensity effects on plasma lactate, ventilatory threshold, and endurance. *Med. Sci. Sports Exerc.*, 21:563, 1989.
2. Andrew, G.M., et al.: Effect of athletic training on exercise cardiac output. *J. Appl. Physiol.*, 21:603, 1966.
3. Asmussen, E., and Nielsen, M.: Experiments on nervous factors controlling respiration and circulation during exercise employing blocking of blood flow. *Acta Physiol. Scand.*, 60:103, 1964.
4. Åstrand, P.O.: *Experimental Studies of Physical Working Capacity in Relation to Sex and Age.* Copenhagen, Munksgaard, 1952.
5. Beaver, W.L., and Wasserman, K.: Transients in ventilation at the start and end of exercise. *J. Appl. Physiol.*, 25:330, 1968.
6. Brooks, G.A.: Anaerobic threshold: review of the concept and directions for future research. *Med. Sci. Sports. Exerc.*, 17:22, 1983.
7. Bye, P.T.P., et al.: Ventilatory muscles during exercise in air and oxygen in normal men. *J. Appl. Physiol.*, 56:464, 1984.
8. Costill, D.L.: Physiology of marathon running. *JAMA*, 221:1024, 1972.
9. Costill, D.L., et al.: Leg muscle pH following sprint running. *Med. Sci. Sports Exerc.*, 15:325, 1983.
10. Coyle, E.F., et al.: Blood lactate threshold in some well-trained ischemic heart disease patients. *J. Appl. Physiol.*, 54:18, 1983.
11. Craig, A.B., Jr.: Principles and problems of underwater diving. *Phys. Sportsmed.*, 8(3):72, 1980.
12. Craig, A.B., Jr.: Summary of 58 cases of loss of consciousness during underwater swimming and diving. *Med. Sci. Sports*, 8:171, 1976.
13. Davis, J.A.: Anaerobic threshold: review of the concept and directions for future research. *Med. Sci. Sports Exerc.*, 17:6, 1985.
14. Davis, C., et al.: Effect of 40 weeks of endurance training on anaerobic threshold. *Int. J. Sports Med.*, 3:208, 1982.
15. Davis, J.A., et al.: Anaerobic threshold alterations caused by endurance training in middle aged men. *J. Appl. Physiol.*, 46:1039, 1979.
16. Davis, J.A., et al.: Anaerobic threshold and maximal aerobic power for three modes of exercise. *J. Appl. Physiol.*, 41:544, 1976.
17. Dempsey, J.A., et al.: Exercise-induced alveolar hypoxemia in healthy human subjects at sea level. *J. Physiol.* (London), 355:161, 1984.
18. Farrell, P.A., et al.: Plasma lactate accumulation and distance running performance. *Med. Sci. Sports*, 11:338, 1979.
19. Fay, L., et al.: Physiological parameters related to distance running performance in female athletes. *Med. Sci. Sports Exerc.*, 21:319, 1989.
20. Fringer, M.N., and Stull, G.A.: Changes in cardiorespiratory parameters during periods of training and detraining in young adult females. *Med. Sci. Sports*, 6:20, 1974.
21. Grimby, G.: Respiration in exercise. *Med. Sci. Sports*, 1:9, 1969.

22. Hagberg, J.M., and Coyle, E.F.: Physiological determinants of endurance performance as studied in competitive racewalkers. *Med. Sci. Sports Exerc.*, 15:287, 1983.
23. Henritze, J., et al.: Effects of training at and above the lactate threshold and maximal oxygen uptake. *Eur. J. Appl. Physiol.*, 54:84, 1985.
24. Hermansen, L.: Effect of metabolic changes on force generation in skeletal muscle during maximal exercise. In *Human Muscle Fatigue: Physiological Mechanisms.* London, Pitman Medical, 1981.
25. Hermansen, L.: Lactate production during exercise. In *Muscle Metabolism During Exercise.* Edited by B. Pernow and B. Saltin. New York, Plenum, 1971.
26. Holloszy, J.O, and Coyle, E.F.: Adaptations of skeletal muscle to endurance exercise and their metabolic consequences. *J. Appl. Physiol.*, 56:834, 1984.
27. Holmér, I., et al.: Hemodynamic and respiratory responses compared in swimming and running. *J. Appl. Physiol.*, 37:48, 1974.
28. Hopkins, S.R., and McKenzie, D.C.: Hypoxic ventilatory response and arterial desaturation during heavy work. *J. Appl. Physiol.*, 67:1119, 1989.
29. Jirka, Z., and Adamus, M.: Changes of ventilation equivalents in young people in the course of three years of training. *J. Sports Med.*, 5:1, 1965.
30. Jacobs, I.: Blood lactate: implications for training and sports performance. *Sports Med.*, 3:10, 1986.
31. Jones, N.L.: Dyspnea in exercise. *Med. Sci. Sports Exerc.*, 16:14, 1984.
32. Jones, N.L., and Ehrsam, R.E.: The anaerobic threshold. In *Exercise and Sport Sciences Reviews.* Vol. 10. Edited by R.L. Terjung. Philadelphia, Franklin Institute, 1982.
33. Jones, N.L., et al.: *Clinical Exercise Testing.* Toronto, W.B. Saunders, 1975.
34. Karlsson, J., et al.: Relevance of muscle fibre type to fatigue in short intense and prolonged exercise in man. In *Human Muscle Fatigue: Physiological Mechanisms.* London, Pitman Medical, 1981.
34a. Kohrt, W.M., et al.: Longitudinal assessment of responses by triathletes to swimming, cycling, and running. *Med. Sci. Sports Exerc.*, 21:569, 1989.
35. Koyal, S.N., et al.: Ventilatory responses to the metabolic acidosis of treadmill and cycle ergometry. *J. Appl. Physiol.*, 40:864, 1976.
36. Kumagi, S., et al.: Relationship of the anaerobic threshold with the 5 km, 10 km, and 10 mile races. *Eur. J. Appl. Physiol.*, 49:13, 1982.
37. LaFontaine, T.P., et al.: The maximal steady state versus selected running events. *Med. Sci. Sports Exerc.*, 13:190, 1981.
38. Levison, H., and Cherniack, R.: Ventilatory cost of exercise in chronic obstructive pulmonary disease. *J. Appl. Physiol.*, 25:21, 1968.
39. Magel, J.R., et al.: Specificity of swim training on maximum oxygen uptake. *J. Appl. Physiol.*, 38:151, 1975.
40. Mahler, D.A., et al.: Ventilatory responses at rest and during exercise in marathon runners. *J. Appl. Physiol.*, 52:388, 1982.
41. Mahler, M.: Neural and humoral signals for pulmonary ventilation arising in exercising muscle. *Med. Sci. Sports*, 11:191, 1979.
42. Mahon, A.D., and Vaccaro, P.: Ventilatory threshold and $\dot{V}O_2$ max changes in children following endurance training. *Med. Sci. Sports Exerc.*, 21:425, 1989.
43. Martin, B.J., and Stager, J.M.: Ventilatory endurance in athletes and non-athletes. *Med. Sci. Sports Exerc.*, 13:21, 1981.
44. Martin, B., et al.: Exercise performance after ventilatory work. *J. Appl. Physiol.*, 52:1581, 1982.
45. McArdle, W.D., et al.: Metabolic and cardiorespiratory response during free swimming and treadmill walking. *J. Appl. Physiol.*, 30:733, 1971.
46. McArdle, W.D., et al.: Metabolic and cardiovascular adjustment to work in air and water at 18, 25 and 33°C. *J. Appl. Physiol.*, 40:85, 1976.
47. Michelson, T.C., and Hagerman, F.C.: Anaerobic threshold measurements of elite oarsmen. *Med. Sci. Sports Exerc.*, 14:44, 1982.
48. Nadel, J.A., and Comroe, J.H.: Acute effects of inhalation of cigarette smoke on airway resistance. *J. Appl. Physiol.*, 16:713, 1961.
49. Nakamura, M., et al.: Acute effects of cigarette smoke inhalation on peripheral airways in dogs. *J. Appl. Physiol.*, 58:27, 1985.

50. Osnes, J.B., and Hermansen, L.: Acid-base balance after maximal exercise of short duration. *J. Appl. Physiol.*, 32:59, 1972.

51. Péronnet, F., et al.: Correlation between the ventilatory threshold and endurance capability in marathon runners. *Med. Sci. Sports Exerc.*, 19:610, 1987.

52. Powers, S.K., et al.: Incidence of exercise induced hypoxemia in elite endurance athletes at sea level. *Eur. J. Appl. Physiol.*, 58:298, 1988.

53. Powers, S.K., et al.: Effects of incomplete pulmonary gas exchange on $\dot{V}O_2$ max. *J. Appl. Physiol.*, 66:2491, 1989.

54. Rasmussen, R., et al.: Pulmonary ventilation, blood gases and blood pH after training of the arms and the legs. *J. Appl. Physiol.*, 38:250, 1975.

55. Rode, A., and Shephard, R.J.: The influence of cigarette smoking upon the oxygen cost of breathing in near-maximal exercise. *Med. Sci. Sports*, 3:51, 1971.

56. Rotstein, A., and Sagiv, M.: Acute effect of cigarette smoking on physiologic response to graded exercise. *Int. J. Sports Med.*, 7:322, 1986.

57. Rowland, T.W., and Green, G.M.: Physiological responses to treadmill exercise in females: adult-child differences. *Med. Sci. Sports Exerc.*, 20:474, 1988.

58. Sjödin, B., et al.: The physiological background on onset blood lactate accumulation (OBLA). In *Proceedings of International Symposium of Sports Biology*. Edited by P.V. Komi. Champaign, IL, 1982.

59. Sutton, J.R., and Jones, N.L.: Control of pulmonary ventilation during exercise and mediators in the blood: CO_2 and hydrogen ion. *Med. Sci. Sports*, 11:198, 1979.

60. Tanaka, K., et al.: Relationship of anaerobic threshold and onset of blood lactate accumulation with endurance performance. *Eur. J. Appl. Physiol.*, 52:51, 1983.

61. Tanaka, K., et al.: A longitudinal assessment of anaerobic threshold and distance-running performance. *Med. Sci. Sports Exerc.*, 16:278, 1984.

62. Taylor, D.J., et al.: Energetics of human muscle: exercise-induced ATP depletion. *Magn. Reson. Med.*, 3:44, 1986.

63. Tzankoff, S.P., et al.: Physiological adjustments to work in older men as affected by physical training. *J. Appl. Physiol.*, 33:346, 1972.

64. Wasserman, K., and Whipp, B.J.: Coupling of ventilation in pulmonary gas exchange during nonsteady-state work in men. *J. Appl. Physiol.*, 54:587, 1983.

65. Wasserman, K., et al.: Anaerobic threshold and respiratory gas exchange during exercise. *J. Appl. Physiol.*, 35:236, 1973.

66. Wasserman, K., et al.: Cardiodynamic hyperpnea: hyperpnea secondary to cardiac output increase. *J. Appl. Physiol.*, 36:457, 1974.

67. Wasserman, K., et al.: Respiratory physiology of exercise: metabolism, gas exchange and ventilatory control. In Respiratory Physiology III. Vol. 23. *International Review of Physiology*. Edited by J.G. Widdicombe. Baltimore, University Park Press, 1981.

68. Weltman, A.: The lactate threshold and endurance performance. *Adv. Sports Med. Fitness*, 2:91, 1989.

69. Weltman, A., et al.: Reliability and validity of a continuous incremental treadmill protocol for the determination of lactate threshold, fixed blood lactate concentrations and $\dot{V}O_2$ max. *Int. J. Sports Med.*, 11:26, 1990.

70. Whipp, B.J., and Davis, J.A.: Peripheral chemoreceptors and exercise hyperpnea. *Med. Sci. Sports*, 11:204, 1979.

71. Whipp, B.J., and Wasserman, K.: The effect of work intensity on the transient respiratory response immediately following exercise. *Med. Sci. Sports*, 5:14, 1973.

72. Williams, J.H., et al.: Hemoglobin desaturation in highly trained athletes during heavy exercise. *Med. Sci. Sports Exerc.*, 18:168, 1986.

73. Yoshida, T., et al.: Blood lactate parameters related to aerobic capacity and endurance performance. *Eur. J. Appl. Physiol.*, 56:7, 1987.

15

THE CARDIOVASCULAR SYSTEM

The **cardiovascular system** serves to integrate the body as a unit and provides the muscles with a continuous stream of nutrients and oxygen so that a high energy output can be maintained for a considerable time period. Conversely, by-products of metabolism are rapidly removed by the circulation from the site of energy release.

In Chapters 15, 16, and 17, we explore the process of circulation, especially its role in oxygen delivery during exercise, and examine the basic differences in cardiovascular function between physically trained and untrained individuals. Oxygen transport, coupled with the capability of specific muscles to generate ATP aerobically, ultimately sets the maximum level of aerobic energy release during strenuous physical activity.

COMPONENTS OF THE CARDIOVASCULAR SYSTEM

The cardiovascular system is a continuous vascular circuit that consists of a pump, a high-pressure distribution circuit, exchange vessels, and a low-pressure collection and return circuit. A schematic view of this system is presented in Figure 15–1.

THE HEART

The heart provides the impetus for blood flow. It is situated in the midcenter of the chest cavity with about two thirds of its mass to the left of the body's midline. Although this four-chambered muscular organ weighs less than a pound, it beats so steadily and powerfully that the force generated during its 40 million beats per year could lift its owner 100 miles above the earth. Even for a person of average fitness, the maximum output of blood from this remarkable organ is greater than the fluid output from a household faucet turned wide open.

The heart muscle, or **myocardium,** is a form of striated muscle similar to skeletal muscle. The individual fibers, however, are multinucleated cells interconnected in a latticework fashion. Consequently, when one cell is stimulated or depolarized, the action potential speeds through the myocardium to all cells, causing the heart to function as a unit.

Figure 15–2 shows the details of the heart as a pump. Functionally, the heart may be viewed as two separate pumps. The hollow chambers that comprise the right side of the heart (right heart) perform two important functions: (1) receive blood returning from all parts of the body, and (2) pump blood to the lungs for aeration by way of the pulmonary circulation. The left heart receives oxygenated blood from the lungs and pumps it into the

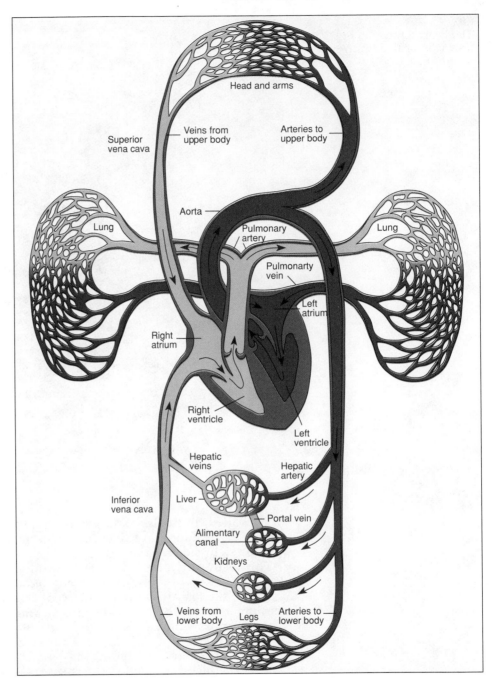

FIG. 15–1.
Schematic view of the cardiovascular system that consists of the heart and the pulmonary and systemic vascular circuits. The dark shading indicates the oxygen-rich arterial blood, whereas the deoxygenated venous blood is somewhat paler. In the pulmonary circuit, the situation is reversed, and oxygenated blood returns to the heart in the right and left pulmonary veins.

FIG. 15–2.
The heart. Direction of blood
flow is indicated by arrows.

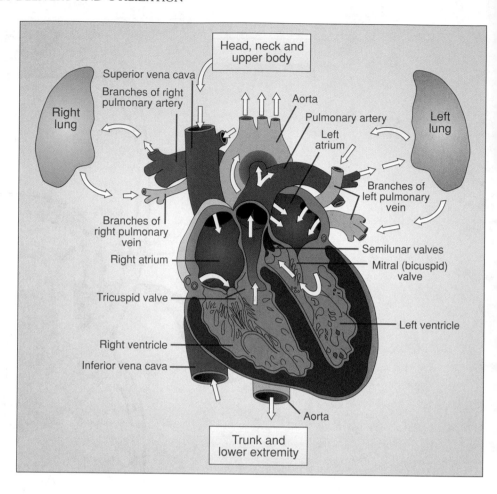

thick-walled, muscular **aorta** for distribution throughout the body in the systemic circulation. A thick, solid muscular wall or **septum** separates the left and right sides of the heart.

The **atrioventricular valves** situated in the heart provide for a one-way passage of blood from the right atrium to the right ventricle (**tricuspid valve**) and from the left atrium to the left ventricle (**mitral or bicuspid valve**). The **semilunar valves** located in the arterial wall just outside the heart prevent blood from flowing back into the heart between contractions.

The relatively thin-walled, saclike atrial chambers serve as primer pumps to receive and store blood during the period of ventricular contraction. About 70% of the blood that returns to the atria flows directly into the ventricle before the atria contract. The simultaneous contraction of both atria then forces the remaining blood into their respective ventricles directly below. Almost immediately following atrial contraction, the ventricles contract and force blood into the arterial system.

As ventricular pressure builds, the atrioventricular valves snap closed. The heart valves remain closed for 0.02 to 0.06 seconds. This brief interval of rising ventricular tension during which the heart volume and fiber length remain unchanged represents the heart's **isovolumetric contraction period.** Blood is ejected from the heart when the ventricular pressure exceeds arterial pressure. By the nature of the spiral and circular arrangement of bands of cardiac muscle, blood is virtually "wrung out" of the heart with each contraction.

THE ARTERIAL SYSTEM

The Arteries

The arteries are the high-pressure tubing that conducts oxygen-rich blood to the tissues. As depicted in Figure 15–3, the arteries are composed of layers of connective tissue and smooth muscle. The walls of these vessels are so thick that no gaseous exchange takes place between arterial blood and the surrounding tissues. Blood pumped from the left ventricle into the highly muscular yet elastic **aorta** is eventually distributed throughout the body via smaller arterial branches called **arterioles.** The walls of arterioles are composed of circular layers of smooth muscle that either constrict or relax to regulate peripheral blood flow. As is discussed in a following section, it is the capacity of these "resistance vessels" to alter dramatically their internal diameter that provides a rapid and effective means for regulating blood flow

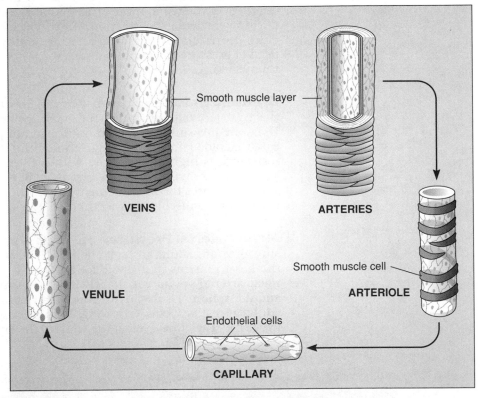

FIG. 15–3.
The walls of blood vessels. A single layer of endothelial cells lines each of the blood vessels. The walls of arteries are surrounded by fibrous tissue and wrapped in several layers of smooth muscle cells. The arterioles are similar but are sheathed in a single layer of muscle cells, whereas the capillaries consist only of a layer of endothelial cells. In the venule, the endothelial cells are sheathed in fibrous tissue whereas the veins also possess a layer of smooth muscle.

through the vascular circuit. This redistribution function is especially important during exercise because blood can be diverted to working muscles from areas that can temporarily compromise their blood supply.

Blood Pressure

A surge of blood enters the aorta with each contraction of the left ventricle. Because the peripheral vessels do not permit blood to be "run-off" from the arterial system as rapidly as it is ejected from the heart, a portion of the blood pumped from the heart is "stored" in the aorta. This creates pressure within the entire arterial system and causes a pressure wave to travel down the aorta to the remote branches of the arterial tree. This stretch and subsequent recoil of the arterial wall during a cardiac cycle can be felt readily as the characteristic "pulse" in any superficial artery of the body. In healthy individuals, the pulse rate and heart rate are identical.

Systolic Blood Pressure. At rest, the highest pressure generated by the heart is usually about 120 mm Hg during the contraction, or **systole,** of the left ventricle. The point of reference for this measurement is usually the brachial artery at the level of the right atrium. **Systolic pressure** provides an estimate of the work of the heart and of the strain against the arterial walls during ventricular contraction. As the heart relaxes and the aortic valves close, the natural elastic recoil of the arterial system provides for a continuous head of pressure to maintain a steady flow of blood into the periphery, until the next surge of blood.

Diastolic Blood Pressure. During **diastole,** or the relaxation phase of the cardiac cycle, arterial blood pressure decreases to 70 to 80 mm Hg. **Diastolic pressure** provides an indication of **peripheral resistance** or of the ease that blood flows from the arterioles into the capillaries. When peripheral resistance is high, the pressure within the arteries after systole is not dissipated and remains elevated for a large portion of the cardiac cycle. Figure 15–4 illustrates the measurement of systolic and diastolic blood pressure by the common auscultation method.

Mean Arterial Pressure. The average systolic and diastolic blood pressure for young adults at rest is about 120 and 80 mm Hg, respectively. Because the heart remains in diastole longer than it does in systole, the average or **mean arterial pressure** is slightly less than simply the average of the systolic and diastolic pressures. Thus, the mean arterial pressure of healthy young adults at rest is about 96 mm Hg. This pressure represents the average force exerted by the blood against the walls of the arteries during the entire cardiac cycle.

THE CAPILLARIES

The arterioles continue to branch and form smaller and less muscular vessels called **metarterioles.** These end in a network of microscopic blood vessels called **capillaries.** These vessels generally contain about 5% of the total blood volume. As was shown in Figure 15–3, the capillary wall consists of only a single layer of endothelial cells. Some capillaries are so narrow that they

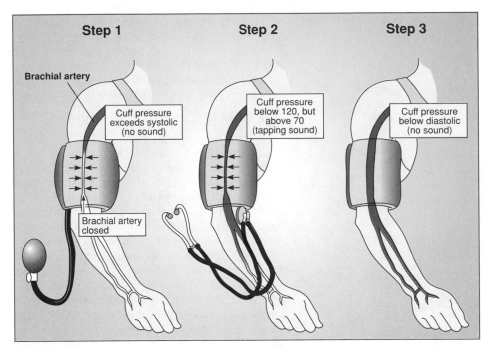

FIG. 15—4.
Measurement of blood pressure at the brachial artery by the **auscultatory method.**
Step 1. A pressure cuff, referred to as a **sphygmomanometer,** is inflated so its
pressure exceeds the **systolic pressure** or highest pressure within the artery. Blood
flow is occluded and a brachial pulse (at the elbow fossa) cannot be palpated or
heard (auscultated). Note the restriction of blood through the brachial artery. Step
2. The pressure within the cuff is reduced by small increments and the examiner
listens until a faint sound occurs. This represents blood flowing through the
brachial artery. The pressure exerted on the walls of the artery when the first soft
tapping sounds occur is called the systolic pressure. Step 3. As the pressure in the
cuff is lowered further, distinct sounds continue to be heard as blood flows
through the artery for longer portions of the cardiac cycle. The pressure in the
artery when the sounds disappear is the **diastolic pressure.**

provide room for only one blood cell to squeeze through, single file. In many
instances, the proliferation of capillaries is so extensive that the walls of
these blood vessels actually abut the membranes of the surrounding cells.
It is likely that the capillary density of human skeletal muscle is between
2000 and 3000 capillaries per square millimeter of tissue.[31] This density is
even greater in heart muscle; thus, no cell lies further away than 0.008 mm
from its nearest capillary.

The diameter of the capillary opening is controlled by a ring of smooth
muscle, the **precapillary sphincter,** that encircles the vessel at its origin. The
action of this sphincter is extremely important in exercise; it provides a
local means for regulating capillary blood flow within a specific tissue to
meet its metabolic requirements.

Branching of the microcirculation results in an increase in the cross-
sectional area of these peripheral vessels that is about 800 times greater than
that of the 1-inch diameter aorta. Because the velocity of blood flow is
inversely proportional to the cross section of the vasculature, there is a

progressive decrease in velocity as blood moves toward and into the capillaries. Thus, about 1.5 seconds are required for a blood cell to pass through an average capillary. The total surface of the capillary walls is more than 100 times greater than the external surface of an average male adult. When this tremendous surface is combined with a slow rate of blood flow, an extremely effective means exists for exchange between the blood and tissues.

THE VEINS

The continuity of the vascular system is maintained as the capillaries feed deoxygenated blood at almost a trickle into the venules or small veins where they merge. Blood flow then increases somewhat because the cross-sectional area of the venous system is now less than that of the capillaries. The smaller veins in the lower portion of the body eventually empty into the body's largest vein, the **inferior vena cava** that travels through the abdominal and chest cavities toward the heart. Venous blood coming from the head, neck, and shoulder regions empties into the **superior vena cava** and moves downward to join the inferior vena cava at heart level. This mixture of blood from the upper and lower body then enters the right atrium where it descends into the right ventricle and is pumped through the pulmonary artery to the lungs. Gas exchange takes place in the alveolar-capillary network of the lungs, and the blood returns in the pulmonary veins to the left side of the heart to once again begin its passage around the body.

As illustrated in Figure 15–5, blood pressure and blood flow vary considerably in the systemic circulation. In the aorta and the large arteries, blood pressure fluctuates between 120 and 80 mm Hg during the cardiac cycle. The pressure then falls in direct proportion to the resistance encountered in the vascular circuit. The blood at the arteriole end of the capillaries exerts

FIG. 15–5.
Blood flow and blood pressure in the systemic circulatory system. The fall in pressure is in direct relation to the resistance offered in each portion of the vascular tree.

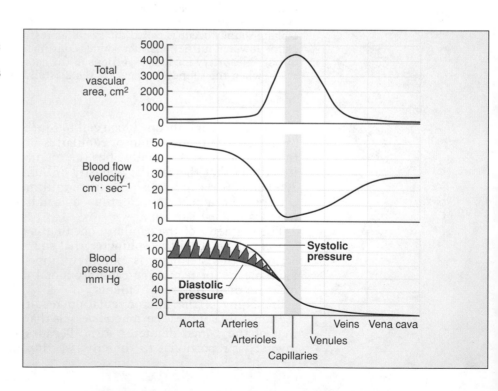

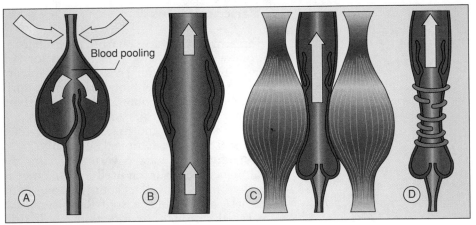

FIG. 15–6.
Valves in the veins (A) prevent returning flow but (B) do not hinder the normal flow very much. Blood can be pushed through veins (C) by nearby active muscle, or (D) by the action of smooth muscle bands. (From Elias, H., and Pauly, J.E.: Human Microanatomy. Philadelphia, F.A. Davis, 1966.)

an average pressure of only 30 mm Hg. As blood enters the venules, the impetus for blood flow is almost entirely lost. By the time blood reaches the right atrium, the pressure has fallen to approximately zero. Because the venous system operates under relatively low pressure, the walls of the veins are much thinner and less muscular than the thicker-walled and less distensible arteries (see Fig. 15–3).

Venous Return

The low pressure of venous blood poses a special problem that is partly solved by a unique characteristic of veins. Figure 15–6 shows that thin, membranous, flaplike **valves** spaced at short intervals within the vein permit a one-way blood flow back to the heart. Because venous blood is under relatively low pressure, veins are easily compressed by the smallest muscular contractions or even by minor pressure changes within the chest cavity during the act of breathing. This alternate compression and relaxation of the veins, as well as the one-way action of their valves, provides a "milking" action similar to the action of the heart. Compression of the veins imparts considerable energy for blood flow, whereas the "diastole" of these vessels enables them to refill as blood moves toward the heart. If valves were not present in these vessels, the blood would tend to pool, as it sometimes does in the veins of the extremities, and people would faint every time they stood up because of a reduction in cerebral blood flow.

The veins are **not** merely passive conduits. At rest, the venous system normally contains about 65% of the total blood volume; hence, the veins are considered capacitance vessels and serve as **blood reservoirs.** A slight increase in the tension or tone of the smooth muscle layer alters the diameter of the venous tree and produces a significant and rapid redistribution of blood from the peripheral venous circulation toward the central blood volume returning to the heart. This gives the venous system an important role as an active blood reservoir to either retard or deliver blood to the systemic circulation.

Varicose Veins

Sometimes the valves within a vein become defective and fail to maintain the one-way flow of blood. This condition is called **varicose veins** and usually occurs in the superficial veins of the lower extremities owing to the force of gravity that retards blood flow in an upright posture.[10a] Blood gathers in these veins, they become excessively distended and painful, and circulation from the affected area is actually impaired. In severe cases, the venous wall becomes inflamed and may degenerate—a condition called phlebitis—and the vessel must be removed surgically.

Individuals who have varicose veins should probably avoid excessive straining exercises often used in resistance training. During such sustained, nonrhythmic muscular contractions, both the muscle and ventilatory "pumps" are unable to contribute significantly to venous return. The increased abdominal pressure associated with straining also impedes venous return. All of these factors act to cause blood to pool in the veins of the lower body and could aggravate an existing varicose-vein condition. Although exercise cannot prevent the occurrence of varicose veins, regular and rhythmic physical activity may minimize complications because muscle action keeps blood moving toward the heart.[48]

Venous Pooling

The rhythmic action of muscular contraction and consequent compression of the vascular tree is so important to venous return that many people faint when forced to maintain an upright posture without movement. The classic "tilt table" experiment demonstrates this point. The subject is strapped to a table that can pivot to different positions. The heart rate and blood pressure are stable as long as the subject remains horizontal. Once the table is tilted vertically, an uninterrupted column of blood exists from the subject's heart to toes. This creates a hydrostatic force of about 80 to 100 mm Hg and causes blood to pool in the lower extremities. This results in a backup of fluid in the capillary bed that then seeps into the surrounding tissues and causes swelling, or **edema.** Consequently, venous return is reduced and blood pressure declines; at the same time, heart rate accelerates and venoconstriction occurs to counter the effects of venous pooling. If the upright position is still maintained, the subject eventually faints due to insufficient cerebral blood supply. Tilting the person either horizontally or head down immediately restores circulation and consciousness is quickly regained.

The Active Cool Down

The preceding discussion of venous pooling can be used to justify the action of those who continue to walk about or jog at a slow pace after strenuous exercise. Such moderate exercise, or "cooling down," would certainly facilitate blood flow through the vascular circuit (including the heart) during recovery. Recall from Chapter 7 that this "active recovery" also aids in removing lactic acid from the blood, and there is some indication that the continuation of mild exercise into recovery may blunt any possible deleterious effects on cardiac function of elevated catecholamines, epinephrine, and norepinephrine that follow exercise.[10] The pressurized suits worn by test pilots and special support stockings also aid in reducing the hydrostatic shift of blood to the veins of the lower extremities in the upright position. A similar supportive effect can be achieved in upright exercise in a swimming pool, because the external support of the water facilitates the return of blood to the heart.

HYPERTENSION

For individuals whose arteries have become "hardened" because fatty materials have deposited within their walls (or because the vessel's connective tissue layer has thickened) or whose arterial system offers excessive resistance to blood flow in the periphery due to nervous strain or kidney malfunction, systolic pressure at rest may be as high as 250 or even 300 mm Hg. The diastolic or run-off pressure may also be elevated above 90 mm Hg. Such **high blood pressure,** called **hypertension,** imposes a chronic, excessive strain on the normal functioning of the cardiovascular system.

It has been estimated that one out of every four persons will have abnormally high blood pressure sometime during their lives, with the disease being especially prevalent among black Americans.[43] Presently, more than 30 million Americans have systolic pressures over 140 mm Hg or diastolic pressures over 90 mm Hg. Uncorrected chronic hypertension can lead to heart failure, myocardial infarction, or stroke. Because elevated blood pressure can progress unnoticed for many years, yet can be effectively treated by medications that reduce extracellular fluid volume or peripheral resistance to blood flow, it is prudent to recommend that blood pressure be checked at periodic intervals.

EFFECTS OF EXERCISE TRAINING

Although the degree to which regular exercise can benefit a hypertensive condition is still unclear, **it does appear that both systolic and diastolic blood pressure can be lowered to a modest degree with a program of aerobic-type exercise.** These results have been observed with normotensive and hypertensive subjects at rest.[6,22,24,27,44,49,51] A reduction in mean arterial pressure during submaximal exercise has also been observed in healthy middle-aged men following endurance training.[25,50]

In patients who have documented coronary-artery disease and in "borderline" hypertensive patients, the effects of exercise training on blood pressure may be more impressive.[7,9,36,42] As indicated in Table 15–1, the average resting systolic pressure of seven middle-aged male patients decreased from 139 to 133 mm Hg following 4 to 6 weeks of interval training. In addition, at similar submaximal exercise levels, systolic pressure fell from 173 to 155 mm Hg, whereas diastolic pressure was also reduced from 92 to 79 mm Hg. Consequently, mean arterial blood pressure during exercise was reduced by approximately 14% following training. Similar findings were observed for an apparently healthy yet borderline hypertensive group of 37 middle-aged men following a 6-month exercise program.[7]

The precise mechanism for the exercise-lowering effect on blood pressure is not known, although it may occur because of a reduction of the catecholamines with training.[11] This response would contribute to a decrease in peripheral resistance to blood flow and a subsequent reduction in blood pressure. Exercise training may also facilitate the elimination of sodium by the kidneys to subsequently reduce fluid volume and blood pressure.[30] It should also be noted that not all research supports the wisdom of exercise in treating hypertension. For example, studies with various animal models for hypertension have not demonstrated a dramatic or consistent benefit of

TABLE 15–1.
MEASURES OF BLOOD PRESSURE AT REST AND DURING SUBMAXIMAL EXERCISE
PRIOR TO AND FOLLOWING 4 TO 6 WEEKS OF TRAINING IN SEVEN MIDDLE-AGED PATIENTS
WITH CORONARY HEART DISEASE[a]

	Rest			Submaximal Exercise		
	Mean Value		Difference	Mean Value		Difference
Measure[b]	**Before**	**After**	**(%)**	**Before**	**After**	**(%)**
Systolic blood pressure (mm Hg)	139	133	−4.3	173	155	−10.4
Diastolic blood pressure (mm Hg)	78	73	−6.4	92	79	−14.1
Mean blood pressure (mm Hg)	97	92	−5.2	127	109	−14.3

[a] Modified from Clausen, J.P., et al.: Physical training in the management of coronary artery disease. Circulation 40:143, 1969.
[b] Blood pressure was measured directly by a pressure transducer inserted into the brachial artery.

regular regimens of forced exercise.[17] Even in those studies that show a modest lowering of blood pressure with exercise training, many studies are subject to methodological shortcomings and inadequate research design.[46] The most glaring weakness is usually the lack of an appropriate control group who also have their blood pressure measured but do not exercise. Until the picture is clarified with sound research on humans, a prudent recommendation is to include exercise and weight loss as a first line of defense in most therapeutic programs to manage borderline hypertension.[28a,38] For more severe elevations in blood pressure, a combination of diet, weight loss, exercise, and pharmacologic therapies may be required.

BLOOD PRESSURE AND EXERCISE

STATIC AND DYNAMIC RESISTANCE EXERCISE

Straining-type exercises compress the peripheral arterial system to bring about acute and dramatic increases in resistance to blood flow.[5,33,18] Research has focused on the blood pressure responses during static and dynamic resistance exercise.[15,34] In a study from one of our laboratories, the blood pressure of normotensive subjects was measured directly with a pressure transducer connected to a catheter that was inserted into the femoral artery.[16] Three forms of exercise were evaluated for blood-pressure response: (1) isometric bench press performed at 25, 50, 75, and 100% of the maximal voluntary contraction (MVC) in the same arm and body position as the two other conditions, (2) free weight, bench press performed at 25 and 50% of the isometric MVC, and (3) hydraulic resistance, bench press exercise performed "all-out" for 20 seconds duration at resistance setting 3 (slow-11 reps) and setting 5 (fast-16 reps). The results displayed in Table 15–2 illustrate clearly that all three forms of exercise significantly augment the magnitude of the cardiovascular response and the corresponding workload of the heart. In addition, exercise that employs the larger muscle mass and the greater relative strain elicits the greater response.[15,17,47] The results show that all

TABLE 15–2.
COMPARISON OF PEAK SYSTOLIC AND DIASTOLIC BLOOD PRESSURE DURING ISOMETRIC AND BENCH-PRESS FREE WEIGHT AND HYDRAULIC RESISTANCE EXERCISE

Condition	Isometric[a] (% MVC)				Free Weight Bench Press[b] (% MVC)		Hydraulic Bench Press[c]	
	25	50	75	100	25	50	Slow	Fast
Peak systolic	172	179	200	225	169	232	237	245
Peak diastolic	106	116	135	156	104	154	101	160

Values are means for 7 subjects. Data from references 16 and unpublished data, Department of Exercise Science, University of Massachusetts, Amherst, MA.
[a] Open glottis (no Valsalva's maneuver); average of 2 trials; contraction time 2 to 3 seconds; arm position that of bench-press exercise with hands just slightly above chest.
[b] The weight lifted was either 25 or 50% of previously determined isometric maximum contraction.
[c] Performed on Hydra-Fitness chest-press apparatus at dial setting 3 (slow) and 5 (fast) for 20 seconds of repeated maximal contractions.

three forms of exercise produce dramatic elevations in blood pressure—even with relatively light, isometric exercise that requires only 25% of the maximal effort. This acute cardiovascular strain could be harmful for individuals who have heart and vascular disease, especially those who are untrained in this form of exercise. For these people, more rhythmic forms of moderate exercise are desirable and beneficial.

CHRONIC EFFECTS OF RESISTANCE TRAINING

Resistance training exercises may cause a greater rise in blood pressure compared to lower-intensity dynamic movement, but it does not seem that this form of training causes any long-term increase in resting blood pressure.[8,14,39] It also appears that a regular program of resistance training blunts the blood pressure response. Trained body builders, for example, show smaller increases in systolic and diastolic blood pressure with resistance exercise than both novice and untrained groups.[14] The finding that regular resistance training benefited the resting blood pressure of borderline hypertensive adolescents[20] complements these observations.

STEADY-RATE EXERCISE

In rhythmic muscular activity such as jogging, swimming, and bicycling, the dilation of the blood vessels in the working muscles decreases total peripheral resistance and enhances the flow of blood through large portions of the peripheral vasculature. The alternate contraction and relaxation of the muscles themselves also provide a significant pumping force to propel blood through the vascular circuit and return it to the heart. The increased blood flow during moderate, rhythmic exercise causes systolic pressure to rise rapidly in the first few minutes of exercise. The blood pressure then levels off at 140 to 160 mm Hg.[13] As steady-rate exercise continues, systolic pressure may gradually fall as the arterioles in the muscles continue to dilate and peripheral resistance to blood flow becomes reduced. During this exercise, the diastolic blood pressure remains relatively unchanged.

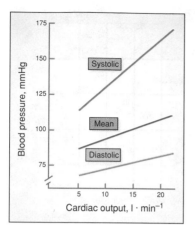

FIG. 15–7.
The relationship between blood flow during exercise and systemic arterial pressures measured at the brachial artery. (From Ekelund, L.G., and Holmgren, A.: Central hemodynamics during exercise. Am. Heart Assoc. Monogr., No. *15*:33, 1967.)

GRADED EXERCISE

The blood pressure response as measured directly by pressure transducers inserted into the brachial artery during progressive exercise of increasing severity is illustrated in Figure 15–7. In this situation, systolic, diastolic, and mean arterial pressure are plotted as a function of the quantity of blood ejected into the arterial circuit each minute, the **cardiac output.**

As can be seen, the various indices of arterial blood pressure increase linearly with cardiac output. The greatest increases in exercise blood pressure are observed during cardiac systole whereas the diastolic pressure increases by only about 12% during the full range of exercise. This response is similar for both physically conditioned and sedentary subjects. During maximum exercise performed by healthy endurance athletes, however, the systolic blood pressure may increase to 200 mm Hg, a response most likely due to the large cardiac outputs of these athletes.[4]

BLOOD PRESSURE IN ARM EXERCISE

As shown in Table 15–3, at a given percentage of the maximal oxygen consumption, systolic and diastolic blood pressures are considerably **higher** when work is performed with the arms than with the legs.[2,40,49a]

It is likely that the smaller muscle mass and vasculature of the arms offer greater resistance to blood flow than the larger muscle mass and vasculature of the legs.[5] Blood flow to the arms during exercise would therefore require a much larger systolic head of pressure. Clearly, this form of exercise represents greater cardiovascular strain because the work of the heart is increased considerably.[35] For individuals who have cardiovascular dysfunction, these observations support the use of exercise that requires large muscle groups, such as walking, bicycling, and running, in contrast to unregulated exercises that engage a rather limited muscle mass such as shoveling, overhead hammering, or even arm ergometry. If a systematic program of arm exercise is utilized, the work loads must be established based on the person's response to this form of exercise and **not** from some exercise stress test prescription that employs bicycling or running. The cardiovascular adjustment to arm exercise is discussed further in Chapter 17.

TABLE 15–3.
COMPARISON OF SYSTOLIC AND DIASTOLIC BLOOD PRESSURE DURING ARM AND LEG EXERCISE AT SIMILAR PERCENTAGES OF THE MAXIMAL OXYGEN INTAKE*

Percent of max $\dot{V}O_2$	Systolic Pressure (mm Hg)		Diastolic Pressure (mm Hg)	
	Arms	**Legs**	**Arms**	**Legs**
25	150	132	90	70
40	165	138	93	71
50	175	144	96	73
75	205	160	103	75

* From Åstrand, P.O., et al.: Intra-arterial blood pressure during exercise with different muscle groups. J. Appl. Physiol., 20:253, 1965.

IN RECOVERY

After a bout of sustained submaximal exercise, systolic blood pressure is temporarily reduced below pre-exercise levels for both normotensive and hypertensive subjects.[23,28,41] This hypotensive response to previous exercise lasts about 2 to 3 hours into recovery. Such findings further support the use of exercise as an important nonpharmacologic line of defense in treating hypertension, and would justify participating in several bouts of light to moderate physical activity interspersed throughout the day.

BODY INVERSION

There is a recent increase in the use of gravity-inversion devices where individuals hang in the upside-down position with the belief that this maneuver can offer relaxation, facilitate a strength-training response, or relieve low back pain. Although it has not yet been demonstrated that inverting the body is of any practical or physiologic benefit, it is now apparent that it can cause a significant elevation in both systolic and diastolic blood pressure. In one study of 50 men and women who had normal blood pressure, body inversion caused systolic blood pressure to rise from an average of 114 mm Hg to nearly 140 mm Hg whereas diastolic pressure increased from 76 to 91 mm Hg.[32] In addition, these changes continued throughout the 3-minute duration of the inversion maneuver. Such responses during inversion at rest in normal individuals, as well as significant increases in intraocular pressure, raise concern about the possible consequences of inversion for hypertensive people and the prudence of recommending exercise in this position without closely monitoring the blood-pressure response.

THE HEART'S BLOOD SUPPLY

Although literally tons of blood may flow through the heart's chambers each day, none of its nourishment passes directly into the myocardium. This is because there are no direct circulatory channels within the heart's chambers leading to its tissues. Instead, the heart muscle has an elaborate circulatory network of its own. As shown in Figure 15–8, these vessels form a visible, crownlike network that arises from the top portion of the heart called the **coronary circulation.**

The openings for the left and right coronary arteries are situated in the aorta just above the semilunar valves at a point where the oxygenated blood leaves the left ventricle. These arteries then curl around the heart's surface; the right coronary supplies predominantly the right atrium and ventricle whereas the greatest volume of blood flows in the left atrium and ventricle and a small part of the right ventricle. These vessels divide and eventually form a dense capillary network within the heart muscle. Blood then leaves the tissues of the left ventricle through the **coronary sinus;** blood from the right ventricle exits via the **anterior cardiac veins,** which empty directly into the right atrium.

With each heart beat, the driving force of the heart pushes a portion of blood into the coronary arteries. At rest, normal blood flow to the myocardium is 200 to 250 ml per minute; this is about 5% of the total cardiac output.

FIG. 15—8.
The coronary circulation.
Arteries are shaded dark, and
veins are unshaded.

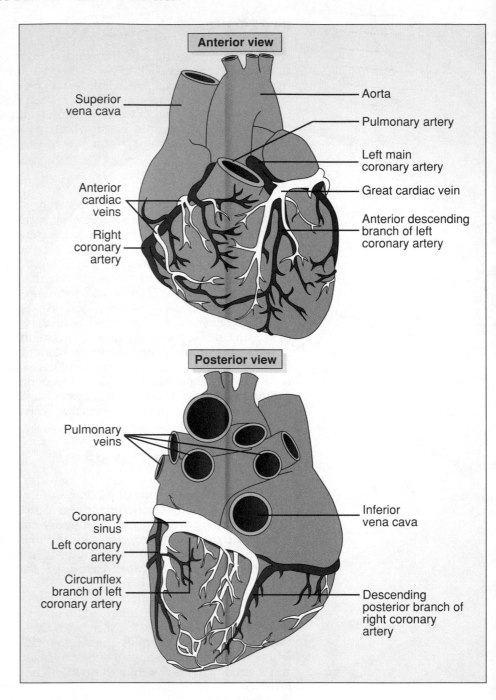

MYOCARDIAL OXYGEN UTILIZATION

At rest, the oxygen utilization of the myocardium is high in relation to its
blood flow. About 70 to 80% of the oxygen is extracted from the blood that
flows in the coronary vessels. This is in contrast to most other tissues at
rest that use as little as one-fourth of the available oxygen. The increased

myocardial oxygen demands during exercise can mainly be met, therefore, by a proportionate increase in coronary blood flow. In vigorous exercise, coronary blood flow may increase 4 to 6 times above the resting level. This is achieved in two ways: (1) increased myocardial metabolism during exercise has a direct effect on the coronary vessels causing them to dilate. For example, hypoxia has an extremely potent effect for increasing blood flow through the myocardium. Adenosine, a byproduct of ATP breakdown, is probably also a major mediator in the autoregulation of myocardial blood flow.[3] In addition to local factors, hormones of the sympathetic nervous system are released during exercise and cause coronary dilatation; and (2) during exercise, the increased aortic pressure forces a proportionately greater quantity of blood into the coronary circulation. The ebb and flow of blood in the coronary vessels fluctuates with each phase of the cardiac cycle. On the average, coronary blood flow is about 2.5 times greater during diastole than during systole.[19]

An adequate oxygen supply is critical for the myocardium, because unlike skeletal muscle, this tissue has an extremely limited ability to generate energy anaerobically. The blood supply to the heart is so profuse that at least one capillary supplies each of the heart's muscle fibers. Impairment in coronary blood flow usually results in chest pains, or **angina pectoris.** These pains become pronounced during exercise. In fact, the stress of exercise is often used to document or evaluate this condition. A blood clot, or **thrombus,** lodged in one of the coronary vessels may severely impair normal heart function. Although this form of "heart attack," or more specifically **myocardial infarction,** may be mild, a more complete blockage causes severe damage to the myocardium and could result in death. A more complete discussion of coronary heart disease, stress testing, and the possible role of exercise as preventive medicine is presented in Chapter 30.

THE RATE-PRESSURE PRODUCT: AN ESTIMATE OF MYOCARDIAL WORK

Myocardial oxygen consumption is determined by interactions between several mechanical factors—most important, the development of tension within the myocardium and its contractility, and heart rate. With increases in each of these factors during exercise, myocardial blood flow increases to balance oxygen supply with demand. One commonly used estimate of **myocardial workload** and resulting oxygen consumption makes use of the product of peak systolic blood pressure (SBP), as measured at the brachial artery, and heart rate (HR). This **index of relative cardiac work,** termed the **double product** or **rate-pressure product** (RPP), is highly related to directly measured myocardial oxygen consumption and coronary blood flow in healthy subjects over a wide range of exercise intensities.[29,37] It is computed as

$$RPP = SBP \times HR$$

The rate-pressure product has been used extensively in exercise studies of coronary heart disease patients to provide a physiologic correlate to either the onset of angina or electrocardiographic abnormalities. Once this is established, various clinical, surgical, or exercise interventions can be evaluated to determine their effect on cardiac performance. For one thing, the well-documented lowering of exercise heart rate and systolic blood pressure

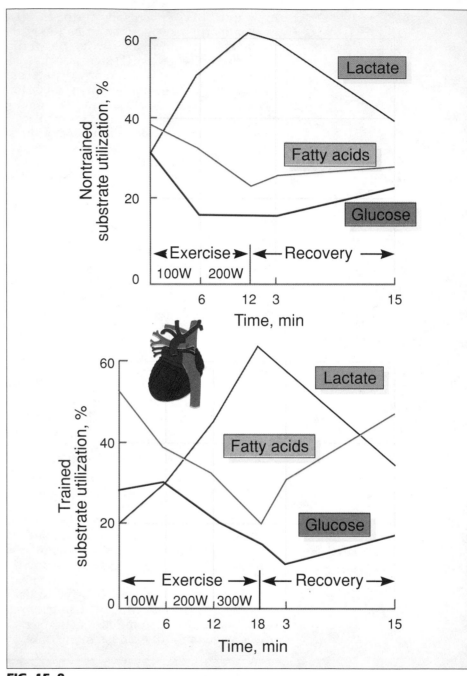

FIG. 15–9.
The relative proportion of substrate used by the heart at various work loads
(W = watts) and during recovery in trained and nontrained subjects. (From Keul,
J., et al.: Medicine and Sport, Vol. 7, Energy Metabolism of Human Muscle. Basel,
Karger, 1972.)

(and hence myocardial oxygen consumption) with training helps explain the improved exercise capacity of cardiac patients with regular exercise. Furthermore, several research studies have shown that prolonged and intense aerobic training can result in higher RPP achieved by cardiac patients.[12,21] In nine patients followed over a 7-year training period the RPP increased by 11.5% before the appearance of ischemic symptoms.[45] Such findings are important because they provide indirect evidence for an improved level of myocardial oxygenation, perhaps due to greater coronary vascularization or reduced obstruction, as part of the training adaptation.

MYOCARDIAL METABOLISM

As with all tissue, the heart utilizes the chemical energy stored in food nutrients to power its work. The heart, however, relies almost totally on energy released in aerobic reactions. In fact, human heart muscle has a 3-fold higher oxidative capacity than skeletal muscle.[26] As such, myocardial fibers have the greatest mitochondrial concentration of all tissues and thus are highly adapted for fat catabolism as a primary source of ATP resynthesis.[52]

Figure 15–9 shows the substrate utilization by the heart at rest and during exercise and recovery. Glucose, fatty acids, and the lactic acid formed in skeletal muscle during glycolysis provide the energy for proper myocardial functioning. At rest, these three substrates are utilized to synthesize ATP. In essence, the heart uses for energy whatever substrate it "sees" on a physiologic level—so during heavy exercise, when the efflux of lactic acid from skeletal muscle into the blood increases significantly, the heart may derive more than 50% of its total energy from the oxidation of circulating lactate. During prolonged submaximal activity that occurs in distance running, skiing, or swimming, the myocardial metabolism of free fatty acids rises to almost 70% of the total energy requirement. This metabolic pattern is similar for trained and untrained subjects, **although the contribution of fats to the total energy requirement is considerably greater among trained endurance athletes.** This difference provides another illustration of the "carbohydrate sparing effect" of training that also occurs in trained skeletal muscle. Such an adaptation helps to preserve the body's glycogen reserves that are crucial to muscle and brain metabolism during prolonged exercise.

■■■■ SUMMARY

1. The striated fibers of the myocardium are interconnected so that portions of the heart contract in a unified manner. Functionally, the heart may be viewed as two separate pumps: One pump receives blood that returns from the body and pumps it to the lungs for aeration (pulmonary circulation), whereas the other receives oxygenated blood from the lungs and pumps it throughout the systemic circulation.
2. Pressure changes created during the cardiac cycle act on the heart's valves to provide for a one-way flow of blood into the vascular circuit.
3. The surge of blood with the contraction of the ventricles (and subsequent run-off during relaxation) creates pressure changes within the arterial vessels. The systolic pressure, or highest pressure generated during the cardiac cycle, occurs during ventricular contraction. The diastolic pressure is the lowest pressure reached before the next ventricle contraction.
4. Hypertension imposes a chronic stress on cardiovascular function. Reg-

ular aerobic training can probably bring about modest reductions in systolic and diastolic blood pressure at rest and during submaximal exercise.

5. Systolic blood pressure increases in proportion to oxygen consumption and cardiac output during graded exercise, whereas diastolic pressure remains relatively unchanged or increases slightly. At the same relative work load, systolic pressures are greater when work is performed with the arms than with the legs.

6. Following exercise blood pressure falls below pre-exercise levels, and may remain lower for several hours.

7. During isometric, free weight, and hydraulic exercise, peak systolic and diastolic blood pressures mirror the hypertensive state and may pose a risk to individuals who have existing hypertension. Regular training with resistance exercise blunts this hypertensive response.

8. The dense capillary network provides a large and effective surface for exchange between the blood and tissues. These vessels adjust blood flow in response to the tissue's metabolic activity.

9. Compression and relaxation of the veins by the action of skeletal muscles impart considerable energy to facilitate venous return. This provides additional justification for the use of active recovery following vigorous exercise.

10. Nerves and hormones act on the smooth muscle layer in the venous walls causing them to constrict or stiffen. This alteration in venous tone can have a profound effect on the distribution of total blood volume.

11. The hemodynamic adjustments with postural changes are of little value in predicting one's exercise capability.

12. At rest, about 80% of the oxygen flowing through the coronary arteries is extracted by the myocardium. This high extraction means that the increased myocardial oxygen demands in exercise can only be met by a proportionate increase in coronary blood flow.

13. Because the myocardium is essentially aerobic tissue, it must continually be supplied with oxygen. Impairment of coronary blood flow causes anginal pains, and blockage of a coronary artery (myocardial infarction) rapidly causes irreversible damage to the heart muscle.

14. The product of heart rate and systolic blood pressure, the rate-pressure product, provides a convenient estimate of the workload on the myocardium. This index has been used to study the effects of exercise training on cardiac performance in heart disease patients.

15. The main substrates used by the heart for energy are glucose, fatty acids, and lactic acid. The percentage utilization of these substrates varies with the severity and duration of exercise.

REFERENCES

1. Armstrong, R.B.: Magnitude and distribution of muscle blood flow in conscious animals during locomotory exercise. *Med. Sci. Sports Exerc.*, 20:S119, 1988.

2. Åstrand, P.O., et al.: Intra-arterial blood pressure during exercise with different muscle groups. *J. Appl. Physiol.*, 20:253, 1965.

3. Berne, R.M.: The role of adenosine in the regulation of coronary blood flow. *Circ. Res.*, 47:807, 1980.

4. Bevegård, S., et al.: Circulatory studies in well-trained athletes at rest and during heavy exercise, with special reference to stroke volume and the influence of body position. *Acta Physiol. Scand.*, 57:26, 1963.

5. Blomqvist, C.G., et al.: Similarity of the hemodynamic responses to static and dynamic exercise of small muscle groups. *Circ. Res.*, 1:87, 1982.

6. Boyer, J.L., and Kasch, F.W.: Exercise therapy in hypertensive men. *JAMA*, 211:1668, 1970.

7. Choquette, G., and Ferguson, R.J.: Blood pressure reduction in "borderline" hypertensives following physical training. *Can. Med. Assoc. J.*, 108:699, 1973.

8. Collander, E.B., et al.: Blood pressure in resistance-trained athletes. *Can. J. Sports Sci.*, 13:31, 1988.

9. Clausen, J.P., et al.: Physical training in management of coronary artery disease. *Circulation*, 30:143, 1969.

10. Dimsdale, J.E., et al.: Postexercise peril: plasma catecholamines and exercise. *JAMA*, 251:630, 1987.

10a. Donaldson, M.C.: Varicose veins in active people. *Phys. Sportsmed.*, 18:46, 1990.

11. Duncan, J.J., et al.: The effects of aerobic exercise on plasma catecholamines and blood pressure in patients with mild hypertension. *JAMA*, 254:2609, 1985.

12. Ehsani, A.A., et al.: Improvement of left ventricular contractile function by exercise training in patients with coronary artery disease. *Circulation*, 74:350, 1986.

13. Ekelund, L.G., and Holmgren, A.: Central hemodynamics during exercise. *Circ. Res.* (Suppl. 1):33, 1967.

14. Fleck, S.J.: Cardiovascular adaptations to resistance training. *Med. Sci. Sports Exerc.*, 20:S146, 1988.

15. Fleck, S.J., and Dean, L.S.: Resistance training experience and the pressor response during resistance training. *J. Appl. Physiol.*, 63:116, 1987.

16. Freedson, P.F., et al.: Intra-arterial blood pressure during free weight and hydraulic resistive exercise. *Med. Sci. Sports Exerc.*, 16:131, 1984.

17. Fregly, M.J.: Effect of an exercise regimen on development of hypertension in rats. *J. Appl. Physiol.*, 56:381, 1984.

18. Gaffney, F.A., et al.: Cardiovascular and metabolic responses to static contraction in man. *Acta Physiol. Scand.*, 138:249, 1990.

19. Gregg, D.E., and Coffman, J.D.: Coronary circulation. In *Blood Vessels and Lymphatics.* Edited by D.S. Abramson. New York, Academic Press, 1962.

20. Hagberg, J.M., et al.: Effect of weight training on blood pressure and hemodynamics in hypertensive adolescents. *J. Pediatr.*, 104:147, 1984.

21. Hagberg, J.M., et al.: Effect of 12 months of intense exercise training on stroke volume in patients with coronary artery disease. *Circulation*, 67:1194, 1983.

22. Hagberg, J.M., et al.: Effect of exercise training on the blood pressure and hemodynamics of adolescent hypertensives. *Am. J. Cardiol.*, 52:763, 1981.

23. Hannum, S.M., and Kasch, F.W.: Acute postexercise blood pressure response of hypertensive and normotensive men. *Scand. J. Sports Sci.*, 3:11, 1981.

24. Hanson, J.S., and Nedde, W.H.: Preliminary observations on physical training for hypertensive males. *Circ. Res.*, 27 (Suppl. 1):49, 1970.

25. Hartley, L.H., et al.: Physical training in sedentary middle-aged and older men. III. Cardiac output and gas exchange at submaximal and maximal exercise. *Scand. J. Clin. Lab. Invest.*, 24:335, 1969.

26. Jansson, E., and Sylven, E.: Myoglobin in human heart and skeletal muscle in relation to oxidative potential as estimated by citrate synthase. *Clin. Physiol.*, 1:596, 1981.

27. Kasch, F.W., and Boyer, L.J.: Changes in maximum work capacity resulting from six months training in patients with ischemic heart disease. *Med. Sci. Sports*, 1:156, 1969.

28. Kaufman, F.L., et al.: Effect of exercise on recovery blood pressure in normotensive and hypertensive subjects. *Med. Sci. Sports Exerc.*, 19:17, 1987.

28a. Kelemen, M.H.: Exercise training combined with antihypertensive drug therapy: effects on lipids, blood pressure, and left ventricular mass. *JAMA*, 263:2766, 1990.

29. Kitamura, K., et al.: Hemodynamic correlates of myocardial oxygen consumption during upright exercise. *J. Appl. Physiol.*, 32:516, 1972.

30. Kiyonaga, A., et al.: Blood pressure and hormonal response to aerobic exercise. *Hypertension*, 17:125, 1985.

31. Krough, A.: *The Anatomy and Physiology of Capillaries.* New Haven, CT, Yale University Press, 1929.

32. Le Marr, J.D., et al: Cardiorespiratory responses to inversion. *Phys. Sportsmed.*, 11:51, 1983.

33. Lind, A.R., et al.: The circulatory effects of sustained voluntary muscle contraction. *Clin Sci.*, 27:229, 1964.

34. MacDougall, J.D., et al.: Arterial blood pressure response to heavy resistance exercise. *J. Appl. Physiol.*, 58:785, 1985.

35. Miles, D.S.: Cardiovascular responses to upper body exercise in normals and cardiac patients. *Med. Sci. Sports Exerc.*, 21:S126, 1989.

36. Nelson, L., et al.: Effect of changing levels of physical activity on blood pressure and haemodynamics in essential hypertension. *Lancet*, 2:473, 1986.

37. Nelson, R.R., et al.: Hemodynamic predictors of myocardial oxygen consumption during static and dynamic exercise. *Circulation*, 50:1179, 1974.

38. Paffenbarger, R.S.: Energy imbalance and hypertension risk. In *Diet and Exercise: Synergism in Health Maintenance*. Edited by P.L. White and T. Mondeika. Chicago, American Medical Association, 1982.

39. Pearson, A.C., et al.: Left ventricular diastolic function in weight lifters. *Am. J. Cardiol.*, 58:1254, 1986.

40. Pendergast, D.R.: Cardiovascular, respiratory, and metabolic responses to upper body exercise. *Med. Sci. Sports Exerc.*, 21:S121, 1989.

41. Raglin, J.S., and Morgan, W.P.: Influence of exercise and quiet rest on state anxiety and blood pressure. *Med. Sci. Sports Exerc.*, 19:456, 1987.

42. Redwood, D.R., et al.: Circulatory and symptomatic effects of physical training in patients with coronary-artery disease and angina pectoris. *N. Engl. J. Med.*, 286:959, 1972.

43. Report on Secretary's Task Force on Black and Minority Health, U.S. Dept. of Health and Human Services publication 0–174–719. Government Printing Office, August 1985.

44. Rissl, J., et al.: Hemodynamic effect of physical training in essential hypertension. *Acta Cardiol.*, 32:121, 1977.

45. Rogers, M.A., et al.: The effects of 7 years of intense exercise training on patients with coronary artery disease. *J. Am. Coll. Cardiol.*, 10:321, 1987.

46. Seals, D.R., and Hagberg, J.M.: The effect of exercise training on human hypertension. *Med. Sci. Sports Exerc.*, 16:207, 1984.

47. Seals, D.R., et al.: Increased cardiovascular response to static contraction of larger muscle groups. *J. Appl. Physiol.*, 54:434, 1983.

48. Smith, M.L., et al.: Effect of muscle tension on the cardiovascular responses to lower body negative pressure in man. *Med. Sci. Sports Exerc.*, 19:436, 1987.

49. Terjung, R.I., et al.: Cardiovascular adaption to twelve minutes of mild daily exercise in middle-aged sedentary men. *J. Am. Geriatr. Soc.*, 21:164, 1973.

49a. Toner, M.M., et al.: Cardiovascular adjustment to exercise distributed between the upper and lower body. *Med. Sci. Sports Exerc.* (in press).

50. Tzankoff, S.P., et al.: Physiological adjustments to work in older men as effected by physical training. *J. Appl. Physiol.*, 33:346, 1972.

51. Urata, H., et al.: Antihypertensive and volume-depleting effects of mild exercise on essential hypertension. *Hypertension*, 9:245, 1987.

52. Vary, T.C., et al.: Control of energy metabolism of heart muscle. *Annu. Rev. Physiol.*, 43:419, 1981.

16

CARDIOVASCULAR REGULATION AND INTEGRATION

In humans, a "closed" circulatory system has evolved in which the blood cells remain trapped within the confines of a continuous vascular circuit. The potential for the vessels of the skin, viscera, and skeletal muscles to conduct blood exceeds by 3- to 4-fold the capacity of the normal heart to pump blood.[16] Consequently, complex mechanisms continually interact to maintain systemic blood pressure and blood flow to various tissues under diverse conditions. Nerves and chemicals regulate both the speed of the pumping heart and the internal opening of various blood vessels. This regulation provides for rapid and effective control of the heart as well as for the distribution of blood throughout the body. When a person is resting comfortably, about 5% of the 5 liters of blood pumped each minute from the heart goes to the skin. This is in contrast to exercise performed in a hot, humid environment in which as much as 20% of the total blood flow is diverted to the body's surface for the purpose of heat dissipation. This "shunting" with appropriate maintenance of blood pressure can occur only within a closed vascular system that has the capability for immediate redistribution of blood depending on the body's metabolic and physiologic needs.[17]

REGULATION OF HEART RATE

Cardiac muscle is unique because it has the capability of maintaining its own rhythm. If left to this inherent rhythmicity, the heart would beat steadily between 70 and 80 times each minute. Nerves, however, that go directly to the heart as well as chemicals that circulate in the blood, can rapidly change the heart rate. These extrinsic controls of cardiac function cause the heart to speed up in "anticipation," even before the start of exercise. To a large extent, extrinsic regulation provides for heart rates that may be as slow as 30 beats per minute at rest in highly trained endurance athletes, and as fast as 220 beats per minute in maximum exercise.

INTRINSIC REGULATION OF HEART RATE

Situated within the posterior wall of the right atrium is a mass of specialized muscle tissue called the **sinoatrial node** or **S-A node.** This node spontaneously depolarizes and repolarizes to provide the "innate" stimulus to the heart. For this reason, the S-A node is referred to as the "pacemaker." The normal route for the transmission of the impulse across the myocardium is shown in Figure 16–1.

FIG. 16–1.
Excitation and conduction of cardiac impulse from the S-A node.

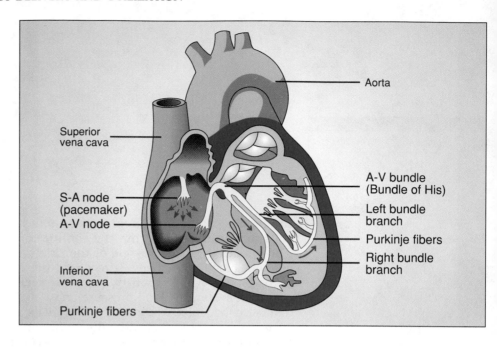

The Heart's Electric Impulse

Rhythms originating at the S-A node spread across the atria to another small knot of tissue, the **atrioventricular node** or **A-V node.** Here the impulse is delayed about 0.10 second to provide sufficient time for the atria to contract and force blood into the ventricles. The A-V node gives rise to the **A-V bundle** (bundle of His) that transmits the impulse rapidly through the ventricles over specialized conducting fibers referred to as the **Purkinje system.** These fibers form distinct branches that penetrate the right and left ventricles. Each ventricular cell is stimulated within about 0.06 second from the passage of the impulse into the ventricles; this permits a unified and simultaneous contraction of the entire musculature of both ventricles. The transmission of the cardiac impulse can be summarized as follows:

S-A node → Atria → A-V node → A-V bundle
(Purkinje fibers) → Ventricles

Electrocardiography: The ECG

Like all nerve and muscle tissue, the outer surface of the myocardial cells is electrically more positive than the inside. This polarity is reversed just prior to contraction when the heart is stimulated and the inside of the cell actually becomes more positive than the outside. During the diastolic phase of the cardiac cycle, the membranes repolarize and the resting membrane potential is re-established.

The electrical activity about the heart creates an electrical field throughout the body. Because the salty body fluids provide an excellent conducting medium, the sequence of electrical events prior to and during each cardiac cycle can be detected as voltage changes by electrodes placed on the skin's surface. The graphic record of the heart's electric activity is called the **elec-**

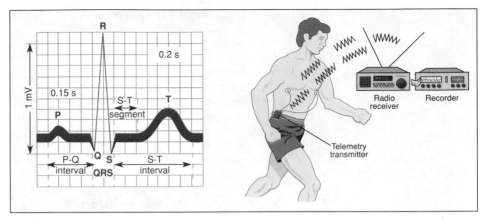

FIG. 16–2.
The normal electrocardiogram. At right is an example of bipolar chest leads and radio telemetry routinely used to obtain the exercise ECG.

trocardiogram, or simply **ECG.** A characteristic, normal electrocardiogram is presented in Figure 16–2.

The **P wave** represents the depolarization of the atria. It lasts about 0.15 seconds and heralds atrial contraction. The P wave is then followed by the relatively large **QRS complex.** This reflects the electrical changes caused by the depolarization of the ventricles; at this point, the ventricles contract. Atrial repolarization that follows the P wave produces a wave so small that it is usually obscured by the large QRS complex. The **T wave** represents repolarization of the ventricles. This occurs during ventricular diastole.

The heart's relatively long period of depolarization of approximately 0.20 to 0.30 second is required before it can receive another impulse and contract again. This "rest" or **refractory period** serves an important function because it provides sufficient time for ventricular filling between beats.

The electrocardiogram serves useful purposes for the cardiologist and exercise specialist. For one thing, it provides an effective means for monitoring heart rate objectively during exercise. Radio telemetry makes it possible to transmit the ECG while the person is free to perform various types of exercise including football, weight lifting, basketball, ice hockey, dancing, and even swimming.

Electrocardiography also serves as an extremely valuable tool for uncovering abnormalities in heart function, especially those related to cardiac rhythm, electrical conduction, myocardial oxygen supply, and actual tissue damage (see Chapter 30).

EXTRINSIC REGULATION OF HEART RATE

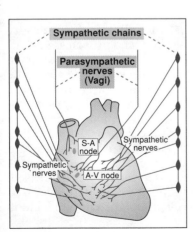

FIG. 16–3.
The heart's distribution of sympathetic and parasympathetic nerves.

Neural influences are superimposed on the inherent rhythmicity and conductivity of the myocardium. These influences originate in the cardiovascular center in the medulla and are transmitted through the sympathetic and parasympathetic components of the autonomic nervous system. As shown in Figure 16–3, the atria are supplied with large numbers of sympathetic and parasympathetic neurons, whereas the ventricles receive sympathetic fibers almost exclusively.

Sympathetic Influence

Stimulation of the sympathetic cardioaccelerator nerves releases epinephrine and norepinephrine. Collectively, these neural hormones are called **catecholamines.** They act to accelerate the depolarization of the sinus node that causes the heart to beat faster. This acceleration in heart rate is termed **tachycardia.** The catecholamines also significantly increase myocardial contractility. It has been estimated that maximum sympathetic stimulation nearly doubles the force of ventricular contraction. Epinephrine released from the medullary portion of the adrenal glands in response to a general sympathetic activation also produces a similar though slower-acting effect on cardiac function.

Parasympathetic Influence

Acetylcholine, the hormone of the parasympathetic nervous system, retards the rate of sinus discharge and slows the heart. This slowing of heart rate is termed **bradycardia.** The effect is largely mediated through the action of the **vagus nerve** whose cell bodies originate in the cardioinhibitory center in the medulla. Vagal stimulation has essentially no effect on myocardial contractility.

Training Effects

Exercise training creates an imbalance between the tonic activity of the sympathetic accelerator and parasympathetic depressor neurons in favor of greater vagal dominance. This is mediated primarily by an increase in parasympathetic activity[10] and perhaps a decrease in sympathetic discharge.[18,23] In addition, training may also decrease the intrinsic rate of firing of the S-A node.[2,9,19] These adaptations account for the significant bradycardia often observed in highly conditioned endurance athletes or in sedentary subjects following aerobic training.[7,8,24]

Peripheral Input

The cardiovascular center in the ventrolateral medulla receives sensory input from peripheral receptors in blood vessels, joints, and muscles. Stimuli from these mechanical and chemical receptors that monitor the state of active muscle modify either vagal or sympathetic outflow to bring about the appropriate cardiovascular response. Receptors in the aortic arch and carotid sinus respond to changes in arterial blood pressure. As blood pressure increases, the stretch of the arterial vessels activates these **baroreceptors** to bring about a reflex slowing of the heart as well as a compensatory dilation of the peripheral vasculature. This causes blood pressure to decrease toward more normal levels. To some degree, this particular feedback mechanism is over-ridden during exercise or its threshold increased or sensitivity decreased, because heart rate and blood pressure are both increased considerably. More than likely, the baroreceptors act as a brake to prevent abnormally high blood pressure levels in exercise.

Carotid Artery Palpation

In some individuals, strong external pressure against the carotid artery produces a slowing effect on the heart rate. This effect is probably mediated by direct stimulation of the baroreceptors at the bifurcation of the carotid artery.[25] Of interest to exercise specialists has been the potential for bradycardia produced by **carotid artery palpation** because this is commonly used for determining pulse rate immediately after exercise. An accurate measure of heart rate is important for training when specific "target" heart rates are assigned to regulate training intensity (see Chapter 21). If the method used to monitor the heart rate consistently gave low values (as could be the case with carotid artery palpation), the person would be pushed to a higher work level. This method would certainly be undesirable in exercise prescription for cardiac patients.

Research in the late 1970s showed that carotid artery palpation significantly slowed the postexercise heart rate and occasionally produced electrocardiographic abnormalities.[26] Several more recent reports, however, have shown rather convincingly that for both healthy adults and cardiac patients, carotid artery palpation caused little or no alteration in heart rate at rest or during exercise and recovery.[4,5,13,20] It appears that palpation of the carotid artery is an appropriate technique to gauge heart rate. It should be kept in mind that various forms of vascular disease affect the sensitivity of the carotid sinus. Under such conditions, palpation could give a falsely low reading for the heart rate. When this is a concern, an excellent substitute method is to determine pulse rate at the radial artery (thumb side of wrist) or temporal artery (side of head at temple) because palpation of these vessels causes no change in heart rate.

Cortical Input

Impulses from the cerebral cortex pass via small afferent nerves through the cardiovascular center in the medulla. Consequently, variations in one's emotional state significantly affect cardiovascular responses and make it difficult to obtain "true" resting values for heart rate and blood pressure. Cerebral impulses also cause the heart rate to rise rapidly and considerably in anticipation of exercise. This **anticipatory heart rate** is probably the result of both an increase in sympathetic discharge and reduction of vagal tone.

The extent of the anticipatory response is clearly demonstrated in Figure 16–4. The heart rate of trained sprint runners was telemetered at rest, at the starting commands, and during a 60-, 220-, and 440-yard race.[12] Heart rate averaged 148 beats per minute at the starting commands in anticipation of the 60-yard sprint; this represented 74% of the total heart rate adjustment to the run! The magnitude of the anticipatory heart rate was greatest in the short sprint events and successively lower prior to the longer sprint distances.

As shown in Table 16–1, this pattern of anticipation was also demonstrated in events of longer duration. For example, the anticipatory heart rate of four athletes trained for the 880-yard run averaged 122 beats per minute, whereas heart rates during the starting commands of the 1-mile and 2-mile runs averaged 118 and 108 beats, respectively. This finding represented only 33% of the total heart rate adjustment to the 2-mile run. The untrained subjects also had an anticipatory increase in heart rate prior to the start of exercise. For these subjects, the pattern of anticipation in relation to the distance to be run was not as clear as in the runners trained for specific events. More than likely, the rather specific nature of the anticipatory heart rate with trained athletes reflects genetic differences between competitive runners and untrained subjects as well as specific long-term conditioning factors resulting from training and perception of effort. The initial neural outflow in anticipation of exercise would be desirable prior to intense activity of

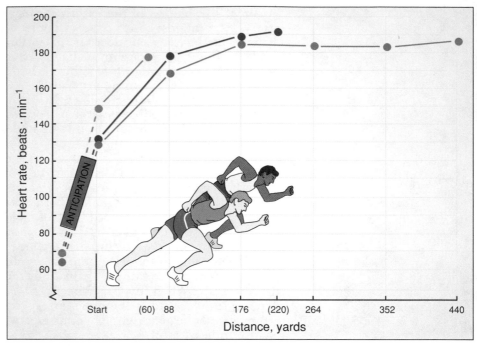

FIG. 16–4.
Heart rate response of runners trained to participate in sprint runs. (Modified from McArdle, W.D., et al. Telemetered cardiac response to selected running events. J. Appl. Physiol., *23*:566, 1967.)

short duration (such as sprinting) to provide for the rapid mobilization of bodily reserves. On the other hand, this mechanism for "revving the body's engine" might be wasteful in events of longer duration.

In essence, the heart is "turned on" during exercise by an increase in sympathetic and decrease in parasympathetic activity combined with input from the central command in the brain.

It seems clear from the preceding observations that a large portion of the

TABLE 16–1.
RESTING, ANTICIPATORY, AND MAXIMUM EXERCISE HEART RATE IN COMPETITIVE RUNNERS AND UNTRAINED SUBJECTS DURING ALL-OUT RUNNING*

Event	N		Rest		Anticipatory		Exercise	
	Trained	Untrained	Trained	Untrained	Trained	Untrained	Trained	Untrained
60 yard	5	4	67	69	148	124	177	162
220 yard	5	4	67	67	130	115	191	186
440 yard	4	4	63	68	129	118	187	189
880 yard	4	4	62	70	122	129	186	194
1 mile	4	4	58	64	118	128	195	198
2 mile	4	4	59	74	108	109	206	199
Values are averages								

* From McArdle, W.D., et al.: Telemetered cardiac response to selected running events. J. Appl. Physiol., 23:566, 1967.

heart rate adjustment to exercise reflects cortical input that occurs during the initial stages of the activity. More than likely, considerable accelerator input is also provided by the activation of receptors in joints and muscles as the exercise begins.[15] Even in the so-called nonsprint events, heart rate is approximately 180 beats per minute within 30 seconds of 1- and 2-mile runs. Further increases in heart rate are gradual, with several plateaus being reached during the run. Almost identical results have been reported for telemetered heart rates during competitive swimming events; only the maximum exercise heart rates were lower in swimming.[11]

It is now apparent that individuals can show voluntary control over heart rate, even during relatively moderate to heavy submaximal exercise. In one study,[14] subjects who saw a beat-by-beat display of the heart rate and were instructed to slow heart rate had a 22% less heart rate increase than controls with no instruction or visual display of heart rate. The precise benefits to exercise of this ability to exert central control over the cardiovascular adjustment to exercise awaits further research.

DISTRIBUTION OF BLOOD

EFFECT OF EXERCISE

Increased energy expenditure usually requires rapid adjustments in blood flow that affect the entire cardiovascular system. For example, nerves and local metabolic conditions act on the smooth muscular bands of arteriole walls and cause them to alter their internal diameter almost instantaneously. In addition, stimulation of nerves to the venous capacitance vessels causes them to "stiffen." Such **venoconstriction** permits large quantities of blood to move from peripheral veins into the central circulation. **The capability of large portions of the vasculature to either constrict or dilate provides a rapid redistribution of blood to meet the tissue's metabolic requirements while maintaining an appropriate blood pressure throughout the entire system.**

During exercise, the vascular portion of active muscles is increased considerably by the dilation of local arterioles. Concurrently, other vessels that can temporarily compromise their blood supply constrict or "shut down." Kidney function vividly illustrates this regulatory capacity for adjusting regional blood flow. Renal blood flow at rest is normally about 1100 ml per minute; this blood flow is about 20% of the cardiac output. In maximal exercise, renal blood flow may be reduced to only 250 ml per minute or about 1% of the exercise cardiac output.[6]

REGULATION OF BLOOD FLOW

Blood flows through the vascular circuit in general accord with the physical laws of hydrodynamics as applied to rigid, cylindric vessels. Because blood is not a homogeneous fluid and the blood vessels are not rigid tubes, these relationships are true mainly in a qualitative sense.

The volume of flow in any vessel is (1) directly proportional to the pressure gradient between the two ends of the vessels and not to the absolute pressure within the vessel, and (2) inversely related to the resistance encountered to the flow. Resistance, which is the force impeding blood flow, is caused by friction between the blood and the internal vascular wall. This resistance is determined by three factors: the thickness or **viscosity** of the blood, the **length** of the conducting tube, and, most important, the **diameter** of the blood vessel. The relationship between pressure, resistance, and flow can be expressed by an equation referred to as **Poiseuille's law** in which:

$$\text{Flow} = \frac{\text{Pressure gradient} \times \text{Vessel radius}^4}{\text{Vessel length} \times \text{Viscosity}}$$

In the body, the viscosity of the blood and the length of the transport vessel remain relatively constant under most circumstances. Therefore, the most important factor that affects blood flow is the **diameter** of the conducting tube. In fact, the resistance to flow changes with the vessel diameter raised to the fourth power; if the diameter is reduced by one-half, flow through the vessel decreases 16 times! Conversely, doubling the vessel's diameter increases the volume 16-fold. If the pressure within the vascular circuit were to remain relatively constant, a considerable alteration in blood flow would be achieved with only a small change in vessel diameter. **In a physiologic sense, constriction and dilation provide the most important mechanisms for regulating regional blood flow.**

Local Factors

At rest, only 1 of every 30 to 40 capillaries is actually open in muscle tissue.[27] The opening of dormant capillaries in exercise serves three important functions: (1) It provides for a significant increase in muscle blood flow. (2) Because more channels are open, the increased blood volume can be delivered with only a minimal increase in the velocity of flow. (3) The enhanced vascularization increases the effective surface for exchange between the blood and the muscle cells.

Local factors related to the level of tissue metabolism act directly on the smooth muscle bands of the small arterioles and precapillary sphincters to cause vasodilatation. The response is almost instantaneous and finely adjusted to the tissue's metabolic needs.[3] In fact, in muscle, the local vascular dilatation is proportional to the force of a contraction.[1] Local regulation even provides for adequate regional blood flow in patients in whom the nerves to the blood vessels have been surgically removed.

A decrease in a tissue's oxygen supply produces a potent local stimulus for vasodilatation in skeletal and cardiac muscle. Furthermore, local increases in temperature, carbon dioxide, acidity, adenosine, and the ions of magnesium and potassium enhance regional blood flow. These **autoregulatory mechanisms** for blood flow make sense from a physiologic standpoint because they reflect elevated tissue metabolism and an increased need for oxygen. The most effective immediate step for increasing a tissue's oxygen supply is rapid and local vasodilatation.

Neural Factors

Superimposed on the vasoregulation afforded by local factors is a central vascular control mediated by the sympathetic, and to a minor degree, the parasympathetic portions of the autonomic nervous system. For example, muscles contain small sensory nerve fibers that are highly sensitive to substances released in local tissue during exercise. When these fibers are stim-

FIG. 16–5.
Schematic view of vascular regulation via sympathetic outflow.

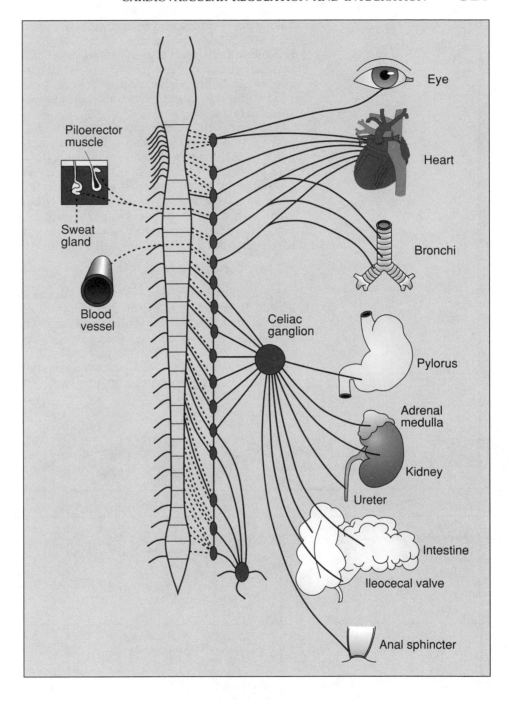

ulated, they provide input to the central nervous system to bring about an appropriate cardiovascular response. With central regulation, blood flow in one area cannot dominate when a concurrent oxygen need exists in other more "needy" tissues. In exercise, for example, blood flow through the skin and kidneys is temporarily reduced, whereas the vessels dilate to the active muscles. This vascular response occurs even in anticipation of exercise.

Figure 16–5 is a schematic view of the distribution of sympathetic outflow.

These nerve fibers end in the muscular layers of small arteries, arterioles, and precapillary sphincters.

Norepinephrine acts as a general vasoconstrictor and is released at certain sympathetic nerve endings. These sympathetic constrictor fibers are called **adrenergic fibers.** Other sympathetic neurons in skeletal and heart muscle release acetylcholine; these are the **cholinergic fibers** and their action is vasodilatation.

The sympathetic nervous system consists of **both** adrenergic constrictor and cholinergic dilator fibers. The constrictor nerves are constantly active; consequently some blood vessels are always in a state of constriction. The relative degree of this constrictor activity is referred to as **vasomotor tone.** Dilatation of blood vessels under the influence of adrenergic neurons is due more to a reduction in vasomotor tone than to an increase in the action of either sympathetic or parasympathetic dilator fibers. In addition, whatever sympathetically-activated vasoconstriction is present in active tissue is rapidly over-ridden by the powerful vasodilation induced by metabolism.[22]

Hormonal Factors

Sympathetic nerves also terminate in the medullary portion of the adrenal glands. In response to sympathetic activation, this glandular tissue secretes large quantities of epinephrine and a smaller amount of norepinephrine into the blood. These hormones then act as chemical messengers to bring about a generalized constrictor response, except in the blood vessels of the heart and skeletal muscles. During exercise, the hormonal control of regional blood flow is relatively minor compared to the more local, rapid, and powerful sympathetic neural drive.

TABLE 16–2.
SUMMARY OF INTEGRATED CHEMICAL, NEURAL, AND HORMONAL ADJUSTMENTS PRIOR TO AND DURING EXERCISE

Condition	Activator	Response
Preexercise "anticipatory" response	Activation of motor cortex and higher areas of brain causes increase in sympathetic outflow and reciprocal inhibition of parasympathetic activity.	Acceleration of heart rate; increased myocardial contractility; vasodilatation in skeletal and heart muscle (cholinergic fibers); vasoconstriction in other areas, especially skin, gut, spleen, liver, and kidneys (adrenergic fibers); increase in arterial blood pressure.
Exercise	Continued sympathetic cholinergic outflow; alterations in local metabolic conditions due to hypoxia, ↓ pH, ↑ Pco_2, ↑ ADP, ↑ Mg^{++}, ↑ Ca^{++}, and ↑ temperature.	Further dilation of muscle vasculature.
	Continued sympathetic adrenergic outflow in conjunction with epinephrine and norepinephrine from the adrenal medullae.	Concomitant constriction of vasculature in inactive tissues to maintain adequate perfusion pressure throughout arterial system. Venous vessels stiffen to reduce their capacity. This venoconstriction facilitates venous return and maintains the central blood volume.

INTEGRATED RESPONSE IN EXERCISE

The chemical, neural, and hormonal adjustments prior to and during exercise are summarized in Table 16–2. At the onset of exercise (and even before exercise begins), cardiovascular changes are initiated from nerve centers above the medullary region. These adjustments significantly increase the rate and pumping strength of the heart, as well as predictable alterations in regional blood flow that are proportional to exercise severity. As exercise continues, sympathetic cholinergic outflow plus local metabolic factors that act on chemosensitive nerves as well as directly on the blood vessels cause dilatation of resistance vessels in active muscles. This reduced peripheral resistance permits the active areas to accommodate greater blood flow. As exercise continues, there are further constrictor adjustments in less active tissues; thus, an adequate perfusion pressure is maintained even with the large dilatation of the muscle's vasculature. This constrictor action provides for the appropriate redistribution of blood to meet the metabolic requirements of working muscles.

Factors that affect venous return are equally as important as those that regulate arterial blood flow. The action of the muscle and ventilatory pumps and the stiffening of the veins themselves (probably mediated by sympathetic activity) immediately increase the return of blood to the right ventricle. As cardiac output increases, venous tone also increases proportionally in both working and nonworking muscles.[21] With these adjustments, the balance is maintained between cardiac output and venous return. The factors that affect blood flow in the venous system are particularly important in upright exercise where gravity tends to counter the venous pressure in the extremities.

SUMMARY

1. The cardiovascular system provides for rapid regulation of heart rate as well as for effective distribution of blood in the vascular circuit while maintaining blood pressure in response to the body's metabolic and physiologic needs.
2. The cardiac rhythm is initiated at the S-A node. The impulse then travels across the atria to the A-V node where it is delayed and then rapidly spreads across the large ventricular mass. With this normal conduction pattern, the atria and ventricles contract effectively to provide impetus for blood flow.
3. The electrocardiogram provides a record of the sequence of the heart's electrical events during the cardiac cycle. Electrocardiography is important for detecting various abnormalities in heart function at rest and during exercise.
4. The sympathetic catecholamines, epinephrine and norepinephrine, act to accelerate heart rate and increase myocardial contractility. The parasympathetic neurotransmitter, acetylcholine, acts through the vagus nerve to slow the heart.
5. The heart is "turned on" in the transition from rest to exercise by an increase in sympathetic and decrease in parasympathetic activity integrated with input from the central command in the brain.
6. Neural and hormonal extrinsic factors modify the heart's inherent rhythmicity, enabling it to speed up rapidly in anticipation of exercise and to increase to 200 beats per minute or higher in maximum exercise.

7. Palpation of the carotid artery is generally an appropriate means for determining the actual heart rate during and immediately after exercise.

8. A large part of the heart rate adjustment to exercise is probably due to cortical influence prior to and during the initial stages of the activity.

9. Nerves, hormones, and local metabolic factors act on the smooth muscle bands in various blood vessels. This causes them to alter their internal diameter to regulate blood flow. Adrenergic sympathetic fibers release norepinephrine that causes vasoconstriction; cholinergic sympathetic neurons secrete acetylcholine that brings about vasodilation.

REFERENCES

1. Bevegård, B.S., and Shepherd, J.T.: Regulation of the circulation during exercise in man. *Physiol. Rev.*, 47:178, 1967.
2. Bolter, C., et al.: Intrinsic rate and cholinergic sensitivity of isolated atria from trained and sedentary rats. *Proc. Soc. Exp. Biol. Med.*, 144:364, 1973.
3. Corcondilas, A., et al.: Effect of a brief contraction of forearm muscles on forearm blood flow. *J. Appl. Physiol.*, 19:1942, 1964.
4. Couldry, W.C., et al.: Carotid vs radial pulse counts. *Phys. Sportsmed.*, 10:67, 1982.
5. Gardner, G.W., et al.: Use of carotid pulse for heart rate monitoring. *Med. Sci. Sports*, 11:111, 1979.
6. Grimby, G., et al.: Cardiac output during submaximal and maximal exercise in active middle-aged athletes. *J. Appl. Physiol.*, 21:1150, 1966.
7. Hanne-Paparo, N., and Kellerman, J.J.: Long-term Holter ECG monitoring of athletes. *Med. Sci. Sports Exerc.*, 13:294, 1981.
8. Kanakis, C., et al.: Left ventricular responses to strenuous endurance training and related training frequencies. *J. Cardiac. Rehab.*, 2:141, 1982.
9. Katona, P.G., et al.: Sympathetic and parasympathetic cardiac control in athletes and non-athletes at rest. *J. Appl. Physiol.*, 52:1652, 1982.
10. Kenney, W.L.: Parasympathetic control of resting heart rate: relationship to aerobic power. *Med. Sci. Sports Exerc.*, 17:451, 1985.
11. Magel, J.R., et al.: Telemetered heart rate response to selected competitive swimming events. *J. Appl. Physiol.*, 26:764, 1969.
12. McArdle, W.D., et al.: Telemetered cardiac response to selected running events. *J. Appl. Physiol.*, 23:566, 1967.
13. Oldridge, N.B., et al.: Carotid palpation, coronary heart disease, and exercise rehabilitation. *Med. Sci. Sports Exerc.*, 13:6, 1981.
14. Perski, A., et al.: Central control of cardiovascular adjustments to exercise. *J. Appl. Physiol.*, 58:431, 1985.
15. Petro, J.K., et al.: Instantaneous cardiac acceleration in a man induced by voluntary muscle contractions. *J. Appl. Physiol.* 29:794, 1970.
16. Rowell, L.B.: General principles of vascular control. In *Human Circulation: Regulation During Physical Stress*. New York, Oxford University Press, 1986.
17. Rowell, L.B.: Human cardiovascular adjustments to exercise and thermal stress. *Physiol. Rev.*, 54:75, 1974.
18. Scheuer, J., and Tipton, C.M.: Cardiovascular adaptations to physical training. *Annu. Rev. Physiol.*, 39:221, 1977.
19. Scheuer, J., et al.: Experimental observations on the effects of physical training upon intrinsic cardiac physiology and biochemistry. *Am. J. Cardiol.*, 33:744, 1974.
20. Sedlock, D.A., et al.: Accuracy of subject-palpated carotid pulse after exercise. *Phys. Sportsmed.*, 11:106, 1983.
21. Shepherd, J.T.: Behavior of resistance and capacity vessels in human limbs during exercise. *Circ. Res.*, 20(Suppl. I):70, 1967.
22. Shepherd, J.T., et al.: Static (isometric) exercise. Retrospection and introspection. *Circ. Res.*, 48(Suppl. I):179, 1981.
23. Smith, M.L., et al.: Exercise training bradycardia: the role of autonomic balance. *Med. Sci. Sports Exerc.*, 21:44, 1989.

24. Upton, S.J., et al.: Comparative physiological profiles among young and middle-aged female distance runners. *Med. Sci. Sports Exerc.,* 16:67, 1984.
25. Weiss, S., and Baker, J.P.: The carotid sinus reflex in health and disease. *Medicine,* 12:297, 1933.
26. White, J.R.: EKG changes using carotid artery for heart rate monitoring. *Med. Sci. Sports,* 9:88, 1977.
27. Zweifach, B.J.: The microcirculation of the blood. *Sci. Am.,* Jan. 1959, p. 54.

17 FUNCTIONAL CAPACITY OF THE CARDIOVASCULAR SYSTEM

MEASUREMENT OF CARDIAC OUTPUT

Cardiac output is the primary indicator of the functional capacity of the circulation to meet the demands of physical activity. Output from the heart, as with any pump, is determined by its rate of pumping (**heart rate**) and by the quantity of blood ejected with each stroke (**stroke volume**). Thus, cardiac output is computed as:

$$\text{Cardiac output} = \text{Heart rate} \times \text{Stroke volume}$$

The output from a hose, pump, or faucet can easily be determined. One need only open the valve, collect and measure the volume of fluid ejected, and record the time. This, however, is not the case with the measurement of cardiac output. Even if such a **direct** technique were applied, the disruption of the main output vessel in a closed circulatory system would in itself dramatically alter the output. With advances in biomedical engineering, however, electromagnetic and ultrasonic flowmeters can be surgically implanted around a main artery in the vascular circuit. For obvious reasons, this technique is usually limited to animal research and has little application for use in a typical exercise setting with healthy humans. The direct Fick, indicator dilution, and CO_2 rebreathing methods are commonly used in human measurement.

DIRECT FICK METHOD

Cardiac output can be computed if one knows a person's oxygen consumption during a minute and the average difference between the oxygen content of arterial and mixed venous blood (a-$\bar{v}$ O_2 difference). The question then to be answered is: How much blood must have circulated during the minute to account for the observed oxygen consumption, given the observed a-$\bar{v}$ O_2 difference? The formula that expresses the relationship between cardiac output, oxygen consumption, and a-$\bar{v}$ O_2 difference embodies the principle set forth by Fick in 1870 and is termed the **Fick equation.**

$$\begin{array}{c}\text{Cardiac} \\ \text{output} \\ \text{(ml} \cdot \text{min}^{-1})\end{array} = \frac{O_2 \text{ consumption (ml} \cdot \text{min}^{-1})}{\text{a-}\bar{v} \ O_2 \text{ difference (ml per 100 ml blood)}} \times 100$$

FIG. 17–1.
Application of the Fick principle for determining cardiac output ($\dot{Q}$).

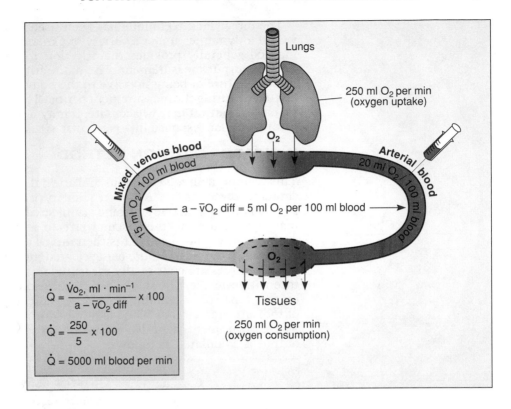

The Fick principle for determining cardiac output is illustrated in Figure 17–1. In this example, 250 ml of oxygen are consumed during a minute at rest, and the a-$\bar{v}$ O_2 difference during this time averages 5 ml of oxygen per 100 ml of blood. These values are substituted in the Fick equation:

$$\text{Cardiac output (ml} \cdot \text{min}^{-1}) = \frac{250 \text{ ml O}_2}{5 \text{ ml O}_2} \times 100 = \frac{5000 \text{ ml}}{\text{blood}}$$

Although the Fick principle is straightforward, the actual measurement of cardiac output by this technique is complex and is usually limited to a clinical setting where the benefits of measurement exceed any potential risk. The measurement of oxygen consumption involves the methods of open-circuit spirometry summarized in Chapter 8. A more difficult aspect is obtaining the a-$\bar{v}$ O_2 difference. A representative sample of arterial blood can be obtained from any convenient systemic artery such as the femoral, radial, or brachial artery. Although these arteries are easily located, the actual arterial puncture can be traumatic to the patient. Sampling of mixed venous blood presents additional difficulties because the blood in each vein only reflects the metabolic activity of the specific area it drains. To obtain an accurate estimate of the average oxygen content of venous blood, it is necessary to sample from an anatomic "mixing chamber" such as the right atrium, right ventricle, or even the pulmonary artery. This is achieved by threading a small flexible tube or catheter through the antecubital vein in the arm, up into the superior vena cava, and into the right heart. Arterial and mixed venous blood are then sampled during the same period of measurement as oxygen consumption.

The direct Fick technique has been used in numerous studies of cardiovascular dynamics under a variety of experimental conditions. In fact, this method generally provides the criterion to validate other techniques for cardiac output measurement. The main criticism of the method is that by its very nature of being **invasive** to the body, cardiovascular dynamics may be altered during the measurement period. Thus, although the obtained value for cardiac output may be accurate, it may not reflect the person's "normal" cardiovascular response in a particular situation.

INDICATOR DILUTION METHOD

This technique involves venous and arterial punctures, but does not require cardiac catheterization. A known quantity of an inert dye such as indocyanine green or of a radioactive substance is injected into a large vein. The indicator material remains in the vascular stream and is usually bound to plasma proteins or red blood cells. It is then mixed as the blood travels to the lungs and back to the heart before being ejected into the systemic circuit. Arterial blood samples are continually measured with a radioactive counter or photosensitive device. The area under the dilution-concentration curve obtained by this repetitive sampling procedure indicates the average concentration of indicator material as blood is pumped from the heart. From the dilution of a known quantity of dye in an unknown quantity of blood, the cardiac output is calculated as follows:

$$\text{Cardiac output} = \frac{\text{Quantity of dye injected}}{\text{Average conc. dye in blood for duration of curve} \times \text{Duration of curve}}$$

CO$_2$ REBREATHING METHOD

Cardiac output can be determined from values of carbon dioxide substituted in the Fick equation. By using a rapid carbon dioxide gas analyzer and making certain reasonable assumptions, it is possible to obtain valid estimates of venous and arterial carbon dioxide levels. The technique is noninvasive or "bloodless," and requires a breath-by-breath analysis of carbon dioxide.[5,40]

Once venous and arterial carbon dioxide concentrations are estimated, cardiac output is calculated in accordance with the Fick principle as follows:

$$\text{Cardiac output} = \frac{\text{Carbon dioxide production}}{\bar{\text{v}}\text{-a CO}_2 \text{ difference}} \times 100$$

The advantages seem obvious for the CO$_2$ rebreathing method over the direct Fick and indicator dilution techniques. The method is bloodless, involves minimal interference with the subject, and does not require close medical supervision. Because the method is noninvasive, it may provide more accurate estimates of the cardiovascular dynamics during exercise than would be obtained by more invasive techniques.[26] One limitation of the method requires the subject to exercise at a steady metabolic rate. This may place some restrictions on its use during maximal and "supermaximal" exercise or during the transition from rest to exercise.

CARDIAC OUTPUT AT REST

Cardiac output shows considerable intra-individual variation at rest. It is affected by emotional conditions that alter cortical outflow to the cardio-accelerator nerves as well as to nerves that act on the resistance and capacitance vessels. On the average, the entire blood volume of approximately 5 liters is pumped from the left ventricle each minute. This value is similar for both trained and untrained subjects.

Untrained

For the average person, a 5-liter cardiac output is usually sustained with a heart rate of about 70 beats per minute. Substituting this heart rate value in the cardiac output equation, the calculated stroke volume of the heart equals 71 ml per beat. Stroke volumes for females usually average 25% below values for men and are 50 to 70 ml per beat at rest. This "gender difference" is essentially due to the smaller body size of the average woman as compared to the average man.

Endurance Athletes

Endurance training causes the sinus node of the heart to come under greater influence of acetylcholine, the parasympathetic hormone that has a slowing effect on heart rate.[64] This effect is probably accompanied by a concomitant reduction in resting sympathetic activity. This training adaptation partially explains the relatively low resting heart rates of many male and female endurance athletes. Their heart rates generally average about 50 beats per minute at rest, although heart rates below 40 beats per minute have been reported for apparently healthy endurance athletes. Extreme bradycardia at rest is not necessarily a general phenomenon with well-trained athletes. For example, resting heart rates of 64 to 76 beats per minute have been observed for Jim Ryun, former world-record-holder in the 1-mile run.[19] Because the resting cardiac output of endurance-trained athletes also averages 5 liters per minute, blood is circulated with the proportionately larger stroke volume of 100 ml per beat. Average values for cardiac output, heart rate, and stroke volume of trained and untrained people at rest are summarized as follows:

	Rest

	Cardiac output	=	Heart rate	×	Stroke volume
Untrained:	5000 ml	=	70 b · min^{-1}	×	71 ml
Trained:	5000 ml	=	50 b · min^{-1}	×	100 ml

Although these calculations are straightforward, the underlying physiologic mechanisms are not fully understood. It is not clear whether the bradycardia that occurs with endurance training "causes" a larger stroke volume or vice versa, because the myocardium itself is strengthened through aerobic exercise. Two factors are probably operative with training: (1) endurance training increases vagal tone that slows the heart, and (2) the heart muscle strengthened through training is capable of a more forceful stroke with each contraction.

CARDIAC OUTPUT DURING EXERCISE

Blood flow increases in proportion to the intensity of exercise. In progressing from rest to steady-rate exercise, cardiac output undergoes a rapid increase followed by a gradual rise until a plateau is reached. At this point, blood flow is presumably sufficient to meet the metabolic requirements of exercise.

In relatively sedentary, college-aged males, cardiac output during **strenuous** exercise increases by about four times the resting level to an average maximum of 20 to 22 liters of blood per minute. Maximum heart rate for these young adults usually averages about 195 beats per minute. Consequently, the stroke volume is generally 103 to 113 ml of blood per beat during maximal exercise. In contrast, world-class endurance athletes have maximum cardiac outputs of 35 to 40 liters per minute. This is even more impressive if one considers that the trained person may have a slightly lower maximum heart rate than the sedentary person of similar age.[42,45] Thus, **the endurance athlete achieves a large cardiac output compared with his or her sedentary counterpart due to a considerably larger stroke volume.** For example, the cardiac output of an Olympic medal winner in cross-country skiing increased almost 8 times above rest to 40 liters per minute in maximum work with an accompanying stroke volume of 210 ml per beat. This is nearly twice the volume of blood pumped per beat in comparison to the maximum stroke volume of healthy, sedentary people of similar age. Lest we become unduly impressed with our functional significance as a species, cardiac outputs of $600 \text{ l} \cdot \text{min}^{-1}$ (with accompanying $150 \text{ ml} \cdot \text{kg}^{-1} \cdot \text{min}^{-1}$ max $\dot{V}_{O_2}$) have been reported for thoroughbred racehorses![73]

The functional capacity of the heart during maximum exercise in trained and untrained men is summarized as follows:

Maximum Exercise		

	Cardiac output	=	Heart rate	×	Stroke volume
Untrained:	22,000 ml	=	$195 \text{ b} \cdot \text{min}^{-1}$	×	113 ml
Trained:	35,000 ml	=	$195 \text{ b} \cdot \text{min}^{-1}$	×	179 ml

STROKE VOLUME IN EXERCISE: TRAINING EFFECTS

Figure 17–2 illustrates the stroke volume response for two groups of men during upright exercise of increasing severity. One group of six highly trained endurance athletes had trained for several years; the other group consisted of three sedentary college students. The students' exercise responses were evaluated before and after a 55-day training program designed to improve aerobic fitness.[61]

From these data, several important conclusions can be drawn: (1) The endurance athlete's heart has a considerably larger stroke volume during rest and exercise than an untrained person of the same age. (2) For both trained and untrained individuals, the greatest increase in stroke volume in upright exercise occurs in the transition from rest to moderate exercise. As exercise becomes more intense, there are only small increases in stroke

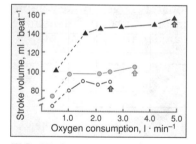

FIG. 17–2.
Stroke volume in relation to oxygen consumption during upright exercise in endurance athletes (▲) and sedentary college students prior to (0) and following (●) 55 days of aerobic training; (○, ▲, ▲ = maximal values). (From Saltin, B.: Physiological effects of physical conditioning. Med. Sci. Sports, *1*:50, 1969.)

volume. (3) Maximum stroke volume is reached at 40 to 50% of the maximal oxygen consumption; this usually represents a heart rate of 110 to 120 beats per minute in young adults. Stroke volume does not decrease at the more intense exercise levels. This suggests that even at rapid heart rates there is still adequate time for the ventricles to fill during diastole, so there is no diminution in the stroke volume. (4) For untrained individuals, there is only a small increase in stroke volume in the transition from rest to exercise. For these individuals, the major increase in cardiac output is brought about by an acceleration in heart rate. For the trained endurance athletes, **both** heart rate and stroke volume increase to augment cardiac output, with the increase in the athlete's stroke volume being generally 50 to 60% above resting values. For previously sedentary subjects, 8 weeks of aerobic training substantially increases stroke volume, but these values are still well below values observed for elite athletes. The degree to which this difference reflects prolonged training, genetics, or a combination of both has yet to be determined.

Stroke Volume and max $\dot{V}O_2$

The importance of stroke volume in differentiating people who have very high and very low values for max $\dot{V}O_2$ is amplified in Table 17–1.

These data were obtained from three groups: athletes, healthy but sedentary men, and patients who had mitral stenosis, a valvular disease of the heart that results in inadequate emptying of the left ventricle. The differences in max $\dot{V}O_2$ between groups are closely related to differences in maximal stroke volume. Patients who had mitral stenosis had an aerobic capacity and maximum stroke volume that was half that of the sedentary subject's. This relationship was also apparent in comparisons between healthy subjects. The maximal oxygen uptakes of the athletes averaged 62% larger than those of the sedentary group. This was paralleled by a 60% larger stroke volume. Because the maximal heart rates of both groups were similar, differences in cardiac output (and max $\dot{V}O_2$) were almost entirely due to differences in maximal stroke volume.

STROKE VOLUME: SYSTOLIC EMPTYING
VERSUS DIASTOLIC FILLING

Essentially, two physiologic mechanisms regulate stroke volume. The first is intrinsic to the myocardium and requires enhanced cardiac filling that is

TABLE 17–1.
MAXIMAL VALUES OF OXYGEN UPTAKE, HEART RATE, STROKE VOLUME, AND CARDIAC OUTPUT IN THREE GROUPS HAVING VERY LOW, NORMAL, AND HIGH MAX $\dot{V}O_2$*

Group	max $\dot{V}O_2$ $(l \cdot min^{-1})$	Max Heart Rate $(beats \cdot min^{-1})$	Max Stroke Volume (ml)	Max Cardiac Output $(l \cdot min^{-1})$
Mitral stenosis	1.6	190	50	9.5
Sedentary	3.2	200	100	20.0
Athlete	5.2	190	160	30.4

* Modified from Rowell, L.B.: Circulation. Med. Sci. Sports, 1:15, 1969.

followed by a more forceful contraction. The second mechanism is governed by neurohormonal influence. It involves normal ventricular filling that is accompanied by an increased stroke due to a forceful systolic ejection that brings about a greater cardiac emptying.

Enhanced Diastolic Filling

Any factor that increases venous return (**preload**) or slows the heart causes greater ventricular filling during the diastolic phase of the cardiac cycle. An increase in **end-diastolic volume** stretches the myocardial fibers and causes a powerful ejection stroke as the heart contracts. As a result, the normal stroke volume is expelled plus the additional blood that entered the ventricles and stretched the myocardium.

The relationship between the force of contraction and the resting length of muscle fibers was described by two physiologists, Frank and Starling, in the early 1900s. The improved contractility probably results from a more optimum arrangement of myofilaments as the muscle stretches. This phenomenon as applied to the myocardium has been termed **Starling's law of the heart**.

For many years it was taught that the Frank-Starling mechanism provided the "modus operandi" for **all** increases in stroke volume during exercise. Physiologists believed that the enhanced venous return in exercise caused a greater cardiac filling, so that the ventricles were stretched in diastole and subsequently responded with a more forceful ejection. In all likelihood, this is the pattern of response for stroke volume in transition from rest to exercise or as a person moves from the upright to the recumbent position. Enhanced diastolic filling probably also occurs in activities such as swimming, where the body's horizontal position optimizes the flow of blood into the heart.

From the data in Table 17–2, it is clear that body position has a significant effect on circulatory dynamics.[7] Cardiac output and stroke volume are high-

TABLE 17–2.
THE EFFECT OF BODY POSITION ON CARDIAC OUTPUT, STROKE VOLUME, AND HEART RATE AT REST AND DURING EXERCISE (ACTIVE SUBJECTS)*

	Rest		Moderate Exercise		Strenuous Exercise	
	Supine	Upright	Supine	Upright	Supine	Upright
Cardiac output, l · min⁻¹	9.2	6.6	19.0	16.9	26.3	24.5
Stroke volume, ml	141	103	163	149	164	155
Heart rate, beats · min⁻¹	65	64	115	112	160	159
Oxygen consumption, ml · min⁻¹	345	384	1769	1864	3364	3387

* Data from Bevegård, S., et al.: Circulatory studies in well-trained athletes at rest and during heavy exercise, with special reference to stroke volume and the influence of body position. Acta Physiol. Scand., 57:26, 1963.

est and most stable in the horizontal position. In this position, the stroke volume is nearly maximum at rest and increases only slightly during exercise. In contrast, the force of gravity in the upright position acts to counter the return flow of blood to the heart; this results in diminished stroke volume and cardiac output. This postural effect is apparent in comparing circulatory dynamics at rest in the upright and supine positions. As the intensity of upright exercise increases, however, stroke volume increases and approaches the maximum stroke volume in the supine position.

Greater Systolic Emptying

In most forms of upright exercise, the heart does not fill to an extent that would cause a significant increase in cardiac volume. Actually, some reports even indicate a decrease in the diastolic size of the heart during such exercise.[59] Although research findings are not always consistent on this topic, more than likely in graded upright exercise both the sedentary and trained heart progressively increases stroke volume mainly by means of a more complete emptying during systole, despite an increasing systolic pressure or **afterload.**[47]

A greater systolic ejection without an accompanying increase in end-diastolic volume is possible because the heart possesses a **functional residual volume.** At rest in the upright position, 40 to 50% of the total end-diastolic blood volume remains in the left ventricle after a contraction; this amounts to approximately 50 to 70 ml of blood. Myocardial contractile force is enhanced in exercise by the sympathetic hormones epinephrine and norepinephrine that produce augmented stroke power and greater systolic emptying of the heart. In addition, endurance training enhances the contractile state of the myocardium itself and improves its capability for achieving a large stroke volume.

HEART RATE DURING EXERCISE: TRAINING EFFECTS

The large stroke volume of topflight endurance-trained athletes and the increases in stroke volume of sedentary subjects following aerobic training are usually accompanied by a proportionate **heart rate reduction** during submaximal exercise. The relationship between heart rate and oxygen consumption is shown in Figure 17–3. As in Figure 17–2 for stroke volume, comparisons are made between athletes and sedentary students before and after training.

The lines relating heart rate and oxygen consumption are essentially linear for both groups throughout the major portion of the work range. Whereas the untrained students' heart rates accelerated rapidly as exercise severity increased, the heart rates of the athletes accelerated to a much lesser extent; that is, the **slope** or rate of change of the lines differed considerably. Consequently, an athlete (or trained student) who has good cardiovascular response to exercise will do more work and achieve a higher oxygen consumption before reaching a particular submaximal heart rate than a sedentary student. At an oxygen consumption of 2.0 liters per minute, the heart rates of the athletes averaged 70 beats per minute lower than those of the sedentary students! Following 55 days of training, this difference in submaximal heart rate was reduced to about 40 beats per minute. In each instance, the cardiac output was approximately the same—**the difference was the stroke volume.**

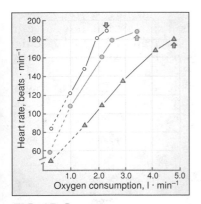

FIG. 17–3.
Heart rate in relation to oxygen consumption during upright exercise in endurance athletes (▲) and sedentary college students prior to (0) and following (●) 55 days of aerobic training; (△, ⬤, ⬤ = maximal values). (From Saltin, B.: Physiological effects of physical conditioning. Med. Sci. Sports, *1*:50, 1969.)

DISTRIBUTION OF CARDIAC OUTPUT

The blood flow to specific tissues is generally proportional to their metabolic activity. Blood flow to the kidneys, skin, and splanchnic areas, however, can also vary with the physiologic function of these tissues in a specific circumstance.

BLOOD FLOW AT REST

At rest in a comfortable environment, the 5-liter cardiac output is distributed in roughly the proportions shown in Table 17–3. About one-fifth of the cardiac output is directed to muscle tissue whereas the major portion of blood flows to the digestive tract, liver, spleen, brain, and kidneys.

BLOOD FLOW DURING EXERCISE

Table 17–4 shows the percentage distribution of the cardiac output during light, moderate, and strenuous exercise. Although regional blood flow during physical activity varies considerably depending on environmental conditions, level of fatigue, and the type of exercise, **the major portion of the exercise cardiac output is diverted to the working muscles.** At rest, about 4 to 7 ml of blood are delivered each minute to every 100 g of muscle. This output increases steadily until at maximum exertion, muscle blood flow may be as high as 50 to 75 ml per 100 g of tissue, although peak values in a limited amount of active 100 g of muscle may reach 300 to 400 ml · min^{-1}.[52] This represents about 85% of the total cardiac output.

Redistribution of Blood

The increase in muscle blood flow in exercise is due largely to increased cardiac output. Muscle blood flow, however, is disproportionately large in relation to blood flow in other tissues. Owing to neural and hormonal vascular regulation and the local metabolic conditions of the muscles themselves, blood is redistributed and directed through working muscles from areas that can temporarily tolerate a reduction in normal blood flow. This **shunting** of blood from specific tissues occurs primarily during maximum exercise. Blood flow to the skin increases during light and moderate exercise so the metabolic

TABLE 17–3.
RELATIVE DISTRIBUTION OF A FIVE-LITER CARDIAC OUTPUT

Organ	Percentage	Volume Per Minute (ml)
Hepatic-splanchnic	27	1350
Kidneys	22	1100
Muscles	20	1000
Brain	14	700
Skin	6	300
Heart	4	200
Other	7	350
TOTAL	100	5000

TABLE 17–4.
DISTRIBUTION OF CARDIAC OUTPUT DURING LIGHT, MODERATE, AND STRENUOUS EXERCISE, AS WELL AS THE OXYGEN EXTRACTION IN THESE VARIOUS TISSUES AT REST*

Tissue	Resting a-v̄ O_2 Difference (ml O_2 per 100 ml Blood)	Exercise Blood Flow, ml · min^{-1}		
		Light	Moderate	Maximum
Splanchnic	4.1	1100 (12%)	600 (3%)	300 (1%)
Renal	1.3	900 (10%)	600 (3%)	250 (1%)
Cerebral	6.3	750 (8%)	750 (4%)	750 (3%)
Coronary	14.0	350 (4%)	750 (4%)	1000 (4%)
Muscle	8.4	4500 (47%)	12,500 (71%)	22,000 (88%)
Skin	1.0	1500 (15%)	1900 (12%)	600 (2%)
Other		400 (4%)	400 (3%)	100 (1%)
		9500	17,500	25,000

* Modified from Anderson, K.L.: The cardiovascular system in exercise. *In* Exercise Physiology. Edited by H.B. Falls. New York, Academic Press, 1968.

heat generated in muscle can be dissipated at the skin's surface. During intense work of short duration, however, this tissue temporarily restricts its blood flow, even if the exercise is performed in a hot environment.[55] In some instances, blood flow is reduced by as much as four-fifths of an organ's blood supply at rest! The kidneys and splanchnic tissues, for example, utilize only 10 to 25% of the oxygen available in their blood supply. Consequently, a considerable reduction in blood flow to these tissues can be tolerated before oxygen demand exceeds supply and function is compromised.[53] With reduced blood flow, the energy needs of the tissues are maintained by increased extraction of oxygen from the available blood supply. A substantial reduction in blood flow to the visceral organs can be sustained for more than an hour during heavy exercise. Redistribution of blood from these tissues occurs even without an increase in cardiac output; this "frees" as much as 600 ml of oxygen per minute for use by the working muscles.[54] Prolonged reduction in blood flow to the liver and kidneys, however, may have its consequences which partially account for the fatigue eventually observed in continuous, submaximal exercise.

Blood Flow to the Heart and Brain

Some tissues cannot compromise their blood supply (Table 17–4). The myocardium normally uses about 75% of the oxygen in the blood flowing through the coronary circulation at rest. With this limited margin of safety, the increased myocardial oxygen needs in exercise can be met mainly by an increase in coronary blood flow. Thus, a four- to five-fold increase in cardiac output is accompanied by a similar increase in coronary circulation; in maximum exercise this amounts to about 1 liter of blood per minute. Recent evidence indicates that cerebral blood flow increases by about 30% with exercise compared to rest.[71]

CARDIAC OUTPUT AND OXYGEN TRANSPORT

REST

Each 100 ml of arterial blood carries about 20 ml of oxygen or 200 ml of oxygen per liter of blood (see Chapter 13). The oxygen-carrying capacity of blood normally varies only slightly because the hemoglobin content of the blood fluctuates little with one's state of training. Because about 5 liters of blood are circulated each minute at rest for trained and untrained adults, potentially 1000 ml of oxygen are available to the body (5 liters blood × 200 ml O_2). Because the oxygen consumption at rest averages only 250 ml per minute, about 750 ml of oxygen return unused to the heart. This, however, is not an unnecessary waste of cardiac output. On the contrary, this extra oxygen above the resting needs represents oxygen in reserve—a margin of safety that can be released immediately should a tissue's metabolic needs suddenly increase.

EXERCISE

A person with a maximum heart rate of 200 beats per minute and a stroke volume of 80 ml per beat generates a maximum cardiac output of 16 liters (200 × 80 ml). Even during maximum exercise, the saturation of hemoglobin with oxygen is nearly complete, so each liter of arterial blood carries about 200 ml of oxygen. Consequently, 3200 ml of oxygen are circulated each minute via a 16-liter cardiac output (16 liters × 200 ml O_2). If all the oxygen could be extracted from this 16-liter cardiac output as it traveled through the body, the greatest possible max $\dot{V}O_2$ would be 3200 ml. However, this is only theoretical because the oxygen needs of certain tissues such as the brain do not increase greatly with exercise, yet these tissues require a rich supply of blood.

Any increase in the maximum cardiac output directly affects a person's capacity to circulate oxygen. Based on the preceding example, if the heart's stroke volume was increased from 80 to 200 ml per beat while the maximum heart rate remained unchanged at 200 beats per minute, maximum cardiac output would be dramatically increased to 40 liters of blood per minute. This means that the quantity of oxygen circulated in maximum exercise each minute would have increased approximately 2.5 times from 3200 to 8000 ml. **An increase in maximum cardiac output clearly results in a proportionate increase in the potential for aerobic metabolism.**

Close Association Between max $\dot{Q}$ and max $\dot{V}O_2$

Figure 17–4 shows the relationship between maximum cardiac output and the capacity for achieving a high level of aerobic metabolism. Included are values for the sedentary and untrained as well as elite endurance athletes. The relationship is unmistakable. A low aerobic capacity is clearly associated with a low maximum cardiac output, whereas the ability to generate a 5- or 6-liter max $\dot{V}O_2$ is always accompanied by a 30- to 40-liter cardiac output.

Figure 17–5 further amplifies the important role of cardiac output in sustaining aerobic metabolism. For both trained athletes and students, the cardiac output increases **linearly** with oxygen consumption throughout the major portion of the work range. This relationship between blood flow and

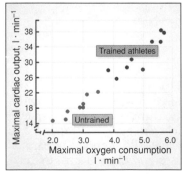

FIG. 17–4.
Relationship between maximal cardiac output and maximal aerobic power.

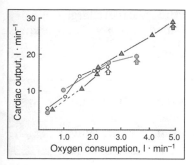

FIG. 17–5.
Cardiac output in relation to oxygen consumption during upright exercise in endurance athletes (▲) and sedentary college students prior to (0) and following (●) 55 days of aerobic training; (⬠, ◆, ◆ = maximal values). (From Saltin, B.: Physiological effects of physical conditioning. Med. Sci. Sports, *1*:50, 1969.)

oxygen consumption also has been demonstrated in growing children. During growth from puberty to adolescence, the oxygen cost of exercise increases as body mass increases; this, in turn, is closely matched by an increase in the heart's stroke volume with a proportionate increase in cardiac output.[18] Each 1-liter increase in oxygen consumption above rest is generally accompanied by a 5- to 6-liter increase in blood flow and this relationship is essentially the same regardless of the type of exercise performed.[39] **Over a wide range of dynamic exercise, there is a tight linkage between systemic blood flow (cardiac output) and the level of oxygen consumption.** The distinguishing feature for the pre-adolescent and adult endurance athlete is a high level of oxygen consumption and cardiac output capacity.[68] The 35% increase in max V_{O_2} noted in Figure 17–5 for the students after 55 days of training was accompanied by an almost proportionate increase in maximum cardiac output.

Differences in Cardiac Output Between Men and Women

The response pattern of cardiac output during exercise is similar between boys and girls and men and women. Both teenage and adult females, however, have a 5 to 10% **larger** cardiac output at any level of **submaximal** oxygen uptake than males.[4,66] This apparent gender difference in cardiac output in submaximal exercise may be due to the lower hemoglobin content of the blood of women, which is about 10% below the values for men.[2] Consequently, within limits, a small decrease in the blood's oxygen-carrying capacity owing to lower hemoglobin is compensated for by a proportionate increase in cardiac output in submaximal exercise.

Training and Submaximal Cardiac Output

Several reports have demonstrated that training, although improving the maximal cardiac output, also tends to **reduce** the minute volume of the heart during moderate exercise.[1,29] In one study, the average cardiac output of young men after 16 weeks of training was reduced by 1.5 and 1.1 liters per minute at a 1.0- and 2.0-liter submaximal oxygen uptake, respectively.[25] As expected, the maximal cardiac output for these men increased 8% from 22.4 to 24.2 liters per minute. With the reduction in submaximal cardiac output, the exercise oxygen requirement was met by a corresponding increase in oxygen extraction in the active muscles. This was presumably the result of an enhanced ability of the trained muscles to generate ATP aerobically and to function at a lower partial pressure of oxygen.

EXTRACTION OF OXYGEN: THE A-$\bar{V}$ O_2 DIFFERENCE

If blood flow was the only means for increasing a tissue's oxygen supply, then cardiac output would have to increase from 5 liters per minute at rest to 100 liters per minute in maximum exercise to achieve a 20-fold increase in oxygen consumption—an increase in oxygen consumption that is not uncommon among trained people. Fortunately, such a large cardiac output is unnecessary during exercise because hemoglobin releases a considerable "extra" quantity of oxygen from the blood that perfuses the active tissues. Consequently, two mechanisms are available to increase the capacity for oxygen consumption. The **first** is to speed up the rate of blood flow, that is, increase cardiac output; the **second** is to utilize the relatively large quantity of oxygen already carried by the blood, that is, expand the a-$\overline{v}$ O_2 difference. The important relationship between cardiac output, a-$\overline{v}$ O_2 difference, and maximal aerobic power is summarized in the following equation:

$$\text{Maximal oxygen consumption} = \begin{matrix} \text{Maximal} \\ \text{cardiac} \\ \text{output} \end{matrix} \times \begin{matrix} \text{Maximal} \\ \text{a-}\overline{v}\ O_2 \\ \text{difference} \end{matrix}$$

a-$\overline{v}$ O_2 DIFFERENCE AT REST

At rest, an average of 5 ml of oxygen is utilized from the 20 ml of oxygen in each 100 ml of arterial blood as it passes through the capillaries. Thus, 75% of the blood's original oxygen load still remains bound to the hemoglobin.

a-$\overline{v}$ O_2 DIFFERENCE IN EXERCISE

Figure 17–6 is a comparison of the relationship between oxygen extraction (a-$\overline{v}$ O_2 difference) and exercise intensity for trained athletes and untrained students. For the students, a-$\overline{v}$ O_2 difference increases steadily during light and moderate exercise and reaches a maximum value of about 15 ml of oxygen per 100 ml of blood. Following 55 days of training, the student's maximum capability for oxygen extraction increased about 11% to 17 ml of oxygen. This means that about 85% of the oxygen was extracted from arterial blood during heavy exercise. Actually, even more oxygen is released in the working muscles because the value for a-$\overline{v}$ O_2 difference reflects an **average** based on calculations from mixed venous blood. This contains blood returning from tissues whose oxygen utilization during exercise is not nearly as high as that of active muscle.

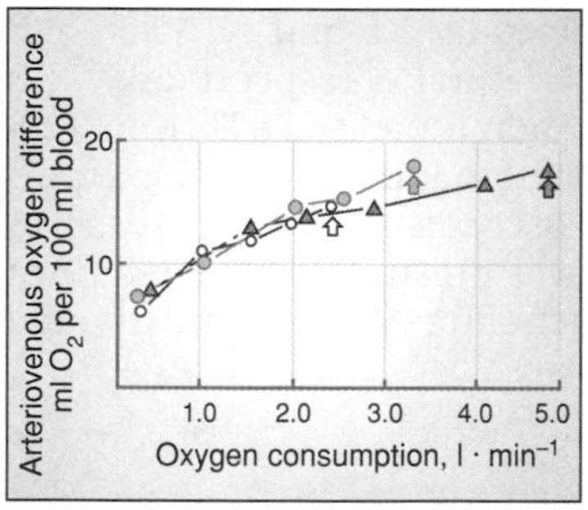

The post-training value for the maximal a-$\overline{v}$ O_2 difference for the students is identical to that achieved by the endurance athletes. Obviously, the rather large difference in max $\dot{V}O_2$ that still exists between the athletes and students is due to the lower cardiac output capacity of the students.

The heart muscle of patients who have advanced coronary artery disease often shows an impaired capacity to perform work or to improve with regular exercise. As a result, training adaptations are negligible in maximal stroke volume and cardiac output. For these patients, however, improvements in exercise tolerance and aerobic capacity are still possible because regular exercise increases the skeletal muscles' ability to receive and utilize oxygen. This contributes to expanding the a-$\overline{v}$ O_2 difference and enables the patients to work at higher levels or at a particular submaximal level with a lower cardiac output.[8,22] A reduced submaximal exercise cardiac output reduces the workload of the heart; naturally, this benefits patients who suffer from exertional angina.

FACTORS THAT AFFECT a-$\overline{v}$ O_2 DIFFERENCE IN EXERCISE

The maximal a-$\overline{v}$ O_2 difference attained during exercise is influenced to some degree by one's capacity to divert a large portion of the cardiac output to

FIG. 17–6.
a-$\overline{v}$ O_2 difference in relation to oxygen consumption during upright exercise in endurance athletes (▲) and sedentary college students prior to (0) and following (●) 55 days of training; (△, ▲, ▲ = maximal values). (From Saltin, B.: Physiological effects of physical conditioning. Med. Sci. Sports, *1*:50, 1969.)

working muscles. As mentioned previously, certain tissues can temporarily compromise their blood supply considerably during exercise for purposes of shunting blood and increasing the quantity of oxygen available for muscle metabolism. This redirection of the central circulation is facilitated by exercise training.

At the local level, the microcirculation of skeletal muscle is also enhanced with aerobic training. Several studies with humans and animals have demonstrated a greater capillary density in specific muscles trained by endurance exercise.[14,62] Muscle biopsies from the **quadriceps femoris** showed a significantly larger ratio of capillaries to muscle fibers in trained than in sedentary men. An increase in the capillary-to-fiber ratio would be a positive adaptation to provide a greater interface for the exchange of nutrients and metabolic gases in exercise.

Another important factor that determines the capacity for oxygen extraction is the ability of individual muscle cells to generate energy aerobically. Aerobic training improves the metabolic capacity of the specific cells trained by exercise. The mitochondria enlarge and even increase in number, as does the quantity of enzymes for aerobic energy transfer.[32,30] All of the local improvements within the muscle ultimately result in an enhanced capacity for the aerobic production of ATP, and the ability of the cell to generate a steady rate of aerobic metabolism without an increase in blood lactate.[15]

CARDIOVASCULAR ADJUSTMENTS TO ARM EXERCISE

The highest oxygen consumption achieved by men and women during arm exercise is generally about 70 to 80% of the max $\dot{V}O_2$ during leg exercise.[41,44,63] Similarly, the maximal values for heart rate and pulmonary ventilation are lower with arm exercise.[65,77] These differences in physiologic response are probably due to the relatively small muscle mass of the upper body used in arm ergometry.[9] In submaximal exercise, however, the response pattern is somewhat reversed. Figure 17–7 shows that for any level of exercise the oxygen consumption is higher when exercising the arms compared to the legs.[30,46,74,76] This difference is small during light work but becomes larger as the intensity of effort increases. This response is generally attributed to a lower mechanical efficiency in arm exercise owing to the static muscular contractions in this form of work (which do not contribute to the external work accomplished). In addition, extra musculature is required to stabilize the torso during arm crank exercise.[11,65]

For any level of oxygen consumption or percent of maximal oxygen consumption, the physiologic strain is greater in arm exercise.[75] Heart rate, ventilation, and perception of effort are generally **higher** when the arms are used.[20] This was also the case for blood pressure in comparisons of arm and leg exercise (Chapter 15). Even when the position of arms is varied either at, above, or below heart level, the differences are not reduced between arm and leg work.[17]

By understanding the differences in physiologic response between arm and leg exercise, the physician and exercise specialist can formulate prudent exercise programs using both forms of work. **The important point is that greater metabolic and physiologic strain accompanies a standard exercise load with the arms.** For this reason, exercise prescriptions based on running and bicycling **cannot** be applied to arm exercise.[27] Furthermore, because the correlation between max $\dot{V}O_2$ for arm and leg exercise is low,[27,38] it is not possible to accurately predict one's capacity for arm exercise from a test

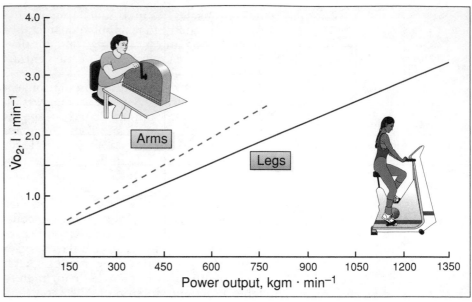

FIG. 17–7.
Arm exercise requires greater oxygen consumption compared to leg exercise at any power output level. (Average data for men and women. From Laboratory of Applied Physiology, Queens College, N.Y.)

using the legs. This further substantiates the concept of specificity for aerobic fitness.

CARDIAC HYPERTROPHY AND THE "ATHLETE'S HEART"

A moderate increase in heart size is often observed in men and women in response to exercise training,[16,24,34,48] independent of age.[3,10] This cardiac **hypertrophy** can be viewed as a fundamental biologic adaptation of muscle to an increased work load.[28] In such situations, a greater synthesis of cellular protein occurs with a concomitant reduction in protein breakdown. This accelerated protein synthesis is due largely to an increase in the muscle's content of RNA. Individual myofibrils thicken; at the same time, the number of these contractile filaments within the muscle fiber increases.[50]

The heart volume determined by x-ray examination of 30 young girl swimmers was generally much larger than expected for their body size.[3] In terms of functional significance, the heart volume correlated well with the girls' maximal oxygen uptake. Additional evidence is presented in Table 17–5 to support the case that in healthy individuals, some cardiac hypertrophy results from training.

In sedentary men, the average heart volume was about 800 ml; this volume increased in athletes in relation to the aerobic nature of the sport, so that for endurance athletes the average heart volumes were about 25% larger than those of sedentary men. The degree to which the relatively large heart volumes of some endurance athletes reflect genetic endowment or training adaptations, or both, has yet to be determined. Furthermore, the duration of training may be a factor in affecting cardiac size and structure. Several

TABLE 17–5.
HEART VOLUMES ESTIMATED FROM AREAS OF HEART SHADOW
DETERMINED BY X-RAY EXAMINATION*

Sports Category	Number of Subjects	Mean Volume (ml)	Range
Normal	67	790	490–1080
Wrestlers and high jumpers	30	782	610–920
Swimmers, soccer players, and tennis players	86	876	605–1130
Skiers, long-distance runners, and swimmers	66	923	645–1180
Professional cyclists	18	1104	880–1460

* From Anderson, K.L.: The cardiovascular system in exercise. In Exercise Physiology. Edited by H.B. Falls. New York, Academic Press, 1968.

studies have reported no changes in cardiac dimensions with short-term training, even though there was significant improvement in both max $\dot{V}_{O_2}$ and submaximal exercise heart rate.[49,72] When left ventricular mass increases with endurance training, this hypertrophy does not appear to be a permanent adaptation. Rather, heart size returns toward pretraining levels when the training intensity is reduced[33] with apparently no deleterious effects.[21]

Considerable cardiac hypertrophy is also seen in certain diseases. In chronic hypertension, for example, the heart must continually work against an excessive resistance to the flow of blood. As as result, the heart muscle stretches and, in accordance with the Frank-Starling mechanism, generates compensatory force to overcome the added vascular resistance. In addition to this ventricular stretching or dilation, the individual muscle cells hypertrophy to adjust to the increased chronic myocardial work imposed by vascular disease.[6] As the untreated hypertension progresses, the myocardial fibers are eventually stretched beyond their optimal length, and the dilated, hypertrophied heart weakens and eventually fails. To the pathologist, the "hypertrophied" heart of the cardiac patient is enlarged, flabby, and functionally inadequate to deliver even enough blood to meet the minimal resting requirements.

FUNCTIONAL VS PATHOLOGIC HYPERTROPHY.

At times, the cardiac hypertrophy in response to chronic pathologic states has been confused with the moderate compensatory growth of the myocardium with endurance training. Although the stress of exercise requires that myocardial fibers generate increased tension, **a critical requirement for initiating compensatory hypertrophy,** the application of this overload differs considerably from that of the chronic resistance imposed by vascular disease. For one thing, during exercise training the myocardial overload is only temporary so a "recuperative" time is available during nonexercise periods. Also, if compensatory heart growth does occur with training, it is not accompanied by a dilation and weakening of the ventricles, a frequent response to chronic hypertension. It is true that the hearts of elite, endurance-trained athletes are usually larger than the hearts of untrained counterparts—**but heart size is generally within the upper range of normal limits. The "athlete's heart" is a muscular heart capable of generating a relatively large stroke volume. This enhanced stroke capacity is generally the result of a more forceful systolic ejection and greater ventricular emptying.**

SPECIFIC NATURE OF TRAINING HYPERTROPHY

The ultrasonic techniques of **echocardiography** have been used to evaluate the structural characteristics of the hearts of athletes and to determine if different patterns of cardiac hypertrophy and enlargement, which may manifest themselves in childhood, were associated with different types of physical conditioning.[12,43,56] Male competitive swimmers, long-distance runners, wrestlers, and shot putters were studied during their competitive seasons and compared to 16 untrained, healthy college men. The swimmers and runners were considered representative of athletes participating in "isotonic" or endurance events; the wrestlers and shot putters represented the "isometric" or resistance-trained athletes. It is clear from the results in Table 17–6 that the structural characteristics of the hearts of apparently healthy athletes differ considerably from those of normal individuals. Also, the pattern of these differences appears to depend on the nature of the exercise conditioning. For example, left ventricular volume and mass were 181 ml and 308 g, respectively for the swimmers, and 160 ml and 302 g for the runners; the nonathletic controls averaged 101 ml for ventricular volume and 211 g for ventricular mass. Ventricular wall thickness was normal for the endurance athletes. This enlarged ventricular cavity with normal wall thickness has also been observed for other groups of male and female athletes.[12,36,58] An intraventricular enlargement could be brought about by the reduction in heart rate and increase in central blood volume observed with endurance training.[13] This, in effect, would dilate the left ventricular cavity in a way similar to water being added to a rubber balloon.[49] In contrast, the athletes involved in resistance exercises had normal left ventricular volumes

TABLE 17–6.
COMPARATIVE AVERAGE CARDIAC DIMENSIONS IN COLLEGE ATHLETES, WORLD-CLASS ATHLETES, AND NORMAL SUBJECTS[a]

Dimension[b]	College Runners (N = 15)	College Swimmers (N = 15)	World Class Runners (N = 10)	College Wrestlers (N = 12)	World Class Shot Putters (N = 4)	Normals (N = 16)
LVID	54	51	48–59[c]	48	43–52[c]	46
LVV, ml	160	181	154	110	122	101
SV, ml	116	—[d]	113	75	68	—[d]
LV wall, mm	11.3	10.6	10.8	13.7	13.8	10.3
Septum, mm	10.9	10.7	10.9	13.0	13.5	10.3
LV mass, g	302	308	283	330	348	211

[a] From Morganroth, J., et al.: Comparative left ventricular dimensions in trained athletes. Ann. Intern. Med. 82:521, 1975.
[b] VID, left ventricular internal dimension at end diastole; LVV, left ventricular volume; SV, stroke volume; LV wall, posterobasal left ventricular wall thickness; Septum, ventricular septal thickness; LV mass, left ventricular mass.
[c] Range.
[d] Values not reported.

but the largest intraventricular septum, ventricular wall thickness, and ventricular mass with little difference in cavity dimensions compared to normal people. This specific enlargement does not appear to be caused by use of anabolic steroids which often is prevalent among resistance trained athletes.[60]

These structural and dimensional differences may largely reflect specific training demands.[23a,67] For example, the training overload for endurance athletes often requires the maintenance of a relatively large cardiac output for many hours each week. As shown in Table 17–6, the myocardial internal dimensions with this training are certainly in keeping with the development of a large stroke volume while optimizing the heart's mechanical performance. On the other hand, athletes who engage in straining-type, isometric exercise are not subjected to a volume overload, but rather to acute episodes of elevated arterial pressure caused by static muscular contractions. Consequently, this added workload of the left ventricle is compensated for by an increase in ventricular wall thickness with probably little effect on the size of the left ventricular cavity.[49,60] More than likely, considerable variability exists among individuals in terms of the heart's structural response to different forms of training.

In one of our studies of world-class body builders at the 1984 Mr. Universe contest, measurements were made of resting and submaximal hemodynamics.[51] No body builder had hypertension at rest, and blood pressure was normal during a progressive treadmill exercise test. By standard evaluation criteria, however, all body builders examined by a cardiologist had abnormal resting electrocardiograms and nine of eleven body builders had left ventricular hypertrophy. The implications are unknown of these apparent differences in training response on myocardial blood supply and long-term cardiovascular health.

OTHER TRAINING ADAPTATIONS

Although considerable debate in this area exists, endurance training may also improve the vascularization of the myocardium,[31,69] especially at the arteriole level.[23,70] In addition, several experimenters have reported an increase in mitochondrial mass and cellular concentration of respiratory enzymes in the hearts of animals trained by forced running or swimming.[64] In some instances, it appeared that new components were actually being added to the existing mitochondria.[37] Of what significance these vascular and cellular adaptations are to the functional capacity of the heart during exercise has yet to be determined because it is not believed that the healthy untrained heart suffers from an oxygen lack during maximum exercise. These training changes may, however, enable myocardial tissue to function at a lower percentage of its total oxidative capacity during exercise. In addition, they may provide some protection from the degenerative process of heart disease.

■ SUMMARY

1. Cardiac output reflects the functional capacity of the circulatory system. The two factors determining the heart's output capacity are heart rate and stroke volume. The relationship is:

Cardiac output = Heart rate × Stroke volume

2. Several invasive and noninvasive methods are available to measure cardiac output. Each has its specific advantages and disadvantages for use with humans, especially during exercise.

3. Cardiac output increases in proportion to the severity of exercise, from about 5 liters per minute at rest to a maximum of 20 to 25 liters per minute and 35 to 40 liters per minute in college-aged men and elite male endurance athletes, respectively. These differences in maximum cardiac output are due entirely to the large stroke volumes of the athletes.

4. During upright exercise, stroke volume increases during the transition from rest to light exercise with maximum values reached at about 45% of max $\dot{V}_{O_2}$. Thereafter, cardiac output is increased by increases in heart rate.

5. Increases in stroke volume in upright exercise are generally the result of a more complete systolic emptying rather than a greater filling of the ventricles during diastole. Systolic ejection is augmented by sympathetic hormones. Endurance training improves myocardial strength, which contributes to stroke power during systole.

6. Heart rate and oxygen consumption are linearly related in trained and untrained individuals throughout the major portion of the exercise range. With endurance training, this line shifts significantly to the right due to improvements in the heart's stroke volume. Consequently, heart rate becomes significantly reduced at any submaximal exercise level.

7. Blood flow to specific tissues is generally regulated in proportion to their metabolic activity. This causes the major portion of exercise cardiac output to be diverted to the working muscles. In addition, a significant quantity of blood is shunted to the muscles from the kidneys and splanchnic regions which can temporarily compromise their blood supply.

8. The maximal oxygen uptake is determined by the maximum cardiac output and the maximum a-$\bar{v}$ O_2 difference. Large cardiac outputs clearly differentiate endurance athletes from untrained counterparts. The ability to generate a large a-$\bar{v}$ O_2 difference is also enhanced with training.

9. Cardiac hypertrophy is a fundamental adaptation to the increased work load imposed by exercise training. It results in a stronger heart that can generate a relatively large stroke volume. There is no scientific evidence that a normal heart is harmed by exercise training.

10. The pattern of structural and dimensional changes in the left ventricle appears to vary with specific forms of exercise training.

REFERENCES

1. Andrew, G.M., et al.: Effect of athletic training on exercise cardiac output. *J. Appl. Physiol.*, 21:503, 1966.

2. Åstrand, P.O., et al.: Cardiac output during submaximal and maximal work. *J. Appl. Physiol.*, 19:268, 1964.

3. Åstrand, P.O., et al.: Girl swimmers—with special references to respiratory and circulatory adaption and gynaecological and psychiatric aspects. *Acta Paediatr.*, 147(Suppl.):5, 1963.

4. Bar-Or, O., et al.: Cardiac output of 10- to 13-year old boys and girls during submaximal exercise. *J. Appl. Physiol.*, 30:219, 1971.

5. Beekman, R.H., et al.: Validity of CO_2-rebreathing cardiac output during rest and exercise in young adults. *Med. Sci. Sports Exerc.*, 16:306, 1984.

6. Bennet, D.H., et al.: Echocardiographic left ventricular dimensions in pressure and volume overload. Their use in assessing aortic stenosis. *Br. Heart J.*, 37:971, 1975.

7. Bevegård, S., et al.: Circulatory studies in well-trained athletes at rest and during heavy

exercise, with special reference to stroke volume and the influence of body position. *Acta Physiol. Scand.*, 57:26, 1963.

8. Blomqvist, C., and Lewis, S.: Physiological effects of training: general circulatory adjustments. In *Physical Conditioning and Cardiovascular Rehabilitation*. Edited by L. Cohen et al. New York, Wiley, 1981.

9. Blomqvist, C.G., et al.: Similarity of the hemodynamic response to static and dynamic exercise of small muscle groups. *Circ. Res.*, 48(Suppl. I):87, 1982.

10. Child, J.S., et al.: Cardiac hypertrophy and function in master endurance runners and sprinters. *J. Appl. Physiol.*, 57:176, 1984.

11. Clausen, J., and Trap-Jensen, J.: Heart rate and arterial blood pressure during exercise in patients with angina pectoris. *Circulation*, 53:436, 1976.

12. Cohen, J.L., and Segal, K.R.: Left ventricular hypertrophy in athletes: an exercise echocardiographic study. *Med. Sci. Sports Exerc.*, 17:695, 1985.

13. Convertino, V.A., et al.: Exercise training-induced hypervolemia: role of plasma albumin, renin, and vasopressin. *J. Appl. Physiol.*, 48:665, 1980.

14. Coyle, E.F., et al.: Time course of loss of adaptations after stopping prolonged intense endurance training. *J. Appl. Physiol.*, 57:1857, 1984.

15. Coyle, E.F., et al.: Blood lactate threshold in some well trained ischemic heart disease patients. *J. Appl. Physiol.*, 54:18, 1983.

16. Cox, M.L., et al.: Exercise training-induced alterations of cardiac morphology. *J. Appl. Physiol.*, 61:926, 1986.

17. Cummins, T.D., and Gladden, L.B.: Responses to submaximal and maximal arm cycling above, at, and below heart level. *Med. Sci. Sports Exerc.*, 15:295, 1983.

18. Cunningham, D.A., et al.: Development of cardiorespiratory function in circumpubertal boys: a longitudinal study. *J. Appl. Physiol.*, 56:302, 1984.

19. Daniels, J.T.: Running with Jim Ryun: a five-year study. *Phys. Sportsmed.*, 2:62, 1974.

20. DeBusk, R.F., et al.: Cardiovascular responses to dynamic and static effort soon after myocardial infarction: application to occupational work assessment. *Circulation*, 58:368, 1978.

21. Dickhuth, H.H., et al.: The long-term involution of physiological cardiomegaly and cardiac hypertrophy. *Med. Sci. Sports Exerc.*, 21:244, 1989.

22. Dressendorfer, R., et al.: Therapeutic effects of exercise training in angina patients. In *Physical Conditioning and Cardiomuscular Rehabilitation*. Edited by L. Cohen et al. New York, John Wiley and Sons, 1981.

23. Eckstein, R.W.: Effect of exercise and coronary artery narrowing on coronary collateral circulation. *Circ. Res.*, 5:230, 1957.

23a. Effron, M.B.: Effects of resistive training on left ventricular function. *Med. Sci. Sports Exerc.*, 21:694, 1989.

24. Ehsani, A.A., et al.: Rapid changes in left ventricular dimensions and mass in response to physical conditioning and deconditioning. *Am. J. Cardiol.*, 42:52, 1978.

25. Ekblom, B., et al.: Effect of training on circulatory response to exercise. *J. Appl. Physiol.*, 24:518, 1968.

26. Ferguson, R.J., et al.: Comparison of cardiac outputs determined by CO_2 rebreathing and dye dilution method. *J. Appl. Physiol.*, 25:450, 1968.

27. Franklin, B.A., et al.: Aerobic requirements of arm ergometry: implications for exercise testing and training. *Phys. Sportsmed.*, 11:81, 1983.

28. Goldberg, A.L.: Mechanism of work-induced hypertrophy of skeletal muscle. *Med. Sci. Sports*, 7:185, 1975.

29. Hanson, J.S., et al.: Long-term physical training and cardiovascular dynamics in middle-aged men. *Circulation*, 38:783, 1968.

30. Hagerman, F.C.: A comparison of energy expenditure during rowing and cycling ergometry. *Med. Sci. Sports Exerc.*, 20:479, 1988.

31. Heaton, W.H., et al.: Beneficial effect of physical training on blood flow to myocardium perfused by chronic collaterals in the exercising dog. *Circulation*, 57:575, 1978.

32. Hickson, R.C.: Skeletal muscle cytochrome c and myoglobin, endurance, and frequency of training. *J. Appl. Physiol.*, 51:746, 1981.

33. Hickson, R.C., et al.: Reduced training intensities and loss of aerobic power, endurance, and cardiac growth. *J. Appl. Physiol.*, 58:492, 1985.

34. Hickson, R.C., et al.: Reduced training duration effects on aerobic power, endurance, and cardiac growth. *J. Appl. Physiol.*, 53:225, 1982.

35. Holloszy, J.O., and Coyle, E.F.: Adaptations of skeletal muscle to endurance training and their metabolic consequences. *J. Appl. Physiol.*, 56:831, 1984.

36. Ikaheimo, M.J., et al.: Noninvasive evaluation of the athletic heart: sprinters versus endurance runners. *Am. J. Cardiol.*, 44:24, 1979.

37. Laugens, R.P., and Gomez-Dumm, L.A.: Fine structure of myocardial mitochondria in rats after exercise for one-half to two hours. *Circ. Res.*, 11:271, 1967.

38. Lazarus, B., et al.: Comparison of the reproducibility of arm and leg exercise tests in men with angina pectoria. *Am. J. Cardiol.*, 47:1074, 1981.

39. Lewis, S.F., et al: Cardiovascular responses to exercise as a function of absolute and relative work load. *J. Appl. Physiol.*, 54:1314, 1983.

40. Magel, J.R., and Andersen, K.L.: Cardiac output in muscular exercise measured by the CO_2 rebreathing technique. In *Ergometry in Cardiology*. Edited by H. Denolin et al. Boehringer, Mannheim GmbH, 1968.

41. Magel, J.R., et al.: Metabolic and cardiovascular adjustment to arm training. *J. Appl. Physiol.*, 45:75, 1978.

42. McArdle, W.D., et al.: Specificity of run training on $\dot{V}O_2$ max and heart rate changes during running and swimming. *Med. Sci. Sports*, 10:16, 1978.

43. Morganroth, J., et al.: Comparative left ventricular dimensions in trained athletes. *Ann. Intern. Med.*, 82:521, 1975.

44. Miles, D.S., et al.: Cardiovascular responses to upper body exercise in normals and cardiac patients. *Med. Sci. Sports Exerc.*, 21:S126, 1989.

45. Pechar, G.S., et al.: Specificity of cardio-respiratory adaptation to bicycle and treadmill training. *J. Appl. Physiol.*, 36:753, 1974.

46. Pendergast, D.R.: Cardiovascular, respiratory, and metabolic responses to upper body exercise. *Med. Sci. Sports Exerc.*, 21:S121, 1989.

47. Poliner, L.R., et al.: Left ventricular performance in normal subjects: a comparison of the responses to exercise in upright and supine positions. *Circulation*, 62:528, 1980.

48. Rerysch, S.K., et al.: Effects of exercise training on left ventricular function in normal subjects: a longitudinal study by radionuclide angiography. *Am. J. Cardiol.*, 45:244, 1980.

49. Ricci, G., et al.: Left ventricular size following endurance, sprint, and strength training. *Med. Sci. Sports Exerc.*, 14:344, 1982.

50. Richter, G.W., and Kellner, A.: Hypertrophy of the human heart at the level of fine structure: an analysis of two postulates. *J. Cell Biol.*, 18:195, 1965.

51. Rippe, J., et al.: Resting cardiovascular examination and response to submaximal treadmill exercise in world champion body builders. *Med. Sci. Sports Exerc.*, 17:283, 1985.

52. Rowell, L.B.: Muscle blood flow in humans: how high can it go? *Med. Sci. Sports Exerc.*, 20:S97, 1988.

53. Rowell, L.B.: Human cardiovascular adjustments to exercise and thermal stress. *Physiol. Rev.*, 54:75, 1974.

54. Rowell, L.B.: Vascular blood flow and metabolism during exercise. In *Frontiers of Fitness*. Edited by R.J. Shephard. Springfield, IL, C.C Thomas, 1971.

55. Rowell, L.B., et al.: Reductions in cardiac output, central blood volume, and stroke volume with thermal stress in normal men during exercise. *J. Clin. Invest.*, 45:1801, 1966.

56. Rowland, T.W., et al.: "Athletes heart" in prepubertal children. *Pediatrics*, 79:800, 1987.

57. Rubal, J.B., et al.: Effects of physical conditioning on the heart size and wall thickness of college women. *Med. Sci. Sports Exerc.*, 19:423, 1987.

58. Rubal, J.B., et al.: Echocardiographic examination of woman softball champions. *Med. Sci. Sports. Exerc.*, 13:176, 1981.

59. Rushmer, R.F.: *Cardiovascular Dynamics*. Philadelphia, W.B. Saunders, 1976.

60. Salke, R.C., et al.: Left ventricular size and function in body builders using anabolic steroids. *Med. Sci. Sports Exerc.*, 17:701, 1985.

61. Saltin, B.: Physiological effects of physical conditioning. *Med. Sci. Sports*, 1:50, 1969.

62. Saltin, B., et al.: Fiber types and metabolic potentials of skeletal muscles in sedentary men and endurance runners. *Ann. N.Y. Acad. Sci.*, 301:3, 1977.

63. Sawka, M.N.: Physiology of upper body exercise. *Exerc. Sport Sci. Rev.*, 14:175, 1986.

64. Scheuer, J., and Tipton, C.M.: Cardiovascular adaptations to training. *Annu. Rev. Physiol.*, 39:221, 1977.

65. Schwade, J., et al.: A comparison of the response to arm and leg work in patients with ischemic heart disease. *Am. Heart J.*, 94:203, 1977.

66. Seely, J.E., et al.: Heart and lung function at rest and during exercise in adolescence. *J. Appl. Physiol.*, 36:34, 1974.
67. Snoeckx, L.H.E.H., et al.: Echocardiographic dimensions in athletes in relation to their training programs. *Med. Sci. Sports Exerc.*, 14:428, 1982.
68. Soto, K.I., et al.: Cardiac output in preadolescent competitive swimmers and in untrained normal children. *J. Sports Med.*, 23:291, 1983.
69. Spear, K.L., et al.: Coronary blood flow in physically trained rats. *Circ. Res.*, 12:135, 1978.
70. Stevenson, J.A.F., et al.: Effect of exercise on coronary tree size in the rat. *Circ. Res.*, 15:265, 1964.
71. Thomas, S.N., et al.: Cerebral blood flow during submaximal and maximal dynamic exercise in humans. *J. Appl. Physiol.*, 67:744, 1989.
72. Thompson, P.D., et al.: Cardiac dimensions and performance after either arm or leg endurance training. *Med. Sci. Sports Exerc.*, 13:303, 1981.
73. Thorston, J., et al.: Effects of training and detraining on oxygen uptake, cardiac output, blood gas tensions, pH, and lactate concentrations during and after exercise in the horse. In *Equine Exercise Physiology*. Edited by D.H. Snow et al. Cambridge, Granta Editions, 1983.
74. Toner, M.M., et al.: Cardiorespiratory responses to exercise distributed between the upper and lower body. *J. Appl. Physiol.*, 54:1403, 1983.
75. Toner, M.M., et al.: Cardiovascular adjustment to exercise distributed between the upper and lower body. *Med. Sci. Sports Exerc.*, in press.
76. Vander, L.B., et al.: Cardiorespiratory responses to arm and leg ergometry in women. *Phys. Sportsmed.*, 12:101, 1984.
77. Vokac, Z., et al.: Oxygen uptake/heart rate relationship in leg and arm exercise, sitting and standing. *J. Appl. Physiol.*, 39:54, 1975.

SKELETAL MUSCLE: STRUCTURE AND FUNCTION

Human movement depends on transforming the chemical energy in ATP into mechanical energy. This specific energy transformation is achieved through the action of skeletal muscles. Muscular forces acting on the body's bony lever system cause one or more bones to move about their joint axis; this enables a person to propel an object, move the body itself, or do both simultaneously. In the sections that follow, we present the architectural organization of **skeletal muscle** and focus on its gross and microscopic structure. The discussion includes the sequence of chemical and mechanical events in muscular contraction and relaxation, as well as the differences in muscle fiber characteristics between sedentary and highly trained people.

GROSS STRUCTURE OF SKELETAL MUSCLE

Each of the more than 430 voluntary muscles in the body contains various wrappings of fibrous connective tissue. Figure 18–1 shows the cross section of a muscle that consists of thousands of cylindric muscle cells called **fibers.** These long, slender multinucleated fibers (whose number is probably fixed by the second trimester of fetal development) lie parallel to each other, and the force of contraction is along the long axis of the fiber.

Each fiber is wrapped and separated from its neighboring fibers by a fine layer of connective tissue, the **endomysium.** Another layer of connective tissue, the **perimysium,** surrounds a bundle of up to 150 fibers called a **fasciculus.** Surrounding the entire muscle is a fascia of fibrous connective tissue known as the **epimysium.** This protective sheath is tapered at its distal end as it blends into and joins the intramuscular tissue sheaths to form the dense, strong connective tissue of the **tendons.** The tendons connect both ends of the muscle to the outermost covering of the skeleton, the **periosteum.** Thus, the force of muscular contraction is transmitted directly from the muscle's connective tissue harness to the tendons, which in turn pull on the bone at their points of attachment. The region where the tendon joins a relatively stable skeletal part is the **origin** of the muscle; the point of attachment to the moving bone is the **insertion.** The origin is generally at the proximal or fixed end of the lever system or nearest the body's midline, whereas the insertion is the distal or movable attachment.

Beneath the endomysium and surrounding each muscle fiber is the **sarcolemma.** This thin, elastic membrane encloses the fiber's cellular contents. The aqueous protoplasm or **sarcoplasm** of the cell contains the contractile proteins, enzymes, fat and glycogen particles, the nuclei, and various specialized cellular organelles. Embedded within the sarcoplasm is an extensive interconnecting network of tubular channels and vesicles known as the

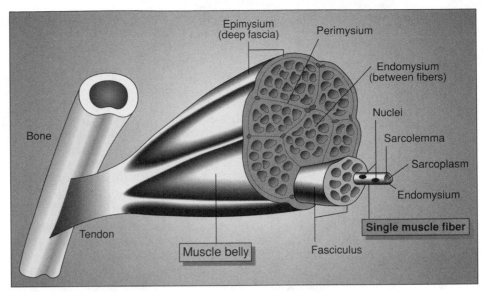

FIG. 18–1.
Cross section of a muscle and the arrangement of connective tissue wrappings. The individual fibers are covered by the endomysium. Groups of fibers called fasciculi are surrounded by the perimysium, and the entire muscle is wrapped in a fibrous sheath of connective tissue, the epimysium. The sarcolemma is a thin, elastic membrane that covers the surface of each muscle fiber.

sarcoplasmic reticulum. This highly specialized system provides the cell with structural integrity and also serves important functions in muscular contraction.

CHEMICAL COMPOSITION

Approximately 75% of skeletal muscle is water, 20% is protein, and the remaining 5% is made up of inorganic salts and other substances that include high-energy phosphates, urea, lactic acid, the minerals calcium, magnesium, and phosphorous, various enzymes and pigments, ions of sodium, potassium, and chloride, and amino acids, fats, and carbohydrates.

The most abundant muscle proteins, in relation to the muscle's total protein content, are **myosin, actin,** and **tropomyosin.** Also, about 700 mg of the conjugated protein **myoglobin** are incorporated into each 100 g of muscle tissue. The specifics of myoglobin function were presented in Chapter 13.

BLOOD SUPPLY

During exercise that requires an oxygen uptake of 4.0 liters per minute, the muscle's oxygen consumption increases nearly 70 times to about 11 ml per 100 g per minute or a total of about 3400 ml per minute. To accommodate this large oxygen requirement of exercising muscles, the local vascular bed must channel large quantities of blood through the active tissues. In rhythmic exercise such as running, swimming, or cycling, the blood flow fluctuates; it decreases during the muscle's contraction phase and increases during the relaxation period. This provides a "milking action" that facilitates blood flow through the muscles and back to the heart. Complementing this pulsatile flow is the rapid dilatation of previously dormant capillaries so that in strenuous exercise, more than 4000 capillaries may be delivering blood to each square millimeter of muscle cross section.

Straining-type activities present a somewhat different picture.[16a] When a muscle contracts to about 60% of its force-generating capacity, blood flow to the muscle is occluded due to elevated intramuscular pressure. With a sustained static or isometric contraction, the compressive force of the contraction can actually stop the flow of blood. Under such conditions, energy for continued muscular effort is generated mainly from the stored phosphagens and through the anaerobic reactions of glycolysis.

Capillarization of Muscle

One factor often proposed for the improved exercise capacity with training is an **increase in capillary density of the trained muscles.** Aside from its role in delivering oxygen, nutrients, and hormones, the capillary circulation also provides the means for removing heat and metabolic by-products from the active tissues. All of these functions would be enhanced by a higher capillary density in muscle tissue.

Several investigations show favorable effects of endurance training on the capillarization of skeletal muscle.[2,8] In one study using the electron microscope,[8] the number of capillaries per muscle (as well as the capillaries per square millimeter of muscle tissue) averaged about 40% greater in endurance athletes than in untrained counterparts. This was almost identical to the 41% difference in maximal oxygen uptake between the two groups. One research group cited unpublished observations that skeletal muscle capillaries "can be easily increased, and that the increase is closely related to the activity level of the muscle."[37] They also reported a high positive relationship for both men and women between maximal oxygen uptake and the average number of muscle capillaries. The functional significance of this relationship is that increased capillarization enhances the oxygenation of the entire muscle cell. This would be beneficial during strenuous exercise that requires a high level of steady-rate aerobic metabolism.

ULTRASTRUCTURE OF SKELETAL MUSCLE

The ultrastructure or microscopic anatomy of skeletal muscle has been revealed with the aid of electron microscopy, x-ray diffraction, and histochemical staining techniques. Figure 18–2 shows the different levels of subcellular organization within skeletal muscle fibers. Each muscle fiber is composed of smaller functional units that lie parallel to the long axis of the fiber. These **fibrils** or **myofibrils** are approximately 1 micron (μm) in diameter (1 μm = 1/1000 mm) and are composed of even smaller subunits, the **filaments** or **myofilaments,** that also lie parallel to the long axis of the myofibril. The myofilaments consist mainly of two proteins, **actin** and **myosin,** that account for about 84% of the myofibrillar complex. Six other proteins have also been identified that have either a structural function or a significant effect on the interaction of protein filaments during contraction. These are (1) **tropomyosin,** located along the actin filaments (5%); (2) **troponin,** also located in the actin filaments (3%); (3) **α-actinin,** distributed in the region of the Z band (7%); (4) **β-actinin,** found in the actin filaments (1%); (5) **M protein,** identified in the region of the M lines within the sarcomere (<1%);

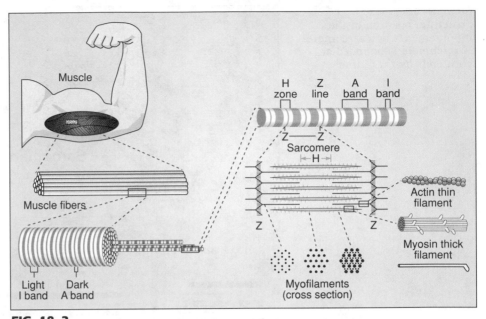

FIG. 18–2.
Microscopic organization of skeletal muscle. The whole muscle is composed of fibers; these in turn are made up of myofibrils, of which the actin and myosin protein filaments are a part. If viewed under a microscope, magnification would be approximately × 205,000. (From Vander, A.J., et al. Human Physiology, 3rd ed. New York, McGraw-Hill, 1985.)

and (6) **C protein** (<1%), thought to maintain the structural integrity of the sarcomere.

THE SARCOMERE

At low magnification, the alternating light and dark bands along the length of the muscle fiber give it its characteristic **striated** appearance. Figure 18–3 illustrates the structural details of this cross-striation pattern within a myofibril. The lighter area is referred to as the I band, whereas the darker zone is known as the A band. The Z line bisects the I band and adheres to the sarcolemma to give stability to the entire structure.* The repeating unit between two Z lines is called **the sarcomere, which is the functional unit of the muscle cell.** The actin and myosin filaments within the sarcomere are primarily involved in the mechanical process of muscular contraction.

The position of the thin actin and thicker myosin proteins in the sarcomere results in an overlap of the two filaments. The center of the A band contains the H zone, a region of lower optical density due to the absence of actin filaments in this area. The central portion of the H zone is bisected by the M line, which delineates the sarcomere's center. The M line consists of the protein structures that support the arrangement of the myosin filaments.

* The bands are named according to their optical properties. When a light passes through the I band, its velocity is the same in all directions (isotropic). Polarized light passing through the A band does not scatter equally (anisotropic). The letter Z is from the German *zwischen* which means "between."

FIG. 18–3.
Structural position of the myofilaments in a sarcomere. A sarcomere is bounded at both ends by the Z line.

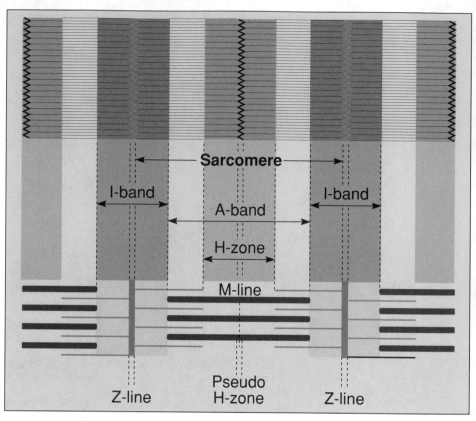

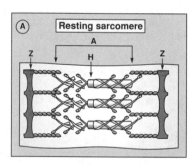

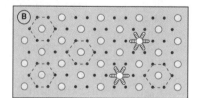

FIG. 18–4.
A, Ultrastructure of actin-myosin orientation within a resting sarcomere. B, Representation of electron micrograph through a cross section of myofibrils in a single muscle fiber. Note the hexagonal orientation of the smaller actin and larger myosin filaments, as well as an example of the cross-bridges that extend from a thick to the thin filament.

ACTIN-MYOSIN ORIENTATION

The top portion of Figure 18–4 illustrates the actin-myosin orientation within a sarcomere at resting length. The bottom portion of the figure shows the hexagonal arrangement of actin and myosin filaments. A thick filament [150 angstroms (Å) in diameter and 1.5 μm long] is bordered by six thinner filaments, each about 50 Å in diameter and 1 μm long. Three thick filaments surround each thin filament. This muscular substructure is extremely impressive. For example, a myofibril 1 μm in diameter contains about 450 thick filaments in the center of the sarcomere and 900 thin filaments at each end of the sarcomere. A single muscle fiber 100 μm in diameter and 1 cm long contains about 8000 myofibrils, each myofibril consisting of 4500 sarcomeres. This results in a total of 16 billion thick and 64 billion thin filaments in a single fiber![43]

Figure 18–5 is a detailed illustration of the spatial orientation of the various proteins that comprise the contractile filaments. Projections or "cross-bridges" spiral about the myosin filament at the region where the filaments of actin and myosin overlap. These cross-bridges are repeated at intervals of about 450 Å along the filament. Their globular "lollipoplike" heads extend perpendicularly to interact with the thinner strands of actin; this is the structural and functional link between the myofilaments.

Tropomyosin and troponin are two other important constituents of the actin helix structure. These proteins appear to regulate the make-and-break contacts between the myofilaments during contraction. Tropomyosin is dis-

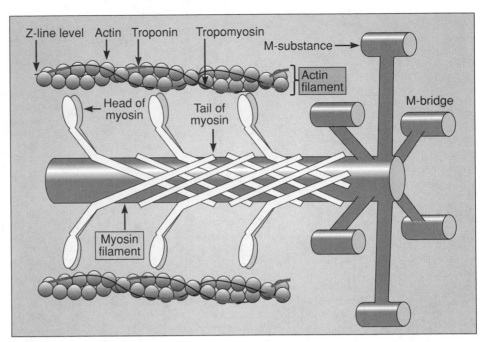

FIG. 18–5.
Details of the thick and thin protein filaments, including tropomyosin, troponin, and the M line. The myosin ATPase is located on the globular heads of the myosin; this "active" head frees the energy from ATP to be used in muscle contraction. (From Edington, D.W., and Edgerton, V.R.: The Biology of Physical Activity. Boston, Houghton Mifflin, 1976.)

tributed along the length of the actin filament in a groove formed by the double helix. It is believed to inhibit actin and myosin interaction or coupling and prevent a permanent bonding of these filaments. Troponin, which is embedded at fairly regular intervals along the actin strands, has a high affinity for calcium ions (Ca^{++}). This mineral plays a crucial role in muscle function.[29] It is the action of Ca^{++} and troponin that triggers the myofibrils to interact and slide past each other. When the fiber is stimulated, the troponin molecules appear to undergo a conformational change that in some way "tugs" on the tropomyosin protein strand. This moves the tropomyosin deeper into the groove between the two actin strands. This action "uncovers" the active sites of the actin and allows contraction to proceed.

The M line consists of transversely and longitudinally oriented proteins that serve to maintain the proper orientation of the thick filament within a sarcomere. As can be observed in Figure 18–5, the perpendicularly oriented M bridges connect with six adjacent thick (myosin) filaments in a hexagonal pattern.

INTRACELLULAR TUBULE SYSTEMS

Figure 18–6 illustrates the tubule system within a muscle fiber. An extensive network of interconnecting tubular channels, the **sarcoplasmic reticulum,** lies parallel to the myofibrils. The lateral end of each tubule terminates in a saclike vesicle that stores Ca^{++}. Another network of tubules known as the transverse tubule system or **T-system** runs perpendicular to the myofibril.

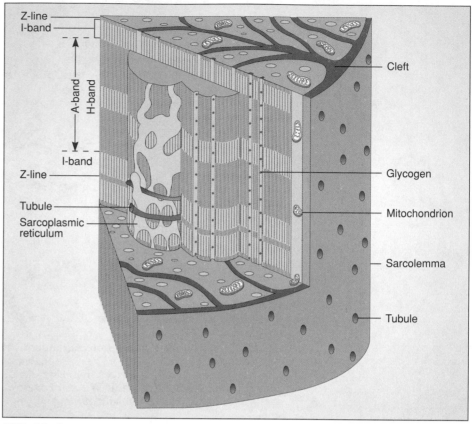

FIG. 18–6.
Three-dimensional view of sarcoplasmic reticulum and T-tubule system within the muscle fiber. (From Graham, H.: How is muscle turned on and off? Sci. Am., 222:84, 1970.)

The T tubules are situated between the lateral-most portion of two sarco-plasmic channels with the vesicles of these structures abutting the T tubule. This repeating pattern of two vesicles and T tubules in the region of each Z line is known as a **triad.** There are two triads in each sarcomere, and the pattern is repeated regularly throughout the length of the myofibril.

The T tubules pass through the fiber and open externally from the inside of the muscle cell. **The triad and T-tubule system appear to function as a microtransportation or plumbing network for spreading the action potential (wave of depolarization) from the fiber's outer membrane inward to the deep regions of the cell.** During this depolarization process, calcium ions are released from the triad sacs and diffuse a short distance to the filaments, presumably to "activate" the actin filaments. Contraction is initiated when the cross-bridges of the myosin filaments are attracted to the active sites on the actin filaments. When electrical excitation ceases, there is a decrease in free calcium concentration in the cytoplasm; this is associated with the relaxation of the muscle.

CHEMICAL AND MECHANICAL EVENTS DURING CONTRACTION AND RELAXATION

The electron microscope has helped unravel many secrets of cellular structure that have led to the formulation of reasonable hypotheses concerning the chemical and mechanical events during muscular contraction and relaxation. Although many gaps remain, there is considerable evidence to support a **"sliding-filament theory"** of muscle contraction that fits nicely with the detailed ultrastructure of muscle discussed previously.

SLIDING-FILAMENT THEORY

The sliding-filament theory proposes that a muscle shortens or lengthens because the thick and thin myofilaments slide past each other without the filaments themselves changing length. This causes a major change in the relative size of the various zones and bands within a sarcomere. Figure 18–7 illustrates that the thin actin myofilaments slide past the myosin myofilaments and move into the region of the A band during contraction (and move out in relaxation). The major structural rearrangement during

FIG. 18–7.
Structural rearrangement of actin and myosin filaments at rest and during contraction.

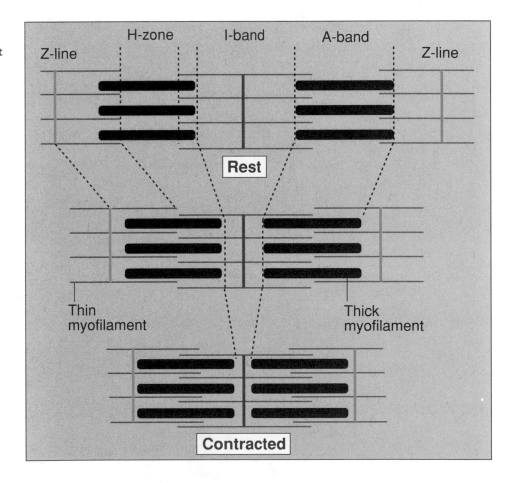

contraction, therefore, occurs in the region of the I band, which decreases markedly. The Z bands are essentially pulled toward the center of each sarcomere. There is no change in the width of the A band, although the H zone can disappear when the actin filaments are in contact at the center of the sarcomere. In an isometric muscular contraction, force is generated while the fiber's length remains relatively unchanged and the relative spacing of I and A bands stays constant; in this situation, the same molecular groups react with one other repeatedly. In an eccentric contraction in which force is generated while the muscle lengthens, the A band becomes broader.

Mechanical Action of the Cross-Bridges

The globular head of the myosin cross-bridge provides the mechanical means for the actin and myosin filaments to slide past each other. Figure 18–8 shows schematically the oscillating to-and-fro nature of the cross-bridges, which move in a way somewhat similar to the action of oars in water. Unlike oars, however, the cross-bridges do not all move in a synchronous manner. During contraction, each cross-bridge undergoes many repeated but independent cycles of movement. Thus, at any one time, only about 50% of the bridges are in contact with the thin actin filaments to

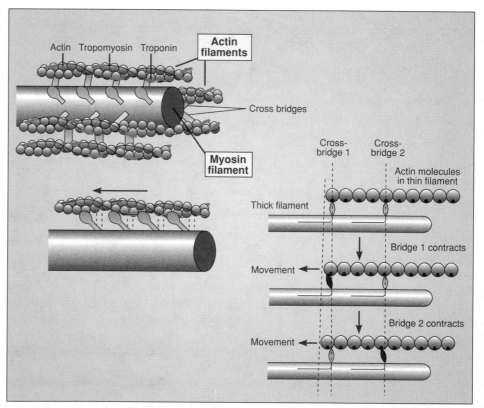

FIG. 18–8.
Relative positioning of actin and myosin filaments during the oscillating movement of the cross-bridges. The action of each bridge contributes a small displacement of movement. For clarity, one of the actin strands is omitted from the left-hand figure.

form the protein complex **actomyosin,** which has contractile properties; the others are at some other position in their vibrating cycle.

As illustrated in the right side of Figure 18–8, each action of a cross-bridge contributes only a small longitudinal displacement in terms of the total sliding action of the filaments. This process has been likened to the action of a person climbing a rope. The arms and legs represent the cross-bridges. Climbing is accomplished by first reaching with the arms, then grabbing, pulling and breaking contact, and then repeating this process over and over throughout the climb.

Link Between Actin, Myosin, and ATP

The interaction and movement of the protein filaments during muscular contraction necessitate that the myosin cross-bridges continually undergo oscillatory movements by combining, detaching, and recombining to new sites along the actin strands.

The detachment of the myosin cross-bridges from the actin filament is brought about when the ATP molecule is joined to the actomyosin complex. This reaction enables the myosin cross-bridge to return to its original state so it is available to bind a new active site on the actin. The dissociation of actomyosin occurs in the following way:

$$\text{Actomyosin} + \text{ATP} \rightarrow \text{Actin} + \text{Myosin} - \text{ATP}$$

ATP also serves an important function in the contraction process. Energy is provided for cross-bridge movement when the terminal phosphate is split from ATP. One of the reacting sites on the globular head of the myosin cross-bridge binds to the reactive site on actin. The other myosin active site acts as the enzyme **myofibrilar adenosinetriphosphatase,** or more commonly, **myosin ATPase.** This enzyme splits ATP so that its energy can be used for muscle contraction. The rate of ATP splitting is relatively slow if myosin and actin remain apart; when they join, however, the reactive rate of myosin ATPase increases considerably. It is believed that energy released from ATP splitting somehow activates the cross-bridges, causing them to oscillate. It is possible that this energy transfer process causes a conformational change in the shape of the globular head of the myosin cross-bridge so that it interacts with the appropriate actin molecule.

Fast-twitch muscle fibers with the ability for rapid and powerful contraction possess a relatively high activity level of myosin ATPase. It is tempting to speculate that specific forms of speed and power training modify enzymatic activity in a manner that facilitates the sequence of events in muscular contraction. More will be said shortly concerning fiber types and training effects.

EXCITATION-CONTRACTION COUPLING

Excitation-contraction is the physiologic mechanism whereby an electric discharge at the muscle initiates the chemical events that lead to contraction.

In the resting state, a muscle's Ca^{++} concentration is relatively low. When a muscle fiber is stimulated to contract, there is an immediate increase in intracellular Ca^{++}. This is initiated by the arrival of the action potential at the transverse tubules that causes Ca^{++} to be released from the lateral sacs of the sarcoplasmic reticulum. The inhibitory action of troponin that prevents actin-myosin interaction is released when Ca^{++} ions bind rapidly with troponin in the actin filaments. In a sense, the muscle is now "turned on."

Actin + Myosin ATPase → Actomyosin ATPase

When the active sites on the actin and myosin are joined, myosin ATPase is activated, which in turn splits ATP. During this process, the transfer of energy causes movement of the myosin cross-bridges and the muscle generates tension.

Actomyosin ATPase → Actomyosin + ADP + P + Energy

The cross-bridges uncouple from actin when ATP **binds** to the myosin bridge. Coupling and uncoupling continue as long as the Ca^{++} concentration remains at a sufficient level to inhibit the troponin-tropomyosin system. When the nerve stimulus to the muscle is removed, Ca^{++} moves back into the lateral sacs of the sarcoplasmic reticulum. This restores the inhibitory action of the troponin-tropomyosin, and actin and myosin remain separated as long as ATP is present. (In **rigor mortis,** the muscles become stiff and rigid soon after death. This occurs because ATP is no longer available in the muscle cells. Without ATP, the myosin cross-bridges and actin remain attached and cannot be pulled apart.) Figure 18–9 illustrates the interaction between the actin and myosin filaments, Ca^{++}, and ATP in a relaxed and contracted muscle.

RELAXATION

When a muscle is no longer stimulated, the flow of Ca^{++} ceases and troponin is free once again to inhibit actin-myosin interaction. During recovery, Ca^{++} is actively pumped into the sarcoplasmic reticulum where it concentrates in the lateral vesicles. The retrieval of Ca^{++} from the troponin-tropomyosin proteins "turns off" the active sites on the actin filament. This deactivation accomplishes two things: (1) It prevents any mechanical link between the myosin cross-bridges and the actin filaments; (2) It reduces the activity of myosin ATPase so there is no more ATP splitting. The muscle's relaxation is brought about by the return of the actin and myosin filaments to their original state.

SEQUENCE OF EVENTS IN MUSCULAR CONTRACTION

The following is a list of the main events in muscular contraction and relaxation. The sequence begins with the initiation of an action potential by the motor nerve. This impulse is then propagated over the entire surface of the muscle fiber as the cell membrane becomes depolarized: (1) The muscle action potential depolarizes the transverse tubules at the A-I junction of the sarcomere; (2) The depolarization of the transverse or T tubules causes Ca^{++} to be released from the lateral sacs of the sarcoplasmic reticulum; (3) Ca^{++} ions bind to troponin-tropomyosin in the actin filaments. This releases the inhibition that prevented actin from combining with myosin; (4) Actin com-

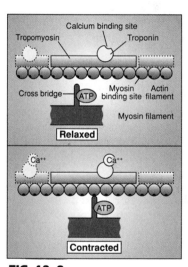

FIG. 18–9.
Interaction between the actin-myosin filaments, Ca^{++}, and ATP in relaxed and contracted muscle. In the relaxed state, troponin and tropomyosin interact with actin, preventing the coupling of the myosin cross bridge to actin. During contraction, the cross bridge couples with actin due to the binding of Ca^{++} with troponin-tropomyosin. (From Vander, A.J., et al.: Human Physiology, 3rd ed. New York, McGraw-Hill Book Company, 1985.)

bines with myosin-ATP. Actin also activates the myosin ATPase, which then splits ATP. Tension is created because the energy from this reaction is used to produce movement of the myosin cross-bridge; (5) ATP binds to the myosin bridge. This breaks the actin-myosin bond and allows the cross-bridge to dissociate from actin. This leads to a relative movement or sliding of the thick and thin filaments past each other and the muscle shortens; (6) Cross-bridge activation continues as long as the concentration of Ca^{++} remains high enough (due to membrane depolarization) to inhibit the action of the troponin-tropomyosin system; (7) When the muscle is no longer stimulated, the concentration of Ca^{++} ions rapidly decreases as they move back into the lateral sacs of the sarcoplasmic reticulum by an energy process that splits ATP; (8) The removal of Ca^{++} ions restores the inhibitory action of troponin-tropomyosin. In the presence of ATP, actin and myosin remain in the dissociated, relaxed state.

MUSCLE FIBER TYPE

Skeletal muscle is not simply a homogeneous group of fibers with similar metabolic and functional properties. Although considerable confusion has existed concerning the method and terminology for classifying human skeletal muscle, **two** distinct fiber types have been identified and classified by their **contractile** and **metabolic** characteristics.[4]

Fast-Twitch Fibers

Fast-twitch muscle fibers have a high capability for the electrochemical transmission of action potentials, a high activity level of myosin ATPase, a rapid level of calcium release and uptake by the sarcoplasmic reticulum, and a high rate of cross-bridge turnover, all of which relate to their ability to generate energy rapidly for quick, forceful contractions. Recall that it is myosin ATPase that splits ATP to provide energy for muscle contraction. In fact, the fast twitch fiber's intrinsic speed of contraction and tension development is two to three times as fast as that of fibers classified as slow-twitch fibers (next section). The fast-twitch fibers rely largely on a well-developed, short-term glycolytic system for energy transfer. They have been labeled **FG fibers** to signify their fast-glycogenolytic capabilities. **Fast-twitch fibers are generally activated in short-term, sprint activities as well as other forceful muscular contractions that depend almost entirely on anaerobic metabolism for energy.**[11,21,27] The metabolic and contractile capacities of these fibers are also important in the stop-and-go or change-of-pace sports such as basketball or field hockey, which at times require rapid energy that only the anaerobic metabolic pathways supply.

Slow-Twitch Fibers

Slow-twitch fibers generate energy for ATP resynthesis predominantly by means of the relatively long-term system of aerobic energy transfer. They are distinguished by a low activity level of myosin ATPase, a slow speed of contraction, and a glycolytic capacity less well developed than their fast-twitch counterparts. The slow-twitch fibers, however, contain relatively large and numerous mitochondria. It is this concentration of mitochondria combined with high levels of myoglobin that give the slow-twitch fibers their characteristic red pigmentation. Accompanying this enhanced metabolic machinery is a high concentration of mitochondrial enzymes required to sustain aerobic metabolism.[15,16,17,24] Thus, **slow twitch fibers are fatigue-resistant and well suited for prolonged aerobic exercise.** These fibers have been labeled **SO fibers** to describe their slow contraction speed and great reliance on oxidative metabolism. Unlike the FG fibers that fatigue readily, the SO fibers (more precisely, motor units) are adapted for prolonged work and are recruited for aerobic activities.[18,26] In fact, studies of muscle glycogen depletion indicate that in prolonged, moderate exercise there is almost exclusive reliance on the slow-twitch muscle fibers. Even after 9 to 12 hours of exercise the limited glycogen that is available is found mostly in the "unused" fast-twitch fibers.[35] It also appears that the capacity for blood flow through muscle is determined by differences in the oxidative capacity of the two fiber types, with the slow-twitch fibers receiving proportionately more blood during exercise than their fast-twitch counterparts.[39]

Many researchers classify slow-twitch fibers as **type I,** whereas the fast-twitch fibers (and proposed subdivisions) are categorized as **type II.** When a person exercises at near maximum aerobic and anaerobic levels, as in middle-distance running or swimming, or in sports such as basketball, field hockey, or soccer, which require a blend of aerobic and anaerobic energy, both types of muscle fibers are activated.[21,35]

Fast-Twitch Subdivisions

Subdivisions of the fast-twitch fiber are present in humans. The **type IIa** fiber is considered intermediate in that its fast contraction speed is combined with a moderately well developed capacity for both aerobic (high level of the aerobic enzyme succinic dehydrogenase or SDH) and anaerobic (high level of the anaerobic enzyme phosphofructokinase or PFK) energy transfer. These are the fast-oxidative-glycolytic or FOG fibers. Another subdivision, the **type IIb** fiber possesses the greatest anaerobic potential and is the "true" fast-glycolytic or FG fiber. The **type IIc** fiber is normally a rare and undifferentiated fiber that may be involved in re-innervation or motor unit transformation.[30]

DIFFERENCES BETWEEN ATHLETIC GROUPS

Several interesting observations can be made concerning muscle fiber types and the possible influence of specific training on fiber composition and metabolic capacity. For one thing, sedentary men and women as well as young children possess 45 to 55% slow-twitch fibers.[6,13] For fast-twitch fibers, the percentage is probably equally distributed between the **a** and **b** subdivisions. Although there are no gender differences in fiber distribution, the individual variation is large, especially among men. Generally the trend in one's muscle fiber type distribution is consistent throughout the body's major muscle groups. There is some debate as to whether the relative proportion of fiber type is altered by aging.[3,31] Certain patterns of fiber distribution are readily apparent among highly proficient athletes.[40] Successful endurance athletes generally demonstrate a predominance of slow-twitch fibers in the muscles activated in their sport. For successful sprint athletes, the fast-twitch muscle fiber predominates. This is shown in Figure 18–10 for top Scandinavian competitors who represent different sports. Athletic groups with the highest aerobic and endurance capacities, such as distance runners and cross-country skiers, also have the highest relative number of slow-twitch fibers, often as high as 90%! Weight lifters, ice-hockey players,

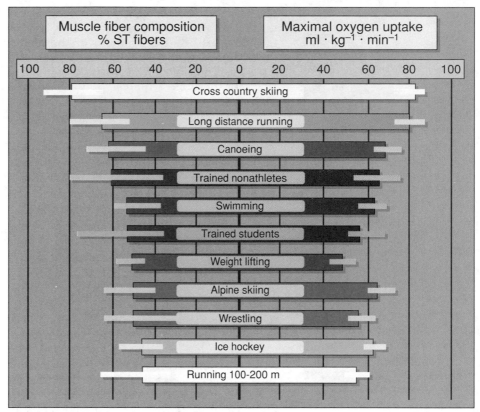

FIG. 18–10.
Muscle fiber composition (percent slow-twitch fibers) (left side) and maximal oxygen uptake (right side) in athletes representing different sports. The outer, lightly shaded bars denote the range. (From Bergh, U., et al.: Maximal oxygen uptake and muscle fiber types in trained and untrained humans. Med. Sci. Sports, *10:*151, 1978.)

and sprinters, on the other hand, tend to have more fast-twitch fibers and a relatively lower max $\dot{V}_{O_2}$.[7] As might be expected, men and women who perform in middle-distance events have an approximately equal percentage of the two types of muscle fibers.[37] This distribution also occurs for power athletes such as throwers, jumpers, and high jumpers.[12] These relatively clear-cut distinctions between performance and muscle fiber composition are for **elite** athletes who have achieved prominence in a specific sport category. A person's fiber composition, however, is clearly not the sole determinant of performance. Several researchers have shown that for a particular group, either trained or untrained, knowledge of a person's predominant fiber type is of limited value in predicting the outcome of specific exercise performances.[9,30] This is not surprising because performance capacity is the end result of the blending of many physiologic, biochemical, neurologic, and biomechanical "support systems"—and is not simply determined by a single factor such as muscle fiber type.[28]

In terms of muscle size, endurance athletes exhibit slow-twitch fibers of relatively normal size.[21] Weightlifters and other power athletes, on the other hand, show a definite enlargement, especially in the fast-twitch fibers.[41,42]

These fibers may be 45% larger than those of endurance athletes or of sedentary people of the same age.[14] This is because power and strength training induce a definite enlargement of the fiber's contractile apparatus—specifically the actin and myosin filaments as well as its total glycogen content.[32,33] The basic distinction between the genders is the generally larger muscle fibers of the male athletes.[12]

CAN FIBER TYPE BE CHANGED?

To determine whether the fiber composition characteristics of specific athletic groups are due to training or natural endowment (that is, can fiber composition be changed?), six men participated in a 5-month program of aerobic bicycle training.[17] Muscle biopsies from the lateral portion of the quadriceps before and after training indicated **no change** in fiber composition, although all men improved considerably in work capacity and aerobic power. Similar observations have been reported for the fiber composition of subjects after endurance or sprint training programs,[36] or after a period of weight training. These data are often used to support the argument that a fast-contracting fiber before training will still be a fast-contracting fiber after training with the same holding true for the slow-twitch fibers.

Additional studies with both humans and animals, however, suggest the possibility of changes in biochemical-physiologic properties of muscle fibers with a progressive transformation in fiber type with specific and chronic training.[1,5,19,34] In one study of 18 weeks of "aerobic" and 11 weeks of "anaerobic" training in four athletes,[25] the anaerobic training caused an **increase** in the percentage of type IIC fibers and a **decrease** in the percentage of type I fibers; the opposite was observed in the aerobic phase of the training sequence. Similarly, recent research has shown a 23% increase in the percentage of fast-twitch fibers and a commensurate decrease in slow-twitch fiber percentage after only 6 weeks of sprint training.[23] These findings suggest that specific training (and perhaps inactivity) may induce an **actual conversion** of type I to type II fibers (or vice versa).[38,40] It is clear that more research needs to be done in this intriguing area before definitive statements can be made concerning the fixed nature of a muscle's fiber composition. **At present, it appears that some transformation in muscle fiber type with chronic activity is in fact possible.**

METABOLIC ADAPTATIONS Are Real and Significant

In summary, considerable variation in fiber type distribution is noted from muscle to muscle and from person to person. These characteristics appear to be determined largely by genetic code with the major direction of a muscle's fiber composition probably being fixed before birth or early in life.[21,24] Whether this status can be modified with prolonged training is still open to question. It also seems likely that, at elite levels of certain sports performances, a particular fiber distribution is "required" for success. Although this suggests an obvious genetic predetermination, it is well documented that specific training significantly enhances aerobic and possibly anaerobic power of both fiber types.[18,24] In fact, enhancement of the oxidative capacity of fast-twitch fibers with high-intensity endurance training brings them to a level at which they are almost as well equipped for oxidative metabolism as are the slow-twitch fibers of untrained subjects.[10,25] This training adaptation in young and mature adults is brought about by the well-documented increase in mitochondrial size and number and the accompanying enhancement in

TABLE 18–1.
EFFECTS OF SPECIFIC FORMS OF TRAINING ON SKELETAL MUSCLE*

	Slow-Twitch Fibers		Fast-Twitch Fibers	
	Type of Training			
Muscle Factor	Strength	Endurance	Strength	Endurance
Percent composition	0 or ?	0 or ?	0 or ?	0 or ?
Size	+	0 or +	+ +	0
Contractile property	0	0	0	0
Oxidative capacity	0	+ +	0	+
Anaerobic capacity	? or +	0	? or +	0
Glycogen content	0	+ +	0	+ +
Fat oxidation	0	+ +	0	+
Capillary density	?	+	?	? or +
Blood flow during work	?	? or +	?	?

0 = no change; ? = unknown; + = moderate increase; + + = large increase.

* Modified from Gollnick, P.D., and Sembrowich, W.L.: Adaptations in human skeletal muscle as a result of training. In Exercise in Cardiovascular Health and Disease. Edited by E.A. Amsterdam et al. New York, Yorke Medical Books, 1977.

the quantity of enzymes relevant to Krebs cycle and electron transport function.[22] Individuals who have adapted to endurance training may show some conversion in the type IIb fiber to the more aerobic type IIa fiber.[10] This is accompanied by a large increase in mitochondrial content and aerobic enzyme levels in these specific fibers.

The fact that **only** the specific muscles (more precisely, muscle fibers) used in training show adaptation to exercise certainly explains why highly trained athletes who switch to a sport requiring different muscle groups feel essentially untrained for the new activity. Within this framework, swimmers or canoeists will not necessarily transfer their upper body "fitness" to a running sport.

The changes that occur in skeletal muscle from specific training are summarized in Table 18–1.[20] It should be noted that both fiber types are involved in most activities; it is just that certain activities require activation of a much greater proportion of one fiber type over another.

SUMMARY

1. Skeletal muscle is encased in various wrappings of connective tissue. These eventually blend into and join the tendinous attachment to bone. This harness enables muscles to act on the bony levers to transform the chemical energy of ATP into mechanical energy and motion.
2. Seventy-five percent of skeletal muscle is water, 20% is protein, and the remainder consists of inorganic salts, enzymes, pigments, fats, and carbohydrates.
3. In vigorous exercise, the muscle's oxygen consumption increases nearly 70 times above the resting level. Supporting this metabolic requirement are immediate adjustments and longer term training adaptations in the local vascular bed.
4. The sarcomere is the functional unit of the muscle cell. It contains the

contractile proteins actin and myosin. There are 4500 sarcomeres and a total of 16 billion thick (myosin) and 64 billion thin (actin) filaments in an average-sized fiber.

5. Projections or "cross-bridges" provide the structural link between the thin and thick contractile filaments. Tropomyosin and troponin, two proteins of the myofibrillar complex, regulate the make-and-break contacts between the filaments during contraction. Tropomyosin inhibits actin and myosin interaction; troponin with calcium triggers the myofibrils to interact and slide past each other.

6. The triad and T-tubule system serve as a microtransportation network for spreading the action potential from the fiber's outer membrane inward to deep regions of the cell. Contraction occurs when calcium activates actin, causing the myosin cross-bridges to attach to active sites on the actin filaments. Relaxation occurs when calcium concentration decreases.

7. The "sliding filament theory" proposes that a muscle shortens or lengthens because the protein filaments slide past each other without changing their length. Excitation-contraction coupling is the mechanism by which electrochemical and mechanical events are linked to achieve muscular contraction.

8. Two types of muscle fibers can be classified by their contractile and metabolic characteristics: (1) fast-twitch fibers, in which energy is predominantly generated anaerobically and rapidly for a quick, powerful contraction (these are labeled FG fibers to signify their fast speed of contraction and high glycolytic properties), and (2) slow-twitch fibers that contract relatively slowly and generate energy for ATP resynthesis predominantly via aerobic metabolism. These are called *SO* fibers to denote their slow contraction speed and reliance on oxidative metabolism. An intermediate, fast-oxidative-glycolytic (FOG) fiber is also present.

9. The percentage distribution of fiber type differs significantly among people. This distribution is probably largely determined by genetic code although there is indication that some modification may take place with physical training.

10. Both types of fibers can be markedly improved in metabolic capacity by specific endurance and power training.

REFERENCES

1. Aitken, J.C., et al.: The effects of high intensity training upon respiratory gas exchanges during fixed term maximal incremental exercise in man. *Eur. J. Appl. Physiol.*, 58:717, 1989.

2. Andersen, P., and Henrickson, J.: Capillary supply of the quadriceps femoris muscle of man: adaptive response to exercise. *J. Physiol. (London)*, 270:677, 1977.

3. Aniansson, A., et al.: Muscle morphology, enzymatic activity, and muscle strength in elderly men: A follow-up study. *Muscle Nerve*, 9:585, 1986.

4. Armstrong, R.B.: Muscle fiber recruitment patterns and their metabolic correlates. In *Exercise, Nutrition, and Energy Metabolism.* Edited by E.S. Horton and R.L. Terjung, New York, Macmillan, 1988.

5. Baldwin, K.M., et al.: Biochemical properties of overloaded fast-twitch skeletal muscle. *J. Appl. Physiol.*, 52:457, 1982.

6. Bell, R.D., et al.: Muscle fiber types and morphometric analysis of skeletal muscle in six-year-old children. *Med. Sci. Sports*, 12:28, 1980.

7. Bergh, U., et al.: Maximal oxygen uptake and muscle fiber types in trained and untrained humans. *Med. Sci. Sports,* 10:151, 1978.
8. Brodal, P., et al: Capillary supply of skeletal muscle fibers in untrained and endurance trained men. *Acta Physiol. Scand. (Suppl.)* Vol 440, 1976.
9. Campbell, C.J., et al.: Muscle fiber composition and performance capacities of women. *Med. Sci. Sports,* 11:260, 1979.
10. Chi, M.M.-Y., et al.: Effects of detraining on enzymes of energy metabolism in individual human muscle fibers. *Am. J. Physiol.,* 244(Cell Physiol. 13):276, 1983.
11. Clarkson, P.M., et al.: Plantar flexion fatigue and muscle fiber type in power and endurance athletes. *Med. Sci. Sports Exerc.,* 12:262, 1980.
12. Costill, D.L., et al.: Skeletal muscle enzyme and fiber composition in male and female track athletes. *J. Appl. Physiol.,* 40:149, 1976.
13. Dudley, G.A., et al.: Muscle fiber composition and blood ammonia levels after intense exercise in humans. *J. Appl. Physiol.,* 54:582, 1983.
14. Edström, L., and Ekblom, B.: Differences in sizes of red and white muscle fibers in vastus lateralis of musculus quadriceps of normal individuals and athletes: relation to physical performance. *Scand. J. Clin. Lab. Invest.,* 30:175, 1972.
15. Essen, B., et al.: Metabolic characteristics of fiber types in human skeletal muscles. *Acta Physiol. Scand.,* 95:153, 1975.
16. Faulkner, J.A., et al.: Contractile properties of isolated human muscle preparation. *Clin. Sci.,* 57:20, 1979.
16a. Gaffney, F.A.: Cardiovascular and metabolic responses to static contraction in man. *Acta Physiol. Scand.,* 138:249, 1990.
17. Gollnick, P.D.: Effects on enzyme activity and fiber composition of human skeletal muscle. *J. Appl. Physiol.,* 34:107, 1973.
18. Gollnick, P.D., and Hermansen, L.: Biochemical adaptations to exercise. Anaerobic metabolism. In *Exercise and Sport Sciences Reviews.* Edited by J.H. Wilmore. New York, Academic Press, 1973.
19. Gollnick, P.D., and Saltin, B.: Significance of skeletal muscle oxidative enzyme enhancement with endurance training. *Clin. Physiol.,* 2:1, 1982.
20. Gollnick, P.D., and Sembrowich, W.L.: Adaptations in human skeletal muscle as a result of training. In *Exercise in Cardiovascular Health and Disease.* Edited by E.A. Amsterdam, et al. New York, Yorke Medical Books, 1977.
21. Gollnick, P.D., et al.: Fiber number and size in overloaded chicken anterior latissimus dorsi muscle. *J. Appl. Physiol.,* 54:1292, 1983.
22. Holloszy, J.O., and Coyle, E.F.: Adaptations of skeletal muscle to endurance training and their metabolic consequences. *J. Appl. Physiol.,* 56:831, 1984.
23. Jacobs, I.: Sprint training effects on muscle myoglobin, enzymes, fiber types, and blood lactate. *Med. Sci. Sports Exerc.,* 19:368, 1987.
24. Jansson, E., and Kaijser, L.: Muscle adaptation to extreme endurance training in man. *Acta Physiol. Scand.,* 100:315, 1977.
25. Jansson, E., et al.: Changes in muscle fiber type distribution in man after physical training. *Acta Physiol. Scand.,* 104:235, 1978.
26. Karlsson, J., and Jacobs, I: Onset of blood lactate accumulation during muscular exercise as a threshold concept. I. Theoretical considerations. *Int. J. Sports Med.,* 3:190, 1982.
27. Karlsson, J.B., et al.: LDH isozymes in skeletal muscles of endurance and strength trained athletes. *Acta Physiol. Scand.,* 93:150, 1975.
28. Klausen, K., et al.: Adaptative changes in work capacity, skeletal muscle capillarization and enzyme levels during training and detraining. *Acta Physiol. Scand.,* 113:9, 1981.
29. Klug, G.A., and Tibbits, G.F.: The effect of activity on calcium mediated events in striated muscle. In *Exercise and Sport Sciences Reviews.* Vol. 16. Edited by K.B. Pandolf. New York, Macmillan, 1988.
30. Komi, P.V., and Karlsson, J.: Skeletal muscle fiber types, enzyme activities and physical performance in young males and females. *Acta Physiol. Scand.,* 103:210, 1978.
31. Larsson, L.: Histochemical characteristics of human skeletal muscle during aging. *Acta Physiol. Scand.,* 117:469, 1983.
32. MacDougall, J.D., et al.: Biochemical adaptation of human skeletal muscle to heavy resistance training and immobilization. *J. Appl. Physiol.,* 43:700, 1977.

33. MacDougall, J.D., et al.: Mitrochondrial volume density in human skeletal muscle following heavy resistance training. *Med. Sci. Sports,* 11:164, 1979.

34. Pette, Q., and Vrbova, G.: Invited review: Neural control of phenotypic expression in mammalian muscle fiber. *Muscle Nerv.,* 8:676, 1985.

35. Saltin, B.: Muscle fiber recruitment and metabolism in prolonged exhaustive dynamic exercise. In *Human Muscle Fatigue: Physiological Mechanisms.* London, Pitman Medical, 1981.

36. Saltin, B., et al.: The nature of the training response; peripheral and central adaptations to one-legged exercise. *Acta Physiol. Scand.,* 96:289, 1976.

37. Saltin, B., et al.: Fiber types and metabolic potentials of skeletal muscles in sedentary man and endurance runners. *Ann. N.Y. Acad. Sci.,* 301:3, 1977.

38. Simoneau, J.-A., et al.: Human skeletal muscle fiber type alteration with high-intensity intermittent training. *Eur. J. Appl. Physiol.,* 54:240, 1985.

39. Terjung, R.L., and Engbretson, B.M.: Blood flow to different rat skeletal muscle fiber type sections during isometric contractions in situ. *Med. Sci. Sports Exerc.,* 20:S124, 1988.

40. Tesch, P.A., and Karlsson, J.: Muscle fiber type and size in trained and untrained muscles of elite athletes. *J. Appl. Physiol.,* 59:1716, 1985.

41. Tesch, P.A., and Larsson, L.: Muscle hypertrophy in body builders. *Eur. J. Appl. Physiol.,* 49:301, 1982.

42. Thorstensson, A.: Muscle strength, fiber types and enzyme activities in man. *Acta Physiol. Scand.* (Suppl.) Vol. 443, 1976.

43. Vander, A.J., et al.: *Human Physiology: The Mechanisms of Body Function.* 3rd ed. New York, McGraw-Hill, 1985.

19

NEURAL CONTROL OF HUMAN MOVEMENT

The correct application of force in relatively complex, learned movements like a tennis serve or the shot put depends on a series of coordinated neuromuscular patterns and not **just** on the strength of the muscle groups recruited for the activity. Such movements are regulated by neural control mechanisms linked together by pathways in the central nervous system. This neural circuitry in the brain and spinal cord is somewhat analogous to a modern computer system, although the integrative and organizational structure of the nervous system is far more highly advanced and specialized. In response to changing internal and external stimuli, bits of sensory input are automatically and rapidly transmitted for processing by the neural control mechanisms. The input is properly organized, routed, and retransmitted to the effector organs, the muscles.

In the sections that follow, we present a general outline describing the neural control of human movement. This outline includes (1) the structural organization for motor control, (2) neuromuscular transmission, (3) the motor unit—the functional unit of neuromuscular activity, and (4) sensory input for muscular activity. For a more thorough discussion of neuromotor regulation and skill acquisition, texts dealing with neuroanatomy, neurophysiology, motor-skill learning, and motor control should be consulted.

ORGANIZATION OF THE NEUROMOTOR SYSTEM

CENTRAL ORGANIZATION

Figure 19–1 presents a general schema for the various subdivisions of the nervous system that regulate and program the sequential patterns required for motor control. Each block represents a major neural subdivision. The arrows show the direction of nerve impulses along the motor (black) or **efferent** and sensory (red) or **afferent** pathways. The circular insert illustrates the basic control mechanism of the motor system. In this closed loop system, the afferent neurons transmit sensory impulses from muscle and also interface with motor neurons that send impulses to muscle.

Tracts of nerve tissue descend from the brain and terminate at neurons in the spinal cord. Two major pathways serve this function, the **pyramidal** and **extrapyramidal** tracts.

Pyramidal Tract

Nerves in the pyramidal or **corticospinal** tract transmit their impulses downward through the spinal cord. By means of direct routes and intercon-

FIG. 19–1.
Simplified block diagram of the nervous system related to motor control. The interconnections between the various subdivisions are indicated by motor (black) and sensory (red) motoneurons. The circular insert represents the closed loop that links each muscle with the spinal cord. The arrows show the direction of the neural impulses between subdivisions.

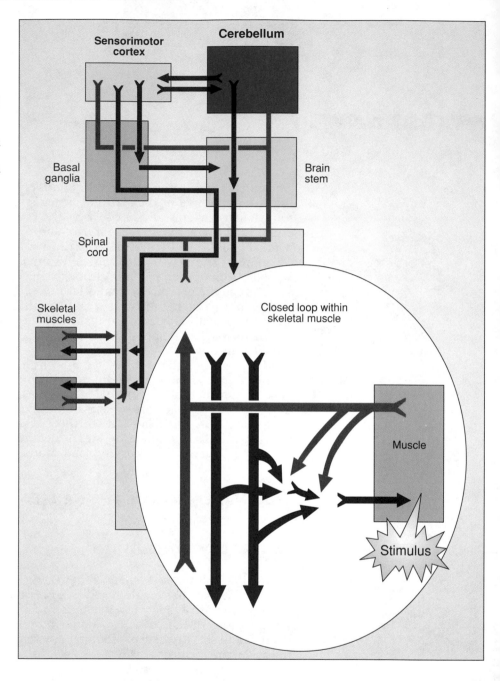

necting neurons in the spinal cord, these nerves eventually excite **alpha (α) motoneurons** that control the various skeletal muscles.

Extrapyramidal Tract

Each of the descending extrapyramidal nerve tracts is named for its area of origin and final connection. For example, nerves in the **vestibulospinal** tract originate in the vestibular nucleus of the cerebellum, whereas the

rubrospinal tract nerves originate in the red nucleus of the mid-brain, and the **reticulospinal** tract nerves have their origin in the reticular formation, a mass of nervous tissue that passes through the brain stem. The neurons of the extrapyramidal tract essentially control posture and provide a continual background level of neuromuscular tone. This is in contrast to the discrete movements stimulated by the nerves in the pyramidal tract.

The extrapyramidal nerves originate in the brain stem and connect at all levels of the spinal cord. The **reticular formation** provides important interconnections between the spinal cord, cerebral cortex, basal ganglia, and cerebellum. It integrates various input that flows through it. This input originates from the stretching of sensors in joints and muscles, or from pain receptors in the skin, visual signals from the eye, or auditory impulses from the ear. Once activated, the reticular system produces either an inhibitory or facilitory effect on other neurons.

The reticular inhibitory center transmits impulses that inhibit neurons to the antigravity muscles involved in postural control. Excitation of the facilitory sensory neurons arouses the reticular nerve cells. This causes excitation in the cerebral cortex, and signals are transmitted back to the reticular system to maintain an appropriate level of cortical arousal. Superimposed on this feedback system is another feedback network that transmits impulses through the spinal cord to the muscles. For example, if the neural outflow goes to the postural muscles, the tension of these muscles then becomes increased. This increased "neuromuscular tone" also stimulates the muscle's own set of sensory modulators, the spindles, to redirect excitatory impulses back to the central nervous system to maintain the excitatory level of the reticular formation. Such a system of multiple feedback control is one of the most complex aspects of the nervous system.

The **basal ganglia** are made up of masses of nerves that receive descending connections from the cortex and pass them through to the brain stem. Although the exact functions of the basal ganglia are unknown, it is believed that this nerve tissue is involved in the coordination of movement at both the volitional and subconscious levels.

The **cerebellum,** located behind the brain stem, functions by means of intricate feedback circuits to monitor and coordinate other areas of the brain involved in motor control. It receives signals concerning motor output from the cortex and sensory information from receptors in muscles, tendons, joints, and skin, as well as from visual, auditory, and vestibular end-organs. The cerebellum influences all motor centers from the cortex down to the spinal cord. It provides a special damping function for movement that would otherwise be jerky. **This specialized brain tissue is the major comparing, evaluating, and integrating center for postural adjustments, locomotion, maintenance of equilibrium, perceptions of speed of body movement, and many other reflex functions related to movement. In essence, it provides the "fine tuning" for muscular activity.**

THE REFLEX ARC

The diagram in Figure 19–2 shows a typical neural arrangement for a **reflex arc** in one of the 31 spinal cord segments. Sensory input is transmitted from the receptor by sensory (afferent) nerves that enter the spinal cord through the dorsal or sensory root. These nerves interconnect or synapse in the cord through **interneurons** that serve as relay stations to distribute information to various levels of the cord. The impulse is then passed over the **motor root pathway** via anterior motoneurons to the effector organ, the muscles.

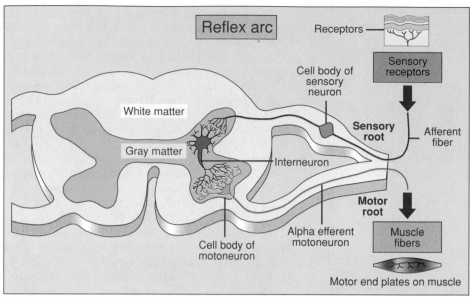

FIG. 19–2.
Afferent and efferent motoneurons in a spinal cord segment. The shaded or gray matter contains the neuron cell bodies; the white matter is made up of longitudinal columns of nerve fibers. Stimulation of a single alpha motoneuron could affect as many as 3000 muscle fibers. The motoneuron and the fibers it innervates are collectively referred to as a motor unit. Only one side of the spinal-nerve complex is shown.

The operation of the reflex arc becomes evident when one unknowingly touches a hot object. Pain receptors in the fingers are stimulated and send sensory information rapidly over afferent fibers to the spinal cord. Here, the efferent or motor fibers are activated to bring about the appropriate muscular response, and the hand is rapidly pulled away. Concurrently, the signal is transmitted up the cord to sensory areas in the brain in which the sensation of pain is actually "felt." These various levels of operation for sensory input, processing, and motor output, including the reflex action just described, account for the removal of the hand from the hot object before the pain is actually perceived. Many muscle functions are controlled by reflex actions in the spinal cord and other subconscious areas of the central nervous system.

NERVE SUPPLY TO MUSCLE

One nerve or its terminal branches innervates at least one of the approximately 250 million muscle fibers in the human body. Because there are only about 420,000 motor nerves, this means that a single nerve usually supplies many individual muscle fibers. **The ratio of muscle to nerve is generally related to a muscle's particular movement function.** The delicate and precise work of the eye muscles, for example, requires a neuron control fewer than 10 muscle fibers. For less complex movements of the big muscles, a motoneuron may innervate as many as 2000 or 3000 fibers. In terms of muscular activity, the spinal cord is the major processing center for motor control. The next sections take a closer look at how information processed

FIG. 19–3.
The anterior motoneuron.

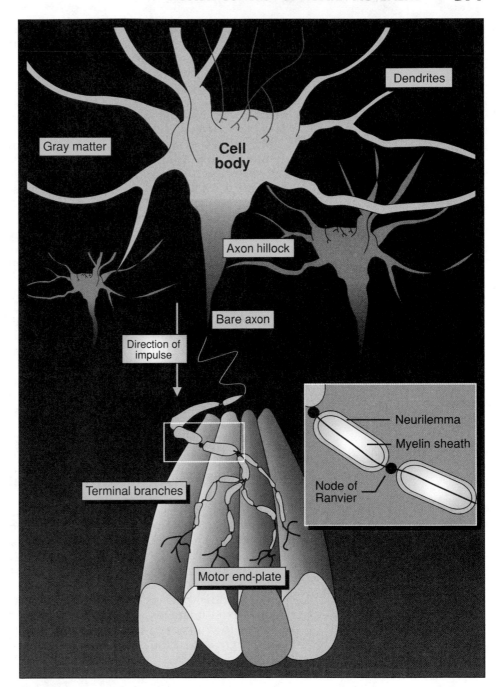

in the central nervous system is delivered to the muscles to bring about an appropriate motor response.

THE ANTERIOR MOTONEURON

The anterior motoneuron illustrated in Figure 19–3 consists of a **cell body, axon,** and **dendrites.** Its unique design enables it to transmit an electrochemical nerve impulse from the spinal cord to the muscle. The cell body houses the control center—the structures involved with replication and transmission of the genetic code. This part of the motoneuron is located within the gray matter of the spinal cord. The axon extends from the cord to deliver the impulse to the muscle; the dendrites are the short neural branches that receive impulses through numerous connections and conduct them toward the cell body. Nerve cells conduct impulses in one direction— down the axon away from the point of stimulation.

The larger nerve fibers are encased in a **myelin sheath,** a lipid-protein membrane that wraps around the axon over most of its length. A specialized cell known as a **Schwann cell** encases the bare axon and then spirals around it, sometimes up to 100 times in the biggest fibers. Myelin forms a large part of this sheath and insulates the axon. A thinner membrane, the **neurilemma,** covers the myelin sheath. The Schwann cells and myelin are interrupted every one or two millimeters along the axon's length at the **nodes of Ranvier.** Whereas the myelin sheath insulates the axon to the flow of ions, the nodes of Ranvier permit depolarization of the axon to occur. This alternating sequence of myelin sheath and nodes of Ranvier permits impulses to "jump" from node to node as the electrical current travels toward the terminal branches at the **motor end-plate.** This means of conduction is responsible for the higher transmission velocity in myelinated than in unmyelinated fibers. In fact, the speed of conduction in a nerve fiber is proportional to its diameter and to the thickness of the myelin sheath.

The anterior motoneurons are also known as **type A α nerve fibers.** Their diameter is large, ranging from about 8 to 20 μm. Other smaller type A fibers are known as **gamma (γ) efferent motoneurons.** They have a diameter no larger than about 10 μm and a conduction velocity about one half that of the larger α fibers. As is discussed in the section on proprioception, the γ fibers connect with special stretch sensors in skeletal muscle that facilitate detecting minute changes in the length of muscle fibers.

NEUROMUSCULAR JUNCTION (MOTOR ENDPLATE)

The interface between the end of a myelinated motoneuron and a muscle fiber is known as the **neuromuscular junction** or **motor end-plate.** Its function is to transmit the nerve impulse to the muscle. For each skeletal muscle fiber, there is usually only one neuromuscular junction. Figure 19–4 illustrates the details of the neuromuscular junction based on electron-microscopic studies.

The terminal portion of the axon below the myelin sheath forms several smaller axon branches whose endings are the **presynaptic terminals.** They lie close to, but not in contact with, the sarcolemma of the muscle fiber. The invaginated region of the **postsynaptic membrane** or **synaptic gutter** has many infoldings that increase its surface area. The region between the synaptic gutter and presynaptic terminal of the axon is called the **synaptic cleft.** The transmission of the neural impulse takes place in this region.

The neurotransmitter responsible for changing a basically electrical neural impulse into a chemical stimulus at the motor endplate is **acetylcholine.** It is released from small, saclike vesicles within the terminal axons. When an impulse arrives at the neuromuscular junction, acetylcholine is released into the synaptic cleft and combines with a transmitter-receptor complex in the postsynaptic membrane. This complex increases the postsynaptic membrane's permeability to sodium and potassium ions that ultimately causes

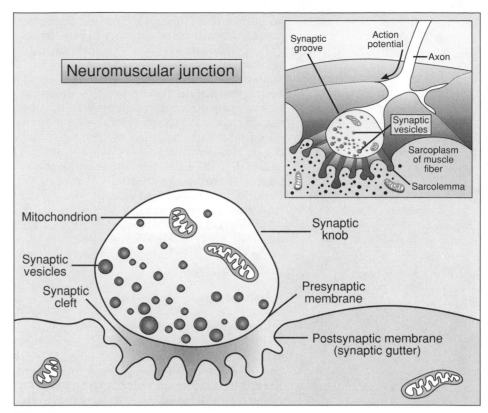

FIG. 19—4.
Microanatomy of the neuromuscular junction. Insert displays the details of the pre- and postsynaptic contact area between the motoneuron and the muscle fiber it innervates. (From Akert, K., et al.: Freeze etching and cytochemistry of vesicles and membrane complexes in synapses of the central nervous system. In Structures and Functions of Synapses. Edited by G.D. Pappas and D.P. Purpua. New York, Raven Press, 1972.)

the membrane to become depolarized. The impulse then begins to spread over the entire muscle fiber.

Within about 5 milliseconds after acetylcholine is released from the synaptic vesicles, it is destroyed by the enzyme **cholinesterase,** which is concentrated at the borders of the synaptic cleft. The destruction of acetylcholine by cholinesterase allows the postsynaptic membrane to repolarize. Acetic acid and choline, the by-products of cholinesterase action, can be taken up by the axon and resynthesized to acetylcholine so the entire process can begin again with the arrival of another nerve impulse.

Excitation

Excitation occurs **only** at the neuromuscular junction. The change in the electrical properties of the postsynaptic membrane elicits an **end-plate potential** that spreads from the motor end-plate to the extrajunctional sarcolemma. It then causes an **action potential** to travel the length of the fiber and down the T tubules. Once this occurs, the contractile machinery of the muscle fiber is primed for its major function—to contract.

Acetylcholine is probably one of the interneuronal excitatory transmitter substances. When it is released from the synaptic vesicles, it excites the postsynaptic membrane of its neighboring neuron. This changes the membrane permeability and permits sodium ions to diffuse rapidly to the inside of a neuron. An action potential is generated if the change in microvoltage is sufficient to reach the **threshold for excitation.** This change in membrane potential at the junction between two neurons (which increases the positive charges inside the cell) is referred to as the **excitatory postsynaptic potential** or **EPSP.** If the EPSP is subthreshold, the neuron does not discharge, but its resting membrane potential is lowered and its tendency to "fire" is temporarily increased. The neuron fires when many subthreshold excitatory impulses arrive in rapid succession. In the terms of neural activity, this condition is known as **temporal summation. Spatial summation** occurs when different presynaptic terminals on the same neuron are stimulated at the same time. Their individual effects are "summed" and an action potential is initiated.

Inhibition

Some presynaptic terminals set up inhibitory impulses. The inhibitory transmitter substance increases the permeability of the postsynaptic membrane to potassium and chloride ions. This produces an increase in the membrane's electrical potential, creating an **inhibitory postsynaptic potential** or **IPSP.** The IPSP hyperpolarizes the neuron, making it more difficult to fire. No action potential is generated if a motoneuron is subjected to both excitatory and inhibitory influences and if the IPSP is large. The reflex to pull one's hand away when removing a splinter, for example, can usually be overridden (inhibited) so the hand can be steadied to expedite this rather painful task.

The exact neurochemical that provokes an IPSP is unknown, although gamma aminobutyric acid (GABA) and the protein glycine are thought to be involved in the inhibitory process. Neural inhibition serves protective functions and also reduces the input of "unwanted" stimuli so that smooth, purposeful responses can occur. Removing inhibitory influences is important under certain exercise conditions. In all-out strength and power activities, for example, the ability to "disinhibit" and maximally activate all motoneurons required for a movement may be crucial to topflight performance.[17] This enhanced disinhibition could lead to full activation of muscle groups during all-out contraction, and may account for the rapid and highly specific strength increases noted in the early stages of a strength development program.[8,9,13,18] In fact, significant improvements in strength have been noted without increases in muscle size.[22] Central nervous system excitation (neuronal facilitation) is perhaps the mechanism by which intense concentration or "psyching" enhances maximal performance. The "psychologic" influence on strength performance is discussed in Chapter 22.

MOTOR UNITS

Although each muscle fiber generally receives only one nerve fiber, a motor nerve may innervate many muscle fibers. This is because the terminal end of an axon forms numerous branches. **The anterior motoneuron and the**

specific muscle fibers it innervates are called a motor unit. This is the functional unit of neuromuscular control. Some motor units contain up to 3,000 muscle fibers whereas others contain relatively few. For example, the first dorsal interosseous muscle of the finger contains 120 motor units that control 41,000 fibers; the medial gastrocnemius muscle (calf) has 580 motor units and 1,030,000 muscle fibers. The ratio of muscle fibers per motor unit is therefore 340 for the finger muscle and 1900 for the gastrocnemius muscle.[4]

All-Or-None Principle

If the stimulus is strong enough to trigger an action potential in the motoneuron, **all** of the accompanying muscle fibers in the motor unit are stimulated to contract synchronously. There is no such thing as a strong or weak contraction from a motor unit—either the impulse is strong enough to elicit a contraction or it is not. Once the neuron is "fired" and the impulse reaches the neuromuscular junction, the muscle cells always contract. This is the principle of "all-or-none" in relation to the normal action of skeletal muscle.

Gradation of Force

How then is the force of contraction varied from slight to maximal? This occurs in two ways: (1) increasing the number of motor units recruited for the activity, and (2) increasing their frequency of discharge. Clearly, if all motor units are active, the force generated is considerable compared to that generated by the activation of only a few. Also, if repetitive stimuli reach a muscle before it has relaxed, the total tension produced is increased. By blending these two factors, recruitment of motor units and the rate of their firing, optimal patterns of neural discharge permit a wide variety of graded contractions.

Characteristics of Motor Units

Motor units are comprised of fibers of one specific fiber type or subdivision of a particular fiber type that have the same metabolic profile.[14] Consequently, these units can be classified into one of three categories depending on their speed of contraction, the amount of force they generate, and the relative fatigability of the fibers. Figure 19–5 illustrates these characteristics for the three categories of motor units: (1) fast twitch, high force, and high fatigue (type IIb); (2) fast twitch, moderate force, and fatigue resistant (type IIa); (3) slow twitch, low tension, and fatigue resistant (type I).

The fast-twitch fibers are innervated by relatively large motoneurons with fast conduction velocities. This motor unit contains between 300 and 500 muscle fibers. These units reach greater peak tension, and develop it nearly twice as fast as slow-twitch motor units. The slow-twitch motor units are innervated by small motoneurons with slow conduction velocities. These units are much more fatigue-resistant than fast-twitch units. It should be recalled from our discussion of muscle fiber type, however, that the particular metabolic characteristics of **all** fibers can be modified by specific endurance training. **With prolonged training, some fast-twitch units can become almost as fatigue-resistant as the slow-twitch units.**

There is some evidence that the particular neurons themselves have a trophic or stimulating effect on the muscle fibers they innervate in a way

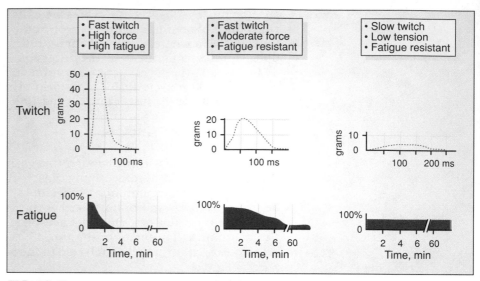

FIG. 19–5.
Speed, force, and fatigue characteristics of motor units. Motoneurons that fire rapidly with short bursts are termed phasic; those that fire slowly but continuously are tonic. (Modified from Edington, D.W., and Edgerton, V.R.: The Biology of Physical Activity. Boston, Houghton Mifflin Co., 1976.)

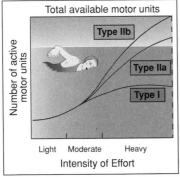

FIG. 19–6.
Recruitment of slow-and fast-twitch muscle fibers (motor units) in relation to the intensity of exercise. As the effort becomes more intense, progressively more fast-twitch fibers are recruited. (Based on the findings of Edgerton, V.R.: Mammalian muscle fiber types and their adaptability. Am. Zoology, *18*:113, 1978.)

that influences the fibers' growth and development.[7] Innervating fast-twitch fibers with the neuron from a slow-twitch motor unit, for example, eventually alters the twitch characteristics of the fast fibers. If this neurotrophic effect is as great as believed, then the myoneural junction takes on much greater significance than only being the site of muscle depolarization. In addition, chronically stimulating a fast-twitch motor unit in a frequency and intensity similar to slow-twitch fiber activity transforms the basic property of the muscle fibers.[15] This suggests a certain plasticity in fiber type that can indeed be altered with chronic use.

Firing Pattern

As shown in Figure 19–6, not all of the motor units in a muscle fire at the same time. If they did, it would be virtually impossible to control the force of a contraction. This is easy to demonstrate if one considers the tremendous gradation of forces and speeds that muscles generate. For example, when lifting a barbell, specific muscles contract to move the limb and weight at some particular speed of movement under a given rate of tension development. If the weight is not too heavy, it can be lifted at a number of speeds. If a heavier weight is used, the speed options decrease accordingly. With a light object like a pencil, the proper force is generated by the muscles to lift the pencil, regardless of how fast or slow the arm is moved. **From the standpoint of neural control, the fast- and slow-twitch motor units are selectively recruited and modulated in their firing pattern to produce the desired response.** In muscular activity requiring contractions of increasing force the slow-twitch fibers or, more specifically, motor units with the lowest functional threshold, are selectively recruited predominantly during lighter effort. This is followed by activation of the more powerful, higher threshold, fast-twitch units when peak force is required.[5] During sustained activities such as jogging or cycling on a level grade or during slow swimming or lifting a light weight at a slow speed, motor units of slow-twitch fibers are selectively recruited; for rapid, powerful movements the fast-twitch fibers, especially the type IIb, come into play.[3,20] More than likely, as a runner reaches a hill during a distance race some fast-twitch units are also activated so that a constant pace is maintained over varying terrain.

The differential control of the motor unit firing pattern is probably the major factor that distinguishes not only skilled from unskilled performances, but also specific athletic groups. For example, weight lifters generally demonstrate a synchronous pattern of motor-unit firing (i.e., many motor units recruited simultaneously during lifting), whereas the firing pattern of endurance athletes is asynchronous (i.e., some units fire while others recover).[21] As discussed previously, the compositional characteristics of a muscle in terms of its specific motor units (muscle fibers) contribute to the performance characteristics of various athletes.[2,6,19] In addition, the synchronous firing of a muscle's fast-twitch fibers certainly aids the weight lifter to generate force quickly for the desired lift. For the endurance athlete, on the other hand, the asynchronous firing of predominantly slow-twitch, fatigue-resistant units provides a built-in recuperative period.

Neuromuscular Fatigue

Muscular fatigue is the result of many factors, each related to the specific demands of the exercise that produces it.[12] These factors can interact in a manner that ultimately affects either contraction or excitation or both. As was shown in Chapter 1, a significant reduction in muscle glycogen is related to fatigue during prolonged submaximal exercise. This "nutrient fatigue" occurs even though sufficient oxygen is available to generate energy through aerobic pathways. Muscle fatigue in short-term maximal exercise is associated with oxygen lack and an increased level of blood and muscle lactic acid, and the accompanying dramatic increase in H^+ concentration in the exercised muscle.[1,11,16] This anaerobic condition may cause drastic intracellular changes within the active muscles. These could include an interference in the contractile mechanism, a depletion of stored high-energy phosphates, an impaired energy transfer via glycolysis owing to reduced activity of key enzymes, a disturbance in the tubular system for transmitting the impulse throughout the cell, and ionic imbalances.[10] Certainly a change in Ca^{++} distribution could alter the activity of the myofilaments and impair muscular performance. This would cause fatigue even though nerve impulses continue to bombard the muscle fiber. Fatigue also can be demonstrated at the neuromuscular junction when an action potential fails to cross from the motoneuron to the muscle fiber. The precise mechanism is unknown for this aspect of "neural fatigue."

As muscle function becomes impaired during prolonged submaximal exercise, additional motor-unit recruitment takes place to maintain the required force output for the particular activity. In all-out exercise, when all motor units are presumably maximally activated, fatigue is accompanied by a decrease in neural activity (as measured by the electromyogram). The fact that neural activity decreases supports the argument that this form of fatigue is partially caused by a failure in neural or myoneural transmission.

RECEPTORS IN MUSCLES, JOINTS, AND TENDONS: THE PROPRIOCEPTORS

Specialized sensory receptors in the muscles, joints, and ligaments are sensitive to stretch, tension, and pressure. These end-organs, known as **proprioceptors,** rapidly relay information concerning muscular dynamics and limb movement to conscious and unconscious portions of the central nervous system for processing. Thus, the progress of any movement or sequence of movements is continually charted to provide the basis for modifying subsequent motor behavior.

MUSCLE SPINDLES

The **muscle spindles** provide sensory information concerning changes in the length and tension of muscle fibers. Their main function is to respond to stretch on a muscle and, through reflex action, to initiate a stronger contraction to reduce this stretch.

Structural Organization

As shown in Figure 19–7, the spindle is fusiform in shape and is attached in parallel to the regular or **extrafusal fibers** of the muscle. Consequently,

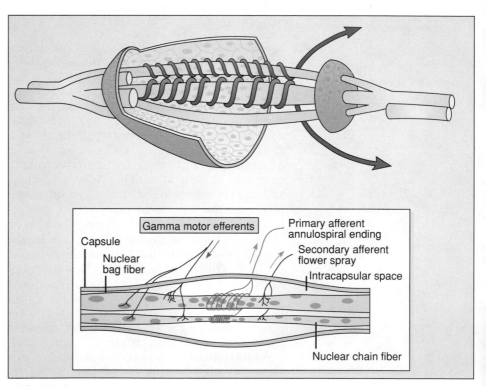

FIG. 19–7.
Structural organization of the muscle spindle. The insert shows an enlarged view of the equatorial region of the spindle. (Modified from Schade, J.P., and Ford, D.H.: Basic Neurology; 2nd ed. Amsterdam, Elsevier/North-Holland Biomedical Press, 1973.)

when the muscle is stretched so is the spindle. The number of spindles contained per gram of muscle varies widely depending on the muscle group. On a relative basis, there are more spindles in muscles requiring complex movements than in those requiring gross movements. The spindle is covered by a sheath of connective tissue. Within the spindle, there are two types of specialized muscle fibers called **intrafusal fibers.** One type, known as **nuclear bag fibers,** is fairly large and has numerous nuclei centrally packed throughout its diameter. There are usually two bag fibers per spindle. The other type of intrafusal fiber contains many nuclei along its length. These are the **nuclear chain fibers** attached to the surface of the longer nuclear bag fibers. There are usually four to five chain fibers in each spindle. The ends of the intrafusal fibers are striated (they contain actin and myosin) and are capable of contracting.

Three different nerve fibers service the spindles; two are afferent or sensory, one is efferent or motor. A primary afferent nerve fiber is entwined about the midregion of the bag fiber. This is the **annulospiral nerve fiber** that responds directly to the stretch of the spindle; its frequency of firing is proportional to the degree of stretch. A second group of smaller, sensory nerve fibers, known as **flower-spray endings,** make connections mainly on the chain fibers, although there are also attachments to the bag fibers. These endings are less sensitive to stretch than are the annulospiral fibers. Activation of the annulospiral and flower-spray sensors relays impulses through sensory roots into the cord to cause a reflex activation of the motoneurons to the muscle. This causes the muscle to contract more forcefully and to shorten; this in turn causes the stretch stimulus to be removed from the spindles.

The third type of spindle nerve fiber has a motor function. These are the thin **gamma efferent fibers** that innervate the contractile, striated ends of the intrafusal fibers. These fibers, activated by higher centers in the brain, provide the mechanism for maintaining the spindle at peak operation at all muscle lengths. Stimulation of the gamma efferents causes the ends of the spindles to contract, thereby regulating their length and sensitivity independent of the overall length of the muscle itself. This activation mechanism prepares the spindle for other lengthening reactions that are about to occur, even though the muscle may already be shortened to a new length. These adjustments enable the spindle to monitor continuously the length of the muscles in which they are located.

The Stretch Reflex

The functional significance of the muscle spindle is its ability to detect, respond to, and control changes in the length of extrafusal muscle fibers. This is important in the regulation of movement and maintenance of posture. Postural muscles are continuously bombarded by neural input; they must maintain their readiness to respond to voluntary movements or to maintain some degree of constant activity to counter the pull of gravity for upright posture. The stretch reflex is fundamental to achieving this end.

There are three main components to the stretch reflex: (1) the muscle spindle, the sensory receptor within the muscle that responds to stretch, (2) an afferent nerve fiber that carries the sensory impulse from the spindle to the spinal cord, and (3) an efferent motor neuron in the spinal cord that signals the muscle to contract.

Figure 19–8 illustrates schematically the neural pathways involved in this basic reflex. In part A, the biceps muscle is contracted to maintain the bony

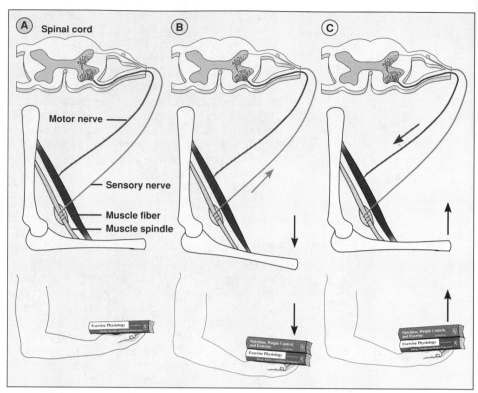

FIG. 19–8.
Schematic representation of the stretch reflex. Because the spindles are arranged in parallel with the extrafusal fibers, they become stretched when these fibers are stretched. This causes a stretch of intrafusal fibers, causing the spindle sensory receptors to fire. This discharge reflexly stimulates the alpha motoneurons to cause the extrafusal fibers to contract. This takes the stretch off of the intrafusal fibers and silences the spindle afferents. In the diagram, the stretch reflex acts as a self-regulating mechanism to maintain the relative constancy of limb position. (From Merton, P.O.: How we control the action of our muscles. Sci. Am., *226*:30, 1972.)

lever at a 90° angle while holding a 1.0-kg book. If the book is suddenly increased twofold in weight (part B), the muscle is stretched. This causes the spindles' sensory endings to direct impulses through the dorsal root into the spinal cord where they directly activate the motoneuron. The returning motor impulses (part C) cause the muscle to contract more forcefully to return the limb to its original position. During this reflex process, interneurons in the cord are also activated to facilitate the appropriate movement response. Excitatory impulses are conveyed to muscles that support the desired movement (**synergistic** muscles), whereas inhibitory impulses flow to the neurons of muscles that are **antagonists** of the movement. In this way, the stretch reflex acts as a self-regulating or compensating mechanism; it enables the muscle to adjust automatically to differences in load (and length) without immediately processing information through higher centers.

GOLGI TENDON ORGANS

Unlike the muscle spindles that lie parallel to the extrafusal muscle fibers, the **Golgi tendon organs** are connected in series to as many as 25 extrafusal

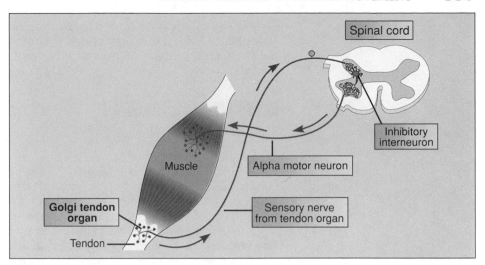

FIG. 19–9.
The Golgi tendon organ. Excessive tension or stretch on a muscle activates the Golgi receptors, which rapidly brings about a reflex inhibition of the muscles they supply. In this way, the Golgi tendon organs function as a protective sensory mechanism.

fibers. These sensory receptors are also located in the ligaments of joints and are mainly responsible for detecting differences in muscle tension rather than length. As shown in Figure 19–9, the Golgi tendon organs respond as a feedback monitor to discharge impulses under one of two conditions: (1) in response to tension created in the muscle when it shortens, and (2) in response to tension when the muscle is passively stretched.

When stimulated by excessive tension or stretch, the Golgi receptors conduct their signals rapidly to bring about a **reflex inhibition** of the muscles they supply. This occurs because of the over-riding influence of the inhibitory spinal interneuron on the motoneurons supplying the muscle. Thus, the Golgi tendon organ functions as a protective sensory mechanism. If the change in tension or stretch is too great, the sensor's discharge increases; this further depresses the activity of the motoneurons and reduces the tension generated in the muscle fibers. If the muscle contraction produces little tension, Golgi receptors are only weakly activated and exert little influence. **The ultimate function of the Golgi tendon organs is to protect the muscle and its connective tissue harness from injury due to an excessive load.**

PACINIAN CORPUSCLES

Pacinian corpuscles are small, ellipsoidal bodies located close to the Golgi tendon organs. These small, onion-like sensory receptors that are sensitive to quick movement and deep pressure are embedded in a single, nonmyelinated nerve fiber. Deformation or compression of the capsule by a mechanical stimulus transmits pressure to the sensory nerve endings within its core. This produces a change in the electric potential of the nerve ending. If this **generator potential** is of sufficient magnitude, a sensory signal is established and propagated down the myelinated axon leaving the corpuscle.

Pacinian corpuscles are "fast-adapting" mechanical sensors because they discharge a few impulses at the onset of a steady stimulus and then remain electrically silent or may discharge a second volley of impulses when the stimulus is removed. Consequently, they detect changes in movement or pressure, rather than how much movement occurred or how much pressure was applied.

▉ SUMMARY

1. Human movement is finely regulated by neural control mechanisms located in the central nervous system. In response to internal and external stimuli, bits of sensory input are automatically and rapidly routed, organized, and transmitted to the effector organs, the muscles.

2. Tracts of nerve tissue descend from the brain to influence neurons in the spinal cord. Neurons in the extrapyramidal tract control posture and provide a continual background level of neuromuscular tone; the pyramidal tract neurons provide for discrete muscular movement.

3. The cerebellum is the major comparing, evaluating, and integrating center that provides the "fine-tuning" for muscular activity.

4. Many muscular functions are controlled in the spinal cord and other subconscious areas of the central nervous system. The reflex arc is the basic mechanism for processing these automatic muscular movements.

5. The number of muscle fibers in a motor unit depends on the muscle's movement function. Intricate movement patterns require a small fiber-to-neuron ratio whereas for gross movements, a single neuron may innervate several thousand muscle fibers.

6. The anterior motoneuron (cell body, axon, and dendrites) transmits the electrochemical nerve impulse from the spinal cord to the muscle. The dendrites receive impulses and conduct them toward the cell body, whereas the axon transmits the impulse in one direction only—down the axon to the muscle.

7. The neuromuscular junction is the interface between the motoneuron and the muscle fiber. Acetylcholine is released here to provide the chemical stimulus that activates muscles.

8. Both excitatory and inhibitory impulses continually bombard the synaptic junctions between neurons. These alter a neuron's threshold for excitation by either increasing or decreasing its tendency to "fire." In all-out power exercise, a high degree of disinhibition is beneficial because it enables maximal activation of a muscle's motor units.

9. There are three types of motor units depending on speed of contraction, force generated, and fatigability: (1) fast twitch, high force, and high fatigue, (2) fast twitch, moderate force, and fatigue resistant, and (3) slow twitch, low tension, fatigue resistant.

10. Gradation of muscle force is accomplished through an interaction of factors that regulate the number and type of motor units recruited as well as their frequency of discharge. Light exercise is accomplished through predominant recruitment of slow-twitch units followed by activation of the fast-twitch units when a more powerful force is required.

11. Alterations in motor unit recruitment and firing pattern probably explain a large portion of the strength improvement with training, especially during the early stages of training.

12. Special sensory receptors in muscles, tendons, and joints relay information concerning muscular dynamics and limb movement to specific portions of the central nervous system. This provides important sensory feedback during physical activity.

REFERENCES

1. Costill, D.L., et al.: Leg muscle pH following sprint running. *Med. Sci. Sports Exerc.*, 15:325, 1983.
2. Costill, D.L., et al.: Skeletal muscle enzyme and fiber composition in male and female track athletes. *J. Appl. Physiol.*, 40:149, 1976.
3. Edgerton, V.R.: Mammalian muscle fiber types and their adaptability. *Am. Zoology*, 18:113, 1978.
4. Feinstein, B., et al.: Morphologic studies of motor units in normal human muscle. *Acta Anat. (Basel)*, 23:127, 1955.
5. Freund, H.J.: Motor unit and muscle activity in voluntary motor control. *Physiol. Rev.*, 63:387, 1983.
6. Gollnick, P.D., and Sembrowich, W.L.: Adaptations in human skeletal muscle as a result of training. In *Exercise in Cardiovascular Health and Disease.* Edited by E.A. Amsterdam, et al. New York, Yorke Medical Books, 1977.
7. Gutman, E.: Neurotrophic relations. *Ann. Rev. Physiol.*, 38:177, 1976.
8. Häkkinen, K., and Komi, P.V.: Electromyographic changes during strength training and detraining. *Med. Sci. Sports Exerc.*, 15:455, 1983.
9. Häkkinen, K., et al.: Effect of combined concentric and eccentric strength training and detraining on force-time, muscle fiber, and metabolic characteristics of leg extensor muscles. *Scand. J. Sports Sci.*, 3:50, 1981.
10. Hermansen, L.: Effect of metabolic changes on force generation in skeletal muscle during maximal exercise. In *Human Muscle Fatigue: Physiological Mechanisms.* London, Pitman Medical, 1981.
11. Hermansen, L., and Osnes, J.B.: Blood and muscle pH after maximal exercise in man. *J. Appl. Physiol.*, 32:304, 1972.
12. Maclaren, D.P.M., et al.: A review of metabolic and physiological factors in fatigue. In *Exercise and Sport Sciences Reviews.* Vol. 17. Edited by K.B. Pandolf. Baltimore, Williams & Wilkins, 1989.
13. Moratini, T., and DeVries, H.: Neural factors versus hypertrophy in the time course of muscle strength gain. *Am. J. Phys. Med.*, 58:115, 1979.
14. Nemete, P., et al.: Comparison of enzyme activities among single muscle fibers within defined motor units. *J. Physiol.*, 311:489, 1985.
15. Pette, D., and Vrbova, G.: Invited review: Neural control of phenotypic expression in mammalian muscle fibers. *Muscle Nerve*, 8:676, 1985.
16. Sahlin, K.: Intracellular pH and energy metabolism in skeletal muscle of man. *Acta Physiol. Scand.* 455(Suppl.):1, 1978.
17. Sale, D.G.: Influence of exercise and training on motor unit activation. In *Exercise and Sport Sciences Reviews.* Vol. 15. Edited by K.B. Pandolf. New York, Macmillan, 1987.
18. Sale, D.G., et al.: Neural adaptation to resistance training. *Med. Sci. Sports Exerc.*, 20:S135, 1988.
19. Saltin, B., et al.: Fiber types and metabolic potentials of skeletal muscles in sedentary man and endurance runners. *Ann. N.Y. Acad. Sci.*, 301:3, 1977.
20. Spectar, S.A., et al.: Muscle architecture and force-velocity characteristics of cat soleus and medial gastrocnemius: implications for motor control. *J. Neurobiol.*, 44:951, 1980.
21. Stepanov, A.S., and Burlakov, M.L.: Electrophysiological investigation of fatigue in muscular activity. *Sechenov Physiol. J. USSR*, 47:43, 1961.
22. Tesch, P.A., et al: Effects of strength training on G tolerance. *Aviat. Space Environ. Med.*, 54:691, 1984.

20 THE ENDOCRINE SYSTEM AND EXERCISE

The endocrine system helps to integrate and control bodily functions and thus provides stability or **homeostasis** in the internal environment. Hormones affect almost all aspects of human function. They regulate growth, development, and reproduction and augment the body's capability for dealing with both physical and psychologic stressors. They maintain internal equilibrium by adjusting electrolyte and acid-base balance and influence the specific metabolic mixture of fuels used to power all biologic work. The sections that follow provide a general review of various aspects of the endocrine system, its functions during rest, its contributions during physical activity, and its role in regulating energy metabolism.

ENDOCRINE SYSTEM OVERVIEW

The endocrine organs are quite small compared to other organs of the body; combined, they weigh only about 0.5 kg. Figure 20–1 shows the location of the major endocrine organs. They include the pituitary, thyroid, parathyroid, adrenal, pineal, and thymus glands. Several other body organs contain discrete areas of endocrine tissue that also produce hormones. These include the pancreas, gonads (ovaries and testes), and hypothalamus (also a major organ of the nervous system). Consequently, the hypothalamus is considered a **neuroendocrine organ.** Other organs and tissues also produce hormones. Pockets of hormone-producing cells are found in the walls of the small intestines, stomach, kidneys, and heart, although these organs have little to do with hormone production per se.

ENDOCRINE SYSTEM ORGANIZATION

THE HOST ORGANS

The endocrine system consists of a host organ (gland), minute quantities of chemical messengers (hormones), and a target or receptor organ. Glands can be classified as either **endocrine** or **exocrine.** Some glands can serve both functions.

Endocrine glands have no ducts and have often been called "ductless glands." They secrete their substances directly into the extracellular spaces around the gland. The hormones then diffuse into the blood for transport throughout the body. It is customary to use the term endocrine synonomously

FIG. 20–1.
Location of the major hormone producing endocrine organs.

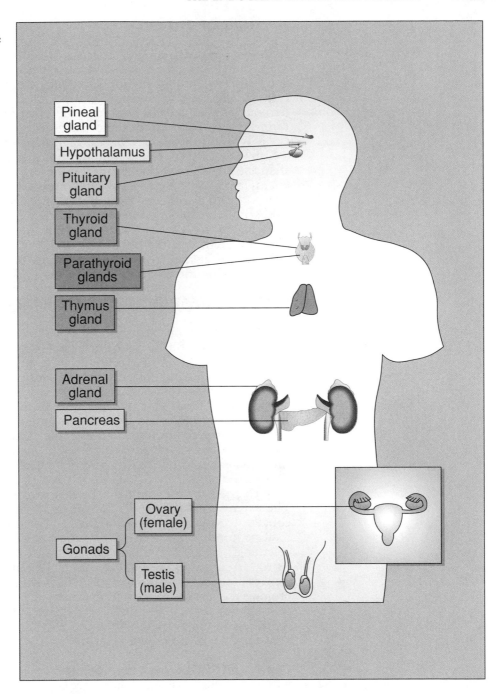

Pineal gland

Hypothalamus

Pituitary gland

Thyroid gland

Parathyroid glands

Thymus gland

Adrenal gland

Pancreas

Ovary (female)

Gonads

Testis (male)

with "hormone secreting." **Exocrine glands,** on the other hand, have secretory ducts that lead directly to a specific compartment or surface where the secreted material is needed. Almost all exocrine glands are under nervous system control. Examples of exocrine glands include sweat glands and glands of the upper digestive tract.

NATURE OF HORMONES

Hormones are chemical substances synthesized by a specific host gland and are secreted into the blood and carried throughout the body. Hormones generally fall into two distinct "chemical" categories: (1) hormones derived from **steroid** compounds, and (2) hormones that are amino acid or **polypeptide** derivatives. The steroid hormones are synthesized from circulating cholesterol by the adrenal cortex and the gonads, whereas other glands produce the polypeptide hormones that range in size from small to very large proteins.

Table 20–1 lists eight different hormones that are produced by organs other than the major endocrine glands. Of these, **prostaglandins** constitute a third chemical class of hormones. They are biologically active **lipids** found in nearly all cell membranes. Another example is **erythropoietin,** a glycoprotein that stimulates production of red blood cells by the bone marrow.

Although most hormones circulate in the blood as messengers that can affect different organs at a distance from the specific gland, other hormones, such as prostaglandins and the gastrointestinal hormone **gastrin,** exert local effects and are metabolized within a limited anatomic region.

HORMONE-TARGET CELL SPECIFICITY

The major function of hormones is to alter the **rates** of specific cellular reactions of specific "target cells." This is accomplished by altering the rate of synthesis of intracellular proteins, by changing the rate of enzyme activity, by modifying cell membrane transport, or by induction of secretory activity.

A target cell's ability to respond to a hormone depends largely on the presence of specific protein receptors on its membrane or in its interior to which that hormone can bind in a complementary way. Receptors for **adrenocorticotropic hormone (ACTH),** for example, are normally found only on certain cells of the adrenal cortex. In contrast, receptors for **thyroxine,** the principal hormone that stimulates cellular metabolism, are found on all cells of the body.

Hormone-receptor binding is the first step in determining hormone action. The extent of a target cell's activation by hormone-receptor interaction is equally dependent on three additional factors: blood hormone levels, the relative number of target cell receptors for that hormone, and the affinity or strength of the union between hormone and receptor. Receptors are dynamic structures. **Up-regulation** is the dynamic state in which target cells form more receptors in response to increasing hormone levels. In contrast, prolonged exposure to high hormone concentrations can desensitize target cells so they begin to respond less vigorously to hormonal stimulation. This **down-regulation** probably involves a loss in the number of receptors to prevent target cells from over-reacting to persistently high hormone levels.

Cyclic AMP: The Intracellular Messenger. The binding of a hormone with a receptor can change the target organ cell's permeability to a particular metabolite, or even change the ability of the cell to manufacture intracellular substances (primarily proteins). This will ultimately affect the level of cellular function. In many cases, the binding hormone acts as a "first messenger" to react with the enzyme **adenyl cyclase,** which is present in the cell membrane. This causes the formation of a compound called cyclic 3', 5'-adenosine monophosphate, or **cyclic AMP.** The cyclic AMP then acts as a "second messenger" to exert its influence on cellular function as a mediator-helping system with at least nine different hormone-receptor cells.[51]

TABLE 20–1.
HORMONES PRODUCED BY ORGANS OTHER THAN THE MAJOR ENDOCRINE ORGANS

Hormone	Composition	Source and Stimulus for Secretion	Target and Outcome
Prostaglandins	20-carbon fatty acid synthesized from arachidonic acid	Source: plasma membrane of different body cells Stimulus: local irritation, different hormones	Target: multiple sites Outcome: controls local hormone response; stimulates arterioles to increase blood pressure; increases uterine contractions, HCl and pepsin secretion in stomach, platelet aggregation, blood clotting, constriction of bronchioles, inflammation, pain, and fever
Gastrin	Peptide	Source: stomach Stimulus: food	Target: stomach Outcome: release of HCl
Enterogastrin	Peptide	Source: duodenum . Stimulus: food (especially fats)	Target: stomach Outcome: inhibits HCl secretion and gastrointestinal mobility
Secretin	Peptide	Source: duodenum Stimulus: food	Target: pancreas Outcome: release of bicarbonate-rich juice Target: liver Outcome: release bile Target: stomach Outcome: inhibits secretion
Cholecystokinin	Peptide	Source: duodenum Stimulus: food	Target: pancreas Outcome: release of bicarbonate-rich juice Target: gall bladder Outcome: expulsion of bile Target: sphincter of Oddi Outcome: relaxes sphincter and allows bile to enter duodenum
Erythropoietin	Glycoprotein	Source: kidney* Stimulus: hypoxia	Target: bone marrow Outcome: production of red blood cells
Active vitamin D₃	Steroid	Source: kidney activates vitamin D made by epidermal skin cells Stimulus: parathyroid hormone	Target: intestines Outcome: active transport of dietary Ca^{++} across intestinal membranes
Atrial natriuretic hormone	Peptide	Source: atrium of heart Stimulus: atrial stretching	Target: kidney Outcome: inhibits Na^+ reabsorption and renin release Target: adrenal cortex Outcome: inhibits secretion of aldosterone

* The kidneys release an enzyme that modifies a circulating blood protein to produce erythropoietin.

The sequence of biochemical reactions set into motion by cyclic AMP depends on the target cell type, the specific protein enzymes it contains, and the specific hormone that acts as first messenger. In thyroid cells, for example, the cyclic AMP generated in response to the binding of **thyroid-stimulating hormone** promotes the synthesis of the hormone **thyroxine.** In bone and muscle cells, on the other hand, the cyclic AMP produced by the action of **growth-hormone** binding activates anabolic reactions in which amino acids are built into tissue proteins.

Hormones can increase an enzyme's activity in one of several ways. One way is simply to stimulate increased production of the enzyme; a second way is by the process of **allosteric modulation.** In this process, the hormone combines with the enzyme to change its shape and ability to act. Allosteric interactions can either increase or decrease enzyme activity, depending on how they alter the enzyme's ability to interact with a substrate. A third way that hormones modify enzyme action is to make active many inactive forms of a particular enzyme. Thus, the total number of active enzymes can be increased by converting some inactive enzyme molecules into active ones.

In addition to altering enzyme activity, hormones can either facilitate or inhibit the transport of substances into cells. Sometimes these actions occur simultaneously to modulate precisely the amount of a substance within the system. **Insulin,** for example, facilitates the transport of glucose through the cell membrane into the cell. This is done in combination with extracellular glucose and a glucose carrier contained within the cell membrane. In contrast, other hormones like epinephrine act in an opposing fashion to inhibit the amount of glucose that enters the cell.

The alteration of enzyme-mediated membrane transport and enzyme synthesis constitutes the major mechanisms of hormone action. In addition to this primary action, hormones also exert secondary effects on the different cells. These secondary effects that in some cases are quite indirect, can be potent. The release of insulin, for example, increases glucose uptake by muscle cells that ultimately leads to the increased synthesis of muscle glycogen. Such a dramatic influence on the metabolism of glucose within a cell has important implications in terms of fuel homeostasis during exercise. For those individuals who may have insulin insufficiency, the effects on exercise metabolism could be quite debilitating. An inadequate cellular uptake of glucose, caused by insulin deficiency, increases the levels of circulating blood glucose and other metabolites that ultimately spill over into the urine. This condition of insulin deficiency is known as **diabetes mellitus.**

FACTORS THAT DETERMINE HORMONE LEVELS

The secretion of hormones rarely occurs at a constant rate. Like the activity of the nervous system, hormone secretion must be adjusted rapidly to meet the immediate demands of changing bodily functions. For this reason, all protein hormone secretions are pulsatile in nature. The concentration of a particular hormone in the blood is a function of the quantity of hormone synthesized in the host gland and the amount released into the blood. Over any extended period of time, hormone synthesis tends to equal hormone release. For a short period, however, it is possible for hormone release to exceed its synthesis. The plasma concentration of a hormone is referred to as the "secreted amount." In reality, this is actually the sum of both synthesis and release by the host gland, as well as its rate of uptake by the receptor tissues, and its rate of removal from the blood by the liver and kidneys. In most cases, the rate of removal, usually measured in the urine, is equal to the rate of release, although this assumption is not always true.

Endocrine glands are stimulated three ways: hormonally, humorally, and neurally.

Hormonal Stimulation. Various hormones influence the secretion of other hormones. For example, the release of most anterior pituitary hormones is regulated by releasing-inhibiting hormones produced by the hypothalamus. Many anterior pituitary hormones, in turn, stimulate other endocrine organs to release their hormones into the blood. As the levels of hormones produced by the final target glands increase in the blood, they provide feedback to inhibit the release of anterior pituitary hormones, and thus their own release.

Humoral Stimulation. Changing levels of certain ions and nutrients in blood, bile, and other body fluids also stimulate hormone release. Such stimuli are referred to as humoral stimuli to distinguish them from hormonal stimuli, which are also fluid-borne chemicals. The release of insulin by pancreatic cells, for example, is prompted by increasing blood sugar (the humoral agent). Because insulin promotes glucose entry into tissue cells, blood sugar levels soon decline, thus ending the initiative for insulin release.

Neural Stimulation. In some cases nerve fibers stimulate hormone release. The classic example of neural stimulation is sympathetic activation of the adrenal medulla to release the catecholamines, epinephrine and norepinephrine, during periods of stress. In certain cases the nervous system overrides normal endocrine control in an effort to maintain homeostasis. Blood sugar levels, for example, are normally maintained within the range of 80 to 120 mg per 100 ml of blood by the action of insulin and other hormones. During exercise, however, activation of the hypothalamus and sympathetic nervous system blunts the release of insulin to cause blood sugar levels to rise much higher to meet the exercise demands and ensure that muscle and nervous tissue will have sufficient carbohydrate fuel.

RESTING AND EXERCISE-INDUCED ENDOCRINE SECRETIONS

Table 20–2 lists the different endocrine host organs, specific hormones secreted, control factors, hormone function, effects of hypo- and hypersecretion, and the influence of exercise on hormone output. In the following sections we will review these important hormones, their function during rest and exercise, and the particular host gland-hormone response to physical training.

Anterior Pituitary Hormones

Figure 20–2 shows the pituitary gland, its secretions, and various target glands and other hormone secretions. The pituitary gland, also called the **hypophysis,** is located beneath the base of the brain. It secretes at least six different polypeptide hormones and influences the secretion of several others. Because of its widespread influence, the anterior pituitary gland was often called the "**master gland.**" It is now known, however, that the hypothalamus actually controls anterior pituitary activity; thus the hypothalamus should truly have the title "master gland." In addition to the hormones displayed in Figure 20–2, the pituitary also secretes a large molecule called **pro-opiomelanocortin** or simply **POMC.** POMC is a large precursor molecule from which other active molecules are split by enzymatic cleavage. POMC is the source of adrenocorticotropic hormone and some of the naturally produced opiates such as **beta-endorphin** (see p. 411).

Growth Hormone. Growth hormone (**GH** or **somatotropin**) has widespread physiologic activity because it promotes cell division and cellular proliferation throughout the body. In the adult, GH facilitates protein synthesis. This action is accomplished by increasing amino acid transport through cell membranes, stimulating an increase in RNA formation, or ac-

TABLE 20–2.
ENDOCRINE GLANDS, THEIR SECRETIONS, FUNCTIONS, CONTROL FACTORS, EFFECTS OF HYPO- AND HYPERSECRETION, AND THE EFFECTS OF EXERCISE ON HORMONE OUTPUT

Host Gland	Hormone	Hormone Effects	Control of Hormone Secretion	Effects of *Hypo-* and *Hyper*secretion	Exercise Effects on Hormone Secretion
Anterior pituitary	Growth hormone (GH; somatotropin)	Stimulates tissue growth; mobilizes fatty acids for energy; inhibits CHO metabolism	Hypothalamic releasing factor (GHRF)	*Hypo-* dwarfism in children; *Hyper-* gigantism in children; acromegly in adults	↑ with increasing exercise
	Thyrotropin (TSH)	Stimulates production and release of thyroxine from thyroid gland	Hypothalamic TSH-releasing factor; thyroxine	*Hypo-* cretinism in children (stunted growth, mental retardation); myxedema in adults (low BMR, constipation, dry skin, puffy eyes, edema, lethargy); *Hyper-* Graves' disease (autoimmune disease—elevated BMR, weight loss, irregular heartbeat), heart disease	↑ with increasing exercise
	Corticotropin (ACTH)	Stimulates production and release of cortisol, aldosterone, and other adrenal hormones	Hypothalamic ACTH-releasing factor; cortisol	*Hypo-* rarely seen; *Hyper-* Cushing's disease	?
	Gonadotropic (FSH and LH)	FSH works with LH to stimulate production of estrogen by ovaries; LH works with FSH to stimulate production of estrogen and progesterone by ovaries and testosterone by male testes	Hypothalamic FSH and LH releasing factor; female—estrogen and progesterone; male—testosterone	*Hypo-* failure of sexual maturation; *Hyper-* none	No change
	Prolactin (PRL)	Inhibits testosterone; mobilizes fatty acids	Hypothalamic PRL-inhibiting factor	*Hypo-* poor milk production in nursing women; *Hyper-* galactorrhea, cessation of menses in females, impotence in males	↑ with increasing exercise
	Endorphins	Blocks pain; promotes euphoria; affects feeding and female menstrual cycle	Stress—physical/emotional (may be intensity related)	Unknown	↑ with long-duration exercise
Posterior pituitary	Vasopressin (ADH)	Controls water excretion by kidneys	Hypothalamic secretory neurons	*Hypo-* diabetes; *Hyper-* unknown	↑ with increasing exercise
	Oxytocin	Stimulates muscles in uterus and breasts; important in birthing and lactation	Hypothalamic secretory neurons	Unknown	?
Adrenal cortex	Cortisol Corticosterone	Promotes use of fatty acids and protein catabolism; conserves blood sugar/insulin antagonist; has anti-inflammatory effects with epinephrine	ACTH; stress	*Hypo-* Addison's disease (weight loss; glucose and sodium levels drop and potassium levels rise, resulting in hypotension and dehydration); *Hyper-* Cushing's disease (persistent hyperglycemia, dramatic losses in muscle and bone protein, and water and salt retention leading to hypertension)	↑ in heavy exercise only

Host Gland	Hormone	Hormone Effects	Control of Hormone Secretion	Effects of *Hypo-* and *Hypersecretion*	Exercise Effects on Hormone Secretion
	Aldosterone	Promotes retention of sodium, potassium, and water by the kidneys	Angiotensin and plasma potassium concentration; renin	*Hypo-* Addison's disease; *Hyper-* aldosteronism (excessive sodium and water retention and accelerated excretion of potassium)	↑ with increasing exercise
Adrenal medulla	Epinephrine Norepinephrine	Facilitates sympathetic activity, increases cardiac output, regulates blood vessels, increases glycogen catabolism and fatty acid release	Stress stimulated hypothalamic sympathetic nerves	*Hypo-* unimportant; *Hyper-* hypertension, increased metabolism	Epinephrine, ↑ in heavy exercise Norepinephrine, ↑ with increasing exercise
Thyroid	Thyroxine T_4 Triiodothyronine T_3	Stimulates metabolic rate; regulates cell growth and activity	TSH; whole body metabolism	*Hypo-* decreased BMR, body temperature, cold intolerance, decreased appetite, weight gain, decreased glucose metabolism, elevated cholesterol, decreased protein synthesis, hypotension, muscle cramps, growth retardation, depressed ovarian function; *Hyper-* increased BMR, temperature, heat intolerance, increased appetite, weight loss, hypertension, enhanced catabolism of glucose, fats, and protein, loss of muscle, muscle atrophy, depressed ovarian function	↑ with increasing exercise
Pancreas	Insulin	Promotes CHO transport into cells; increases CHO catabolism and decreases blood glucose; promotes fatty acid and amino acid transport into cells	Plasma glucose levels	*Hypo-* diabetes; *Hyper-* hypoglycemia, anxiety, nervousness, weakness	↓ with increasing exercise
	Glucagon	Promotes release of glucose from liver to blood; increases fat metabolism, reduces amino acid levels	Plasma glucose levels	*Hypo-* chronic hypoglycemia, low circulating amino acids; *Hyper-* hyperglycemia	↑ with increasing exercise
Parathyroid	Parathormone	Raises blood calcium; lowers blood phosphate	Plasma calcium concentration	*Hypo-* hypocalcemia, respiratory paralysis, uncontrolled spasms and convulsions; *Hyper-* hypercalcemia, extreme leaching of calcium from bones, depression of nervous system activity, muscle weakness, formation of kidney stones	↑ with long-term exercise
Ovaries	Estrogen Progesterone	Controls menstrual cycle; increases fat deposition; promotes female sex characteristics	FSH, LH	*Hypo* (estrogen)- *Hyper* (progesterone)- masculinization or virilization	↑ with exercise; depends on menstrual phase

TABLE 20–2 (cont.).
ENDOCRINE GLANDS, THEIR SECRETIONS, FUNCTIONS, CONTROL FACTORS, EFFECTS OF HYPO- AND HYPERSECRETION AND THE EFFECTS OF EXERCISE ON HORMONE OUTPUT

Host Gland	Hormone	Hormone Effects	Control of Hormone Secretion	Effects of *Hypo-* and *Hypersecretion*	Exercise Effects on Hormone Secretion
Testes	Testosterone	Controls muscle size; increases RBC; decreases body fat; promotes male sex characteristics	LH	*Hypo-* feminization; *Hyper-* masculinization or virilization	↑ with exercise
Kidney	Renin	Stimulates aldosterone secretion	Plasma sodium concentration	*Hypo-* hypertension; *Hyper-* hypotension	↑ with increasing exercise

↑ = increases

tivating cellular ribosomes that increase protein synthesis.[41,98] The release of GH also results in a decrease in the rate of carbohydrate utilization with a subsequent increase in the mobilization and use of fats as an energy source. Regardless of training status, women appear to have higher levels of GH at rest compared to men; this difference disappears during prolonged exercise.[24]

Studies on exercise-induced production of GH have revealed a delay of a few minutes in GH secretion after exercise starts.[64,70] With successively increasing exercise levels, there is a sharp rise in GH production and total secretion.[40,41] This would certainly be a beneficial response for muscle, bone, and connective tissue growth, as well as for optimizing the metabolic mixture during exercise. The precise relationship between GH synthesis and exercise intensity and duration has not been established, nor has the stimulus been identified for increased GH production with exercise. Concurrent measurements of circulating lactate, alanine, pyruvate, blood glucose, and body temperature reveal that none of these factors are responsible for regulating the pattern of GH secretion.[64] Thus, it is most probable that **neural factors** primarily control GH secretion.

Growth hormone secretion during rest is influenced by a GH-releasing factor that acts directly on the anterior pituitary gland. In fact, each of the

FIG. 20–2.
The pituitary gland and its secretions.

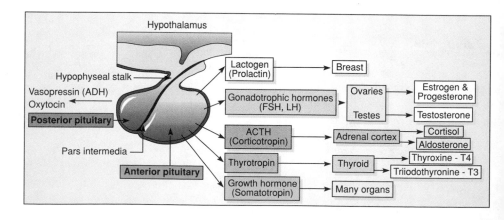

primary pituitary hormones has its own hypothalamic releasing hormone, sometimes called a **releasing factor.** These releasing hormones are controlled by neural input to the hypothalamus by factors such as anxiety, stress, and exercise.

Growth Hormone, Exercise, and Tissue Synthesis. The exact mechanism is not entirely clear by which GH and exercise interact to bring about increases in protein synthesis, cartilage formation, skeletal growth, and cell proliferation.[22,25,65,74,82] One hypothesis suggests that exercise directly stimulates GH production and the pulsatile pattern of its release that in turn stimulates anabolic processes.[13] It has been shown that exercise is directly associated with doubling of both GH pulse frequency and amplitude. Furthermore, exercise stimulates the production of endogenous opiates that facilitate GH release by inhibiting the liver's production of **somatostatin,** a hormone that blunts the release of GH.[14] Figure 20–3 shows a proposed plan for classifying the overall metabolic actions of GH. In terms of exercise modulation, GH stimulates fat breakdown and release from adipose tissue while inhibiting glucose uptake by the cells, thus maintaining blood sugar

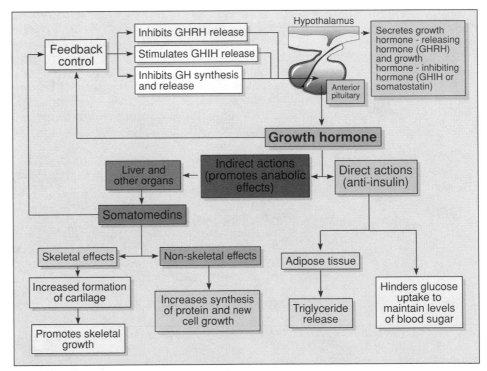

FIG. 20–3.
Overview plan for classifying the actions of growth hormone (GH). GH stimulates the breakdown and release of triglycerides from adipose tissue and hinders glucose uptake by the cells to maintain relatively high levels of blood sugar. GH is said to exert an anti-insulin effect because its action is opposite that of insulin. The indirect anabolic effects of GH are mediated through somatomedins. Elevated levels of GH and somatomedins feed back to promote GHIH release and depress GHRH release by the hypothalamus to further inhibit its release by the anterior pituitary.

at fairly high levels. This sparing of glucose would certainly contribute to one's ability to perform endurance exercise.

Intensity and duration of effort are important factors that affect GH secretion during exercise. When exercised to exhaustion, both fit and unfit subjects show similar increases in GH concentration; unfit subjects, however, continue to maintain high levels of GH for several hours into recovery.[94] In a standard submaximal bout of exercise, the GH response is greater in the unfit compared to trained counterparts.[95] Because this absolute work level represents a greater demand on the less fit, it appears that the release of GH is in some way related to the relative strenuousness of effort.

Thyrotropin. Thyrotropin, sometimes called **thyroid-stimulating hormone (TSH)**, controls the amount of hormone secreted by the thyroid gland. This hormone acts to maintain the growth and development of the thyroid gland as well as to increase the activity rate of the thyroid cell. In light of the important role of thyroid hormone in regulating cellular metabolism, it is not surprising that TSH output from the pituitary increases during exercise, although this effect may not be consistent.[99]

Corticotropin. Corticotropin, or **adrenocorticotropic hormone (ACTH)**, regulates the output of the hormones secreted by the adrenal cortex in a manner similar to the way TSH controls the secretion of thyroid hormone. ACTH acts throughout the body to directly enhance fat mobilization from adipose tissue, increase the rate of gluconeogenesis, and stimulate protein catabolism. Owing to difficulty in assay methods and the rapid disappearance of this hormone from the blood, clear evidence is scarce concerning the response of ACTH during exercise.[41] Because ACTH is directly responsible for the synthesis of cortisol (a hormone of the adrenal cortex gland), and cortisol is increased during exercise, the inference is that ACTH output is augmented during exercise.

Prolactin. The hormone prolactin **(PRL)** initiates and supports milk secretion from mammary glands. Like GH, the output of PRL increases with higher levels of exercise but to a lesser degree.[84] PRL levels return toward baseline within 45 minutes after cessation of exercise,[23] and its effects appear to be transient. Owing to its important role in female sexual function, it is possible that repeated exercise-induced PRL release may inhibit the ovaries and contribute to the alterations in menstrual cycle often observed among athletic females.[12,16,77] Increases in PRL are greater in women who run without wearing a bra;[86] the release of this hormone is also enhanced by either fasting or consuming a high-fat diet.[45,61] Recent evidence also shows that PRL increases in males following an acute, maximal bout of exercise.[31]

Gonadotropic Hormones. Gonadotropic hormones stimulate the male and female sex organs to grow and secrete their hormones at a faster rate. The two gonadotropic hormones are **follicle-stimulating hormone (FSH)** and **luteinizing hormone (LH)**. In the female, FSH initiates growth of follicles in the ovaries and stimulates the ovaries to secrete estrogens, one type of female sex hormone. Luteinizing hormone works with FSH to cause estrogen secretion as well as rupture of the follicle to allow the ovum to pass through the fallopian tube for fertilization. In the male, FSH stimulates growth of the germinal epithelium in the testes to promote the development of sperm. Luteinizing hormone also stimulates the testes to secrete the male hormone **testosterone.**

Reports of exercise-associated alterations in FSH and LH are inconsistent[11,30,55,103] and are confounded by the nature of gonadotropin release. Because LH is normally released in a pulsatile manner in response to many factors, it is difficult to separate any specific exercise-related change from the normal pattern of pulsatile release. Anxiety can also either raise or lower LH levels via the action of norepinephrine, which is influenced by stress.[66] Interestingly, recent findings show that LH rises even before exercise starts and reaches its peak after exercise is completed. This pre-exercise increase in LH was synchronous with the release of testosterone and suggests that gonadotropin stimulation is not responsible for exercise-induced increases in testosterone.[31]

Posterior Pituitary Hormones

The posterior pituitary gland shown in Figure 20–2 is often called the **neurohypophysis** and stores two hormones, **antidiuretic hormone (ADH or vasopressin)** and **oxytocin.** The posterior pituitary gland is really an outgrowth of the hypothalamus and resembles true neural tissue. This gland does not synthesize its hormones. Rather, they are produced in the hypothalamus and secreted to the neurohypophysis where they are then released as needed to the general circulation. In fact, if the posterior pituitary is damaged or surgically removed, the production of ADH and oxytocin would not be dramatically affected. The release of hormones from the neurohypophysis is controlled by nervous stimulation.

The major function of the polypeptide ADH is to control the excretion of water by the kidneys. Essentially, ADH limits the production of large volumes of urine to conserve water. The role of oxytocin is to stimulate the contraction of muscle in the uterus and to stimulate ejection of milk from the breasts during lactation; thus, oxytocin is important during birthing and nursing.

Exercise is a potent stimulus for ADH secretion,[28,95,105] which explains the increased water retention of the kidneys during and after severe exercise when dehydration occurs.[27,32] This hormonal response would certainly be helpful in aiding the body to conserve fluids, especially during hot-weather exercise. It is believed that sweating stimulates ADH secretion, which in turn causes faster reabsorption of water in the kidney tubules. In contrast, ADH release is decreased and urine volume increased proportionately in response to a fluid overload.

Thyroid Hormones

The thyroid gland is located in the neck just below the larynx. This gland is under the influence of **thyroid stimulating hormone (TSH)** produced by the anterior portion of the pituitary gland. The thyroid gland secretes two amino acid-iodine bound thyroid hormones known as **thyroxine** (T_4), and **triiodothyronine** (T_3). Thyroxine is secreted in greater quantity than T_3. Although less abundant, T_3 acts several times as rapidly as T_4. Most receptor cells for T_4 can metabolize T_4 to T_3, the active form of thyroid hormone.

An increase in T_4 secretion raises the metabolic rate of all cells. This stimulation of metabolism is thought to be a consequence of the elevation of enzymatic activities by T_4.[110] Thyroxine, for example, can raise the basal metabolic rate by as much as four times. Because of this influence, deviations in the BMR are often used as an index of thyroid gland abnormalities (see Chapter 9).

Thyroid hormone increases carbohydrate and fat metabolism. Consequently, a person with elevated thyroxine production will have an increased metabolic rate and potentially can lose weight rapidly. If thyroid activity is low, on the other hand, the person will usually gain weight. **However, fewer than 3% of obese people have abnormal thyroid function.**[8,21] In terms of nervous system function, T_3 release facilitates reflex activity, and a decrease in T_4 production can make people sluggish, so they sometimes sleep as much as 10 to 15 hours a day.

Synthesis of the thyroid hormones appears to be under the regulation of TSH, which in turn is directly influenced by the rate of whole body metabolism. If the body's metabolism falls to some critical value, hypothalamic release of TSH is stimulated, thyroid output increases, and the resting metabolic rate becomes elevated. Conversely, an increase in whole body metabolism causes a decrease in TSH production and a corresponding decrease in the metabolic rate. This exquisite "feedback" system is illustrated in Figure 20–4. During exercise, blood levels of free thyroxine concentration (that is, thyroxine not bound to plasma proteins) is increased by about 35%.[44,99] This increase could be the result of an exercise-induced hyperthermia that alters binding for several hormones including T_4. Whereas liver concentrations of T_4 increase during exercise, muscle concentrations appear unchanged;[111] the importance of these transient alterations remains unknown.

Adrenal Glands

The adrenal glands are flattened, cap-like tissues located just above each kidney (Fig. 20–5). There are two distinct parts to the adrenal gland: the **medulla** (inner portion) and the **cortex** (outer portion). Each part secretes

FIG. 20–4.

Feedback system for control of thyroid hormone.

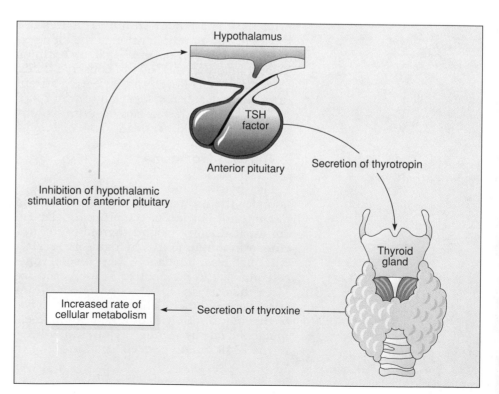

FIG. 20–5.
The adrenal gland and its secretions.

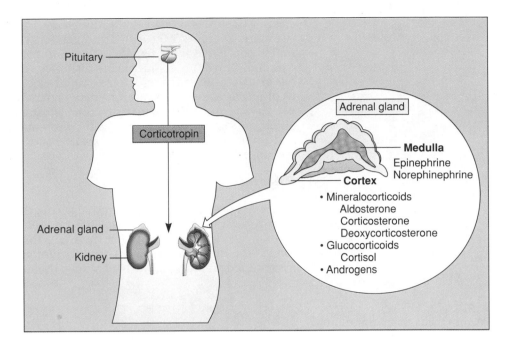

different types of hormones, and essentially they can be considered two distinct glands.

The adrenal medulla is part of the sympathetic nervous system. It acts to prolong and augment sympathetic effects by secreting two hormones, **epinephrine** and **norepinephrine,** collectively called **catecholamines.** Adrenal medulla secretions are under the direct influence of the hypothalamus. An outflow of neural impulses from the hypothalamus stimulates the adrenal medulla to increase its output of catecholamines. These hormones affect the heart, blood vessels, and glands in essentially the same way as direct stimulation by the sympathetic nervous system.

The adrenal cortex, stimulated by corticotropin from the pituitary, secretes what are known as **adrenocortical hormones.** These steroid hormones can be categorized by function into one of three groups: **mineralocorticoids, glucocorticoids,** and **androgens;** each is produced in a different zone or layer of the adrenal cortex.

Adrenal Medulla Hormones

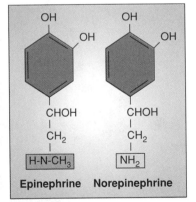

FIG. 20–6.
Chemical structure of epinephrine and norepinephrine. Norepinephrine is both a hormone and a precursor of epinephrine. Norepinephrine is also considered a neurotransmitter when it is released by sympathetic nerve endings. The structural differences between these hormones are shown by the shaded areas.

Figure 20–6 shows the chemical structure of epinephrine and norepinephrine. Norepinephrine is both a hormone in its own right and a precursor of epinephrine. It is also considered a neurotransmitter when released by sympathetic nerve endings. Norepinephrine levels are increased 2- to 6-fold by light to maximal exercise. More than likely, this secretion is from sympathetic postganglionic nerve endings and is related to the cardiovascular and metabolic adjustments of the working tissues. Epinephrine output from adrenal medulla secretion increases with exercise, and the magnitude of this increase is related to the intensity of effort.[37,41,98] The duration of exercise is also important in determining the magnitude of the catecholamine response. It has been shown, for example, that there is an inverse relationship between plasma epinephrine and norepinephrine and mileage run.[108] A number of other factors also determine catecholamine response to exercise, including age (plasma catecholamine secretion is greater in older subjects at the same exercise intensity[39]) and gender (greater epinephrine secretion in males compared with females who perform the same relative intensity of exercise, but the norepinephrine response was the same[96]). Recent research has also demonstrated selective sympathetic neural activation that is specific to different tissues and thus not necessarily whole-body dependent.[101] Sympathetic activity affects many factors; these include blood-flow distribution, enhanced cardiac contractility, substrate mobilization, liver glycogenolysis, and adipose tissue lipolysis. All of these actions are beneficial to the exercise response.

Adrenocortical Hormones

Mineralocorticoids. As the name suggests, the role of the mineralocorticoids is to regulate the mineral salts sodium and potassium in the body's extracellular fluid spaces. Although there are three mineralocorticoids, **aldosterone** is the most physiologically important and comprises almost 95% of all mineralocorticoids. Figure 20–7 shows the major controlling mechanism for

FIG. 20–7.
Major mechanisms controlling aldosterone release from the adrenal cortex.

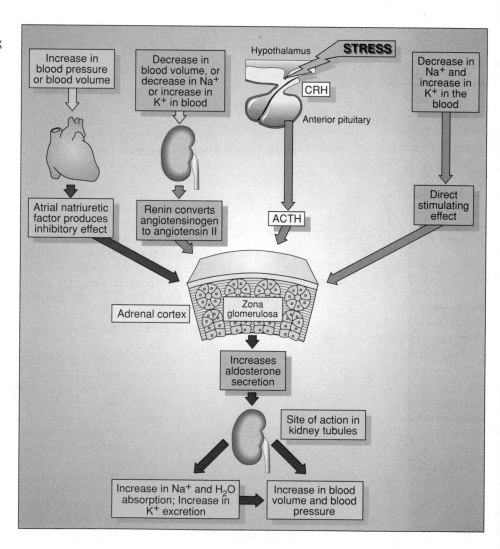

aldosterone release from the adrenal cortex. As can be seen, aldosterone acts by regulating sodium reabsorption in the distal tubules of the kidneys. When large quantities of aldosterone are secreted, the sodium ions that enter the kidney are reabsorbed from the tubules back into the blood along with increased fluid. Consequently, little sodium passes into the urine. In contrast, sodium is literally poured into the urine with cessation of aldosterone secretion. Thus, aldosterone secretion is essential for controlling total sodium concentration as well as the body's extracellular fluid volume. Mineral balance is important for nerve transmission and muscle function—neuromuscular activity would be impossible without the regulation of sodium and potassium. Because the kidneys exchange either potassium or a hydrogen ion for each sodium ion that is reabsorbed, aldosterone also contributes to the homeostatic maintenance of serum potassium and pH.

An increase in aldosterone secretion produces a corresponding rise in extracellular fluid as water is reabsorbed from the kidneys. This increases blood volume, and is accompanied by a concomitant increase in cardiac output and arterial blood pressure. During exercise, the outflow from the sympathetic nervous system constricts blood vessels to the kidneys. In turn, this reduction in blood flow stimulates the kidneys to release the enzyme **renin** into the blood. Increased renin stimulates the production of still another kidney hormone, **angiotensin,** that stimulates the adrenal cortex to secrete aldosterone.[32] Aldosterone secretion rises progressively during exercise, with peak plasma aldosterone levels reaching as high as six times the resting level.[75] There is a strong relationship between renin release via beta-adrenergic stimulation and aldosterone secretion.[93] During rest, the renin-angiotensin mechanism, stimulated by a decrease in blood pressure in the afferent arterioles of the kidneys, controls aldosterone secretion.

Glucocorticoids. **Cortisol,** also called hydrocortisone, is the major glucocorticoid secreted by the adrenal cortex. The most important function of cortisol relates to its effects on energy metabolism. Cortisol does the following: (1) stimulates protein breakdown to the amino acid constituents in all cells of the body except the liver. These "liberated" amino acids circulate to the liver where they are synthesized to glucose in the process of gluconeogenesis; (2) supports the action of other hormones, primarily glucagon and growth hormone, in the gluconeogenic process; and (3) serves as an insulin antagonist by inhibiting glucose uptake and oxidation.

Increased cortisol production results in a decrease of protein in tissue other than the liver by suppressing the rate of protein formation and stimulating protein catabolism. This results in an increase in liver and plasma protein, much of which serves the process of glucose synthesis. The excess protein can also be used for various bodily functions including growth. Prolonged, high concentrations of cortisol in the blood ultimately lead to tissue-wasting and create a negative nitrogen balance in the body.

Cortisol secretion also accelerates fat mobilization and utilization for energy. This occurs during starvation as well as exercise. With rapid and large increases in cortisol output, the liver splits the mobilized fat into ketoacids. Above normal concentrations of ketoacids in the extracellular fluid can lead to the potentially dangerous medical condition of **ketosis** (a form of acidosis). Ketosis, stimulated by increased cortisol secretion, is a common condition of individuals who consume very low-carbohydrate, low-calorie (ketogenic) weight-loss diets.

The effects of glucocorticoid secretion are ideally suited to meet severe stress situations. In fact, emotional stress or the stress encountered in physical exertion induce neural responses from the periphery to the hypothalamus. The hypothalamus then secretes **corticotropin releasing factor** which stimulates the anterior pituitary to release ACTH. In turn, ACTH causes the adrenal cortex to release cortisol into the circulation.

There is considerable variability in the cortisol response to exercise, depending on many factors that include exercise intensity and duration, fitness level, meal status, and even circadian rhythm.[20,93] Despite a few studies to the contrary,[69,105] most research indicates that cortisol output increases with exercise intensity.[83] Extremely high levels of cortisol have been observed in subjects following long-duration exercise including a marathon.[19,29,94] Even at lower work rates, plasma cortisol rises if the exercise period is sufficiently long.[20] Cortisol response during exercise is best studied by observing cortisol turnover; that is, the difference between production and removal. When this approach is used, research shows that highly-trained runners are in a state of hypercortisolism, which is particularly heightened prior to competition.[73] Cortisol levels also remain elevated for as long as 2 hours after exercise. These responses are probably related to cortisol's antistress functions.

Androgens. The sex steroids are produced in the ovaries and adrenals for females, and in the testes and adrenals for males. There are no distinctly singular "male" or "female" hormones but rather general differences in concentrations between genders. The ovaries are the primary source of **estradiol** and luteal phase **progesterone;** the adrenals are the chief source of **dehydroepiandrosterone (DHEA)** and its sulfate DHEAS. The ovaries and adrenals are the main source of **androstenedione** and **testosterone.** The male testes produce testosterone. The ovaries of females also secrete small quantities of testosterone; conversely, for normal men testosterone can be converted to estrogen in peripheral tissues.

In the female, abnormal adrenal androgen secretion produces masculinizing effects. Plasma testosterone concentration in females, while about one-tenth that of males, nevertheless has been shown to increase with exercise. Exercise also elevates estradiol and progesterone.[62] For untrained males, both resistance training and moderate aerobic exercise appear to significantly increase both serum and free testosterone levels after about 15 to 20 minutes compared to resting values.[31,44,103] For prolonged, strenuous exercise such as marathon running, there is some indication that the level of testosterone falls below that observed at rest. The mechanism for the alterations in testosterone levels with exercise cannot be presently explained. It appears, however, that it is not associated with any increased production of LH.[103]

Pancreatic Hormones

The pancreas is a gland about 14 cm long; it weighs about 60 g and lies just below the stomach. It is composed of two different types of tissues, acini and islets of Langerhans (Fig. 20–8). The islets of Langerhans are endocrine tissues composed of two types of cells—alpha and beta cells that secrete glucagon and insulin, respectively. When blood sugar levels are high the pancreas releases insulin. Insulin stimulates glucose uptake by cells and glycogen formation in the liver, which subsequently lowers blood sugar levels. In essence, insulin exerts a **hypoglycemic effect.** The acini portion of the pancreas is exocrine in function and secretes digestive enzymes.

Insulin. Insulin, first discovered in 1921, has many different bodily functions, all of which are directly related to cellular metabolism. Insulin's major

FIG. 20—8.
The pancreas and its
secretions.

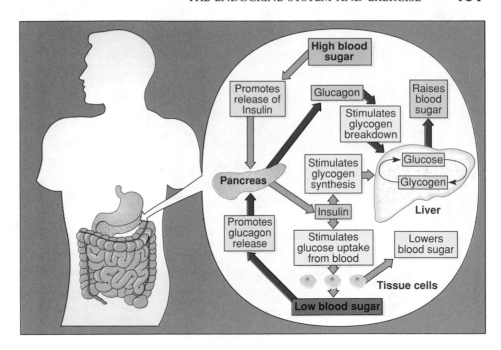

function is to regulate total body glucose metabolism in all tissues except the brain. It does this by increasing the rate of glucose transport through the membranes of muscle and adipose tissue cells. In the total absence of insulin, only trace amounts of glucose can be transported into cells. Insulin is essentially the mediator of **"facilitated diffusion,"** whereby glucose in the presence of insulin combines with a glucose carrier for transport into the cells. In this way, insulin actually controls the rate of cellular glucose metabolism. Glucose and fat oxidation are maintained by increasing the cellular uptake of glucose. Any glucose not immediately catabolized for energy is stored as glycogen for later use. The effects of insulin on metabolism are shown in Figure 20–9.

Increased insulin-mediated glucose uptake results in glycogen synthesis that leads to a decrease in the circulating levels of blood-borne glucose. Conversely, with diminished insulin secretion the level of blood glucose increases and sometimes rises from a normal level of 80 to 95 mg $\cdot$ 100 ml^{-1} blood to a high of 350 mg $\cdot$ 100 ml^{-1}.[9] Glucose will eventually spill into the urine unless these high levels of blood glucose are lowered.

Insulin also has a pronounced effect on fat metabolism. As glucose levels in the blood rise, as would normally occur after a meal, insulin is released into the blood to cause some of the glucose to be transported into fat cells. Much of this glucose is synthesized to fat. Thus, with a well-regulated insulin response, carbohydrates are utilized preferentially for energy and excess carbohydrate is stored as fat. In the absence of insulin, fatty acids are predominantly mobilized and utilized in place of sugar. Insulin will also cause protein deposition in cells. This is accomplished by one or all of the following: (1) increasing the levels of RNA in cells, (2) increasing the rate of amino acid transport through cell membranes, or (3) increasing the formation of proteins by ribosomes.

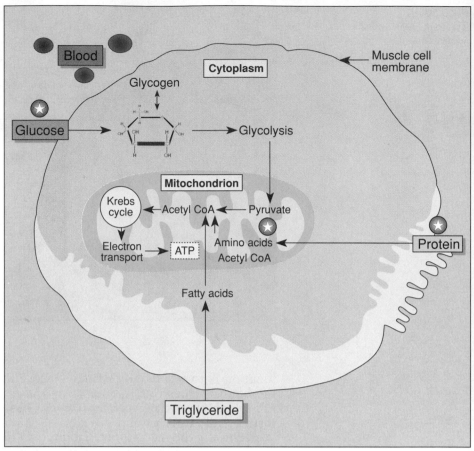

FIG. 20–9.
Primary function of insulin in the body. The circled stars show the location in metabolism where insulin exerts its influence.

Glucose-Insulin Interaction. Insulin secretion is directly controlled by the level of blood glucose that passes through the pancreas. An increase in circulating blood glucose stimulates insulin secretion, while a reduction in blood glucose leads to a decreased insulin response. Insulin secretion is also sensitive to plasma amino acid levels; elevated amino acid levels cause an increase in insulin secretion. The insulin-glucose mechanism provides feedback for controlling the body's levels of insulin and blood glucose. A rise in blood glucose stimulates insulin secretion; insulin then induces glucose entry into cells, which lowers blood glucose and reduces the stimulus to the pancreas for insulin secretion.

During exercise of increasing intensity and duration, the circulating levels of blood glucose and insulin progressively decrease (Fig. 20–10) as a result of the inhibitory effect of increased alpha-adrenergic input to the beta cells.[8,43] This directly enhances hepatic glucose output and sensitizes the liver to the effects of glucagon and epinephrine. These actions help to maintain blood glucose. **The catecholamine suppression of insulin, which is proportional to the intensity of exercise, helps to explain why the hypoglycemia with pre-exercise glucose feedings is not observed when glucose is consumed**

FIG. 20–10.
Plasma insulin levels during exercise and recovery. (From the Applied Physiology Laboratory. University of Michigan.)

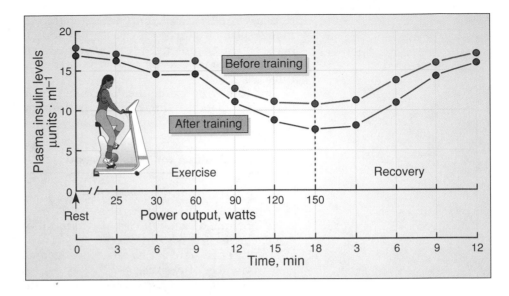

during exercise. As insulin output becomes reduced in exercise of long duration, progressively more energy is derived from the mobilization and metabolism of free fatty acids.[2]

Diabetes. The syndrome of **diabetes mellitus** consists of different subgroups of disorders with different pathophysiologies. The two largest subgroups are labeled **Type I** and **Type II.**

Type I. Type I, or **insulin-dependent diabetes,** formerly called juvenile-onset diabetes, generally occurs in younger individuals. It is associated with an absolute deficiency of insulin and often of other pancreatic hormones. Compared to the Type II subgroup, these patients generally have a more severe abnormality of glucose homeostasis, the effects of exercise on the metabolic state are more pronounced, and the management of exercise-related problems more difficult.

Type II. Type II, or **non-insulin-dependent diabetes,** formerly called maturity-onset diabetes, tends to occur in older individuals. It is associated with significant resistance to insulin's actions, an abnormal but relatively well maintained insulin secretion, and normal to elevated plasma insulin levels.

In diabetes, adequate amounts of glucose fail to enter the cells; this causes blood glucose levels to increase to abnormally high levels. The increased glucose is passed into the kidney tubules and voided in the urine. The glucose particles in plasma filtrate cause an increase in osmotic pressure in the kidney tubules that also diminishes the reabsorption of water. As a result, the diabetic person loses large amounts of both water and glucose from the body. With decreased glucose uptake by the cells, the diabetic person relies heavily on fat metabolism for energy. This produces an excess of ketoacids and results in acidosis. In extreme cases of metabolic acidosis the blood pH falls to as low as 7.0, eventually resulting in a diabetic coma. Also associated with diabetes are such medical conditions as arteriosclerosis, small blood vessel and nerve disease, and susceptibility to infection.[63]

Recommendations for Exercise. Hypoglycemia in Type I diabetes is the most common disturbance of glucose homeostasis during exercise. It is particularly intense in patients who undergo intensive insulin therapy to normalize plasma glucose levels throughout the day. Usually hypoglycemia occurs during a prolonged session of moderate exercise when hepatic glucose production cannot keep pace with the increased use of glucose by exercising muscle.

The following guidelines are useful for the well-controlled, insulin-dependent diabetic who wishes to perform prolonged and strenuous exercise with minimal risk of hypoglycemia:[102]

Exercise Guidelines for Type I Diabetic

- Ingest 15 to 30 g of carbohydrate for every 30 minutes of intense exercise.
- Consume a snack of carbohydrate following the exercise period.
- Decrease the insulin dose:
  a. Intermediate-acting insulin—decrease the dose by 30 to 35% on the day of exercise.
  b. Intermediate- and short-acting insulin—omit the dose if it precedes exercise.
  c. Multiple doses of short-acting insulin—reduce the dose prior to exercise by 30% and supplement carbohydrate intake.
  d. Continuous subcutaneous insulin infusion—eliminate mealtime bolus or insulin increment that precedes or follows exercise.
- Avoid exercising for 1 hour those muscles in which short-acting insulin was injected.
- Avoid exercise in the late evening.

Glucagon. Glucagon, termed the "insulin antagonist" hormone, is secreted from the alpha cells of the islets of Langerhans. In contrast to insulin, glucagon's major function in the body is to raise the level of blood glucose and exert a hyperglycemic effect. Glucagon increases cellular availability of glucose by stimulating both glycogenolysis and gluconeogenesis in the liver. The glucose generated in this process is then released into the bloodstream. In contrast, insulin performs the opposite function by increasing the transport of glucose from the blood into the cells.[100]

Glucagon exerts its effect by activating the enzyme adenyl cyclase, which stimulates cyclic AMP in the liver. This process ultimately stimulates the breakdown of liver glycogen to glucose. Glucagon also exerts a stimulating effect on the gluconeogenic process in the liver by promoting amino acid uptake by the liver.

Like its antagonist insulin, glucagon is controlled by circulating levels of blood glucose. When glucose concentration falls, as occurs in prolonged exercise, the alpha cells of the pancreas are stimulated to secrete glucagon. Glucagon, in turn, causes an almost instantaneous release of glucose from the liver.[38] The glucagon mechanism is active during heavy, long-duration exercise and starvation. In both conditions there can be marked decreases in blood glucose, and the body's carbohydrate reserves are reduced dramatically.

In contrast to insulin secretion, glucagon release does not appear to be mediated by the autonomic nervous system because neither alpha nor beta blockade influences glucagon secretion during prolonged exercise.[42] Also, there are no gender differences in the glucagon response to exercise when working at the same percentage of aerobic capacity.[96] The stimulating effects of hepatic glucose output by glucagon may be important to glucose balance during exercise. However, the increase in glucagon concentration is considerably delayed following the onset of exercise in humans. This suggests that glucagon might not be important in the early regulation of hepatic glycogenolysis.[7]

Other Glands and Hormones

Other important hormones regulate bodily functions. The liver, for example, secretes the hormone **somatomedin** that affects the growth of muscle, cartilage, and other tissues. The mucosal lining of the small intestines secretes **secretin, gastrin,** and **cholecystokinin** to promote, regulate, and coordinate activities involved in the digestive process. The parathyroid gland secretes **parathormone,** a hormone that promotes calcium absorption from the intestines, kidneys, and bones. This provides for the critical regulation of blood calcium. Limited evidence suggests that prolonged physical activity has a powerful effect in stimulating the secretion of parathormone.[71] The hypothalamus itself is an important endocrine gland that directly secretes several stimulating or releasing hormones. **Somatoliberin,** for example, which is released from the hypothalamus, stimulates release of somatotropin from the anterior pituitary gland.

The **gonads** also serve important endocrine functions. The interstitial cells of the testes secrete the androgen testosterone which is believed responsible for increases in muscle mass and decreases in body fat. The female gonads (ovaries) secrete estrogens (mainly **estradiol**) and progesterone. These hormones are responsible for the onset and continuation of menses and for promoting female sex characteristics.

EFFECTS OF EXERCISE TRAINING ON ENDOCRINE FUNCTION

Table 20–3 lists the different hormones and their general response pattern to exercise training. These "training responses" are often based on limited evidence. Because of the complex interactions between endocrine secretions and the nervous system, research is especially sparse with respect to multiple hormone secretions and changes consequent to physical training. Except for a few isolated studies, the present state of knowledge is basically limited to research on single hormone responses to different forms of exercise training.[98]

Anterior Pituitary Hormones

Growth Hormone. Because GH stimulates lipolysis and inhibits carbohydrate metabolism, it is an attractive hypothesis that trained individuals will have enhanced GH secretion in comparison to untrained counterparts. This response for GH to exercise training, however, is not well documented. Limited research suggests that resting levels of GH are unaffected by training. In fact, there is evidence to suggest that endurance-trained individuals have **less** of a rise in blood levels of GH during exercise than untrained counterparts. This is probably an adaptation of reduced psychologic stress to exercise as training progresses.[10]

TABLE 20-3.
HORMONES AND THEIR RESPONSE TO EXERCISE TRAINING

Hormone	Training Response
Hypothalamus-Pituitary Hormones	
Growth hormone	No effect on resting values; trained tend to have less dramatic rise during exercise
Thyrotropin	No known training effect
ACTH	Trained have increased exercise values
Prolactin	Some evidence that training lowers resting values
FSH, LH, and	Trained females have depressed values
Testosterone	Trained males have depressed testosterone, with probably no change in LH and FSH (Testosterone levels may increase in males with long-term strength training)
Posterior Pituitary Hormones	
Vasopressin (ADH)	Some evidence that training results in slight reductions in ADH at a given workload
Oxytocin	No research information available
Thyroid Hormones	
Thyroxine	Reduced concentration of total T_3 and an increased free thyroxine at rest. Increased
Triiodothyronine	turnover of T_3 and T_4 during exercise
Adrenal Hormones	
Aldosterone	No significant training adaptation
Cortisol	Trained exhibit slight elevations during exercise
Epinephrine	Decrease in secretion at rest and same exercise intensity after training
Norepinephrine	
Pancreatic Hormones	
Insulin	Training increases sensitivity to insulin; normal decrease in insulin during exercise is greatly reduced in response to training
Glucagon	Smaller increase in glucose levels during exercise at both absolute and relative workloads
Kidney Hormones	
Renin	No apparent training effect
Angiotensin	

ACTH. Corticotropin is a potent stimulator of adrenal hormone secretion and also directly increases fat mobilization for energy. Physical training appears to increase levels of ACTH during exercise. Also, because ACTH is a stimulator of cortisol, the effects of ACTH with training would be to increase the responsiveness of the adrenal gland so it will result in a greater sparing of glucose. This would certainly be beneficial to long-term exercise performance.

Prolactin. Long-term training changes in prolactin that are not mediated by alterations in sympathetic activity or other multiple hormone effects are poorly documented.[40,41] In one report, the resting prolactin levels of male runners were significantly lower than the prolactin levels of nonrunners.[109]

FSH, LH, and Testosterone. Research suggests that females with a long history of exercise participation have altered levels of FSH and LH at different times in their menstrual cycle.[16] Alterations in these hormone levels are usually responsible for menstrual dysfunction. Follicle-stimulating hormone levels, for example, are depressed in chronically trained females throughout an abbreviated anovulatory menstrual cycle, whereas LH and progesterone concentrations are elevated in the follicular phase of the cycle.[78]

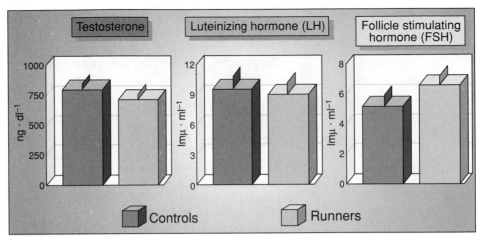

FIG. 20–11.
Comparison of testosterone, LH, and FSH levels between runners and controls. There are significantly reduced testosterone levels in trained runners with no significant difference in LH and FSH. (From Wheeler, G.D., et al.: Reduced serum testosterone and prolactin levels in male distance runners. JAMA, 252:514, 1984.)

Although the exact factors are unknown that cause alterations in reproductive system function in women athletes, several considerations have emerged, including acute and chronic effects of exercise, energy drain, weight loss, dietary changes, alterations in lean-to-fat ratio, physical and emotional stress of competition, and alterations in clearance rates of gonadal steroid hormones.

In males, research has shown that endurance training may have chronic effects on pituitary-gonadal function,[109] including levels of testosterone and prolactin.[52,58a] Figure 20–11 shows the results of comparisons between 46 male runners (average weekly distance = 64 km) and 18 nonrunners matched in age, stature, and body mass. The runners had depressed testosterone levels with no significant differences in the levels of LH and FSH compared to the nonrunners. The reduced testosterone concentration in males is similar to and may be the parallel of sex-steroid reductions observed in females who undergo marathon training and its associated effect of lowering one's level of body fat.[16,91] Both increased metabolic clearance and lower production rates of testosterone are present in the highly trained athlete.[109] Because both LH and FSH are not different between the trained and the untrained, this suggests that impaired gonadotropin release from the anterior pituitary is not responsible for the reduced testosterone levels observed in the trained state.

Resistance training may offer a different picture because significant increases in serum testosterone, LH, and FSH levels have been noted for elite male athletes who trained over a 2-year period. The researchers speculated that such increases create more optimal conditions to facilitate muscular adaptations in response to intense programs of muscular overload.[54]

Posterior Pituitary Hormones

Antidiuretic Hormone. Studies of ADH show no difference between trained and untrained subjects following exercise to exhaustion, or in subjects who undergo exercise at 65% of maximum following training.[27,78] There is some evidence, however, that ADH concentration is reduced with training in response to exercise at the same absolute exercise intensity.[104]

Oxytocin. We are unaware of research on training-induced changes in this hormone.

Thyroid Hormones

Thyroxine. Repeated bouts of exercise result in a coordinated pituitary-thyroid response consistent with an increased turnover of thyroid hormones. A repeated increase in thyroxine turnover is usually associated with excessive hormonal action that ultimately leads to hyperthyroidism. There is no evidence, however, that this condition occurs in highly trained individuals.[41] Differences in BMR and body temperature, for example, are rarely aberrant in the trained. If thyroid function were abnormal, one would certainly expect an abnormal response in these measures. Thus, it appears that the greater thyroxine turnover found with physical training occurs through a mechanism that differs from the "normal" actions of this hormone. The importance of changes in thyroxine levels in the adaptive process of training has not been established.

Research on women who engage in marathon training has revealed some interesting results regarding thyroid turnover.[18] Going from a baseline of relatively sedentary living to training 48 km a week resulted in a mild thyroid impairment as reflected by a decrease in the levels of T_3 and T_4. Increasing training to 80 km a week was associated with a significant increase in these hormones. In this study there was a suggestion that perhaps changes in body composition that accompany training might play a role in any exercise-induced change in thyroid function in females. Six months of resistance training in men slightly reduced the concentrations of T_4 and serum free T_4, with no change in thyrotropin. The authors concluded that the small changes were of no clinical or physiologic significance.[85]

Adrenal Hormones

Aldosterone. The response of the renin-angiotensin-aldosterone system during exercise is well suited for homeostatic control.[106] These responses are apparently transient as research shows that systematic exercise training does not affect resting levels of these hormones or their response to exercise.[47] There has been some evidence that renin activity during rest may be lower in trained subjects, but there are a host of variables other than exercise training per se that could explain these results.[49]

Cortisol. Cortisol levels are elevated during intense exercise, while changes during light exercise are associated with the psychologic stress-induced cortisol production that is most apparent in the untrained. For example, it appears that plasma cortisol levels tend to increase less in trained compared to sedentary subjects during the same moderate exercise performance.[89] With repeated bouts of high intensity exercise and correspondingly high cortisol output, adrenal gland enlargement can occur due to both cellular hypertrophy and hyperplasia.[90]

Epinephrine and Norepinephrine. Because sympathetic nerve endings, including those to the adrenal gland, secrete both epinephrine and norepinephrine, it is appropriate to discuss the "sympathoadrenal" response to exercise and training, rather than simply the adrenal gland response. About 75% of the secretion of the adrenal medulla is epinephrine, while most of the production of norepinephrine is by the sympathetic nerve endings in blood vessels and other tissues.

Increases in norepinephrine concentration have been found even at fairly light exercise levels, with dramatic 2- to 6-fold increases at maximal work-loads. In contrast, epinephrine levels are modest during light exercise but increase rapidly during intense exercise.[10,37] This suggests that repeated, heavy exercise is required to produce massive involvement of the adrenals. **The sympathoadrenal response to exercise is related to the relative rather than to the absolute amount of exercise performed.** It follows, therefore, that the sympathoadrenal response, especially for norepinephrine, to a given sub-maximum workload would be lower in the trained than the untrained in-dividual.[35,53] This reduction in blood catecholamine levels with training is also observed for the resting state.[15] Such an adaptation to training is a favorable response because it contributes to a lowering of myocardial oxygen demands both at rest and during submaximal exercise, as well as during other forms of stress.[72]

Figure 20–12 illustrates the reduction in epinephrine and norepinephrine response to standard heavy exercise during physical training. As can be seen, the concentration of both hormones falls dramatically during the first several weeks of training. The appearance of bradycardia and smaller rise in blood

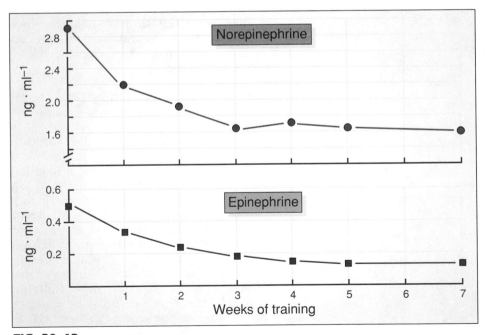

FIG. 20–12.
Week-by-week changes in catecholamines during a 5-min exercise bout at 243 watts for six male subjects. Training consisted of running and stationary cycling for 6 days a week. (From Winder, W.W., et al.: Time course of sympathoadrenal adaptations to endurance exercise training in man. J. Appl. Physiol., 45:370, 1978.)

pressure during submaximal exercise with training is probably the most familiar reflection of this sympathoadrenal response.

Pancreatic Hormones

Insulin and Glucagon. Because mild exercise alters circulating blood glucose levels, it is reasonable to expect a corresponding exercise-glucostatic response. This response includes immediate sympathetic discharge, an elevation of plasma glucagon levels, and a decrease in plasma insulin. The effect of increased sympathetic and glucagon output is to increase blood glucose by liver glycogenolysis. Moreover, glucagon also provides an alternative glucose supply by stimulating the process of gluconeogenesis. Concurrently, the decrease in insulin release appears to facilitate both of these processes that are geared to maintain adequate levels of blood glucose.

Studies of training-induced changes in insulin and glucagon reveal several interesting findings. For one thing, it appears that as a consequence of endurance training, the levels of insulin and glucagon in the blood are maintained closer to resting values in response to exercise. This seemingly blunted hormonal response to exercise training is accomplished in one of several ways. It is possible, for example, that physical training actually increases a person's sensitivity to insulin. Consequently, less insulin would be required to regulate blood glucose after training than before.[33a,36,80] This **improved insulin sensitivity** is probably related to the binding capacity of insulin to receptor sites of individual muscle cells. There is also an increase in hepatic insulin sensitivity.[33] During heavy exercise, the trained person is also able to spare carbohydrate and use fat as a fuel; thus, there would be a lowered insulin requirement. **In essence, the trained state requires less insulin at any stage from rest through light to moderate intensities of exercise.** In addition, a smaller insulin output is required to clear any excess glucose from the circulation. This is characteristic of an enhanced insulin effectiveness with training.

Exercise Training in Diabetes. Despite the clear association between physical training and improved insulin sensitivity, the clinical utility of physical training in improving glucose control in Type I diabetes has not been clearly demonstrated. Studies suggest that improved overall glycemic control in these patients is difficult to demonstrate during physical training, despite sustained improvements in insulin sensitivity.[58]

As a consequence of obesity, many overweight individuals have a reduced glucose tolerance that results in hyperinsulinemia and a generalized insulin resistance. Many of these individuals develop adult-onset diabetes. In this situation, exercise training can often reduce resting plasma insulin levels and lower insulin output during a glucose tolerance test, both of which are evidence for improved insulin sensitivity.[5] **Improved insulin sensitivity with exercise training is important "therapy" that can ultimately result in a lowered insulin requirement for those individuals who are insulin-dependent.** Some diabetics, however, must be cautious because exercise can actually trigger a dual response: an enhanced glucose uptake and a greater insulin supply due to the rapid circulation that accompanies exercise.[67] Such a response would actually worsen the imbalance between glucose supply and utilization and could result in serious complications from hypoglycemia. In these cases, the site of insulin injection has proven to be important in terms of insulin action in the exercising diabetic.[5,107]

Diabetics can safely engage in regular heavy exercise, but they must be cautious and proceed under medical supervision to appropriately monitor blood glucose and insulin.

SPECIAL ENDOCRINE CONSIDERATIONS

OPIOID PEPTIDES AND EXERCISE

In the 1970s, scientists studying the pain-relieving effects of opioid peptides on brain function (e.g., morphine) discovered that these substances had neurotransmitter properties that probably acted on specific opioid receptor sites in the brain. With this finding came the realization that perhaps the brain itself produced endogenous opioid-like, mood-altering substances.[56,81]

Evidence for the existence of endogenous substances that behaved like opiates was first provided in the mid-1970s with the isolation and purification of two opioid pentapeptides, **methionine** and **leucine enkephalin** (enkephalin is Greek meaning "in the brain.")[50,97] These opioids are part of a larger proopiocortin precursor molecule produced in the anterior part of the pituitary gland.[46] Several other opioid substances have been identified; these include **beta-lipotrophin, beta-endorphin,** and most recently, **dynorphin,** the most potent of the opioid peptides. Also originating from this pro-opiocortin precursor molecule is the pituitary hormone, ACTH. The effects of the various endogenously produced opioids are widespread, ranging from their function as a neurohormone to that of a neurotransmitter. It is generally accepted that endogenous opiates produce a strong inhibition of hormonal release from the posterior pituitary, especially the release of LH and FSH.[50] On the other hand, release of GH and PRL are stimulated by the opioid peptides.

Research shows that the serum concentrations of endogenous opioids, primarily beta-endorphin and/or beta-lipotrophin, increase in response to exercise[46,48,68,76] with a significant variation in the magnitude of the response noted among individuals.[34] The elevation of beta-endorphin in exercising men and women has ranged to as high as five times the resting level,[87] with even higher values probably occurring in the brain itself.[56] While the precise physiologic significance of the various endogenous opioid peptides' response to exercise remains unclear, several important effects must be considered. The most notable postulated endorphin effect has been its role in triggering a so-called "exercise high"—a state of euphoria when the exerciser experiences an exhilaration and other noted mood-improving and relaxing benefits[87] as the duration of moderate to intense exercise progresses. Endorphin secretion has also been implicated in increased pain tolerance, improved appetite control, and a reduction in anxiety, tension, anger, and confusion.[81,112]

Endorphins have been implicated in the regulation of the female menstrual cycle, as well as in influencing the response of numerous other hormones, including GH, ACTH, prolactin, catecholamines, and cortisol.[77] The endogenous opioids are also involved in regulating and modulating secretion of the hormones LH and FSH. This regulation is thought to be inhibitory in nature and, therefore, may possibly play a key role in menstrual cycle disturbances such as delay in menarche, dysfunctional uterine bleeding, secondary amenorrhea, and inadequacy of the luteal phase.[3]

The effect of training on endorphin response is controversial because only limited data are available.[1,26,59,79] In one study, no significant changes were found in beta-endorphin response to prolonged exercise following 8 weeks of endurance training. In contrast, another study showed that physical training augmented the responses of beta-endorphins and beta-lipotrophins to exercise.[26] Recent studies indicate that different types of training may bring about differing effects on the release of these hormones during exercise, with anaerobic or sprint-type training producing the largest hormonal increase.[68] This suggests that anaerobic factors may affect the pattern of hormonal release. It is tempting to hypothesize that with physical training, the individual becomes more sensitive to opioid effects so that it would take less of a concentration of the hormone to induce a specific effect. In this sense, regular exercise could be viewed as a type of "positive addiction." An increased opioid sensitivity would be similar to the insulin response to physical training. It also appears that the opiates produced in the body during exercise are degraded more slowly in the blood of trained individuals compared to the untrained.[60] Certainly, this slower rate of disposal would facilitate a given opiate response and might augment one's tolerance for extended exercise. More research will certainly be forthcoming in this exciting field of study.

SUMMARY

1. The endocrine system consists of a host organ, a transmitted substance called a hormone, and a target or receptor organ. Hormones are either steroids or amino acid (polypeptide) derivatives.
2. The major function of hormones is to alter the rates at which cellular reactions take place. This function is achieved through hormones that act at specific receptor sites by either enhancing or inhibiting enzyme function.
3. Hormone concentration in the blood is determined by the amount of hormone synthesized, the amount released, the amount taken up by the target organ, and the rate of hormone removal from the blood.
4. The anterior pituitary is responsible for the secretion of at least six hormones: prolactin, the gonadotropic hormones FSH and LH, corticotropin, thyrotropin, and growth hormone.
5. Growth hormone promotes cell division and cellular proliferation; thyrotropin controls the amount of hormone secreted by the thyroid gland; ACTH regulates the output of the hormones secreted by the adrenal cortex; prolactin is important in reproduction and in the development of secondary sex characteristics of females; and FSH and LH stimulate the ovaries to secrete estrogen in females and sperm in males.
6. The posterior pituitary secretes the hormone vasopressin (ADH), which controls water excretion by the kidneys. It also secretes oxytocin, which is an important hormone in birthing and milk secretion.
7. Thyroxine increases the metabolic rate of all cells and increases the breakdown of carbohydrate and fat in energy metabolism.
8. The inner (medulla) and the outer (cortex) portions of the adrenal gland secrete two different types of hormones. The medulla secretes the catecholamines, epinephrine and norepinephrine. The adrenal cortex secretes mineralocorticoids, which regulate extracellular sodium and potassium, glucocorticoids, which stimulate gluconeogenesis and serve as an insulin antagonist, and androgens, which control male secondary sex characteristics.

9. The main function of insulin, secreted from the pancreas, is to increase the rate of glucose transport into cells and thereby control the rate of carbohydrate metabolism in the body. Insulin also has a pronounced effect on fat metabolism. Diminished insulin production results in the condition known as diabetes. The pancreas also secretes glucagon, an insulin antagonist that acts to raise the level of blood sugar.

10. Exercise training has differential effects on resting and exercise-induced hormone production and release. Trained persons exhibit elevated hormone response during exercise for ACTH and cortisol, depressed values for GH, prolactin, FSH, LH, testosterone, ADH, thyroxine, and insulin, and no training response for aldosterone, renin, and angiotensin.

11. Exercise-induced elevation of beta-endorphins has been associated with euphoria, increased pain tolerance, the "exercise high," and altered menstrual function.

REFERENCES

1. Aforzo, G.A., et al.: In vivo opioid receptor occupation in the rat brain following exercise. *Med. Sci. Sports Exerc.*, 18:380, 1986.
2. Ahlborg, G., et al.: Influence of lactate infusion on glucose and FFA metabolism in man. *Scand. J. Clin. Lab. Invest.*, 36:193, 1976.
3. Baker, E.R.: Menstrual dysfunction and hormonal status in athletic women: a review. *Fertil. Steril.*, 36:691, 1981.
4. Baker, E.R., et al.: Female runners and secondary amenorrhea: correlation with age, parity, mileage, and plasma hormone and sex-hormone binding globulin concentrations. *Fertil. Steril.*, 36:183, 1981.
5. Berger, M., et al.: Pharmacokinetics of subcutaneously injected tritiated insulin: effects of exercise. *Diabetes*, 28 (Suppl. 1), 1979.
6. Berger, D., et al.: The effect of adrenergic receptor blockage on the exercise-induced rise in pancreatic polypeptide in man. *J. Clin. Endocrinol. Metab.*, 50:33, 1980.
7. Bjorkman, O., et al.: Influence of hypoglucagonemia on splanchnic glucose output during leg exercise in man. *Clin. Physiol.*, 1:43, 1981.
8. Björntrop, P., et al.: Physical training in human hyperplastic obesity. IV. Effects on the hormonal system. *Metabolism*, 26:319, 1977.
9. Björntrop, P.: The effects of exercise on plasma insulin. *Int. J. Sports Med.*, 2:125, 1981.
10. Bloom, S.R., et al.: Differences in the metabolic and hormonal responses to exercise between racing cyclists and untrained individuals. *J. Physiol.*, 258:1, 1976.
11. Bonen, A., et al.: Effects of exercise on serum concentrations of FSH, LH, Progesterone and Estradiol. *Eur. J. Appl. Physiol.*, 42:15, 1979.
12. Bonen, A., et al.: Effects of exercise and training on menstrual cycle hormones. *Aust. J. Sports Med.*, 10:39, 1978.
13. Borer, K.T.: Exercise-induced facilitation of pulsatile growth hormone (GH) secretion and somatic growth. In *Hormones and Sport*. Vol. 55. Edited by Z. Laron and A.D. Rogol. Serono Symposia Publications. New York, Raven Press, 1989.
14. Borer, K.T., and Pearson, W.: Adrenergic restraint of somatostatin action promotes pulsatile growth hormone secretion in hamsters (Abstr.). Program of the 68th Annual Meeting of the Endocrine Society. Cornell University, NY, 1986.
15. Bove, A.A., et al.: Increased conjugated dopamine in plasma after exercise training. *J. Lab. Clin. Med.*, 104:77, 1984.
16. Boyden, T.W., et al.: Prolactin responses, menstrual cycles and body composition of women runners. *J. Clin. Endocrinol. Metab.*, 58:711, 1982.
17. Boyden, T.W., et al.: Evidence for mild thyroidal impairment in women undergoing endurance training. *J. Clin. Endocrinol. Metab.*, 54:53, 1982.
18. Boyden, T.W., et al.: Thyroidal changes associated with endurance training in women. *Med. Sci. Sports Exerc.*, 16:234, 1984.
19. Brandenberger, G.: Cortisol responses to exercise and other interactions with diurnal

secretory peaks. In: *Exercise Endocrinology*. Edited by F. Fotherby and S.B. Pal. Berlin, Walter de Gruyter, 1985.

20. Brandenberger, G., and Follenius, M.: Influence of timing and intensity of muscular exercise on temporal patterns of plasma cortisol levels. *J. Clin. Endocrinol. Metab.*, 54:592, 1982.
21. Bray, G.: *The Obese Paitent*. Vol. 9. Major Problems in Internal Medicine. Philadelphia, W.B. Saunders, 1976.
22. Brillon, D., et al.: Cholinergic but not serotonergic mediation of exercise-induced growth hormone secretion. *Endocrine Research*, 12:137, 1986.
23. Brisson, D., et al.: Exercise-induced dissociation of the blood prolactin response in young women according to their sports habits. *Horm. Metab. Res.*, 12:201, 1980.
24. Bunt, J.C., et al.: Sex and training differences in human growth hormone levels during prolonged exercise. *J. Appl. Physiol.*, 61:1796, 1986.
25. Cappa, M.R., et al.: Modification of somatostatinergic tone induced by physical exercise. In *Hormones in Sport*. Vol. 55. Edited by Z. Laron and A.D. Rogol. Serono Symposia Publications. New York, Raven Press, 1989.
26. Carr, D.B., et al.: Physical conditioning facilitates exercise induced secretion of beta-endorphins and beta-lipotrophin in women. *N. Engl. J. Med.*, 305:560, 1981.
27. Convertino, V.A., et al.: Exercise induced hypervolemia: role of plasma albumin, renin and vasopressin. *J. Appl. Physiol.*, 48:665, 1980.
28. Convertino, V.A., et al.: Plasma volume, osmolarity, vasopressin and renin activity during graded exercise in man. *J. Appl. Physiol.*, 50:123, 1981.
29. Cook, N.J., et al.: Changes in adrenal and testicular activity monitored by salivary sampling in males throughout marathon runs. *J. Appl. Physiol.*, 55:634, 1986.
30. Cumming, D.C., and Rebar, R.W.: Exercise and reproductive function in women. *Prog. Clin. Biol. Res.*, 117:113, 1983.
31. Cumming, D.C., et al.: Reproductive hormone increases in response to acute exercise in men. *Med. Sci. Sports Exerc.*, 18:369, 1986.
32. De Souza, M.J., et al.: Menstrual status and plasma vasopressin, renin activity, aldosterone and exercise responses. *J. Appl. Physiol.*, 67:736, 1989.
33. Devlin, J.T., et al.: Enhanced peripheral and splanchnic insulin sensitivity in NIDDM men after a single bout of exercise. *Diabetes*, 36:434, 1987.
33a. Dolkas, C.B., et al.: Effect of body weight gain on insulin sensitivity after retirement from exercise training. *J. Appl. Physiol.*, 68:520, 1990.
34. Donevan, R.H., and Andrew, G.M.: Plasma B-endorphin immunoreactivity during graded cycle ergometry. *Med. Sci. Sports Exerc.*, 19:229, 1987.
35. Duncan, J.J., et al.: The effects of aerobic exercise on plasma catecholamines and blood pressure in patients with mild hypertension. *JAMA*, 254:2609, 1985.
36. Farrell, P.A.: Decreased insulin response to sustained hyperglycemia in exercise trained rats. *Med. Sci. Sports Exerc.*, 22:469, 1988.
37. Farrell, P.A., et al.: Enkephalins, catecholamines, and psychological mood alterations: effects of prolonged exercise. *Med. Sci. Sports Exerc.*, 19:347, 1987.
38. Felig, P., et al.: Plasma glucagon levels in exercising man. *N. Engl. J. Med.*, 287:184, 1972.
39. Felig, J.L., et al.: Age related agumentation of plasma catecholamines during dynamic exercise in healthy males. *J. Appl. Physiol.*, 59:1033, 1985.
40. Galbo, H.: Endocrinology and metabolism in exercise. *Int. J. Sports Med.*, 2:203, 1981.
41. Galbo, H.: *Hormonal and Metabolic Adaptation to Exercise*. New York, G.T. Verlag, 1983.
42. Galbo, H., et al.: Glucagon and plasma catecholamine responses to graded and prolonged exercise in man. *J. Appl. Physiol.*, 38:70, 1975.
43. Galbo, H., et al.: Catecholamines and pancreatic hormones during autonomic blockade in exercising man. *Acta. Physiol. Scand.* 101:428, 1977.
44. Galbo, H., et al.: Thyroid and testicular hormone response to graded and prolonged exercise in man. *Eur. J. Appl. Physiol.*, 36:101, 1977.
45. Galbo, H, et al.: The effect of fasting on the hormonal response to graded exercise. *J. Clin. Endocrinol. Metab.*, 52:1106, 1981.
46. Gambert, S.R., et al.: Exercise and endogenous opioids. *N. Engl. J. Med.*, 305:1590, 1981.
47. Geyssant, A., et al.: Plasma vasopressin, renin activity, and aldosterone: effects of exercise and training. *Eur. J. Appl. Physiol.*, 46:21, 1980.

48. Goldfarb, A.H., et al.: Serum β-endorphin levels during a graded exercise test to exhaustion. *Med. Sci. Sports Exerc.*, 19:78, 1987.

49. Greenleaf, J.E., et al.: Exercise training hypotension: implications for plasma volume, renin, and vasopressin. *J. App. Physiol.*, 51:298, 1981.

50. Grossman, A., and Sutton, J.R.: Endorphins: What are they? How are they measured? What is their role in exercise? *Med. Sci. Sports Exerc.*, 17:74, 1985.

51. Guyton, A.: *Physiology of the Human Body*, 6th ed. Philadelphia, W.B. Saunders, 1984.

52. Hackney, A.C., et al.: Reproductive hormonal profiles of endurance-trained and untrained males. *Med. Sci. Sports Exerc.*, 20:60, 1988.

53. Haggendal, L., et al.: Arterial noradrenaline concentration during exercise in relation to the relative work levels. *Scand. J. Lab. Invest.*, 26:337, 1970.

54. Häkkinen, K., et al.: Neuromuscular and hormonal adaptations in athletes to strength training in two years. *J. Appl. Physiol.*, 65:2406, 1988.

55. Hale, R.W., et al.: A marathon: The immediate effect on female runners luteinizing hormone, follicle-stimulating hormone, prolactin, testosterone, and cortisol levels. *Am. J. Obstet. Gynecol.*, 146:550, 1983.

56. Harbor, V.J., and Sutton, J.R.: Endorphins and exercise. *Sports Med.*, 1:154, 1984.

57. Heyward, V.H., et al.: Gender differences in strength. *Res. Q. Exerc. Sport*, 57:154, 1986.

58. Holm, G., and Stromblad, G.: Type I diabetes and physical exercise. *Acta Med. Scand.*, 671(suppl): 95, 1983.

58a. Houmard, J.A., et al.: Testosterone, cortisol, and creatine kinase levels in male distance runners during reduced training. *Int. J. Sports Med.*, 11:41, 1990.

59. Howlett, T.A., et al.: Release of beta-endorphin and met-enkephalin during exercise in normal women: Response to training. *Br. Med. J.*, 288:1950, 1984.

60. Jaskowski, M.A., et al.: Enkephalin metabolism: effect of acute exercise stress and cardiovascular fitness. *Med. Sci. Sports Exerc.*, 21:154, 1989.

61. Johannessen, A., et al.: Prolactin, growth hormone, thyrotropin, 3,5,3-triiodothyronine, and thyroxine response to exercise after fat and carbohydrate enriched diet. *J. Clin. Endocrinol. Metab.*, 52:56, 1981.

62. Jurkowski, J., et al.: Ovarian hormone response to exercise. *J. Appl. Physiol.*, 44:109, 1978.

63. Kannel, W.B., and McGee, D.L.: Diabetes and cardiovascular risk factors: the Framingham study. *Circulation*, 59:8, 1979.

64. Karagiorgos, A., et al.: Growth hormone response to continuous and intermittent exercise. *Med. Sci. Sports*, 11:302, 1979.

65. Keizer, H.A., et al.: Effect of a 3 month endurance training program on metabolic and multiple hormonal responses to exercise. *Int. J. Sports Med.*, 8(Suppl. 3):150, 1987.

66. Knobil, E.: The neuroendocrine control of the menstrual cycle. *Recent Prog. Horm. Res.*, 36:53, 1980.

67. Koivisto, V.A., and Felig, P.: Effects of leg exercise on insulin absorption in diabetic patients. *N. Engl. J. Med.*, 298:79, 1978.

68. Kraemer, W.J., et al.: Training responses of plasma beta-endorphin, adrenocorticotropin, and cortisol. *Med. Sci. Sports Exerc.*, 21:146, 1989.

69. Kuoppasalmi, K., et al.: Plasma cortisol, androstenedione, testosterone and luteinizing hormone in running exercise of different intensities. *Scand. J. Clin. Lab. Invest.*, 40:403, 1980.

70. Lassarre, C., et al.: Kinetics of human growth hormone during submaximum exercise. *J. Appl. Physiol.*, 37:826, 1974.

71. Ljunghall, S., et al.: Increase in serum parathyroid hormone levels after prolonged physical exercise. *Med. Sci. Sports Exerc.*, 20:122, 1988.

72. Lockette, W., et al.: Endurance training and human α_2-adrenergic receptors on platelets. *Med. Sci. Sports Exerc.*, 19:7, 1987.

73. Luger, A., et al.: Acute hypothalamic-pituitary-adrenal responses to the stress of treadmill exercise: Physiologic adaptations to physical training, *N. Engl. J. Med.*, 316:1309, 1987.

74. MacIntyre, J.G.: Growth hormone and athletes. *Sports Med.*, 4:129, 1987.

75. Maher, J.T., et al.: Aldosterone dynamics during graded exercise at sea level and high altitude. *J. Appl. Physiol.*, 39:18, 1975.

76. Mahler, D.A., et al.: β-endorphin activity and hypercapnic ventilatory responsiveness after marathon running. *J. Appl. Physiol.*, 66:2431, 1989.

77. McArthur, J.W.: Endorphins and exercise in females: possible connection with reproductive dysfunction. *Med. Sci. Sports Exerc.*, 17:82, 1985.
78. Melin, B., et al.: Plasma AVD, neurophysin, renin activity and aldosterone during submaximal exercise performed until exhaustion in trained and untrained men. *Eur. J. Appl. Physiol.*, 44:141, 1980.
79. Metzger, J.M., and Stein, E.A.: Beta-endorphin and sprint training. *Life Sci.*, 34:1541, 1984.
80. Mikines, K.J., et al.: Insulin action and insulin secretion; effects of different levels of physical activity. *Can. J. Sports Sci.*, 12:113, 1987.
81. Morgan, W.P.: Affective beneficence of vigorous physical activity. *Med. Sci. Sports Exerc.*, 17:94, 1985.
82. Naveri, H., et al.: Metabolic and hormonal changes in moderate and intense long-term running exercises. *Int. J. Sports Med.*, 6:276, 1985.
83. Newmark, S.R., et al.: Adrenalcortical response to marathon running. *J. Clin. Endocrinol. Metab.*, 42:393, 1976.
84. Noel, G.L., et al.: Human prolactin and growth hormone release during surgery and other conditions of stress. *J. Clin. Endocrinol. Metab.*, 35:840, 1972.
85. Pakarinen, A., et al.: Serum thyroid hormones, thyrotropin and thyroxine binding globulin during prolonged strength training. *Eur. J. Appl. Physiol.*, 57:394, 1988.
86. Prior, J.C., et al.: Prolactin changes with exercise vary with breast motion: Analysis of running versus cycling. *Fertil. Steril.*, 36:268, 1981.
87. Rahkila, P., et al.: Response of plasma endorphins to running exercises in male and female endurance athletes. *Med. Sci. Sports Exerc.*, 19:451, 1987.
88. Rogol, A.D., et al.: Pulsatile secretion of gonadotropins and prolactin in endurance-trained men: relation to endogenous opiate system. *J. Androl.*, 5:21, 1984.
89. Shephard, R.J., and Sidney, K.H.: Effects of physical exercise on plasma growth hormone and cortisol levels in human subjects. In *Exercise and Sport Sciences Reviews*. Vol. 3. Edited by J.H. Wilmore. New York, Academic Press, 1975.
90. Song, M.K., et al.: The mode of adrenal gland enlargement in the rat in response to exercise training. *Pflugers Arch.*, 339:59, 1973.
91. Strauss, R.H., et al.: Weight loss in amateur wrestlers and its effects on serum testosterone levels. *JAMA*, 254:3337, 1985.
92. Sutton, J.R.: Hormonal and metabolic responses to exercise in subjects of high and low work capacities. *Med. Sci. Sports*, 10:1, 1978.
93. Sutton, J.R., and Farrell, P.: Endocrine responses to prolonged exercise. In *Exercise Science and Sports Medicine*. Vol. 1. Edited by D.R. Lamb, and R. Murray. Indianapolis, Benchmark Press, 1988.
94. Sutton, J.R., et al.: The hormonal response to physical exercise. *Aust. Ann. Med.*, 18:84, 1969.
95. Sutton, J.R., et al.: Plasma vasopressin, catecholamines and lactate during exhaustive exercise at extreme simulated altitude: "Operation Everest II." *Can. J. Appl. Sports Sci.*, 11:43P, 1986.
96. Tarnopolsky, L., et al.: Gender differences in hormonal and metabolic responses to prolonged exercise in males and females. *J. Appl. Physiol.*, 68:650, 1990.
97. Terenius, L., and Wahlstrom, A.: Inhibitors of narcotic receptor binding in brain extracts and cerebrospinal fluid. *Acta Pharmacol. Toxicol.* (Copenh.), 35 (Suppl. 87):55, 1974.
98. Terjung, R.: Endocrine Response to Exercise. In *Exercise and Sport Sciences Reviews*. Vol. 7. Edited by R.S. Hutton and D.I. Miller. Philadelphia, Franklin Institute Press, 1979.
99. Terjung, R., and Tipton, C.M.: Plasma thyroxine and thyroid-stimulating hormone levels during submaximal exercise in humans. *Am. J. Physiol.*, 220:1840, 1971.
100. Unger, R.H., et al.: Insulin, glucagon and somatostatin secretion in the regulation of metabolism. *Ann. Rev. Physiol.*, 40:307, 1978.
101. Victor, R.G., et al.: Differential control of heart rate and sympathetic nerve activity during dynamic exercise. *J. Clin. Invest.*, 79:508–516, 1987.
102. Vitug, A., et al.: Exercise and Type I diabetes mellitus. In *Exercise and Sport Sciences Reviews*. Vol. 16. Edited by K.B. Pandolf. New York, Macmillan, 1988.
103. Vogel, R.B., et al.: Increase of free and total testosterone during submaximal exercise in normal males. *Med. Sci. Sports Exerc.*, 17:119, 1985.
104. Wade, C.E.: Response, regulation, and actions of vasopressin during exercise: a review. *Med. Sci. Sports Exerc.*, 16:506, 1984.

105. Wade, C.E., and Claybaugh, J.R.: Plasma renin activity, vasopressin concentration, and urinary excretory responses to exercise in men. *J. Appl. Physiol.*, 49:930, 1980.
106. Wade, C.E., et al.: Plasma aldosterone and renal function in runners during a 20-day road race. *Eur. J. Appl. Physiol.*, 54:456, 1985.
107. Wahren, J.: Glucose turnover during exercise in healthy man and in patients with diabetes mellitus. *Diabetes*, 28:82, 1979.
108. Ward, M.M., et al.: Exercise and plasma catecholamine release. In *Exercise Endocrinology.* Edited by K. Fotherby, and S.P. Pal. Berlin, Walter de Gruyter, 1985.
109. Wheeler, G.D.: Reduced serum testosterone and prolactin levels in male distance runners. *JAMA*, 252:514, 1984.
110. Winder, W.W.: Time course of the T_3- and T_4-induced increase in rat soleus muscle mitochondria. *Am. J. Physiol. Cell Physiol.*, 5:C132, 1979.
111. Winder, W.W., and Heninger, V.: Effect of exercise on tissue levels of thyroid hormones in the rat. *Amer. J. Physiol.*, 221:1139, 1971.
112. Yates, A., et al.: Running—an analogue of anorexia. *N. Engl. J. Med.*, 308:251, 1983.

APPLIED EXERCISE PHYSIOLOGY

ENHANCEMENT OF ENERGY CAPACITY

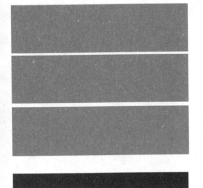

In many cases, exercise training is more art than science. The success of different conditioning programs is usually evaluated by individual achievements or won-loss records rather than by scientific inquiry and discovery. Too often, coaches of sports such as basketball or soccer place considerable importance on the development of cardiovascular or aerobic capacity and devote little time to various phases of vigorous anaerobic conditioning. While these sports require a relatively steady release of aerobic energy, there are crucial situations that demand all-out effort. If the relative capacity of the athlete's anaerobic energy transfer system is poor, the player may be unable to perform at full potential. Training the anaerobic capacity of endurance athletes, on the other hand, would be wasteful because the contribution of anaerobic energy to successful performance is minimal. Rather, these activities demand a well-conditioned heart and vascular system capable of circulating large quantities of blood as well as a high capacity of muscle cells to generate ATP aerobically. At the other extreme, one's capacity for aerobic metabolism contributes little to overall success in sprint activities and sports such as football. Here, performance largely depends on muscular strength and power where energy is generated primarily from reactions that do not utilize oxygen.

With a clear understanding of energy transfer and the effects of specific training on the systems of energy delivery and utilization, it should be possible to construct a sound training program to achieve optimum performance. In Chapters 21 and 22 we discuss the basis of training for aerobic and anaerobic power and muscular strength, the physiologic consequences of such training, and the important factors that affect training success. In Chapter 23, we take a closer look at factors purported to increase human exercise performance.

21

TRAINING FOR ANAEROBIC AND AEROBIC POWER

Throughout this book we have stressed that different activities, depending on their duration and intensity, require the activation of specific energy systems. This is shown in Figure 21–1 in which exercise is broadly classified in terms of duration and predominant energy pathways.

We realize that it is difficult to place certain activities in one category. For example, as a person increases aerobic fitness, an activity previously classified as anaerobic might be reclassified as aerobic. In many cases, all three energy transfer systems, the ATP-CP system, the glycolytic or lactic acid system, and the aerobic system, operate at different times during exercise. Their relative contributions to the energy continuum are directly related to the duration and intensity (power output) of the specific activity.

Brief power activities that last up to about 6 seconds rely almost exclusively on "immediate" energy generated from the breakdown of the stored intramuscular phosphates, ATP and CP. Consequently, power athletes like sprinters must gear training to improve the capacity of this energy transfer system. As all-out exercise progresses to 60 seconds and power output becomes somewhat reduced, the major portion of energy is still generated through anaerobic pathways. These metabolic reactions involve the short-term energy system of glycolysis and subsequent lactic acid formation. As exercise intensity diminishes somewhat, and duration extends to 2 to 4 minutes, reliance on energy from phosphate stores and anaerobic glycolysis decreases, whereas the aerobic production of ATP becomes increasingly more important. Prolonged exercise progresses on a "pay-as-you-go" basis with more than 99% of the energy requirement being generated by aerobic reactions. Clearly, an efficient training program is one that allocates a proportionate commitment to training the specific energy systems involved in the activity. In the sections that follow, we discuss anaerobic and aerobic conditioning with special emphasis on principles, methods, and short- and long-term adaptations. The approach to physiologic conditioning is basically the same for men and women within a broad age range; both respond and adapt to training in essentially the same manner.[19,24,78] Furthermore, as long as accepted principles of exercise training are followed, similar fitness improvements occur in exercise programs offered to the public compared to rigidly controlled training experiments carried out under laboratory conditions.[79]

PRINCIPLES OF TRAINING

The major objective in training is to cause biologic adaptations to improve performance in specific tasks. This requires adherence to carefully planned and executed activities. Attention is focused on factors such as frequency

FIG. 21–1.
Classification of activities based on duration of performance and the predominant intracellular energy pathways.

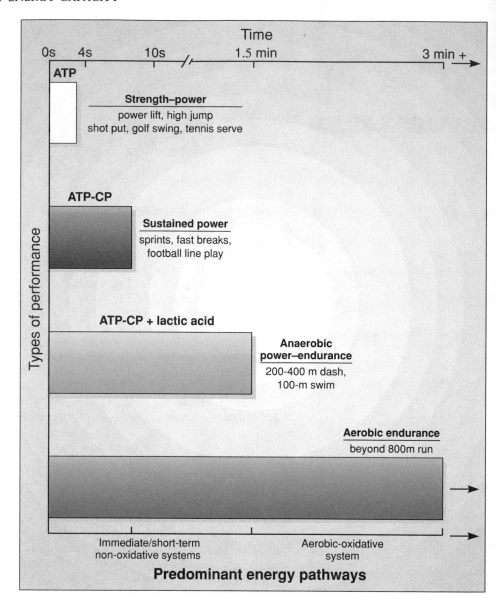

and length of workouts, type of training, speed, intensity, duration, and repetition of the activity, and appropriate competition. Although these factors vary depending on the performance goal, it is possible to identify several principles of physiologic conditioning common to the performance classifications shown in Figure 21–1.

OVERLOAD PRINCIPLE

A specific exercise **overload** must be applied to enhance physiologic improvement effectively and to bring about a training change. By exercising at a level above normal, a variety of training adaptations enable the body to function more efficiently. The appropriate overload for each person can be achieved by manipulating combinations of training **frequency, intensity, mode,** and **duration.**

This concept of individualized and progressive overload applies to the athlete, the sedentary person, the disabled, and even the cardiac patient. In fact, an increasing number in this latter group use an appropriately formulated exercise rehabilitation program to walk, jog, and eventually run and complete marathons![59]

SPECIFICITY PRINCIPLE

When applied to training, specificity refers to adaptations in the **metabolic** and **physiologic** systems depending on the type of overload imposed. It is known that a specific exercise stress such as strength–power training induces specific strength–power adaptations, and that specific aerobic or cardiovascular exercise elicits specific endurance-training adaptations with essentially no interchange between strength and aerobic training.[43,50,52,97,98] The specificity principle, however, goes beyond this because development of aerobic fitness for swimming,[66] bicycling,[77,100], running[68] or arm exercise[63,67] is most effectively achieved when the exerciser trains the specific muscles involved in the desired performance. **Specific exercise elicits specific adaptations creating specific training effects.**

Specificity of max $\dot{V}O_2$

Table 21–1 shows the results of a study in one of our laboratories designed to investigate specificity of endurance swim training on improvements in maximal aerobic power.[66] Fifteen men trained 1 hour a day, 3 days a week, for 10 weeks. Before and after the training, the max $\dot{V}O_2$ of all subjects was measured during both treadmill running and swimming. A large amount of

TABLE 21–1.
EFFECTS OF 10 WEEKS OF INTERVAL SWIM TRAINING ON CHANGES IN MAX $\dot{V}O_2$ AND ENDURANCE PERFORMANCE AS MEASURED DURING RUNNING AND SWIMMING*

Subjects	Measure	Running Test			Swimming Test		
		Pre-Training	Post-Training	% Change	Pre-Training	Post-Training	% Change
Swim Training	**Max $\dot{V}O_2$**						
	$l \cdot min^{-1}$	4.05	4.11	+1.5	3.44	3.82	+11.0
N = 15	$ml \cdot kg^{-1} \cdot min^{-1}$	54.9	55.7	+1.5	46.6	51.8	+11.0
	Max work time, min	19.6	20.5	+4.6	11.9	15.9	+34.0
Nontraining Controls	**Max $\dot{V}O_2$**						
	$l \cdot min^{-1}$	4.12	4.18	+1.5	3.51	3.40	3.1
N = 15	$ml \cdot kg^{-1} \cdot min^{-1}$	55.1	55.5	+0.7	46.8	45.0	−3.8
	Max work time, min	20.7	19.7	−4.8	11.5	11.5	0

* From Magel, J.R., et al.: Specificity of swim training on maximum oxygen uptake. J. Appl. Physiol., 38:151, 1975.

specificity was noted for the improvement in max $\dot{V}O_2$ with swim training. Because vigorous training exercise such as swimming provides a general overload to the central circulation, we had expected at least some improvement or "transfer" in aerobic power from swimming to running. This was not the case, however. In fact, if only treadmill running had been used to evaluate swim training effects, we would have mistakenly concluded that there was no training effect!

Based on available research,[62a,67,68,77] it is reasonable to advise that in training for specific aerobic activities like cycling, swimming, rowing, or running, the overload must engage the appropriate muscles required by the activity as well as provide an exercise stress for the central cardiovascular system.[30,39,84] Little improvement occurs when aerobic capacity is measured by a dissimilar exercise, yet improvements are significant when the test exercise is the same exercise used in training.[32] Thus, one can appreciate how difficult it is to be in "good shape" for diverse forms of aerobic exercise.

It is also noteworthy from Table 21–1, that although swimming max $\dot{V}O_2$ improved 11% with swim training, the maximum work time increased 34% during the swim test. Maximum $\dot{V}O_2$ improvements probably reach a peak in training, and thereafter, improvements in performance are supported by other mechanisms only partly related to the capacity of the oxygen transport system. These adaptations are most likely peripheral in nature.

Whereas the increase in aerobic fitness with training is highly specific, it appears that improvements in cardiac performance (e.g., ventricular contractility) are rather general in nature. Training studies indicate that improved heart function induced by one form of exercise has been observed with exercise of an untrained limb.[96] This indicates that the heart muscle per se can be conditioned under a variety of exercise modes.

Specificity of Local Changes

In endurance training, the overload of specific muscle groups enhances work performance and aerobic power by facilitating both oxygen transport and utilization at the local level.[39,50,84] The oxidative capacity of the vastus lateralis muscle, for example, is greater in well-trained cyclists than in endurance runners,[37] and is improved significantly following training on a bicycle ergometer.[38] Such adaptations would certainly increase the capacity of the trained muscles to generate ATP aerobically. The specificity of aerobic improvement may also result from greater regional blood flow in active tissues due either to increased microcirculation or to more effective distribution of cardiac output, or both. Regardless of the mechanism, such adaptations would occur only in the specifically trained muscles and would only be seen when these muscles were activated.

INDIVIDUAL DIFFERENCES PRINCIPLE

Many factors contribute to individual variation in training response. The person's relative fitness level at the start of training is important. It is unrealistic to expect different people to be in the same "state" of training at the same time. Consequently, it is counter-productive to insist that all performers of the same team (or even the same event) train the same way or at the same relative or absolute work rate. It is also unrealistic to expect all individuals to respond to a given training dosage in precisely the same manner. As discussed in Chapter 11, genetic factors clearly interact to influence the training response. **Training benefits are optimized when programs are planned to meet the individual needs and capacities of the participants.**

REVERSIBILITY PRINCIPLE

Detraining occurs rapidly when a person stops exercising. After only 1 or 2 weeks of detraining, significant reductions in both metabolic and working capacity can be measured, and many of the training improvements are lost within several months. Table 21–2 shows the physiologic and metabolic consequences of various durations of detraining as evaluated in several studies.[87] The research of one group of workers is especially interesting. For five subjects confined to bed for 20 consecutive days, max $\dot{V}O_2$ decreased 25% accompanied by a similar decrement in maximal stroke volume and cardiac output. This finding corresponded approximately to a 1% decrease in physiologic function each day. Additionally, the number of capillaries within trained muscle decreases between 14 and 25% within 3 weeks of detraining.[86]

The important point is that even among highly trained athletes the beneficial effects of exercise training are **transient** and **reversible**.[10,16,72] For this reason, most athletes begin a reconditioning program several months prior to the start of the competitive season or maintain some moderate level of sport-specific exercise to perhaps slow down the rate of deconditioning.[73] Many ex-athletes are in poorer physiologic condition several years after they retire from active participation than is the business executive who exercises on a regular basis.

TABLE 21–2.
CHANGES IN PHYSIOLOGIC AND METABOLIC VALUES RESULTING FROM VARIOUS DURATIONS OF DETRAINING

Study*	N	Sex	Duration (Days)	Variable	Pre-Detraining Average	Post-Detraining Average	Percent Change
87	5	M	20 (bedrest)	Max $\dot{V}O_2$, $l \cdot min^{-1}$	3.3	2.4	−27
				Stroke volume, $ml \cdot beat^{-1}$	116	88	−24
				Cardiac output, $l \cdot min^{-1}$	20	14.8	−26
22	7	F	84	Max $\dot{V}O_2$, $ml \cdot kg^{-1} \cdot min^{-1}$	47.8	40.4	−15.5
				$\dot{V}_E$ max, $l \cdot min^{-1}$	77.5	69.5	−10.3
				O_2 pulse, $ml \cdot beat^{-1}$	12.7	10.9	−14.2
69	17	M	70	Sum of 3-min recovery heart rate	190	237	−24.7
65	9	M	35	CP, $mmols \cdot g$ wet wt^{-1}	17.9	13.0	−27.4
				ATP, $mmols \cdot g$ wet wt^{-1}	5.97	5.08	−14.9
				Glycogen, $mmols \cdot g$ wet wt^{-1}	113.9	57.4	−49.6
				Elbow extension strength, ft-lb	39.0	25.5	−34.6
16	6	M	56	Max $\dot{V}O_2$, $l \cdot min^{-1}$	4.22	3.67	−14
	1	F		Max $\dot{V}O_2$, $ml \cdot kg \cdot min^{-1}$	62.1	53.2	−14
				HR max, $l \cdot min^{-1}$	187	199	+6
				Stroke volume, $ml \cdot beat^{-1}$	148	127	−14
				Cardiac output, $l \cdot min^{-1}$	27.8	25.2	−9
				Max a-$\bar{v}$ O_2 diff, $ml \cdot 100$ ml^{-1}	15.1	14.5	−19
				Citrate synthase, $mol \cdot kg$ protein$^{-1} \cdot h^{-1}$	10.0	6.0	−40.6
				SDH $mol \cdot kg$ protein$^{-1} \cdot h^{-1}$	4.43	2.73	−38.4

* Number in parentheses refers to study in reference list.

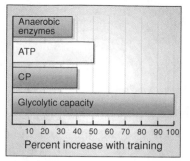

FIG. 21–2.
Increases in the anaerobic
functions of skeletal muscle as
a result of heavy physical
training.

PHYSIOLOGIC CONSEQUENCES OF TRAINING

Many of the biologic changes that accompany training have been presented in other sections throughout this book. We now summarize the various training adaptations outlined in Table 21–3.

ANAEROBIC SYSTEM CHANGES

Figure 21–2 summarizes the metabolic adaptations in **anaerobic function** that accompany strenuous physical training.

In keeping with the concept of specificity of training, activities that demand a high level of anaerobic metabolism bring about specific changes in the immediate and short-term energy systems, without a concomitant increase in aerobic functions. Specifically, the metabolic changes that occur with sprint and power-type training include:

TABLE 21–3.
TYPICAL METABOLIC AND PHYSIOLOGIC VALUES FOR HEALTHY TRAINED AND UNTRAINED MEN[a]

Variable	Untrained	Trained	Percent Difference[b]
Glycogen, mmol · g wet muscle^{-1}	85.0	120	41
Number of mitochondria, mmol^3	0.59	1.20	103
Mitochondrial volume, % muscle cell	2.15	8.00	272
Resting ATP, mmol · g wet muscle^{-1}	3.0	6.0	100
Resting CP, mmol · g wet muscle^{-1}	11.0	18.0	64
Resting creatine, mmol · g wet muscle^{-1}	10.7	14.5	35
Glycolytic enzymes			
Phosphofructokinase, mmol · g wet muscle^{-1}	50.0	50.0	0
Phosphorylase, mmol · g wet muscle^{-1}	4–6	6–9	60
Aerobic enzymes			
Succinate dehydrogenase, mmol · kg wet muscle^{-1}	5–10	15–20	133
Max lactic acid, mmol · kg wet muscle^{-1}	110	150	36
Muscle fibers			
Fast twitch, %	50	20–30	−50
Slow twitch, %	50	60	20
Max stroke volume, ml · beat^{-1}	120	180	50
Max cardiac output, l · min^{-1}	20	30–40	75
Resting heart rate, beats · min^{-1}	70	40	−43
Max heart rate, beats · min^{-1}	190	180	−5
Max a-v̄ O_2 diff, ml · 100 ml^{-1}	14.5	16.0	10
Max V̇O_2, ml · kg^{-1} · min^{-1}	30–40	65–80	107
Heart volume, l	7.5	9.5	27
Blood volume, l	4.7	6.0	28
V̇$_E$ max, l · min^{-1}	110	190	73
Percent body fat	15	11	−27

[a] In some cases, approximate values are used. In all cases, the trained values represent data from endurance athletes. Caution is advised in assuming that the percent differences between trained and untrained are necessarily the result of training, because genetic differences between individuals probably exert a strong influence on many of these factors.

[b] Computed as the percent that the trained differs from the corresponding value for the untrained.

1. **Increases in resting levels of anaerobic substrates.**[51,58,65] As determined from muscle biopsies taken before and after resistance training (Table 21–4), 28% improvement in strength was accompanied by significant increases in the muscle's resting levels of ATP, CP, free creatine, and glycogen.
2. **Increases in the quantity and activity of key enzymes that control the anaerobic phase of glucose breakdown.**[25,53,97] These changes are not of the magnitude observed for oxidative enzymes with aerobic training. The most dramatic alterations in anaerobic enzyme function and increase in fiber size occur in the fast-twitch muscle fibers.
3. **Increases in capacity for levels of blood lactate during all-out exercise that follow anaerobic training.**[53] This is probably due to enhanced levels of glycogen and glycolytic enzymes, as well as to improved motivation and "pain" tolerance to fatiguing exercise. Increased anaerobic exercise capacity is also observed concurrent with metabolic adaptations.[17,20,65]

AEROBIC SYSTEM CHANGES

Aerobic overload training significantly improves a variety of functional capacities related to oxygen transport and use.[49] The most notable adaptations accompanying aerobic training include the following:

1. **Mitochondria from trained skeletel muscle have a greatly increased capacity to generate ATP aerobically by oxidative phosphorylation.**[50]
2. **Associated with the increased capacity for mitochondrial oxygen uptake is an increase in both the size and number of mitochondria and a potential twofold increase in the level of aerobic system enzymes.**[5,60] These changes may be important in sustaining a high percentage of aerobic capacity during prolonged exercise. To a large extent, this training response is independent of the aging process.[75,102]
3. **Skeletal muscle myoglobin content of animals increases by as much as 80%.**[76] As a result, the quantity of oxygen within the cell at any one time is increased, which probably facilitates oxygen diffusion to the mitochondria. It has not been established whether exercise training induces similar increases of myoglobin in humans.

TABLE 21–4.
CHANGES IN RESTING CONCENTRATIONS OF CP, CREATINE, ATP, AND GLYCOGEN FOLLOWING 5 MONTHS OF HEAVY-RESISTANCE WEIGHT TRAINING IN NINE MALE SUBJECTS[a]

Variable[b]	Control	Post-Training	Percent Difference[c]
CP	17.07	17.94	+5.1
Creatine	10.74	14.52	+35.2
ATP	5.07	5.97	+17.8
Glycogen	86.28	113.90	+32.0

[a] From MacDougall, J.D., et al.: Biochemical adaptation of human skeletal muscle to heavy resistance training and immobilization. J. Appl. Physiol., 43:700, 1977.
[b] All values are expressed in mmol per gram of wet muscle tissue.
[c] All percent differences are statistically significant.

4. **There is an increase in the trained muscle's capacity to mobilize and oxidize fat.**[21,83] This occurs by an increase in blood flow within muscle and in the activity of fat-mobilizing and fat-metabolizing enzymes. At any submaximal work rate, a trained person uses more free fatty acids for energy than an untrained counterpart. This is beneficial to endurance athletes because it conserves the carbohydrate stores so important in prolonged exercise.

5. **Trained muscle exhibits a greater capability to oxidize carbohydrate.**[35] Consequently, large quantities of pyruvic acid move through the aerobic energy pathways. This is consistent with the increased oxidative capacity of the mitochondria and increased glycogen storage within the trained muscles.

6. **Aerobic training produces metabolic adaptations in the different types of muscle fibers.** It is generally believed that the basic fiber type does not "change" to any great extent, but instead, all fibers develop their already existing aerobic potential.[37,97]

7. **There may also be selective hypertrophy of different muscle fibers to specific overload training.**[37,98] Highly trained endurance athletes show larger slow-twitch fibers than fast-twitch fibers in the same muscle. Conversely, for athletes trained in predominantly anaerobic activities, the fast-twitch fibers occupy a greater cross-sectional area.

RELATED CARDIOVASCULAR AND RESPIRATORY CHANGES

Because the cardiovascular and respiratory systems are intimately linked with aerobic processes, related changes occur that are both functional and dimensional.

Heart Size

The weight and volume of the heart generally **increase** with long-term aerobic training.[71] A mild cardiac hypertrophy is a normal training adaptation characterized by an increase in the size of the left ventricular cavity as well as by a thickening of its walls. This cardiac enlargement returns to control levels with reduced training intensity.[48]

Blood Volume

Plasma volume and total hemoglobin tend to **increase** with endurance training.[61] This adaptation may enhance circulatory and thermoregulatory dynamics to facilitate oxygen delivery capacity during exercise.[12]

Heart Rate

Resting and submaximal exercise heart rate **decrease** during aerobic training, especially for previously sedentary individuals. Consequently, heart rate changes provide a convenient index to measure training improvement.

Stroke Volume

The heart's stroke volume **increases** significantly at rest and during exercise. Large stroke volumes are evident among endurance athletes and generally result from a large ventricular volume accompanied by enhanced myocardial contractility.

Cardiac Output

The most significant change in cardiovascular function with aerobic training is the **increase** in maximum cardiac output. Because the maximal heart rate may even decrease slightly, the heart's increased outflow capacity results directly from improved stroke volume. A large cardiac output is a major factor that distinguishes champion endurance athletes from well-trained and untrained individuals.

Oxygen Extraction

Training produces significant **increases** in the amount of oxygen extracted from the circulating blood.[67,85] An increase in the arteriovenous oxygen difference results from a more effective distribution of the cardiac output to working muscles, as well as of enhanced capacity of the trained muscle cells to extract and utilize oxygen. Among individuals 60 years of age and older, the significant increase in max $\dot{V}O_2$ with regular training appears to be mediated more through an increase in the maximal arteriovenous oxygen differences and less through an increase in maximum cardiac output and stroke volume.[88] For women, training adaptations appear to be independent of menopausal status.[15]

Blood Flow and Distribution

There is some indication that trained individuals perform **submaximal** exercise with a relatively **lower** cardiac output than untrained counterparts. This is probably the result of specific cellular changes with training. As the cell's ability to deliver, extract, and utilize oxygen increases, less regional blood flow is required to meet the muscle's oxygen needs. This decrease in muscle blood flow would "free" blood that could now be delivered to nonworking but important tissues such as the skin, liver, and kidneys. This could be of considerable benefit in sustaining prolonged submaximal exercise.

Of course, aerobic training causes large **increases** in total muscle blood flow during **maximal** exercise due to: (1) improvements in maximal cardiac output, and (2) redistribution of blood from nonworking areas that temporarily compromise their blood flow in response to all-out effort.

Blood Pressure

Regular aerobic training tends to **reduce** both systolic and diastolic blood pressure during rest and submaximal exercise. The largest decreases occur in systolic pressure and are most apparent in hypertensive subjects.

Respiratory Function

Increased breathing volumes accompany improvements in max $\dot{V}O_2$. Higher maximum ventilation is due to **increases** in both tidal volume and breathing frequency. In submaximal exercise, the trained person ventilates **less** than before training. This adaptation should be helpful in prolonged exercise because increased ventilatory economy means more oxygen availability to the working muscles.

PRACTICAL IMPLICATIONS

A schema is proposed to summarize adaptive changes in active muscle that accompany changes in max $\dot{V}_{O_2}$ with endurance training. As shown in Figure 21–3, maximal oxygen uptake increases about 15 to 30% in the first 3 months of intensive training and may rise as much as 50% over a 2-year period. When training stops, the aerobic capacity returns toward the pretraining level. The picture is even more impressive for the aerobic enzymes of the Krebs cycle and electron transport system. These enzymes that facilitate carbohydrate and fat breakdown increase rapidly and substantially throughout the training period in both fiber types and subdivisions. Conversely, a large amount of this metabolic adaptation is lost within a few weeks after training ceases.[10,14,16,62] The number of muscle capillaries increases throughout training.[2,7] This adaptation in blood supply is probably lost at a relatively slow rate with detraining.

As illustrated in Figure 21–3, intensive training lasting longer than 6 months causes an increase in the mitochondrial respiratory capacity of the trained muscles. This "local" metabolic improvement greatly outstrips the body's ability to circulate, deliver, and use oxygen during intense exercise (as demonstrated by the increase in max $\dot{V}_{O_2}$). In this phase of training, however, a muscle's lactate level (lower production or greater removal rate)

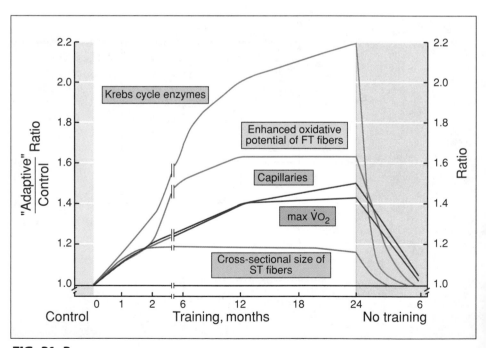

FIG. 21–3.
A schematic, and in part, hypothetical summary of some of the adaptations taking place in active muscle with endurance training. The graph is based on longitudinal and cross-sectional studies of humans. (From Saltin, B., et al.: Fiber types and metabolic potentials of skeletal muscles in sedentary man and endurance runners. *Ann. N.Y. Acad. Sci.*, 301:3, 1977.)

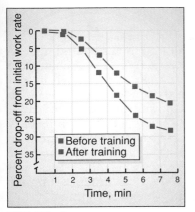

FIG. 21–4.
Percent drop-off from initial work rate before and after 10 weeks of endurance cycle training. (From Applied Physiology Laboratory, University of Michigan.)

may be much lower than that observed in submaximal exercise of similar relative intensity prior to training. **These cellular adjustments may account for a trained person being able to do prolonged steady-rate work at a larger percentage of max $\dot{V}O_2$.** To a large extent, endurance for sustained exercise may be more closely related to the oxidative capacity of mitochondria within specific muscles than to whole body oxygen consumption as reflected by the max $\dot{V}O_2$. In addition, the local metabolic and circulatory adaptations with long-term training may partially explain the lowered rate of glycogen utilization and concomitant increase in mitochondrial capability for free fatty acid metabolism in exercise.[36] This glycogen-sparing effect is important in prolonged strenuous exercise.

OTHER TRAINING-INDUCED CHANGES

Performance Changes

Enhanced exercise performance usually accompanies the physiologic adjustments with training. For example, a training study consisted of long-distance cycling for 40 to 60 minutes, 4 days per week for 10 weeks at an intensity of 85% of max $\dot{V}O_2$. The performance test consisted of trying to maintain a constant work rate of 265 watts for 8 minutes. As shown in Figure 21–4, there was significantly less drop-off from the prescribed initial pace following training.

Body Composition Changes

For the obese and borderline obese, regular endurance exercise causes a reduction in body weight accompanied by a decrease in body fat. Increases in lean body weight often accompany resistance training programs. If weight loss is attempted by **diet alone,** a significant reduction in lean body tissue also accompanies fat loss. When exercise is used alone or combined with diet, however, more of the weight lost is fat because exercise appears to have a conserving effect on the body's lean tissues.[4,6a,103]

Body Heat Transfer

Well-hydrated, trained individuals exercise more comfortably in hot environments because of a larger blood volume and a more responsive heat regulatory mechanism. Trained individuals dissipate heat faster and more economically and thereby cool the body more effectively. This means that the metabolic heat generated by exercise causes less heat strain and thus prolongs exercise tolerance.

FACTORS THAT AFFECT TRAINING

An appropriate exercise prescription must consider the major factors that relate to training improvement: (1) initial fitness level, (2) intensity of exercise, (3) duration of exercise, (4) frequency of exercise, and (5) type of exercise.

INITIAL FITNESS LEVEL

The amount of training improvement depends on one's initial fitness level.[90,91] If someone rates low at the start, there is room for considerable improvement. If capacity is already high, then naturally there will be relatively little improvement. In studies of sedentary, middle-aged men with heart disease, for example, maximal aerobic power improved by 50%, whereas for the same type of training in normally active, healthy adults only a 10 to 15% improvement occurred.[80] Of course, a 5% improvement in physiologic function for an elite athlete is just as important as a 40% increase for a sedentary person. As a general guideline, however, aerobic fitness improvements of 5 to 25% can be expected from systematic endurance training. These improvements are noted within the first 3 weeks of training.[46]

EXERCISE INTENSITY

Training-induced physiologic changes depend **primarily** on the intensity of the overload. Exercise intensity reflects both the caloric cost of the work and the specific energy systems activated. Intensity can be applied either on an absolute or relative basis.

An example of an absolute training intensity would be to have all individuals do the same work at the same rate, such as 200 watts on the bicycle ergometer or expend 300 kcal in a 30-minute exercise session. With everyone doing the same amount of work, a considerable exercise stress for one person might be below the training threshold intensity for another more highly conditioned person. For this reason, training is usually assigned based on the **relative stress** placed on a person's physiologic systems. Relative intensity is usually assigned as some percentage of maximum function, for example, max $\dot{V}O_2$, maximum heart rate, or maximum working capacity. The general practice for establishing aerobic training intensity is to either directly measure or estimate the person's max $\dot{V}O_2$ or maximum heart rate, and then assign a work schedule that corresponds to some **percentage** of these maximums.

Although establishing training intensity from measures of oxygen consumption is reasonably accurate, it is impractical without sophisticated equipment. An effective alternative is to use **heart rate** to classify exercise in terms of relative intensity and to establish a training protocol. This practice is based on the fact that percent max $\dot{V}O_2$ and percent maximum heart rate (HR max) are related in a predictable way, regardless of gender, fitness level, or age. Selected values for percent max $\dot{V}O_2$ and corresponding percentages of HR max obtained from several sources are presented in Table 21–5.[3,95] The error in estimating percent max $\dot{V}O_2$ from percent HR max, or vice versa, is about ± 8%. Because of this intrinsic relationship, it is only necessary to monitor heart rate to estimate percent max $\dot{V}O_2$. The relationship between percent max $\dot{V}O_2$ and percent HR max is essentially the same for either arm or leg exercise for both normal people and cardiac patients.[29,40] **The important point is that maximum heart rate is significantly lower in arm exercise, and this difference must be considered in formulating the exercise prescription for different exercise modes** (*see Chapter 17*).

Training at a Percentage of Maximum Heart Rate

As a general rule, aerobic capacity improves if exercise is of sufficient intensity to increase heart rate to about 70% of maximum. This is equivalent

TABLE 21–5. RELATION BETWEEN PERCENT MAX $\dot{V}_{O_2}$ AND PERCENT MAX HEART RATE	
Percent Max HR	**Percent Max $\dot{V}_{O_2}$**
50	28
60	42
70	56
80	70
90	83
100	100

to about 50 to 55% of the maximum aerobic capacity or, for college-aged men and women, to a heart rate of 130 to 140 beats per minute with leg exercise as in cycling, walking, and running. This intensity appears to be the **minimal stimulus** required to provide training improvements.

An alternate and equally effective method to establish the training threshold is to exercise at a heart rate about 60% of the difference between resting and maximal.[57] This approach to determining the threshold training heart rate tends to give a somewhat higher value compared to the heart rate computed simply as 70% of HR_{max}. This is calculated as:

$$HR_{threshold} = HR_{rest} + 0.60 (HR_{max} - HR_{rest})$$

Clearly, exercise need not be strenuous to obtain positive results. An exercise heart rate of 70% maximum represents moderate exercise that can be continued for a long time with little or no discomfort. This training level is frequently referred to as "conversational exercise" because it is sufficiently intense to stimulate a training effect, yet not so strenuous that it limits a person from talking during the workout. It is unnecessary to exercise above this heart rate to improve physiologic capacity. Figure 21–5 shows that as cardiovascular fitness improves and the person becomes "trained," the exercise heart rate at a given oxygen consumption is reduced. It is common for submaximal heart rate to be lowered by 10 to 20 beats per minute during an aerobic conditioning program. To keep pace, the work rate must be increased periodically to achieve the threshold heart rate or whatever target heart rate has been established. A person who began training by walking would have to walk more briskly; this would be gradually replaced by jogging for periods of the workout; eventually, continuous running would be required to achieve the same relative strenuousness at the desired target heart rate. If the progression in exercise intensity is not matched to training improvements, the exercise program essentially becomes a "maintenance" program for aerobic fitness.[46]

Is Strenuous Training More Effective?

Generally, the greater the relative training intensity above threshold, the greater is the training improvement.[78,91] However, this is only true within certain limits. Although there may be a minimal threshold intensity below which a training effect does not occur, there may also be a ceiling "threshold" above which there are no further gains. The lower and upper limits may depend on the participant's initial capacity and state of training. For people in relatively poor condition, the training threshold may be closer to 60% of HR max that corresponds to about 45% max $\dot{V}_{O_2}$;[31] individuals at higher

FIG. 21–5.

HR-$\dot{V}_{O_2}$ line for a 20-year-old woman before and after a 10-week aerobic conditioning program.

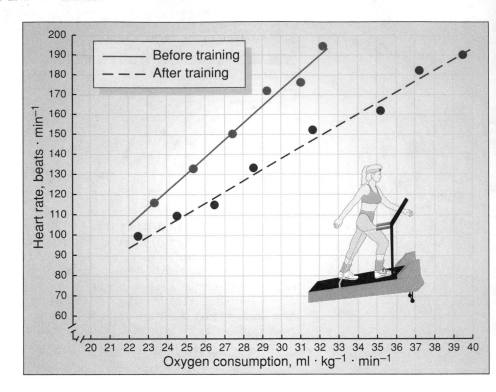

fitness levels generally have a higher threshold level. The ceiling for training intensity is unknown, although 85% max $\dot{V}_{O_2}$ (corresponding to 90% HR max) is thought to be the upper limit—at present, no definitive research is available to either prove or disprove this notion.

The "Training-Sensitive Zone"

Maximum heart rate for a specific activity can usually be determined immediately after 2 to 4 minutes of all-out exercise in that form of work. This level of exercise, however, requires considerable motivation and certainly is not advisable for adults without medical clearance or those predisposed to coronary heart disease. Consequently, people should consider themselves "average" and use the age-adjusted maximum heart rates presented in Figure 21–6. Although all people of a particular age do not possess the same maximum heart rate, the loss in accuracy due to individual variation (generally ± 10 beats per minute is about one standard deviation at any age-predicted heart rate) is usually of little significance in establishing an effective training program for healthy people. Maximum heart rate is established as 220 minus the person's age, with values being independent of race or gender in children and adults.[54,64] Although this represents a convenient "rule of thumb," this is only an estimate. Within normal variation, about 95% (± 2 standard deviations) of 40-year-old men and women will have a maximum heart rate between 160 and 200 beats per minute. Figure 21–6 also shows the "training-sensitive zone" in relation to age. Conditioning of the aerobic systems should occur as long as the exercise heart rate is maintained within this zone.

If a 40-year-old woman wishes to train at moderate intensity, yet still be

FIG. 21–6.
Maximal heart rates and training-sensitive zones for use in aerobic training programs for people of different ages.

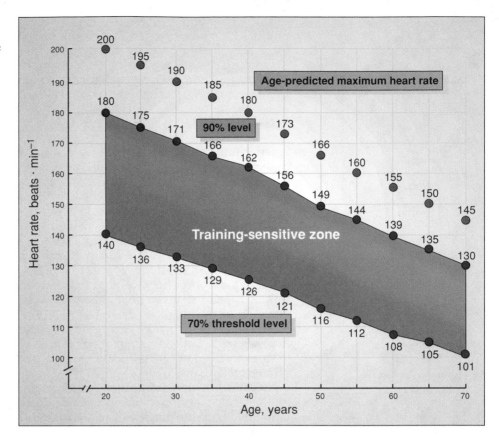

at or above the threshold level, a training heart rate would be selected equal to 70% of the age-predicted maximum heart rate, i.e., a target exercise heart rate of 126 beats per minute (0.70 × 180). Then, by trial and error (using progressive increments in light to moderate exercise), the woman could arrive at a walking, jogging, or cycling load that produced the desired target heart rate. If the woman wished to increase her training intensity to 85% of maximum, her exercise heart rate would have to be increased to 153 beats per minute (0.85 × 180).

Running Versus Swimming and Other Forms of Arm Exercise

An adjustment should be made in estimating maximum heart rate if swimming or other forms of arm exercise are used for training.[30] Maximum heart rate in these exercise modes averages about 10 to 13 beats per minute lower than running in both trained and untrained subjects.[66,68,99] This difference is probably the result of a relatively smaller muscle mass being activated in upper body exercise such as arm cranking and swimming compared to running or cycling. With swimming, the horizontal body position and the cooling effect of the water may also contribute to the lower maximum heart rate. To establish the appropriate exercise intensity for swimming and arm work, this difference of about 13 beats per minute should be subtracted from the age-predicted HR max in Figure 21–6. Consequently, a 30-year-old person wishing to swim at 70% HR max would select a swimming speed that produced a heart rate of 124 beats per minute [(0.70) × (190 − 13)]. This more accurately represents the appropriate threshold training heart rate for swimming. If this is not done, a prescription for arm exercise based on a percentage of the HR max for leg exercise results in an **overestimation** of the appropriate training heart rate.

TRAINING DURATION

A threshold duration per workout has **not** been identified for optimal cardiovascular improvement. Such a threshold probably depends on many factors; these include the total work done, exercise intensity, training frequency, and initial fitness level. Whereas 3- to 5-minute periods of daily exercise produce training effects in some poorly conditioned people, 20 to 30 minutes of exercise per session is more optimal (yet still practical in terms of time) if the intensity is at 70% HR max. With high-intensity training, significant improvements occur with 10- to 15-minute exercise periods per workout. Conversely, a 45-minute continuous exercise period may be required to produce a training effect when exercise intensity is below the threshold heart rate. **It appears that the lower intensity of exercise is offset by the increased duration of training.**

How long does it take before improvements are noted? This, of course, depends on the specific systems affected. Adaptations in cardiovascular fitness and aerobic capacity occur rapidly, and significant improvements are often noted within 1 or 2 weeks.[23,45] Figure 21–7 shows absolute and percentage improvements in max $\dot{V}_{O_2}$ for subjects who trained 6 days a week for 10 weeks. Training consisted of 30 minutes of bicycling 3 days a week combined with running for up to 40 minutes on alternate days. As illustrated, there was continuous week-by-week improvement in aerobic capacity. This suggests that for relatively sedentary people, training improvements occur rapidly and continue in a relatively steady fashion. Of course, these adaptive responses eventually begin to level off as the person approaches their "genetically determined" maximum. The exact duration until this leveling-off occurs is unknown, particularly for those who undergo high-intensity training. It no doubt varies depending on the particular physiologic and metabolic systems affected. It should be kept in mind that "more" is not necessarily better in terms of aerobic training because the number of running injuries reported by nonathletic adults increases dramatically as there are increases in frequency and duration of exercise.[55,94] For both men and women, for example, the only variable consistently associated with running injuries was the number of miles run per week.[82] For preadolescent children, running excessive distances may place undue strain on the articular cartilage. For susceptible children, this strain could result in injury to the bone's growth plate (epiphysis) with subsequent adverse effects on growth.[9]

TRAINING FREQUENCY

Is it better to work out 2 or 5 days a week if the duration and intensity are the same for each training session? Unfortunately, a precise answer is unavailable. Although some investigators report that training frequency is an important factor in causing cardiovascular improvements,[33,81] others maintain that this factor is considerably less important than either exercise intensity or duration.[20,78,91] Several studies using interval training showed that 2-days-per week training resulted in changes in max $\dot{V}_{O_2}$ similar to those

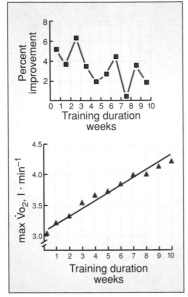

FIG. 21–7.

Improvements in max $\dot{V}_{O_2}$ over 8 weeks of high-intensity endurance training. The upper graph shows the week-by-week percent improvement. (From Hickson, R.C., et al.: Linear increases in aerobic power induced by a program of endurance exercise. J. Appl. Physiol., *42*:373, 1977.)

observed with 5-days-per-week training.[26,27] In other studies where total work was held constant,[42,93] there were no differences in max $\dot{V}O_2$ improvements when training was 2 versus 4, or 3 versus 5 days a week. As was the case with training duration, if the training is performed at a lower intensity, more frequent training is beneficial.

It seems that an extra investment of time may not be that profitable in terms of producing beneficial changes in physiologic function. On the other hand, if exercise is used as a means for weight control, strong consideration should be given to exercising 5 to 7 days a week, because this frequency of exercise represents a considerable caloric expenditure when compared with training only 2 days a week. **To bring about meaningful weight loss through exercise, it is recommended that each exercise session last at least 60 minutes and be of sufficient intensity to expend at least 300 kcal.** Training fewer than 2 days a week generally does not produce adequate changes in either anaerobic or aerobic capacity or body composition.[1]

Typical training programs are conducted 3 days a week with a rest day spaced between 2 workout days. A meaningful question, however, is whether training could occur on consecutive days and still produce equally effective results? In an experiment concerned with this exact question, improvements in max $\dot{V}O_2$ were nearly **identical** regardless of the sequence of the 3-day-a-week training schedule.[70] This result suggests that perhaps the stimulus for aerobic training is closely tied to the intensity and total work accomplished and **not** to the sequence of training.

EXERCISE MODE

If intensity, duration, and frequency are held constant, training improvements are similar, regardless of training mode—as long as the exercise involves large muscle groups activated in a rhythmic, aerobic nature. Bicycling, walking, running, swimming, and rope skipping are all excellent exercises to stress the aerobic system.[81,101] Of course, based on the specificity concept, the magnitude of training changes may vary considerably depending on the **mode of testing.** Individuals trained on a bicycle show greater improvements when tested on the bicycle than on the treadmill.[77,100] Likewise, those trained by running show the greatest improvements when measured with an exercise involving the leg muscles rather than muscles of the upper body.[68]

MAINTENANCE OF FITNESS

An interesting question concerns the optimal frequency and level of exercise required to maintain the aerobic improvements attained through training. In one study, healthy young adults achieved a 25% improvement in max $\dot{V}O_2$ after 10 weeks of interval bicycling and running for 40 minutes, 6 days a week.[44,56] They were then placed into one of two groups that continued to exercise for an additional 15 weeks at the same intensity and duration but at a reduced **frequency** of either 4 or 2 days a week. For both groups, the gains in aerobic capacity were maintained despite the reduced training frequency.

The effect of reduced training **duration** on the maintenance of improved aerobic fitness has been studied.[47] Following the same protocol as outlined above for the initial 10 weeks of training, the subjects continued to maintain the intensity and frequency of training for an additional 15 weeks, but they reduced training duration from the original 40-minute sessions to either 26 or 13 minutes per day. Almost all of the max $\dot{V}O_2$ and performance increases were maintained despite this reduction in training duration by as much as two thirds.

These results indicate that improvements in fitness may involve somewhat different training requirements and adaptions compared to fitness maintenance. **If exercise intensity is maintained, the frequency and duration of physical activity required to maintain a certain level of aerobic fitness is less than that required to improve it.** When the **intensity** of training is reduced and the frequency and duration held constant, even a one-third reduction in work rate causes the max $\dot{V}_{O_2}$ to decline.[48] This strongly suggests that training intensity plays a principal role in maintaining the increase in aerobic power achieved through training.

METHODS OF TRAINING

Each year improvements in performance are noted in almost all athletic activities. These advances are generally attributed to increased opportunities for participation, so that individuals with "natural endowment" are more likely to be exposed to particular sports. Also contributing to superior performances are improved nutrition and health care, better athletic equipment, and a more systematic and scientific approach to athletic training and conditioning. In the following sections, general guidelines for both anaerobic and aerobic training are presented with particular emphasis on three general training classifications: (1) **interval training,** (2) **continuous training,** and (3) **fartlek training.**

ANAEROBIC TRAINING

We showed in Figure 21–1 that the capacity to perform and persist in all-out exercise for brief periods of time up to 60 seconds is largely dependent on ATP generated by the immediate and short-term anaerobic energy systems.

The Phosphate Pool

Sports such as football, weightlifting, and various other brief sprint activities rely almost exclusively on energy derived from the muscle's phosphate pool.

The phosphate pool can be overloaded by engaging specific muscles in repeated **maximum** bursts of effort for 5 to 10 seconds. Because the high-energy phosphates supply energy for intense, intermittent exercise, only small amounts of lactic acid are produced and recovery is rapid. Thus, a subsequent exercise bout can begin after a 30- to 60-second rest period. This use of brief all-out work periods interspersed with recovery represents a specific application of **interval training,** a technique useful for anaerobic conditioning.

In training to enhance the ATP-CP energy capacity of specific muscles, the activities selected must engage the muscles at the appropriate speed of movement for which the athlete desires improved anaerobic power. Not only does this enhance the anaerobic metabolic capacity of the specifically trained muscle fibers, but it also facilitates the recruitment of the appropriate motor units used in the actual movement.

Lactic Acid

As the duration of all-out effort extends beyond 10 seconds, dependence on anaerobic energy from the phosphates decreases while the quantity of anaerobic energy generated in glycolysis increases. To improve capability for anaerobic energy release via the short-term lactic acid energy system, the physiologic conditioning program must overload this aspect of energy metabolism.

Heavy anaerobic training is psychologically taxing and requires considerable motivation. Repeat bouts of up to 1 minute of maximum running, swimming, or cycling, stopped about 30 seconds before subjective feelings of exhaustion, cause lactic acid to increase to near maximum levels. Each exercise bout should be repeated after 3 to 5 minutes of recovery. Several repeats cause a "lactate stacking" that results in higher levels of lactic acid than just one bout of all-out effort to the point of voluntary exhaustion.[41] Of course, it is critical to use the specific muscle groups that require this enhanced anaerobic capacity. A backstroke swimmer should train by swimming backstroke, a cyclist must bicycle, and the basketball, hockey, or soccer player should rapidly perform various movements and direction changes similar to those required by the demands of the sport.

The time necessary for recovery can be considerable when exercise involves a significant anaerobic component. For this reason, anaerobic power training should occur at the end of the conditioning session. Otherwise, fatigue would carry over and perhaps hinder the efficiency of subsequent aerobic training.

AEROBIC TRAINING

Figure 21–8 indicates two important factors in formulating an aerobic training program. For one thing, the training must be geared to provide a sufficient cardiovascular overload to stimulate increases in stroke volume and cardiac output. This central overload should be accomplished with the appropriate muscle groups to concurrently enhance the local circulation and "metabolic machinery" within the specific muscles. This essentially embodies the specificity principle as applied to aerobic training. Simply stated, runners should run, cyclists should bicycle, and swimmers should swim.

Brief bouts of repeated exercise (interval training) as well as continuous, long-duration work (continuous training) enhance aerobic capacity, provided the exercise is sufficiently intense to overload the aerobic system. Interval training, continuous training, and fartlek training are three common methods to improve aerobic fitness.

Interval Training

Many elite athletes attribute their success to interval training.[28] With the correct spacing of exercise and rest periods, a tremendous amount of work can be accomplished that would not normally be completed in a workout in which the exercise was performed continuously. Repeated exercise bouts (with rest periods or **relief intervals**) can vary from a few seconds to several minutes or more depending on the desired outcome.[18] The interval training prescription can be modified in terms of intensity and duration of the exercise interval, the length and type of relief interval, the number of work intervals (**repetitions**), and the number of repetition blocks (**sets**) per workout. Adjustment of any or all of these can easily be made to meet the specific requirements for different performances. This offers flexible options for de-

FIG. 21–8.
The two major goals of aerobic conditioning.

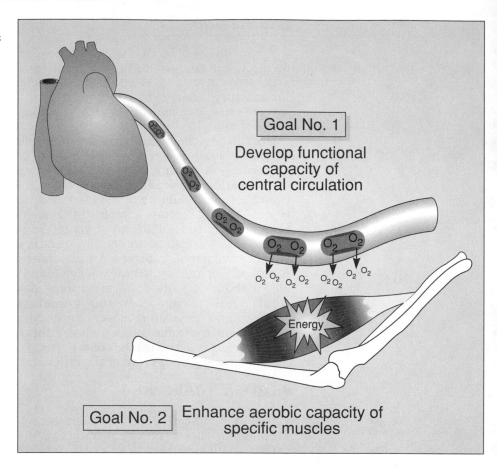

Goal No. 1

Develop functional capacity of central circulation

Energy

Goal No. 2 Enhance aerobic capacity of specific muscles

veloping the anaerobic and aerobic energy transfer systems. A longer work interval engages the aerobic system, whereas shorter exercise intervals place greater overload on the anaerobic energy system.

One value of interval training is that it permits high-intensity, intermittent exercise for a relatively long period. In essence, a person can reach max $\dot{V}_{O_2}$ repeatedly in a training session. For example, few people can maintain a 4-minute mile pace for longer than a minute, let alone complete a mile within 4 minutes. If running intervals, however, were limited to only 15 seconds followed by 30 seconds of recovery, it would not be exceedingly difficult to maintain these exercise–rest intervals and complete the mile in 4 minutes of actual running! Although this does not suggest a world-class performance, the point is that a significant quantity of normally exhausting work at high levels of both aerobic and anaerobic metabolism can be achieved with the proper spacing of rest and work intervals.

In the example of a continuous run at a 4-minute mile pace, a large portion of energy would be supplied through anaerobic glycolysis. Within a minute or two, lactic acid levels would rise precipitously and the runner would become exhausted. With interval training, repeated work bouts of about 15 seconds would permit a severe load to be imposed on the aerobic energy system of specific muscles without an appreciable buildup of lactic acid. Fatigue incurred during the work interval would be minor and recovery could take place quickly. The work interval could then begin after only a brief rest

TABLE 21–6.
GUIDELINES FOR DETERMINING INTERVAL-TRAINING WORK RATE FOR RUNNING AND SWIMMING DIFFERENT DISTANCES*

Interval Training Distances (Yards)		Work Rate for Each Exercise Interval or Repeat
Run	Swim	
55	15	1.5 ⎧ seconds *slower* than best
110	25	3 ⎨ times from running
		(or swimming) start
220	55	5 ⎩ for each distance
440	110	1 to 4 seconds *faster* than the average 440-yard run or 110-yard swim times recorded during a mile run or 440-yard swim
660–1320	165–320	3 to 4 seconds *slower* than the average 440-yard run or 100-yard swim times recorded during a mile run or 440-yard swim

* From Fox, E.L., and Mathers, D.K.: Interval Training. Philadelphia, W.B. Saunders Co., 1974.

period, and a high level of aerobic metabolism would be sustained (see Chap. 7).

A practical system for determining interval training work rates is presented in Table 21–6.

1. **Exercise interval:** Generally 1.5 to 5.0 seconds are **added** to the exerciser's "best time" for training distances between 55 and 220 yards for running and 15 to 55 yards for swimming.[28,101] If a person can run 60 yards from a running start in 8 seconds, the training time for each repeat would therefore be 8 + 1.5 or 9.5 seconds. For interval training distances of 110 and 220 yards, 3 and 5 seconds are added, respectively, to the best running times. This particular application of interval training is suited for training the immediate, anaerobic power energy system.

    For training distances of 440 yards running or 110 yards swimming, the work rate is determined by **subtracting** 1 to 4 seconds from the best 440-yard part of a mile run or 110-yard part of a 440-yard swim. If a person runs a 7-minute mile (averaging 105 seconds per 440 yards), the interval time for each 440-yard repeat would be between 104 seconds (105 − 1) and 101 seconds (105 − 4). For training intervals beyond 440 yards, 3 to 4 seconds are **added** for each 440-yard portion of the interval distance. In running an interval of 880 yards, the 7-minute miler would thus run each interval at about 216 seconds [(105 + 3) × (2) = 216].

2. **Relief interval:** The relief interval can be either passive (rest-relief) or active (work-relief). The recommended duration of relief is usually expressed as a ratio of work duration to recovery duration. The ratio of 1 to 3 is generally recommended for training the immediate energy system. Thus, for a sprinter who runs 10-second intervals, the relief interval is usually about 30 seconds. For training the short-term glycolytic energy system, the relief interval is twice as long as the work interval or a ratio of 1 to 2. These specific ratios of work-to-relief for anaerobic training supposedly ensure sufficient restoration of the phosphagen pools or lactic

acid removal so that the next work bout can proceed without undue fatigue.

For training the long-term aerobic energy system, the work-recovery interval ratio is usually 1 to 1 or 1 to 1.5. During a 60- to 90-second exercise interval, for example, oxygen consumption is insufficient to meet the energy requirements of the exercise. The recommended recovery interval enables the succeeding exercise interval to begin before recovery is complete. This ensures that the circulatory and aerobic metabolic stress reach near peak levels even though the exercise intervals are relatively short. With longer periods of intermittent exercise there is sufficient time for metabolic and circulatory adjustments; thus, the duration of the rest interval is not as crucial.

Continuous Training

Continuous or **long slow distance** training involves steady-paced exercise performed at either moderate or high aerobic intensity (60 to 80% max $\dot{V}_{O_2}$) for a sustained duration. The exact pace can vary, but it must be of sufficient threshold intensity to ensure physiologic adaptation. The method to establish this threshold so that the person might exercise in the "training-sensitive zone" was outlined previously. Continuous training for an hour or longer is popular among joggers and other fitness enthusiasts as well as competitive endurance athletes. It is not uncommon for distance runners to train twice a day on a yearly basis and run between 100 and 150 miles each week. In one report,[13] a man training for the 52.5 mile "ultramarathon" ran twice a day, 20 miles in the morning and 13 miles in the evening, interspersed with an occasional 30- to 60-mile nonstop run at a 7- to 8-minute per mile pace. With this schedule, he ran more than 800 miles each month and totaled 9600 miles per year! The precise effects and benefits of such considerable training are unknown.

By its nature, continuous exercise training is submaximum and, therefore, can be engaged in for considerable time in relative comfort. Because of the potential hazards of high-intensity interval training for coronary-prone individuals (as well as an almost total dedication required by such strenuous exercise), continuous training is particularly suitable for those just beginning an exercise program. When applied in athletic training, continuous training is really "over-distance" training, with most athletes covering between two to five times the actual distance of their racing event. It is believed that over-distance training produces the largest aerobic adaptations in both the central circulation and peripheral tissues. Overload is generally accomplished by increasing exercise duration, although the work rate increases progressively as training improvements are achieved.

One of the advantages of continuous training is that it permits training at nearly the same intensity as actual competition. Because the recruitment of appropriate motor units depends on work rate, continuous training may be best suited for the endurance athlete in terms of adaptations at the cellular level.[25,34] This is in contrast to interval training that may place a disproportionate stress on the fast-twitch muscle fibers; these are **not** the fibers predominantly recruited in endurance competition!

Fartlek Training

Fartlek is a Swedish word meaning "speed play." This training method, introduced to the U.S. in the 1940s, is a relatively "unscientific" adaptation of interval and continuous training that is well suited for exercising out-of-doors over natural terrain. With this system, alternate running is done at both fast and slow speeds on both a level and hilly course.

In contrast to the precise exercise prescription in interval training, fartlek training does not require systematic manipulation of the work and relief intervals. Instead, the performer determines the training scheme based on "how it feels" at the time. If done properly, this system develops one or all of the energy systems. An added advantage is the flexibility it affords in determining the extent of training. Although lacking the systematic and quantified base of interval and continuous training, fartlek training is ideally suited for general conditioning or off-season training and for maintaining a certain "freedom" and variety in workouts.

At present, there is insufficient evidence for the superiority of any specific training method for improving aerobic capacity. Each training procedure results in success; they probably can be used interchangeably and certainly should be used to modify training and achieve a more pleasing psychologic set.

INITIATING A TRAINING PROGRAM

Dropout rates of 50% or higher within 6 months to a year are commonly reported for men and women who embark on an exercise program.[74] Undoubtedly, adherence could be greatly improved with a more thoughtful and systematic approach in formulating and delivering the exercise program.

AIMS AND OBJECTIVES

The most important aspect in formulating a training program is to identify specific aims and achievement objectives. This is equally important for both the Olympic decathlon hopeful and the person simply wishing to "get-in-shape." Whether the goal is a 4-minute mile, a sub-3-hour marathon, or a non-stop jog or run for 60 minutes, motivation, willingness, and dedication during training are enhanced if individuals know their performance goals and the specific directions to meet their training objectives.

Specific goals are usually expressed in performance terms, such as being able to complete a marathon. It is also possible to state goals in physiologic terms, for example, achieving a certain increase in max $\dot{V}_{O_2}$ or a decrease in heart rate on a step test, etc.

ASSESSMENT OF INDIVIDUAL STATUS

Measuring initial status prior to a conditioning program depends on available facilities and technical expertise, as well as the exerciser's training goals. For competitive athletes, a complete assessment is provided with the help of films, coaches' expert opinion and field and laboratory tests.[93] For others, adequate fitness assessment may consist of simple tests easily administered at home.[92] Regardless of the method or extent of assessment, we believe it is prudent for previously sedentary individuals to have medical clearance prior to initiating an exercise program. For some over the age of 35, this may involve a supervised exercise stress test (Chapter 30). As a minimum for adults considering an exercise program, we advise a physical examination that includes assessment of the cardiovascular system, blood pressure and blood chemistry, muscles, and joints. The assessment of fitness status involves four major areas: (1) muscular strength, (2) joint flexibility, (3) body composition (percent fat, lean body weight, and desirable body weight), and (4) cardiovascular-respiratory functional capacity.

THE PROGRAM

Training should begin at a relatively moderate intensity and gradually increase to the target level. **The quickest way to extinguish enthusiasm for exercise is to prescribe too much too soon!** Several minutes of mild calisthenics or jogging in place generally provide a sufficient warm-up so that the heart and circulation are not suddenly taxed.[6] A 5-minute "cool-down" that consists of exercise of gradually lessening intensity is generally advocated before exercise is stopped. This makes sense because it reduces the chances of abrupt physiologic alterations, especially those involving cardiovascular dynamics, which are more likely with the sudden cessation of strenuous exercise.

Provision should be made for progressive improvements, adequate use of facilities, variety, some competition, and periodic evaluation. It is desirable to sketch out the whole program at the beginning, with full realization that changes and modifications are necessary to meet individual needs and rates of change. With a well-planned yet flexible program, both the exercise specialist and participant know exactly what to expect and in what fitness areas to concentrate. This does not mean the program must be complex or excessively time-consuming. In fact, in one school district, simply adding between 5 and 14 minutes of jogging to the daily physical education program significantly improved the students' running endurance by 18% at the end of the semester.[11]

LEADERSHIP

We maintain that individuals trained in the science of exercise are most qualified to establish, lead, and direct individual and community fitness and training programs. Although it is often argued that practical experience is the best teacher, it is our belief that knowing the **why** of training and exercise is fundamental in establishing a well-rounded, scientifically sound program. It is no longer sufficient to have been a "letter winner" or former athlete to qualify for a job in the "fitness marketplace." Programs of certification are now available for qualification as exercise leaders and program directors, laboratory stress-test coordinators or exercise technicians, athletic trainers, physical educators, and coaches.

SUMMARY

1. Activities can be classified in terms of their predominant activation of a specific system of energy transfer. An effective training program is one that allocates a proportionate time commitment to training the specific energy system(s) involved in the activity.
2. Proper physical conditioning is based on sound principles that produce optimum improvements. Of crucial importance are the overload principle, the specificity of exercise principle, the individual difference principle, and the reversibility principle.

3. Exercise training brings about specific metabolic and physiologic adaptations that involve subtle cellular as well as gross physiologic changes. Anaerobic training increases resting levels of anaerobic substrates and key glycolytic enzymes. This is usually accompanied by concomitant increases in all-out exercise performance. Aerobic training changes include increases in mitochondrial size and number as well as the activity of aerobic enzymes, increased capillarization of the trained muscle, and enhanced oxidation of fat and carbohydrate. These adaptations are geared to a greater aerobic production of ATP.

4. Aerobic training also brings about both functional and dimensional changes in the cardiovascular system. These include decreases in resting and submaximal exercise heart rate, enhanced stroke volume and cardiac output, and an expanded $\mathrm{a}\text{-}\bar{\mathrm{v}}\ \mathrm{O}_2$ difference.

5. The major factors that affect training improvement are initial fitness level, frequency of training, exercise intensity, duration of exercise, and type (mode) of training. Of these, intensity is most crucial.

6. Training intensity can be applied either on an absolute basis in terms of exercise load, or relative to an individual's physiologic response. It is practical and effective to set exercise intensity to a percent of a person's maximum heart rate response. Training levels that correspond to 70 to 90% of maximum heart rate are most desirable for inducing aerobic fitness changes.

7. Training duration and intensity are intimately related. Nevertheless, 30 minutes per session seems to be desirable in terms of exercise duration. Extending the duration can compensate somewhat for a reduced intensity.

8. Frequency for optimum aerobic training appears to be a minimum of 3 days per week. Optimal frequency levels have not been established.

9. If intensity, duration, and frequency are held constant, training improvements are similar regardless of training mode, as long as large muscle groups are exercised.

10. The level of physical activity required to maintain an improved level of aerobic fitness is less than that required to improve it.

11. Methods of training differ markedly. Interval, continuous, and fartlek training can be used effectively for conditioning the different energy systems. Interval training seems to be most desirable for promoting changes in the anaerobic energy systems.

12. Aerobic training must be geared to enhance both circulatory function and the metabolic capacity of the specific muscles. Peripheral adaptations may have a profound effect on enhancing exercise performance.

13. When initiating a training program, care must be given to establish individual aims and objectives, assessment of performance and physiologic status design of the program, and appropriate leadership.

REFERENCES

1. American College of Sports Medicine: The Recommended Quantity and Quality of Exercise for Developing and Maintaining Fitness in Healthy Adults. *Sports Med. Bull.*, 13:1, 1978.
2. Anderson, P., and Henrickson, J.: Capillary supply of quadriceps femoris muscle of man: adaptive response to exercise. *J. Physiol. (London)*, 270:677, 1977.
3. Åstrand, P.O., and Rodahl, K.: *Textbook of Work Physiology*. 3rd ed. New York, McGraw-Hill, 1986.

4. Ballor, D.L., et al.: Resistance weight training during caloric restriction enhances lean body weight maintenance. *Am. J. Clin. Nutr.*, 47:19, 1988.

5. Barnard, R. J., et al.: Effects of exercise of skeletal muscle: I. Biochemical and histochemical properties. *J. Appl. Physiol.*, 28:762, 1970.

6. Barnard, R.J., et al.: Ischemic response to sudden strenuous exercise in healthy men. *Circulation*, 48:936, 1973.

6a. Bouchard, C., et al.: Long-term exercise training with constant energy intake. I: Effect on body composition and selected metabolic variables. *Int. J. Obesity.* 14:57, 1990.

7. Brodahl, P., et al.: Capillary supply of skeletal muscle fibers in untrained and endurance-trained men. *Am J. Physiol.*, 232:705, 1977.

8. Burke, E. J., and Franks, B. D.: Changes in $\dot{V}O_2$ max resulting from bicycle training at different intensities holding total mechanical work constant. *Res. Q.* 46:31, 1975.

9. Caine, D.J., and Lindner, D.J.: Growth plate injury: a threat to young distance runners. *Phys. Sportsmed.*, 12:119, 1984.

10. Chi, M.-Y., et al.: Effects of detraining on enzymes of energy metabolism in individual human muscle fibers. *Am. J. Physiol.*, 244 (Cell Physiol. 13):276, 1983.

11. Cooper, K.H., et al.: An aerobics conditioning program for Fort Worth, Texas school district. *Res. Q.*, 46:345, 1975.

12. Convertino, V.A.: Heart rate and sweat rate responses associated with exercise-induced hypervolemia. *Med. Sci. Sports Exerc.*, 15:77, 1982.

13. Costill, D., and Fox, E.: The ultra-marathoner. *Distance Running News*, 3:4, 1968.

14. Costill, D.L., et al.: Metabolic characteristics of skeletal muscle during detraining from competitive swimming. *Med. Sci. Sports Exerc.*, 17:339, 1985.

15. Cowan, M.M., and Gregory, L.W.: Responses of pre- and post-menopausal females to aerobic conditioning. *Med. Sci. Sports Exerc.*, 17:138, 1985.

16. Coyle, E.F., et al.: Time course of loss of adaptations after stopping prolonged intense endurance training. *J. Appl. Physiol.*, 57:1857, 1984.

17. Cunningham, D.A., and Faulkner, J.A.: The effect of training on aerobic and anaerobic metabolism during a short exhaustive run. *Med. Sci. Sports*, 1:65, 1969.

18. Daniels, J., and Scardinia, N.: Interval training and performance. *Sports Medicine*, 1:327, 1984.

19. Daniels, W.L.: Physiological effects of a military training program on male and female cadets. *Aviat. Space Environ. Med.*, 50:562, 1979.

20. Davies, C.T.M., and Knibbs, A.V.: The training stimulus: the effects of intensity, duration and frequency of effort on maximum aerobic power output. *Int. Z. Angew. Physiol.*, 29:299, 1971.

21. Donovan, C.M., and Brooks, G.A.: Endurance training affects lactate clearance, not lactate production. *Am. J. Physiol.*, 244 (Endocrinol. Metab. 7):83, 1983.

22. Drinkwater, B., and Horvath, S.: Detraining effects on young women. *Med. Sci. Sports*, 4:91, 1972.

23. Durnin, J.V.G.A., et al.: Effects of a short period of training of varying severity on some measurements of physical fitness. *J. Appl. Physiol.*, 15:161, 1960.

24. Eddy, D.O., et al.: The effects of continuous and interval training in women and men. *Eur. J. Appl. Physiol.*, 37:83, 1977.

25. Fouriner, M., et al.: Skeletal muscle adaptation in adolescent boys: sprint and endurance training and detraining. *Med. Sci. Sports Exerc.*, 14:453, 1982.

26. Fox, E.L., and Matthews, D.K.: *Interval Training: Conditioning for Sports and General Fitness.* Philadelphia, W.B. Saunders, 1974.

27. Fox, E.L., et al.: Intensity and distance of interval training programs and changes in aerobic power. *Med. Sci. Sports*, 5:18, 1973.

28. Fox, E.L., et al.: Frequency and duration of interval training programs and changes in aerobic power. *J. Appl. Physiol.*, 38:481, 1975.

29. Franklin, B.A.: Aerobic requirements of arm ergometry: implications for testing and training. *Phys. Sportsmed.*, 11:81, 1983.

30. Franklin, B.A.: Aerobic exercise training programs for the upper body. *Med. Sci. Sports Exerc.*, 21:S141, 1989.

31. Gaesser, G.A., and Rich, G.A.: Effects of high- and low-intensity exercises on aerobic capacity and blood lipids. *Med. Sci. Sports Exerc.*, 16:269, 1984.

32. Gergley, T.J., et al.: Specificity of arm training on aerobic power during swimming and running. *Med. Sci. Sports Exerc.*, 16:349, 1984.
33. Gettman, L.R., et al.: Physiological responses of men to 1, 3, and 5 day per week programs *Res. Q.*, 47:638, 1976.
34. Gillespie, A.C.: Enzyme adaptations in rat skeletal muscle after two intensities of treadmill training. *Med. Sci. Sports Exerc.*, 14:461, 1982.
35. Gollnick, P., and Hermansen, L.: Biochemical adaptation to exercise: anaerobic metabolism. In *Exercise and Sport Sciences Reviews.* Vol. 1. Edited by J. Wilmore. New York, Academic Press, 1973.
36. Gollnick, P., and Saltin, B.: Significance of skeletal muscle oxidative enzyme enhancement with endurance training. *Clin. Physiol.*, 2:1, 1983.
37. Gollnick, P., et al.: Enzyme activity and fiber composition in skeletal muscle of untrained men. *J. Appl. Physiol.*, 33:312, 1972.
38. Gollnick, P., et al.: Effects of training on enzyme activity and fiber composition of human skeletal muscle. *J. Appl. Physiol.*, 34:107, 1973.
39. Gregg, S.G., et al.: Interactive effects of anemia and muscle oxidative capacity on exercise endurance. *J. Appl. Physiol.*, 67:765, 1989.
40. Hellerstein, H.K., and Franklin, B.A.: Exercise testing and prescription. In *Rehabilitation of the Coronary Patient.* Edited by N.K. Wenger and H.K. Hellerstein. New York, Wiley, 1978.
41. Hermansen, L.: Lactate production during exercise. In *Muscle Metabolism During Exercise.* Edited by B. Pernow and B. Saltin. New York, Plenum Press, 1971.
42. Hill, J.S.: The effects of frequency of exercise on cardiorespiratory fitness of adult men. Unpublished M.S. thesis. London, Ontario, University of Western Ontario, 1969.
43. Hickson, R.C.: Interference of strength development by simultaneously training for strength and endurance. *Eur. J. Appl. Physiol.*, 45:255, 1980.
44. Hickson, R.C., and Rosenkoetter, M.A.: Reduced training frequencies and maintenance of aerobic power. *Med. Sci. Sports Exerc.*, 13:13, 1981.
45. Hickson, R.C., et al.: Linear increases in aerobic power induced by a strenuous program of endurance exercise. *J. Appl. Physiol.*, 42:373, 1977.
46. Hickson, R.C., et al.: Time course of the adaptive responses of aerobic power and heart rate to training. *Med. Sci. Sports Exerc.*, 13:17, 1981.
47. Hickson, R.C., et al.: Reduced training duration effects on aerobic power, endurance, and cardiac growth. *J. Appl. Physiol.*, 53:255, 1982.
48. Hickson, R.C., et al.: Reduced training intensities and loss of aerobic power, endurance, and cardiac growth. *J. Appl. Physiol.*, 58:492, 1985.
49. Holloszy, J.O.: Metabolic consequences of endurance exercise training. In *Exercise, Nutrition, and Energy Metabolism.* Edited by E.S. Horton, and R.L. Terjung. New York, Macmillan, 1988.
50. Holloszy, J.O., and Coyle, E.F.: Adaptations of skeletal muscle to endurance exercise and their metabolic consequences. *J. Appl. Physiol.*, 56:831, 1984.
51. Houston, M.E., and Thompson, J.A.: The response of endurance adapted adults to intense anaerobic training. *Eur. J. Appl. Physiol.*, 36:207, 1977.
52. Hurley, B.F., et al.: Effects of high intensity strength training on cardiovascular function. *Med. Sci. Sports Exerc.*, 16:483, 1984.
53. Jacobs, I.: Sprint training effects on muscle myoglobin, enzymes, fiber types, and blood lactate. *Med. Sci. Sports Exerc.*, 19:368, 1987.
54. James, F., et al.: Responses of normal children and young adults to controlled bicycle exercise. *Circulation*, 61:902, 1980.
55. James, S.: Injuries to runners. *Am. J. Sports Med.*, 6:43, 1978.
56. Kanakis, C., et al.: Left ventricular responses to strenuous endurance training and reduced training frequencies. *J. Cardiac. Rehab.*, 2:141, 1982.
57. Karvonen, M.J., et al.: The effects of training on heart rate. A longitudinal study. *Ann. Med. Exp. Biol. Fenn.*, 35:305, 1957.
58. Karlsson, J., et al.: Muscle lactate, ATP, and CP levels during exercise after physical training in man. *J. Appl. Physiol.*, 33:199, 1972.
59. Kavanagh, T., et al.: Characteristics of postcoronary marathon runners. *Ann. N.Y. Acad. Sci.*, 301:455, 1977.

60. Kiessling, K.: Effects of physical training on ultrastructural features in human skeletal muscle. In *Muscle Metabolism During Exercise.* Edited by B. Pernow and B. Saltin. New York, Plenum Press, 1971.

61. Kjellberg, S., et al.: Increase of the amount of hemoglobin and blood volume in connection with physical training. *Acta Physiol. Scand.*, 19:146, 1949.

62. Klausen, K., et al.: Adaptative changes in work capacity, skeletal muscle capillarization, and enzyme levels during training and detraining. *Acta Physiol. Scand.*, 113:9, 1981.

62a. Kohrt, W.M., et al.: Longitudinal assessment of responses of triathletes to swimming, cycling, and running. *Med. Sci. Sports Exerc.*, 21:569, 1989.

63. Loftin, M., et al.: Effect of arm training on central and peripheral circulatory function. *Med. Sci. Sports Exerc.*, 20:136, 1988.

64. Londeree, B.R., and Moeschberger, M.L.: Effect of age and other factors on maximal heart rate. *Res. Q. Exerc. Sport*, 53:297, 1982.

65. MacDougall, J.D., et al.: Biochemical adaptation of human skeletal muscle to heavy resistance training and immobilization. *J. Appl. Physiol.*, 43:700, 1977.

66. Magel, J.R., et al.: Specificity of swim training on maximum oxygen uptake. *J. Appl. Physiol.*, 38:151, 1975.

67. Magel, J.R., et al.: Metabolic and cardiovascular adjustment to arm training. *J. Appl. Physiol.*, 45:75, 1978.

68. McArdle, W.D., et al.: Specificity of run training on $\dot{V}O_2$ max and heart rate changes during running and swimming. *Med. Sci. Sports*, 10:16, 1978.

69. Michael, E., et al.: Physiological changes of teenage girls during five months of detraining. *Med. Sci. Sports*, 4:214, 1972.

70. Moffatt, R.: Placement of tri-weekly training sessions: importance regarding enhancement of aerobic capacity. *Res. Q.*, 48:583, 1977.

71. Morganroth, J., et al.: Comparative left ventricular dimensions in trained athletes. *Ann. Intern. Med.*, 82:521, 1975.

72. Murase, Y., et al.: Longitudinal study of aerobic power in superior junior athletes. *Med. Sci. Sports Exerc.*, 13:180, 1981.

73. Neufer, P.D., et al.: Effect of reduced training on muscular strength and endurance in competitive swimmers. *Med. Sci. Sports Exerc.*, 19:486, 1987.

74. Oldridge, N.B.: Compliance and exercise in primary and secondary prevention of coronary heart disease: A review. *Prev. Med.*, 11:56, 1982.

75. Örlander, J., and Aniansson, A.: Effects of physical training on skeletal muscle metabolism and ultrastructure in 70- to 75-year-old men. *Acta Physiol. Scand.*, 109:149, 1980.

76. Pattengale, P.K., and Holloszy, J.O.: Augmentation of skeletal muscle myoglobin by programs of treadmill running. *Am. J. Physiol.*, 213:783, 1967.

77. Pechar, G.S., et al.: Specificity of cardiorespiratory adaptation to bicycle and treadmill training. *J. Appl. Physiol.*, 36:753, 1974.

78. Pollock, M.L.: The quantification of endurance training programs. In *Exercise Sport Sciences Reviews.* Vol. 1. Edited by J. Wilmore, New York, Academic Press, 1973.

79. Pollock, M.L.: Effects of a YMCA starter fitness program. *Phys. Sportsmed.*, 10:97, 1982.

80. Pollock, M.L., et al.: Effects of mode of training on cardiovascular function and body composition of adult men. *Med. Sci. Sports*, 7:139, 1975.

81. Pollock, M.L., et al.: Effects of frequency and duration of training on attrition and incidence of injury. *Med. Sci. Sports*, 9:31, 1977.

82. Powell, K.E., et al.: An epidemiological perspective on the causes of running injuries. *Phys. Sportsmed.*, 14:100, 1986.

83. Riviere, D., et al.: Lipolytic response of fat cells to catecholamines in sedentary and exercise-trained women. *J. Appl. Physiol.*, 66:330, 1989.

84. Roca, J., et al.: Evidence for tissue diffusion limitation of $\dot{V}O_2$ max in normal humans. *J. Appl. Physiol.*, 67:291, 1989.

85. Rowell, L.: Human cardiovascular adjustments to exercise and thermal stress. *Physiol. Rev.*, 54:75, 1974.

86. Saltin, B., and Rowell, L.B. Functional adaptations to physical activity and inactivity. *Fed. Proc.*, 39:1506, 1980.

87. Saltin, B., et al.: Response to exercise after bed rest and after training. *Circulation*, 38(Suppl. 7), 1968.

88. Seals, D.R., et al.: Endurance training in older men and women. I. Cardiovascular responses to exercise. *J. Appl. Physiol.*, 57:1024, 1984.
89. Seals, D.R., et al.: Endurance training in older men and women. II. Blood lactate response to submaximal exercise. *J. Appl. Physiol.*, 57:1030, 1984.
90. Sharkey, B.J.: Intensity and duration of training and the development of cardiorespiratory endurance. *Med. Sci. Sports*, 2:197, 1970.
91. Shephard, R.J.: Intensity, duration, and frequency of exercise as determinants of the response to a training regime. *Int. Z. Angew. Physiol.*, 26:272, 1968.
92. Shephard, R.J., et al.: Development of the Canadian home fitness test. *Can. Med. Assoc.*, 114:675, 1976.
93. Sidney, K. H., et al.: In *Training: Scientific Basis and Application.* Edited by A.W. Taylor. Springfield, IL, Charles C Thomas, 1972.
94. Stanish, W.D.: Overuse injuries in athletes; a perspective. *Med. Sci. Sports Exerc.*, 16:1, 1984.
95. Taylor, H.L., et al.: Exercise Tests: a summary of procedures and concepts of stress testing for cardiovascular diagnosis and function evaluation. In *Measurement in Exercise Electrocardiography.* Edited by H. Blackburn. Springfield, IL, Charles C Thomas, 1969.
96. Thompson, P.D., et al.: Cardiac dimensions and performance after either arm or leg endurance training. *Med. Sci. Sports Exerc.*, 13:303, 1981.
97. Thorstensson, A., et al.: Enzyme activities and muscle strength after sprint training in man. *Acta Physiol. Scand.*, 94:313, 1975.
98. Thorstensson, A., et al.: Effect of strengh training on enzyme activities and fiber characteristics in human skeletal muscle. *Acta Physiol. Scand.*, 96:392, 1976.
99. Vander, L.B., et al.: Cardiorespiratory responses to arm and leg ergometry in women. *Phys. Sportsmed.*, 12:101, 1984.
100. Wilmore, J., et al.: Physiological alterations consequent to 20-week conditioning programs of bicycling, tennis, and jogging. *Med. Sci. Sports*, 12:1, 1980.
101. Wilt, F.: Training for competitive running. In *Exercise Physiology.* Edited by H. Falls. New York, Academic Press, 1968.
102. Young, J.C., et al.: Maintenance of the adaptation of skeletal muscle mitochondria to exercise in old rats. *Med. Sci. Sports Exerc.*, 15:243, 1983.
103. Zuti, W.B., and Golding, L.A.: Comparing diet and exercise as weight reduction tools. *Phys. Sportsmed.*, 4:49, 1976.

MUSCULAR STRENGTH: TRAINING MUSCLES TO BECOME STRONGER

P A R T

1

Measuring and Improving Muscular Strength

In the 1940s and 1950s, specific strength or "weight lifting" exercises were used predominantly by body builders, competitive weight lifters, and some wrestlers. Most other athletes refrained from weight lifting for fear that such exercises would slow them down and increase muscle size to the point where they would lose joint flexibility and become **muscle-bound!** This myth was essentially dispelled by subsequent research in the late 1950s and 1960s that showed that exercises that strengthen muscles do not reduce movement speed or flexibility. In fact, the opposite was usually the case because elite weight lifters and body builders demonstrated exceptional joint flexibility and were certainly not limited in general movement speed. In longitudinal experiments with untrained healthy subjects, heavy-resistance exercises have increased both speed and power of muscular effort. Certainly, these effects would not be detrimental to sports performance.

In the sections that follow, we explore the underlying rationale, the process, and the physiologic adjustments that occur as muscles are trained to become stronger. Discussion centers on the various ways muscular strength is measured, strength differences between men and women, and the various resistance training programs designed to increase muscular strength.

MEASUREMENT OF MUSCULAR STRENGTH

Muscular strength, or more precisely, **the maximum force or tension generated by a muscle** (or muscle groups), is generally measured by one of four methods: (1) tensiometry, (2) dynamometry, (3) one-repetition maximum or 1-RM, and the newest approach, (4) computer-assisted force and work output determinations.

CABLE TENSIOMETRY

Figure 22-1 shows a cable tensiometer and its use for measuring muscular force during knee extension. As the force on the cable increases, the riser is depressed over which the cable passes. This deflects the pointer and in-

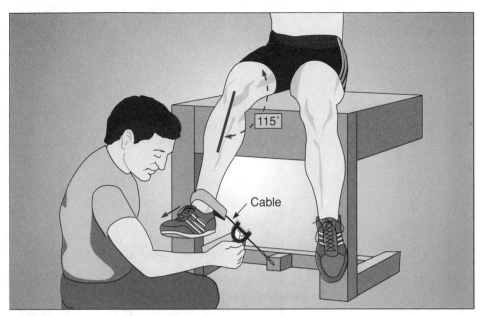

FIG. 22–1.
Measurement of knee extension static force using the cable tensiometer.

dicates the subject's strength score. This instrument measures muscular force during a static or isometric contraction where there is essentially no change in the muscle's external length. This application of the tensiometer is considerably different from its original use for measuring the tension of the steel cables linking various parts of an airplane.

The tensiometer is lightweight, portable, durable, easy to use, and has the advantage of versatility for recording force measurements at virtually all angles in the range of motion of a specific joint. Cable tension strength-test batteries have been developed to measure the static force of essentially all major muscle groups.[17,18,21] These tests are excellent for isolating and evaluating strength impairment in specific muscles weakened as a result of disease or injury. The muscle can be isolated and evaluated at a specific joint angle. This can then be objectively reproduced on repeated measurement to determine the status of specific muscle groups at the beginning of a therapeutic exercise program and during rehabilitation. In addition, since more than one muscle group is usually activated in a particular movement, the tensiometer can be applied in many phases of the movement. This may give a clearer picture of "strength" (or weakness) than do standard weight-lifting tests.

DYNAMOMETRY

Hand-grip and back-lift dynamometers used for strength measurement are shown in Figure 22-2. Both devices operate on the principle of compression. When an external force is applied to the dynamometer, a steel spring is compressed and moves a pointer. By knowing how much force is required to move the pointer a particular distance, one can then determine exactly how much external "static" force has been applied to the dynamometer.

FIG. 22–2.
Hand-grip and back-leg lift
dynamometers.

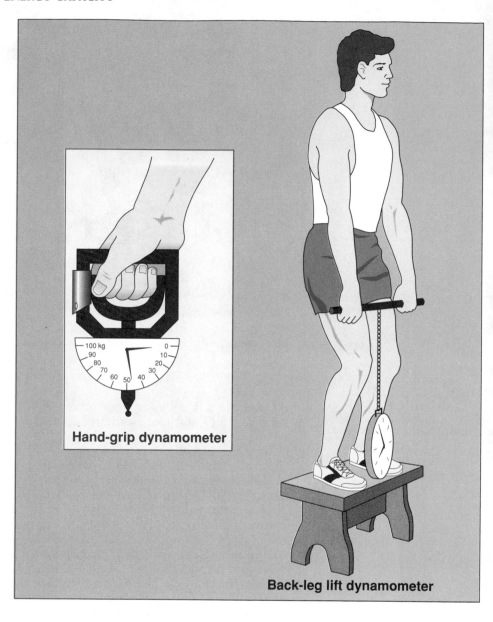

Hand-grip dynamometer

Back-leg lift dynamometer

ONE-REPETITION MAXIMUM (1-RM)

A dynamic method of measuring muscular strength makes use of the one-repetition maximum or 1-RM method. This refers to the maximum amount of weight lifted **one time** with correct form during the performance of a predetermined weight-lifting exercise. To test 1-RM for any particular muscle group or groups such as forearm flexors, leg extensors, or shoulders, a suitable starting weight is selected close to but below the subject's maximum lifting capacity. If one repetition is completed, weight is added to the exercise device until maximum lift capacity is achieved. Depending on the muscle group evaluated, the weight increments are usually either 5, 2 or 1 kg during the period of measurement.

COMPUTER-ASSISTED, ELECTROMECHANICAL, AND ISOKINETIC METHODS

The emergence of microprocessor technology has made possible a rapid way to quantify accurately muscular forces generated during a variety of movements. Sensitive instruments are currently available to measure force, acceleration, and velocity of body segments in various movement patterns.

Force platforms can be used to measure the external application of muscular force by a limb such as in jumping (Chapter 11, p. 201 for example). Other electromechanical devices measure the forces generated during all phases of an exercise movement such as cycling or during a supine bench press or leg press.[69]

An **isokinetic dynamometer** is an electromechanical instrument that contains a speed-controlling mechanism that acclerates to a preset speed when any force is applied. Once this constant speed is attained, the isokinetic loading mechanism accommodates automatically to provide a counterforce in relation to the force generated by the muscle. **Thus, maximum force (or any percentage of maximum effort) can be applied during all phases of the movement at a constant velocity.** A load cell inside the dynamometer continuously monitors the immediate level of applied force (and matching resistance) and feeds this information into an appropriate recorder. An electronic integrator placed in series with the recorder provides a readout of the average force generated for a given time. The voltage output from the integrator can be interfaced directly with a computer to provide almost instantaneous readouts of average force and total work.

The interface of computer technology with mechanical devices provides the exercise scientist with valuable data for purposes of testing, evaluation, and prescription. Such advances in technology, however, are not accepted universally by many who still consider a maximum lift (1-RM) as the best criterion of overall muscular strength. The counter argument of such dogma is that the dynamics of muscle strength involve considerably more than just the final outcome of a 1-RM. Even if two people have a 1-RM score of 100 kg, for example, the force-time curves could be quite dissimilar. Such differences in force dynamics (e.g., time to peak tension) may reflect an entirely different underlying neuromuscular physiology. Without the aid of the newer technologies, it simply would not be possible to study many of the exciting areas of muscle dynamics.

CATEGORIES OF RESISTANCE TRAINING EQUIPMENT

There are currently three basic categories of resistive exercise equipment used to train muscles. **Category I** is the more standard-type weight lifting equipment (free weights and barbells) that controls neither speed nor resistance through the range of motion. **Category II** has two subdivisions: IIA provides for a controlled or constant speed and variable resistance (true isokinetic equipment which imparts the speed), and IIB, that also provides a constant speed and variable resistance by means of a hydraulic device where the individual controls the speed. In **Category III**, speed is variable and resistance is constant (some cam devices and concentric-eccentric devices). There is currently no machine whereby muscles contract under conditions of true constant speed and true constant resistance. Surely, the newer generation of microprocessors, interfaced with mechanical devices to evaluate specific movement patterns, will help provide answers to many questions concerning the dynamics of force generation by muscles.

STRENGTH TESTING CONSIDERATIONS

The following considerations are important when individuals are tested for "strength," whether by cable tensiometry, dynamometry, 1-RM, or computer-assisted methods. This will ensure that all subjects are treated "equally" so that fair comparisons can be made among subjects.

1. Standardized instructions should be given prior to testing.
2. If a "warm-up" is given, it should be of uniform duration and intensity.
3. The subject must have adequate practice prior to the actual test to minimize a "learning" component that could compromise initial results.
4. Care must be taken to ensure that the angle of measurement on the limb or the test device is consistent among subjects.
5. A minimum number of trials (repetitions) should be determined a priori in order to establish a criterion score. For example, if five repetitions of a test are administered, what score should be used to represent the individual's score on the test? Is the "highest" score best, or should an average be used? There appears to be no clear answer on this point. It depends on the nature of the test. In most cases, however, an average of several trials will provide a more representative score of an individual's strength or power performance than a single trial. Stated another way, a single score is usually less reliable than an average of several scores.
6. Select tests that result in known reliability of measurement. This is a crucial aspect of testing that is often overlooked. If a test is created without determining the variability of the subjects' responses, the resultant test scores may not truly reflect an individual's performance, especially when evaluating the effects of a strength improvement protocol. Unreliability of measurement per se can mask the individual's performance on the test.
7. Be prepared to consider individual differences in such factors as body size and composition when evaluating strength scores between individuals and groups. For example, is it fair to compare the "strength" of a 120-kg defensive lineman with the "strength" of a 62-kg distance runner? Unfortunately, there is no clear-cut answer to this dilemma, but in the next section we do present some alternatives.

STRENGTH DIFFERENCES BETWEEN MEN AND WOMEN

Three basic approaches have been used to determine whether true "sex differences" exist between men and women in terms of muscular strength. Strength has been evaluated (1) in relation to muscle cross-sectional area, (2) on an **absolute** basis as total force exerted, and (3) as **relative** strength, that is, strength in relation to body mass or lean body mass.

STRENGTH IN RELATION TO MUSCLE CROSS SECTION

Human skeletal muscle can generate approximately 3 to 8 kg of force per cm² of muscle cross section regardless of gender. In the body, however, this force-output capacity varies depending on the arrangement of the bony levers and muscle architecture. Figure 22-3 presents a comparison of the arm flexor strength of men and women in relation to the cross-sectional area of muscle. Clearly, the greatest force is exerted by individuals with the largest muscle cross section. The linear relationship between strength and muscle size,

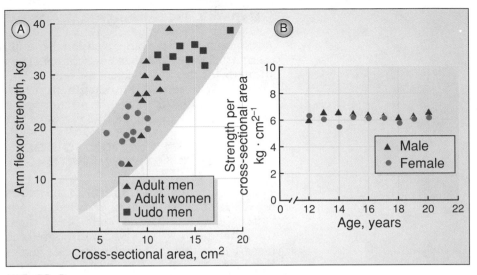

FIG. 22–3.
A. Variability of upper arm flexor strength of men and women as a function of upper arm flexor cross-sectional area. Absolute strength was computed as the measured strength × 4.90. B. Strength per unit cross-sectional area of muscle in males and females 12 to 20 years of age. (From Ikai, M., and Fukunaga, T.: Calculation of muscle strength per unit cross-sectional area of human muscle by means of ultrasonic measurements. Arbeitsphysiologie, 26:26, 1968.) For more recent data see ref 43a.

however, indicates little difference in arm flexor strength for the same size muscle in men and women. This is further demonstrated in the insert graph when the strength of males and females is expressed per unit area of muscle cross section.[72]

ABSOLUTE MUSCULAR STRENGTH

When strength is compared on an absolute score basis (that is, **total** force in pounds or kilograms), men are usually considerably stronger than women for all muscle groups tested. These strength differences between the genders are particularly apparent in comparisons of upper body strength; females are about 50% weaker than males in contrast to lower body strength, where they are about 30% weaker.[63,99] This gender characterization of muscular strength is true regardless of the device used to measure strength. The exceptions may occur for some strength-trained female track and field athletes and bodybuilders who significantly increase the strength of specific muscle groups by resistance exercises.

Isometric Measurements

In a review of seven studies comparing the isometric strengths of men and women,[85] the absolute strength of the upper extremities measured in 11 different positions in women was about 56% less than that of their male counterparts; for five other tests of leg strength, the women's strength scores averaged about 72% of the maximum values recorded by men.

Dynamic Measurements

Prior experiments have evaluated the absolute weight-lifting and dynamometric strength of men and women. The results were clear—the absolute strength for the men was about one third greater than that for women for each measurement.[137] Similar findings have been reported by others[117,118] and indicate that the dynamic strength of women averaged about 70% that of men, with a range of 59 to 84%, depending on the muscle groups tested.

STRENGTH IN RELATION TO BODY MASS AND LEAN BODY MASS, BODY FAT, AND GIRTH

There is controversy concerning the true extent of gender differences in muscular strength because of different approaches that are used to express "strength." The traditional relative method creates a ratio score by dividing the individual's strength score by a reference measurement such as body mass, lean body mass, total or muscle cross-section area, limb volume, fatfold thickness, or girth. For example, when body mass or lean body mass is used to "equate" genders in strength by creating a strength score as a ratio of strength per unit of body mass or lean body mass, the large absolute strength differences between men and women are reduced considerably. In fact, when this method is used to compute leg-press strength per kilogram of lean mass, females are considered "stronger" than males.[137] **We must point out, however, that this statistical method of attempting to equate for gender differences in body size may not truly "equalize" the men and women in terms of the underlying physiology.** The ratio method only creates a score that presumes to eradicate apparent gender differences. In fact, a recent study has concluded that men and women did not differ significantly in either upper or lower body strength when gender differences were "corrected" by using the traditional ratio method that incorporates lean body mass, local girth, and estimated local fat content.[63]

As was the case for aerobic capacity discussed in Chapter 11, one way to fairly evaluate gender differences in a criterion trait such as muscular strength (or aerobic capacity) is to compare males and females who do **not** differ in such body size variables as body mass or lean body mass. Consequently, there is no need to create a ratio score because the males and females are of equal body size. With this approach, five measures of muscular strength were assessed using concentric (shortening) contractions for the bench press and squat (1-RM) and isokinetic dynamometry for assessing the maximum force output for knee flexion and extension and seated shoulder press for men and women.[80] Figure 22-4 shows that when males and females were matched for body mass, the gender differences were largest for the sedentary group (44.0% for the shoulders and 20.3% for knee flexion) compared to the trained group (33.0% for the bench press and 10.7% for knee extension). The percentage differences were lower for the sedentary and trained groups when they were matched for LBM. The largest gender differences for the sedentary group occurred for the shoulder press (39.4%) and bench press (31.2%); the corresponding gender differences for the trained group were 30.6% (shoulder press) and 35.4% (chest press).

These results differ markedly from prior studies that have used the traditional approach of attempting to equate males and females by generating a ratio score to express muscular performance. The traditional approach would provide a strong argument that few gender differences exist in muscle quality, at least as reflected by force output capacity. In contrast, when males

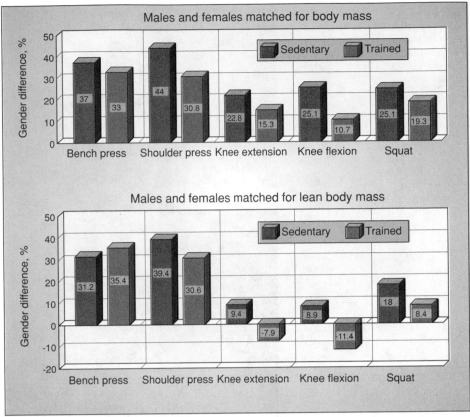

FIG. 22–4.

Males and females matched for body mass and lean body mass in relation to five measures of muscular strength. Above the zero line is the percent that males exceed females. (Data courtesy of Keller, B.: The influence of body size variables on gender differences in strength and maximum aerobic capacity. Unpublished doctoral dissertation. University of Massachusetts, 1989.)

and females are matched before testing for body size and composition and training status, the results reveal that males are still "stronger," especially in measures of upper body strength.

Until more research is available, however, it would be unfair to conclude that gender differences in selected aspects of muscular strength are truly biologic in nature. A more complete answer awaits further research.

TRAINING MUSCLES TO BECOME STRONGER

As a general rule for men and women, a muscle worked close to its force-generating capacity will increase in strength.[91] The overload can be applied with standard weight-plate equipment, pulleys or springs, immovable bars, or a variety of isokinetic devices. The important point is, that strength improvements are generally governed by the **intensity** of overload (level of tension placed on muscle) and not by the specific type. It is just that certain methods lend themselves to a precise and systematic overload application.

FIG. 22–5.
A, Concentric (shortening) contraction; B, Eccentric (lengthening) contraction; and C, Isometric (static) contraction.

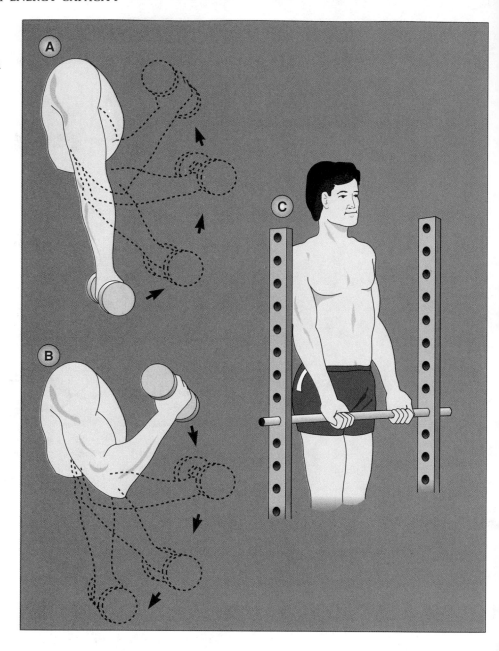

Progressive resistance weight training, isometric training, and **isokinetic training** are three common exercise systems used for training muscles to become stronger. These systems rely, to varying degrees, upon the three types of muscular contractions illustrated in Figure 22-5A-C.

Types of Contraction

When movement of the skeleton takes place, the action is considered **dynamic.** Concentric and eccentric contractions occur during dynamic activity. With a **concentric** muscular contraction (Fig. 22-5A), the muscle shortens as it develops tension to overcome the external resistance to cause movement. This type of contraction is common to weight lifting as well as to most other sports activities. When a muscle contracts **eccentrically** (Fig. 22-5*B*), the external resistance exceeds the muscle force and the muscle lengthens while developing tension. This usually occurs as a muscle acts to oppose the force of gravity. In weight lifting, muscles frequently contract eccentrically as the exerciser slowly returns the weight to the starting position to begin again the next concentric contraction. These combinations of concentric and eccentric contractions in weight lifting have frequently, but imprecisely been termed **isotonic exercise** (from the Greek **iso,** the same or equal; **tonos,** tension or strain). Although the absolute weight lifted or external resistance in a specific exercise remains constant, the bony lever system presents the muscle with an ever-changing resistance through the full range of motion.

In Figure 22-5*C*, the object is too heavy to move and the muscle contracts **isometrically.** Considerable force may be generated with no noticeable shortening of the muscle fibers or skeletal movement.

Resistance Plus Aerobic Training Equals Less Strength Improvement

Concurrent maintenance of resistance and aerobic training programs reduces the magnitude of strength development.[35,65] This may partly explain why power athletes and body builders often refrain from endurance activities while resistance training. Such findings should not deter those who desire a well-rounded conditioning program that offers the specific fitness and health benefits available from both forms of training.

RESISTANCE TRAINING FOR CHILDREN

Relatively little is known concerning the benefits and possible risks of resistance training in preadolescents. Because the skeletal system is in the formative stage, obvious concern arises as to the potential for injury of heavy muscular overload in growing children. Furthermore, because the hormonal profile is still in progressive development, especially for the tissue-building hormone testosterone, one might question whether resistance training could induce significant strength improvements at a relatively young age. The limited available evidence indicates that closely supervised resistance training programs that employ concentric muscular contractions with high repetitions and relatively low resistance can significantly improve the muscular strength of children with no adverse effect on bone or muscle.[106,116,136] Certainly, more studies are warranted before definitive statements can be made regarding the benefit-to-risk ratio and long-term effect on growth and development of regular and more stressful muscular overload.

PROGRESSIVE RESISTANCE WEIGHT TRAINING

Probably the most popular form of resistance training involves weight lifting. With this method, exercises are designed to strengthen specific muscles by causing them to overcome a fixed resistance, usually in the form of a barbell, dumbbell, or weight machine.

Progressive Resistance Exercise (PRE)

Working in a rehabilitation setting following World War II, researchers devised a method of weight training to improve the strength capacity of previously injured limbs.[29] Their method involved three sets of exercises—each set consisting of ten repetitions done consecutively without resting. The first set was done with one half of the maximum weight that could be lifted 10 times or $\frac{1}{2}$ 10-RM; the second set was done with $\frac{3}{4}$ 10-RM, whereas the final 10-RM set was done with maximum weight. As patients trained, the exercised limbs became stronger, and it was necessary to increase periodically the 10-RM resistance so that continued strength improvements would occur. Even when the progression of exercise intensity was reversed so that the 10-RM was performed first, similar improvements were observed.

This technique of **progressive resistance exercise** or **PRE** is a practical application of the overload principle and forms the basis of most resistance training programs.

Variations of PRE

Variations of PRE for weight training have been studied to determine the optimal number of sets and repetitions, and the frequency and relative intensity of training required to improve muscular strength.[20] The findings can be summarized as follows: (1) Performing an exercise between 3-RM and 9-RM is the most effective number of repetitions for increasing muscular strength. (2) PRE training once weekly with 1-RM for one set increases strength significantly after the first week of training and each week up to at least the sixth week. (3) No particular sequence of PRE training with different percentages of 10-RM is more effective for strength improvement, as long as one set of 10-RM is performed each training session. (4) Performing one set of an exercise is less effective for increasing strength than performing two or three sets, and there is some indication that three sets are more effective than two sets. (5) The optimum number of training days per week with PRE is unknown for improving muscular strength. Significant increases have occurred training 1 day weekly to 5 days per week for beginners. (6) When PRE training uses several different exercises, training 4 or 5 days a week may be less effective for increasing dynamic strength than training 2 or 3 times per week. The more frequent resistance training may prevent sufficient recuperation between training sessions. This possibly could retard progress in neuromuscular adaptation and strength development. (7) There is some indication that for a given resistance or load, a fast rate of movement may generate a better strength improvement than lifting at a slower rate. Furthermore, neither free weights (barbells or dumbbells) nor concentric-eccentric-type weight machines are inherently superior to the other in terms of developing muscular strength.[61,68]

Practical Recommendations for Initiating a Weight-Training Program

In the beginning stages of a program, maximum lifts should be avoided. Excessive weight contributes little to strength development and greatly increases the chances for muscle or joint injury. In fact, a load that is equal to 60 to 80% of a muscle's force-generating capacity (1-RM) is sufficient to increase strength. This generally permits the completion of about ten repetitions of a particular exercise. Using a lighter weight (and thus more repetitions) may be more effective when starting a weight-training program. Experience has shown that beginners should initially attempt to complete 12 to 15 repetitions. This will not place an excessive strain on the muscles during the beginning phase of the muscular conditioning program. A heavier weight should be used if the weight selected for the 12 repetitions feels too easy. The weight is too heavy if the exerciser cannot do 12 repetitions. This is a trial-and-error process and may take several exercise sessions before a proper starting weight is selected. After a week or two of training, when the muscles have adapted and the correct movements are learned, the number of repetitions can be reduced to six to eight. Each time this new target number of repetitions is reached, more weight is added. This is progressive resistance training—as the muscles become stronger, the weight is adjusted and a heavier load is attempted. In general, the exercise sequence should proceed from larger to smaller muscle groups in an attempt to reduce early exhaustion of larger muscles.

The Lower Back

Many orthopedists consider muscular weakness, especially in the abdominal region, and poor joint flexibility in the back and legs, prime factors related to the **low back pain syndrome.** Both strengthening and flexibility exercises are commonly prescribed for the prevention and rehabilitation of chronic low back strain.

The proper application of resistance training provides an excellent means for strengthening the abdomen and lumbar extensor muscles of the lower back through their full range of motion.[41,105] These muscles provide the necessary support and protection for the spine. If done improperly, however, with heavy weight and hips thrust forward and back arched, weight lifting can impose a considerable strain on the lower spine. Pressing and curling exercises, if performed with excessive hyperextension or arch to the back may also create unusually high shearing stresses on the lumbar vertebrae. This spinal pressure can trigger low back pain. Proper execution should not be sacrificed to lift a heavier load or "squeeze out" an additional repetition. The extra weight lifted through improper technique will not facilitate strengthening the desired muscle groups and may precipitate an injury.

Wearing a weight-lifting "belt" during heavy lifts (squats, dead lifts, clean and jerk maneuvers) significantly reduces intra-abdominal pressure compared to not wearing the belt.[59] This may reduce the potentially injurious compressive forces on spinal discs during heavy lifting as in most Olympic and power lifting events. The researchers concluded that a belt should always be employed for near maximal or maximal lifts and that someone who normally wears a belt should be extremely cautious about lifting without one. A further recommendation is that at least some of the training be done without a belt to help strengthen the deep abdominal muscles and develop a pattern of muscle recruitment needed to generate high intra-abdominal pressures when a belt is not worn.

ISOMETRIC STRENGTH TRAINING

Research in Germany showed that a weekly increase in isometric strength of about 5% of the initial strength level could be achieved by performing a daily single, maximum isometric contraction of only 1-second duration or a 6-second contraction at two thirds maximum.[62] Repeating this contraction 5 to 10 times daily produced greater increases in isometric strength. These results have been observed for subjects who differed in initial strength and age.[25,98]

Although isometric exercise is effective in providing muscular overload and improving strength, its use may be limited for sports training. For one thing, it is difficult to evaluate progress in training. Because there is no movement, it is impractical to measure force output or whether a person's strength is actually improving. Also, the development of isometric strength is highly **specific.** Thus, a muscle trained isometrically demonstrates improved strength mainly when the muscle contracts isometrically, especially at the joint angle and body position at which the strength was developed.[81,86]

If isometric training is used to develop "strengths" in a particular movement, it is probably necessary to train isometrically at many points through the range of motion. This can become time-consuming, especially with ready availability of conventional weight training and, more recently, isokinetic and hydraulic methods. The isometric method, however, does seem to be beneficial in muscle testing and rehabilitation. With isometric techniques,[19–21] specific muscle weakness can be detected and strengthening exercises performed at the appropriate joint angle.

WHICH ARE BETTER, STATIC OR DYNAMIC METHODS?

It is generally true that both static and dynamic training methods produce significant increases in muscular strength.[17] The resistance training method selected must be determined by the individual's specific needs.

Specificity of the Training Response

The isometrically trained muscle is stronger when measured isometrically; the dynamically-trained muscle is stronger when evaluated during resistance lifting. Muscles trained in a particular limited range of motion demonstrate the greatest strength improvement when measured in that specific range of motion. This **angle-specific training response** is observed for both dynamic[51] and static exercise.[81] This **specificity of resistance training** is probably explained because **improvements in a muscle's force production capacity with resistance training are related to a blend of favorable adaptations that occur both in the muscle itself as well as in the neural organization and excitability for a particular voluntary movement.**[37,92,111] Likewise, all-out muscular effort depends not only on local factors such as muscle fiber type and cross section, but also on neural factors that determine effective recruitment and the synchronization of firing of the appropriate motor units.[55,98,110]

A 3-month study with young adult men and women emphasized the highly specific nature of the adaptations to resistance training.[34] One group trained the adductor pollicis muscle isometrically with 10 daily contractions of 5-seconds duration at a frequency of one contraction per minute. The other group trained the same muscle dynamically with 10 daily series of 10 contractions moving a weight that represented one third maximal strength. The non-trained muscle served as the control. To eliminate any training influences on psychologic factors and central nervous system adaptations, the trained muscle was evaluated with a supramaximal electrical stimulation applied to the motor nerve. The results indicated that both groups improved in maximal force capacity and in the peak rate of force development with training. The improvement in maximal force for the isometrically trained group, however, was nearly twice that of the group that trained dynamically.

Conversely, improvement in the speed of force development was about 70% greater in the group that trained with dynamic contractions. Such findings provide strong evidence that resistance training per se is not a single entity; rather, improvements in the contractile properties (maximal force, velocity of shortening, and rate of tension development) of the muscle itself can vary in a manner that is highly specific with the type of contraction used in training. Consequently, although it is true that both static and dynamic training methods produce significant increases in muscular strength, no one system can really be considered "superior" to the other. The crucial consideration is the intended purpose for the newly acquired strength.

Practical Implications

The complex interaction between the nervous and muscular systems provides some explanation for the observation that the leg muscles, when strengthened in an activity like squats or deep knee bends, do not usually show improved force capability when used in another leg movement such as jumping. Also, a muscle group strengthened with weights does not generate an equal improvement in force when measured isometrically. Consequently, strengthening muscles for use in a specific activity such as golf, rowing, swimming, or football requires more than just identifying and overloading muscles involved in the movement. It requires that training be specific with regard to the exact movements involved. Training the muscles of the leg to become stronger by weight lifting does not necessarily mean that the performance of **all** subsequent leg movements will be improved. There is little transfer of newly acquired strength to other types of movements, even though the **same** muscles are involved.[15,27,34,131] This is clearly shown in a recent study in which a 227% increase in leg extension strength occurred through a standard program of weight training. When peak torque of the leg extensors was evaluated with an isokinetic dynamometer, however, there was only a 10 to 17% improvement![43] **To improve a specific performance by the strengthened musculature, the muscles must be trained with movements as close as possible to the desired movement or actual skill.**

Within this framework of training specificity, isokinetic-type methods show great promise for effective strength improvement at different speeds of movement.

Specificity Not Always Observed

Even though the results of most training studies support the concept of specificity, the exceptions are frequently overlooked. In one study subjects who trained with weights improved equally on isotonic and isokinetic tests.[40] Similarly, subjects who trained isometrically or isokinetically failed to show superiority on either of the specific tests,[82] and those who trained eccentrically or concentrically revealed comparable gains on both test modes.[75,76,102]

In a recent training study,[68] subjects in a free weight group (FW) trained 3 times a week for 12 weeks with eccentric and concentric actions; a second group trained with concentric-only contractions using hydraulic resistance (HY), and a control group did not train. Training with FW and HY included five sets of supine bench press and upright squat at an intensity of 1 to 6 RM, plus five supplementary exercises at 5 to 10 RM for a total of 20 sets per session for approximately 50 minutes. Testing before and after training included the 1-RM bench press and squat performed with concentric and eccentric contractions, and without prestretch (concentric only) using standard free weight testing procedures.

Figure 22-6 (top) shows that the FW group made large increases in concentric squat strength (+35%), but these increases were not significantly different from those observed in the group who trained using only concentric muscular contractions (+39%). When squat strength was assessed with eccentric and concentric conditions (bottom panel), both groups improved approximately 30%. Although the specificity principle in resistance training may encompass angular or length specificity, velocity specificity, and task or movement pattern specificity,[77,125] the principle may not be totally valid for contraction type or test mode specificity.[75,102,103] It is likely that exercise training with free weights and hydraulic devices was not distinct enough in

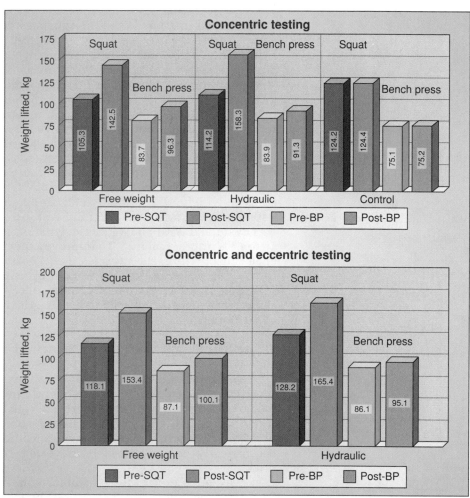

FIG. 22–6.
Effects of 12 weeks of resistance training with concentric resistance and combined eccentric + concentric resistance on (Top panel) pre- and post-test eccentric squat (SQT) and bench press (BP) contractions, and (Bottom panel) pre- and post-test concentric SQT and BP contractions. (From Hortobágyi, T., and Katch, F.I.: Role of concentric force in limiting improvement in muscular strength. J. Appl. Physiol., 68:650, 1990.)

terms of muscular contraction to elicit specific training adaptations. Unlike isokinetic and hydraulic routines, however, free-weight exercise includes both eccentric and concentric actions. During an exercise that incorporates both eccentric and concentric muscle actions, the limiting factor in overcoming resistance is the concentric and not the eccentric muscle force. Thus, despite differences between the contraction modes used for free-weight and hydraulic exercise, FW- and HY-trained subjects may have still used the same training stimulus as the concentric load. It is therefore a distinct possibility that any form of concentric or combined concentric plus eccentric resistance training should result in comparable gains in muscle strength as long as similar movement patterns and velocities are employed for training and testing.[92a]

It is also possible that the general principle of specificity may not be applicable to different contraction modes under conditions of fatigue. To evaluate this, endurance athletes and power athletes were tested under four conditions of fatigue that included the following: (1) 49 repetitions of maximal knee extension interspersed with 5-second isometric contractions, (2) 49 repetitions of maximal knee extension with lengthening (eccentric) contractions, (3) 49 maximal knee extensions with shortening (concentric) contractions, and (4) 1 minute of continuous, maximal vertical jumps.[67a] Prior to and following each of the fatiguing sessions, the maximum knee extension force was measured by isometric, lengthening, and shortening contractions with the same apparatus that was used to induce fatigue. The graphs in Figure 22-7 show that the power athletes fatigued significantly more than the endurance athletes. However, there were no significant differences among the **changes** in the three measures of force output (strength) following the four fatigue protocols. These data suggest that the failure of the contractile mechanism within the muscle is independent of the type of contraction that is used to induce the fatigue.

ISOKINETIC RESISTANCE TRAINING

Isokinetic resistance training attempts to combine the best features of both isometrics and weight training to provide muscular overload at a constant preset speed while the muscle mobilizes its variable force-generating capacity through the full range of movement. Any effort encounters an opposing force relative to the force being applied; this is an **accommodating resistance exercise** accomplished with the aid of a mechanical device. Theoretically, isokinetic-type training would make it possible to activate the largest number of motor units and consistently overload muscles to achieve their maximum tension-developing or force output capacity at every point in the range of motion, even at the relatively "weaker" joint angles.

Isokinetics Versus Standard Weight Lifting

An important distinction can be made between a muscle overloaded isokinetically and one overloaded with a standard weight-lifting exercise. Figure 22-8 shows the relationship between maximal force generated by the knee extensor muscles at various knee joint angles. As with all joints of the body, the muscular force exerted against an external resistance varies with the bony lever configuration as the joint moves through its normal range of motion. Therefore, when training with weights, the resistance is usually fixed at the greatest load that allows completion of the movement. Consequently, the resistance can be no greater than the maximum strength of

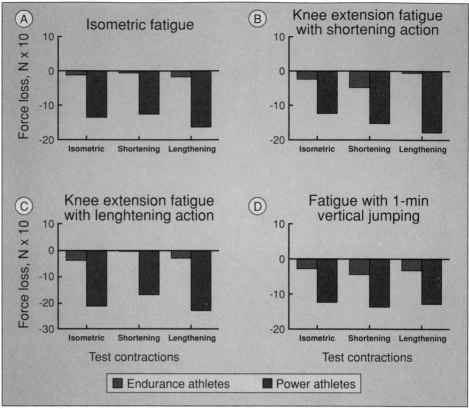

FIG. 22–7.
Effects of four different fatigue conditions on the forces exerted with isometric shortening, and lengthening contractions in 12 endurance and 12 power athletes. **A.** Isometric fatigue. **B.** Knee extension fatigue with shortening contractions. **C.** Knee extension fatigue with lengthening contractions. **D.** Fatigue induced by 1 minute of continuous vertical jumping. (Data courtesy of Hortobágyi, T.: Specificity of muscular fatigue and force enhancing mechanisms with isometric, shortening and lengthening muscle actions, and myotatic stretch in power and endurance athletes. Unpublished doctoral dissertation. University of Massachusetts, Amherst, 1990.)

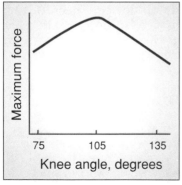

FIG. 22–8.
Peak force generated by the knee extensor muscles in relation to the joint angle of the knee.

the weakest muscle (or joint position) in the range of motion. Otherwise, the movement would not be completed. Weight lifters frequently refer to this point in a range of motion as the "sticking point." A main limitation of weight-lifting exercise is that the force generated by the muscles during a contraction is not maximum throughout **all** phases of the movement. To help overcome this problem, several equipment manufacturers have devised what they term a **variable resistance** machine that utilizes an irregularly shaped metal cam or other device to adjust the resistance in accordance with the lever characteristics of a particular joint movement. Essentially, this equipment still represents a classic mode of weight lifting, except that the resistance offered to the muscle is, in theory, relatively constant throughout the full range of motion. The machine does not control the speed of movement and the design of the mechanical device that is supposed to achieve a variable resistance is based on the average physical dimensions of a population with little means to adjust for individual variation in body structure. Furthermore, there is little evidence that such cam devices actually compensate fully to account for individual differences in mechanics and force application at all phases of the particular movement on a given exercise device.[58] This is not the case in an isokinetically loaded muscle. The desired speed of movement occurs quickly and the muscle is able to generate peak power output throughout the range of motion at a specific, but controlled velocity of contraction.

Experiments With Isokinetic Exercise and Training

Experiments using isokinetic exercise have been designed to explore the force–velocity relationships in various exercises and to relate this to the muscle's fiber composition.[26,131] Figure 22-9 displays the progressive decline in peak torque output in relation to increasing angular velocity of the knee extensor muscles in two groups of subjects who differed in sports training and muscle fiber composition. In the experiments that involved movement at 180 degrees per second, the power athletes, elite Swedish track and field sprinters and jumpers produced significantly higher torque values per kilogram of body mass than the other group of athletes, that included downhill skiers, competition walkers, and cross-country runners. At this angular velocity, the decrement in maximal torque was equal to about 55% of the maximal isometric torque.

The two curves in Figure 22-9 can also be distinguished in terms of peak torque depending on the group's muscle fiber composition. At zero velocity

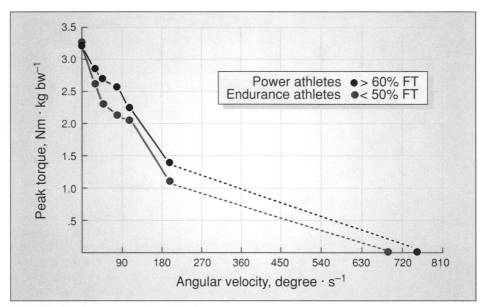

FIG. 22–9.
Peak torque output per kg of body weight expressed as a function of increasing angular velocity in two groups of athletes with different muscle fiber composition. The torque-velocity curves were extrapolated to the approximated values for maximal velocity of knee extension. (From Thorstensson, A.: Muscle strength, fiber types, and enzyme activities in man. Acta Physiol. Scand. (Suppl.) 443, 1976.)

(isometric contraction), peak force was the same for athletes with relatively high or low percentages of fast-twitch fibers; this indicated that both fast- and slow-twitch motor units were activated in maximal isometric knee extension. **As movement velocity increased, greater torque was achieved by individuals with a higher percentage of fast-twitch fibers.** This suggests that a high percentage of fast-twitch fibers is desirable for power activities in which success is largely influenced by one's ability to generate large muscular torque in rapid velocities of movement.[26,132]

Fast-Versus Slow-Speed Isokinetic Training

Studies have compared strength and power improvement with isokinetic training at slow and fast limb speeds. Such studies have provided further support for the principle of specificity as applied to both performance and training response. For example, several research studies support the contention that gains in "strength" from slow-speed training are highly specific to the angular velocity of movements used in training. Exercising at fast speeds, on the other hand, gives a more general improvement because increases in power output were noted at **both** fast and slow velocities of movement.[104] In fact, muscular hypertrophy was noted only during fast-speed training and this occurred only in the fast contracting type II muscle fibers.[27] This hypertrophy possibly accounts for a more general strength improvement with fast-speed training. In the aforementioned studies, the greatest improvements in muscular performance occurred at the specific angular velocity and movement pattern at which training took place. No definitive conclusion can be drawn whether fast or slow velocity training is significantly more effective for improving muscular force and power throughout the entire velocity spectrum.

The fundamental basis for isokinetic training is attractive because muscles can be overloaded through a full range of motion at speeds somewhat similar to, but considerably slower than, specific sport and physical activities. It should be remembered that while isokinetic-type training occurs at a relatively slow but constant, preselected angular velocity of movement, there are few physical activities that are performed at a constant, relatively slow speed of joint motion. Also, the present generation of isokinetic-type dynamometers cannot provide effective overload for eccentric muscular contractions that seem to be important for limb deceleration and "braking" control of human movements.

PLYOMETRIC TRAINING

Athletes involved in sports that require specific powerful movements—such as football, volleyball, sprinting, and basketball—have recently used a special form of exercise training drills termed **plyometrics** or explosive jump training. With plyometric exercise, movements are structured to make use of the inherent stretch-recoil characteristics of skeletal muscle as well as the modulation of a muscle's response via the stretch or myotatic reflex. Essentially, with plyometric exercise, overload is applied to skeletal muscle in a manner that rapidly places the muscle on stretch (eccentric or lengthening phase) immediately prior to the concentric phase or contraction. This rapid lengthening phase probably facilitates a subsequent more rapid and forceful movement and hence augments the speed-power benefits of training.[56]

In its practical application, a plyometric drill utilizes one's body mass and the force of gravity to provide the all-important rapid prestretch or "cocking" phase to "activate" the muscle's natural elastic recoil elements. This then augments the subsequent concentric contraction in the opposite direction. An example of a natural form of plyometric movement is an eccentric pre-stretch brought on by rapidly lowering and dropping the arms to the side prior to vertical jumping. Specific plyometric drills for the lower body include a standing jump, multiple jumps, repetitive jumping in place, depth-jumps or drop jumping from a height, single and double leg jumps, and various modifications. It is believed that repetitions of these exercises will provide the proper neural and muscular training to enhance the power performance of specific muscles.[16]

Testimonials abound as to the benefits of plyometric training, but carefully controlled evaluation is lacking of both benefits and possible orthopedic risks of such exercise.[9a] At this point, research is required to establish the appropriate role, if any, of plyometric drills in a total strength-power training program.

PHYSICAL TESTING IN THE OCCUPATIONAL SETTING: THE ROLE OF SPECIFICITY

The high degree of specificity within various components of physical performance and physiologic function such as aerobic fitness and muscular strength and power, as well as the specific nature of the training response, casts serious doubt on the assumption that broad constructs of physical fitness exist to any significant extent. Clearly, there is no one measure of muscular strength and no one measure of aerobic fitness. An individual possesses an array of muscular "strengths" and "powers." Often, these expressions of physiologic function and performance are poorly related to each other, if at all. Likewise, a person has diverse capabilities for expressing aerobic function, depending on the muscle mass used. Certainly, within the occupational setting, the use of a 12-minute run to infer aerobic capacity for firefighting or lumbering (both requiring significant upper body aerobic function), or the use of static grip or leg strength to evaluate the diverse strengths and powers required in such occupations is physiologically naive in light of current knowledge in exercise physiology. In the occupational setting it is imperative to apply measurement that most closely resembles the actual content of the job, not only in terms of specific tasks, but also in a manner that faithfully reflects the intensity, duration, and pace of the job.

ELECTROMYOGRAPHY (EMG) DURING MAXIMAL BALLISTIC CONTRACTIONS

The EMG signal provides a convenient methodology to study the intricacy of muscle neurophysiology during various types of contractions. The EMG is influenced by both the quality and quantity of electrical activity generated by the muscles. In isometric contractions, for example, the EMG signal is proportional to the amount of force generated by the muscle. In dynamic-type contractions, the situation is more complex because of the changing force-torque characteristics during contraction at different angular velocities of movement. During rapid, ballistic-type movements, the EMG is characterized by alternating bursts of electrical activity in the agonist and antagonistic muscles. The EMG pattern is triphasic in nature; the first burst of electrical activity occurs in the agonist, followed by signals from the antagonist (during which time the agonist is electrically silent), and then another burst of activity in the agonist. Each phase of the EMG is associated with certain aspects of the contraction. The first burst of the agonist is believed to create the propulsive force to set the limb in motion, the antagonist's burst stops the limb, and the agonist's second burst is involved with the final positioning of the limb. In our studies of professional baseball pitchers and champion body builders, there were striking differences between the groups in the triphasic pattern elicited during maximal speed, unloaded arm flexion. Figure 22-10 displays a comparison using the integrated EMG signal. For the 19 pitchers, the second burst occurs sooner (probably a pro-

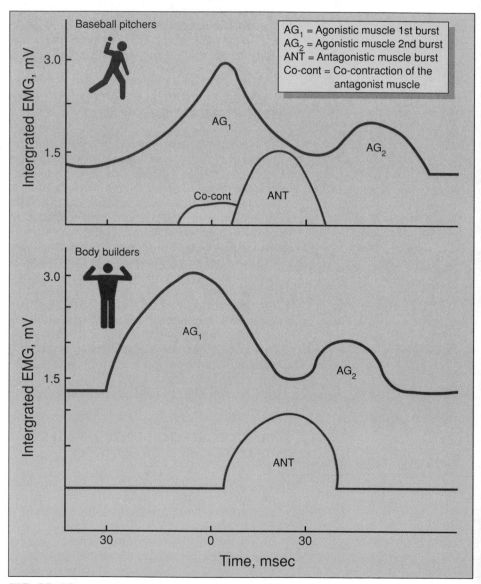

FIG. 22–10.
Comparison of the triphasic EMG pattern during rapid arm flexion in professional baseball pitchers and champion body builders. (Data courtesy of Dr. Pierre Lagasse, Human Motor Performance Research Laboratory, Laval University, Quebec City, Canada).

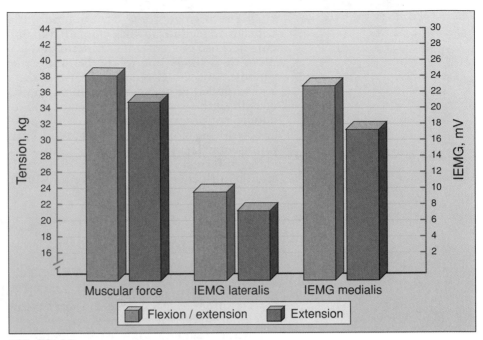

FIG. 22–11.
Increase in maximal knee extension tension output (kg) and integrated EMG (IEMG) activity (mV) during knee extension and knee flexion/extension. (From Lagasse, P., Normand, M., and Larue, J.: Neuromuscular facilitation of muscle tension output by reciprocal muscle work. Annals of the French-Canadian Association for the Advancement of Sciences, 50:222, 1983.)

tective mechanism to slow limb speed), and the amplitude is less. For the 11 body builders, the first agonist burst occurs rapidly and there is a distinct "delay" before the antagonist fires. This is probably related to a training adaptation of rapid limb acceleration during repetitive efforts of maximal effort in weight lifting over many years of training.

EMG During Concentric Bidirectional Muscle Work

A series of experiments has demonstrated the beneficial effect of bidirectional (reciprocal) concentric muscle exercise over conventional, unidirectional concentric contractions.[87] There were two treatment conditions: (1) from a fully flexed position of the right knee, subjects executed a maximal knee extension while seated in an exercise device that permitted concentric-only contractions; (2) from a fully extended position, the leg was flexed through the full range of motion and then extended to the original starting position. All contractions during flexion and extension were performed at maximal speed. The order of administration of treatment conditions was balanced across subjects to attenuate sequence effects. Figure 22-11 shows the results for maximal tension output (determined by strain gauge with simultaneous recording of joint displacement) and integrated EMG from the vastus lateralis and vastus medialis during the two treatment conditions. When the maximal flexion movement preceded extension, tension output increased significantly by 11.4%, and the EMG activity from the quadriceps increased significantly by 31.2% (vastus medialis) and 42.9% (vastus lateralis). The augmentation of tension output and greater EMG activity during concentric reciprocal work compared to unidirectional, concentric work occurs because of greater recruitment and activation of motor units which is more than likely due to the facilitatory effects of muscle proprioceptors (spindles and Golgi tendon organs). If the right and left limbs were flexed and extended in alternating fashion, the tension output and EMG activity of both limbs would be even greater due to the facilitative influence of the double reciprocal contractions. An appropriate term for such facilitatory effects during bidirectional, double concentric contractions is **quadruple neuromuscular facilitation (QNF).** *[108] Studies of QNF should provide considerable information regarding the basic patterns of neuromuscular control and the influence of such contractions on subsequent athletic performance.

SUMMARY

1. The most common methods for measuring muscular strength are: (1) tensiometry, (2) dynamometry, (3) 1-RM testing with weights, and (4) computer-assisted force and work-output determinations including isokinetic-type measurement.
2. Human skeletal muscle can generate about 3 to 8 kg of force per square centimeter of muscle cross section, regardless of gender. On an absolute basis, however, men are usually stronger than women.
3. Two methods have been used to evaluate gender differences in muscular strength. The traditional method creates a ratio score for muscular strength [e.g., strength per unit body size (body mass, lean body mass, limb volume, girth)]. When body size and composition are considered in this manner, the large strength differences between men and women are reduced considerably, and in some cases women score higher than men.
4. When men and women are compared for muscular strength by matching body size and training status, strength differences still persist between genders.
5. Muscles become stronger in response to overload training. Overload is created by either increasing the load, increasing the speed of muscular contraction, or by a combination of the above.
6. A load that represents 60 to 80% of a muscle's force-generating capacity is usually sufficient overload to produce strength gains.
7. The three major systems for developing strength are progressive resistance weight training, isometrics, and isokinetic-type training. Each system results in strength gains that are highly specific to the type of training. Isokinetic-type training, because of the possibility for generating maximum force throughout the full range of motion at different velocities of limb movement, appears to offer a unique method for resistance training. For all resistance training, appropriate attention must be given to proper technique and safety.
8. The magnitude of strength improvement is somewhat blunted if strength and aerobic training are performed concurrently.
9. Based on limited data, closely supervised resistance training programs

* Term suggested by Professor Pierre Lagasse, Human Motor Performance Research Laboratory, Laval University, Quebec City, Canada.

using relatively moderate levels of concentric exercise significantly improve the muscular strength of children with no adverse effect on bone or muscle.

10. Plyometric training drills attempt to utilize the inherent stretch-recoil characteristics of the neuromuscular system to facilitate the development of muscular power of specific muscles. Determining both the risks and benefits of such training awaits further research.

11. The high degree of specificity of physiologic and performance measures, as well as their response to training, casts doubt on the wisdom of using broad or general fitness measures to infer one's ability to perform specific tasks or occupations.

12. EMG activity during rapid, ballistic movements can be characterized by a distinct triphasic pattern that may differ among individuals depending on prior athletic training and method of strength acquisition.

13. Muscle tension output and EMG activity are augmented during concentric, bidirectional contractions compared to unilateral contractions because of neuromuscular facilitation and subsequent recruitment of additional motor units.

P A R T 2

Adaptations with Resistance Training

Figure 22-12 displays six factors that have an impact on the development and maintenance of muscle mass. Without doubt, genetics provides the governing frame of reference that influences the effect of each of the other factors on the ultimate outcome of increased muscle mass and strength. Muscle activity contributes little to tissue growth without appropriate nutrition to provide essential building blocks. Similarly, specific hormones and patterns of nervous system innervation are crucial for patterning the appropriate training response. Without tension overload, however, each of the other factors is relatively ineffective to produce the desired training response.

FIG. 22–12.
Six factors that have an impact on total muscle mass.

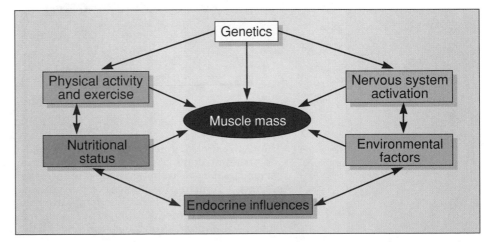

FACTORS THAT MODIFY THE EXPRESSION OF HUMAN STRENGTH

As shown in Figure 22-13, factors broadly characterized as psychologic, or neural, and muscular influence the expression of human strength. Many of these factors are modified by a program of systematic resistance training, whereas others appear to be training-resistant; these are probably determined by natural endowment or are fixed early in life.

PSYCHOLOGIC/NEURAL FACTORS

In a series of unique experiments conducted at George Williams College,[73] a group of college men and women was subjected to various treatments designed to evaluate "psychologic" influences on human strength. Arm strength was measured under normal conditions, immediately after a loud noise or while the subject screamed loudly at the time of exertion, under the influence of alcohol and amphetamines or "pep pills," and under hypnosis in which they were told they would be considerably stronger than usual and should have no fear of injury. Each of these factors generally increased

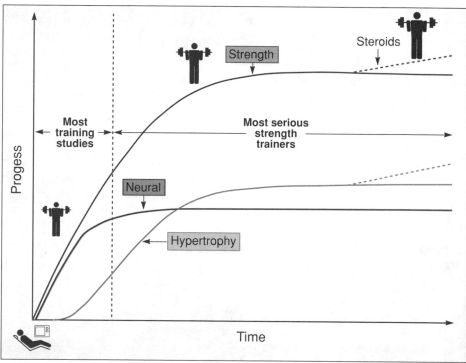

FIG. 22–13.
The relative roles of neural and muscular adaptation to strength training. Neural adaptation predominates in the early phase of training. This phase also encompasses most training studies. In intermediate and advanced training, progress is limited to the extent of muscular adaptation that can be achieved, notably hypertrophy—hence the temptation to use anabolic steroids when it becomes difficult to induce hypertrophy by training alone. (From Sale, D.G.: Neural adaptation to resistance training. Med. Sci. Sports Exerc., 20:135, 1988.)

TABLE 22–1. PHYSIOLOGIC ADAPTATIONS THAT OCCUR IN RESPONSE TO RESISTANCE TRAINING

System/Variable	Response
Muscle Fibers	
Number	Equivocal
Size	Increase
Type	Probably no change
Capillary Desity	
In bodybuilders	No change
In power lifters	Decrease
Mitochondrial Volume Density	Decrease
Twitch Contraction Time Enzymes	Decrease
Creatine phosphokinase	Increase
Myokinase	Increase
Enzymes of Glycolysis	
Phosphofructokinase	Increase
Lactate dehydrogenase	No change
Aerobic Metabolism Enzymes	
Carbohydrate metabolism	Increase
Triglycerides	Not known
Intramuscular Fuel Stores	
Adenosine triphosphate	Increase
Phosphocreatine	Increase
Glycogen	Increase
Triglycerides	Not known
max $\dot{V}O_2$	
Circuit resistance training	Increase
Heavy resistance training	No change
Connective Tissue	
Ligament strength	Increase
Tendon strength	Increase
Collagen content of muscle	No change
Bone	
Mineral content	Increase
Cross-sectional area	No change

(Modified from Fleck, S.J., and Kramer, W.J.: Resistance training: physiological responses and adaptations (Part 2 of 4). Phys. Sportsmed., 16:108, 1988.)

strength above normal levels. In fact, the greatest increments were observed under hypnosis, the most "mental" of all treatments!

The investigators speculated that strength improvements under the various experimental treatments were due to a temporary modification in central nervous system function. It was argued that most people normally operate at a level of neural inhibition, perhaps via protective reflex mechanisms, that prevents them from expressing their true strength capacity. This capacity is largely established by the cross section and fiber type of the muscle and by the mechanical arrangement of bone and muscle. Neuromuscular inhibition could be the result of unpleasant past experiences with exercise, an overly protective home environment, or a fear of injury. Regardless of the reason, the person is usually unable to express maximum strength capability. During the excitement of intense competition, or under the influence of disinhibitory drugs or hypnotic suggestion, however, inhibition is removed, motoneurons are optimally recruited, and an apparent "supermaximal" performance is attained.

An enhanced arousal level and accompanying disinhibition (or neural facilitation) could lead to full activation of muscle groups.[57] This may explain the so-called unexplainable feats of strength of men and women in emergency situations. In all likelihood, under such circumstances, the person is now able to achieve true maximum strength. In fact, drugs, loud noises, and hypnotic suggestion are not the only factors that might improve strength performance. Highly trained athletes in many sports create an almost self-hypnotic state by intensely concentrating or "psyching" prior to competition. In the first few days or weeks of resistance training, changes in neural facilitation more fully activate prime movers in a specific muscular action. This probably accounts for the early and rapid strength increase that is not associated with an increase in muscle size and cross-sectional area.[39,109]

MUSCULAR FACTORS

Although psychologic inhibition and learning factors greatly modify one's ability to express muscular strength in the early phase of training, the ultimate limit for strength is determined by anatomic and physiologic factors within the muscle. These factors are not immutable and can be modified with appropriate training procedures (Table 22-1). The changes in resistance-trained muscles are generally limited to adaptations in the contractile mechanisms and are usually accompanied by substantial increases in ability to exert force and apply power through a given range of movement.

Muscular Hypertrophy

Increases in skeletal muscle size with resistance training can be viewed as a fundamental biologic adaptation to an increased work load. This compensatory adjustment ultimately leads to an increase in the muscle's capacity to generate tension. It should be kept in mind that an increase in muscle size is not necessarily a prerequisite for improving strength and power with training.[27,137] This is probably related to the important neurologic factors involved in the expression of human strength. The later, slower strength improvements, however, generally coincide with alterations in muscle architecture.[55,98]

Muscular growth in response to overload training occurs primarily from an enlargement or **hypertrophy** of individual muscle fibers.[46,57,95] The fast-twitch muscle fibers of weight lifters, for example, were 45% larger than those of healthy sedentary people and endurance athletes.[38] The process of hypertrophy is directly related to the synthesis of cellular material, particularly the protein filaments that constitute the contractile elements. Within the cell, myofibrils thicken and increase in number and additional sarcomeres are formed by an acceleration of protein systhesis and corresponding decreases in protein breakdown.[3,39,45,48,95] It appears that the primary requirement for initiating muscular hypertrophy is an increase in the tension or force the muscle must generate. Also observed with resistance training are significant increases in local stores of ATP, CP, and glycogen.[89] Undoubtedly, this increase of intramuscular anaerobic energy stores contributes to the rapid rate of energy transfer required in this form of exercise.

The increase in total contractile protein and energy-generating compounds with heavy resistance training occurs without parallel increases in capillarization or in the total volume of mitochondria or mitochondrial enzymes within the muscle cells.[113,127,130] Thus the ratio of mitochondrial volume and enzyme concentration to myofibrillar (contractile protein) volume is actually **reduced** in resistance-trained muscle.[2] Although this training response is apparently not harmful to the performance of strength and power athletes due to the anaerobic nature of such effort, it may actually be detrimental to endurance performance by decreasing the fiber's aerobic potential per unit of muscle mass.

Aside from enlarging existing muscle fibers, tension overload may also stimulate an increase in bone mineral content as well as a proliferation of connective tissue and satellite cells that surround the individual muscle fibers.[45] This thickens and strengthens the muscle's connective tissue harness.[36,123] Muscular overload also improves the structural and functional integrity of both tendons and ligaments.[120,134] These adaptations may provide some protection from joint and muscle injury; this supports the use of resistance exercise in preventive and rehabilitative strength programs for athletes.

Figure 22-14 shows the changes in muscle fiber size that accompany exercise-induced hypertrophy. The top figure compares the exercised and unexercised soleus muscle of a rat. The hypertrophied muscle is on the right. The bottom figures are typical cross sections of untrained and hypertrophied muscles. Not only was the average diameter of the hypertrophied muscle larger by about 30%, but there was also a 46% increase in the number of nuclei present within the cells. These compensatory changes with intense muscular overload are related to a marked increase in DNA synthesis as well as to a proliferation of small, mononucleated satellite cells located under the basement membrane adjacent to the muscle fibers.[45]

Hyperplasia: Are New Muscle Fibers Made?

Whether the actual number of muscle cells increases (**hyperplasia**) with training is a question that is frequently raised. If this does take place, to what extent does it contribute to muscular enlargement in humans? Researchers have reported that some muscle fibers from trained animals undergo a process of **longitudinal splitting**,[49,50,67] or the development of new muscle fibers from satellite cells.[112] With longitudinal splitting, the split fibers become two individual daughter cells through a process of lateral budding.

One of the problems often encountered with animal research is generalizing the findings to humans. For example, most animals do not undergo the massive hypertrophy observed in humans with resistance training. Thus,

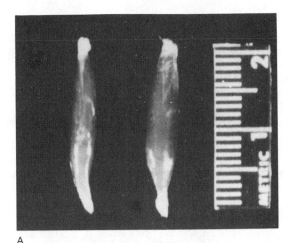

A

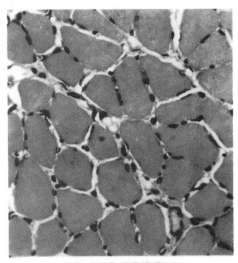

CONTROL   HYPERTROPHY

B

FIG. 22–14.
A, Control and hypertrophied rat soleus muscle. *B,* Cross sections of control and hypertrophied muscles shown in *A.* The average diameter for 50 fibers of the hypertrophied muscle was 24 to 34% greater than in controls; the average number of nuclei in the hypertrophied muscle was 40 to 52% greater than in the control. (From Goldberg, A.L., et al.: Mechanism of work-induced hypertrophy of skeletal muscle. Med. Sci. Sports, 3:185, 1975.)

for various animal species, muscle cell proliferation may be an important compensatory adjustment to overload. Whether this takes place in humans is currently an area of debate. Cross-sectional studies of body builders with relatively large limb circumference and muscle mass have failed to show that these athletes possessed a significant hypertrophy of individual muscle fibers.[90,91,129] This certainly leaves open the possibility of hyperplasia in humans with resistance training;[3,84,126] it suggests either an inherited difference in fiber number or that muscle cells may adapt differently to the high volume, high intensity training used by body builders compared to the typical low-repetition, heavy-load system favored by weight and power athletes. Even if hyperplasia is replicated in other human studies (and even if the response is a positive adjustment), **the greatest contribution to muscular size with overload training is made by the enlargement of existing individual muscle cells.**[92]

Changes in Muscle Fiber Composition

The effects of 8 weeks of progressive resistance exercise on muscle fiber size and composition were evaluated for the leg extensor muscles of 14 men who performed three sets of 6-RM leg squats 3 times a week.[131] Muscle biopsies were taken from the vastus lateralis before and after training. The results for muscle fiber type were clear; there was **no change** with resistance training in the percentage distribution of fast- and slow-twitch muscle fibers as indicated by the activity level of myofibrillar ATPase. This was consistent with previous studies using both resistance and endurance-type training,[47] and suggested that relatively short-term resistance training in adults does not alter the basic fiber composition of skeletal muscle. It is still open to question whether specific training early in life or for the prolonged time periods engaged in by Olympic-caliber athletes can cause a change in the inherent twitch (speed of shortening) characteristics of muscle fibers. Although research indicates a possible potential for progressive fiber type transformation with prolonged, specific training (see Chapter 18), the position presently taken is that the predominant muscle fiber distribution is established early in life and is largely determined by genetic factors.[38,47]

Although basic fiber type probably does not change dramatically during life,[52] other characteristics of specific fibers can undergo change with training. In the resistance training experiment just described, there were significant increases in the volume of fast-twitch fibers in the leg extensor muscles. This is clearly shown in Figure 22-15, in which the relative areas of the fast- and slow-twitch fibers are presented for each subject before and after training. Progressive resistance training, especially as performed among power and Olympic-type lifters, produces a significant hypertrophy, predominently of the fast-twitch fibers.[128,130] This makes sense within the framework of exercise specificity because the fast-twitch motor units are recruited in near-maximal, low-volume, high-resistance exercise that involves powerful contractions and requires anaerobic energy transfer. With the relatively lower resistance but higher volume training performed by body builders, a tendency has been noted toward hypertrophy of the slow-twitch fibers. Recent evidence suggests possible racial influences on muscle fiber composition with black males possessing a relatively larger percentage of type II fibers compared to Caucasian males.[7] If these findings are consistent, they would suggest that black individuals are well suited, in terms of muscle fiber type, for sport events requiring muscular strength and power and energy-transfer capacity for short duration.

NEUROMUSCULAR FACTORS

In previous sections, we have presented evidence that neuromuscular factors play an intimate role in the expression of muscular strength. What is less well researched are the changes in neuromuscular control consequent to muscular overload by resistance training. One study focused on the adaptations and neuromuscular patterning with resistance training.[107] The experiment investigated the effects of relatively slow-speed, isokinetic-type training in sedentary males and females on the latencies and amplitudes of

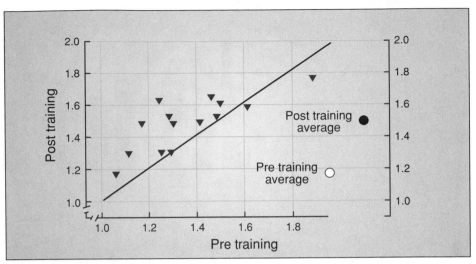

FIG. 22–15.
Individual changes for 14 male subjects in the ratio of the area of fast- to slow-twitch fibers following 8 weeks of strength training. Open circle on right indicates average pretraining FT/ST area ratio while dark circle is the post-training average. (From Thorstensson, A.: Muscle strength, fiber types, and enzyme activities in man. Acta Physiol., 443 (Suppl.) 1976.)

the muscular response (M) and patellar tendon reflex response (T) that accompanies increases in torque output. Figure 22-16 shows the results following 10 weeks of slow-speed, hydraulic resistance training that consisted of five training sessions per week. For training, subjects performed two sets of 15 knee extension concentric contractions at an angular velocity of 90° per second. The top part of the figure displays the improvements in peak torque output measured at 3 velocities of movement (90°, 120°, and 180° per second) that ranged from 48 to 54% for males and 67 to 70% for females. The differences for a nontraining control group were non-significant. The bottom part of the figure demonstrates that the T-reflex latencies were unaltered for males and females as a result of training. The T reflex was elicited by percussion of the patellar tendon and measured by EMG surface electrodes placed over the vastus medialis. The M response was measured by stimulation of 1 ms pulses at 0.2 Hz frequency of the femoral nerve. The M-response latencies decreased significantly in the males and females, whereas the amplitude of the M response increased approximately two-fold. The changes in the T reflex and M response that accompanied the substantial increases in maximal torque output were attributed to enhancement of alpha motor neuron excitability and, therefore, **greater recruitment of motor units** and improvement in the elastic properties of the trained muscles.

COMPARATIVE RESPONSES OF MEN AND WOMEN TO STRENGTH TRAINING

In today's society, women are successfully participating in all sports and physical activities. One area that women had generally shied away from was resistance training. Many women feared that such exercise would develop overly enlarged muscles similar to those observed for men engaged in heavy

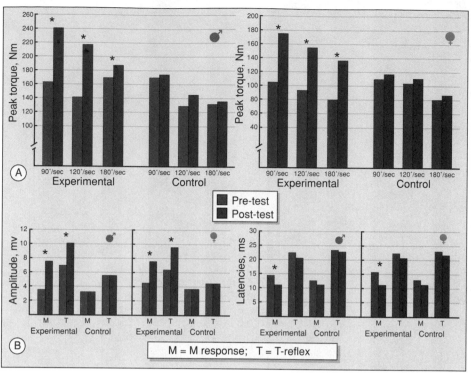

FIG. 22–16.
Effects of slow-speed resistance training on neuromuscular factors in males and females. A, Changes in peak torque (Nm) measured concentrically at three velocities of movement. B, Changes in the amplitude and latency of the vastus medialis (M) response and patellar tendon (T) reflex. The stars indicate significant differences between pre- and post-test values (P < 0.001). (From Roy, M.A., et al.: Changes in alphamotoneuron excitability following isokinetic training in sedentary males and females. Can. J. Appl. Sport Sci., 9:20P, 1984.)

weight-lifting programs. This was unfortunate because both men and women often lack sufficient strength to successfully perform activities such as tennis, golf, skiing, dance, and gymnastics. Appropriate forms of muscular overload rapidly improve strength to levels required to learn basic sport skills. A proper program of resistance training improves muscular strength and may have a favorable effect on body composition.[10,137]

MUSCULAR STRENGTH AND HYPERTROPHY

The basic gender difference in response to resistance training appears to be the degree of muscle hypertrophy. Despite similar strength improvements, increases in muscle girth are generally less for women. Researchers have speculated that this was due to differences in hormonal levels between the genders, especially the 20 to 30 times higher testosterone level in men which exerts a strong anabolic or tissue-building effect. It should also be noted that testosterone levels are on a continuum between men and women, with some females normally possessing high levels of this hormone.

Recent experiments using computed axial tomography (CAT) scans to directly evaluate muscle cross-sectional area suggest that the hypertrophic response to resistance training is **similar** for men and women. Absolute changes in size are certainly greater for men (because the initial muscle mass of the male is greater), but the enlargement of muscle on a percentage basis was the same between genders.[28] This ability for women to achieve substantial muscular hypertrophy with long-term resistance training[120a] has also been suggested in comparisons between elite male and female bodybuilders.[3] More research is needed before definitive statements can be made concerning similarities and differences in the resistance training responses of men and women. The limited data from relatively short-term studies do suggest that women can utilize conventional resistance-training exercise without developing overly large muscles.

IS MUSCULAR STRENGTH RELATED TO BONE DENSITY?

Several early studies reported a poor association between muscular strength and bone density. However, strength was assessed for the arm muscles using isometric methods. One could argue that such procedures failed to evaluate the true extent of the relationship between strength and bone density because isometric strength is generally not related to dynamic measures of strength that include the larger muscles of the trunk and extremities. A recent experiment in one of our laboratories has documented differences in maximum flexion and extension dynamic strength in postmenopausal women with and without osteoporosis.[74,122] Osteoporosis was defined as bone mineral density (BMD) greater than 1 standard deviation below age-adjusted normal levels measured by dual-photon absorptiometry in the lumbar spine and neck of the femur. Muscular strength was measured during five maximum chest press, lat pull, and knee and shoulder flexion and extension concentric contractions at three velocities of movement using an Omnitron hydraulic dynamometer. Figure 22-17 shows the results for chest flexion and extension

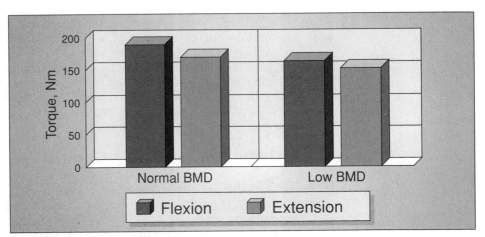

FIG. 22–17.
Comparison of chest press extension and flexion strength in age and weight matched postmenopausal women with normal and low bone mineral density (BMD). Strength differences were statistically significant between and within the groups. (From Stock, J.L., et al.: Dynamic muscle strength is decreased in postmenopausal women with low bone density. J. Bone Min. Res. 2;338, 1987, and Janney, C., et al.: Maximum muscular strength differs in postmenopausal women with and without osteoporosis. Med. Sci. Sports Exerc., 19:S61, 1987).

strength for the women with normal BMD and women with low BMD. The results were unequivocal: in 11 of 12 test comparisons for flexion, the women without osteoporosis were stronger by an average of 20%; for the extension tests, 4 of 12 comparisons were higher by 13% in women without osteoporosis than in counterparts with osteoporosis. These results demonstrate that differences in maximum dynamic strength among postmenopausal women may serve a clinically useful role in screening for osteoporosis. It would be valuable to know if the offspring of those with osteoporosis are also "weaker," especially at a younger age. This is a plausible hypothesis because there may be a genetic link in the development of osteoporosis.

METABOLIC STRESS OF RESISTANCE TRAINING

Although the various traditional resistance training methods are effective in enhancing a muscle's force-generating capacity, these exercises provide only minimal stimulus to improve aerobic capacity and reduce body fat.[64,65] In a study of the effects of high intensity, variable-resistance (Nautilus) training on cardiovascular function,[70] no adaptive improvement occurred in max $\dot{V}O_2$ or submaximal exercise heart rate and stroke volume. This lack of improvement was probably due to the relatively low metabolic cost of the resistance-training exercises. For example, data obtained during standard isometric and weight-lifting exercises with young adult men indicated that such exercise would be classified as light to moderate in terms of heart rate (generally <130 beats per minute) and oxygen consumption (about 3 to 4 times rest).[93] In all instances, resting levels were reached within 1 or 2 minutes of recovery. For both young adults and the elderly, conflicting reports emerge as to the potential benefits of resistance training in relation to the blood lipid profile and other heart-disease risk factors.[54,71,83]

Undoubtedly, the stress on specific muscles is considerable in resistance training. Owing to the brief activation period and the relatively small muscle mass used, however, the cardiovascular and aerobic metabolic demands are small compared to those of vigorous walking or running, swimming, and cycling, or any other activity utilizing large muscles. Although a person may spend an hour or more completing a strength-training workout, the total time spent in actual exercise is relatively small, usually no more than 6 or 7 minutes per hour! Clearly, traditional resistance training exercise should not make up the major portion of a program designed for cardiovascular overload and weight control.

CIRCUIT RESISTANCE TRAINING

By modifying the standard approach to resistance training it is possible to increase the caloric cost of exercise and bring about improvements in more than one aspect of fitness.[53,138,139] The approach, called **circuit resistance training** or **CRT,** de-emphasizes the heavy local muscle overload of standard resistance training to provide for a more general conditioning to improve body composition, muscular strength and endurance, and some cardiovascular fitness.[8,44,78,96] With this approach, a person lifts a weight that is 40 to 55% of I-RM. The weight is then lifted as many times as possible for 30 seconds. After a 15-second rest, the participant moves to the next resistance-exercise station and so on until the circuit is completed. Between 8 to 15 exercise stations are usually used. (A modification that appears to provide similar energy expenditures during CRT is to employ exercise-to-rest ratios of 1:1 with either 15- or 30-second exercise periods.[9]) The circuit is repeated several times to allow for 30–50 minutes of continuous exercise. As strength increases, the weight lifted at each station also increases.

This modification of standard resistance training is an attractive alternative for those fitness enthusiasts desiring a generalized conditioning program. Medically supervised programs of CRT may also be effective for coronary-prone, cardiac, and spinal-cord-injured patients who desire to improve fitness through resistance-training activities.[12,23,79a,119] It may also be a good supplemental off-season conditioning for athletes involved in sports requiring a high level of strength, power, and muscular endurance.

Specificity of Aerobic Improvement. CRT is generally believed to be less effective for improving aerobic fitness than bicycle or run training.[44] This form of training, however, usually involves substantial upper body exercise, yet the assessment of aerobic benefits of such training has utilized treadmill or bicycle tests, which predominantly activate the lower body musculature. Recently, the effects of CRT on aerobic function have been assessed with both treadmill running and arm-crank ergometry.[60] In keeping with the training specificity principle, aerobic capacity increased 7.8% with treadmill testing and 21.1% with arm cranking! These findings were impressive because they were obtained with a group of borderline hypertensives without negative effects. In fact, training produced a significant increase in muscular strength, a decrease in blood pressure, and a modest improvement in body composition.

CALORIC COST FOR DIFFERENT METHODS OF RESISTANCE EXERCISE

A number of studies have evaluated the caloric cost for different forms of resistance exercise. Table 22-2 displays the results for caloric expenditure expressed per minute for exercise performed using free weights, Nautilus (eccentric), Universal Gym (concentric/eccentric), Cybex (isokinetic), and Hydra-Fitness (hydraulic-concentric).

The results show that caloric expenditure for the hydraulic exercises av-

---

TABLE 22–2.
COMPARISON OF ENERGY EXPENDITURE FOR DIFFERENT MODES OF RESISTANCE EXERCISE*

Mode	Gender	$kJ \cdot min^{-1}$	$kcal \cdot min^{-1}$
Nautilus, circuit	M	29.7	7.1
	F	24.3	5.8
Nautilus, circuit	M	22.6	5.4
Universal, circuit	M	33.1	7.9
	F	28.5	6.8
Isokinetic, slow	M	40.2	9.6
Isokinetic, fast	M	41.4	9.9
Isometric, free-weight	M	25.1	6.0
Hydra-Fitness, circuit	M	37.7	9.0
Walking on level	M	22.6	5.4

* Based on a body weight of 68 kg. (Data from Katch, F.I., et al.: Evaluation of acute cardiorespiratory responses to hydraulic resistance exercise. Med. Sci. Sports Exerc., 17:168, 1985.)

eraged 9.0 kcal · min⁻¹ (37.7 kJ); this was approximately 35% higher than exercise with free weights, 29.4% greater than the average caloric cost based on studies using Nautilus, and 11.5% higher than circuit exercise with Universal equipment. The energy expenditure values for the hydraulic exercise averaged 8.9% less than slow- and fast-speed isokinetic circuit exercise. As a frame of comparison, the caloric cost is also presented for walking at a normal pace on the level.

MUSCLE SORENESS AND STIFFNESS

Following an extended layoff from exercise, most of us have experienced soreness and stiffness in the exercised muscles and joints. A temporary soreness may persist for several hours immediately after unaccustomed exercise, whereas a residual soreness may appear later and last for 3 to 4 days. Any one of at least four factors may be the causative agent: (1) minute tears in the muscle tissue itself, (2) osmotic pressure changes that causes retention of fluids in the surrounding tissues, (3) muscle spasms, and (4) overstretching and perhaps tearing of portions of the muscle's connective tissue harness.

Predominantly With Eccentric Contractions

Although the precise cause of muscle soreness remains unknown, the degree of discomfort depends to a large extent on the intensity and duration of effort[133] and the type of exercise performed. **Eccentric and to some extent isometric muscular contractions generally cause the greatest postexercise discomfort.**[114,115,121,124,135] When muscle soreness was rated by subjects immediately after exercise, and after 24, 48, and 72 hours postexercise, soreness was greater when the exercise involved repeated eccentric contractions than when it involved concentric and isometric contractions. This effect did not relate to lactate build up because level running (concentric contractions) produced no residual soreness despite significant elevations in blood lactate.[114] Downhill running (eccentric contractions), on the other hand, caused moderate-to-severe **delayed onset muscle soreness (DOMS)** with no elevation in lactate during or after exercise.

Actual Cellular Damage

In a study that used downhill running to study DOMS,[13] there was no question that running downhill at −10° slope for 30 minutes produced significant DOMS at 42 hours postexercise. There were also corresponding increases for the serum levels of the muscles' specific enzyme, creatine kinase (CK), and myoglobin (Mb), both of which are commonly used markers of muscle injury. The subjects were then retested on the same exercise after 3, 6 and 9 weeks. Figure 22-18 shows the perceived soreness rating for the leg muscles as a function of time postexercise for the three study durations. For the 3- and 6-week comparisons, the differences between bouts were significant with diminished DOMS noted in the second trial (open squares). The pattern of results for CK and Mb was similar to the perception of muscle soreness. It is interesting to note that peak soreness ratings achieved at 48 hours did not correlate with the absolute or relative changes in CK or Mb. That is, individuals reporting the greatest soreness did not necessarily have the highest CK and Mb values. The researchers agree with others[4,42,66] that

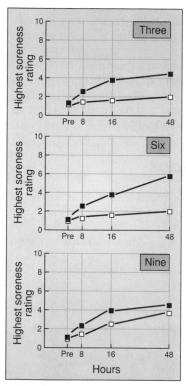

FIG. 22–18.
Highest soreness rating prior to and 6, 18, and 48 hours following bout 1 (closed square) and bout 2 (open square) performed either 3, 6, or 9 weeks later. Similar results were obtained for creatine kinase and myoglobin. (From Byrnes, W.C., et al.: Delayed onset muscle soreness following repeated bouts of downhill running. J. Appl. Physiol. 59:710, 1985.)

the first bout of repetitive, high-intensity exercise may disrupt the integrity of the cellular environment and produce temporary ultrastructural muscle damage in a pool of stress-susceptible or degenerating muscle fibers. This damage becomes more extensive several days after exercise than in the immediate postexercise period.[42,101] Cellular damage and accompanying soreness from eccentric exercise is associated with mitochondrial swelling as well as an increase in the serum level of muscle enzymes and myoglobin.[94] Of interest is that a single bout of exercise has a significant prophylactic effect on the development of muscle soreness in subsequent exercise,[13,100] and this effect appears to last for up to 6 weeks. Such results support the wisdom of initiating a training program with light exercise to protect against the muscle soreness that is almost sure to follow an initial heavy exercise bout that contains an eccentric component. However, even prior lower-intensity exercise of specific muscles does not provide complete protection from subsequent soreness with more intense exercise.[22]

Table 22-3 displays the results for muscle soreness and CK activity following an exercise circuit of either concentric only or concentric and eccentric muscular contractions.[14] Group 1 performed three sets of 8 exercises (concentric-eccentric) at 60% of 1 RM on Universal gym equipment; one set was equal to 20 seconds of exercise followed by 40 seconds of rest; total exercise time was 24 minutes. Group 2 followed the same exercise protocol but exercised maximally for each repetition on hydraulic resistance devices

TABLE 22–3.
ACUTE EFFECTS OF CONCENTRIC ONLY AND CONCENTRIC-ECCENTRIC EXERCISE ON DOMS AT 25 HOURS POST-EXERCISE*

	Soreness Ratings			Soreness Ratings	
Site	Concentric $\bar{X}$	Concentric-Eccentric $\bar{X}$	Site	Concentric $\bar{X}$	Concentric-Eccentric $\bar{X}$
Chest	2.3	5.1	Forearm (front)	1.7	3.4
Back (upper)	2.6	2.8	Forearm (back)	1.7	2.9
Shoulders (front)	2.2	3.6	Back (lower)	1.7	2.9
Shoulders (back)	1.9	3.6	Buttocks	1.8	2.5
Biceps (mid)	1.9	4.3	Quads (mid)	2.0	4.1
Biceps (lower)	1.8	3.5	Quads (lower)	2.1	3.8
Triceps (mid)	1.9	3.4	Hamstrings (mid)	2.1	3.5
Triceps (lower)	1.9	3.0	Hamstrings (lower)	2.1	3.0

	CK Activity (mU · ml^{-1})	
Sample time	Concentric $\bar{X}$	Concentric-Eccentric $\bar{X}$
Pre	86.7	126.9
5 hr post	344.8	232.0
10 hr post	394.3	368.5
25 hr post	288.0	482.2
Absolute increase	319.3	399.9
Relative increase	435.6	355.4

* All differences between groups were statistically significant. (From Byrnes, W.C.: Muscle soreness following resistance exercise with and without eccentric contractions. Res. Q. Exerc. Sport, 56:283, 1985.)

(Hydra-Fitness) that utilized concentric only contractions. Blood samples and ratings of perceived muscle soreness were obtained before exercise and at 5, 10, and 25 hours postexercise. The major difference in soreness ratings between exercise groups occurred at 25 hours postexercise; the concentric-eccentric workout produced significantly higher perceived ratings of soreness for the major muscle groups exercised. For serum CK, the magnitude of the increase was the same between groups from 5 to 25 hours postexercise. Whereas both modes of exercise resulted in elevations in serum CK, the concentric-only muscular contractions did not produce significant DOMS. This finding of no DOMS with concentric-only exercise has recently been reported by others.[24]

EXPERIMENTAL STUDIES

The Spasm Hypothesis

Several experiments provide evidence to support the spasm hypothesis of muscle soreness. In one study,[30] soreness was produced in both arms by repeated wrist hyperextensions against a 4.3-kg resistance. The muscles of the nondominant arm were stretched by the static technique. This was beneficial because significantly greater levels of soreness occurred 24 and 48 hours after exercise in the dominant arm that was not stretched. In another study,[31] the resting EMG activity of the anterior lower leg muscles was evaluated as a consequence of static stretching procedures in subjects with shin splints, a musculoskeletal disorder involving severe pain in the anterior portion of the lower leg. Static stretching procedures brought about marked reductions in EMG activity. Symptomatic relief with stretching was also accompanied by a decrease in the amplitude of the EMG tracings. It is noteworthy, however, that postexercise stretching has not consistently been shown to alleviate delayed muscle soreness.[11]

The Tear Theory

As already discussed, the tear theory of muscle soreness proposes that minute tears or ruptures of individual fibers cause the delayed soreness.[33,133] Because eccentric contractions can place greater strain on connective tissue and muscle fibers compared to concentric muscle action, it is possible that these contractions increase the likelihood of structural changes in muscle fibers with a resulting enzyme efflux.

Excess Metabolite Theory

A competing **excess metabolite theory** proposes that prolonged exercise that follows a layoff causes an accumulation of metabolites in the muscle. This accumulation triggers osmotic changes in the cellular environment and fluid is retained. The edema caused by increased osmotic pressure excites sensory nerve endings and causes pain. This explanation is also inadequate.[6] It should be recalled that muscle soreness was much greater following eccentric exercise than following static or concentric contractions. The metabolic stress of concentric (positive) work, however, is usually about 5 to 7 times greater than that of eccentric work. Consequently, one would expect the metabolite buildup and accompanying soreness to be greater in the concentric exercise and not in the eccentric exercise, as is generally reported.

Tonic muscular spasms may also cause the temporary and residual soreness that follows exercise. It is argued that exercise above some minimal level produces a diminished oxygen supply due to inadequate blood flow in the active muscles. This ischemia can produce pain and initiate a reflex contraction of the muscle. This situation further prolongs the ischemia and sets up repeated cycles of pain, ischemia, and spasm. **Static stretching** is often recommended to alleviate some forms of muscle soreness. With this form of stretching, a body position is held that stabilizes the joint in a position that places the sore muscle at the greatest possible length. The recommended duration of the stretch is 2 minutes, followed by a 1-minute rest, and then another 2 minutes of stretching. This static stretch results in the least possible reflex stimulation (and thus contraction) to the particular muscle by the stretch reflex. A **bouncing stretch** may have an undesirable effect because the sore and rapidly stretched muscle would be reflexly stimulated to contract, and the spasms, pain, and soreness would be prolonged.[32]

Connective Tissue Damage

In other studies, muscle soreness was induced by repeated bouts of arm-curl, weight-lifting exercise, and bench stepping. Several measures were employed to evaluate the degree of soreness and to shed light on possible causative mechanisms. Surface electrodes recorded the electric activity of the muscles for durations up to 48 hours. Urine samples were obtained at selected intervals to detect postexercise **myoglobinuria,** because the presence of myoglobin in the urine is an indicator of trauma to muscle fibers. To evaluate connective tissue damage, urinary levels of **hydroxyproline** were measured. This assay is useful for determining specific breakdown products of connective tissue and for evaluating collagen metabolism. Subjects also rated the degree of muscle soreness by a subjective rating scale. Analysis of these results showed no relationship between muscle pain and EMG activity in the sore muscles. When static stretching procedures were used to alleviate the soreness, there was no change in EMG activity, although pain was reduced somewhat for 1 to 2 minutes. In fact, slowly flexing and extending the arm relieved the pain to the same extent as the static stretching procedures!

The results for postexercise myoglobinuria showed that myoglobin release was unrelated to the development of muscle soreness. Elevated myoglobin levels were found in the urine of 7 of 8 subjects in the arm-soreness experiments and also after cycling, even though no soreness occurred following cycling at heavy loads. The presence of myoglobinuria, therefore, must have been a normal consequence of the exercise and was not necessarily due to the soreness per se.

Interesting results were observed for the hydroxyproline excretion rates in relation to muscle soreness produced by bench stepping. As a result of stepping, all subjects reported soreness in both the concentric (quadriceps) and eccentric (gastrocnemius) working muscles. There were statistically significant increases in the 48-hour postexercise hydroxyproline levels and in the total 4-day mean excretion levels for the exercise as compared to the nonexercise condition. The significant increase in hydroxyproline 48 hours after exercise, coupled with the fact that subjects complained of pain in the tendons of the eccentrically exercised muscles, suggests that **connective** tissue damage in and around the muscles or an imbalance of collagen metabolism (a degradation process) is somehow involved in exercise-induced muscle soreness.

SUMMARY

1. One's capacity for muscular strength is largely determined by physiologic factors such as size and type of muscle fibers, as well as by the anatomic-lever arrangement of bone and muscle. This strength capacity is probably greatly affected by neural influences from the central nervous system that activate the prime movers in a specific muscular action.

2. Genetic, exercise, nutritional, hormonal, environmental, and neural factors interact to regulate skeletal muscle mass and corresponding strength development.

3. Increases in strength with resistance training are due to improved capacity for motor unit recruitment, changes in the efficiency of the firing pattern of motor nerves, and significant alterations in the contractile elements within the muscle cell itself.

4. As muscles are overloaded and become stronger, they normally hypertrophy or grow larger. This process involves increased protein synthesis with resulting myofibril thickening, proliferation of connective tissue cells, and an increase in the number of satellite cells that surround each fiber.

5. Muscular hypertrophy generally involves structural changes within the contractile mechanism of individual fibers, especially that of the fast-twitch fibers as well as increases in anaerobic energy stores. If new muscle fibers actually develop, their contribution to muscular enlargement is as yet undetermined.

6. Heavy resistance training does not bring about adaptations in cellular components that would contribute to enhanced local aerobic energy transfer.

7. In short-term training studies, strength improvements for women on a percentage basis are similar to those of men. Recent research also indicates that women are capable of significant relative increases in muscle mass as a result of resistance training.

8. Conventional resistance-training exercises per se contribute little to cardiovascular-aerobic fitness. Due to their relatively low caloric cost, they would not be effective major activities in weight-reducing programs.

9. By use of lower resistance and higher repetitions, circuit resistance training offers an effective alternative for combining the muscle-training benefits of resistance exercise with the cardiovascular benefits of more continuous dynamic exercise.

10. There are several possible explanations for the muscle soreness and stiffness that accompany strength training: (1) minute tears in the muscle tissue, (2) osmotic pressure changes and water retention, (3) muscle spasms, and (4) overstretching and perhaps muscle and connective tissue tears. There is significant experimental evidence to support the argument for muscle tears and connective tissue damage.

11. Significantly greater delayed onset muscle soreness (DOMS) occurs in eccentric muscular contractions evaluated by perceived ratings of soreness compared to concentric only or isometric exercise, even though serum markers of muscle damage (e.g., creatine kinase and myoglobin) are elevated following all three modes of exercise.

12. A single bout of exercise has a significant protective effect on the development of muscle soreness and damage in subsequent exercise. This supports the wisdom of gradual progression when beginning an exercise program.

REFERENCES

1. Abraham, W.M.: Factors in delayed muscle soreness. *Med. Sci. Sports*, 9:11, 1977.
2. Alway, S.E., et al.: Functional and structural adaptations in skeletal muscle of trained athletes. *J. Appl. Physiol.*, 64:1114, 1988.
3. Alway, S.E., et al.: Contrasts in muscle and myofibers of elite male and female body builders. *J. Appl. Physiol.*, 67:24, 1989.
4. Armstrong, R.R.: Eccentric exercise induced damage to rat skeletal muscle. *J. Appl. Physiol.*, 54:80, 1983.
5. Ashmore, C.R., and Summers, P.J.: Stretch-induced growth of chicken wing muscles: myofibrillar proliferation. *Am. J. Physiol.*, 240(Cell Physiol. 10):93, 1981.
6. Assmussen, E.: Observations on experimental muscle soreness. *Acta Rheumatol. Scand.*, 1:109, 1956.
7. Ama, P.F.M., et al.: Skeletal muscle characteristics in sedentary Black and Caucasian males. *J. Appl. Physiol.*, 61:1758, 1986.
8. Ballor, D.L., et al.: Metabolic response during hydraulic resistance exercise. *Med. Sci. Sports Exerc.*, 19:363, 1987.
9. Ballor, D.L., et al.: Physiological response to nine different exercise:rest protocols. *Med. Sci. Sports Exerc.*, 21:90, 1989.
9a. Boocock, M.G., et al.: Changes in stature following drop jumping and post-exercise gravity inversion. Med. Sci. Sports Exerc., 22:385, 1990.
10. Brown, R.D., and Harrison, J.A.: The effects of a strength training program on the strength and self-confidence of two female age groups. *Res. Q. Exerc. Sport*, 57:315, 1986.
11. Buroker, K.C., and Schwane, J.A.: Does post-exercise stretching alleviate delayed muscle soreness? *Phys. Sportsmed.*, 17:65, 1989.
12. Butler, R.M., et al.: The cardiovascular response to circuit weight training in patients with cardiac disease. *J. Cardiopulmonary Rehabil.*, 7:402, 1987.
13. Byrnes, W.C., et al.: Delayed onset muscle soreness following repeated bouts of downhill running. *J. Appl. Physiol.*, 59:710, 1985.
14. Byrnes, W.C.: Muscle soreness following resistance exercise with and without eccentric contractions. *Res. Q. Exer. Sport*, 56:283, 1985.
15. Caiozzo, V.J., et al.: Training-induced alterations in the in-vivo force-velocity relationship of human muscle. *J. Appl. Physiol.*, 51:750, 1981.
16. Cho, D.A.: Plyometric exercise. *NSCA Journal*, 5:56, 1984.
17. Clarke, D.H.: Adaptations in strength and muscular endurance resulting from exercise. In *Exercise and Sport Science Reviews*. Vol. 1. Edited by J.H. Wilmore. New York, Academic Press, 1973.
18. Clarke, D.H., and Clarke, H.H.: *Research Processes in Physical Education, Recreation, and Health*. Englewood Cliffs, NJ, Prentice-Hall, 1984.
19. Clarke, H.H.: Objective strength test of affected muscle groups in orthopedic disabilities. *Res. Q.*, 19:118, 1948.
20. Clarke, H.H.: Development of muscular strength and endurance. *Phys. Fitness Res. Dig.*, President's Council on Physical Fitness in Sports. Washington, DC, U.S. Government Printing Office, Jan., 1974.
21. Clarke, H.H., et al.: New objective strength tests of muscle groups by cable tension methods. *Res. Q.*, 23:136, 1952.
22. Clarkson, P.M., and Tremblay, I.: Exercise induced muscle damage, repair, and adaptation in humans. *J. Appl. Physiol.*, 65:1, 1988.
23. Cooney, M.M., and Walker, J.B.: Hydraulic resistance exercise benefits cardiovascular fitness in spinal cord injured. *Med. Sci. Sports Exerc.*, 18:522, 1986.
24. Cote, C., et al.: Isokinetic strength training protocols: do they induce skeletal muscle fiber hypertrophy? *Arch. Phys. Med. Rehab.* 69:281, 1988.

25. Cotten, D.: Relationship of the duration of sustained voluntary isometric contraction to changes in endurance and strength. *Res. Q.*, 33:366, 1967.
26. Coyle, E.F., et al.: Leg extension power and muscle fiber composition. *Med. Sci. Sports,* 11:12, 1979.
27. Coyle, E.F., et al.: Specificity of power improvements through slow and fast isokinetic training. *J. Appl. Physiol.*, 51:1437, 1981.
28. Cureton, K.J., et al.: Muscle hypertrophy in men and women. *Med. Sci. Sports Exerc.,* 20:338, 1988.
29. DeLorme, T.L., and Watkins, A.L.: *Progressive Resistance Exercise.* New York, Appleton-Century-Crofts, 1951.
30. DeVries, H.A.: Electromyographic observations of the effects of static stretching upon muscular distress. *Res. Q.*, 32:468, 1961.
31. DeVries, H.A.: Prevention of muscular distress after exercise. *Res. Q.*, 32:177, 1961.
32. DeVries, H.A.: Evaluation of static stretching procedures for improvement of flexibility. *Res. Q.*, 33:222, 1961.
33. Dressendorfer, R.H., and Wade, C.E.: The muscular overuse syndrome in long-distance runners. *Phys. Sportsmed.*, 11:116, 1983.
34. Duchateau, J., and Hainaut, K.: Isometric or dynamic training: differentiated effects on mechanical properties of human muscle. *J. Appl. Physiol.*, 56:296, 1984.
35. Dudley, G., and Djamil, R.: Incompatibility of endurance and strength training modes of exercise. *J. Appl. Physiol.*, 59:1446, 1985.
36. Edgerton, V.R.: Exercise and the growth and development of muscle tissue. In *Physical Activity; Human Growth and Development.* Edited by G.L. Rarick. New York, Academic Press, 1973.
37. Edgerton, V.R.: Mammalian muscle fiber types and their adaptability. *Am. Zool.*, 18:113, 1978.
38. Edstrom, L., and Ekblom, B.: Differences in sizes of red and white muscle fibers in vastus lateralis of musculus quadriceps femoris of normal individuals and athletes. *Scand. J. Clin. Lab. Invest.*, 30:175, 1972.
39. Edström L., and Grimby, L.: Effects of exercise on the motor unit. *Muscle Nerve*, 9:104, 1986.
40. Fahey, T.D., and C.H. Brown.: The effects of an anabolic steroid on the strength, body composition, and endurance of college males when accompanied by a weight training program. *Med. Sci. Sports,* 5:272, 1973.
41. Foster, D.N., et al.: Effect of training frequency on lumbar extension strength. *Med. Sci. Sports Exerc.*, 21:S88, 1989.
42. Friden, J., et al.: Myofibrillar damage following eccentric exercise in man. *Int. J. Sports Med.*, 4:170, 1983.
43. Frontera, W.R., et al.: Strength conditioning in older men: skeletal muscle hypertrophy and improved function. *J. Appl. Physiol.*, 64:1038, 1988.
43a. Fukunaga, T., et al.: Physiological cross-section area of human leg muscles based on magnetic resonance imaging. *J. Orthopaedic Res.*, in press.
44. Gettman, L.R.: Physiologic effects on adult men of circuit strength training and jogging. *Arch. Phys. Med. Rehab.*, 60:115, 1979.
45. Goldberg, A.L., et al.: Mechanism of work-induced hypertrophy of skeletal muscle. *Med. Sci. Sports,* 7:185, 1975.
46. Gollnick, P.D.: Fiber number and size in overloaded chicken anterior latissimus dorsi muscle. *J. Appl. Physiol.*, 54:1292, 1983.
47. Gollnick, P.D., et al.: Effect of training on enzyme activity and fiber composition of human skeletal muscle. *J. Appl. Physiol.*, 34:107, 1973.
48. Gollnick, P.D., et al.: Muscular enlargement and number of fibers in skeletal muscles of rats. *J. Appl. Physiol.*, 50:936, 1981.
49. Gonyea, W.J.: Role of exercise in inducing increases in skeletal muscle fiber number. *J. Appl. Physiol.*, 48:421, 1980.
50. Gonyea, W.J., et al.: Exercise induced increases in muscle fiber number. *Eur. J. Appl. Physiol.*, 55:137, 1986.
51. Graves, J.E., et al.: Specificity of limited range of motion variable resistance training. *Med. Sci. Sports Exerc.*, 21:84, 1989.

52. Grimby, G., et al.: Muscle morphology and function in 67- to 81-year-old men and women (abstract). *Med. Sci. Sports*, 12:96, 1980.

53. Haennel, R., et al.: Effects of hydraulic circuit training on cardiovascular function. *Med. Sci. Sports Exerc.*, 21:605, 1989.

54. Hagberg, J., et al.: Cardiovascular responses of 70- to 79-yr old men and women to exercise training. *J. Appl. Physiol.*, 66:2589, 1989.

55. Häkkinen, K., and Komi, P.V.: Electromyographic changes during strength training and detraining. *Med. Sci. Sports Exerc.*, 15:455, 1983.

56. Häkkinen, K., et al.: Effect of explosive type strength training on isometric force- and relaxation-time, electromyographic and muscle fiber characteristics of leg extensor muscles. *Acta Physiol. Scand.*, 125:587, 1985.

57. Häkkinen, K., et al.: Neuromuscular and hormonal adaptations in athletes to strength training in two years. *J. Appl. Physiol.*, 65:2406, 1988.

58. Harmon, E.A.: Resistive torque analysis of 5 Nautilus exercise machines. *Med. Sci. Sports Exerc.*, 15:113, 1983.

59. Harmon, E.A., et al.: Effect of a belt on intra-abdominal pressure during weight lifting. *Med. Sci. Sports Exerc.*, 21:186, 1989.

60. Harris, K.A., and Holly, R.G.: Physiological response to circuit weight training in borderline hypertensive subjects. *Med. Sci. Sports Exerc.*, 19:246, 1987.

61. Hay, J.G., et al.: Effects of lifting rate on elbow torques exerted during arm curl exercise. *Med. Sci. Sports Exerc.*, 15:63, 1983.

62. Hettinger, T.L., and Muller, E.A.: Muskelleistung und muskeltraining. *Int. Z. Angew. Physiol.*, 15:111, 1953.

63. Heyward, V.H., et al.: Gender differences in strength. *Res. Q. Exerc. Sport*, 57:154, 1986.

64. Hickson, R.C.: Interference of strength development by simultaneously training for strength and endurance. *Eur. J. Appl. Physiol.*, 45:255, 1980.

65. Hickson, R.C., et al.: Strength training effects on aerobic power and short-term endurance. *Med. Sci. Sports Exerc.*, 12:336, 1980.

66. Hikida, R.S., et al.: Muscle fiber necrosis associated with marathon running. *J. Neurol. Sci.*, 59:185, 1983.

67. Ho, K.W., et al.: Skeletal muscle fiber splitting with weight-lifting exercises in rats. *Am. J. Anat.*, 157:433, 1980.

67a. Hortobágyi, T.: Specificity of muscular fatigue and force enhancing mechanisms with isometric, shortening, and lengthening muscle actions, and myotatic stretch in power and endurance athletes. Unpublished doctoral dissertation. University of Massachusetts, Amherst, 1990.

68. Hortobágyi, T., and Katch, F.I.: Role of concentric force in limiting improvement in muscular strength. *J. Appl. Physiol.*, 68:650, 1990.

69. Hortobágyi, T., et al.: Interrelationships among various measures of upper body strength assessed by different contraction modes. *Eur. J. Appl. Physiol.*, 58:749, 1989.

70. Hurley, B.F., et al.: Effects of high intensity strength training on cardiovascular function. *Med. Sci. Sports Exerc.*, 16:483, 1984.

71. Hurley, B.F., et al.: Resistance training can reduce coronary risk factors without altering VO_2 max or percent body fat. *Med. Sci. Sports Exerc.*, 20:150, 1988.

72. Ikai, M., and Fukunaga, T.: Calculation of muscle strength per unit cross-sectional area of a human muscle by means of ultrasonic measurements. *Int. Z. Angew. Physiol.*, 26:26, 1968.

73. Ikai, M., and Steinhaus, A.H.: Some factors modifying the expression of human strength. *J. Appl. Physiol.*, 16:157, 1961.

74. Janney, C., et al.: Maximum muscular strength differs in postmenopausal women with and without osteoporosis. *Med. Sci. Sports Exerc.* (abstract), 19:561, 1987.

75. Johnson, B.L., et al.: A comparison of concentric and eccentric muscle training. *Med. Sci. Sports*, 8:35, 1976.

76. Jones, D.A., and Rutherford, O.M.: Human muscle strength training: the effects of three different regimes and the nature of the resultant changes. *J. Physiol.*, 391:1, 1987.

77. Kanehisa, H., and Miyashita, M.: Specificity of velocity in strength training. *Eur. J. Appl. Physiol.*, 52:104, 1983.

78. Katch, F.I., et al.: Evaluation of acute cardiorespiratory responses to hydraulic resistance exercise. *Med. Sci. Sports Exerc.*, 17:168, 1985.

79. Karlsson, J.B., et al.: LDH isozymes in skeletal muscles of endurance and strength trained athletes. *Acta Physiol. Scand.*, 93:150, 1975.

79a. Kelemen, M.H.: Resistance training safety and assessment guidelines for cardiac and coronary prone patients. *Med. Sci. Sports Exerc.*, 21:675, 1989.

80. Keller, B.: The influence of body size variables on gender differences in strength and maximum aerobic capacity. Unpublished doctoral dissertation. University of Massachusetts, Amherst, 1989.

81. Kitai, T.A., et al.: Specificity of joint angle in isometric training. *Eur. J. Appl. Physiol.*, 58:744, 1989.

82. Knapik, J.J., et al.: Angular specificity and test mode specificity of isometric and isokinetic strength training. *J. Orthop. Sports Phys. Ther.*, 5:58, 1983.

83. Kokkinos, P.F., et al.: Effects of low- and high-repetition resistance training on lipoprotein-lipid profiles. *Med. Sci. Sports Exerc.*, 20:50, 1988.

84. Larsson, L., and Tesch, P.A.: Motor unit fiber density in extremely hypertrophied skeletal muscles in man. Electrophysiological signs of muscle fiber hypertrophy. *Eur. J. Appl. Physiol.*, 55:130, 1986.

85. Laubach, L.L.: Comparative muscular strength of men and women: a review of the literature. *Aviat. Space Environ. Med.*, 47:534, 1976.

86. Lindh, M.: Increase of muscle strength from isometric quadriceps exercises at different knee angles. *Scand. J. Rehabil. Med.*, 11:36, 1979.

87. LaGasse, P., et al.: Neuromuscular facilitation of muscle tension output by reciprocal muscle work. *Ann. French-Canadian Assoc. Adv. Sciences*, 50:222, 1983.

88. MacDougall, J.D.: Mitochondrial volume density in human skeletal muscle following heavy resistance training. *Med. Sci. Sports*, 11:164, 1979.

89. MacDougall, J.D.: Morphological changes in human skeletal muscle following strength training and immobilization. In: *Human Muscle Power*. Edited by N.L. Jones et al. Champaign, IL, Human Kinetics, 1986.

90. MacDougall, J.D., et al.: Effects of strength training and immobilization of human muscle fibers. *Eur. J. Appl. Physiol.*, 43:25, 1980.

91. MacDougall, J.D., et al.: Muscle ultrastructural characteristics of the elite powerlifters and bodybuilders. *Med. Sci. Sports*, 2:131, 1980.

92. MacDougall, J.D., et al.: Muscle fiber number in biceps brachii in bodybuilders and control subjects. *J. Appl.Physiol.*, 57:1399, 1984.

92a. Manning, R.M., et al: Constant vs variable resistance knee extension training. *Med. Sci. Sports Exerc.*, 22:397, 1990.

93. McArdle, W.D., and Foglia, G.F.: Energy cost and cardiorespiratory stress of isometric and weight training exercises. *J. Sports Med. Phys. Fitness*, 9:23, 1969.

94. McCully, K.K. and Faulkner, J.A. Injury to skeletal muscle fibers of mice following lengthening contractions. *J. Appl. Physiol.*, 59:119, 1985.

95. McDonnagh, M.J., and Davies, C.T.M.: Adaptative response to mammalian skeletal muscle to exercise with high load. *Eur. J. Appl. Physiol.*, 52:139, 1984.

96. Marcinik, E.J., et al.: Aerobic/calisthenic and aerobic/circuit weight training for Navy men: a comparative study. *Med. Sci. Sports Exerc.*, 17:482, 1985.

97. Morehouse, C.A.: Development and maintenance of isometric strength of subjects with diverse initial strength. *Res. Q.*, 38:449, 1967.

98. Moritani, T., and DeVries, H.: Neural factors versus hypertrophy in the time course of muscle strength gain. *Am. J. Phys. Med.*, 58:115, 1979.

99. Morrow, J.R., Jr., and Hosler, W.W.: Strength comparisons in untrained men and trained women. *Med. Sci. Sports Exerc.*, 13:194, 1981.

100. Newham, D.J., et al.: Repeated high force eccentric exercise; effects on muscle pain and damage. *J. Appl. Physiol.*, 63:1381, 1987.

101. Paul, G.L.: Serum and urinary markers of skeletal muscle tissue damage after weight lifting exercise. *Eur. J. Appl. Physiol.*, 58:786, 1989.

102. Pavone, E., and Moffat, M.: Isometric torque of the quadriceps femoris after concentric, eccentric and isometric training. *Arch. Phys. Med. Rehabil.*, 66:168, 1985.

103. Petersen, S.R., et al.: The influence of isokinetic concentric resistance training on concentric and eccentric torque outputs and cross-sectional area of the quadriceps femoris. *Can. J. Sport Sci.*, 13:76, 1988.

104. Pipes, T.V., and Wilmore, J.H.: Isokinetic vs. isotonic strength training in adult men. *Med. Sci. Sports*, 7:262, 1975.

105. Pollock, M.L., et al.: Effect of resistance training on lumbar extension strength. *Am. J. Sports Med.*, 17:624, 1989.

106. Rians, C.B., et al.: Strength training for prepubescent males: is it safe? *Am. J. Sports Med.*, 15:483, 1987.

107. Roy, M.A., et al.: Changes in alpha motoneuron excitability following isokinetic training in sedentary males and females. *Can. J. Appl. Sport Sci.*, 9:20P, 1984.

108. Roy, M.A., et al.: Proprioceptive facilitation of muscle tension during unilateral and bilateral knee extension. *Int. J. Sports Med.*, in press.

109. Sale, D.G.: Influence of exercise and training on motor unit activation. In *Exercise and Sport Sciences Reviews*. Vol. 15. Edited by K.B. Pandolf. New York, Macmillan, 1987.

110. Sale, D.G.: Neural adaptations to resistance training. *Med. Sci. Sports Exerc.*, 20:S135, 1988.

111. Sale, D.G., et al.: Effect of strength training upon motoneuron excitability in man. *Med. Sci. Sports Exerc.*, 15:57, 1983.

112. Salleo, A., et al.: New muscle fiber production during compensatory hypertrophy. *Med. Sci. Sports*, 12:268, 1980.

113. Schantz, P.G., and Källman, M.: NADH shuttle enzymes and cytochrome b_5 reductase in human skeletal muscle: effect of strength training. *J. Appl. Physiol.*, 67:123, 1989.

114. Schwane, J.A., et al.: Delayed-onset muscular soreness and plasma CPK and LDH activities after downhill running. *Med. Sci. Sports Exerc.*, 15:51, 1983.

115. Schwane, J.A., et al.: Effects of training on delayed muscle soreness and serum creatine kinase activity after running. *Med. Sci. Sports Exerc.*, 19:584, 1987.

116. Sewall, L., and Michelli, L.J.: Strength training for children. *J. Pediatr. Orthop.*, 6:143, 1986.

117. Snook, S.H., et al.: Maximum weights and work loads acceptable to male industrial workers. *Am. Ind. Hyg. Assoc. J.*, 31:579, 1970.

118. Snook, S.H., and Ciriello, V.M.: Maximum weights and work loads acceptable to female workers. *J. Occup. Med.*, 16:527, 1974.

119. Sparling, P.S., and Cantwell, J.A.: Strength training guidelines for cardiac patients. *Phys. Sportsmed.*, 17:191, 1989.

120. Staff, P.H.: The effects of physical activity on joints, cartilage, tendons, and ligaments. *Scand. J. Soc. Med.*, 29(Suppl.):59, 1982.

120a. Starron, R.S., et al.: Muscle hypertrophy and fast fiber type conversions in heavy resistance-trained women. *Eur. J. Appl. Physiol.*, 60:71, 1990.

121. Stauber, W.T.: Eccentric action of muscles: physiology, injury, and adaptation. In *Exercise and Sport Sciences Reviews*. Vol. 17. Edited by K.B. Pandolf. New York, Macmillan, 1989.

122. Stock, J.L., et al.: Dynamic muscle strength is decreased in post-menopausal women with low bone density. *J. Bone Min. Res.*, 2:338, 1987.

123. Stone, M.H.: Implications for connective tissue and bone alterations resulting from resistance exercise training. *Med. Sci. Sports Exerc.*, 20:S162, 1988.

124. Talag, T.S.: Residual muscular soreness influenced by concentric, eccentric, and static contractions. *Res. Q.*, 44:458, 1973.

125. Ter Haar Romeny, B.M., et al.: Relation between location of a motor unit in the human biceps brachii and its critical firing levels for different tasks. *Exp. Neurol.*, 85:631, 1984.

126. Tesch, P.A.: Skeletal muscle adaptations consequent to long-term heavy resistance exercise. *Med. Sci. Sports Exerc.*, 20:S132, 1988.

127. Tesch, P.A.: Enzyme activities of FT and ST muscle fibers in heavy-resistance trained athletes. *J. Appl. Physiol.*, 83:67, 1989.

128. Tesch, P.A., and Karlsson, J.: Muscle fiber types and size in trained and untrained muscles of elite athletes. *J. Appl. Physiol.*, 59:1716, 1985.

129. Tesch, P.A., and Larsson, L.: Muscle hypertrophy in body builders. *Eur. J. Appl. Physiol.*, 49:301, 1982.

130. Tesch, P.A., et al.: Muscle capillary supply and fiber type characteristics in weight and power lifters. *J. Appl. Physiol.*, 56:35, 1984.

131. Thorstensson, A.: Muscle strength, fiber types and enzyme activities in man. *Acta Physiol. Scand.*, Suppl. 443, 1976.

132. Thorstensson, A., et al.: Force-velocity relations and fiber composition in human knee extensor muscles. *J. Appl. Physiol.*, 40:12, 1976.

133. Tiidus, P.M., and Ianuzzo, C.D.: Effects of intensity and duration of muscular exercise on delayed soreness and serum enzyme activities. *Med. Sci. Sports Exerc.*, 15:461, 1983.

134. Tipton, C.M., et al.: The influence of physical activity on ligaments and tendons. *Med. Sci. Sports*, 7:165, 1975.

135. Triffletti, P., et al.: Creatine kinase and muscle soreness after repeated isometric exercise. *Med. Sci. Sports Exerc.*, 20:242, 1988.

136. Weltman, A., et al.: The effects of hydraulic resistance strength training in pre-pubertal males. *Med. Sci. Sports Exerc.*, 18:629, 1986.

137. Wilmore, J.H.: Alterations in strength, body composition and anthrometric measurements consequent to a 10-week weight training program. *Med. Sci. Sports*, 6:133, 1974.

138. Wilmore, J.H., et al.: Energy cost of circuit weight training. *Med. Sci. Sports*, 10:75, 1978.

139. Wilmore, J.H., et al.: Physiological alterations consequent to circuit weight training. *Med. Sci. Sports*, 10:79, 1978.

23 SPECIAL AIDS TO PERFORMANCE AND CONDITIONING

Coaches and athletes are continually searching for ways to gain the competitive "edge" and improve athletic performance. It is not surprising, therefore, that a variety of **ergogenic*** substances and procedures are used routinely at almost all competitive levels. Drugs are used most often by college and professional athletes, whereas nutrition supplementation and warm-up procedures are common to individuals who train for fitness and sport activities.

Considerable literature exists on the topic of ergogenic aids and athletic performance. It includes studies of the potential performance benefits of alcohol, amphetamines, epinephrine, aspartates, red cell reinfusion, caffeine, steroids, protein, phosphates, oxygen-rich breathing mixtures, gelatin, lecithin, wheat-germ oil, vitamins, sugar, ionized air, music, hypnosis, and even marijuana and cocaine! Only a few of these alleged aids, however, are used routinely by athletes and only a few cause real controversy. Of specific interest is the use of anabolic steroids and amphetamines and the unique procedure of "blood doping." Because warm-up, oxygen administration, and nutritional supplementation are in common use, we also include these in our discussion of the practical implications of ergogenic aids for human exercise physiology and performance.

PHARMACOLOGIC AGENTS

Many male and female athletes use a variety of pharmacologic agents in the belief that a specific drug will have a positive influence on skill, strength, or endurance. In our drug-oriented, competitive culture, it is not surprising to find drug use for ergogenic purposes on the upswing among high school and even junior high school athletes. When winning becomes all important, one can do little to prevent the use **and** abuse of drugs by athletes, even if little "hard" scientific evidence of a direct relationship exists between drug use and improved athletic performance. It seems ironic that athletes go to great lengths to promote all aspects of their health; they train hard, they eat well-balanced meals, they seek and receive medical advice for various injuries (no matter how minor), yet they ingest synthetic agents, many of which can precipitate side effects ranging from nausea, hair loss, itching, and nervous irritability, to severe consequences such as sterility, liver disease, drug addiction, and even death caused by liver and blood cancer.

We will now take a closer look at two categories of drugs often used by athletes—anabolic steroids and amphetamines.

---

* Ergogenic relates to the application of a nutritional, physical, mechanical, psychologic, or pharmacologic procedure or aid to improve physical work capacity or athletic performance.

497

ANABOLIC STEROIDS

Anabolic steroids came into prominence for medical use in the early 1950s to treat patients who were deficient in natural androgens or had muscle-wasting diseases.[28] They have now become an integral part of the high-technology scene of competitive sports.

Structure and Action

An anabolic steroid is a drug that functions in a manner similar to that of the principal male hormone testosterone. By binding with special receptor sites on muscle and various other tissues, this hormone greatly contributes to the male secondary sex characteristics and to the gender differences in muscle mass and strength that begin to develop at the onset of puberty. The hormone's androgenic or masculinizing effects can be minimized by synthetically manipulating the chemical structure of the steroid so that the cell's anabolic tissue-building, nitrogen-retaining process is emphasized for purposes of promoting increased muscular growth. Nevertheless, the masculinizing effect is still noticeable, especially when the drug is used by females. Athletes who take these drugs do so usually during the active years of their athletic careers, often taking a progressively increasing combined steroid dose in both oral and injectable form—called "stacking"—far in excess of the recommended medical dose.[16,85] The dosage is then progressively reduced in the months prior to competition to reduce chances of detection.

Steroids are most frequently taken in conjunction with a strength development program and augmented protein intake. The aim is to improve performance in sports that require strength, speed, and power. Federal authorities conservatively estimate that the emerging illegal trafficking in steroids exceeds 100 million dollars yearly and is growing rapidly. The fact that many athletes get their steroids on the "black market" raises the fear that misinformed individuals may take a massive and prolonged dose of the drug without any medical monitoring for possible harmful alterations in physiologic function. Particularly worrisome is steroid use among young boys and girls[128] and its accompanying effects including extreme virilization and premature cessation of bone growth. A recent study estimated that 1 of 15 high school students or about 500,000 adolescents have used steroids.[15] Improved athletic performance was the most common reason cited, although 25% said the main reason was an enhanced appearance.

Are They Effective?

The use of anabolic steroids by **men and women** in sports has generated considerable controversy. Testimonials abound as to the muscular size and performance benefits from steroid use. This situation has been further aggravated by the disqualification and suspension of numerous elite amateur and professional athletes for steroid use. Aside from the moral and ethical issues, this is unfortunate because of the conflicting scientific data as to whether they even exert a positive influence on growth and performance in normal, healthy athletes.

Much of the confusion as to the ergogenic effectiveness of anabolic steroids has been due to variations in experimental design, poor controls, and differences in specific drugs, dosages, treatment duration, training intensity, measurement techniques, previous experience as subjects, individual variation in response, and nutritional supplementation.[40,104] There is also speculation that the relatively small residual androgenic action of the anabolic steroid acts via the central nervous system to facilitate improvements by making the athlete more **aggressive, competitive,** and **fatigue resistant** so that he or she trains harder for a longer period of time,[20,68,105,110] or gains the impression that augmented training effects are taking place while using the drug.[29]

Based on research with animals,[93] it has been suggested that treatment with anabolic steroids, when combined with exercise and adequate protein intake, may stimulate the complex process of protein synthesis and increase the content of skeletal muscle protein (myosin, myofibrillar, and sarcoplasmic factors). Animal studies, however, have shown no benefit from steroid treatment on the leg muscle weight of rats subjected to functional overload by surgical removal of the synergistic muscle.[73] The researchers concluded that treatment with sex steroids does not act in a synergistic manner with functional overload to stimulate muscular development. The situation with humans is unclear because some studies show augmented body mass gains with steroid use in men who exercise, whereas others show no effects on muscular strength and power, or body composition, even when calorie and protein intake were sufficient to support an anabolic effect.[51,64,126] When body mass gain does occur with steroid use, it is unclear as to the compositional nature of this gain in terms of muscle and water.

Are There Risks?

In our opinion, the infrequent but distinct possibility of harmful side effects greatly out-weighs any potential performance benefits to be gained from steroid treatment. In addition, it now appears that high dosages of anabolic steroids lead to a long-lasting impairment of normal testosterone endocrine function. In fact, a study of 5 male power athletes showed that the cessation of 26 weeks of steroid administration brought about a reduction in serum testosterone to levels that were less than half that seen at the beginning of the study; this effect lasted throughout the 12–16 week follow-up period. Other accompanying hormonal alterations in males included a 7-fold increased concentration of circulating estradiol, a major female hormone, during steroid administration.[2] This level was representative of average values for normal females and possibly explains the **gynecomastia** (palpable breast tissue) often reported among male users of anabolic steroids.[42] As part of a long-range educational program, the American College of Sports Medicine (ACSM) has taken a stand on the use and abuse of anabolic-androgenic steroids in sports.[3] We endorse their position paper which follows.

ACSM Position Statement on Anabolic Steroids

Based on a comprehensive survey of the world literature and a careful analysis of the claims made for and against the efficacy of anabolic-androgenic steroids in improving human physical performance, it is the position of the American College of Sports Medicine that:

1. Anabolic-androgenic steroids in the presence of an adequate diet and training can contribute to increases in body weight, often in the lean mass compartment.
2. The gains in muscular strength achieved through high-intensity exercise and proper diet can occur by the increased use of anabolic-androgenic steroids in some individuals.

3. Anabolic-androgenic steroids do not increase aerobic power or capacity for muscular exercise.
4. Anabolic-androgenic steroids have been associated with adverse effects on the liver, cardiovascular system, reproductive system, and psychologic status in therapeutic trials and in limited research on athletes. Until further research is completed, the potential hazards of the use of the anabolic-androgenic steroids in athletes must include those found in therapeutic trials.
5. The use of anabolic-androgenic steroids by athletes is contrary to the rules and ethical principles of athletic competition as set forth by many of the sports governing bodies. The American College of Sports Medicine supports these ethical principles and deplores the use of anabolic-androgenic steroids by athletes.

Steroid Use and Life-Threatening Disease

In the ACSM position statement, evidence was presented concerning a possible link between anabolic steroid use and alterations in normal liver function. Table 23–1 amplifies the observations for liver disease.[57] We present these data not as a scare tactic, but to emphasize the potentially serious side effects, even when the drug is authorized by a physician in the recommended dosage. It should be noted that although the duration of drug treatment for patients is greater than usually used by athletes, some athletes take steroids on and off for a period of years, with the dosage often greatly exceeding that typically prescribed for therapeutic purposes (50 to 200 mg per day versus the usual dosage of 5 to 20 mg). Recent preliminary data suggest that anabolic-androgenic steroid use as practiced by many athletes may interfere with the responsiveness of the body's immune system.[18] The precise clinical signif-

TABLE 23–1.
ORAL ANDROGENIC-ANABOLIC STEROIDS ASSOCIATED WITH DETRIMENTAL SIDE EFFECTS[a]

Chemical Name	Trade Name	Daily Dosage (mg)	Duration in Affected Patients (Months)	Complications
Oxymetholone	Ora-Testryl Adroyd Anapolon Anadrol 50	10–250	10–51	**Peliosis hepatis[b]**
Methyltestosterone	Oreton Methyl Metandren Android	20–50	1–165	**Hepatoma[c]**
Stanazolol	Winstrol	15	18	**Hepatoma[c]**
Methandrostenolone	Dianabol	10–15	12–18	**Hepatoma[c]**
Fluoxymesterone	Halotestin	15–80	4–16	**Peliosis hepatis[b]**
Norethandrolone	Nilevar	20–30	1.5–9	**Peliosis hepatis[b]**

[a] Data from Johnson, F.L.: The association of oral androgenic-anabolic steroids and life threatening disease. Med. Sci. Sports, 7:284, 1975.
[b] Severe liver malfunction
[c] Liver cancer

icance of these findings awaits further research. We believe that any potential gain in exercise performance from anabolic steroids is simply outweighed by the chances of developing severe and harmful side effects.

Steroid Use and Plasma Lipoproteins

Reports indicate that steroid use produces a rapid and profound lowering of high-density lipoprotein cholesterol and elevations in both low-density lipoprotein cholesterol and total cholesterol in healthy, trained men and women.[25,53,81,112] In one study of weight lifters who used anabolic steroids, the average value for HDL cholesterol was 26 mg $\cdot$ dl^{-1} compared to 50 mg $\cdot$ dl^{-1} for the weight lifters not on this drug![60] Because a low level of this particular lipoprotein is related to an increased incidence of coronary artery disease, the use of steroids is particularly troublesome and clinically inadvisable for use by healthy men and women.

Steroid Use by Females

There is little information on the use of anabolic agents by female athletes, but hearsay reports and "off-the-record" statements by coaches, physicians, and the athletes themselves give every reason to conclude that this form of drug abuse is on the upswing. In addition to the broad range of side effects discussed previously, women, especially those who have not fully matured, are susceptible to specific dangers. These include masculinization, disruption of normal growth pattern by premature closure of the plates for bone growth (also for boys), voice changes, dramatic increase in sebaceous gland size, acne, hirsutism, and enlargement of the clitoris. The long-term effects on reproductive function are unknown, but anabolic steroids may be harmful in this area. Their ability to interfere with the menstrual cycle has been well documented.[3]

GROWTH HORMONE: THE NEXT MAGIC PILL?

Physicians and pharmacologists are now predicting that the anabolic steroid will soon become obsolete as a training aid, and that it will be replaced by human growth hormone (GH), also known as somatotrophic hormone. This hormone is produced by the adenohypophysis of the pituitary gland and is intimately involved in tissue-building processes and normal human growth. Medically, it is given to children who are deficient in GH to help them achieve near-normal size. The use of GH is attractive to the athlete because it stimulates amino acid uptake and protein synthesis by muscle, the breakdown of fat, and decreases the quantity of carbohydrate used by the body.[69] In a recent double-blind study, 6 well-trained men were administered either biosynthetic GH or a placebo while maintaining a high-protein diet. During 6 weeks of standard resistance training with GH, percent body fat decreased and fat-free weight increased significantly. No changes in body composition were noted for the group who trained with the placebo.[30]

Because GH occurs naturally in the body, there is as yet no foolproof way to detect its use as an ergogenic aid. At present, the hormone is expensive and usually only available to the athlete on the black market and often in a highly adulterated form. With the techniques of gene splicing, it will soon be possible to mass-produce a synthetic form of GH at a reasonable price. Undoubtedly, as athletes begin to use this hormone in the hope of attaining a competitive edge, we will see an increased incidence of gigantism in children and acromegalic syndrome—coarsening of the skin, thickening of bones, and overgrowth of soft tissue—in adults.[92]

AMPHETAMINES

Amphetamines or "pep pills" are a group of pharmacologic compounds that exert a powerful stimulating effect on central nervous system function. Amphetamine (Benzedrine) and dextroamphetamine sulfate (Dexedrine) are the compounds used most frequently. Amphetamines are sympathomimetic in that their action mimics that of the sympathetic hormones epinephrine and norepinephrine. Consequently, they cause a rise in blood pressure, pulse rate, cardiac output, breathing rate, metabolism, and blood-sugar level. Five to 20 mg of amphetamine usually exerts its effect for 30 to 90 minutes after ingestion, although its influence can persist for much longer. Aside from bringing about an aroused level of sympathetic function, amphetamines are supposed to increase alertness and wakefulness as well as the capacity to perform increased amounts of work; this is achieved by depressing the sensation of muscle fatigue.[20] It is not surprising, therefore, that athletes frequently use amphetamines with the hope of gaining an ergogenic edge.

Dangers of Amphetamines

The use of amphetamines in athletics is ill-advised for the following medical reasons:

1. **Continual use** can lead to either physiologic or emotional drug dependency. This often brings about a cyclical dependency on "uppers" (amphetamines) or "downers" (barbiturates)—the barbiturates are taken to reduce or tranquilize the "hyper" state brought on by amphetamines.
2. **General side effects** of amphetamines are headaches, tremulousness, agitation, fever, dizziness, and confusion—all of which can have a negative effect on sports performance requiring reaction, judgment, and a high level of steadiness and mental concentration.
3. **Larger doses** are eventually required to achieve the same effect because individual tolerances to the drug increase with prolonged use; this may aggravate and even precipitate certain cardiovascular disorders.
4. **Agents that inhibit or suppresss** the body's normal mechanisms for perceiving and responding to pain, fatigue, or heat stress can severely jeopardize the health and safety of the athlete.
5. **The effects of prolonged intake** of high doses of amphetamines are unknown.

Amphetamine Use and Athletic Performance

Table 23–2 summarizes the results of seven experiments that dealt with the effects of amphetamines on athletic performance. In almost all instances, amphetamines had little or no affect on exercise performances or on simple psychomotor skills.

The major reason athletes take amphetamines is to get "up" for the event and to keep up and be psychologically ready to compete. The day or evening before a contest, however, competitors are often nervous and irritable and have difficulty relaxing. Under these circumstances, a barbiturate is used to induce sleep. The athlete then regains the "hyper" condition by popping an "upper." Not only is this cycle of depressant-to-stimulant undesirable and

TABLE 23–2.
SUMMARY OF RESULTS ON THE USE OF AMPHETAMINES AND ATHLETIC PERFORMANCE

Study*	Dose (mg)	Type of Experiment	Effect of Amphetamines
(61)	10–20	2 exhaustive treadmill runs with 10 min rest between runs	None
		Consecutive 100 yd swims with 10 min rest intervals	None
		220–440 yd swims	None
		220 yd track runs for time	None
		100 yd to 2 mile track runs	None
(41)	10	Bench stepping to fatigue carrying weights equal to $\frac{1}{3}$ body mass, 3 times with 3 min rest intervals	None
(48)	5	100 yd swim for speed	None
(46)	15	All-out treadmill runs	None
(129)	10	Stationary cycling at work rates of 275–2215 kgm · min⁻¹ for 25–35 min followed by treadmill run to exhaustion	None on submaximal or maximal oxygen consumption, heart rate, ventilation volume, or blood lactic acid; work time on the bicycle and treadmill increased significantly
(83)	20	Reaction and movement time to a visual stimulus	None; subjective feelings of alertness or lethargy unrelated to reaction or movement time
(75)	5	Psychomotor performance during a simulated airplane flight	Enhanced performance and lessened fatigue, but if preceded by secobarbital (barbiturate), decreased performance

* The number in parentheses refers to the specific reference listed at the end of the chapter.

potentially dangerous, but the stimulant does not act in its normal manner after a barbiturate. Knowledgeable and prudent people urge that amphetamines be banned from sport competition. The International Olympic Committee, the American Medical Association, and most athletic governing groups have rules to disqualify athletes using amphetamines. Ironically, most research has indicated that amphetamines do **not** enhance physical performance. Perhaps their greatest influence is in the psychologic realm, in which athletes are easily convinced that any supplement will bring on a superior performance. A placebo containing an inert substance often produces identical results!

CAFFEINE

A possible exception to the general rule against stimulants is caffeine.[8,54,76] Caffeine is one of a group of compounds called methylxanthines found naturally in coffee beans, tea leaves, chocolate, cocoa beans, and cola nuts, and is often added to carbonated beverages and nonprescription medicines. Depending on preparation, one cup of brewed coffee contains between 50 and 150 mg of caffeine, tea between 10 and 50 mg, and caffeinated soft drinks about 50 mg. It is absorbed rapidly into the body and reaches peak concentrations in the blood approximately 1 hour after ingestion to exert an influence on the nervous, cardiovascular, and muscular systems.

Although all studies do not support the ergogenic benefits of caffeine, it has been shown that consuming the amount of caffeine commonly found in 2.5 cups of regularly percolated coffee (330 mg) 60 minutes before exercising significantly extended endurance in moderately strenuous exercise.[22] With caffeine, subjects were able to perform an average of 90.2 minutes of exercise compared to 75.5 minutes during a decaffeinated exercise treatment. Even though values for heart rate and oxygen consumption during the two trials were similar, the caffeine also made the work feel easier. During exercise prior to which caffeine had been ingested, the plasma glycerol and free fatty acid levels and the respiratory exchange ratio indicated a high level of fat metabolism and a corresponding reduced rate of carbohydrate oxidation. It is likely that this ergogenic effect of caffeine is due to the **facilitated use of fat** as a fuel for exercise, perhaps mediated via catecholamine release, thus sparing the body's limited carbohydrate reserves.[54,65] Certainly, conserving muscle and liver glycogen would be of considerable benefit in prolonged exercise in which glycogen depletion is intimately related to diminished endurance capacity. It is likely that a lessening of the subjective ratings of effort was due to a central analgesic effect of caffeine, or its effect on neuronal excitability, possibly through a lowering of the threshold for motor-unit recruitment and nerve transmission.[22] It is also possible that caffeine has a favorable effect on maximal aerobic exercise; a small increase in max $\dot{V}O_2$ has been reported following ingestion of a beverage containing 350 mg of caffeine 1 hour prior to exercise,[111] and a high dose of caffeine (10 to 15 mg per kg of body mass) may also provide ergogenic benefits during incremental exercise.[76]

Endurance Effects Are Often Inconsistent

The effect of caffeine ingestion on the mobilization of free fatty acids is significantly blunted, and the metabolic mixture unaltered during prolonged submaximal exercise in individuals who maintain a high carbohydrate intake.[113] This influence of prior nutrition may partly account for the wide and often inconsistent variation in individual response to exercise following caffeine ingestion. Individual sensitivity and tolerance due to patterns of caffeine consumption are probably also important factors that affect the ergogenic nature of this drug, because benefits are not consistently noted,[19,127] especially among habitual caffeine users.[109]

Effects on Muscle

Caffeine may also act directly on muscle to enhance its capacity for exercise.[67,106] By means of a double-blind research design, both voluntary and electrically stimulated muscle contractions were evaluated under normal conditions and following the oral administration of 500 mg of caffeine. Electrical stimulation of the motor nerve enabled the researchers to look at the **direct effects** of caffeine on skeletal muscle. The results showed no effect of caffeine on maximal force developed by the muscle with either voluntary or electrically stimulated contractions. For submaximal work, however, caffeine ingestion produced an increase in the force developed for a given low frequency of electrical stimulation before and after muscular fatigue. These data certainly suggest a specific ergogenic effect of caffeine that appears to be mediated by direct action on muscle. One hypothesis is that caffeine acts on the sarcoplasmic reticulum to increase its permeability to calcium, thus making it readily available for the contraction process.

Warning

The use of caffeine is not without its hazard for certain individuals, especially those who normally avoid this drug.[86] Caffeine is a stimulant to the central nervous system; it can cause headache, insomnia, and nervous irritability, as well as produce extra beats of the left ventricle. Evidence also indicates that coffee consumption in excess of 2 to 3 cups per day is linked to lipoprotein profiles suggestive of increased cardiovascular risk.[123] From a point of temperature regulation, caffeine causes a potent diuretic effect that could be significant during long-term exercise in a hot environment.

PANGAMIC ACID

Pangamic acid, commonly known as "vitamin" B_{15}, has been widely touted among athletes for its alleged ergogenic benefits in aerobic exercise. Confusion exists as to the exact chemical structure of B_{15}. Products sold as pangamic acid contain a variety of different chemicals, although the original compound was a mixture of calcium gluconate and N, N–dimethylglycine in a 60–40 ratio, respectively.

The proponents of pangamic acid argue that studies conducted in Russia indicate that this compound increases the cell's efficiency to use oxygen, reduces lactic acid buildup, and thus enhances endurance. As with many proposed ergogenic aids, testimonials from athletes abound as to its effectiveness as a training aid and performance enhancer. On careful scrutiny of the early studies of pangamic acid, however, it is difficult to interpret the validity of the findings in light of the significant limitations in research design. With research conducted in this country, the results have generally been contradictory. In one study,[84] the ingestion of pangamic acid was associated with improved exercise performance. Several other studies, however, utilizing an appropriate research design were unable to show any benefit of this compound on aerobic capacity, endurance performance, or circulating levels of blood glucose and blood lactate.[43,47] Within the framework of the available, although limited, research, there appears to be no physiologic basis to ingest pangamic acid as an ergogenic aid. From a nutritional perspective, pangamic acid has no vitamin or pro-vitamin properties, and appears to serve no particular need for the body. Concern has been expressed that synthetic mixtures sold as B_{15} may be harmful to humans.[49] In fact, the Food and Drug Administration states that it is illegal to sell this compound as a dietary supplement or drug.

BUFFERING SOLUTIONS

Short-term exhaustive exercise is accompanied by dramatic alterations in the balance of the intra- and extracellular environment because the muscle cells rely predominantly on anaerobic means for energy transfer. As a result, significant quantities of lactic acid accumulate with a concurrent fall in intracellular pH.[24,107] The increase in acidity may ultimately inhibit the energy transfer and contractile capabilities of the active muscle fibers.[50,70,80] In the blood this acidosis is reflected by increases in both H^+ and lactate.

A major line of defense against an increase in intracellular H^+ concentration is the bicarbonate aspect of the body's buffering system (Chapter 14).

If extracellular bicarbonate is maintained at a high level, H^+ will leave the cell more rapidly, thus reducing intracellular acidosis.[71] This fact has led to speculation that an increase in the body's bicarbonate reserve prior to exercise might enhance short-term anaerobic exercise by delaying the fall in intracellular pH associated with this type of effort. The research in this area, however, has produced conflicting results,[26,52,59,63,72] perhaps owing to variations in the pre-exercise dose of sodium bicarbonate and the type of exercise used to evaluate the possible ergogenic benefits of alkalosis. To remove some of these previous limitations, one study investigated the effects of acute metabolic alkalosis induced by ingesting a solution of sodium bicarbonate on short-term, all-out exercise in which fatigue was associated with significant increases in anaerobic metabolites.[117] Six trained middle-distance runners were evaluated during an 880-m race under control conditions and following induced alkalosis by ingesting a sodium bicarbonate solution (300 mg per kg body mass) or a placebo of similar concentration of calcium carbonate. As shown in Table 23–3, ingestion of an alkaline drink raised a subject's pH and standard bicarbonate level prior to exercise. In addition, subjects ran an average of 2.9 seconds faster under alkalosis and achieved higher post-exercise values for blood lactate, pH, and extracellular H^+ concentration than under the control or placebo condition.

These findings suggest that the ergogenic effect of pre-exercise alkalosis is due to an elevated level of anaerobic energy transfer during exercise. It is possible that the increased extracellular buffering provided by ingesting sodium bicarbonate facilitates H^+ efflux from the working muscle cells. This would delay the fall in intracellular pH and its subsequent negative effects

TABLE 23–3.
PERFORMANCE TIME AND ACID-BASE PROFILES FOR SUBJECTS UNDER CONTROL, PLACEBO, AND INDUCED PRE-EXERCISE ALKALOSIS CONDITIONS PRIOR TO AND FOLLOWING AN 800-M RACE

Variable	Condition	Pre-Treatment	Pre-Exercise	Post-Exercise
pH	Control	7.40	7.39	7.07
	Placebo	7.39	7.40	7.09
	Alkalosis	7.40	7.49**	7.18*
Lactate $(mmol \cdot 1^{-1})$	Control	1.21	1.15	12.62
	Placebo	1.38	1.23	13.32
	Alkalosis	1.29	1.31	14.29*
Standard $[HCO^{3-}]$ $(meq \cdot 1^{-1})$	Control	25.8	24.5	9.9
	Placebo	25.6	26.2	11.0
	Alkalosis	25.2	33.5**	14.3*

	Control	Placebo	Alkalosis
Performance time (min:sec)	2:05.8	2:05.1	2:02.9***

* Alkalosis values significantly higher than placebo and control values post-exercise.
** Pre-exercise values significantly higher than pre-treatment values.
*** Alkalosis time significantly faster than control and placebo times.
(From Wilkes, D., et al.: Effects of induced metabolic alkalosis on 800-m racing time. Med. Sci. Sports Exerc., 15:277, 1983.)

on muscle function. It is significant to note that an improvement of 2.9 seconds in 800-m race time is rather dramatic. At race pace this transposes to a distance of 19 meters which in most 800-m races would bring a last place finisher to first place!

Effect Related to Dosage and Degree of Anaerobiosis. An interaction exists between bicarbonate dosage and the cumulative anaerobic nature of the exercise.[52] Heavier dosages of bicarbonate (at least 0.3 g per kg body mass) may facilitate H^+ efflux from the cell and significantly enhance a single 1- to 2-minute bout of maximal exertion[13,45,117] as well as longer-term arm or leg exercise that produces exhaustion within 6 to 8 minutes.[91] All-out effort of less than 1 minute may only be improved when exercise is repetitive in nature. This is also true for lower dosages of bicarbonate.[26] Such intermittent anaerobic exercise would produce a high intracellular concentration of H^+ and perhaps generate a sufficient gradient to facilitate H^+ efflux from the cell for effective buffering. Undoubtedly, more research is needed concerning the ergogenic benefits and possible dangers of acute induced metabolic alkalosis. Many individuals who bicarbonate-load often experience urgent gastrointestinal distress in the form of abdominal cramps and diarrhea about 1 hour after ingestion. For the susceptible athlete such a side effect could negate any potential ergogenic benefit.

TEMPOROMANDIBULAR JOINT (TMJ) REPOSITIONING

During the past 15 years, a number of reports in the popular and dental literature have claimed either real or potential benefits to exercise performance with temporomandibular joint repositioning by means of a specially designed, custom-fitted bite splint that optimizes the alignment of the upper and lower jaw.[55,122] At the same time, the sports performance benefits of such jaw appliances have been extolled by World Class athletes in a variety of sports that include football, tennis, baseball, boxing, luge, field events, and sprint and distance running.[77] The proponents of this form of "sports dentistry" argue that improper TMJ alignment negatively affects the skeletal frame in general and the relatively large neural and vascular component in the jaw region in particular. Consequently, correct positioning and proper stability of the mandible will align the cervical vertebrae and alleviate negative neural and vascular input from the orofacial area to the brain. This adjustment will ultimately translate into optimal exercise performance.

Whereas the practice of TMJ repositioning in sports has increased, research in this area is sparse. Most claims for improved exercise performance are based on case study, anecdotal "success stories," and subjective subject response and evaluation. Even when objective measurement has been made, the research design was often flawed and poorly controlled. In one of our laboratories we evaluated the effects of a mandibular orthopedic repositioning appliance (MORA) on maximum and submaximum physiologic and performance measures in young adult men and women with documented TMJ malalignment.[74] The subjects were randomly assigned to each of four conditions: (1) normal, without a bite splint, (2) with a placebo splint with no bite surface so as to maintain normal jaw position, (3) with a MORA that optimized jaw position, and (4) with a MORA that magnified the subject's normal degree of malocclusion. To assure a double-blind nature to the study each of the three bite splints was coded for a particular subject with the code maintained by the dentists. Because the dentists did not take part in the laboratory measurements, neither the subjects nor the staff had knowledge of what particular bite splint was being used during testing. Data analysis revealed that the use of a MORA device had **no effect** on measures of visual reaction and movement time, muscular strength of the grip, elbow flexors, and leg extensors, submaximal and maximal oxygen consumption, perception of effort, anaerobic power output, running economy, and all-out working capacity in both arm and leg exercise on a cycle ergometer. In fact, placing the jaw in a less-than-optimal position, greater than the subject's normal degree of malocclusion, had no deleterious effect on any performance measures. Clearly, these findings were contrary to that predicted by advocates of mandibular repositioning appliances and in agreement with several studies showing no effect of TMJ repositioning on isometric and isokinetic measures of muscular strength and power.[78,99,130] Such findings support the contention that the beneficial effects of short-term TMJ repositioning on exercise performance noted in previous articles are the result of inadequacies in research design rather than any real ergogenic benefits of a MORA device.

RED BLOOD CELL REINFUSION—BLOOD DOPING

Red blood cell reinfusion, often called induced erythrocythemia, blood boosting, or "blood doping," came into public prominence as a possible ergogenic technique during the 1976 Montreal Olympics when a champion endurance athlete was alleged to have used this in preparation for his eventual gold medal endurance run.

How It Works

With this procedure, usually between 1 to 4 units (1 unit = 450 ml of blood) of a person's blood (autologous) are withdrawn, the plasma is removed and immediately reinfused, and the packed red cells are placed in frozen storage. (With homologous transfusion a type-matched donor's blood is infused.) To prevent a dramatic reduction in blood-cell concentration, each unit of blood is withdrawn over a 3 to 8 week period because it generally takes this long for the person to reestablish normal red blood cell levels. The stored blood cells are then reinfused 1 to 7 days before an endurance event. As a result, the red blood cell count and hemoglobin level of the blood is often elevated some 8 to 20%. This hemoconcentration translates to an average increase in hemoglobin for men from a normal of 15 g per 100 ml of blood to 19 g per 100 ml (or from hematocrits of 40 to 60%). These hematologic characteristics remain elevated for at least 14 days.[44] It is theorized that the added blood volume contributes to a larger maximal cardiac output and that the red blood cell packing increases the blood's oxygen-carrying capacity and thus the quantity of oxygen available to the working muscles. This would be beneficial to the endurance athlete, especially the marathoner, for whom oxygen transport is often a limiting factor in exercise.

Usually 900 to 1800 ml of freeze-preserved autologous blood is the amount infused to bring about ergogenic benefits. For each infusion of 500 ml of whole blood or its equivalent of 275 ml of packed red cells, about 100 ml of oxygen are theoretically added to the total oxygen-carrying capacity of the blood. (This is because each 100 ml of whole blood carries about 20 ml of oxygen.) Because an athlete's total blood volume circulates five or six times each minute in all-out exercise, the potential "extra" oxygen available to the tissues from red cell reinfusion is about 0.5 liters.

It is also possible that blood doping could have effects opposite to those intended. A large infusion of red blood cells (and resulting increase in cellular concentration) could increase blood viscosity and bring about a **decrease** in cardiac output, a **decrease** in blood flow velocity, and a **reduction** in peripheral oxygen content—all of which would **reduce** aerobic capacity. Certainly any increase in blood viscosity or "thickness" could also compromise blood flow through the atherosclerotic vessels of those with significant coronary artery disease.

Does It Work?

A theoretical basis for blood doping exists, and there is experimental evidence to justify this procedure.[98] Much of the early conflict as to the ergogenic benefits has arisen from poor experimental design, inconsistent criteria for performance, variations in techniques for storing blood, and in the timing and quantity of blood withdrawn and replaced. The early research in this area noted significant and rapid increases in max $\dot{V}O_2$ following the infusion of whole blood.[33] Other experiments, however, showed essentially no improvement in endurance performance after red cell reinfusion.[119,120] In commenting on the reported 23% overnight increase in performance and 9% increase in maximal oxygen uptake after blood doping,[35] as well as the favorable results of other studies, researchers point out that control groups were not used, both subjects and investigators generally had knowledge of the specific conditions under which performance measures were made, and that subsequent work tests performed after blood withdrawal may have biased the results owing to a training effect. These researchers failed to acknowledge the possibility that the differences in results with blood doping might be due to the quantity of blood reinfused, the techniques for blood storage, and the duration between withdrawal and reinfusion. Subsequent investigations, however, including a study by one of the past critics of this technique, support the findings of the early studies and showed physiologic and performance improvements with red blood cell reinfusion.[14,89,90,103,127]

The differences among the various research studies are based largely on blood storage methods. Frozen red blood cells can be stored in excess of 6 weeks without significant loss of cells compared to conventional storage at 4°C (used in earlier studies); at 4°C substantial hemolysis occurs after only 3 weeks. This is important because, as shown in Figure 23–1, it usually takes a person 5 to 6 weeks to re-establish the blood cells lost following the withdrawal of 2 units of whole blood.[44]

The procedure of red blood cell reinfusion has a significant effect in elevating hematologic characteristics for both men and women. This, in turn, translates to a 5 to 13% increase in aerobic capacity, a reduced submaximal heart rate and blood lactate for a standard exercise task, and an augmented endurance performance both at altitude and sea level. In addition thermoregulatory benefits (reduced body heat storage and improved sweating responses) during exercise in the heat are derived from pre-exercise red blood cell infusion.[96,97] It is possible that the increased oxygen content of arterial blood in the infused state "frees" relatively more blood to be delivered to the skin for heat dissipation during exercise heat stress. Table 23–4 illustrates hematologic, physiologic, and performance responses for five adult men dur-

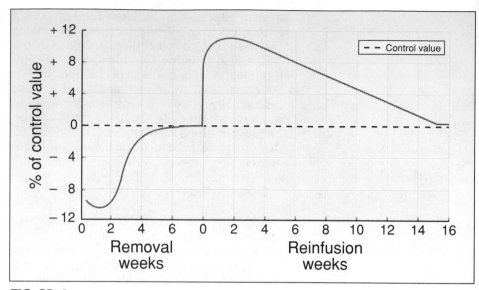

FIG. 23–1.
Time course of hematology changes after the removal and reinfusion of 900 ml of freeze-preserved blood. (From Gledhill, N.: Blood doping and related issues: a brief review. Med. Sci. Sports Exerc., 14:183, 1982.)

ing submaximal and maximal exercise before and 24 hours after the comparatively large infusion of 750 ml of packed red blood cells. These response patterns are generally representative of the more recent research in this area.

A NEW TWIST—HORMONAL BLOOD BOOSTING!

In an attempt to eliminate the somewhat cumbersome and lengthy process of blood doping, endurance athletes are now experimenting with **erythropoietin,** a hormone normally produced by the kidneys. This hormone stim-

TABLE 23–4.
PHYSIOLOGIC, PERFORMANCE, AND HEMATOLOGIC CHARACTERISTICS PRIOR TO AND 24 HOURS AFTER THE REINFUSION OF 750 ML OF PACKED RED BLOOD CELLS

	Pre-Infusion	Post-Infusion	Difference	Difference, %
Hemoglobin, g · 100 ml blood^{-1}	13.8	17.6	3.8*	+27.5*
Hematocrit[a], %	43.3	54.8	11.5*	+26.5*
Submaximal $\dot{V}O_2$, l · min^{-1}	1.60	1.59	−0.01	−0.6
Submaximal HR, b · min^{-1}	127.4	109.2	18.2	−14.3*
Max $\dot{V}O_2$, l · min^{-1}	3.28	3.70	0.42*	+12.8*
Max HR, b · min^{-1}	181.6	180.0	−1.6	−0.9
Treadmill Run Time, sec	793	918	125*	+15.8

[a] Hematocrit is presented as the percent (%) of 100 ml of whole blood occupied by the red blood cells.
* Difference is statistically significant.
(From Robertson, R.J., et al.: Effect of induced erythrocythemia on hypoxia tolerance during exercise. J. Appl. Physiol., 53:490, 1982.)

ulates the bone marrow to produce red blood cells. From a medical standpoint, it has proved quite useful in combating the anemia often observed in patients with severe kidney disease.[39] Normally, when the blood count is low or when the pressure of oxygen in arterial blood decreases (as in severe lung disease or on ascent to high altitude), this hormone is released to stimulate red blood cell production. Unfortunately, if administered exogeneously in an unregulated and unmonitored fashion (simply injecting the hormone requires much less sophistication than the procedures for blood doping), hematocrits can increase to dangerous levels, in excess of 60%. Such a significant increase in hemoconcentration, and thus blood viscosity, greatly increases the likelihood for stroke, heart attack, heart failure, and pulmonary edema.

WARM-UP (PRELIMINARY EXERCISE)

Engaging in some type of physical activity or warm-up prior to vigorous exercise is generally accepted as a valid procedure by coaches, trainers, and athletes at all levels of competition. The underlying belief is that this preliminary exercise aids the performer in preparing either physiologically or psychologically for an event and may reduce the chances of joint and muscle injury. With animals, greater forces and increases in muscle length were required to injure a "warmed-up" muscle compared to a muscle in the "cold" condition.[94] It was suggested that the warming-up process stretches the muscle-tendon unit and subsequently allows for greater length and less tension at any given load on the unit.

The warm up is generally classified under one of two categories, although overlap often exists. These are (1) **general warm-up,** involving calisthenics, stretching, and general body movements or "loosening-up" exercises generally unrelated to the specific neuromuscular action of the anticipated performance, and (2) **specific warm-up,** which provides a skill rehearsal in the actual activity for which the participant is preparing. Swinging a golf club, throwing a baseball or football, tennis practice, and preliminary lead-up in the high jump or pole vault are examples of specific warm-up.

PSYCHOLOGIC CONSIDERATIONS

Competitors at all levels often consider that some prior activity prepares them mentally for their event, so that their concentration and "psyche" become clearly focused on the upcoming performance. Some evidence supports the contention that a specific warm-up related to the activity itself improves the necessary skill and coordination. Consequently, sports that require accuracy, timing, and precise movements generally benefit from some type of specific or "formal" preliminary practice.

There is also the notion that prior exercise, especially before a strenuous effort, gradually prepares a person to go "all out" without fear of injury. A good example is the ritual warm-up of baseball pitchers. Is it conceivable that a pitcher would ever enter a game, throwing at competitive speeds, without previously warming up? Would any athlete begin competition without first engaging in a particular form, intensity, or duration of warm-up? Although in most instances the answer is a definite "no," it would be nearly impossible to design an experiment with topflight athletes to resolve whether warm-up is really necessary and, in fact, whether it improves subsequent performance.

In certain situations, peak performance is expected as soon as play begins, and there is little time for warming up. For example, when a reserve player goes into the last few minutes of a game there is no time for stretching, vigorous calisthenics, or taking practice shots; the player is expected to go all out with no warm-up, except that done before the game or at intermission. Are more injuries recorded in such cases? Is physical performance such as shooting, rebounding, or basketball defense, for example, poorer during the first few minutes of this "unwarmed" condition than it is following a performance preceded by a warm-up?

Psychologic factors such as an athlete's ingrained belief in the importance of warming up establish a definite bias in comparing performance in the "no warm-up" condition.[62] It is difficult to obtain maximum effort with no warm-up if a subject believes warm-up is important. In this regard, some researchers have hypnotized their subjects to neutralize preconceived notions about warm-up.

PHYSIOLOGIC CONSIDERATIONS

On purely physiologic grounds, there are five possible mechanisms by which warm-up should improve performance owing to subsequent increases in blood flow and muscle and core temperature:[7,32] (1) increased speed of contraction and relaxation of muscles; (2) greater mechanical efficiency because of lowered viscous resistance within the muscles; (3) facilitated oxygen utilization by the muscles because hemoglobin releases oxygen more readily at higher temperatures; (4) facilitated nerve transmission and muscle metabolism at higher temperatures; a specific warm-up may also facilitate the recruitment of motor units required in a subsequent all-out activity; and (5) increased blood flow through active tissues as the local vascular bed dilates with higher muscle temperatures.

EFFECTS ON PERFORMANCE

There is little concrete evidence that warm-up per se directly affects subsequent exercise performance. That is not to say that warm-up is unimportant for such purposes. Rather, there is simply little justification from laboratory studies to support such practices. Because of the strong psychologic component and possible physical benefits of warming up, however, whether it be passive (massage, heat applications, and diathermy), general (calisthenics, jogging), or specific (practice of the actual movements), we recommend that such procedures be continued. Until there is substantial evidence justifying its elimination, a brief warm-up is certainly a comfortable way to lead up to more vigorous exercise. **The warm-up should be gradual and sufficient to increase muscle and core temperature without causing fatigue or reducing energy stores.** This consideration is highly individualized; adequate warm-up in terms of intensity and duration for an Olympic swimmer would totally exhaust the average recreational swimmer. To reap the possible benefits from increased body temperature, the actual event or activity should begin within several minutes from the end of the warm-up. In warming up, the specific muscles should be used in a way that mimics the anticipated activity and brings about a full range of joint motion.

SUDDEN STRENUOUS EXERCISE

Several studies have been done to evaluate the effects of preliminary exercise on cardiovascular response to sudden, strenuous exercise. The findings provide an essentially different physiologic framework for justifying warm-up that is important to those involved in adult fitness and cardiac rehabilitation, as well as in occupations and sports requiring a sudden burst of high-intensity exercise.

In one study, 44 men, free from overt symptoms of coronary heart disease, ran on a treadmill at an intense workload for 10 to 15 seconds without prior warm-up.[5] Evaluation of the post-exercise ECG revealed that 70% of the subjects displayed abnormal electrocardiographic changes that could be attributed to inadequate oxygen supply to the heart muscle. These changes were not related to age or fitness level. To evaluate the effect of a warm-up, 22 of the men jogged in place at moderate intensity (heart rate about 145 beats per min) for 2 minutes prior to the treadmill run. With warm-up, 10 men who had previously shown abnormal ECG responses to the treadmill run now had normal tracings and 10 men had improved their S-T segment changes, whereas only two subjects still showed significant S-T changes. The blood pressure response also improved with warm-up. For seven subjects with no warm-up, systolic blood pressure averaged 168 mm Hg immediately after the treadmill run. This was reduced to 140 mm Hg with the 2-minute jog-in-place warm-up.

In an extension of this research,[6] the same pattern of blood pressure and ECG response was obtained with regard to the beneficial warm-up effects on sudden strenuous exercise; warm-up exercise that preceded sudden exertion either eliminated or reduced the ischemic response of the myocardium. These observations indicate that the adaptation of coronary blood flow to a sudden and vigorous cardiac work load is not instantaneous and that transient myocardial ischemia (poor oxygen supply) may occur in apparently healthy and fit individuals. **The effect of prior warm-up (at least 2 minutes of easy jogging) on the electrocardiogram and blood pressure appears to be significant in establishing a more favorable relationship between myocardial oxygen supply and demand.**

Although warm-up preceding strenuous exercise is probably a prudent practice for all people, it is most important for those who have cardiovascular problems that limit the heart's oxygen supply. Brief, prior exercise probably provides for more optimal blood pressure and hormonal adjustment at the onset of subsequent strenuous exercise. This warm-up would serve two purposes: (1) reduce the myocardial work load and thus the myocardial oxygen requirement, and (2) provide adequate coronary blood flow in sudden, high-intensity exercise.

OXYGEN INHALATION (HYPEROXIA)

It is common to observe athletes breathing oxygen-enriched or hyperoxic gas mixtures during times out, at half-time, or following strenuous exercise. The belief is that this procedure significantly enhances the blood's oxygen-carrying capacity and thus facilitates oxygen transport to the exercising muscles. The fact is, however, that when healthy people breathe ambient air at sea level, the hemoglobin in arterial blood leaving the lungs is about 95 to 98% saturated with oxygen. Thus, breathing high concentrations of oxygen could increase oxygen transport by hemoglobin to only a small extent, i.e., about 1 ml of extra oxygen for every 100 ml of whole blood. The oxygen dissolved in plasma when breathing a hyperoxic mixture would also increase slightly from its normal quantity of 0.3 ml to about 0.7 ml per 100 ml of blood. Thus, the blood's oxygen-carrying capacity under hyperoxic conditions would be increased potentially by about 1.4 ml of oxygen for every 100 ml of blood—1.0 ml extra attached to hemoglobin and 0.4 ml extra dissolved in plasma.

PRE-EXERCISE OXYGEN BREATHING

A 70-kg person has about 5000 ml of blood. A hyperoxic breathing mixture could therefore potentially add or "store" about 70 ml of oxygen in the total blood volume (5000 ml blood × 1.4 ml "extra" O_2 per 100 ml blood or 1.4 ml O_2 × 50). Thus, despite the potential psychologic benefit of the athlete believing that pre-exercise oxygen breathing helps performance, this procedure might confer only a slight performance advantage owing to the oxygen per se. This, however, could occur only if the subsequent exercise took place almost immediately after oxygen administration and if ambient air was not breathed in the interval between hyperoxic breathing and exercise. The half-back who breathes oxygen on the sideline before returning to the game or the swimmer who takes a few breaths of oxygen before moving to the blocks for the starting instructions does not really gain the competitive edge due to physiologic benefits. This is especially ironic in football, because the energy to power each play is generated almost totally by metabolic reactions that do not utilize oxygen! The positive psychologic influence of oxygen breathing should not be discounted, however, for it may provide a useful rationale for continuing this practice.

OXYGEN BREATHING DURING EXERCISE

There is considerable evidence that breathing hyperoxic gas during submaximal and maximal aerobic exercise enhances physical performance of an aerobic nature.[114,124,125] Oxygen breathing during both light and heavy exercise has resulted in reduced blood lactic acid levels, heart rates, ventilation volumes, and a significant increase in maximal oxygen consumption.[1,17,34]

In one study,[116] subjects performed a 6.5-minute endurance ride on a bicycle ergometer at a work level equivalent to 115% of max $\dot{V}O_2$ while breathing either room air or 100% oxygen. To mask a subject's knowledge of the breathing mixture, both air and oxygen were supplied from tanks of compressed gas. Figure 23–2A gives the details of the endurance ride showing there was superiority in endurance (with less drop-off in pedal revolutions) while breathing pure oxygen. Figure 23–2B shows the oxygen uptake curves during the endurance ride breathing oxygen and room air. The results indicated that oxygen uptake was higher in the 100% oxygen condition with a correspondingly faster rate of oxygen uptake in the early stages of work. Because the available evidence does not show that hyperoxic gas mixtures increase the maximal cardiac output,[114] the increase in max $\dot{V}O_2$ must be due to an expanded a-$\bar{v}O_2$ difference. This may be partially explained by the fact that even a small increase in hemoglobin saturation during hyperoxia, as well as additional oxygen dissolved in the plasma, increases total oxygen availability during strenuous exercise where the total blood volume is circulated 4 to 7 times each minute. The increase in partial pressure of oxygen in solution breathing hyperoxic gas also facilitates its diffusion across the tissue-capillary membrane to the mitochondria. This may account for its more rapid rate of utilization in the beginning phase of exercise.

When considering the increase in arterial oxygen content when breathing

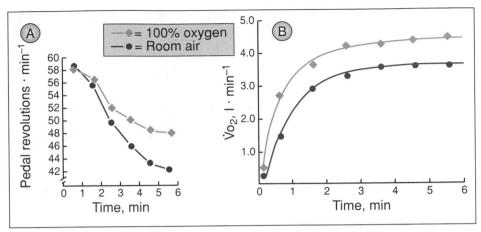

FIG. 23–2.

A, Superiority of endurance (measured by pedal revolutions each minute) breathing pure oxygen versus room air. B, Oxygen uptake curves during the endurance rides. (Data from Weltman, A., et al.: Effects of increasing oxygen availability on bicycle ergometer endurance performance. Ergonomics, 21:427, 1978.)

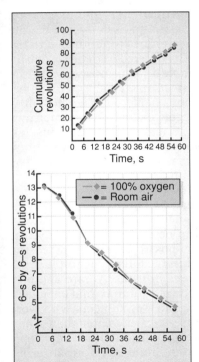

FIG. 23–3.

Absolute and cumulative (top) endurance scores for 1 minute of all-out exercise following oxygen or room air inhalation during recovery from a previous all-out work bout. (From Weltman, A., et al.: Exercise recovery, lactate removal, and subsequent high intensity exercise performance. Res. Q. 48:786, 1977.)

hyperoxic gas that is reported in the literature, however, it appears that the maximum increase does not exceed 10%. In terms of exercise performance, the reduction in pulmonary ventilation commonly observed breathing hyperoxic gas would reduce the oxygen cost of breathing and theoretically liberate significant oxygen for use by the working muscles. Evidence also suggests that hyperoxia increases local muscular performance in static and dynamic movements in which the central circulation does not appear to be a limiting factor. The proposed mechanism for this ergogenic benefit is that the local high oxygen pressure enhances the rate of energy release in the active muscle.[36,37]

Although breathing hyperoxic mixtures appears to offer positive ergogenic benefits **during** endurance performance, its practical application in sports seems limited. Even if an appropriate breathing system could be devised, its "legality" during actual competition is unlikely.

OXYGEN BREATHING DURING RECOVERY

Figure 23–3 illustrates the effects of breathing hyperoxic gas during recovery from strenuous exercise on subsequent exercise performance.[115] Following one minute of all-out exercise on a bicycle ergometer, subjects recovered passively (quiet sitting) or actively (light pedaling) while breathing room air or 100% oxygen for either 10 or 20 minutes. They then repeated the all-out bicycle ride. There were no significant differences in the 6-second revolutions and cumulative revolutions (note the top graph) for the 1-minute ride after breathing room air or pure oxygen during recovery from previous all-out exercise. There were also no significant differences when comparing blood lactate measurements during the 10- and 20-minute recovery periods breathing room air or oxygen. This finding indicated that oxygen inhalation did not preferentially alter lactate removal. In summary, research does not support the use of hyperoxic breathing mixtures as an ergogenic aid during recovery, nor as an adjuvant to improved performance following different durations of recovery from previous exercise.

MODIFICATION OF CARBOHYDRATE INTAKE

Carbohydrate loading is one of the more popular methods of nutritional modification used by endurance athletes to improve performance. Although the judicious adherence to this dietary technique can significantly improve specific performances, there are also some negative aspects that could prove detrimental.

NUTRIENT-RELATED FATIGUE IN PROLONGED EXERCISE

During intense aerobic exercise the energy requirement is supplied largely by the glycogen stored in the liver and the exercising muscles. If intense steady-rate exercise continues and the body's glycogen reserves become reduced, a progressively greater percentage of the energy for exercise must be supplied through the metabolism of fat. This food nutrient is mobilized from storage sites such as the adipose tissue and the liver and is delivered via the circulation to the working muscles. If, however, exercise is performed to the point at which muscle glycogen becomes severely lowered, fatigue can easily occur, even though sufficient oxygen is available to the muscles and the potential energy from stored fat remains almost unlimited.[31,101] This is because the glycogen stored in the muscles becomes depleted. If a solution of glucose and water is ingested at the point of fatigue, exercise can sometimes be prolonged for an additional time, but for all practical purposes the muscles' "fuel tank" reads empty and continued energy production is severely limited. This occurrence, accompanied by sensations of fatigue and muscle pain, has been termed "hitting the wall" by marathon runners.[79]

In the late 1930s, scientists observed that endurance activities could be markedly improved by consuming a carbohydrate-rich diet. Conversely, if the diet consisted predominantly of fat, endurance capacity was drastically reduced. Because of this important relationship between diet and physical performance, researchers have evaluated several possible ways of increasing the body's glycogen reserves. As was discussed in Chapter 1, a simple modification of the diet greatly alters the body stores of carbohydrate and profoundly affects subsequent performance in prolonged submaximal exercise.[9] In one series of experiments the endurance capacity of subjects who were fed a high-carbohydrate diet was more than three times greater than when the same subjects were on a high-fat diet. The recognition of the important role of carbohydrate as an energy substrate in endurance performance has led researchers to focus on additional means to elevate the body's glycogen reserves.

GLYCOGEN SUPERCOMPENSATION

Research has shown that a particular combination of diet and exercise results in a significant "packing" of muscle glycogen. This procedure is termed **carbohydrate loading** or **glycogen supercompensation** and is commonly used by endurance athletes. The end result of this specific dietary modification is an even greater increase in muscle glycogen than occurs with the high-carbohydrate diet discussed previously. In fact, as many as 4 or 5 g of glycogen are "packed" into each 100 g of muscle.[66]

TABLE 23–5.
TWO-STAGE DIETARY
PLAN FOR INCREASING
MUSCLE GLYCOGEN
STORAGE

Stage 1—Depletion

Day 1: Exhausting exercise performed to deplete muscle glycogen in specific muscles

Days 2, 3, 4: Low carbohydrate food intake (high percentage of protein and fat in the daily diet)

Stage 2—Carbohydrate Loading

Days 5, 6, 7: High carbohydrate food intake (normal percentage of protein in the daily diet)

Competition Day

Follow high-carbohydrate pre-event meal outlined in Chapter 3.

Classic Loading Procedure

The classic procedure for achieving the supercompensation effect outlined in Table 23–5 is to reduce the muscle's glycogen content with prolonged steady-rate exercise about 6 days prior to competition. Because glycogen supercompensation occurs **only** in those specific muscles exercised, athletes should be sure to engage the muscles involved in their sport. In preparation for a marathon, a 15- or 20-mile run is usually necessary, whereas for swimming and bicycling, moderately intense submaximal exercise, also for 90 minutes, is required. Then the athlete maintains a low-carbohydrate diet (about 60 to 100 g per day) for several days to further deplete glycogen stores.* During this time moderate training is continued. Then, at least 3 days before the competition, the athlete switches to a high-carbohydrate diet (400 to 600 g per day) and maintains it up to and including the pre-event meal. The 3-day requirement for a high-carbohydrate intake is important because it generally takes this long for full restoration of muscle glycogen following severe depletion.[82] Of course, adequate daily protein, minerals and vitamins, and abundant water must also be part of the supercompensation diet.

One should keep in mind that the potential benefits from carbohydrate loading apply to fairly intense aerobic activities of a prolonged nature. In most instances of sports competition and intense training, a daily diet containing about 60 to 70% of its calories as carbohydrate provides adequate muscle and liver glycogen reserves. Because most competitive endurance athletes normally maintain this carbohydrate intake or higher, their levels of muscle glycogen are usually about twice those of untrained individuals. For them, the supercompensation effect would be relatively small. With heavy training, however, some athletes may not upgrade their daily caloric and carbohydrate intakes to meet energy demands, and consequently they may experience a state of chronic muscular fatigue and staleness.[27]

REPLENISHING CARBOHYDRATE

The rapid replenishment of glycogen following exercise is largely influenced by the amount of carbohydrate consumed, with 70% of the total calories from carbohydrate being a practical recommendation. The form in which carbohydrate is ingested also appears to affect its rate of replenishment. Carbohydrate in simple form facilitates glycogen storage to a greater extent than the more complex variety.[88] Whether the carbohydrate is administered as a liquid or a solid does not make a difference.[87] Glucose and sucrose are the simple sugars of choice because they produce a faster rate of glycogen resynthesis compared to fructose, at least within the first 6 hours following exhaustive exercise.[12] These feedings were most effective at a dose of 0.7 g per kg of body mass taken every second hour for the 6 hours following exercise.

It is not necessary to supercompensate fully for all competitions. For exercise that lasts less than 60 minutes the normal carbohydrate intake and glycogen reserves are adequate. Carbohydrate loading and associated high levels of muscle and liver glycogen, for example, were of no benefit to the subsequent performance of trained runners in a 20.9-km (13-mile) run compared to a run following a low carbohydrate diet.[102] For activities of relatively

* Although the precise mechanism is poorly understood, this glycogen depletion causes the formation of intermediate forms of the glycogen-storing enzyme **glycogen synthetase** in the muscle cell.

short duration, it probably would be beneficial to increase the carbohydrate percentage in the diet for 1 or 2 days before the event. This ensures that glycogen stores are not low enough to limit performance.

If, after weighing all the pros and cons, an athlete decides to supercompensate, it should be tried in stages during training. For example, the athlete should start with a long run followed by a high-carbohydrate diet. A detailed log should be kept of what is done and what happens. Subjective feelings should be noted during both exercise depletion and supercompensation phases. If the feelings are positive, then the athlete should try the entire series of depletion, low-carbohydrate diet, and high-carbohydrate diet—but stay on the low-carbohydrate diet for only one day. If there are no adverse effects, then the low carbohydrate diet should be gradually extended to a maximum of 3 to 4 days.

SAMPLE DIETS FOR ACHIEVING THE SUPERCOMPENSATION EFFECT

Table 23–6 provides an example of meal plans that can be used during carbohydrate depletion (Stage 1) and carbohydrate loading (Stage 2) preceding the endurance event.

TABLE 23–6.
SAMPLE MEAL PLAN FOR CARBOHYDRATE DEPLETION AND CARBOHYDRATE LOADING PRECEDING ENDURANCE EVENT*

Meal	Stage 1 Depletion	Stage 2 Carbohydrate Loading
Breakfast	½ cup fruit juice 2 eggs 1 slice whole-wheat toast 1 glass whole milk	1 cup fruit juice 1 bowl hot or cold cereal 1 to 2 muffins 1 tbsp butter coffee (cream/sugar)
Lunch	6-oz hamburger 2 slices bread salad 1 tbsp mayonnaise and salad dressing 1 glass whole milk	2–3-oz hamburger with bun 1 cup juice 1 orange 1 tbsp mayonnaise pie or cake
Snack	1 cup yogurt	1 cup yogurt, fruit, or cookies

Dinner	2 to 3 pieces chicken, fried
	1 baked potato with sour
	  cream
	½ cup vegetables
	iced tea (no sugar)
	2 tbsp butter
		1–1½ pieces chicken, baked
		1 baked potato with sour
		  cream
		1 cup vegetables
		½ cup sweetened pineapple
		iced tea (sugar)
		1 tbsp butter

Snack	1 glass whole milk	1 glass chocolate milk with 4 cookies

* During Stage 1, the intake of carbohydrate is approximately 100 grams or 400 calories; in Stage 2, the carbohydrate intake is increased to 400 to 625 grams or about 1600 to 2500 calories.

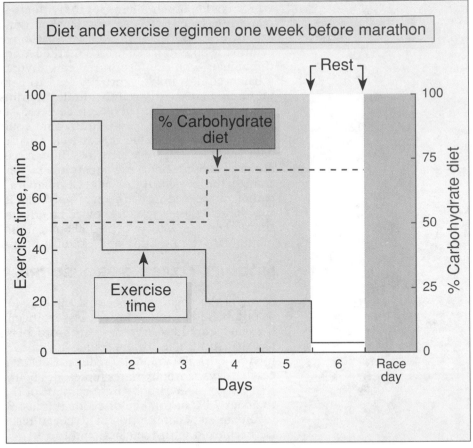

FIG. 23–4.
Recommended combination of diet and exercise for overloading muscle glycogen stores during the week before an important contest. Exercise is gradually reduced during the week, and the carbohydrate content of the diet is increased for the last three days. (From Sherman, W.M. et al. Effect of exercise-diet manipulation on muscle glycogen and its subsequent utilization during performance. *Int. J. Sports Med.*, 2:114, 1981.)

NEGATIVE ASPECTS

The additional 2.7 g of water stored with each gram of muscle glycogen makes this a heavy fuel compared to an equal quantity of calories stored as fat. Body mass is increased and may make the athlete feel "too heavy" and uncomfortable; any extra weight load also directly adds to the energy cost of activities such as running. Extra weight may actually negate the potential benefits to be derived from the extra glycogen storage. On the positive side, the water liberated during glycogen breakdown is available for temperature regulation during exercise in the heat.

The classic model for supercompensation may be potentially hazardous to individuals who have predispositions to specific health problems. A severe and chronic carbohydrate overload interspersed with periods of high fat or high protein intake can increase blood cholesterol and urea nitrogen levels which could pose problems for individuals susceptible to adult diabetes and heart disease, or for those who have certain muscle enzyme deficiencies or kidney disease.[4,56] During the low-carbohydrate phase of the loading procedure the potential exists for a marked ketosis that is often observed among individuals who exercise in the carbohydrate-depleted state. Failure to eat a balanced diet may eventually lead to deficiencies of some minerals and vitamins, particularly water-soluble vitamins; this may require some dietary supplementation. The glycogen-depleted state certainly reduces one's capability to engage in hard training that could result in an actual detraining effect. It is also possible that the elimination of dietary carbohydrate for 3 days could set the stage for a loss of lean tissue as the muscles' amino acids are used as gluconeogenic substrates to maintain blood-glucose levels. Although the possibility of adverse changes in the electrocardiogram with carbohydrate loading has been raised, studies of marathoners who undertook this diet-exercise regimen showed no effects on the ECG pattern.[10] Athletes should become well informed about carbohydrate loading before trying to manipulate their dietary and exercise habits to achieve glycogen loading.

MODIFIED LOADING PROCEDURE

Many of the negative aspects of the classic glycogen-loading sequence can be eliminated by following the less-stringent modified dietary protocol outlined in Figure 23–4.[102] This 6-day protocol is achieved without prior exercise to exhaustion. The athlete trains at about 75% of max $\dot{V}_{O_2}$ (~85% max HR) for 1.5 hours and gradually reduces (tapers) the duration of exercise on successive days. Carbohydrates represent about 50% of total calories during the first 3 days. The carbohydrate content of the diet is concurrently increased to about 70% of total calories for the last 3 days before competition. This results in an accumulation of glycogen reserves to about the same point as that achieved with the classic protocol.[100,102]

SUMMARY

1. Ergogenic aids are substances or procedures that are thought to improve physical work capacity, physiologic function, or athletic performance.
2. One of the most common pharmacologic agents used as ergogenic aids is anabolic steroids. An anabolic steroid is a drug that functions in a manner similar to the hormone testosterone. Although research findings are often inconsistent and the precise mechanism unclear, steroids may help to increase muscle size, strength, and power in some individuals.
3. Administration of growth hormone appears to augment increases in muscle mass when combined with resistance training. As with anabolic steroids, significant potential health risks will probably affect those who abuse this chemical.
4. There is little evidence that amphetamines or "pep pills" aid exercise performance or psychomotor skills, other than what can be derived from a simple placebo effect. The side effects of amphetamines include drug dependency, headache, dizziness, confusion and upset stomach.
5. In some individuals, caffeine ingestion exerts an ergogenic effect by extending endurance in moderately intense aerobic exercise. This is achieved by increasing the utilization of fat for energy and conserving the body's glycogen reserves. These effects are less apparent for indi-

viduals who maintain a high-carbohydrate diet, as well as for those who are habitual users of caffeine.

6. Relatively concentrated buffering solutions consumed prior to exercise exert a significant benefit on all-out anaerobic performance. Neither repositioning of the TMJ nor ingesting pangamic acid offers any ergogenic benefit.

7. Red blood cell reinfusion or "blood doping" involves the drawing, storage, and reinfusion several weeks later of concentrated red blood cells. It is believed that the added blood volume contributes to a larger maximum cardiac output and an increase in the blood's oxygen-carrying capacity and hence max $\dot{V}_{O_2}$. Studies applying appropriate methods of blood storage and research design support the ergogenic benefits of this process for both aerobic exercise and thermoregulation.

8. There is a physiologic argument that warm-up enhances exercise performance. This includes the possible ergogenic effects of warm-up on muscular speed and efficiency, enhanced oxygen delivery and utilization, and facilitated transmission of nerve impulses. Research to support the benefits of warm-up beyond a strong psychologic component, however, is limited. Likewise, there is little evidence justifying its elimination if the performer feels that warm-up is important.

9. There is evidence to support the usefulness of moderate warm-up prior to sudden strenuous exercise. It reduces cardiac work load and may provide for adequate coronary blood flow. In this instance, warm-up may prevent transient myocardial ischemia and potentially dangerous side effects.

10. Breathing 100% oxygen during performance extends endurance by increasing oxygen uptake, reducing blood lactic acid production, and lowering ventilation rate. Breathing hyperoxic mixtures prior to or following exercise is of no ergogenic benefit.

11. Techniques for carbohydrate loading are generally effective for augmenting endurance in prolonged submaximal exercise. Because of potential negative effects, athletes should become well informed about these nutritional modifications before trying to manipulate their diets to achieve glycogen loading. A modification of the classic loading procedure provides for essentially the same high-level of glycogen storage without dramatic alterations in diet and exercise routine.

12. Carbohydrate in simple form (glucose and sucrose) facilitates glycogen storage to a greater extent than the more complex variety. The quantity of carbohydrate consumed in the replenishment phase should represent at least 70% of total calories.

REFERENCES

1. Adams, R.P., and Welch, H.G.: Oxygen uptake, acid-base status, and performance with varied inspired oxygen fractions. *J. Appl. Physiol.*, 49:863, 1980.
2. Alén, M., et al.: Response of serum hormones to androgen administration in power athletes. *Med. Sci. Sports Exerc.*, 17:354, 1985.
3. American College of Sports Medicine. The use of anabolic-androgenic steroids in sports. *Sports Med. Bull.*, 19:13, 1984.
4. American Medical Association Council on Food and Nutrition. *JAMA*, 224:1418, 1973.
5. Barnard, R.J., et al.: Cardiovascular responses to sudden strenuous exercise: heart rate, blood pressure, and ECG. *J. Appl. Physiol.*, 34:883, 1973.
6. Barnard, R.J., et al.: Ischemic response to sudden strenuous exercise in healthy men. *Circulation*, 48:936, 1973.

7. Bergh, U., and Ekblom, B.: Physical performance and peak aerobic power at different body temperatures. *J. Appl. Physiol.*, 46:885, 1979.
8. Berglund, B., and Hemmingsson, P.: Effects of caffeine ingestion on exercise performance at low and high altitudes in cross-country skiers. *Int. J. Sports Med.*, 3:234, 1982.
9. Bergstrom, J., et al.: Diet, muscle glycogen and physical performance. *Acta Physiol. Scand.*, 71:140, 1967.
10. Blair, S.: Blood lipid and ECG responses to carbohydrate loading. *Phys. Sportsmed.*, 8:68, 1980.
11. Blom, P.C.S.: Exhaustive running: inappropriate as a stimulus of muscle glycogen supercompensation. *Med. Sci. Sports Exerc.*, 19:398, 1987.
12. Blom, P.C.S., et al.: Effect of different post-exercise sugar diets on the rate of muscle glycogen synthesis. *Med. Sci. Sports Exerc.*, 19:491, 1987.
13. Bouissou, P., et al.: Metabolic and blood catecholamine responses to exercise during alkalosis. *Med. Sci. Sports Exerc.*, 20:228, 1988.
14. Brien, A.J., and Simon, T.L.: The effects of red blood cell infusion on 10-km race time. *JAMA*, 257:2761, 1987.
15. Buckley, W.E., et al.: Estimated prevalence of anabolic steroid use among male high school seniors. *JAMA*, 260:3441, 1988.
16. Burkett, L.N., and Falduto, M.T.: Steroid use by athletes in a metropolitan area. *Phys. Sportsmed.*, 12:69, 1984.
17. Byrnes, W.C., et al.: Submaximal exercise quantified as percent of normoxic and hyperoxic maximum oxygen uptakes. *Med. Sci. Sports Exerc.*, 16:572, 1984.
18. Calabrese, L.H., et al.: The effects of anabolic steroids and strength training on the human immune response. *Med. Sci. Sports Exerc.*, 21:386, 1989.
19. Casal, D.C., and Leon, A.S.: Failure of caffeine to affect substrate utilization during prolonged exercise. *Med. Sci. Sports Exerc.*, 17:174, 1985.
20. Chandler, J.V., and Blair, S.N.: The effect of amphetamines on selected physiological components related to athletic success. *Med. Sci. Sports*, 12:65, 1980.
21. Clausen, J.P.: Circulatory adjustments to dynamic exercise and effects of physical training in normal subjects and in patients with coronary artery disease. *Prog. Cardiovasc. Dis.*, 18:459, 1976.
22. Costill, D.L., et al.: Effects of caffeine ingestion on metabolism and exercise performance. *Med. Sci. Sports*, 10:155, 1978.
23. Costill, D.L., et al.: The role of dietary carbohydrates in muscle glycogen resynthesis after strenuous running. *Am. J. Clin. Nutr.*, 9:1831, 1981.
24. Costill, D.L., et al.: Leg muscle pH following sprint running. *Med. Sci. Sports Exerc.*, 15:325, 1983.
25. Costill, D.L., et al.: Anabolic steroid use among athletes: changes in HDL-C levels. *Phys. Sportsmed.*, 12:112, 1984.
26. Costill, D.L., et al.: Acid-base balance during repeated bouts of exercise: influence of HCO_3. *Int. J. Sports Med.*, 5:228, 1984.
27. Costill, D.L., et al.: Effects of repeated days of intensified training on muscle glycogen and swimming performance. *Med. Sci. Sports Exerc.*, 20:249, 1988.
28. Cowart, V.: Steroids in sports: After four decades, time to return to the genie's bottle? *JAMA*, 257:421, 1987.
29. Crist, D.M., et al.: Effects of androgenic anabolic steroids on neuromuscular power and body composition. *J. Appl. Physiol.*, 54:366, 1983.
30. Crist, D.M., et al.: Body composition response to exogenous GH during training in highly conditioned adults. *J. Appl. Physiol.*, 65:579, 1988.
31. Davies, C.T.M., and Thompson, M.W.: Aerobic performance of female marathon and ultramarathon athletes. *Eur. J. Appl. Physiol.*, 41:233, 1979.
32. DeVries, H.A.: *Physiology of Exercise for Physical Education and Athletics.* 2nd ed., Dubuque, IA, W.C. Brown, 1986.
33. Ekblom, B.: Response to exercise after blood loss and reinfusion. *J. Appl. Physiol.*, 33:175, 1972.
34. Ekblom, B., et al.: Effect of changes in arterial oxygen content on circulation and physical performance. *J. Appl. Physiol.*, 39:71, 1975.
35. Ekblom, B., et al.: Central circulation during exercise after venesection and reinfusion of red blood cells. *J. Appl. Physiol.*, 40:379, 1976.

36. Eiken, O., et al.: Human skeletal muscle function and metabolism during intense exercise at high O_2 and N_2 pressures. *J. Appl. Physiol.*, 63:571, 1987.

37. Eiken, O., and Tesch, P.A.: Effects of hyperoxia and hypoxia on dynamic and sustained static performance of the human quadriceps muscle. *Acta Physiol. Scand.*, 122:629, 1984.

38. Erickson, M.A., et al.: Effects of caffeine, fructose, and glucose ingestion on muscle glycogen utilization during exercise. *Med. Sci. Sports Exerc.*, 19:579, 1987.

39. Eschbach, J.W., et al.: Correction of the anemia of end-stage renal disease with recombinant human erythropoietin. Results of a combined Phase I and II clinical trials. *N. Engl. J. Med.*, 316:73, 1987.

40. Fahey, T.D., and Brown, C.H.: The effects of anabolic steroids on the strength, body composition and endurance of college males when accompanied by a weight training program. *Med. Sci. Sports*, 5:272, 1973.

41. Foltz, E.E., et al.: The influence of amphetamine (Benzedrine) sulfate and caffeine on the performance of rapidly exhausting work by untrained subjects. *J. Lab. Clin. Med.*, 28:601, 1943.

42. Friedl, K.E., and Yesalis, C.E.: Self-treatment of gynecomastia in body builders who use anabolic steroids. *Phys. Sportsmed.*, 17:67, 1989.

43. Girondola, R.N., et al.: Effects of pangamic acid (B_{15}) ingestion on metabolic response to exercise. *Biochem. Med.*, 24:218, 1980.

44. Gledhill, N.: Blood doping and related issues: a brief review. *Med. Sci. Sports Exerc.*, 14:183, 1982.

45. Goldfinch, J., et al.: Induced metabolic alkalosis and its effects on 400-m racing time. *Eur. J. Appl. Physiol.*, 57:45, 1988.

46. Golding, L.A., and Barnard, R.J.: The effects of d-amphetamine sulfate on physical performance. *J. Sports Med. Phys. Fitness*, 3:221, 1963.

47. Gray, M.E., and Titlow, L.W.: B_{15}: myth or miracle? *Phys. Sportsmed.*, 10(1):107, 1982.

48. Haldi, J., and Wynn, W.: Action of drugs on efficiency of swimmers. *Res. Q*, 17:96, 1959.

49. Herbert, V.: Pangamic acid ("vitamin B_{15}"). *Am. J. Clin. Nutr.*, 32:1534, 1979.

50. Hermansen, L.: Effect of metabolic changes on force generation in skeletal muscle during maximal exercise. In *Human Muscle Fatigue: Physiological Mechanisms.* Edited by R. Porter, and J. Whelan. CIBA Foundation, London, Pitman Medical, 1981.

51. Hervey, G.R., et al.: Effects of methandienone on the performance and body composition of men undergoing athletic training. *Clin. Sci.*, 60:457, 1981.

52. Horswill, C.A., et al.: Influence of sodium bicarbonate on sprint performance: relationship to dosage. *Med. Sci. Sports Exerc.*, 20:566, 1988.

53. Hurley, B.F., et al.: High density lipoprotein cholesterol in body builders vs. power-lifters. Negative effects of androgen use. *JAMA*, 252:4, 1984.

54. Ivy, J.L., et al.: Influence of caffeine and carbohydrate feedings on endurance performance. *Med. Sci. Sports*, 11:6, 1979.

55. Jakush, J.: Divergent views: can dental therapy enhance athletic performance? *J. Am. Dent. Assoc.*, 104:292, 1982.

56. Jetté, M., et al.: The nutritional and metabolic effects of a carbohydrate-rich diet in a glycogen supercompensation training regimen. *Am J. Clin. Nutr.*, 31:2140, 1978.

57. Johnson, F.L.: The association of oral androgenic steroids and life-threatening disease. *Med. Sci. Sports*, 7:284, 1975.

58. Johnson, W.R., and Black, D.H.: Comparison of effects of certain blood alkalinizers and glucose upon competitive endurance performance. *J. Appl. Physiol.*, 5:557, 1953.

59. Jones, N.L., et al.: Effect of pH on cardiorespiratory and metabolic response to exercise. *J. Appl. Physiol.*, 43:959, 1977.

60. Kantor, M.A., et al.: Androgens reduce HDL_2-cholesterol and increase hepatic triglyceride lipase activity. *Med. Sci. Sports Exerc.*, 17:462, 1985.

61. Karpovich, P.V.: Effect of amphetamine sulfate on athletic performance. *JAMA*, 170:558, 1959.

62. Karpovich, P.V.: *Encyclopedia of Sport Sciences and Medicine.* New York, Macmillan, 1971.

63. Katz, A., et al.: Maximal exercise tolerance after induced alkalosis. *J. Sports Med.*, 5:107, 1984.

64. Lamb, D.R.: Anabolic steroids and athletic performance. In *Hormones and Sport.* Edited by Z. Laron, and A. Rogol. Rome, Serono, 1989.

65. LeBlanc, J., et al.: Enhanced metabolic response to caffeine in exercise-trained human subjects. *J. Appl. Physiol.*, 59:832, 1985.
66. Londeree, B.: To supercompensate or not? *Runners World*, 9:26, 1974.
67. Lopes, J.M., et al.: Effect of caffeine on skeletal muscle function before and after fatigue. *J. Appl. Physiol.*, 54:1303, 1983.
68. MacDougall, D.: Anabolic steroids. *Phys. Sportsmed.*, 11:95, 1983.
69. MacIntyre, J.G.: Growth hormone and athletes. *Sports Med.*, 4:129, 1987.
70. MacLaren, D.P.M., et al.: A review of metabolic and physiological factors in fatigue. In *Exercise and Sport Sciences Reviews*. Vol 17. Edited by K.B. Pandolf. New York, Macmillan, 1989.
71. Mainwood, G.W., and Worsley-Brown, P.: The effects of extracellular pH and buffer concentration on the efflux of lactate from frog sartorius muscle. *J. Physiol. (London)*, 250:1, 1975.
72. Margaria, R.: Effect of alkalosis on performance in supramaximal exercise. *Int. Z. Agnew. Physiol.*, 29:215,1971.
73. Max, S.R., and Rance, N.E.: No effect of sex steroids on compensatory muscle hypertrophy. *J. Appl. Physiol.*, 56:1589, 1984.
74. McArdle, W.D., et al.: Temporomandibular joint repositioning and exercise performance: a double-blind study. *Med. Sci. Sports Exerc.*, 16:228, 1984.
75. McKenzie, R.E., and Elliot, L.L.: Effects of secobarbital and D-amphetamine on performance during a simulated air mission. *Aerosp. Med.*, 36:774, 1965.
76. McNaughton, L.: Two levels of caffeine ingestion on blood lactate and free fatty acid responses during incremental exercise. *Res. Q. Exerc. Sport.*, 58:255, 1987.
77. Moore, M.: Corrective mouth guards: performance aids or expensive placebos? *Phys. Sportsmed.*, 9:127, 1981.
78. Moore, T.J., et al.: Temporomandibular orthopedic repositioning appliance and its effect on power production in conditioned athletes. *Phys. Sportsmed.*, 14:137, 1986.
79. Morril, S.: How to beat the wall: the myth and reality behind the marathoner's notorious 20-mile juncture. *Runner*, 3:43, 1981.
80. Parkhouse, W.S., and McKenzie, D.C.: Possible contribution of skeletal muscle buffers to enhanced anaerobic performance: a brief review. *Med. Sci. Sports Exerc.*, 16:328, 1984.
81. Peterson, G.E., and Fahey, T.D.: HDL-C in five elite athletes using anabolic-androgenic steroids. *Phys. Sportsmed.*, 12:120, 1984.
82. Piehl, K.: Time course for refilling of glycogen stores in human muscle fibers following exercise-induced glycogen depletion. *Acta Physiol. Scand.*, 90:297, 1974.
83. Pierson, W.R., et al.: Some psychological effects of the administration of amphetamine sulfate and meprobamate on speed of movement and reaction time. *Med. Sci. Sports*, 1:61, 1961.
84. Pipes, T.V.: The effects of pangamic acid on performance in trained athletes. *Med. Sci. Sports*, 12(Abstract):98, 1980.
85. Pope, H.G., Jr., and Katz, D.L.: Affective and psychotic symptoms associated with anabolic steroid use. *Am. J. Psychiatry*, 145:487, 1988.
86. Rappaport, J.L., and Kruesi, M.J.P.: Behavior and nutrition: a mini review. *Contemp. Nutr.*, 8, 1983.
87. Reed, M.J., et al.: Muscle glycogen storage post-exercise: effect of mode of carbohydrate administration. *J. Appl. Physiol.*, 66:720, 1989.
88. Roberts, K.M., et al.: The effect of simple and complex carbohydrate diets on skeletal muscle glycogen and lipoprotein lipase of marathon runners. *Clinical Physiol.*, 5:41, 1985.
89. Robertson, R.J., et al.: Effect of induced erythrocythemia on hypoxia tolerance during physical exercise. *J. Appl. Physiol.*, 53:490, 1982.
90. Robertson, R.J., et al.: Hemoglobin concentration and aerobic work capacity in women following induced erythrocythemia. *J. Appl. Physiol.*, 57:568, 1984.
91. Robertson, R.J., et al.: Effect of induced alkalosis on physical work capacity during arm and leg exercise. *Ergonomics*, 30:19, 1987.
92. Rogol, A.D.: Growth hormone: physiology, therapeutic use, and potential for abuse. In *Exercise and Sport Sciences Reviews*. Vol. 17. Edited by K.B. Pandolf. New York, Macmillan, 1989.
93. Rogozkin, V.: Metabolic effects of anabolic steroids on skeletal muscle. *Med. Sci. Sports*, 11:160, 1979.

94. Safran, M.R., et al.: The role of warm up in muscular injury prevention. *Am. J. Sports Med.*, 16:123, 1988.

95. Sawka, M.N., et al.: Thermoregulatory and blood responses during exercise at graded hypohydration levels. *J. Appl. Physiol.*, 59:1394, 1984.

96. Sawka, M.N., et al.: Influence of polycythemia on blood volume and thermoregulation during exercise heat stress. *J. Appl. Physiol.*, 62:912, 1987.

97. Sawka, M.N., et al.: Polycythemia and hydration: effects on thermoregulation on blood volume during exercise-heat stress, *Am. J. Physiol.*, 255:R456, 1988.

98. Sawka, M.N., and Young, A.J.: Acute polycythemia and human performance during exercise and exposure to extreme environments. *Exercise and Sport Sciences Reviews.* Vol. 17. Edited by K.B. Pandolf. New York, Macmillan, 1989.

99. Schubert, M.M., et al: Changes in shoulder and leg strength in athletes wearing mandibular orthopedic repositioning appliances. *J. Am. Dent. Assoc.*, 108:334, 1984.

100. Sherman, W.M.: Carbohydrates, muscle glycogen, and muscle glycogen super-compensation. In *Ergogenic Aids in Sports.* Edited by M.H. Williams. Champaign, IL, Human Kinetics, 1983.

101. Sherman, W.M., and Costill, D.L.: The marathon: dietary manipulation to optimize performance. *Am. J. Sports Med.*, 12:44, 1984.

102. Sherman, W.M., et al.: Effect of exercise-diet manipulation on muscle glycogen and its subsequent utilization during performance. *Int. J. Sports Med.*, 2:114, 1981.

103. Spiret, L.L., et al.: Effect of graded erythrocythemia on cardiovascular and metabolic responses to exercise. *J. Appl. Physiol.*, 61:1942, 1986.

104. Stamford, B., and Moffatt, R.: Anabolic steroids: effectiveness as an ergogenic aid to experienced weight trainers. *J. Sports Med.*, 14:191, 1974.

105. Strauss, R.H.: Side effects of anabolic steroids in weight-trained men. *Phys. Sportsmed.*, 11:87, 1983.

106. Supinski, G.S., et al.: Caffeine effect on respiratory muscle endurance and sense of effort during loaded breathing, *J. Appl. Physiol.*, 60:2040, 1986.

107. Sutton, J.R., et al.: Effect of pH on muscle glycolysis during exercise. *Clin. Sci.*, 61:331, 1981.

108. Tan, M.H., et al.: Muscle glycogen repletion after exercise in trained normal and diabetic rats. *J. Appl. Physiol.*, 57:1404, 1984.

109. Tarnopolsky, M.A., et al.: Physiological responses to caffeine during endurance running in habitual caffeine users. *Med. Sci. Sports Exerc.*, 21:418, 1989.

110. Taylor, W.N.: Synthetic anabolic—androgenic steroids: a plea for controlled substance status. *Phys. Sportsmed.*, 15:140, 1987.

111. Toner, M.M., et al.: Metabolic and cardiovascular responses to exercise with caffeine. *Ergonomics*, 25:1175, 1982.

112. Webb, O.L., et al.: Severe depression of high density lipoprotein cholesterol levels in weight lifters and body builders by self-administered exogenous testosterone and anabolic-androgenic steroids. *Metabolism*, 33:11, 1984.

113. Weir, J., et al.: A high carbohydrate diet negates the metabolic effects of caffeine during exercise. *Med. Sci. Sports Exerc.*, 19:100, 1987.

114. Welch, H.G.: Hyperoxia and human performance: a brief review. *Med. Sci. Sports Exerc.*, 14:253, 1982.

115. Weltman, A.L., et al.: Exercise recovery, lactate removal, and subsequent high intensity exercise performance. *Res. Q.*, 48:786, 1977.

116. Weltman, A.L., et al.: Effects of increasing oxygen availability on bicycle ergometer endurance performance. *Ergonomics*, 21:427, 1978.

117. Wilkes, D., et al.: Effect of acute induced metabolic alkalosis on 800-m racing time. *Med. Sci. Sports Exerc.*, 15:277, 1983.

118. Williams, M.H.: *Nutritional Aspects of Human Physical and Athletic Performance.* 2nd ed., Springfield, IL, Charles C Thomas, 1985.

119. Williams, M.H., et al.: Effect of blood reinjection upon endurance capacity and heart rate. *Med. Sci. Sports*, 5:181, 1973.

120. Williams, M.H., et al.: The effect of blood infusion upon endurance capacity and ratings of perceived exertion. *Med. Sci. Sports*, 10:13, 1978.

121. Williams, M.H., et al.: The effect of induced erythrocythemia upon 5-mile treadmill run time. *Med. Sci. Sports Exerc.*, 13:169, 1981.

122. Williams, M.O., et al.: The effect of mandibular position on appendage muscle strength. *J. Prosthet. Dent.*, 49:560, 1983.

123. Williams, P.T., et al.: Coffee intake and elevated cholesterol and apolipoprotein β levels in men. *JAMA*, 253:1407, 1985.

124. Wilson, B.A., et al.: Effects of hyperoxic gas mixtures on energy metabolism during prolonged work. *J. Appl. Physiol.*, 30:267, 1975.

125. Wilson, G.D., and Welch, H.G.: Effects of varying concentrations of N_2/O_2 and He/O_2 on exercise tolerance in man. *Med. Sci. Sports*, 12:380, 1980.

126. Wilson, J.D.: Androgen abuse by athletes. *Endocrine Rev.*, 9:181, 1988.

127. Winder, W.W.: Effect of intravenous caffeine on liver glycogenolysis during prolonged exercise. *Med. Sci. Sports Exerc.*, 18:192, 1986.

128. Windsor, R., and Dumitrv, D.: Prevalence of anabolic steroid use by male and female adolescents. *Med. Sci. Sports Exerc.*, 21:494, 1989.

129. Wyndham, C.H., et al.: Physiological effects of the amphetamines during exercise. *S. Afr. Med. J.*, 45:247, 1971.

130. Yates, J.W., et al.: Effect of a mandibular orthopedic repositioning appliance on muscular strength. *J. Am. Dent. Assoc.*, 108:331, 1984.

WORK PERFORMANCE AND ENVIRONMENTAL STRESS

Our main concern so far has been to focus on the physiologic and metabolic adjustments that enable humans to generate energy for exercise in "normal" environments. In this context, the stress on the organism is largely that imposed by the specific form of work, such as walking or running, bicycling, or swimming in relatively warm water. In many instances, however, the environment compounds the stress of exercise. Sport activities are often held at altitudes that impair the normal oxygenation of blood flowing through the lungs. Above a certain elevation, the capacity to generate aerobic energy for exercise is severely limited. At the other extreme, to explore beneath the water's surface poses a different challenge. In this case, we must bring our sea level environment with us. This usually takes the form of air compressed in a scuba tank carried on the diver's back. For some sports enthusiasts, however, no external assistance is provided, and the length of an underwater excursion is generally limited by the quantity of air inhaled into the lungs just prior to the dive. In both breathhold diving and diving with scuba, unique challenges and dangers are provided by the environment. These are often independent of the stress of exercise.

Exercising in a hot, humid environment or under conditions of extreme cold can also impose severe thermal stress. The total effect of each environmental stressor is clearly determined by the degree to which it deviates from neutral conditions as well as by the duration of the exposure. In addition, several environmental stressors operating at the same time (e.g., extreme cold exposure at high altitude) may exceed and override the simple additive effects of each stressor were it imposed by itself.

In the three chapters that follow, we explore the specific problems encountered at altitude and during exercise in hot and cold environments. We also discuss the immediate physiologic adjustments and long-term adaptations as the body strives to maintain internal consistency despite an environmental challenge. The chapter on sport diving considers the unique problems associated with this increasingly popular form of sport and recreation.

EXERCISE AT MEDIUM
AND HIGH ALTITUDE

More than 40 million people live, work and recreate at terrestrial elevations between 3048 meters (m) and 5486 m (10,000 to 18,000 ft) above sea level. In terms of the earth's topography, these elevations encompass the range of what is generally considered **high altitude.** Although high-altitude natives are self-sustaining, some of them living in permanent settlements as high as 5486 m in the Andes and Himalaya mountains, prolonged exposure of an unacclimatized person to such an altitude may cause death from hypoxia even if the person remains inactive. The physiologic challenge of even moderately high or **medium altitude** becomes readily apparent during physical activity.[53] This is particularly true for newcomers who have not had time to acclimatize to the decreased partial pressure of oxygen at high elevations.

THE STRESS OF ALTITUDE

The challenge of altitude is essentially the decreased partial pressure of ambient oxygen (PO_2) and not the reduced total barometric pressure per se, or of the relative concentrations of the inspired gases. Figure 24–1 illustrates the barometric pressure, the pressures of the respired gases, and the percent saturation of hemoglobin at various terrestrial elevations.[8,35]

The density of air decreases progressively as one ascends above sea level. For example, the barometric pressure at sea level averages 760 mm Hg whereas at 3048 m the barometer reads 510 mm Hg, and at an elevation of 5486 m, the pressure of a column of air at the earth's surface is about half its pressure at sea level. Although dry ambient air at sea level and altitude contains 20.9% oxygen, the PO_2 or density of oxygen molecules is lowered in direct proportion to the fall in barometric pressure upon ascending to higher elevations ($PO_2 = 0.209 \times$ barometric pressure). Thus, ambient PO_2 at sea level averages about 150 mm Hg, but it is only 107 mm Hg at 3048 m. At the summit of Mt. Everest (8848 m; 29,028 ft) the pressure of ambient air is about 250 mm Hg with a concomitant alveolar oxygen pressure of 25 mm Hg, or about 30% of the oxygen available at sea level.[59] **It is this reduction in PO_2 (and accompanying arterial hypoxia) that precipitates the immediate physiologic adjustments to altitude as well as the long-term process of acclimatization.**

OXYGEN LOADING AT ALTITUDE

Due to the s-shaped nature of the oxyhemoglobin dissociation curve (Chapter 13, Fig. 13–3), only a small change in the percent saturation of hemoglobin is observed with decreasing PO_2 until an altitude of about 3048 m. At 1981

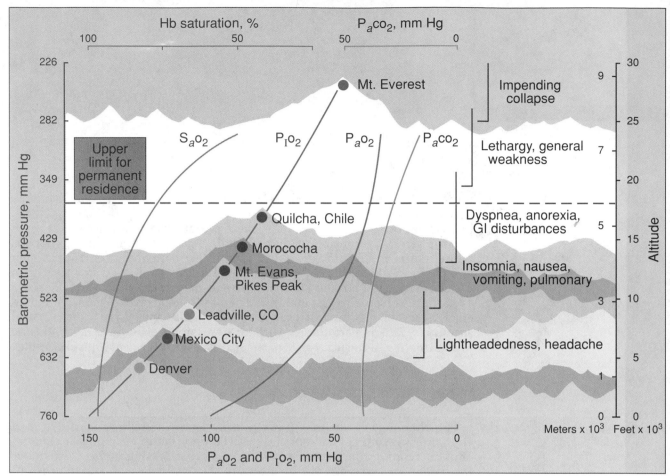

FIG. 24–1.
Changes in environmental and physiologic variables with altitude. (P_aO_2 = partial pressure of arterial oxygen; P_aCO_2 = partial pressure of arterial carbon dioxide; P_{IO_2} = partial pressure of oxygen in inspired air; S_aO_2 = oxygen saturation of hemoglobin) (From Kollias, J., and Buskirk, E.: Exercise and altitude. In Science and Medicine of Exercise and Sport. Edited by W. Johnson and E. Buskirk. New York, Harper and Row, 1974.)

m (6500 ft), for example, the alveolar oxygen partial pressure is reduced from its sea level value of 100 mm Hg to 78 mm Hg, yet hemoglobin is still about 90% saturated with oxygen. Although this relatively small decrease in the amount of oxygen carried by the blood may have little effect on an individual at rest or even during mild exercise, vigorous aerobic activities are altitude sensitive. Research indicates that impairment in aerobic capacity may be linearly related to altitude (about 10% per 1000 m); measurable negative effects on max $\dot{V}O_2$ are noted at altitudes as low as 1219 m (4023 ft),[53] although the altitude-related decrements among individuals are quite variable.[62] The relatively poor performance of men and women in middle distance and distance running and swimming during the 1968 Olympics in Mexico City (altitude 2300 m; 7546 ft) were attributed to the small reduction in oxygen transport at this altitude.[10,18] **No world records were established in events lasting longer than 2.5 minutes.**

At higher elevations such as in the Andes and Himalayas, the reduced loading of hemoglobin with oxygen is readily apparent, and sustained physical activity is difficult. An acute exposure to an altitude of 4300 m, for example, causes a 32% reduction in aerobic capacity compared to sea level values.[4,63] At altitudes above 5182 m (17,000 ft), permanent living is nearly impossible, and mountain climbing at these high elevations is usually done with the aid of oxygen equipment.[40] At 5486 m, for example, arterial P_{O_2} is about 38 mm Hg and hemoglobin is 73% saturated. Because this oxygen partial pressure is on the steep portion of the oxyhemoglobin dissociation curve, any further increase in altitude brings about a relatively large decrease in hemoglobin saturation and oxygen transport capacity. There are reports of acclimatized mountaineers who lived for weeks at 6706 m (28,000 ft) breathing only ambient air.[28] In fact, members of two Swiss expeditions to Mt. Everest remained at the summit for 2 hours without using oxygen equipment.[39] This is an impressive feat considering that arterial oxygen tension is only about 28 mm Hg with a corresponding 58% arterial blood oxygen saturation;[61] an unacclimatized person under these conditions would become unconscious within 20 to 30 seconds. For acclimatized men at simulated extreme altitudes that approach the summit of Mr. Everest, max $\dot{V}_{O_2}$ was reduced about 70% from 4.13 $l \cdot min^{-1}$ to 1.17 $l \cdot min^{-1}$ or 49.1 $ml \cdot kg^{-1} \cdot min^{-1}$ to 15.3 $ml \cdot kg^{-1} \cdot min^{-1}$.[16] These values represent the sea level aerobic capacity of a 70- to 80 year-old man! Although such performances are clearly the exception and not the rule, they do demonstrate the enormous adaptative capability of humans to work and survive without external support at extreme altitudes.

ACCLIMATIZATION

During the many years that mountaineers have attempted to climb the world's highest peaks, it has been well known that weeks are required for sea-level residents to adjust to successively higher elevations. **The adaptive responses that improve one's tolerance to altitude hypoxia are broadly termed acclimatization. Each adjustment to a higher altitude is progressive, and full acclimatization requires time. Successful adjustment to medium altitude represents only partial adjustment to a higher elevation.** Residents of moderate altitudes however, show less decrement in physiologic and exercise performance compared to lowlanders when both groups travel to a high altitude.[38]

As summarized in Table 24–1, certain compensatory responses to altitude occur almost immediately, whereas other physiologic and metabolic adaptations take weeks or even months. Although the intensity of the response is largely altitude-dependent, considerable **individual variability** exists for both the rate and success of an individual's acclimatization.[41]

IMMEDIATE RESPONSES TO ALTITUDE

Upon arrival at elevations of about 2300 m and higher, rapid physiologic adjustments occur to compensate for the thinner air and accompanying reduced alveolar oxygen pressure. The most important of these responses are

TABLE 24—1.
IMMEDIATE AND LONGER-TERM ADJUSTMENTS TO ALTITUDE HYPOXIA

System	Immediate	Longer Term
Pulmonary Acid-Base	• Hyperventilation • Body fluids become more alkaline due to reduction to CO_2 with hyperventilation	• Hyperventilation • Excretion of base via the kidneys and concomitant reduction in alkaline reserve
Cardiovascular	• Increase in submaximal heart rate • Increase in submaximal cardiac output • Stroke volume remains the same or is slightly lowered • Maximum heart rate remains the same or is slightly lowered • Maximum cardiac output remains the same or is slightly lowered	• Submaximal heart rate remains elevated • Submaximal cardiac output falls to or below sea-level values • Stroke volume is lowered • Maximum heart rate is lowered • Maximum cardiac output is lowered
Hematologic		• Decrease in plasma volume • Increased hematocrit • Increased hemoglobin concentration • Increased total number of red blood cells
Local		• Possible increased capillarization of skeletal muscle • Increased red-blood-cell 2,3-DPG • Increased mitochondria • Increased aerobic enzymes

(1) an increase in the respiratory drive, which results in hyperventilation, and (2) an increase in blood flow at rest and during submaximal exercise.

Hyperventilation

Probably the most important and clear-cut immediate response of the native lowlander to altitude exposure is hyperventilation brought on by the reduced arterial P_{O_2}.[13,50] Once initiated, this "hypoxic drive" increases during the first few weeks and may remain elevated for a year or longer during prolonged altitude residence.[34,39]

Special receptors sensitive to reduced oxygen pressure are located in the aortic arch and at the branching of the carotid arteries in the neck (Chapter 14). Any significant reduction in arterial P_{O_2} (e.g., which occurs at an altitude above 2000 m) progressively stimulates these chemoreceptors. This, in turn, modifies inspiratory activity to increase alveolar ventilation and cause alveolar oxygen concentration to increase toward the level in ambient air—the greater the hyperventilation, the more closely alveolar air resembles inspired air. The increase in alveolar P_{O_2} with hyperventilation facilitates oxygen loading in the lungs and provides, both at rest and during exercise, the rapid first line of defense against the stress of reduced ambient P_{O_2}.[56] Studies indicate that mountaineers who respond with a strong, hypoxic ventilatory drive may perform exercise tasks better at extreme altitudes and can reach a higher altitude compared to other climbers who show a somewhat blunted ventilatory response when exposed to a reduced ambient P_{O_2}.[50]

Increased Cardiovascular Response

In the early stages of altitude adaptation, submaximal heart rate and cardiac output may increase 50% above sea-level values,[31,57] whereas the heart's stroke volume remains essentially unchanged. Because the oxygen cost of work at altitude is essentially no different compared to sea level,[41] the increase in submaximal blood flow partially compensates for the reduced oxygen in arterial blood. For example, a 10% increase in cardiac output at rest or in moderate exercise offsets a 10% reduction in arterial oxygen saturation, at least in terms of the total oxygen circulated through the body.

The effects of **acute altitude exposure** on the metabolic and cardiorespiratory response to submaximal and all-out bicycle exercise in young men are shown in Table 24–2. Physiologic measures were obtained at sea level and during a brief exposure to a simulated altitude of 4000 m (13,124 ft).[54] Even with the increase in pulmonary ventilation during submaximal exercise at "altitude" as compared to sea level, arterial oxygen saturation was reduced from 96% to about 70% at all work levels. In submaximal exercise, however, the blood's reduced oxygen content was entirely compensated for by an increased cardiac output. The increase in blood flow was due to an elevated heart rate because the heart's stroke volume remained unchanged during altitude exposure. With this circulatory adjustment, oxygen consumption was nearly identical in submaximal work at sea level and at altitude. The greatest effect of altitude on aerobic metabolism was observed during maximal exercise, when the max $\dot{V}O_2$ was reduced to 72% of sea-level values. At this exercise intensity, the ventilatory and circulatory adjustments to acute altitude exposure cannot compensate for the lower oxygen content of arterial blood.

TABLE 24–2. CARDIORESPIRATORY AND METABOLIC RESPONSE DURING SUBMAXIMAL AND MAXIMAL EXERCISE AT SEA LEVEL AND SIMULATED ALTITUDE OF 4000 M (13,115 FT) IN SIX YOUNG MEN.*

Exercise Level	$\dot{V}O_2$ ($l \cdot min^{-1}$)		$\dot{V}_E$ ($l \cdot min^{-1}$ BTPS)		Arterial Saturation (%)	
ALTITUDE, m	0	4000	0	4000	0	4000
600 kg – m·min^{-1}	1.50	1.56	39.6	53.7	96	71
900 kg – m·min^{-1}	2.17	2.23	59.0	93.7	95	69
Maximum	3.46	2.50	123.5	118.0	94	70

Exercise Level	$\dot{Q}$ ($l \cdot min^{-1}$)		H R ($b \cdot min^{-1}$)		S V (ml)		$a - \bar{v}O_2$ DIFF (ml $O_2 \cdot$ 100 ml^{-1})	
ALTITUDE, m	0	4000	0	4000	0	4000	0	4000
600 kg – m·min^{-1}	13.0	16.7	115	148	122	113	10.8	9.4
900 kg – m·min^{-1}	19.2	21.6	154	176	125	123	11.4	10.4
Maximum	23.7	23.2	186	184	127	126	14.6	10.8

* From Stenberg, J., et al.: Hemodynamic response to work at simulated altitude. 4000 m. J. Appl. Physiol., 21:1589, 1966.

Acute Mountain Sickness

Despite the body's rapid defense against the stress of altitude, many people experience the acute discomfort of **mountain sickness** during the first few days at altitudes of about 3048 m or higher. This is apparent among people who ascend rapidly to high altitudes without benefiting from a gradual and progressive acclimatization to lower altitudes. These symptoms usually include headache (most frequent symptom probably due to dilation of cerebral blood vessels), dizziness, nausea, vomiting, dimness of vision, insomnia, and generalized weakness.[29] Appetite suppression can be severe during the early stages of high altitude stay. This may result in an average reduction in energy intake of about 40% and an accompanying loss of body mass. Diets low in salt and high in carbohydrates appear to be well tolerated by subjects during the early stay at high altitude and ameliorate the ill effects of altitude exposure. For one thing, the energy liberated per unit of oxygen consumed is greater in carbohydrate breakdown than in fat breakdown (5.0 kcal vs. 4.7 kcal per liter O_2). Also, high levels of circulating fats following a high-fat meal may reduce arterial oxygen saturation. This improvement in arterial oxygen transport with a high-carbohydrate diet tends to (1) enhance altitude tolerance, (2) reduce the severity of mountain sickness, and (3) lessen the physical performance decrements during the early stages of altitude exposure. For people suffering the effects of mountain sickness, even moderate exercise can be intolerable. With acclimatization, symptoms subside and many disappear. Concurrently, a person's ability to exercise improves and considerably more work can be accomplished. Mountain sickness can usually be prevented by acclimatizing slowly to moderate altitudes below 3048 m, followed by a slow progression to higher altitudes. Exercise should also be minimized during the first days of altitude exposure.

For unknown reasons, about 2% of the sojourners to altitudes above 10,000 ft (3048 m) experience a severe aspect of acute mountain sickness termed **high-altitude pulmonary edema** or **HAPE.** In this life-threatening condition fluid accumulates in the lungs and brain.[49] To prevent severe disability or even death, the best treatment for this malady is immediate descent to lower altitude.

Fluid Loss

Because the air in mountainous regions is usually cool and dry, considerable body water can be lost through evaporation as air is warmed and moistened in the respiratory passages. This fluid loss often leads to moderate dehydration and accompanying symptoms of dryness of the lips, mouth, and throat. This is particularly true for active people for whom the daily total sweat loss and pulmonary ventilation (and hence water loss) are large. For these active people, body mass should be checked frequently and easy access to water provided at all times.

LONGER-TERM ADJUSTMENTS TO ALTITUDE

Hyperventilation and increased submaximal cardiac output provide a rapid and relatively effective counter to the acute challenge of altitude. Concurrently, other slower acting physiologic and metabolic adjustments occur during a prolonged altitude stay.[64] The most important of these involve (1) maintenance of the acid-base balance of body fluids altered by hyperventilation, (2) increased formation of hemoglobin and red blood cells, and (3) changes in local circulation and cellular function. All of these adaptations generally reduce distress and improve tolerance to the relative hypoxia of medium and high altitudes.

Acid-Base Readjustment

Although hyperventilation at altitude favorably increases alveolar oxygen concentration, it has the opposite effect on carbon dioxide. Because ambient air contains essentially no carbon dioxide, the increased breathing at altitude tends to "wash out" or dilute this gas in the alveoli. This creates a larger-than-normal gradient for the diffusion of carbon dioxide from the blood to the lungs, and arterial carbon dioxide is reduced considerably. Upon exposure to an altitude of 3048 m, for example, alveolar P_{CO_2} falls to approximately 24 mm Hg. This is in contrast to the P_{CO_2} of 40 mm Hg usually maintained at sea level. During a prolonged stay at higher altitudes, the pressure of alveolar carbon dioxide falls as low as 10 mm Hg.

The loss of carbon dioxide from the body's fluids in hypoxic environments causes a physiologic disequilibrium. The largest quantity of carbon dioxide is normally carried as carbonic acid (Chapter 13). This relatively weak acid readily ionizes to H^+ and HCO_3^-, which are then transported to the lungs by the venous circulation. In the pulmonary capillaries, carbon dioxide and water re-form, and carbon dioxide diffuses into the alveoli. A decrease in carbon dioxide, as would occur in hyperventilation, causes the pH to rise and the blood becomes more alkaline.

Because hyperventilation is a normal response to altitude, adjustments must be made during acclimatization to minimize the accompanying "side effects" that disrupt acid-base balance. This control of respiratory alkalosis is accomplished slowly by the kidneys that excrete base (HCO_3^-) through the renal tubules. In turn, the restoration of a normal pH increases the responsiveness of the respiratory center, and ventilation increases further to adjust to altitude hypoxia.

Reduced Buffering Capacity

The establishment of acid-base equilibrium with acclimatization occurs at the expense of a loss of alkaline reserve. Thus, although the pathways of anaerobic metabolism are unaffected at altitude, the blood's buffering capacity for acids is gradually decreased, and the critical limit is lowered for the accumulation of acid metabolites.[30,34] A general depression in maximum lactate concentration is especially apparent in all-out exercise at altitudes above 4000 m.[42]

Hematologic Changes

The most important long-term adaptation to altitude is an increase in the blood's oxygen-carrying capacity. Two factors are responsible: (1) an initial decrease in plasma volume, which is then followed by (2) a rapid formation of erythrocytes and hemoglobin.

1. **Decrease in plasma volume:** Red blood cells become more concentrated during the first few days at altitude because of a decrease in plasma volume.[3,7,25] After a week at 2300 m, for example, the plasma volume decreased by about 8% whereas the concentrations of red blood cells (hematocrit) and hemoglobin increased 4 and 10%, respectively. A week

at 4300 m (14,108 ft) caused a 16 to 25% decrease in plasma volume, whereas hematocrit rose about 6% and hemoglobin increases 20%.[7,25] This rather rapid adjustment in plasma volume and accompanying hemoconcentration causes the oxygen content of arterial blood to increase significantly above values observed immediately on ascent to altitude.

2. **Increase in red cell mass:** The reduced arterial oxygen pressure also stimulates an increase in the total number of red blood cells, a process termed **polycythemia.** This response is mediated by an erythrocyte-stimulating factor, erythropoietin, released from the kidneys and other tissues within 15 hours after altitude ascent.[1] In the weeks that follow, the production of erythrocytes in the marrow of the long bones increases considerably and remains elevated during residence at altitude.[16,21,46] A typical miner in the Andes, for example, has 38% more circulating red blood cells than his low-altitude counterpart. In some apparently healthy high-altitude natives, the red cell count may be more than 50% greater than normal[37]— 8 million cells per cubic millimeter compared to 5.3 million for the native lowlander! In fact, during a 1973 Mt. Everest Expedition, a 40% increase in blood hemoglobin concentration and a 66% increase in hematocrit were noted for members acclimatized at 6500 m.[9] This probably approaches the limit of a beneficial rise in red blood cell concentration. Any further erythrocyte packing would increase the blood's viscosity and probably restrict oxygen diffusion and blood flow through the body.

In effect, polycythemia directly translates into a large increase in the blood's capacity to transport oxygen. The oxygen-carrying capacity of blood for high-altitude residents of Peru is 28% above sea-level averages.[29] For well-acclimatized mountaineers, the oxygen-carrying capacity is 25 to 31 ml of oxygen per 100 ml of blood compared to about 19.7 ml for lowland residents.[40,60] Thus, even with the reduced saturation of hemoglobin at altitude, the actual quantity of oxygen in arterial blood closely approaches or even equals sea-level values.

The general trend for increases in hemoglobin and hematocrit during altitude acclimatization is illustrated in Figure 24–2A. These data were obtained for eight young women at the University of Missouri (altitude 213 m) who lived and worked for 10 weeks at the 4267-m summit of Pikes Peak.[25] Because previous work of the researchers showed markedly fewer hematologic changes during acclimatization in women than in men, possibly due to inadequate iron intake, each woman received iron supplementation prior to, during, and on return from altitude.

Upon reaching Pikes Peak, red blood cell concentration increased rapidly. This increase was caused by a reduction in plasma volume during the first 24 hours at altitude. In the month that followed, hemoglobin concentration and hematocrit continued to increase and then stabilized for the remainder of the stay. Prealtitude values were established 2 weeks after the women returned to Missouri.

As shown in Figure 24–2B, iron supplementation increased the prealtitude values for hematocrit and hemoglobin. This finding is not surprising. As pointed out in Chapter 2, young women frequently suffer from mild iron insufficiency with depressed iron reserves. When the acclimatization curves for the iron-supplemented women were compared with those of another group of women not given additional iron, there was a greater hematocrit increase in the group given supplements. Thus, at least in these women, iron supplementation enhanced the rate of hematocrit increase at altitude to a level similar for men at the same location. These findings indicate that

FIG. 24–2.
A, Effects of altitude on hemoglobin and hematocrit levels of 8 young women prior to, during, and 2 weeks after exposure to altitude of 4267 m. (From Hannon, J.P., et al.: Effects of altitude acclimatization on blood composition of women. J. Appl. Physiol., 26:540, 1968.) B, hematocrit response of young women receiving supplemental iron (+ Fe) prior to and during altitude exposure compared to groups of male and female subjects receiving no supplemental iron. (Courtesy of Dr. J.P. Hannon.)

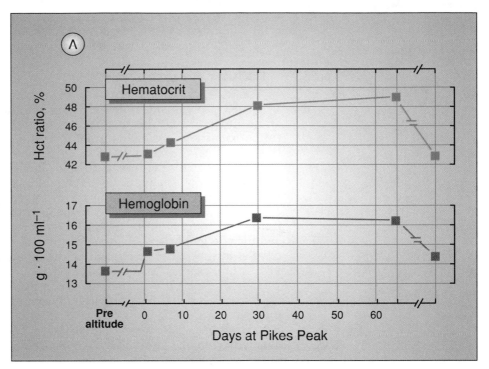

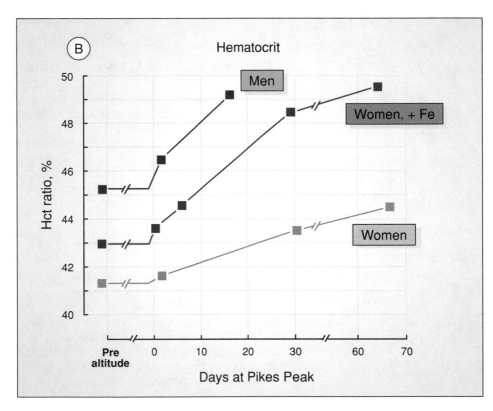

athletes who have borderline iron stores may not respond to the acclimatization process as well as individuals who arrive at a specific altitude with adequate iron reserves to sustain an increase in red blood cell production.

Cellular Adaptations

A subject of considerable debate is whether significant local circulatory and cellular adaptations to extremes of hypoxia take place that could maximize oxidative function in humans.[14,15,45] Capillaries are more concentrated in the skeletal muscle of animals born and raised at high altitude compared to sea level.[58] This modification in local circulation reduces the distance for oxygen diffusion between the blood and tissues. Also, muscle biopsies from humans living at altitude indicate an increase in myoglobin by as much as 16% after acclimatization. This was complemented by an increase in the number of mitochondria and in the concentration of enzymes required for aerobic energy transfer.[45] Such adaptations would increase the "storage" of oxygen in specific muscles and facilitate intracellular oxygen delivery and utilization, especially at low tissue P_{O_2}.

High-altitude natives also benefit from a slight shift to the right of the oxyhemoglobin dissociation curve. A decrease in the oxygen affinity of hemoglobin at altitude favors the release of available oxygen to the tissues for a given drop in P_{O_2}. This adaptation occurs because of an increase in the concentration of 2, 3-diphosphoglycerate in red blood cells (see Chapter 13) during long-term altitude residence.[17,36] An increase in 2,3-DPG coupled with an increased quantity of circulating hemoglobin places the long-term altitude resident in a more favorable physiologic position for supplying oxygen to active tissue during strenuous exercise.

Changes in Body Mass and Composition

During long-term exposure to high altitude a significant loss is generally observed in body mass. Recent studies observed 6 men who participated in progressive decompression over 40 days to an ambient pressure of 249 mm Hg in a hyperbaric chamber in a simulated ascent to the top of Mt. Everest. Daily caloric intake decreased by 43% during the exposure period. This represented a loss of 7.4 kg of body mass, predominantly derived from the muscle component of fat-free mass.[47]

TIME REQUIRED FOR ACCLIMATIZATION

In general, the length of the acclimatization period depends on the altitude. Acclimation at one altitude ensures only partial adjustment to a higher elevation.[38] As a broad guideline, about 2 weeks are required to adapt to altitudes up to 2300 m. Thereafter, for each 610-m increase in altitude, an additional week is required for full adaptation up to an altitude of 4572 m.

For athletes desiring to compete at altitude, intense training should commence as soon as possible during the acclimatization period. This will minimize any detraining effects because it is difficult to engage in hard training in the early days of one's altitude stay.[35] **The benefits of acclimatization are probably lost within 2 or 3 weeks after returning to sea level.**

METABOLIC, PHYSIOLOGIC, AND EXERCISE CAPACITIES AT ALTITUDE

The stress of high altitude imposes significant restrictions on work capacity and physiologic function. Even at lower altitudes, the body's adjustments do not fully compensate for the reduced oxygen pressure, and there is a compromise in both physiologic function and exercise performance. In fact, certain circulatory parameters, especially stroke volume and maximum heart rate, are altered in a direction that contributes to a reduced capacity for oxygen transport.[13a,48]

MAXIMUM AEROBIC CAPACITY

Aerobic capacity shows a progressive and somewhat linear decrease with increases in altitude.[43,53] One can generally expect a 1.5 to 3.5% reduction in max $\dot{V}O_2$ for every 305 m (1000 ft) above 1524 m (5000 ft) altitude.[7,19,43] At 6248 m (20,500 ft), the max $\dot{V}O_2$ is approximately half the value at sea level whereas the maximum aerobic power of a relatively fit man atop Mt. Everest would be about 1000 ml of oxygen per minute.[39,64] This value corresponds to a work rate of only about 50 watts on a bicycle ergometer.

The degree of physical conditioning prior to altitude exposure offers little protection because the percentage reduction in max $\dot{V}O_2$ is equal in both trained and untrained individuals. For well-conditioned individuals, however, a particular work task at altitude still provides relatively less stress because it can be performed at a lower percentage of the trained person's maximum aerobic capacity.

CIRCULATORY FACTORS

Even after several months of acclimatization, maximum aerobic power still remains significantly below sea-level values.[19,23,34] This occurs because the benefits of acclimatization are offset by a reduction in circulatory efficiency in moderate and strenuous exercise.[3,24,33]

Submaximal Exercise

Although the immediate altitude response involves an increase in submaximal cardiac output (Table 24–2), the exercise cardiac output in the days and weeks of acclimatization that follow is reduced and does not improve with longer altitude exposure.[3,31] This is due mainly to a **decrease in stroke volume** as the altitude stay progresses. With reduced cardiac output, oxygen consumption is maintained by an expanded a-$\bar{v}O_2$ difference. To some extent the decrease in stroke volume observed at altitude is offset by an increase in submaximal heart rate. Even at altitudes as high as Mt. Everest, the rate and contractile function of the heart are maintained in reasonable fashion despite the considerable level of chronic hypoxis.[44]

Maximal Exercise

A reduction in maximum cardiac output also occurs after about a week at altitudes above 3048 m.[23,43,57] This reduction in blood flow during all-out exercise, that persists through one's altitude stay, is generally the combined effect of a decrease in maximum heart rate[34,48] and stroke volume.[24,31 – 33,43,48] The reduced stroke volume may be the result of a decreased plasma and blood volume and an increased total peripheral resistance. Myocardial hypoxia has not been demonstrated with measurements of ECG,[26,48] or coronary blood flow during vigorous exercise at high altitudes. The reduction in maximum heart rate at altitude may be influenced by enhanced parasympathetic tone induced by prolonged altitude exposure.[27]

PERFORMANCE MEASURES

In order to perform aerobic exercise at the same relative intensity at altitude as at sea level, the pace must be slowed. If it is not, a larger portion of the energy for exercise is provided by anaerobic metabolism and fatigue develops.[61a] For example, a 2 to 13% decrement in performance is observed for fit subjects in the 1- and 3-mile runs at the medium altitude of 2300 m.[19] This agrees with the 7.2% increase in 2-mile run times reported for highly trained middle-distance runners at the same altitude.[2] Even after 29 days of acclimatization, significant decrements in 3-mile running performance at high altitude are noted as compared to run times near sea level.[42]

ALTITUDE TRAINING AND SEA-LEVEL PERFORMANCE

It is clear that altitude acclimatization improves one's capacity to work at altitude, especially at high altitude. What is not clear, however, is the effect of prior altitude exposure and altitude training on aerobic power and endurance performance immediately on return to sea level. Certainly, possible adaptations in local circulation and cellular function as well as the compensatory increase in ventilation and the blood's oxygen-carrying capacity should facilitate sea-level performance. Also, if tissue hypoxia is an important training stimulus, altitude **and** training should act synergistically so that the total effect exceeds that of similar training at sea level. Unfortunately, much of the previous altitude research has not been designed to evaluate adequately this possibility. Often, the activity level of the subjects is poorly controlled, so it is difficult to determine whether an improved max $\dot{V}_{O_2}$ or performance score on return from altitude represents a training effect, an altitude effect, or synergism between altitude and training.[7,19]

MAX $\dot{V}_{O_2}$ ON RETURN TO SEA LEVEL

When max $\dot{V}_{O_2}$ is used as the criterion, sea-level performance is not significantly improved after living at altitude.[4,6,19,38a] For example, there was no significant change in the original 25% reduction in maximal aerobic power in young runners during 18 days at 3100 m.[23] On return to sea level, the max $\dot{V}_{O_2}$ was about the same as the prealtitude measures. These findings have been duplicated at even higher altitudes.[6] Highly trained varsity trackmen were flown to Nunoa, Peru, altitude 4000 m (13,000 ft). They continued to train and acclimatize for 40 to 57 days. After the initial 3 days at altitude, max $\dot{V}_{O_2}$ was reduced 29% below sea-level values; after 48 days, it was still 26% lower. Running performance after acclimatization was measured in the 440-yard, 880-yard, and 1- and 2-mile runs during a "track meet" with the altitude natives. Run times were still considerably slower than prealtitude times, especially in the longer runs. Furthermore, when the athletes returned to sea level, max $\dot{V}O_2$ and running performance were generally no different than they had been previously. On no occasion did a runner improve his previous prealtitude run time! In fact, running times in the longer events remained about 5% below prealtitude trials. Recent studies have shown that simulated altitude training provides no additional benefit to sea level performance compared to similar training at sea level. Of course, the altitude-trained group showed significantly higher physical performance at altitude compared to their sea level counterparts.[55]

Even in those studies showing a small improvement in max $\dot{V}O_2$ at altitude and upon return to sea level,[14,18,32] the change is generally attributed to an increase in physical activity (that is, training and/or repeated testing) during the altitude exposure.

Some of the physiologic changes that occur during prolonged altitude exposure may actually negate adaptations that possibly could improve exercise performance upon return to sea level. The residual effects of a loss in muscle mass and a reduced maximum heart rate and stroke volume frequently observed at altitude certainly would not enhance sea-level performance. Any reduction in maximum cardiac output would offset the benefits derived from the blood's greater oxygen-carrying capacity. Although circulatory function does return to normal after a few weeks at sea level, so also do the potentially positive hematologic changes.[25] Within a physiologic context, it appears that the controversial use of blood doping mimics the hematologic benefits of a stay at altitude without the potential negative effects of altitude on cardiovascular dynamics.[20]

CAN TRAINING BE MAINTAINED AT ALTITUDE?

Exposure to altitudes of 2300 m and higher makes it nearly impossible for athletes to train at the same intensity as that engaged in at sea level. Table 24–3 shows this reduction in training intensity in relation to sea-level standards for six competitive college athletes.[35]

At successively higher elevations, the absolute training intensity was progressively below what was possible at a lower altitude. At 4000 m, for example, the runners were only able to train at 39% of the sea level max $\dot{V}O_2$ compared to an intensity of 78% when training at sea level. This reduction in training intensity is possibly of such magnitude that an athlete may be unable to maintain peak condition for sea-level competition. Elite athletes might benefit by periodically returning from altitude to sea level for intensive

TABLE 24–3.
EFFECT OF ALTITUDE ON TRAINING INTENSITY OF SIX COLLEGIATE ATHLETES*

	Altitude (m)			
	300	**2300**	**3100**	**4000**
Intensity of Workout (% of max $\dot{V}O_2$ at 200 m)	78	60	56	39

* From Kollias, J., and Buskirk, E.R.: Exercise and altitude. IN Science and Medicine of Exercise and Sports, 2nd ed. Edited by W.R. Johnson, and E. R. Buskirk. New York, Harper and Row, 1974.

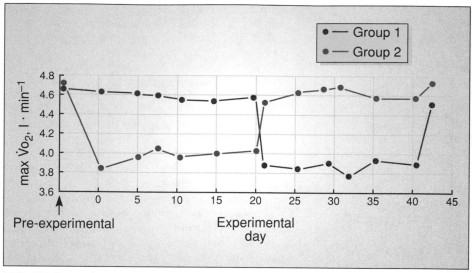

FIG. 24–3.
Maximal oxygen uptakes of two equivalent groups during training for 3 weeks at altitude and 3 weeks at sea level. Group 1 trained first at sea level and then continued training for 3 weeks at altitude. For group 2, the procedure was reversed as they trained first at altitude and then at sea level. (From Adams, W.C., et al.: Effects of equivalent sea-level and altitude training on $\dot{V}O_2$ max and running performance. J. Appl. Physiol., 39:262, 1975.)

training. This would help offset any "detraining" that takes place during a prolonged altitude stay. Such a procedure would not interfere with acclimatization and might even be beneficial to altitude performance.[12] If this approach is impractical, athletes should be sure to include speed work in their altitude training program to maintain muscle power.[5]

IS ALTITUDE TRAINING MORE EFFECTIVE THAN SEA-LEVEL TRAINING?

Equivalent groups were used to determine whether altitude training is more effective than sea-level training. Six highly trained middle-distance runners were trained at sea level for 3 weeks at 75% of sea level max $\dot{V}O_2$, whereas another group of six runners trained an equivalent distance at the same percentage of the max $\dot{V}O_2$ at 2300 m. The groups then exchanged training sites and continued training for 3 weeks at an intensity similar to that of the preceding group. Initially, 2-mile run times were 7.2% slower at altitude than at sea level. This improved about 2.0% for both groups after altitude training, but post-altitude performance at sea level was unchanged when compared to prealtitude sea-level runs. As shown in Figure 24–3, max $\dot{V}O_2$ for both groups at altitude was reduced initially by about 17.4%. This improved only slightly after 20 days of altitude training. When the runners were then measured at sea level, aerobic power was 2.8% below prealtitude sea-level values! Clearly, for these well-conditioned middle-distance runners, there was no synergistic effect of hard aerobic training at medium altitude over equivalently severe training at sea level.

SUMMARY

1. The progressive reduction in ambient P_{O_2} upon ascent to altitude eventually leads to inadequate oxygenation of hemoglobin. This produces noticeable performance decrements in aerobic activities at altitudes of 2000 m and higher. Short-term sprint performances that depend on energy from the stored phosphagens are not adversely affected at altitude.

2. The reduced P_{O_2} and accompanying hypoxia at altitude stimulate physiologic responses and adjustments that improve one's altitude tolerance during rest and exercise. The primary immediate responses are relative hyperventilation and increased submaximal cardiac output via an elevated heart rate.

3. The longer term acclimatization process involves physiologic and metabolic adjustments that greatly improve tolerance to altitude hypoxia. The main adjustments involve (1) re-establishing the acid-base balance of the body fluids, (2) increased formation of hemoglobin and red blood cells, and (3) possible changes in local circulation and cellular function. Adjustments (2) and (3) significantly facilitate oxygen transport and utilization.

4. The rate of altitude acclimatization depends on the altitude. Noticeable improvements are generally observed within several days, and the major adjustments require about 2 weeks, although 4 to 6 weeks may be required to acclimatize to relatively high altitudes.

5. For acclimatized men at a simulated altitude that approaches the summit of Mt. Everest alveolar P_{O_2} is about 25 mm Hg. This results in a 70% reduction in max $\dot{V}_{O_2}$ to about 15 ml $O_2 \cdot kg^{-1} \cdot min^{-1}$. Unacclimatized individuals at this altitude would become unconscious within 30 seconds.

6. Acclimatization does not fully compensate for the stress of altitude. Even after acclimatization, the max $\dot{V}_{O_2}$ is decreased about 2% for every 300 m above 1500 m. This is paralleled by a drop in performance in endurance-related activities.

7. The inability to achieve sea-level max $\dot{V}_{O_2}$ values at altitude is partially explained by the fact that the beneficial effects of acclimatization are somewhat offset by altitude-related decrements in physiologic function. The latter involve mainly a reduction in maximum heart rate and stroke volume.

8. Although acclimatization to altitude would certainly seem to enhance aerobic power and endurance performance upon return to sea level, research results generally do not support this contention.

9. Training at altitude provides no additional benefit to sea level performance compared to equivalent training at sea level.

REFERENCES

1. Abbrecht, P.H., and Littell, J.K.: Plasma erythropoietin in men and mice during acclimatization to different altitudes. *J. Appl. Physiol.*, 32:54, 1972.
2. Adams, W.C., et al.: Effects of equivalent sea-level and altitude training on $\dot{V}_{O_2}$ max and running performance. *J. Appl. Physiol.*, 39:262, 1975.
3. Alexander, J.K., et al.: Reduction of stroke volume during exercise in man following ascent to 3,100 m altitude. *J. Appl. Physiol.*, 23:849, 1967.
4. Balke, B., et al.: Effects of altitude acclimatization on work capacity. *Fed. Proc.*, 15:7, 1966.

5. Balke, B., et al.: Variation in altitude and its effects on exercise performance. In *Exercise Physiology*. Edited by H.B. Falls. New York, Academic Press, 1968.

6. Buskirk, E.R., et al.: Maximal performance at altitude and on return from altitude in conditioned runners. *J. Appl. Physiol.* 23:259, 1967.

7. Buskirk, E.R., et al.: Physiology and performance of track athletes at various altitudes in the United States and Peru. In *The International Symposium on the Effects of Altitude on Physical Performance*. Edited by R.F. Goddard. Chicago, The Athletic Institute, 1967.

8. Buskirk, E.R.: Work and fatigue in high altitude. In *Physiology of Work Capacity and Fatigue*. Edited by E. Simonsen. Springfield, IL, Charles C Thomas, 1971.

9. Cerretelli, P.: Limiting factors to oxygen transport on Mount Everest. *J. Appl. Physiol.*, 40:658, 1976.

10. Craig, A.B., Jr.: Olympics 1968: a post-mortem. *Med. Sci. Sports*, 1:177, 1969.

11. Cymerman, A., et al.: Operation Everest II: maximal oxygen uptake at extreme altitude. *J. Appl. Physiol.*, 66:2446, 1989.

12. Daniels, J., and Oldridge N.: The effects of alternate exposure to altitude and sea level on world-class middle-distance runners. *Med. Sci. Sports*, 2:107, 1970.

13. Dempsey, J.A.: Effects of acute through life-long hypoxic exposure on exercise pulmonary gas exchange. *Resp. Physiol.*, 13:62, 1971.

13a. Ferretti, G., et al.: Oxygen transport system before and after exposure to chronic hypoxia. *Int. J. Sports Med.*, 11:S15, 1990.

14. Gold, A.J., et al.: Effects of altitude stress on mitochondrial function. *Am. J. Physiol.*, 224:946, 1973.

15. Green, H.J., et al.: Operation Everest II: adaptations in human skeletal muscle. *J. Appl. Physiol.*, 66:2454, 1989.

16. Groves, B.M., et al.: Operation Everest II: elevated high-altitude pulmonary resistance unresponsive to oxygen. *J. Appl. Physiol.* 63:521, 1987.

17. Eaton, J.W., et al.: Role of red cell 2,3-diphosphoglycerate (DPG) in adaptation of men to altitude. *J. Lab. Clin. Med.*, 73:603, 1969.

18. Faulkner, J.A., et al.: Effects of training at moderate altitude on physical performance capacity. *J. Appl. Physiol.*, 23:85, 1967.

19. Faulkner, J.A., et al.: Maximum aerobic capacity and running performance at altitude. *J. Appl. Physiol.*, 24:685, 1968.

20. Gledhill, N.: Blood doping and related issues: a brief review. *Med. Sci. Sports Exerc.*, 14:183. 1982.

21. Gordon, A.S.: Hemopoietin. *Physiol. Rev.*, 39:1, 1959.

22. Grover, R.F., et al.: Decreased stroke volume during maximum exercise in man at high altitude. *Fed. Proc.*, 26:655, 1967.

23. Grover, R.F., and Reeves, J.T.: Exercise performance of athletes at sea level and 3,100 meters altitude. In *The Effects of Altitude on Physical Performance*. Edited by R.F. Goddard. Chicago, IL, Athletic Institute, 1967.

24. Grover, R.F., et al.: Alterations in coronary circulation of man following ascent to 3,100 m altitude. *J. Appl. Physiol.*, 4:832, 1976.

25. Hannon, J.P., et al.: Effects of altitude acclimatization on blood composition of women. *J. Appl. Physiol.*, 26:540, 1969.

26. Harris, C.W., and Hansen, J.E.: Electrocardiographic changes during exposure to high altitude. *Am. J. Cardiol.*, 18:183, 1966.

27. Hartley, L.H., et al.: Reduction of maximal exercise heart rate at altitude and its reversal with atropine. *J. Appl. Physiol.*, 36:362, 1976.

28. Hunt, J., and Hillary, E.: The Conquest of Everest. New York, E.P. Dutton Co., 1954.

29. Hurtado, A.: Animals in high altitudes: resident man. In *Handbook of Physiology*. Edited by D.B. Dill et al. Baltimore, Williams & Wilkins, 1964.

30. Hurtado, A., et al.: Mechanisms of natural acclimatization. Studies on the native resident of Morococha, Peru, at an altitude of 14,900 ft. *Technical Documentary Report No. SAM-TDR–56–1*, Washington, DC, USAF School of Aerospace Medicine, 1956.

31. Klausen, K.: Cardiac output in man in rest and work during and after acclimatization to 3800 m. *J. Appl. Physiol.*, 21:609, 1969.

32. Klausen, K., et al.: Effect of high altitude on maximal working capacity. *J. Appl. Physiol.*, 21:1191, 1966.

33. Klausen, K.: Exercise under hypoxic conditions. *Med. Sci. Sports*, 1:43, 1969.

34. Klausen, K., et al.: Exercise at ambient and high oxygen pressure at high altitude and at sea level. *J. Appl. Physiol.*, 29:456, 1970.
35. Kollias, J., and Buskirk, E.R.: Exercise at altitude. In *Science and Medicine of Exercise and Sports.* Edited by W.R. Johnson, and E.R. Buskirk. New York, Harper and Row, 1974.
36. Lenfant, C.P., et al.: Effect of chronic hypoxic hypoxia on the O_2-Hb dissociation curve and respiratory gas transport in man. *Resp. Physiol.*, 7:7, 1969.
37. Manier, G., et al.: Pulmonary gas exchange in Andean natives with excessive polycythemia—effect of hemodilution. *J. Appl. Physiol.*, 65:2107, 1988.
38. Maresch, C.M., et al.: Maximal exercise during hypobaric hypoxia (447 Torr) in moderate-altitude natives. *Med. Sci. Sports Exerc.*, 15:360, 1983.
38a. Mizuno, M., et al.: Limb skeletal muscle adaptation in athletes after training at altitude. *J. Appl. Physiol.*, 68:496, 1990.
39. Pugh, L.C.G.E.: Muscular exercise on Mount Everest. *J. Physiol* (London), 141:233, 1958.
40. Pugh, L.C.G.E.: Physiological and medical aspects of the Himalayan Scientific and Mountaineering Expedition, 1960–61. *Brit. Med. J.* 2:621, 1962.
41. Pugh, L.C.G.E.: Muscular exercise at great altitudes. *J. Appl. Physiol.*, 19:431, 1964.
42. Pugh, L.C.G.E.: Athletes at altitude. *J. Physiol.* (London), 192:619, 1967.
43. Pugh, L.C.G.E.: Animals in high altitudes: an above 5000 meters-mountain exploration. In *Handbook of Physiology.* Edited by D.B. Dill, et al. Baltimore, Williams & Wilkins, 1964.
44. Reeves, J.T.: Operation Everest II; preservation of cardiac function at extreme altitude. *J. Appl. Physiol.*, 63:531, 1987.
45. Reynafarje, C.: Myoglobin content and enzymatic activity of muscle and altitude adaptation. *J. Appl. Physiol.*, 17:301, 1962.
46. Reynafarje, C.: Hematologic changes during rest and physical activity in man at high altitude. In *The Physiological Effects of High Altitude.* Edited by W.H. Weihe. New York, Macmillan, 1964.
47. Rose, M.S., et al.: Operation Everest II: nutrition and body composition. *J. Appl. Physiol.*, 65:2545, 1988.
48. Saltin, B., et al.: Maximal oxygen uptake and cardiac output after 2 weeks at 4300 m. *J. Appl. Physiol.*, 25:400, 1968.
49. Schoene, R.B.: High-altitude pulmonary edena: the disguised killer. *Phys. Sportsmed.*, 16:103, 1988.
50. Schoene, R.B., et al.: Relationship of hypoxic ventilatory response to exercise performance on Mount Everest. *J. Appl. Physiol.*, 56:1478, 1984.
51. Smith, M.H., and Sharkey, B.S.: Altitude training: who benefits? *Phys. Sportsmed.*, 12:48, 1984.
52. Sorensen, S.C., and Severinghaus, J.: Respiratory sensitivity to acute hypoxia in man born at sea level living at high altitude. *J. Appl. Physiol.*, 25:211, 1968.
53. Squires, R. W., and Buskirk, E.R.: Aerobic capacity during acute exposure to simulated altitude, 914 to 2,286 meters. *Med. Sci. Sports Exerc.*, 14:36, 1982.
54. Stenberg, J., et al.: Hemodynamic response to work at simulated altitude, 4,000 m. *J. Appl. Physiol.*, 21:1589, 1966.
55. Terrados, N., et al.: Effects of training at simulated altitude on performance and muscle metabolic capacity in competitive road cyclists. *Eur. J. Appl. Physiol.*, 57:203, 1988.
56. Torre-Bueno, J.R., et al.: Diffusion limitation in normal humans during exercise at sea-level and simulated altitude. *J. Appl. Physiol.*, 58:989, 1985.
57. Vogel, J.A., et al.: Cardiovascular responses in man during exhaustive work at sea level and high altitude. *J. Appl. Physiol.*, 23:531, 1967.
58. Valdivia, E.: Total capillary bed in striated muscle of guinea pigs native to Peruvian mountains. *Am. J. Physiol.*, 194:585, 1958.
59. West, J.B.: Do climbs to extreme altitudes cause brain damage? *Lancet*, 2:387, 1986.
60. West, J.B., et al.: Arterial oxygen saturation during exercise at high altitude. *J. Appl. Physiol.*, 17:617, 1962.
61. West, J.B., et al.: Pulmonary gas exchange on the summit of Mount Everest. *J. Appl. Physiol.*, 55:678, 1983.
61a. Young, A.J.: Energy substrate utilization during exercise in extreme environments. In *Exercise and Sport Sciences Reviews.* Vol. 18. Edited by K.B. Pandolf. Baltimore, Williams & Wilkins, 1990.

62. Young, A.J., et al.: The influence of cardiorespiratory fitness on the decrement in maximal aerobic power at high altitude. *Eur. J. Appl. Physiol.*, 54:12, 1985.

63. Young, A.J., and Young, P.M. Human acclimatization to high terrestrial altitude. In *Human Performance Physiology and Environmental Medicine at Terrestrial Extremes.* Edited by K.B. Pandolf, et al. Indianapolis, Benchmark Press, 1988.

64. Young, P.M., et al. Altitude acclimatization attenuates plasma ammonia accumulation during submaximal exercise. *J. Appl. Physiol.*, 63:758, 1987.

Mechanisms of Thermoregulation

The requirements for **thermoregulation** can be considerable; the price for failure is death. A drop in deep body temperature of 10° C and an increase of only 5°C can be tolerated. This has been vividly illustrated by the fact that between 1960 and 1983, 70 football players have died as a direct result of excessive heat stress during practice or actual competition. Heat injury is also an unfortunate common occurrence in a variety of longer duration athletic events. Such tragedies are avoidable with the proper understanding of thermoregulation and the best ways to support these mechanisms. A major part of this responsibility rests with the people who organize and guide sport and physical activity programs.

THERMAL BALANCE

As shown in Figure 25–1, body temperature or, more specifically, the temperature of the deeper tissues or **core,** is in dynamic equilibrium as a result of a balance between factors that add and subtract body heat. This balance is maintained by the integration of mechanisms that alter heat transfer to the periphery or **shell,** regulate evaporative cooling, and vary the body's rate of heat production. If heat gain outstrips heat loss, as can readily occur in vigorous exercise in a warm environment, core temperature rises; in the cold, on the other hand, heat loss often exceeds heat production and core temperature falls.

Heat is gained directly from the reactions of energy metabolism. When muscles become active, their heat contribution is tremendous. From shivering alone, the total metabolic rate can increase 3- to 5-fold.[57] During sustained vigorous exercise, the metabolic rate increases to 20 to 25 times above the basal level; this theoretically can increase core temperature by about 1° C or 1.8° F every 5 minutes! Heat is also absorbed from the environment by solar radiation and from objects that are warmer than the body. Heat is lost

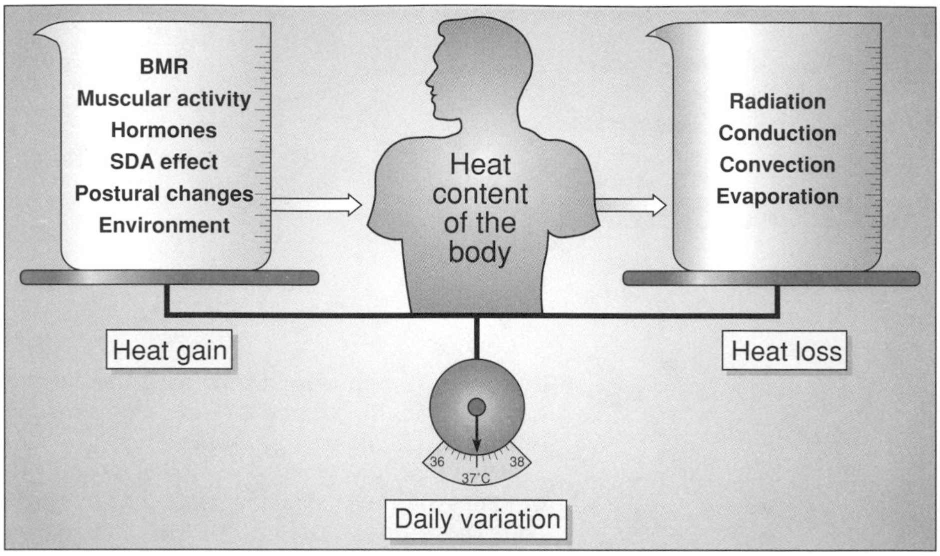

FIG. 25–1.
Factors that contribute to heat gain and heat loss so core temperature is balanced at approximately 37° C.

by the physical mechanisms of radiation, conduction, and convection, and the vaporization of water from the skin and respiratory passages.

Circulatory adjustments provide the "fine tuning" for temperature regulation. Heat is conserved by rapidly shunting blood deep to the cranial, thoracic, and abdominal cavities and portions of the muscle mass. This optimizes the insulation from subcutaneous fat and other portions of the body's shell. Conversely, when internal heat becomes excessive, peripheral vessels dilate and warm blood is channeled to the cooler periphery. The drive for thermal balance is so strong that it may elicit a sweat rate of 3.5 liters per hour in exercise in the heat or an oxygen consumption of 1000 ml per minute brought on by shivering in severe cold.

HYPOTHALAMIC REGULATION OF TEMPERATURE

The **hypothalamus** contains the coordinating center for the various processes of temperature regulation. This group of specialized neurons at the floor of the brain acts as a "thermostat" (usually set and carefully regulated at 37° C ± 1° C) that makes thermoregulatory adjustments to deviations from a temperature norm. Unlike our home thermostat, however, the hypothalamus cannot "turn off" the heat; it can only initiate responses to protect the body from a buildup or loss of heat.

Heat-regulating mechanisms are activated in two ways: (1) by thermal receptors in the skin that provide input to the central control center, and (2) by direct stimulation of the hypothalamus through changes in blood temperature perfusing these areas.

Peripheral thermal receptors, or sensors responsive to rapid changes in heat and cold, are distributed predominantly as free nerve endings in the skin. The cutaneous cold receptors are generally toward the skin surface and are more abundant than the deeper heat receptors. They play an important role in initiating the regulatory response to a cold environment. The cutaneous thermal receptors act as an "early warning system" that relays sensory information to the hypothalamus and cortex to bring about appropriate heat-conserving or heat-dissipating adjustments and to cause the individual consciously to seek relief from a thermal challenge.

A **central regulatory center** plays the most important role in maintaining thermal balance. In addition to peripheral input, cells in the anterior portion of the hypothalamus are themselves capable of detecting changes in blood temperature. These cells then activate other hypothalamic regions to initiate coordinated responses for heat conservation (posterior hypothalamus) or heat loss (anterior hypothalamus). In contrast to the importance of peripheral receptors in detecting cold, **body warmth is monitored mainly by the temperature of the blood perfusing the hypothalamus.**

THERMOREGULATION IN COLD STRESS: HEAT CONSERVATION AND HEAT PRODUCTION

Normally, the gradient for heat transfer is from the body to the environment, and core temperature is maintained without excessive physiologic strain. In extreme cold, however, excessive heat loss can occur, especially at rest. In this situation, heat production is increased and heat loss is retarded as adjustments are made to prevent a fall in internal temperature.

VASCULAR ADJUSTMENTS

Stimulation of cutaneous cold receptors causes constriction of peripheral blood vessels, which immediately reduces the flow of warm blood to the body's cooler surface and redirects it to the warmer core. Consequently, skin temperature falls toward the ambient temperature, and the insulatory benefits of skin and subcutaneous fat are used to their maximal advantage. A fat person, therefore, can derive great benefits from this heat-conserving mechanism when exposed to cold stress.[70,79,91] For a thinly-clad person who is not overfat, the regulation of cutaneous blood flow generally provides for effective thermoregulation at ambient temperatures between 25 and 29° C (77 to 84° F).

MUSCULAR ACTIVITY

Although significant metabolic heat is generated through shivering, the greatest contribution of muscle to defense against cold occurs during physical activity. Exercise energy metabolism can sustain a constant core temperature in air temperature as low as −30° C (−22° F) without the need for heavy, restrictive clothing. It should be noted, however, that the thermoregulatory defense against cold is mediated by internal temperature and **not** by the heat production in the body per se.[81] Thus, shivering is observed even during exercise if the core temperature is low. As a result, exercise oxygen consumption is proportionally higher (due directly to shivering) in cold stress than it is during the same exercise in a warmer environment.[32,70]

HORMONAL OUTPUT

During cold exposure, increased heat production is due partially to the action of the two hormones of the adrenal medulla, epinephrine and norepinephrine.

It is also possible that prolonged cold stress increases the release of the thyroid hormone, thyroxine, that leads to sustained elevation in resting metabolsim.

THERMOREGULATION IN HEAT STRESS: HEAT LOSS

The mechanisms for thermoregulation are primarily geared to protect against overheating.[106] This is particularly important during exercise in hot weather, when an inherent competition exists between mechanisms that maintain a large muscle blood flow and those that provide for adequate thermoregulation. Figure 25–2 illustrates the potential avenues for heat exchange in an exercising human. **Body heat may be lost by radiation, conduction, convection, and evaporation.**

HEAT LOSS BY RADIATION

Objects are continually emitting electromagnetic heat waves. Because our bodies are usually warmer than the environment, the net exchange of radiant heat energy is through the air to the solid, cooler objects in the environment. This form of heat transfer does not require molecular contact with the warmer object and is essentially the means by which the sun's rays warm the earth. A person can remain warm by absorbing radiant heat energy from direct

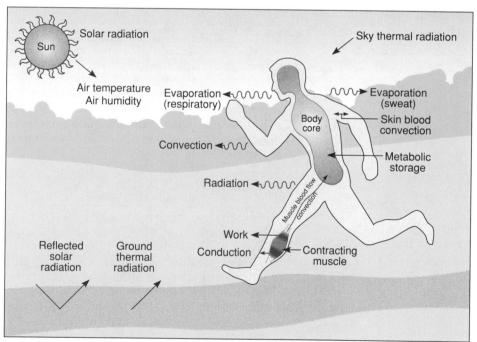

FIG. 25–2.
Heat production within working muscle and its subsequent transfer to the core and skin. Under appropriate environmental conditions, the excess body heat will be dissipated to the environment. (From Gisolfi, C.V., and Wenger, C.B.: Temperature regulation during exercise: old concepts, new ideas. In Exercise and Sport Sciences Reviews. Vol. 12. Edited by R.L. Terjung. New York, Macmillan Publishing Co., 1984.)

sunlight (or reflected from the snow, sand, or water), even in subfreezing temperatures. When the temperature of objects in the environment exceeds skin temperature, radiant heat energy is absorbed from the surroundings. Under these conditions, the only avenue for heat loss is from evaporative cooling.

HEAT LOSS BY CONDUCTION

This process of heat exchange involves the direct transfer of heat through a liquid, solid, or gas from one molecule to another. Although most of the body heat is transported to the shell by the circulation, a small amount continually moves by conduction directly through the deep tissues to the cooler surface. Here, heat loss by conduction involves the warming of air molecules and cooler surfaces in contact with the skin.

The rate of conductive heat loss depends on the temperature gradient between the skin and surrounding surfaces and their thermal qualities. For example, heat loss in water can be considerable.[32] This is clearly illustrated by placing one hand in water at room temperature. The hand in water feels much colder than the hand in air—even though the water is the same temperature as the air. **This occurs because water can absorb several thousand times more heat than air and conduct it away from the warm body.** For this reason, sitting in an indoor swimming pool is more uncomfortable than sitting on the pool deck, even though the air and water are the same temperature.

HEAT LOSS BY CONVECTION

The effectiveness of heat loss by conduction depends on how rapidly the air (or water) adjacent to the body is exchanged once it becomes warmed. If air movement or convection is slow, the air next to the skin is warmed and acts as a zone of insulation. This minimizes further conductive heat loss. Conversely, if the warmer air surrounding the body is continually replaced by cooler air (as occurs on a breezy day or in a room with a fan or during running), heat loss increases as convective currents carry the heat away. Air currents at 4 miles per hour are about twice as effective for cooling as air currents at 1 mile per hour. This is the basis of the **Wind Chill Index** (see p. 573), which gives the equivalent still-air temperature for a particular ambient temperature at different wind velocities. In water, convection is also an important factor because heat is lost more rapidly via convection while swimming than while lying motionless in the water.[79]

HEAT LOSS BY EVAPORATION

Evaporation provides the major physiologic defense against overheating. Heat is continually transferred to the environment as water is vaporized from the respiratory passages and skin surface. For each liter of water that vaporizes, 580 kcal are extracted from the body and transferred to the environment.

Approximately 2 to 4 million sweat glands are distributed throughout the surface of the body. In response to heat stress, these **eccrine glands,** controlled by cholinergic sympathetic nerve fibers, secrete large quantities of weak saline solution (hypotonic—0.2 to 0.4% NaCl). When sweat comes in contact with the skin, a cooling effect occurs as sweat evaporates. The cooled skin in turn serves to cool the blood that has been shunted from the interior to the surface. In addition to heat loss through sweating, about 350 ml of water seep through the skin each day and evaporate to the environment. Also, about 300 ml of water vaporize from the moist mucous membranes of the respiratory passages. This is seen as "foggy breath" in very cold weather.

Heat Loss at High Ambient Temperatures

As ambient temperature increases, the effectiveness of heat loss decreases by conduction, convection, and radiation. When ambient temperature exceeds body temperature, heat is actually gained by these mechanisms of thermal transfer. In such environments (or when conduction, convection, and radiation are inadequate to dissipate a large metabolic heat load), the **only means** for heat dissipation is by sweat evaporation and the small contribution to cooling provided by the vaporization of water from the respiratory tract. In fact, the rate of sweating increases directly with the ambient temperature.[85]

Heat Loss in High Humidity

The total sweat vaporized from the skin depends on three factors: (1) the surface exposed to the environment, (2) the temperature and humidity of the ambient air, and (3) the convective air currents about the body. **By far, relative humidity is the most important factor that determines the effectiveness of evaporative heat loss.** Relative humidity is defined as the ratio of water in ambient air to the total quantity of moisture that can be carried in air at a particular ambient temperature, expressed as a percentage. For example, 40% relative humidity means that ambient air contains only 40% of the air's moisture-carrying capacity at the specific temperature. When humidity is high the ambient vapor pressure approaches that of the moist skin (about 40 mm Hg) and evaporation is greatly reduced. Thus, this avenue for heat loss is essentially closed, even though large quantities of sweat bead on the skin and eventually roll off. This form of sweating represents a useless water loss that can lead to a dangerous state of dehydration and overheating.

Evaporative cooling is also thwarted by continually drying the skin with a towel before sweat has a chance to evaporate. **Sweat per se does not cool the skin; evaporation cools the skin.** As long as the humidity is low, relatively high environmental temperatures are tolerated. For this reason, hot, dry desert climates are more comfortable than cooler but more humid tropical climates.

INTEGRATION OF HEAT-DISSIPATING MECHANISMS

The mechanisms for heat loss are the same whether the heat load is imposed internally (metabolic heat) or externally (environmental heat).

Circulation

The circulatory system serves as the "workhorse" in maintaining thermal homeostasis. At rest, the heart rate and cardiac output increase while superficial venous and arterial blood vessels dilate to divert warm blood to the body shell. This is seen as a flushed or reddened face on a hot day or during vigorous exercise. With extreme heat stress, 15 to 25% of the cardiac output passes through the skin. This greatly increases the thermal conductance of peripheral tissues and favors radiative heat loss to the environment, especially from the hands, forehead, forearms, ears, and tibial areas.

Evaporation

Sweating begins within 1.5 seconds after the start of vigorous exercise,[13] and, after about 30 minutes, reaches an equilibrium that is in direct relation to the exercise load.[53] An effective heat defense is established when evaporative cooling is combined with a large cutaneous blood flow. The cooled peripheral blood then returns to the deeper tissues to pick up additional heat.

Hormonal Adjustments

Because both water and electrolytes are lost through sweating, hormonal adjustments are initiated in heat stress as the body attempts to conserve salts and fluid.[36,50] The pituitary gland releases vasopressin or **antidiuretic hormone** (ADH) to increase water reabsorption from the kidneys. This causes the urine to become more concentrated during heat stress. Concurrently, during repeated days of exercise in the heat,[108] or with just a single bout of exercise,[30] the sodium-conserving hormone **aldosterone** is released from the adrenal cortex. This hormone acts on the renal tubules to increase the reabsorption of sodium. Through as yet undefined mechanisms, the sodium concentration in sweat is also decreased during repeated heat exposure to aid in conserving electrolytes.

EFFECTS OF CLOTHING ON THERMOREGULATION

Clothing insulates the body from its surroundings. It may reduce radiant heat gain in a hot environment or retard conductive and convective heat loss in the cold.

COLD-WEATHER CLOTHING

In providing insulation from the cold, the mesh of the cloth fibers traps air that then becomes warm. Because both cloth and air are poor heat conductors, a barrier to heat loss is established; **the thicker the zone of trapped air next to the skin, the more effective is the insulation.** For this reason, several layers of light clothing, or garments lined with animal fur, feathers, or synthetic fabrics (with numerous layers of trapped air), provide much greater insulation than a single bulky layer of winter clothing. The clothing layer against the skin must also be effective in transporting moisture away from the body's surface to the next clothing layer, the insulating layer, for subsequent evaporation. Wool or synthetics such as polypropylene that insulate well and dry quickly can serve this purpose. A wool cap contributes considerably to heat conservation because nearly 30 to 40% of body heat is lost through the highly vascularized head region that represents only about 8% of the body's surface area. Conversely, cooling the head during exercise in hot weather is quite effective in reducing symptoms of thermal discomfort. When clothing becomes wet, through either external moisture or condensation from sweating, it loses nearly 90% of its insulating properties and actually facilitates heat transfer from the body because water conducts heat much faster than air.

When working in cold air, the problem is usually not one of adequate insulation but rather the dissipation of metabolic heat through a thick air-clothing barrier.[19] Cross-country skiers alleviate this problem by removing layers of clothing as the body becomes warm. In this way, core temperature is maintained without reliance on evaporative cooling. **The ideal winter garment in dry weather is impermeable to air movement but permits the escape of water vapor from the skin through the clothing if sweating should occur.**

WARM-WEATHER CLOTHING

Dry clothing, no matter how light, retards heat exchange compared to the same clothing soaking wet. The practice of switching to a dry tennis, basketball, or football uniform in hot weather makes little sense from the standpoint of temperture regulation. **Evaporative heat loss occurs only when the clothing becomes wet throughout.** A dry uniform simply prolongs the time lag between sweating and cooling.

Different materials absorb water at different rates. Cottons and linens absorb moisture readily. On the other hand, heavy "sweat shirts" and clothing made of rubber or plastic produce high relative humidity close to the skin and retard the vaporization of moisture from the skin surface; this significantly inhibits or even prevents evaporative cooling. Warm-weather clothing should be loose fitting to permit the free circulation of air between the skin and environment to promote water movement away from the skin. Color is also important because dark colors absorb light and add to the radiant heat gain whereas light colors reflect heat rays.

FOOTBALL UNIFORMS

Of all athletic uniforms and equipment, those used in football present the most significant barrier to heat dissipation.[49,69] Even with loose-fitting porous jerseys, the wrappings, the padding (with its plastic covering), the helmet, and other objects of "armor" effectively seal off 50% of the body surface from the benefits of evaporative cooling. To this is added the metabolic cost of carrying the 6 or 7 kg of equipment, frequently over a relatively hot artificial playing surface. This situation is further magnified by the large size of these athletes, especially the offensive and defensive linemen who possess a relatively small surface area-to-mass ratio and a higher percentage of body fat than teammates at other positions.[115]

The metabolic and thermal stress provided by the football uniform is shown in Figure 25–3. In this experiment, 9 men were tested at 25.6° C (78°F) and 35% relative humidity while running for 30 minutes.

In one test, the men wore only shorts; in another, they wore the complete football uniform including helmet and plastic padding. In a third series of measures, they wore shorts and carried a backpack containing 6.2 kg, exactly the weight of the uniform and equipment.

The effects of the uniform on heat dissipation are clear. Rectal temperature and skin temperature were significantly higher in both exercise and recovery than either of the other exercise conditions. The temperature directly beneath the padding averaged only 1° C less than rectal temperature. This meant that subcutaneous blood in these areas was cooled only about one fifth as much as skin surface directly exposed to the environment. Because rectal temperature remained elevated in recovery with uniforms, it appears that a

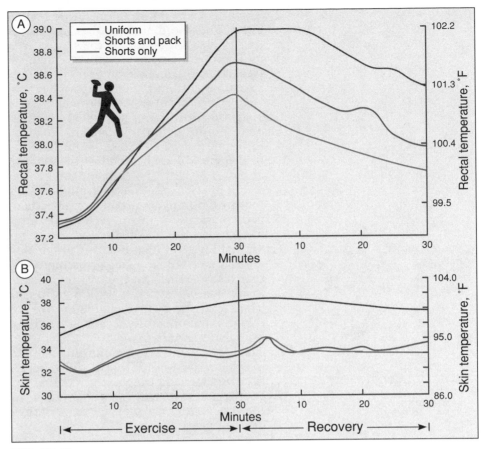

FIG. 25–3.
Effects of full football uniform and its equivalent weight on (A) rectal temperature and (B) skin temperature during exercise. Subjects ran at 9.6 km · h^{-1} for 30 minutes at 25.6° C and 35% relative humidity. Owing to its effect in retarding evaporative cooling, the uniform caused the largest heat stress as indicated by significant elevations in rectal and skin temperatures. (From Mathews, D.K., et al.: Physiological responses during exercise and recovery in a football uniform. J. Appl. Physiol., 26:611, 1969.)

"rest" period is of limited value in normalizing thermal regulation unless the uniform is removed.

As shown by the blue line, a large portion of the heat load was provided simply by the weight of the uniform. Skin temperatures, however, were much cooler and sweat rates less when the uniform was not worn. Without the uniform, evaporation from the skin was relatively free, whereas the uniform insulated the athlete and reduced the evaporative surface.

SUMMARY

1. Humans can tolerate only relatively small variations in internal temperature. Consequently, exposure to heat or cold stress initiates thermoregulatory mechanisms that generate and conserve heat at low ambient temperatures and dissipate heat at high temperatures.

2. The "thermostat" for temperature regulation is located in the hypothalamus. This coordinating center initiates adjustments in response to input from thermal receptors in the skin as well as changes in the temperature of blood perfusing hypothalamic regions.

3. Heat conservation in cold stress is achieved by vascular adjustments that shunt blood from the cooler periphery to the warmer deep tissues of the body's core. If this action is ineffective, shivering is initiated to provide a significant input of metabolic heat. Hormones that cause a sustained elevation in resting metabolism are also released.

4. In response to heat stress, warm blood is diverted from the body's core to the shell. Body heat is lost by radiation, conduction, convection, and evaporation. At high ambient temperatures and during exercise, evaporation provides the major physiologic defense against overheating.

5. In warm, humid environments, the effectiveness of evaporative heat loss is dramatically reduced. This makes a person especially susceptible to a dangerous state of dehydration and spiraling core temperature.

6. Several layers of light clothing provide a relatively thick zone of trapped air against the skin. This gives more effective insulation from the cold than a single thick layer of winter clothing. When clothing becomes wet, insulation is lost and heat flow from a body is greatly facilitated.

7. The metabolic heat generated during vigorous exercise generally maintains core temperature in cold air environments, even if the person wears only little clothing.

8. The ideal warm-weather clothing is light-weight, loose-fitting, and light in color. Even wearing this clothing, heat loss is retarded until the clothing becomes wet and evaporative cooling can proceed.

9. Football uniforms impose a significant barrier to heat dissipation because they effectively seal off about 50% of the body's surface from the benefits of evaporative cooling.

P A R T
2

Thermoregulation and Environmental Stress During Exercise

EXERCISE IN THE HEAT

When exercising in the heat, the body is faced with two competitive demands: (1) the muscles require oxygen to sustain energy metabolism and, equally important, (2) metabolic heat must be transported by the blood from the deep tissues to the periphery. Consequently, this blood cannot deliver its oxygen to the working muscles.

In addition to vascular adjustments, the dissipation of metabolic heat during exercise in hot weather almost totally depends on the refrigeration mechanism of evaporative cooling. A price is paid, however, as demands are placed on the body's fluid reserves and a relative state of dehydration frequently occurs. Excessive sweating leads to more serious fluid loss and an accompanying reduction in plasma volume. This can cause circulatory failure, and core temperature may rise to lethal levels.

CIRCULATORY ADJUSTMENTS

For reasons not fully understood, the stroke volume of the heart is usually lower during exercise in the heat,[96,121] and is compensated for by a proportionate increase in heart rate at all levels of submaximal exercise. Consequently, cardiac outputs in submaximal exercise are similar in hot and cool environments.[101,102] In maximal exercise, however, this reflex compensatory increase in heart rate is insufficient to offset the stroke volume decrease and maximal cardiac output and associated aerobic power are reduced.

Constriction and Dilation

In the heat, adequate cutaneous and muscle blood flow are achieved at the expense of other tissues that can temporarily compromise their blood supply.[99] For example, vasodilatation of the subcutaneous vessels is rapidly countered by compensatory constriction of the splanchnic vascular bed and renal tissues. Such a significant and often prolonged reduction in blood flow to these tissues may account for the large number of liver and kidney complications noted with exertional heat stress.

Maintenance of Blood Pressure

Aside from redirecting blood to areas in great need, vasoconstriction in the viscera serves to increase total vascular resistance. In this way, arterial blood pressure is maintained during work in the heat. During heavy exercise in the heat with its accompanying dehydration, relatively less blood is shunted to peripheral areas for heat dissipation.[47] This probably reflects the body's attempt to maintain cardiac output in the face of a diminishing plasma volume caused by sweating. In strenuous exercise, circulatory regulation and muscle blood flow probably take precedence over temperature regulation.[80]

Even when submaximal exercise is well tolerated in the heat, the work is generally accomplished with a greater dependence on anaerobic metabolism than in cooler conditions.[123a] This results in the early accumulation of lactic acid[20] and encroachment on glycogen stores during submaximal exercise.[44] An increased lactic acid level is probably due to (1) decreased lactate uptake by the liver because significant reductions in hepatic blood flow occur during exercise in the heat, and (2) reduced muscle circulation because large quantities of blood are shunted to the periphery for heat dissipation. Both of these factors may be responsible for early fatigue during only moderate exercise in the heat.

CORE TEMPERATURE DURING EXERCISE

The heat generated by exercising muscles can raise body temperature to fever levels that would incapacitate a person if caused by external heat stress alone.[6] In fact, champion distance runners show no ill effects from rectal temperatures as high as 41° C (105.8° F) recorded at the end of a 3-mile race.[63]

Higher Core Temperatures Are Regulated. Within limits, the increase in core temperature in exercise does not reflect a failure of the heat-

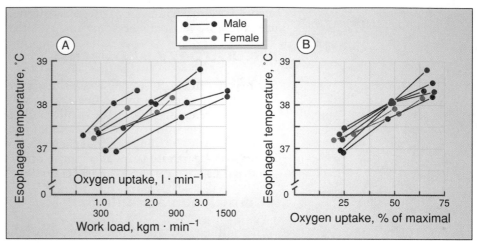

FIG. 25–4.
Relationship between esophageal temperature and (A) oxygen consumption-work intensity and (B) oxygen uptake as a percent of each person's max $\dot{V}O_2$ (From Saltin, B., and Hermansen, L.: Esophageal, rectal, and muscle temperature during exercise. J. Appl. Physiol., 21:1757, 1966.)

dissipating mechanisms. To the contrary, it is a well regulated response that even occurs during exercise in the cold.[40,92] More than likely, a modest rise in core temperature reflects a favorable adaptative response that creates an optimal thermal environment for physiologic and metabolic function.

Figure 25–4A illustrates the relationship between esophageal temperature and oxygen consumption for five men and two women of varying fitness levels during exercise of increasing severity. For all subjects, body temperature increases to a higher level as the exercise becomes more intense. Wide variability in temperature response exists, however, between subjects. When temperature is plotted in relation to oxygen consumption expressed as a percentage of each person's max $\dot{V}O_2$ (Fig. 25–4B), the lines move closer together. **It is the relative work load (i.e., the percentage of one's capacity) that determines the change in core temperature with exercise.**[7,41,54] In general, work in a comfortable environment at 50% of max $\dot{V}O_2$ increases temperture to a new steady level of about 37.3° C (99° F), whereas work at 75% of maximum elevates core temperature to 38.5° C (101° F), regardless of the absolute level of oxygen consumption. Thus, at the same percentage of max $\dot{V}O_2$, a fit person generates more energy in exercise yet still has about the same core temperature as a less fit counterpart.[103] The extra metabolic heat is dissipated in a larger sweat output by the person working at the higher absolute work load. Of course, at the same workload, the trained person exercises with a lower core temperature.

WATER LOSS IN THE HEAT—DEHYDRATION

In a few hours of hard exercise in the heat, water loss or dehydration can reach proportions that impede heat dissipation and severely compromise cardiovascular function and work capacity.

Magnitude of Fluid Loss

For an acclimatized person, water loss by sweating may reach a peak of about 3 liters per hour during severe work and average nearly 12 liters (26 lb) on a daily basis. Furthermore, several hours of intense sweating can cause sweat-gland fatigue that ultimately leads to the inability to regulate core temperature. Elite marathon runners frequently experience fluid losses in excess of 5 liters during competition.[92] For these athletes, this loss represents 6 to 10% of their body mass. For a slower marathon or ultramarathon, the average fluid loss probably rarely exceeds 500 ml per hour.[86]

Distance running is not the only sport in which there are large sweat outputs. Football, basketball, and hockey players may lose similar large quantities of fluid in the course of a contest. It has also been reported that high school wrestlers lose 9 to 13% of their preseason body mass prior to certification; the greatest portion of this weight loss comes from voluntarily reducing water intake and excessive sweating just prior to the weigh-in.[124] **In the desire to "make weight," high school and collegiate wrestlers usually compete in a dehydrated state.[125]**

Significant Consequences

As dehydration progresses and plasma volume drops, sweating is reduced and thermoregulation becomes progressively more difficult. A pre-exercise dehydration equivalent to 5% of body mass significantly increases rectal temperature and heart rate, and decreases sweat rate for both men and women who exercise in either a comfortable or hot environment compared to exercise when they are normally hydrated.[104] The elevated heart rate in the dehydrated state is attributed to a reduced central blood volume that leads to a lower ventricular filling pressure and stroke volume. An elevated core temperature is related to a reduction in both sweating and blood flow to the skin.[46,82]

Fluid loss is particularly apparent during exercise in hot, humid environments. In this situation, the effectiveness of evaporative cooling is thwarted by the high vapor pressure of ambient air. As illustrated in Figure 25–5, there is a linear relationship between sweat rate, both at rest and during exercise, and the air's moisture content.[63] Ironically, the excessive output of sweat in high humidity contributes little to cooling, because evaporation is at a minimum. In this regard, clothing that retards the rapid diffusion and subsequent evaporation of sweat creates an extremely humid microclimate that envelops a large portion of the body. Such clothing "encourages" dehydration and overheating.

Physiologic and Performance Decrements

A fluid loss equivalent to as little as 1% of body mass is associated with a significant increase in rectal temperature compared to the same exercise with normal hydration.[20,40] When water loss reaches 4 to 5% of body mass (a value commonly seen among high school wrestlers[112]), a definite impairment is noted in physical work capacity[94,102] and physiologic function.[16,21] Although wrestlers have about 5 hours between weigh-in and match time, evidence is not conclusive that this time is adequate to assure complete rehydration and electrolyte balance at the time of competition.[120] Because a large portion of water loss through sweating comes from the blood,[2,25,27] circulatory capacity is adversely affected as sweat loss progresses (if water

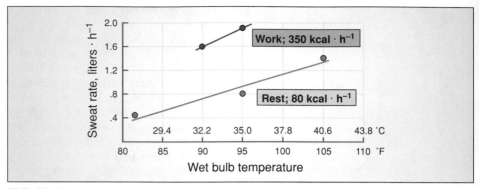

FIG. 25–5.
The effect of humidity (wet-bulb temperature) on sweat rate at rest and during work in the heat. Ambient temperature (dry-bulb) at all times was 43.3° C (110° F). (From Lampietro, P.F.: Exercise in hot environments. In Frontiers of Fitness. Edited by R.J. Shephard. Springfield, Ill., Charles C Thomas, 1971.)

is not continually replenished). This is manifested by a decrease in circulating blood volume, a fall in stroke volume and a compensatory increase in heart rate, and a general deterioration in circulatory and thermoregulatory efficiency during exercise. In terms of performance, a 48% reduction in walking endurance was noted when subjects were dehydrated to 4.3% of body mass; concurrently, max $\dot{V}O_2$ decreased by 22%.[33] In these same experiments, endurance performance and max $\dot{V}O_2$ were reduced by 22% and 10%, respectively, when dehydration averaged only 1.9% of body mass. **Clearly, dehydration reduces the capability of the circulatory and temperature-regulating systems to meet the metabolic and thermal stress of exercise.**

Use of Diuretics

A greater percentage of water is drawn from the plasma if body water is lost as a result of diuretic-induced dehydration.[21] Thus, wrestlers who use these drugs to lose body water rapidly to "make weight" are at a distinct disadvantage because of a disproportionate reduction in plasma volume and its negative effects on thermoregulation and cardiovascular function. Chemicals to induce vomiting and diarrhea are also used by athletes to cause weight loss. Not only does this abuse lead to dehydration, but it may also cause excessive potassium loss with accompanying muscle weakness. The use of such drugs would clearly give the competitive "edge" to the opponent. In addition, drugs that induce a diuretic effect may also cause a marked impairment in neuromuscular function that is not noted when comparable fluid loss is brought on by exercise.[17]

WATER REPLACEMENT

The primary aim of fluid replacement is to maintain plasma volume so that circulation and sweating can progress at optimal levels. Prevention of dehydration and its consequences, especially hyperthermia, can only be achieved with an adequate and strictly adhered to water replacement schedule.[18,52,53] This may be "easier said than done," because some coaches and athletes feel that ingesting water hinders performance. Also, in some sports, water intake is actually prohibited. In international marathon competition prior to the 1976 Olympics, drinking fluids was prohibited during the first 10 km of this 42 km race! For wrestlers, dehydration is a "way of life," because young boys and men lose considerable weight to wrestle in a lower-weight class. This may also be the case for young and older ballet dancers, who have a continual preoccupation with body mass regulation so they appear thin. The enlightened exercise specialist must be keenly aware of the importance of proper hydration for thermoregulation, exercise performance, and safety.

Periodic application of cold towels to the forehead and abdomen during exercise or a cold shower prior to exercise in the heat have been proposed as practical approaches to reducing heat stress.[42] These remedies may result in a slower exercise heart rate, lower rectal temperature, and reduced fluid loss. It is assumed that these "cold treatments" cool the blood in the periphery and facilitate heat transfer from the body surface. It has been shown, however, that the spraying of 50 ml of 30°C water on the skin every 10 minutes during a 2-hour treadmill run was of no thermoregulatory benefit compared to the same exercise without skin wetting.[12] **It must be stressed that the most effective defense against heat stress is adequate hydration.** This is achieved by balancing water loss with water intake, not by pouring water over the head or body. There is simply no evidence to indicate that restricting fluid intake during training in some way makes an athlete better able to adjust to subsequent work in the heat. **A well-hydrated athlete always functions at a higher level than one who exercises in a dehydrated state.**

Practical Recommendations for Fluid Replacement

Ingestion of "extra" water or **hyperhydration** prior to exercise in the heat provides some protection because it delays the development of dehydration, increases sweating during exercise, and brings about a smaller rise in core temperature.[53,75,82] In this regard, it would be wise to consume 400 to 600 ml (13 to 20 oz.) of cold water 10 to 20 minutes **before** exercising in the heat. This procedure, however, does not replace the need for continual fluid replacement and is not as effective in maintaining thermal balance as consuming an equal volume of water during the exercise.[52] **In activities like distance running, matching fluid loss with fluid intake may be virtually impossible because only about 800 ml of fluid can be emptied from the stomach each hour during vigorous exercise.** This is insufficient to match a water loss that may average nearly 2 liters per hour. Consequently, athletes must be carefully monitored during exercise even if they are permitted free access to water.

Gastric Emptying.   Cold fluids (5° C; 41°F) are emptied from the stomach at a faster rate than fluids at body temperature.[26] The volume of fluid in the stomach is also of importance because gastric emptying speeds up for each 100 ml increase in gastric volume up to 600 ml. A volume of about 250 ml (8.5 oz) ingested at 10- to 15-minute intervals is probably a realistic goal because larger volumes tend to produce feelings of a "full stomach." Thus, to obtain a high rate of fluid absorption, the stomach should remain partially filled and the fluid ingested should be relatively cold. Within a broad range of submaximal intensities, the movement of water from the stomach is not retarded by physical activity.[45] Gastric emptying is actually facilitated while running compared to rest, perhaps the result of increased mechanical movement of fluid during the run.[84]

Of importance is the observation that gastric emptying is **retarded** when the ingested fluid contains concentrated simple sugars, whether in the form of glucose, fructose, or sucrose![31] A 10% glucose solution, for example, cuts the rate of gastric emptying by one-half and a 40% sugar solution is emptied from the stomach at a rate that is only one fifth that of plain water! With intense exercise, even a small amount of simple sugar blocks fluid movement from the stomach into the intestinal tract, and the addition of salt to the fluid further depresses emptying of the stomach. From a practical standpoint, during exercise in the heat, when the need for water greatly exceeds the need for carbohydrate supplementation, concentrated sugar solutions may hinder water replenishment.

Other Considerations. Although gastric emptying may be influenced by the **osmolality** (number of particles in solution) of the ingested fluid, the precise relationship between the rate of fluid movement out of the stomach and its subsequent absorption in the intestines remains unclear. Impairment in physiologic function and thermoregulation, when consuming beverages of moderate carbohydrate and electrolyte content, has not been consistently observed.[18a,35,78,89] In fact, consuming glucose polymer-containing drinks (see next section) of 7.5% concentration in small amounts every 15 minutes during exercise does not appear to inhibit gastric emptying.[76] This means that the endurance benefits of regular sugar feedings during exercise can be attained without compromise in the all-important aspect of fluid replacement. More concentrated beverages provide a different picture, however. Recent findings indicate that while 12 and 18% polymerized carbohydrate solutions deliver more carbohydrate compared to a less-concentrated drink, fluid replacement was impaired.[77] **This suggests that "athletic drinks" with sugar concentrations above 7 to 8% could pose significant adverse consequences for thermoregulation during prolonged exercise in the heat.**

On the other hand, during prolonged exercise in a cool environment, fluid loss from sweating may not be great. Here, a reduction in gastric emptying and fluid uptake can be tolerated and a strong sugar solution may be beneficial (15 to 30 g per 100 ml water). One should keep in mind, however, that it may take 20 to 30 minutes for the carbohydrate to reach the muscles after it enters the stomach. The possible "trade-off" between the composition of the fluid ingested and the rate of gastric emptying must be evaluated on the basis of environmental and metabolic demands. **In terms of survival, fluid replacement is primary during prolonged exercise in the heat.**

Glucose Polymers. Because both fluid and carbohydrate depletion can limit exercise performance, it is encouraging to note that the negative effects of simple sugar solutions on gastric emptying can be greatly reduced by using a solution of maltodextrins, a polymerized form of glucose[48,107] Polymerization of glucose maximizes carbohydrate content while minimizing osmolality.[43] For example, a glucose polymer solution has one fifth the osmolality of a solution of simple sugars. In effect, a 5% solution of polymerized glucose gives the athlete water and carbohydrate at a more rapid rate compared to a drink of similar carbohydrate content composed of mono- and disaccharides.

Adequacy of Rehydration

Changes in body mass should be used to indicate water loss in exercise and the adequacy of rehydration during and following exercise. Recommendations for fluid intake with body mass loss during exercise are presented

TABLE 25–1.
RECOMMENDED FLUID AVAILABILITY AND INTAKE FOR A STRENUOUS 90-MINUTE ATHLETIC PRACTICE*

Weight Loss		Minutes Between Water Breaks	Fluid Per Break		Fluid Availability for an 11-Member Squad	
lb	(kg)		oz	(ml)	Gallons	(Liters)
8	(3.6)	No practice	—		—	
7½	(3.4)	recommended	—		—	
7	(3.2)	10	8–10	(266)	6½–8	(27.4)
6½	(3.0)	10	8–9	(251)	6½–7	(25.5)
6	(2.7)	10	8–9	(251)	6½–7	(25.5)
5½	(2.5)	15	10–12	(325)	5½–6½	(22.7)
5	(2.3)	15	10–11	(311)	5½–6	(21.8)
4½	(2.1)	15	9–10	(281)	5 –5½	(19.9)
4	(1.8)	15	8–9	(251)	4½–5	(18.0)
3½	(1.6)	20	10–11	(311)	4 –4½	(16.1)
3	(1.4)	20	9–10	(281)	3½–4	(14.2)
2½	(1.1)	20	7–8	(222)	3	(11.4)
2	(0.9)	30	8	(237)	2½	(9.5)
1½	(0.7)	30	6	(177)	1½	(5.7)
1	(0.5)	45	6	(177)	1	(3.8)
½	(0.2)	60	6	(177)	½	(1.9)

* Based on an 80% replacement of weight loss.

in Table 25–1. Although these standards were developed for a 90-minute football practice, they are easily adapted to most exercise situations.

Coaches can have their athletes "weigh-in" before and after practice and insist that weight loss be minimized by periodic water breaks during activity. Water must be available and consumed during practice and competition. Because the thirst mechanism is generally an imprecise guide to water needs,[2,97] athletes must be urged to rehydrate themselves. In fact, if rehydration were left entirely to the person's thirst, it could take several days to re-establish fluid balance after severe dehydration.

ELECTROLYTE REPLACEMENT

There is no evidence that electrolyte intake during exercise in the heat improves performance or reduces physiologic strain including muscle cramps. Because sweat is hypotonic to the body fluids, it is much more of an immediate concern to replace water than ions. For a fluid loss of less than 2.7 kg (6 lb) in adults, electrolytes are readily replenished by adding a slight amount of salt to the food when the need exists. For example, the addition of the electrolytes sodium and potassium chloride to the drinking water was of minimum value for men and women who were dehydrated due to sweating by about 3% of body mass on 5 successive days, but who were permitted food and water ad libitum during each daily recovery period.[28] Furthermore, sodium losses during endurance exercise are generally balanced by the sodium-conservation mechanisms of the kidney, which are activated in such exercise.[64]

A Small Amount of Salt May Be Beneficial. Studies with animals indicate that a small amount of electrolytes added to the drinking water brings about a more complete rehydration compared to water alone.[87] In humans, similar benefits seem to exist.[18a,88] It appears that pure water absorbed from the gut rapidly dilutes the plasma concentration of sodium, which stimulates urine production and blunts the sodium-dependent stimulation of thirst. Maintaining plasma concentration of sodium by adding a small amount of this electrolyte to the ingested fluid may sustain the thirst drive and more rapidly restore lost plasma volume.

With prolonged exercise in the heat, sweat loss may deplete the body of 13 to 17 g of salt (2.3 to 3.4 g per liter of sweat) per day. This amount is about 8 g in excess of that provided in the daily diet. In this case, salt supplements may be necessary, i.e., about one third of a teaspoon of table salt added to a liter of water. It is doubtful whether potassium supplements are needed because the potassium loss through sweating is negligible except under the most extreme conditions.[29,39] In this case, potassium loss can be replaced by increasing the intake of potassium-rich foods like citrus fruits and bananas. A glass of orange juice or tomato juice replaces almost all of the potassium, calcium, and magnesium excreted in 2 to 3 liters of sweat. **For all but unusual cases, dietary modifications and electrolyte conservation by the kidneys adequately compensate for electrolyte loss through sweating.**

FACTORS THAT MODIFY HEAT TOLERANCE

Acclimatization

Tasks that are relatively easy when performed in cool weather become extremely taxing if attempted on the first hot day of spring. The early stages of spring training are often the most hazardous in terms of heat injury, because thermoregulatory mechanisms are not adjusted to the dual challenge of exercise and heat. Repeated exposure of men and women to hot environments, especially when combined with exercise, results in improved capacity for exercise and less discomfort upon heat exposure.[34,59,104] The physiologic adaptive changes that improve heat tolerance are collectively termed **heat acclimatization.**[119]

As depicted in Figure 25–6, the major acclimatization occurs during the first week of heat exposure and is essentially complete by the end of 10 days.[22,68,111] Only 2 to 4 hours of daily heat exposure are required. In practical terms, the first several exercise sessions in the heat should be light and last about 15 to 20 minutes. Thereafter, exercise sessions can increase in duration and intensity.

Table 25–2 summarizes the main physiologic adjustments during heat acclimatization. With acclimatization, larger quantities of blood are shunted to cutaneous vessels and heat moves from the core to the periphery. A more effective distribution of cardiac output is also achieved so that blood pressure remains more stable during exercise. This "circulatory acclimatization" is complemented by a lowered threshold for sweating. Consequently, the cooling process is initiated before temperature increases too markedly. After 10 days of heat exposure, the capacity for sweating is nearly doubled, the sweat becomes more dilute (less salt is lost), and is more evenly distributed on the skin surface.[30] These adjustments in circulation and evaporative cooling enable the heat-acclimatized person to exercise with a lower skin and core temperature and heart rate than an unacclimatized person.[1,11] This lower core temperature might result in less need for blood flow to the skin, thus freeing a greater portion of the cardiac output for distribution to the working muscles. It is important to note that unless the person is well hydrated the acclimatization process is blunted. Also, the major benefits of acclimati-

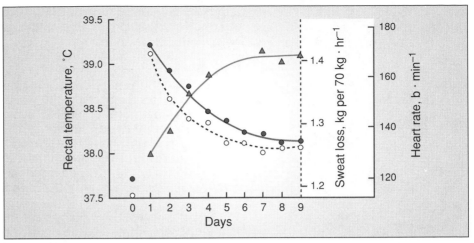

FIG. 25–6.
Average rectal temperature ●, heart rate ○, and sweat loss ▲ during 100-minute daily heat-exercise exposure for 9 consecutive days. On day 0, the men walked on a treadmill at 300 kcal· h⁻¹ in a cool climate. Thereafter, daily exercise was performed in the heat at 48.9° C (26.7° C wet-bulb). (From Lind, A.R.,and Bass, D.E.: Optimal exposure time for development of acclimatization to heat. Fed. Proc., 22:704, 1963.)

zation are lost within 2 to 3 weeks after returning to a more temperate environment.[122]

Training

In a cool environment, exercise-induced "internal" heat stress brings about adjustments in peripheral circulation and evaporative cooling similar to those observed at hot ambient temperatures. This is noted during an 8- to 12-week training period at an exercise intensity that exceeds 50% of the person's aerobic capacity. For this reason, well-conditioned men and women generally respond more effectively to a severe heat stress than their sedentary coun-

TABLE 25–2.
PHYSIOLOGIC ADJUSTMENTS DURING HEAT ACCLIMATIZATION

Acclimatization Response	Effect
● Improved cutaneous blood flow	Transports metabolic heat from deep tissues to the body's shell
● Effective distribution of cardiac output	Appropriate circulation to skin and muscles to meet demands of metabolism and thermoregulation; greater stability in blood pressure during exercise
● Lowered threshold for start of sweating	Evaporative cooling begins early in exercise
● More effective distribution of sweat over skin surface	Optimum use of effective surface for evaporative cooling
● Increased sweat output	Maximizes evaporative cooling
● Lowered salt concentration of sweat	Dilute sweat preserves electrolytes in extracellular fluid

terparts.[5,41] For one thing, training increases the sensitivity and capacity of the sweating response so that sweating begins at a lower body temperature and larger volumes of more dilute sweat are produced. This seems to be brought on by intrinsic adaptations in the sweat glands themselves.[15] This beneficial thermoregulatory response is also associated with significant increase in plasma volume noted with endurance-type training.[23,24] This added fluid could provide for the needs of the sweat glands during heat stress while, at the same time, maintaining an adequate circulating blood volume to meet both the cardiovascular and thermoregulatory demands of exercise. Thus, the trained person stores less heat during the thermal transient phase of exercise and arrives at a thermal steady state sooner and at a lower core temperature than does an untrained counterpart. **This training advantage for thermoregulation occurs only if the individual is fully hydrated during exercise.**[105]

As might be expected, however, this form of heat conditioning is much less effective than acclimatization derived from similar exercise training in the heat.[109,110] **Full heat acclimatization cannot be achieved without actual exposure to heat stress.** Athletes who train and compete in hot weather have a distinct advantage over athletes who train in cool climates but periodically compete in hot weather.

Age

Debate exists as to whether the ability to tolerate and acclimatize to moderate heat stress appreciably deteriorates with age.[37,65,90,96,100] In one experiment, two groups of men and women aged 60 to 93 years were exposed to 70 minutes of heat stress; they progressively exercised at intensities ranging from 2 to 5 times the resting metabolism.[59] The relationship between heart rate and work intensity during work in the heat for these subjects and for young men and women is shown in Figure 25–7.

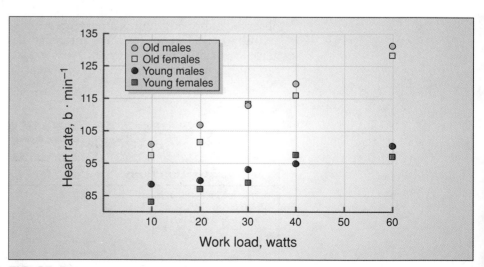

FIG. 25–7.
Heart rate during moderate exercise in the heat for young and old men and women. Dry-bulb ambient temperature was 33.5° C and wet-bulb, 28.5° C. (Adapted from Henshel, A.: The environment and performance. In Physiology of Work Capacity and Fatigue. Edited by E. Simonsen. Springfield, Charles C Thomas, 1971.)

As expected, heart rates were higher for the generally less fit elderly subjects than for the young adults of the same gender. The heat, however, imposed no greater physiologic strain upon the older subjects because their body temperature increased an average of 0.3°C compared to 0.2°C for the younger group. The elderly subjects were also tested in the spring and fall to evaluate the extent of natural heat acclimatization during the summer months. After the summer, pulse rates during the standard thermal-work stress were significantly lower for the men and women.

It has been suggested that age is a limiting factor during vigorous exercise in excessive heat,[37,56] especially when the humidity is low.[4] This age effect is attributed to both an apparent delayed onset in sweating with advancing years as well as an inadequate sweating response due either to a limitation in sweat gland output or to a dehydration-limited response if fluid replacement is insufficient.[65] These conclusions are not supported, however, by findings from competitive runners,[95] because no differentiation could be made between young and middle-aged men in their capacity for temperature regulation during marathon running. For physically trained men in their fifth decade of life, little impairment in thermoregulatory function was noted in comparison to young men.[90] Supporting these findings is a report that the capacity for sweating is fully adequate to regulate body temperature during desert walks for men aged 58 to 84 years.[38] Research that controls for such factors as body composition and fitness level, in addition to chronologic age, is required before the true effect of aging on the thermoregulatory process is known.

Gender

Early comparisons of thermoregulation of men and women during exercise indicated that men showed a greater heat tolerance. A major flaw in this research, however, was that the women were consistently exercising at much higher intensities in relation to their aerobic capacity. When this factor was controlled, and men and women of equal fitness were compared, the thermoregulatory differences between the genders were less pronounced.[51,55,61]

Sweating. **The distinct difference in thermoregulation between men and women is in sweating.** Women are less prolific sweaters than men. Women start to sweat at higher skin and core temperatures; they also produce less sweat than men for a comparable heat-exercise load,[39] even after acclimatization comparable to that achieved by men.[116] This difference in sweating response occurs even though women possess more heat-activated sweat glands per unit skin area than men.[9,15]

Evaporative Versus Circulatory Cooling. Despite a lower sweat output, women show heat tolerance similar to that in men of equal aerobic fitness at the same exercise level.[8,10,51] Women probably rely more on circulatory mechanisms for heat dissipation whereas men make greater use of evaporative cooling. Clearly, the production of less sweat to maintain thermal balance would provide significant protection from dehydration for women during work at high ambient temperatures.

Ratio of Surface Area to Volume. A favorable dimensional characteristic for heat dissipation possessed by most women compared to men is a relatively large surface area-to-mass ratio; that is, per unit of body mass the female has a larger surface exposed to the environment. Consequently, under identical conditions of heat exposure, women would tend to cool at a faster rate than men through a smaller body mass across a relatively large surface area.

Effect of Menstruation. The sweating response and the ability of women to exercise in the heat is not related to the **menstrual cycle.**[118] No differences have been noted for untrained, acclimatized young women in skin temperature, core temperature, pulmonary ventilation, oxygen consumption, or sweat rate with exercise during different phases of menstruation.

The available research indicates that women can tolerate the physiologic and thermal stress of exercise at least as well as men of comparable fitness and level of acclimatization and can acclimatize to a similar degree.[3,117]

Fatness

Obesity is a liability when working in the heat. Because the specific heat of fat is greater than for muscle tissue, excess fat increases the insulatory quality of the body shell and retards conduction of heat to the periphery. The large, overfat person also has a relatively small body surface for the evaporation of sweat compared to a leaner, smaller person.

In addition to interfering with heat exchange, excess fat directly adds to the metabolic cost of activities in which the body mass must be moved. When this effect is compounded by the additional weight of equipment such as football gear, intense competition, and a hot, humid environment, the overfat person is at a distinct disadvantage in terms of heat regulation and physical performance. In fact, fatal heat stroke occurs 3.5 times more frequently in young adults who are excessively overweight than in individuals whose body mass is within reasonable limits.[58]

COMPLICATIONS FROM EXCESSIVE HEAT STRESS

Heat Illness

If the normal signs of heat stress—thirst, tiredness, grogginess, and visual disturbances—are not heeded, cardiovascular compensation begins to fail and a series of disabling complications termed **heat illness** can result. The major forms of heat illness in order of increasing severity are heat cramps, heat exhaustion, and heat stroke. There is often no clear-cut demarcation between these maladies because symptoms often overlap.[62] When heat illness does occur, however, immediate action must be taken to relieve the heat stress and rehydrate the person until medical help arrives.

Heat Cramps. Heat cramps, or involuntary muscle spasms, occur during or after intense physical activity and are usually observed in the specific muscles exercised. This form of heat illness is probably due to an imbalance in the body's fluid and electrolyte concentrations. During heat exposure, salts can be lost as a result of sweating. Muscle pain and spasm may occur if these electrolytes are not replenished. (The spasms occur most commonly in the muscles of the abdomen and extremities.) With heat cramps, body temperature is not necessarily elevated. Prevention can usually be assured by providing copious amounts of water and by increasing the daily intake of salt several days preceding the period of heat stress; the easiest way is to add a bit more salt to the foods at mealtime.

Heat Exhaustion. This condition usually develops in unacclimatized people and is often reported during the first heat wave of the summer or during the first hard training session on a hot day. Exercise-induced heat exhaustion is believed to be caused by ineffective circulatory adjustments compounded by a depletion of extracellular fluid, especially blood volume owing to excessive sweating.[62] Blood usually pools in the dilated peripheral vessels. This drastically reduces the central blood volume necessary to maintain cardiac output. Heat exhaustion is usually characterized by a weak, rapid pulse, low blood pressure in the upright position, headache, dizziness and general weakness. Sweating may be reduced somewhat, but body temperature is not elevated to dangerous levels (less than 104°F or 40°C). A person experiencing these symptoms should stop exercising and move to a cooler environment; fluids should be administered, usually via intravenous therapy.

Heat Stroke. This is the most serious and complex of the heat-stress maladies. **It requires immediate medical attention.** Heat stroke is essentially a failure of the heat-regulating mechanisms brought on by excessively high body temperatures. When thermoregulation fails, sweating usually ceases, the skin becomes dry and hot, body temperature rises to dangerous levels (40°C and higher), and excessive strain is placed on the circulatory system. The complexity of the problem of emergency hyperthermia is confounded because the symptoms are often subtle. With intense exercise, usually in young, highly motivated individuals, sweating may be present, but heat gain by the body greatly outstrips the avenues for heat loss. If left untreated, the disability progresses and death ensues due to circulatory collapse and eventual damage to the central nervous system. **Heat stroke is a medical emergency!** While awaiting medical treatment, one must take aggressive steps rapidly to lower the elevated core temperature because mortality is related to both magnitude and duration of hyperthermia. These include alcohol rubs, ice packs, and whole-body immersion in cold or even ice water.

Don't Rely on Oral Temperature. Oral temperature is often a highly inaccurate measure of body temperature after strenuous exercise. In one report,[98] large and consistent differences were noted between the oral and rectal temperatures—the average rectal temperature was 103.5°F while oral temperature was a normal 98°F following a 14-mile race in a tropical climate!

HOW HOT IS "TOO HOT"?

By now it should be clear that factors other than air temperature determine the physiologic strain imposed by heat. These factors include individual variations in body size and fatness, state of training and acclimatization, and external factors such as convective air currents, radiant heat gain, intensity of exercise, amount, type, and color of clothing, and most important, the relative humidity. In fact, several football deaths have been reported when the air temperature was below 75°F (23.9°C) but the relative humidity exceeded 95%.

The most effective way to control heat stress injuries is to prevent their occurrence. For one thing, acclimatization greatly reduces the chance for heat injury. Another consideration is to evaluate the environment in terms of its potential heat challenge. This can be done with the use of a **heat stress index;** this was developed by the military and was derived from measures of ambient temperature, relative humidity, and radiant heat.[74] The index, termed the **wet bulb-globe temperature** or **WB-GT** index, is calculated as:

$$\text{WB-GT} = 0.1 \times \text{DBT} + 0.7 \times \text{WBT} + 0.2 \times \text{GT}$$

where DBT is the dry-bulb or air temperature recorded by an ordinary mercury thermometer used for recording air temperature.

WBT is the wet-bulb temperature recorded by a similar thermometer except that a wet wick surrounds the mercury bulb that is exposed to rapid air movement. When the relative humidity is high, little evaporative cooling occurs from the wetted bulb, so the temperature of this thermometer is similar to that of the dry-bulb. On a dry day, however, significant evaporation occurs from the wetted bulb and the difference between the two thermometer readings is maximized. A small difference between readings indicates a high relative humidity, whereas a large difference indicates little air moisture and a high rate of evaporation.

GT is the globe temperature recorded by a thermometer whose bulb is enclosed in a metal sphere painted black. The black globe absorbs radiant energy from the surroundings to provide a measure of this important source of heat gain.

Table 25–3 (top) presents WB-GT guidelines that can be applied to athletic activities to reduce the chance of heat injury. These standards apply to lightly clothed humans, but do not take into account the specific heat load imposed by uniforms or equipment such as that used in football. For this activity, the lower end of each temperature range serves as a more prudent guide.

A simple indication of the ambient heat load can also be obtained from the wet-bulb thermometer. This reading evaluates both temperature and humidity. The thermometer is relatively inexpensive and can be purchased at most industrial supply companies. The bottom of Table 25–3 presents recommendations based on wet-bulb temperature.

TABLE 25–3.
WB-GT FOR OUTDOOR ACTIVITIES AND WET-BULB TEMPERATURE (WBT) GUIDE*

WB-GT Range		Recommendations
°F	°C	
80–84	26.5–28.8	• Use discretion, especially if unconditioned or unacclimatized
85–87	29.5–30.5	• Avoid strenuous activity in the sun
88	31.2	• Avoid exercise training

WBT Range		Recommendations
°F	°C	
60	15.5	• No prevention necessary
61–65	16.2–18.4	• Alert all participants to problems of heat stress and importance of adequate hydration
66–70	18.8–21.1	• Insist that appropriate quantity of fluid be ingested
71–75	21.6–23.8	• Rest periods and water breaks every 20 to 30 minutes; limits placed on intense activity
76–79	24.5–26.1	• Practice curtailed and modified considerably
80	26.5	• Practice cancelled

* Modified from Murphy, R.J., and Ashe, W.F.: Prevention of heat illness in football players. JAMA, 194:650, 1965.

SUMMARY

1. Core temperature normally increases during exercise with the magnitude of the rise in temperature determined by the relative stress of a particular work load. This well-regulated temperature adjustment probably creates a favorable environment for physiologic and metabolic function.
2. Sweating places demands on the body's fluid reserves and creates a relative state of dehydration. If sweating is excessive and fluids are not continually replaced, blood volume falls and core temperature may rise to lethal levels.
3. Exercise in hot, humid environments poses a great challenge to temperature regulation because the large sweat loss in high humidity contributes little to evaporative cooling.
4. Fluid loss in excess of 4 to 5% body mass significantly impedes heat dissipation and compromises cardiovascular function and work capacity.
5. The primary aim of fluid replacement is to maintain plasma volume so that circulation and sweating progress at optimal levels. For the ideal replacement schedule during exercise, fluid intake should match fluid loss. This can be monitored effectively by changes in body mass.
6. Although about 800 ml of water can be absorbed from the digestive tract each hour, several factors affect this rate of absorption: (1) colder fluids are emptied from the stomach more rapidly than fluids at body temperature, (2) when the stomach is partially filled with fluid, the rate of gastric emptying is increased, and (3) concentrated carbohydrate solutions impair gastric emptying and fluid replacement.
7. Beverages of moderate glucose-polymer content do not impair physiologic function and thermoregulation.
8. Electrolytes lost through sweating are readily replaced by adding a small amount of salt to the food in the daily diet. Salt tablets are not recommended.
9. Repeated heat stress initiates thermoregulatory adjustments that result in improved exercise capacity and less discomfort on subsequent heat exposure. This heat acclimatization brings about a favorable distribution of cardiac output and a greatly increased capacity for sweating. Full acclimatization generally occurs in about 10 days of heat exposure.
10. The ability to tolerate and acclimatize to moderate heat stress probably does not appreciably deteriorate with age.
11. Women seem to be at least as efficient in temperature regulation as men because they produce less sweat while maintaining the same temperature.
12. Various practical heat stress indices make use of ambient temperature, radiant heat, and relative humidity to evaluate the potential heat challenge of an environment to an exercising subject.
13. The major forms of heat illness are heat cramps, heat exhaustion, and heat stroke. Heat stroke is by far the most serious and complex of these maladies.

EXERCISE IN THE COLD

Water is an excellent medium in which to study physiologic adjustment to cold because body heat is lost about two to four times as fast in cool water as in air at the same temperature. Shivering is frequently observed if people remain inactive in a pool or ocean environment because of a large conductive heat loss. Even with moderate exercise in cold water, the metabolic heat generated is often insufficient to counter the large thermal drain. This is especially true during swimming because heat transfer by convection markedly increases with the rapid movement of water past the skin surface.

In light and moderate exercise in cold water, oxygen consumption is higher and body temperature lower compared to identical exercise in warmer water.[32,72] For example, swimming at a submaximal pace in a flume at 18°C (64°F) requires about 500 ml of oxygen more per minute than swimming at the same speed in 26°C (79°F) water.[79] This additional oxygen uptake is directly related to the added energy cost of shivering as the body attempts to combat heat loss.

BODY FATNESS, EXERCISE, AND COLD STRESS

Differences between individuals in the body's total fat content have a significant effect on physiologic function in the cold both at rest and during exercise.[71,72,79] Successful Channel swimmers, for example, usually possess a relatively large amount of subcutaneous fat compared to highly trained non-distance swimmers.[91] This greatly increases their effective insulation when peripheral blood is redirected to the body's core in cold water. With this advantage, these athletes can swim in cool ocean waters with almost no fall in core temperature. For leaner swimmers, however, heat generated in exercise is insufficient to counter the heat drain to the water and, as such, the body cools.

To a large extent, the stress from "cold" is highly relative. The physiologic strain imposed by the cold depends not only on the environmental temperature per se, but also on the level of metabolism and the resistance to heat flow provided by body fat,[113] especially in water colder than 25°C.[93,114] A fat person who is comfortable resting immersed to the neck in 26°C water may sweat about the forehead during vigorous exercise. For this person, 18°C may be a more favorable water temperature for heavy exercise. For the lean person, on the other hand, water at 18°C is debilitating both at rest and during exercise. Thus, there is an optimum water temperature for each person and for each activity. For most of us, water temperatures between 26°C and 30°C provide for effective heat dissipation in exercise, yet are not excessively cold so that work capacity is compromised. With short-term intense exercise even cooler water may be optimal, especially for fatter people.

ACCLIMATIZATION TO THE COLD

Humans possess much less capacity for adaptation to prolonged cold exposure than to prolonged exposure to heat.[123] Indeed, the basic response of Eskimos and Lapps is to avoid the cold or minimize its effect. For example, the clothing of these cold weather inhabitants provides a near-tropical microclimate, and the temperature inside an igloo is generally maintained at about 21°C (70°F).

Some indication of cold adaptation has been provided from studies of the Ama, the women divers of Korea and southern Japan.[60] These women can tolerate daily prolonged exposure to diving for food in cold water, which in winter is about 10°C (50°F). In addition to an apparent psychologic "toughness," the capability of these women to tolerate extreme cold has been attributed to an elevated resting metabolism. In winter, this is increased by about 25% compared to nondiving women of the same community. Interestingly, the body fat of these women is no greater than their nondiving female counterparts. It is possible, therefore, that circulatory adaptations also aid these divers by retarding heat transfer from the core to the skin.

A type of general cold adaptation appears to occur following regular and prolonged cold-air exposure. As a result, heat loss is not compensated for by increased heat production and individuals "regulate" at a **lower** rectal temperature in response to cold.[67]

Some peripheral adaptations also reflect a form of acclimation with severe local cold exposure.[66,67a,83] Repeated cold exposure to the hands or feet brings about an increased blood flow through these areas when they are subjected to cold stress. This is readily apparent in cold-water fisherman who handle nets and fish in extreme cold. Although this local adaptation would actually result in a loss of body heat from the periphery, it does represent a form of "self-defense" because a vigorous circulation in the exposed areas aids in preventing tissue damage due to hypothermia. Although not specifically a form of cold acclimatization, improved fitness, as reflected by a high aerobic capacity, enhances a person's thermoregulatory defense against cold stress. This manifests itself in a larger shivering response as well as an earlier or more sensitive onset of this response with cold exposure.[14]

HOW COLD IS "TOO COLD"?

Cold injuries from overexposure are on the upswing because of increased interest in outdoor winter activities such as ice skating, cross-country skiing, snowmobiling, and jogging. Owing to the pronounced peripheral vasoconstriction during severe cold exposure, the temperature of the skin and the extremities may fall to dangerous levels. Early warning signs of cold injury include a tingling and numbness in the fingers and toes, or a burning sensation of the nose and ears. If these signs are not heeded, overexposure can lead to tissue damage in the form of frostbite; in extreme cases the damage is irreversible and the tissue must be surgically removed.

The Wind Chill Index.   One dilemma in evaluating the thermal quality of an environment is that ambient temperature alone is not always a valid indication of "coldness." All too often we have experienced the chilling winds of a spring day, even though the temperature was well above freezing. On the other hand, a calm subfreezing day may feel quite comfortable. **The important factor is the wind—on a windy day air currents magnify heat loss as the warmer insulating air layer surrounding the body is continually replaced by cooler ambient air.**

The cooling effect of wind is clearly shown in the Wind Chill Index presented in Figure 25–8. This figure illustrates the effects of wind velocity on bare skin for different temperatures and velocities. For example, a 30° F reading is equivalent to 0° F when the wind speed is 25 mph, while a 10° F reading is equivalent to −29° F when the wind speed is 25 mph. In addition, if a person runs, skis, or skates into the wind the effective cooling from the wind is increased in direct relation to the exerciser's velocity. Thus, running at 8 mph into a 12 mph headwind is the equivalent of a 20 mph wind speed. Conversely, running at 8 mph with a 12 mph wind at your back creates a relative wind speed of only 4 mph. In the lightly shaded zone on the left there is relatively little danger from cold exposure for a properly clothed person. In contrast, in the medium-shaded zone, which generally begins at about 22° F, there is increasing danger to exposed flesh, especially the ears, nose, and fingers. In the heavily shaded zone on the right the equivalent

FIG. 25—8.
The Wind Chill Index.

	Ambient temperature, °F**													
	40	35	30	25	20	15	10	5	0	−5	−10	−15	−20	−25	−30
	Equivalent temperature, °F													
Calm	40	35	30	25	20	15	10	5	0	−5	−10	−15	−20	−25	−30
5	37	33	27	21	16	12	6	1	−5	−11	−15	−20	−26	−31	−35
10	28	21	16	9	4	−2	−9	−15	−21	−27	−33	−38	−46	−52	−58
15	22	16	11	1	−5	−11	−18	−25	−36	−40	−45	−51	−58	−65	−70
20	18	12	3	−4	−10	−17	−25	−32	−39	−46	−53	−60	−67	−76	−81
25	16	7	0	−7	−15	−22	−29	−37	−44	−52	−59	−67	−74	−83	−89
30	13	5	−2	−11	−18	−26	−33	−41	−48	−56	−63	−70	−79	−87	−94
35	11	3	−4	−13	−20	−27	−35	−43	−49	−60	−67	−72	−82	−90	−98
***40**	10	1	−6	−15	−21	−29	−37	−45	−53	−62	−69	−76	−85	−94	−101

(Wind speed, mph on left axis; Calm / 5 / 10 / 15 / 20 / 25 / 30 / 35 / 40* on right axis)

☐ Little danger   ☐ Danger   ☐ Great danger

* Convective heat loss at wind speeds above 40 mph has little additional effect on body cooling.
** °C = 0.556 (°F −32)

temperatures pose a serious danger to the freezing of exposed flesh within a matter of minutes.

THE RESPIRATORY TRACT DURING COLD-WEATHER EXERCISE

Cold ambient air generally does not pose a special danger in terms of damage to the respiratory passages. Even in extreme cold, the incoming air is warmed to between 26.5° C and 32.2° C by the time it reaches the bronchi,[19] although values as low as 20° C have been observed while breathing large volumes of cold, dry air.[73] When the incoming breath of air is warmed, its capacity to hold moisture increases greatly and humidification occurs at the expense of water from the respiratory passages. Thus, significant amounts of water and heat can be lost from the respiratory tract, especially during exercise when ventilatory volumes are quite large. This often contributes to frequent complaints during exercise in the cold. A general dehydration can accompany exercise, as well as a dryness of the mouth, a burning sensation in the throat, and an irritation of the respiratory passages. These symptoms can be greatly reduced by wearing a scarf or mask-type "balaclava" that covers the nose and mouth and that traps the water in the exhaled air. This action subsequently warms and moistens the next breath of incoming air.

SUMMARY

1. Because heat conduction in water is about 25 times greater than in air, immersion in water of only 28 to 30° C can provide a considerable thermal stress and bring about thermoregulatory adjustments in a relatively short period of time.

2. Heat flux to a cold environment is offset to some degree by heat production associated with shivering and muscular exercise. For the average man, shivering can increase the metabolic rate about 3 to 6 times above the resting level.
3. Subcutaneous fat provides excellent body insulation against cold stress. It greatly enhances the effectiveness of vasomotor adjustments and enables fat individuals to retain a large percentage of metabolic heat. This is apparent in cold water where relatively fat individuals show proportionately smaller thermal and cardiovascular adjustments and greater exercise tolerance in comparison to leaner counterparts.
4. The body is much less capable of adapting physiologically to prolonged cold stress than to prolonged heat exposure. In most instances, appropriate clothing enables humans to tolerate even the coldest climates on earth.
5. The "coldness" of an environment is influenced by the ambient temperature and the wind. The cooling effect of the wind can be determined by the Wind Chill Index.
6. Although considerable water can be lost from the respiratory passages when exercising on a cold day, the temperature of the inspired ambient air generally does not pose a danger to the respiratory tract.

REFERENCES

1. Adams, W.C., et al.: Thermoregulation during marathon running in cool, moderate, and hot environments. *J. Appl. Physiol.*, 38:1030, 1975.
2. Adolph, E.F.: *Physiology of Man in the Desert.* New York, Interscience, 1947.
3. American College of Sports Medicine: American College of Sports Medicine position stand on prevention of thermal injuries during distance running. *Sports Med. Bull.*, 19:8, 1984.
4. Anderson, R.K., and Kenney, W.L.: Effect of age on heat-activated sweat gland density and flow during exercise in dry heat. *J. Appl. Physiol.*, 63:1089, 1987.
5. Armstrong, L.E., and Pandolf, K.B.: Physical training, cardiorespiratory physical fitness and exercise-heat tolerance. In *Human Performance Physiology and Environmental Medicine at Terrestrial Extremes.* Edited by K.B. Pandolf et al. Indianapolis, Benchmark Press, 1988.
6. Asmussen, E., and Bøje, O.: Body temperature in muscular work. *Acta Physiol. Scand.*, 10:1, 1945.
7. Åstrand, I.: Aerobic work capacity in men and women with special reference to age. *Acta Physiol. Scand.*, 49(Suppl. 169):1, 1960.
8. Avellini, B.A., et al.: Physiological responses of physically fit men and women to acclimation to humid heat. *J. Appl. Physiol.*, 49:254, 1980.
9. Bar-Or, O., et al.: Distribution of heat-activated sweat glands in obese and lean men and women. *Hum. Biol.*, 40:235, 1968.
10. Bar-Or, O., et al.: Heat tolerance of exercising obese and lean women. *J. Appl. Physiol.*, 26:403, 1969.
11. Bass, D.E., et al.: Mechanisms of acclimatization to heat in man. *Medicine*, 34:323, 1955.
12. Bassett, D.R. Jr., et al.: Thermoregulatory responses to skin wetting during prolonged treadmill running. *Med. Sci. Sports Exerc.*, 19:28, 1987.
13. Beaumont, W. van, and Bullard, R.W.: Sweating: its rapid response to muscular work. *Science*, 141:643, 1963.
14. Bittel, J.H.M., et al.: Physical fitness and thermoregulatory reactions in a cold environment. *J. Appl. Physiol.*, 65:1984, 1988.
15. Buono, M.J., and Sjoholm, N.T.: Effect of physical training on peripheral sweat production. *J. Appl. Physiol.*, 65:811, 1988.
16. Buskirk, E.R., et al.: Work performance after dehydration: effects of physical conditioning and heat acclimatization. *J. Appl. Physiol.*, 12:189, 1958.
17. Caldwell, J.E., et al.: Diuretic therapy, physical performance, and neuromuscular function. *Phys. Sportsmed.*, 12:73, 1984.

18. Candas, V., et al.: Hydration during exercise: effects on thermal and cardiovascular adjustments. *Eur. J. Appl. Physiol.*, 55:113, 1986.

18a. Carter, J.E., and Gisolfi, C.V.: Fluid replacement during and after exercise in the heat. *Med. Sci. Sports Exerc.*, 21:532, 1989.

19. Claremont, A.D.: Taking winter in stride requires proper attire. *Phys. Sportsmed.*, 4:65, 1976.

20. Claremont, A.D., et al.: Comparison of metabolic, temperature, heart rate and ventilatory responses to exercise at extreme ambient temperature (0° and 35° C). *Med. Sci. Sports*, 7:150, 1975

21. Claremont, A.D., et al.: Heat tolerance following diuretic induced dehydration. *Med. Sci. Sports*, 8:239, 1976.

22. Cleland, T.S., et al.: Acclimatization of women to heat after training. *Int. Z. Angew. Physiol.*, 27:15, 1969.

23. Convertino, V.A.: Heart rate and sweat rate responses associated with exercise-induced hypervolemia. *Med. Sci. Sports Exerc.*, 15:77, 1983.

24. Convertino, V.A., et al.: Plasma volume, renin, and vasopressin responses to graded exercise after training. *J. Appl. Physiol.*, 54:508, 1983.

25. Costill, D.L., and Fink, W.J.: Plasma volume changes following exercise and thermal dehydration. *J. Appl. Physiol.*, 37:521, 1974.

26. Costill, D.L., and Saltin, B.: Factors limiting gastric emptying during rest and exercise. *J. Appl. Physiol.*, 37:679, 1974.

27. Costill, D.L., and Sparks, K.E.: Rapid fluid replacement following thermal dehydration. *J. Appl. Physiol.*, 34:299, 1973.

28. Costill, D.L., et al.: Water and electrolyte replacement during repeated days of work in the heat. *Aviat. Space Environ. Med.*, 46:795, 1975.

29. Costill, D.L., et al.: Muscle water and electrolytes following varied levels of dehydration in man. *J. Appl. Physiol.*, 40:6, 1976.

30. Costill, D.L., et al.: Exercise induced sodium conservation: changes in plasma renin and aldosterone. *Med. Sci. Sports*, 8:209, 1976.

31. Coyle, E.F.: Gastric emptying rates for selected athletic drinks. *Res. Q.*, 49:119, 1978.

32. Craig, A.B., Jr., and Dvorak, M.: Thermal regulation of man exercising during water immersion. *J. Appl. Physiol.*, 25:28, 1968.

33. Craig, F.N., and Cummings, E.G.: Dehydration and muscular work. *J. Appl. Physiol.*, 21:670, 1966.

34. Davies, C.T.M.: Effect of acclimatization to heat on the regulation of sweating during moderate and severe exercise. *J. Appl. Physiol.*, 50:741, 1981.

35. Davis, J.M., et al.: Carbohydrate-electrolyte drinks: effects on endurance cycling in the heat. *Am. J. Clin. Nutr.*, 48:1023, 1988.

36. DeSouza, M.J., et al.: Menstrual status and plasma vasopressin, renin and aldosterone exercise responses. *J. Appl. Physiol.*, 67:736, 1989.

37. Dill, D.B., and Consolazio, C.F.: Responses to exercise as related to age and environmental temperature. *J. Appl. Physiol.*, 17:64, 1962.

38. Dill, D.B., et al.: Cardiovascular responses and temperature in relation to age. *Aust. J. Sports Med.*, 7:99, 1975.

39. Dill, D.B., et al.: Capacity of young males and females for running in desert heat. *Med. Sci. Sports*, 9:137, 1977.

40. Drinkwater, B.L.: Thermoregulatory response of women to intermittent work in the heat. *J. Appl. Physiol.*, 41:57, 1976.

41. Drinkwater, B.L., et al.: Aerobic power as a factor in women's response to work in hot environments. *J. Appl. Physiol.*, 41:815, 1976.

42. Falls, H.B., and Humphrey, L.D.: Cold water application effects on responses to heat stress during exercise. *Res Q.*, 42:21, 1971.

43. Fink, W.J.: Fluid intake for maximizing athletic performance. In *Nutrition and Athletic Performance.* Edited by W. Haskell, et al. Palo Alto, CA, Bull, 1982.

44. Fink, W.J., et al.: Leg muscle metabolism during exercise in the heat and cold. *Eur. J. Appl. Physiol.*, 34:183, 1975.

45. Fordtran, J.S., and Saltin, B.: Gastric emptying and intestinal absorption during prolonged severe exercise. *J. Appl. Physiol.*, 23:331, 1967.

46. Fortney, S.M., et al.: Effect of blood pressure on sweating rate and blood lipids in exercising humans. *J. Appl. Physiol.*, 51:1594, 1981.

47. Fortney, S.M., et al.: Effect of hyperosmolarity on control of blood flow and sweating. *J. Appl. Physiol.*, 57:1668, 1984.

48. Foster, C., et al.: Gastric emptying characteristics of glucose and glucose polymer solutions. *Res. Q.*, 51:299, 1980.

49. Fox, E., et al.: Effects of football equipment on thermal balance and energy cost during exercise. *Res. Q.*, 37:322, 1966.

50. Francesconi, R.P.: Endocrine logical responses to exercise in stressful environments. In *Exercise and Sport Sciences Reviews.* Vol. 16. Edited by K.B. Pandolf. New York. Macmillan, 1988.

51. Frye, A.J., and Kamon, E.: Responses to dry heat of men and women with similar capacities. *J. Appl. Physiol.*, 50:65, 1981.

52. Gisolfi, C.V., and Copping, J.R.: Thermal effects of prolonged treadmill exercise in the heat. *Med. Sci. Sports*, 6:108, 1974.

53. Greenleaf, J.E.: Hyperthermia in exercise. In *International Review of Physiology: Environmental Physiology* III. Vol. 20. Edited by D. Robertshaw. Baltimore, MD, University Park Press, 1979.

54. Greenleaf, J.E., et al.: Maximal oxygen uptake, sweating and tolerance to exercise in the heat. *Int. J. Biometeorol.*, 16:375, 1972.

55. Haymes, E.M.: Physiological responses of female athletes to heat stress: a review. *Phys. Sportsmed.*, 12:45, 1984.

56. Hellon, R.F., et al.: The physiological reactions of men of two age groups to a hot environment. *J. Physiol.* (London), 133:118, 1956.

57. Hemingway, A.: Shivering. *Physiol. Rev.*, 43:397, 1963.

58. Henshel, A.: Obesity as an occupational hazard. *Can. J. Public Health*, 58:491, 1967.

59. Henshel, A.: The environment and performance. In *Physiology of Work Capacity and Fatigue.* Edited by E. Simonsen. Springfield, IL, Charles C Thomas, 1971.

60. Hong, S.K., and Rahn, H.: The diving women of Korea and Japan. *Sci. Am.*, 216:34, 1967.

61. Horstman, D.H., and Christensen, E.: Acclimatization to dry heat: active men vs. active women. *J. Appl. Physiol.*, 52:825, 1982.

62. Hubbard, R.W., et al.: The heat illnesses: biochemical, ultrastructural and fluid-electrolyte considerations. In *Human Performance, Physiology and Environmental Medicine at Terrestrial Extremes,* Edited by K.B. Pandolf et al. Indianapolis, IN, Benchmark Press, 1988.

63. Iampietro, P.F.: Exercise in hot environments. In *Frontiers of Fitness.* Edited by R.J. Shephard. Springfield, IL, Charles C Thomas, 1971.

64. Irving, R.A., et al. The immediate and delayed effects of marathon running on renal function. *J. Urol.*, 136:1176, 1986.

65. Kenney, W.L., and Anderson, R.K.: Response of older and younger women in dry and humid heat without fluid replacement. *Med. Sci. Sports Exerc.*, 20:155, 1988.

66. LeBlanc, J.: Local adaptation to cold of Gaspé fisherman. *J. Appl. Physiol.*, 17:950, 1962.

67. LeBlanc, J.: Factors affecting cold acclimation and thermogenesis in man. *Med. Sci. Sports Exerc.*, 20:S193, 1988.

67a. Leftheriotis, G., et al.: Finger and forearm vasodilatory changes after local acclimation. *Eur. J. Appl. Physiol.*, 60:49, 1990.

68. Lind, A.R., and Bass, D.E.: Optimal exposure time for development of acclimatization to heat. *Fed. Proc.*, 22:704, 1963.

69. Mathews, D.K., et al.: Physiological responses during exercise and recovery in a football uniform. *J. Appl. Physiol.*, 26:611, 1969.

70. McArdle, W.D., et al.: Metabolic and cardiovascular adjustment to work in air and water at 18, 25, 33° C. *J. Appl. Physiol.*, 40:85, 1976.

71. McArdle, W.D., et al.: Thermal adjustment to cold-water exposure in resting men and women. *J. Appl. Physiol.*, 56:1565, 1984.

72. McArdle, W.D., et al.: Thermal adjustment to cold-water exposure in exercising men and women. *J. Appl. Physiol.*, 56:1572, 1984.

73. McFadden, E.R., Jr.: Respiratory heat and water exchange: physiological and clinical implications. *J. Appl. Physiol.*, 54:331, 1984.

74. Minard, D., et al.: Prevention of heat casualties. *JAMA*, 165:1813, 1957.

75. Miroff, S.V., and Bass, D.E.: Effects of overhydration on man's physiological responses to work in the heat. *J. Appl. Physiol.*, 20:267, 1965.
76. Mitchell, J.B., et al.: Effects of carbohydrate ingestion on gastric emptying and exercise performance. *Med. Sci. Sports Exerc.*, 20:110, 1988.
77. Mitchell, J.B., et al.: Gastric emptying: influence of prolonged exercise and carbohydrate concentration. *Med. Sci. Sports Exerc.*, 21:269, 1989.
78. Murray, R. et al.: The effect of fluid and carbohydrate feedings during intermittent cycling exercise. *Med. Sci. Sports Exerc.*, 19:597, 1987.
79. Nadel, E.R. Thermal and energetic exchanges during swimming. In *Problems With Temperature Regulation During Exercise.* New York, Academic Press, 1977.
80. Nadel, E.R.: Circulatory regulation during exercise in different ambient temperatures, *J. Appl. Physiol.*, 46:430, 1979.
81. Nadel, E.R., et al.: Thermoregulatory shivering during exercise. *Life Sci.*, 13:983, 1973.
82. Nadel, E.R., et al.: Effect of hydration state on circulatory and thermal regulation. *J. Appl. Physiol.*, 49:751, 1980.
83. Nelms, J.D., and Soper, J.G.: Cold vasodilatation and cold acclimatization in the hands of British fish filleters. *J. Appl. Physiol.*, 17:444, 1962.
84. Neuffer, P.D.: Effects of exercise and carbohydrate composition on gastric emptying. *Med. Sci. Sports Exerc.*, 18:658, 1986.
85. Nielsen, B., and Nielsen, M.: On the regulation of sweat secretion in exercise. *Acta Physiol. Scand.*, 64:314, 1965.
86. Noakes, T.D., et al.: The danger of an inadequate water intake during prolonged exercise. A novel concept re-visited. *Eur. J. Appl. Physiol.*, 57:210, 1988.
87. Nose, H., et al.: Recovery of blood volume and osmolality after thermal dehydration in rats. *Am. J. Physiol.*, 251:R492, 1986.
88. Nose, H., et al.: Role of osmolality and plasma volume during rehydration in humans. *J. Appl. Physiol.*, 65:325, 1988.
89. Owen, M.D., et al. Effects of carbohydrate ingestion on thermoregulation, gastric emptying, and plasma volume during exercise in the heat. *Med. Sci. Sports Exerc.*, 18:568, 1981.
90. Pandolf, K.B., et al.: Thermoregulatory responses of middle-aged and young men during dry-heat acclimatization. *J. Appl. Physiol.*, 65:65, 1988.
91. Pugh, L.C.G.E.: A physiological study of channel swimming. *J. Clin. Invest.*, 37:538, 1960.
92. Pugh, L.C.G.E., et al.: Rectal temperatures, weight losses and sweat rates in marathon running. *J. Appl. Physiol.*, 21:1251, 1966.
93. Rennie, D.W.: Tissue heat transfer in water: lessons from the Korean divers. *Med. Sci. Sports Exerc.*, 20:S177, 1988.
94. Ribisl, P.M., and Herbert, W.G.: Effects of rapid weight reduction and subsequent rehydration upon the physical working capacity of wrestlers. *Res. Q.*, 41:536, 1970.
95. Robinson, S.: Training, acclimatization and heat tolerance. *Can. Med. Assoc. J.*, 96:795, 1967.
96. Robinson, S., et al.: Acclimatization of older men to work in the heat. *J. Appl. Physiol.*, 20:583, 1965.
97. Rolls, B.J., et al.: Thirst following water deprivation in humans. *Am. J. Physiol.*, 239(Regulatory Integrative Comp. Physiol. 8):476, 1980.
98. Rozycki, T.J.: Oral and rectal temperatures in runners. *Phys. Sportsmed.*, 12:105, 1984.
99. Rowell, L.B.: Hepatic clearance of idocyanine green in man under thermal and exercise stresses. *J. Appl. Physiol.*, 20:384, 1965.
100. Rowell, L.B.: Human cardiovascular adjustment to exercise and thermal stress. *Physiol. Rev.*, 54:75, 1974.
101. Rowell, L.B., et al.: Splanchnic blood flow and metabolism in heat-stressed man. *J. Appl. Physiol.*, 24:475, 1968.
102. Saltin, B.: Circulatory response to submaximal and maximal exercise after thermal dehydration. *J. Appl. Physiol.*, 19:1125, 1964.
103. Saltin, B., and Hermansen, L.: Esophageal, rectal and muscle temperature during exercise. *J. Appl. Physiol.*, 21:1757, 1966.
104. Sawka, M.N., et al.: Hypohydration and exercise: effects of heat acclimatization, gender, and environment. *J. Appl. Physiol.*, 55:1147, 1983.
105. Sawka, M.N., et al.: Influence of hydration level and body fluids on exercise performance in the heat. *JAMA*, 252:1165, 1984.

106. Sawka, M.N., and Wegner, C.B.: Physiological responses to acute-exercise heat stress. In *Human Performance Physiology and Environmental Medicine at Terrestrial Extremes.* Edited by K.B. Pandolf et al. Indianapolis, IN, Benchmark Press, 1988.

107. Seiple, R.S., et al.: Gastric-emptying characteristics of two glucose polymer-electrolyte solutions. *Med. Sci. Sports Exerc.,* 15:366, 1983.

108. Smiles, K.A., and Robinson, S.: Sodium ion conservation during acclimatization of men to work in the heat. *J. Appl. Physiol.,* 31:63, 1971.

109. Strydom, N.B., et al.: Comparison of oral and rectal temperatures during work in the heat. *J. Appl. Physiol.,* 8:406, 1956.

110. Strydom, N.B., et al.: Acclimatization to humid heat and the role of physical conditioning. *J. Appl. Physiol.,* 21:636, 1966.

111. Taylor, H.L., et al.: Cardiovascular adjustments of man at rest and work during exposure to dry heat. *Am. J. Physiol.,* 139:583, 1955.

112. Tipton, C.M., and Tcheng, T.K.: Iowa wrestling study: weight loss in high school students. *JAMA,* 214:1269, 1970.

113. Toner, M.M., et al.: Thermal responses during arm and leg and combined arm-leg exercise in water. *J. Appl. Physiol.,* 56:1355, 1984.

114. Toner, M.M., and McArdle, W.D.: Physiological adjustments of a man to cold. In *Human Performance Physiology and Environmental Medicine at Terrestrial Extremes.* Edited by K.B. Pandolf et al. Indianapolis, IN Benchmark Press, 1988.

115. Wailgum, T.D., and Paolone, A.M.: Heat tolerance of college football lineman and backs. *Phys. Sportsmed.,* 12:81, 1984.

116. Wells, C.L.: Sexual differences in heat stress response. *Phys. Sportsmed.,* 5:79, 1977.

117. Wells, C.L.: Responses of physically active and acclimatized men and women to exercise in a desert environment. *Med. Sci. Sports,* 12:9, 1980.

118. Wells, C.L., and Horvath, S.M.: Responses to exercise in a hot environment as related to the menstrual cycle. *J. Appl. Physiol.,* 36:299, 1974.

119. Wegner, C.B.: Human heat acclimatization. In *Human Performance Physiology and Environmental Medicine at Terrestrial Extremes.* Edited by K.B. Pandolf et al. Indianapolis, IN, Benchmark Press, 1988.

120. Widerman, P.M., and Hagen, R.D.: Body weight loss in a wrestler preparing for competition: a case study. *Med. Sci. Sports Exerc.,* 14:413, 1982.

121. Williams, C.G., et al.: Circulatory and metabolic reactions to work in heat. *J. Appl. Physiol.,* 17:625, 1962.

122. Williams C.G., et al.: Rate of loss of acclimatization in summer and winter. *J. Appl. Physiol.,* 22:21, 1967.

123. Young, A.J.: Human adaptation to cold. In *Human Performance Physiology and Environmental Medicine at Terrestrial Extremes.* Edited by K.B. Pandolf et al. Indianapolis, IN, Benchmark Press, 1988.

123a. Young, A.J.: Energy substrate utilization during exercise in extreme environments. In *Exercise and Sport Sciences Reviews.* Vol. 18. Edited by K.B. Pandolf and J.O. Holloszy. Baltimore, Williams & Wilkins, 1990.

124. Zambraski, E.J., et al.: Iowa wrestling study: urinary profiles of state finalists prior to competition. *Med. Sci. Sports,* 6:129, 1974.

125. Zambraski, E.J., et al.: Iowa wrestling study: weight loss and urinary profiles of collegiate wrestlers. *Med. Sci Sports,* 8:105,1976.

It is estimated that there are 4 million scuba divers in the United States and an additional 200,000 divers are trained each year.[1,14] With this interest in diving, pool directors and physical educators are being called upon to initiate, teach, and supervise instructional programs in diving. In the following sections, we outline general principles of diving, as well as potential dangers as a person descends and ascends in the water. Unquestionably, safe diving requires thorough knowledge of diving physics and physiology. Within this framework, emphasis is placed on the relationship between diving depth, pressure, and gas volume, as well as on the potentially toxic effects of various gases respired in diving.[5]

PRESSURE-VOLUME RELATIONSHIPS AND DIVING DEPTH

DIVING DEPTH AND PRESSURE

Because water is essentially noncompressible, the water pressure against the diver's body increases directly with the depth of the dive. This pressure is the result of two forces: (1) the weight of the column of water directly above, and (2) the weight of the atmosphere at the surface. As shown in Table 26–1, **a column of fresh water exerts a force of one sea level atmosphere** or **760 mm Hg (14.7 lb per square inch) for each 33 ft (10 m) one descends below the water's surface.** (Because salt water is more dense than fresh water, a depth of about 32 ft corresponds to the pressure equivalent of **one atmosphere** in ocean diving.) Consequently, a person diving to a depth of 33 ft is exposed to a pressure of 2 atmospheres—1 atmosphere due to the weight of the ambient air at the surface and the other due to the weight of the column of water itself. Diving from sea level to a depth of 66 ft (20 m) exposes the diver to an absolute pressure equal to 3 atmospheres; at 99 ft (30 m) the pressure is 4 atmospheres, and so on. Clearly, considerable pressure is exerted at relatively short distances below the surface.

Because the tissues of the body are largely water, they too are incompressible and thus are not especially susceptible to the increased external pressure in diving. The body, however, also contains air-filled cavities—notably the lungs, respiratory passages, and sinus and middle-ear spaces. Volume and pressure in these spaces is greatly modified by any increase or decrease in diving depth. **Pain, injury, and even death can occur unless adjustments are made to equalize the rapid and significant pressure changes in diving.**

TABLE 26–1.
RELATIONSHIP OF DEPTH IN WATER TO PRESSURE AND VOLUME

Depth (ft)	Depth (m)	Pressure (Atmosphere)	Pressure (mm Hg)	Hypothetical Lung Volume (ml)	Inspired Air P_{O_2}	Inspired Air P_{N_2}
Sea level		1	760	6000	159	600
33	10	2	1520	3000	318	1201
66	20	3	2280	2000	477	1802
99	30	4	3040	1500	636	2402
133	40	5	3800	1200	795	3003
166	50	6	4560	1000	954	3604
200	60	7	5320	857	1113	4204
300	90	10	7600	600	1590	6006
400	120	13	9880	461	2068	7808
500	150	16	12160	375	2545	9610
600	180	19	14440	316	3022	11412

DIVING DEPTH AND GAS VOLUME

In accordance with **Boyle's law,** the volume of any gas varies inversely with the pressure on it. Thus, if the pressure is doubled, the volume is halved; conversely, reducing the pressure by half causes the volume of any gas to expand to twice its previous size.

As shown in Table 26–1 and illustrated in Figure 26–1, if divers fill their lungs with 6 liters of air at sea level and descend to a depth of 10 m, the lung volume is compressed to 3 liters; diving an additional 10 m to a depth of 20 m (where the external pressure is now 3 atmospheres), the original 6-liter lung volume is reduced one third to 2 liters. In fact, at a depth of 91 m (300 ft), the lung volume is compressed to about 0.6 liter simply due to the force of water acting against the air-filled chest cavity. For most of us, any further increase in diving depth would cause the air volume in the respiratory tract to become so small as to produce serious damage to the chest wall and lung tissue. When the diver returns to the surface the air volume reexpands to its original 6-liter volume. Of great significance to the scuba diver who breathes pressurized air is the fact that 6 liters of air in the lungs at a depth of 10 m will expand to 12 liters when brought to the surface. This same 6-liter volume at 50 m will occupy 36 liters at sea-level pressure! **If this "extra" air is not permitted to escape through the nose or mouth during ascent, the lung tissue will rupture under the force of expanding gases.**

SNORKELING AND BREATHHOLD DIVING

Swimming at the surface of the water with fins, mask, and snorkel is a common form of recreation and sport. This "skin diving" is popular for spearfishing and exploring shallow areas of clear water. The snorkel is a J-shaped tube that allows the swimmer to breathe continually with the face immersed in water. The swimmer periodically takes a full breath of air and dives beneath the water for a closer look. After about 30 seconds, the carbon dioxide level in the blood increases, the diver senses the need to breathe and quickly surfaces. This activity is basically an extension of swimming and is limited entirely by the swimmer's breathhold ability.

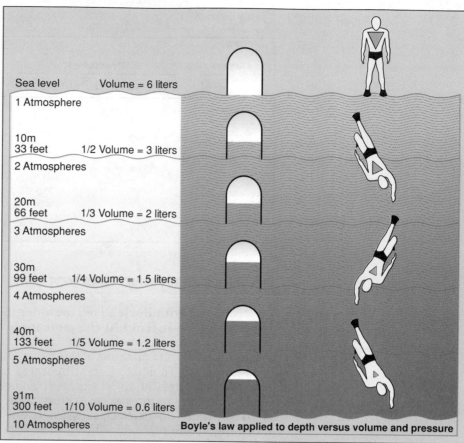

Sea level   Volume = 6 liters

1 Atmosphere

10m
33 feet   1/2 Volume = 3 liters

2 Atmospheres

20m
66 feet   1/3 Volume = 2 liters

3 Atmospheres

30m
99 feet   1/4 Volume = 1.5 liters

4 Atmospheres

40m
133 feet   1/5 Volume = 1.2 liters

5 Atmospheres

91m
300 feet   1/10 Volume = 0.6 liters

10 Atmospheres   **Boyle's law applied to depth versus volume and pressure**

FIG. 26–1.
Any gas volume varies inversely with the pressure on it. A 6-liter volume, whether in an open bell or in the flexible chest cavity, is compressed to 3 liters in 33 feet of water. This occurs because the external pressure has been doubled. At 99 feet or 4 atmospheres, the gas is reduced to 25% of its original volume or 1.5 liters.

LIMITS TO SNORKEL SIZE

Novice skin divers often speculate that if only the snorkel were longer, it would certainly be possible to swim deeper in the water and still breathe ambient air through the top of the snorkel. Some novices even believe they could sit at the pool bottom and breathe through a garden hose extending up to the pool deck! Although the idea of a longer snorkel is intriguing, two factors must be considered: (1) the increased water pressure on the chest cavity as one descends beneath the water, and (2) the increase in pulmonary dead space brought on by enlarging the snorkel.

Inspiratory Capacity in Relation to Depth

When breathing through a snorkel, the diver must inspire air at atmospheric pressure. At a depth of about 3 ft (1 m), the compressive force of water against the chest cavity is so large that the inspiratory muscles are usually unable to overcome external pressure and expand thoracic dimensions. Consequently, inspiration is impossible unless air is supplied at a pressure sufficient to counter the compressive force of water at the particular depth. This is the basic principle of scuba that is discussed fully in a later section.

Snorkel Size in Relation to Pulmonary Dead Space

Not all of the inspired air enters the alveoli. About 150 ml of each breath fills the nose, mouth, and other nondiffusable portions of the respiratory tract. The snorkel adds to the volume of this anatomic dead space.

Ideally, the snorkel should be about 15 in. (38 cm) long with an inside diameter of five-eighths to three-quarters inch.[16] Any further increase in snorkel size significantly increases the dead space and encroaches on one's capacity for alveolar ventilation.

BREATHHOLD DIVING

If a person takes a full inspiration of ambient air, about 1 liter of oxygen is brought into the lungs. If the breath is then held, 600 to 700 ml of oxygen in the lungs can be utilized to sustain metabolism before the partial pressures of arterial oxygen and carbon dioxide signal the need for renewed breathing. With some training, most of us can breathhold for about 1 minute. During this time, arterial P_{O_2} drops to about 60 mm Hg whereas P_{CO_2} rises to 50 mm Hg. During exercise, breathhold time is greatly reduced because oxygen consumption and carbon dioxide production increase directly with the severity of the activity. **Breathhold diving, preceded by hyperventilation, may extend the breathhold period yet significantly increase the danger of the dive.**

Hyperventilation and Breathhold Diving: Blackout

The problem of sudden loss of consciousness or **blackout** is unique to skin diving and usually occurs in divers who attempt to extend the duration of a dive beyond reasonable limits. The cause is probably either a critical reduction in oxygen, an increase in carbon dioxide, or the combined effects of both.

The breakpoint for breathhold usually corresponds to an increase in arterial P_{CO_2} to about 50 mm Hg. For some people, however, it is possible to "ignore" this stimulus and continue to breathhold until carbon dioxide reaches levels that cause severe disorientation and even blackout.[3] With hyperventilation prior to breathhold, arterial P_{CO_2} may decrease from its normal value of 40 mm Hg to 20 or 15 mm Hg. This elimination of carbon dioxide significantly extends the duration of breathhold until the arterial P_{CO_2} increases to a level to stimulate ventilation. For example, the longest breathhold recorded while breathing air without prior hyperventilation is 270 seconds. Breathholds of 15 to 20 minutes have been reported with hyperventilation followed by several breaths of pure oxygen.[12]

The attempt to extend breathhold time in diving, however, is not without serious risks.[3,4,11] If a skin diver hyperventilates at the surface prior to a dive, arterial P_{CO_2} is reduced and the potential length of the breathhold is extended. The diver now takes a full inhalation and descends beneath the water. Alveolar oxygen continually moves into the blood to be delivered to the working muscles. Owing to the previous hyperventilation, carbon dioxide levels remain low and the diver is essentially "free" from the urge to breathe. Concurrently, as the diver goes deeper, the external water pressure compresses the thorax. This increased pressure maintains a relatively high P_{O_2} within the alveoli. Thus, even though the absolute quantity of oxygen in the lungs is lowered (as oxygen moves into the blood during the dive), adequate pressure is maintained to load hemoglobin as the dive progresses. Now, as the diver senses the need to breathe and begins the ascent, significant reversals in pressure occur. As water pressure on the thorax decreases and lung volume expands, the partial pressure of alveolar oxygen is reduced proportionately. In fact, as the diver nears the surface, the alveolar P_{O_2} may be so low that dissolved oxygen actually leaves the blood and flows into the alveoli! In this situation, the diver may suddenly lose consciousness before reaching the surface.

Hyperventilation Should Not Be Used in the Water

Two other physiologic effects must be considered in addition to the dangerous reduction in P_{O_2} with prolonged breathhold.

1. **A normal quantity of arterial carbon dioxide is necessary to maintain the acid-base balance of the blood.** This is mediated by the release of H^+ as carbonic acid is formed from the union of carbon dioxide and water. By reducing the blood's carbon dioxide content through hyperventilation, the H^+ concentration is lowered, causing the blood to become more **alkaline.**
2. **Arterial P_{CO_2} also provides a continuous stimulus for the dilation of small arteries in the brain.** The elimination of carbon dioxide during hyperventilation can reduce cerebral blood flow and cause dizziness or even loss of consciousness. This would obviously create a dangerous situation in the water.

Depth Limits with Breathhold Diving: Thoracic Squeeze

The body's air cavities are subjected to tremendous compressive forces as the skin diver progresses deeper beneath the water. In general, if the lung volume is compressed below 1.0 to 1.5 liters (to the residual lung volume), internal and external pressures are unable to equalize and **lung squeeze** occurs. This effect of excessive hydrostatic pressure can cause extensive damage to respiratory tissue.

There is considerable variability among individuals as to the safe depth for breathhold diving without danger of lung squeeze. This critical depth is generally determined by the ratio of the diver's total lung volume to residual lung volume; this ratio usually averages about 4:1 at the surface. **There is no danger from lung squeeze if the lung volume remains greater than the residual volume.** This is because sufficient air remains in the lungs and rigid respiratory passages to equalize pressure and prevent damage due to compression. For most of us, this critical depth is usually about 30 m. If the lung volume during a dive is reduced below residual volume—that is, if the ratio of total to residual volume is less than 1.00—the pulmonary air pressure becomes less than the external water pressure. This unequalized pressure creates a relative vacuum within the lungs that can cause serious damage to alveolar tissue. In severe cases of lung squeeze, blood is literally sucked from the pulmonary capillaries through the alveoli into the lungs. In this situation, the diver actually drowns in his own blood. Further increases in depth cause compression fractures of the ribs as the chest cavity begins to cave in from the external pressure.

Other Problems

If internal and external pressures are unable to equalize, problems other than lung squeeze limit the depth of any dive. For example, if air at ambient pressure remains trapped in the middle ear (owing to inflamed tissue or a mucous plug) and cannot equilibrate with air from the lungs, the hydrostatic pressure will cause the eardrum to move inward and eventually to rupture. This can occur even at relatively shallow depths.

The sinuses are also a source of difficulty for skin divers. Air that is compressed in the lungs by the external force of water attempts to move into the paranasal sinuses. Inflamed and irritated sinuses (due to infection) have extremely narrow openings and often are unable to equilibrate with small pressure change. This creates a relative vacuum in these cavities and distorts the shape of the involved tissue causing intense sinus pain. In severe cases, fluid and blood move into the sinuses to fill the vacuum.

SCUBA DIVING

In our discussion of snorkeling, it was noted that at depths below 1 m, the inspiratory muscles are unable to overcome the compressive force of a column of water against the chest cavity. To counteract this external force, air under pressure must be supplied from an external source so that inspiratory action is possible. The **self-contained underwater breathing apparatus** or **scuba** is the most common apparatus used for this purpose. The scuba system is strapped to the diver's chest or back and consists of a tank of compressed air and a regulator with a hose and mouthpiece or full face mask. Two basic scuba designs are used: (1) the popular open-circuit system, and (2) the closed-circuit system that is useful for military operations and special applications that require mixed gases.

OPEN-CIRCUIT SCUBA

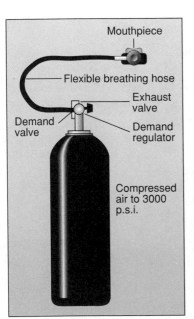

FIG. 26–2.
Open-circuit scuba system.

The typical open-circuit scuba system is illustrated in Figure 26–2. **This is the only form of scuba that should be used in sport diving!** For most diving purposes, the tanks are constructed of aluminum and contain 1000 to 2000 liters of air compressed to pressures of about 3000 pounds per square inch (p.s.i.). One tank supplies enough air for $\frac{1}{2}$ to 1 hour at moderate depths. The compressed air flows through a regulator valve that reduces the tank pressure to the "ambient" pressure at a particular depth. A slight negative pressure is created as the diver begins inspiration. This causes the demand valve to open and release air to the diver at a pressure about equal to the external pressure of the water. On exhalation, the inspiratory valves close and the exhaled air is discharged into the water.

Open-circuit scuba is not without its drawbacks. Because expired air generally contains about 16 to 17% oxygen, the open-circuit system is wasteful because about 75% of the total oxygen in the tank is ultimately exhaled into the water. In addition, a significant mass of air must be supplied to the diver at increased depths to provide the appropriate tidal volume for adequate pulmonary ventilation. As an extreme example, the equivalent of 50 liters of air at sea level must be pumped to a diver at 300 ft (90 m) to provide a 5-liter volume! This effect of pressure greatly limits the length of time one can remain at a great depth before the air is depleted in the scuba tank.

Figure 26–3 shows the theoretical air time limit for a diver doing similar work at various depths. These theoretical times are based on a completely filled standard tank and a 60-ft per minute rate of ascent and descent. For example, a single aluminium tank with 80 cubic ft of air compressed to 3000 p.s.i. normally supplies an 80 minute dive near the surface. At a depth of 10 m, however, this tank supplies enough air for about 40 minutes, whereas at 3 atmospheres or 20 m, the diving time is reduced by one third to about 27 minutes. Of course, these time limits will vary due to the effects of body weight, type of activity, level of fitness, and experience of the diver, all of which impact on ventilatory volumes.

CLOSED-CIRCUIT SCUBA

The closed-circuit underwater breathing apparatus operates in the same manner as the closed-circuit spirometer described in Chapter 8. A small cylinder of pure oxygen feeds into a bellows or bag from which the diver breathes. This breathing bag acts as a pressure regulator. Appropriate valves in the breathing mask direct the exhaled gas through a cannister containing soda lime that absorbs carbon dioxide and passes the carbon dioxide-free gas back to the diver. The cylinder of oxygen replenishes the oxygen used in metabolism. Consequently, oxygen is continually rebreathed, and the only gas eliminated from the tank is that used to supply the metabolic requirements of the dive. With only a small cylinder of oxygen, the diver can remain submerged for several hours. The system also provides for an almost completely silent and "bubble-free" operation in contrast to open-circuit diving systems.

Two main problems exist with closed-circuit scuba: First, serious injury

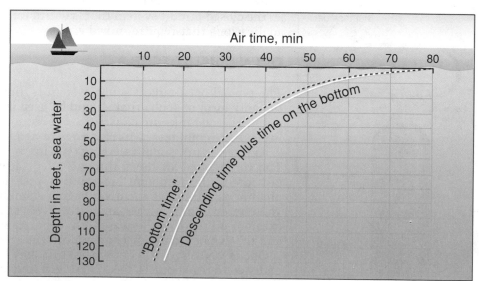

FIG. 26–3.
Theoretical air time for a single tank containing 80 cubic feet of air. The solid line includes the time spent while descending (at a rate of 60 ft · min⁻¹) plus the time on the bottom; the dashed line represents only "bottom time."

can occur should the carbon dioxide output exceed its rate of absorption or should the absorbent fail altogether. With a faulty rebreathing system, the diver may not receive warning symptoms and can drown as a result of being anesthetized by carbon dioxide. Second, high concentrations of oxygen, especially when breathed under pressure, have a variety of adverse effects on physiologic functions. In fact, at greater than 2 atmospheres of pressure, oxygen becomes a deadly poison. This phenomenon is discussed more fully in the section on "oxygen poisoning."

SPECIAL PROBLEMS BREATHING GASES AT HIGH PRESSURES

Underwater breathing systems must supply air or other gas mixtures at sufficient pressure to overcome the force of water against the diver's chest. At 20 m or 3 atmospheres of pressure, for example, the respired gas must be delivered at approximately 2280 mm Hg (3 × 760 mm Hg), whereas at 60 m the gas is delivered at a pressure of 5320 mm Hg. In the sections that follow, we consider the specific dynamics of breathing gases at high pressures and examine their effects on bodily functions as well as on the physical responses of gas to abrupt pressure changes. Figure 26–4 summarizes the main hazards of scuba diving that result from improper equalization of pressure within the body's air spaces (and diving mask) in response to changes in external pressure.

AIR EMBOLISM

An air volume breathed underwater expands in direct proportion to the reduction in external pressure as the diver ascends toward the surface. Air breathed at a depth of 10 m doubles in volume if brought to the surface. This expanding air vents freely if normal breathing continues during the ascent. If a diver takes a full breath at this depth but fails to exhale while ascending to the surface, the progressive and rapid expansion of gas eventually ruptures the lungs before the surface is reached. The potential for **lung burst** is quite real; many beginning divers react to a perceived danger by filling their lungs and then "holding their breath" while rapidly swimming to the surface. This particular diving hazard is not necessarily related to deep dives. Accidents caused by breathhold ascent with scuba frequently occur near the surface where the changes in pressure have the greatest effect on the expanding lung volume. **A full inhalation in 6 ft of water can cause serious overdistension of lung tissue during ascent if the diver does not exhale on the way up.** In fact, fatal **air embolism** has occurred in swimming pools as shallow as 8 ft for an inexperienced diver using scuba.[8]

If the expansion of air in the respiratory tract causes lung tissue to rupture, air bubbles may be forced into the pulmonary venous system. These bubbles or **emboli** are then carried to the heart and passed into the systemic circulation. Because the diver is usually in the head-up, vertical position on ascent, the bubbles move upward in the body. Eventually, they become lodged in the small arterioles or capillaries and can restrict the blood supply to vital tissue. General symptoms of air embolism include confusion, weakness, dizziness, and blurred vision. Severe blockage of pulmonary, coronary, and cerebral circulation causes collapse, unconsciousness, and frequently death. The only effective treatment for air embolism is rapid decompression to reduce the size of the bubbles and to force them into solution in order to

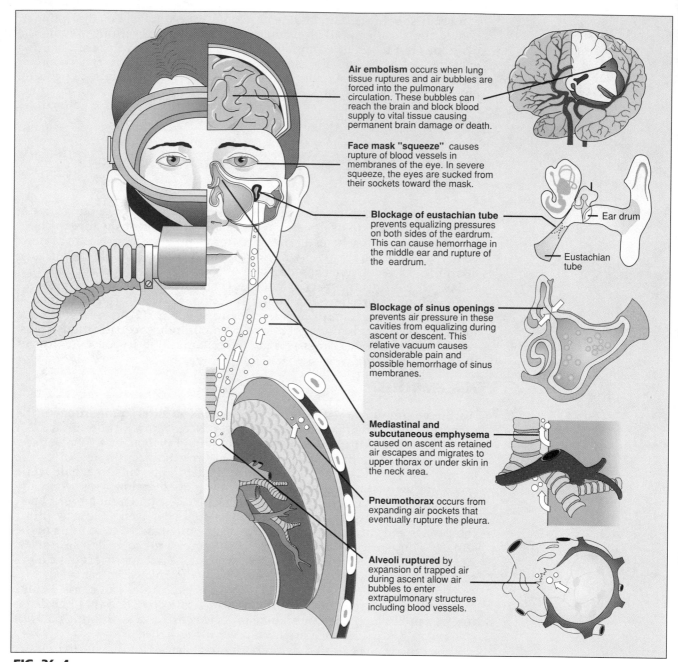

FIG. 26—4.
Hazards in scuba diving due to inability to equalize internal and external pressure.

open the plugged vessels. Even with rapid, expert treatment, approximately 16% of air embolism victims die.

PNEUMOTHORAX: LUNG COLLAPSE

Often, when lung tissue ruptures and air is forced through the alveoli, it migrates laterally to burst through the pleural sac covering the lungs.[8] This causes an air pocket to form in the chest cavity outside the lungs between the chest wall and lung tissue. Continued expansion of this trapped air during ascent causes the ruptured lung to collapse and the heart and other organs to be pushed out of their normal positions. Treatment for this condition of **pneumothorax** often requires surgical intervention in which a syringe is used to "extract" the air bubble.

To eliminate the danger of air embolism and pneumothorax, all divers must be instructed to ascend slowly and breathe normally when using scuba gear. It is imperative that the diver's lungs be free of any disease that could lead to the trapping of air and make it difficult for alveolar pressure to equalize during ascent.

MASK SQUEEZE

Prior to a dive, air in the face mask or goggles is at the same pressure as ambient air at the surfac. A considerable pressure differential develops between the inside and outside of the mask as the diver progresses beneath the water. This creates a relative vacuum within the mask. For example, wearing swimming goggles to protect the eyes from chemical irritants and to improve vision beneath the water can cause the eyes to bulge or "squeeze" from their sockets, which leads to capillary rupture and hemorrhage of the eyes and surrounding soft tissue at depths below 6 to 8 ft—a depth which represents the deep end of most pools.[6] This is because most goggles are made of rigid materials so that the only means to equalize pressure during surface diving is through the movement of the eye and surrounding soft tissue into the air space between the eye and the goggles. As many of the newer pools with separate diving wells reach depths of 14 ft (4.3 m), the use of goggles could pose a special problem to swimmers who frequently dive to the bottom to explore or retrieve a lost object. During breathhold diving, however, with a face mask that covers both the eyes and nose, the pressure in the mask can equalize to the ambient hydrostatic pressure because the nasal passages are connected to other air compartments, especially to the relatively large volume of air in the lungs. With scuba, the inspired air is automatically adjusted to the outside water pressure. Thus, periodically exhaling through the nose into the mask balances the pressures on both sides of the face mask. In breathhold diving, air in the lungs becomes compressed and can be passed through the nose to equalize the mask pressure.

AEROTITIS: MIDDLE-EAR SQUEEZE

Another significant problem encountered by the diver is equalizing the air space within the **eustachian tubes,** the passages that connect the middle ear with the back of the throat.[1] These mucus-lined passages are relatively narrow and generally resist air flow. In healthy individuals, the tubes are clear so that changes in external pressure against the eardrum can be equalized by the same pressure transmitted from within the lungs through the eustachian tubes. In both skin and scuba diving (as well as in air travel in nonpressurized aircraft), middle-ear pressure can usually be equalized by blowing gently against closed nostrils. Swallowing, yawning, or moving the jaws from side to side is also helpful in "popping the ears."

With an upper respiratory infection, the eustachian tube membranes swell and produce mucus that may plug these air passages. The greatest difficulty is usually experienced in equalizing middle-ear pressure during descent because a change in pressure against the outer surface of the eardrum is not readily met by an equal force from the interior. In diving, the pressure changes are considerable compared to those experienced in air travel. The diver can suffer severe pain in only a few feet of water because the eardrum becomes stretched and moves inward toward the plugged canal. With further pressure disequilibrium, the relative vacuum in the middle ear causes tissue to hemorrhage. When the eustachian tubes are totally blocked, the eardrums rupture and water rushes into the middle ear, which is at a considerably lower pressure.

Never Dive with Earplugs!

Earplugs should never be worn while diving. The external water pressure will push the ear plug deep into the external ear canal. If a pocket of ambient air is trapped between the plug and eardrum, the eardrum may rupture outward during descent. People who have respiratory disease, perforated eardrums, or temporary blockage of the eustachian tubes due to infection should not dive. In the latter case, diving can be resumed again when the infection subsides and the ear canals clear.

Aerosinusitis

Inflamed and congested sinuses prevent the air pressure in these cavities from equalizing during descent or ascent. If sinus air pressure does not equalize during descent, the air in these spaces remains at atmospheric pressure while external pressure increases. This relative vacuum creates "sinus squeeze," which causes bleeding in the sinus membranes as blood moves in to equalize the pressure differential.

NITROGEN NARCOSIS: RAPTURE OF THE DEEP

During the dive, the total pressure of the respired gas increases in direct proportion with the depth of the dive. Likewise, the partial pressure of each gas in the breathing mixture increases so that at 10 m the partial pressure of nitrogen is double the sea-level value, i.e., 1200 mm Hg. With each additional 10 m, the partial pressure of nitrogen increases by 600 mm Hg—at a depth of 60 m, the inspired P_{N_2} is about 4200 mm Hg. Thus, at each successive depth, a gradient exists for the net flow of nitrogen across the alveolar membrane into the blood and eventually into the tissues, with which it equilibrates. At 20 m, for example, all tissues will eventually contain about three times as much nitrogen as they did before the dive. Factors such as tissue perfusion, solubility coefficients of tissue, and temperature all influence the uptake of nitrogen at the tissue level.

An increase in the pressure and quantity of dissolved nitrogen causes physical and mental reactions characterized by a general state of euphoria not unlike alcohol intoxication. This has been termed **rapture of the deep.** In fact, this effect at a depth of 30 m is likened to the feelings one often experiences after drinking a martini on an empty stomach; at 60 m the feelings are similar to the effects of two or three martinis. Eventually, high nitrogen levels produce a numbing or anesthetic effect on the central nervous system, the symptoms of which are collectively termed **nitrogen narcosis.** This affects thought processes to the extent that a diver may feel that the scuba gear is unnecessary and may actually remove it and swim deeper instead of toward the surface.

As nitrogen diffuses slowly into the body tissues, the narcosis effect depends not only on depth but also on the duration of the dive. Although great individual variation in sensitivity exists, a mild narcosis usually appears after an hour or more at a depth of 30 to 40 m. **This is usually the maximum recommended depth range for recreational scuba divers.** The treatment for nitrogen narcosis simply requires that the diver ascend to a shallower depth—recovery is usually immediate and complete.[16]

DECOMPRESSION SICKNESS: THE BENDS

Decompression sickness or the "bends" occurs when dissolved nitrogen moves out of solution and forms bubbles in body tissues and fluids. It is caused by a diver ascending to the surface too rapidly after a deep, prolonged dive which is often made possible with the use of double and triple air tanks. Because nitrogen reaches equilibrium slowly in many tissues, especially fatty tissues, it leaves the body slowly. This means that women, who usually possess a greater percentage of body fat than men, may possibly be at greater risk of decompression sickness.[9] The relationship of body fat to nitrogen elimination from the body is illustrated for two dogs in Figure 26–5. The fact that dog B was fat and dog A was lean is reflected by the large difference in nitrogen content in proportion to their body mass.

Nitrogen Elimination

If the diver ascends at a prescribed, relatively slow rate, all of the body's excess dissolved nitrogen has sufficient time to diffuse from the tissues into

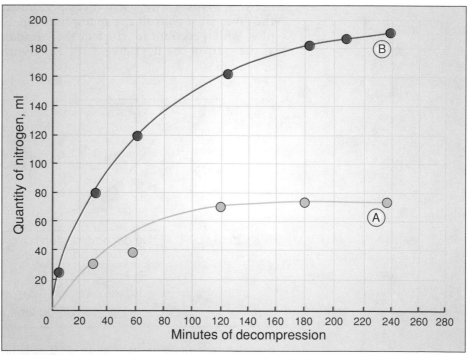

FIG. 26–5.
Elimination of nitrogen from the body tissues during decompression of a relatively lean dog (A) and one that was high in body fat (B). (Courtesy of Dr. A.R. Behnke)

the blood and be eliminated through the lungs. If the ascent is too rapid and the external pressure against the diver's body is reduced dramatically, the excess nitrogen begins to separate from the dissolved state and eventually forms bubbles in the tissues. This effect is not unlike the formation of carbon dioxide bubbles in a carbonated beverage when the cap is removed. As long as the cap is in place, the gas is under pressure and remains dissolved. When the cap is removed, the pressure is suddenly reduced and bubbles form.

Zero Decompression Limits

Diving at 30 m for up to 30 minutes, for example, is the upper limit before sufficient nitrogen is dissolved to pose danger from the bends. About 18-minutes duration is the limit at 40 m, whereas almost an hour can be spent at 20 m without danger of decompression sickness. If the diver exceeds the depth-duration recommendations for compressed-air diving shown in Figure 26–6, the ascent to the surface must be accomplished in stages at a specified rate to specific depths in a stepwise fashion. Such pauses or decompression stops give sufficient time for excess nitrogen to diffuse from the tissues via the blood to exit through the lungs without bubbles forming.[16] For example, a dive to 30 m for 50 minutes requires one 2-minute decompression stop at 6 m (20 ft) and another 24-minute stop at 3 m (10 ft).

For the sport diver, a conservative approach is recommended and the diver should not exceed a 20- to 25-m depth with 30 m considered as maximum. Furthermore, the diver should never approach the time limits indicated by the decompression tables during single or repetitive dives.[7] The recommendations in Figure 26–6 assume a single dive with a minimum of 12 hours between dives. For repeated dives performed within 12 hours of each other the diver must consult the appropriate repetitive dive decompression schedules, which account for the fact that residual nitrogen remains in the body at the start of the next dive if it occurs within the 12-hour period.

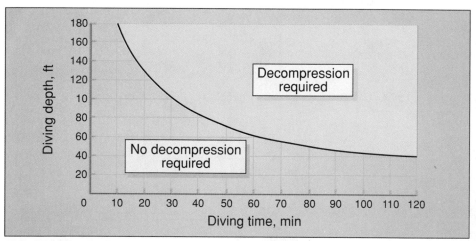

FIG. 26–6.
Zero decompression limits. Any single dive falling on the left side of the curve requires no decompression as long as the rate of ascent does not exceed 60 ft per minute (m = ft × 0.3048). Dives on the right side of the line require a decompression period, the specifics of which can be obtained from Navy Standard Decompression Tables.[13]

Consequences of Inadequate Decompression

The evidence is overwhelming that bubbles within the vascular circuit initiate the complications responsible for decompression injury. With the exception of bubbles forming in central nervous tissue, the primary bubbles occur in the venous and arterial vascular bed.[2] Symptoms of **decompression sickness** usually appear within 4 to 6 hours after the dive. When decompression procedures have been severely violated, for example, when a diver runs out of air and ascends rapidly, symptoms may appear immediately and progress to paralysis within minutes.[15] The most common symptoms of inadequate decompression are dizziness, itchy skin, and pain in the legs and arms, especially in the "tight" tissues such as ligaments and tendons. The degree of injury depends on the size of the bubbles and where they are formed. Bubbles in the lungs can cause choking and asphyxia, whereas bubbles in the brain and coronary arteries block blood flow. The bubbles deprive these vital tissues of oxygen and nutrients and cause subsequent cellular damage and death. Central nervous system "bends" is relatively common and especially serious because it can cause permanent neural damage if treatment is delayed.

Treatment for the bends involves the lengthy process of recompression in a small chamber to force the nitrogen gas back into solution. This step is followed by slow decompression so that the expanding gas has sufficient time to leave the body as the diver is brought back to the "surface."

For the sport diver, the chances are slim of having a recompression chamber nearby. This makes it imperative that the recommendations for diving depth and duration not be exceeded.

OXYGEN POISONING

In general, when the inspired P_{O_2} exceeds 2 atmospheres or 1520 mm Hg, the diver becomes highly susceptible to **oxygen poisoning**.[10] For this reason, closed-circuit scuba systems that use pure oxygen place severe restrictions on both the depth and length of a dive. At depths greater than 25 ft (7.6 m), the use of pure oxygen is not recommended except in extraordinary circumstances (Table 26–2).

High oxygen pressures affect bodily functions in several ways: (1) a high P_{O_2} directly irritates respiratory passages and eventually induces bronchopneumonia if exposure is continued; (2) at pressures of about 2 atmospheres, oxygen tends to constrict cerebral blood vessels and has a profound effect on central nervous system function; and (3) high pressures of inspired oxygen may also cause dysfunction by affecting carbon dioxide elimination. Specifically, a high inspired P_{O_2} may force sufficient oxygen into solution in the plasma to supply the metabolic needs of the diver. Oxygen is therefore returned in combination with hemoglobin so that carbon dioxide elimination via hemoglobin is greatly reduced.

The treatment for oxygen poisoning consists of breathing air at sea-level pressure.

CARBON MONOXIDE POISONING

Carbon monoxide is a lethal gas that combines with hemoglobin 200 times more readily than oxygen. Thus, tissue hypoxia can easily occur with only a small quantity of carbon monoxide in the inspired mixture. This becomes

TABLE 26.2.
UNITED STATES NAVY RECOMMENDED DEPTH-TIME LIMITS BREATHING PURE OXYGEN DURING WORKING DIVES

Normal Operations*		
Depth (ft)	(m)	Time (min)
10	3.0	240
15	4.6	150
20	6.1	110
25	7.6	75

Exceptional Operations		
Depth (ft)	(m)	Time (min)
30	9.2	45
35	10.7	20
40	12.2	10

* No symptoms of oxygen poisoning were noted at these depths and durations.

a problem in deep dives because the partial pressures of all gases in the breathing mixture (including the impurities) increase greatly.

Contaminants from automotive and industrial exhausts, including carbon monoxide and oxides of sulfur, may be high in urban areas. For this reason, it is not advisable to fill scuba tanks when an air pollution alert is in effect. Aside from the contaminants present in free air, carbon monoxide and oil impurities can also be added during the operation of the gasoline or diesel engine compressor. This potential source of contamination can be eliminated by placing the engine's exhaust downstream from the air intake that provides the compressed air.

SUMMARY

1. The underwater diver is exposed to high pressures and to the possibility of rapid changes in pressures. The diver can suffer severe injury and even death unless adjustments are made to equalize pressures in the body's air-filled cavities.
2. Snorkel size is limited by two factors: (1) the underwater depth at which the skin diver can generate sufficient inspiratory force to breathe, and (2) the additional pulmonary dead space created by the snorkel's volume.
3. Hyperventilation increases breathhold time but can also contribute to underwater blackout. This extremely dangerous consequence of hyperventilation is probably brought on by a critical reduction in oxygen, an abnormal increase in carbon dioxide, acid-base imbalance, or the combined effects of all these factors.
4. The maximum depth for breathhold diving is generally determined by the point at which the diver's lung volume is compressed to residual volume. Below this critical depth, internal and external pressures are unable to equalize and lung squeeze results.
5. Because scuba systems supply breathing mixtures at great depths and pressures, specific hazards result from improper equalization of pressure in the lungs, sinus, and middle-ear spaces. The most significant dangers are from air embolism, pneumothorax, mask and middle-ear squeeze, and aerosinusitis.
6. Gases breathed at high pressures move across the pulmonary membrane and eventually dissolve and equilibrate in all fluids and tissues of the body. High tissue pressures of oxygen and nitrogen have profound effects on physiologic function. Because of these problems, the maximum recommended depth for breathing compressed air is generally about 30 m.
7. As a scuba diver ascends toward the surface, a gradient is created for the net flow of nitrogen from the body fluids into the lungs. If the ascent is too rapid, the excess nitrogen that cannot exit through the lungs escapes from the dissolved state and forms bubbles in the tissues. This extremely painful condition is termed the "bends."

REFERENCES

1. Becker, G.D.: Barotrauma resulting from scuba diving: an otolaryngological perspective. *Phys. Sportsmed.*, 13:113, 1985.
2. Behnke, A.R.: Decompression sickness: advances and interpretations. *Aerosp. Med.*, 42:255, 1971.

3. Craig, A.B., Jr.: Causes of loss of consciousness during underwater swimming. *J. Appl. Physiol.*, 16:583, 1961.
4. Craig, A.B., Jr.: Summary of 58 cases of loss of consciousness during underwater swimming and diving. *Med. Sci. Sports*, 8:171, 1976.
5. Craig, A.B., Jr.: Principles and problems of underwater diving. *Phys. Sportsmed.*, 8:72, 1980.
6. Craig, A.B., Jr.: Physics and physiology of swimming goggles. *Phys. Sportsmed.*, 12:107, 1984.
7. Davis, J.C.: Decompression sickness in sport scuba diving. *Phys. Sportsmed.*, 16(2):108, 1988.
8. Kindwall, E.P.: Medical aspects of sport scuba diving. *Aqua Notes*, 3:1, 1976.
9. Kizer, K.W.: Women and diving. *Phys. Sportmed.*, 9:84, 1981.
10. Lambertsen, C.J.: Effects of oxygen at high partial pressure. In *Handbook of Physiology*. Edited by W.O. Fenn and H. Rahn. Washington, DC, American Physiological Society, 1965.
11. Lanphier, E.H., and Rahn, H.: Alveolar gas exchange during breathhold diving. *J. Appl. Physiol.*, 18:471, 1963.
12. Lin, Y.-C.: Breath-hold diving in terrestrial mammals. In *Exercise and Sport Sciences Reviews*. Vol. 10. Edited by R.L. Terjung. Philadelphia, Franklin Institute, 1982.
13. Miller, J.W. (ed).: *NOAA Diving Manual*. Washington, DC: U.S. Department of Commerce, U.S. Government Printing Office, 1979.
14. Strauss, R.H.: Diving medicine. *Am. Rev. Respir. Dis.*, 119:1001, 1979.
15. Strauss, R.H., and Yount, D.E.: Decompression sickness. *Am. Sci.*, 65:598, 1977.
16. *The NOAA Diving Manual*. U.S. Department of Commerce, Washington, DC, National Oceanic and Atmospheric Administration, 1976.

BODY COMPOSITION, ENERGY BALANCE, AND WEIGHT CONTROL

An accurate appraisal of body composition provides an important basis to formulate an intelligent program of total fitness. The frequently used standard—the height-weight tables—is of limited value in evaluating physique, since it is well established that overweight and overfat are not synonymous. This point is clearly illustrated with athletes, many of whom are muscular and exceed some average weight for their age and height, but otherwise are lean in terms of body composition. For such people, a weight loss program is unnecessary and may even be detrimental to sports performance. In contrast, comprehensive dietary management programs are surely needed by the 55 million dieters in the United States. In 1989, they spent in excess of 30 billion dollars on 54 million diet books and for services and products at 1500 weight control clinics in the hope of trying to permanently reduce excess fat! By 1995, spending will inflate to a projected 51 billion dollars as dieters intensify their efforts to achieve permanent weight loss. Also of importance is the need for effective programs of weight control among adults who suffer the insidious consequences of physical inactivity and overeating. For these people, body fat exceeds even the most liberal limits for normalcy. To this end, exercise plays a crucial role and, in our opinion, is the key ingredient in the formula to attain long-term success.

In this section we discuss body composition—its assessment, and differences between men and women, trained and untrained. We also deal with topics relevant to obesity and weight-control programs that use diet and exercise.

The evaluation of body composition permits quantification of the major structural components of the body—muscle, bone, and fat. Although height-weight* tables are often used to assess the extent of "overweightness" based on age and "frame size," **such tables do not provide reliable information about the relative composition or quality of an individual's body.** They are essentially based on statistics of the average ranges of body mass for people aged 25 to 59 years where the **mortality rate** is lowest, without regard to specific causes of death or quality of health prior to death. Someone may weigh much more than the average weight-for-height standards based on life insurance company statistics, yet still be "underfat" in terms of the body's total quantity of fat. The "extra" weight could simply be additional muscle mass. According to tables, assuming a "large" frame size, the desirable body mass range for a professional football player who is 188 cm tall and weighs 116 kg is 78 to 88 kg. The average weight for young adult males who are 188 cm tall is 85 kg. Using either set of criteria, the player is clearly "overweight" and by conventional standards should reduce his body mass at least 28 kg just to achieve the upper limit of the desirable weight range! He must reduce an additional 3 kg to match his "average" American male counterpart.

If the player followed these guidelines, it is a good bet he would no longer be playing football and might even jeopardize his overall health by undertaking a crash or bizarre diet regimen that prohibited the proper intake of essential nutrients. Although many larger-sized persons are indeed "overweight," they are not necessarily too fat and may not necessarily need to reduce. The total fat content of the football player was only 12.7% of his body weight compared with 15.0% body fat typically reported for young male nonathletes.[54] This player's body fat was below that normally found in the general population, even though he weighed much more than the average.

Such observations were first noted by Navy physician Dr. Albert Behnke in the early 1940s on 25 football players, 17 of whom were found unfit for military service because they were overweight and presumably too fat or obese.[6] A careful evaluation of each player's body composition, however, revealed that the so-called excess weight was due primarily to muscular hypertrophy. **The term overweight refers only to body mass in excess of some standard, usually the mean body mass for a given stature.** Being above some "average," "ideal," or "desirable" body mass based on height-weight tables should not necessarily dictate whether or not someone goes on a reducing regimen. A more desirable alternative is to determine the body

---

* The proper scientific terms (SI units) are stature (cm) and mass (kg). We acknowledge the distinction and have changed weight to mass and height to stature as long as it preserves readability.

composition by one of several laboratory or field techniques. In this chapter, we review various methods for the assessment of body composition.

COMPOSITION OF THE HUMAN BODY

The three major structural components of the human body include muscle, fat, and bone. Because there are marked gender differences in body composition, a convenient basis for evaluation and comparison is to employ the concept proposed by Behnke of the **reference man** and **reference woman.**[4] Figure 27–1 depicts the gross composition for a reference man and woman in terms of muscle, fat, and bone. This theoretical model is based on the average physical dimensions obtained from detailed measurements of thousands of individuals from large-scale anthropometric surveys.

The reference man is taller, heavier, his skeleton weighs more, and he has a larger muscle mass and lower total fat content than the reference female. These gender differences exist even when the amount of fat, muscle, and bone are expressed as a percentage of body mass. It is not known how much of this difference in body fat is biologic or how much is behavioral, owing perhaps to the more sedentary lifestyle of the average female. More than likely, hormonal differences play an important role. The concept of reference standards does not mean that men and women should strive to achieve the body composition of the reference models or that the reference man and woman are in fact "average." The models are useful as a frame of reference for statistical comparison and interpretation with data from other studies.

ESSENTIAL AND STORAGE FAT

The total amount of body fat exists in two depots or storage sites. The first depot, termed **essential fat,** is the fat stored in the marrow of bones and in the heart, lungs, liver, spleen, kidneys, intestines, muscles, and lipid-rich tissues throughout the central nervous system. This fat is required for normal physiologic functioning. In the female, essential fat also includes **sex-specific or sex-characteristic fat.** It is not at all clear whether this fat depot is expendable or serves as reserve storage.

The other major fat depot, **storage fat,** consists of fat that accumulates in adipose tissue. This nutritional reserve includes the fatty tissues that protect the various internal organs from trauma and the larger subcutaneous fat volume deposited beneath the skin surface. Although the proportional distribution of storage fat in males and females is similar (12% in males, 15% in females), the total percentage of essential fat in females, **which includes the sex-specific fat,** is 4 times higher than in males. More than likely, the additional essential fat is biologically important for child-bearing and other hormone-related functions.

Figure 27–2 illustrates a proposed theoretical model of body fat distribution for Behnke's reference female. Note that for the woman, as part of the 5 to 9% sex-specific reserve storage fat, the breasts contribute approximately 4.4% of the total mass of body fat, or no more than 12.5% to the quantity of sex-specific fat.[61] We interpret this to mean there must be other substantial sex-specific depots in the female such as the pelvic, buttock, and thigh regions that contribute quantitatively to female body-fat stores.[83,84]

Fat-free Mass and Lean Body Mass. The terms fat-free mass and lean body mass are often considered interchangeable when they should not be.

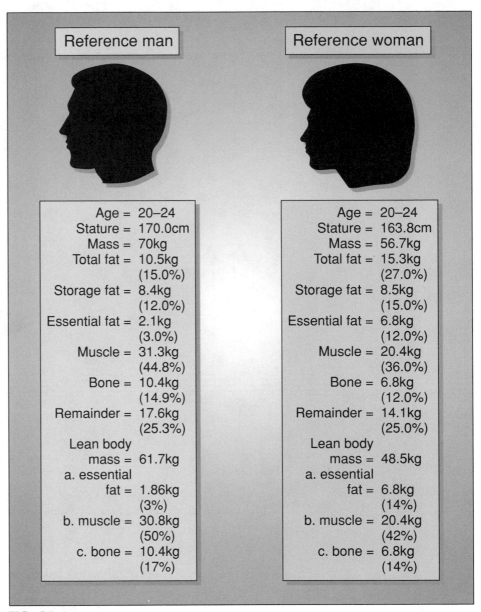

Reference man

Age = 20–24
Stature = 170.0cm
Mass = 70kg
Total fat = 10.5kg
(15.0%)
Storage fat = 8.4kg
(12.0%)
Essential fat = 2.1kg
(3.0%)
Muscle = 31.3kg
(44.8%)
Bone = 10.4kg
(14.9%)
Remainder = 17.6kg
(25.3%)
Lean body
mass = 61.7kg
a. essential
fat = 1.86kg
(3%)
b. muscle = 30.8kg
(50%)
c. bone = 10.4kg
(17%)

Reference woman

Age = 20–24
Stature = 163.8cm
Mass = 56.7kg
Total fat = 15.3kg
(27.0%)
Storage fat = 8.5kg
(15.0%)
Essential fat = 6.8kg
(12.0%)
Muscle = 20.4kg
(36.0%)
Bone = 6.8kg
(12.0%)
Remainder = 14.1kg
(25.0%)
Lean body
mass = 48.5kg
a. essential
fat = 6.8kg
(14%)
b. muscle = 20.4kg
(42%)
c. bone = 6.8kg
(14%)

FIG. 27–1.
Behnke's theoretical model for a reference man and woman. The mean values for 13 circumference measures and 8 skeletal diameters, including proportionality constants for the reference man and woman and comparisons with various groups, are presented in various publications.[5,8,51]

The lean body mass contains a small percentage of essential fat stores (perhaps as much as 3%), chiefly within the central nervous system, marrow of bones, and internal organs. In contrast, use of the term "fat-free" mass refers to the body mass devoid of all extractable fat. Behnke points out that the fat-free mass is an in vitro entity, and is the appropriate term with regard to carcass analysis.[5] Behnke views the lean body mass, on the other hand, as an in vivo

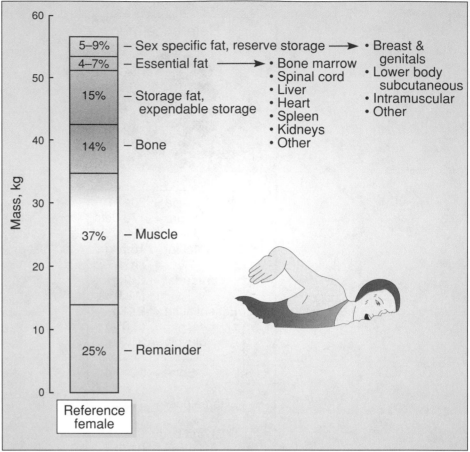

FIG. 27–2.
Theoretical model of body fat distribution for a reference female whose mass is 56.7 kg (stature is 163.8 cm), and body fat is 23.6% of body mass. (From Katch, V.L., et al.: Contribution of breast volume and weight to body fat distribution in females. *Am. J. Phys. Anthropol.,* 53:93, 1980.)

entity that remains relatively constant throughout the active adult life span with respect to water, organic matter, and minerals. **In normally hydrated, healthy adults, the only difference between the fat-free mass and lean body mass is the "essential" lipid-rich stores in bone marrow, brain, spinal cord, and internal organs.** Thus, in the calculation of lean body mass (LBM), even though "total" body fat is subtracted (LBM = Body mass − Fat mass), the small quantity of essential fat would still be present.

As shown in Figure 27–1, the lean body mass in males and the minimal body mass in females (see p. 603) is composed chiefly of essential fat (plus sex-specific reserve fat for females), muscle, and bone. The density of the reference man with 12% storage fat is 1.070 g · cc^{-1}; this is in addition to the approximately 3% essential fat. The density of the lean body mass is 1.094 g · cc^{-1}. If the "total" fat of the reference man is 15.0% (storage plus essential fat), then the density of the hypothetical "fat-free" body attains the upper limit of 1.100 g · cc^{-1}.

For the reference female, the average body density is 1.040 g · cc^{-1} at a body fat percentage of 26%; of this, about 12% is considered to be essential. The density of the minimal body mass of 48.5 kg is 1.072 g · cc^{-1}. In actual practice, densities of 1.075 to 1.080 g · cc^{-1} are rarely exceeded by the leanest, healthy, athlete, and normally hydrated female.

MINIMAL STANDARDS FOR LEANNESS

There seems to be a biologic lower limit beyond which a person's body mass cannot be reduced without impairing health status. This lower limit in men is the lean body mass and is calculated as body mass minus the mass of storage fat. For the reference man, the lean mass is equivalent to 61.7 kg; this includes approximately 3% or 2.1 kg essential fat. According to Behnke,[5] this amount of fat presumably is a lower limit, and any encroachment into this reserve may impair normal physiologic function or capacity for endurance-type exercise.

Low values of body fat have been reported for champion male athletes, chiefly world-class marathon runners, and some conscientious objectors who voluntarily reduced their body fat stores during a prolonged experiment with semistarvation.[62] The low fat levels of marathon runners, which ranged from 1 to 8% of body mass, probably reflect an adaptation to the severe training requirements of prolonged distance running. A minimal fat level permits a more effective gradient for the rapid transfer of metabolic heat produced during prolonged, high-intensity exercise.

Considerable individual differences also are found in the lean body mass of different athletes, with values ranging from a low of 48.1 kg in some jockeys to 112.5 kg in an All-Pro football defensive lineman and a high of 116.1 kg in an Olympic champion discus thrower. Table 27–1 presents data on the physique status and body composition of selected professional athletes who could be classified as "underfat" and "overweight." There are striking differences in body size between these groups, as well as in the relative body fat, lean body mass, the lean-to-fat ratio, and various girth measures. The defensive backs and offensive backs were "underfat" in relation to the reference man (or any other nonathletic standard), whereas the linemen and shot putters were clearly "overweight" for their stature (mean body mass relative to stature, or mass per unit size, was greater than 126 which represents the 90th percentile for nonathletic males).

Minimal Body Mass

In contrast to the lower limit of body mass of males that includes 3 to 4% essential fat, the lower limit of body mass for the reference female includes 12 to 14% essential fat. This theoretical limit is termed **minimal body mass,** and for the reference woman it is equivalent to 48.5 kg. In general, the leanest women in the population do not have body fat levels below about 10 to 12% of body mass.[51] This probably represents the lower limit of fatness for most women in good health. **This concept of minimal body mass in females that incorporates about 12% essential fat is equivalent to lean body mass in males that includes 3% essential fat.**

It should be emphasized that the concept of female minimal body mass is based on theoretical considerations, with little actual data. In carefully conducted experiments, however, values lower than 10% body fat are rarely reported. Data from female distance runners constitute an exception,[104] in which a value of 5.9% body fat was reported for one runner who weighed

TABLE 27–1.
PHYSIQUE AND BODY COMPOSITION OF "UNDERFAT" PROFESSIONAL FOOTBALL PLAYERS AND "OVER-WEIGHT" OFFENSIVE AND DEFENSIVE PROFESSIONAL FOOTBALL LINEMEN AND SHOT-PUTTERS

Variable	Defensive Backs (All-Pro)				Offensive Back (All-Pro, N = 1)	Offensive Lineman (Dallas, 1977, N = 10)	Defensive Lineman (Dallas, 1977, N = 5)	Shot-Putters (Olympics, N = 13)
	1	2	3	4				
Age, years	27.1	30.2	29.4	24.0	32	—	—	24.0
Stature, cm	184.7	181.9	187.2	181.5	184.7	193.8	197.6	187.0
Mass, kg	87.9	87.1	88.4	88.9	90.6	116.0	116.5	112.3
Relative fat, %	3.9	3.8	3.8	2.5	1.4	18.6	13.2	14.8
Absolute fat, kg	3.4	3.3	3.4	2.2	1.3	21.6	15.4	16.6
Lean body mass, kg	84.5	83.8	85.0	86.7	89.3	94.4	101.1	95.7
Lean/fat ratio	24.85	25.39	25.00	39.41	68.69	4.37	6.57	5.77
Girths, cm								
Shoulders	122.1	119.0	120.5	117.2	121.8	129.5	122.5	133.3
Chest	101.6	101.0	99.5	107.5	102.0	116.5	109.9	118.5
Abdomen, average	81.8	85.5	81.0	82.6	81.7	102.0	97.0	100.3
Buttocks	98.0	99.0	101.9	102.0	96.5	112.8	111.5	112.3
Thigh	61.0	61.0	58.5	64.0	63.2	66.2	69.3	69.4
Knee	39.5	41.3	41.1	38.0	41.0	44.8	45.8	42.9
Calf	37.6	38.8	38.8	37.8	41.3	43.5	42.4	43.6
Ankle	21.8	23.1	23.5	22.4	22.7	25.8	25.7	24.7
Forearm	31.8	29.1	31.1	31.8	33.5	33.5	34.8	33.7
Biceps	38.0	35.8	37.1	37.7	40.4	41.5	41.7	42.2
Wrist	18.5	17.2	17.4	17.5	18.0	19.3	19.3	18.9
D*	6.51	6.50	6.50	6.64	6.62	7.34	7.20	7.40
3F†	6.54	6.56	6.52	6.58	6.64	7.35	7.28	7.35
us-W‡	113.5	115.4	111.5	118.2	117.0	138.0	134.1	141.9

* D = sum of 11 girths/100.

† 3F = 3 x $\sqrt{\text{(mass, kg/stature, dm)}}$.

‡ us-W = unit size-Weight (mean body mass relative to mean stature). For the reference man and woman, us-W = 100. For males, us-W = 183.6 W/h$^{1.7}$, where W = mass, kg and h = stature, dm. For females, us-W = 204.2 W/h$^{1.7}$. The 90th percentile for us-W is 126 for males and 123 for females.

(From Katch, F.I., and Katch, V.L.: The body composition profile: techniques of measurement and applications. Clin. Sports Med. 3:31, 1984.)

52.6 kg. Although some measurement error is to be expected, this value is well below the limits for minimal weight specified by the Behnke model. Corroboration of such findings with a larger sample will probably result in modification of the lower limits of minimal body mass.

Calculation of Minimal Mass. Behnke has proposed a relatively simple method for estimating a woman's minimal body mass based on bone diameter measurements.[4] If body mass is lower than the computed minimal body mass, then the woman is clearly underweight and should not reduce body mass further without medical supervision. The following equation is used to calculate minimal body mass:

$$\text{Minimal body mass} = (D/33.5)^2 \times h_{dm} \times 0.111$$

where D is the sum of eight bone diameters, h_{dm} is stature in decimeters, and 33.5 and 0.111 are constants.

TABLE 27–2. PROCEDURE FOR COMPUTING A WOMAN'S MINIMAL BODY MASS

Diameter	Measurement, cm
Biacromial	34.4
Chest	23.8
Bi-Iliac	22.7
Bitrochanteric	29.8
Knees[a]	16.1
Ankles[a]	11.5
Elbows[a]	11.1
Wrists[a]	10.0
Sum =	159.4

[a] Sum of right and left sides

Step 1. Compute D, which is the sum of the 8 diameters. Note that the last 4 measurements represent the sum of the right and left sides.

Step 2. Substitute in the equation for minimal body mass:

$$\begin{aligned} \text{Minimal mass} &= (D/33.5)^2 \times h_{dm} \times 0.111 \\ &= (159.4/33.5)^2 \times 16.67 \\ &\quad \times 0.111 \\ &= 41.9 \text{ kg} \end{aligned}$$

The example shown in Table 27–2 illustrates the computation of minimal body mass in a young, thin-appearing female who weighed 38.7 kg (85.3 lb) and was 65.6 in. tall (166.7 cm or 16.67 dm). The eight bone diameter measurements were taken according to standard procedures.[4] Clearly, by Behnke's standards, this woman would be classified as **underweight** because her body mass of 38.7 kg is about 8% below her recommended minimal body mass.

UNDERWEIGHT AND THIN

The terms underweight and thin are not necessarily synonymous. In fact, in some cases they describe physical conditions that differ considerably. Measurements in our laboratories have focused on the structural characteristics of apparently "thin" females.[59] Subjects were initially screened subjectively as appearing thin or "skinny." The 26 women then underwent a thorough anthropometric evaluation that included the measurement of fatfolds, circumferences, and bone diameters, and the determination of body fat and lean body mass by hydrostatic weighing. (These techniques are discussed in detail beginning on page 607.)

The results were unexpected because the percent body fat of these women averaged 18.2%, which is about 7 percentage units below the average value of 25% body fat typically reported for college-aged women. The other striking finding was the lack of significant differences in four trunk and four extremity bone diameter measurements between the thin-appearing women and 174 women who averaged 25.6% fat and 31 women who averaged 31.4% fat. Appearing thin or skinny does not necessarily mean that skeletal frame size is diminutive or that the body's total fat content is excessively low, as would be the case for the lower limits for minimal body mass and essential fat proposed by Behnke.

We believe that three criteria can be used to designate a female who is underweight: (1) body mass lower than minimal body mass calculated from skeletal measurements as outlined in Table 27–2; (2) body mass lower than the 20th percentile by stature, and (3) percent body fat lower than 17%.

LEANNESS, EXERCISE, AND MENSTRUAL IRREGULARITY

Researchers have suggested that physically active females in general, as well as athletes in specific sports with a low percentage of body fat, increase their chances of either a delayed onset of menstruation, an irregular menstrual cycle (oligomenorrhea), or a complete cessation of the menses (amenorrhea).[69a] In support of this position are studies of female ballet dancers who, as a group, are quite lean and report a greater incidence of menstrual irregularities, eating disorders, and a higher mean age at menarche compared to age-matched, non-dance females.[29,99] When menstrual function is irregular or absent in premenopausal women, they face an increased risk of musculoskeletal injury when they participate in vigorous exercise.[66] One third to one half of female athletes in certain sports are believed to have some menstrual irregularity.[98] The speculation is that in some way the body "senses" when energy reserves are inadequate to sustain a pregnancy and thus ceases ovulation to prevent conception. Often cited is the maintenance of at least 17% body fat as the "critical level" for the onset of menstruation and 22% fat as the level required to maintain a normal cycle. It is argued by some that hormonal and metabolic disturbances are triggered that affect the menses if body fat falls below these levels.[28,29]

Leanness Is Not the Only Factor

Although the lean-to-fat ratio does appear to be important for normal menstrual function (perhaps through the role of peripheral fat in the conversion of androgens to estrogens), other factors must be considered. **This is because there are many active females who are significantly below the supposed critical level of 17% body fat who have normal menstrual cycles and who maintain a high level of physiologic and performance capacity.**[13,20,21,86] On the other hand, there are amenorrheic athletes with average levels of body fat. In addition, when the menstrual cycle returns to normal, it is not always associated with an increase in body mass or body fat. These observations have led researchers to focus on the role of physiologic and psychologic stress, including that provided by exercise training, in influencing normal menstrual function.[11,12,73,87,88] More recently, nutritional inadequacy and energy imbalance among athletes with irregular menstrual function have also been identified as possible predisposing factors.[13,44,86]

There are currently many outstanding female distance runners, gymnasts, and body builders in highly structured and vigorous training programs who have normal menses during intensive training and competition and who compete at a body fat level below 17%. In a study from one of our laboratories,[56] 30 female athletes and 30 nonathletes, all below 20% body fat, were compared for menstrual cycle regularity. Four of the athletes and 3 nonathletes who ranged from 11 to 15% body fat had regular cycles, whereas 7 athletes and 2 nonathletes had irregular cycles or were amenorrheic. For the total sample, 14 athletes and 21 nonathletes had regular cycles. These data corroborate other findings, and support the contention that the hypothesis is simply not valid of a critical fat level of 17 to 22% based on estimates of body fat determined from regression equations that use stature and body mass.[69]

Other Factors Involved

The complex interplay of physical, nutritional, genetic, hormonal, regional fat distribution, psychologic, and environmental factors on menstrual function must be considered. Research shows that for active women, an intense bout of exercise triggers the release of an array of hormones, some of which have antireproductive properties.[17] It remains to be determined whether regular bouts of heavy exercise have a cumulative effect sufficient to alter a normal menses. It is of interest to note that when young amenorrheic ballet dancers received injuries that prevented them from exercising, normal menstruation resumed, although body weight remained unchanged.[99]

It appears that exercise-associated disturbances in menstrual function can be reversed with changes in lifestyle without serious consequences.[80,99] If a critical fat level does exist, it is probably specific for each woman and probably changes throughout life.

In light of what is presently known on this topic, approximately 13 to 17% body fat should be regarded as an upper bound estimate of some yet to be determined minimal level of fatness associated with the regularity of menstrual function. The effects and risks of sustained amenorrhea on the reproductive system are still to be determined. Failure to menstruate or cessation of the normal cycle should be evaluated by a gynecologist/endocrinologist because it may reflect a significant medical condition such as pituitary or thyroid gland malfunction or premature menopause.[88a] Additional studies are needed to define the lower limits of body fatness compatible with regularity of menstrual cycle status, and to determine if a low level of body fat per se acts to modify hormonal regulation of ovulatory patterns. Recent studies demonstrate that the "height-weight" based regression equations to predict body fatness should not be employed to provide answers to such questions.[56,69]

DELAYED ONSET OF MENSTRUATION AND CANCER RISK

Researchers now suggest that the delayed onset of menarche generally observed in chronically active young females[34a,71,91] may provide a positive health benefit. Female college athletes who had started training in high school or earlier show a lower lifetime occurrence of cancer of the breast and reproductive organs, as well as nonreproductive-system cancers compared to their less active counterparts.[30,32] It is speculated that the mechanism for the lower cancer risk is linked to the production of less total estrogen or a less potent form of estrogen over the lifetime of an athlete with a delayed onset of menstruation.[31,36] Lower body fat levels may also be involved because fatty tissues convert androgens to estrogen.

COMMON TECHNIQUES FOR ASSESSING BODY COMPOSITION

Two general procedures are used to evaluate body composition: (1) direct, i.e., chemical analysis of the animal carcass and human cadaver, and (2) indirect, i.e., hydrostatic weighing and skinfold and girth measurements. Other indirect techniques such as isotope dilution, neutron activation analysis, potassium-40 counting, dual-photon absorptiometry, and infrared interactance are discussed in detail in various reviews and technical publications.[9,14,36a,70,105] Although the direct methods provide the theoretical validity for the indirect procedures, it is the indirect, noninvasive techniques that enable the exercise specialist to assess the fat and non-fat components of living persons.

The following section focuses on indirect procedures for assessing body composition. The first describes the application of Archimedes' principle of **hydrostatic weighing.** With this method, percent body fat is computed from body density (the ratio of body mass to body volume). The other procedures involve the **prediction** of body fat from fatfold and girth measurements, x-ray, total body electrical conductivity, ultrasound, computerized tomography, and magnetic resonance imaging.

ARCHIMEDES' PRINCIPLE: HYDROSTATIC WEIGHING

The Greek mathematician Archimedes is attributed with discovering the physical principle that serves as the basis of body composition evaluation. When King Hieron asked Archimedes to determine the gold content of his crown, he was in fact asking for an analysis of a two-component system of gold and some other metal that was supposedly diluting the crown's gold content. In solving this problem, Archimedes reasoned that the volume of water that overflowed his bath when he entered was equal to the volume of his submerged body. He also reasoned that an object either submerged or floating in water is buoyed up by a counterforce that equals the weight of water displaced. This buoyant force helps support the submerged object against the downward pull of gravity. Thus, an object is said to lose weight in water. **Because the object's loss of mass in water equals the mass of the volume of water it displaces, its specific gravity* can be defined as the ratio of its mass in air divided by its loss of mass in water.**

$$\frac{\text{Specific}}{\text{gravity}} = \frac{\text{Mass in air}}{\text{Loss of mass in water}}$$
$$\text{(Mass in Air} - \text{Mass in Water)}$$

Archimedes used quantities of silver and gold each having the same mass as the crown. When the silver and gold were submerged in a container filled with water, each caused a different volume to overflow. When the crown was submerged, it displaced more water than the gold but less than the silver. Thus, Archimedes deduced that the King's gold crown had been adulterated and was indeed composed of both gold **and** silver.

Validity of Hydrostatic Weighing for Estimating Body Fat

There is both direct and indirect experimental evidence that establishes the validity of hydrostatic weighing for estimating body fat content. In Behnke's early studies of Navy divers, 64 subjects were placed into two groups based on their body density. The mean difference between the groups in body mass and body volume was 12.4 kg and 13.29 liters, respectively. The ratio of these average differences (Δ weight $\div$ Δ volume), was 0.933 $g \cdot ml^{-1}$, a value within the densitometric range of 0.92 to 0.96 $g \cdot cc^{-1}$ for human adipose tissue. Thus, the difference in body mass between the high- and low-density groups was equivalent to the density of adipose tissue. When the density of a group of heavy but lean pro football players was determined by hydrostatic weighing, the players had an average density of 1.080 $g \cdot cc^{-1}$, and an average lean body mass that was 20 kg higher than that of the divers. As Behnke has stated,[4] "Here indeed was a presumptive demonstration that fat could be separated from bone and muscle in vivo or 'the silver from the gold' by application of a principle renowned in antiquity."

The upper and lower limits of body density in the population are approximately 0.93 $g \cdot cc^{-1}$ in the very obese and 1.10 $g \cdot cc^{-1}$ in the leanest males. This coincides nicely with the 1.10 density of fat-free and 0.93 for fat samples of homogenized whole body tissues in small mammals at 37° C.[63]

Possible Limitations

It must be pointed out that the validity of body density for estimating total body fat is based on several assumptions, some of which have not been verified in humans. Although 19th-century German anatomists had conducted experiments on the gross composition of various body tissues, only 6 adult human bodies had been analyzed for skin, muscle, adipose tissue,

---

* The specific gravity can be thought of as an object's degree of "heaviness" in relation to its volume. Objects of the same volume may, of course, vary considerably in density, density being defined as mass per unit volume (density = mass $\div$ volume). The volume of 1 g of water occupies exactly one cubic centimeter at a temperature of 39.2° F (4° C). Its density would be 1 g per cubic centimeter, or 1 $g \cdot cc^{-1}$. Because the density of water is greatest at 4° C, increasing the temperature would increase the volume of 1 g of water and decrease its density. It is necessary, therefore, to correct the volume of an object weighed in water in terms of the water's density at the weighing temperature. The temperature effect distinguishes density from specific gravity.

bone, and organ components prior to more recent cadaver analysis studies. In 1984, anatomic data were presented from analysis of 12 embalmed (6 male, 6 female) and 13 unembalmed (6 male, 7 female) cadavers ranging in age from 55 to 94 years.[23]

The time-consuming and meticulous nature of the cadaver experiments revealed that several of the basic assumptions were not consistent with the two-component model (fat and fat-free compartments) for body composition assessment. The basic assumption could not be affirmed that the densities of the fat-free components (bone, mineral, muscle, water) are constant. For example, the range for muscle as a percent of the fat-free mass was 41.9% (female) to 59.4% (male); similarly, the range for bone was 17.4% (male) to 25.7% (female). Also, other experiments have shown large individual variation in bone density in living humans.[2] The evidence from the cadaver studies on the tissue components and the studies on bone density indicate that body density determined by hydrostatic weighing may not be the ideal "gold standard" by which to estimate the true extent of total body fat and lean body mass. Hopefully, more cadaver studies over a broad age range in "healthy" males and females will be undertaken to help assess the theoretical implications raised by the latest cadaver studies.[23]

Computing Body Density

According to Archimedes' principle, if an object weighs 75 kg in air and 3 kg when submerged in water, the loss of weight in water of 72 kg is equal to the weight of the displaced water. Because the density of water at any temperature is known, the volume of water displaced easily can be computed. In the example, 72 kg of water is equal to 72 liters or 72,000 cubic centimeters (1 g of water = 1 cc in volume). If the water temperature is 4° C, there would be no correction factor. Refer to Appendix E for the correction factors at higher temperatures. The density of the person, computed as mass ÷ volume, would be 75,000 g ÷ 72,000 cc, or 1.0416 g · cc^{-1}. The laboratory procedures for measuring body density are discussed on pages (611 to 615). Once the body density is known, the next step is to convert the density value to an estimate of percentage of body fat.

Computing Percent Body Fat

The percentage of fat in the body can be determined from a simple equation that incorporates density.* The simplified "Siri equation" is obtained by substituting 0.90 g · cc^{-1} and 1.10 g · cc^{-1} for the densities of fat and lean tissue, respectively.

$$D = \frac{F + L}{(F/f) + (L/1)}$$

Because the density of the whole system equals the sum of its parts, F + L = 1.00.

---

* An algebraic expression that incorporates the densities of the fat and lean tissues can be represented by D, the density of the whole system, where F and L are the fat and lean tissue components, each with densities f and 1, respectively. The object is to solve for F.

$$D = \frac{1}{(F/f) + (L/1)}$$

By rearranging terms, the proportional contribution of F becomes:

$$F = \frac{1}{D} \times \frac{f \times 1}{(1-f)} - \frac{f}{(1-f)}$$

Its derivation by Berkeley scientist Dr. Willaim Siri[89] is:

Percent body fat $= \dfrac{495}{\textbf{Density}} - 450$

This equation was derived from a two-compartment model of the body consisting of fat and fat free tissue. Fat extracted from adipose tissue has a density of 0.90 g · cc^{-1} at 36° C, whereas lean tissue has a density of approximately 1.10 g · cc^{-1}. The assumption of early research in this area was that each of these densities remains relatively constant even with large individual variations in total body fat. In addition, it is assumed that the lean tissue components of bone and muscle are essentially at the same density among different individuals.

In the previous example, where mass and volume were equal to 75 kg and 72 liters, respectively, the density of 1.0416 g · cc^{-1} when converted to percent fat by the Siri equation equals 25.2%.

$$\text{Percent fat} = \frac{495}{1.0416} - 450$$
$$= 25.2$$

Several formulas other than Siri's equation have also been devised to estimate body fat from density.[16] The basic difference in calculating body fat between the formulas is generally less than about 1% within a range of body fat of 4 to 30%.

Possible Limitations for Density Assumptions

It should be noted that the generalized density values for lean and fat tissues of 1.10 and 0.90 g · cc^{-1}, respectively, are average values for young and middle-aged adults. The assumed densities for the components of the fat-free mass at 37° C are: water, 0.993 g · cc^{-1}; mineral, 3.000 g · cc^{-1}; protein, 1.340 g · cc^{-1}. Although it is assumed that these values are constants, this may not be the case. For example, the density of the lean body mass for blacks was estimated to be significantly **greater** than a corresponding group of whites (1.113 g · cc^{-1} *vs* 1.100 g · cc^{-1}).[85] The existing equations to calculate body composition from body density in whites, therefore, would tend to **overestimate** the lean body mass when applied to blacks.

In addition to racial differences, applying constant density values of the various tissues to growing children or aging adults could also add uncertainty in predicting body composition.[102] For example, the density of the skeleton (as well as water content and potassium concentration) is probably in continual change during the growth period as well as during the well documented demineralization of osteoporosis with aging. This would make the actual density of the lean tissue of young children and the elderly lower than the assumed constant of $1.10 \text{ g} \cdot \text{cc}^{-1}$. As a result, an **overestimation** of relative body fat would occur with densitometry for these individuals. This is one main reason why some researchers have not attempted to convert body density to body fat in these diverse groups, although others have adjusted for such factors in computing percent body fat from body density in pre-pubertal children.[67] With highly trained and select groups of athletes such as football players, the density of the lean component could theoretically exceed $1.10 \text{ g} \cdot \text{cc}^{-1}$. This would cause an **underestimation** of relative fat and would account for the negative values for percent body fat reported for some of these athletes.[1] A significantly large residual volume may have also caused the discrepancy in these results. In our own experience in measuring several thousand high-quality white and black performers (Olympic athletes, professional football, baseball, basketball, and soccer players, as well as champion body builders and gymnasts), we have never recorded a density value greater than $1.10 \text{ g} \cdot \text{cc}^{-1}$ when residual volume is measured directly and found to be in the range appropriate for body size and age.

Computing the Mass of Fat

The mass of the body's fat content is computed by multiplying percent fat by body mass.

$$\textbf{Fat mass} = \textbf{(Percent fat/100)} \times \textbf{Body mass}$$
$$= .252 \times 75 \text{ kg}$$
$$= 18.9 \text{ kg}$$

Further calculations could be done to subdivide the total fat mass for this person into essential and storage fat. For a female with about 12% essential fat, there would be 9.0 kg of essential fat (0.12×75 kg) and a remainder of 9.9 kg of storage fat (0.132×75 kg); for a male with 3% essential fat and 22.2% storage fat (based on a percent fat of 25.2), the corresponding values would be 2.3 kg for essential fat and 16.5 kg for storage fat. Clearly, if a man and a woman have the same percent body fat, the man could be considered "fatter" because a larger percentage of his body fat is in the form of storage fat. Because each gram of fat contains about 9 kcal (9000 kcal per kg), a rough approximation can also be made of the potential energy stored in each fat depot. For storage fat, the values would be 81,000 kcal for the female and 148,500 kcal for the male; for essential fat, the values would be 89,100 kcal for the woman and 20,700 kcal for the man.

Computing Lean Body Mass

Lean body mass is calculated by subtracting the mass of fat from body mass.

$$\textbf{Lean body mass} = \textbf{Body mass} - \textbf{Fat mass}$$
$$= 75 \text{ kg} - 18.9 \text{ kg}$$
$$= 56.1 \text{ kg}$$

MEASUREMENT OF BODY VOLUME

The principle discovered by Archimedes is applied to the measurement of body volume in one of two ways: (1) water displacement, or (2) underwater weighing. Body volume must be measured accurately because small variations in volume can have a substantial effect on the density calculation and, hence, on the computed values for percent fat and lean body mass.

Water Displacement

The volume of an object submerged in water can be measured by the corresponding rise in the level of water within a container. With this technique, the rise of water is measured in a thin tube secured to the side of a tank. This finely calibrated tube permits accurate volume measurements. When the volume of the body is measured in this manner, the volume of air remaining in the subject's lungs during submersion must be considered. This residual volume is usually determined before the subject enters the tank and is subtracted from the total body volume determined by water displacement.

Bottle Buoyancy

A new method has recently been introduced that is inexpensive, simple to administer, and has the added advantage of not requiring a scale or other measurement device to record the underwater weight.[60] The procedure involves the direct measurement of water volume by applying the buoyancy principle that states that 1 liter of displaced water has a buoyant force exerted that is equivalent to 1 kg at 4° C. The subject simply holds onto a container that is partially filled with water and submerges underwater. The underwater weight is equivalent to the weight of the volume of water added back to the container when the subject neither sinks nor floats, but is suspended just beneath the water surface.

The results are shown in Figure 27–3 of a recent experiment confirming that the measurement of total body volume by the buoyancy method (bottle) was as reliable and produces individual scores that are as dependable as those obtained by conventional underwater weighing (chair method) explained in the next section.

Underwater Weighing

With this procedure, body volume is computed as the difference between body mass measured in air and mass measured during water submersion. **Body volume is equal to the loss of mass in water with the appropriate temperature correction for the density of water.** Figure 27–4 illustrates the procedure to measure body volume by underwater weighing.

The subject's body mass is first determined in air on a balance scale accurate to ± 50 g. A diver's belt is usually secured around the waist of fatter-appearing subjects to ensure that they do not float upward during submersion. (The underwater mass of this belt and chair is determined beforehand and is subtracted from the subject's total mass under water.) The subject, who wears a thin nylon swim suit, sits in a light-weight, plastic tubular chair suspended from the scale and submerged beneath the surface of the water. A swimming pool can serve the same purpose as the tank, and the scale and chair assembly can be suspended from a support at the side of the pool. In the tank, water temperature is maintained at about 95° F, which is close to the subject's skin temperature. Water temperature is recorded to correct for the density of water at the weighing temperature.

The subject makes a forced maximal exhalation as the head is lowered

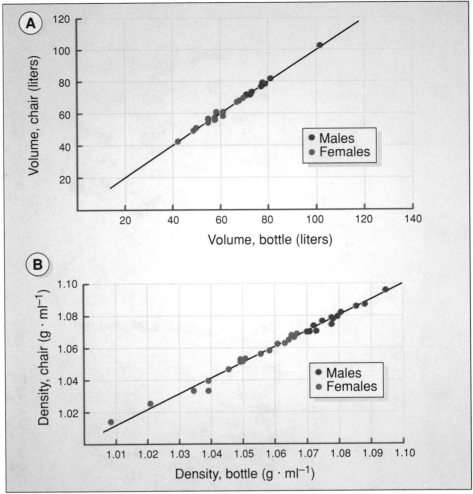

FIG. 27–3.
Results for (A) body volume and (B) body density by the bottle buoyancy method compared to traditional hydrostatic weighing (chair) in males and females. With the bottle buoyancy method, many people can be measured simultaneously in a swimming pool. (From Katch, F.I., et al.: Reliability and validity of a new method for the measurement of total body volume. *Res. Q. Exerc. Sport*, 60:286, 1989.)

under water. The breath is held for about 5 to 10 seconds while mass under water is recorded on a sensitive scale accurate to ± 10 g or on a force transducer system with digital readout. The underwater weighing procedure is repeated 8 to 12 times, because subjects "learn" to expel more air from their lungs with each additional underwater trial. We use an average of the last two or three weighings because these trials represent the subject's "true" weight under water with minimal intraindividual variation.[46] As was the case with the water displacement method, residual volume (preferably measured, not predicted to eliminate a potential source of significant error[74]) must be subtracted from total body volume determinations. Reproducibility

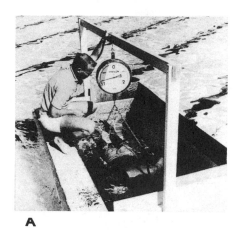

A

B

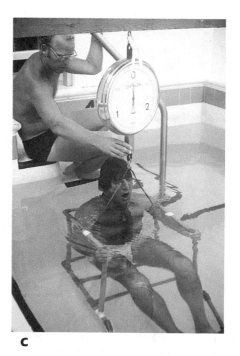

C

D

FIG. 27–4.
Measuring body volume by the procedure of underwater weighing. Prone and supine methods can be used for underwater weighing with no difference in results, [37] and residual lung volume can be measured either before, during, or after the underwater weighing. (A) Prone, (B) seated in a swimming pool, (C) seated in a therapy pool, and (D) upright in a stainless steel tank with plexiglass front in the laboratory. For any of the methods, a snorkel with nose clip can be used by subjects who are apprehensive about submersion.

of body volume scores measured several times on the same day or on consecutive days is always high, with the test-retest reliability coefficient usually above r = 0.94.[57] Normal fluctuations in body mass (chiefly body water) related to the menstrual cycle are not large enough to affect body density and body fat assessed by hydrostatic weighing.[19] In the small number of females who experience perceptible changes in body mass(>1.0 kg) during the menstrual cycle, however, this alteration in total body water will affect body density and introduce some error in computing percent body fat.[18]

Calculation of Body Composition from Body Mass, Body Volume, and Residual Lung Volume

Data from measurements on two professional football players who were "All-Pro" and played on Super Bowl teams are shown in Table 27–3.

The conventional formula for density is mass ÷ volume, where density is expressed in $g \cdot cc^{-1}$, and mass is in kilograms and volume is in liters. The difference between M_a and M_a is equal to the body volume when the appropriate water temperature correction (D_w) is applied. Because the air remaining in the lungs and any other gas remaining in the "spaces" of the body (abdominal viscera, sinuses) at the time of underwater weighing contribute to buoyancy, these volumes must be considered. In most subjects, abdominal gas and sinus air volumes are generally small and thus are ignored. **The residual lung volume is large and variable; it must be measured and subtracted from the total body volume.**

Whereas the residual lung volume tends to be slightly smaller when measured with the subject immersed in water compared to air,[76] probably due to the compressing force of water against the thoracic cavity, the effect is small on computed body fat.[81] Residual lung volume can be measured in air prior to underwater weighing without loss in accuracy compared to the simultaneous measurement of residual volume at the time of underwater weighing.[37]

TABLE 27–3.
MEASUREMENTS OF TWO PROFESSIONAL FOOTBALL PLAYERS USING THE UNDERWATER WEIGHING METHOD.

Variable	Symbol	Defensive Lineman	Running Back
Body mass, kg	M_a	121.73	97.37
Net underwater mass, kg	M_w	7.39	6.52
Water temperature correction	D_w	0.99336	0.99336
Residual lung volume, liters	RV	1.213	1.374
Total body volume, liters	TBV	113.89	90.08
Body density, $g \cdot cc^{-1}$	D_b	1.0688	1.0809
Body Composition			
Relative percent body fat, %*	% Fat	13.1	8.0
Absolute body fat, kg	FM	15.9	7.2
Lean body mass, kg	LBM	105.8	90.2

* Siri equation.[89]

The formula for calculating the density of the body (D_b) is:

$$D_b = \frac{Mass}{Volume} = \frac{M_a}{\dfrac{(M_a - M_w)}{D_w} - RV}$$

For ease in computation, the formula can be rewritten as:

$$D_b = M_a \times D_w / (M_a - M_w - RV \times D_w)$$

The lower part of Table 27–3 presents the body composition results based on the computation of body density.

FATFOLD AND CIRCUMFERENCE MEASUREMENTS

Hydrostatic weighing and water displacement are two widely used indirect methods currently available for assessing total body volume. When proper laboratory facilities are unavailable, alternative but simple procedures can be used to predict body fatness. Two of these procedures, the measurement of **subcutaneous fatfolds** and of **girths** or **circumferences,** require relatively inexpensive equipment.

Measurement of Subcutaneous Fatfolds

The rationale for fatfold measurements to estimate total body fat is that a relationship exists between the fat located in the depots directly beneath the skin and internal fat, and body density.[77]

The Caliper. By 1930, a special pincer-type caliper was used to measure subcutaneous fat at selected sites on the body with relative accuracy. The caliper works on the same principle as the micrometer used to measure distance between two points. The procedure for measuring fatfold thickness is to grasp firmly with the thumb and forefingers a fold of skin and subcutaneous fat, pulling it away from the underlying muscular tissue following the natural contour of the fatfold. Constant tension of $10 \text{ g} \cdot \text{mm}^2$ is exerted by the pincer jaws of the calipers at their point of contact with the skin. The thickness of the double layer of skin and subcutaneous tissue is then read directly from the caliper dial and recorded in millimeters within two seconds after applying the full force of the caliper. This is done to avoid the complications of fat fold compression when taking the measurements.[3]

The Sites. The most common areas for taking fatfold measurements are at the triceps and subscapula, and at the suprailiac, abdominal, and upper thigh sites. All measures are taken on the right side of the body with the subject standing. A minimum of two or three measurements are made at each site, and the average value is used as the fatfold score. When fatfolds are measured for research purposes, the investigator has usually had considerable experience and is consistent in duplicating values for the same subject made on the same day, consecutive days, or even weeks apart. Figure 27–5 shows the anatomic location of the five most frequently measured

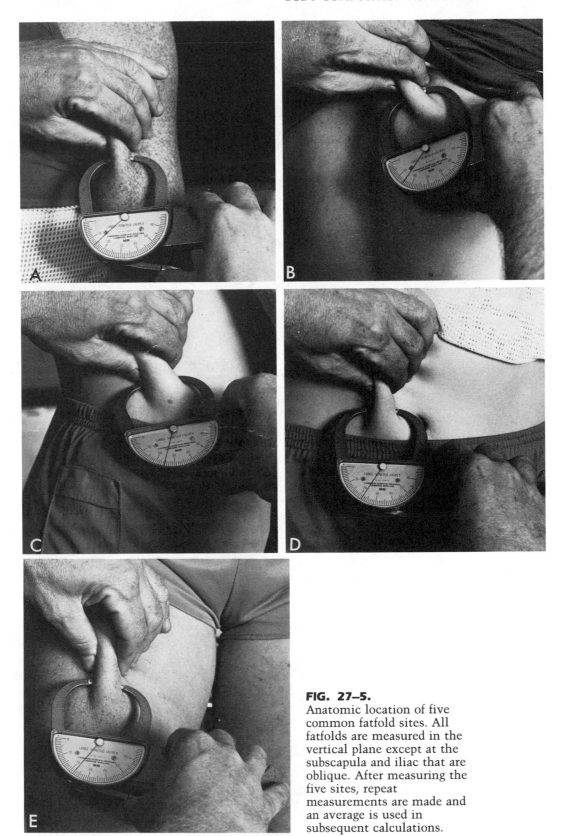

FIG. 27–5.
Anatomic location of five common fatfold sites. All fatfolds are measured in the vertical plane except at the subscapula and iliac that are oblique. After measuring the five sites, repeat measurements are made and an average is used in subsequent calculations.

fatfold sites: (A) triceps—vertical fold measured at the midline of the upper arm halfway between the tip of the shoulder and the tip of the elbow; (B) subscapula—oblique fold measured just below the bottom tip of the scapula; (C) suprailiac—slightly oblique fold measured just above the hip bone. The fold is lifted to follow the natural diagonal line at this point; (D) abdomen—vertical fold measured 1 inch to the right of the umbilicus; and (E) thigh—vertical fold measured at the midline of the thigh, two thirds of the distance from the knee cap to the hip. Fatfolds are also sometimes taken on the medial, lateral, and posterior calf and on the anterior chest wall at the level of the armpit. However, these measurements are often difficult to take depending on the degree of fatness of the individual.

Usefulness of Fatfold Scores

Fatfold measurements can provide fairly consistent and meaningful information concerning body fat and its distribution.[35] There are basically two ways to use fatfolds. The first is to sum the scores as an indication of the relative degree of fatness among individuals. The "sum of fatfolds", for example, can also be used to reflect changes in fatness "before" and "after" a physical conditioning regimen. Changes in individual fatfold values as well as the total score can then be evaluated on either an absolute or percentage basis.

From the fatfold data shown in Table 27–4, obtained from a 22-year-old female college student prior to and following a 16-week exercise program, the following observations can be made:

1. The largest changes in fatfold thickness occurred at the suprailiac and abdominal sites.
2. The triceps showed the largest decrease and the subscapular the smallest decrease when changes were expressed as percentages.
3. The total reduction in subcutaneous fatfolds at the five sites was 16.6 mm or 12.6% of the "before" condition.

The second way to use fatfolds is in conjunction with mathematical equations designed to **predict** body density or percent body fat. These equations are "population specific,"[40,50,78] and predict fatness fairly accurately for subjects similar in age, gender, state of training, fatness, and probably race, to those from which the equations were derived. **When these criteria are met, the predicted value of fatness for an individual is usually within 3 to 5% of**

TABLE 27–4.
CHANGES IN SELECTED FATFOLDS FOR A YOUNG WOMAN DURING A 16-WEEK EXERCISE PROGRAM

Fatfolds, mm	Before	After	Absolute Change	Percent Change
Triceps	22.5	19.4	−3.1	−13.8
Subscapular	19.0	17.0	−2.0	−10.5
Suprailiac	34.5	30.2	−4.3	−12.8
Abdomen	33.7	29.4	−4.3	−12.8
Thigh	21.6	18.7	−2.9	−13.4
Sum	131.3	114.7	−16.6	−12.6

the body fat based on body density measurements determined by hydrostatic weighing or water displacement.

As a result of several experiments conducted in our laboratories, equations have been developed to predict body fat from triceps and subscapula fatfolds.[53,54] The equations most useful for predicting total body fat in young women and men are:

Young women, ages 17 to 26 years
% body fat = .55(A) + .31(B) + 6.13
Where A = triceps fatfold, mm
B = subscapula fatfold, mm

Young men, ages 17 to 26 years
% body fat = .43(A) + .58(B) + 1.47
Where A = triceps fatfold, mm
B = subscapula fatfold, mm

Using the fatfold data of the young woman who participated in the 16-week physical conditioning program, we can compute her percent body fat prior to and following the conditioning program using the equation for young women. Substituting the pretraining values for triceps (22.5 mm) and subscapula (19.0 mm) fatfolds, percent body fat equals 24.4%.

% body fat = .55(A) + .31(B) + 6.13
= .55(22.5) + .31(19.0) + 6.13
= 12.38 + 5.89 + 6.13
= 24.4%

Substituting the values for triceps (19.4 mm) and subscapula (17.0 mm) fatfolds, the post-training value for percent body fat is computed in a similar fashion.

% body fat = .55(19.4) + .31(17.0) + 6.13
= 10.67 + 5.27 + 6.13
= 22.1%

The determination of percent body fat prior to and following a conditioning and weight control program provides a convenient means to evaluate alterations in body composition that often are independent of changes in body mass.

For young adults, approximately one half of the body's total fat is subcutaneous fat and the remainder is internal or organ fat. With advancing age, a proportionately greater quantity of fat is deposited internally in comparison to subcutaneous fat. Thus, the same fatfold score reflects a greater total percent body fat as one gets older. **For this reason, age-adjusted or generalized equations should be used when predicting body fat from fatfolds or girths (see next section) for older adult men and women.**[78,95]

Although the use of fatfolds to predict percent body fat has been widely used in the allied health professions, a major drawback is that the person taking the measurements must have considerable experience with the proper techniques to obtain consistent values. With extremely obese people the thickness of the fatfold often exceeds the width of the caliper's jaws. The particular caliper used may also contribute to the measurement error.[68] Because there are no standards by which to compare the results between different investigators from diverse geographic regions, it is difficult to determine which sets of fatfold data are the best to use. Thus, prediction equations developed by a particular researcher (that may be highly valid for the sample measured) may result in large errors in prediction when another person takes the fatfolds. The error in predicting body fat could be plus or minus 200% or higher![50]

Even when a different approach is used for body fat prediction that incorporates surface area with fatfolds,[58] there is still the basic limitation imposed by technique. The surface area formula method described below, may provide valid estimates of body fat for different populations. The basic equation is:

$$\% \text{ Fat} = \frac{\Sigma \text{ fatfolds}}{3F \times k(sf)}$$

where Σ fatfolds is the sum of two or more fatfold sites, depending on availability; $3F = 3\sqrt{\text{mass/stature}}$ (where mass is in kg and stature is in dm); and $k(sf) = \Sigma$ fatfolds/$3F \times \%$ fat. The % fat is based on a criterion method such as underwater weighing, and is based on the observed average for a group or population (of a particular age, gender, state of training, or sport). Thus, different $k(sf)$ constants are required for diverse populations. For example, consider the mean values for a population that represents relatively sedentary young men: stature = 18.42 dm; mass = 72.16 kg; fatfolds = 67.3 mm; % fat = 15.3. The $k(sf)$ constant is calculated as:

$$k(sf) = \frac{67.3}{3\sqrt{3.9175} \times 15.3} = 0.741$$

By first calculating a $k(sf)$ constant based on group data, the percent body fat for any similar individual can be computed with the basic equation. If mass = 74.0 kg, stature = 17.52 dm, and the sum of 5 fatfolds = 57.5 mm, then percent fat for this particular young man is computed as follows:

$$\% \text{ fat} = \frac{\Sigma \text{ fatfolds}}{3F \times k(sf)}$$

$$\% \text{ fat} = \frac{57.5}{6.166 \times 0.741}$$

$$\% \text{ fat} = 12.6\%$$

This was the method used to estimate the body fat content of the two champion body builders whose data are displayed in Figure 28–1 (Chapter 28). For the reference group of body builders, 3F was 6.666, percent fat was 6.4%, the sum of 5 fatfolds was 30.4 mm, and k was 0.71818.

Measurement of Girths

A linen or plastic measuring tape should be applied lightly to the skin surface so that the tape is taut but not tight. This procedure avoids skin compression that produces lower than normal scores. Duplicate measurements should be taken at each site and the average used. The anatomic landmarks for the various girths for young and older men and women are: (1) abdomen—one inch above the umbilicus; (2) buttocks—maximum protrusion with the heels together; (3) right thigh—upper thigh just below the buttocks; (4) right upper arm—arm straight, palm up, and extended in front of the body, measured at the midpoint between the shoulder and the elbow; (5) right forearm—maximum girth with the arm extended in front of the body with palm up; and (6) right calf—widest girth midway between the ankle and knee.

Different prediction equations have been developed for different genders and age groups.[54,88,89] The equations developed for these subgroups, although cross-validated on different samples with good results,[55] nevertheless seem to be population specific and should not be used to predict fatness in individuals (1) who appear very thin or very fat, and (2) who have been involved for a number of years in strenuous sports or resistance training programs.

Usefulness of Girth Measurements

The girth-based prediction equations are most useful in ranking or ordering individuals within a group according to their relative fatness. **If one uses the equations and constants presented in Appendix F for young and older men and women, the error in predicting an individual's body fat is generally between ±2.5 to 4.0%.** These relatively low prediction errors make the equations particularly useful to those without access to laboratory facilities, because the measurements are easy to take and a tape measure is inexpensive. Recently, specific equations utilizing girth measures have been developed to estimate the body composition of obese adult females and males.[41a,100] **In addition to predicting percent body fat, the use of girths is also well suited in the obese for determining patterns of fat distribution on the body as well as changes in body fat following weight loss.**[42,97]

Predicting Body Fat from Girths

The various sites measured in individuals of different genders and ages are indicated at the beginning of Appendix F. From the appropriate tables in the Appendix, one can substitute the corresponding constants, A, B, and C in the formula shown at the bottom of each table. One addition and two substraction steps are required. The following five-step example shows how to calculate percent fat, fat mass, and lean body mass for a 21-year-old man who weighed 174 lb (79.1 kg):

Step 1. The upper arm, abdomen, and right forearm girths were measured with a cloth tape and recorded to the nearest one-fourth inch.
   Upper arm = 11.5 in. (29.21 cm)
   Abdomen = 31.0 in. (78.74 cm)
   Right forearm = 10.75 in. (27.30 cm)

Step 2. The three constants A, B, and C corresponding to the three girth measures were determined from Appendix F.
   Constant A, corresponding to 11.5 in.   = 42.56
   Constant B, corresponding to 31.0 in.   = 40.68
   Constant C, corresponding to 10.75 in. = 58.37

Step 3. Percent body fat was computed by substituting the appropriate constants in the formula shown at the bottom of Chart 1 in Appendix F.

   % fat = Constant A + Constant B − Constant C − 10.2
      = 42.56 + 40.68 − 58.37 − 10.2
      = 83.24 − 58.37 − 10.2
      = 24.87 − 10.2
      = 14.7%

Step 4. Fat mass = % fat/100 × Body mass
   Fat mass = 14.7/100 × 174 lb
      = 0.147 × 174 lb
      = 25.6 lb

Step 5. LBM = Body mass − Fat mass
LBM = 174 lb − 25.6 lb
= 148.4 lb

Target Body Fat from Changes in Abdominal Girth

A new method has recently been devised calculated as a target percent body fat based on the difference between an initial value for abdominal girth and a calculated "target" abdominal girth formulated from a desired level of percent body fat.[60a] The method differs from the traditional approach that first determines percent body fat, and then the individual attempts to achieve a desired change in body mass or composition. With the new approach, the question is asked, "How much does the abdominal girth need to be reduced to achieve a desired percent body fat"?

In the following example, excess abdominal girth (measured at the level of the umbilicus) is calculated based on a desired level of fatness that represents the 50th percentile for the population:

Step 1. A target abdominal girth expressed in cm is computed as the product of (body mass, kg/stature, m) and a constant Q (at the 50th percentile for percent body fat: Q for males is 12.36 and 14.25 for females). For example, for a male with an abdominal girth of 89.7 cm whose body mass is 85.5 kg and stature is 1.876 m, (85.5/1.876) = 6.751. This value is then multiplied by Q (12.36) to yield the target abdominal girth of 83.4 cm.

Step 2. Excess abdominal girth is computed as the measured abdominal girth (89.7 cm in the above example) minus the target abdominal girth of 83.4 cm from Step 1. The difference of 6.3 cm (89.7 cm − 83.4 cm) is the excess girth. This person would then attempt to reduce abdominal girth by 6.3 cm. When this goal is reached, percent body fat will correspond to approximately the 50th percentile.

An important consideration with this new approach is to decide on the target or desired level of percent body fat. If different percentile values for body fat are desired, then different Q values must be used in Step 1. (See reference 60a for specific Q values.) Because this is a new approach to the quantification of determining excess body fat, future research studies must still refine and extend the applicability of the different Q constants that are used in Step 1.

APPLICATION OF SURFACE ANTHROPOMETRY: THE BODY PROFILE

A matrix of 11 girths can be integrated into a muscular and non-muscular "body profile" that provides a quantitative assessment of body shape.[5,7,51,59a] Appendix I presents a step-by-step procedure for computing and constructing the body profile that includes examples for an adult man over a 28 year span, Amherst College students contrasted between 1882 and 1886, and two children at ages 66 and 101 months. If the anthropometric proportions of the individual conformed to group symmetry, all of the deviation values on the body profile would fall within ±2% units of the vertical or zero deviation reference line.

The practical application of the body profile method of analysis allows one to quantify the relative proportions of the body's girth dimensions and chart the **changes** in the physical dimensions of individuals due to such factors as acute and chronic training, dietary intervention, or the influence of aging.[52] The body profile method also permits quantification of differences in physique status between athletes in different sports (e.g., male gymnasts vs. male long distance swimmers) or within the same sport (football defensive lineman vs. quarterbacks, or small or large body builders).

BIOELECTRIC IMPEDANCE ANALYSIS (BIA)

The principle of body impedance is based on the concept that electric flow is facilitated through hydrated fat-free tissue and extracellular water compared to fat tissue due to the greater electrolyte content (and thus lower electric resistance) of the fat-free component. Consequently, impedance to the flow of electric current will be directly related to the level of body fat. With the BIA technique, electrodes are placed on the hands and feet, a painless, localized electric signal is introduced, and the impedance or resistance to current flow is determined. The impedance value is then converted to body density, which in turn is converted to percent body fat by use of either the Brožek or the Siri equation (see p. 610).

One factor affecting the accuracy of this technique is the maintenance of normal hydration levels, as either dehydration or overhydration affects the normal concentrations of electrolytes in the body.[43,48,72] This will affect current flow independent of real changes in body fat. More specifically, a loss of body water will decrease the impedance measure to yield a lower % fat, whereas overhydration will produce the opposite effect. Even under conditions of normal hydration the resulting prediction of body fat may be questionable in relation to values obtained from hydrostatic weighing.[38] In fact, the technique may be less accurate than the various anthropometric methods that use girths and fatfolds to predict body fat.[41] Skin temperature, which is influenced by ambient conditions, also affects whole-body resistance and thus the prediction of body fat by BIA. Predicted body fat is significantly lower in a warm environment compared to a cold one.[22] Conflicting evidence relates to whether the technique may be able to detect small changes in body composition during weight loss compared to densitometry.[26,82] **At best, the technique of BIA is another noninvasive, indirect, relatively easy means for providing a general assessment of body composition, providing measurements are made under conditions strictly standardized for both ambient temperature and level of hydration.**[2a]

ULTRASOUND ASSESSMENT OF FAT

A lightweight, portable ultrasound meter is commercially available to measure the distance between the skin and fat-muscle layer and fat-muscle layer and bone. The ultrasound meter operates by emitting high frequency sound waves that penetrate the skin surface. The sound waves pass through adipose tissue until the muscle layer is reached; here the sound waves are then reflected from the fat-muscle interface to produce an echo that returns to the ultrasound unit. The time for sound wave transmission through the tissues and back to the receiver is converted to a distance score and displayed on a light emitting diode scale.

The reliability of repeat measurements in the lying and standing positions is high, as is the association between the two methods of obtaining measurements of surface fat thickness.[47] The ultrasound technique may be especially useful with the obese, for whom variation in compression of subcutaneous body fat is greatest.[64] The use of ultrasound for "mapping" of muscle and fat thickness at different body regions, as well as quantifying changes in the topographical fat pattern, is a valuable adjunct for the assessment of body composition.

ARM X-RAY ASSESSMENT OF FAT

The arm radiograph is a useful method for analysis of body composition that permits quantification of reliable and valid estimates of body fat.[49,61a] The objective of the x-ray assessment is to provide a direct and unambiguous estimate of body fat content. The thickness of the fat layers at points A, B, and C shown in Figure 27–6 are transformed into a value of body fat for the individual by use of the fatfold-surface area equation on page 620. The only difference is that the fat thickness from the roentgenogram is substituted for fatfolds. The k constant is derived in the same manner, and differs among individuals depending on age, level of fitness assessed by max $\dot{V}_{O_2}$, and race.[56] The validity of the x-ray procedure was determined by comparing values of fatness based on the roentgenogram and densitometry. The relationship between the two methods of estimating body fat was high (correlation of r = ~0.90 in young and older white and black men and women). For an individual, the conversion of the x-ray widths of fat to total body fat percentage was accurate to within ±3% units of body fat determined hydrostatically. This degree of accuracy is generally higher than for estimates of body fat derived from fatfolds or girths. The applications of the x-ray procedure also include assessment of muscle hypertrophy[27] that is useful in cross-sectional and longitudinal studies of body composition and aging, as well as clinical evaluation of nutritional status.

COMPUTERIZED TOMOGRAPHY (CT) ASSESSMENT OF FAT

The CT scanning procedure can provide cross-sectional images for any part of the body. By use of appropriate computer software, the scan provides pictorial and quantitative information for the distribution of the various tissue components within the scanned area.[86a,93] The top left portion of Figure 27–7 shows a cross section of the mid thigh. The picture elements (pixels) are plotted in the histogram, with the low density adipose tissue and the higher density muscle tissue forming two distinct peaks. The CT scans B and C show an anterior view of both upper legs and the cross section at mid thigh in a professional walker who completed an 11,200 mile walk through the 50 United States in 50 weeks.[92] A comparison of CT scans prior to and after completion of the walk showed a significant increase in the total cross-section of muscle area and decrease in peripheral subcutaneous fat in the mid-thigh region.

The future is indeed bright for the quantitative assessment of body composition including regional fat distribution as well as fat within the intraabdominal cavity by use of the newer, emerging technologies. The MRI (magnetic resonance imaging) scan illustrated in the bottom right portion of Figure 27–7 should prove to be invaluable for partitioning of the various tissues, and for analyzing changes in abdominal tissue components during aging or exercise training and nutritional supplementation or dietary restriction.

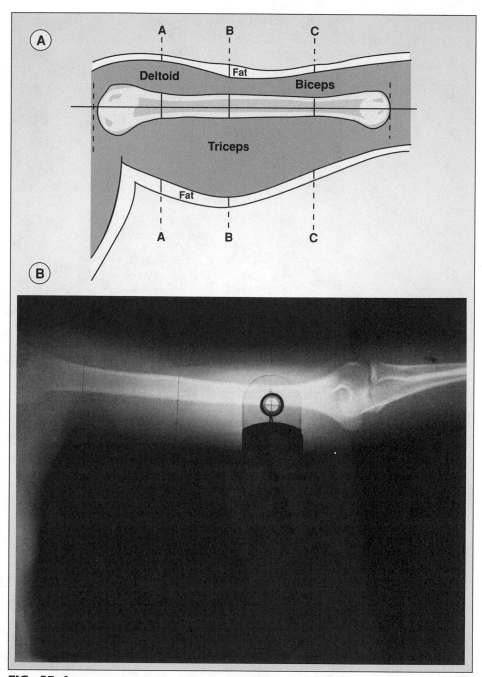

FIG. 27—6.
A, Schematic drawing of a radiograph of the arm of a 24-year-old female. The six fat widths are represented by the vertical lines drawn perpendicular to the long axis of the humerus at points A, B, and C. There is a demarcation on the radiograph between fat, muscle, and bone that permits an accurate assessment of radiographic widths. Total body fat determined from this radiograph was 23.6%; by underwater weighing, body fat was 23.3%. B, The technician is using a digitizer to calculate fat width on the radiograph. The information from the digitizer is processed by a computer to a printer and graphics plotter. (Line art and photo courtesy of Dr. A.R. Behnke and F. Katch.)

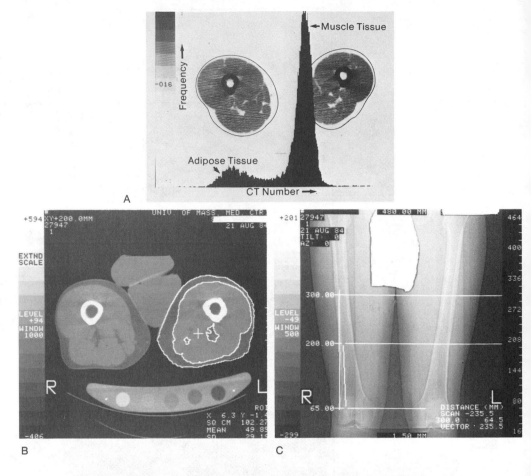

FIG. 27–7.
CT and MRI scans. A, Plot of pixel elements that illustrates extent of adipose tissue and muscle tissue in the cross section of the thigh. B, C, Cross section at mid thigh and anterior view of both upper legs prior to an 11,200-mile walk around the United States. D, MRI scan of mid-thigh of middle-aged adult during 13-week resistance-training program. (A, courtesy of Dr. Steven Heymsfeld, M.D. B, C, courtesy of the University of Massachusetts Medical School, Department of Radiology, Worcester, MA. D, courtesy of Radiology Department, Bay State Medical Center, Springfield, MA.)

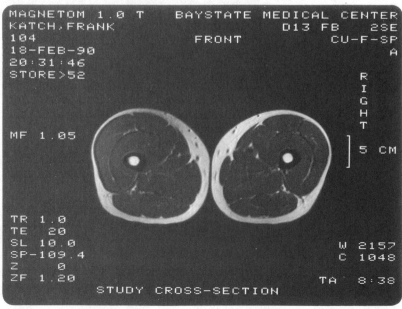

WHAT IS AVERAGE FOR PERCENT BODY FAT?

Considerable data are available concerning the average body composition for various groups of men and women of different ages and fitness levels. The best that can be done is to present the mean values of body fat for young and older men and women in Table 27–5 including plus or minus one unit of variation, the standard deviation. This gives some idea of the spread about the average body fat in each of the studies for 68% of the people in the samples measured.

It is evident from the mean values in Table 27–5 that with increasing age body fat increases in both genders. This average change does not necessarily mean the trend should be interpreted as being "normal," because studies have shown that participation in vigorous physical activities after age 35 can retard the "average increase in body fatness.[45,65,96] Higher relative fat values could be due in part to the skeleton as it becomes demineralized and porous, thereby reducing the body density because of the decrease in bone density. Adaptation to a more sedentary life style and a concomitant reduction in the level of daily physical activity could also cause the relative increases in body fat. This would occur even if the daily caloric consumption remained the same.

DESIRABLE BODY MASS

Although large quantities of body fat are undesirable from a health perspective, precise statements cannot be made for an optimum level of body fat or body mass for a particular individual. More than likely, this optimum varies from person to person and is greatly influenced by a variety of genetic factors. **Based on data from active young adults and competitive athletes, however, it does appear that it would be desirable to maintain body fat at about 15% of body mass, or less for men, and 25% or less for women.** This "optimal" or desirable body mass can be computed (based on a desired body fat level) as follows:

$$\text{Desirable body mass} = \frac{\text{Lean body mass}}{1.00 - \text{Desired \% fat}}$$

Suppose a 91 kg (200 lb) man who has 20% body fat wishes to know the mass he should attain so his new lower body mass would contain 10% body fat. The computations would be:

$$\text{Fat mass} = 91 \text{ kg} \times .20 = 18.2 \text{ kg}$$
$$\text{LBM} = 91 \text{ kg} - 18.2 \text{ kg} = 72.8 \text{ kg}$$
$$\textbf{Desirable body mass} = \frac{72.8 \text{ kg}}{1.00 - .10}$$
$$= \frac{72.8 \text{ kg}}{.90}$$
$$= 80.9 \text{ kg}$$
$$\textbf{Desirable fat loss} = \textbf{Current body mass} - \textbf{Desirable body mass}$$
$$= 91 \text{ kg} - 80.9 \text{ kg}$$
$$= 10.1 \text{ kg } (22.2 \text{ lb})$$

TABLE 27–5.
AVERAGE VALUES OF PERCENT BODY FAT FOR YOUNGER AND OLDER WOMEN AND MEN FROM SELECTED STUDIES

Study	Reference	Age Range	Stature cm	Mass kg	% Fat[a]	68% Variation Limits
Younger Women						
North Carolina, 1962	(90)	17–25	165.0	55.5	22.9	17.5–28.5
New York, 1962	(106)	16–30	167.5	59.0	28.7	24.6–32.9
California, 1968	(53)	19–23	165.9	58.4	21.9	17.0–26.9
California, 1970	(103)	17–29	164.9	58.6	25.5	21.0–30.1
Air Force, 1972	(24)	17–22	164.1	55.8	28.7	22.3–35.3
New York, 1973	(54)	17–26	160.4	59.0	26.2	23.4–33.3
North Carolina, 1975	e		166.1	57.5	24.6	—
Massachusetts, 1990	b	17–30	165.3	57.6	21.9	16.7–27.1
Older Women						
Minnesota, 1953	c	31–45	163.3	60.7	28.9	25.1–32.8
		43–68	160.0	60.9	34.2	28.0–40.5
New York, 1963	(107)	30–40	164.9	59.6	28.6	22.1–35.3
		40–50	163.1	56.4	34.4	29.5–39.5
North Carolina, 1975	e	33–50	—	—	29.7	23.1–36.5
Massachusetts, 1990	b	31–50	165.2	58.9	25.1	19.2–31.0
Younger Men						
Minnesota, 1951	(15)	17–26	177.8	69.1	11.8	5.9–11.8
Colorado, 1956	(79)	17–25	172.4	68.3	13.5	8.3–18.8
Indiana, 1966	d	18–23	180.1	75.5	12.6	8.7–16.5
California, 1968	g	16–31	175.7	74.1	15.2	6.3–24.2
New York, 1973	(54)	17–26	176.4	71.4	15.0	8.9–21.1
Texas, 1977	(39)	18–24	179.9	74.6	13.4	7.4–19.4
Massachusetts, 1990	b	17–30	178.2	76.2	12.9	7.8–18.0
Older Men						
Indiana, 1966	d	24–38	179.0	76.6	17.8	11.3–24.3
		40–48	177.0	80.5	22.3	16.3–28.3
North Carolina, 1976	f	27–50	—	—	23.7	17.9–30.1
Texas, 1977	(39)	27–59	180.0	85.3	27.1	23.7–30.5
Massachusetts, 1990	b	31–50	177.1	77.5	19.8	13.2–26.4

[a] % body fat was computed from body density measured by hydrostatic weighing, where % Fat = 495/density − 450.
[b] Unpublished data, Katch, F.I. and A.R. Behnke. Anthropometric-arm radiographic assessment of body composition, muscularity and frame size. U.S. Army Medical Research and Development Command. Fort Detrick, Frederick, MD. No. DAMD17–80–C–0108.
[c] Chen, K.P.: *J. Formosan Med. Assoc.,* 52:271, 1953.
[d] Myhre, L.G., and Kessler, W.V.: *J. Appl. Physiol.,* 21:1251, 1966.
[e] Pollock, M.L., et al.: *J. Appl. Physiol.,* 38:745, 1975.
[f] Pollock, M.L., et al.: *J. Appl. Physiol.,* 40:300, 1976.
[g] Wilmore, J.A., and Behnke, A.R.: *J. Appl. Physiol.,* 25:349, 1968.

If this man lost 10.1 kg of body fat, his new body mass of 80.9 kg would have a fat content equal to 10% of body mass. These calculations assume no change in LBM with weight loss.

 SUMMARY

1. Standard "height-weight" tables reveal little about an individual's body composition, which, at any given body mass and stature, may vary considerably. One can be overweight but not overfat.

2. Total body fat consists of essential fat and storage fat. Essential fat is the fat present in bone marrow, nerve tissue, and the various organs; it is generally required for normal physiologic function. Storage fat is the energy reserve that accumulates mainly as adipose tissue beneath the skin.

3. True gender differences appear to exist for quantities of essential fat. Although storage fat values for men and women average 12 to 15% of body mass, respectively, the essential fat differences are large and amount to 3% for men and 12% for women. This difference is probably related to child-bearing and hormonal functions.

4. It appears that a person cannot reduce below the essential fat level and still maintain good health.

5. Menstrual irregularities and cessation of menstruation occur among certain groups of female athletes, especially among women who train hard and maintain low levels of body fat. The precise interaction between the physiologic and psychologic stress of intense, regular training, hormonal balance, energy and nutrient intake, and body fat requires further study.

6. A positive aspect may exist between the delayed onset of menarche observed in chronically active young females and health because such individuals show a lower lifetime occurrence of reproductive organ and other cancers.

7. The two most popular indirect methods for body composition assessment are hydrostatic weighing and prediction methods from fatfolds and circumferences. Hydrostatic weighing involves the determination of body density and the subsequent estimation of percent body fat. This assumes a constant density for human fat and fat-free tissues. Lean body mass is calculated by subtracting fat mass from body mass.

8. Part of the error inherent in predicting body fat from whole body density lies in the assumptions concerning the density of the body's fat and nonfat components. The differences in body density probably vary and may differ from assumed constants in relation to race, age, and athletic experience. At this point in time, firm data are lacking for improving the theoretically based constants.

9. The prediction methods to assess body composition employ equations developed from relationships between selected fatfolds and girths and body density and percent fat. These equations are "population specific" because they are most accurate with subjects similar to those from which the equations were derived. A surface area formula based on stature, body mass, and fatfolds also can be used to estimate body composition in diverse population groups of different ages.

REFERENCES

1. Adams, J., et al.: Total body fat content in a group of professional football players. *Can. J. Appl. Sport Sci.*, 7:36, 1982.

2. Bakker, H.K., and Strvikenkamp, R.S.: Biological variability and lean body mass estimates. *Hum. Biol.*, 53:181, 1977.

2a. Baumgartner, R.N., et al.: Bioelectric impedence for body composition. In *Exercise and Sport Sciences Reviews*, Vol. 18. Edited by K.B. Pandolf, and J.O. Holloszy. Baltimore, Williams & Wilkens, 1990.

3. Becque, M.D., et al.: Time course of skin-plus-fat compression in males and females. *Hum. Biol.*, 58:33, 1984.

4. Behnke, A.R., and Wilmore, J.H.: *Evaluation and Regulation of Body Build and Composition.* Englewood Cliffs, NJ, Prentice-Hall, 1974.

5. Behnke, A.R.: New concepts in height-weight relationships. In *Obesity*. Edited by N. Wilson. Philadelphia, F.A. Davis, 1969.
6. Behnke, A.R., et al.: The specific gravity of healthy men. *JAMA*, 118:495, 1942.
7. Behnke, A.R., et al.: Quantification of body weight and configuration from anthropometric measurements. *Hum. Biol.*, 31:213, 1959.
8. Behnke, A.R., et al.: Routine anthropometry and arm radiography in assessment of nutritional status: its potential. *J. Parent. Enteral. Nutr.*, 2:532, 1978.
9. *Body Composition in Animals and Man*. Washington, DC, National Academy of Sciences, Publication 1598, 1968.
10. Boling, E.A., and Lipkind, J.B.: Body composition and serum electrolyte concentrations. *J. Appl. Physiol.*, 18:943, 1963.
11. Bonen, A., et al.: Profiles of selected hormones during menstrual cycles of teenage athletes. *J. Appl. Physiol*, 50:545, 1981.
12. Boyden, T.: Prolactin responses, menstrual cycles and body composition of women runners. *J. Clin. Endocrinol. Metab.*, 54:712, 1982.
13. Brooks-Gunn, J., et al.: The relation of eating problems and amenorrhea in ballet dancers. *Med. Sci. Sports Exerc.*, 19:41, 1987.
14. Brožek, J. (Ed.): *Human Body Composition. Approaches and Applications*. Oxford, Pergamon Press, 1965.
15. Brožek, J., and Keys, A.: The evaluation of leanness-fatness in man: norms and interrelationships. *Br. J. Nutr.*, 36:32, 1951.
16. Brožek, J., et al.: Densitometric analysis of body composition: revision of some quantitative assumptions. *Ann. N.Y. Acad. Sci.*, 110:113, 1963.
17. Bullen, B.A., et al.: Endurance training effects on plasma hormonal responsiveness and sex hormone excretion. *J. Appl. Physiol.*, 56:1453, 1984.
18. Bunt, J.C., et al.: Impact of total body water fluctuations on estimation of body fat from body density. *Med. Sci. Sports Exerc.*, 21:96, 1989.
19. Byrd, P.J., and Thomas, T.R.: Hydrostatic weighing during different stages of the menstrual cycle. *Res. Q. Exerc. Sport*, 54:296, 1983.
20. Calabrese, L.H., et al.: Menstrual abnormalities, nutritional patterns, and body composition in female classical ballet dancers. *Phys. Sportsmed.*, 11:86, 1983.
21. Carlberg, K.A., et al.: Body composition of oligo/amenorrheic athletes. *Med. Sci. Sports Exerc.*, 15:215, 1983.
22. Caton, J.R., et al.: Body composition by bioelectrical impedence: effect of skin temperature. *Med. Sci. Sports Exerc.*, 20:489, 1988.
23. Clarys, J.P., et al.: Gross tissue weights in the human body by cadaver dissection. *Hum. Biol.*, 56:459, 1984.
24. Clauser, C.E., et al.: Anthropometry of air force women. AMRL-TR-70-5. Ohio, Wright Patterson Air Force Base, 1972.
25. Dale, E.D., et al.: Menstrual dysfunction in distance runners. *Obstet. Gynecol.*, 54:47, 1979.
26. Durenberg, P., et al.: Changes in fat-free mass during weight loss measured by bioelectrical impedence and by densitometry. *Am. J. Clin. Nutr.*, 49:33, 1989.
27. Drumm, S., et al.: Changes in body composition, anthropometry, and arm radiography following 10 weeks of hydraulic resistive circuit exercise. *Med. Sci. Sports Exerc.*, 16:184, 1984.
28. Freedson, P.S., et al.: Physique, body composition, and psychological characteristics of competitive female body builders. *Phys. Sportsmed.*, 11:85, 1983.
29. Frisch, R.E., et al.: Delayed menarche and amenorrhea in ballet dancers. *N. Engl. J. Med.*, 303:17, 1980.
30. Frisch, R.E., et al.: Lower lifetime occurance of breast cancer and cancers of the reproductive system among former college athletes. *Am. J. Clin. Nutr.*, 45:328, 1987.
31. Frisch, R.E., et al.: Lower prevalence of breast cancer and cancers of the reproductive system among former college athletes compared to non-athletes. *Br. J. Cancer*, 52:885, 1985.
32. Frisch, R.E., et al.: Lower prevalence of non-reproductive cancers among female former college athletes. *Med. Sci. Sports Exerc.*, 21:250, 1989.
33. Graver, W.O., et al.: Quantification of body fat distribution in the abdomen using computed tomography. *Am. J. Clin. Nutr.*, 39:631, 1984.

34. Hamilton, L. H., et al.: The role of selectivity in the pathogenesis of eating problems in ballet dancers. *Med. Sci. Sports Exerc.*, 20:560, 1988.

34a. Hata, E., and Aoki, K.: Age at menarche and select menstrual characteristics in young Japanese athletes. *Res. Q. Exer. Sports*, 61:178, 1990.

35. Hayes, P.A., et al.: Sub-cutaneous fat thickness measured by magnetic resonance imaging, ultrasound, and calipers. *Med. Sci. Sports Exerc.*, 20:303, 1988.

36. Henderson, B.E., et al.: Do regular ovulatory cycles increase breast cancer risk? *Cancer*, 56:1206, 1985.

36a. Heymfield, S.B., et al.: Appendicular skeletal muscle mass: measurement by dual-photon absorptiometry. 52:214, 1990.

37. Hsieh, S., et al.: Measurement of residual volume sitting and lying in air and water (and during underwater weighing) and its effects on computed body density. *Med. Sci. Sports Exerc.*, 17:204, 1985.

38. Hutcheson, L., et al.: Body impedence analysis and body water loss. *Res. Q. Exerc. Sport.*, 59:359, 1988.

39. Jackson, A.S., and Pollock, M.L.: Prediction accuracy of body density, lean body weight, and total body volume equations. *Med. Sci. Sports*, 9:197, 1977.

40. Jackson, A.S., et al.: Generalized equations for predicting body density of women. *Med. Sci. Sports*, 12:175, 1980.

41. Jackson, A.S., et al.: Reliability and validity of bioelectrical impedence in determining body composition. *J. Appl. Physiol.* 64:529, 1988.

41a. Jensen, R.L., and F.I. Katch: Cross validation of body composition prediction equations in obese, adult males. *Am. J. Hum. Biol.*, (In Press, 1991.)

42. Johnston, F.E.: Body fat deposition in adult obese women. I. Patterns of fat distribution. *Am. J. Clin. Nutr.*, 47:225, 1988.

43. Kahled, M.A., et al.: Electrical impedence in assessing human body composition: the BIA Method. *Am. J. Clin. Nutr.*, 47:789, 1988.

44. Kaiserauer, S., et al.: Nutritional, physiological, and menstrual status of distance runners. *Med. Sci. Sports Exerc.*, 21:120, 1989.

45. Kasch, F.W., and Wallace, J.P.: Physiological variables during 10 years of endurance exercise. *Med. Sci. Sports*, 8:5, 1976.

46. Katch, F.I.: Practice curves and errors of measurement in estimating underwater weight by hydrostatic weighing. *Med. Sci. Sports*, 1:212, 1969.

47. Katch, F.I.: Reliability and individual differences in ultrasound assessment of subcutaneous fat: effects of body position. *Hum. Biol.*, 55:789, 1983.

48. Katch, F.I.: Assessment of lean body tissues by radiography and bioelectrical impedence. In *Body Composition in Youth and Adults.* Columbus, OH, Ross Laboratories, 1985.

49. Katch, F.I., and Behnke, A.R.: Arm x-ray assessment of body fat in men and women. *Med. Sci. Sports Exerc.*, 16:316, 1984.

50. Katch, F.I., and Katch, V.L.: Measurement and prediction errors in body composition assessment and the search for the perfect prediction equation. *Res. Q. Exerc. Sport*, 51:249, 1980.

51. Katch, F.I., and Katch, V.L.: The body composition profile: techniques of measurement and applications. *Clin. Sports Med.*, 3:31, 1984.

52. Katch, F.I., and Katch, V.L.: Computer technology to evaluate body composition, nutrition, and exercise. *Prev. Med.*, 12:619, 1983.

53. Katch, F.I., and Michael, E.D.: Prediction of body density from skinfold and girth measurements of college females. *J. Appl. Physiol.*, 25:92, 1968.

54. Katch, F.I., and McArdle, W.D.: Prediction of body density from simple anthropometric measurements in college-age men and women. *Hum. Biol.*, 45:445, 1973.

55. Katch, F.I., and McArdle, W.D.: Validity of body composition prediction equations for college men and women. *Am. J. Clin. Nutr.*, 28:105, 1975.

56. Katch, F.I., and Spiak, D.L.: Validity of the Mellits and Cheek method for body-fat estimation in relation to menstrual cycle status in athletes and non-athletes below 22 percent fat. *Ann. Hum. Biol.*, 11:389, 1984.

57. Katch, F.I., et al.: Estimation of body volume by underwater weighing: description of a simple method. *J. Appl. Physiol.*, 23:811, 1967.

58. Katch, F.I., et al.: Estimation of body fat from skinfolds and surface area. *Hum. Biol.*, 51:411, 1979.

59. Katch, F.I., et al.: The underweight female. *Phys. Sportsmed.*, 8:55, 1980.

59a. Katch, F.I., et al.: The ponderal somatogram: evaluation of body size and shape from anthropometric girths and stature. *Hum. Biol.*, 59:439, 1987.

60. Katch, F.I., et al.: Reliability and validity of a new method for the measurement of total body volume. *Res. Q. Exerc. Sport.*, 60:286, 1989.

60a. Katch, F.I., et al.: A new approach for estimating excess body fat from changes in abdominal girth. *Am. J. Hum. Biol.*, 2:125,1990.

60b. Katch, F.I., and T. Hortobagyi: Validity of surface anthropometry to estimate upper-arm muscularity, including changes with body mass loss. *Am. J. Clin. Nutr.*, 52:591, 1990.

61. Katch, V.L., et al.: Contribution of breast volume and weight to body fat distribution in females. *Am. J. Phys. Anthropol.*, 53:93, 1980.

62. Keys, A., et al.: *The Biology of Human Starvation.* Minneapolis, University of Minnesota Press, 1950.

63. Kodama, A.A.: In vivo and in vitro determinations of body fat and body water in the hamster. *J. Appl. Physiol.*, 31:218,1971.

64. Kuczmarski, R.J., et al.: Ultrasonic assessment of body composition in obese adults: overcoming the limitations of the skinfold caliper. *Am. J. Clin. Nutr.*, 45:717, 1987.

65. Lewis, S., et al.: Body composition of middle-aged female endurance athletes. In *Biomechanics of Sports and Kinanthropometry.* Vol. 6. Edited by F. Landry, and W.A.R. Orban. Miami, FL, Symposia Specialists, 1978.

66. Lloyd, T., et al.: Women athletes with menstrual irregularities have increased musculoskeletal injuries. *Med. Sci. Sports Exerc.*, 18:374, 1986.

67. Lohman, T.G., et al.: Bone mineral measurements and their relation to body density in children and adults. *Hum. Biol.*, 56:667, 1984.

68. Lohman, T.G., et al.: Methodological factors and the prediction of body fat in female athletes. *Med. Sci. Sports Exerc.*, 16:92, 1984.

69. Loucks, A.B., et al.: Menstrual status and validation of body fat prediction in athletes. *Hum. Biol.*, 56:383, 1984.

69a. Loucks, A.B.: Effects of exercise training on the menstrual cycle: existence and mechanisms. *Med. Sci. Sports Exerc.*, 22:275, 1990.

70. Lukaski, H.C., et al.: A comparison of methods of assessment of body composition involving neutron activation analysis of total body nitrogen. *Metabolism*, 30:777, 1981.

71. Malina, R.M.: Menarche in athletes: a synthesis and hypothesis. *Ann. Human Biol.*, 1:1, 1983.

72. Malina, R.M.: Bioelectric methods for estimating body composition: an overview and discussion. *Hum. Biol.*, 59:329, 1987.

73. Martin, B.J.: Is athletic amenorrhea specific to runners? *Am. J. Obst. Gynecol.*, 143:859, 1982.

74. Morrow, J.R., Jr.: Accuracy of measured and predicted residual lung volume on body density measurement. *Med. Sci. Sports Exerc.*, 18:647, 1986.

75. Mueller, W.H., and Malina, R.M.: Relative reliability of circumferences and skinfold measures of body fat distribution. *Am. J. Phys. Anthropol.*, 72:437, 1987.

76. Ostrove, S.M., and Vaccaro, P.: Effect of immersion on RV in young women: implications for measurement of body density. *Int. J. Sports Med.*, 3:220, 1982.

77. Pascale, L.R., et al.: Correlations between thickness of skin-folds and body density in 88 soldiers. *Hum. Biol.*, 28:165, 1956.

78. Pollock, M.L., and Jackson, A.S.: Measurement of cardiorespiratory fitness and body composition in the clinical setting: *Compr. Ther.*, 6:12, 1980.

79. Pollock, M.L., et al.: Body composition of elite class distance runners. *Ann. N.Y. Acad. Sci.*, 301:361, 1977.

80. Rebar, R.W., et al.: Reproductive function in women athletes (editorial). *JAMA*, 246:1590, 1981.

81. Robertson, C.H., et al.: Lung volumes in men immersed to the neck: dilution and plethysmographic techniques. *J. Appl. Physiol.*, 44:679, 1978.

82. Ross, R., et al.: Sensitivity of bioelectrical impedence to detect changes in body composition. *J. Appl. Physiol.*, 67:1643, 1989.

83. Rubeffe-Scrive, M.L., et al.: Fat cell metabolism in different regions in women: effect on menstrual cycle, pregnancy, and lactation. *J. Clin. Invest.*, 75: 1973, 1985.

84. Savard, R., et al.: The effect of exercise training on regional adipose tissue metabolism in pregnant rats. *Am. J. Physiol.*, 250:R837, 1986.

85. Schutte, J.E., et al.: Density of lean body mass is greater in Blacks than Whites. *J. Appl. Physiol.*, 56:1647, 1984.

86. Schweiger, V., et al.: Caloric intake, stress, and menstrual function in atheletes. *Fertil. Steril.*, 49:447, 1988.

86a. Feidell, J.C., et al.: Imaging techniques for measuring adipose-tissue distribution—a comparison between computed tomography and 1.5-T magnetic resonance. *Am. J. Clin. Nutr.*, 51:953, 1990.

87. Shangold, M.M.: Sports and menstrual function. *Phys. Sportsmed.*, 8:66, 1980.

88. Shangold, M.M.: Do women's sports lead to menstrual problems? *Contemp. Obst. Gynecol.*, 17:52, 1981.

88a. Shangold, M.M., et al.: Evaluation and management of menstrual dysfunction in athletes. *JAMA*, 262:1665, 1990.

89. Siri, W.E.: Gross composition of the body. In *Advances in Biological and Medical Physics.* Vol. 4. Edited by J.H. Lawrence, and C.A. Tobias. New York, Academic Press, 1956.

90. Sloan, A.W., et al.: Estimation of body fat in young women. *J. Appl. Physiol.*, 17:967, 1962.

91. Stager, J.M., and Hatler, L.K.: Menarche in athletes: the influence of genetics and prepubertal training. *Med. Sci. Sports Exerc.*, 20:369, 1988.

92. Sweetgall, R., et al.: *Fitness Walking*, New York, Putnam, 1985.

93. Tokunaga, K.: A novel technique for the determination of body fat by computed tomography. *Int. J. Obesity*, 7:437, 1983.

94. Tran, Z.V., and Weltman, A.: Predicting body composition of men from girth measurements. *Hum. Biol.*, 60:167, 1988.

95. Tran, Z.V., and Weltman, A.: Generalized equation for predicting body density of women from girth measurements. *Med. Sci. Sports Exerc.*, 21:101, 1989.

96. Vaccaro, P., et al.: Body composition and physiological responses of Masters female swimmers 20 to 70 years of age. *Res. Q. Exerc. Sport*, 55:278, 1984.

97. Wadden, T. A., et al.: Body fat deposition in adult obese women. II. Changes in fat distribution accompanying weight reduction. *Am. J. Clin. Nutr.*, 47:229, 1988.

98. Wakat, D.K., et al.: Reproductive system function in women cross-country runners. *Med. Sci. Sports Exerc.*, 14:263, 1982.

99. Warren, M.P.: The effects of exercise on pubertal progression and reproductive function. *J. Clin. Endocrinol. Metab.*, 51:1150, 1980.

100. Weltman, A., et al.: Accurate assessment of body composition in obese females. *Am. J. Clin. Nutr.*, 48:1179, 1988.

101. Wilmore, J.H.: The use of actual, predicted and constant residual volumes in the assessment of body composition by underwater weighing. *Med. Sci. Sports*, 1:87, 1969.

102. Wilmore, J.H.: Body composition in sport and exercise: directions for future research. *Med. Sci. Sports Exerc.*, 15:21, 1983.

103. Wilmore, J.H., and Behnke, A.R.: An anthropometric estimation of body density and lean body weight in young women. *Am. J. Clin. Nutr.*, 23:267, 1970.

104. Wilmore, J.H., and Brown, C.H.: Physiological profiles of women distance runners. *Med. Sci. Sports*, 6:178, 1974.

105. Yang, M-V.: Body composition and resting metabolic rate in obesity. In *Obesity and Weight Control.* Edited by R.T. Frankle, and M-V. Yang. Rockville, MD, Aspen Press, 1988.

106. Young, C.M., et al.: Predicting specific gravity and body fatness in young women. *J. Am. Diet. Assoc.*, 40:102, 1962.

107. Young, C.M., et al.: Body composition of "older" women. *J. Am. Diet Assoc.*, 43:344, 1963.

The evaluation of body composition provides an excellent opportunity to partition the gross size of a person into two major structural components—body fat and lean body mass. As discussed in Chapter 27, the physique of the adult man differs considerably from that of the woman. Differences in physique are also pronounced when comparisons are made among sports participants of the same gender such as Olympic competitors, track and field specialists, wrestlers, and football players, and these differences are noted even in comparisons between highly proficient adolescent competitors.[45] Aside from describing physique in relation to sports category and level of competition, some experiments have determined the effects of different forms of sports training and exercise on the body composition of men and women. In this chapter, we take a closer look at the physiques of champion athletes in various sports and evaluate the relative effects of diet and exercise in modifying body composition.

PHYSIQUE OF CHAMPION ATHLETES

Considerable research has dealt with the anthropometric methodology of quantifying physique status. Visual appraisal is often used to describe individuals as small, medium, or large, or thin (**endomorphy**), muscular (**mesomorphy**), or fat (**ectomorphy**). A basic shortcoming of this approach, termed **somatotyping,** is that visual inspection does little more than describe body shape in terms of a particular category such as thin or muscular. Visual inspection does not really quantify the various body dimensions in terms of how large the chest or shoulders are or how much the biceps are developed (or underdeveloped) in relation to the thighs or calves. In this section, we focus on body composition and quantification of physique by use of anthropometry in Olympic competitors, endurance runners, collegiate and professional football players, triathletes, high school wrestlers, champion male and female body builders, and diverse subgroups using the body profile technique discussed in Chapter 27 and detailed in Appendix I.

OLYMPIC SPECIALISTS

Early studies of Olympic competitors revealed that physique was related to a high level of achievement in certain sports.[14,29] Tables 28–1 and 28–2 list the anthropometric characteristics of male and female competitors who participated in the 1964 Tokyo and 1968 Mexico City Olympics.[15,20] Lean body mass and percent body fat were estimated from bone diameter measures and

TABLE 28–1.
AGE, BODY SIZE, AND BODY COMPOSITION OF MALE ATHLETES IN SELECTED EVENTS WHO COMPETED IN THE TOKYO AND MEXICO CITY OLYMPICS[a]

Event	Specialty	Olympics	N	Age years	Stature cm	Mass kg	LBM[b] kg	Body Fat %
Sprinters	100–200 m; 4 × 100 m; 110-m hurdles	Toyko	172	24.9	178.4	72.2	64.9	10.1
		Mexico City	79	23.9	175.4	68.4	62.8	8.2
Long-distance runners	3000, 5000, 10,000 m	Tokyo	99	27.3	173.6	62.4	61.5	1.4
		Mexico City	34	25.3	171.9	59.8	60.1	−0.5
Marathoners	42.2 km	Tokyo	74	28.3	170.3	60.8	59.2	2.7
		Mexico City	20	26.4	168.7	56.6	58.1	2.7
Decathlon		Tokyo	26	26.3	183.2	83.5	68.5	18.0
		Mexico City	8	25.1	181.3	77.5	67.1	13.4
Jumpers	High, long, triple jump	Tokyo	89	25.3	181.5	73.2	67.2	8.2
		Mexico City	14	23.5	182.8	73.2	68.2	6.8
Weight throwers	Shot, discus, hammer	Tokyo	79	27.6	187.3	101.4	71.6	29.4
		Mexico City	9	27.3	186.1	102.3	70.7	30.9
Swimmers	Free, breast, back, butterfly, medley	Tokyo	450	20.4	178.7	74.1	65.1	12.1
		Mexico City	66	19.2	179.3	72.1	65.6	9.0
Basketball		Tokyo	186	25.3	189.4	84.3	73.2	13.2
		Mexico City	63	24.0	189.1	79.7	73.0	8.4
Gymnastics	All events	Tokyo	122	26.0	167.2	63.3	57.0	9.9
		Mexico City	28	23.6	167.4	61.5	57.2	7.0
Wrestling	Bantam and featherweight	Tokyo	29	27.3	163.3	62.3	54.4	12.7
		Mexico City	32	22.5	166.1	57.0	56.3	1.2
Rowing	Single and double skulls; pairs, fours, eights	Tokyo	357	25.0	186.0	82.2	70.6	14.1
		Mexico City	85	24.3	185.1	82.6	69.9	15.4

[a] Adapted from De Garay, et al.: Genetic and Anthropological Studies of Olympic Athletes. New York, Academic Press, 1974, and from Hirata, K.: Physique and age of Tokyo Olympic champions. J. Sports Med. Phys. Fitness, 6:207, 1966.
[b] Calculated by Behnke's method[6]: LBM = $h^2 \times 0.204$, where h = stature, dm.
[c] Body fat = (Body mass − LBM)/Body mass × 100

TABLE 28–2.
AGE, BODY SIZE, AND BODY COMPOSITION OF FEMALE ATHLETES WHO COMPETED IN THE TOKYO AND MEXICO CITY OLYMPICS IN SELECTED EVENTS[a]

Event	Specialty	Olympics	N	Age year	Stature cm	Mass kg	LBM[b] kg	Body Fat %
Sprinters	100–200 m; 100 m hurdles	Tokyo	85	22.7	166.0	56.6	49.6	12.4
		Mexico City	28	20.7	165.0	56.8	49.0	13.7
Jumpers	High, long, triple jump	Tokyo	56	23.6	169.5	60.2	51.7	14.1
		Mexico City	12	21.5	169.4	56.4	51.7	8.4
Weight throwers	Shot, discus, hammer	Tokyo	37	26.2	170.4	79.0	52.3	33.8
		Mexico City	9	19.9	170.9	73.5	52.6	28.5
Swimmers	Free, breast, back, butterfly, medley	Tokyo	272	18.6	166.3	59.7	49.8	16.6
		Mexico City	28	16.3	164.4	56.9	48.6	14.5
Diving	Spring, high	Tokyo	65	18.5	160.9	54.1	46.6	13.9
		Mexico City	7	21.1	160.4	52.3	46.3	11.5
Gymnastics	All events	Tokyo	102	22.7	157.0	52.0	44.4	14.7
		Mexico City	21	17.8	156.9	49.8	44.3	11.0

[a] Adapted from De Garay, et al.: Genetic and Anthropological Studies of Olympic Athletes. New York, Academic Press, 1974 and from Hirata, K.: Physique and age of Tokyo Olympic champions. J. Sports Med. Phys. Fitness, 6:207, 1966.
[b] Calculated by Behnke's method[6]: LBM (lean body mass) = $h^2 \times 0.18$, where h = stature, dm.
[c] Body fat = (Body mass − LBM)/Body mass × 100

stature and show relative differences in body composition among the different specialists.[6]

Gender Differences

For the men, basketball players, rowers, and weight throwers were the tallest and heaviest competitors; they also possessed the largest amount of lean body mass and percent body fat. For example, weight throwers in both Olympiads averaged 30% body fat, whereas 94 marathon and 133 long-distance runners averaged an exceptionally low 1.6% body fat. The biggest discrepancy in body composition within a particular sports category was noted between the Tokyo wrestlers, who averaged 12.7% body fat, and the wrestlers in Mexico City, who averaged only 1.2% body fat. This difference is even more remarkable because the age, stature, and lean body mass of both groups of wrestlers were very similar.

For female Olympic athletes, the most striking observation is their relatively low body fat values. Except for the weight throwers, who averaged about 31% body fat, the other sports groups were very close to the average body fat of 13.1% for all 676 female participants in both Olympics.

Racial Differences

Racial differences in physique have been observed among Olympic competitors and this may be significant in terms of athletic performance.[43] Black sprinters and high jumpers, for example, have longer limbs and narrower hips than their white counterparts. From a mechanical perspective, a black sprinter with leg and arm size identical to that of a white sprinter would have a lighter, shorter, and slimmer body to propel. This might confer a more favorable power to body mass ratio for black athletes at any given size compared to white competitors. A relatively greater power output would be advantageous in jumping events and in all sprint running events where generating rapid energy for short periods is crucial for successful performance. This advantage may be diminished somewhat in the various throwing events. In contrast to whites and blacks, Asian athletes have short legs relative to their upper body size. This might confer some advantage in the very short and longer distance races, as well as in weight lifting. In fact, successful weight lifters of all races have relatively short arms and legs for their stature compared to other groups of athletes.

FEMALE LONG-DISTANCE RUNNERS

Table 28–3 presents data for body mass, stature, and body composition of 11 female long-distance runners of national and international caliber.[46] The runners averaged 15.2% body fat (determined by underwater weighing), which is similar to reports of high school cross-country runners[9] and considerably lower than the average value of about 26% reported for sedentary females of the same age, stature, and body mass.[24] Compared to other female athletes, the runners have a lower average fat value than collegiate basketball players (20.9%),[41] competitive gymnasts (15.5%),[40] younger distance runners (18%),[29] swimmers (20.1%),[23] or tennis players (22.8%).[23]

Interestingly, the average body fat for these runners is the same as the 15% average generally reported for males, and is close to the quantity of

TABLE 28–3.
BODY COMPOSITION OF FEMALE LONG-DISTANCE RUNNERS[a]

Subjects	Age year	Stature cm	Mass kg	LBM kg	Body Fat kg	Body Fat %
1[b]	24	172.7	52.6	49.5	3.1	5.9
2[c]	26	159.8	71.5	46.2	25.3	35.4
3[d]	28	162.6	50.7	47.6	3.1	6.1
4	31	171.5	52.0	47.3	4.7	9.0
5	33	176.5	61.2	50.8	10.4	17.0
6	34	166.4	52.9	44.8	8.1	15.2
7	35	168.4	55.0	48.7	6.3	11.6
8	36	164.5	53.1	44.3	8.8	16.6
9	36	182.9	61.5	50.4	11.1	18.1
10	36	182.9	65.4	55.7	9.7	14.8
11	37	154.9	53.6	44.0	9.6	18.0
Average	32.4	169.4	57.2	48.1	9.1	15.2

[a] From Wilmore, J.H., and Brown, C.H.: Physiological profiles of women distance runners. Med. Sci. Sports, 6:178, 1974.
[b] World's best time in marathon (2:49:40) as of 1974.
[c] World's best time in 50-mile run (7:05:31); established 18 months after the body composition evaluation.
[d] Noted U.S. distance runner. Five consecutive national and international crosscountry championships.

essential fat proposed by Behnke in his model for the "reference woman." In fact, the body fat of several of the female distance runners listed in Table 28–3 was within the range of values reported for topflight male distance runners. Based on Behnke's reference woman the leanest women in the population have essential fat equal to about 12 to 14% of body mass. This apparent discrepancy between the estimated fat content of some distance runners and Behnke's theoretical lower limit for body fat in women requires further study. It is also difficult to explain the relatively high body fat content of one of the best runners and indicates that other factors override the obvious limitations to distance running imposed by excess fat.

MALE LONG-DISTANCE RUNNERS

Table 28–4 presents the body composition of male middle and long-distance runners and eight elite marathoners. This high caliber group included Prefontaine (former American record holder in the 800- and 1500-m runs) and Shorter (1976 Olympic Gold Medalist). For comparative purposes, data are presented for a typical sample of untrained college-aged men.

For both groups of runners, body fat values were extremely low, considering that the quantity of essential fat is about 3% of body mass. Clearly, these endurance runners are at the lower end of the lean-to-fat continuum for topflight athletes. Apparently, this physique characteristic is a prerequisite for success in distance running. This makes sense for several reasons. First, the body's ability to dissipate heat during running is of primary importance in maintaining thermal balance during competition. Excess fat thwarts heat dissipation. Second, excess body fat is "dead weight" that adds directly to the energy cost of running.

In terms of body structure, elite male distance runners generally have smaller girths and bone diameters than untrained males.[14] One could con-

TABLE 28–4.
ANTHROPOMETRIC CHARACTERISTICS OF ELITE MALE DISTANCE RUNNERS

Variable	Middle-Long Distance[a] (N = 11)	Marathon[a] (N = 8)	College Men[b] (N = 54)
Stature, cm	176.0	176.8	176.4
Mass, kg	63.1	62.1	71.4
Body fat, %	5.0	4.3	15.3
Body fat, kg	3.2	2.7	10.9
LBM, kg	59.9	59.4	60.5

[a] Data from Pollock, M.L., et al.: Body composition of elite class distance runners. Ann. N.Y. Acad. Sci., 301:361, 1977.
[b] Data from Katch, F.I., and McArdle, W.D.: Prediction of body density from simple anthropometric measurements in college-age men and women. Hum. Biol., 45:445, 1973.

sider that these structural differences, especially the bone diameters, are "genetic." **The best long-distance runners inherit a physique that is slight of build, not only in terms of stature but also in skeletal dimensions. When this physique is blended with a lean body composition, highly developed aerobic system, and the proper psychologic attitude for prolonged, intensive training, the proper ingredients certainly exist for a champion!**

TRIATHLETES

The triathlon combines endurance performance in three continuous activities—swimming, bicycling, and running. At the upper extreme of triathlon requirements is the ultra endurance Ironman competition in which each performer is required to swim 3.9 km (2.4 miles), bicycle 180.2 km (112 miles), and run a standard marathon of 42.2 km (26.2 miles). Average training for the serious triathlete involves about 3 hours per day covering 280 miles per week in swimming (7.2 miles; 30:00 pace), bicycling (227 miles; 18.6 mph), and running (45 miles; 7:42 pace).[21] Data for 6 males and 3 females who participated in the 1982 Ironman Triathlon showed that body fat values ranged between 5.0 and 11.3% for males and 7.4 and 17.2% for females. For the top 15 male finishers, body fat averaged 7.1% with a corresponding max $\dot{V}O_2$ of 72.0 ml·kg^{-1}·min^{-1}. From these data, it appears that the body fat and aerobic capacity of triathletes are similar to values reported for other endurance athletes in single endurance sports.[34] A subsequent study of 14 triathletes training for the 1984 Ironman competition concluded that the physique of both male and female triathlete is most similar to that of elite cyclists.[35] For aerobic capacity, the values for men were similar to those for trained swimmers, and for the females, the aerobic capacity was at the upper range of data reported for elite endurance-trained runners.

FOOTBALL PLAYERS

The first detailed body composition analyses of football players vividly demonstrated the inadequacy of determining a person's "optimal" body mass from "height-weight" standards.[46] Football players as a group had a body fat content that averaged only 10.4% of body mass and lean body mass averaged 81.3 kg (179.3 lb). Certainly the men were heavy, but they were **not** fat. The heaviest lineman weighed 118 kg (260 lb, 17.4% body fat, 215 lb lean body mass), whereas the fattest lineman had 23.2% body fat at a body mass of 115.4 kg (252 lb). The player with the least fat was a defensive back; for his body mass of 82.3 kg (182 lb), body fat was 3.3% with a lean body mass of 79.6 kg (175 lb).

Table 28–5 presents a clearer picture of the average values for body mass, stature, percent body fat, and lean body mass of college and professional players grouped by position.[47,49] The pro, older group consists of 25 players from the 1942 Washington Redskins, the first professional players measured

TABLE 28–5.
COMPARISON OF BODY COMPOSITION BETWEEN COLLEGIATE AND PROFESSIONAL FOOTBALL PLAYERS GROUPED BY POSITION[a]

Position	Level	N	Stature (cm)	Mass (kg)	Body Fat (%)	LBM (kg)
Defensive backs	St Cloud[b]	15	178.3	77.3	11.5	68.4
	U Mass[c]	12	179.9	83.1	8.8	76.8
	USC[d]	15	183.0	83.7	9.6	75.7
	Pro, current[e]	26	182.5	84.8	9.6	76.7
	Pro, older[g]	25	183.0	91.2	10.7	81.4
Offensive backs and receivers	St Cloud	15	179.7	79.8	12.4	69.6
	U Mass	29	181.8	84.1	9.5	76.4
	USC	18	185.6	86.1	9.9	77.6
	Pro, current	40	183.8	90.7	9.4	81.9
	Pro, older	25	183.0	91.7	10.0	87.5
Linebackers	St Cloud	7	180.1	87.2	13.4	75.4
	U Mass	17	186.1	97.1	13.1	84.2
	USC	17	185.6	98.8	13.2	85.8
	Pro, current	28	188.6	102.2	14.0	87.6
Offensive linemen and tight ends	St Cloud	13	186.0	99.2	19.1	79.8
	U Mass	23	187.5	107.6	19.5	86.6
	USC	25	191.1	106.5	15.3	90.3
	Pro, current	38	193.0	112.6	15.6	94.7
Defensive linemen	St Cloud	15	186.6	97.8	18.5	79.3
	U Mass	8	188.8	114.3	19.5	91.9
	USC	13	191.1	109.3	14.7	93.2
	Pro, current	32	192.4	117.1	18.2	95.8
	Pro, older	25	185.7	97.1	14.0	83.5
Total	St Cloud	65	182.5	88.0	15.0	74.2
	U Mass	91	184.9	97.3	13.9	83.2
	USC	88	186.6	96.6	11.4	84.6
	Pro, current	164	188.1	101.5	13.4	87.3
	Pro, older	25	183.1	91.2	10.4	81.3
	Dallas-Jets[f]	107	188.2	100.4	12.6	87.7

[a] Grouping according to reference 49.
[b] Data from Wickkiser, J.D., and Kelly, J.M.: The body composition of a college football team. Med. Sci. Sports, 7:199, 1975.
[c] U. Mass data courtesy of Coach Robert Stull and F. Katch. University of Massachusetts. Data collected during spring practice, 1985; % fat by densitometry.
[d] USC data courtesy of Dr. Robert Girandola, University of Southern California, Los Angeles, 1978.
[e] Data from Wilmore, J.H., et al.: Football pros' strengths—and CV weakness—charted. Phys. Sportsmed., 4:45, 1976.
[f] Data from Katch, F.I., and Katch, V.L.: Body composition of the Dallas Cowboys and New York Jets Football teams. Unpublished data, 1978.
[g] Data from Dr. A.R. Behnke and reference 46.

for body composition using densitometry to assess fat content. The pro, current group consists of 164 players from 14 teams in the National Football League (NFL) (69% were veterans, 31% were rookies). The third group is 107 members of the 1976–1978 Dallas Cowboys and New York Jets football teams. For comparison, three groups of collegiate players are represented: (1) the St. Cloud State College, Minnesota, players who were candidates for spring practice, (2) players from the University of Massachusetts (U. Mass.) who were also candidates for spring practice, and (3) teams from the University of Southern California (USC), 1973 to 1977, who were National Champions and participants in two Rose Bowls.

One would generally expect modern day professional players to be larger in body size at each position than a representative collegiate group. Although this was generally true for the comparison with the St. Cloud and U. Mass. players, the USC players were similar in physique to the professionals. With the exception of defensive linemen, the USC players at each position had almost the same body fat content, although they tended to weigh less than the current players at each position. For the all-important component of lean body mass, the USC players were no more than 4.4 kg lighter than the "pros" at each position. In fact, the average defensive lineman in the NFL outweighed his USC counterpart in lean tissue by only 1.8 kg. The total body mass of the pro linemen, however, was significantly heavier than that of the USC counterparts. This difference occurred because the professional linemen possessed 18.2% body fat whereas collegians were leaner at 14.7%.

It is of interest to note that as a group, the pro players of 50 years ago were lowest in terms of total body fat (10.4%), and were shorter in stature and lighter in total body mass and lean body mass than their modern counterparts. The exceptions were the defensive and offensive backs and receivers; they were almost identical in body size and composition to current players. The biggest differences in physique status were for the defensive lineman (current players were taller by 6.7 cm, heavier by 20 kg, and with 12.3 kg more lean body mass and 4.2 percentage points more percent body fat). Obviously, "bigness" was not yet an important factor in the line play of the 1940s.

In comparing the modern pro players with their collegiate contemporaries, one can conclude that at the highest levels of collegiate competition, the body size and composition of college and professional players are similar.

HIGH SCHOOL WRESTLERS

Wrestlers represent a unique group of athletes who undergo both severe training and acute weight loss. Despite warnings from medical and professional groups regarding rapid weight loss,[1,3] most high school and college wrestlers lose considerable weight a few days before or on the day of competition. This is done with the hope of gaining a competitive advantage by wrestling at a lower weight category. This process of "making weight" usually occurs by combining food restriction and dehydration, either through fluid deprivation or exercising in a hot environment while wearing plastic or rubber garments. To reduce the possibility of injury from acute weight loss and dehydration, the American College of Sports Medicine recommends that each wrestler's body composition be assessed several weeks prior to the competitive season to determine an acceptable minimal wrestling weight.[2] **A lower limit of 5% body fat is proposed as the lowest acceptable level for safe wrestling competition.** The monitoring of relative leanness may also be appropriate for competitors below the high school level; research indicates that prepubescent wrestlers possess significantly less body fat than normally active boys.[38]

Physical Characteristics of High School Wrestlers

The physical characteristics of three groups of high school wrestlers are presented in Table 28–6.[12,22] The "certified" Iowa and Minnesota wrestlers were assigned to wrestle at one of 12 different weight categories; the "champion" wrestlers competed in the State or Conference finals. Except for age and fatfolds, there was little difference in the physical characteristics of the Iowa and Minnesota certified and champion wrestlers. As reflected by the fatfold measures, however, the champions were considerably leaner than their less successful teammates. Because differences in body mass were small among groups, the elite wrestlers actually competed at a heavier lean body mass. This may have contributed greatly to their success in a particular weight class. The last column in Table 28–6 presents recent additional data on 409 Nebraska high school wrestlers.[22] Their anthropometric characteristics are most similar to the Iowa and Minnesota certified wrestlers. Their relative percent body fat determined densitometrically was 11.0 ± 4.0% (range from 1.5 to 26.0%). Their minimal wrestling weight at 5% body fat averaged 59.1 kg.

Weight Loss Recommendations for Wrestlers

Prior to the season's start, suppose a high school wrestler who wishes to compete in the lowest possible weight category compatible with good health

TABLE 28–6.
ANTHROPOMETRIC COMPARISONS BETWEEN CERTIFIED AND CHAMPION IOWA AND MINNESOTA HIGH SCHOOL WRESTLERS[a] AND COMPARISON TO NEBRASKA WRESTLERS[b]

| Measurement | Certified Wrestlers | | Champion Wrestlers | | Nebraska Wrestlers |
	Iowa (N = 484)	Minn (N = 245)	Iowa (N = 382)	Minn (N = 164)	(N = 409)
Age, years	15.9	16.8	17.8	17.4	16.4
Stature, cm	169.9	172.0	171.7	172.5	171.0
Body mass, kg	64.3	65.3	64.6	64.7	63.2
Chest diameter, cm	26.8	26.5	27.7	26.6	27.9
Chest depth, cm	19.0	16.8	19.2	17.3	18.9
Bitrochanteric diameter, cm	31.0	31.4	31.1	31.5	31.0
Ankles diameter, cm	14.3	14.3	14.0	14.3	13.8
Fatfolds, mm					
Scapula	8.4	7.9	6.4	6.8	8.8
Triceps	8.6	9.1	6.0	5.6	8.9
Suprailiac	13.3	12.3	9.1	7.5	11.8
Abdominal	13.1	12.2	8.6	8.3	11.6
Thigh	10.8	13.6	7.7	8.3	9.4
Sum of 5 fatfolds	54.2	55.1	37.8	36.5	50.5

[a] From Clarke, K.S.: Predicting certified weight of young wrestlers: a field study of the Tcheng-Tipton method. *Med. Sci. Sports,* 6:52, 1974.
[b] From Housh, T.J., et al. Validity of anthropometric estimations of body composition in high school wrestlers. *Res. Q. Exec. Sport,* 60:239, 1989.

and performance weighs 70 kg with 15% body fat. A prudent recommendation for weight loss would be to reduce body fat content no lower than 5% of body mass.[44] The proposed weight loss would therefore be 10% of his body mass or about 7 kg, resulting in a final body mass of 63 kg. With increased training and moderate caloric restriction, a weight loss of 1 kg per week is a reasonable objective.[42] Thus, in about 7 weeks, the desired weight loss would be achieved and the wrestler could effectively compete in the 145-lb (65.8 kg) weight class.

WEIGHT LIFTERS AND BODY BUILDERS

Men

Individuals who engage in resistance training often exhibit remarkable muscular development, although quantification has been limited of this hypertrophy. Data from the 1968 Olympic weightlifting team reveals that for these men the estimated average lean body mass accounted for nearly 90% of their total body mass compared to between 80 and 85% usually observed for non-weightlifting counterparts. Bodybuilders are a group of athletes who lift weights solely to improve body configuration and form, with little concern for subsequent athletic performance. This group differs from the other types of weight lifters who train for the purpose of increasing the amount of weight lifted with less concern for muscle size and definition.

Excess muscle development and level of lean body mass has been quantified in competitive body builders, Olympic weight lifters, and power weight lifters.[28] For percent body fat, the values were quite similar averaging 9.3%, 9.1%, and 10.8% for the body builders, power weight lifters, and Olympic weight lifters, respectively. This degree of leanness existed even though each group of athletes was classified as being between 14 to 19% "overweight" from the "height-weight" tables! Also, no differences between groups were present in skeletal frame size, lean body mass, fatfolds, and bone diameter measurements. The only differences in anthropometry between the groups were for the shoulders, chest, biceps relaxed and flexed, and forearm girths, with the body builders being larger at each site. Estimation of excess muscle revealed that the body builders possessed nearly 16 kg of excess muscle, while power weight lifters had 15 kg and Olympic weight lifters had 13 kg.

Women

Body building for females gained widespread popularity throughout the United States during the late 1970s as women aggressively pursued this previously male-dominated sport. As more women undertook the vigorous demands of resistance training, competition became more intense and the level of achievement increased markedly. Because success in bodybuilding is based on a "slim" and lean appearance, complemented by a well-defined yet enlarged musculature, interesting questions are raised with regard to body composition. How lean are such competitors, and is their presumably low level of body fat accompanied by a relatively large muscle mass?

A study of the body composition of 10 competitive female body builders revealed that these athletes were quite lean, averaging 13.2% body fat (range from 8.0 to 18.3%) with an average lean body mass of 46.6 kg.[17] Their body mass averaged 53.8 kg, stature 160.8 cm, and age 27.1 years. With the exception of champion gymnasts who also average about 13% body fat, the body builders were shorter in stature by 3 to 4%, lower in body mass by 4 to 5%, and possessed 7 to 10% less total mass of body fat compared with other top female athletes. The most striking compositional characteristic of the female body builders was their dramatically large lean-to-fat ratio of 7:1 (weight of the lean mass relative to the weight of the fat mass) in comparison to 4.3:1 for other female athletic groups. This occurred without the use of steroids as reported by the women in a questionnaire. It is interesting to note that menstrual function was reported as normal by eight of ten body builders in the aforementioned study, despite their relatively low level of body fat.

Males vs Females

Table 28–7 compares the body composition and girths of male and female body builders, including excess muscle. The latter was defined as the difference between scale mass and mass-for-height taken from the Metropolitan Life Insurance tables. The men are 14.8 kg (18%) overweight, and the women are 1.2 kg (12%) overweight. Obviously for these athletes, excess mass is primarily lean body mass and excess muscle.

The comparison of the girth data makes it possible to compare individuals (or groups) who differ in overall body size. The analysis shows that gender

TABLE 28–7.
BODY COMPOSITION AND ANTHROPOMETRIC GIRTHS OF MALE AND FEMALE BODY BUILDERS

Gender	Age (year)	Mass (kg)	Stature (cm)	Fat (%)	LBM (kg)	Excess Mass[a] (kg)
Male[b] (N = 18)	27.0	82.4	177.1	9.3	74.6	14.8
Female[c] (N = 10)	27.0	53.8	160.8	13.2	46.6	1.2

Body Part (cm)	Males		Females		% Difference (males vs. females)	
	Raw	Adjusted[d]	Raw	Adjusted[d]	Raw	Adjusted
Shoulders	123.1	37.1	101.7	36.7	17.4	1.1
Chest	106.4	32.1	90.6	32.7	14.9	− 1.9
Waist	82.0	24.7	64.5	23.3	21.3	5.7
Abdomen	82.3	24.8	67.7	25.1	15.3	− 1.2
Hips	95.6	28.8	87.0	31.4	9.0	− 9.0
Biceps relaxed	35.9	10.8	25.8	9.3	28.1	13.9
Biceps flexed	40.4	12.2	28.9	10.4	28.5	14.8
Forearm	30.7	9.2	24.0	8.7	21.8	5.4
Wrist	17.4	5.2	15.1	5.4	13.2	− 3.8
Thigh	59.6	17.9	53.0	19.1	11.1	− 6.7
Calf	37.3	11.2	32.4	11.7	13.1	− 4.5
Ankle	22.8	6.9	26.3	7.3	11.0	− 5.8

[a] Body mass minus body mass estimated from "height-weight" tables.
[b] Reference 28.
[c] Reference 17.
[d] Calculated as $G_i/\sqrt{\text{mass}}$, kg/stature, dm$^{0.7}$), where G_i equals any one of the girths. The term (mass/stature$^{0.7}$) is a frame structure estimate of perimetric (girth) size. The adjusted values are the perimetric equivalent adjusted girths due to gender differences because they are corrected for whatever differences may exist due to differences in body size.

differences in girths, when scaled to body size (referred to as "adjusted" in the table), are not as different as would be expected if only comparing the raw girths. In fact, relative to their body size, the females are actually **larger** than the male body builders in 7 of the 12 body areas. **This suggests that women can probably alter their muscle size to almost the same extent as males, at least when scaled to a body size factor.** The fact that the women are larger in the hips probably relates to greater fat stores in this location.

COMPARATIVE DATA: ANALYSIS OF MUSCULAR AND NONMUSCULAR COMPONENTS

The body profile technique introduced in Chapter 27 provides a unique way to partition the anthropometric aspects of body size and shape into muscular and nonmuscular components. A major advantage with this approach is that diversity in overall body dimensions can be readily compared among individuals and groups. The technique also permits the easy tracking of progress (or lack thereof) in the various body components that result from different modes of training, weight loss and gain, growth, and aging. Table 28–8 presents the percentage deviations for the muscular and nonmuscular components for 10 different groups. The percentage-deviation scores are directly comparable because the data analysis used the Behnke reference man and woman as standards of comparison, and adjustments were made among the groups to equalize them in terms of stature.

TABLE 28–8.
PERCENTAGE DEVIATION FOR THE MUSCULAR AND NONMUSCULAR COMPONENTS AMONG INDIVIDUALS AND GROUPS THAT DIFFER MARKEDLY IN BODY SIZE AND SHAPE. (GROUPS 1 TO 10 ARE IDENTIFIED IN FOOTNOTE*)

Reference Data Group	Group									
	1	**2**	**3**	**4**	**5**	**6**	**7**	**8**	**9**	**10**
Age, years	21.8	32.1	27.1	18–22	17.2	17.2	17.0	17.4	17.2	21.2
Body mass, kg	51.4	84.2	53.4	78.5	69.3	58.2	43.4	141.8	59.1	64.6
Stature, cm	166.4	170.6	160.6	180.3	178.1	165.0	161.2	189.7	171.0	172.9
Muscular Girths					**Percentage Deviation**					
Shoulders	10.6	28.2	12.9	1.7	2.7	−1.0	0.7	−6.4	—	—
Chest	8.6	49.2	19.4	6.4	2.2	1.4	3.8	4.1	−1.2	2.0
Biceps	−9.2	77.9	14.9	2.3	−2.4	−2.9	5.4	3.1	−10.4	17.3
Forearm	−1.8	43.7	9.1	3.3	1.6	−0.8	2.0	−15.3	2.6	4.0
Thigh	0.0	17.9	−1.6	6.3	2.5	−2.2	−20.2	26.1	−4.1	−2.4
Calf	9.8	20.9	3.4	4.7	4.4	2.0	1.4	11.0	2.1	2.0
Nonmuscular Girths										
Abdomen	−1.2	−28.7	−7.0	−2.6	−4.5	0.2	−7.8	31.9	−5.9	−7.6
Hips	−3.7	−30.1	−11.0	−3.1	−0.4	−2.0	−1.4	3.2	2.4	−2.5
Wrist	−5.8	−26.4	−8.3	−7.8	−4.7	−3.9	16.4	−36.4	3.4	−7.0
Knee	−4.9	−29.0	−12.1	−3.8	−1.0	4.3	7.6	−12.2	8.7	0.6
Ankle	0.6	−27.7	−6.1	−5.9	2.6	5.8	11.1	−4.7	—	—

* The numbers for the groups refer to the following: 1 = professional ballet dancers; 2 = champion male body builders; 3 = champion female body builders; 4 = male college students; 5 = senior high school males; 6 = senior high school females; 7 = smallest boy in a 12th grade class; 8 = largest boy in a 12th grade class; 9 = entering class of Amherst College in 1882; 10 = graduating class of Amherst College in 1886. More detailed anthropometric information about Groups 1 to 10, including original references, appears in reference 27.

Professional Ballet Dancers

The subjects were 10 prima ballerinas from the San Francisco Ballet Company. Not unexpectedly, six of the 11 percentage deviations were negative; the exceptions were the shoulders, chest, thigh, calf, and ankles.

Champion Male and Female Body Builders

The excessive muscular development for the champion group of 11 Mr. Universe body builders (Table 28–8) is evident for the six muscular sites, especially the biceps—77.9%—the largest deviation of any body site among the groups. Expressed as the ponderal equivalent for the biceps, the 77.9% deviation is equivalent to 127.9 kg. This means that for a body builder who weighed 84.2 kg, the mass of the biceps would be the projected size of a biceps in an untrained individual weighing 127.9 kg. The ratio of the muscular to nonmuscular girths was 1.396; this would verify the existence of excess muscle. In contrast, the muscular/nonmuscular ratio for the 10 champion female body builders (group 3) was considerably lower at 1.097. Although visual appraisal of a female body builder gives the appearance that the muscular girths predominate, the body profile analysis suggests that the eye may be unable to detect excess muscle as well as it can perceive general refinement and muscular shaping in different body parts. Stated differently, female body builders as a group were more evenly proportional between the muscular and nonmuscular components.

College Age Males

Group 4 was a representative sample of 100 males from Eastern Oregon College. They were all nonathletes.

Twelfth Grade Caucasian Males and Females

Groups 5 and 6 were part of a large-scale multiracial nutrition and anthropometric survey of ninth- through twelfth-grade students at Berkeley High School, Berkeley, California. Despite substantial differences in mass and stature at identical ages, the percentage deviations were surprisingly similar for the muscular and nonmuscular components. The absolute values for the various body parts, however, were larger for the males by approximately the same difference as body mass.

Smallest and Largest Boy in the Twelfth Grade

Groups 7 and 8 represent the smallest and largest boy, respectively, of the same age in the 450-member twelfth-grade class at Berkeley High School. The comparison is dramatic; the difference between the boys in body mass was 98.4 kg (217 lb) and 28.5 cm in stature (almost one foot)!

Table 28–9 compares the ponderal (mass) equivalents for the muscular and non-muscular body components for these two boys. The ponderal values that are listed are the mass equivalents for a person possessing a specific girth. For example, the ponderal equivalent of 44.7 kg for the diminutive boy is the projected mass based on that particular girth. Note that the smallest boy weighed 43.4 kg, so his shoulder mass was certainly of "normal" development. This percentage deviation for the shoulder of 0.7 listed in Table 28–8 (Group 7) is much smaller than the −6.4 percent deviation for the

TABLE 28–9.
PONDERAL EQUIVALENT (P-E) MUSCULAR AND NONMUSCULAR COMPONENTS FOR THE SMALLEST AND LARGEST BOY IN A TWELFTH-GRADE CLASS.*

Variable	Smallest Boy Mass = 43.4 kg Stature = 161.2 cm		Largest Boy Mass = 141.8 kg Stature = 189.7 cm	
	Girth (cm)	P-E (kg)	Girth (cm)	P-E (kg)
Muscular				
Shoulders	92.3	44.7	144.2	128.5
Chest	74.7	42.7	126.0	142.9
Biceps	25.6	42.0	43.3	141.6
Forearm	22.1	43.5	33.3	116.3
Thigh	40.6	35.4	82.8	173.2
Calf	29.9	45.0	50.7	152.4
Nonmuscular				
Abdomen	60.9	38.9	123.4	188.0
Hips	75.0	41.6	130.0	147.0
Wrist	15.1	49.1	18.9	90.6
Knee	30.7	45.4	47.0	125.1
Ankle	19.2	46.9	30.1	135.8

* Data for Groups 7 and 8 in Table 28–8.

largest boy (Group 8). His ponderal equivalent based on the shoulder measurement is 128.5 kg (Table 28–9), yet his body mass is only 13.3 kg greater. Viewed somewhat differently, the largest boy in the class had a shoulder ponderal equivalent of someone who weighed 128.5 kg. It certainly would be fair to conclude that despite the boy's size, his shoulders were underdeveloped, as were his forearms (116.3 kg), wrists (90.6 kg), knees (125.1 kg), and ankles (135.8 kg).

For the nonmuscular abdominal component, the ponderal equivalent for the diminutive boy was also small (38.9 kg) in relation to his scale weight (43.4 kg). For the largest boy, the abdominal component was excessively large; it corresponded to a projected size for someone who would be expected to weigh 46.2 kg more or 188.0 kg (415 lb)!

It may be that in relatively small children or large adolescents, the occurrence of a large positive deviation for the ponderal equivalent for the abdomen is a "signal" that coincides with the onset of late-adolescent or even adult-onset obesity. If this is true, then determination of the abdominal ponderal equivalent at an early age with subsequent follow-up could be useful as a clinical marker for preventive purposes.

Entering and Graduating Classes of Amherst College (1882–1886)

This remarkable data set for groups 9 and 10 was retrieved from the archival records of the pioneer exercise scientist-anthropometrist, Dr. Edward Hitchcock. Note the dramatic increase in biceps girth and decreases in the nonmuscular abdomen and hip regions: these changes coincided with the start of daily resistance training.

EFFECTS OF EXERCISE TRAINING ON BODY COMPOSITION

In this section, we consider the effectiveness of walking, jogging, cycling, and resistance training in modifying the body composition of young and middle-aged adults. These findings have special meaning to health professionals and others involved in leading exercise programs because of the important role exercise can play in effective weight control.

TEN-WEEK JOGGING PROGRAM

Table 28–10 shows changes in physique for men aged 17 to 59 years who jogged 3 days a week for 10 weeks.[48] The average distance run by the end of 10 weeks was 84.4 km (51.8 miles), or about 2.8 km or 1.7 miles a day. Body composition changes did occur but they were relatively small. Because lean body mass did not change, the decrease in body mass was due to a reduction in percent body fat from pretest (18.9%) to post-test (17.8%) values, which represented a fat loss of 1.07 kg. The reduction in individual fatfold values paralleled the decrease in body fat.

It is possible that this small reduction in body fat was due to (1) the relatively short duration of the jogging program, and (2) the average body fat of the group prior to training was only slightly greater than the average for college-aged men. As such, these men were not overfat and did not need to greatly reduce body size.

WALKING-RUNNING FOR DIFFERENT DURATIONS

The duration of exercise has an effect on fat loss with training. Table 28–11 shows the changes in body fat (predicted from fatfolds) for three groups

TABLE 28–10.
BODY COMPOSITION CHANGES RESULTING FROM A 10-WEEK JOGGING PROGRAM[a]

Variable	Pretraining	Post Training	Difference
Body mass, kg	79.59	78.58	− 1.01[b]
Body fat, %	18.88	17.77	− 1.11[b]
Body fat, kg	15.03	13.96	− 1.07
LBM, kg	64.56	64.62	0.06
Fatfolds, mm			
Triceps	11.5	11.1	− 0.4
Scapula	16.3	15.1	− 1.2[b]
Suprailiac	24.9	24.4	− 0.5
Mid axillary	17.3	14.3	− 3.0[b]
Abdominal	24.4	23.5	− 0.9
Thigh	16.9	16.0	− 0.9[b]
Chest	12.7	11.5	− 1.2[b]
Girths, cm			
Waist	84.8	84.4	− 0.8[b]
Abdomen	88.2	87.7	− 0.5

[a] From Wilmore, J.H., et al.: Body composition changes with a 10-week program of jogging. Med. Sci. Sports, 2:113, 1970.
[b] Statistically significant.

TABLE 28–11.
EFFECTS OF THREE TRAINING DURATIONS OF WALKING AND RUNNING ON BODY COMPOSITION CHANGES*

| | | | | | Training Group | | | |
| | Control (N = 16) | | 15 Minute (N = 14) | | 30 Minute (N = 17) | | 45 Minute (N = 12) | |
Variable	PRE	POST	PRE	POST	PRE	POST	PRE	POST
Body mass, kg	72.1	73.2	76.9	76.3	80.6	78.9	70.9	69.9
Body fat, %	12.5	13.0	13.7	13.2	14.2	13.6	13.2	12.0
Sum of fatfolds, mm	73.8	79.6	83.0	77.0	90.0	83.8	77.5	67.0
Waist girth, cm	82.7	84.9	84.3	82.8	88.2	86.1	83.6	81.8
Distance run per workout (miles)	week	4		1.56		2.89		4.13
		8		1.54		2.95		4.46
		13		1.79		3.19		4.82
		17		1.75		3.24		5.06
Total time of exercise (min:s)	week	4		14:58		30:25		41:18
		8		14:11		28:40		42:48
		13		15:51		29:43		43:19
		17		14:53		30:12		42:27
Training heart rate (beats · min⁻¹)	week	4		179		175		174
		8		179		174		169
		13		182		175		177
		17		180		175		175
Intensity (% of max HR)	week	4		89.4		83.8		84.5
		8		89.8		73.4		81.0
		13		94.0		90.1		89.5
		17		92.5		90.2		88.1

* From Milesis, C.A., et al.: Effects of different durations of physical training on cardiorespiratory function, body composition, and serum lipids. Res. Q., 47:716, 1976.

of men who trained walking and running for either 15, 30, or 45 minutes per workout.[32] Also included are the distance run and the total duration of the weekly workouts, training heart rate, body mass, the sum of fatfolds (chest, axilla, triceps, abdomen, suprailiac, and anterior thigh), and waist girth.

Compared to the control group, that remained unchanged over the 20-week training period, the three exercise groups significantly decreased their body fat, fatfolds, and waist girth. Body mass was also significantly lowered with exercise except for the 15-minute group whose mass remained stable. When comparisons were made between the three groups, the 45-minute training group lost a greater percentage of body fat then either the 30- or 15-minute exercise groups. **This was attributed to the greater calorie-burning effect of the longer exercise period.** For individuals desiring to use walking as the sole means for exercise training and weight loss, hand, wrist, and ankle weights can be used to increase exercise intensity and calorie output as the program progresses.[18]

TWO-YEAR PROGRAM OF CALISTHENICS AND JOGGING

In a 2-year calisthenics and jogging program the body composition of seven middle-aged men was evaluated.[11] Comparative data were also presented for six controls measured at the same 6-month intervals, but who did not take part in the exercise program. The exercisers participated in a supervised program 3 days a week. Initially, they walked and jogged for 10 minutes; thereafter, they jogged for 30 to 35 minutes. The average distance covered increased from 2.4 to 12.1 km per week, and the total mileage run per subject after 2 years of training averaged 1188 km or 738 miles.

Compared with control subjects whose body composition remained relatively constant during the 2-year period, the exercisers after the first year significantly reduced their body mass (5.7%), sum of fatfolds (27.4%), and girth measurements (3.1%). Thereafter, there was little further change in body mass and body composition. **These findings show that calisthenics and jogging can significantly alter the physique of previously sedentary 40- to 60-year-old men.** The changes were paralleled by a 25% improvement in aerobic capacity.

WALKING, RUNNING, AND BICYCLING

When the relative training effects of walking, running, or bicycling on body composition are evaluated, each mode of exercise is found to reduce significantly body mass, body fat, fatfold thickness, and girths. **In addition, there is generally no selective effect of running, walking, or bicycling; each training mode is equally effective in altering body composition.**

RESISTANCE TRAINING

Resistance training has been considered as a possible modality to induce favorable changes in body composition.[7,13] Because the caloric expenditure of circuit resistance training averages about 9 kcal per minute, this form of exercise can potentially produce a substantial caloric output during a typical 30 to 60-minute workout.[5,26] Table 28–12 presents the results of a study of the body composition changes of initially obese females who trained 3 days a week for 8 weeks without modifying daily caloric intake.[4] The exercise regimen consisted of an eight-station routine performed on a multistation hydraulic apparatus. Subjects performed 3 sets of 10 repetitions of the bench press, inverse leg press, lateral-pull down, biceps curl, triceps extension, calf raise, leg extension, and hamstring curl. Strength, evaluated by the 1-RM bench press (Chapter 22), improved by 5.0 kg from 35 kg to 40 kg.

Considering the relatively brief duration of training, there was an impressive 4.9% increase in biceps girth; this was more than likely due to the 6.0% increase in muscle plus bone cross-sectional area from radiographic examination of the upper arm and a corresponding decrease of 5.3% in the cross-sectional area for fat. Body composition changed favorably: percent fat decreased by 1.2% fat units (−3.4%), fat mass decreased by 0.6 kg (−2.3%), and lean body mass increased by 1.1 kg (2.3%).

FREQUENCY OF TRAINING

Research has been conducted to investigate optimal training frequency.[36] Training consisted of either running or walking and was conducted for 30 to 47 minutes a day for 20 weeks; exercise intensity was always maintained at 80 to 95% of maximum heart rate. Training 2 days a week did not significantly change body mass, fatfolds, or percent body fat but training 3 and 4 days a week did. Subjects who trained 4 days a week reduced their body

TABLE 28–12.
BODY COMPOSITION AND MEASUREMENTS IN 10 OBESE FEMALES BEFORE AND AFTER RESISTANCE TRAINING

Variable	Mean Value		Change (%)
	Before Training	After Training	
Body Composition			
Body mass, kg	73.9	74.3	0.5
Body fat, %[a]	35.1	33.9	−3.4[d]
Fat mass, kg	26.2	25.6	−2.3
LBM, kg	47.7	48.8	2.3[d]
Anthropometry[b]			
Waist girth, cm	80.5	81.1	0.7
Thigh girth, cm	64.4	65.4	1.6
Calf girth, cm	38.5	38.8	0.8
Biceps girth, cm	32.8	34.4	4.9[d]
Forearm girth, cm	26.8	27.6	3.0
Sum 5 fatfolds, mm	170.1	170.6	0.3
Arm Radiography[c]			
Arm area, cm²	255.7	261.8	2.4
Muscle + bone area, cm²	173.8	184.3	6.0[d]
Fat area, cm²	81.8	77.5	−5.3[d]

[a] Densitometry.
[b] Measurement procedures discussed in Chapter 27 and Appendix I.
[c] Measurement procedures shown in Figure 27–6.
[d] Changes were statistically significant compared to control subjects matched for age, stature, body mass, LBM, and % body fat. (From Ballor, D.L., et al.: Resistance weight training during caloric restriction enhances lean body weight maintenance. Am. J. Clin. Nutr., 47:19, 1988.)

mass and fatfolds significantly more than the 3-day per week group. Reductions in percent body fat, however, were similar for both groups.

Within the framework of these findings, at least 3 days of training per week are required to change body composition through exercise; more frequent training may even be more effective. More than likely, such effects result from the added caloric stress provided by the extra training. In addition, the calorie-burning effect of each exercise session should reach a threshold of about 300 kcal.[37] This is generally achieved with 20 to 30 minutes of moderate to vigorous running, swimming, bicycling, circuit resistance training, or walking programs of at least 60 minutes duration.

EFFECTS OF DIET AND EXERCISE ON BODY COMPOSITION DURING WEIGHT LOSS

The addition of exercise to a program of weight control can favorably modify the composition of the weight lost in the direction of greater fat loss.[4,6a,30,31,51] To evaluate this possibility, a caloric deficit of 500 calories per day was maintained by each of three groups of adult women during a 16-week period of weight loss.[51] The diet group reduced daily food intake by 500 kcal, whereas women in the exercise group increased their energy output by 500 kcal through participation in a supervised walking and exercise

FIG. 28–1.
Changes in body composition with exercise and diet in obese females. (Ballor, D.L., et al.: Resistance weight training during caloric restriction enhances lean body mass maintenance. *Am. J. Clin. Nutr.*, 47:19, 1988.)

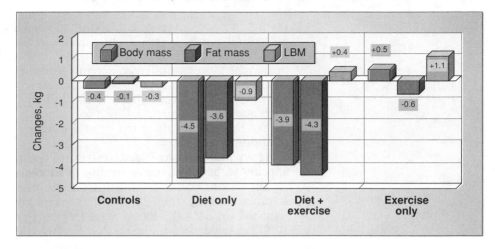

program 5 days a week. The women using diet plus exercise created their daily 500 kcal deficit by reducing food intake by 250 kcal and increasing energy output by 250 kcal through exercise. In terms of weight loss, there was no significant difference between the three groups because each group lost approximately 5 kg. This finding shows that as long as a caloric deficit is created body mass will be reduced, regardless of the method used to create the imbalance. **In terms of reducing body fat, however, combining diet and exercise was the most effective approach to weight loss.** The most interesting observation concerned lean body mass. Although the exercise and combination groups increased their lean body mass by 0.9 and 0.5 respectively, the dieters **lost** 1.1 kg of lean tissue!

Figure 28–1 displays the body composition changes in a recent experiment in which 40 obese women were placed in one of four groups: (1) control, no exercise and no diet; (2) diet only without exercise (DO); (3) diet plus resistance exercise (D + E), and (4) resistance exercise only without diet (EO). The women trained 3 days a week for 8 weeks. They performed 10 repetitions for each of 3 sets of eight strength exercises. Body mass decreased significantly for DO (−4.5 kg) and D + E (−3.9 kg) compared to EO (+0.5 kg) and controls (−0.4 kg). Interestingly, lean body mass increased significantly for EO (1.1 kg) relative to DO (−0.9 kg) and controls (−0.3 kg), and lean tissue was spared for D + E (+0.4 kg) compared to the loss in lean body mass for DO (−0.9 kg). The authors concluded that augmenting a caloric-restriction program with resistance exercise training preserves the lean body mass in comparison to a diet program without such exercise.[4]

When body mass is reduced by diet only, there is a greater loss of lean tissue and less fat than when a similar loss in body mass occurs with the appropriate use of exercise.[8,10,33]

GAINING WEIGHT

Gaining weight for athletes poses a unique problem that is not easily resolved. Weight gain per se is a relatively easy and often enjoyable task brought about by imbalancing the body's energy balance in favor of a greater caloric intake. For a sedentary person, an excess intake of 3500 kcal results in a body mass gain of about 0.5 kg. This is because the excess calories are stored as body fat. Weight gain for athletes, however, should be in the lean tissues specifically muscle mass. It is generally agreed that this form of weight gain can only be accomplished if an increased caloric intake is accompanied by an appropriate exercise program.[42] Although endurance exercise can increase lean body mass slightly,[37] the body composition change is frequently accompanied by a loss in body mass due to fat loss. This loss is probably the result of the calorie-burning and appetite-depressing effects of endurance exercise.

Heavy muscular overload (resistance training) supported by a prudent diet appears to be an effective means to increase muscle mass and strength. If all the "extra" calories consumed were used for muscle growth during resistance training, then 2000 to 2500 extra kcal from a well-balanced diet are required for each 0.5 kg increase in lean tissue. In a practical sense, 700 to 1000 kcal added to the daily diet supply the nutrients to support a weekly 0.5- to 1.0-kg gain in lean tissue as well as the energy requirements of the training.[16] This ideal situation presupposes that all extra calories are used to synthesize lean tissue. Variation from this ideal depends on many factors including the type, intensity, and frequency of training and the hormonal characteristics of the athlete. Athletes who have a relatively high androgen-estrogen ratio probably increase lean tissue to a greater extent than a similarly trained athlete with lower androgen production.[19] One means to verify whether the combination of training and increased food intake is increasing lean tissue (and not body fat) is to monitor regularly body mass and body fat. This can be accomplished in the laboratory with hydrostatic weighing or appropriate anthropometric measurements.

SUMMARY

1. Body composition assessment reveals that athletes generally have physique characteristics unique to their specific sport. For example, field-event athletes have relatively large quantities of lean tissue and a high percent body fat, whereas long-distance runners have the least amount of lean tissue and fat mass.
2. Physique characteristics blended with highly developed physiologic support systems provide important ingredients for a champion performance.
3. Body composition analysis of football players reveals they are among the heaviest and leanest of all sportsmen. At the highest levels of competition, collegiate and professional football players are similar in body size and composition.
4. Wrestlers should be discouraged from drastically reducing body fat, especially if it brings them below 5% body fat.
5. The body profile technique provides a practical method to subdivide anthropometric dimensions into muscular and non-muscular components, and to monitor progress resulting from training, diet, growth, and aging.
6. Continuous moderate to high intensity exercise, can be an effective for weight reduction. The greater the caloric expenditure, the greater the potential for body fat loss. This effect is independent of mode of exercise, as long as there is a sufficient caloric deficit caused by the exercise. At least 3 training days a week are required, but seven days are preferred.
7. Thirty minutes of moderately strenuous running, bicycling, circuit resistance exercise, or swimming, or at least 60 minutes of walking will

stimulate fat loss. This threshold level of exercise generally represents about a 300-kcal increase in daily energy expenditure.

8. A combination of diet and exercise offers the greatest flexibility for achieving a negative caloric balance and desirable body fat loss. The inclusion of exercise with diet in a weight-loss program provides protection against an excessive loss of lean tissue. This results in greater fat loss than would be achieved by diet alone.

9. For athletes, weight gain should be in the form of lean body tissue, i.e., muscle mass. Increased caloric intake plus resistance training increases muscle mass and strength effectively. Ideally, 700 to 1000 extra kcal per day support a weekly 0.5- to 1.0-kg gain in lean tissue and training energy requirements. Realistically, individual physiologic variations and training factors also affect weight gain. For this reason, changes in body mass and body fat should be monitored on a regular basis.

REFERENCES

1. Abraham, S., et al.: Weight and height of adults 18 to 74 years of age. United States Vital and Health Statistics, Series H, No. 211. Washington, DC, U.S. Government Printing Office, 1979.

2. American College of Sports Medicine: Weight loss in wrestlers. *Med. Sci. Sports*, 8:11, 1976.

3. American Medical Association Committee on the Medical Aspects of Sports: Wrestling and weight control. *JAMA*, 201:541, 1967.

4. Ballor, D.L., et al.: Resistance weight training during caloric restriction enhances lean body weight maintenance. *Am. J. Clin. Nutr.*, 47:19, 1988.

5. Ballor, D.L., et al.: Physiologic responses to nine different work:rest protocols during hydraulic resistive exercises. *Med. Sci. Sports Exerc.*, 21:90, 1989.

6. Behnke, A.R., and Wilmore, J.H.: *Evaluation and Regulation of Body Build and Composition*. Englewood Cliffs, NJ, Prentice Hall, 1974.

6a. Bouchard, C., et al.: Long-term exercise training with constant energy intake. 1: Effect on body composition and selected metabolic variables. *Int. J. Obesity*, 14:57, 1990.

7. Brown, C.H., and Wilmore, J.H.: The effects of maximal resistance training on the strength and body composition of women athletes. *Med. Sci. Sports*, 6:174, 1974.

8. Buono, M.J., et al.: Effects of a diet and exercise program on blood lipids, cardiorespiratory function, and body composition in obese women. *Med. Sci. Sports Exerc.*, 17:189, 1985.

9. Butts, N.K.: Physiological profile of high school female cross-country runners. *Phys. Sportsmed.*, 10:103, 1983.

10. Butts, N.K., et al.: Effects of a 12-week weight training program on the body composition of women over and under forty years of age. *Med. Sci. Sports Exerc.*, 17:267, 1985.

11. Carter, J.E.L., and Phillips, W.H.: Structural changes in exercising middle-aged males during a 2-year period. *J. Appl. Physiol.*, 27:787, 1969.

12. Clarke, K.C.: Predicting certified weight of young wrestlers: a field study of the Tcheng-Tipton Method. *Med. Sci. Sports*, 6:52, 1974.

13. Coleman, A.E.: Nautilus vs. Universal gym strength training in adult males. *Am. Corr. Ther. J.*, 31:103, 1977.

14. Cureton, T.K.: *Physical Fitness of Champion Athletes*. Urbana, IL, University of Illinois Press, 1951.

15. DeGaray, A.L., et al.: *Genetic and Anthropological Studies of Olympic Athletes*. New York, Academic Press, 1974.

16. Fahey, T.D., and Brown, C.: The effects of an anabolic steroid on the strength, body composition, and endurance of college males when accompanied by a weight training program. *Med. Sci. Sports*, 5:272, 1973.

17. Freedson, P.S., et al.: Physique, body composition, and psychological characteristics of competitive female bodybuilders. *Phys. Sportmed.*, 11:85, 1983.

18. Graves, J.E., et al.: Physiological responses to walking with hand weights, wrist weights and ankle weights. *Med. Sci. Sports Exerc.*, 20:265, 1988.
19. Harris, D.V.: The female athlete: strength, endurance and performance. In *Toward an Understanding of Human Performance.* Edited by E.J. Burke. Ithaca, NY, Mouvement Publications, 1977.
20. Hirata, K.: Physique and age of Tokyo Olympic champions. *J. Sports Med. Phys. Fitness,* 6:207, 1966.
21. Holly, R.G., et al.: Triathlete characterization and response to prolonged competition. *Med. Sci. Sports Exerc.*, 18:123, 1986.
22. Housh, T.J., et al.: Validity of anthropometric estimations of body composition in high school wrestlers. *Res. Q. Exerc. Sport.*, 60:239, 1989.
23. Katch, F.I., et al.: Effects of physical training on the body composition and diet of females. *Res. Q.*, 40:99, 1969.
24. Katch, F.I., and McArdle, W.D.: Prediction of body density from simple anthropometric measurements in college-age men and women. *Hum. Biol.*, 45:445, 1973.
25. Katch, F.I., and Katch, V.L.: The body composition profile: techniques of measurement and applications. *Clin. Sports Med.*, 3:31, 1984.
26. Katch, F.I., et al.: Evaluation of acute cardiorespiratory responses to hydraulic resistance exercise. *Med. Sci. Sports Exerc.*, 17:168, 1985.
27. Katch, F.I., et al.: The ponderal somatogram: evaluation of body size and shape from anthropometric girths and stature. *Hum. Biol.*, 59:439, 1987.
28. Katch, V.L., et al.: Muscular development and lean body weight in body-builders and weight lifters. *Med. Sci. Sports*, 12:340, 1980.
29. Kohlraush, W.: Zusammenhang von Korperform und Leistung. Ergebnisse der anthropometrischen Messungen an der Athletern der Amsterdamer Olympiade. *Int. Z. Angrew. Physiol.*, 2:187, 1970.
30. Konstantin, N.P., et al: Effects of dieting and exercise on lean body mass, oxygen uptake, and strength. *Med. Sci. Sports Exerc.*, 17:446, 1985.
31. McMurray, R.G., et al.: Responses of endurance trained subjects to caloric deficits induced by diet or exercise. *Med. Sci. Sports Exerc.*, 17:574, 1985.
32. Milesis, C.A., et al.: Effects of different durations of physical training on cardiorespiratory function, body composition, and serum lipids. *Res. Q.* 47:716, 1976.
33. Moyer, C.L., et al.: Body composition changes in obese women on a very low calorie diet with and without exercise. *Med. Sci. Sports Exerc.*, 17:292, 1985.
34. O'Toole, M.L., et al.: The ultraendurance triathlete: a physiological profile. *Med. Sci. Sports Exerc.*, 19:45, 1987.
35. O'Toole, M.L.: Training for ultraendurance triathletes. *Med. Sci. Sports Exerc.*, 21:209, 1989.
36. Pollock, M.L., et al.: Frequency of training as a determinant for improvement in cardiovascular function and body composition of middle-aged men. *Arch. Phys. Med. Rehabil.*, 56:141, 1975.
37. Pollock, M.L., and Jackson, A.: Body composition: measurement and changes resulting from physical training. In *Toward an Understanding of Human Performance.* Edited by E.J. Burke. Ithaca, NY, Mouvement Publications, 1977.
38. Sady, S.P., et al.: Physiological characteristics of high-ability prepubescent wrestlers. *Med. Sci. Sports Exerc.*, 16:72, 1984.
39. Sady, S.P., and Freedson, P.S.: Body composition and structural comparisons of female and male athletes. *Clin. Sports Med.*, 3:755, 1984.
40. Sinning, W.E., and Lindberg, G.D.: Physical characteristics of college age women gymnasts. *Res. Q.* 43:226, 1972.
41. Sinning, W.E.: Body composition, cardiorespiratory function, and rule changes in women's basketball. *Res. Q.*, 44:313, 1973.
42. Smith, N.J.: Gaining and losing weight in athletics. *JAMA*, 236:149, 1976.
43. Tanner, J.M.: *The Physique of the Olympic Athlete.* London, Allen and Unwin, 1964.
44. Tcheng, T., and Tipton, C.M.: Iowa wrestling study: anthropometric measurements and the prediction of a "minimal" body weight for high school wrestlers. *Med. Sci. Sports*, 5:1, 1973.
45. Thorland, W.G., et al.: Body composition and somatotype characteristics of Junior Olympic Athletes. *Med. Sci. Sports Exerc.*, 13:332, 1981.

46. Welham, W.C., and Behnke A.R.: The specific gravity of healthy men. *JAMA*, 118:498, 1942.

47. Wickkiser, J.D., and Kelly, J.M.: The body composition of a college football team. *Med. Sci. Sports*, 7:199, 1975.

48. Wilmore, J.H., et al.: Body composition changes with a 10-week program of jogging. *Med. Sci. Sports*, 2:113, 1970.

49. Wilmore, J.H., and Haskell, W.L.: Body composition and endurance capacity of professional football players. *J. Appl. Physiol.*, 33:564, 1972.

50. Wilmore, J.H., and Brown, C.H.: Physiological profiles of women distance runners. *Med. Sci. Sports*, 6:178, 1974.

51. Zuti, W.B., and Golding, L.A.: Comparing diet and exercise as weight reduction tools. *Phys. Sportsmed*; 4:49, 1976.

CHAPTER 29

OBESITY AND WEIGHT CONTROL

PART 1

Obesity

It is indeed unfortunate that in our modern era of technologic and scientific achievement, in which man has walked on the moon, developed surgical procedures to prolong and enhance the quality of life, and discovered many of the secrets of molecular interaction, there is no adequate explanation for a seemingly simple question: "Why do people become too fat, and what can be done to prevent it?" Despite increasing attempts at weight loss, more Americans are overweight than a generation ago. This is particularly true for an increasing number of children. More specifically, about 50 million men and 60 million women between the ages of 18 and 79 are "too fat" and need to reduce excess weight.[1] This amounts to about 377 million kg of excess fat for men and 667 million kg for women, or a total of 1044 million kg (2297 million lb) for the United States adult population! If the overfat men and women dieted by consuming 600 fewer calories each day to reduce to a "normal" value of body fat (achievable in 68 days for men and 101 days for women), the reduced caloric intake would equal 5.7 trillion calories. Translating this into fossil fuel energy and considering such factors as the energy required to plant, cultivate, harvest, feed, process, transport, wholesale, retail, acquire, store, and cook the food, the annual energy savings would be equal to that required to supply the residential electric demands of Boston, Chicago, San Francisco, and Washington, DC, or 1.3 billion gallons of gasoline to fuel 900,000 autos per year.[64]

Often a long-term process. Obesity often begins early in childhood and, if this occurs, the chances for adult obesity are three times greater compared to children of normal body mass.[56,58,128] Simply stated, a child generally does not "grow out of" an obesity problem.[35] Excessive fatness also develops slowly during the adult years with ages 25 to 44 being the danger years. Middle-aged men and women invariably weigh more than their college-aged counterparts of the same stature. There may also be racial differences in adipose tissue metabolism between male black Africans and Caucasians who are compared at similar levels of total body fat and fat distribution.[4] In the Western world, the average 35-year-old male gains between 0.2 and 0.8 kg of fat each year until the sixth decade of life, despite a progressive decrease in food intake! In one longitudinal study, the fat content of 27 adult men increased an average of 6.5 kg over a 12-year period from age 32 to 44. This was equal to the group's total gain in body mass over the duration of the study.[36] Women are the biggest weight gainers with about 14% putting on more than 30 pounds between the ages of 25 and 34. It is unknown the extent to which this "creeping obesity" during adulthood reflects a normal biologic pattern.

Not necessarily overeating. Until recently, the major cause of obesity was believed to be overeating. If obesity were truly a unitary disorder, and gluttony and overindulgence were the only factors associated with an increase in body fat, the easiest way to permanently reduce would surely be to cut back on food. Of course, if it were that simple, obesity would soon be eliminated as a major health problem. There are obviously other factors operative such as genetic, environmental, social and perhaps racial influences.[25,26,59,128,146] Research also suggests that differences in specific factors may predispose a person to excessive weight gain.[27,90a,123] These include: eating patterns, environment, food packaging, body image, and biochemical differences related to resting metabolic rate, dietary-induced thermogenesis, level of spontaneous activity or "fidgeting," basal body temperature, levels of cellular adenosine triphosphatase, lipoprotein lipase, and other enzymes, and metabolically active brown adipose tissue.

It is difficult to partition the cause(s) of obesity into distinct categories because the cause(s) probably overlap. It seems fairly certain that the treatment procedures devised so far, whether they are diets, surgery, drugs, psychologic methods, or exercise, either alone or in combination, have not been particularly successful in solving the problem on a **long-term** basis. There is optimism, nonetheless, that as researchers continue to investigate the many facets of obesity, as well as to test and quantify various treatment modalities, significant progress will be made to conquer this major health problem.

Physical activity: an important component. Observations of older men and women who maintain active lifestyles suggest that the pattern of fat gain in adulthood can be attenuated significantly.[24,153,154] **In fact, increases in body fat may be more a function of activity level than age.** For both young and middle-aged men who exercised regularly, the time spent in activity was **inversely** related to their body fat level.[65,105] Surprisingly, no relationship emerged between body fat and caloric intake. This suggested that the greater body fat observed among the active middle-aged men compared to their younger counterparts was the consequence of less-vigorous training and **not** greater food intake. Such findings indicate that the trend for increases in body fat with aging can be somewhat blunted with increases in daily physical activity.

Evidence has been presented recently that reduced energy expenditure, measured by the doubly labeled water method,[75] is a crucial factor in the magnitude of weight gain in infants born to overweight mothers, compared to infants born to lean mothers.[125] The infants drank two stable isotopes; one was labeled with deuterium and the other with ^{18}O. Total energy expenditure was estimated by subsequent analysis of urine for 7 consecutive days. Up to 3 months of age, the infants could not be distinguished in terms of anthropometry (fatfolds and body mass index), postprandial metabolic rate, respiratory quotient, or metabolizable energy intake. From age 3 months to 1 year, however, the total energy expenditure of the infants who later became overweight was significantly lower by 21% compared to the infants that maintained a normal weight gain. In another study,[123] the 24-hour energy expenditure was measured in young adult American Indians and correlated with the rate of change in body mass over a 2-year period. The estimated risk of gaining more than 7.5 kg in body mass was 4 times higher in people with a low compared to a high 24-hour energy expenditure. The difference in 24-hour energy expenditure, as well as resting metabolic rate,[23] was also prevalent in families, and is referred to as familial aggregation. Furthermore, for the obese person, there is little if any potentiating effect of exercise on the thermic effects of food,[133] or the thermic effects of a mixed meal following 30 minutes of exercise.[135] These blunted responses to thermogenesis are probably due to the obese state per se and not the absolute quantity of body fat.[133a]

This section deals with various aspects of obesity, including: (1) measurement and risk factors, (2) comparison of fat cell size and number in normal and obese persons before and after weight gain and reduction, (3) development of adipose cellularity in animals and humans, and (4) modification of adipose cellularity by diet and exercise.

HEALTH RISKS OF OBESITY

It is difficult to determine quantitatively the importance of excess body fat as a risk to good health. To a large extent, what is considered "fat" and "thin" in our society has more to do with fashion than with distinct medical risks. Nearly 37 million Americans, however, weigh 20% or more above their desirable body weight, and more than 12 million of these men and women are severely obese and at high risk of developing a variety of diseases related to their obesity.

THE CULPRIT: BODY WEIGHT OR BODY FAT?

There is some confusion as to whether being overweight without being overfat is related to excess risk for cardiovascular disease. Some studies have indicated that blood pressure and cholesterol were significantly greater in men who were significantly overweight and not obese compared to normal-weight, nonobese men.[157] These men also appeared to be at greater risk for premature heart disease.[124] When more rigorous standards for determining body fat levels are used, however, recent evidence indicates that it is **excess fat** and not body mass per se that drives the relationship between body mass and increased risk for cardiovascular disease.[134] Such findings point to the need for further clarification of the role of body composition in one's health risk profile.

Although it has been argued that a moderate excess in body fat is not, in itself, harmful,[88] a report from the National Institutes of Health[110] concluded that obesity should be viewed as a disease, because there are multiple biologic hazards at surprisingly low levels of excess fat that represent only 5 to 10 lb above "desirable body weight." An 8-year study of nearly 116,000 female nurses concluded that all but the thinnest women were at increased risk for heart attack and chest pains.[101a] Women of average weight had 30% more heart attacks than the thinnest women, while the risk for the moderately overweight was 80% higher. **In fact, it is now argued rather convincingly that obesity is an independent and powerful heart disease risk that may be equal to that of smoking, elevated blood lipids, and hypertension.**[71]

EXCESS FAT RELATES TO HEALTH RISK

Although there is little agreement as to the exact cause(s) of obesity, there is considerable information regarding the associations between excessive body fatness and a number of health risks.[8,27] What is not clear is whether obesity causes the risks or simply is a by-product of a particular medical condition. **Clearly, obesity is associated with multiple atherogenic traits, and an excessive fat accumulation contributes to an increased risk of disease.**[122] The following are health-related correlates of obesity: (1) impairment of cardiac function due to an increase in the heart's mechanical work,[3] and to autonomic and left ventricular dysfunction,[1,130] (2) hypertension and stroke,[9,51,143] (3) diabetes, as about 80% of adult-onset diabetics are overweight,[144] (4) renal disease,[161] (5) gallbladder disease,[101] (6) pulmonary diseases and impaired function due to the added effort to move the chest wall,[11,34] (7) problems in administration of anesthetics during surgery,[155] (8) osteoarthritis, degenerative joint disease, and gout,[149] (9) several types of cancer[136] (e.g., a severely obese postmenopausal woman has a five-fold greater than normal risk of developing cancer of the uterine lining),[22] (10) abnormal plasma lipid and lipoprotein concentrations,[131] (11) menstrual irregularities,[110] and (12) an enormous psychologic burden.[27]

Weight loss with accompanying fat reduction often normalizes serum cholesterol and triglyceride and has a beneficial effect on blood pressure. In fact, the normally observed relationship between age and blood pressure is partially explained by the tendency to gain weight with age.

CRITERIA FOR OBESITY: HOW FAT IS TOO FAT?

A person's fat content is generally evaluated in terms of the percent of body mass that is fat (percent body fat) or in relation to the size and number of individual fat cells.

PERCENT BODY FAT

Obesity can be defined as excessive enlargement of the body's total quantity of fat, or the excess storage of energy in adipose tissue. Where do we draw the line, however, between what is considered normal and what is obese? In Chapter 27, we suggested a "normal" range of body fat in adult men and women—an "average" value for body fat plus or minus one unit of variation. For men and women aged 17 to 50, this variation unit is approximately 5% body fat. Using this statistical boundary, "overfatness" would then correspond to body fat that exceeds the average value plus 5% units for body fat. For example, in young men whose body fat averages 15% of body mass, the borderline for obesity would be 20% body fat. For older men whose average fatness is approximately 25%, obesity would be defined as a body fat content in excess of 30%. For young women aged 17 to 27, obesity would correspond to a body fat content above 30%; for older women, aged 27 to 50, the borderline between the average and obesity would be about 37% body fat. We emphasize, however, that just because the "average" value for percent body fat tends to increase with age, this should **not** dictate that people should expect to get fatter as they grow older. **We believe that one criterion for what is considered "too fat" should be that established for younger men and women—above 20% for men and above 30% for women.** In this way, "average" population values do not become the reference standard, and should not subsequently be accepted as "normal."

> ### Standards for Overfatness
> Men—above 20%;
> Women—above 30%

It should be kept in mind that there is a gradation in obesity that progresses from the upper limit of normal—20% for men and 30% for women—to as high as 50 to 70% of body mass in massively obese people. This group includes people who weigh in the range of 170 to 250 kg or higher. In this situation, body fat often exceeds lean body mass and obesity may be life-threatening.

REGIONAL FAT DISTRIBUTION

It is apparent that the patterning of adipose tissue distribution, independent of total body fat, alters the health risk of obesity.[34a,55,116a] For example, ratios of waist-to-hip girth that exceed 0.90 are associated with an increased risk of death from coronary artery disease as well as a variety of illnesses, most notably diabetes, elevated triglycerides, hypertension, and general overall mortality.[16,92,94] This may be because excess fat in the abdominal area (**central or android-type obesity**) is more active metabolically than fat located in the hips and thighs (**peripheral or gynoid-type obesity**), and thus more capable of entering into processes related to heart disease.[97,109] As a frame of reference, the waist-to-hip ratio for the Behnke reference female is 0.696 (65.6 cm ÷ 94.2 cm). To some extent, one's pattern of fat distribution is inherited[51] and is probably governed by the regional activity of lipoprotein lipase, the rate-limiting enzyme for triglyceride uptake by the fat cell.

FAT CELL SIZE AND NUMBER: HYPERTROPHY VERSUS HYPERPLASIA

Another way to determine and classify obesity is to measure the size and number of fat cells. Adipose tissue increases in two ways: Existing fat cells are enlarged or filled with more fat—a process called **fat cell hypertrophy**, or the total number of fat cells is increased—a process called **fat cell hyperplasia.**

A variety of techniques are used to study adipose cellularity in both humans and animals. One technique involves sucking small fragments of subcutaneous tissue into a syringe with a needle inserted directly into a fat depot. These tissue fragments are usually sampled from the back of the arm at the triceps, the subscapular region, the buttocks, and the lower abdomen. The tissue is then treated chemically so the fat cells can be isolated and counted.

Once the number of fat cells is determined for a known weight of fat tissue, the average quantity of fat per cell is determined by dividing the quantity of fat in the sample by the total number of fat cells present. If total body fat is known, a good estimate can then be made of the total number of fat cells in the body. If an individual weighs 88 kg, and 13% body fat is determined by the underwater weighing method outlined in Chapter 27, total fat mass equals 11.4 kg (0.13 × 88 kg). The total number of fat cells in the body is determined by dividing 11.4 kg by the average content of fat per cell. For example, if the average fat cell contained 0.60μg of fat, then there would be 19 billion (11.4 kg ÷ 0.6 μg) fat cells in the body.

$$\text{Total cell number} = \frac{\text{Mass of body fat}}{\text{Fat content per cell}}$$

In one of our laboratories, needle biopsy and photomicrographic techniques have been used to extract and measure the average size of fat cells at selected sites in the body. After the fat sample is obtained, it is prepared for sizing by appropriate biochemical methods and photographed for later projection on a large screen that permits measuring cell diameters with a light-emitting pen that interfaces with a computer.[95] At least 200 cells are measured per site. Figure 29–1 depicts fat cells from the abdominal area in an endurance-trained athlete and one of this textbook's authors. Once the mean diameter

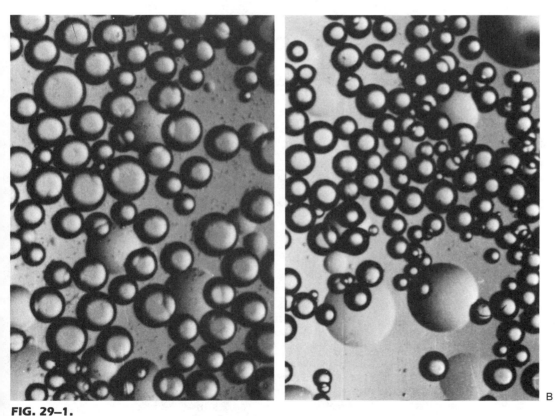

B

FIG. 29–1.
Photomicrograph of fat cells biopsied from the buttocks region. The larger fat cells (A) were from an untrained, middle-aged professor. The smaller cells (B) were taken from an endurance-trained, experienced marathoner. The large, spherical structures in the background are lipid droplets. (Courtesy of the Muscle Biochemistry Laboratory, Department of Exercise Science, University of Massachusetts.)

of the cells is known, the fat content per cell is determined by the appropriate volume conversion, $\pi \times$ cell diameter. For the middle-aged author whose total mass of body fat was 17.02 kg (89.1 kg and 19.1% body fat) with 0.73 μg of lipid per cell, the total number of cells was estimated to be 23.3 billion (17.02 kg ÷ 0.73 μg).

VARIETY OF HUMAN OBESITIES

It is easy enough to establish standards for obesity, but it is not easy to distinguish between the gradations of obesity. Attempts have been made to describe obesity types based on the amount, distribution, and texture of fat tissue.[12,39,156] The idea of a classification schema is appealing because it permits quantification of the phenotype and, in addition, allows for the evaluation of a variety of hormonal and biochemical correlates of obesity.[85,86,92] Figure 29–2 presents six phenotypic patterns that have been observed in female obesity (comparable data are unavailable for males). In addition, the photographic inserts complement several of the outline patterns and show examples of severe, intractable obesity. The two panels at the bottom right show the effects of a 35-kg weight loss over 15 months in a 49-year-old woman. Most of the weight loss was from the large abdominal panniculus that was partially removed surgically. Note that even with a relatively large decrease in overall body mass, the pattern of the phenotype remains relatively invariant. Considerably more basic and applied research in the area of obesity classification is needed in males and females of all ages, particularly in conjunction with anthropometric, metabolic, and biochemical studies.

ADIPOSE CELLULARITY

COMPARISON OF FAT CELLULARITY IN NORMAL AND OBESE INDIVIDUALS

Several comparative studies of adipose cellularity in obese and nonobese humans show conclusively that fat accumulation in the obese occurs either by storing larger quantities of fat in existing adipose cells (hypertrophy), new fat cell formation (hyperplasia), or by both hypertrophy and hyperplasia.

Figure 29–3 compares body mass, total fat content, and cellularity in 25 subjects, 20 of whom were clinically classified as obese. The body mass of the obese subjects averaged more than twice that of the nonobese, and their total fat content was nearly three times larger than that of the leaner group. In terms of cellularity, the average fat content per cell was about 35% greater in the obese, whereas the total number of fat cells was approximately three times greater than that of the nonobese (75 compared to 27 billion). **These results illustrate dramatically that the major structural difference in adipose tissue cellularity between the obese and nonobese is cell number.**

The importance of fat cell number in obesity can be further illustrated by relating total fat content to both cell size and cell number. Most available research suggests that fat cells may reach some biologic upper limit. Once this size is reached, cell number becomes the key factor in determining any further extent of obesity. Even if fat cells could double in size, this would still not account for the large difference in the total fat content of the obese

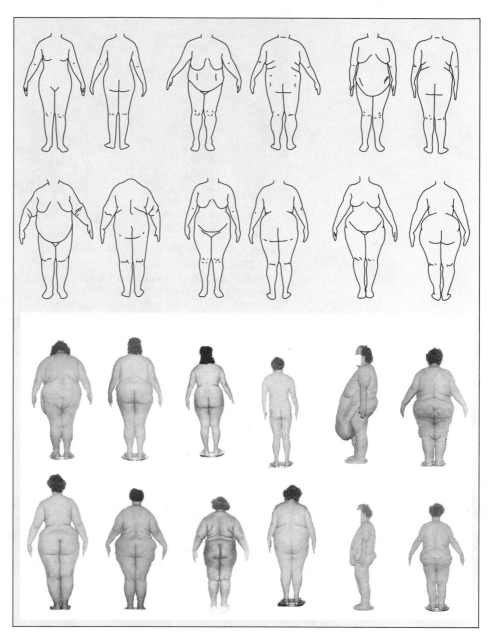

FIG. 29–2.
Diagrams of patterns of female body form that illustrate the varieties of human obesity. The details of the rating profile used with this classification are presented in reference 38. The photographic inserts show examples of several of the body patterns, as well as intractable obesity where the abdominal paniculus was estimated to weigh over 35 kg prior to weight loss (lower right panels). (The photographs and outline patterns are courtesy of Leela S. Craig, M.D.)

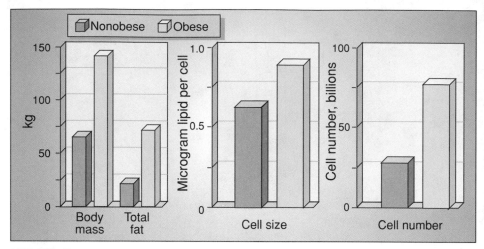

FIG. 29–3.
Comparison of body mass, total body fat, cell size, and cell number in obese and nonobese subjects. (Modified from Hirsch, J., and Knittle, J.: Cellularity of obese and non-obese human adipose tissue. Fed. Proc., 29:1518, 1970).

as compared to normal individuals. The excessive quantity of adipose tissue in obesity must, therefore, occur by the process of fat cell hyperplasia. For comparison, a nonobese person has approximately 25 to 30 billion fat cells, whereas the number of fat cells in the "extremely obese" may be as high as 260 billion.[18]

Table 29–1 presents data on fat cell size and number, total body fat, and percentage body fat in men and women by 5-year age intervals from 20 to 50 years. The interesting observation in this sample of 217 subjects was the

TABLE 29–1.
COMPARISON OF ADIPOCYTE SIZE AND NUMBER, AND BODY COMPOSITION IN MEN AND WOMEN BY 5-YEAR AGE GROUPS[c]

Age Group	Gender	Adipocyte Size µg lipid per cell	Adipocyte Number, billions	Total Body Fat, kg	% Body Fat[a]
20–24	M	.37[b]	28.83	9.4[b]	13.0[b]
	F	.47	33.48	15.1	25.0
25–29	M	.37	34.62	13.0	17.5[b]
	F	.41	38.37	14.2	24.7
30–34	M	.42	34.85	13.4	17.8[b]
	F	.47	32.87	15.7	26.4
35–39	M	.41	42.08	16.6	21.8[b]
	F	.44	40.28	17.3	27.9
40–44	M	.39	45.57	18.9	22.7[b]
	F	.46	43.34	19.7	31.5
45–50	M	.51	50.07	22.0[b]	26.3
	F	.49	39.33	17.1	29.0

[a] Determined by underwater weighing.
[b] Differences between the genders statistically significant.
[c] From Chumlea, W.C., et al.: Adipocytes and adiposity in adults. Am J. Clin. Nutr. 34:1798, 1981.

lack of gender differences among the variables, except the 20- to 24-year age group where the women had significantly larger adipocytes and more total body fat than the men. On a percentage basis, the women had the greatest amount of body fat in every age group except at ages 45 to 50. Longitudinal data, where the same subjects are remeasured over time, will confirm whether or not adipocyte number can increase as people get older (as it appears to do in these cross-sectional data). If this is true, then the notion that cell number becomes fixed early in life requires re-examination.

EFFECTS OF WEIGHT REDUCTION

When obese adults reduce body size, there is a decrease in fat cell size but no change in cell number. If normal body mass and body fatness are achieved, then individual fat cells shrink and actually become smaller in size than the fat cells of nonobese individuals. Figure 29–4 depicts the results of weight reduction on adipose cellularity in obese subjects.

In this study, 19 obese adults reduced their body mass from 149 to 113 kg by the end of the first stage of the experiment. The average number of fat cells before weight reduction was approximately 75 billion and remained essentially unchanged even after reducing by 46 kg (101 lb). Fat cell size, on the other hand, was reduced from 0.9 to 0.6μg of fat per cell, a decrease of 33%. When subjects attained normal body mass by reducing 28 more kg, cell number again remained unchanged but cell size continued to shrink to

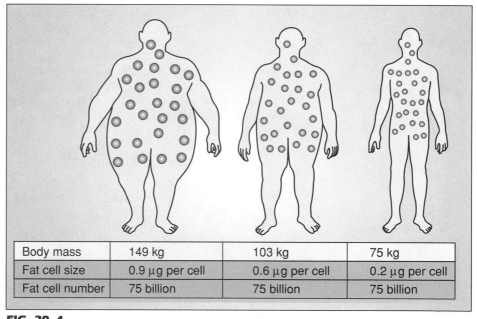

Body mass	149 kg	103 kg	75 kg
Fat cell size	0.9 μg per cell	0.6 μg per cell	0.2 μg per cell
Fat cell number	75 billion	75 billion	75 billion

FIG. 29–4.
Changes in adipose cellularity with weight reduction in obese subjects. (Data from Hirsch, J.: Adipose cellularity in relation to human obesity. In Advances in Internal Medicine, Vol. 17. Edited by G. H. Stollerman. Chicago, Year Book, 1971.)

about one third the size of the fat cells of normal, nonobese subjects. **The major structural change in adipose cellularity following weight loss in adults is a shrinkage in fat cell size with no change in cell number.**

These findings suggest that formerly obese persons are not really "cured" of their obesity, at least in terms of the total number of fat cells present. There is no doubt that formerly obese patients have an extremely difficult time maintaining their new body size.[74,147]

EFFECTS OF WEIGHT GAIN

An interesting series of studies was conducted on the development of obesity.[138] Adult male volunteers with an initial average body fat content of 15% deliberately increased their caloric intake three times above normal to about 7000 kcal per day for a period of 40 weeks. For a typical subject, body mass increased 25%, and body fat doubled from 14.6 to 28.2% of body mass. Consequently, of the 12.7 kg gained during the period of overeating, 10.5 kg were caused by increased deposition of body fat.

In a similar experiment with nonobese subjects who had no previous personal or family history of obesity, voluntary overeating produced an average increase in body mass of 16.4 kg.[132] In comparing cell size and number before and after the 4-month experiment, the average fat cell increased substantially in size with no change in cell number. When caloric intake was reduced and subjects achieved their normal weight, total body fat declined and the fat cells reverted to their original size. **These results indicate that when adults get fatter as a result of overeating, they are filling or enlarging existing adipose cells rather than creating new ones.**

There is some indication that **extreme** fat development in adults may modify adipose cellularity.[17,67] This is because there is an upper limit to fat cell size beyond which hypertrophy fails to occur; this limit is reached when the cell contains about 1.0 μg of lipid per cell (normal is about 0.5 to 0.6 μg). In the massively obese (60% body fat; about 170% of normal weight), almost all the fat cells have attained their hypertrophic limit and more cells may be recruited from the preadipocyte pool to increase cell number. Thus, in maturity-onset severe obesity, in which the already fat adult becomes fatter, hypercellularity may accompany the greatly increasing size of the existing fat cells!

FAT CELL DEVELOPMENT

The development of adipose tissue during growth has been studied in both animals and humans.

ANIMAL STUDIES

The most extensive studies of adipose cellularity have been conducted with rats because these mammals have a relatively short life span, and various diets and exercise regimens can be studied easily during the growth cycle.

Figure 29–5 illustrates the general upward-trending curves for body mass, fat mass, and fat cell size and number in rats during the first 5 months of life. Note that both the number and size of fat cells increased during weeks 6 through 15. As the animals became heavier and total body fat increased,

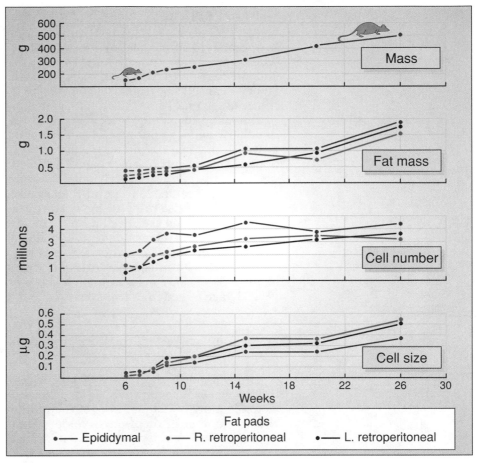

FIG. 29–5.
Changes in body mass, fat mass, cell number, and cell size during the first 5 months of growth in rats. (From Hirsch, J., and Han, P. W.: Cellularity of rat adipose tissue: effects of growth, starvation, and obesity, J. Lipid Res., 10:77, 1969).

there was a corresponding increase only in the **size** of the fat cells. The additional increase in body fat occurred because existing cells became filled with fat, not because new fat cells developed.

In contrast to this traditional view regarding the development of fat cells, some investigators argue that under certain conditions adult adipose cell number is not permanently fixed.[50] When adult rats were fed highly palatable diets so they gained both weight and fat significantly, cell size increased to a maximum level; the further increase in the mass of adipose tissue occurred by an increase in adipocyte **number** that persisted even after weight loss. The researchers postulated that cell size may not be infinitely expandable and that once maximum cell size is reached, and food continues to be available, a response may be triggered to proliferate new fat cells. The increased cell number at this point would constitute a failure of adipocyte regulation that leads to obesity.

HUMAN STUDIES

In one study,[68] adipose cellularity was established for 34 infants and children who ranged in age from a few days to age 13. Fat cell size in newborn infants and children up to the age of 1 year was about one fourth the size of adult fat cells. Fat cell size tripled during the first 6 years with little further increase in size to age 13. Although several studies have determined adipose cellularity in childhood and adolescence,[32,62] there still is a scarcity of data on changes in adipose cell size during this growth period. We may reasonably assume, however, that cell size increases during this period because cell size in adulthood is significantly larger than cell size at age 13 or in late adolescence.

Cell number increases fairly rapidly during the first year of life, being about three times greater at this point than at birth. It is believed that most of the fat cells that exist prior to birth are formed during the last trimester of pregnancy. Beyond age 1, cell number increases gradually to the age of about 10. Like cell size, there is significant cell hyperplasia during the growth spurt in adolescence until adulthood; thereafter, there is generally little further increase in cell number. In terms of body fat, the percentage increases from about 16% of body mass at birth to between 24 to 30% of body mass at 1 year. By age 6, body fat decreases to about 14% of body mass.

CAN ADIPOSE TISSUE CELLULARITY BE MODIFIED?

Although the precise causes for fat cell development are poorly understood, it does appear that certain practices can affect fat cellularity. In humans, for example, nutritional practices of the mother during pregnancy may modify the body composition of the developing fetus. A weight gain by the mother in excess of 18 kg was associated with a significantly larger fatfold thickness of the offspring than for a woman who followed a recommended weight gain during pregnancy.[152] Bottle feeding and the early introduction of solid food may also be associated with the development of obesity. Conversely, breast feeding that allows the infant to set the limits to food consumption, and a delayed introduction to solid food, may prevent overfeeding, the development of poor eating habits, and subsequent obesity.[90]

Research in animals suggests that alterations in fat cell size and number can be achieved in one of two ways: (1) modification of early nutrition, and (2) exercise.

INFLUENCE OF NUTRITION

In one study,[87] rats were distributed at birth so that some mothers had large litters of 22 animals and others smaller litters of four animals. After weaning at 21 days, both groups of animals had unlimited access to food. Six animals in each group were then sacrificed at 5, 10, 15, and 20 weeks of age. At weaning and after each subsequent 5-week period, the body mass of animals from large (calorically deprived) litters was significantly lower than the other group. This suggested that early nutritional deprivation caused a permanent stunting of growth, even though both groups of animals had free access to food after weaning.

In both groups of animals, fat mass increased from weaning to 20 weeks of age; the dramatic differences occurred in animals reared in small litters, especially at weeks 15 and 20. These relative differences were larger than

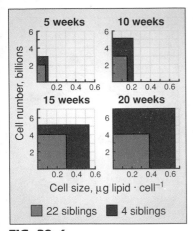

FIG. 29–6.
Changes in cell size and number in animals raised in small and large litters. (Adapted from Knittle, J., and Hirsch, J.: Effect of early nutrition on the development of rat epididymal fat pads: cellularity and metabolism. J. Clin. Invest., *47*:2091, 1968.)

the differences observed in body mass for the same period. Figure 29–6 shows that for cell size and number, the nutritionally deprived animals had fewer and smaller fat cells at all age intervals than animals reared in small litters.

At 5 and 10 weeks of age, the height of the rectangles is greater than the base, indicating that the proliferation of fat cells made a greater contribution to adipose mass than to cell size. For the 15- and 20-week periods, the shape of the bar approaches a square, indicating that cell size played an increasingly important role in the development of adipose tissue. An interesting comparison is seen between animals raised in large and small litters. The purple area of the bar represents the fat depot of the overnourished animals raised in small litters; the orange area represents the undernourished animals, with the difference in shading illustrating the difference in depot size between the two groups. By 15 weeks of age, the underfed group reached a definite plateau in fat cell number. Cell number continued to increase in the overfed, small-litter animals. In both groups, cell size increased progressively during the experimental period. **These data certainly suggest that there may be a critical time during the growth period when a permanent modification in adipose tissue cellularity can occur.**

Although extrapolation from animal experiments to humans is difficult, some striking similarities are worth noting in adipose tissue development between humans and rats. As shown in a previous section, extremely obese humans show a large **number** of fat cells and, to a lesser extent, an increased **size** of individual fat cells. When obese adults lose body mass, the number of fat cells remains unaltered and the decrease in total body fat is achieved almost exclusively by a reduction in fat cell size. The same process seems to take place in studies of adult rats. When adult rats are deprived of food, the observed decrease in body mass is only temporary and is rapidly reinstated upon refeeding. In such animals, the weight loss is attributable to a decrease in fat cell size with no corresponding change in cell number. Overfeeding of adult animals produces an increase in total body fat, but as with humans, this increase is usually brought about by "stuffing" of cells with fat rather than by an increase in cell number. The exception occurs when adipocyte size reaches a maximum; further increases in adipose mass occur by additional proliferation of fat cells.[50,67] Furthermore, when the fat content of adult humans is reduced, cell size shrinks accordingly only to expand again when the body's content of fat is restored. The number of existing or newly added fat cells remains constant.

INFLUENCE OF PHYSICAL ACTIVITY

Figure 29–7 summarizes the results of studies with young rats with free access to food who were forced to swim in plastic barrels 6 days a week for 14 to 16 weeks.[115] The exercise sessions were gradually lengthened until the animals were swimming for 360 minutes. Two adult groups of rats remained sedentary; one group had free access to food and water; the other group was food restricted to maintain body mass at the same level as the exercise group.

The results were convincing; animals given unlimited food but forced to exercise for 15 weeks gained weight more slowly and had a lower final body mass than sedentary, freely eating rats. Because both groups consumed the same number of calories each day, the lower rate of weight gain in the exercisers could be attributed to the increased caloric requirements of the exercise. It was also shown that the total fat content of the nonexercise group was about four times higher than the fat content of the freely eating exercise

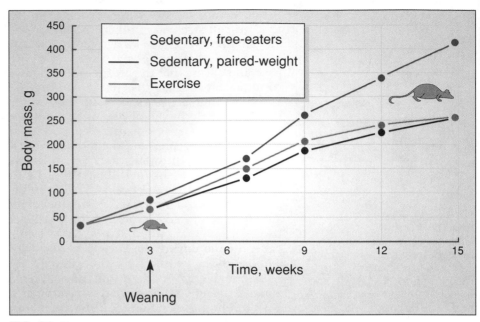

FIG. 29–7.
Effects of exercise and food restriction on body weight of rats. (From Oscai, L., et al.: Effects of exercise and of food restriction on adipose tissue cellularity. J. Lipid Res., *13*:588, 1972.)

group. The exercise intervention program during the growth period resulted in a significant reduction in total body fat due to a decrease in both cell size and number.

The total fat content of the sedentary, food-restricted group was lower than for the sedentary animals who could eat ad libitum. Reducing food intake resulted in a reduction in cell size and cell number. When the body fat of the food restricted and exercised animals was compared, the exercisers had fewer fat cells and less fat per cell, even though the final body mass was approximately equal in both groups. The results demonstrate that exercise performed early during the growth period depresses the growth of new fat cells.

In a followup experiment,[116] the fat-retarding effects of exercise or diet early in an animal's life were studied to determine whether either would reduce fat accumulation in adulthood. Three groups of animals were used: an exercise group, a sedentary group with free access to food and water, and a sedentary group with restricted food intake. Exercise and food restriction were terminated after 28 weeks. Several animals from each group were then sacrificed and the groups compared for growth, body fat, and adipose cell size and number. The remaining animals were subjected to 34 weeks of sedentary living without exercise and were allowed unlimited food and water. The animals were then sacrificed and the groups compared for body mass, cell size, and cell number. The data in Figure 29–8 show that the exercised animals had lower body masses at 28 weeks of age than the free-eaters.

During the next 34 weeks of inactivity, the previously exercised animals continued to maintain a lower body mass than the sedentary animals. Thus, the 28-week exercise program performed earlier in life caused a reduction

FIG. 29–8.

Effects of 28 weeks of exercise and food restriction on body mass in rats followed by no exercise with unlimited access to food. (Data from Oscai, L., et al.: Exercise or food restriction: effect on adipose tissue cellularity. Am. J. Physiol., 277:901, 1974.)

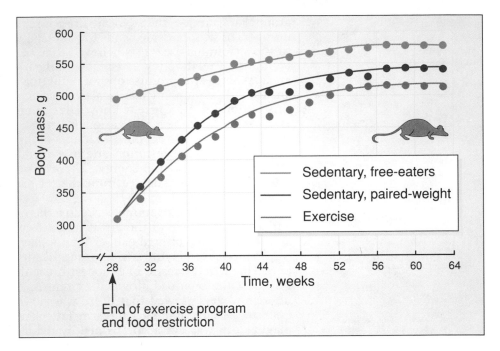

in body mass that was still evident at the end of the experiment. Comparing cell size and number at the end of the training period revealed that the exercisers had **fewer and smaller fat cells** than either sedentary group of animals. These results were in agreement with the previous experiment.

The exercise group had a lower final body mass and reduced total body fat content than their sedentary counterparts, as well as significantly fewer fat cells in later life than animals in the other groups. Twenty-six weeks of exercise, begun early in life and then terminated, retarded the expansion and proliferation of fat cells during adulthood—even though the exercise period was followed by 34 weeks of inactivity.

If these findings can be applied to humans, it is possible that the introduction of caloric restriction or exercise during the early stages of growth may aid in controlling the proliferation of new fat cells and the filling up of previously dormant ones. Programs of exercise and weight control begun later in life and maintained thereafter can be effective in lowering the body's total quantity of fat. As far as we know, however, it is only cell size and not cell number that can be reduced. If exercise or dietary intervention is discontinued, then the existing adipose tissue mass is likely to increase again by expansion of the cellular volume. **Early prevention of obesity through exercise and food restraint, rather than correction of obesity once it is present, may be the most effective method to curb the grossly "overfat" condition so common in teenagers and adults.**

SPOT REDUCTION: DOES IT WORK?

Spot reduction is localized exercise that is supposed to reduce fat stores in the exercised areas. The underlying basis for this notion is that by exercising a specific body area, more fat will be selectively reduced from that area than if exercise of the same caloric intensity was performed by a different muscle group. For example, the advocate of spot reduction would recommend large numbers of situps or side bends for a person who has excessive fat deposited in the abdominal area. It is believed that an increase in a muscle's activity facilitates a relatively greater fat mobilization from these specific storage areas. Whereas the promise of spot reduction with exercise is attractive from an aesthetic standpoint, a critical evaluation of the research does **not** support this notion.[61,91,113] Current knowledge of energy supply indicates that exercise stimulates the mobilization of fatty acids through hormones delivered through the blood to act on the fat depots throughout the body.[140] The areas of greatest fat concentration and/or enzyme activity probably supply the greatest amount of this energy. **There is no evidence that fatty acids are released to a greater degree from the fat pads directly over the exercising muscle.**

In an attempt to examine critically the claims for spot reduction, comparisons were made of the girths and subcutaneous fat stores of the right and left forearms of high-caliber tennis players.[61] As expected, the girth of the dominant or playing arm was significantly larger than the nondominant arm. This was the result of a modest muscular hypertrophy associated with the muscular overload provided by tennis. Measurements of fatfold thickness, however, showed that there was no difference between arms in the quantity of subcutaneous forearm fat. Clearly, prolonged exercise of the playing arm was not accompanied by reduced deposits specifically in that arm.

In another experiment,[79] fat biopsies were taken at the abdominal, subscapular, and buttock sites before and after a 27-day period of progressively increased situp exercise training. The experimental subjects performed an average of 185 situps daily in comparison to a control group who did not exercise. Situp intensity increased from 140 situps at the end of the first week to 336 situps on day 27. Following the situp training regimen, the fat cells in the abdominal region did not become significantly smaller than adipose cells in the relatively unexercised buttocks or subscapular regions. There were no significant changes in fatfolds, girths, or total fat content assessed by underwater weighing.

There is no doubt that the negative caloric balance created through regular exercise can significantly contribute to a reduction in total body fat. This fat, however, is **not** reduced selectively from the exercised areas, but rather from total body fat reserves and usually from the areas of greatest fat concentration.

WHERE ON THE BODY DOES FAT REDUCTION OCCUR?

An often asked question concerning weight loss is, "Where on the body do changes occur when weight is lost?" To help answer this question, changes in body composition were evaluated in 26 initially obese females at successive 2.3-kg (5 lb) increments of weight loss.[77,84] Weight loss was induced by caloric restriction and a 3-day-per-week, 45-minute-per-session exercise program for 14 weeks duration. The supervised program included 15 minutes of general calisthenics and 30 minutes of walking and jogging performed at an exercise intensity recommended for producing training effects.[76] Body composition assessment included 5 fatfolds, 12 girths, and densitometry to determine absolute and relative body fat and lean body mass. Figure 29–9 displays the changes in body composition, fatfolds, and girths for the three subgroups that reduced body mass by 2.3, 4.5, and 9.1 kg. The salient findings are that a 4.5-kg loss in body mass produced approximately twice the amount

FIG. 29–9.
Changes in body composition (top), fatfolds (middle), and girths (bottom) with specified amounts of weight loss. Data from reference 84. The abbreviations for the girths are Shoul = shoulders, Umbil = umbilicus abdomen, Fore = forearm. The anatomic sites are those described in Appendix F.

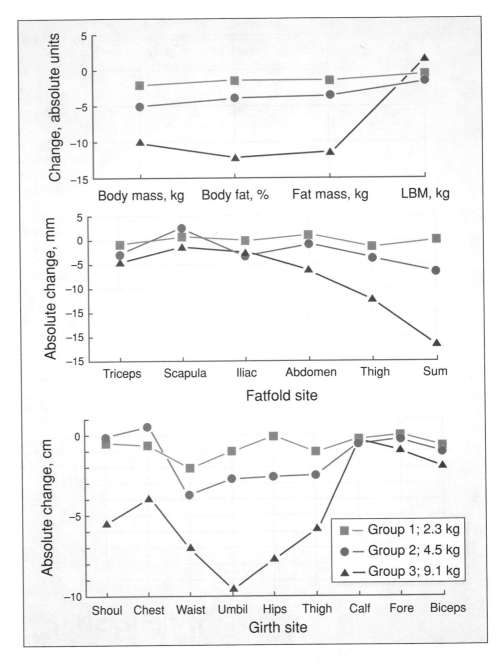

of change in overall body composition compared to changes observed at 2.3 kg. When weight loss doubled from 4.5 to 9.1 kg, the corresponding change in body composition was almost three times as much. This occurred for changes in fat mass, sum of 5 fatfolds, and sum of 11 girths. It is noteworthy that there was an approximately 2-fold greater change in fatfolds and girths in the trunk region compared to the extremities. In addition, the proportional changes in body composition and anthropometry apparently mirror changes in body mass loss up to 4.5 kg. Thereafter, the relative change in composition becomes more pronounced and is apparently not related to the amount of additional weight loss up to 9.1 kg. Additional studies are needed to quantify the magnitude of change in overall body composition with additional losses in body mass.

▬▬▬ SUMMARY

1. Obesity is usually defined in terms of excessive quantities of total body fat. There is probably no biologic reason for men and women to get fatter as they grow older. Therefore, the standards for overfatness for adult men and women should probably be that established for younger adults, namely, men—above 20%, women—above 30% body fat.
2. The location of adipose tissue should also be considered; fat distributed in the abdominal region (android-type obesity) poses a greater health risk compared to fat deposited at the thigh and buttocks (gynoid-type obesity).
3. Another classification for obesity is based on the size and number of fat cells. Before adulthood, body fat increases in two ways: by enlargement of individual fat cells, termed fat cell hypertrophy, or by an increase in the total number of fat cells, termed fat cell hyperplasia. Fat cells probably reach some biologic upper limit in size, so that cell number becomes the key factor determining the extent of obesity.
4. The number of fat cells becomes stable sometime before adulthood; any weight gain or loss thereafter is usually related to a change in the size of the individual cells. In extreme cases, fat cell number may increase in adults once the hypertrophic limit is reached for cell size.
5. Increases in the number of fat cells appear to involve three critical periods: the last trimester of pregnancy, the first year of life, and the adolescent growth spurt prior to adulthood.
6. Fat cell development in animals is influenced by dietary restriction and exercise. These effects are most prominent during early growth when the rate of fat cell division can be retarded.
7. Selective reduction of fat at specific body areas by "spot exercise" does not work.

P A R T

2 **Weight Control**

The body mass of most adults fluctuates only slightly during the year, even though the annual food intake averages close to 900 kg. The relative constancy in body mass is impressive when one considers that a slight but consistent increase in food intake can cause body mass to increase substantially if there is no compensatory alteration in daily energy expenditure. For example, for a person who originally weighed 120 kg (263 lb) an intake of only 218 extra calories a day for 22 years could result in a body mass of 340 kg or 748 lb.[57] The fact, however, that the mass of most adults fluctuates only slightly illustrates the body's exquisite regulatory control in balancing daily calorie intake with daily energy expenditure. **If the total calories ingested as food exceed the daily energy expenditure, the excess calories are stored as fat in adipose tissue.**

The composition of the diet also influences the efficiency at which the body converts and stores excess calories as fat.[137] Only about 3% of the calories in ingested fat is required to convert these excess calories to stored body fat, whereas 25% of the calories in carbohydrate are "burned" in the conversion process. Simply stated, it is easier for the body to make fat from dietary fat than from an equivalent caloric excess in the form of carbohydrate. Consequently, shifting the diet's composition toward higher carbohydrate would result in less fat gain should a caloric excess occur. **To prevent an increase in body mass and fat because of a caloric disequilibrium, an effective program of weight control must establish a balance between energy input and energy output.**

ENERGY BALANCE: INPUT VERSUS OUTPUT

A review of the scientific literature dealing with body mass loss in obese persons reveals that initial success in modifying body composition has little relation to the long-term effects of maintaining desired body size and shape. This has been demonstrated in studies concerned with follow-up measurements of participants in weight reduction programs where caloric intake was carefully regulated and monitored. In an early survey of studies dealing with weight control management during a 10-year period, the dropout rate varied from 20 to 80%.[147] Of those who remained in a program, no more than 25% lost as much as 9 kg and only 5% lost 18 kg or more. Such statistics indicate that long-term success in maintaining a reduced body mass is extremely difficult; it is particularly difficult in the relaxed atmosphere of one's home where there is ready access to food.

The energy balance equation states that body mass remains constant when caloric intake equals caloric expenditure. Any chronic imbalance on the energy output or input side of the equation causes the body mass to change.

> Energy input = Energy output → **Stable body mass**
> Energy input > Energy output → **Increase in body mass**
> Energy input < Energy output → **Decrease in body mass**

There are three ways to "unbalance" the energy balance equation: (1) reduce caloric intake below daily energy requirements, (2) maintain regular food intake and increase energy expenditure through additional physical activity above daily energy requirements, and (3) combine methods (1) and (2) by decreasing daily food intake and increasing daily energy expenditure.

When considering the sensitivity of the energy balance equation in regulating overall energy balance, we note that if caloric intake exceeds output by 100 kcal per day, the surplus number of calories consumed in a year would be 365 days × 100 kcal or 36,500 kcal. Because 0.45 kg of body fat contains about 3500 kcal (each pound of adipose tissue is about 87% fat or 395 g × 9 kcal · g^{-1} = 3555 kcal per pound), this is equivalent to a gain of 4.7 kg of fat in one year. On the other hand, if daily food intake is reduced by just

100 kcal and energy expenditure is increased 100 kcal by jogging 1 mile each day, then the caloric deficit is equivalent to a reduction of about 9.5 kg or 21 lb of fat in one year!

DIETING FOR WEIGHT CONTROL

Dieting for weight loss creates a disequilibrium in the energy balance equation by reducing energy intake. If an obese woman who consumes 2800 calories daily and maintains body mass at 79.4 kg wishes to lose weight, she must maintain her regular level of activity but reduce daily food intake to 1800 calories to create a 1000-calorie deficit. In 7 days, the caloric deficit created would equal 7000 calories, the caloric equivalent of 0.91 kg of body fat. Actually, considerably more than 0.91 kg would be lost during the first week of caloric restriction because the body's carbohydrate stores would be used up first. This stored nutrient contains fewer calories per gram and much more water than fat. **For this reason, short periods of caloric restriction often prove encouraging to the dieter but result in a large percentage of water and carbohydrate loss per unit of body mass reduction, with only a minimal decrease in body fat. Then, as weight loss continues, a larger proportion of body fat is used for energy to supply the caloric deficit created by food restriction.** To reduce fat by another 1.4 kg, the reduced caloric intake of 1800 calories would have to be maintained for another 10.5 days, and body fat theoretically would be reduced at a rate of 0.45 kg every 3.5 days. Although the mathematics of weight loss through caloric restriction are straightforward, they depend upon several basic assumptions that, if violated, reduce the effectiveness of weight loss through dieting.

Many professional organizations have voiced strong opposition to various dietary practices, in particular, extremes of fasting as well as the low-carbohydrate, high-fat, and high-protein diets.[6,111,112] This is particularly troublesome to those who work in the areas of sports medicine and athletics, where reports consistently document the use of bizarre and often pathogenic weight control behaviors among athletes.[19,127,149a] In the following sections we review several of the more popular dietary approaches to weight loss.

KETOGENIC DIETS

Ketogenic diets emphasize carbohydrate restriction while ignoring the total caloric content of the diet. Advocates of these diets maintain that with minimal carbohydrates for energy, the body must metabolize fat. A high fat/protein diet generates excess ketone bodies (by-products of incomplete fat breakdown) to supposedly suppress appetite and cause urinary loss of these unused calories. This accounts for significant weight loss despite a moderately high-caloric intake. It is argued that this caloric loss is so great that dieters can eat all they wish, as long as carbohydrates are restricted. At best, however, the calories lost by urinary excretion of ketones would equal only 100 to 150 calories a day.[7] This amount would account for only a small weight loss of approximately 0.45 kg per month—not very appealing when the major proportion of the daily calorie intake amounts to a fat intake that may be as high as 60 to 70% of the food consumed. Also, any initial weight loss on this diet may be due largely to dehydration brought about by sodium loss and extra solute load on the kidneys that increases the excretion of urinary water. Such water loss is of no lasting significance in a program designed to reduce body fat. When compared to a standard, well-balanced, low-calorie diet, the ketogenic diet shows no advantage in facilitating a loss in body fat.[100] Low carbohydrate diets have the potential for causing the body to lose significant amounts of muscle tissue because the body uses its protein as a primary fuel to maintain blood glucose. This is certainly an undesirable side effect for a diet designed to bring about fat loss.

High-fat, low-carbohydrate diets are also potentially hazardous in several ways. The diet can raise serum uric acid levels, lower potassium levels that can facilitate undesirable cardiac arrhythmias, cause acidosis, aggravate kidney problems owing to the extra solute burden placed on the renal system, elevate blood lipids and thus increase a primary heart disease risk factor, deplete glycogen reserves and contribute to a fatigued state, and cause a relative dehydration. The diet is definitely contraindicated during pregnancy because adequate carbohydrate metabolism is essential for proper fetal development.

HIGH-PROTEIN DIETS

Other dietary plans, specifically the various modifications of a high-protein diet, can be potentially lethal. The high-protein diet has been extrolled as the "last chance diet" for the obese. It is argued that protein diets cause suppression of appetite through the body's excessive reliance on fat mobilization and the formation of ketone bodies, although this effect has yet to be supported with careful research. It is also argued that the elevated calorie-burning thermic effect of dietary protein, as well as its relatively low coefficient of digestability, ultimately reduce the net calories available from this food compared to a well-balanced meal of equal caloric value. The calorigenic effect of protein ingestion is due largely to digestive processes as well as the extra energy required by the liver to assimilate amino acids. Although this point may have some validity, many other factors must be considered in formulating a sound program for weight loss—not to mention the potentially harmful strain on kidney and liver function and accompanying dehydration, electrolyte imbalance, and lean-tissue loss resulting from diets that are excessively high in protein. When the protein is in liquid form, the "miracle liquid" is made palatable with artificial flavoring and often includes a blend of ground-up animal hooves and horns, and pigskin mixed in a broth with enzymes and tenderizers to "predigest" it. This protein blend is often not of the highest quality in terms of the amino-acid mixture and is generally lacking in required vitamins and minerals.

According to the Federal Drug Administration, this particular brand of "protein elixir" and others like it were associated with 58 deaths.[40,102] Sixteen of the victims were obese women who lost an average of 37.6 kg within 2 to 8 months. None of the women had a previous history of heart disease; they all died suddenly while on the diet or shortly thereafter. Formal complaints were received from 165 people who reported a variety of side effects that included hair loss, nausea, headaches, constipation, neural disorders, bad breath, faintness, muscle weakness, decreased libido, and gastrointestinal disorders.

STARVATION DIETS

A starvation diet or **therapeutic fast** may be recommended in cases of severe obesity where body fat exceeds 40 to 50% of body mass.[2] Such diets are usually prescribed for up to 3 months, but only as a "last resort" prior to undertaking more extreme medical approaches that include various surgical treatments.[45,46,49] The "very-low-caloric-diet" (VLCD) approach to weight loss is predicated on the hope that abstinence from food may break established dietary habits and, in turn, improve the long-term prospects for successful weight loss.[73] Such low-calorie diets may also produce a significant depression of appetite that could help the obese person comply with the diet plan.[159] This form of dieting must be closely supervised, usually in a hospital setting.

If a person fasts even for several days and then attempts to exercise, deterioration in performance and fatigue are likely to occur. Because adequate carbohydrates are not consumed in the starvation diet, the glycogen-storage depots in the liver and muscles are reduced to low levels; this may cause impairment in most tasks that require a sustained muscular effort.

Daily medications are usually prescribed and include calcium carbonate or antihistamines for nausea; bicarbonate of soda and potassium chloride to maintain consistency of body fluids; mouthwash and sugar-free chewing gum for bad breath (owing to high levels of fat metabolism) that is present as long as fasting persists; and various bath oils for dry skin. **Clearly, for most individuals, starvation is not an "ultimate diet" or proper approach to weight control.** Furthermore, the success rate of prolonged fasting is poor.[74] And because the lean tissue lost through dieting may not all be regained as weight is gained back, the "new" body mass may now possess a greater percentage of body fat! There is also the possibility that lean tissue loss may occur disproportionally from critical organs such as the heart.[158] Table 29–2 summarizes the principles and main advantages and disadvantages of some of the popular dietary approaches to weight loss.

Although most diets produce a weight loss during the first several weeks, most of the weight lost is body water. In addition, loss of lean tissue is usually significant with dieting alone,[92a] especially in the early phase of very-low-calorie dieting.[20] Unless a person can maintain a reduced caloric intake for a considerable time, the weight is eventually regained. The net result is a return to original body size, often at the expense of feelings of hunger and other psychologic stresses while the diet plan is actually followed. Anyone who has seriously tried to maintain a diet knows the difficulties encountered. Although it is certainly possible to lose significant weight through dieting, few people have enough self-control to stick with a diet plan long enough to change body size successfully and permanently.

A review of the scientific literature that deals with weight loss in obese subjects reveals that people who are initially successful in modifying their body composition are usually unsuccessful in permanently maintaining their desired body size and shape.[73] This has been pointed out in numerous studies that deal with follow-up measurements of patients who have participated in weight-reduction programs in which caloric intake was carefully regulated and monitored. Over a 7.3-year follow-up period in 121 patients, much of the reduced weight was maintained for the first 12 to 18 months. The tendency to regain weight was independent of length of fast (up to 2 months), extent of weight loss (up to 41.4 kg), or age at onset of obesity. Return to original weight occurred in 50% within 2 to 3 years, and only seven patients remained at their reduced weights. Such statistics, which are illustrated in Figure 29–10, are rather discouraging and indicate that the long-term maintenance of a particular low-calorie diet is extremely difficult.

SETPOINT THEORY: A CASE AGAINST DIETING

When reviewing the scientific literature on the success of weight loss through dieting, one is forced to conclude that on a long-term basis, dieting for many

TABLE 29–2.
SOME POPULAR WEIGHT LOSS METHODS

Method	Principle	Advantages	Disadvantages	Comments
Surgical procedures	Alteration of the gastro-intestinal tract changes capacity or amount of absorptive surface	Caloric restriction is less necessary	Risks of surgery and post-surgical complications include death	Radical procedures include stapling of the stomach and removal of a section of the small intestine (a jejunoileal bypass)
Fasting	No energy input assures negative energy balance	Weight loss is rapid (which may be a disadvantage) Exposure to temptation is reduced	Ketogenic A large portion of weight lost is from lean body mass Nutrients are lacking	Medical supervision is mandatory and hospitalization is recommended
Protein-sparing modified fast	Same as fasting except protein or protein with carbohydrate intake assumedly helps preserve lean body mass	Same as in fasting	Ketogenic Nutrients are lacking Some unconfirmed deaths have been reported, possibly from potassium depletion	Medical supervision is mandatory Popular presentation was made in Linn's *The Last Chance Diet*
One-food-centered diets	Low-caloric intake favors negative energy balance	Being easy to follow has initial psychologic appeal	Being too restrictive means nutrients are probably lacking Repetitious nature may cause boredom	No food or food combination is known to "burn off" fat Examples include the grapefruit diet and the egg diet
Low-carbohydrate/high-fat diets	Increased ketone excretion removes energy-containing substances from the body Fat intake is often voluntarily decreased; a low caloric diet results	Inclusion of rich foods may have psychologic appeal Initial rapid loss of water may be an incentive	Ketogenic High-fat intake is contraindicated for heart and diabetes patients Nutrients are often lacking	Popular versions have been offered by Taller and Atkins; some have been called the "Mayo," "Drinking Man's," and "Air Force" diets
Low-carbohydrate/high-protein diets	Low-caloric intake favors negative energy balance		Expense and repetitious nature may make it difficult to sustain	If meat is emphasized, the diet becomes one that is high in fat The Pennington diet is an example
High-carbohydrate/low-fat diets	Low-caloric intake favors negative energy balance	Wise food selections can make the diet nutritionally sound	Initial water retention may be discouraging	The Pritikin diet is an example

Modified and reprinted by permission from Reed, P. B.: Nutrition: An Applied Science. Copyright © 1980 by West Publishing Co. All Rights Reserved.

people just does not work. One can crash off large amounts of body mass in a relatively short time period by simply not eating, but this success, however, is short-lived and eventually the urge to eat wins out and body mass is regained. The reason for this failure, it is argued, lies in "setpoints" that differ from what the dieter would like to have. The proponents of a **setpoint theory** argue that the body has an internal control mechanism, a setpoint, probably located deep within the brain's lateral hypothalamus, that drives the body to maintain a particular level of body fat.[14,81,82] In a practical sense, this would be the body mass one would achieve when one is not counting calories. The problem is that we all have different setpoints, and various factors such as the drugs fenfluramine, amphetamine, and nicotine,

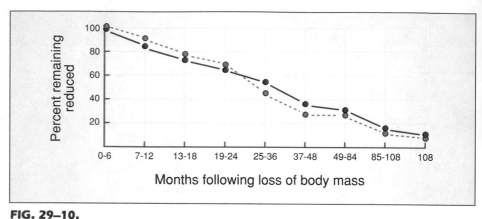

FIG. 29–10.
Percent of patients remaining at reduced weights at various time intervals
following accomplished weight loss. Solid line represents 60 subjects with obesity
onset before age 21; broken line, 42 subjects with obesity onset after age 21.
(From Johnson, D., and Drenick, E. J.: Therapeutic fasting in morbid obesity.
Arch. Intern. Med., *137*:1381, 1977.)

as well as exercise, lower the particular setting—whereas dieting has no
effect. Each time we manage to reduce our fat level below our "natural"
setpoint, the body makes internal adjustments to resist this change and
conserve or replenish body fat. In fact, even when a person attempts to gain
weight above one's normal level by means of overfeeding, the body resists
this change through an increase in the resting metabolic rate.[42,93,160]

Resting Metabolism is Lowered

**One well-documented change that occurs during weight loss through diet-
ing is the dramatic and sustained reduction in resting metabolic rate.**[47,108]
In fact, the decrease in resting metabolism is often greater than the decrease
attributable to either body mass or lean body mass loss. For example, severe
caloric restriction depresses resting metabolism by as much as 45%![145] This
calorie-sparing effect may even become more apparent with repeated bouts
of dieting (so common in the general population and among athletes as well),
so the depression of resting metabolism is enhanced with each subsequent
attempt to reduce caloric intake. This greatly conserves energy and causes
the diet to become progressively less effective despite surprisingly low caloric
intakes. As a result, a plateau in weight loss is reached and further decreases
in weight are considerably less than predicted from the mathematics of the
restricted food intake. When the rewards of one's efforts are no longer apparent, the dieter usually quits and reverts to the previous eating behaviors.
Figure 29–11 displays the results from one study of six obese men in which
body mass, resting oxygen consumption (minimal energy requirements), and
caloric intake were carefully monitored for 31 consecutive days. The subjects
consumed 3500 kcal per day for the first 7 days of the experiment. For the
remaining 24 days the daily caloric intake was reduced markedly to 450
kcal.

During the prediet period, body mass and resting oxygen consumption
remained stable. For this group, 3500 kcal a day was just adequate to equal
the daily energy expenditure. When the subjects switched to the low-calorie

FIG. 29–11.

Effects of two levels of caloric intake on changes in body mass and resting oxygen consumption. (Adapted from Bray, G.: Effect of caloric restriction on energy expenditure in obese subjects. Lancet, 2:397, 1969.)

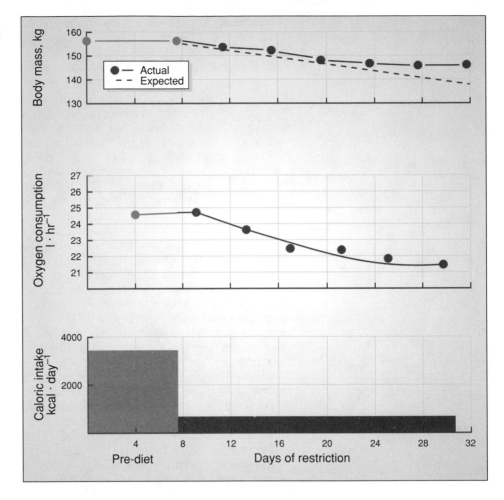

semistarvation diet, however, both body mass and resting metabolism declined. Interestingly, the percentage decline in resting energy expenditure was greater than the decrease in body mass. The dashed line represents the expected weight loss for this 450-kcal diet. The decline in resting energy metabolism actually conserved energy and caused the diet to be less effective. More than half the total weight loss occurred within the first 8 days of the 24-day diet, with the remaining weight loss occurring during the final 16 days. This slowing up of the theoretical weight-loss curve often leaves the dieter frustrated and discouraged. Dieters are anxious for weight loss to occur rapidly and according to an expected schedule.

Classic starvation studies in the 1940s to help the military plan the repatriation of war prisoners showed that even when an extreme diet ends and food intake is increased above the daily energy output, the dieter usually remains preoccupied with food. In fact, binge eating and psychologic distress continued until the original weight and fat level were attained.[83] The fact that the body itself demands a certain amount of adipose tissue, as proposed by the setpoint theory, was further substantiated in a series of experiments in 1960 in which sedentary prisoners gained weight by increasing their food intake. As body mass and fat increased concurrently, the resting metabolic rate increased so it took an astounding 7000 kcal per day for weight gain to continue!

Weight Cycling: Going Noplace Fast

The futility of repeated cycles of weight loss and weight gain, the so-called yo-yo effect, is shown in food efficiency studies in which efficiency is evaluated by the ratio of body mass change to ingested calories.[21,30,96] In general, weight gain occurs more easily with repeated cycles of body mass loss. In one study, animals required twice the time to lose the same weight during a second period of caloric restriction and only one-third the time to regain it![29] Furthermore, if the body is unable to regain its lean tissue at the rate it was lost through dieting, then fat may be replaced instead.

Recent research provides further disconcerting news for those desiring a permanent fat loss. It now appears that when obese people lose body mass the fat cells increase their level of lipoprotein lipase, the enzyme that facilitates fat synthesis and storage.[82a] The effect of these high enzyme levels is to make it easier for the formerly obese to regain fat—and the fatter people are before the weight loss the more lipoprotein lipase they produce when they reduce. In essence, these data suggest that the fatter people are originally the more vigorously their bodies attempt to regain the lost weight. If further research supports these observations, the findings will provide a biologic mechanism to explain the great difficulty the obese encounter in maintaining weight loss.

Although the setpoint theory may be unwelcome news for those who possess a setpoint that is tuned "too high," the good news, according to the setpoint advocates, is that sustained, vigorous exercise may lower the setpoint towards a more desirable level.[44,166] Concurrently, regular exercise increases lean body mass, raises basal metabolism, and brings about enzymatic changes that facilitate fat metabolism in the tissues, all of which would make dieting more effective.[44,98,108,156a] For overweight men and women who exercise regularly, food intake tends to drop initially despite the increase in caloric output, and body fat decreases. Eventually, as an active lifestyle is maintained, caloric intake balances the daily energy requirements so body mass is stabilized at a new, lower level.

EXERCISE FOR WEIGHT CONTROL

The relative contributions of how physical inactivity and excessive caloric intake are linked to obesity are not clear.

Not simply a problem of gluttony. In the past, it was generally accepted that the obese condition was the result of excessive food intake. Clearly then, the effective approach to weight control would be some form of caloric restriction by dieting. **This view of obesity is overly simplistic as available evidence indicates that excess weight gain throughout life often closely parallels reduced physical activity rather than an increased caloric intake.**[119] In fact, among active young and middle-aged endurance-trained men, body fat was inversely related to energy expenditure; no relationship existed between body fat and food consumption.[105] Among the physically active, those who eat the most often weigh the least and are the most fit![139] In the United States, the caloric intake per person has steadily decreased over the past 80 years, yet body mass and body fat have slowly increased. Americans now eat 5 to 10% fewer calories than they did 20 years ago, yet they weigh an average of 2.3 kg more. Certainly, if dieting were effective this reduction in caloric intake should bring the national body mass to a lower, not higher level!

Furthermore, obese infants do not characteristically consume more calories than the recommended dietary standards.[13] Infant offspring of obese parents often display lower resting metabolic rates and less spontaneous movements than infants of normal-weight parents. This finding suggests the possibility that such subdued movement patterns are abnormal and may reflect an inherited characteristic.[70] Time-in-motion photography used to document activity patterns of elementary-school children clearly showed that overweight school children were considerably less active than their normal-weight peers, and that excess weight was not related to food intake. The caloric intake of obese high-school girls and boys is actually below their nonobese peers.[33,63,72,74a,126] This observation that fat people often eat the same or even less than thinner ones[141] is also true for adults over a broad age range as they become less active and slowly begin to add weight.[48,163] **In fact, one is hard pressed for evidence that groups of overweight individuals actually eat more, on the average, than people of normal weight.**[80] Consequently, to further reduce caloric intake in an effort to cause weight loss would seem neither prudent nor appropriate as the **only** method to combat the overfat condition.

Increased energy output worth considering. **It is increasingly clear that people who maintain a physically active lifestyle or who become involved in endurance exercise programs maintain a desirable level of body composition.** Evidence is accumulating to support the contention that exercise may be more effective than dieting for long-term weight control.[41,107] Within this framework, a strong case can be made for habitual, vigorous physical activity for individuals of all ages. Increased caloric output through endurance-type exercise provides a significant option for unbalancing the energy balance equation to bring about weight loss and a desirable modification in body composition.[38,104,162] Only recently, however, has this approach to weight control come to prominence.[52] Two arguments have generally been raised against the exercise approach. One is the belief that exercise inevitably causes an increase in appetite so that any caloric deficit is rapidly made up by a proportionate increase in food intake. The second argument is that the calorie-burning effects of exercise are so small that a reasonable use of exercise would make only a "dent" in the body's fat reserves compared to the approach using starvation or semistarvation. We shall take a closer look at these two misconceptions.

MISCONCEPTION 1. EXERCISE AND FOOD INTAKE

To some extent, regular physical activity appears to contribute to the normal functioning of the brain's feeding control mechanisms. A delicate balance between energy expenditure and food intake is often not maintained in sedentary people. This lack of precision in regulating food intake at the lower end of the physical activity spectrum may account for the "creeping obesity" observed in highly mechanized and technically advanced societies. On the other hand, for individuals who exercise on a regular basis, appetite control falls within a "reactive zone", whereby it is simpler to match food intake to the daily level of energy expenditure.

In considering the effects of exercise on appetite and food consumption, a distinction should be made between the type and duration of the exercise. People such as lumberjacks, farm laborers, and endurance athletes who perform hard, physical labor for prolonged periods, consume about twice the daily calories (4000 to 6000 kcal) than more sedentary individuals (2000 to 3000 kcal). Under such circumstances, the high caloric intakes are required to balance the extremely high daily caloric requirements, especially for many athletes who spend considerable time each day in vigorous physical training. Endurance athletes like marathon runners, cross-country skiers, and cyclists consume about 4000 to 5000 kcal each day, yet are among the leanest people in the world! Obviously, this extreme caloric intake is required just to meet the energy requirements of training.

When considering the dietary intake for people who train for relatively short periods, the apparent compensatory appetite-stimulating effect of exercise is noticeably reduced or manifested only immediately following exercise.[148,164,165] This lack of increased caloric intake with moderate training is shown for college women whose daily caloric intake was evaluated before and during a season of competitive swimming and tennis.[78] Swim workouts were conducted daily for 2 hours from January to May. Total distance swum each day was 2000 to 4000 m during January and March and 1000 to 2000 m in April and May. The workouts for the tennis players consisted of 10 minutes of rope-skipping, 45 minutes of organized practice and games, followed by a half-mile jog. Rope-skipping and jogging were discontinued after January. Average daily caloric intake was assessed for each woman before and during the 5-month training and competitive season by a 7-day dietary inventory.

The average caloric intake of the swimmers shown in Table 29–3 remained about 15% higher than for the tennis players during the training and competitive seasons. Within each group, there were insignificant changes in caloric intake or food composition before and after the experiment. In terms of body composition, there was no change for both groups in body mass, percent body fat, and lean body mass.

MISCONCEPTION 2. CALORIC STRESS OF PHYSICAL ACTIVITY

A common misconception about the role of increased exercise in programs of weight reduction concerns the calorigenic effects of regular exercise. It has been argued that an inordinate amount of exercise must be performed

TABLE 29–3.
AVERAGE KCAL INTAKE OF SWIMMERS AND TENNIS PLAYERS BEFORE AND DURING A 5-MONTH TRAINING AND COMPETITIVE SEASON*

Group	Calories		Protein (g)		Fats (g)		Carbohydrates (g)	
	Before	During	Before	During	Before	During	Before	During
Swimmers	2091	2065	80.8	71.8	92.3	90.0	247.9	240.6
Tennis players	1811	1797	78.1	74.2	78.9	78.8	192.6	195.2

* From Katch, F.I., et al.: Effects of physical training on the body composition and diet of females. *Res. Q.*, 40:99, 1969.

just to lose 1 lb (0.45 kg) of body fat, as for example, chopping wood for 10 hours, playing golf for 20 hours, performing mild calisthenic exercises for 22 hours, playing ping-pong for 28 hours or volleyball for 32 hours. From a different perspective, however, golf played only 2 hours (about 350 kcal per day), 2 days per week (700 kcal) would take about 5 weeks to reduce 1 lb of fat. Assuming one could play year round, devoting 2 days a week to this form of exercise would result in a 4.5 kg loss of fat during the year, provided that the food intake remained fairly constant. **The calorie-expending effects of exercise are cumulative. A caloric deficit of 3500 kcal is equivalent to a 0.45 kg body fat loss, whether the deficit occurs rapidly or systematically over time.**

In determining the caloric cost of various physical activities, the assumption is made that the energy cost is fairly constant among people of a particular body size. In Chapter 8, we noted that the energy cost data for most physical activities were averages often based on few observations. A wide range of values is therefore possible because of individual differences in performance style and technique, and environmental factors that include terrain, temperature, and wind resistance, as well as the intensity of participation.

The values of energy expenditure for physical activities presented in Appendix C should not be considered absolute. These are "average" values, applicable under "average" conditions when applied to the "average" person of a given body mass. These gross values, however, do provide a relative approximation of energy expenditure for establishing the appropriate caloric cost of different physical activities. With submaximal exercise as performed by most people who use exercise for weight control, the contribution of recovery metabolism—the so-called "afterglow"—to the total energy expenditure is probably quite small.[28]

EXERCISE IS EFFECTIVE

Regular aerobic exercise, even without dietary restriction, brings about favorable changes in body mass and body composition.[26a,43] Even performance of conventional resistance training exercise combined with caloric restriction results in the maintenance of lean body mass compared to a program of weight loss that relies exclusively on dieting.[10] The effectiveness of an exercise program for weight loss is linked to the degree of obesity at the start. As a general rule, persons who are obese lose weight and fat more readily than their normal counterparts.[60,106] In addition, exercise provides significant positive "spin off" because it alters body composition (reduced fat and maintenance or even small increase in lean tissue) in such a way that the resting level of energy expenditure is maintained or even increased in both lean and moderately obese individuals.[20,151,151a] This reduces the body's tendency to store calories.

When considering exercise for weight control, factors such as frequency, intensity, and duration, as well as the specific form of exercise must be considered. Because 0.45 kg of body fat contains approximately 3500 kcal, the exercise program must establish this negative caloric balance to bring about a 0.45 kg fat loss. Continuous, big muscle aerobic activities that have a moderate to high caloric cost such as walking, running, rope skipping, cycling, and swimming are ideal. Many recreational sports and games are also effective in reducing body mass, although precise quantification of energy expenditure is difficult during such activities. These rhythmic exercises burn considerable calories, stimulate lipid metabolism, reduce body fat, establish favorable blood pressure responses, and generally promote cardiovascular fitness. In addition, there is generally no selective effect of running, walking, or bicycling; each is equally effective in altering body composition, provided the duration, frequency, and intensity of exercise are similar.[121] An extra 300 kcal daily caloric output with moderate jogging for 30 minutes causes a 0.45 kg fat loss in about 12 days. Over a year's time, this theoretically represents a total caloric deficit equivalent to approximately 13.6 kg of body fat.

A DOSE-RESPONSE RELATIONSHIP

Generally, the total energy expended is the most important factor that influences the effectiveness of the exercise program for weight loss. A direct dose-response relationship has been demonstrated between weight loss and the time spent in exercise.[61] Thus, an overfat person who starts out at a light exercise intensity with slow walking can accrue a considerable caloric expenditure simply by extending the duration of the exercise. This effect of duration offsets the inability (and inadvisability) of the previously sedentary, obese person beginning a program at high exercise intensities. Also, the energy cost of weight-bearing exercise such as walking is proportional to body mass; thus, the overweight person expends considerably more calories to perform the same task than someone of normal body mass.

The importance of exercise duration for weight loss was illustrated in a study of three groups of men who exercised for 20 weeks by walking and running for either 15, 30, or 45 minutes per session.[106] Compared to a sedentary control group, the three exercise groups significantly decreased their body fat, fatfolds, and waist girth. When comparisons were made between the three groups, the 45-minute training group lost more body fat than either the 30- or 15-minute group. This was attributed to the greater caloric expenditure of the longer exercise period.

START SLOWLY

The initial stage of an exercise program for a previously sedentary, overfat person should be developmental in nature and should not include a high total energy output. During this time, the individual should be urged to adopt longterm goals, personal discipline, and a restructuring of both eating and exercise behaviors.[104] It is often counterproductive to include unduly rapid training progressions because many obese men and women initially show psychologic resistance to physical training.[53] During the first few weeks, slow walking is replaced by intervals of walking and jogging that eventually lead to continuous jogging. At least 6 to 8 weeks are required before observable changes occur. Behavioral approaches should also be applied to cause meaningful lifestyle changes in physical activity. For example, walking or bicycling can replace the use of the auto, stair climbing can replace the elevator, and manual tools can replace power tools.

Table 29–4 shows the effectiveness of regular exercise for weight loss. In this study six sedentary, obese young men exercised 5 days a week for 16 weeks doing 90 minutes of walking on a motor-driven treadmill at each session.[99] The men lost an average of almost 6 kg of body fat, representing a decrease in percent body fat from 23.5 to 18.6%. In addition, physiologic fitness and work capacity improved, as did the level of high-density lipoprotein (15.6%) and the high/low density lipoprotein ratio (25.9%).

TABLE 29—4.
CHANGES IN BODY COMPOSITION AND BLOOD LIPIDS IN SIX OBESE YOUNG ADULT MEN WITH A SIXTEEN-WEEK WALKING PROGRAM[a]

Variable	Pre-Training[b]	Post-Training	Difference
Body mass, kg	99.1	93.4	−5.7[c]
Body density, $g \cdot ml^{-1}$	1.044	1.056	+0.012[c]
Body fat, %	23.5	18.6	−4.9[c]
Fat mass, kg	23.3	17.4	−5.9[c]
Lean body mass, kg	75.8	76.0	+0.2
Sum of fatfolds, mm	142.9	104.8	−38.1[c]
HDL cholesterol, $mg \cdot 100\ ml^{-1}$	32	37	+5.0[c]
HDL/LDL cholesterol	0.27	0.34	+0.07[c]

[a] From Leon, A.S., et al.: Effects of a vigorous walking program on body composition, and carbohydrate and lipid metabolism of obese young men. Amer. J. Clin. Nutr., 33:1776, 1979.
[b] Values are means.
[c] Statistically significant.

REGULARITY IS THE KEY

Exercise frequency is important when prescribing considering exercise for weight reduction. In a summary of six studies that investigated optimal training frequency,[121] it was observed that training 2 days a week did not change body mass, fatfolds, or percent body fat. Training 3 and 4 days a week, however, had a significant effect. Subjects who trained 4 days a week reduced their body mass and fatfolds significantly more than the 3 days per week group. Reductions in percent body fat, however, were similar for both groups.

Within the framework of available research, it appears that at least 3 days of training per week are required to bring about changes in body composition through exercise.[6] More frequent exercise may even be more effective. More than likely, this effect is the direct result of the added caloric output provided by the extra exercise. **Although it is difficult to precisely determine a threshhold energy expenditure for weight reduction and fat loss, it is generally recommended that the calorie-burning effect of each exercise session should be at least 300 kcal.** This can be achieved with 30 minutes of moderate to vigorous running, swimming, or bicycling, or walking for at least 60 minutes.

DIET PLUS EXERCISE: THE IDEAL COMBINATION

For moderately obese children and adults, combinations of regular aerobic exercise and diet offer considerably more flexibility in achieving a negative caloric balance and accompanying fat loss than either exercise alone or diet alone.[5,31,54] In fact, the addition of exercise to the program of weight control may facilitate a more permanent fat loss than would total reliance on caloric restriction.[15,37,103,118,142]

How can an obese person, utilizing exercise and diet while attempting to maintain a prudent weight loss of about 0.45 kg a week, reduce body mass by 9.1 kg (20 lb)? Under these conditions, even under the best circumstances, 20 weeks would be required to achieve a 9.1 kg fat loss. With this goal, the average weekly deficit would have to be 3500 calories whereas the daily deficit must be 500 calories.

One-half hour of moderate exercise (about 350 "extra" calories) performed 3 days a week adds 1050 calories to the weekly caloric deficit. Consequently, the weekly caloric intake would have to be reduced by only 2400 calories instead of 3500 calories to lose the desired 0.45 kg of fat each week. If the number of exercise days is increased from 3 to 5, food intake need only be reduced by 250 calories each day. If the duration of the 5-day per week workouts were prolonged from 30 minutes to 1 hour, then no reduction in food intake would be necessary for weight loss to occur, because the required 3500 caloric deficit would have been created entirely through exercise.

If the intensity of the 1-hour exercise performed 5 days a week was then increased by only 10% (cycling at 22 mph instead of 20 mph; running a mile in 9 minutes instead of 10 minutes), the number of calories burned each week through exercise would increase by an additional 350 kcal (3500 kcal/ week × 10%). This new weekly deficit of 3850 kcal or 550 kcal per day would actually permit the "dieter" to **increase** the daily food intake by 50 kcal and still lose 0.45 kg of fat each week!

Clearly, physical activity can be used effectively by itself, or in combination with mild dietary restriction, to create an effective loss of body fat.[26a] Either approach is likely to produce fewer feelings of intense hunger and other psychologic stresses that occur with a program of weight loss that relies exclusively on caloric restriction. Perhaps of equal or greater significance is that the use of both aerobic and resistance exercise in a weight-reducing program provides protection against a loss in lean tissue usually observed when weight loss is achieved by diet alone.[66,87,117,150,167] The preservation of the lean tissue mass is partly due to aerobic exercise training that enhances the mobilization and breakdown of fat from the body's adipose tissue depots. In addition, vigorous exercise tends to increase the rate of protein building in skeletal muscle, while at the same time retarding its rate of breakdown. **These protein-sparing effects of exercise will cause more of the weight loss to be fat loss.**

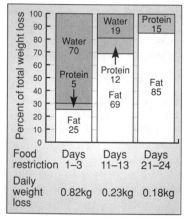

FIG. 29–12.
Percentage composition of weight loss at the start, middle, and end of 24 days of food restriction (1000 kcal per day) plus enforced exercise of 2.5 hours per day. (From Grande, F.: Nutrition and energy balance in body composition studies. In Techniques for Measuring Body Composition. Washington, D.C. National Academy of Sciences–National Research Council, 1961.)

FACTORS THAT AFFECT WEIGHT LOSS

When caloric intake is below the daily energy requirement, the initial decrease in body mass occurs primarily from water loss and corresponding depletion of the body's carbohydrate reserves; with further weight loss, a larger proportion of body fat and protein is metabolized to supply the caloric deficit created by restricting food intake or increasing physical activity.

EARLY WEIGHT LOSS IS LARGELY WATER

Figure 29–12 shows the percentage composition of the average daily weight loss for water, protein, and fat during 24 days on a 1000-kcal-per-day carbohydrate diet and 2.5 hours of prescribed exercise. In the first 3 days of caloric deficit, 70% of the weight loss could be attributed to water loss. The contribution of water to weight loss became progressively less as the caloric deficit continued and represented only 19% of the weight loss during days 11 to 13. At the same time, fat loss accelerated from 25 to 69% during this period. From day 21 to day 24, 85% of the weight loss was due to a reduction in body fat with no corresponding increase in water loss. The contribution of protein to weight loss increased from 5% initially to 12% during days 11 to 13 and to 15% by the end of the observation period.

MAINTAIN ADEQUATE HYDRATION

Figure 29–13 shows the relationship between the proportion of water, protein, and fat lost and the amount of water consumed during the first few days of caloric restriction. The left bar is the same as that shown in Figure 29–12 for days 1 to 3. In this experiment, subjects had unlimited access to water. The middle bar represents the compositional loss in body mass for six men who also subsisted on the 1000-kcal carbohydrate diet, but with daily water intake restricted to 1800 ml. The bar at the right displays the average weight loss for six men who consumed an identical diet, but whose daily water intake was further reduced to 900 ml. Water restriction during the first 3 days significantly increased the proportion of water loss and decreased the proportion of fat loss, especially in subjects who consumed the least water. Although more total weight was lost when daily water intake was reduced to 900 ml per day, this additional loss in body mass could be attributed to water loss. **Regardless of whether or not fluid intake was restricted, the total quantity of fat lost was essentially the same.**

LONGER-TERM DEFICIT PROMOTES FAT LOSS

Figure 29–14 illustrates the important concept that the caloric equivalent of weight loss increases as a function of increasing the duration of caloric restriction. **This is the major reason why it is crucial to maintain a sustained caloric deficit for extended periods because shorter periods cause a larger percentage of water and carbohydrate loss per unit of weight reduction with only a minimal decrease in body fat.** It should be noted that the caloric equivalent per kg of weight loss more than doubles after 2 months as compared to the first 4 or 5 days.

The results of the prior studies that have evaluated various approaches to establish a caloric imbalance can be summarized as follows:

1. Exercise combined with dietary restriction is a more effective approach for achieving a long-term negative caloric balance as compared with exercise or diet alone.
2. During the first few days of weight reduction, the rapid weight loss is due primarily to a loss in body water and carbohydrates; longer periods of weight reduction are associated with a substantially greater loss of fat per unit of weight loss.
3. Water intake should not be restricted when beginning weight reduction because this can precipitate dehydration but no additional fat loss.
4. Undesirable psychologic and medically related problems may occur with prolonged caloric restriction maintained below minimal energy requirements.
5. Weight loss by diet alone causes a significant loss of muscle mass. Exercise protects against lean tissue losses; thus, more of the weight lost is fat.

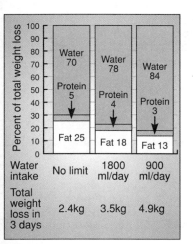

FIG. 29–13.
Percentage composition of weight loss during the first 3 days on a 1000-kcal carbohydrate diet with water intake unrestricted and reduced to 1800 ml and 900 ml per day. (From Grande, F.: Nutrition of energy balance in body composition studies. In Techniques for Measuring Body Composition. Washington, D.C. National Academy of Sciences–National Research Council, 1961.)

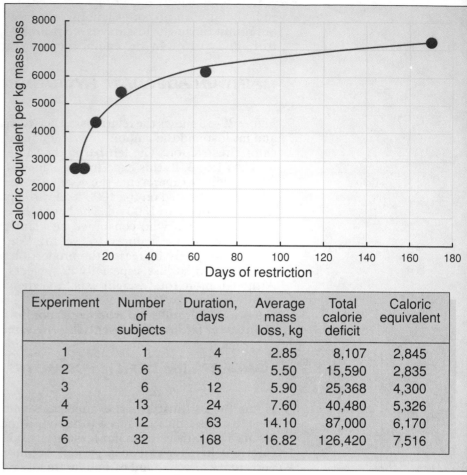

FIG. 29–14.
Caloric equivalent of weight loss in relation to the duration of calorie restriction. Each datum point represents an experiment summarized in the accompanying legend. (Adapted from Grande, F.: Nutrition and energy balance in body composition studies. In Techniques for Measuring Body Composition. Washington, D.C. National Academy of Sciences–National Research Council, 1961.)

SUMMARY

1. There are three ways to unbalance the energy balance equation and bring about weight loss: (1) reduce caloric intake below daily energy expenditure, (2) maintain regular food intake and increase energy output, and (3) combine methods 1 and 2 by decreasing food intake and increasing energy expenditure.
2. Long-term weight control through dietary restriction is generally successful less than 20% of the time.
3. A caloric deficit of 3500 kcal created either through diet or exercise is the equivalent to the calories in 1 lb (0.45 kg) of adipose tissue.
4. Dieting for weight loss can be effective if done properly. The disadvantages

of semistarvation, however, are significant and include a loss of lean body tissue, lethargy, possible malnutrition and metabolic disorders, and a decrease in the basal energy expenditure. Some of these factors actually conserve energy and cause the diet to be less effective.

5. Repeated cycles of weight loss–weight gain cause the body to increase its ability to conserve energy. This ultimately leads to a greater difficulty in achieving weight loss with subsequent dieting and makes the process easier for regaining the lost weight.

6. The calorie-expending effects of exercise are cumulative, so that a little exercise performed routinely has a dramatic effect over time. The role of exercise in appetite suppression or stimulation is unclear. Over time, most athletes eventually consume enough calories to counterbalance caloric expenditure—but many of these athletes are the leanest in the world.

7. Combinations of exercise and diet offer a flexible and effective approach to weight control. Exercise enhances the mobilization and utilization of fat, thus increasing fat mass loss, and at the same time retards lean tissue loss.

8. The rapid weight loss during the first few days of caloric deficit is due primarily to a loss in body water and carbohydrates. Continued weight reduction is associated with a greater loss of fat per unit of weight loss.

REFERENCES

1. Abraham, S., and Johnson, C.L.: Prevalence of severe obesity in adults in the United States. *Am. J. Clin. Nutr.*, 33:364, 1980.

2. Alban, H.J., et al.: Metabolic response to low- and very-low-calorie diets. *Am. J. Clin. Nutr.*, 49:745, 1989.

3. Alexander, J.K., and Peterson, K.L.: Cardiovascular effects of weight reduction. *Circulation*, 45:310, 1972.

4. Ama, P.F.M., et al.: Fat distribution and adipose tissue metabolism in non-obese male black Africans and caucasian subjects. *Int. J. Obesity*, 10:503, 1986.

5. American College of Sports Medicine: Position statement on the recommended quantity and quality of exercise for developing and maintaining fitness in healthy adults. *Med. Sci. Sports*, 10:7, 1978.

6. American College of Sports Medicine: Position statement on proper and improper weight loss programs. *Med. Sci. Sports Exerc.*, 15:9, 1983.

7. American Medical Association: A critique of low-carbohydrate ketogenic weight and reduction regimens (a review of Dr. Atkins' diet revolution). *JAMA*, 224:1418, 1973.

8. Angel, A., and Roncari, D.A.K.: Medical complications of obesity. *Can. Med. Assoc. J.*, 119:1408, 1978.

9. Backman, L., et al.: Cardiovascular function in extreme obesity. *Acta Med. Scand.*, 193:437, 1972.

10. Ballor, C.L., et al.: Resistance weight training during caloric restriction enhances lean body weight maintenance. *Am. J. Clin. Nutr.*, 47:19, 1988.

11. Barlett, H.L., and Buskirk, E.R.: Body composition and the expiratory reserve volume in lean and obese men and women. *Int. J. Obesity*, 7:339, 1983.

12. Bayley, N., and Bayer, L.M.: Assessment of somatic androgyny. *Am. J. Phys. Anthrop.*, 4:433, 1946.

13. Beaton, G., et al.: An examination of factors believed to be associated with infantile obesity. *Am. J. Clin. Nutr.*, 32:1997, 1979.

14. Bennett, W., and Gurin, J.: Do diets really work? *Science*, 82, 3:42, 1982.

15. Björntorp, P.: Interrelation of physical activity and nutrition on obesity. In *Diet and Exercise: Synergism in Health Maintenance*. Edited by P.L. White, and T. Mondeika. Chicago, American Medical Association, 1982.

16. Björntorp, P.: Adipose tissue in obesity. In *Recent Advances in Obesity Research.* Vol. 4. Edited by J. Hirsch, and T.B. Van Itallie. London, Libby, 1985.

17. Björntorp, P.: Fat cells in obesity. In *Handbook of Eating Disorders.* Edited by K.D. Brownell, and J.P. Foreyt. New York, Basic Books, 1986.

18. Björntorp, P., et al.: Effect of an energy reduced dietary regimen in relation to adipose tissue cellularity in obese women. *Am. J. Clin. Nutr.,* 28:445, 1975.

19. Black, D.R., and Burckes-Miller, M.E.: Male and female college athletes: use of anorexia nervosa and bulimia nervosa weight loss methods. *Res. Q. Exer. Sport,* 59:252, 1988.

20. Blackburn, G.L., et al.: The very-low-calorie-diet: a weight reduction technique. In *Handbook of Eating Disorders.* Edited by K.D. Brownell, and J.P. Foreyt. New York: Basic Books, 1986.

21. Blackburn, G.L., et al.: Weight cycling: the experience of human dieters. *Am. J. Clin. Nutr.,* 49:1105, 1989.

22. Blitzer, P.H., et al.: Association between teenage obesity and cancer in 56,111 women. *Prev. Med.,* 5:20, 1976.

23. Bogardus, C., et al.: Familiar dependence of the resting metabolic rate. *N. Engl. J. Med.,* 315:96, 1986.

24. Bortz, W.M.: Disuse and aging. *JAMA,* 248:1203, 1982.

25. Bouchard, C., and Perusse, L.: Heredity and body fat. *Ann. Rev. Nutr.,* 8:259, 1988.

26. Bouchard, C., et al.: Inheritance of the amount and distribution of human body fat. *Int. J. Obesity,* 72:205, 1988.

26a. Bouchard, C., et al.: Long-term exercise training with constant energy intake. 1: Effect on body composition and selected metabolic variables. *Int. J. Obesity,* 14:57, 1990.

27. Bray, G.A.: Effects of obesity on health and happiness. In *Handbook of Eating Disorders.* Edited by K.D. Brownell, and J.B. Foreyt. New York, Basic Books, 1986.

28. Brehm, B.A., and Gutin, B.: Recovery energy expenditure for steady state exercise in runners and nonexercisers. *Med. Sci. Sports Exerc.,* 18:205, 1986.

29. Brownell, K.D., et al.: The effects of repeated cycles of weight loss and regain in rats. *Physiol. Behav.,* 38:459, 1986.

30. Brownell, K.D., et al.: Weight regulation practices in athletes: analysis of metabolic and health effects. *Med. Sci. Sports Exerc.,* 19:546, 1987.

31. Brownell, K.D., and Kaye, K.S.: A school-based behavioral modification, nutrition education, and physical activity program for obese children. *Am. J. Clin. Nutr.,* 35:277, 1982.

32. Brook, C.G.D., and Lloyd, J.K.: Adipose cell size and glucose tolerance in obese children and effects of diet. *Arch. Dis. Child.,* 48:301, 1973.

33. Bullen, B.A., et al.: Physical activity of obese and non-obese adolescent girls appraised by motion picture sampling. *Am. J. Clin. Nutr.,* 14:211, 1964.

34. Burwell, C.S., et al.: Extreme obesity associated with alveolar hypoventilation—a Pickwickian syndrome. *Am. J. Med.,* 21:811, 1956.

34a. Campaigne, B.N.: Body fat distribution: metabolic consequences and implications for weight loss. *Med. Sci. Sports Exerc.,* 22:291, 1990.

35. Charney, H.C., et al.: Childhood antecedents of adult obesity. *N. Engl. J. Med.,* 295:6, 1976.

36. Chien, S., et al.: Longitudinal measurements of blood volume and essential body mass in human subjects. *J. Appl. Physiol.,* 39:818, 1975.

37. Colvin, R.H., and Olson, S.B.: A descriptive analysis of men and women who have lost significant weight and are highly successful at maintaining the loss. *Addictive Behaviors,* 8:287, 1983.

38. Craig, B.W., et al.: Effects of stopping training on size and response to insulin of fat cells in female rats. *J. Appl. Physiol.,* 54:571, 1983.

39. Craig, L.S., and Bayer, L.M.: Androgynic phenotypes in obese women. *Am. J. Phys. Anthrop.,* 26:23, 1967.

40. Cyborski, C.K.: Deaths associated with the protein-sparing fast. *JAMA,* 239:971, 1978.

41. Dahlkoetter, J.A.: Obesity and the unbalanced energy equation: exercise versus eating habit change. *J. Consult. Clin. Psychol.,* 47:898, 1979.

42. Danforth, E., Jr.: Nutritionally induced alterations in metabolism. In *Assessment of Energy Metabolism in Health and Disease.* Columbus, OH, Ross Laboratories, 1980.

43. Desperés, J.P., et al.: Effects of aerobic training on fat distribution in male subjects. *Med. Sci. Sports Exerc.,* 17:113, 1985.

44. Donohoe, C.T., Jr., et al.: Metabolic consequences of dieting and exercise in the treatment of obesity. *J. Consult. Clin. Psychol.*, 52:829, 1984.

45. Drenick, E.J.: Weight reduction by prolonged fasting. In *Obesity in Perspective*. Edited by G.A. Bray, Bethesda, MD., United States Department of Health, Education and Welfare, 1975.

46. Drenick, E.J., and Johnson, D.: Therapeutic fasting in morbid obesity—long term follow-up. *Arch. Intern. Med.*, 137:1381, 1977.

47. Elliot, D.L., et al.: Sustained depression of the resting metabolic rate after massive weight loss. *Am. J. Clin. Nutr.*, 49:93, 1989.

48. Epstein, L.H., and Wing, R.R.: Aerobic exercise and weight. *Addictive Behaviors*, 5:371, 1980.

49. Faloon, W.W. (Ed.): Conference on jejunoileostomy for obesity. *Am. J. Clin. Nutr.*, 30:1, 1977.

50. Faust, I.M., et al.: Diet induced adipocyte number increase in adult rats: a new model of obesity. *Am. J. Physiol.*, 235:279, 1978.

51. Forbes, G.B., et al.: Genetic factors in abdominal obesity, a risk factor for stroke. *N. Engl. J. Med.*, 318:1070, 1988.

52. Foreyt, J.P., et al.: Weight control and nutrition education programs in occupational settings. *Public Health Reports*, 95:127, 1980.

53. Foss, M.L., et al.: Physical training program for rehabilitating extremely obese patients. *Arch. Intern. Med.*, 137:1381, 1977.

54. Franklin, B.A., and Rubenfine, M.: Losing weight through exercise. *JAMA*, 244:377, 1980.

55. Freedman, D.S., et al.: Relation of body fat distribution to hyperinsulinemia in children and adolescents: the Bogalusa Heart Study. *Am. J. Clin. Nutr.*, 46:403, 1987.

56. Garn, S.M.: Continuities and changes in fatness from infancy through adulthood. *Curr. Probl. Pediatr.*, 15:1, 1985.

57. Garrow, J.S.: Problems in measuring human energy balance. In *Assessment of Energy Metabolism in Health and Disease*. Columbus, Ohio, Ross Laboratories, 1980.

58. Gartmaker, S.L., et al.: Increasing pediatric obesity in the United States. *Am. J. Dis. Child*, 141:535, 1987.

59. Greenwood, M.R.C., and Pitmann-Waller, V.: Weight control: a complex, various, and controversial problem. In *Obesity and Weight Control*. Edited by R.I. Frankle, and M-U. Yang. Maryland, Aspen, 1988.

60. Gettman, L.R., et al.: Physiological responses of young men to 1, 3, and 5 day per week training programs. *Res. Q.*, 47:638, 1976.

61. Gwinup, G., et al.: Thickness of subcutaneous fat and activity of underlying muscles. *Ann. Intern. Med*, 74:408, 1971.

62. Hagar, A., et al.: Adipose tissue cellularity in obese school girls before and after dietary treatment. *Am. J. Clin. Nutr.*, 31:68, 1978.

63. Hampton, M.C., et al.: Caloric and nutrient intake of teenagers. *J. Am. Diet. Assoc.*, 50:385, 1987.

64. Hannon, B.M., and Lohman, T.G.: The energy cost of overweight in the United States. *Am. J. Public Health*, 68:765, 1978.

65. Heath, G.W., et al.: A physiological comparison of younger and older endurance-trained athletes. *J. Appl. Physiol.*, 51:634, 1981.

66. Hill, J.O., et al.: Effects of exercise and food restriction on body composition and metabolic rates in obese women. *Am. J. Clin. Nutr.*, 46:622, 1987.

67. Hirsch, J., and Batchelor, B.R.: Adipose tissue cellularity in human obesity. *Clin. Endocrinol. Metab.*, 5:299, 1976.

68. Hirsch, J., and Knittle, J.: Cellularity of obese and non-obese human adipose tissue. *Fed. Prod.*, 29:1518, 1970.

69. Hirsch, J., and Leibel, R.L.: New light on obesity. *N. Engl. J. Med.*, 318:509, 1988.

70. Hollenberg, C.H.: Human obesity—a survey and suggestions. *Can. Med. Assoc. J.*, 199:1383, 1978.

71. Hubert, H.B., et al.: Obesity as an independent risk factor for cardiovascular disease: a 26-year follow-up of participants in the Framingham Heart Study. *Circulation*, 67:968, 1983.

72. Huenemann, R.: Food habits of obese and non-obese adolescents. *Postgrad. Med.*, 51:99, 1972.

73. Innes, J.A., et al.: Long term follow-up of therapeutic starvation. *Br. Med. J.*, 2:356, 1974.
74. Johnson, D., and Drenick, E.J.: Therapeutic fasting in morbid obesity. *Arch. Int. Med.*, 137:1381, 1977.
74a. Johnson, M.L., et al.: Relative importance of inactivity and overeating in energy balance in obese high school girls. *Amer. J. Clin. Nutr.*, 44:779, 1986.
75. Jones, P.J., et al.: Validation of doubly labeled water for assessing energy expenditure in infants. *Pediatr. Res.*, 21:242, 1987.
76. Karvonen, M., et al.: The effects of training heart rate: a longitudinal study. *Ann. Med. Exptl. Biol. Fem.*, 35:307, 1957.
77. Katch, F.I., and Katch, V.L.: Computer technology to evaluate body composition, nutrition, and exercise. *Prev. Med.*, 12:619, 1984.
78. Katch, F.I., et al.: Effects of physical training on the body composition and diet of females. *Res. Q.*, 40:99, 1969.
79. Katch, F.I., et al.: Effects of situp exercise training on adipose cell size and adiposity. *Res. Q. Exerc. Sport*, 55:242, 1984.
80. Keen, H., et al.: Nutrient intake, adiposity, and diabetes. *Br. Med. J.*, 1:655, 1979.
81. Kessey, R.E.: A set-point theory of obesity. In *Handbook of Eating Disorders*. Edited by K.D. Brownell, and J.P. Foreyt. New York, Basic Books, 1986.
82. Keesey, R.E., and Corbett, S.W.: Metabolic defense of the body weight set point. In *Eating and Its Disorders*. Edited by A.J. Stunkard, and E. Stellar. New York, Raven Press, 1984.
82a. Kern, P.A., et al.: The effects of weight loss on the activity and expression of adipose-tissue lipoprotein lipase in very obese humans. *N. Engl. J. Med.*, 322:1053, 1990.
83. Keys, A., et al.: *The Biology of Human Starvation*. Minneapolis, University of Minnesota Press, 1970.
84. King, M.A., and Katch, F.I.: Changes in body density, fatfolds and girths at 2.3 kg increments of weight loss. *Hum. Biol.*, 58:709, 1986.
85. Kissebah, A.H., et al.: Relationship of body fat distribution to glucose tolerance and clinical diabetes in obese women. *Clin. Res.*, 38:520A, 1980.
86. Kissebah, A.H., et al.: Relationship of body fat distribution to metabolic complications of obesity. *J. Clin. Endocr. Metab.*, 54:154, 1982.
87. Knittle, J., and Hirsch, J.: Effect of early nutrition on the development of rat epididymal fat pads; cellularity and metabolism. *J. Clin. Invest.*, 47:2901, 1968.
88. Kohrs, M.B., and Kamath, S.K. (Eds.): Symposium on Metabolism and Aging. *Am. J. Clin. Nutr.*, 35(Suppl.):1, 1982.
89. Konstantin, N., et al.: Effects of dieting and exercise on lean body mass, oxygen uptake, and strength. *Med. Sci. Sports Exerc.*, 17:466, 1985.
90. Kramer, M.S.: Do breast feeding and delayed introduction to solid food protect against subsequent obesity? *J. Pediatrics*, 98:883, 1981.
90a. Krieger, D.R., and Landsberg, L.: Role of hormones in the etiology and pathogenesis of obesity. In *Obesity and Weight Control*. Edited by R.T. Frankel, and M-U. Yang. Maryland, Aspen, 1988.
91. Krotkiewski, M., et al.: The effect of unilateral isokinetic strength training on local adipose and muscle tissue morphology, thickness and enzymes. *Eur. J. Appl. Physiol.*, 22:221, 1979.
92. Krotkiewski, M., et al.: Impact of obesity on metabolism in men and women: importance of regional adipose tissue distribution. *J. Clin. Invest.*, 72:1150, 1983.
92a. Krotkiewski, M., et al.: Increased muscle dynamic endurance associated with weight reduction on a very-low-calorie diet. *Am. J. Clin. Nutr.*, 51:321, 1990.
93. Lambert, O., and Hansen, E.S.: Effects of excessive caloric intake and caloric restriction on body weight and energy expenditure at rest and light exercise. *Acta Physiol. Scand.*, 114:135, 1982.
94. Larsson, B., et al.: Abdominal adipose tissue distribution, obesity, and risk of cardiovascular disease and death: 13 year follow-up of participants in the study of newborn in 1913. *Br. Med. J.*, 288:1401, 1984.
95. LaVau, J.: Reliable photomicrographic method of determining fat cell size and number: application to dietary obesity. *Proc. Soc. Exp. Biol. Med.*, 156:251, 1977.
96. Leibel, R.L., and Hirsh, J.: Diminished energy requirements in reduced-obese patients. *Metabolism*, 33:164, 1984.

97. Leibel, R.L., et al.: Alterations in adipocyte free fatty acid reesterfication associated with obesity and weight reduction in man. *Am. J. Clin. Nutr.*, 42:198, 1985.
98. Lennon, D., et al.: Diet and exercise training effects on resting metabolic rate. *Int. J. Obesity*, 9:39, 1985.
99. Leon, A.S., et al.: Effects of a vigorous walking program on body composition, and carbohydrate and lipid metabolism of obese young men. *Am. J. Clin. Nutr.*, 32:1776, 1979.
100. Lewis, S.B., et al.: Effect of a diet composition on metabolic adaptations to hypocaloric nutrition: comparison of high carbohydrate and high fat isocaloric diets. *Am. J. Clin. Nutr.*, 30:160, 1977.
101. Mabee, T.M., et al.: The mechanism of increased gallstone formation in obese human subjects. *Surgery*, 79:460, 1976.
101a. Manson, J.E., et al.: A prospective study of obesity and risk of coronary heart disease in women. *N. Engl. J. Med.*, 322:822, 1990.
102. Marliss, B.: Protein diets for obesity: metabolic and clinical aspects. *Can. Med. Assoc. J.*, 119:1413, 1978.
103. McArdle, W.D., and Magel, J.R.: Weight management: diet and exercise. In *The Medical Aspects of Clinical Nutrition*. Edited by J. Bland, and N. Shealy. New Canaan, CT, Keats, 1983.
104. McArdle, W.D., and Toner, M.M.: Application of exercise for weight control: the exercise prescription. In *Obesity and Weight Control*. Edited by R.T. Frankle, and M-U. Yang. Rockville, MD, Aspen, 1988.
105. Meredith, C.N., et al.: Body composition and aerobic capacity in young and middle-aged endurance-trained men. *Med. Sci. Sports Exerc.*, 19:557, 1987.
106. Milesis, C.A., et al.: Effects of different durations of physical training on cardiorespiratory function, body composition and serum lipids. *Res. Q.*, 47:716, 1976.
107. Miller, P.M., and Sims, K.L.: Evaluation and component analysis of a comprehensive weight control program. *Int. J. Obesity*, 5:57, 1981.
108. Molé, P.A., et al.: Exercise reverses depressed metabolic rate produced by severe caloric restriction. *Med. Sci. Sports Exerc.*, 21:29, 1989.
109. Mueller, W.H.: The genetics of human fatness. *Yearbook Phys. Anthropol.*, 26:215, 1983.
110. National Institutes of Health: Health implications of obesity, NIH Consensus Development Conference Statement, Vol. 5, No. 9. U.S. Government Printing Office, 1985.
111. Newmark, S.R., and Williamson, B.: Survey of very low calorie weight reduction diets. I. Novelty diets. *Arch. Intern. Med.*, 143:1195, 1983.
112. Newmark, S.R., and Williamson, B.: Survey of very low calorie weight reduction diets. II. Total fasting, protein-sparing modified fasts, chemically defined diets. *Arch. Intern. Med.*, 143:1423, 1983.
113. Noland, M., and Kearney, J.T.: Anthropometric and densitometric responses of women to specific and general exercise. *Res. Q.*, 49:322, 1978.
114. Oscai, L.B., and Holloszy, J.O.: Effects of weight changes produced by exercise, food restriction, or overeating on body composition. *J. Clin. Invest.*, 48:2124, 1969.
115. Oscai, L., et al.: Effects of exercise and of food restriction on adipose tissue cellularity. *J. Lipid Res.*, 13:588, 1972.
116. Oscai, L., et al.: Exercise or food restriction: effect on adipose tissue cellularity. *Am. J. Physiol.*, 227:901, 1974.
116a. Ostlund, R.E. et al.: The ratio of waist-to-hip circumference, plasma insulin level, and glucose intolerance as independent predictors of the HDL_2 cholesterol level in older adults. *N. Engl. J. Med.*, 332:229, 1990.
117. Pavlov, K.N., et al.: The effects of dieting and exercise on lean body mass, oxygen uptake, and strength. *Med. Sci. Sports Exerc.*, 17:446, 1985.
118. Pavlov, K.N., et al.: Exercise as an adjunct to weight loss and maintenance in moderately obese subjects. *Am. J. Clin. Nutr.*, 49:1115, 1989.
119. Pi-Sunyer, F.X.: Exercise in the treatment of obesity. In *Obesity and Weight Control*. Edited by R.T. Frankle, and M-U. Yang. Rockville, MD, Aspen, 1988.
120. Poehlman, E.T.: Exercise and its influence on resting energy metabolism in man. *Med. Sci. Sports Exerc.*, 5:515, 1989.
121. Pollock, M.L., et al.: Effects of mode of training on cardiovascular function and body composition of adult men. *Med. Sci. Sports*, 7:139, 1975.
122. Rabkin, S.W., et al.: Relation of body weight to the development of ischemic heart disease

in a cohort of young North American men after a 26 year observation period: the Manitoba study. *Am. J. Cardiol.*, 39:452, 1977.

123. Ravussin, E., et al.: Reduced rate of energy expenditure as a risk factor for body-weight gain. *N. Engl. J. Med.*, 318:467, 1988.

124. Rissanen, V.: Coronary and aortic atherosclerosis in relation to body-build factors. *Ann. Clin. Res.*, 7:402, 1975.

125. Roberts, S.B., et al.: Energy expenditure and intake in infants born to lean and overweight mothers. *N. Engl. J. Med.*, 318:461, 1988.

126. Rolland-Cachera, M.F., and Bellisle, F.: No correlation between adiposity and food intake: why are working class children fatter? *Am. J. Clin. Nutr.*, 44:779, 1986.

127. Rosen, L.W., and Hough, D.O.: Pathogenic weight-control behaviors of female college gymnasts. *Phys. Sportsmed.*, 16:141, 1988.

128. Ross, J.G., and Pate, R.R.: The National Children and Youth Fitness Study. II: a summary of findings. *JOPERD*, 58:51, 1987.

129. Ross, C.E., and Mirowsky, J.: Social epidemiology of obesity: A substantive and methodological investigation. *J. Health and Social Behavior*, 24:288, 1983.

130. Rossi, M., et al.: Cardiac autonomic dysfunction in obese subjects. *Clin. Sci.*, 76:567, 1989.

131. Rossner, S., and Hallberg, D.: Serum lipoproteins in massive obesity. *Acta Med. Scand.*, 204:103, 1978.

132. Salans, L.B., et al.: Experimental obesity in man: cellular character of the adipose tissue. *J. Clin. Invest.*, 50:1005, 1971.

133. Segal, K.R., and Pi-Sunyer, F.X.: Exercise, resting metabolic rate, and thermogenesis. *Diab./Metab. Rev.*, 2:19, 1986.

133a. Segal, K.R., et al.: Thermic effects of food and exercise in lean and obese men of similar lean body mass. *Am. J. Physiol.*, 252 (Endocrinol. Metab. 19): E110, 1987.

134. Segal, K.R., et al.: Body composition, not body weight, is related to cardiovascular disease risk factors and sex hormone levels in man. *J. Clin. Invest.*, 80:1050, 1987.

135. Segal, K.R., et al.: Impact of body fat mass and percent fat on metabolic rate and thermogenesis in men. *Am. J. Physiol.*, 256 (Endocrinol. Metab. 19): E573, 1989.

136. Simopoulos, A.P.: Obesity and carcinogenesis: historical perspective. *Am. J. Clin. Nutr.*, 45:271, 1987.

137. Sims, E.A.H., and Danforth, E., Jr.: Expenditure and storage of energy in man (perspective). *J. Clin. Invest.*, 79:1019, 1987.

138. Sims, E.A.H., and Horton, E.S.: Endocrine and metabolic adaptation to obesity and starvation. *Am. J. Clin. Nutr.*, 21:1455, 1968.

139. Slattery, M.L., and Jacobs, D.R., Jr.: The interrelationships of physical activity, physical fitness, and body measurements. *Med. Sci. Sports Exerc.*, 19:564, 1987.

140. Sonka, J.: Effects of diet or diet and exercise in weight reducing regimens. In *Nutrition, Physical Fitness, and Health.* Edited by J. Parizkova, and V. Rogozkin. Baltimore, MD, University Park Press, 1978.

141. Spitzer, L., and Rodin, J.: Human eating behavior: A critical review of studies in normal weight and overweight individuals. *Appetite*, 2:293, 1981.

142. Stalonas, P.M., et al.: Behavior modification for obesity: the evaluation of exercise, contingency management and program adherence. *J. Consult. Clin. Psychol.*, 46:463, 1978.

143. Stamler, R., et al.: Weight and blood pressure. Findings in hypertension screening of 1 million Americans. *JAMA*, 240:1607, 1978.

144. Stern, J.S., et al.: Pancreatic insulin release and peripheral tissue resistance in Zucker obese rats fed high and low carbohydrate diets. *Am. J. Physiol.*, 228:543, 1975.

145. Stordy, B.J., et al.: Weight gain, thermic effect of glucose and resting metabolic rate during recovery from anorexia nervosa. *Am. J. Clin. Nutr.*, 30:138, 1977.

146. Stunkard, A.J.: An adoption study of human obesity. *N. Engl. J. Med.*, 314:193, 1986.

147. Stunkard, A.J., and McLaren-Hume, M.: The results of treatment of obesity: a review of the literature and report of a series. *Arch. Intern. Med.*, 103:79, 1959.

148. Thompson, D.A., et al.: Acute effects of exercise intensity on appetite in young men. *Med. Sci. Sports Exerc.*, 20:227, 1988.

149. Thorn, G.W., et al.: *Harrison's Principles of Internal Medicine.* 8th ed. New York, McGraw-Hill, 1977.

149a. Thornton, J.S.: Feast or famine: eating disorders in athletes. *Phys. Sportsmed.*, 18:116, 1990.

150. Trembly, A., et al.: The effects of exercise training on energy balance and adipose tissue morphology and metabolism. *Sports Med.*, 2:223, 1985.

151. Trembly, A., et al.: The effect of exercise training on resting metabolic rate in lean and moderately obese individual. *Int. J. Obesity*, 10:511, 1986.

151a. Trembly, A., et al.: Exercise training with constant energy intake. 2: Effect on glucose metabolism and resting energy expenditure. *Int. J. Obesity*, 14:75, 1990.

152. Udall, J.G., et al.: Interaction of maternal and neonatal obesity. *Pediatrics*, 62:17, 1978.

153. Vaccaro, P., et al.: Physiological characteristics of masters female distance runners. *Phys. Sportsmed.*, 9:105, 1981.

154. Vaccaro, P., et al.: Body composition and physiological responses of Masters female swimmers 20 to 70 years of age. *Res. Q. Exerc. Sport*, 55:278, 1984.

155. Warner, W.W.: The obese patient and anesthesia. *JAMA*, 205:102, 1968.

156. Vague, J., et al.: The various forms of obesity. *Triangle. Sandoz J. Med. Sci.*, 13:41, 1974.

156a. VanDale, D., and Saris, W.H.M. Repetitive weight loss and weight regain: effects on weight reduction, resting metabolic rate, and lipolytic activity before and after exercise and/or diet treatment. *Am. J. Clin. Nutr.*, 49:409, 1989.

157. Van Itallie, T.B.: Health implications of overweight and obesity in the United States. *Ann. Intern. Med.*, 103:983, 1985.

158. Wadden, T.A.: Very low calorie diets: Their efficacy, safety, and future. *Ann. Intern. Med.*, 99:675, 1983.

159. Wadden, T.A., et al.: Less food, less hunger: reports of appetite and symptoms in a controlled study of a protein-sparing modified fast. *Int. J. Obesity*, 11:239, 1987.

160. Welle, S.L.: Some metabolic effects of overeating in man. *Am. J. Clin. Nutr.*, 44:718, 1986.

161. Weisinger, J.R., et al.: The nephrotic syndrome: a complication of massive obesity. *Ann. Intern. Med.*, 50:233, 1974.

162. Wilmore, J.H.: Appetite and body composition consequent to physical activity. *Res. Q. Exerc. Sport*, 54:415, 1983.

163. Wolley, S.C., et al.: Theoretical, practical, and social issues in behavioral treatments of obesity. *J. Appl. Behav. Anal.*, 12:3, 1979.

164. Woo, R.: Voluntary food intake during prolonged exercise in obese women. *Am. J. Clin. Nutr.*, 36:478, 1982.

165. Woo, R., et al.: Effect of exercise on spontaneous caloric intake in obesity. *Am. J. Clin. Nutr.*, 36:470, 1982.

166. Young, J.C., et al.: Prior exercise potentiates the thermic effect of a carbohydrate load. *Metabolism*, 35:1048, 1986.

167. Zuti, W.B., and Golding, L.A.: Comparing diet and exercise as weight reduction tools. *Phys. Sportsmed.*, 4:49, 1976.

PHYSICAL ACTIVITY, HEALTH, AND AGING

The elderly represent the fastest growing segment of the American population, with the average life expectancy for men and women rapidly approaching 80 years. Aside from merely extending life, it is crucial to more fully understand the factors that can modify and upgrade one's quality of life. In this chapter, we explore physical activity as a health-related behavior with specific reference to aging and cardiovascular disease. We also present concepts about physical activity epidemiology and their application to health and longevity.

PHYSICAL ACTIVITY EPIDEMIOLOGY

Epidemiology is concerned with quantifying the rate of health-related events that occur within a group with the ultimate goal of generalizing this information to a larger population in hopes of controlling or modifying health problems.[75,125] The field of **physical activity epidemiology** takes on added importance when epidemiology is linked to studying the association of physical activity as a health-related behavior with disease and other outcomes.[24]

TERMINOLOGY

In applications of physical activity epidemiology, specific definitions are used to characterize the activity behavioral patterns of the people under investigation. These definitions take on a fuller meaning when evaluated within the context of health and disease:[25]

> **Physical activity**—any body movement produced by muscles that results in increased energy expenditure
> **Exercise**—physical activity that is planned, structured, repetitive, and purposeful
> **Physical fitness**—a set of attributes that relate to the ability of people to perform physical activity
> **Health**—physical, mental, and social well-being, not simply the absence of disease
> **Health-related physical fitness**—components of physical fitness that are associated with some aspect of health. These important factors in health-related fitness include cardiorespiratory endurance, muscular endurance, muscular strength, body composition, and joint flexibility
> **Longevity**—length of life

Within this framework, physical activity is a generic term with exercise as the major component. Similarly, the definition of health focuses on the broad spectrum of well-being that ranges from complete absence of health (death) to the highest levels of functional capability. Although such definitions often challenge our ability to measure and quantify, they do offer a broader frame of reference to understand both physical activity and exercise as they relate to a healthy life-style.

ASSESSMENT OF AND PARTICIPATION IN PHYSICAL ACTIVITY

More than 30 different methods have been used to assess physical activity, including direct and indirect calorimetry, self-reporting, job classifications, survey procedures, physiologic markers, behavioral observations, mechanical or electronic monitors, and dietary measures. Each approach has unique advantages and disadvantages depending on the situation and the population studied. It is difficult to obtain valid estimates of physical activity in large populations because such studies, by necessity, make use of self-reports of daily activity and exercise participation.

The current status of participation in physical activity of adult North Americans is not encouraging. According to data from the U.S. National Center for Health Statistics on the physical activity of noninstitutionalized adults aged 18 years and older, only 8.1% of men and 7% of women reported that they engaged in regular vigorous exercise. Regular but less-intense activity was done by 36.2% of the men and 31.5% of the women, indicating that only 44.3% males and 38.5% of females engage in some regular physical activity.[26]

Figure 30–1 illustrates the findings for exercise participation from a recent large-scale study of over 15,000 adults enrolled in exercise programs that included a variety of aerobic and muscle-strengthening activities.[17] Participants' ages ranged from 19 to 64 years and were grouped into 5 age categories. With increasing age there was a progressive and predictable decline in par-

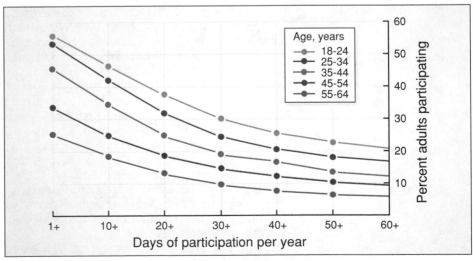

FIG. 30–1.
Percentage of adults in the United States grouped by age, who participate in fitness activities. (Courtesy of C. Brooks, Department of Sports Management and Communication, University of Michigan, Ann Arbor, 1989.)

ticipation in fitness activities. The percentage of the group that participated regularly in exercise was quite small, especially for the older individuals. Such findings are complemented by a report that revealed only 14% of American adults expended more than 1600 kcal per week in leisure-time activities, and only 10% participated in vigorous activity on a regular basis.[18]

It appears that at best no more than 20%, and possibly less than 10%, of the adults in the United States obtain sufficient regular and vigorous physical activity to impart discernible health and fitness benefits. Clearly, there is a real need to improve the physical activity profile of the U.S. population as a whole. With this in mind, the Public Health Service has published a report generally referred to as the "1990 Objectives" that outlines national objectives for health promotion and disease prevention.[26] In the area of physical fitness and exercise, 11 objectives pertain to such topics as physical activity habits, awareness of the fitness benefits of physical activity, worksite programs, and the possible health benefits of exercise. By focusing on the interaction between various government, medical, and recreational agencies, one of the 1990 Objectives is to increase to 60% the number of adults who participate in regular vigorous physical activity.[100] If this ambitious goal is to be achieved, new strategies must be applied that encourage women, the elderly, persons in the lower socioeconomic strata, and minorities to adopt exercise as a way of life.

IS EXERCISE SAFE?

Many people have raised this question largely because of several well-publicized reports of sudden death during exercise.[48,72,117,127] In actuality, sudden death rates during exercise have **declined** over the past 20 years even though there has been an overall increase in exercise participation. In one report of cardiovascular episodes during 65 months, 2935 exercisers recorded 374,798 hours of exercise that included 2,726,272 km of running and walking. There were no deaths during this time and only 2 nonfatal cardiovascular complications. The rates were 3 complications per 100,000 hours of exercise for men and 2 per 100,000 hours for women. For apparently healthy adults, the potential benefits of regular physical activity far outweigh any potential cardiovascular complications. Perhaps not surprising, the most prevalent exercise complications are musculoskeletal in nature. In a longitudinal study of aerobic dance injuries for 351 participants and 60 instructors at 6 dance facilities, 327 medical complaints were reported during nearly 30,000 hours of activity.[47] Only 84 of the injuries resulted in disability (2.8 per 1000 person hours of participation), and just 2.1% of the injuries required medical attention. Age does not appear to be a factor in the occurrence of orthopedic problems related to moderate exercise.[79]

Although exercise is safe for the majority of adults, there is still a need for further studies of the potential risks, especially if we are to encourage markedly increased rates of participation consistent with the 1990 Objectives. Furthermore, a clearer knowledge of injury rates, types of injuries, and conditions under which they occur will enhance the development of appropriate preventive strategies.

One of the recurring themes in physical activity epidemiology is the impact of aging on the development of health risk. In the following sections we explore several aspects of the aging process with reference to exercise and its relation to the development of cardiovascular disease, and exercise performance.

AGING AND PHYSIOLOGIC FUNCTION

Physiologic and performance measures generally improve rapidly during childhood and reach a maximum between the late teens and 30 years of age. Functional capacity then declines with age. Although all measures decline with age, not all decline at the same rate. Nerve conduction velocity, for example, declines only 10 to 15% from 30 to 80 years of age, whereas resting cardiac index (ratio of cardiac output to surface area) declines 20 to 30%; maximum breathing capacity at age 80 is about 40% that of a 30-year-old. In addition, some functions (such as ventilation) that show only a moderate aging effect at rest may show dramatic changes under the stress of exercise. Because long-term exercise studies on the same subjects are scarce, it is not known the extent to which regular exercise participation can change the actual rate of decline in physiologic function or "override" deterioration in function that normally occurs with increasing age.[121] What is known is that there is an alarmingly large number of individuals with such poor functional capacity that they cannot do relatively simple physical tasks without assistance.[32]

MUSCULAR STRENGTH

Maximum strength of men and women is generally achieved between the ages of 20 and 30 years,[28,90] at a time when muscular cross-sectional area is usually the largest. Thereafter, there is a progressive decline in strength for most muscle groups.

There is a decline of at least 16.5% in muscular strength after the third decade of life.[40] This strength loss is directly related to limited mobility and physical performance as well as to increases in the incidence of accidents suffered by those with muscle weakness. The loss of hand grip strength in males by age 65 is about 20% compared to values for 20 year olds; for females, the loss over the same 45 years ranges from 2 to 20%.[13,23] The lesser decline in females could be attributable to a relatively lower peak strength in the younger years due to less occupational use of the hands. Other data from individuals beyond age 65 suggest that the loss of strength further accelerates with aging, with the overall age-related strength loss ranging from 24 to 45%.[74,92]

Decrease in Muscle Mass

A reduced muscle mass is a primary factor responsible for the age-associated loss of strength that reflects a loss of total muscle protein brought about by inactivity, aging, or both.[62] The loss of muscle volume may be due to a reduced fiber size, particularly in the fast-twitch (type II) fibers.[4,5,129] A reduced size of these specific fibers would result in a proportionate increase in slow-twitch (type I) area. There may also be an actual reduction in the total number of muscle fibers.[78] Electromyographic studies indicate there is a loss of functioning motor neurons in the elderly.[19] This could account for a loss of muscle fibers because denervation leads to fiber atrophy and eventual replacement by connective tissue.

Strength Trainability Among the Elderly

Habitual physical training facilitates protein retention and can delay the decrement in lean body mass and strength with aging.[85] In a recent study,[45]

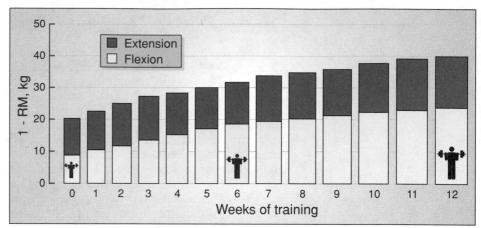

FIG. 30–2.
Weekly measurements of dynamic muscle strength (1-RM) of left knee extension (blue) and flexion (yellow). (From Frontera, W.R., et al.: Strength conditioning in older man: skeletal muscle hypertrophy and improved function. *J. Appl. Physiol., 64:*1038, 1988.)

healthy men between the ages of 60 and 72 years were trained for 12 weeks with a standard resistance training program that used loads equivalent to 80% of the 1-RM. As shown in Figure 30–2, muscular strength increased progressively throughout the training. By week 12, knee flexion strength had increased by 107% and knee extension strength by 227%! This rate of strength improvement (about 5% per training session) was similar to increases previously reported for young adults. In addition, these dramatic strength improvements were accompanied by significant hypertrophy of both fiber types. **Such findings clearly indicate an impressive plasticity in physiologic, structural, and performance characteristics, and that marked and rapid improvement can be achieved with vigorous training, at least into the 7th decade of life.** Improvement in muscular strength by resistance training may be the best way to prevent or reduce the incidence of injury with older individuals.[121]

Table 30–1 summarizes relevant literature on changes in muscle strength and hypertrophy in older adults who participated in various programs of resistance training. In most instances, training responses were positive and strength improvements ranged between 1.9 and 72% in individuals beyond age 60. Possible mechanisms to explain these often impressive training responses include neural (recruitment and innervation) and muscular (hypertrophy) factors previously discussed in Chapter 22. As with younger adults, the number of sets and repetitions, and the intensity, duration, and frequency of training, are crucial in determining strength adaptations in the elderly.

NEURAL FUNCTION

The cumulative effects of aging on central nervous system function are exhibited by a 37% decline in the number of spinal cord axons, a 10% decline in nerve conduction velocity, and a significant loss in the elastic properties of connective tissue.[8,116] These changes may partially explain the age-related

TABLE 30–1.
RESULTS FROM STRENGTH TRAINING STUDIES IN OLDER INDIVIDUALS

STUDY	AGE (yr)	GENDER	TRAINING TYPE	STRENGTH GAIN (%)	HYPERTROPHY (%)
1	62–84	M + F	RT	57	NM
2	18–26	M	RT	30	9
	67–72	M	RT	22	NC
3	69–74	M	CT	9–22	NC
4	22–65	M	CT	3–8	19–20
5	63–84	F	RT	7–13	10
6	60–72	M	RT	9–19	9–12
1	62–84	M + F	Iso	45.8	NM
7	41–80	M	Iso	12–24	NM
8	20–26	F	Iso	95	NM
	65–73	F	Iso	72	NM
9	55–78	M + F	CV	1.9–13.4	NM
10	51–87	M	CV(6 wk)	6.4	NC
		M	CV(42 wk)	11.9	1.0
11	65	M + F	CV	5–13	NM
12	65	M	Hdr	15–132	NC

RT = resistance training, CT = circuit training, Iso = isometric training, CV = cardiovascular endurance training, Hdr = hydraulic resistance training, NC = no change, NM = not measured.

1. Perkins, L.C., and Kaiser, H: *Phys. Ther. Rev.,* 41:633, 1961.
2. Moritaini, T., and deVries, HA: *J. Gerontol.,* 35:672, 1980.
3. Aniansson, A., and Gustafsson, E.: *Clin. Physiol.,* 1:87, 1981.
4. Larsson, L.: *Med. Sci. Sports Exerc.,* 19:203, 1982.
5. Aniansson, A., et al.: *Arch. Gerontol. Geriat.,* 3:229, 1984.
6. Frontera, W.F., et al.: *J. Appl. Physiol.,* 64:1038, 1988.
7. Liemohm, W.P.: *Int. J. Aging Hum. Dev.,* 6:347, 1975.
8. Kauffman, T.L.: *Arch. Phys. Med. Rehab.,* 66:223, 1985.
9. Barry, A.J., et al.: *J. Gerontol.,* 21:192, 1966.
10. deVries, H.A.: *J. Gerontol.,* 25:325, 1970.
11. Sidney, L.H., et al.: *Am. J. Clin. Nutr.,* 30:326, 1977.
12. Becque, M.D.: Ph.D. Dissertation, University of Michigan, 1989.

decrement in neuromuscular performance as assessed by both simple and complex reaction and movement times.[29,119] When reaction time is partitioned into central processing time and muscle contraction time, the central processing time is affected most by the aging process. Thus, aging affects the ability to detect a stimulus and process the information to produce a response. Because reflexes, such as the knee jerk reflex, do not involve processing in the brain, they are less affected by the aging process than voluntary responses, i.e., reaction and movement times.[30] As shown in Figure 30–3, movement times for simple and complex tasks were significantly slower for older subjects than for younger counterparts of the same activity level. **In all instances, however, the active groups (be they young or old) moved significantly faster than a corresponding age group that was less active.** These observations suggest that an active lifestyle may significantly and positively affect neuromuscular functioning at any age. Older men who have remained active for 20 years or longer have reaction times that are equal to or faster than inactive men in their 20s.[120] It is tempting to speculate that the biologic

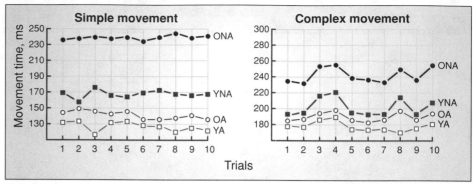

FIG. 30–3.
Simple and complex movement time in young active (YA), old active (OA), young nonactive (YNA), and old nonactive (ONA) subjects. Note that in both simple and complex movement tasks, the movement time is slower for the old and young nonactive individuals than in the young and old active individuals. (From Spirduso, W.W.: Reaction and movement time as a function of age and physical activity level. J. Gerontol., *30:*435, 1975.)

aging of certain neuromuscular functions can be somewhat slowed by regular participation in physical activity.

PULMONARY FUNCTION

Both static and dynamic measures of lung function generally deteriorate with age.[69,91] How regular exercise throughout one's lifetime can override this "aging" of the pulmonary system is unknown, although older endurance-trained athletes have greater pulmonary functional capacity than sedentary peers. For example, athletes over the age of 60 have values for vital capacity, total lung capacity, residual volume, maximum voluntary ventilation, $FEV_{1.0}$, and $FEV_{1.0}/FVC$ consistently larger than expected based on their body size, and significantly larger than those of sedentary, age-matched, healthy counterparts.[53] Such findings are encouraging; they indicate that a lifetime of regular physical activity may retard the decline in pulmonary function associated with aging.

CARDIOVASCULAR FUNCTION

Aerobic Capacity

It has been estimated from cross-sectional data that the decline in max $\dot{V}O_2$ is approximately $0.4 \text{ ml} \cdot \text{kg}^{-1} \cdot \text{min}^{-1}$ each year.[35] This estimate may be somewhat high because a clear difference exists in the rate of decline in max $\dot{V}O_2$ with aging between sedentary versus active individuals.[65a] **Sedentary individuals have nearly a two-fold faster rate of decline in max $\dot{V}O_2$ as they age.**[21,22]

Many confounding factors influence the age-related decline in max $\dot{V}O_2$. Heredity undoubtedly plays an important role, as does the well-documented decrease in muscle mass.[41] It may be that the normal decline in aerobic capacity follows a two-component curve: one portion of the curve represents a faster rate of decline in sedentary adults from age 20 to 30 years compared to adults who are physically active; thereafter, a slower rate of decline is observed for both groups.[23]

Central and Peripheral Function

For both the active and sedentary person, the decline in aerobic capacity is influenced by the various age-related decrements in both central and peripheral physiologic functions linked to oxygen transport and utilization.[103]

Heart Rate One well-documented change in cardiovascular function with age is a decline in the maximal heart rate.[51] This apparent age-effect is progressive with advancing years and occurs to the same extent in both men and women. A rough approximation of the maximal heart rate is expressed by the following relationship:

$$\text{max HR (beats} \cdot \text{min}^{-1}) = 220 - \text{age (years)}$$

Cardiac Output As a consequence of a lower maximum heart rate, maximum cardiac output may be reduced with age. Also contributing to this reduced blood flow capacity in some individuals is a reduction in the heart's stroke volume that may reflect changes in myocardial contractility, although it does appear that contractile function can be well maintained.[134] Consequently, in the absence of disease, a decrease in maximum cardiac output does not always occur. The healthy elderly may compensate for a diminished heart rate response with an increased cardiac filling (end diastolic volume) that leads to a subsequent increase in stroke volume by the Frank-Starling mechanism.[106,112]

Local Factors Other age-related changes in the cardiovascular system include a decrease in muscle mass and a reduction in peripheral blood flow capacity. This reduction may be due to a decrease in the capillary-to-muscle fiber ratio and a reduction in arterial cross-sectional area.[98]

Whether the changes in cardiovascular function are a direct result of the aging process per se or of a lack of habitual physical activity has not been determined. **In fact, sedentary living may bring about losses in functional capacity that are as great as the effects of aging itself.** Results from training studies indicate a high degree of trainability among older men and women with adaptations similar to younger individuals.[53,54,114] Both low and high intensity regular exercise enables older individuals to retain cardiovascular functioning much above age-paired sedentary subjects.[6,21,57,86] When previously active middle-aged men followed a regular endurance exercise program over a 10-year-period, the usual 9 to 15% decline in work capacity and maximal aerobic power was forestalled.[65] In fact, at age 55, these active men had maintained the same values for blood pressure, body weight, and max $\dot{V}O_2$ as at age 45.

Exceptional Performance Capacity

Further evidence for the dramatic effects of exercise training on the preservation of cardiovascular function throughout life can be observed by comparing the performance times on endurance-type events for individuals of different ages. Figure 30–4 presents a plot of age-group, world-record mar-

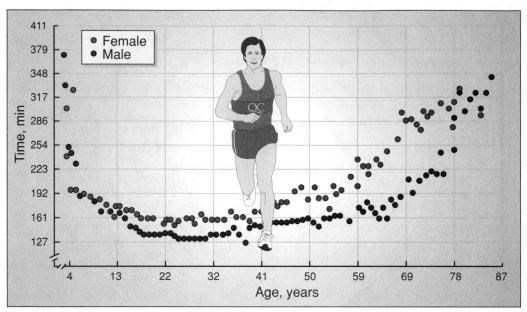

FIG. 30–4.
Plot of world record times for the marathon for men and women of different ages.

athon times for males and females starting at age 4 for males and at age 5 for females up to age 86 for males and age 80 for females. The world record of 2 hr 7 min 11 sec by Carlos Lopes (aged 37) of Portugal (recorded in April 1985) corresponds to a running speed of 4 min 51 sec per mile (12.4 mph). As can be seen, with the very young and as people get older, endurance performance sharply decreases. Whereas from age 30 to 86 the decrease in performance is progressive and significant, the 86-year-old world record time for the male of 340.2 minutes corresponds to a 12.9 min per mile pace, and for the 80-year-old female the world record time is 328.6 minutes and corresponds to a 12.5 min per mile pace. This rate of running for 26.2 miles is truly remarkable for individuals in their eighth decade of life and attests to the tremendous cardiovascular capabilities of older individuals who continue to train on a regular basis as they grow older.

BODY COMPOSITION

After age 35, men and women tend to progressively add weight until the fifth or sixth decade of life.[97] Figure 30–5 clearly shows this trend in percent body fat for men and women of different ages.

After age 60, total body mass is reduced despite increasing body fat. This is partly explained because in the upper age group many of the grossly overweight people have died and there are just not many heavy subjects to be measured. Also, lean body mass does tend to decrease with age. This is largely due to the aging skeleton that becomes demineralized and porous; concurrently, there is a reduction in total muscle mass. The extent is unknown of how regular physical activity can retard these changes in body density.

A major limitation of age-trend studies is that the same subjects are not followed over time, but rather different subjects in different age categories

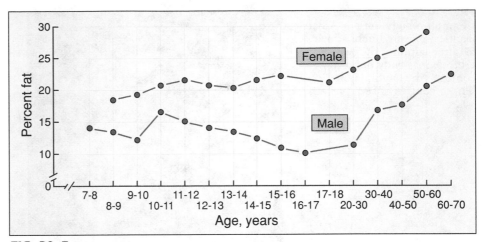

FIG. 30–5.
Body fat changes as a function of age. (From Parizkova, J.: Body composition and exercise during growth and development. In Physical Activity: Human Growth and Development. Edited by G.L. Rarick. New York, Academic Press, 1974.)

are evaluated at the same time. From these **cross-sectional** data, one attempts to generalize as to expected age-related changes for an individual. Sometimes these generalizations are misleading. For example, today's 70 and 80-year-olds are generally shorter than 20-year-old college students. This observation does not necessarily mean that we get shorter as we grow older (although this does happen to some extent). Rather, the young adults of this generation are better nourished than their 80-year-old counterparts were at the age of 20. In terms of body fat changes, the limited **longitudinal** data from the same subjects tend to support the trends noted in cross-sectional studies. When the fat content of 27 adult men was studied over a 12-year span from age 32 to 44 years, fat weight increased an average of 6.5 kg which was equal to the total gain in body mass over the 12 years.[27]

Although it is common to see that most "normal" individuals grow fatter with increasing age, individuals who engage in heavy resistance training increase their lean body mass and decrease body fat. Figure 30–6 shows an example of an "older" resistance trained athlete and demonstrates quite dramatically that it is possible to maintain and even gain muscle mass with increasing age. Indeed, it is a universal observation that individuals who engage in heavy resistance training seem to defy certain aspects of the normal aging process. In this sport there is certainly great truth to the statement, "use it or lose it."

BONE MASS

Osteoporosis is a major problem in aging. This condition results in a loss of bone mass, increased bone porosity, and a decrease in the thickness of bone cortex. For people over 60 years of age, these alterations in aging bone can reduce the bone mass by 30 to 50%.[11] These changes in bone density may invalidate the equations that underly the use of hydrostatic weighing for body fat prediction and may result in gross inaccuracies in predicting the body fat content of older individuals. As mentioned in Chapter 2, regular

FIG. 30—6.
Champion body builder Albert
Beckles, pictured here at age
52 after winning the world
professional body building
contest, Mr. Olympia, in 1984.
(Photo courtesy of M. Neveux)

weight-bearing exercise may not only retard bone loss, but can actually
increase bone mass in the elderly, even in postmenopausal women.[34,71,102]

TRAINABILITY AND AGE

**Regular vigorous physical activity produces physiologic improvements
regardless of age.** Of course, the magnitude of the changes depends on several
factors that include initial fitness status, age, and the specific type of
training.[99] With regard to the age factor, it has generally been held that older
individuals are not able to improve their strength and endurance capacity
to the same extent as younger people.[110] The reasons for this decreased
"trainability" were not well understood, although it was attributed to a
general decline in neuromuscular function and the age-related impairment
in the cell's capability for protein synthesis and chemical regulation.
Figure 30–7 shows the classic view of the improvement to be expected
from physical conditioning for people of different ages in relation to their
initial fitness level at the start of training. Essentially, when a person, young
or old, has a relatively high functional capacity at the start of training, there
is less room for improvement compared to someone who starts at a lower
level and has considerable room for improvement. At the same time, ability
to improve may be age-related; older persons have often shown less improvement when they begin to train later in life than younger counterparts

FIG. 30–7.
Traditional theoretical representation of the improvement that might be expected with conditioning relative to age and initial level of functioning.

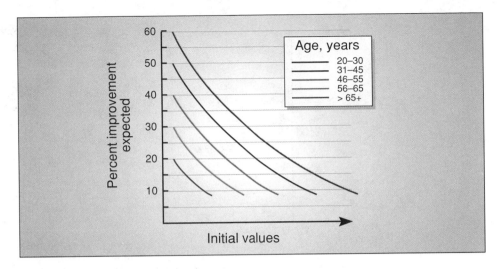

who start training at the same initial level of fitness. As mentioned in the beginning sections of this chapter, however, **large and rapid improvement in physiologic function can occur in the healthy elderly, often at the rate and magnitude recorded for younger individuals.**[45,107,114,126] This occurs with relatively intense training that is continuously adjusted to keep pace with training improvements. Recent research indicates that healthy older adults show no negative metabolic or hormonal responses or maladaptations to regular exercise that would contraindicate participation in a standard exercise-training program.[52]

EXERCISE AND LONGEVITY

Because older fit individuals have many of the functional characteristics of younger people, one could argue that improved physical fitness may help retard the aging process and thus offer some protection to health in later life.

In one of the first studies of the possible beneficial influence of sport and regular exercise on prolonging life,[58] former Harvard oarsmen exceeded their predicted longevity by 5.1 years per man. Earlier studies showed similar but more modest results.[3] These studies, however, were plagued with methodologic problems, including inadequate record keeping, small sample size, improper statistical procedures for estimating expected longevity, and an inability to account for other important factors such as socioeconomic background, body type, cigarette smoking, and family background. One group of researchers attempted to overcome many of these limitations in their study of the diseases and longevity of former college athletes.[88] Because collegiate athletes usually have a longer involvement in habitual physical activity prior to entering college than nonathletes, and because they may remain more physically active after college,[89] this seemed to be an excellent group to study to provide insight concerning exercise and longevity.

Figure 30–8 shows that essentially no difference existed in the longevity of the ex-athletes compared to their nonathletic counterparts. Some degree of equality in genetic background existed between the groups because the

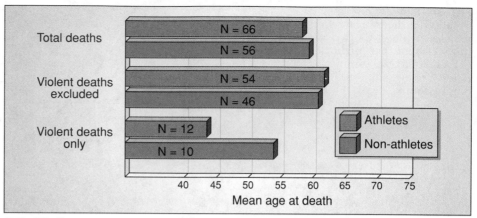

FIG. 30–8.

Age at death of athletes and nonathletes. None of the differences between the groups is statistically significant. (From Montoye, H.J., et al.: The Longevity and Morbidity of College Athletes. Indianapolis, Ind., Phi Epsilon Kappa, 1957.)

average age at death was also similar of grandparents, parents, and siblings of exathletes and nonathletes. These and more recent findings[16] suggest that participation in athletics only as a young adult does not necessarily ensure increased longevity. **It appears, however, that regular physical activity throughout life offers protection in terms of health and longevity.**[12,96]

ENHANCED QUALITY TO A LONGER LIFE: A STUDY OF HARVARD ALUMNI

Research concerning the lifestyles and exercise habits of 17,000 Harvard alumni who entered college between 1916 and 1950 gives strong evidence that only **moderate** aerobic exercise, equivalent to jogging about 3 miles a day, promotes good health and may actually add years to life.[96] The results of these long-term studies can be summarized as follows:

> Regular exercise countered the life-shortening effects of cigarette smoking and excess body mass. Even for people with high blood pressure, those who exercised regularly reduced their death rate by one-half. Genetic tendencies toward an early death were countered by regular exercise. For individuals who had one or both parents die before age 65 (a significant health risk), a lifestyle that included regular exercise reduced the risk of death by 25%. A 50% reduction in mortality rate was observed for those whose parents lived beyond 65 years.

As shown in Figure 30–9, the person who exercised more had an improved health profile. For example, the mortality rates were 21% lower for men who walked 9 or more miles a week than men who walked 3 miles or less. Performing the equivalent of light sport activity increased life expectancy over men who remained sedentary. From the perspective of energy expenditure, the life expectancy of Harvard alumni increased steadily from an exercise energy expenditure of 500 kcal per week to 3500 kcal; this was equivalent to 6 to 8 hours of strenuous weekly exercise. In addition, active

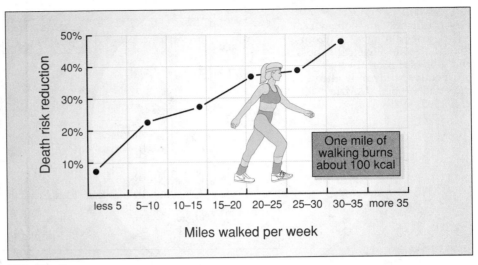

FIG. 30–9.
Reduced risk of death with regular exercise. (From Paffenbarger, R.S., Jr., et al.: Physical activity, all-cause mortality, and longevity of college alumni. *N. Engl. J. Med., 314*:605, 1986.)

men lived an average of 1 to 2 years longer than sedentary classmates. Beyond weekly exercise of 3500 kcal, there were **no additional** health or longevity benefits. When exercise was carried to extremes, the men had higher death rates than their more moderately active colleagues.

EPIDEMIOLOGIC EVIDENCE

A critique of 43 studies of the relationship between physical inactivity and coronary heart disease concluded that lack of regular activity contributes to the heart disease process in a cause-and-effect manner, with the sedentary person almost twice as likely to develop heart disease as the most active individual.[101] **The strength of this protective association was essentially the same as that observed for hypertension, cigarette smoking, and high serum cholesterol.** In the researchers' opinion, this placed physical inactivity as the greater heart disease risk considering that more people lead sedentary lifestyles than possess one or more of the other risks. Although vigorous exercise does provide a small risk of sudden death during the activity, the significant longer-term health benefits of regular activity far outweigh any potential acute risk.[117]

From available data, it appears that if life-extending benefits of exercise exist, they are more associated with the prevention of early mortality than an improvement in overall life span. While the maximum life span may not be extended greatly, more active people tend to survive to that "ripe old age."[60,101] That only moderate exercise is needed to achieve these benefits is further good news.

IMPROVED FITNESS: A LITTLE GOES A LONG WAY

A recent study of more than 13,000 men and women followed for an average of 8 years indicates that even modest amounts of exercise substantially reduce

FIG. 30–10.
Physical fitness and longevity: a little goes a long way. (From Blair, S., et al.: Physical fitness and all-cause mortality: a prospective study of healthy men and women. *JAMA,* 262:2395, 1989.)

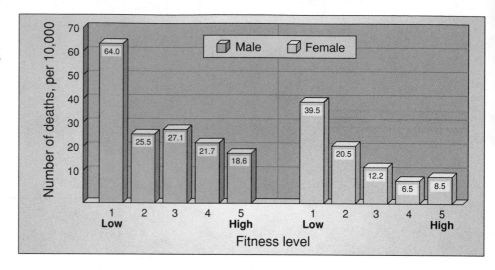

the risk of dying from heart disease, cancer, and other causes.[12] This is one of the few studies that has looked directly at fitness performance rather than verbal or written reports of regular physical activity habits. To isolate the effect of fitness per se, the study considered such factors as smoking, cholesterol and blood sugar levels, blood pressure, and family history of CHD. Based on age-adjusted death rates per 10,000 person-years, Figure 30–10 illustrates that death rates of the least fit group were more than 3 times that of the most fit. The most striking finding was that the greatest health benefit occurred in the group rated just above those in the most sedentary category. For men, the drop in death rate from the least fit category to the next fit was more than 38 (64.0 vs. 25.5 deaths per 10,000 person-years), whereas the drop in mortality from moving to the second group up to the most fit category was only 7. Similar benefits were found for women. The amount of exercise required to move from the most sedentary category to the next most fit—the jump that showed the greatest health benefits—occurred for moderate intensity exercise such as walking briskly for 30 minutes several times a week.

CORONARY HEART DISEASE

Coronary heart disease (CHD) generally involves degenerative changes in the intima or inner lining of the larger arteries that supply the heart muscle.

CHANGES ON THE CELLULAR LEVEL

The action and chemical modification of various compounds, including the cholesterol in low-density lipoproteins, initiate a complex process that ultimately causes bulging lesions in the arterial wall. These initially take the form of fatty streaks, the first signs of **atherosclerosis.** With further damage and proliferation of underlying cells the vessel becomes progressively congested with either lipid-filled plaques or fibrous scar tissue, or both. This

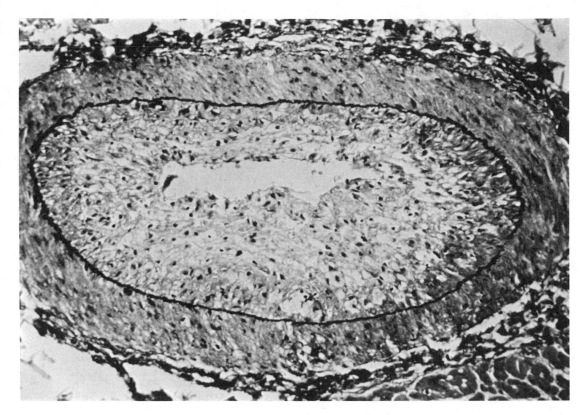

FIG. 30–11.
Deterioration of a normal coronary artery is seen as atherosclerosis develops with
the beginning of deposits of fatty substances that roughen the vessel's center. The
clot then forms and plugs the artery, depriving the heart muscle of vital blood.
The result is a myocardial infarction or heart attack.

change progressively reduces the capacity for blood flow and causes the
myocardium to become **ischemic**—that is, poorly supplied with oxygen due
to reduced blood flow. Figure 30–11 shows the progressive occlusion of an
artery with a buildup of calcified fatty substances in the process of atherosclerosis. In this degenerative process that often begins early in life,[93,123]
the roughened, hardened lining of the coronary artery frequently causes the
slowly flowing blood to clot. This blood clot or **thrombus** may plug one of
the smaller coronary vessels. In such cases, a portion of the heart muscle
dies and the person is said to have suffered a heart attack or **myocardial
infarction.** If the blockage is not too severe but blood flow is still reduced
below the heart's requirement, the person may experience temporary chest
pains termed **angina pectoris.** These pains are felt usually during exertion,
because this causes the greatest demand for myocardial blood flow. Such
anginal attacks provide painful and dramatic evidence of the importance of
adequate oxygen supply to this vital organ.

STILL AN EPIDEMIC

Coronary heart disease has reached epidemic proportions throughout the
United States and most technologically advanced societies. It is responsible

for more than 1.5 million myocardial infarctions and causes more than 550,000 deaths annually in the United States. While deaths from CHD have declined more than a third since 1970, it still remains the **leading cause of death** in the U.S. and other developed countries. Beginning at about age 40 for men and age 60 for women, CHD is the single largest cause of death in the Western world. For example, about twice as many people die from CHD as from cancer. As depicted in Figure 30–12 after age 35 in males and age 45 in females, the chance of dying from CHD increases progressively and dramatically. Between ages 55 and 65, about 13 of every 100 men, and about 6 of every 100 women die from coronary artery disease.

Almost all people show some evidence of coronary artery disease, and it can be severe in seemingly healthy young adults. Actually, the disease probably starts early in life because fatty streaks are common in the coronary arteries of children by the age of 5 years! There seems to be little harm, however, unless a marked narrowing of the arteries is present. **At rest, the heart's blood supply becomes deficient only when obstruction of the coronary vessels reaches 80%. In fact, at least 50 to 70% occlusion must take place before the disease can be clinically detected.** Generally, death from CHD occurs when there is advanced obstruction in several major vessels supplying the myocardium.

CORONARY HEART DISEASE RISK FACTORS

Significant information has been provided on the natural history and dynamics of heart disease. Various personal characteristics and environmental factors have been identified over the past 30 years that appear to play causitive roles in making individuals more susceptible to CHD. The following is a list of the more frequently implicated **risk factors** that can be used to identify those at high or low risk for CHD, even in childhood:[93] (1) age and gender;

FIG. 30–12.
The chance of a single individual dying from coronary atherosclerosis. (Data from the American Heart Association.)

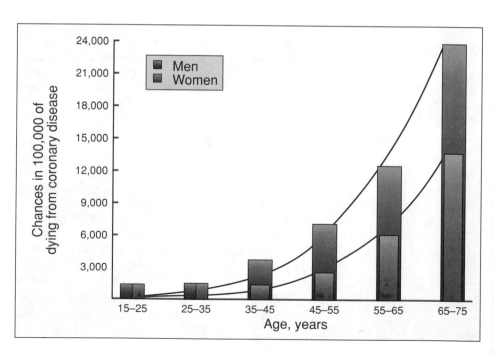

(2) elevated blood lipids; (3) hypertension; (4) cigarette smoking; (5) physical inactivity; (6) obesity; (7) diabetes mellitus; (8) diet; (9) heredity; (10) personality and behavior patterns; (11) high uric acid levels; (12) pulmonary function abnormalities; (13) race; (14) electrocardiographic abnormalities during rest and exercise; (15) family history; and (16) tension and stress.

It is difficult to determine quantitatively the importance of any single CHD risk factor in comparison to any other because many of the factors are interrelated. For example, blood lipid abnormalities, diabetes, heredity, and obesity often go hand-in-hand. One research study reported that men living in Ireland consumed more saturated fat than their blood brothers who lived in the United States, yet the former had a much lower incidence of CHD.[130] This protection was attributed to higher physical activity levels for those living in Ireland. Similar findings of high saturated fat intake, high physical activity, and low incidence of CHD have been reported for Masai tribesmen of East Africa and farm laborers in Georgia.[80,83]

In the following sections, we discuss blood lipid abnormalities, obesity, cigarette smoking, and physical inactivity as potent yet modifiable CHD risk factors that are influenced by lifestyle and health habits. Hypertension, another important risk factor, was discussed in Chapter 15. These factors were chosen because they have received the greatest public attention. All are powerful and consistent predictors in assessing one's chances for developing heart disease. It must be noted, however, that although these factors are closely associated with CHD, the associations do not necessarily infer causality. In many instances, it still remains to be shown that risk factor modification offers effective protection from the disease. Until definite proof is demonstrated, however, it seems logical to assume that elimination or reduction of one or more risk factors will cause a corresponding decrease in the probability of contracting CHD.[124a]

Blood Lipid Abnormalities

Table 30–2 presents levels of serum cholesterol above which young and older adults should seek advice on treatment. In general, a cholesterol of 200 mg · dl^{-1} or lower is considered desirable.[50] A cholesterol value of 230 mg · dl^{-1} increases the risk of heart attack to about twice that of a person with 180 mg · dl^{-1}, and a value of 300 mg · dl^{-1} increases the risk 4-fold. An increased lipid level in the blood is termed **hyperlipidemia.** Cholesterol and triglycerides are the two most common lipids associated with this CHD

TABLE 30–2.
DESIRABLE LEVEL OF TOTAL CHOLESTEROL (mg · dl^{-1})
AND LEVELS ABOVE WHICH ADULTS SHOULD RECEIVE TREATMENT

Age	Goal	Moderate Risk (75th percentile)	High Risk (90th percentile)
20–29	<180	>200	>220
30–39	<200	>220	>240
40 and over	<200	>240	>260

(Adapted from National Institutes of Health Consensus Development Conference Statement: Lowering Blood Cholesterol to Prevent Heart Disease, *JAMA*, 253:2080, 1985.)

risk. These fats do not circulate freely in the blood plasma, but rather are transported in combination with a carrier protein to form a **lipoprotein.** Table 30–3 lists the four different lipoproteins, their approximate density, and percent composition in the blood. Serum cholesterol represents a composite of the total cholesterol contained in the different lipoproteins. Although it is proper to refer to hyperlipidemia, it is more meaningful to evaluate and discuss the different types of **hyperlipoproteinemia.**

The precise mechanism by which elevated blood lipids affect the development of CHD is complex and not fully understood. Nevertheless, the overwhelming evidence seems to link high levels of blood lipids, especially low density lipoproteins, with increased incidence of CHD. In many cases, these elevated lipids are related to consuming diets high in saturated fats and cholesterol.

The distribution of cholesterol among the various types of lipoproteins may be a more powerful predictor of heart disease than simply the total quantity of plasma lipids. Specifically, a high level of high-density lipoproteins (HDL, which comprise the smallest portion of lipoproteins, carry about 20% of the total cholesterol, and contain the largest quantity of protein) is associated with a lower heart disease risk, even among individuals with total cholesterol below 200 mg · dl^{-1}; elevated levels of the low-density lipoproteins (LDL and VLDL) represent an increased risk. An effective way to evaluate lipoprotein status is to divide total cholesterol by HDL cholesterol. A ratio greater than 4.5 indicates a high heart disease risk, while a ratio of 3.5 or lower may be optimal.[81]

Although controversy exists as to the precise role of lipoproteins in heart disease, it is generally believed that the LDL and VLDL are means for transporting fat throughout the body for delivery to the cells, including those of the smooth muscle walls of the arteries. Here it ultimately becomes involved in the artery-narrowing process of atherosclerosis. Whereas LDL is targeted for peripheral tissue and is associated with arterial damage, HDL reflects the removal aspect of lipid dynamics by promoting the movement of cholesterol from peripheral tissues (including arterial walls) for transport to the liver where it is excreted through bile synthesis.

Research is currently progressing to clarify the mechanism by which HDL (and its subfractions) is protective and what factors can raise them. It is encouraging from an exercise perspective that HDL levels are elevated in endurance athletes and may be favorably altered in sedentary people who engage in either vigorous aerobic training[82,128] or more moderate levels of

TABLE 30–3.
APPROXIMATE COMPOSITION OF LIPOPROTEINS IN THE BLOOD

	Chylomicrons	Very Low Density Lipoproteins (VLDL: Prebeta)	Low-Density Lipoproteins (LDL: Beta)	High-Density Lipoproteins (HDL: Alpha)
Density, g.cc^{-1}	0.95	0.95–1.006	1.006–1.019	1.063–1.210
Protein, %	0.5–1.0	5–15	25	45–55
Lipid, %	99	95	75	50
Cholesterol, %	2–5	10–20	40–45	18
Triglyceride, %	85	50–70	5–10	2
Phospholipid, %	3–6	10–20	20–25	30

regular exercise.[31,137] These favorable alterations in the blood lipid profile take place independently of changes in body mass,[128,131] and may be related to an enhanced triglyceride clearance from plasma in response to regular exercise.[109] Even among trained endurance athletes, however, the variability is quite high in HDL cholesterol. Some runners have HDL cholesterol values near the median value for the general population.[108] Moreover, no single factor—nutritional, body compositional, or related to the level of training—distinguished runners with higher HDL values from those with lower HDL values. This suggests that genetic factors may be the most important influence on blood lipid status.

Obesity

Although excess body fat has received great notoriety as a CHD risk factor, its relationship is often co-dependent with such factors as hypertension, diabetes mellitus, and cigarette smoking[61] Autopsy studies have not revealed a strong association between body fatness per se and the degree of atherosclerosis.[118] Research indicates, however, that overfat individuals are often hypertensive and have elevated serum lipid levels[61] Weight loss, and accompanying fat reduction whether through diet or exercise, generally normalize cholesterol and triglyceride levels and have a beneficial effect on blood pressure and adult-onset diabetes. Although being too fat may not be a primary CHD risk factor, its role cannot be denied as a secondary and contributing factor in heart disease and various other disease processes (see Chapter 29).

Cigarette Smoking

Cigarette smoking may be one of the best predictors of CHD and the risk is directly related to the quantity of cigarettes smoked. The probability of death from heart disease for smokers is almost twice as great as for non-smokers.[77a] The increase in death rate from heart disease among women in this country almost parallels their increased consumption of cigarettes.[136] Surprisingly, this CHD risk is associated with more deaths than the excess mortality rate of cigarette smokers due to lung cancer!

It is generally observed that the smoking risk acts independently of other risk factors. At the same time, however, if other risk factors are present, cigarette smoking accentuates their influence. If smoking is stopped, the risk of CHD often returns to that of nonsmokers.[43] Cigarette smoking may increase heart disease risk through its effect on serum lipoproteins; individuals who smoke have lower levels of HDL cholesterol compared to nonsmokers. The good news is that if smoking is stopped the HDL can return to normal levels.[122]

Physical Inactivity

In terms of CHD, the general consensus among the more recent investigations is that a consistent relationship emerges between a sedentary lifestyle and an increased chance for heart disease. The relative risk of fatal CHD among the sedentary is about twice that of more active individuals. For both men and women, the maintenance of physical fitness throughout life also provides significant protection in terms of both risk factors and the occurrence of actual disease.[1,12,36] It could be argued that one's level of physical fitness is more related to genetic factors and somewhat less to exercise patterns in daily life. Because fitness level is strongly related to the person's reported activity level, however, regular physical activity may be even more important than genetics in determining fitness and consequent health benefits.

Mechanisms of Protection   It has been suggested that many of the conclusions concerning the singular beneficial effects of exercise "outrun" the facts.[59] **Although the data on physical inactivity and CHD generally fall short of critical "proof," there is certainly no evidence that the prudent use of exercise is harmful.** In fact, the major bulk of research on animals and humans indicates that regular aerobic exercise may operate against CHD in a variety of beneficial ways to:

1. Improve myocardial circulation and metabolism to protect the heart from hypoxic stress; this includes enhanced vascularization, as well as modest increases in cardiac glycogen stores and glycolytic capacity that could be beneficial when the heart's oxygen supply is compromised.[70,76,77,111]
2. Enhance the mechanical or contractile properties of the myocardium to enable the conditioned heart to maintain or increase contractility during a specific challenge.
3. Possibly establish more favorable blood clotting characteristics and other hemostatic mechanisms.[14,39]
4. Normalizes the blood lipid profile.[5,38,113]
5. Favorably alter heart rate and blood pressure so the work of the myocardium is significantly reduced at rest and during exercise.[94]
6. Achieve a more desirable body composition.[61,66]
7. Establish a more favorable neural-hormonal balance to conserve oxygen for the myocardium.[101]
8. Provide a favorable outlet and response pattern to psychologic stress and tensions.[33]

INTERACTION OF CHD RISK FACTORS

Many risk factors are associated with each other as well as with CHD itself.[64] Figure 30–13 shows the effect of combining risk factors on the incidence of CHD. Those with no risk factors have a CHD incidence of about 60 per 100,000 persons. With one risk factor the risk is almost doubled, whereas possessing 3 risk factors increases the incidence rate almost 10-fold! Clearly, the relationship of risk factors to CHD incidence is simply not additive but rather related in an exponential manner. **The interaction of risk factors most definitely magnifies their singular effect.** An important consideration is that many of the risk factors have a common root in health behavior patterns, and consequently can be influenced by similar, and in some cases, identical interventions. For example, regular physical activity can exert a positive influence on the risk factors of obesity, hypertension, and elevated lipid profiles.

RISK FACTORS IN CHILDREN

Several studies have documented multiple CHD risk factors in young children.[35a,49,135] The prevalence of risk factors for active and apparently healthy

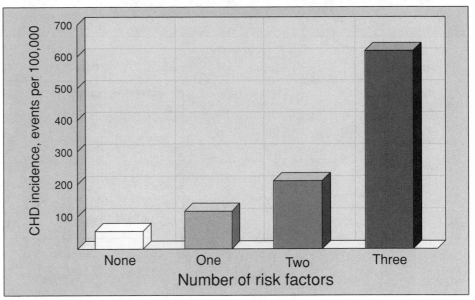

FIG. 30–13.
Relationship between a combination of abnormal risk factors (cholesterol > 250 mg · dl^{-1}; systolic blood pressure > 160 mmHg; smoking > 1 pack a day), and incidence of coronary heart disease. (Adapted from Kannel, W.B., and Gordon, T.: *The Framingham Study: An epidemiological investigation of cardiovascular disease.* Section 30. Washington, DC, Public Health Service, NIH, DHEW Publication #74–599, Feb. 1974.)

boys and girls aged 7 to 12 years is shown in Table 30–4. Obesity and a family history of heart disease were the two most frequently occurring risk factors.[49] A relatively large percentage of children also showed abnormally high blood lipids. Of the total group, 65% had at least one or two risk factors whereas 31% had three or more! As is generally the case with adults, the association between body fatness and serum lipid levels becomes readily apparent in subjects classified as obese; the fattest children generally have the highest levels of cholesterol and triglycerides.

Because of the prevalence of CHD risk factors in some children, plus autopsy observations of young adults and children, it seems likely that heart disease has its origins in childhood. School-based programs aimed at reducing risk factors in adolescents are effective for increasing knowledge about risk factors and daily level of physical activity compared to students not receiving such a program.[68] It remains to be shown whether the benefits from such "risk intervention" will improve the child's health outlook and can be carried into adulthood to improve overall health. Certainly, if regular physical activity can modify or at least stabilize the risk factor profile, and if this protects against CHD, then all children should be encouraged and taught to pursue active lifestyles. **It is our strong belief that such behaviors should be taught in the elementary school grades, and that physical education should be a required course taught on a daily basis.**

The aforementioned studies were conducted on normal, healthy children, and the results are probably predictive of the general population of children. It is possible, however, that risk factor development would be different for obese children, or children who already have one or more primary risk factors.

TABLE 30–4.
PREVALENCE OF CHD RISK FACTORS IN BOYS AND GIRLS AGED 7 TO 12*

Risk Factor	Prevalence		Total	N	Percent in Total Sample
	Male	**Female**			
Obesity (>20% body fat)	10	4	14	47	30
Low work capacity (<31 ml · kg^{-1} · min^{-1})	3	1	4	34	12
Elevated blood lipids					
Cholesterol (>200 mg%)	1	3	4	38	10
Triglycerides (>100 mg%)	4	3	7	38	18
Lipoprotein classification					
Type II	1	1	2	38	5
Type IV	4	3	7	38	18
Family history of CHD	7	5	12	47	26

* From Gilliam, T., et al.: Prevalence of coronary heart disease risk factors in active children, 7 to 12 years of age. Med. Sci. Sports, 9:21, 1977.

In an evaluation of CHD risk factors in a sample of 62 obese children, aged 10 to 15, only one child had just one risk factor[9]! Fourteen percent had two risk factors, 30% had three risk factors, 29% had four risk factors, 18% had five risk factors, and five children or 8% of the obese group had six risk factors! Some of these obese children were then enrolled in a 20-week experiment to evaluate the effects of diet + behavior therapy or exercise + diet + behavior therapy on CHD risk. Results showed no differences in multiple risk reduction for either the control or diet + behavior group. In contrast, the exercise + diet + behavior therapy group experienced a dramatic reduction in the multiple risk factor profile shown in Figure 30–14. Such findings are encouraging because they show that a moderate program of food restriction, exercise, and behavior change can significantly decrease the CHD risk of obese adolescents. By far, regular, moderately intense exercise was the more potent intervention that caused changes in the risk factor profile.

CALCULATING THE RISK

Many attempts have been made to quantify an individual's susceptibility for CHD; this has led to the development of CHD risk inventories, commonly called "risk appraisals." Most risk inventories assign different point values to different aspects of a person's lifestyle. Usually the point assignments are rather arbitrary and are not based on conclusive research that shows actual risk (mortality or morbidity) quantitatively related to a given lifestyle trait. Thus, such risk inventories should be used and interpreted with caution, and should be used only as a general screen.

In Table 30–5, we present a popular risk inventory, RISKO, developed by the Michigan Heart Association. To determine your own risk profile, study each risk factor and the accompanying numerical value that best describes your status. Find the box applicable to you and circle the number in it. For example, if you are 19 years old, circle the number one in the box labeled 10 to 20 years. After checking all the rows, add the circled numbers. The

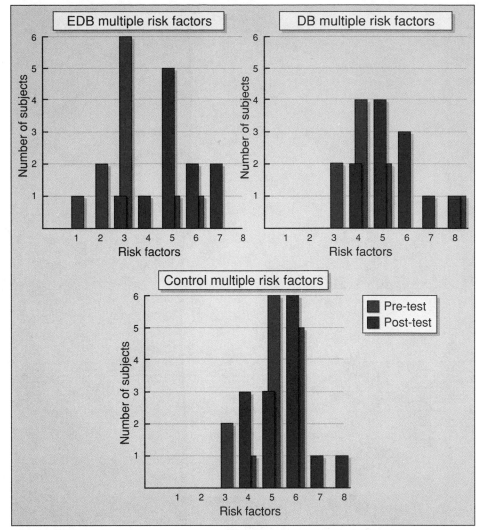

FIG. 30–14.
Multiple coronary heart disease risk factors for obese adolescents before and after treatment. DB = diet + behavior change group; EDB = exercise + diet + behavior change group. (From Becque, M.D., et al.: Coronary risk incidence of obese adolescents: reduction by exercise plus diet intervention. Pediatrics, *81*:605, 1988.)

total number of points is your risk score. Refer to Table 30–5 for your **Relative Risk Category.**

EXERCISE STRESS TESTING

Without doubt, vigorous aerobic exercise increases the functional capacity of the circulatory system. In addition, it is now generally accepted that this type of exercise can serve important protective and rehabilitative functions in the battle against CHD. The potential therapeutic and fitness benefits of

RELATIVE RISK CATEGORY

Score	Relative Risk Category
6–11	Risk well below average
12–17	Risk below average
18–24	Average risk
25–31	Moderate risk
32–40	High risk
41–62	Very high risk, see your physician

TABLE 30–5.
CHD RISK APPRAISAL, RISKO

AGE	10 to 20 **1**	21 to 30 **2**	31 to 40 **3**	41 to 50 **4**	51 to 60 **6**	61 to 70 and over **8**
HEREDITY	No known history of heart disease **1**	1 relative with cardio-vascular disease over 60 **2**	2 relatives with cardio-vascular disease over 60 **3**	1 relative with cardio-vascular disease under 60 **4**	2 relatives with cardio-vascular disease under 60 **6**	3 relatives with cardio-vascular disease under 60 **7**
WEIGHT	More than 5 lb below standard weight **0**	−5 to +5 lb standard weight **1**	6–20 lb over weight **2**	21–35 lb over weight **3**	36–50 lb over weight **5**	51–65 lb over weight **7**
TOBACCO SMOKING	Non-user **0**	Cigar and/or pipe **1**	10 cigarettes or less a day **2**	20 cigarettes a day **4**	30 cigarettes a day **6**	40 cigarettes a day or more **10**
EXERCISE	Intensive occupational and recreational exertion **1**	Moderate occupational and recreational exertion **2**	Sedentary work and intense recreational exertion **3**	Sedentary occupational and moderate recreational exertion **5**	Sedentary work and light recreational exertion **6**	Complete lack of all exercise **8**
CHOLESTEROL OR FAT % IN DIET	Cholesterol below 180 mg/dl Diet contains no animal or solid fats **1**	Cholesterol 181–205 mg/dl Diet contains 10% animal or solid fats **2**	Cholesterol 206–230 mg/dl Diet contains 20% animal or solid fats **3**	Cholesterol 231–255 mg/dl Diet contains 30% animal or solid fats **4**	Cholesterol 256–280 mg/dl Diet contains 40% animal or solid fats **5**	Cholesterol 281–300 mg/dl Diet contains 50% animal or solid fats **7**
BLOOD PRESSURE	100 upper reading **1**	120 upper reading **2**	140 upper reading **3**	160 upper reading **4**	180 upper reading **6**	200 or over upper reading **8**
GENDER	Female under 40 **1**	Female 40–50 **2**	Female over 50 **3**	Male **4**	Stocky male **6**	Bald stocky male **7**

Explanation of variables: *Heredity*—count parents, brothers, and sisters who have had a heart attack or stroke; *Smoking*—if you inhale deeply and smoke a cigarette way down, add one point to your score. Do not subtract because you think you do not inhale or smoke only a half inch on a cigarette; *Exercise*—lower your score one point if you exercise regularly and frequently; *Cholesterol/Saturated Fat Intake*—a cholesterol blood level is best. If you have not had a blood test recently, then estimate honestly the percentage of solid fats you eat. These are usually of animal origin—lard, cream, butter, and beef and lamb fat. If you eat much saturated fat, your cholesterol level will probably be high; *Blood Pressure*—if you have no recent reading but have passed an insurance or general medical examination, chances are you have a systolic blood pressure level of 140 or less; *Gender*—this takes into account the fact that men have from 6 to 10 times more heart attacks than women of child-bearing age. (Adapted from the Michigan Heart Association.)

exercise, however, should be viewed in perspective. For a sedentary person with significant, undetected coronary heart disease, a sudden burst of strenuous exercise with its concomitant catecholamine release could place an inordinate strain on the cardiovascular system. This risk can be reduced considerably with proper medical evaluation that should minimally include a thorough personal health and family history, and physical examination. The physical examination should emphasize signs and symptoms of cardiovascular disease and include blood pressure, resting 12-lead ECG, cardiac murmurs and dysrhythmias, edema, blood analysis, and chronic lung disease. For many people, an important part of the medical evaluation is the **exercise stress test.**

The term "stress test" is generally used to describe the systematic use of exercise for two main purposes: (1) for ECG observations, and (2) to evaluate the adequacy of physiologic adjustments to metabolic demands that exceed the resting requirement. The test can be a single-stage test such as the **Master "2-step"** bench stepping test where work load remains constant throughout the exercise period. Multistage bicycle and treadmill tests, however, are the most popular means for stress testing. These tests are graded in terms of physical work. They include several levels of 3 to 5 minutes of submaximal exercise and may often bring the person to a self-imposed fatigue level called the **maximum physical working capacity** *or* **PWC max.** The submaximal work levels allow work to be increased in small increments until ischemic manifestations (e.g., anginal pain or ECG abnormalities) make their appearance. If heart disease is detected, this provides for a more precise manipulation of work load and gives a reliable and quantitative index of the person's functional impairment. This enhances the accuracy of diagnosis and subsequent exercise prescription. For most screening purposes, the test need not be maximal, but it is desirable to bring a person to at least 85% of the age-predicted maximal heart rate.

A resting electrocardiogram should precede the exercise test. The ECG establishes that the person can engage safely in graded exercise, and also provides the important baseline to compare exercise results. For diagnostic purposes, the stress test provides the means for obtaining important electrocardiographic information. The exercise ECG often picks up signs of subtle coronary disease that include myocardial ischemia and cardiac rhythm disorders. Stress testing is limited, however, by its inability to show the extent and specific location of the disease. Also, 25 to 40% of the people with relatively advanced CHD demonstrate **false-negative** readings (normal stress test but abnormal coronary angiogram; see p. 729).

For the person with documented cardiovascular disease, the exercise stress test provides an objective and reliable means to define the patient's limitations and capacity for exercise and to evaluate specific therapy. This aspect of stress testing also provides the basis for prudent exercise prescription for healthy and coronary-prone adults as well as for patients with CHD. With proper medical clearance, it is not out of the question for individuals with documented CHD to train safely for and complete marathons.[67]

WHY USE STRESS TESTS?

There are at least six reasons to include stress testing in an overall CHD evaluation:

1. **To establish, from ECG observations, a diagnosis of overt heart disease and also to screen for possible "silent" coronary disease in seemingly**

healthy men and women. Approximately 30% of the people with confirmed coronary artery disease have normal resting electrocardiograms. During relatively intense exercise, however, about 80% of these abnormalities are uncovered.

2. **To reproduce and assess exercise-related chest symptoms.** In many instances, individuals over the age of 40 suffer chest or related pain in the left shoulder or arm on physical exertion. Proper electrocardiographic analysis during an exercise stress test helps identify myocardial abnormalities and provides a more precise diagnosis of exercise-induced pain.

3. **To screen candidates for preventive and cardiac rehabilitative exercise programs.** Stress test results can then be used to design an exercise program that is within the person's current functional capacity and health status, with emphasis on intensity, frequency, duration, and type of exercise. Repeated testing aids in evaluating progress in the exercise intervention program as well as in determining the need for safe program modification.

4. **To detect an abnormal blood pressure response.** It is not uncommon to find individuals with a normal resting blood pressure who show higher than normal increases in systolic blood pressure with exercise. This exercise hypertension may signify developing cardiovascular complications.

5. **To monitor responses to various therapeutic interventions (drug, surgical, and dietary) designed to improve cardiovascular functioning.** For example, the success of coronary bypass surgery can be detected by a patient's adjustment to exercise and ability to successfully reach a target exercise heart rate without complications.

6. **To define the functional aerobic capacity and evaluate its degree of deviation from normal standards.**

WHO SHOULD BE STRESS-TESTED?

Table 30–6 shows a classification system by age and health status for screening and supervisory procedures for use in conjunction with an exercise stress test. The prudent rules are:

1. **If a person is less than 35 years of age and has no previous history of cardiovascular disease and no known primary risk factors (and has had a medical evaluation within the past 2 years), it is generally acceptable to begin an exercise program without special medical clearance.** These people may also be stress tested for purposes of functional evaluation and for preparing the exercise prescription by a trained exercise specialist.

2. **If a person is younger than 35 years of age but has evidence of CHD or a significant combination of risk factors, he or she should be medically cleared prior to embarking on an exercise program.** This should include a physician supervised graded exercise test.

3. **For all adults above 35 years of age, medical evaluation is advised prior to any major increase in exercise habits.** This medical evaluation should include an ECG monitored before, during, and in recovery from a graded exercise test and supervised by a physician.

These standards conform to policies and practices of the American College of Sports Medicine and the American Medical Association.[2]

EXERCISE-INDUCED INDICATORS OF CHD

Several clues to CHD become apparent during exercise because exercise creates the greatest demand on coronary blood flow.

TABLE 30–6.
CLASSIFICATION BY AGE AND HEALTH STATUS FOR MEN AND WOMEN WHO REQUIRE DIFFERENT SCREENING AND SUPERVISORY PROCEDURES PRIOR TO AND DURING A STRESS TEST

Age	Patient Health Status	Evaluation and Required Medical Clearance	12-Lead Resting ECG	Personnel Involved During Stress Test
<35	No known primary CHD risk factors;[a] may have secondary CHD risk factors[b]	During past 2 years, signed statement	Not required	No test required
35–40	No known primary CHD risk factors; may have secondary CHD risk factors	During past 2 years, signed statement	Required	Exercise technician; exercise physiologist; M.D. in area
>40	No known primary CHD risk factors; may have secondary CHD risk factors	During past 2 years, signed statement	Required	Exercise technician; exercise physiologist; M.D. in area
Any age	Documented CHD: hypertension; suspected CHD	During past 2 years, signed statement	Required	Exercise technician; exercise physiologist; M.D. conducting test

[a] Primary risk factors: Hypertension, hyperlipidemia, and cigarette smoking.
[b] Secondary risk factors: Family history, obesity, physical inactivity, diabetes mellitus, and hyperglycemia.

Angina Pectoris

Approximately 30% of the initial manifestations of CHD take the form of chest-related pain called angina pectoris. This is a temporary but painful condition that indicates that coronary blood flow (oxygen supply) has momentarily reached a critically low level. This myocardial ischemia (usually the result of restricted coronary circulation brought about by coronary atherosclerosis) stimulates sensory nerves in the walls of the coronary arteries and myocardium itself. The resulting pain or discomfort is generally felt in the upper chest region, although it is frequently characterized by a sensation of pressure or constriction in the left shoulder, neck, jaw, or left arm. Many people also report sensations of "being smothered" during an angina episode. In addition to pain, cardiac performance is also impaired with angina. This depressed myocardial function is accompanied by reduced cardiac output, reduced stroke volume, and generally impaired contractility of the left ventricle. After a few minutes of rest, the pain usually subsides with no permanent damage being done to the heart muscle.

Electrocardiographic Disorders

Alterations in the heart's normal pattern of electric activity are often indicative of insufficient oxygen supply to the myocardium. These electric "clues," however, are rarely observed until the metabolic (and blood flow) requirements are increased above the resting level.

The ECG tracing illustrated in Figure 30–15 is a plot of the dynamic electric activity of the heart muscle in millivolts (mV). Normal ECG paper is divided into small 1-mm squares and larger 5-mm squares. Horizontally, each small square is equivalent to 0.04 seconds (with standard paper speed of 25 mm/sec); each large square represents 0.2 seconds. On the vertical axis each small square indicates a 0.1-mV deflection with a calibration of 10 mm/mV. A

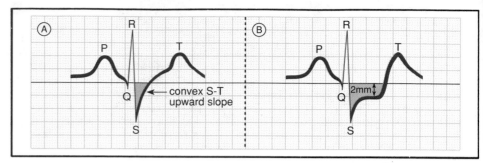

FIG. 30–15.
A tracing that illustrates the normal ECG complex. The arrow points to the slightly convex and upward sloping S-T segment. B, Tracing shows an abnormal horizontal S-T-segment depression (shaded area) of 2 mm measured from a stable baseline.

normal single heart beat or **cardiac cycle** consists of five major electric waves: P, Q, R, S, and T (Fig. 30–15A). The P wave is the electric impulse or wave of depolarization associated with atrial contraction. The Q, R, and S waves are considered as a unit that represents the depolarization and subsequent contraction of the ventricles. Collectively this is known as the QRS complex. The T wave is generated by the repolarization of the ventricles. Although it is not known why the S-T segment becomes depressed, this deviation from normal is closely correlated with other indicators of CHD including coronary artery narrowing. Those individuals with significant S-T-segment depression usually have severe and extensive obstruction of the coronary arteries. This generally involves a reduction of more than 70% of the normal opening of one or more coronary vessels. In addition, the amount of S-T segment depression is related directly to the chances of dying from CHD. In one study, persons with 1- to 2-mm S-T segment depression during exercise had a 4.6-fold increase in mortality from CHD, whereas those with more than 2-mm depression had a 19.1-fold greater chance of dying from CHD.[104]

Cardiac Rhythm Abnormalities

Although exercise can induce abnormalities in the S-T segment of the electrocardiogram, it also provides an effective way to observe both normal and abnormal cardiac rhythm. One significant alteration in cardiac rhythm (**arrythmia**) with exercise is the occurrence of **premature ventricular contractions** or **PVCs.** In this situation, the ventricles demonstrate disorganized electric activity. They are not stimulated by the normal passage of the wave of depolarization through the atrioventricular node. Rather, portions of the ventricle become spontaneously depolarized. This shows up on the electrocardiogram as an "extra-ventricular beat" or QRS complex that occurs without being preceded by a P wave that indicates atrial depolarization.

Premature ventricular contractions in exercise generally herald the presence of severe ischemic atherosclerotic heart disease, often involving two or more major coronary vessels. This specific electric instability of the myocardium has a greater predictive value than S-T segment depression for the diagnosis of CHD. The incidence of sudden death owing to **ventricular fibrillation** is generally about six to ten times as high in a group with frequent PVCs as in heart disease patients without this abnormality. With fibrillation, the ventricles are unable to contract in a unified manner. As a result, blood is not pumped effectively and cardiac output falls dramatically.

Other Indices of CHD

Two useful non-electrocardiographic indices of possible CHD are the blood pressure and heart rate responses to exercise. During a graded exercise test, there is a normal, progressive increase in systolic blood pressure from about 120 mm Hg to 160 to 190 mm Hg at peak exercise. The change in diastolic pressure is generally less than 10 mm Hg. For some individuals, however, strenuous exercise may cause the systolic blood pressure to rise well above 200 mm Hg, whereas the diastolic pressure can increase to 100 to 150 mm Hg. This abnormal **hypertensive response** can be a significant clue to cardiovascular disease.

The inability of blood pressure to increase with exercise can also reflect cardiovascular malfunction. For example, failure of the systolic blood pressure to increase by at least 20 or 30 mm Hg during graded exercise (**hypotensive response**) may reflect diminished cardiac reserve. A rapid, large increase in heart rate (tachycardia) early in exercise is often a harbinger of cardiac problems. Likewise, abnormally low exercise heart rates may reflect unhealthy function of the heart's sinus node. Also, the inability of the heart rate to increase during exercise, especially when accompanied by extreme fatigue, may be indicative of cardiac strain and heart disease.

STRESS TEST PROTOCOLS

There are many different stress test protocols. In a national survey of 1400 exercise stress test centers,[125] the treadmill was the most common mode of testing. Seventy-one percent of the facilities reported using the treadmill, 17% used the bicycle ergometer, whereas only 12% tested with step tests. For the treadmill, the Bruce protocol was used 65.5% of the time, whereas about 10% of the tests utilized the protocol designed by Balke.[7]

The Balke and Bruce Treadmill Tests

The Balke and Bruce tests (see Chapt. 11, p. 215 for descriptions) are two popular treadmill protocols. Each test has advantages and disadvantages. For example, the Bruce test has a more abrupt increase in work loads between stages. Although this may be beneficial in terms of sensitivity (see p. 729) to ischemic ECG responses, it also means that the subjects must be able to tolerate the increased work. Both the Bruce and Balke protocol start at relatively high levels of exercise, especially for cardiac patients. Because of this, both tests have undergone modifications. For the Bruce protocol, lower exercise levels were added, whereas for the Balke test, a preliminary 2 to 3 minute stage (2 mph, 0% grade) was included.

It is sometimes difficult to decide which treadmill protocol to use for testing. Choice of protocol must be made based on the particular population studied, health status, age, and fitness status. In general, a graded stress test should start at a low level and have increments in exercise intensity approximately every 2 to 3 minutes. It is also advisable to use a warm-up, either separately or incorporated into the test itself. The total duration of the test should be at least 8 minutes. On the other hand, a test that is much longer than 15 minutes is not needed as most important cardiac and physiologic data can be obtained within this time period.

Bicycle Ergometer Tests

Bicycle ergometers are also a good means for exercise stress testing. In contrast to the treadmill, power output on the ergometer is independent of the person's body mass and is easily calculated and regulated. Also, the bike is portable, safe, and relatively inexpensive. There are two different types of bicycle ergometers: (1) electronically braked ergometers and (2) mass-loaded, friction-type ergometers. With electronically braked ergometers power output remains fixed within a range for a specified rate of pedaling. With mechanically loaded ergometers, power output is directly related to the frictional resistance and rate of pedaling.

The same general guidelines used for treadmill testing apply for graded exercise tests on the bicycle ergometer. Most bicycle ergometer rates of power output are expressed in $kgm \cdot min^{-1}$ or watts ($1 \, W = 6.12 \, kgm \cdot min^{-1}$). Bicycle ergometer test protocols have 2- to 4- minute stages of graded exercise with an initial resistance between zero and 15 or 30 watts; power output is generally increased by 15 to 30 watts per stage. Pedaling rate for mechanically braked ergometers is usually set at 50 or 60 revolutions per min.

Arm Ergometers

Arm cranking also can be used successfully for stress testing. Arm exercise results in lower max $\dot{V}_{O_2}$ values by as much as 20 to 30%; heart rate max is generally 10 to 15 beats per minute lower; and blood pressure is difficult to measure. In addition, blood pressure, heart rate, and oxygen consumption values are higher during submaximal arm cranking compared to the same exercise level performed with the legs. Nevertheless, it is possible to use the same type of graded protocols developed for leg cycling tests when an evaluation is desired of the physiologic response to arm exercise. Of course, the starting frictional resistance is less and the incremental power outputs must be adjusted accordingly.

Other Exercise Diagnostic Tests

Although the conventional exercise ECG stress test has been widely used, it is not without limitations. It is estimated that **false-negatives** occur 25% of the time and **false-positives** 15% of the time.[37] These limitations have led to the development of two other noninvasive tests that use radionuclide methods for detecting CHD. One of the methods, thallium-201 (^{201}Ti) imaging, assesses myocardial perfusion, and the other, radionuclide angiocardiography, assesses ventricular function. Both of these techniques can be used during exercise.[42,44]

IS STRESS TESTING SAFE?

The yearly death rate is 2 to 12% for men who show clinical evidence of heart disease. This broad variation in expected mortality rate is directly related to the number and severity of diseased coronary arteries. In general, if a person has had a previous myocardial infarction or episode of angina, the risk of having another episode or dying is increased 10 to 30 times above normal. This raises some question as to the advisability of stress testing adults, especially patients who have had myocardial infarctions or who have demonstrated other heart disease symptoms or significant risk profiles.

One research group reported that in about 170,000 submaximum and maximum stress tests only 16 high-risk but apparently healthy patients suffered coronary episodes during testing.[105,124] This represented about one person per 10,000 or about 0.01% of the total group. Other researchers determined that the risk of having a coronary episode for apparently healthy middle-aged adults was about 1 in 3000 during a maximum stress test.[84] This amounts to a risk that is 6 to 12 times normal for most middle-aged adults. For patients with documented heart disease, the risk in stress testing may increase to as much as 30 to 60 times normal. In one study, five cases of cardiac arrest were reported following exercise in men with prior myocardial infarction or angina pectoris,[63] while a 1982 report of more than 9000 stress tests showed no cardiovascular episodes for subjects with heart disease risks.[115]

HOW PREDICTIVE ARE STRESS TESTS?

The value of stress testing depends directly on how well results on a particular test predict the existence of heart disease. The **sensitivity** of a stress test refers to the percentage of people with disease who have an abnormal test result. No test is ever perfect for diagnosing heart disease. There are four possible outcomes from a graded exercise stress test:

1. **True-Positive:** In this condition, the test results are correct in diagnosing a person with heart disease (the stress test is a success).
2. **False-Negative:** In this condition, the test results are normal yet the person actually has heart disease (the test is unsuccessful in correctly making a diagnosis—a person with heart disease goes undiagnosed).
3. **True-Negative:** In this condition, the test results are normal and the person does not have heart disease (the test is a success).
4. **False-Positive:** In this condition, the test results are abnormal yet the person has no heart disease (the test is unsuccessful—a normal person is diagnosed with heart disease).

Both a false-negative and a false-positive test can have dramatic ramifications, especially a false-negative result. Whenever a stress test indicates the presence of heart disease, secondary tests are performed to confirm the diagnosis. Similarly, just because results from graded stress tests are normal does not necessarily rule out the occurrence of heart disease; the predictive value of an abnormal test is much greater than the predictive value of a normal stress test.[133]

GUIDELINES FOR STRESS TESTING

The following guidelines should be used for stopping a stress test. Each of these symptoms generally indicates extreme cardiovascular strain that could be dangerous to the patient.

1. **Repeated presence of premature ventricular contractions (PVCs).**
2. **Progressive angina pain regardless of the presence or absence of ECG abnormalities consistent with ischemia.**
3. **Electrocardiographic changes that include the presence of S-T segment depression of 2 mm or more, continuous bigeminal or trigeminal**

ectopic ventricular complexes, and evidence of atrioventricular conduction disturbances (AV block).

4. An extremely rapid increase in heart rate that may reflect a severely compromised cardiovascular response.

5. Failure of heart rate or blood pressure to increase with progressive exercise or a progressive drop in systolic blood pressure with increasing work load.

6. An increase in diastolic pressure of 20 mm Hg or more, or a rise above 110 mm Hg.

7. Headache, blurred vision, pale, clammy skin, or extreme fatigue.

8. Marked dyspnea (breathlessness) or cyanosis.

9. Dizziness or near fainting.

10. Nausea.

Persons who exhibit these responses require further medical evaluation and should be excluded from unsupervised exercise programs pending such a study. Those patients who complete the stress test without significant ECG responses or other evidence of CHD can be medically cleared for unsupervised exercise that does not exceed the intensity of exercise reached during the test.

EXERCISE PRESCRIPTION

Heart rate and oxygen uptake data obtained during the stress test are used to formulate the exercise prescription; this is an individualized exercise program based on the person's current fitness and health status, with emphasis on intensity, frequency, duration, and type of exercise. This is important because many people who start exercising do not recognize their limitations and may exercise above a prudent level. Even group exercise programs that require medical clearance are limited because members exercise at about the same work level (walk, jog, or swim at the same speed) with little attention paid to individual differences in fitness.

Practical Illustration

Figure 30–16 illustrates a practical approach that permits the functional translation of treadmill or cycle ergometer exercise test responses to the exercise prescription.[41a] The results depicted are for a male cardiac patient and were generated from an algorithm that used exercise test responses from the Bruce treadmill protocol for level-ground ambulation. During the Bruce test, the heart rate (A) was plotted as a function of time. A mathematical line of "best fit" was then applied to the heart rate data, and line B was drawn through the data points. A target zone for heart rate was calculated as 60 to 75% based on the maximum heart rate of 167 beats per minute (shaded portion represented as C). The individualized prescription is then detailed in terms of pace (14.0 to 15.4 min per mile; D) and METS (3.9 to 5.9; E). The acceptable range of exercise intensity in area C, based on heart rate response during the graded exercise test, includes the following recreational activities; bicycling, touch football, canoeing, alpine skiing, aerobics, volleyball, tennis and badminton, swimming, skating, and waterskiing. This quantitative method of exercise prescription may improve the specificity and precision of exercise prescription for the healthy previously sedentary individual as well as the patient with known cardiovascular disease.

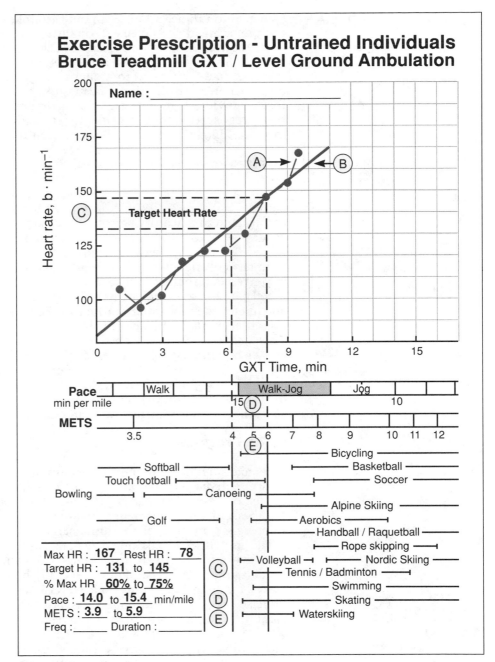

FIG. 30–16.
Exercise prescription based on the functional translation algorithm for level ground ambulation. (Used with permission from Dr. C. Foster, Department of Medicine, Cardiovascular Disease Section, Sinai Samaritan Medical Center, Milwaukee, WI).

In addition to heart rate as an indicator of exercise intensity, one can also use the **rating of perceived exertion (RPE).** During the stress test, with each progression in intensity, subjects are asked to rate on a numerical scale how they feel in relation to their level of exertion (see Fig. 30–17). The levels of exercise that correspond to higher levels of energy expenditure and physiologic strain result in higher ratings of perceived exertion. It is possible to teach individuals to exercise at a particular RPE based on their subjective feeling of exertion that coincides nicely with objective measures of exercise intensity. In this sense, the axion "listen to your body" is appropriate.

Improvements in CHD Patients

With a properly prescribed and monitored exercise program many cardiac patients can expect to safely improve their functional capacity to a degree comparable to a healthy person of the same age.[132] In some instances, even clinical symptoms such as ECG abnormalities are improved or even eliminated. This occurs because many individuals respond to exercise training with physiologic adjustments that actually reduce the work of the heart at any given external work load. For example, reduced exercise heart rate and blood pressure (two major determinants of myocardial oxygen consumption) and improved myocardial contractility reduce myocardial effort. This delays the onset of anginal pain allowing work of greater intensity and duration. This seems to be a primary benefit for angina patients, because the threshold at which pain develops is often at the same level of heart rate and systolic pressure before and after training. **For individuals whose occupations predominantly require arm work, this musculature should be exercised in training because many of the benefits of physical conditioning are highly specific and generally not transferable from one muscle group to another.**

The Program

Exercise programs for preventive purposes as well as for rehabilitation are most effective when they are individualized. The exercises prescribed usually consist of a slow but steady program of general movements that stimulate cardiovascular improvement such as walking, jogging, running, cycling, rope skipping, swimming, or rhythmic calisthenics.

The guidelines for making decisions concerning frequency, duration, and intensity of training were discussed in Chapter 21. **For physical conditioning of adult groups, exercise should generally be performed at least three times a week for 20 to 40 minutes each exercise session.** Longer and more frequent exercise will confer greater benefits, but the relatively small extra improvement may not be worth the time invested. Big muscle endurance activities should be performed at 50 to 75% of the person's working capacity or oxygen intake capacity as measured on the stress test. This "target zone" puts individuals above the threshold level for a training effect, yet ensures that they are not unduly stressed by the training program. Ideally, the personalized exercise prescription should include a recommendation for weight loss and dietary modification (if necessary), as well as warm-up and cool-down exercises and a developmental strength program. Some heart disease patients show a reduced heart rate response to exercise with a corresponding reduction in maximum heart rate.[102] For these individuals, the use of target heart rates based on an age-predicted maximum for the general population will grossly overestimate the appropriate training intensity. These observations argue

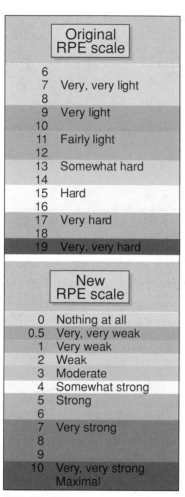

FIG. 30–17.
Scales used for ratings of perceived exertion (RPE). Original scale (6–19) above and revised briefer scale (1–10) below. These scales are related to physiologic stress and can be used to establish exercise intensity for the purpose of training (From Borg, G.A.: *Med. Sci. Sports Exerc.,* 14:377, 1982.)

TABLE 30–7.
TYPES OF EXERCISE PROGRAMS

Type	Participants	Entry MET Level	Supervision
A. Unsupervised	Asymptomatic	8+	None
B. Supervised			
1. Inpatient	All symtomatics—post MI, post op, pul dis	3	Supervised ambulatory therapy
2. Outpatient	All symtomatics—post MI, post operative, pulmonary disease	3+	Exercise specialist, physician on call
3. In-Home	Symptomatic + asymptomatic	>3–5	Unsupervised; periodic hospital re-evaluation
4. Community	Symptomatic + asymptomatic, 6–8 wk post infarct, 4–8 wk post operative	>5	Exercise program director + exercise specialist

for exercise testing each patient to a symptom-limited maximum and then formulating the exercise prescription based on the actual heart rate data.

Exercise Programs: Level of Supervision

The American College of Sports Medicine has categorized several types of exercise programs with specific criteria for entry and supervision as shown in Table 30–7. They are classified as **unsupervised** and **supervised** with 4 subdivisions of the supervised category. Unsupervised programs are for asymptomatic participants of any age with functional capacities of at least 8 METS with no known major risk factors. The supervised exercise programs are geared to people with different and specific needs. These include asymptomatic physically active or inactive persons of any age with CHD risk factors but no known disease (B-4) and symptomatic individuals (B1–3) including those with recent onset of CHD and those who report a change in disease status.

■■■■   SUMMARY

1. Physiologic and performance capability generally declines after about 30 years of age. The rates of decline in the various functions differ and are significantly influenced by many factors, including level of physical activity. Regular physical training enables older persons to retain higher levels of functional capacity, especially cardiovascular function.
2. Regardless of age, regular vigorous physical activity produces measurable physiologic improvements. The magnitude of these improvements depends on many factors that include initial fitness status, age, and type and amount of training.
3. Participation in vigorous activity early in life probably contributes little to increased longevity or health in later life. However, a physically active lifestyle throughout life confers significant health benefits.

4. Coronary heart disease is the single largest cause of death in the Western world. The pathogenesis of this disease involves degenerative changes in the inner lining of the arterial wall that results in progressive occlusion. The mechanism is not fully understood for the development of coronary atherosclerosis.

5. Numerous factors have been identified that make individuals more susceptible to developing CHD. The major risk factors are age and gender, elevated blood lipids, hypertension, cigarette smoking, obesity, physical inactivity, diet, heredity, and ECG abnormalities during rest and exercise.

6. Many CHD risk factors can be modified by proper programs of nutrition, exercise, and weight control. This "risk intervention" generally improves an individual's health outlook.

7. Stress testing provides the means to evaluate objectively physiologic capacity and observe heart function during exercise. The stress test often detects subtle signs of coronary artery disease that would normally remain undetected under resting conditions. These CHD indicators include angina pectoris, cardiac rhythm disorders, ECG abnormalities, and an abnormal blood pressure response.

8. Stress testing is relatively safe provided appropriate guidelines are followed. Graded exercise testing provides valuable information for clearing individuals to begin exercising and also for individualizing the exercise prescription.

9. No stress test is perfect for diagnosing heart disease. The four possible outcomes of a stress test are: True-Positive (the test is a success); False-Negative (a person with heart disease is not diagnosed); True-Negative (the test is a success); and False-Positive (a healthy person is diagnosed with heart disease).

10. Cardiac patients can often improve their functional capacity to the same extent as a healthy person of the same age.

REFERENCES

1. Abbott, R.F., et al.: Cardiovascular risk factors and graded treadmill exercise endurance in healthy adults: The Framingham Offspring Study. *Am. J. Cardiol.*, 63:342, 1989.

2. American College of Sports Medicine: *Guidelines for Exercise, Testing and Prescription.* 3rd ed. Philadelphia, Lea & Febiger, 1986.

3. Andersen, W.G.: Further studies on the longevity of Yale athletes. *Med. Times*, 44:75, 1916.

4. Aniansson, A., et al.: Muscle morphology, enzyme activity and muscle strength in elderly men and women. *Clin. Physiol.*, 1:73, 1981.

5. Aniansson, A., et al.: Muscle morphology, enzmatic activity, and muscle strength in elderly men: A follow-up study. *Muscle Nerve*, 9:585, 1986.

6. Badenhop, D.T., et al.: Physiological adjustments to higher- or lower-intensity exercise in elders. *Med. Sci. Sports Exerc.*, 15:496, 1983.

7. Balke, B., et al.: An experimental study of physical fitness of Air Force personnel. *U.S. Armed Forces Med. J.*, 20:745, 1965.

8. Barrett, J.H.: *Gerontological Psychology.* Springfield, IL, Charles C Thomas, 1972.

9. Becque, M.D.: Effects of 12 weeks of hydraulic resistance training on body composition, strength, and walking performance of older males. Unpublished Ph.D. dissertation, University of Michigan, Ann Arbor, 1989.

10. Becque, M.D., et al.: Coronary risk incidence of obese adolescents: Reduction by exercise plus diet intervention. *Pediatrics*, 81:605, 1988.

11. Behnke, A.R., and Wilmore, J.H.: *Evaluation and Regulation of Body Build and Composition.* Englewood Cliffs, NJ, Prentice-Hall, 1974.

12. Blair, S.N., et al.: Physical fitness and all-cause mortality: a prospective study of healthy men and women. *JAMA,* 262:2395, 1989.

13. Bookwalter, K.W.: Grip strength norms for males. *Res Q.,* 21:249, 1950.

14. Bourey, R.E., and Santoro, S.A.: Interactions of exercise, coagulation, platelets, and fibrinolysis—a brief review. *Med. Sci. Sports Exerc.,* 20:439, 1988.

15. Brewer, V., et al.: Role of exercise in prevention of involutional bone loss. *Med. Sci. Sports Exerc.,* 15:445, 1983.

16. Brill, P.A., et al.: The impact of athleticism on exercise habits, physical fitness, and coronary heart disease risk factors in middle-aged men. *Res. Q. Exer. Sport,* 60:209, 1989.

17. Brooks C.M.: Adult participation in physical activities requiring moderate to high levels of energy expenditure. *Phys. Sportsmed.,* 15:118, 1987.

18. Brooks, C.M.: Leisure time physical activity assessment of American adults through analysis of time diaries collected in 1981. *Am. J. Public Health,* 77:455, 1987.

19. Brown, W.F.: A method for estimating the number of motor units in thenar muscles and the change in motor unit count with aging. *J. Neurol. Neurosurg. Psychiatry,* 35:845, 1972.

20. Bruce, R.A., et al: Maximal oxygen uptake and nomographic assessment of functional aortic impairment in cardiovascular disease. *Am. Heart J.,* 85:545, 1973.

21. Bruce, R.A.: Exercise, functional aerobic capacity and aging—another viewpoint. *Med. Sci. Sports Exerc.,* 16:8, 1984.

22. Bruce, R.A., et al.: Longitudinal comparisons of responses to maximal exercise. In *Environmental Stress: Individual Adaptations.* Edited by L.A. Folinsbee, et al. New York, Academic Press, 1978.

23. Buskirk, E.R., and Hodgson, J.L.: Age and aerobic power: The rate of change in men and women. *Fed. Proc.,* 46:1824, 1987.

24. Casperson, C.J.: Physical activity epidemiology: concepts, methods, and applications to exercise science. In *Exercise and Sport Sciences Reviews.* Vol. 17. Edited by K.B. Pandolf. New York, Macmillan, 1989.

25. Caspersen C.J., et al.: Physical activity, exercise and physical fitness: definitions and distinctions for health-related research. *Public Health Rep.,* 100:126, 1985.

26. Caspersen C.J., et al.: Status of the 1990 physical fitness and exercise objectives—evidence from NHIS 1985. *Public Health Rep.,* 101:587, 1986.

27. Chien, S., et al.: Longitudinal measurements of blood volume and essential body mass in human subjects. *J. Appl. Physiol.,* 39:818, 1975.

28. Clarke, D.H.: Adaptations in strength and muscular endurance resulting from exercise. In *Exercise and Sport Science Reviews.* Vol. 1. Edited by J. Wilmore. New York, Academic Press, 1973.

29. Clarkson, P.M.: The effect of age and activity level on simple and fractionated response time. *Eur. J. Appl. Physiol.,* 40:17, 1978.

30. Clarkson, P.M.: The relationship of age and level of physical activity with the fractionated components of patellar reflex time. *J. Gerontol.,* 3:650, 1978.

31. Cook, T.C., et al.: Chronic low level physical activity as a determinant of high density lipoprotein cholesterol and subfractions. *Med. Sci. Sports Exerc.,* 18:653, 1986.

32. Cornoni-Huntley, J., et al.: Established Populations for Epidemiologic Studies of the Elderly: Resource Data Book. NIH Pub. No. 86–2443, National Institutes of Aging, Public Health Service, 1986.

33. Crews, D.J., and Landers, D.M.: A meta-analytic review of aerobic fitness and reactivity to psychosocial stressors. *Med. Sci. Sports Exerc.,* 19:S114, 1987.

34. Dalsky, G.P., et al.: Weight-bearing exercise, training and lumbar bone mineral content in post-menopausal women. *Ann. Intern. Med.,* 108:824, 1988.

35. Dehn, M.M., et al.: Longitudinal variations in maximal oxygen intake with age and activity. *J. Appl. Physiol.,* 33:805, 1972.

35a. Després, J-P., et al. Physical activity and coronary heart disease risk factors during childhood and adolescense. In *Exercise and Sport Sciences Reviews.* Vol. 18. Edited by K.B. Pandolf, and J.O. Holloszy. Baltimore, Williams & Wilkins, 1990.

36. Ekelund, L., et al. Physical fitness as a predictor of cardiovascular mortality in asymp-

tomatic North American men: the Lipid Research Clinics mortality follow-up study. *N. Engl. J. Med.*, 319:1379, 1988.

37. Epstein, S.E.: Implications of probability analysis on the strategy used for non-invasive detection of coronary artery disease. *Am. J. Cardiol.*, 46:491, 1980.
38. Fang, C-L., et al.: Exercise modality and selected coronary risk factors: a multivariate approach. *Med. Sci. Sports Exerc.*, 20:455, 1988.
39. Ferguson, E.W., et al.: Effects of exercise and conditioning on clotting and fibrinolytic activity in men. *J. Appl. Physiol.*, 62:1416, 1987.
40. Fisher, M.B., and Birren, J.E.: Age and strength. *J. Appl Psychol.*, 31:490, 1947.
41. Fleg, J.L., and Lakatta, E.G.: Role of muscle mass in the age-associated decrease in $\dot{V}O_2$ max. *J. Appl. Physiol.*, 65:1147, 1988.
41a. Foster, C.: Translation of exercise test responses to exercise prescription. In *Cardiac Rehabilitation and Clinical Exercise Programs: Theory and Practice*. Edited by N.B. Oldridge, et al. Ithaca, NY, Movement Publications, 1989.
42. Franklin, B.A., et al.: Additional diagnostic test: special populations. In *Resource Manual for Guidelines for Exercise Testing and Prescription*. Edited by S. Blair, et al., Philadelphia, Lea & Febiger, 1988.
43. Friedman, G.D., and Siegelaub, M.S.: Changes after quitting cigarette smoking. *Circulation*, 61:617, 1980.
44. Froelicher, V.F.: *Exercise Testing and Training*. New York, LeJacq, Publishing Co., 1983.
45. Frontera, W.R., et al.: Strength conditioning in older men: skeletal muscle hypertrophy and improved function. *J. Appl. Physiol.*, 64:1038, 1988.
46. Garn, S.M.: Bone loss and aging. In *The Physiology and Pathology of Aging*. Edited by R. Goldman, and M. Rockstein. New York, Academic Press, 1975.
47. Garrick, J.G., et al.: The epidemiology of aerobic dance injuries. *Am. J. Sports Med.*, 14:67, 1986.
48. Gibbons, L.W., et al.: The acute cardiac risk of strenuous exercise. *JAMA*, 224:1799, 1980.
49. Gilliam, T., et al.: Prevalence of coronary heart disease risk factors in active children, 7–12 years of age. *Med. Sci. Sports*, 9:21, 1977.
50. Goodman, D.S.: Report of National Cholesterol Education Program expert panel on detection, evaluation, and treatment of high cholesterol in adults. *Arch. Intern. Med.*, 148:36, 1988.
51. Hagberg, J.M.: A hemodynamic comparison of young and old endurance athletes during exercise. *J. Appl. Physiol.*, 58:2041, 1985.
52. Hagberg, J.M., et al.: Metabolic responses to exercise in young and older athletes and sedentary men. *J. Appl. Physiol.*, 65:900, 1988.
53. Hagberg, J.M., et al.: Pulmonary function in young and older athletes and untrained men. *J. Appl. Physiol.*, 65:101, 1988.
54. Hagberg, J., et al.: Cardiovascular responses of 70- to 79-yr old men and women to exercise training. *J. Appl. Physiol.*, 66:2589, 1989.
55. Hanson, P.: Clinical exercise testing. In *Resource Manual for Guidelines for Exercise Testing and Prescription*. Edited by S. Blair, et al., Philadelphia, Lea & Febiger, 1988.
56. Haskell, W.L.: Cardiovascular complications during exercise training of cardiac patients. *Circulation*, 47:920, 1978.
57. Higginbotham, M.B., et al.: Physiologic basis for the age-related decline in aerobic work capacity. *Am. J. Cardiol.*, 57:1374, 1986.
58. Hill, A.B.: Cricket and its relation to the duration of life. *Lancet*, 2:949, 1927.
59. Holloszy, J.O.: Exercise, health and aging: a need for more information. *Med. Sci. Sports Exerc.*, 15:1, 1983.
60. Holloszy, J.O., et al.: Effect of voluntary exercise on longevity of rats. *J. Appl. Physiol.*, 59:826, 1985.
61. Hubert, H.A., et al.: Obesity as an independent risk factor for cardiovascular disease. A 26-year follow-up of participants in the Framingham heart study. *Circulation*, 67:968, 1983.
62. Imamura, K., et al.: Human major psoas muscle and scarospinalis muscle in relation to age: a study by computed tomography. *J. Gerontology*, 33:678, 1983.
63. Irving, J.B., and Bruce, R.A.: Exertional hypotension and postexertional ventricular fibrillation in stress testing. *Am. J. Cardiol.*, 39:849, 1977.
64. Kannel, W.B., and Gordon, T.: The Framingham study: *An epidemiological investigation*

of cardiovascular disease. Section 30. Washington, DC, Public Health Service, DHEW No. (NIH) 74–599, 1974.

65. Kasch, F.W.: The effects of exercise on the aging process. *Phys. Sportsmed.,* 4:64, 1976.

65a. Kasch, F.W., et al.: The effect of physical activity and inactivity on aerobic power in older men (a longitudinal study). *Phys. Sports Med.,* 18:73, 1990.

66. Katch, F.I., and Katch, V.L.: Optimal health and body composition. In *Women and Exercise: Physiology and Sports Medicine.* Edited by M. Shangold, and G. Mirkin. Philadelphia, F.A. Davis, 1988.

67. Kavanagh, T., et al.: Marathon running after myocardial infarction. *JAMA,* 229:1602, 1974.

68. Killen, J.D., et al. Cardiovascular disease risk reduction for tenth graders. A multiple-factor school-based program. *JAMA,* 260:1728, 1988.

69. Kory, R.C., et al.: The Veterans Administration—Army cooperative study of pulmonary function. *Am. J. Med.,* 30:243, 1961.

70. Kramsch, D.M., et al.: Reduction in coronary atherosclerosis by moderate conditioning exercise in monkeys on an atherogenic diet. *N. Engl. J. Med.,* 305:1483, 1981.

71. Krolner B., et al.: Bone mass of the axial and the appendicular skeleton in women with Colles' fracture: Its relation to physical activity. *Clin. Physiol.,* 2:147, 1982.

72. Kuller, L.H., et al.: Sudden death and the decline in coronary heart disease mortality. *J. Chron. Dis.,* 39:1001, 1986.

73. Kutal, I., et al.: Muscle strength and lean body mass in old men of different physical activity. *J. Appl. Physiol.,* 29:168, 1970.

74. Larsson, L., et al.: Muscle strength and speed of movement in relation to age and muscle morphology. *J. Appl. Physiol,* 46:451, 1979.

75. Last, J.M. (Ed.): *A Dictionary of Epidemiology.* New York, Oxford University Press, 1983.

76. Laughlin, M.H.: Effects of training on coronary transport capacity. *J. Appl. Physiol.,* 58:468, 1985.

77. Laughlin, M.H., et al.: Exercise training increases coronary transport reserve in miniature swine. *J. Appl. Physiol.,* 67:1140, 1989.

77a. Levy, D., et al.: Stratifying the patient at risk from coronary disease: New insights from the Framingham Heart Study. *Am. Heart J.,* 119:712, 1990.

78. Lexell, J., et al.: Distribution of different fiber types in human skeletal muscle. 3. Effects of aging in m. vastus lateralis studies in whole muscle cross-sections. *Muscle Nerve,* 6:588, 1983.

79. Macera, C.A., et al.: Age, physical activity, physical fitness, body composition, and incidence of orthopedic problems. *Res. Q. Exerc. Sport,* 60:225, 1989.

80. Mann, G.V., et al: Physical fitness and immunity to heart disease in Masai. *Lancet,* 2:1308, 1965.

81. Manninen, V., et al.: Lipid alterations and decline in the incidence of coronary heart disease in the Helsinki Study. *JAMA,* 260:641, 1988.

82. Martin, R.P., et al: Blood chemistry and lipid profiles of elite distance runners. *Ann. N.Y. Acad. Sci.,* 301:346, 1977.

83. McDonough, J.R., et al.: Coronary heart disease among Negroes and Whites in Evans County, Georgia. *J. Chronic Dis.,* 18:443, 1965.

84. McDonough, J.R., and Bruce, R.A.: Maximal exercise testing in assessing cardiovascular function. *J. S.C. Med. Assoc.,* 65:26, 1969.

85. Meredith, C.N., et al.: Body composition and aerobic capacity in young and middle-aged endurance-trained men. *Med. Sci. Sports Exerc.,* 19:557, 1987.

86. Meredith, C.N., et al.: Peripheral effects of endurance training in young and old subjects. *J. Appl. Physiol.,* 66:2844, 1989.

87. Miller, A.J., and Miller, N.E.: Plasma high-density lipoprotein concentration and development of ischemic heart disease. *Lancet,* 1:16, 1975.

88. Montoye, H.J., et al.: *The Longevity and Morbidity of College Athletes.* Indianapolis, Phi Epsilon Kappa, 1957.

89. Montoye, H.J.: *Physical Activity and Health: An Epidemiologic Study of an Entire Community.* Englewood Cliffs, NJ, Prentice-Hall, 1975.

90. Montoye, H.J., and Lamphiear, D.E.: Grip and arm strength in males and females, age 10 to 69. *Res. Q.,* 48:109, 1977.

91. Morris, J.F., et al.: Spirometric standards for healthy nonsmoking adults. *Am. Rev. Respir. Dis.,* 103:57, 1971.

92. Murray, M.P., and Gardner, G.M.: Strength of isometric and isokinetic contractions. *Phys. Ther.*, 60:412, 1980.

93. Newnan, W.B. III, et al.: Relation of serum lipoprotein levels and systolic blood pressure to early atherosclerosis: the Bogalusa Study. *N. Engl. J. Med.*, 314:138, 1986.

94. Paffenbarger, R.S., Jr., et al.: Physical activity, all-cause mortality, and longevity of college alumni. *N. Engl. J. Med.*, 314:605, 1986.

95. Paffenbarger, R.S., Jr., et al.: Physical activity as an index of heart attack risk in college alumni. *Am. J. Epidemiol.*, 108:161, 1978.

96. Paffenbarger, R.S., Jr., et al.: Physical activity, all-cause mortality, and longevity of college alumni. *N. Engl. J. Med.*, 314:605, 1986.

97. Parizkova, J.: Body composition and exercise during growth and development. In *Physical Activity: Human Growth and Development.* Edited by G. L. Rarick. New York, Academic Press, 1974.

98. Parizkova, J., et al.: Body composition, aerobic capacity and density of muscle capillaries of young and old men. *J. Appl. Physiol.*, 31:323, 1971.

99. Pollock, M.L., et al.: Effects of walking on body composition and cardiovascular function of middle-aged men. *J. Appl. Physiol.*, 30:126, 1971.

100. Powell, K.E., et al.: Objective for the nation: assessing the role physical education must play. *J. Physical Educ.*, 55:18, 1984.

101. Powell, K.E., et al.: Physical activity and the incidence of coronary heart disease. *Ann. Rev. Public Health.*, 8:253, 1987.

102. Powles, A.C.P., et al.: Reduced heart rate response to exercise in ischemic heart disease: the fallacy of the target heart rate in exercise testing. *Med. Sci. Sports*, 11:227, 1979.

102a. Rikli, R.E., and B.G. Mcmanis. Effects of exercise on bone mineral content in postmenopausal women. *Res. Q. Exer. Sport*, 61:243, 1990.

103. Rivera, A.M., et al.: Physiological factors associated with lower maximal oxygen consumption of master runners. *J. Appl. Physiol.*, 66:949, 1989.

104. Robb, G.P., and Marks, H.H.: Latent coronary artery disease determination, its presence and severity by the exercise electrocardiogram. *Am. J. Cardiol.*, 13:603, 1964.

105. Rochmis, P., and Blackburn, H.: Exercise tests. A survey of procedures, safety and litigation experience in approximately 170,000 tests. *JAMA*, 217:1060, 1971.

106. Rodenheffer, R.J., et al.: Exercise cardiac output is maintained with advancing age in healthy human subjects: cardiac dilation and increased stroke volume compensate for diminished heart rate. *Circulation*, 69:203, 1984.

107. Rogers, M.A., et al.: Effect of 6d of exercise training on responses to maximal and submaximal exercise in middle-aged man. *Med. Sci. Sports Exerc.*, 20:260, 1988.

108. Sady, S., et al.: Training, diet and physical characteristics of distance runners with low or high concentrations of high density lipoprotein cholesterol. *Atherosclerosis*, 53:273, 1984.

109. Sady, S.P., et al.: Elevated high-density lipoprotein cholesterol in endurance athletes is related to enhanced plasma triglyceride clearance. *Metabolism*, 37:568, 1988.

110. Saltin, B., et al.: Physical training in sedentary middle-aged and older men. II. Oxygen uptake, heart rate, and blood lactate concentration at submaximal and maximal exercise. *Scand. J. Clin. Lab. Invest.*, 24:323, 1969.

111. Scheuer, J.: Effects of physical training on myocardial vascularization and perfusion. *Circulation*, 66:491, 1982.

112. Schulman, S.P., and Gerstenblith, G.: Cardiovascular changes with aging. The response to exercise. *J. Cardiopulmonary Rehab.*, 9:12, 1989.

113. Shulhan, D., et al.: Phasic cardiac selectivity to psychological stress as a function of aerobic fitness level. *Psychophysiology*, 23:562, 1986.

114. Seals, D.R., et al.: Endurance training in older men and women. I. Cardiovascular responses to exercise. *J. Appl. Physiol.*, 57:1024, 1984.

115. Sheffield, T., et al.: Safety of exercise testing volunteer subjects: The Lipid Research Clinics Prevalence Study experience. *J. Cardiac Rehab.*, 2:395, 1982.

116. Shock, N.W.: Physical activity and the rate of aging. *Can. Med. Assoc. J.*, 96:836, 1967.

117. Siscovick, D.S., et al.: The incidence of primary cardiac arrest during vigorous exercise. *N. Engl. J. Med.*, 311:874, 1984.

118. Spain, D.W., et al.: Weight, body type, and prevalence of atherosclerotic heart disease in males. *Am. J. Med. Sci.*, 245:63, 1963.

119. Spirduso, W.W.: Reaction and movement time as a function of age and physical activity level. *J. Gerontol.*, 30:435, 1975.
120. Spirduso, W.W., and Clifford, P.: Replication of age and physical activity effects on reaction and movement time. *J. Gerontol.*, 33:26, 1978.
121. Stamford, B.A.: Exercise and the elderly. In *Exercise and Sport Sciences Reviews*. Edited by K.B. Pandolf. Vol. 16. New York, Macmillan, 1988.
122. Stamford, B.A., et al.: Effects of smoking cessation on weight gain, metabolic rate, caloric consumption and blood lipids. *Am. J. Clin. Nutr.*, 43:486, 1986.
123. Strong, W.B.: Atherosclerosis: Its pediatric roots. In *Prevention of Coronary Heart Disease*. Edited by N.M. Kaplin, and J. Stamler. Philadelphia, W.B. Saunders, 1983.
124. Stuart, R.J., and Ellestad, M.H.: National survey of exercise stress test facilities. *Chest*, 77:94, 1980.
124a. Sytkowski, P.A., et al.: Changes in risk factors and the decline in mortality from cardiovascular disease. *N. Engl. J. Med.*, 322:1635, 1990.
125. Thacker, S.B., and Berkelman, R.L.: Public health surveillance in the United States. *Epidemiol. Rev.*, 10:164, 1988.
126. Thomas, S.G., et al.: Determinants of the training response in elderly men. *Med. Sci. Sports Exerc.*, 17:667, 1985.
127. Thompson, P.D., et al.: Incidence of death during jogging in Rhode Island from 1975 through 1980. *JAMA*, 245:2535, 1982.
128. Thompson, P.D., et al.: Modest changes in high-density lipoprotein concentration with prolonged exercise training. *Circulation*, 78:25,1988.
129. Tomonaga, M.: Histochemical and ultrastructural changes in senile human skeletal muscle. *J. Am. Geriatr. Soc.*, 25:125–131, 1977.
130. Trulson, M.F., et al.: Comparisons of siblings in Boston and Ireland. *J. Am. Diet. Assoc.*, 45:225, 1964.
131. Tran, Z.V., and Weltman, A.: Differential effects of exercise on serum lipids and lipoprotein levels seen with changes in body weight: a meta analysis. *JAMA*, 254:919, 1985.
132. Van Camp, S.P., and Peterson, R.A.: Cardiovascular complications of outpatient cardiac rehabilitation programs. *JAMA*, 256:1160, 1986.
133. Vecchio, T.H.: Predictive value of single diagnostic test in unselected populations. *N. Engl. J. Med.*, 274:1171, 1966.
134. Weisfeldt, M.L., and Gerstenblith, G.: Cardiovascular aging and adaptation to disease. In *The Heart*. Edited by J. W. Hurst. New York, Macmillan, 1986.
135. Wilmore, J.H., and McNamera, J.J.: Prevalence of coronary heart disease risk factors in boys 8 to 12 years of age. *J. Ped.*, 84:527, 1974.
136. Willett, W.C., et al.: Relative and absolute excess risks of coronary heart disease among women who smoke cigarettes. *N. Engl. J. Med.*, 317:1303, 1987.
137. Williams, J., et al.: The effects of running mileage and duration on plasma lipoprotein levels. *JAMA*, 247:2634, 1982.

APPENDICES

THE METRIC SYSTEM AND SI UNITS

This appendix has two parts. Part 1 deals with the metric system, and Part 2 discusses le Système international d'Unités (SI units).

P A R T

1 The Metric System

Most measurements in science are expressed in terms of the metric system. This system uses units that are related to one another by some power of 10. The prefix "centi" means one hundredth, "milli" means one thousandth, whereas the prefix "kilo" is derived from a word that means one thousand. In the following sections, we show the relationship between metric units and English units of measurement that are relevant to the material presented in this book.

UNITS OF LENGTH		
Metric Unit	**Equivalent Metric Unit**	**Equivalent English Unit**
meter (m)	100 cm 1000 mm	39.37 in. (3.28 ft; 1.09 yd)
centimeter (cm)	0.01 m 10 mm	0.3937 in.
millimeter (mm)	0.001 m 0.1 cm	0.03937 in.

UNITS OF WEIGHT		
Metric Unit	**Equivalent Metric Unit**	**Equivalent English Unit**
kilogram (kg)	1000 g 1,000,000 mg	35.3 oz (2.2046 lb)
gram (g)	0.001 kg 1000 mg	0.0353 oz
milligram (mg)	0.000001 kg 0.001 g	0.0000353 oz

UNITS OF VOLUME

Metric Unit	Equivalent Metric Unit	Equivalent English Unit
liter (l)	1000 ml	1.057 qt
milliliter (ml) or cubic centimeter (cc)	0.001 l	0.001057 qt

TEMPERATURE

To convert a Fahrenheit temperature to Celsius:
$$°C = (°F - 32)/1.8$$

To convert a Celsius temperature to Fahrenheit:
$$°F = (1.8 × °C) + 32$$

With the Fahrenheit scale, the freezing point of water is 32° F and the boiling point 212° F. On the Celsius scale, the freezing point of water is 0°C and the boiling point is 100°C.

UNITS OF SPEED

MPH	KM · HR^{-1}	M · S^{-1}	MPH	KM · HR^{-1}	M · S^{-1}
1	1.6	0.47	11	17.7	5.17
2	3.2	0.94	12	19.3	5.64
3	4.8	1.41	13	20.9	6.11
4	6.4	1.88	14	22.5	6.58
5	8.0	2.35	15	24.1	7.05
6	9.6	2.82	16	25.8	7.52
7	11.2	3.29	17	27.4	7.99
8	12.8	3.76	18	29.0	8.46
9	14.4	4.23	19	30.6	8.93
10	16.0	4.70	20	32.2	9.40

[handwritten in margin: 1 watt = 0.1433 kcal · min^{-1}]

COMMON EXPRESSIONS OF WORK, ENERGY, AND POWER

Watts	Kilocalories (kcal)	Foot-Pounds (ft-lb)
1 watt = 0.73756 ft-lb · s^{-1}	1 kcal = 3086 ft-lb	1 ft-lb = 3.2389 × 10^{-3} kcal
1 watt = 0.01433 kcal · min^{-1}	1 kcal = 426.8 kg-m	1 ft-lb = 0.13825 kg-m
1 watt = 1.341 × 10^{-3} hp or 0.0013 hp	1 kcal = 3087.4 ft-lb	1 ft-lb = 5.050 × 10^{-3} hp · h^{-1}
1 watt = 6.12 kg-m · min^{-1}	1 kcal = 1.5593 × 10^{-3} hp · h^{-1}	

DEFINITIONS OF COMMON SI UNITS

Degree Celsius (°C)	The degree Celsius is equivalent to K − 273.15.
Radian (rad)	The radian is the plane angle between two radii of a circle which subtend on the circumference an arc equal in length to the radius.
Joule (J)	The joule is the work done when the point of application of a force of one newton is displaced through a distance of one meter in the direction of the force. 1 J = 1 Nm.
Kelvin (K)	The kelvin is the fraction 1/273.16 of the thermodynamic temperature of the triple point of water.
Kilogram (kg)	The kilogram is a unit of mass equal to the mass of the international prototype of the kilogram
Meter (m)	The meter is the length equal to 1,650,763.73 wavelengths in vacuum of the radiation that corresponds to the transition between the levels $2p_{10}$ and $5d_5$ of the krypton-86 atom.
Newton (N)	The newton is that force that when applied to a mass of one kilogram, gives it an acceleration of one meter per second squared. 1 N = 1 kg · m/s².
Pascal (pa)	The pascal is the pressure produced by a force of one newton applied, with uniform distribution, over an area of one square meter. 1 Pa = 1 N/m².
Second (s)	The second is the duration of 9,192,631,770 periods of the radiation that corresponds to the transition between the two hyperfine levels of the ground state of the cesium-133 atom.
Watt (W)	The watt is the power that in one second gives rise to the energy of one joule. 1 W = 1 J/s.

TABLE A–1
CONVERSION FACTORS FOR USE IN THE EXERCISE SCIENCES

To Convert	Into	Multiply By
A		
ampere-hours	coulombs	3,600.0
ampere-hours	faradays	0.03731
angstrom units	inches	3.937×10^{-9}
angstrom units	meters	1×10^{-10}
angstrom units	microns	1×10^{-4}
B		
Btu	ergs	1.0550×10^{10}
Btu	foot-lb	778.3
Btu	gram-calories	252.0
Btu	horsepower-hours	3.931×10^{-4}
Btu	joules	1,054.8
Btu	kilogram-calories	0.2520
Btu	kilogram-meters	107.5
Btu	kilowatt-hours	2.928×10^{-4}
Btu/hour	foot-pounds/second	0.2162
Btu/hour	gram-cal/second	0.0700
Btu/hour	horsepower-hours	3.929×10^{-4}
Btu/hour	watts	0.2931
Btu/min	foot-lb/second	12.96
Btu/min	horsepower	0.02356
Btu/min	kilowatts	0.01757
Btu/min	watts	17.57
Btu/sq ft/min	watts/sq in	0.1221

To Convert	Into	Multiply By
C		
calories, gram (mean)	Btu (mean)	3.9685×10^{-3}
centigrade	Fahrenheit	$(C° \times 9/5) + 32$
centigrams	grams	0.01
centiliters	ounces fluid (U.S.)	0.3382
centiliters	cubic inches	0.6103
centiliters	drams	2.705
centiliters	liters	0.01
centimeters	feet	3.281×10^{-2}
centimeters	inches	0.3937
centimeters	kilometers	1×10^{-5}
centimeters	meters	0.01
centimeters	miles	6.214×10^{-6}
centimeters	millimeters	10.0
centimeters	mils	393.7
centimeters	yards	1.094×10^{-2}
centimeter-dynes	cm-grams	1.020×10^{-3}
centimeter-dynes	meter-kg	1.020×10^{-8}
centimeter-dynes	pound-feet	7.376×10^{-8}
centimeter-grams	cm-dynes	980.7
centimeter-grams	meter-kg	1×10^{-5}
centimeter-grams	pound-feet	7.233×10^{-6}
centimeters of mercury	atmospheres	0.01316
centimeters of mercury	feet of water	0.4461
centimeters of mercury	kg/sq meter	136.0
centimeters of mercury	pounds/sq ft	27.85
centimeters of mercury	pounds/sq in	0.1934
centimeters/second	feet/min	1.1969
centimeters/second	feet/sec	0.03281
centimeters/second	kilometers/hour	0.036
centimeters/second	knots	0.1943
centimeters/second	meters/min	0.6
centimeters/second	miles/hour	0.02237
centimeters/second	miles/min	3.728×10^{-4}
centimeters/second/second	feet/sec/sec	0.03281
centimeters/second/second	km/hour/second	0.036
centimeters/second/second	meters/second/second	0.01
centimeters/second/second	miles/hour/second	0.02237
cubic centimeters	cu feet	3.531×10^{-5}
cubic centimeters	cu inches	0.06102
cubic centimeters	cu meters	1×10^{-6}
cubic centimeters	cu yards	1.308×10^{-6}
cubic centimeters	gallons (U.S. liq.)	2.642×10^{-4}
cubic centimeters	liters	0.001
cubic centimeters	pints (U.S. liq.)	2.113×10^{-3}
cubic centimeters	quarts (U.S. liq.)	1.057×10^{-3}
cubic feet	bushels (dry)	0.8036
cubic feet	cu cm	28,320.0
cubic feet	cu inches	1,728.0
cubic feet	cu meters	0.02832
cubic feet	cu yards	0.03704
cubic feet	gallons (U.S. liq.)	7.48052
cubic feet	liters	28.32
cubic feet	pints (U.S. liq.)	59.84
cubic feet	quarts (U.S. liq.)	29.92
cubic feet/min	cu cm/second	472.0
cubic feet/min	gallons/second	0.1247
cubic feet/min	liters/second	0.4720
cubic feet/min	pounds of water/min	62.43
cubic feet/second	million gal/day	0.646317
cubic feet/second	gallons/min	448.831
cubic inches	cu cm	16.39

To Convert	Into	Multiply By
cubic inches	cu feet	5.787×10^{-4}
cubic inches	cu meters	1.639×10^{-5}
cubic inches	cu yards	2.143×10^{-5}
cubic inches	gallons	4.329×10^{-3}
cubic inches	liters	0.01639
cubic inches	mil-feet	1.061×10^{5}
cubic inches	pints (U.S. liq.)	0.03463
cubic inches	quarts (U.S. liq.)	0.01732
cubic meters	bushels (dry)	28.38
cubic meters	cu cm	1×10^{6}
cubic meters	cu feet	35.31
cubic meters	cu inches	61,023.0
cubic meters	cu yards	1.308
cubic meters	gallons (U.S. liq.)	264.2
cubic meters	liters	1,000.0
cubic meters	pints (U.S. liq.)	2,113.0
cubic meters	quarts (U.S. liq.)	1,057.0
cubic yards	cu cm	7.646×10^{5}
cubic yards	cu feet	27.0
cubic yards	cu inches	46,656.0
cubic yards	cu meters	0.7646
cubic yards	gallons (U.S. liq.)	202.0
cubic yards	liters	764.6
cubic yards	pints (U.S. liq.)	1,615.9
cubic yards	quarts (U.S. liq.)	807.9
cubic yards/min	cubic ft/second	0.45
cubic yards/min	gallons/second	3.367
cubic yards/min	liters/second	12.74
D		
days	seconds	86,400.0
decigrams	grams	0.1
deciliters	liters	0.1
decimeters	meters	0.1
degrees (angle)	quadrants	0.01111
degrees (angle)	radians	0.01745
degrees (angle)	seconds	3,600.0
degrees/second	radians/second	0.01745
degrees/second	revolutions/min	0.1667
degrees/second	revolutions/second	2.778×10^{-3}
dekagrams	grams	10.0
dekaliters	liters	10.0
dekameters	meters	10.0
drams	grams	1.7718
drams	ounces	0.0625
dyne/cm	ergs/sq millimeter	0.01
dyne/sq cm	atmospheres	9.869×10^{-7}
dyne/sq cm	inches of mercury at 0°C	2.953×10^{-5}
dyne/sq cm	inches of water at 4°C	4.015×10^{-4}
dynes	grams	1.020×10^{-3}
dynes	joules/cm	1×10^{-7}
dynes	jouls/meter (newtons)	1×10^{-5}
dynes	kilograms	1.020×10^{-6}
dynes	poundals	7.233×10^{-5}
dynes	pounds	2.248×10^{-6}
dynes/sq cm	bars	1×10^{-6}
E		
ergs	Btu	9.480×10^{-11}
ergs	dyne-centimeters	1.0
ergs	foot-pounds	7.3670×10^{-8}
ergs	gram-calories	0.2389×10^{-7}

To Convert	Into	Multiply By
ergs	grams-cm	1.020×10^{-3}
ergs	horsepower hours	3.7250×10^{-14}
ergs	joules	1×10^{-7}
ergs	kg-calories	2.389×10^{-11}
ergs	kg-meters	1.020×10^{-8}
ergs	kilowatt-hours	0.2778×10^{-13}
ergs	watt-hours	0.2778×10^{-10}
ergs/second	Btu/min	$5,688 \times 10^{-9}$
ergs/second	ft-lb/min	4.427×10^{-6}
ergs/second	ft-lb/second	7.3756×10^{-8}
ergs/second	horsepower	1.341×10^{-10}
ergs/second	kg-calories/min	1.433×10^{-9}
ergs/second	kilowatts	1×10^{-10}
F		
feet	centimeters	30.48
feet	kilometers	3.048×10^{-4}
feet	meters	0.3048
feet	miles (naut.)	1.645×10^{-4}
feet	miles (stat.)	1.894×10^{-4}
feet	millimeters	304.8
feet	mils	1.2×10^{4}
feet of water	atmospheres	0.02950
feet of water	inches of mercury	0.8826
feet of water	kg/sq cm	0.03048
feet of water	kg/sq meter	304.8
feet of water	pounds/sq ft	62.43
feet of water	pounds/sq in	0.4335
feet/min	cm/second	0.5080
feet/min	feet/second	0.01667
feet/min	km/hour	0.01829
feet/min	meters/min	0.3048
feet/min	miles/hour	0.01136
feet/second	cm/second	30.48
feet/second	km/hour	1.097
feet/second	knots	0.5921
feet/second	meters/min	18.29
feet/second	miles/hour	0.6818
feet/second	miles/min	0.01136
feet/second/second	cm/sec	30.48
feet/second/second	km/hour/second	1.097
feet/second/second	meters/sec/sec	0.3048
feet/second/second	miles/hour/second	0.6818
feet/100 feet	percent grade	1.0
foot-candles	lumen/sq meter	10.764
foot-pounds	Btu	1.286×10^{-3}
foot-pounds	ergs	1.356×10^{7}
foot-pounds	gram-calories	0.3238
foot-pounds	hp-hours	5.050×10^{-7}
foot-pounds	joules	1.356
foot-pounds	kg-calories	3.24×10^{-4}
foot-pounds	kg-meters	0.1383
foot-pounds	kilowatt-hours	3.766×10^{-7}
foot-pounds/min	Btu/min	1.286×10^{-3}
foot-pounds/min	foot-pounds/sec	0.01667
foot-pounds/min	horsepower	3.030×10^{-5}
foot-pounds/min	kg-calories/min	3.24×10^{-4}
foot-pounds/min	kilowatts	2.260×10^{-5}
foot-pounds/second	Btu/hour	4.6263
foot-pounds/second	Btu/min	0.07717
foot-pounds/second	horsepower	0.818×10^{-3}

To Convert	Into	Multiply By
foot-pounds/second	kg-calories/min	1.01945
foot-pounds/second	kilowatts	1.356×10^{-3}
G		
gallons	cu cm	3,785.0
gallons	cu feet	0.1337
gallons	cu inches	231.0
gallons	cu meters	3.785×10^{-3}
gallons	cu yards	4.951×10^{-3}
gallons	liters	3.785
gallons (liq. British imp.)	gallons (U.S. liq.)	1.20095
gallons (U.S.)	gallons (imp.)	0.83267
gallons of water	pounds of water	8.3453
gallons/min	cu ft/sec	2.228×10^{-3}
gallons/min	liters/sec	0.06308
gallons/min	cu ft/hour	8.0208
grams	dynes	980.7
grams	grains	15.43
grams	joules/cm	9.807×10^{-5}
grams	joules/meter (newtons)	9.807×10^{-3}
grams	kilograms	0.001
grams	milligrams	1,000.0
grams	ounces (avoirdupois)	0.03527
grams	ounces (troy)	0.03215
grams	poundals	0.07093
grams	pounds	2.205×10^{-3}
grams/cm	pounds/inch	5.600×10^{-3}
grams/cu cm	pounds/cu ft	62.43
grams/cu cm	pounds/cu in	0.03613
grams/cu cm	pounds/mil-foot	3.405×10^{-7}
grams/liter	grains/gal	58.417
grams/liter	pounds/1000 gal	8.345
grams/liter	pounds/cu ft	0.062427
grams/liter	parts/million	1,000.0
grams/sq cm	pounds/sq ft	2.0481
gram-calories	Btu	3.9683×10^{-3}
gram-calories	ergs	4.1868×10^{7}
gram-calories	foot-pounds	3.0880
gram-calories	horsepower-hours	1.5596×10^{-6}
gram-calories	kilowatt-hours	1.1630×10^{-6}
gram-calories	watt-hours	1.1630×10^{-3}
gram-calories/second	Btu/hour	14.286
gram-centimeters	Btu	9.297×10^{-8}
gram-centimeters	ergs	980.7
gram-centimeters	joules	9.807×10^{-5}
gram-centimeters	kg-cal	2.343×10^{-8}
gram-centimeters	kg-meters	1×10^{-5}
H		
horsepower	Btu/min	42.44
horsepower	foot-lb/min	33,000.0
horsepower	foot-lb/sec	550.0
horsepower (metric) (542.5 ft lb/second)	horsepower (550 ft lb/sec)	0.9863
horsepower (550 ft lb/second)	horsepower (metric) (542.5 ft lb/sec)	1.014
horsepower	kg-calories/min	10.68
horsepower	kilowatts	0.7457
horsepower	watts	745.7
horsepower (boiler)	Btu/hour	33,479.0
horsepower (boiler)	kilowatts	9.803
horsepower-hours	Btu	2,547.0

To Convert	Into	Multiply By
horsepower-hours	ergs	2.6845×10^{13}
horsepower-hours	foot-lb	1.98×10^{6}
horsepower-hours	gram-calories	641,190.0
horsepower-hours	joules	2.684×10^{6}
horsepower-hours	kg-calories	641.1
horsepower-hours	kg-meters	2.737×10^{5}
horsepower-hours	kilowatt-hours	0.7457
hours	days	4.167×10^{-2}
hours	weeks	5.952×10^{-3}
I		
inches	centimeters	2.540
inches	meters	2.540×10^{-2}
inches	miles	1.578×10^{-5}
inches	millimeters	25.40
inches	mils	1,000.0
inches	yards	2.778×10^{-2}
inches of mercury	atmospheres	0.03342
inches of mercury	feet of water	1.133
inches of mercury	kg/sq cm	0.03453
inches of mercury	kg/sq meter	345.3
inches of mercury	pounds/sq ft	70.73
inches of mercury	pounds/sq in	0.4912
inches of water (at 4°C)	atmospheres	2.458×10^{-3}
inches of water (at 4°C)	inches of mercury	0.07355
inches of water (at 4°C)	kg/sq cm	2.540×10^{-3}
inches of water (at 4°C)	ounces/sq in	0.5781
inches of water (at 4°C)	pounds/sq ft	5.204
inches of water (at 4°C)	pounds/sq in	0.03613
J		
joules	Btu	9.480×10^{-4}
joules	ergs	1×10^{7}
joules	foot-pounds	0.7376
joules	kg-calories	2.389×10^{-4}
joules	kg-meters	0.1020
joules	watt-hours	2.778×10^{-4}
joules/cm	grams	1.020×10^{4}
joules/cm	dynes	1×10^{7}
joules/cm	joules/meter (newtons)	100.0
joules/cm	poundals	723.3
joules/cm	pounds	22.48
K		
kilograms	dynes	980,665.0
kilograms	grams	1,000.0
kilograms	joules/cm	0.09807
kilograms	joules/meter (newtons)	9.807
kilograms	poundals	70.93
kilograms	pounds	2.205
kilograms	tons (long)	9.842×10^{-4}
kilograms	tons (short)	1.102×10^{-3}
kilograms/cu meter	grams/cu cm	0.001
kilograms/cu meter	pounds/cu ft	0.06243
kilograms/cu meter	pounds/cu in	3.613×10^{-5}
kilograms/cu meter	pounds/mil-foot	3.405×10^{-10}
kilograms/meter	pounds/ft	0.6720
kilograms/sq cm	dynes	980,665.0
kilograms/sq cm	atmospheres	0.9678
kilograms/sq cm	feet of water	32.81
kilograms/sq cm	inches of mercury	28.96
kilograms/sq cm	pounds/sq ft	2,048.0

To Convert	Into	Multiply By
kilograms/sq cm	pounds/sq in	14.22
kilograms/sq meter	atmospheres	9.678×10^{-5}
kilograms/sq meter	bars	98.07×10^{-6}
kilograms/sq meter	feet of water	3.281×10^{-3}
kilograms/sq meter	inches of mercury	2.896×10^{-3}
kilograms/sq meter	pounds/sq ft	0.2048
kilograms/sq meter	pounds/sq in	1.422×10^{-3}
kilograms/sq mm	kg/sq meter	1×10^{6}
kilogram-calories	Btu	3.968
kilogram-calories	foot-pounds	3,088.0
kilogram-calories	hp-hours	1.560×10^{-3}
kilogram-calories	joules	4,186.0
kilogram-calories	kg-meters	426.9
kilogram-calories	kilojoules	4.186
kilogram-calories	kilowatt-hours	1.163×10^{-3}
kilogram-meters	Btu	9.294×10^{-3}
kilogram-meters	ergs	9.804×10^{7}
kilogram-meters	foot-pounds	7.233
kilogram-meters	joules	9.804
kilogram-meters	kg-caloies	2.342×10^{-3}
kilogram-meters	kilowatt-hours	2.723×10^{-6}
kilolines	maxwells	1,000.0
kiloliters	liters	1,000.0
kilometers	centimeters	1×10^{5}
kilometers	feet	3,281.0
kilometers	inches	3.937×10^{4}
kilometers	meters	1,000.0
kilometers	miles	0.6214
kilometers	millimeters	1×10^{6}
kilometers	yards	1,094.0
kilometers/hour	cm/second	27.78
kilometers/hour	feet/min	54.68
kilometers/hour	feet/second	0.9113
kilometers/hour	knots	0.5396
kilometers/hour	meters/min	16.67
kilometers/hour	miles/hour	0.6214
kilometers/hour/second	cm/second/second	27.78
kilometers/hour/second	ft/second/second	0.9113
kilometers/hour/second	meters/second/second	0.2778
kilometers/hour/second	miles/hour/second	0.6214
kilowatts	Btu/min	56.92
kilowatts	foot-lb/min	4.426×10^{4}
kilowatts	foot-lb/second	737.6
kilowatts	horsepower	1.341
kilowatts	kg-calories/min	14.34
kilowatts	watts	1,000.0
kilowatt-hr	Btu	3,413.0
kilowatt-hr	ergs	3.600×10^{13}
kilowatt-hr	foot-lb	2.655×10^{6}
kilowatt-hr	gram-calories	859,850.0
kilowatt-hr	horsepower-hours	1.341
kilowatt-hr	joules	3.6×10^{6}
kilowatt-hr	kg-calories	860.5
kilowatt-hr	kg-meters	3.671×10^{5}
kilowatt-hr	pounds of water evaporated from and at 212°F	3.53
kilowatt-hr	pounds of water raised from 62° to 212°F	22.75
L		
liters	bushels (U.S. dry)	0.02838
liters	cu cm	1,000.0

To Convert	Into	Multiply By
liters	cu feet	0.03531
liters	cu inches	61.02
liters	cu meters	0.001
liters	cu yards	1.308×10^{-3}
liters	gallons (U.S. liq.)	0.2642
liters	pints (U.S. liq.)	2.113
liters	quarts (U.S. liq.)	1.057
liters/min	cu ft/second	5.886×10^{-4}
liters/min	gals/second	4.403×10^{-3}
M		
meters	centimeters	100.0
meters	feet	3.281
meters	inches	39.37
meters	kilometers	0.001
meters	miles (nautical)	5.396×10^{-4}
meters	miles (statute)	6.214×10^{-4}
meters	millimeters	1,000.0
meters	yards	1.094
meters/min	cms/second	1.667
meters/min	feet/min	3.281
meters/min	feet/second	0.05468
meters/min	km/hour	0.06
meters/min	knots	0.03238
meters/min	miles/hour	0.03728
meters/second	feet/min	196.8
meters/second	feet/second	3.281
meters/second	kilometers/hour	3.6
meters/second	kilometers/min	0.06
meters/second	miles/hour	2.237
meters/second	miles/min	0.03728
meters/second/second	cm/second/second	100.0
meters/second/second	ft/second/second	3.281
meters/second/second	km/hour/second	3.6
meters/second/second	miles/hour/second	2.237
meter-kilograms	cm-dynes	9.807×10^{7}
meter-kilograms	cm-grams	1×10^{5}
meter-kilograms	pound-feet	7.233
miles (nautical)	feet	6,080.27
miles (nautical)	kilometers	1.853
miles (nautical)	meters	1,853.0
miles (nautical)	miles (statute)	1.1516
miles (nautical)	yards	2,027.0
miles (statute)	centimeters	1.609×10^{5}
miles (statute)	feet	5,280.0
miles (statute)	inches	6.336×10^{4}
miles (statute)	kilometers	1.609
miles (statute)	meters	1,609.0
miles (statute)	miles (nautical)	0.8684
miles (statute)	yards	1,760.0
miles/hour	cm/second	44.70
miles/hour	feet/min	88.0
miles/hour	feet/second	1.467
miles/hour	km/hour	1.609
miles/hour	km/min	0.02682
miles/hour	knots	0.8684
miles/hour	meters/min	26.82
miles/hour	miles/min	0.1667
miles/hour/second	cm/second/second	44.70
miles/hour/second	feet/second/second	1.467
miles/hour/second	km/hour/second	1.609
miles/hour/second	meters/second/second	0.4470

To Convert	Into	Multiply By
miles/min	cm/second	2,682.0
miles/min	feet/second	88.0
miles/min	km/min	1.609
miles/min	knots/min	0.8684
miles/min	miles/hour	60.0
mil-feet	cu inches	9.425×10^{-6}
milliers	kilograms	1,000.0
millimicrons	meters	1×10^{-9}
milligrams	grains	0.01543236
milligrams	grams	0.001
milligrams/liter	parts/million	1.0
millihenries	henries	0.001
milliliters	liters	0.001
millimeters	centimeters	0.1
millimeters	feet	3.281×10^{-3}
millimeters	inches	0.03937
millimeters	kilometers	1×10^{-6}
millimeters	meters	0.001
millimeters	miles	6.214×10^{-7}
millimeters	mils	39.37
millimeters	yards	1.094×10^{-3}
minutes (angles)	degrees	0.01667
minutes (angles)	quadrants	1.852×10^{-4}
minutes (angles)	radians	2.909×10^{-4}
minutes (angles)	seconds	60.0

N

newtons	dynes	1×10^{5}

O

ohms (international)	ohms (absolute)	1.005
ohms	megohms	1×10^{-6}
ohms	microhms	1×10^{6}
ounces	drams	16.0
ounces	grains	437.5
ounces	grams	28.349527
ounces	pounds	0.0625
ounces	ounces (troy)	0.9115
ounces	tons (long)	2.790×10^{-5}
ounces	tons (metric)	2.835×10^{-5}
ounces (fluid)	cu inches	1.805
ounces (fluid)	liters	0.02957
ounces (troy)	grains	480.0
ounces (troy)	grams	31.103481
ounces (troy)	ounces (avoirdupois)	1.09714
ounces (troy)	pennyweights (troy)	20.0
ounces (troy)	pounds (troy)	0.08333
ounces/sq inch	dynes/sq cm	4,309.0
ounces/sq inch	pounds/sq in	0.0625

P

pints (liquid)	cu cm	473.2
pints (liquid)	cu feet	0.01671
pints (liquid)	cu inches	28.87
pints (liquid)	cu meters	4.732×10^{-4}
pints (liquid)	cu yards	6.189×10^{-4}
pints (liquid)	gallons	0.125
pints (liquid)	liters	0.4732
pints (liquid)	quarts (liquid)	0.5
pounds (avoirdupois)	ounces (troy)	14.5833
pounds	drams	256.0
pounds	dynes	44.4823×10^{4}

To Convert	Into	Multiply By
pounds	grains	7,000.0
pounds	grams	453.5924
pounds	joules/cm	0.04448
pounds	joules/meter (newtons)	4.448
pounds	kilograms	0.4536
pounds	ounces	16.0
pounds	ounces (troy)	14.5833
pounds	poundals	32.17
pounds	pounds (troy)	1.21528
pounds	tons (short)	0.0005
pounds of water	cu inches	27.68
pounds of water	gallons	0.1198
Q		
quarts (dry)	cu inches	67.20
quarts (liquid)	cu cm	946.4
quarts (liquid)	cu feet	0.03342
quarts (liquid)	cu inches	57.75
quarts (liquid)	cu meters	9.464×10^{-4}
quarts (liquid)	cu yards	1.238×10^{-3}
quarts (liquid)	gallons	0.25
quarts (liquid)	liters	0.9463
R		
radians	degrees	57.30
revolutions	degrees	360.0
revolutions	quadrants	4.0
revolutions	radians	6.283
revolutions/min	degrees/second	6.0
revolutions/min	radians/second	0.1047
revolutions/min	revolutions/second	0.01667
revolutions/min/min	radians/second/second	1.745×10^{-3}
revolutions/min/min	rev/min/second	0.01667
revolutions/min/min	revolutions/second/second	2.778×10^{-4}
revolutions/second	degrees/second	360.0
revolutions/second	radians/second	6.283
revolutions/second	rev/min	60.0
revolutions/second/second	radians/second/second	6.283
revolutions/second/second	rev/min/min	3,600.0
revolutions/second/second	rev/min/second	60.0
S		
seconds (angle)	degrees	2.778×10^{-4}
seconds (angle)	minutes	0.01667
seconds (angle)	quadrants	3.087×10^{-6}
seconds (angle)	radians	4.848×10^{-6}
square centimeters	circular mils	1.973×10^{5}
square centimeters	square feet	1.076×10^{-3}
square centimeters	square inches	0.1550
square centimeters	square meters	0.0001
square centimeters	sq miles	3.861×10^{-11}
square centimeters	sq millimeters	100.0
square centimeters	sq yards	1.196×10^{-4}
square feet	acres	2.296×10^{-5}
square feet	circular mils	1.833×10^{8}
square feet	sq cm	929.0
square feet	sq inches	144.0
square feet	sq meters	0.09290
square feet	sq miles	3.587×10^{-8}
square feet	sq millimetes	9.290×10^{4}
square feet	sq yards	0.1111
square inches	circular mils	1.273×10^{6}

To Convert	Into	Multiply By
square inches	sq cm	6.452
square inches	sq feet	6.944×10^{-3}
square inches	sq millimeters	645.2
square inches	sq mils	1×10^{6}
square inches	sq yards	7.716×10^{-4}
square kilometers	acres	247.1
square kilometers	sq cm	1×10^{10}
square kilometers	sq ft	10.76×10^{6}
square kilometers	sq inches	1.550×10^{9}
square kilometers	sq meters	1×10^{6}
square kilometers	sq miles	0.3861
square kilometers	sq yards	1.196×10^{6}
square meters	acres	2.471×10^{-4}
square meters	sq cm	1×10^{4}
square meters	sq feet	10.76
square meters	sq inches	1,550.0
square meters	sq miles	3.861×10^{-7}
square meters	sq millimeters	1×10^{6}
square meters	sq yards	1.196
square miles	acres	640.0
square miles	sq feet	27.88×10^{6}
square miles	sq km	2.590
square miles	sq meters	2.590×10^{6}
square miles	sq yards	3.098×10^{6}
square millimeters	circular mils	1,973.0
square millimeters	sq cm	0.01
square millimeters	sq feet	1.076×10^{-5}
square millimeters	sq inches	1.550×10^{-3}
square yards	acres	2.066×10^{-4}
square yards	sq cm	8,361.0
square yards	sq feet	9.0
square yards	sq inches	1,296.0
square yards	sq meters	0.8361
square yards	sq miles	3.228×10^{-7}
square yards	sq millimeters	8.361×10^{5}

T

To Convert	Into	Multiply By
temperature (°F) + 460	absolute temperature (°F)	1.0
temperature (°F) − 32	temperature (°C)	5/9
temperature (°F) + 460	absolute temperature (°F)	1.0
temperature (°F) − 32	temperature (°C)	5.9
tons (metric)	kilograms	1,000.0
tons (metric)	pounds	2,205.0

W

To Convert	Into	Multiply By
watts	Btu/hour	3.4129
watts	Btu/min	0.05688
watts	ergs/second	107.0
watts	foot-lb/min	44.27
watts	foot-lb/second	0.7378
watts	horsepower	1.341×10^{-3}
watts	horsepower (metric)	1.360×10^{-3}
watts	kg-calories/min	0.01433
watts	kilowatts	0.001
watts (absolute)	Btu (mean)/min	0.056884
watts (absolute)	joules/sec	1.0
watt-hours	Btu	3.413
watt-hours	ergs	3.60×10^{10}

To Convert	Into	Multiply By
watt-hours	foot-pounds	2,656.0
watt-hours	gram-calories	859.85
watt-hours	horsepower-hours	1.341×10^{-3}
watt-hours	kilogram-calories	0.8605
watt-hours	kilogram-meters	367.2
watt-hours	kilowatt-hours	0.001
watts (international)	watt (absolute)	1.0002
Y		
yards	centimeters	91.44
yards	kilometers	9.144×10^{-4}
yards	meters	0.9144
yards	miles (nautical)	4.934×10^{-4}
yards	miles (statute)	5.682×10^{-4}
yards	millimeters	914.4

TERMINOLOGY AND UNITS OF MEASUREMENT

The American College of Sports Medicine suggests that the following terminology and units of measurement be used in scientific endeavors to promote consistency and clarity of communication, and to avoid ambiguity. The terms defined below utilize the units of measurement of the Système International d'Unités (SI).

Exercise: Any and all activity involving generation of force by the activated muscle(s) which results in a disruption of a homeostatic state. In dynamic exercise, the muscle may perform shortening (concentric) contractions or be overcome by external resistance and perform lengthening (eccentric) contractions. When muscle force results in no movement, the contraction should be termed static or isometric.

Exercise Intensity: A specific level of maintenance of muscular activity that can be quantified in terms of power (energy expenditure or work performed per unit of time), isometric force sustained, or velocity of progression.

Endurance: The time limit of a person's ability to maintain either a specific isometric force or a specific power level involving combinations of concentric or eccentric muscular contractions.

Mass: A quantity of matter of an object, a direct measure of the object's inertia (note: mass = weight/acceleration due to gravity; units: gram or kilogram).

Weight: The force with which a quantity of matter is attracted toward Earth by normal acceleration of gravity (traditional unit: kilogram of weight).

Energy: The capability of producing force, performing work, or generating heat (unit: joule or kilojoule).

Force: That which changes or tends to change the state of rest or motion in matter (unit: newton).

Speed: Total distance travelled per unit of time (units: meter per second).

Velocity: Displacement per unit of time. A vector quantity requiring that direction be stated or strongly implied (units: meter per second or kilometer per hour).

Work: Force expressed through a distance but with no limitation on time (unit: joule or kilojoule). Quantities of energy and heat expressed independently of time should also be presented in joules. The term "work" should *not* be employed synonymously with muscular exercise.

Power: The rate of performing work; the derivative of work with respect to time; the product of force and velocity (unit: watt). Other related processes such as energy release and heat transfer should, when expressed per unit of time, be quantified and presented in watts.

Torque: The effectiveness of a force to produce rotation about an axis (unit: newton meter).

Volume: A space occupied, for example, by a quantity of fluid gas (unit: liter or milliliter). Gas volumes should be indicated as ATPS, BTPS, or STPD.

Amount of a substance: The amount of a substance is frequently expressed in moles. A mole is the quantity of a chemical substance that has a weight in mass units (e.g., grams) numerically equal to the molecular weight or that in the case of a gas has a volume occupied by such a weight under specified conditions. One mole of a respiratory gas is equal to 22.4 liters at STPD.

P A R T

2 SI Units

The uniform system of reporting numerical values is known as le Système International d'Unités, or its abbreviation, SI. The SI was developed through International cooperation to create a universally acceptable system for units of measurement. The SI ensures that units of measurement are uniform in concept and style. The SI system permits quantities in common use to be more easily compared. Many scientific organizations endorse the concept of the SI, and leading journals in the exercise and sport sciences now require that laboratory data be presented in SI units. The information in this appendix has been summarized from a detailed description about the SI published in the following article:

> Young, D.S.: Implementation of SI units for clinical laboratory data. Style specifications and conversion tables. *Ann. Intern. Med.*, 106:114, 1987.

TABLE A–2
BASE UNITS OF SI NOMENCLATURE

Physical Quantity	Base Unit	SI Symbol
1. Length	meter	m
2. Mass	kilogram	kg
3. Time	second	s
4. Amount of substance	mole	mol
5. Thermodynamic temperature	kelvin	K
6. Electric current	ampere	A
7. Luminous intensity	candela	cd

TABLE A–3
GENERAL SI STYLE GUIDELINES

Guidelines	Example	Incorrect Style	Correct Style
Use lowercase for symbols or abbreviations	kilogram	Kg	kg
Exceptions:			
	kelvin	k	K
	ampere	a	A
	liter	l	L
Symbols are not followed by period	meter	m.	m
	mole	mol.	mol
Exception: end of sentence			
Do not pluralize symbols	kilograms	kgs	kg
	meters	ms	m
Names and symbols are not to be combined	force	kilogram $\cdot$ meter $\cdot$ s^{-2}	kg $\cdot$ m $\cdot$ s^{-2}
			kg $\cdot$ m/s^2
When numbers are printed symbols are preferred		100 meters	100 m
		2 moles	2 mol
Space between number and symbol		50ml	50 mL
The product of units is indicated by a dot above the line		kg $\times$ m/s^2	kg $\cdot$ m $\cdot$ s^{-2}
			kg $\cdot$ m/s^2
Only one solidus (/) per expression		mmol/L/s	mmol/(L $\cdot$ s)
Place zero before decimal		.01	0.01
Decimal numbers are preferable to fractions		$\frac{3}{4}$	0.75
		75%	0.75
Spaces are used to separate long numbers		1,500,000	1 500 000
Exception: optional with four-digit number		1,000	1000 or 1 000

For SI units in exercise physiology, the term body weight is properly referred to as mass (kg), height should be referred to as stature (m), second is s, minute is min, hour is h, week is wk, month is mo, year is y, day is d, gram is g, liter is L, hertz is Hz, joule is J, kilocalorie is kcal, ohm is Ω, pascal is Pa, revolutions per minute is rpm, volt is V, and watt is W. These abbreviations or symbols are used for the singular or plural form.

Table A-4 lists common values in clinical hematology and clinical chemistry. The table lists the present reference intervals, the present unit, the conversion factor, the SI reference intervals, the SI unit symbol, the significant digits, and the suggested minimum increment.

TABLE A–4
SI CONVERSION TABLE FOR COMMON VALUES IN CLINICAL HEMATOLOGY AND CLINICAL CHEMISTRY

Component	Present Reference Intervals (Examples)	Present Unit	Conversion Factor	SI Reference Intervals	SI Unit Symbol	Significant Digits	Suggested Minimum Increment
Hemoglobin (B)							
Mass concentration							
—female	12.0–15.0	g/dL	10	120–150	g/L	XXX	1 g/L
—male	13.6–17.2	g/dL	10	136–172	g/L	XXX	1 g/L
Substance conc. Hb [Fe]							
—female	12.0–15.0	g/dL	0.6206	7.45–9.30	mmol/L	XX.XX	0.05 mmol/L
—male	13.6–17.2	g/dL	0.6206	8.45–10.65	mmol/L	XX.XX	0.05 mmol/L
Alkaline phosphatase (S)	30–120	U/L	0.01667	0.5–2.0	μkat/L	X.X	0.1 μkat/L
Amino acid nitrogen (P)	4.0–6.0	mg/dL	0.7139	2.9–4.3	mmol/L	X.X	0.1 mmol/L

Component	Present Reference Intervals (Examples)	Present Unit	Conversion Factor	SI Reference Intervals	SI Unit Symbol	Signif-icant Digits	Suggested Minimum Increment
Amino acid nitrogen (U)	50–200	mg/24 h	0.07139	3.6–14.3	mmol/d	X.X	0.1 mmol/d
Androstenedione (S)							
—male > 18 years	0.2–3.0	µg/L	3.492	0.5–10.5	mmol/L	XX.X	0.5 nmol/L
—female > 18 years	0.8–3.0	µg/L	3.492	3.0–10.5	nmol/L	XX.X	0.5 nmol/L
Bilirubin, total (S)	0.1–1.0	mg/dL	17.10	2–18	µmol/L	XX	2 µmol/L
Calcium (S)							
—male	8.8–10.3	mg/dL	0.2495	2.20–2.58	mmol/L	X.XX	0.02 mmol/L
—female <50 y	8.8–10.0	mg/dL	0.2495	2.20–2.50	mmol/L	X.XX	0.02 mmol/L
—female >50 y	8.8–10.2	mg/dL	0.2495	2.20–2.56	mmol/L	X.XX	0.02 mmol/L
Calcium (U), normal diet	<250	mg/24 h	0.02495	<6.2	mmol/d	X.X	0.1 mmol/d
Cholesterol (P)							
—<29 years	<200	mg/dL	0.02586	<5.20	mmol/L	X.XX	0.05 mmol/L
—30–39 years	<225	mg/dL	0.02586	<5.85	mmol/L	X.XX	0.05 mmol/L
—40–49 years	<245	mg/dL	0.02586	<6.35	mmol/L	X.XX	0.05 mmol/L
—>50 years	<265	mg/dL	0.02586	<6.85	mmol/L	X.XX	0.05 mmol/L
Ferritin (S)	18–300	ng/mL	1.00	18–300	µg/L	XX0	10 µg/L
Glucose (P)—fasting	70–110	mg/dL	0.05551	3.9–6.1	mmol/L	XX.X	0.1 mmol/L
Hemoglobin (B)							
—male	14.0–18.0	g/dL	10.0	140–180	g/L	XXX	1 g/L
—female	11.5–15.5	g/dL	10.0	115–155	g/L	XXX	1 g/L
Insulin (P,S)	5–20	µU/mL	7.175	35–145	pmol/L	XXX	5 pmol/L
	5–20	mU/L	7.175	35–145	pmol/L	XXX	5 pmol/L
	0.20–0.84	µg/mL	172.2	35–145	pmol/L	XXX	5 pmol/L
Iron (S)							
—male	80–180	µg/dL	0.1791	14–32	µmol/L	XX	1 µmol/L
—female	60–160	µg/dL	0.1791	11–29	µmol/L	XX	1 µmol/L
Lipoproteins (P)							
Low density [LDL]— as cholesterol	50–190	mg/dL	0.02586	1.30–4.90	mmol/L	X.XX	0.05 mmol/L
High density [HDL]— as cholesterol							
Male	30–70	mg/dL	0.02586	0.80–1.80	mmol/L	X.XX	0.05 mmol/L
Female	30–90	mg/dL	0.02586	0.80–2.35	mmol/L	X.XX	0.05 mmol/L
Testosterone (P)							
—female	0.6	ng/mL	3.467	2.0	nmol/L	XX.X	0.5 nmol/L
—male	4.6–8.0	ng/mL	3.467	14.0–28.0	nmol/L	XX.X	0.5 nmol/L
Thyroid tests:							
Thyroid stimulating hormone [TSH] (S)	2–11	µU/mL	1.00	2–11	mU/L	XX	1 mU/L
Thyroxine [T$_4$] (S)	4.0–11.0	µg/dL	12.87	51–142	nmol/L	XXX	1 nmol/L
Thyroxine binding globulin [TBG] (S)—[as thyroxine]	12.0–28.0	µg/dL	12.87	150–360	nmol/L	XX0	1 nmol/L
Thyroxine, free (S)	0.8–2.8	ng/dL	12.87	10–36	pmol/L	XX	1 pmol/L
Triiodothyronine [T$_3$] (S)	75–220	ng/dL	0.01536	1.2–3.4	nmol/L	X.X	0.1 nmol/L
T$_3$ uptake (S)	25–35	%	0.01	0.25–0.35	1	0.XX	0.01
Tolbutamide (P)— therapeutic	50–120	mg/L	3.699	180–450	µmol/L	XX0	10 µmol/L
Transferrin (S)	170–370	mg/dL	0.01	1.70–3.70	g/L	X.XX	0.01 g/L
Triglycerides (P) [as triolein]	<160	mg/dL	0.01129	<1.80	mmol/L	X.XX	0.02 mmol/L
Vitamin A [retinol] (P,S)	10–50	µg/dL	0.03491	0.35–1.75	µmol/L	X.XX	0.05 µmol/L
Vitamin B$_1$ [thiamine hydrochloride] (U)	60–500	µg/24 h	0.002965	0.18–1.48	µmol/d	X.XX	0.01 µmol/d
Vitamin B$_2$ [riboflavin] (S)	2.6–3.7	µg/dL	26.57	70–100	nmol/L	XXX	5 nmol/L

Component	Present Reference Intervals (Examples)	Present Unit	Conversion Factor	SI Reference Intervals	SI Unit Symbol	Significant Digits	Suggested Minimum Increment
Vitamin B_6 [pyridoxal] (B)	20–90	ng/mL	5.982	120–540	nmol/L	XXX	5 nmol/L
Vitamin B_{12} [cyano-cobalamin] (P,S)	200–1000	pg/mL ng/dL	0.7378 7.378	150–750	pmol/L pmol/L	XX0	10 pmol/L
Vitamin C (ascorbic acid)	0.6–2.00	mg/dL	56.78	30–110	μmol/L	X0	10 μmol/L
Vitamin D_3 [cholecalciferol] (P)	24–40	μg/mL	2.599	60–105	nmol/L	XXX	5 nmol/L
25 OH-cholecalciferol	18–36	ng/mL	2.496	45–90	nmol/L	XXX	5 mmol/L
Vitamin E [alpha-tocopherol] (P, S)	0.78–1.25	mg/dL	23.22	18–29	μmol/L	XX	1 μmol/L

NUTRITIVE VALUES FOR COMMON FOODS, ALCOHOLIC AND NONALCOHOLIC BEVERAGES, AND SPECIALITY AND FAST-FOOD ITEMS

This appendix has three parts. Part 1 lists nutritive values for common foods, Part 2 lists nutritive values for alcoholic and nonalcoholic beverages, and Part 3 presents nutritive values for specialty and fast-food items. The nutritive values of foods and alcoholic and nonalcoholic beverages are expressed in 1-ounce (28.4 g) portions so comparisons can readily be made among and between the different food categories. Thus, for example, the protein content of 1.55 g for 1 ounce of banana nut bread can be compared directly to the protein content of 6.28 g for 1 ounce of processed American cheese.

NUTRITIVE VALUES FOR COMMON FOODS*

The foods are grouped into categories and are listed in alphabetical order within each category. The categories include breads, cakes and pies, cookies, candy bars, chocolate, desserts, cereals, cheese, fish, fruits, meats, eggs, dairy products, and vegetables. An additional section labeled Variety consists of food items such as soups, sandwichs, salad dressings, oils, some condiments, and other "goodies." The nutritive value for each food is expressed per ounce or 28.4 g of that food item. The specific values for each food include the caloric content (kcal) for 1 ounce, protein, total fat, carbohydrate, calcium, iron, vitamin B_1, vitamin B_2, fiber content, and cholesterol.

---

* The information about the nutritive value of the foods was taken from a variety of sources. This includes primarily data from Watt, B.K., and Merrill, A.L.: *Composition of Foods—Raw, Processed and Prepared.* U.S. Department of Agriculture, Washington, DC, 1963, Adams, C., and Richardson, M.: *Nutritive Value of Foods.* Home and Garden Bulletin No. 72, rev, Government Printing Office, Washington, DC, 1981, and Pennington, J.A.T., and Church, H.N.: *Food Values of Portions Commonly Used.* 14th ed. New York, Harper & Row, 1985. Other sources included a comprehensive database on the Cyber mainframe computer at the University of Massachusetts, the public relations departments of manufacturers, and journal articles that evaluated specific food items.

Breads	kcal	Protein (g)	Fat (g)	CHO (g)	Ca (mg)	Fe (mg)	B₁ (mg)	B₂ (mg)	Fiber (g)	Cholesterol (mg)
Banana nut	91	1.55	4.00	12.7	10.0	0.470	0.054	0.046	0.66	18.3
Boston brown—canned	60	1.26	0.39	13.2	25.8	0.567	0.038	0.025	1.34	1.9
Cornmeal muffin—recipe	91	1.89	3.15	13.2	41.6	0.567	0.069	0.069	1.00	14.5
Cracked wheat—slice	74	2.63	0.99	14.2	18.1	0.755	0.108	0.108	1.50	0
Cracked wheat—toast	88	3.00	1.17	16.9	21.6	0.899	0.100	0.128	1.82	0
French—chunk	81	2.67	1.10	14.3	31.6	0.875	0.130	0.097	0.57	0
Italian	78	2.55	0.25	16.0	4.7	0.756	0.116	0.066	0.47	0
Mixed grain—slice	74	2.27	1.05	13.6	30.6	0.907	0.113	0.113	1.78	0
Mixed grain—toast	80	2.47	1.15	14.8	33.3	0.986	0.099	0.123	1.97	0
Oatmeal—slice	74	2.37	1.25	13.6	17.0	0.794	0.130	0.075	1.10	0
Oatmeal—toast	80	2.47	1.36	14.8	18.5	0.863	0.110	0.081	1.20	0
Pita pocket	78	2.94	0.42	15.6	23.2	0.685	0.129	0.061	0.45	0
Raisin—slice	77	2.15	1.12	15.0	28.4	0.879	0.093	0.176	0.68	0
Raisin—toasted	92	2.57	1.34	17.6	33.8	1.080	0.081	0.209	0.81	0
Rye—light—piece	74	2.40	1.04	13.6	22.7	0.771	0.116	0.090	1.87	0
Rye—light—toast	84	2.73	1.18	15.5	25.8	0.876	0.107	0.103	2.15	0
Pumpernickel—slice	71	2.60	0.98	13.6	20.4	0.777	0.097	0.147	1.67	0
Pumpernickel—toast	78	2.86	1.08	15.0	22.5	0.857	0.088	0.162	1.87	0
Vienna—slice	79	2.72	1.10	14.4	31.2	0.873	0.130	0.100	0.91	0
White—slice	76	2.35	1.10	13.8	35.7	0.806	0.133	0.088	0.54	0
White—toast	84	2.67	1.26	15.7	40.6	0.915	0.121	0.103	0.64	0
Whole wheat—slice	69	2.84	1.22	12.9	20.2	0.964	0.100	0.059	2.10	0
Whole wheat—toasted	79	3.42	1.47	14.4	22.5	1.090	0.090	0.066	2.74	0
Croutons—dry	105	3.69	1.04	20.5	35.0	1.020	0.099	0.099	0.09	0
Bread crumbs—dry grated	111	3.69	1.42	20.7	34.6	1.160	0.099	0.099	1.15	1.42
Bread crumbs—soft	76	2.35	1.10	13.9	35.9	0.806	0.134	0.088	0.54	0
Bread sticks wo/salt	109	3.40	0.82	21.3	7.9	0.255	0.017	0.020	0.43	0
Bread sticks w/salt	86	2.67	0.89	16.4	13.0	0.243	0.016	0.024	0.41	0

Cakes and Pies	kcal	Protein (g)	Fat (g)	CHO (g)	Ca (mg)	Fe (mg)	B₁ (mg)	B₂ (mg)	Fiber (g)	Cholesterol (mg)
Angel food cake	67	1.71	0.09	15.2	23.5	0.123	0.014	0.057		0
Boston cream pie—⅛	61	0.59	1.89	10.4	6.1	0.142	0.002	0.043		4.7
Carrot cake	103	1.05	5.32	13.2	6.9	0.304	0.030	0.035		14.6
Cheesecake	86	1.54	5.45	8.1	15.9	0.136	0.009	0.037		52.4
Choc cupcake/choc frosting	97	1.24	3.29	16.5	16.9	0.575	0.029	0.041		15.2
Coffee cake	91	1.78	2.70	14.8	17.3	0.480	0.054	0.059		18.5
Dark fruitcake	109	1.32	4.62	16.5	27.0	0.791	0.053	0.053		13.2
Gingerbread cake	91	1.15	2.86	15.2	12.2	0.706	0.042	0.038		7.8
Pound cake	113	1.89	4.72	14.2	18.9	0.472	0.047	0.057		30.2
Sheet cake—plain	104	1.32	3.96	15.8	18.1	0.429	0.046	0.049		20.1
Sheet cake—white frosting	104	0.94	3.28	18.0	14.3	0.281	0.030	0.037		16.4
Sponge cake	83	2.01	1.27	16.0	10.8	0.524	0.043	0.046		58.8
White cake/coconut	109	1.30	4.05	17.0	13.7	0.454	0.041	0.053		1.2
White cake/white frosting	104	1.20	3.59	16.8	13.2	0.399	0.080	0.052		1.2
Yellow cake/choc	101	1.03	4.48	15.9	9.5	0.509	0.020	0.058		15.6
Apple pie	73	0.66	3.14	10.7	5.0	0.300	0.031	0.023		0
Apple pie—fried	85	0.73	4.67	10.7	4.0	0.312	0.030	0.020		4.7
Banana cream pie	46	0.90	1.85	6.7	21.0	0.156	0.022	0.042		2.2
Blueberry pie	68	0.72	3.05	9.9	4.7	0.377	0.031	0.025		0
Boston cream pie	61	0.59	1.89	10.4	6.1	0.142	0.002	0.043		4.7
Cherry pie	74	0.77	3.19	10.9	6.6	0.569	0.034	0.025		0
Cherry pie—fried	83	0.68	4.74	10.7	3.7	0.233	0.020	0.020		4.3
Chocolate cream pie	50	1.20	2.04	6.9	25.9	0.175	0.024	0.049		2.4
Coconut cream pie	57	1.03	2.79	7.2	24.0	0.198	0.021	0.042		2.5
Coconut custard pie	66	1.69	3.85	6.3	25.0	0.304	0.029	0.055		31.4
Cream pie	85	0.56	4.29	11.0	8.6	0.205	0.011	0.028		1.5

Cakes and Pies (cont.)	kcal	Protein (g)	Fat (g)	CHO (g)	Ca (mg)	Fe (mg)	B₁ (mg)	B₂ (mg)	Fiber (g)	Cholesterol (mg)
Custard pie	55	1.43	2.65	6.3	23.1	0.269	0.026	0.050		27.6
Lemon meringue pie	72	0.95	2.90	10.7	5.1	0.283	0.020	0.028		27.7
Mincemeat pie	70	0.65	2.13	12.8	6.9	0.360	0.028	0.024		0
Peach pie	73	0.63	3.14	10.9	4.8	0.340	0.031	0.028		0
Pecan pie	120	1.30	4.87	18.9	7.2	0.380	0.045	0.034		28.1
Pumpkin pie	52	1.28	2.23	7.3	30.0	0.373	0.019	0.042		15.5
Strawberry chiffon pie	65	0.85	3.46	8.0	7.7	0.254	0.022	0.023		7.1

Cookies	kcal	Protein (g)	Fat (g)	CHO (g)	Ca (mg)	Fe (mg)	B₁ (mg)	B₂ (mg)	Fiber (g)	Cholesterol (mg)
Animal cookies	120	1.90	2.89	22.0	3.0	0.918	0.080	0.130		0.1
Butter cookies	130	1.76	4.82	20.2	36.3	0.170	0.011	0.017		4.1
Brownies w/nuts	135	1.84	8.93	15.6	12.8	0.567	0.070	0.070		25.5
Fig bars	106	1.02	1.93	21.4	20.2	0.689	0.039	0.037		13.7
Lady fingers	102	2.19	2.19	18.3	11.6	0.515	0.019	0.039		101.0
Oatmeal raisin cookies	134	1.64	5.45	19.6	9.8	0.600	0.049	0.044		1.1
Peanut butter cookies	145	2.36	8.27	16.5	12.4	0.650	0.041	0.041		13.0
Sandwich type cookies	138	1.42	5.67	20.6	8.5	0.992	0.064	0.050		0
Shortbread cookies	137	1.77	7.09	17.7	11.5	0.709	0.089	0.080		23.9
Sugar cookies	139	1.18	7.09	18.3	29.5	0.532	0.053	0.035		17.1
Vanilla wafers	131	1.42	4.96	20.6	11.3	0.567	0.050	0.070		17.7

Candy Bars	kcal	Protein (g)	Fat (g)	CHO (g)	Ca (mg)	Fe (mg)	B₁ (mg)	B₂ (mg)	Fiber (g)	Cholesterol (mg)
Almond Joy	151	1.69	7.82	18.5	2.0	0.778				0
Sugar-coated almonds	146	3.10	9.12	14.6	39.6	0.775	0.042	0.156		0
Bittersweet chocolate	141	1.90	9.73	15.7	13.0	1.040	0.015	0.050		0
Caramel—plain or choc	115	1.00	2.99	22.0	41.9	0.399	0.010	0.050		1.00
Chocolate-coated almonds	161	3.92	12.70	8.0	47.8	1.090	0.052	0.186		0
Chocolate-covered coconut	133	0.91	7.10	17.5	8.4	0.614	0.008	0.016		0
Chocolate-covered mints	116	0.50	2.99	23.0	16.0	0.299	0.010	0.020		0
Chocolate-covered peanuts	159	5.00	11.70	9.8	32.9	0.689	0.086	0.043		0
Chocolate-covered raisins	111	1.06	2.71	20.6	12.2	0.663	0.034	0.025		0
Chocolate fudge	115	0.56	2.78	21.0	22.0	0.299	0.010	0.030		1.00
Chocolate fudge with nuts	114	1.06	4.99	18.8	22.0	0.299	0.016	0.030		7.35
English toffee	195	0.89	16.90	9.8	0	0.177	0.470	0.044		0
Hard candy	109	0	0	27.6	6.0	0.100	0.	0		0
Jelly beans	104	0	0.10	26.4	1.0	0.299	0	0		0
Gum drops	98	0	0.20	24.8	2.0	0.100	0	0		0
Chocolate candy kisses	154	2.10	8.98	15.9	52.9	0.499	0.020	0.080		
Kit Kat	138	1.98	7.25	16.5	42.9	0.369	0.020	0.073		
Krackle	149	2.00	8.09	16.9	50.0	0.400	0.017	0.075		
Malted milk balls	135	2.30	6.99	17.8	62.9					
M&M's plain chocolate	140	1.95	6.08	19.5	46.7	0.449	0.015	0.073		0
M&M's peanut chocolate	144	3.23	7.25	16.5	35.4	0.402	0.016	0.056		0
Mars bar	136	2.27	6.24	17.0	48.2	0.312	0.014	0.093		0
Milk chocolate—plain	145	2.00	8.98	16.0	49.9	0.399	0.020	0.100		5.99
Milk chocolate w/almonds	150	2.90	10.40	15.0	60.9	0.559	0.030	0.130		4.49
Milk chocolate w/peanuts	155	4.89	11.70	10.0	31.9	0.679	0.112	0.065		2.99
Milk chocolate + rice cereal	140	2.00	6.99	18.0	47.9	0.200	0.010	0.080		5.99
Milky Way	123	1.53	4.25	20.3	40.6	0.232	0.013	0.070		6.61

Candy Bars (cont.)	kcal	Protein (g)	Fat (g)	CHO (g)	Ca (mg)	Fe (mg)	B₁ (mg)	B₂ (mg)	Fiber (g)	Cholesterol (mg)
Mr. Goodbar	151	3.62	9.05	13.9	39.2	0.567	0.030	0.072		4.22
Reese's peanut butter cup	151	3.65	9.07	13.9	21.7	0.430	0.020	0.032		1.62
Snickers—2.2 oz	134	3.08	6.62	17.0	32.4	0.227	0.013	0.050		0
Vanilla fudge	118	0.70	3.15	22.0	29.9	0.030	0.006	0.025		9.98
Vanilla fudge with nuts	122	1.00	5.01	18.3	25.0	0.159	0.017	0.026		8.53

Chocolate	kcal	Protein (g)	Fat (g)	CHO (g)	Ca (mg)	Fe (mg)	B₁ (mg)	B₂ (mg)	Fiber (g)	Cholesterol (mg)
Baking chocolate	145	3.49	15.00	7.5	22.0	1.900	0.015	0.099		0
Bittersweet chocolate	141	1.90	9.73	15.7	13.0	1.040	0.015	0.050		0
Milk chocolate—plain	145	2.00	8.98	16.0	49.9	0.399	0.020	0.100		6.0
Semi-swt chocolate chips	143	1.17	10.20	16.2	8.5	0.967	0.017	0.023		0
Dark chocolate—sweet	150	1.00	9.98	16.0	7.0	0.599	0.010	0.040		0
Choc cupcake/choc frosting	97	1.24	3.29	16.5	16.9	0.575	0.029	0.041		15.2
Chocolate candy kisses	154	2.10	8.98	15.9	52.9	0.499	0.020	0.080		
Chocolate chip cookies	122	1.54	5.94	18.9	11.0	0.540	0.068	0.155		3.4
Chocolate coated almonds	161	3.92	12.70	8.0	47.8	1.090	0.052	0.186		0
Chocolate covered mints	116	0.50	2.99	23.0	16.0	0.299	0.010	0.020		0
Chocolate coated peanuts	159	5.00	11.70	9.8	32.9	0.689	0.086	0.043		0
Chocolate covered raisins	111	1.06	2.71	20.6	12.2	0.663	0.034	0.025		0
Chocolate cream pie	50	1.20	2.04	6.9	25.9	0.175	0.024	0.049		2.4
Chocolate fudge	115	0.56	2.78	21.0	22.0	0.299	0.010	0.030		1.0
Chocolate fudge with nuts	114	1.06	4.99	18.8	22.0	0.299	0.016	0.030		7.4
Cake flour-baked value	103	2.08	0.28	22.4	4.5	1.250	0.154	0.096		0
Reese's peanut butter cup	151	3.65	9.07	13.9	21.7	0.430	0.020	0.032		1.6
Chocolate pudding/recipe	42	0.88	1.25	7.3	27.3	0.142	0.005	0.039		4.3
Chocolate pudding—instant	34	0.85	0.818	5.9	28.4	0.065	0.009	0.039		3.1

Desserts	kcal	Protein (g)	Fat (g)	CHO (g)	Ca (mg)	Fe (mg)	B₁ (mg)	B₂ (mg)	Fiber (g)	Cholesterol (mg)
Apple brown betty	43	0.30	1.60	7.40	5.4	0.130	0.016	0.012		3.8
Apple cobbler	55	0.53	1.74	9.57	8.8	0.206	0.023	0.019		0.3
Apple crisp	53	0.33	1.93	9.09	7.4	0.278	0.018	0.013		0
Apple dumpling	55	0.32	2.37	8.64	7.1	0.253	0.012	0.013		0
Banana nut bread	91	1.60	4.00	12.70	10.0	0.470	0.054	0.046		18.3
Bread + raisin pudding	60	1.20	2.47	8.54	27.7	0.304	0.030	0.048		24.4
Cheesecake	86	1.50	5.45	8.10	15.9	0.136	0.009	0.037		52.4
Cherry cobbler	44	0.53	1.37	7.52	8.5	0.391	0.019	0.021		0.3
Cherry&cream cheese torte	79	1.28	3.96	10.00	28.5	0.266	0.015	0.052		11.2
Vanilla milkshake	32	0.98	0.841	5.09	34.5	0.026	0.013	0.052		3.2
Cream puff w/custard fill	72	1.24	4.54	6.83	16.4	0.276	0.015	0.040		58.8
Choc eclair w/custard fill	79	1.20	4.43	8.96	18.6	0.258	0.019	0.041		50.4
Gelatin salad	17	0.43	0	3.99	0.5	0.024	0.002	0.002		0
Peach cobbler	28	0.50	1.35	7.96	7.6	0.197	0.018	0.017		0.3
Peach crisp	34	0.31	1.06	6.18	4.8	0.203	0.010	0.010		0
Doughnut—cake type	119	1.33	6.75	13.90	13.0	0.454	0.068	0.068		11.3
Doughnut—jelly filled	99	1.48	3.84	13.00	12.2	0.349	0.052	0.044		0
Doughnut—yeast-raised	111	1.89	6.28	12.30	8.0	0.661	0.132	0.057		9.9
Chocolate pudding	42	0.88	1.25	7.28	27.3	0.142	0.005	0.039		4.3
Tapioca pudding	38	1.43	1.44	4.85	29.7	0.120	0.012	0.052		27.3

Desserts (cont.)	kcal	Protein (g)	Fat (g)	CHO (g)	Ca (mg)	Fe (mg)	B₁ (mg)	B₂ (mg)	Fiber (g)	Cholesterol (mg)
Vanilla pudding	32	0.99	1.10	4.50	33.1	0.089	0.009	0.046		4.1
Chocolate pudding— instant	34	0.85	0.82	5.89	28.4	0.065	0.009	0.039		3.1
Rice pudding	33	0.86	0.86	5.80	28.6	0.107	0.021	0.039		3.2
Butterscotch pudding pop	47	1.19	1.29	7.80	37.8	0.020	0.015	0.055		0.5
Chocolate pudding pop	49	1.34	1.34	8.20	42.8	0.179	0.015	0.055		0.5
Vanilla pudding pop	46	1.19	1.29	7.80	37.8	0.020	0.015	0.055		0.5

Cereals (without milk)	kcal	Protein (g)	Fat (g)	CHO (g)	Ca (mg)	Fe (mg)	B₁ (mg)	B₂ (mg)	Fiber (g)	Cholesterol (mg)
Alpha Bits	111	2.20	0.60	24.6	7.99	1.80	0.399	0.399	0.650	
Apple Jacks	110	1.50	0.10	25.7	2.99	4.49	0.399	0.399	0.200	
Bran Buds	73	3.95	0.68	21.6	18.90	4.52	0.371	0.439	7.860	
Bran Chex	90	2.95	0.81	22.6	16.80	4.51	0.347	0.150	5.200	
Buc Wheats	110	2.00	1.00	24.0	59.90	8.09	0.674	0.764	2.000	
C.W. Post—plain	126	2.54	4.44	20.3	13.70	4.50	0.380	0.438	0.643	
All-Bran	70	3.99	0.50	21.0	23.00	4.49	0.369	0.429	8.490	
C.W. Post w/raisins	123	2.45	4.05	20.3	14.00	4.51	0.358	0.413	0.660	
Cap'n Crunch	120	1.46	2.60	22.9	4.60	7.53	0.506	0.544	0.709	
Cap'n Crunchberries	118	1.46	2.35	23.0	8.91	7.32	0.478	0.543	0.324	
Cap'n Crunch—pnut butter	125	2.03	3.64	21.5	5.67	7.37	0.486	0.567	0.324	
Cheerios	110	4.24	1.77	19.4	47.30	4.44	0.394	0.394	3.000	
Cocoa Krispies	109	1.50	0.39	25.2	4.73	1.81	0.394	0.394	0.354	
Cocoa Pebbles	117	1.35	1.49	24.7	5.40	1.75	0.405	0.405	0.312	
Corn Bran	98	1.97	1.02	23.9	32.30	9.60	0.299	0.551	5.390	
Corn Chex	111	2.00	0.10	24.9	2.99	1.80	0.399	0.070	0.499	
Corn flakes—Kellogg's	110	2.30	0.09	24.4	1.00	1.80	0.367	0.424	0.594	
Corn Flakes—Post Toasties	110	2.30	0.09	24.4	1.00	0.70	0.367	0.424	0.594	
Corn grits—enr yellow dry	105	2.49	0.33	22.5	0.55	1.10	0.182	0.107	3.270	
Corn grits—enr yellow ckd	17	0.41	0.06	3.7	0.12	0.18	0.028	0.018	0.527	
Cracklin' Oat Bran	108	2.60	4.16	19.4	18.90	1.80	0.378	0.425	4.280	
Cream of Rice	15	0.24	0.01	3.3	0.93	0.05	0.012	0	0.163	
Cream of Wheat	16	0.42	0.07	3.4	6.27	1.27	0.028	0.008	0.395	
Crispy Wheat 'n Raisins	99	2.00	0.46	23.1	46.80	4.48	0.396	0.396	1.320	
Farina—cooked	14	0.41	0.02	3.0	0.49	0.14	0.023	0.015	0.389	
40% Bran Flakes— Kelloggs	91	3.60	0.54	22.2	13.80	8.14	0.369	0.430	0.850	
40% Bran Flakes—Post	92	3.20	0.45	22.3	12.70	4.50	0.374	0.435	3.800	
Fortified Oat Flakes	105	5.32	0.41	20.5	40.20	8.09	0.354	0.413	0.827	
Froot Loops	111	1.70	1.00	25.0	2.99	4.49	0.399	0.399	0.299	
Frosted Mini-Wheats	102	2.93	0.27	23.4	9.15	1.83	0.366	0.457	2.160	
Frosted Rice Krispies	109	1.30	0.10	25.7	1.00	1.80	0.399	0.399	0.998	
Fruity Pebbles	115	1.10	1.50	24.4	2.99	1.80	0.399	0.399	0.226	
Fruit & Fiber w/apples	90	2.99	1.00	22.0	9.98	4.49	0.374	0.424	4.190	
Fruit & Fiber w/dates	90	2.99	1.00	21.0	9.98	4.49	0.374	0.424	4.190	
Fruitful Bran	92	2.50	0	22.5	8.34	6.75	0.313	0.354	4.170	
Golden Grahams	109	1.60	1.09	24.1	17.40	4.50	0.363	0.436	1.670	
Granola—Nature Valley	126	2.89	4.92	18.9	17.80	0.95	0.098	0.048	2.960	
Granola—homemade	138	3.49	7.69	15.6	17.70	1.12	0.170	0.072	2.970	
Grape Nuts	100	3.28	0.11	23.2	10.90	1.22	0.398	0.398	1.840	
Grape Nuts Flakes	102	2.99	0.30	23.2	11.00	4.49	0.399	0.399	1.900	
Honey & Nut Corn Flakes	113	1.80	1.50	23.3	2.99	1.80	0.399	0.399	0.299	
Honey Nut Cheerios	107	3.09	0.69	22.8	19.80	4.47	0.344	0.430	0.790	
Honey Bran	96	2.51	0.57	23.2	13.00	4.54	0.405	0.405	3.160	
Honey Comb	111	1.68	0.52	25.3	5.15	1.80	0.387	0.387	0.387	

Cereals (without milk) (cont.)	kcal	Protein (g)	Fat (g)	CHO (g)	Ca (mg)	Fe (mg)	B₁ (mg)	B₂ (mg)	Fiber (g)	Cholesterol (mg)
King Vitamin	115	1.49	1.62	24.0		17.10	0.124	1.430	0.135	
Kix	109	2.49	0.70	23.3	34.80	8.06	0.398	0.398	0.398	
Life	104	5.22	0.52	20.3	99.20	7.47	0.612	0.644	0.902	
Lucky Charms	111	2.57	1.06	23.1	31.90	4.52	0.354	0.443	0.624	
Malt-O-Meal	14	0.43	0.03	3.1	0.59	1.13	0.057	0.028	0.354	
Maypo—cooked	1	0.02	0.01	0.1	0.52	0.04	0.003	0.003	0.012	
Nutri-Grain—barley	106	3.11	0.21	23.4	7.60	1.00	0.346	0.415	1.660	
Nutri-Grain—corn	108	2.30	0.68	24.0	0.68	0.60	0.338	0.405	1.750	
Nutri-Grain—rye	102	2.48	0.21	24.0	5.67	0.80	0.354	0.425	2.160	
Nutri-Grain—wheat	102	2.45	0.32	24.0	7.73	0.80	0.387	0.451	1.800	
Oatmeal—prepared	18	0.73	0.29	3.1	2.42	0.19	0.032	0.006	0.497	
Rolled Oats	109	4.55	1.78	19.0	14.70	1.19	0.206	0.038	3.090	
Instant Oatmeal w/apples	26	0.74	0.30	5.0	30.00	1.15	0.091	0.053	0.552	
Inst Oatmeal w/bran/raisn	23	0.71	0.28	4.4	25.20	1.10	0.081	0.092	0.480	
Inst Oatmeal w/maple	30	0.84	0.35	5.8	29.60	1.16	0.097	0.059	0.530	
Inst Oatmeal w/cinn/spice	31	0.85	0.34	6.2	30.30	1.17	0.099	0.060	0.510	
Inst Oatmeal w/rais/spice	29	0.77	0.32	5.7	29.60	1.18	0.092	0.065	0.556	
100% Bran	77	3.57	1.42	20.7	19.80	3.49	0.687	0.773	8.380	
100% Natural	135	3.02	6.02	18.0	48.90	0.83	0.085	0.150	3.390	
100% Natural—w/apples	130	2.92	5.32	19.0	42.80	0.79	0.090	0.158	1.300	
100% Natural—w/rais/ dates	128	2.89	5.23	18.7	41.20	0.80	0.077	0.165	1.080	
Product 19	108	2.75	0.17	23.5	3.44	18.00	1.460	1.720	0.369	
Puffed Rice	111	1.79	0.20	25.5	2.03	0.30	0.030	0.028	0.227	
Puffed Wheat	104	4.25	0.24	22.4	7.09	1.35	0.047	0.070	5.430	
Quisp	117	1.42	2.08	23.6	8.50	5.96	0.510	0.718	0.378	
Raisin Bran—Kellogg's	91	3.07	0.46	21.4	14.50	13.90	0.293	0.332	3.410	
Raisin Bran—Post	86	2.68	0.55	21.4	13.70	4.56	0.373	0.430	3.190	
Raisins, Rice & Rye	96	1.60	0.06	24.2	6.16	3.45	0.308	0.370	0.308	
Ralston—cooked	15	0.62	0.09	3.2	1.57	0.18	0.022	0.020	0.370	
Rice Chex	112	1.49	1.00	25.2	3.88	1.79	0.400	0.298	1.840	
Rice Krispies	109	1.86	0.20	24.2	3.91	1.76	0.391	0.391	0.312	
Roman Meal—dry	91	4.07	0.60	20.4	18.40	1.31	0.142	0.069	0.905	
Roman Meal—cooked	17	0.77	0.11	3.9	3.45	0.25	0.028	0.014	0.877	
Shredded Wheat	102	3.09	0.71	22.5	11.00	1.20	0.070	0.080	3.100	
Shredded wheat	97	3.06	0.45	16.4	11.20	0.89	0.082	0.075	2.900	
Special K	111	5.58	0.10	21.3	7.97	4.48	0.399	0.399	0.266	
Sugar Corn Pops	108	1.40	0.10	25.6	0.10	1.80	0.399	0.399	0.100	
Sugar Frosted Flakes	108	1.46	0.08	25.7	0.81	1.78	0.405	0.405	0.446	
Sugar Smacks	106	2.00	0.50	24.7	2.99	1.80	0.369	0.429	0.319	
Super Golden Crisp	106	1.80	0.26	25.6	6.01	1.80	0.344	0.430	0.430	
Team	111	1.82	0.48	24.3	4.05	1.73	0.371	0.425	0.270	
Total	105	2.84	0.60	22.3	172.00	18.00	1.460	1.720	2.060	
Trix	108	1.50	0.40	24.9	5.99	4.49	0.399	0.399	0.184	
Wheat & Raisin Chex	97	2.68	0.21	22.6		4.04	0.263	0.315	1.890	
Wheat Chex	104	2.77	0.68	23.3	11.00	4.50	0.370	0.105	2.100	
Wheat germ—toasted	108	8.25	3.09	14.0	12.50	2.19	0.474	0.233	3.910	
Wheat germ w/brn sgr, honey	107	6.19	2.30	17.2	8.98	1.93	0.349	0.180	3.390	
Wheatena—cooked	16	0.58	0.13	3.4	1.28	0.16	0.002	0.006	0.385	
Wheaties	99	2.74	0.51	22.6	43.00	4.50	0.391	0.391	2.540	
Whole wheat berries	16	0.54	0.11	3.2	1.70	0.17	0.023	0.006	0.680	
Whole wheat cereal— cooked	18	0.58	0.11	3.9	1.99	0.18	0.020	0.014	0.457	

Cheese	kcal	Protein (g)	Fat (g)	CHO (g)	Ca (mg)	Fe (mg)	B₁ (mg)	B₂ (mg)	Fiber (g)	Cholesterol (mg)
American—processed	106	6.28	8.84	0.45	174	0.110	0.008	405		27.0
American cheese food—cold pack	94	5.23	6.78	2.36	145	0.240	0.009	274		18.0
American cheese spread	82	5.16	6.00	2.48	159	0.090	0.014	380		16.0
Blue	100	6.09	8.14	0.66	150	0.090	0.008	395		21.0
Brick	105	6.40	8.40	0.79	191	0.130	0.004	159		27.0
Brie	95	5.87	7.84	0.13	52	0.140	0.020	178		28.0
Camembert	85	5.60	6.86	0.13	110	0.094	0.008	236		20.0
Caraway	107	7.13	8.27	0.87	191	0.100	0.009	196		25.0
Cheddar	114	7.05	9.38	0.36	204	0.197	0.008	176		29.9
Cheshire	110	6.60	8.66	1.36	182	0.060	0.013	198		28.9
Colby	112	6.73	9.08	0.73	194	0.216	0.004	171		27.0
Cottage	29	3.54	1.20	0.76	17	0.040	0.006	115		4.2
Cottage—lowfat 2%	26	3.90	0.55	1.03	20	0.045	0.007	115		2.4
Cottage—lowfat 1%	21	3.51	0.29	0.77	18	0.040	0.006	115		1.3
Cottage—dry curd	24	4.89	0.12	0.52	9	0.065	0.007	4		2.0
Cottage—w/fruit	35	2.80	0.96	3.78	14	0.031	0.005	115		3.1
Cream	99	2.10	9.87	0.75	23	0.337	0.005	84		30.9
Edam	101	7.07	7.79	0.40	207	0.125	0.010	274		25.0
Feta	75	4.49	6.19	1.16	140	0.180	0.040	315		25.0
Gorgonzola	111	6.99	8.98	0	149	0.120	0.010	512		25.0
Gruyere	117	8.44	9.05	0.10	286	0.060	0.017	95		30.9
Fontina	110	7.25	8.62	0.44	156	0.060	0.006			32.9
Gjetost	132	2.74	8.32	12.00	113	0.130	0.009	170		25.0
Gouda	101	7.06	7.72	0.63	198	0.070	0.009	232		31.9
Liederkranz	87	4.99	7.99	0	110	0.120	0.010	389		21.0
Limburger	93	5.67	7.59	0.14	141	0.040	0.023	227		26.0
Monterey jack	106	6.93	8.56	0.19	212	0.200	0.004	152		26.0
Mozzarella—skim, low moist	80	7.60	4.67	0.89	207	0.076	0.006	150		15.0
Mozzarella—whl milk, reg	80	5.50	5.75	0.63	147	0.050	0.004	106		22.0
Mozzarella—whl mlk, low moist	90	6.10	7.19	0.43	163	0.060	0.005	119		25.0
Muenster	104	6.40	8.42	0.32	203	0.125	0.004	178		27.0
Neufchatel	74	2.82	6.70	0.83	21	0.080	0.004	113		22.0
Parmesan—hard	111	10.00	7.30	0.91	335	0.230	0.010	453		19.0
Parmesan—grated	129	11.80	8.50	1.06	389	0.270	0.013	527		22.0
Pimento processed	106	6.26	8.82	0.49	174	0.120	0.008	404		27.0
Port du salut	100	6.73	7.99	0.16	84	0.140	0.004	151		34.9
Provolone	100	7.13	7.54	0.61	214	0.146	0.005	248		20.0
Ricotta—part skim	39	3.23	2.25	1.45	77	0.126	0.006	35		8.8
Ricotta—whole milk	49	3.19	3.68	0.86	59	0.108	0.004	24		14.3
Romano	110	9.00	7.63	1.03	301	0.230	0.010	339		28.9
Romano—grated	128	10.50	8.86	1.20	350	0.270	0.013	394		32.9
Roquefort	105	6.10	8.93	0.57	188	0.172	0.010	512		26.0
Swiss	107	8.03	7.79	0.96	272	0.050	0.006	74		26.0
Swiss processed	95	7.00	6.97	0.60	219	0.170	0.004	387		24.0

Fish	kcal	Protein (g)	Fat (g)	CHO (g)	Ca (mg)	Fe (mg)	B₁ (mg)	B₂ (mg)	Fiber (g)	Cholesterol (mg)
Bass—freshwater raw	32	5.36	1.05	0	22.7	0.422	0.028	0.009	0	19.3
Bluefish—baked/broiled	45	7.43	1.42	0	2.6	0.174	0.022	0.030	0	17.9
Bluefish—fried in crumbs	58	6.44	2.78	1.33	2.3	0.151	0.017	0.023	0	17
Bluefish—raw	35	5.67	1.20	0	2.0	0.136	0.016	0.023	0	16.7
Carp—raw	36	5.05	1.59	0	11.6	0.352	0.013	0.011	0	18.7
Catfish—Channel—raw	33	5.16	1.20	0	11.3	0.275	0.013	0.030	0	16.4
Cod—baked w/butter	37	6.46	0.94	0	5.7	0.139	0.025	0.022	0	17.0
Cod—batter-fried	56	5.56	2.92	2.13	22.7	0.142	0.011	0.011	0	15.6

Fish (cont.)	kcal	Protein (g)	Fat (g)	CHO (g)	Ca (mg)	Fe (mg)	B$_1$ (mg)	B$_2$ (mg)	Fiber (g)	Cholesterol (mg)
Cod—baked/broiled	30	6.46	0.24	0	4.0	0.139	0.025	0.022	0	15.6
Cod—poached	29	6.24	0.24	0	4.0	0.139	0.025	0.022	0	15.6
Cod—steamed	29	6.24	0.24	0	4.0	0.139	0.025	0.022	0	15.9
Cod—smoked	22	5.19	0.17	0	4.0	0.113	0.023	0.020	0	14.2
Cod—Atlantic—raw	23	5.05	0.19	0	4.5	0.108	0.022	0.018	0	12.2
Cod liver oil	255	0	28.40	0				0	0	162.0
Eel—smoked	94	5.27	7.88	0.23	26.9	0.198	0.040	0.099	0	19.8
Sole/flounder—bkd w/butter	40	5.34	2.00	0	5.3	0.093	0.023	0.032	0	22.7
Sole/flounder—bkd/broiled	33	6.84	0.43	0	5.3	0.093	0.023	0.032	0	19.3
Sole/flounder—battr-fried	83	4.47	5.10	4.07	16.7	0.239	0.057	0.042	0.01	15.0
Sole/flounder—brd-fried	53	4.96	2.55	2.54	11.3	0.128	0.037	0.034	0	15.0
Sole/flounder—steamed	26	5.67	0.33	0	4.5	0.079	0.017	0.027	0	14.7
Sole/flounder—raw	26	5.33	0.34	0	5.1	0.102	0.025	0.022	0	13.6
Haddock—breaded/fried	58	5.67	3.00	2.33	11.3	0.384	0.020	0.033	0.01	18.3
Haddock—smoked	33	7.14	0.27	0	13.9	0.397	0.013	0.014	0	21.8
Haddock—raw	22	5.36	0.20	0	9.4	0.298	0.010	0.010	0	16.2
Herring—pickled	74	4.03	5.10	2.73	21.8	0.346	0.010	0.039	0	3.7
Herring—smoked/kippered	62	6.97	3.52	0	23.8	0.428	0.036	0.090	0	22.7
Herring—canned w/liquid	59	5.64	3.86	0	41.7	0.879	0.007	0.051	0	27.5
Mackerel—fried	49	7.00	2.35	0	4.3	0.445	0.045	0.116	0	19.8
Mackerel—Atlant—bkd/brld	74	6.78	5.05	0	4.3	0.445	0.045	0.117	0	21.3
Mackerel—Atlantic—raw	58	5.27	3.94	0	3.4	0.462	0.050	0.088	0	19.8
Mackerel—Pacific—raw	45	6.12	2.84	0	2.3	0.567	0.043	0.096	0	22.7
Ocean perch—breaded/fried	62	5.34	3.67	2.33	30.7	0.400	0.033	0.037	0.03	15.3
Northern pike—raw	25	5.47	0.20	0	16.2	0.156	0.017	0.018	0	11.0
Pollock—baked/broiled	28	6.60	0.31	0	19.3	0.149	0.014	0.057	0	19.8
Pollock—poached	36	6.60	0.31	0	17.0	0.149	0.010	0.050	0	19.8
Salmon—broiled/baked	61	7.74	3.10	0	2.0	0.157	0.061	0.048	0	24.7
Coho salmon—steamed/poached	52	7.77	2.14	0	8.22	0.252	0.057	0.031	0	13.9
Smoked salmon—Chinook	33	5.17	1.22	0	3.0	0.240	0.007	0.029	0	6.7
Atlantic salmon—small can	36	5.05	1.62	0	3.1	0.204	0.057	0.097	0	17.0
Pink salmon—raw	33	5.64	0.98	0	11.3	0.218	0.040	0.057	0	14.7
Sardines	59	7.00	3.24	0	108.0	0.826	0.023	0.064	0	40.4
Shad—baked with bacon	57	6.58	3.20	0	6.8	0.170	0.037	0.074	0	17.0
Smelt-rainbow—raw	28	4.99	0.69	0	17.0	0.255	0.016	0.034	0	19.8
Snapper—baked or broiled	36	7.46	0.49	0	11.3	0.068	0.015	0.021	0	13.3
Snapper—raw	28	5.81	0.38	0	9.1	0.051	0.013	0.017	0	10.5
Lemon sole—raw	23	4.85	0.21	0	4.8	0.088	0.026	0.023	0	17.0
Lemon sole—fried w/crumbs	56	4.56	3.12	2.64	26.9	0.176	0.020	0.023	0	18.4
Lemon sole—steamed	26	5.84	0.26	0	6.0	0.147	0.026	0.026	0	17.0
Seatrout/steelhead—raw	30	4.73	1.02	0	4.8	0.077	0.023	0.057	0	23.5
Seatrout/steelhead—cooked	37	6.07	1.42	0	5.7	0.088	0.024	0.064	0	32.3
Swordfish—raw	34	5.61	1.14	0	1.1	0.230	0.010	0.027	0	11.0
Swordfish—broiled/baked	44	7.20	1.46	0	1.7	0.295	0.012	0.033	0	14.2
Trout—baked/broiled	43	7.47	1.22	0	24.3	0.690	0.024	0.064	0	20.7
Tuna—oil pack	56	8.26	2.34	0	3.8	0.395	0.010	0.030	0	5.0
Tuna—water pack	37	8.38	0.14	0	3.4	0.409	0.010	0.033	0	16.0
Tuna—raw	31	6.63	0.27	0	4.5	0.207	0.123	0.013	0	12.8
Whiting—flour/bread-fried	54	5.13	1.56	1.98	11.3	0.198	0.023	0.020	0	18.4

Fruits	kcal	Protein (g)	Fat (g)	CHO (g)	Ca (mg)	Fe (mg)	B₁ (mg)	B₂ (mg)	Fiber (g)	Cholesterol (mg)
Apple w/peel	16	0.055	0.100	4.31	2.05	0.051	0.005	0.004	0.709	0
Apple slices w/peel—frsh	17	0.054	0.100	4.33	2.06	0.052	0.005	0.004	0.709	0
Apple juice—cannd/bottld	13	0.017	0.032	3.32	1.94	0.105	0.006	0.005	0.034	0
Apple juice—frozen conc	47	0.144	0.105	11.60	5.78	0.258	0.003	0.015	0.089	0
Applesauce—sweetened	22	0.052	0.052	5.67	1.11	0.111	0.003	0.008	0.397	0
Apricot—fresh halves	14	0.397	0.110	3.15	4.02	0.154	0.009	0.011	0.538	0
Apricot halves—light syrup	18	0.150	0.013	4.67	3.34	0.110	0.005	0.006	0.319	0
Apricot nectar—canned	16	0.104	0.025	4.08	2.03	0.108	0.003	0.004	0.170	0
Avocado—average	46	0.563	0.340	2.10	3.07	0.284	0.030	0.035	2.720	0
Banana—fresh slices	26	0.293	0.136	6.63	1.74	0.088	0.013	0.028	0.578	0
Blackberries—fresh	15	0.205	0.110	3.62	9.06	0.158	0.008	0.011	1.910	0
Blackberries—frozen	18	0.334	0.122	4.45	8.26	0.173	0.008	0.013	1.460	0
Blueberries—fresh	16	0.190	0.108	4.00	1.76	0.047	0.014	0.014	0.763	0
Blueberries—fzn unsweetnd	14	0.119	0.181	3.46	2.19	0.051	0.009	0.010	0.658	0
Boysenberries—frozen	14	0.314	0.075	3.46	7.73	0.240	0.015	0.010	1.100	0
Sour cherries—frozen	13	0.260	0.124	3.13	3.66	0.150	0.012	0.010	0.384	0
Sweet cherries—fresh	20	0.340	0.272	4.69	4.10	0.110	0.014	0.017	0.430	0
Sweet cherries—frozen	25	0.325	0.037	6.34	3.39	0.099	0.008	0.013	0.224	0
Cranberries—whole—raw	14	0.110	0.057	3.58	2.09	0.057	0.009	0.006	1.190	0
Cranberry/apple juice	19	0.015	0.090	4.82	2.02	0.017	0.001	0.006	0.070	0
Cranberry juice cocktail	16	0.009	0.015	4.03	0.90	0.043	0.002	0.002	0.085	0
Date—whole—each	78	0.557	0.127	20.80	9.22	0.342	0.026	0.028	2.300	0
Figs—medium—fresh	21	0.215	0.085	5.44	10.20	0.102	0.017	0.014	1.050	0
Fig—dried—each	72	0.864	0.330	18.50	40.80	0.634	0.020	0.025	3.140	0
Fruit cocktail—hvy syrup	21	0.111	0.020	5.36	1.78	0.081	0.005	0.005	0.280	0
Fruit cocktail—lite syrup	16	0.114	0.020	4.23	1.80	0.082	0.005	0.005	0.284	0
Grapefruit half—pink/red	9	0.157	0.028	2.18	3.00	0.034	0.010	0.006	0.369	0
Grapefruit half—white	10	0.195	0.029	2.38	3.36	0.017	0.010	0.006	0.368	0
Grapefruit sections—fresh	9	0.179	0.028	2.29	3.33	0.025	0.010	0.006	0.370	0
Grapefruit sections—cnd	17	0.160	0.028	4.38	4.02	0.114	0.010	0.006	0.313	0
Grapefruit juice—fresh	11	0.142	0.029	2.60	2.53	0.056	0.011	0.006	0.113	0
Grapefruit juice—sw	13	0.164	0.026	3.15	2.27	0.102	0.011	0.007	0.076	0
Grapefruit juice—unsw	11	0.148	0.028	2.54	1.95	0.057	0.012	0.006	0.077	0
Grapefruit juice—frzn conc	41	0.548	0.137	9.86	7.67	0.140	0.041	0.022	0.383	0
Grapes—Thompson	20	0.188	0.163	5.03	3.01	0.073	0.026	0.016	0.333	0
Grape juice—bottled/cnd	17	0.158	0.021	4.25	2.47	0.068	0.007	0.010	0.141	0
Grape juice—frzn conc	51	0.184	0.088	12.60	3.68	0.102	0.015	0.026	0.492	0
Grape juice—prep frozn	15	0.053	0.026	3.62	1.13	0.029	0.004	0.007	0.142	0
Kiwi fruit	17	0.280	0.127	4.22	7.46	0.112	0.007	0.015	0.962	0
Lemon—fresh wo/peel	8	0.313	0.083	2.64	7.33	0.171	0.011	0.006	0.582	0
Lemon juice—fresh	7	0.107	0.081	2.45	2.09	0.009	0.008	0.003	0.099	0
Lemon juice—bottled	6	0.114	0.081	1.84	3.02	0.036	0.012	0.003	0.085	0
Lime—fresh	9	0.199	0.055	2.99	9.30	0.169	0.008	0.006	0.228	0
Lime juice—fresh	8	0.124	0.029	2.56	2.54	0.009	0.006	0.003	0.113	0
Lime juice—bottled	6	0.115	0.115	1.84	3.46	0.069	0.009	0.001	0.099	0
Loganberries—fresh	20	0.430	0.088	3.69	8.50	0.181	0.014	0.010	1.760	0
Loganberries—frozen	16	0.430	0.089	3.68	7.33	0.181	0.014	0.010	1.760	0
Mango—fresh—slices	19	0.146	0.077	4.83	2.92	0.360	0.016	0.016	0.997	0
Mango—fresh—whole	19	0.145	0.078	4.82	2.88	0.356	0.016	0.016	1.010	0
Cantaloupe—cubes	10	0.250	0.079	2.37	3.19	0.060	0.006	0.006	0.284	0
Casaba melon—cubes	8	0.255	0.028	1.75	1.50	0.113	0.017	0.006	0.284	0
Honeydew melon—cubes	10	0.128	0.028	2.60	1.67	0.020	0.022	0.005	0.307	0
Nectarine	14	0.267	0.129	3.34	1.25	0.044	0.005	0.012	0.554	0
Melon balls—mixed—frozen	9	0.239	0.023	2.25	2.79	0.008	0.005	0.006	0.295	0
Mixed fruit—dried	69	0.697	0.139	18.20	10.6	0.767	0.012	0.045	1.220	0
Mixed fruit—frozen—thawed	28	0.397	0.052	6.87	2.04	0.079	0.005	0.010	0.386	0
Orange	13	0.266	0.035	3.33	11.30	0.029	0.025	0.011	0.680	0
Orange sections—fresh	13	0.266	0.035	3.34	11.30	0.029	0.025	0.011	0.680	0

Fruits (cont.)	kcal	Protein (g)	Fat (g)	CHO (g)	Ca (mg)	Fe (mg)	B₁ (mg)	B₂ (mg)	Fiber (g)	Cholesterol (mg)
Mandarin oranges— canned	17	0.113	0.011	4.61	2.03	0.101	0.015	0.012	0.478	0
Orange juice—fresh	13	0.199	0.057	2.95	3.09	0.057	0.025	0.008	0.113	0
Orange juice—frozen conc	45	0.679	0.059	10.80	9.05	0.099	0.079	0.018	0.313	0
Orange juice—prep frzn	13	0.191	0.016	3.05	2.50	0.031	0.023	0.005	0.057	0
Papaya—whole fresh	11	0.173	0.040	2.78	6.71	0.028	0.008	0.009	0.482	0
Papaya—slices fresh	12	0.174	0.040	2.77	6.68	0.060	0.008	0.009	0.482	0
Papaya nectar—canned	16	0.049	0.043	4.12	2.72	0.098	0.002	0.001	0.170	0
Peaches—fresh	12	0.199	0.026	3.14	1.30	0.031	0.005	0.012	0.489	0
Peach slices—frozen/ thawd	27	0.177	0.037	6.80	0.91	0.105	0.004	0.010	0.467	0
Peach halves—heavy syrup	21	0.130	0.028	5.67	1.05	0.077	0.003	0.007	0.315	0
Peach halves—light syrup	15	0.126	0.010	4.13	1.05	0.101	0.002	0.007	0.402	0
Peach halves—dried	68	0.020	0.216	17.40	8.07	1.150	0	0.060	2.330	0
Peach nectar—canned	15	0.076	0.006	3.95	1.48	0.054	0	0.004	0.170	0
Pears—Bartlett	17	0.111	0.113	4.29	3.24	0.070	0.006	0.011	0.779	0
Pear halves—heavy syrup	21	0.057	0.036	5.42	1.44	0.061	0.004	0.006	0.395	0
Pear halves—lite syrup	16	0.054	0.007	4.30	1.44	0.082	0.003	0.005	0.395	0
Pear nectar—canned	17	0.030	0.003	4.47	1.25	0.073	0	0.004	0.204	0
Pineapple slices—hvy syrup	22	0.098	0.034	5.72	3.91	0.108	0.025	0.007	0.270	0
Pineapple slices—lite syrp	15	0.103	0.034	3.81	3.91	0.110	0.026	0.007	0.268	0
Pineapple—frozen— sweetened	24	0.113	0.029	6.29	2.55	0.113	0.028	0.009	0.496	0
Pineapple juice—frzn conc	51	0.369	0.029	12.60	11.00	0.255	0.065	0.017	0.255	0
Plums	16	0.223	0.176	3.69	1.29	0.030	0.012	0.027	0.550	0
Plums—cnd—hvy syrup	25	0.102	0.028	6.59	2.53	0.238	0.005	0.010	0.434	0
Plums—cnd—lite syrup	18	0.105	0.029	4.61	2.70	0.243	0.005	0.011	0.444	0
Prunes—dried	68	0.739	0.145	17.80	14.50	0.702	0.023	0.046	2.700	0
Prune juice—bottled	20	0.172	0.009	4.95	3.43	0.334	0.005	0.020	0.310	0
Raisins—seedless	85	0.914	0.130	22.50	13.90	0.594	0.044	0.025	1.670	0
Raspberries—fresh	14	0.256	0.157	3.27	6.22	0.162	0.009	0.026	1.770	0
Raspberries—canned w/liq	26	0.235	0.034	6.62	2.99	0.120	0.006	0.009	1.200	0
Raspberries—frozen	29	0.197	0.044	7.42	4.30	0.184	0.005	0.013	1.300	0
Rhubarb—raw—diced	6	0.253	0.056	1.29	39.50	0.062	0.006	0.009	0.737	0
Rhubarb—cooked w/sugar	33	0.111	0.013	8.85	41.10	0.060	0.005	0.006	0.624	0
Strawberries—fresh	9	0.173	0.105	2.00	4.00	0.108	0.006	0.019	0.736	0
Strawberries—frozen	10	0.120	0.030	2.59	4.38	0.213	0.006	0.010	0.736	0
Tangerine—fresh	13	0.179	0.054	3.17	4.05	0.028	0.030	0.006	0.574	0
Tangerines—cnd—lite syrup	17	0.127	0.028	4.61	2.03	0.105	0.015	0.012	0.453	0
Watermelon	9	0.175	0.121	2.04	2.24	0.048	0.023	0.006	0.114	0

Meats	kcal	Protein (g)	Fat (g)	CHO (g)	Ca (mg)	Fe (mg)	B₁ (mg)	B₂ (mg)	Fiber (g)	Cholesterol (mg)
Beef chuck—pot rstd	108	7.20	8.64	0	3.67	0.840	0.020	0.065	0	29.0
Beef chuck—pot rstd— lean	77	8.80	4.34	0	3.67	1.040	0.024	0.080	0	30.0
Beef round—pot rstd— ln&ft	74	8.44	4.20	0	1.67	0.920	0.020	0.069	0	27.0
Beef round—pot rstd lean	63	8.97	2.74	0	1.33	0.980	0.021	0.074	0	27.0
Ground beef—lean	77	7.00	5.34	0	3.00	0.600	0.013	0.060	0	24.7
Ground beef—regular	82	6.67	5.94	0	3.00	0.700	0.010	0.053	0	25.3
Sirloin steak—lean	57	8.10	2.53	0	2.33	0.700	0.026	0.056	0	21.7
T-bone steak—lean & fat	92	6.80	6.97	0	2.67	0.720	0.026	0.059	0	23.7
Beef—thin-sliced lunchmt	50	7.96	1.09	1.62	2.99	0.759	0.023	0.054	0	12.0

Meats (cont.)	kcal	Protein (g)	Fat (g)	CHO (g)	Ca (mg)	Fe (mg)	B₁ (mg)	B₂ (mg)	Fiber (g)	Cholesterol (mg)
Beef lunchmeat—loaf/roll	87	4.06	7.42	0.82	2.99	0.659	0.030	0.062	0	18.0
Beef rib—oven rstd—lean	68	7.70	3.90	0	3.34	0.740	0.023	0.060	0	22.7
Beef round—oven rstd—lean	54	8.14	2.12	0	1.67	0.834	0.028	0.076	0	23.0
Beef rump roast—lean only	51	8.40	1.89	0	1.13	0.567	0.026	0.049	0	19.6
Beef brains—pan-fried	56	3.57	4.50	0	2.67	0.630	0.037	0.074	0	566.0
Beef heart	47	8.17	1.59	0.12	1.67	2.130	0.040	0.436	0	54.7
Beef kidney	41	7.23	0.97	0.27	5.00	2.070	0.054	1.150	0	110.0
Beef liver—fried	61	7.57	2.27	2.23	3.00	1.780	0.060	1.170	0	137.0
Beef tongue—cooked	80	6.27	5.87	0.09	2.00	0.960	0.009	0.009	0	30.4
Beef tripe—raw	28	4.14	1.12	0	36.00	0.553	0.002	0.047	0	26.9
Beef tripe—pickled	17	3.29	0.40	0	25.00	0.389	0	0.028	0	15.0
Corned beef—canned	71	7.67	4.24	0	5.67	0.590	0.006	0.042	0	24.3
Corned beef hash—canned	49	2.35	1.29	2.82	3.74	0.567	0.016	0.052	0.153	17.0
Beef—dried/cured	47	8.24	1.10	0.44	2.00	1.280	0.050	0.230	0	45.9
Beef & veg stew	26	1.85	1.27	1.74	3.36	0.336	0.017	0.020	0.393	8.2
Beef stew—canned	22	1.64	0.88	2.06	2.66	0.368	0.008	0.014	0.150	1.7
Burrito—beef & bean	63	3.40	2.84	6.48	26.70	0.437	0.042	0.047	0.810	8.4
Tostada w/beans & beef	49	2.72	3.06	2.98	27.50	0.319	0.012	0.036	0.583	9.1
Beef + macaroni + tom	24	1.25	0.73	3.16	3.81	0.300	0.024	0.021	0.291	2.8
Beef enchilada	69	3.09	3.26	3.69	60.20	0.418	0.017	0.039	0.465	8.9
Frankfurter—beef	92	3.20	8.36	0.68	3.48	0.378	0.014	0.029	0	13.4
Frankfurter—beef & pork	91	3.20	8.26	0.73	2.98	0.328	0.056	0.034	0	14.4
Beef pot pie—fr/frozen	52	1.99	2.73	4.77	2.42	0.436	0.022	0.018	0.109	5.0
Beef pot pie—recipe	70	2.84	4.05	5.27	3.92	0.513	0.039	0.039	0.155	5.7
Beef taco	75	4.94	4.80	3.67	30.90	0.469	0.010	0.049	0.407	16.2
Chicken meat—all-fried	62	8.67	2.59	0.48	4.86	0.383	0.024	0.056	0.002	26.5
Chicken meat—all-roasted	54	8.20	2.10	0	4.25	0.342	0.020	0.050	0	25.3
Chicken meat—all-stewed	50	7.74	1.90	0	4.05	0.330	0.014	0.046	0	23.5
Boned chicken w/broth	47	6.17	2.26	0	3.99	0.439	0.004	0.037	0	17.6
Chicken—dark meat—fried	68	8.22	3.30	0.74	5.06	0.423	0.026	0.070	0.002	27.3
Chicken—dark meat—roasted	58	7.76	2.75	0	4.25	0.377	0.020	0.064	0	26.3
Chicken—dark meat—stewed	55	7.37	2.55	0	4.05	0.385	0.016	0.057	0	24.9
Chicken—light meat—fried	54	9.31	1.57	0.12	4.45	0.322	0.020	0.036	0	25.3
Chicken—light meat—roastd	49	8.77	1.28	0	4.25	0.302	0.018	0.033	0	23.9
Chicken—light meat—stewed	45	8.18	1.17	0	3.64	0.265	0.012	0.033	0	21.7
Chick breast—no skin	47	8.80	0.99	0	4.29	0.295	0.020	0.032	0	24.0
Chick breast meat—stewed	43	8.20	0.86	0	3.58	0.250	0.012	0.034	0	21.8
Chick drumstick—battr frd	76	6.22	4.45	2.36	4.73	0.382	0.032	0.061	0.008	24.4
Chick drumstick—roasted	61	7.69	3.16	0	3.27	0.376	0.020	0.061	0	26.2
Chick wing—batter-fried	92	5.64	6.19	3.10	5.79	0.365	0.030	0.043	0.012	22.6
Chick wing—flour-fried	91	7.40	6.28	0.67	4.43	0.354	0.017	0.039	0.009	23.0
Chick wing—roasted	83	7.48	5.53	0	4.17	0.359	0.012	0.037	0	24.2
Chicken gizzards—simmered	44	7.73	1.04	0.32	2.58	1.180	0.008	0.070	0	54.9
Chicken hearts—simmered	52	7.49	2.23	0.03	5.27	2.560	0.017	0.206	0	68.7
Chicken livers—simmered	44	6.90	1.55	0.25	4.05	2.400	0.043	0.496	0	179.0
Chicken roll—light meat	45	5.52	2.08	0.69	11.90	0.274	0.018	0.037	0	13.9
Chicken frankfurter	73	3.67	5.52	1.93	27.00	0.567	0.019	0.033	0	28.4
Chicken a la king	54	3.12	3.93	1.39	14.70	0.289	0.012	0.049	0.154	25.6
Chicken + noodles	43	2.60	2.13	3.07	3.07	0.278	0.006	0.020	0.142	12.2
Chicken chow mein	29	2.60	1.25	1.13	6.58	0.284	0.009	0.026	0.466	8.5

Meats (cont.)	kcal	Protein (g)	Fat (g)	CHO (g)	Ca (mg)	Fe (mg)	B₁ (mg)	B₂ (mg)	Fiber (g)	Cholesterol (mg)
Chicken curry	26	2.21	1.53	0.64	2.52	0.170	0.009	0.019	0.033	5.3
Chicken frankfurter	72	3.67	5.52	1.93	27.00	0.567	0.019	0.033	0	28.4
Chicken pot pie—fr/frozen	53	1.84	2.84	5.08	3.70	0.382	0.020	0.020	0.210	4.9
Chicken roll—light meat	45	5.52	2.08	0.69	11.90	0.274	0.018	0.037	0	13.9
Chicken salad w/celery	97	3.82	8.90	0.47	5.92	0.239	0.012	0.028	0.109	17.3
Chicken patty sandwich	79	4.48	4.06	6.10	7.95	0.338	0.052	0.047	0.244	12.3
Chicken broth—from dry	2	0.16	0.13	0.17	1.74	0.009	0	0.004	0.001	0.1
Chicken broth—from cube	2	0.11	0.04	0.18	0.04	0.014	0.001	0.003	0	0.1
Chicken noodle soup	9	0.48	0.29	1.10	2.00	0.092	0.006	0.007	0.085	0.8
Tostada w/beans/chicken	45	3.50	2.06	3.38	29.30	0.305	0.013	0.034	0.668	9.6
Chicken taco	63	5.60	3.03	3.67	31.60	0.327	0.014	0.043	0.407	16.5
Chicken enchilada	64	3.38	2.48	3.69	60.50	0.359	0.018	0.036	0.465	9.1
Turkey dark meat—roasted	53	8.10	2.05	0	9.11	0.662	0.018	0.070	0	24.0
Turkey white meat—roasted	44	8.48	0.91	0	5.47	0.380	0.017	0.037	0	19.6
Turkey breast—barbecued	40	6.39	1.40	0	2.00	0.120	0.010	0.030	0	16.0
Turkey gizzards	46	8.34	1.10	0.17	4.32	1.540	0.009	0.093	0	65.6
Turkey hearts	50	7.58	1.73	0.58	3.72	1.950	0.019	0.250	0	64.0
Turkey livers	48	6.80	1.69	0.97	3.02	2.210	0.015	0.404	0	177.0
Turkey loaf	31	6.38	0.45	0	2.00	0.113	0.011	0.030	0	11.6
Turkey roll	41	5.27	2.03	0.15	11.40	0.358	0.025	0.064	0	11.9
Turkey bologna	56	3.86	4.27	0.27	23.40	0.432	0.015	0.047	0	28.0
Turkey frankfurter	64	4.05	5.22	0.42	36.50	0.485	0.023	0.050	0	24.6
Turkey ham	36	5.37	1.49	0.42	2.49	0.776	0.020	0.075	0	15.9
Turkey pastrami	37	5.22	1.75	0.43	2.49	0.403	0.022	0.075	0	14.9
Turkey salami	55	4.62	3.89	0.15	5.47	0.463	0.029	0.075	0	22.9
Turkey pot pie—frozen	51	1.80	2.75	4.65	7.79	0.256	0.020	0.020	0.110	2.4

Eggs	kcal	Protein (g)	Fat (g)	CHO (g)	Ca (mg)	Fe (mg)	B₁ (mg)	B₂ (mg)	Fiber (g)	Cholesterol (mg)
Egg white, cooked	13	2.68	0	0.33	3.20	0.008	0.002	0.072	0	0
Egg yolk, cooked	108	4.76	8.73	0.07	44.40	1.620	0.044	0.121	0	355.0
Fried egg, in butter	56	3.54	3.45	0.38	15.70	0.536	0.018	0.148	0	129.0
Hard-cooked egg	40	3.52	2.79	0.34	13.80	0.474	0.014	0.132	0	113.0
Poached egg	40	3.50	2.79	0.34	13.80	0.474	0.014	0.132	0	113.0
Hard-cooked egg	40	3.52	2.79	0.34	13.80	0.474	0.014	0.132	0	113.0
Scrambled egg, milk + butter	40	2.88	2.57	0.61	23.90	0.412	0.013	0.106	0	93.9
Egg raw—large	40	3.52	2.79	0.34	13.80	0.474	0.017	0.139	0	113.0
Egg white—raw	13	2.68	0	0.33	3.20	0.008	0.002	0.075	0	0
Egg yolk, raw	108	4.76	8.73	0.07	44.40	1.620	0.051	0.126	0	355.0
Egg substitute, frozen	45	3.20	3.15	0.91	20.80	0.562	0.034	0.110	0	0.5
Egg substitute, powder	125	15.60	3.66	6.12	90.70	0.879	0.062	0.493	0	162.0

Dairy Products	kcal	Protein (g)	Fat (g)	CHO (g)	Ca (mg)	Fe (mg)	B₁ (mg)	B₂ (mg)	Fiber (g)	Cholesterol (mg)
Milk—1% lowfat	12	0.93	0.30	1.36	34.9	0.014	0.011	0.047	0	1.16
Milk—2% lowfat	14	0.94	0.56	1.36	34.5	0.012	0.011	0.047	0	2.56
Milk—skim	10	0.97	0.05	1.38	34.9	0.012	0.010	0.040	0	0.46
Milk—whole	17	0.93	0.95	1.32	33.8	0.014	0.010	0.046	0	3.83
Buttermilk	12	0.94	0.25	1.35	33.0	0.014	0.010	0.044	0	1.04
Milk—instant nonfat dry	102	9.96	0.21	14.80	349.0	0.088	0.117	0.496	0	5.00
Canned skim milk—evap	22	2.11	0.06	3.22	82.0	0.078	0.013	0.088	0	1.11

Dairy Products (cont.)	kcal	Protein (g)	Fat (g)	CHO (g)	Ca (mg)	Fe (mg)	B₁ (mg)	B₂ (mg)	Fiber (g)	Cholesterol (mg)
Canned whole milk—evap	38	1.91	2.20	2.81	73.9	0.054	0.014	0.090	0	8.33
Carob flavor mix—powder	106	0.47	0.05	26.50		1.300	0.002		4.020	0
Chocolate milk—1%	18	0.92	0.28	2.96	32.5	0.068	0.010	0.047	0.425	0.79
Chocolate milk—2%	20	0.91	0.57	2.95	32.2	0.068	0.010	0.046	0.425	1.93
Chocolate milk—whole	24	0.90	0.96	2.94	31.8	0.068	0.010	0.046	0.425	3.52
Hot cocoa—with whole milk	25	1.03	1.03	2.93	33.8	0.088	0.012	0.049	0.340	3.74
Inst breakfast w/2% milk	25	1.52	0.47	3.50	31.0	0.807	0.040	0.048	0	1.82
Inst breakfast w/1% milk	23	1.51	0.25	3.50	31.3	0.807	0.040	0.048	0	1.00
Inst breakfast w/skim milk	22	1.55	0.04	3.50	31.4	0.804	0.039	0.041	0	0.40
Inst breakfast w/whl milk	28	1.51	0.82	3.47	30.4	0.807	0.040	0.047	0	3.33
Egg nog	38	1.08	2.12	3.84	36.8	0.057	0.010	0.054	0	16.60
Kefir	20	1.13	0.55	1.07	42.6	0.060	0.055	0.054	0	1.22
Malt powder—choc flavored	107	1.49	1.08	24.80	17.6	0.648	0.049	0.057	0.540	1.35
Malted milk powder	117	3.12	2.30	21.50	85.0	0.209	0.143	0.260	0.405	5.40
Malted milk drink—choc	25	1.00	0.95	3.19	32.5	0.064	0.014	0.047	0.043	3.64
Chocolate milkshake	36	0.96	1.05	5.80	32.0	0.088	0.016	0.069	0.035	3.70
Strawberry milkshake	32	0.95	0.80	5.35	32.0	0.030	0.013	0.055	0.024	3.10
Vanilla milkshake	32	0.98	0.84	5.09	34.5	0.026	0.013	0.052	0.019	3.20
Ovaltine powder—choc flvr	102	2.00	0.85	23.70	134.0	6.190	0.719	0.772	0.013	0
Ovaltine powder—malt flvr	104	2.53	0.24	23.60	106.0	5.820	0.772	1.010	0.040	0
Ovaltine drink—choc flvr	24	1.02	0.94	3.12	41.9	0.510	0.067	0.104	0.001	3.53
Ovaltine drink—malt flvr	24	1.06	0.89	3.10	39.7	0.480	0.072	0.124	0.003	3.53
Milk—goat	20	1.00	1.17	1.27	37.9	0.014	0.014	0.039	0	3.25
Milk—sheep	31	1.70	1.99	1.52	54.8	0.028	0.018	0.100	0	
Milk—soybean	9	0.78	0.54	0.51	1.2	0.163	0.046	0.020	0	0
Ice cream—regular-vanilla	57	1.02	3.05	6.76	37.5	0.026	0.011	0.070		12.60
Ice cream—rich-vanilla	67	0.79	4.54	6.13	28.9	0.019	0.008	0.054		16.90
Ice cream—soft-serve	62	1.15	3.69	6.28	38.7	0.070	0.013	0.073		25.00
Creamsicle ice cream bar	44	0.52	1.33	7.56	19.8	0	0.009	0.034		
Drumstick ice cream bar	88	1.23	4.68	10.20	31.7	0.047	0.009	0.043		
Fudgesicle ice cream bar	35	1.48	0.08	7.22	50.0	0.039	0.012	0.070		
Ice milk	40	1.12	1.22	6.28	38.0	0.039	0.016	0.075		3.90
Ice milk—soft-serve—3% fat	36	1.30	0.75	6.22	44.4	0.045	0.019	0.088		2.10
Yogurt—coffee-vanilla	24	1.40	0.35	3.90	48.5	0.020	0.012	0.057		1.42
Yogurt—lowfat with fruit	29	1.24	0.31	5.37	43.0	0.020	0.010	0.050		1.25
Yogurt—lowfat-plain	18	1.49	0.43	2.00	51.8	0.022	0.012	0.060		1.75
Yogurt—nonfat milk	16	1.62	0.05	2.17	56.5	0.025	0.014	0.066		0.50
Yogurt—whole milk	17	0.98	0.92	1.32	34.3	0.014	0.008	0.040		3.68

Vegetables	kcal	Protein (g)	Fat (g)	CHO (g)	Ca (mg)	Fe (mg)	B₁ (mg)	B₂ (mg)	Fiber (g)	Cholesterol (mg)
Broccoli—frzn ckd spears	8	0.879	0.034	1.51	14.5	0.173	0.015	0.023	0.826	0
Brussels sprouts—raw	12	0.960	0.084	2.54	11.6	0.396	0.039	0.026	1.260	0
Brussels sprouts—ckd	11	1.090	0.145	2.45	10.2	0.342	0.030	0.023	1.220	0
Brussels sprouts—frzn, ckd	12	1.030	0.112	2.36	7.0	0.210	0.029	0.032	1.230	0
Cabbage—raw, shredded	7	0.340	0.049	1.52	13.0	0.162	0.015	0.009	0.680	0
Cabbage—cooked	6	0.272	0.070	1.35	9.5	0.110	0.016	0.016	0.661	0
Bok choy—raw, shredded	4	0.425	0.057	0.62	30.0	0.227	0.011	0.020	0.486	
Bok choy—cooked	3	0.442	0.045	0.51	26.3	0.295	0.009	0.018	0.454	
Red cabbage—raw	8	0.393	0.073	1.74	14.6	0.142	0.018	0.009	0.648	0
Red cabbage—cooked	6	0.299	0.057	1.32	10.6	0.102	0.010	0.006	1.567	0
Carrot—whole, raw	12	0.291	0.055	2.87	7.5	0.142	0.028	0.017	1.906	0
Carrot—grated, raw	12	0.289	0.052	2.88	7.7	0.142	0.027	0.016	0.907	0

Vegetables (cont.)	kcal	Protein (g)	Fat (g)	CHO (g)	Ca (mg)	Fe (mg)	B₁ (mg)	B₂ (mg)	Fiber (g)	Cholesterol (mg)
Carrots—sliced, cooked	13	0.309	0.050	2.97	8.7	0.176	0.010	0.016	0.992	0
Carrots—fr/frozen, cooked	10	0.338	0.031	2.34	8.2	0.136	0.008	0.010	1.050	0
Carrots—canned, drained	7	0.183	0.054	1.57	7.4	0.181	0.005	0.009	0.435	0
Carrot juice	11	0.267	0.041	2.63	6.7	0.130	0.026	0.015	0.385	0
Cauliflower—raw	7	0.561	0.051	1.39	7.9	0.164	0.022	0.016	0.720	0
Cauliflower—cooked	7	0.530	0.050	1.31	7.8	0.119	0.018	0.018	0.622	0
Cauliflower—frozen, cooked	5	0.457	0.061	1.06	4.9	0.116	0.010	0.015	0.535	0
Celery—raw—chopped	5	0.189	0.033	1.03	10.4	0.137	0.009	0.009	0.472	0
Swiss chard—raw	5	0.510	0.057	1.06	14.5	0.510	0.011	0.025	0.512	0
Swiss chard—cooked	6	0.533	0.023	1.17	16.5	0.642	0.010	0.024	0.616	0
Collards—fresh	5	0.445	0.062	1.07	33.0	0.299	0.009	0.018	0.590	0
Artichoke hearts—marinated	28	0.680	0.250	2.18	6.5	0.270	0.010	0.029	1.780	0
Asparagus—raw spears	6	0.865	0.063	1.05	6.4	0.193	0.032	0.035	0.395	0
Asparagus—canned spears	5	0.606	0.184	0.70	3.9	0.177	0.017	0.025	0.454	0
Bamboo shoots—sliced—raw	8	0.738	0.088	1.47	3.8	0.143	0.043	0.020	0.738	0
Bamboo shoots—sliced, canned	5	0.489	0.113	0.91	2.2	0.090	0.007	0.007	0.706	0
Bean sprouts—fresh raw	9	0.861	0.050	1.68	3.8	0.258	0.024	0.035	0.736	0
Bean sprouts—boiled	6	0.576	0.025	1.19	3.4	0.185	0.014	0.029	0.572	0
Bean sprouts—stir fried	14	1.220	0.059	3.00	3.7	0.549	0.040	0.050	0.777	0
Black beans—cooked	37	2.500	0.152	6.72	7.8	0.593	0.069	0.017	2.540	0
Green beans—raw uncooked	9	0.515	0.034	2.02	10.6	0.363	0.024	0.030	0.644	0
Green beans—fresh—cooked	10	0.535	0.082	2.24	13.2	0.363	0.021	0.027	0.737	0
Green beans—frzn—cooked	8	0.386	0.038	1.73	12.8	0.233	0.014	0.021	0.880	0
Green beans—canned/draind	6	0.326	0.028	1.28	7.6	0.256	0.004	0.016	0.378	0
Red kidney beans—dry	94	6.690	0.234	16.90	40.5	2.330	0.150	0.062	6.160	0
Lima beans—dry large	96	6.080	0.194	18.00	22.9	2.130	0.144	0.057	8.600	0
Lima beans—fresh-cooked	35	1.930	0.090	6.70	9.0	0.695	0.040	0.027	2.670	0
Lima beans—dry small	98	5.790	0.397	18.20	18.4	2.270	0.136	0.048	8.500	0
Lima beans—canned/drained	27	1.530	0.100	5.20	8.0	0.487	0.010	0.013	2.420	0
Beans w/franks—canned	40	1.900	1.860	4.37	13.6	0.490	0.016	0.016	1.950	1.7
Pork & beans—canned	32	1.500	0.413	5.95	17.4	0.470	0.013	0.017	1.560	1.9
Navy beans—dry, cooked	40	2.460	0.162	7.77	19.9	0.703	0.057	0.017	2.490	0
Pinto beans—dry, cooked	39	2.320	0.148	7.28	13.6	0.741	0.053	0.026	3.230	0
Refried beans—canned	30	1.770	0.303	5.24	13.2	0.500	0.014	0.016	2.470	0
Soybeans—dry	118	10.400	5.650	8.55	78.5	4.450	0.248	0.247	1.570	0
White beans—dry	95	5.990	0.334	17.70	36.0	2.190	0.210	0.059	0.766	0
White beans—dry, cooked	40	2.550	0.182	7.32	20.7	0.808	0.067	0.017	2.230	0
Yellow wax beans—raw	9	0.515	0.034	2.02	10.6	0.294	0.024	0.030	0.644	0
Yellow wax beans—raw	10	0.535	0.082	2.24	13.2	0.363	0.021	0.027	0.726	0
Yellow wax beans—frzn	8	0.386	0.038	1.73	12.8	0.233	0.014	0.021	0.880	0
Beets—cooked	9	0.300	0.014	1.90	3.1	0.176	0.009	0.004	0.539	0
Beets—pickled slices	18	0.227	0.028	4.63	3.1	0.116	0.006	0.014	0.587	0
Broccoli—raw chopped	8	0.844	0.097	1.49	13.5	0.251	0.019	0.034	0.934	0
Broccoli—raw spears	8	0.845	0.098	1.49	13.5	0.250	0.018	0.034	0.935	0
Collards—fresh, cooked	4	0.313	0.043	0.75	22.0	0.116	0.005	0.012	0.798	0
Collards—frozen, cooked	10	0.840	0.115	2.02	59.5	0.317	0.013	0.033	0.794	0
Corn—kernels raw	24	0.913	0.335	5.38	0.6	0.147	0.057	0.017	1.220	0
Corn on cob—cooked	31	0.943	0.363	7.14	0.6	0.173	0.061	0.020	1.190	0
Corn—cooked from frozen	23	0.857	0.020	5.80	0.6	0.085	0.020	0.020	1.190	0
Corn—canned, drained	23	0.743	0.283	5.26	1.4	0.142	0.009	0.014	0.398	0
Corn—canned cream style	21	0.494	0.119	5.14	0.9	0.108	0.007	0.015	0.354	0
Cucumber slices w/peel	4	0.153	0.037	0.82	4.0	0.079	0.009	0.006	0.329	0

Vegetables (cont.)	kcal	Protein (g)	Fat (g)	CHO (g)	Ca (mg)	Fe (mg)	B₁ (mg)	B₂ (mg)	Fiber (g)	Cholesterol (mg)
Eggplant—cooked	8	0.236	0.066	1.88	1.7	0.099	0.022	0.006	1.060	0
Escarole/curly endive—chp	5	0.354	0.057	0.95	14.7	0.235	0.023	0.022	0.369	0
Garbanzo/chickpeas—dry	103	5.470	1.720	17.20	29.9	1.770	0.135	0.060	5.390	0
Garbanzo/chickpeas—cooked	47	2.500	0.735	7.78	13.8	0.819	0.033	0.018	1.920	0
Kale—fresh, chopped	14	0.935	0.198	2.84	38.0	0.482	0.031	0.037	1.650	0
Jerusalem artichoke—raw	22	0.567	0.004	4.95	4.0	0.964	0.057	0.017	0.369	0
Lentils—dry	96	7.960	0.273	16.20	14.6	2.550	0.135	0.069	3.400	0
Lentils—cooked from dry	33	2.560	0.106	5.73	5.3	0.944	0.048	0.020	1.430	0
Kohlrabi—raw slices	8	0.482	0.028	1.76	6.9	0.113	0.014	0.006	0.405	0
Kohlrabi—cooked	8	0.510	0.030	1.96	7.0	0.113	0.011	0.006	0.395	0
Leeks—chopped raw	17	0.425	0.085	4.00	16.7	0.594	0.017	0.008	0.668	0
Leeks—cooked, chopped	9	0.230	0.057	2.16	8.5	0.310	0.007	0.006	0.927	0
Lettuce—butterhead	4	0.367	0.062	0.66	9.5	0.085	0.017	0.017	0.397	0
Lettuce—iceberg	4	0.286	0.054	0.59	5.4	0.142	0.013	0.009	0.347	0
Lettuce—romaine	5	0.459	0.057	0.67	10.2	0.312	0.028	0.028	0.482	0
Mushrooms—raw sliced	7	0.593	0.119	1.32	1.4	0.352	0.029	0.127	0.508	0
Mushrooms—cooked	8	0.614	0.134	1.46	1.7	0.494	0.020	0.085	0.625	0
Mushrooms—canned, drained	7	0.530	0.082	1.40	3.1	0.224	0.017	0.063	0.596	0
Mustard greens—fresh	7	0.764	0.057	1.39	29.4	0.414	0.023	0.031	0.764	0
Mustard greens—ckd	4	0.640	0.068	0.60	21.0	0.316	0.012	0.018	0.587	0
Okra pods—cooked	9	0.530	0.048	2.04	17.9	0.128	0.037	0.016	0.624	0
Okra slices—cooked	11	0.589	0.085	2.32	27.1	0.190	0.028	0.035	0.709	0
Onions—chopped, raw	10	0.335	0.074	2.07	7.1	0.105	0.017	0.003	0.454	0
Onion slices—raw	10	0.335	0.074	2.08	7.2	0.105	0.017	0.003	0.454	0
Onion—dehydrated flakes	91	2.530	0.122	23.70	72.9	0.446	0.022	0.018	2.510	0
Onion rings—frozen, heated	115	1.520	7.570	10.80	8.8	0.482	0.079	0.040	0.240	0
Parsley—freeze dried	81	8.910	1.420	11.90	40.5	15.200	0.304	0.648	22.000	0
Parsley—fresh chopped	9	0.624	0.085	1.96	36.9	1.760	0.023	0.031	1.550	0
Parsnips—sliced raw	21	0.341	0.085	5.09	10.0	0.167	0.026	0.014	1.280	0
Fresh peas—uncooked	23	1.530	0.113	4.10	7.0	0.416	0.075	0.037	1.380	0
Peas—cooked	24	1.520	0.060	4.43	7.8	0.438	0.073	0.042	1.360	0
Peas—frozen, cooked	22	1.460	0.078	4.04	6.7	0.443	0.080	0.050	1.280	0
Peas—edible pods—fresh	12	0.794	0.057	2.15	12.1	0.589	0.043	0.023	0.794	0
Split peas—dry	97	6.970	0.328	17.10	15.5	1.250	0.206	0.061	4.030	0
Peas + carrots—frzn, ckd	14	0.875	0.120	2.87	6.4	0.266	0.064	0.020	1.170	0
Green chili pepper—raw	11	0.567	0.057	2.68	5.0	0.340	0.026	0.026	0.504	0
Red chili peppers—raw/chopped	11	0.567	0.057	2.68	4.9	0.340	0.026	0.026	0.454	0
Jalapeno peppers—cnd/chopped	7	0.225	0.170	1.39	7.5	0.792	0.008	0.014	0.850	0
Baked potato with skin	31	0.653	0.028	7.16	2.8	0.386	0.030	0.009	0.660	0
Baked potato—flesh only	26	0.556	0.029	6.10	1.5	0.100	0.030	0.006	0.436	0
Potato skin—oven baked	56	1.220	0.029	13.20	9.8	1.080	0.035	0.034	1.130	0
Potato + peel—microwaved	30	0.692	0.028	6.83	3.1	0.350	0.034	0.009	0.660	0
Peeled potato—boiled	24	0.485	0.029	5.67	2.1	0.088	0.028	0.005	0.426	0
French fries—oven heated	63	0.980	2.480	9.64	2.3	0.380	0.035	0.009	0.567	0
French fries—frzn—veg oil	90	1.140	4.690	11.20	5.7	0.215	0.050	0.008	0.567	0
Cottage—fried potatoes	62	0.975	2.320	9.64	2.8	0.425	0.034	0.009	0.567	0
Hash-brown potatoes	30	0.343	1.980	3.02	1.1	0.115	0.010	0.003	0.567	0
Mashed potatoes prep/milk	22	0.548	0.166	4.98	7.4	0.077	0.025	0.011	0.405	0.5
Mashed potatoes—mlk & margarine	30	0.533	1.200	4.74	7.3	0.074	0.024	0.014	0.405	0.5
Potato pancakes	88	1.730	4.700	9.85	7.8	0.451	0.039	0.035	0.563	34.7
Potatoes au gratin mix	26	0.653	1.170	3.65	23.5	0.090	0.006	0.023	0.487	1.4
Scalloped potatoes—recipe	24	0.813	1.040	3.05	16.2	0.163	0.020	0.026	0.289	3.4
Potato chips	148	1.590	10.000	14.70	7.0	0.339	0.040	0.006	1.360	0

Vegetables (cont.)	kcal	Protein (g)	Fat (g)	CHO (g)	Ca (mg)	Fe (mg)	B₁ (mg)	B₂ (mg)	Fiber (g)	Cholesterol (mg)
Potato flour	100	2.260	0.226	22.60	9.3	4.880	0.119	0.040	0.317	0
Pumpkin—canned	10	0.311	0.080	2.28	7.4	0.394	0.007	0.015	0.519	0
Red radishes	4	0.170	0.151	1.01	5.7	0.082	0.001	0.013	0.624	0
Rutabaga—cooked cubes	10	0.314	0.053	2.19	12.0	0.133	0.020	0.010	0.434	0
Sauerkraut—canned w/ liqd	5	0.258	0.040	1.21	8.7	0.417	0.006	0.006	0.529	0
Spinach—cooked from fresh	7	0.843	0.074	1.06	38.4	1.010	0.027	0.067	0.702	0
Summer squash—raw slices	6	0.334	0.061	1.23	5.7	0.130	0.018	0.010	0.425	0
Zucchini squash—cooked	5	0.181	0.014	1.11	3.6	0.099	0.012	0.012	0.567	0
Acorn squash—boiled/ mashed	10	0.190	0.023	2.49	7.5	0.159	0.028	0.002	0.680	0
Butternut squash—bkd— cube	12	0.256	0.026	2.97	11.6	0.170	0.020	0.005	0.794	0
Winter squash—boiled	10	0.319	0.050	2.40	5.2	0.136	0.018	0.006	0.794	0
Sweet potato—bkd in skin	29	0.487	0.032	6.89	8.0	0.129	0.020	0.036	0.850	0
Candied sweet potatoes	39	0.246	0.920	7.91	7.3	0.324	0.005	0.012	0.545	0
Tofu (soybean curd)	22	2.290	1.360	0.53	29.7	1.520	0.023	0.015	0.343	0
Tomato—fresh whole	6	0.251	0.060	1.23	2.1	0.136	0.017	0.014	0.415	0
Tomatoes—whole canned	6	0.265	0.070	1.22	7.4	0.171	0.013	0.009	0.298	0
Tomato sauce—canned	9	0.376	0.047	2.04	3.9	0.218	0.019	0.016	0.425	0
Tomato paste—canned	24	1.070	0.249	5.33	9.9	0.848	0.044	0.054	1.210	0
Tomato juice—canned	5	0.215	0.017	1.20	2.6	0.164	0.013	0.009	0.220	0
Turnip cubes—raw	8	0.255	0.028	1.76	8.5	0.085	0.011	0.009	0.587	0
Mixed vegetables—frzn, cooked	17	0.812	0.043	3.70	7.2	0.232	0.020	0.034	1.120	0
Vegetable juice cocktail	6	0.178	0.026	1.29	3.1	0.119	0.012	0.008	0.178	0
Water chestnuts—raw	30	0.398	0.027	6.77	3.2	0.170	0.040	0.057	0.869	0
Watercress—fresh	3	0.650	0.033	0.37	33.4	0.050	0.025	0.033	0.719	0
Lentils—sprouted, raw	30	2.540	0.156	6.30	7.0	0.909	0.065	0.036	1.150	0
Soybeans—mature, raw	37	3.700	1.900	3.17	19.4	0.599	0.096	0.033	0.656	0
White yams—raw	34	0.438	0.049	7.90	8.7	0.153	0.032	0.009	0.822	0

Variety	kcal	Protein (g)	Fat (g)	CHO (g)	Ca (mg)	Fe (mg)	B₁ (mg)	B₂ (mg)	Fiber (g)	Cholesterol (mg)
Almonds—dried, chopped	167	5.65	14.80	5.78	75.5	1.040	0.060	0.220	3.36	0
Almonds—whole, toasted	167	5.77	14.40	6.48	80.0	1.400	0.037	0.170	3.99	0
Gelatin salad/dessert	17	0.43	0	3.99	0.5	0.024	0.002	0.002	0.02	0
Keilbasa sausage	88	3.76	7.70	0.61	12.0	0.414	0.064	0.061	0	18.5
Knockwurst sausage	87	3.37	7.88	0.50	2.9	0.258	0.097	0.040	0	16.3
Crepe, unfilled	49	2.06	1.32	7.07	25.0	0.475	0.045	0.070	0.21	43.0
Pancakes—plain	63	2.10	2.10	9.45	28.4	0.525	0.063	0.074	0.42	16.8
Croissant	117	2.32	6.02	13.40	10.0	1.040	0.085	0.065	0.54	6.5
Patty melt—ground beef/ rye	91	5.09	6.07	3.96	36.5	0.533	0.040	0.072	0.81	17.1
Pizza—cheese—⅛th	69	3.54	2.13	9.21	52.0	0.378	0.080	0.069	0.51	13.2
Pizza—mozzarella—⅛th	80	7.60	4.67	0.89	207.0	0.076	0.006	0.097	0	15.0
Pizza—Canadian bacon— ⅛th	52	6.82	2.36	0.38	3.0	0.229	0.231	0.055	0	16.3
Pizza—pepperoni—⅛th	140	5.94	12.50	0.81	2.8	0.397	0.090	0.070	0	9.8
Pizza—onion—⅛th	10	0.34	0.07	2.07	7.1	1.105	0.017	0.003	0.45	0
Polish sausage	92	3.99	8.13	0.46	3.0	0.409	0.142	0.042	0	20.0
Popcorn—plain, air popped	106	3.5	1.42	21.30	3.5	0.709	0.106	0.035	4.60	0
Popcorn—cooked in oil/ salted	142	2.32	7.99	15.50	7.7	0.696	0.026	0.052	3.09	0
Popcorn—syrup-coated	109	1.6	0.81	24.30	1.6	0.405	0.106	0.016	0.81	0

Variety (cont.)	kcal	Protein (g)	Fat (g)	CHO (g)	Ca (mg)	Fe (mg)	B₁ (mg)	B₂ (mg)	Fiber (g)	Cholesterol (mg)
Quiche lorraine—⅛th	97	2.09	7.73	4.67	34.00	0.226	0.018	0.052	0.09	45.9
Chicken salad w/celery	97	3.82	8.90	0.47	5.92	0.239	0.012	0.028	0.11	17.3
Cole slaw	20	0.36	0.74	3.52	12.80	0.166	0.019	0.017	0.57	2.3
Egg salad	68	2.91	6.01	0.45	14.50	0.525	0.019	0.070	0	97.4
Ham salad spread	62	2.46	4.39	3.01	2.24	0.168	0.123	0.034	0.03	10.4
Macaroni salad—no cheese	75	0.54	6.66	3.52	5.50	0.229	0.020	0.014	0.27	4.9
Potato salad w/mayo + eggs	41	0.76	2.32	3.16	5.44	0.185	0.022	0.017	0.42	19.3
Tuna salad	53	4.55	2.53	2.67	4.84	0.282	0.009	0.019	0.34	3.7
Waldorf salad	85	0.72	8.33	2.62	8.82	0.196	0.020	0.013	0.72	4.3
Lobster meat—cooked	28	5.80	0.17	0.36	17.20	0.111	0.020	0.019	0	20.3
Shrimp—boiled—2 large	28	5.93	0.31	0	11.00	0.876	0.009	0.009	0	55.3
Sunflower seeds—dry	162	6.46	14.00	5.32	32.90	1.920	0.650	0.070	1.97	0
Cream of celery soup	20	0.38	1.27	2.00	9.04	0.141	0.007	0.011	0.09	3.2
Chicken soup—chunky	20	1.43	0.75	1.95	2.71	0.195	0.010	0.020	0.03	3.4
Chicken + dumpling soup	23	1.30	1.28	1.39	3.34	0.144	0.004	0.017	0.10	7.6
Chicken gumbo soup	13	0.60	0.32	1.89	5.53	0.201	0.006	0.009	0.05	0.9
Chick-noodl soup—chunky	14	1.50	0.70	0.24	2.84	0.170	0.009	0.020	0.09	2.1
Chili with beans	32	1.62	1.56	3.38	13.20	0.973	0.014	0.030	0.91	4.8
Clam chowder—New England	19	1.08	0.75	1.90	21.40	0.169	0.008	0.027	0.11	2.5
Minestrone soup—chunky	15	0.60	0.33	2.45	7.20	0.209	0.006	0.014	0.12	0.6
Cream of mushroom soup	29	0.46	2.15	2.10	7.23	0.119	0.007	0.019	0.06	0.4
Mushroom-barley soup	14	0.43	0.51	1.69	2.82	0.113	0.006	0.020	0.17	0
Pea soup—prepared w/milk	27	1.40	0.79	3.59	19.30	0.224	0.017	0.030	0.07	2.0
Cream of potato soup	17	0.39	0.53	2.59	4.52	0.107	0.008	0.008	0.10	1.5
Split pea + ham soup	22	1.31	0.47	3.17	3.90	0.253	0.014	0.011	0.19	0.8
Tomato-beef-noodle soup	32	1.00	0.97	4.78	3.95	0.252	0.019	0.020	0.03	0.9
Tomato bisque prep w/milk	22	0.71	0.75	3.32	21.00	0.099	0.013	0.030	0.01	2.5
Turkey soup—chunky	16	1.23	0.53	1.69	6.00	0.229	0.010	0.029	0.12	1.1
Turkey noodle soup	16	0.88	0.45	1.95	2.60	0.212	0.017	0.014	0.03	1.1
Cream vegetable soup—dry mix	126	2.27	6.84	14.80			1.470	0.127	0.22	1.2
Vegetable soup	16	0.49	0.45	2.77	4.96	0.249	0.012	0.010	0.37	0
Spinach souffle	45	2.29	3.84	0.59	47.90	0.279	0.019	0.064	0.79	38.4
Potato chips—sour crm onion	153	2.40	9.48	14.60	21.00	0.474		0.055	1.35	1.0
Soy sauce	14	1.46	0.02	2.40	4.73	0.567	0.014	0.036	0	0
Garlic cloves	42	1.80	0.14	9.38	51.30	0.482	0.057	0.030	0.47	0
Catsup	30	0.52	0.11	7.17	6.23	0.228	0.026	0.020	0.45	0
Mustard	21	1.34	1.25	1.81	23.80	0.567	0.024	0.057	0.11	0
Alfalfa sprouts	9	1.13	0.20	1.07	9.45	0.272	0.021	0.036	1.03	0
Tang orange juice crystals	13	0.17	0.06	3.06	4.57	0.002	0.023	0.008	0	0
Doirtos—nacho flavor	139	2.20	6.79	18.00	17.00	0.399	0.040	0.030	1.10	0
Doritos—taco flavor	140	2.60	6.59	17.60	44.90	0.699	0.080	0.090	1.10	0
Corn tortilla—fried	82	2.08	2.84	12.30	39.70	0.567	0.047	0.028	2.27	0
Cocoa butter oil	251	0	28.40	0	0	0			0	0
Corn oil	251	0	28.40	0		0.001	0	0	0	0
Cottonseed oil	251	0	28.40	0					0	0
Olive oil	251	0	28.40	0	0.05	0.109	0	0	0	0
Palm oil	251	0	28.40	0	0	0.003	0	0	0	0
Palm kernel oil	251	0	28.40	0	0	0	0	0	0	0
Peanut oil	251	0	28.40	0	0.03	0.008	0	0	0	0
Safflower oil	251	0	28.40	0	0	0	0	0	0	0
Sesame oil	250	0	28.40	0	0	0	0	0	0	0
Soybean oil	251	0	28.40	0	0.01	0.007	0	0	0	0

Variety	kcal	Protein (g)	Fat (g)	CHO (g)	Ca (mg)	Fe (mg)	B₁ (mg)	B₂ (mg)	Fiber (g)	Cholesterol (mg)
Sunflower oil	251	0	28.40	0	0	0	0	0	0	0
Walnut oil	251	0	28.40	0	0	0	0	0	0	0
Wheat germ oil	250	0	28.40	0	0	0	0	0	0	0
Vegetable shortening	251	0	28.40	0	0	0	0	0	0	0
Wheat cracker—thin	124	3.19	4.96	17.70	10.6	1.060	0.142	0.106	1.84	0
Whole wheat crackers	124	3.19	5.32	17.70	10.6	0.850	0.070	0.106	2.94	0
Whole wheat spaghetti	35	1.53	0.11	7.49	4.3	0.244	0.048	0.020	1.04	0
BLT Sandwich—whole wheat	68	2.49	3.77	6.45	11.4	0.570	0.077	0.043	1.55	4.1
BLT sandwich—white	70	2.30	3.77	6.70	18.2	0.489	0.092	0.055	0.43	4.2
Grilled cheese sand/wheat	91	4.22	5.37	7.10	87.4	0.565	0.057	0.076	1.59	11.8
Grilled cheese sand/part WW	95	4.39	5.82	6.59	103.0	0.526	0.067	0.093	0.61	13.3
Grilled cheese sand—white	97	4.22	5.79	6.88	103.0	0.441	0.068	0.092	0.30	13.3
Chicken salad sand—wheat	80	3.00	4.25	8.13	14.7	0.679	0.066	0.046	1.88	6.2
Chicken salad sand—white	85	2.83	4.63	8.05	22.7	0.552	0.080	0.060	0.39	7.1
Corn dog	84	2.55	5.10	6.97	8.7	0.495	0.072	0.043	0.03	9.5
Corned beef & swiss on rye	83	5.26	4.59	4.90	63.8	0.768	0.044	0.078	0.97	16.4
Engl mfn (egg/cheese/bacon)	74	3.70	3.70	6.37	40.5	0.637	0.095	0.103	0.32	43.8
Egg salad sand—wheat	79	2.60	4.56	7.44	17.0	0.744	0.063	0.059	1.67	48.8
Egg salad sand/soft white	83	2.41	4.90	7.25	24.5	0.636	0.075	0.074	0.30	41.6
Ham sandwich—rye bread	59	3.84	2.09	6.10	12.0	0.474	0.182	0.073	1.24	7.1
Ham sandwich—whole wheat	59	3.79	2.04	6.84	12.4	0.617	0.164	0.060	1.54	6.1
Ham sandwich—soft white	61	3.74	2.08	6.60	18.6	0.504	0.186	0.073	0.29	6.7
Ham & swiss sand/on rye	68	4.65	3.23	5.08	63.5	0.389	0.147	0.079	0.99	10.8
Ham & cheese sand—wheat	67	4.20	3.23	5.72	40.5	0.527	0.136	0.066	1.27	9.7
Ham & cheese sand—soft white	69	4.20	3.36	5.43	48.0	0.428	0.152	0.078	0.23	10.6
Ham salad sand—wheat	75	2.45	4.02	7.83	11.5	0.569	0.104	0.045	1.52	5.6
Ham salad sand—white	78	2.25	4.26	7.71	17.5	0.454	0.119	0.057	0.28	6.2
Hotdog/frankfurter & bun	87	2.80	5.10	7.04	19.6	0.570	0.095	0.062	0.40	7.6
Peanut butter & jam—whole wheat	92	3.40	3.85	12.30	15.4	0.751	0.070	0.045	2.29	0
Peanut butter & jam—white	98	3.29	4.14	12.80	23.4	0.632	0.085	0.059	0.85	0
Reuben sandwich—grilled	58	3.43	3.38	3.53	43.6	0.633	0.030	0.052	0.80	10.4
Roast beef sand—whole wheat	64	4.00	2.52	6.71	12.3	0.760	0.061	0.054	1.53	6.2
Roast beef sand—white	67	3.97	2.63	6.46	18.5	0.665	0.072	0.066	0.27	6.9
Tuna salad sand—wheat	72	3.29	3.29	8.03	13.0	0.667	0.057	0.040	1.74	5.5
Tuna salad sand—white	76	3.18	3.47	7.92	19.6	0.555	0.068	0.052	0.44	6.2
Turkey sand—whole wheat	62	4.09	2.33	6.67	11.6	0.554	0.056	0.044	1.53	6.0
Turkey sand—white	64	4.07	2.42	6.41	17.8	0.435	0.066	0.055	0.27	6.7
Turkey & ham sandwich—rye	58	3.74	2.12	6.18	12.3	0.743	0.060	0.079	1.23	8.4
Turkey & ham sand—whole wheat	59	3.71	2.07	6.90	12.6	0.846	0.060	0.064	1.54	7.2
Turkey & ham sand—white	60	3.65	2.10	6.67	18.8	0.762	0.071	0.078	0.28	8.0
Turkey & ham & cheese on rye	68	4.22	3.46	5.04	44.2	0.616	0.050	0.083	0.99	12.1

Variety (cont.)	kcal	Protein (g)	Fat (g)	CHO (g)	Ca (mg)	Fe (mg)	B$_1$ (mg)	B$_2$ (mg)	Fiber (g)	Cholesterol (mg)
Turkey & ham & cheese—wheat	67	4.14	3.25	5.77	40.5	0.716	0.051	0.070	1.27	10.6
Turkey & ham & cheese—white	69	4.13	3.38	5.48	48.30	0.636	0.059	0.082	0.23	11.6
Bordelaise sauce	24	0.33	1.46	1.10	3.80	0.179	0.008	0.010	0.08	3.8
Hot chili sauce, red pepper	5	0.25	0.17	1.10	2.51	0.137	0.003	0.026	2.29	0
Teriyaki sauce	24	1.69	0.09	4.52	6.30	0.488	0.008	0.020	0	0
Bologna—beef	89	3.32	8.04	0.56	3.70	0.394	0.016	0.036	0	16.0
Brotwurst	92	4.05	7.90	0.84	13.80	0.292	0.070	0.064	0	17.8
Liverwurst	93	4.00	8.10	0.63	7.88	1.810	0.077	0.291	0	44.1
Turkey pastrami	37	5.22	1.75	0.43	2.49	0.403	0.022	0.075	0	14.9
Scampi—fried in crumbs	69	6.07	3.49	3.26	19.00	0.357	0.037	0.039	0.04	50.2
Spaghetti meat sauce	30	1.03	1.30	3.72	4.95	0.385	0.028	0.022	0.19	2.3
Tomato soup—canned	19	0.47	0.43	3.75	3.05	0.396	0.020	0.011	0.11	0
Oyster stew	14	0.49	0.89	0.94	4.96	0.227	0.005	0.008	0	3.1
Onion soup—canned	13	0.87	0.40	1.89	6.10	0.156	0.008	0.006	0.11	0
Blue cheese salad dressing	143	1.37	14.80	2.09	22.90	0.057	0.003	0.028	0.02	7.6
Ceasar's salad dressing	126	2.66	12.70	0.67	44.40	0.239	0.007	0.025	0.05	33.9
French dressing	150	0.16	16.00	1.81	3.55	0.113	0	0	0.22	0
Italian dressing—lo cal	15	0.02	1.18	1.37	0.59	0.059	0	0	0.08	1.7
Mayonnaise	203	0.31	22.60	0.77	5.67	0.168	0.005	0.012	0	16.8
Imitation mayonnaise	66	0	5.67	3.78	0	0	0	0	0	7.6
Ranch salad dressing	104	0.86	10.70	1.31	28.40	0.075	0.010	0.040	0	11.1
Russian salad dressing	140	0.45	14.50	3.00	5.44	0.174	0.014	0.014	0.08	18.4
1000 island dressing	107	0.26	10.10	4.30	3.52	0.170	0.006	0.009	0.06	7.3
1000 island dressing—lo cal	45	0.22	3.00	4.58	3.12	1.174	0.006	0.008	0.34	3.4
Vinegar & oil dressing	124	0	14.20	0		0	0	0	0	0
Salami—beef	72	4.17	5.69	0.70	2.47	0.567	0.036	0.073	0	17.3
Salami—pork & beef	72	3.94	5.70	0.64	3.68	0.755	0.068	0.106	0	18.5
Salami—turkey	55	4.62	3.89	0.15	5.47	0.463	0.029	0.075	0	22.9
Salami—dry—beef & pork	120	6.49	9.75	0.74	2.84	0.425	0.170	0.082	0	22.7
Sandwich type cookies	138	1.42	5.67	20.60	8.50	0.992	0.064	0.050	0.28	0
Avocado cheese sand—white	64	2.02	4.00	5.39	43.10	0.418	0.057	0.059	0.98	4.4

P   A   R   T

2

Nutritive Values for Alcoholic and Nonalcoholic Beverages

The nutritive values for alcoholic and nonalcoholic beverages are expressed in 1-ounce (28.4 g) portions. We have also included the nutritive values for the minerals calcium, iron, magnesium, phosphorus, and potassium and the vitamins thiamine, riboflavin, niacin, and cobalamin. The alcoholic beverages contain no cholesterol or fat.

ALCOHOLIC BEVERAGES (1 OUNCE)

| Beverage | kcal | Protein (g) | CHO (g) | Minerals | | | | | Vitamins | | | |
				Ca (mg)	Fe (mg)	Mg (mg)	P (mg)	K (mg)	B₁ (mg)	B₂ (mg)	Niacin (mg)	B₁₂ (mg)
Beer, light	12	0.072	1.1	1.4	0.009	1.830	3.500	7.090	0.002	0.007	0.128	0.005
Beer, regular	8	0.057	0.4	1.4	0.011	1.420	3.440	5.130	0.003	0.008	0.111	0.002
Brandy	69	0	10.6	2.5	0.012		1.010	1.010	0.002	0.002	0.004	0
Champagne	22	0.043	0.6	1.6	0.093	2.400	1.900	22.600	0	0.003	0.019	0
Dessert wine, dry	36	0.057	1.2	2.3	0.068	2.550	2.550	26.200	0.005	0.005	0.060	0
Dessert wine, sweet	44	0.057	3.3	2.3	0.057	2.550	2.640	26.200	0.005	0.005	0.060	0
Gin, rum, vodka, scotch, whiskey, 80 p	64	0	0	0	0.010	0	0	1.010	0	0	0	0
Gin, rum, vodka, scotch, whiskey, 86 p	71	0	0	0	0.012	0	1.160	0.557	0.002	0.002	0.004	0
Gin, rum, vodka, scotch, whiskey, 90 p	74	0	0	0	0.010	0	0	0.860	0	0	0	0
Sherry, dry	28	0.024	0.3	2.1	0.052	1.960	2.600	17.800	0.002	0.002	0.024	0
Sherry, medium	40	0.066	2.3	2.3	0.071	2.270	1.890	23.600	0.002	0.008	0.035	0
Vermouth, dry	34	0.028	1.6	2.0	0.096	1.420	1.890	11.300			0.011	0
Vermouth, sweet	44	0.014	4.5	1.7	0.099	1.130	1.650	8.500			0.011	0
Wine, dry white	19	0.029	0.2	2.6	0.093	2.620	1.670	17.400	0	0.001	0.019	0
Wine, medium white	19	0.028	0.2	2.5	0.085	3.030	3.840	22.600	0.001	0.001	0.019	0
Wine, red	20	0.055	0.5	2.2	0.122	3.600	3.840	31.500	0.001	0.008	0.023	0.004
Wine, rosé	20	0.055	0.4	2.4	0.108	2.740	4.080	28.100	0.001	0.004	0.020	0.002
Creme de menthe	105	0	11.8	0	0.023	0	0	0	0	0	0.001	0
Bloody mary	22	0.153	0.9	1.9	0.105	2.100	4.020	41.400	0.010	0.006	0.123	0
Bourbon and soda	26	0	0	1.0		0.244	0.489	0.489	0	0	0.005	0
Daiquiri	52	0	1.9	0.9	0.043	0.472	1.890	6.140	0.004	0	0.012	0
Manhattan	64	0	0.9	0.5	0.025	0.012	1.990	7.460	0.003	0.001	0.026	0
Martini	63	0	0.1	0.4	0.024	0.405	0.810	5.260	0	0	0.004	0
Pina colada	53	0.120	8.0	2.2	0.062		2.010	20.100	0.008	0.004	0.033	0
Screwdriver	23	0.160	2.5	2.1	0.023	2.260	3.86	43.300	0.018	0.004	0.046	0
Tequila	31	0.099	2.4	1.7	0.077	1.980	2.800	29.300	0.010	0.005	0.054	0
Tom collins	16	0.013	0.4	1.3		0.383	0.128	2.300	0	0	0.004	0
Whiskey sour	42	0	3.7	0.3	0.021	0.267	1.600	5.080	0.003	0.002	0.006	0
Coffee + cream liqueur	93	0.784	5.9	4.2	0.036	0.603	13.900	9.050	0	0.016	0.022	0
Coffee liqueur	95	0	13.3	0.5	0.016	0.545	1.640	8.180	0.001	0.003	0.040	0

Note: Alcoholic beverages contain no fat or cholesterol; light beer contains 0.5 g fiber and regular beer contains 1.2 g fiber. All of the other nonmixed alcoholic beverages have no fiber.

NONALCOHOLIC BEVERAGES (1 OUNCE)

Beverage	kcal	Protein (g)	CHO (g)	Minerals					Vitamins			
				Ca (mg)	Fe (mg)	Mg (mg)	P (mg)	K (mg)	B₁ (mg)	B₂ (mg)	Niacin (mg)	B₁₂ (mg)
Hot cocoa with whole milk	25	1.030	2.9	33.8	0.088	6.350	30.600	54.400	0.012	0.049	0.041	0.099
Cocoa mix + water-diet	7	0.561	1.3	13.3	0.110	4.870	19.800	59.800	0.006	0.030	0.024	0
Coffee-brewed	0	0.016	0.1	0.5	0.113	1.590	0.320	15.400	0	0.002	0.063	0
Coffee-instant dry powder	32	0.157	12.4	47.3	1.260	94.500	78.800	108.000	0	0.016	7.990	0
Cola beverage, regular	12	0	3.0	0.7	0.009	0.230	3.520	0.306	0	0	0	0
Diet cola-w/aspartame	0	0	0	1.0	0.009	0.319	2.400	0	0.001	0.007	0	0
Club soda	0	0	0	1.4	0.012	0.319	0	0.479	0	0	0	0
Cream soda	15	0	3.8	1.5	0.015	0.229	0	0.306	0	0	0	0
Diet soda-avg assorted	0	0	0.0	1.1	0.011	0.200	3.030	0.559	0	0	0	0
Egg nog—commercial	38	1.080	3.8	36.8	0.057	5.250	31.000	46.900	0.010	0.054	0.030	0.127
Five Alive citrus	13	0.135	3.1	1.7	0.021	1.950	2.700	34.000	0.015	0.003	0.060	0
Fruit flavored soda pop	13	0	3.2	1.1	0.020	0.305	0.152	1.520	0	0	0.002	0
Fruit punch drink—canned	13	0.015	3.4	2.1	0.058	0.610	0.305	7.160	0.006	0.006	0.006	0
Gatorade	5	0	1.3	2.8			0	2.840				0
Ginger ale	10	0.008	2.5	0.9	0.051	0.232	0.077	0.387	0	0	0	0
Grape soda, carbonated	12	0	3.2	0.9	0.024	0.305	0	0.229	0	0	0	0
Koolade w/NutraSweet	0	0	0	0	0	0.028	0	0	0	0	0	0
Koolade w/sugar added	12	0	3.0	0	0	0	0	0	0	0	0	0
Lemon-lime soda	12	0	3.0	0.7	0.019	0.154	0.077	0.308	0	0	0.004	0
Lemonade drink from dry	11	0	2.9	7.6	0.016	0.322	3.650	3.540	0		0.004	0
Lemonade frozen conc	51	0.078	13.3	1.9	0.205	1.420	2.460	19.200	0.007	0.027	0.020	0
Limeade frozen conc	53	0.052	14.0	1.4	0.029	7.800	1.690	16.800	0.003	0.003	0.028	0
Chocolate milkshake	36	0.962	5.8	32.0	0.088	4.700	28.900	56.800	0.016	0.069	0.046	0.097
Strawberry milkshake	32	0.952	5.4	32.0	0.030	3.600	28.400	51.700	0.013	0.055	0.050	0.088
Vanilla milkshake	32	0.982	5.1	34.5	0.026	3.500	29.000	49.300	0.013	0.052	0.052	0.101
Orange drink/carbonated	14	0	3.5	1.5	0.018	0.305	0.305	0.686	0	0	0	0
Pepper type soda	12	0	2.9	0.9	0.010	0.077	3.160	0.154	0	0	0	0
Root beer	12	0.008	3.0	1.5	0.014	0.306	0.153	0.230	0	0	0	0
Pineapple grapefruit drink	13	0.068	3.3	2.0	0.087	1.700	1.590	17.500	0.009	0.005	0.076	0
Pineapple orange drink	14	0.352	3.3	1.5	0.076	1.590	1.130	13.200	0.009	0.005	0.059	0
Tonic water/Quinine water	10	0	2.5	0.4	0.019	0.077	0	0.077	0	0	0	0
Tea-brewed	0	0.001	0.1	0	0.006	0.796	0.159	10.500	0	0.004	0.012	0
Herbal tea—brewed	0	0	0	0.6	0.022	0.319	0	2.390	0.003	0.001	0	0
Perrier water	0	0	0	3.8	0	0.148	0	0	0	0	0	0
Poland Springs bottle water	0	0	0	0.4	0.001	0.239	0	0	0	0	0	0

Note: Other nonalcoholic beverages are listed in the sections on fruits and vegetables.

Nutritive Values for Specialty and Fast-Food Items

Nutrient information was kindly provided by the manufacturer or its representative. Unlike Parts 1 and 2, nutritive values are not given for 1-ounce portions but for the actual amounts of the foods as sold commercially. To make a direct comparison of the kcal values and the various nutrients, we recommend that the weight of the food and its nutrients be expressed relative to 1-ounce (28.4 g) portions.

SPECIALTY AND FAST FOOD ITEMS (DASHES INDICATE INFORMATION NOT PROVIDED BY SOURCES.)
ARBY'S

	Mass (g)	kcal	Protein (g)	Fat (g)	CHO (g)	Ca (mg)	Fe (mg)	Vit A (IU)	Vit C (mg)	Vit B₁ (mg)	Vit B₂ (mg)
Bac'n Cheddar Deluxe	225	561	28	34	78	—	—	—	—	—	—
Baked Potato											
Plain	312	290	8	1	0	—	—	—	—	—	—
Beef 'n Cheddar	190	490	24	21	51	—	—	—	—	—	—
Chicken Breast Sandwich	210	592	28	27	57	—	—	—	—	—	—
Chocolate Shake	300	384	9	11	32	—	—	—	—	—	—
French Fries	71	211	2	8	6	—	—	—	—	—	—
Hot Ham 'n Cheese Sandwich	161	353	26	13	50	—	—	—	—	—	—
Jamocha Shake	305	424	8	10	31	—	—	—	—	—	—
Junior Roast Beef	86	218	12	8	22	—	—	—	—	—	—
King Roast Beef	192	467	27	19	49	—	—	—	—	—	—
Potato Cakes	85	201	2	14	13	—	—	—	—	—	—
Regular Roast Beef	147	353	22	15	32	—	—	—	—	—	—
Super Roast Beef	234	501	25	22	40	—	—	—	—	—	—
Superstuffed Potato											
Broccoli and Cheddar	340	541	13	22	24	—	—	—	—	—	—
Superstuffed Potato											
Deluxe	312	648	18	38	72	—	—	—	—	—	—
Superstuffed Potato											
Mushroom and Cheese	300	506	16	22	21	—	—	—	—	—	—
Superstuffed Potato											
Taco	425	619	23	27	145	—	—	—	—	—	—
Turkey Deluxe	197	375	24	17	39	—	—	—	—	—	—
Vanilla Shake	250	295	8	10	30	—	—	—	—	—	—

Source: Arby's Inc. Nutritional information provided by Consumer Affairs, Arby's Inc., Atlanta, GA, 1986.

ARTHUR TREACHER'S

	Mass (g)	kcal	Pro-tein (g)	Fat (g)	CHO (g)	Ca (mg)	Fe (mg)	Vit A (IU)	Vit C (mg)	Vit B₁ (mg)	Vit B₂ (mg)
Chicken	136	369	27.1	21.6	16.5	11.2	0.799	102	1.5	0.07	0.148
Chicken Sandwich	156	41	16.2	19.2	44.0	58.8	1.700	117	19.2	0.17	0.240
Chips	113	276	4.0	13.2	34.9	12.4	0.473	85	5.9	0.17	0.035
Chowder	170	112	4.6	5.4	11.2	61.4	0.092	340	1.7	0.07	0.140
Cole Slaw	85	123	1.0	8.2	11.1	24.0	0.185	170	59.1	0.03	0.025
Fish	147	355	19.2	19.8	25.4	14.9	0.566	111	1.5	0.10	0.075
Fish Sandwich	156	440	16.4	24.0	39.4	88.9	1.490	117	1.6	0.27	0.215
Krunch Pups	57	203	5.4	14.8	12.0	8.0	0.601	43	3.5	0.05	0.052
Lemon Luvs	85	276	2.6	13.9	35.1	9.9	0.851	64	0.9	0.18	0.089
Shrimp	115	381	13.1	24.4	27.2	56.7	0.638	86	1.3	0.08	0.051

Source: Arthur Treacher's Inc., Youngstown, OH. All sampling and analysis was conducted by Warf Institute, Inc., Madison, WI, July, 1977.

BURGER CHEF

	Mass (g)	kcal	Pro-tein (g)	Fat (g)	CHO (g)	Ca (mg)	Fe (mg)	Vit A (IU)	Vit C (mg)	Vit B₁ (mg)	Vit B₂ (mg)
Big Chef	186	542	23	34	35	189	3.4	282	2	0.34	0.35
Cheeseburger	104	304	14	17	24	156	2.0	266	1	0.22	0.23
Double Cheeseburger	145	434	24	26	24	246	3.1	430	1	0.25	0.34
French Fries	68	187	3	9	25	10	0.9	tr	14	0.09	0.05
Hamburger, Regular	91	258	11	13	24	69	1.9	114	1	0.22	0.18
Mariner Platter	373	680	32	24	85	137	4.7	448	24	0.37	0.40
Rancher Platter	316	640	30	38	44	57	5.1	367	24	0.30	0.37
Shake	305	326	11	11	47	411	0.2	10	2	0.11	0.57
Skipper's Treat	179	604	21	37	47	201	2.5	303	1	0.29	0.30
Super Shef	252	600	29	37	39	240	4.2	763	9	0.37	0.43

Source: Burger King Corporation, Miami, FL.

BURGER KING

	Mass (g)	kcal	Protein (g)	Fat (g)	CHO (g)	Ca (mg)	Fe (mg)	Vit A (IU)	Vit C (mg)	Vit B₁ (mg)	Vit B₂ (mg)
Bacon Double Cheeseburger	159	510	31.92	31	27	167.79	3.80	383.68	—	0.31	0.42
Breakfast Croissan'wich Bacon, Egg, Cheese	119	355	14.55	24	20	135.70	2.01	426.11	—	0.32	0.30
Breakfast Croissan'wich Ham, Egg, Cheese	145	386	17.53	20	20	136.46	2.16	426.11	—	0.49	0.32
Breakfast Croissan'wich Sausage, Egg, Cheese	163	538	19.36	41	20	145.22	2.89	426.11	—	0.36	0.32
Cheeseburger	120	317	17.26	15	30	102.17	2.74	341.40	3.09	0.23	0.29
Chicken Sandwich	230	688	26.03	40	56	78.51	3.30	126.13	—	0.45	0.31
Chicken Tenders	95	204	20.24	10	10	18.29	0.67	95.26	—	0.08	0.08
Coffee, Regular	244	2	0	0	0	—	—	—	—	—	—
Double Beef Whopper	351	—	46.36	—	—	91.31	7.31	617.34	13.88	0.34	0.56
Double Beef Whopper w/Cheese	374	—	50.82	—	—	221.95	7.34	1000.64	13.88	0.35	0.63
French Fries, Reg.	74	227	3.00	13	24	—	0.52	—	—	0.10	0.30
French Toast Platter w/Bacon	117	469	11.02	30	41	59.09	2.72	—	—	0.24	0.24
French Toast Platter w/Sausage	158	635	16.08	46	41	69.92	3.71	—	—	0.29	0.29
Ham and Cheese	230	471	23.74	23	44	194.90	3.19	725.02	7.37	0.87	0.42
Hamburger	109	275	15.03	12	29	36.86	2.73	149.76	3.09	0.23	0.25
Hot Chocolate	244	131	1.34	4	22	26.71	0.71	—	—	—	—
Hot Dog	—	291	11.00	17	23	40.00	2.00	0	0	0.04	0.02
Milk 2% lowfat	244	121	8.13	5	12	296.70	—	500.20	2.32	0.10	0.40
Milk Whole	244	157	8.00	9	11	290.36	—	336.72	3.59	0.09	0.39
Onion Rings, Regular	79	274	3.64	16	28	124.01	0.80	—	—	—	—
Orange Juice	183	82	1.00	0	20	—	—	142.74	71.19	0.14	—
Pies Apple	125	305	3.12	12	44	—	1.18	—	4.99	0.27	0.16
Pies Cherry	128	357	3.57	13	55	—	1.12	370.04	7.66	0.24	0.16
Pies Pecan	113	459	4.99	20	64	23.81	1.13	—	—	0.28	0.18
Salad w/1000 Island	176	145	2.46	12	9	42.38	1.44	1659.13	42.55	0.06	0.13
Salad w/Bleu Cheese	176	184	3.43	16	7	66.19	1.33	1638.16	41.70	0.06	0.15
Salad w/Creamy Italian	176	NA	NA	NA	NA	NA	NA	NA	NA	NA	NA
Salad w/French	176	152	2.29	11	13	40.40	1.35	1688.62	43.26	0.06	0.12
Salad w/Golden Italian	176	162	2.24	14	7	39.83	1.26	1597.90	42.32	0.05	0.12
Salad w/House	176	159	2.92	13	8	44.36	1.31	1604.42	41.84	0.06	0.15
Salad w/Reduced Calorie Italian	176	42	2.21	1	7	40.40	1.35	1591.38	42.04	0.05	0.12
Salad—Plain	148	28	2.16	0	5	36.71	1.24	1582.60	41.67	0.06	0.12
Salad Dressings 1000 Island	—	117	—	12	4	—	—	—	—	—	—
Salad Dressings Bleu Cheese	—	156	—	16	2	—	—	—	—	—	—
Salad Dressings Creamy Italian	—	NA	—	NA	NA	—	—	—	—	—	—
Salad Dressings French	—	123	—	11	8	—	—	—	—	—	—
Salad Dressings Golden Italian	—	134	—	14	2	—	—	—	—	—	—
Salad Dressings House	—	130	—	13	3	—	—	—	—	—	—

BURGER KING

	Mass (g)	kcal	Pro- tein (g)	Fat (g)	CHO (g)	Ca (mg)	Fe (mg)	Vit A (IU)	Vit C (mg)	Vit B₁ (mg)	Vit B₂ (mg)
Salad Dressings											
Reduced Calorie Italian	—	14	—	0	2	—	—	—	—	—	—
Scrambled Egg Platter	195	—	14.24	—	—	101.45	2.69	374.90	2.62	0.31	0.35
Scrambled Egg Platter w/Bacon	206	536	18.02	36	33	102.84	2.82	374.90	2.62	0.39	0.38
Scrambled Egg Platter w/Sausage	247	702	21.88	52	33	112.04	3.66	375.01	2.62	0.42	0.40
Shakes, Medium, Chocolate	273	320	8.02	12	46	259.88	1.61	—	—	0.13	0.55
Shakes, Med. Chocolate (added syrup)	284	374	8.33	11	60	247.58	1.55	—	—	0.12	0.51
Shakes, Med. Vanilla	273	321	8.73	10	49	294.52	—	—	—	0.11	0.57
Shakes, Med. Vanilla (added syrup)	284	334	9	10	51	NA	NA	NA	NA	NA	NA
Soft Drinks, Med. 7–Up	366	144	0	0	38	—	—	—	—	—	—
Soft Drinks, Med. Diet Pepsi	366	1	0	0	0	—	—	—	—	—	—
Soft Drinks, Med. Dr. Pepper	366	155	0	0	40	—	—	—	—	—	—
Soft Drinks, Med. Mountain Dew	366	158	0	0	42	—	—	—	—	—	—
Soft Drinks, Med. Pepsi Cola	366	159	0	0	40	—	—	—	—	—	—
Whaler Sandwich	189	488	18.64	27	45	46.46	2.22	35.54	—	0.28	0.21
Whaler w/Cheese	201	530	20.86	30	46	111.78	2.24	227.19	—	0.27	0.24
Whopper Jr. Sandwich	136	322	15.19	17	30	39.97	2.81	296.38	6.24	0.23	0.25
Whopper Jr. w/Cheese	147	364	17.42	20	31	105.29	2.82	488.02	6.24	0.23	0.29
Whopper Sandwich	265	640	26.94	41	42	79.79	4.88	617.57	13.88	0.33	0.41
Whopper w/Cheese	289	723	31.39	48	43	210.40	4.91	1000.64	13.88	0.34	0.48

Source: Burger Chef Systems, Inc., Indianapolis, IN, 1978 (Analyses obtained from USDA Handbook No. 8).

CHURCH'S FRIED CHICKEN

	Mass (g)	Description	kcal	Pro- tein (g)	Fat (g)	CHO (g)	Ca (mg)	Fe (mg)	Vit A (IU)	Vit C (mg)	Vit B₁ (mg)	Vit B₂ (mg)
Catfish	21		66.7	3.9	4.0	3.8	—	—	—	—	—	
Corn	168	1 ear with Butter Oil	236.5	4.2	9.3	32.9	—	—	—	—	—	
French Fries	85	1 reg. w/100 mg. salt added	138	2.1	5.5	20.1	—	—	—	—	—	
Fried Chicken Breast	93		278	21.3	17.3	9.4	—	—	—	—	—	—
Fried Chicken Leg	56		147.1	12.9	8.6	4.5	—	—	—	—	—	—

CHURCH'S FRIED CHICKEN

	Mass (g)	Description	kcal	Pro-tein (g)	Fat (g)	CHO (g)	Ca (mg)	Fe (mg)	Vit A (IU)	Vit C (mg)	Vit B₁ (mg)	Vit B₂ (mg)
Fried Chicken												
Thigh	93		305.8	18.5	21.6	9.2	—	—	—	—	—	—
Fried Chicken												
Wing—Breast	97		—	—	—	—	—	—	—	—	—	—
Hushpuppy	23	1	78.0	1.3	2.9	11.6	—	—	—	—	—	—
Nuggets												
Regular	18		55.1	3.0	3.1	3.7	—	—	—	—	—	—
Spicy	18		51.8	3.1	2.9	3.4	—	—	—	—	—	—

Source: Church's Fried Chicken, Inc., San Antonio, TX. Nutritional information based on Texas Testing Laboratories, Inc., May, 1985., Pioneer Flour Mills, Inc., August, 1985, and Bowes & Church's Food Values of Portions Commonly Used—Pennington & Church, 14th Edition, 1985.

DAIRY QUEEN

	Mass (g)	Description	kcal	Pro-tein (g)	Fat (g)	CHO (g)	Ca (mg)	Fe (mg)	Vit A (IU)	Vit C (mg)	Vit B₁ (mg)	Vit B₂ (mg)
Banana Split	383		540	9	11	103	—	—	—	—	—	—
Big Brazier Deluxe	213		470	28	24	36	111	5.2	—	<2.5	0.34	0.37
Big Brazier Regular	184		184	27	23	37	113	5.2	—	<2.0	0.37	0.39
Big Brazier w/Cheese	213		553	32	30	38	268	5.2	495	<2.3	0.34	0.53
Blizzard Banana Split	—	regular	763	—	—	—	—	—	—	—	—	—
Blizzard Banana Split	—	large	1333	—	—	—	—	—	—	—	—	—
Blizzard Chocolate Sandwich Cookies	—	regular	600	—	—	—	—	—	—	—	—	—
Blizzard Chocolate Sandwich Cookies	—	large	1050	—	—	—	—	—	—	—	—	—
Blizzard German Chocolate	—	regular	794	—	—	—	—	—	—	—	—	—
Blizzard German Chocolate	—	large	1460	—	—	—	—	—	—	—	—	—
Blizzard Heath	—	regular	824	—	—	—	—	—	—	—	—	—
Blizzard Heath	—	large	1212	—	—	—	—	—	—	—	—	—
Blizzard M&M	—	regular	766	—	—	—	—	—	—	—	—	—
Blizzard M&M	—	large	1154	—	—	—	—	—	—	—	—	—
Brazier Cheese Dog	113		330	15	19	24	168	1.6	—	—	—	0.18
Brazier Chili Dog	128		330	13	20	25	86	2.0	—	11.0	0.15	0.23
Brazier Dog	99		273	11	15	23	75	1.5	—	11.0	0.12	0.15
Brazier French Fries	71		200	2	10	25	tr	0.4	tr	3.6	0.06	tr
Brazier French Fries	113		320	3	16	40	tr	0.4	tr	4.8	0.09	0.03
Brazier Onion Rings	85		300	6	17	33	20	0.4	tr	2.4	0.09	tr
Brazier Regular	106		260	13	9	28	70	3.5	—	<1.0	0.28	0.26
Brazier w/Cheese	121		318	18	14	30	163	3.5	—	<1.2	0.29	0.29
Buster Bar	149		460	10	29	41	—	—	—	—	—	—
Chicken Sandwich	220		670	29	41	46	—	—	—	—	—	—
Cone large	213		340	9	10	57	—	—	—	—	—	—
Cone regular	142		240	6	7	38	—	—	—	—	—	—
Cone small	85		140	3	4	22	—	—	—	—	—	—
Dairy Queen Parfait	284		460	10	11	81	300	1.8	400	tr	0.12	0.43
Dilly Bar	85		240	4	15	22	100	0.4	100	tr	0.06	0.17
Dilly Bar	85		210	3	13	21	—	—	—	—	—	—
Dipped Cone large	234		510	9	24	64	—	—	—	—	—	—
Dipped Cone regular	156		340	6	16	42	—	—	—	—	—	—
Dipped Cone small	92		190	3	9	25	—	—	—	—	—	—
Double Delight	255		490	9	20	69	—	—	—	—	—	—

DAIRY QUEEN

	Mass (g)	Description	kcal	Protein (g)	Fat (g)	CHO (g)	Ca (mg)	Fe (mg)	Vit A (IU)	Vit C (mg)	Vit B₁ (mg)	Vit B₂ (mg)
Double Hamburger	210		530	36	28	33	—	—	—	—	—	—
Double w/Cheese	239		650	43	37	34	—	—	—	—	—	—
Chocolate Dipped Cone	234	large	450	10	20	58	300	0.4	400	tr	0.12	0.51
Chocolate Dipped Cone	156	medium	300	7	13	40	200	0.4	300	tr	0.09	0.34
Chocolate Dipped Cone	78	small	150	3	7	20	100	tr	100	tr	0.03	0.17
Chocolate Malt	588	large	840	22	28	125	600	5.4	750	6.0	0.15	0.85
Chocolate Malt	418	medium	600	15	20	89	500	3.6	750	3.6	0.12	0.60
Chocolate Malt	241	small	340	10	11	51	300	1.8	400	2.4	0.06	0.34
Chocolate Sundae	248	large	400	9	9	71	300	1.8	400	tr	0.09	0.43
Chocolate Sundae	184	medium	300	6	7	53	200	1.1	300	tr	0.06	0.26
Chocolate Sundae	106	small	170	4	4	30	100	0.7	100	tr	0.03	0.17
Cone	213	large	340	10	10	52	300	tr	400	tr	0.15	0.43
Cone	142	medium	230	6	7	35	200	tr	300	tr	0.09	0.26
Cone	71	small	110	3	3	18	100	tr	100	tr	0.03	0.14
Float	397		330	6	8	59	200	tr	100	tr	0.12	0.17
Freeze	397		520	11	13	89	300	tr	200	tr	0.15	0.34
Sandwich	60		140	3	4	24	60	0.4	100	tr	0.03	0.14
Fiesta Sundae	269		570	9	22	84	200	tr	200	tr	0.23	0.26
Fish Sandwich	170		400	20	17	41	60	1.1	tr	tr	0.15	0.26
Fish Sandwich w/cheese	177		440	24	21	39	150	0.4	100	tr	0.15	0.26
Float	397		410	5	7	82	—	—	—	—	—	—
Freeze	397		500	9	12	89	—	—	—	—	—	—
French Fries	71		200	2	10	25	—	—	—	—	—	—
French Fries	113	large	320	3	16	40	—	—	—	—	—	—
Frozen Dessert	113		180	4	6	27	—	—	—	—	—	—
Hot Dog	100		280	11	16	21	—	—	—	—	—	—
Hot Dog w/Cheese	114		330	15	21	21	—	—	—	—	—	—
Hot Dog w/Chili	128		320	13	20	23	—	—	—	—	—	—
Hot Fudge Brownie Delight	266		600	9	25	85	—	—	—	—	—	—
Malt large	588		1060	20	25	187	—	—	—	—	—	—
Malt regular	418		760	14	18	134	—	—	—	—	—	—
Malt small	291		520	10	13	91	—	—	—	—	—	—
Mr. Misty	439	large	340	0	0	84	—	—	—	—	—	—
Mr. Misty	330	regular	250	0	0	63	—	—	—	—	—	—
Mr. Misty	248	small	190	0	0	48	—	—	—	—	—	—
Mr. Misty Float	404		440	6	8	85	200	tr	120	tr	0.12	0.17
Mr. Misty Float	411		390	5	7	74	—	—	—	—	—	—
Mr. Misty Freeze	411		500	9	12	91	—	—	—	—	—	—
Mr. Misty Kiss	89		70	0	0	17	—	—	—	—	—	—
Onion Rings	85		280	4	16	31	—	—	—	—	—	—
Parfait	283		430	8	8	76	—	—	—	—	—	—
Peanut Buster Parfait	305		740	16	34	94	—	—	—	—	—	—
Shake large	588		990	19	26	168	—	—	—	—	—	—
Shake regular	418		710	14	19	120	—	—	—	—	—	—
Shake small	291		490	10	13	82	—	—	—	—	—	—
Single Hamburger	148		360	21	16	33	—	—	—	—	—	—
Single w/Cheese	162		410	24	20	33	—	—	—	—	—	—
Strawberry Shortcake	312		540	10	11	100	—	—	—	—	—	—
Sundae large	248		440	8	10	78	—	—	—	—	—	—
Sundae regular	177		310	5	8	56	—	—	—	—	—	—
Sundae small	106		190	3	4	33	—	—	—	—	—	—
Super Brazier	298		783	53	48	35	282	7.3	—	<3.2	0.39	0.69
Super Brazier Chili Dog	210		555	23	33	42	158	4.0	—	18.0	0.42	0.48
Super Brazier Dog	182		518	20	30	41	158	4.3	tr	14.0	0.42	0.44
Super Brazier Dog w/Cheese	203		593	26	36	43	297	4.4	—	14.0	0.43	0.48
Super Hot Dog	175		520	17	27	44	—	—	—	—	—	—
Super Hot Dog w/Cheese	196		580	22	34	45	—	—	—	—	—	—
Super Hot Dog w/Chili	218		570	21	32	47	—	—	—	—	—	—

DAIRY QUEEN

	Mass (g)	Description	kcal	Pro-tein (g)	Fat (g)	CHO (g)	Ca (mg)	Fe (mg)	Vit A (IU)	Vit C (mg)	Vit B₁ (mg)	Vit B₂ (mg)
Triple Hamburger	272		710	51	45	33	—	—	—	—	—	—
Triple w/Cheese	301		820	58	50	34	—	—	—	—	—	—

Source: International Dairy Queen, Inc., Minneapolis, MN, 1982. Nutritional information reviewed and edited by Dr. David J. Aulik in cooperation with Raltech Scientific Services.

JACK IN THE BOX

	Mass (g)	Description	kcal	Pro-tein (g)	Fat (g)	CHO (g)	Ca (mg)	Fe (mg)	Vit A (IU)	Vit C (mg)	Vit B₁ (mg)	Vit B₂ (mg)
1000 Island Dressing	—		250	0	24	9	—	—	—	—	—	—
Apple Turnover	—		410	4	24	45	—	—	—	—	—	—
Bacon	—	2 slices	70	3	6	0	—	—	—	—	—	—
Bacon Cheeseburger Supreme	—		724	34	46	44	—	—	—	—	—	—
Bleu Cheese Dressing	—		210	0	18	11	—	—	—	—	—	—
Breakfast Jack	—		307	18	13	30	—	—	—	—	—	—
Buttermilk House Dressing	—		290	0	29	6	—	—	—	—	—	—
Canadian Crescent	—		452	19	31	25	—	—	—	—	—	—
Cheese Nachos	—		571	15	35	49	—	—	—	—	—	—
Cheeseburger	—		323	16	15	32	—	—	—	—	—	—
Chicken Strips Dinner	—		689	40	30	65	—	—	—	—	—	—
Chicken Supreme	—		601	31	36	39	—	—	—	—	—	—
Chocolate Shake	—		330	11	7	55	—	—	—	—	—	—
Club Pita	—		284	22	8	30	—	—	—	—	—	—
Grape Jelly	—		38	0	0	9	—	—	—	—	—	—
Ham & Swiss Burger	—		638	36	39	37	—	—	—	—	—	—
Hamburger	—		276	13	12	30	—	—	—	—	—	—
Jumbo Jack	—		485	26	26	38	—	—	—	—	—	—
Jumbo Jack w/Cheese	—		630	32	35	45	—	—	—	—	—	—
Ketchup	—		10	0	2	0	—	—	—	—	—	—
Milk	—		137	10	5	14	—	—	—	—	—	—
Moby Jack	—		444	16	25	39	—	—	—	—	—	—
Mushroom Burger	—		477	28	27	30	—	—	—	—	—	—
Onion Rings	—		382	5	23	39	—	—	—	—	—	—
Orange Juice	—		80	1	0	20	—	—	—	—	—	—
Pancake Breakfast	—		630	16	27	79	—	—	—	—	—	—
Pasta Seafood Salad	—		394	15	22	32	—	—	—	—	—	—
Regular French Fries	—		221	2	12	27	—	—	—	—	—	—
Regular Taco	—		191	8	11	16	—	—	—	—	—	—
Sausage Crescent	—		584	22	43	28	—	—	—	—	—	—
Scrambled Eggs Breakfast	—		720	26	44	55	—	—	—	—	—	—
Shrimp Dinner	—		731	22	37	77	—	—	—	—	—	—
Sirloin Steak Dinner	—		699	38	27	75	—	—	—	—	—	—
Strawberry Shake	—		320	10	7	55	—	—	—	—	—	—
Super Taco	—		288	12	17	21	—	—	—	—	—	—
Supreme Crescent	—		547	20	40	27	—	—	—	—	—	—
Supreme Nachos	—		718	23	40	66	—	—	—	—	—	—
Swiss & Bacon Burger	—		643	33	43	31	—	—	—	—	—	—
Taco Salad	—		377	31	24	10	—	—	—	—	—	—
Vanilla Shake	—		320	10	6	57	—	—	—	—	—	—

Source: Jack In The Box; nutritional information provided by Foodmaker, Inc., San Diego, CA.

KENTUCKY FRIED CHICKEN

	Mass (g)	kcal	Protein (g)	Fat (g)	CHO (g)	Ca (mg)	Fe (mg)	Vit A (IU)	Vit C (mg)	Vit B₁ (mg)	Vit B₂ (mg)
9 Pieces	652	1892	152	116	59	—	8.8	—	—	0.49	1.27
Drumstick	54	136	14	8	2	20	0.9	30	0.6	0.04	0.12
Extra Crispy Dinner	437	950	52	54	63	150	3.6	750	27.0	0.38	0.56
Keel	96	283	25	13	6	—	0.9	50	1.2	0.07	0.13
Original Recipe Dinner	425	830	52	46	56	150	4.5	750	27.0	0.38	0.56
Rib	82	241	19	15	8	55	1.0	58	<1.0	0.06	0.14
Thigh	97	276	20	19	12	39	1.4	74	<1.0	0.08	0.24
Wing	45	151	11	10	4	—	0.6	—	<1.0	0.03	0.07

Source: Nutritional Content of Average Serving, Heublein Food Service and Franchising Group, June 1976.

LONG JOHN SILVER'S

	Mass (g)	Description	kcal	Protein (g)	Fat (g)	CHO (g)	Ca (mg)	Fe (mg)	Vit A (IU)	Vit C (mg)	Vit B₁ (mg)	Vit B₂ (mg)
3 Pc. Nugget Dinner	—	6 chicken nuggets, Fryes, slaw	699	23	45	54	—	—	—	—	—	—
Apple Pie	113		280	2	11	43	—	—	—	—	—	—
Barbecue Sauce	34		45	0	0	11	—	—	—	—	—	—
Battered Shrimp Dinner	—	6 battered shrimp, Fryes, slaw	711	17	45	60	—	—	—	—	—	—
Bleu Cheese Dressing	45		225	4	23	3	—	—	—	—	—	—
Breaded Clams	—		465	13	25	46	—	—	—	—	—	—
Breaded Fish Sandwich Platter	—	Fish sandwich, Fryes, slaw	835	30	42	84	—	—	—	—	—	—
Breaded Oysters	—	6 pc.	460	14	19	58	—	—	—	—	—	—
Breaded Shrimp Platter	—	Breaded shrimp, Fryes, slaw, 2 hush puppies	962	20	57	93	—	—	—	—	—	—
Cherry Pie	113		294	3	11	46	—	—	—	—	—	—
Chicken Planks	—	4 pc.	458	27	23	35	—	—	—	—	—	—
Clam Chowder	187		128	7	5	15	—	—	—	—	—	—
Clam Dinner	—	Clams, Fryes, slaw	955	22	58	100	—	—	—	—	—	—
Cole Slaw	—		138	1	8	16	—	—	—	—	—	—
Cole Slaw, drained on fork	98		182	1	15	11	—	—	—	—	—	—
Combo Salad	—	4.25 oz seafood salad, 2 oz salad shrimp, 6 oz lettuce 2.4 oz tomato, 1 pkg crackers	397	27	29	21	—	—	—	—	—	—
Corn on Cob	—	1 pc.	174	5	4	29	—	—	—	—	—	—
Corn on the Cob	150	1 ear	176	5	4	29	—	—	—	—	—	—

LONG JOHN SILVER'S

	Mass (g)	Description	kcal	Protein (g)	Fat (g)	CHO (g)	Ca (mg)	Fe (mg)	Vit A (IU)	Vit C (mg)	Vit B₁ (mg)	Vit B₂ (mg)
Fish & Chicken	—	1 fish, 2 Tender Chicken Planks, Fryes, slaw	935	36	55	73	—	—	—	—	—	—
Fish & Fryes	—	3 fish, fryes	853	43	48	64	—	—	—	—	—	—
Fish & Fryes	—	2 pc fish, fryes	651	30	36	53	—	—	—	—	—	—
Fish & More	—	2 fish, Fryes, slaw, 2 hush puppies	978	34	58	92	—	—	—	—	—	—
Fish w/Batter	—	2 pc	319	19	19	19	—	—	—	—	—	—
Fish w/Batter	—	3 pc	477	28	28	28	—	—	—	—	—	—
Four Nuggets and Fryes	—		427	16	24	39	—	—	—	—	—	—
Fryes	85		247	4	12	31	—	—	—	—	—	—
Fryes	—		275	4	15	32	—	—	—	—	—	—
Honey–Mustard Sauce	35		56	—	—	14	—	—	—	—	—	—
Hush Puppies	47	2 pieces	145	3	7	18	—	—	—	—	—	—
Hush Puppies	—	3 pc	158	1	7	20	—	—	—	—	—	—
Kitchen–Breaded Fish (Three Piece Dinner)	—	3 kitchen breaded fish, Fryes, slaw, 2 hush puppies	940	35	52	84	—	—	—	—	—	—
Kitchen–Breaded Fish (Two Piece Dinner)	—	2 kitchen– breaded fish, Fryes, slaw, 2 hush puppies	818	26	46	76	—	—	—	—	—	—
Lemon Meringue Pie	99		200	2	6	37	—	—	—	—	—	—
Ocean Chef Salad	—	6 oz. lettuce, 1.25 oz shrimp, 2 oz. seafood blend, 2 tomato wedges, ¾ oz cheese	229	27	8	13	—	—	—	—	—	—
Ocean Scallops	—	6 pieces	257	10	12	27	—	—	—	—	—	—
One Fish and Fryes	—		449	16	24	42	—	—	—	—	—	—
One Fish, Two Nuggets, and Fryes	—		539	23	30	46	—	—	—	—	—	—
Oyster Dinner	—	6 oysters, Fryes, slaw	789	17	45	78	—	—	—	—	—	—
Pecan Pie	113		446	5	22	59	—	—	—	—	—	—
Peg Leg w/Batter	—	5 pieces	514	25	33	30	—	—	—	—	—	—
Pumpkin Pie	113		251	4	11	34	—	—	—	—	—	—
Reduced Calorie Italian Dressing	49		20	0	1	3	—	—	—	—	—	—
Scallop Dinner	—	6 scallops, Fryes, slaw	747	17	45	66	—	—	—	—	—	—
Sea Salad Dressing	45		220	4	21	5	—	—	—	—	—	—
Seafood Platter	—	1 fish, 2 battered shrimp, 2 scallops, Fryes, slaw	976	29	58	85	—	—	—	—	—	—

LONG JOHN SILVER'S

	Mass (g)	Description	kcal	Pro-tein (g)	Fat (g)	CHO (g)	Ca (mg)	Fe (mg)	Vit A (IU)	Vit C (mg)	Vit B₁ (mg)	Vit B₂ (mg)
Seafood Salad	—	5.6 oz seafood salad, 6 oz. lettuce, 2.4 oz to-mato	426	19	30	22	—	—	—	—	—	—
Shrimp & Fish Dinner	—	1 fish, 3 battered shrimp, Fryes, slaw, 2 hush pup-pies	917	27	55	80	—	—	—	—	—	—
Shrimp Salad	—	4.5 oz salad shrimp, 6 oz. lettuce, 2.4 oz tomato	203	28	3	16	—	—	—	—	—	—
Shrimp w/Batter	—	5 pieces	269	9	13	31	—	—	—	—	—	—
Sweet–n–Sour Sauce	30		—	—	—	—	—	—	—	—	—	—
Tartar Sauce	30		117	—	11	5	—	—	—	—	—	—
Tender Chicken Plank Dinner	—	3 Chicken Planks, Fryes, slaw	885	32	51	72	—	—	—	—	—	—
Tender Chicken Plank Dinner	—	4 Chicken Planks, Fryes, slaw	1037	41	59	82	—	—	—	—	—	—
Thousand Island Dressing	48		223	—	22	8	—	—	—	—	—	—
Three Piece Fish Dinner	—	3 fish, Fryes, slaw, 2 hush pup-pies	1180	47	70	93	—	—	—	—	—	—
Treasure Chest	—	2 pc Fish, 2 Peg Legs	467	25	29	27	—	—	—	—	—	—
Two Planks and Fryes	—		551	22	28	51	—	—	—	—	—	—

Source: Long John Silver's Seafood Shoppes, sampling and nutrient analysis conducted independently by the Department of Nutrition and Food Science, University of Kentucky, April 10, 1986.

MCDONALD'S

	Mass (g)	Description	kcal	Pro-tein (g)	Fat (g)	CHO (g)	Ca (mg)	Fe (mg)	Vit A (IU)	Vit C (mg)	Vit B₁ (mg)	Vit B₂ (mg)
Apple Pie	85		253	1.9	14.3	29.3	14	0.62	<34	<0.85	0.02	0.02
Barbeque Sauce	32		60	0.4	0.4	13.7	4	0.12	45	<0.64	0.01	0.01
Big Mac	200		570	24.6	35.0	39.2	203	4.90	380	3.00	0.48	0.38
Biscuit with Bacon, Egg and Cheese	145		483	16.5	31.6	33.2	2	2.57	653	1.60	0.30	0.43
Biscuit with Sausage	121		467	12.1	30.9	35.3	82	2.05	61	<1.20	0.56	0.22
Biscuit with Sausage and Egg	175		585	19.8	39.9	36.4	119	3.43	420	<1.75	0.53	0.49
Biscuit, plain	85		330	4.9	18.2	36.6	74	1.30	179	<0.85	0.21	0.15
Cake Cones	115		185	4.3	5.2	30.2	20	<2.00	4	—	4.00	20.00
Carmel Sundae	165		361	7.2	10.0	608.0	200	0.23	279	3.61	0.07	0.31
Cheeseburger	114		318	15.0	16.0	28.5	169	2.84	353	2.05	0.30	0.24
Cherry Pie	88		260	2.0	13.6	32.1	12	0.59	114	<0.88	0.03	0.02
Chocolate Milk Shake	291		383	9.9	9.0	65.5	320	0.84	349	<2.91	0.12	0.44
Chocolaty Chip Cookies	69		342	4.2	16.3	44.8	29	1.56	75.9	1.04	0.12	0.21

MCDONALD'S

	Mass (g)	Description	kcal	Protein (g)	Fat (g)	CHO (g)	Ca (mg)	Fe (mg)	Vit A (IU)	Vit C (mg)	Vit B₁ (mg)	Vit B₂ (mg)
Cones	115		185	4.3	5.2	30.2	183	0.12	218	<1.15	0.06	0.36
Egg McMuffin	138		340	18.5	15.8	31.0	226	2.93	591	<1.38	0.47	0.44
English Muffin w/Butter	63		186	5.0	5.3	29.5	117	1.51	164	0.82	0.28	0.49
Filet–O–Fish	143		435	14.7	25.7	35.9	133	2.47	186	<2.15	0.36	0.23
French Fries	68		220	3.0	11.5	26.1	9	0.61	<17	12.53	0.12	0.02
Hamburger	100		263	12.4	11.3	28.3	84	2.85	100	1.79	0.31	0.22
Hash Brown Potatoes	55		125	1.5	7.0	14.0	5	0.40	<14	4.14	0.06	<0.01
Honey	14		50	0	0	12.4	1	0.02	<14	<0.15	0	0
Hot Fudge Sundae	164		357	7.0	10.8	58.0	215	0.61	230	2.46	0.07	0.31
Hot Mustard Sauce	30		63	0.6	2.1	10.5	8	0.17	9	<0.30	0.01	0
Hotcakes w/Butter & Syrup	214		500	7.9	10.3	93.9	103	2.23	257	4.71	0.26	0.36
McD.L.T.	254		680	30.0	44.0	40.0	230	6.60	508	8.00	0.56	0.46
McDonaldLand Cookies	67		308	4.2	10.8	48.7	12	1.47	<27	<0.94	0.23	0.23
McNuggets	109	6 pieces	323	19.1	21.3	13.7	11	1.25	<109	2.07	0.16	0.14
Quarter Pounder	160		427	24.6	23.5	29.3	98	4.30	128	2.56	0.35	0.32
Quarter Pounder w/Cheese	186		525	29.6	31.6	30.5	255	4.84	614	2.79	0.37	0.41
Sausage	53		210	9.8	18.6	0.6	16	0.82	<32	<0.53	0.27	0.11
Sausage McMuffin with Egg	165		517	22.9	32.9	32.2	196	3.47	660	1.65	0.84	0.50
Scrambled Eggs	98		180	13.2	13.0	2.5	61	2.53	652	1.18	0.08	0.47
Strawberry Milk Shake	290		362	9.0	8.7	62.1	30	<2.00	377	4.06	0.12	0.44
Strawberry Sundae	164		320	6.0	8.7	54.0	174	0.38	230	2.79	0.07	0.30
Sweet & Sour Sauce	32		64	0.2	0.3	15.0	2	0.08	200	<0.30	0.01	<0.01
Vanilla Milk Shake	291		352	9.3	8.4	59.6	329	0.18	349	3.20	0.12	0.70

Source: McDonald's Corporation. Nutritional analysis reported by Hazleton Laboratories, Inc., 1986.

PIZZA HUT

	Mass (g)	kcal	Protein (g)	Fat (g)	CHO (g)	Ca (mg)	Fe (mg)	Vit A (IU)	Vit C (mg)	Vit B₁ (mg)	Vit B₂ (mg)
Thick 'N' Chewy, Beef	—	620	38	20	73	400	7.2	750	<1.2	0.68	0.60
Thick 'N' Chewy, Cheese	—	560	34	14	71	500	5.4	1000	<1.2	0.68	0.68
Thick 'N' Chewy, Pepperoni	—	560	31	18	68	400	5.4	1250	3.6	0.68	0.68
Thick 'N' Chewy, Pork	—	640	36	23	71	400	7.2	750	1.2	0.90	0.77
Thick 'N' Chewy, Supreme	—	640	36	22	74	400	7.2	1000	9.0	0.75	0.85
Thin 'N' Crispy, Beef	—	490	29	19	51	350	6.3	750	<1.2	0.30	0.60
Thin 'N' Crispy, Cheese	—	450	25	15	54	450	4.5	750	<1.2	0.30	0.51
Thin 'N' Crispy, Pepperoni	—	430	23	17	45	300	4.5	1000	<1.2	0.30	0.51
Thin 'N' Crispy, Pork	—	520	27	23	51	350	6.3	1000	<1.2	0.38	0.68
Thin 'N' Crispy, Supreme	—	510	27	21	51	350	7.2	1250	2.4	0.38	0.68

Source: Research 900 and Pizza Hut, Inc., Wichita, KS.

ROY ROGERS

	Mass (g)	Description	kcal	Pro-tein (g)	Fat (g)	CHO (g)	Ca (mg)	Fe (mg)	Vit A (IU)	Vit C (mg)	Vit B₁ (mg)	Vit B₂ (mg)
Apple Danish	71		249	4.5	11.6	31.6	—	—	—	—	—	—
Bacon Cheeseburger	180		581	32.3	39.2	25.0	—	—	—	—	—	—
Biscuit	63		231	4.4	12.1	26.2	—	—	—	—	—	—
Breakfast Crescent Sandwich	127		401	13.3	27.3	25.3	—	—	—	—	—	—
Breakfast Crescent Sandwich w/bacon	133		431	15.4	29.7	25.5	—	—	—	—	—	—
Breakfast Crescent Sandwich w/ham	165		557	19.8	41.7	25.3	—	—	—	—	—	—
Breakfast Crescent Sandwich w/sausage	162		449	19.9	29.4	25.9	—	—	—	—	—	—
Breast & Wing	196		604	43.5	36.5	25.4	—	—	—	—	—	—
Brownie	64		264	3.3	11.4	37.3	—	—	—	—	—	—
Caramel Sundae	145		293	7.0	8.5	51.5	—	—	—	—	—	—
Cheese Danish	71		254	4.9	12.2	31.4	—	—	—	—	—	—
Cheeseburger	173		563	29.5	37.3	27.4	—	—	—	—	—	—
Cherry Danish	71		271	4.4	14.4	31.7	—	—	—	—	—	—
Chicken Breast	144		412	33.0	23.7	16.9	—	—	—	—	—	—
Chocolate Shake	319		358	7.9	10.2	61.3	—	—	—	—	—	—
Cole Slaw	99		110	1.0	6.9	11.0	—	—	—	—	—	—
Crescent Roll	70		287	4.7	17.7	27.2	—	—	—	—	—	—
Egg and Biscuit Platter	165		394	16.9	26.5	21.9	—	—	—	—	—	—
Egg and Biscuit Platter w/bacon	173		435	19.7	29.6	22.1	—	—	—	—	—	—
Egg and Biscuit Platter w/ham	200		442	23.5	28.6	22.5	—	—	—	—	—	—
Egg and Biscuit Platter w/sausage	203		550	23.4	40.9	21.9	—	—	—	—	—	—
French Fries	85		268	3.9	13.5	32.0	—	—	—	—	—	—
Hamburger	143		456	23.8	28.3	65.6	—	—	—	—	—	—
Hot Chocolate	—		123	3.0	2.0	22.0	—	—	—	—	—	—
Hot Fudge Sundae	151		337	6.5	12.5	53.3	—	—	—	—	—	—
Hot Topped Potato plain	227		211	5.9	0.2	47.9	—	—	—	—	—	—
Hot Topped Potato w/bacon 'n cheese	248		397	17.1	21.7	33.3	—	—	—	—	—	—
Hot Topped Potato w/broccoli 'n cheese	312		376	13.7	18.1	39.6	—	—	—	—	—	—
Hot Topped Potato w/oleo	236		274	5.9	7.3	47.9	—	—	—	—	—	—
Hot Topped Potato w/sour cream 'n chives	297		408	7.3	20.9	47.6	—	—	—	—	—	—
Hot Topped Potato w/taco beef 'n cheese	359		463	21.8	21.8	45.0	—	—	—	—	—	—
Large Fries	113		357	5.3	18.4	42.7	—	—	—	—	—	—
Large Roast Beef	182		360	33.9	11.9	29.6	—	—	—	—	—	—
Large Roast Beef w/Cheese	211		467	39.6	20.9	30.3	—	—	—	—	—	—
Leg	53		140	11.5	8.0	5.5	—	—	—	—	—	—
Macaroni	100		186	3.1	10.7	19.4	—	—	—	—	—	—
Milk	—		150	8.0	8.2	11.4	—	—	—	—	—	—
Orange Juice	—		99	1.5	0.2	22.8	—	—	—	—	—	—
Orange Juice	—		136	2.0	0.3	31.3	—	—	—	—	—	—
Pancake Platter (w.syrup, butter)	165		452	7.7	15.2	71.8	—	—	—	—	—	—
Pancake Platter (w.syrup, butter) w/bacon	173		493	10.4	18.3	72.0	—	—	—	—	—	—
Pancake Platter (w.syrup, butter) w/ham	200		506	14.3	17.3	72.4	—	—	—	—	—	—
Pancake Platter (w.syrup, butter) w/sausage	203		608	14.2	29.6	71.8	—	—	—	—	—	—
Potato Salad	100		107	2.0	6.1	10.9	—	—	—	—	—	—
Roast Beef Sandwich	154		317	27.2	10.2	29.1	—	—	—	—	—	—
Roast Beef Sandwich w/Cheese	182		424	32.9	19.2	29.9	—	—	—	—	—	—
RR Bar Burger	208		611	36.1	39.4	28.0	—	—	—	—	—	—
Salad Bar 1,000 Island	—	2 T	160	—	16.0	4.0	—	—	—	—	—	—

ROY ROGERS

	Mass (g)	Description	kcal	Protein (g)	Fat (g)	CHO (g)	Ca (mg)	Fe (mg)	Vit A (IU)	Vit C (mg)	Vit B₁ (mg)	Vit B₂ (mg)
Salad Bar Bacon 'n Tomato	—	2 T	136	—	12.0	6.0	—	—	—	—	—	—
Salad Bar Bacon Bits	—	1 T	24	4.0	1.0	38.0	—	—	—	—	—	—
Salad Bar Blue Cheese Dressing	—	2 T	150	2.0	16.0	2.0	—	—	—	—	—	—
Salad Bar Cheddar Cheese	—	¼ cup	112	5.8	9.0	0.8	—	—	—	—	—	—
Salad Bar Chinese Noodles	—	¼ cup	55	1.5	2.8	6.5	—	—	—	—	—	—
Salad Bar Chopped Eggs	—	2 T	55	4.0	4.0	0.7	—	—	—	—	—	—
Salad Bar Croutons	—	2 T	132	5.5	0	31.0	—	—	—	—	—	—
Salad Bar Cucumbers	—	5–6 slices	4	—	0	1.0	—	—	—	—	—	—
Salad Bar Green Peas	—	¼ cup	7	0.5	0	1.2	—	—	—	—	—	—
Salad Bar Green Peppers	—	2 T	4	0.3	0	1.0	—	—	—	—	—	—
Salad Bar Lettuce	—	1 cup	10	—	0	4.0	—	—	—	—	—	—
Salad Bar Lo–cal Italian	—	2 T	70	—	6.0	2.0	—	—	—	—	—	—
Salad Bar Macaroni Salad	—	2 T	60	1.0	3.6	6.2	—	—	—	—	—	—
Salad Bar Mushrooms	—	¼ cup	5	0.5	0	0.7	—	—	—	—	—	—
Salad Bar Potato Salad	—	2 T	50	1.0	3.0	5.5	—	—	—	—	—	—
Salad Bar Ranch	—	2 T	155	—	14.0	4.0	—	—	—	—	—	—
Salad Bar Shredded Carrots	—	¼ cup	12	0.6	0	24.0	—	—	—	—	—	—
Salad Bar Sliced Beets	—	¼ cup	16	0.5	0	3.8	—	—	—	—	—	—
Salad Bar Sunflower Seeds	—	2 T	101	4.0	9.0	5.0	—	—	—	—	—	—
Salad Bar Tomatoes	—	3 slices	20	0.8	0	4.8	—	—	—	—	—	—
Strawberry Shake	312		315	7.6	10.2	49.4	—	—	—	—	—	—
Strawberry Shortcake	205		447	10.1	19.2	59.3	—	—	—	—	—	—
Strawberry Sundae	142		216	5.7	7.1	33.1	—	—	—	—	—	—
Thigh	98		296	18.4	19.5	11.7	—	—	—	—	—	—
Thigh & Leg	151		436	29.9	27.5	17.2	—	—	—	—	—	—
Vanilla Shake	306		306	8.0	10.7	45.0	—	—	—	—	—	—
Wing	52		192	10.5	12.8	8.5	—	—	—	—	—	—

Source: Roy Rogers Restaurants, Marriott Corporation, Washington, DC. Nutritional data furnished by Lancaster Laboratories, 1985.

TACO BELL

	Mass (g)	kcal	Protein (g)	Fat (g)	CHO (g)	Ca (mg)	Fe (mg)	Vit A (IU)	Vit C (mg)	Vit B₁ (mg)	Vit B₂ (mg)
Bean Burrito	166	343	11	12	48	98	2.8	1657	15.2	0.37	0.22
Beef Burrito	184	466	30	21	37	83	4.6	1675	15.2	0.30	0.39
Beefy Tostada	184	291	19	15	21	208	3.4	3450	12.7	0.16	0.27
Bellbeefer	123	221	15	7	23	40	2.6	2961	10.0	0.15	0.20
Bellbeefer w/Cheese	137	278	19	12	23	147	2.7	3146	10.0	0.16	0.27
Burrito Supreme	225	457	21	22	43	121	3.8	3462	16.0	0.33	0.35
Combination Burrito	175	404	21	16	43	91	3.7	1666	15.2	0.34	0.31
Enchirito	207	454	25	21	42	259	3.8	1178	9.5	0.31	0.37
Pintos 'N' Cheese	158	168	11	5	21	150	2.3	3123	9.3	0.26	0.16
Taco	83	186	15	8	14	120	2.5	120	0.2	0.09	0.16
Tostada	138	179	9	6	25	191	2.3	3152	9.7	0.18	0.15

Sources: Menu Item Portions, July 1976. Taco Bell Co., San Antonio, TX.
Adams CF: *Nutritive Value of American Foods in Common Units*. USDA Agricultural Research Service, Agricultural Handbook No. 456, November 1975.
Church CF, Church HN: *Food Values of Portions Commonly Used*, ed 12. Philadelphia, J. B. Lippincott, 1975.
Valley Baptist Medical Center, Food Service Department: Descriptions of Mexican–American Foods, NASCO. Fort Atkinson, WI.

WENDY'S

	Mass (g)	Description	kcal	Pro-tein (g)	Fat (g)	CHO (g)	Ca (mg)	Fe (mg)	Vit A (IU)	Vit C (mg)	Vit B₁ (mg)	Vit B₂ (mg)
Bacon	18	2 strips	110	5	10	—	—	—	—	—	—	—
Bacon Cheeseburger on white bun	147		460	29	28	23	—	—	—	—	—	—
Breakfast Sandwich	129		370	17	19	33	—	—	—	—	—	—
Chicken Sandwich on multi—grain wheat bun	128		320	25	10	31	—	—	—	—	—	—
Chili	256		260	21	8	26	—	—	—	—	—	—
Danish	85	1 piece	360	6	18	44	—	—	—	—	—	—
Double Hamburger on white bun	197		560	41	34	24	—	—	—	—	—	—
French Fries (salted) (reg)	98		280	4	14	35	—	—	—	—	—	—
French Toast	135	2 slices	400	11	19	45	—	—	—	—	—	—
Frosty Dairy Desert	243		400	8	14	59	—	—	—	—	—	—
Home Fries	103		360	4	22	37	—	—	—	—	—	—
Hot Stuffed Baked Potatoes Bacon & Cheese	350		570	19	30	57	—	—	—	—	—	—
Hot Stuffed Baked Potatoes Broccoli & Cheese	365		500	13	25	54	—	—	—	—	—	—
Hot Stuffed Baked Potatoes Cheese	350		590	17	34	55	—	—	—	—	—	—
Hot Stuffed Baked Potatoes Chili & Cheese	400		510	22	20	63	—	—	—	—	—	—
Hot Stuffed Baked Potatoes Plain	250		250	6	2	52	—	—	—	—	—	—
Hot Stuffed Baked Potatoes Sour Cream & Chives	310		460	6	24	53	—	—	—	—	—	—
Kids' Meal Hamburger	75		220	13	8	11	—	—	—	—	—	—
Multi-grain Wheat Bun	48		135	5	3	23	—	—	—	—	—	—
Omelet #1 Ham & Cheese	114		250	18	17	6	—	—	—	—	—	—
Omelet #2 Ham, Cheese & Mushroom	118		290	18	21	7	—	—	—	—	—	—
Omelet #3 Ham, Cheese, Onion, Green Pepper	128		280	19	19	7	—	—	—	—	—	—
Omelet #4 Mushroom, Onion, Green Pepper	114		210	14	15	7	—	—	—	—	—	—
Orange Juice	180		80	1	—	17	—	—	—	—	—	—
Pick-Up Window Salad	510		110	8	6	5	—	—	—	—	—	—
Sausage	45	1 patty	200	9	18	—	—	—	—	—	—	—
Scrambled Eggs	91		190	14	12	7	—	—	—	—	—	—
Single Hamburger on multi—grain wheat bun	119		340	25	17	20	—	—	—	—	—	—
Single Hamburger on white bun	117		350	21	18	27	—	—	—	—	—	—
Taco Salad	357		390	23	18	36	—	—	—	—	—	—
Toast with Margarine	69	2 slices	250	6	9	35	—	—	—	—	—	—
White Bun	52		160	5	3	28	—	—	—	—	—	—

Source: Wendy's International, Inc., Dublin, OH, 1985. Nutritional information provided by Hazelton Laboratories America, Inc., and U.S. Department of Agriculture Handbook #8.

APPENDIX C

METABOLIC COMPUTATIONS IN OPEN-CIRCUIT SPIROMETRY

STANDARDIZING GAS VOLUMES: ENVIRONMENTAL FACTORS

Gas volumes obtained during physiologic measurements are usually expressed in one of three ways: (1) *ATPS*, (2) *STPD*, or (3) *BTPS*.

ATPS refers to the volume of gas at the specific conditions of measurement, which are therefore at Ambient Temperature (273°K + ambient temperature°C), ambient Pressure, and Saturated with water vapor. Gas volumes collected during open-circuit spirometry and pulmonary function tests are measured initially at ATPS.

The volume of a gas varies, however, depending on its temperature, pressure, and content of water vapor, even though the absolute number of gas molecules remains constant. These environmental influences are summarized as follows:

Temperature: The volume of a gas varies *directly* with temperature. Increasing the temperature causes the molecules to move more rapidly; the gas mixture expands, and the volume increases proportionately (*Charles' Law*).

Pressure: The volume of a gas varies *inversely* with pressure. Increasing the pressure on a gas forces the molecules closer together, causing the volume to decrease in proportion to the increase in pressure (*Boyle's Law*).

Water vapor: The volume of a gas varies depending on its water vapor content. The volume of a gas is greater when the gas is saturated with water vapor than it is when the same gas is dry (i.e., contains no moisture).

These three factors—temperature, pressure, and the relative degree of saturation of the gas with water vapor—must be considered, especially when gas volumes are to be compared under different environmental conditions and used subsequently in metabolic and physiologic calculations. The standards that provide the frame of reference for expressing a volume of gas are either STPD or BTPS.

STPD refers to the volume of a gas expressed under Standard conditions of Temperature (273°K or 0°C), Pressure (760 mm Hg), and Dry (no water vapor). Expressing a gas volume STPD, for example, makes it possible to evaluate and compare the volumes of expired air measured while running in the rain at high altitude, along a beach in the cold of winter, or in a hot desert environment below sea level. *In all metabolic calculations, gas volumes are always expressed at STPD.*

1. To reduce a gas volume to standard temperature (ST), the following formula is applied:

$$\text{Gas volume ST} = V_{\text{ATPS}} \times \frac{273°\text{K}}{273°\text{K} + \text{T}°\text{C}} \tag{1}$$

where $\text{T}°\text{C}$ = temperature of the gas in the measuring device and $273°\text{K}$ = absolute temperature Kelvin, which is equivalent to $0°\text{C}$.

TABLE C–1.
VAPOR PRESSURE (P_{H_2O}) OF WET GAS AT TEMPERATURES NORMALLY ENCOUNTERED IN THE LABORATORY

T(°C)	$P_{H_2O\text{(mm Hg)}}$	T(°C)	$P_{H_2O\text{(mm Hg)}}$
20	17.5	31	33.7
21	18.7	32	35.7
22	19.8	33	37.7
23	21.1	34	39.9
24	22.4	35	42.2
25	23.8	36	44.6
26	25.2	37	47.1
27	26.7	38	49.7
28	28.4	39	52.4
29	30.0	40	55.3
30	31.8		

2. The following equation is used to express a gas volume at standard pressure (SP):

$$\text{Gas volume SP} = V_{\text{ATPS}} \times \frac{P_B}{760 \text{ mm Hg}} \tag{2}$$

where P_B = ambient barometric pressure in mm Hg and 760 = standard barometric pressure at sea level, mm Hg.

3. To reduce a gas to standard dry (SD) conditions, the effects of water vapor pressure at the particular environmental temperature must be subtracted from the volume of gas. Because expired air is 100% saturated with water vapor, it is not necessary to determine its percent saturation from measures of relative humidity. The vapor pressure in moist or completely humidified air at a particular ambient temperature can be obtained in Table C-1 and is expressed in mm Hg. This vapor pressure (P_{H_2O}) is then subtracted from the ambient barometric pressure (P_B) to reduce the gas to standard pressure dry (SPD) as follows:

$$\text{Gas volume SPD} = V_{\text{ATPS}} \times \frac{P_B - P_{H_2O}}{760} \tag{3}$$

By combining equations (1) and (3), any volume of moist air can be converted to STPD as follows:

$$\text{Gas volume STPD} = V_{\text{ATPS}} \left(\frac{273}{273 + \text{T}°\text{C}} \right) \left(\frac{P_B - P_{H_2O}}{760} \right) \tag{4}$$

Fortunately, these computations need not be carried out, because the appropriate *STPD correction factors* have already been calculated for moist gas in the range of temperatures and pressures ordinarily encountered in most laboratories. These factors are presented in Table C-2. Multiplying any gas volume ATPS by the appropriate correction factor gives the same gas volume STPD that would be obtained if values for the ambient temperature, barometric pressure, and water vapor pressure were substituted in equation (4).

The term *BTPS* refers to a volume of a gas expressed at *B*ody *T*emperature (usually 273°K + 37°C or 310°K), ambient *P*ressure (whatever the barometer reads), and *S*aturated with water vapor with a partial pressure of 47 mm Hg at 37°C. Conventionally, pulmonary physiologists express lung volumes such as vital capacity, inspiratory and expiratory capacity, residual lung volume, and the dynamic measures of lung function such as maximum breathing

TABLE C–2.
FACTORS TO REDUCE MOIST GAS TO A DRY GAS VOLUME AT 0° C AND 760 mm Hg

Baro-metric Reading	Temperature, °C																	
	15	16	17	18	19	20	21	22	23	24	25	26	27	28	29	30	31	32
700	0.855	851	847	842	838	834	829	825	821	816	812	807	802	797	793	788	783	778
702	857	853	849	845	840	836	832	827	823	818	814	809	805	800	795	790	785	780
704	860	856	852	847	843	839	834	830	825	821	816	812	807	802	797	792	787	783
706	862	858	854	850	845	841	837	832	828	823	819	814	810	804	800	795	790	785
708	865	861	856	852	848	843	839	834	830	825	821	816	812	807	802	797	792	787
710	867	863	859	855	850	846	842	837	833	828	824	819	814	809	804	799	795	790
712	870	866	861	857	853	848	844	839	836	830	826	821	817	812	807	802	797	792
714	872	868	864	859	855	851	846	842	837	833	828	824	819	814	809	804	799	794
716	875	871	866	862	858	853	849	844	840	835	831	826	822	816	812	807	802	797
718	877	873	869	864	860	856	851	847	842	838	833	828	824	819	814	809	804	799
720	880	876	871	867	863	858	854	849	845	840	836	831	826	821	816	812	807	802
722	882	878	874	869	865	861	856	852	847	843	838	833	829	824	819	814	809	804
724	885	880	876	872	867	863	858	854	849	845	840	835	831	826	821	816	811	806
726	887	883	879	874	870	866	861	856	852	847	843	838	833	829	824	818	813	808
728	890	886	881	877	872	868	863	859	854	850	845	840	836	831	826	821	816	811
730	892	888	884	879	875	871	866	861	857	852	847	843	838	833	828	823	818	813
732	895	890	886	882	877	873	868	864	859	854	850	845	840	836	831	825	820	815
734	897	893	889	884	880	875	871	866	862	857	852	847	843	838	833	828	823	818
736	900	895	891	887	882	878	873	869	864	859	855	850	845	840	835	830	825	820
738	902	898	894	889	885	880	876	871	866	862	857	852	848	843	838	833	828	822
740	905	900	896	892	887	883	878	874	869	864	860	855	850	845	840	835	830	825
742	907	903	898	894	890	885	881	876	871	867	862	857	852	847	842	837	832	827
744	910	906	901	897	892	888	883	878	874	869	864	859	855	850	845	840	834	829
746	912	908	903	899	895	890	886	881	876	872	867	862	857	852	847	842	837	832
748	915	910	906	901	897	892	888	883	879	874	869	864	860	854	850	845	839	834
750	917	913	908	904	900	895	890	886	881	876	872	867	862	857	852	847	842	837
752	920	915	911	906	902	897	893	888	883	879	874	869	864	859	854	849	844	839
754	922	918	913	909	904	900	895	891	886	881	876	872	867	862	857	852	846	841
756	925	920	916	911	907	902	898	893	888	883	879	874	869	864	859	854	849	844
758	927	923	918	914	909	905	900	896	891	886	881	876	872	866	861	856	851	846
760	930	925	921	916	912	907	902	898	893	888	883	879	874	869	864	859	854	848
762	932	928	923	919	914	910	905	900	896	891	886	881	876	871	866	861	856	851
764	936	930	926	921	916	912	907	903	898	893	888	884	879	874	869	864	858	853
766	937	933	928	924	919	915	910	905	900	896	891	886	881	876	871	866	861	855
768	940	935	931	926	922	917	912	908	903	898	893	888	883	878	873	868	863	858
770	942	938	933	928	924	919	915	910	905	901	896	891	886	881	876	871	865	860

capacity at body temperature and moist, or BTPS. The following equation converts a gas volume ATPS to BTPS:

$$\text{Gas volume BTPS} = V_{\text{ATPS}} \left(\frac{P_B - P_{H_2O}}{P_B - 47 \text{ mm Hg}} \right) \left(\frac{310}{273 + T°C} \right) \tag{5}$$

As was the case with the correction to STPD, appropriate BTPS *correction factors* are available for converting a moist gas volume at ambient conditions to a volume BTPS. These BTPS factors for a broad range of ambient temperatures are presented in Table C-3. These factors have been computed assuming a barometric pressure of 760 mm Hg, and small deviations (± 10 mm Hg) from this pressure introduce only a minimal error.

TABLE C–3. BTPS FACTORS			
T(°C)	**BTPS**	**T(°C)**	**BTPS**
20	1.102	29	1.051
21	1.096	30	1.045
22	1.091	31	1.039
23	1.085	32	1.032
24	1.080	33	1.026
25	1.075	34	1.020
26	1.068	35	1.014
27	1.063	36	1.007
28	1.057	37	1.000

* Body temperature, ambient pressure, and saturated with water vapor.

CALCULATION OF OXYGEN CONSUMPTION

In determining oxygen consumption by open-circuit spirometry, we are interested in knowing how much oxygen has been removed from the *inspired air*. Because the composition of inspired air remains relatively constant ($CO_2 = 0.03\%$, $O_2 = 20.93\%$, $N_2 = 79.04\%$), it is possible to determine how much oxygen has been removed from the inspired air by measuring the amount and composition of the expired air. When this is done, the expired air contains more carbon dioxide (usually 2.5 to 5.0%), less oxygen (usually 15.0 to 18.5%), and more nitrogen (usually 79.04 to 79.60%). It should be noted, however, that nitrogen is inert in terms of metabolism; any change in its concentration in expired air reflects the fact that the number of oxygen molecules removed from the inspired air is not replaced by the same number of carbon dioxide molecules produced in metabolism. This results in the volume of expired air (V_E, STPD) being unequal to the inspired volume (V_I, STPD). For example, if the respiratory quotient is less than 1.00 (i.e., less CO_2 produced in relation to O_2 consumed), and 3 liters of air are inspired, *less* than 3 liters of air will be expired. In this case, the nitrogen concentration is higher in the expired air than in the inspired air. This is not to say that nitrogen has been produced, only that nitrogen molecules now represent a larger percentage of V_E compared to V_I. In fact, V_E differs from V_I in direct proportion to the change in nitrogen concentration between the inspired and expired volumes. Thus, V_I can be determined from V_E using the relative change in nitrogen in an equation known as the *Haldane transformation*.

$$V_I, STPD = V_E, STPD \times \frac{\% N_{2_E}}{\% N_{2_I}} \qquad (6)$$

where $\% N_{2_I} = 79.04$ and $\% N_{2_E}$ = percent nitrogen in expired air computed from gas analysis as $[(100 - (\% O_{2_E} + \% CO_2)]$.

The volume of O_2 in the inspired air $(V_{O_{2_I}})$ can then be determined as follows:

$$V_{O_{2_I}} = V_I \times \% O_{2_I} \qquad (7)$$

Substituting equation (6) for V_I,

$$V_{O_{2_I}} = V_E \times \frac{\% N_{2_E}}{79.04\%} \times \% O_{2_I} \qquad (8)$$
$$\text{where } \% O_{2_I} = 20.93\%$$

The amount or volume of oxygen in the expired air $(\dot{V}_{O_{2_E}})$ is computed as

$$V_{O_{2_E}} = V_E \times \% O_{2_E} \qquad (9)$$

where $\% O_{2_E}$ is the fractional concentration of oxygen in expired air determined by gas analysis (chemical or electronic methods).

The amount of O_2 removed from the inspired air each minute $(\dot{V}_{O_2})$ can then be computed as follows:

$$\dot{V}_{O_2} = (\dot{V}_I \times \% O_{2_I}) - (\dot{V}_E \times \% O_{2_E}) \qquad (10)$$

By substitution

$$\dot{V}_{O_2} = \left\{ \left[\left(\dot{V}_E \times \frac{\% N_{2_E}}{79.04\%} \right) \times 20.93\% \right] - (\dot{V}_E \times \% O_{2_E}) \right\} \qquad (11)$$

where $\dot{V}_{O_2}$ = volume of oxygen consumed per minute, expressed in milliliters or liters, and $\dot{V}_E$ = expired air volume per minute expressed in milliliters or liters.

Equation (11) can be simplified to:

$$\dot{V}_{O_2} = \dot{V}_E \left[\left(\frac{\% N_{2_E}}{79.04\%} \times 20.93\% \right) - \% O_{2_E} \right] \qquad (12)$$

The final form of the equation is:

$$\dot{V}_{O_2} = \dot{V}_E [(\% N_{2_E} \times .265) - \% O_{2_E}] \qquad (13)$$

The value obtained within the brackets in equations (12) and (13) is referred to as the *true O_2*; this represents the "oxygen extraction" or, more precisely, the percentage of oxygen consumed for any volume of air *expired*.

Although equation (13) is the equation used most widely to compute oxygen consumption from measures of expired air, it is also possible to calculate $\dot{V}_{O_2}$ from direct measurements of both V_I and V_E. In this case, the Haldane transformation is not used, and oxygen consumption is calculated directly as

$$\dot{V}_{O_2} = (\dot{V}_I \times 20.93) - (\dot{V}_E \times \%O_{2_E}) \tag{14}$$

In situations in which only V_I is measured, the V_E can be calculated from the Haldane transformation as

$$\dot{V}_E = \dot{V}_I \frac{\%N_{2_I}}{\%N_{2_E}}$$

By substitution in equation (14), the computational equation is:

$$\dot{V}_{O_2} = \dot{V}_I \left[\%O_{2_I} - \left(\frac{\%N_{2_I}}{\%N_{2_E}} \times \%O_{2_E} \right) \right] \tag{15}$$

CALCULATION OF CARBON DIOXIDE PRODUCTION

The carbon dioxide production per minute ($\dot{V}_{CO_2}$) is calculated as follows:

$$\dot{V}_{CO_2} = \dot{V}_E(\%CO_{2_E} - \%CO_{2_I}) \tag{16}$$

where $\%CO_{2_E}$ = percent carbon dioxide in expired air determined by gas analysis, and $\%CO_{2_I}$ = percent carbon dioxide in inspired air, which is essentially constant at 0.03%.

The final form of the equation is:

$$\dot{V}_{CO_2} = \dot{V}_E(\%CO_{2_E} - 0.03\%) \tag{17}$$

CALCULATION OF RESPIRATORY QUOTIENT

The respiratory quotient (R.Q.) is calculated in one of two ways:

1. $\text{R.Q.} = \dot{V}_{CO_2}/\dot{V}_{O_2}$ (18)

   or

2. $\text{R.Q.} = \dfrac{\%CO_{2_E} - 0.03\%)}{\text{"true" } O_2}$ (19)

SAMPLE METABOLIC CALCULATIONS

The following data were obtained during the last minute of a steady-rate, 10-minute treadmill run performed at 6 miles per hour at a 5% grade.

$\dot{V}_E$: 62.1 liters, ATPS
Barometric pressure: 750 mm Hg
Temperature: 26°C
$\%O_2$ expired: 16.86 (O_2 analyzer)
$\%CO_2$ expired: 3.60 (CO_2 analyzer)
$\%N_2$ expired: $[100 - (16.86 + 3.60)] = 79.54$

Determine the following:
1. $\dot{V}_E$, STPD
2. $\dot{V}_{O_2}$, STPD
3. $\dot{V}_{CO_2}$ STPD

4. R.Q.
5. kcal $\cdot$ min^{-1}

1. $\dot{V}_E$, STPD (use equation 4 or STPD correction factor in Table C-2).

$$\dot{V}_E, \text{STPD} = \dot{V}_E, \text{ATPS} \left(\frac{273}{273 + T°C}\right)\left(\frac{P_B - P_{H_2O}}{760}\right)$$

$$= 62.1 \left(\frac{273}{299}\right)\left(\frac{750 - 25.2}{760}\right)$$

$$= 62.1 \, (.913 \times .954)$$

$$= 54.07 \text{ liters} \cdot \text{min}^{-1}$$

2. $\dot{V}_{O_2}$, STPD (use equation 13)

$$\dot{V}_{O_2}, \text{STPD} = \dot{V}_E, \text{STPD} \, [(\%N_{2_E} \times .265) - \%O_{2_E}]$$

$$= 54.07 \, [(.7954 \times .265) - .1686]$$

$$= 54.07 \, (.0422)$$

$$= 2.281 \text{ liters} \cdot \text{min}^{-1}$$

3. $\dot{V}_{CO_2}$, STPD (use equation 17)

$$\dot{V}_{CO_2}, \text{STPD} = \dot{V}_E, \text{STPD} \, (CO_{2_E} - 0.03\%)$$

$$= 54.07 \, (.0360 - .0003)$$

$$= 54.07 \, (.0357)$$

$$= 1.930 \text{ liters} \cdot \text{min}^{-1}$$

4. R.Q. (use equation 18 or 19)

$$\text{R.Q.} = \dot{V}_{CO_2}/\dot{V}_{O_2}$$

$$= \frac{1.930 \text{ liters CO}_2/\text{min}}{2.281 \text{ liters O}_2/\text{min}}$$

$$= 0.846$$

or

$$\text{R.Q.} = \frac{(\%CO_{2_E} - 0.03\%)}{\text{``true'' O}_2}$$

$$= \frac{3.60 - .03}{4.22}$$

$$= 0.846$$

Because the exercise was performed in a steady rate of aerobic metabolism, the obtained R.Q. of 0.846 can be applied in Table 8–1 to obtain the appropriate caloric transformation. In this way, the exercise oxygen consumption can be transposed to kcal of energy expended per minute as follows:

5. Energy expenditure (kcal $\cdot$ min^{-1}) = $\dot{V}_{O_2}$ (liters $\cdot$ min^{-1}) $\times$ caloric equivalent per liter O_2 at the given steady-rate R.Q.
Energy expenditure = 2.281×4.862
$$= 11.09 \text{ kcal} \cdot \text{min}^{-1}$$

Assuming that the R.Q. value reflects the nonprotein R.Q., a reasonable estimate of both the percentage and quantity of fat and carbohydrate metabolized during each minute of the run can be obtained from Table 8–1.

Percentage kcal derived from fat = 50.7%

Percentage kcal derived from carbohydrate = 49.3%

Grams of fat utilized = 0.267 g per liter of oxygen or approximately 0.61 g per minute (.267 × 2.281 l O_2)

Grams of carbohydrate utilized = 0.580 g per liter of oxygen or approximately 1.36 g per minute (.580 × 2.281 l O_2)

ENERGY EXPENDITURE IN HOUSEHOLD, RECREATIONAL, AND SPORTS ACTIVITIES (IN KCAL · MIN⁻¹)*

Activity	kcal · min⁻¹ · kg⁻¹	kg lb	50 110	53 117	56 123	59 130	62 137	65 143	68 150
Archery	0.065		3.3	3.4	3.6	3.8	4.0	4.2	4.4
Badminton	0.097		4.9	5.1	5.4	5.7	6.0	6.3	6.6
Bakery, general (F)	0.035		1.8	1.9	2.0	2.1	2.2	2.3	2.4
Basketball	0.138		6.9	7.3	7.7	8.1	8.6	9.0	9.4
Billiards	0.042		2.1	2.2	2.4	2.5	2.6	2.7	2.9
Bookbinding	0.038		1.9	2.0	2.1	2.2	2.4	2.5	2.6
Boxing									
in ring	0.222		6.9	7.3	7.7	8.1	8.6	9.0	9.4
sparring	0.138		11.1	11.8	12.4	13.1	13.8	14.4	15.1
Canoeing									
leisure	0.044		2.2	2.3	2.5	2.6	2.7	2.9	3.0
racing	0.103		5.2	5.5	5.8	6.1	6.4	6.7	7.0
Card playing	0.025		1.3	1.3	1.4	1.5	1.6	1.6	1.7
Carpentry, general	0.052		2.6	2.8	2.9	3.1	3.2	3.4	3.5
Carpet sweeping (F)	0.045		2.3	2.4	2.5	2.7	2.8	2.9	3.1
Carpet sweeping (M)	0.048		2.4	2.5	2.7	2.8	3.0	3.1	3.3
Circuit training									
Hydra-Fitness	0.132		6.6	7.0	7.4	7.8	8.2	8.6	9.0
Universal	0.116		5.8	6.2	6.5	6.9	7.2	7.5	7.9
Nautilus	0.092		4.6	4.9	5.2	5.5	5.8	6.0	6.3
Free Weights	0.086		4.3	4.5	4.8	5.0	5.3	5.5	5.8
Cleaning (F)	0.062		3.1	3.3	3.5	3.7	3.8	4.0	4.2
Cleaning (M)	0.058		2.9	3.1	3.2	3.4	3.6	3.8	3.9
Climbing hills									
with no load	0.121		6.1	6.4	6.8	7.1	7.5	7.9	8.2
with 5-kg load	0.129		6.5	6.8	7.2	7.6	8.0	8.4	8.8
with 10-kg load	0.140		7.0	7.4	7.8	8.3	8.7	9.1	9.5
with 20-kg load	0.147		7.4	7.8	8.2	8.7	9.1	9.6	10.0
Coal mining									
drilling coal, rock	0.094		4.7	5.0	5.3	5.5	5.8	6.1	6.4
erecting supports	0.088		4.4	4.7	4.9	5.2	5.5	5.7	6.0
shoveling coal	0.108		5.4	5.7	6.0	6.4	6.7	7.0	7.3
Cooking (F)	0.045		2.3	2.4	2.5	2.7	2.8	2.9	3.1
Cooking (M)	0.048		2.4	2.5	2.7	2.8	3.0	3.1	3.3

* Data from Bannister, E.W. and Brown, S.R.: The relative energy requirements of physical activity, IN H.B. Falls (ed.): Exercise Physiology. New York, Academic Press, 1968; Howley, E.T. and Glover, M.E.: The caloric costs of running and walking one mile for men and women. Medicine and Science in Sports 6:235, 1974; Passmore, R. and Durnin, J.V.G.A.: Human energy expenditure. Physiological Reviews 35:801, 1955. *Note:* Symbols (M) and (F) denote experiments for males and females, respectively. See p. 168 for instructions on how to use this appendix.

71 157	74 163	77 170	80 176	83 183	86 190	89 196	92 203	95 209	98 216
4.6	4.8	5.0	5.2	5.4	5.6	5.8	6.0	6.2	6.4
6.9	7.2	7.5	7.8	8.1	8.3	8.6	8.9	9.2	9.5
2.5	2.6	2.7	2.8	2.9	3.0	3.1	3.2	3.3	3.4
9.8	10.2	10.6	11.0	11.5	11.9	12.3	12.7	13.1	13.5
3.0	3.1	3.2	3.4	3.5	3.6	3.7	3.9	4.0	4.1
2.7	2.8	2.9	3.0	3.2	3.3	3.4	3.5	3.6	3.7
9.8	10.2	10.6	11.0	11.5	11.9	12.3	12.7	13.1	13.5
15.8	16.4	17.1	17.8	18.4	19.1	19.8	20.4	21.1	21.8
3.1	3.3	3.4	3.5	3.7	3.8	3.9	4.0	4.2	4.3
7.3	7.6	7.9	8.2	8.5	8.9	9.2	9.5	9.8	10.1
1.8	1.9	1.9	2.0	2.1	2.2	2.2	2.3	2.4	2.5
3.7	3.8	4.0	4.2	4.3	4.5	4.6	4.8	4.9	5.1
3.2	3.3	3.5	3.6	3.7	3.9	4.0	4.1	4.3	4.4
3.4	3.6	3.7	3.8	4.0	4.1	4.3	4.4	4.6	4.7
9.4	9.7	10.2	10.5	10.9	11.4	11.7	12.1	12.5	12.9
8.3	8.6	8.9	9.3	9.6	10.0	10.3	10.7	11.0	11.4
6.6	6.8	7.1	7.4	7.7	8.0	8.2	8.5	8.8	9.1
6.1	6.3	6.6	6.8	7.1	7.4	7.6	7.9	8.1	8.4
4.4	4.6	4.8	5.0	5.1	5.3	5.5	5.7	5.9	6.1
4.1	4.3	4.5	4.6	4.8	5.0	5.2	5.3	5.5	5.7
8.6	9.0	9.3	9.7	10.0	10.4	10.8	11.1	11.5	11.9
9.2	9.5	9.9	10.3	10.7	11.1	11.5	11.9	12.3	12.6
9.9	10.4	10.8	11.2	11.6	12.0	12.5	12.9	13.3	13.7
10.4	10.9	11.3	11.8	12.2	12.6	13.1	13.5	14.0	14.4
6.7	7.0	7.2	7.5	7.8	8.1	8.4	8.6	8.9	9.2
6.2	6.5	6.8	7.0	7.3	7.6	7.8	8.1	8.4	8.6
7.7	8.0	8.3	8.6	9.0	9.3	9.6	9.9	10.3	10.6
3.2	3.3	3.5	3.6	3.7	3.9	4.0	4.1	4.3	4.4
3.4	3.6	3.7	3.8	4.0	4.1	4.3	4.4	4.6	4.7

Activity	kcal · min⁻¹ · kg⁻¹	kg lb	50 110	53 117	56 123	59 130	62 137	65 143	68 150
Cricket									
batting	0.083		4.2	4.4	4.6	4.9	5.1	5.4	5.6
bowling	0.090		4.5	4.8	5.0	5.3	5.6	5.9	6.1
Croquet	0.059		3.0	3.1	3.3	3.5	3.7	3.8	4.0
Cycling									
leisure, 5.5 mph	0.064		3.2	3.4	3.6	3.8	4.0	4.2	4.4
leisure, 9.4 mph	0.100		5.0	5.3	5.6	5.9	6.2	6.5	6.8
racing	0.169		8.5	9.0	9.5	10.0	10.5	11.0	11.5
Dancing									
Dancing (F)									
aerobic, medium	0.103		5.2	5.5	5.8	6.1	6.4	6.7	7.0
aerobic, intense	0.135		6.7	7.1	7.5	7.9	8.3	8.7	9.2
ballroom	0.051		2.6	2.7	2.9	3.0	3.2	3.3	3.5
choreographed			8.4	8.9	9.4	9.9	10.4	10.9	11.4
"twist," "lambada"	0.168		5.2	5.5	5.8	6.1	6.4	6.7	7.0
Digging trenches	0.145		7.3	7.7	8.1	8.6	9.0	9.4	9.9
Drawing (standing)	0.036		1.8	1.9	2.0	2.1	2.2	2.3	2.4
Eating (sitting)	0.023		1.2	1.2	1.3	1.4	1.4	1.5	1.6
Electrical work	0.058		2.9	3.1	3.2	3.4	3.6	3.8	3.9
Farming									
barn cleaning	0.135		6.8	7.2	7.6	8.0	8.4	8.8	9.2
driving harvester	0.040		2.0	2.1	2.2	2.4	2.5	2.6	2.7
driving tractor	0.037		1.9	2.0	2.1	2.2	2.3	2.4	2.5
feeding cattle	0.085		4.3	4.5	4.8	5.0	5.3	5.5	5.8
feeding animals	0.065		3.3	3.4	3.6	3.8	4.0	4.2	4.4
forking straw bales	0.138		6.9	7.3	7.7	8.1	8.6	9.0	9.4
milking by hand	0.054		2.7	2.9	3.0	3.2	3.3	3.5	3.7
milking by machine	0.023		1.2	1.2	1.3	1.4	1.4	1.5	1.6
shoveling grain	0.085		4.3	4.5	4.8	5.0	5.3	5.5	5.8
Field hockey	0.134		6.7	7.1	7.5	7.9	8.3	8.7	9.1
Fishing	0.062		3.1	3.3	3.5	3.7	3.8	4.0	4.2
Food shopping (F)	0.062		3.1	3.3	3.5	3.7	3.8	4.0	4.2
Food shopping (M)	0.058		2.9	3.1	3.2	3.4	3.6	3.8	3.9
Football	0.132		6.6	7.0	7.4	7.8	8.2	8.6	9.0
Forestry									
ax chopping, fast	0.297		14.9	15.7	16.6	17.5	18.4	19.3	20.2
ax chopping, slow	0.085		4.3	4.5	4.8	5.0	5.3	5.5	5.8
barking trees	0.123		6.2	6.5	6.9	7.3	7.6	8.0	8.4
carrying logs	0.186		9.3	9.9	10.4	11.0	11.5	12.1	12.6
felling trees	0.132		6.6	7.0	7.4	7.8	8.2	8.6	9.0
hoeing	0.091		4.6	4.8	5.1	5.4	5.6	5.9	6.2
planting by hand	0.109		5.5	5.8	6.1	6.4	6.8	7.1	7.4
sawing by hand	0.122		6.1	6.5	6.8	7.2	7.6	7.9	8.3
sawing, power	0.075		3.8	4.0	4.2	4.4	4.7	4.9	5.1
stacking firewood	0.088		4.4	4.7	4.9	5.2	5.5	5.7	6.0
trimming trees	0.129		6.5	6.8	7.2	7.6	8.0	8.4	8.8
weeding	0.072		3.6	3.8	4.0	4.2	4.5	4.7	4.9
Furriery	0.083		4.2	4.4	4.6	4.9	5.1	5.4	5.6
Gardening									
digging	0.126		6.3	6.7	7.1	7.4	7.8	8.2	8.6
hedging	0.077		3.9	4.1	4.3	4.5	4.8	5.0	5.2
mowing	0.112		5.6	5.9	6.3	6.6	6.9	7.3	7.6
raking	0.054		2.7	2.9	3.0	3.2	3.3	3.5	3.7
Golf	0.085		4.3	4.5	4.8	5.0	5.3	5.5	5.8
Gymnastics	0.066		3.3	3.5	3.7	3.9	4.1	4.3	4.5
Horse-grooming	0.128		6.4	6.8	7.2	7.6	7.9	8.3	8.7

71 157	74 163	77 170	80 176	83 183	86 190	89 196	92 203	95 209	98 216
5.9	6.1	6.4	6.6	6.9	7.1	7.4	7.6	7.9	8.1
6.4	6.7	6.9	7.2	7.5	7.7	8.0	8.3	8.6	8.8
4.2	4.4	4.5	4.7	4.9	5.1	5.3	5.4	5.6	5.8
4.5	4.7	4.9	5.1	5.3	5.5	5.7	5.9	6.1	6.3
7.1	7.4	7.7	8.0	8.3	8.6	8.9	9.2	9.5	9.8
12.0	12.5	13.0	13.5	14.0	14.5	15.0	15.5	16.1	16.6
7.3	7.6	7.9	8.2	8.5	8.9	9.2	9.5	9.8	10.1
9.6	10.0	10.4	10.8	11.2	11.6	12.0	12.4	12.8	13.2
3.6	3.8	3.9	4.1	4.2	4.4	4.5	4.7	4.8	5.0
11.9	12.4	12.9	13.4	13.9	14.4	15.0	15.5	16.0	16.5
7.3	7.6	7.9	8.2	8.5	8.9	9.2	9.5	9.8	10.1
10.3	10.7	11.2	11.6	12.0	12.5	12.9	13.3	13.8	14.2
2.6	2.7	2.8	2.9	3.0	3.1	3.2	3.3	3.4	3.5
1.6	1.7	1.8	1.8	1.9	2.0	2.0	2.1	2.2	2.3
4.1	4.3	4.5	4.6	4.8	5.0	5.2	5.3	5.5	5.7
9.6	10.0	10.4	10.8	11.2	11.6	12.0	12.4	12.8	13.2
2.8	3.0	3.1	3.2	3.3	3.4	3.6	3.7	3.8	3.9
2.6	2.7	2.8	3.0	3.1	3.2	3.3	3.4	3.5	3.6
6.0	6.3	6.5	6.8	7.1	7.3	7.6	7.8	8.1	8.3
4.6	4.8	5.0	5.2	5.4	5.6	5.8	6.0	6.2	6.4
9.8	10.2	10.6	11.0	11.5	11.9	12.3	12.7	13.1	13.5
3.8	4.0	4.2	4.3	4.5	4.6	4.8	5.0	5.1	5.3
1.6	1.7	1.8	1.8	1.9	2.0	2.0	2.1	2.2	2.3
6.0	6.3	6.5	6.8	7.1	7.3	7.6	7.8	8.1	8.3
9.5	9.9	10.3	10.7	11.1	11.5	11.9	12.3	12.7	13.1
4.4	4.6	4.8	5.0	5.1	5.3	5.5	5.7	5.9	6.1
4.4	4.6	4.8	5.0	5.1	5.3	5.5	5.7	5.9	6.1
4.1	4.3	4.5	4.6	4.8	5.0	5.2	5.3	5.5	5.7
9.4	9.8	10.2	10.6	11.0	11.4	11.7	12.1	12.5	12.9
21.1	22.0	22.9	23.8	24.7	25.5	26.4	27.3	28.2	29.1
6.0	6.3	6.5	6.8	7.1	7.3	7.6	7.8	8.1	8.3
8.7	9.1	9.5	9.8	10.2	10.6	10.9	11.3	11.7	12.1
13.2	13.8	14.3	14.9	15.4	16.0	16.6	17.1	17.7	18.2
9.4	9.8	10.2	10.6	11.0	11.4	11.7	12.1	12.5	12.9
6.5	6.7	7.0	7.3	7.6	7.8	8.1	8.4	8.6	8.9
7.7	8.1	8.4	8.7	9.0	9.4	9.7	10.0	10.4	10.7
8.7	9.0	9.4	9.8	10.1	10.5	10.9	11.2	11.6	12.0
5.3	5.6	5.8	6.0	6.2	6.5	6.7	6.9	7.1	7.4
6.2	6.5	6.8	7.0	7.3	7.6	7.8	8.1	8.4	8.6
9.2	9.5	9.9	10.3	10.7	11.1	11.5	11.9	12.3	12.6
5.1	5.3	5.5	5.8	6.0	6.2	6.4	6.6	6.8	7.1
5.9	6.1	6.4	6.6	6.9	7.1	7.4	7.6	7.9	8.1
8.9	9.3	9.7	10.1	10.5	10.8	11.2	11.6	12.0	12.3
5.5	5.7	5.9	6.2	6.4	6.6	6.9	7.1	7.3	7.5
8.0	8.3	8.6	9.0	9.3	9.6	10.0	10.3	10.6	11.0
3.8	4.0	4.2	4.3	4.5	4.6	4.8	5.0	5.1	5.3
6.0	6.3	6.5	6.8	7.1	7.3	7.6	7.8	8.1	8.3
4.7	4.9	5.1	5.3	5.5	5.7	5.9	6.1	6.3	6.5
9.1	9.5	9.9	10.2	10.6	11.0	11.4	11.8	12.2	12.5

Activity	kcal · min⁻¹ · kg⁻¹	kg lb	50 110	53 117	56 123	59 130	62 137	65 143	68 150
Horse-racing									
galloping	0.137		6.9	7.3	7.7	8.1	8.5	8.9	9.3
trotting	0.110		5.5	5.8	6.2	6.5	6.8	7.2	7.5
walking	0.041		2.1	2.2	2.3	2.4	2.5	2.7	2.8
Ironing (F)	0.033		1.7	1.7	1.8	1.9	2.0	2.1	2.2
Ironing (M)	0.064		3.2	3.4	3.6	3.8	4.0	4.2	4.4
Judo	0.195		9.8	10.3	10.9	11.5	12.1	12.7	13.3
Jumping rope									
70 per min	0.162		8.1	8.6	9.1	9.6	10.0	10.5	11.0
80 per min	0.164		8.2	8.7	9.2	9.7	10.2	10.7	11.2
125 per min	0.177		8.9	9.4	9.9	10.4	11.0	11.5	12.0
145 per min	0.197		9.9	10.4	11.0	11.6	12.2	12.8	13.4
Knitting, sewing (F)	0.022		1.1	1.2	1.2	1.3	1.4	1.4	1.5
Knitting, sewing (M)	0.023		1.2	1.2	1.3	1.4	1.4	1.5	1.6
Locksmith	0.057		2.9	3.0	3.2	3.4	3.5	3.7	3.9
Lying at ease	0.022		1.1	1.2	1.2	1.3	1.4	1.4	1.5
Machine-tooling									
machining	0.048		2.4	2.5	2.7	2.8	3.0	3.1	3.3
operating lathe	0.052		2.6	2.8	2.9	3.1	3.2	3.4	3.5
operating punch press	0.088		4.4	4.7	4.9	5.2	5.5	5.7	6.0
tapping and drilling	0.065		3.3	3.4	3.6	3.8	4.0	4.2	4.4
welding	0.052		2.6	2.8	2.9	3.1	3.2	3.4	3.5
working sheet metal	0.048		2.4	2.5	2.7	2.8	3.0	3.1	3.3
Marching, rapid	0.142		7.1	7.5	8.0	8.4	8.8	9.2	9.7
Mopping floor (F)	0.062		3.1	3.3	3.5	3.7	3.8	4.0	4.2
Mopping floor (M)	0.058		2.9	3.1	3.2	3.4	3.6	3.8	3.9
Music playing									
accordion (sitting)	0.032		1.6	1.7	1.8	1.9	2.0	2.1	2.2
cello (sitting)	0.041		2.1	2.2	2.3	2.4	2.5	2.7	2.8
conducting	0.039		2.0	2.1	2.2	2.3	2.4	2.5	2.7
drums (sitting)	0.066		3.3	3.5	3.7	3.9	4.1	4.3	4.5
flute (sitting)	0.035		1.8	1.9	2.0	2.1	2.2	2.3	2.4
horn (sitting)	0.029		1.5	1.5	1.6	1.7	1.8	1.9	2.0
organ (sitting)	0.053		2.7	2.8	3.0	3.1	3.3	3.4	3.6
piano (sitting)	0.040		2.0	2.1	2.2	2.4	2.5	2.6	2.7
trumpet (standing)	0.031		1.6	1.6	1.7	1.8	1.9	2.0	2.1
violin (sitting)	0.045		2.3	2.4	2.5	2.7	2.8	2.9	3.1
woodwind (sitting)	0.032		1.6	1.7	1.8	1.9	2.0	2.1	2.2
Painting, inside	0.034		1.7	1.8	1.9	2.0	2.1	2.2	2.3
Painting, outside	0.077		3.9	4.1	4.3	4.5	4.8	5.0	5.2
Planting seedlings	0.070		3.5	3.7	3.9	4.1	4.3	4.6	4.8
Plastering	0.078		3.9	4.1	4.4	4.6	4.8	5.1	5.3
Printing	0.035		1.8	1.9	2.0	2.1	2.2	2.3	2.4
Racquetball	0.178		8.9	9.4	10.0	10.5	11.0	11.6	12.1
Running, cross-country	0.163		8.2	8.6	9.1	9.6	10.1	10.6	11.1
Running, horizontal									
11 min, 30 s per mile	0.135		6.8	7.2	7.6	8.0	8.4	8.8	9.2
9 min per mile	0.193		9.7	10.2	10.8	11.4	12.0	12.5	13.1
8 min per mile	0.208		10.8	11.3	11.9	12.5	13.1	13.6	14.2
7 min per mile	0.228		12.2	12.7	13.3	13.9	14.5	15.0	15.6
6 min per mile	0.252		13.9	14.4	15.0	15.6	16.2	16.7	17.3
5 min, 30 s per mile	0.289		14.5	15.3	16.2	17.1	17.9	18.8	19.7
Scraping paint	0.063		3.2	3.3	3.5	3.7	3.9	4.1	4.3

71 157	74 163	77 170	80 176	83 183	86 190	89 196	92 203	95 209	98 216
9.7	10.1	10.6	11.0	11.4	11.8	12.2	12.6	13.0	13.4
7.8	8.1	8.5	8.8	9.1	9.5	9.8	10.1	10.5	10.8
2.9	3.0	3.2	3.3	3.4	3.5	3.6	3.8	3.9	4.0
2.3	2.4	2.5	2.6	2.7	2.8	2.9	3.0	3.1	3.2
4.5	4.7	4.9	5.1	5.3	5.5	5.7	5.9	6.1	6.3
13.8	14.4	15.0	15.6	16.2	16.8	17.4	17.9	18.5	19.1
11.5	12.0	12.5	13.0	13.4	13.9	14.4	14.9	15.4	15.9
11.6	12.1	12.6	13.1	13.6	14.1	14.6	14.6	15.6	16.1
12.6	13.1	13.6	14.2	14.7	15.2	15.8	16.3	16.8	17.3
14.0	14.6	15.2	15.8	16.4	16.9	17.5	18.1	18.7	19.3
1.6	1.6	1.7	1.8	1.8	1.9	2.0	2.0	2.1	2.2
1.6	1.7	1.8	1.8	1.9	2.0	2.0	2.1	2.2	2.3
4.0	4.2	4.4	4.6	4.7	4.9	5.1	5.2	5.4	5.6
1.6	1.6	1.7	1.8	1.8	1.9	2.0	2.0	2.1	2.2
3.4	3.6	3.7	3.8	4.0	4.1	4.3	4.4	4.6	4.7
3.7	3.8	4.0	4.2	4.3	4.5	4.6	4.8	4.9	5.1
6.2	6.5	6.8	7.0	7.3	7.6	7.8	8.1	8.4	8.6
4.6	4.8	5.0	5.2	5.4	5.6	5.8	6.0	6.2	6.4
3.7	3.8	4.0	4.2	4.3	4.5	4.6	4.8	4.9	5.1
3.4	3.6	3.7	3.8	4.0	4.1	4.3	4.4	4.6	4.7
10.1	10.5	10.9	11.4	11.8	12.2	12.6	13.1	13.5	13.9
4.4	4.6	4.8	5.0	5.1	5.3	5.5	5.7	5.9	6.1
4.1	4.3	4.5	4.6	4.8	5.0	5.2	5.0	5.5	5.7
2.3	2.4	2.5	2.6	2.7	2.8	2.8	2.9	3.0	3.1
2.9	3.0	3.2	3.3	3.4	3.5	3.6	3.8	3.9	4.0
2.8	2.9	3.0	3.1	3.2	3.4	3.5	3.6	3.7	3.8
4.7	4.9	5.1	5.3	5.5	5.7	5.9	6.1	6.3	6.6
2.5	2.6	2.7	2.8	2.9	3.0	3.1	3.2	3.3	3.4
2.1	2.1	2.2	2.3	2.4	2.5	2.6	2.7	2.8	2.8
3.8	3.9	4.1	4.2	4.4	4.6	4.7	4.9	5.0	5.2
2.8	3.0	3.1	3.2	3.3	3.4	3.6	3.7	3.8	3.9
2.2	2.3	2.4	2.5	2.6	2.7	2.8	2.9	2.9	3.0
3.2	3.3	3.5	3.6	3.7	3.9	4.0	4.1	4.3	4.4
2.3	2.4	2.5	2.6	2.7	2.8	2.8	2.9	3.0	3.1
2.4	2.5	2.6	2.7	2.8	2.9	3.0	3.1	3.2	3.3
5.5	5.7	5.9	6.2	6.4	6.6	6.9	7.1	7.3	7.5
5.0	5.2	5.4	5.6	5.8	6.0	6.2	6.4	6.7	6.9
5.5	5.8	6.0	6.2	6.5	6.7	6.9	7.2	7.4	7.6
2.5	2.6	2.7	2.8	2.9	3.0	3.1	3.2	3.3	3.4
12.6	13.2	13.7	14.2	14.8	15.3	15.8	16.4	16.9	17.4
11.6	12.1	12.6	13.0	13.5	14.0	14.5	15.0	15.5	16.0
9.6	10.0	10.5	10.9	11.3	11.7	12.1	12.5	12.9	13.3
13.7	14.3	14.9	15.4	16.0	16.6	17.2	17.8	18.3	18.9
14.8	15.4	16.0	16.5	17.1	17.7	18.3	18.9	19.4	20.0
16.2	16.8	17.4	17.9	18.5	19.1	19.7	20.3	20.8	21.4
17.9	18.5	19.1	19.6	20.2	20.8	21.4	22.0	22.5	23.1
20.5	21.4	22.3	23.1	24.0	24.9	25.7	26.6	27.5	28.3
4.5	4.7	4.9	5.0	5.2	5.4	5.6	5.8	6.0	6.2

Activity	kcal · min⁻¹ · kg⁻¹	kg 50 lb 110	53 117	56 123	59 130	62 137	65 143	68 150
Scrubbing floors (F)	0.109	5.5	5.8	6.1	6.4	6.8	7.1	7.4
Scrubbing floors (M)	0.108	5.4	5.7	6.0	6.4	6.7	7.0	7.3
Shoe repair, general	0.045	2.3	2.4	2.5	2.7	2.8	2.9	3.1
Sitting quietly	0.021	1.1	1.1	1.2	1.2	1.3	1.4	1.4
Skiing, hard snow								
level, moderate speed	0.119	6.0	6.3	6.7	7.0	7.4	7.7	8.1
level, walking speed	0.143	7.2	7.6	8.0	8.4	8.9	9.3	9.7
uphill, maximum speed	0.274	13.7	14.5	15.3	16.2	17.0	17.8	18.6
Skiing, soft snow								
leisure (F)	0.111	4.9	5.2	5.5	5.8	6.1	6.4	6.7
leisure (M)	0.098	5.6	5.9	6.2	6.5	6.9	7.2	7.5
Skindiving, as frogman								
considerable motion	0.276	13.8	14.6	15.5	16.3	17.1	17.9	18.8
moderate motion	0.206	10.3	10.9	11.5	12.2	12.8	13.4	14.0
Snowshoeing, soft snow	0.166	8.3	8.8	9.3	9.8	10.3	10.8	11.3
Squash	0.212	10.6	11.2	11.9	12.5	13.1	13.8	14.4
Standing quietly (F)	0.025	1.3	1.3	1.4	1.5	1.6	1.6	1.7
Standing quietly (M)	0.027	1.4	1.4	1.5	1.6	1.7	1.8	1.8
Steel mill, working in								
fettling	0.089	4.5	4.7	5.0	5.3	5.5	5.8	6.1
forging	0.100	5.0	5.3	5.6	5.9	6.2	6.5	6.8
hand rolling	0.137	6.9	7.3	7.7	8.1	8.5	8.9	9.3
merchant mill rolling	0.145	7.3	7.7	8.1	8.6	9.0	9.4	9.9
removing slag	0.178	8.9	9.4	10.0	10.5	11.0	11.6	12.1
tending furnace	0.126	6.3	6.7	7.1	7.4	7.8	8.2	8.6
tipping molds	0.092	4.6	4.9	5.2	5.4	5.7	6.0	6.3
Stock clerking	0.054	2.7	2.9	3.0	3.2	3.3	3.5	3.7
Swimming								
back stroke	0.169	8.5	9.0	9.5	10.0	10.5	11.0	11.5
breast stroke	0.162	8.1	8.6	9.1	9.6	10.0	10.5	11.0
crawl, fast	0.156	7.8	8.3	8.7	9.2	9.7	10.1	10.6
crawl, slow	0.128	6.4	6.8	7.2	7.6	7.9	8.3	8.7
side stroke	0.122	6.1	6.5	6.8	7.2	7.6	7.9	8.3
treading, fast	0.170	8.5	9.0	9.5	10.0	10.5	11.1	11.6
treading, normal	0.062	3.1	3.3	3.5	3.7	3.8	4.0	4.2
Table tennis (ping pong)	0.068	3.4	3.6	3.8	4.0	4.2	4.4	4.6
Tailoring								
cutting	0.041	2.1	2.2	2.3	2.4	2.5	2.7	2.8
hand-sewing	0.032	1.6	1.7	1.8	1.9	2.0	2.1	2.2
machine-sewing	0.045	2.3	2.4	2.5	2.7	2.8	2.9	3.1
pressing	0.062	3.1	3.3	3.5	3.7	3.8	4.0	4.2
Tennis	0.109	5.5	5.8	6.1	6.4	6.8	7.1	7.4
Typing								
electric	0.027	1.4	1.4	1.5	1.6	1.7	1.8	1.8
manual	0.031	1.6	1.6	1.7	1.8	1.9	2.0	2.1
Volleyball	0.050	2.5	2.7	2.8	3.0	3.1	3.3	3.4
Walking, normal pace								
asphalt road	0.080	4.0	4.2	4.5	4.7	5.0	5.2	5.4
fields and hillsides	0.082	4.1	4.3	4.6	4.8	5.1	5.3	5.6
grass track	0.081	4.1	4.3	4.5	4.8	5.0	5.3	5.5
plowed field	0.077	3.9	4.1	4.3	4.5	4.8	5.0	5.2
Wallpapering	0.048	2.4	2.5	2.7	2.8	3.0	3.1	3.3
Watch repairing	0.025	1.3	1.3	1.4	1.5	1.6	1.6	1.7
Window cleaning (F)	0.059	3.0	3.1	3.3	3.5	3.7	3.8	4.0
Window cleaning (M)	0.058	2.9	3.1	3.2	3.4	3.6	3.8	3.9
Writing (sitting)	0.029	1.5	1.5	1.6	1.7	1.8	1.9	2.0

71 157	74 163	77 170	80 176	83 183	86 190	89 196	92 203	95 209	98 216
7.7	8.1	8.4	8.7	9.0	9.4	9.7	10.0	10.4	10.7
7.7	8.0	8.3	8.6	9.0	9.3	9.6	9.9	10.3	10.6
3.2	3.3	3.5	3.6	3.7	3.9	4.0	4.1	4.3	4.4
1.5	1.6	1.6	1.7	1.7	1.8	1.9	1.9	2.0	2.1
8.4	8.8	9.2	9.5	9.9	10.2	10.6	10.9	11.3	11.7
10.2	10.6	11.0	11.4	11.9	12.3	12.7	13.2	13.6	14.0
19.5	20.3	21.1	21.9	22.7	23.6	24.4	25.2	26.0	26.9
7.0	7.3	7.5	7.8	8.1	8.4	8.7	9.0	9.3	9.6
7.9	8.2	8.5	8.9	9.2	9.5	9.9	10.2	10.5	10.9
19.6	20.4	21.3	22.1	22.9	23.7	24.6	25.4	26.2	27.0
14.6	15.2	15.9	16.5	17.1	17.7	18.3	19.0	19.6	20.2
11.8	12.3	12.8	13.3	13.8	14.3	14.8	15.3	15.8	16.3
15.1	15.7	16.3	17.0	17.6	18.2	18.9	19.5	20.1	20.8
1.8	1.9	1.9	2.0	2.1	2.2	2.2	2.3	2.4	2.5
1.9	2.0	2.1	2.2	2.2	2.3	2.4	2.5	2.6	2.6
6.3	6.6	6.9	7.1	7.4	7.7	7.9	8.2	8.5	8.7
7.1	7.4	7.7	8.0	8.3	8.6	8.9	9.2	9.5	9.8
9.7	10.1	10.6	11.0	11.4	11.8	12.2	12.6	13.0	13.4
10.3	10.7	11.2	11.6	12.0	12.5	12.9	13.3	13.8	14.2
12.6	13.2	13.7	14.2	14.8	15.3	15.8	16.4	16.9	17.4
8.9	9.3	9.7	10.1	10.5	10.8	11.2	11.6	12.0	12.3
6.5	6.8	7.1	7.4	7.6	7.9	8.2	8.5	8.7	9.0
3.8	4.0	4.2	4.3	4.5	4.6	4.8	5.0	5.1	5.3
12.0	12.5	13.0	13.5	14.0	14.5	15.0	15.5	16.1	16.6
11.5	12.0	12.5	13.0	13.4	13.9	14.4	14.9	15.4	15.9
11.1	11.5	12.0	12.5	12.9	13.4	13.9	14.4	14.8	15.3
9.1	9.5	9.9	10.2	10.6	11.0	11.4	11.8	12.2	12.5
8.7	9.0	9.4	9.8	10.1	10.5	10.9	11.2	11.6	12.0
12.1	12.6	13.1	13.6	14.1	14.6	15.1	15.6	16.2	16.7
4.4	4.6	4.8	5.0	5.1	5.3	5.5	5.7	5.9	6.1
4.8	5.0	5.2	5.4	5.6	5.8	6.1	6.3	6.5	6.7
2.9	3.0	3.2	3.3	3.4	3.5	3.6	3.8	3.9	4.0
2.3	2.4	2.5	2.6	2.7	2.8	2.8	2.9	3.0	3.1
3.2	3.3	3.5	3.6	3.7	3.9	4.0	4.1	4.3	4.4
4.4	4.6	4.8	5.0	5.1	5.3	5.5	5.7	5.9	6.1
7.7	8.1	8.4	8.7	9.0	9.4	9.7	10.0	10.4	10.7
1.9	2.0	2.1	2.2	2.2	2.3	2.4	2.5	2.6	2.6
2.2	2.3	2.4	2.5	2.6	2.7	2.8	2.9	2.9	3.0
3.6	3.7	3.9	4.0	4.2	4.3	4.5	4.6	4.8	4.9
5.7	5.9	6.2	6.4	6.6	6.9	7.1	7.4	7.6	7.8
5.8	6.1	6.3	6.6	6.8	7.1	7.3	7.5	7.8	8.0
5.8	6.0	6.2	6.5	6.7	7.0	7.2	7.5	7.7	7.9
5.5	5.7	5.9	6.2	6.4	6.6	6.9	7.1	7.3	7.5
3.4	3.6	3.7	3.8	4.0	4.1	4.3	4.4	4.6	4.7
1.8	1.9	1.9	2.0	2.1	2.2	2.2	2.3	2.4	2.5
4.2	4.4	4.5	4.7	4.9	5.1	5.3	5.4	5.6	5.8
4.1	4.3	4.5	4.6	4.8	5.0	5.2	5.3	5.5	5.7
2.1	2.1	2.2	2.3	2.4	2.5	2.6	2.7	2.8	2.8

CORRECTION FOR WATER DENSITY AT DIFFERENT TEMPERATURES

Temperature, C°	Density	Temperature, C°	Density
4	1.00000	31	0.99537
10	0.99973	32	0.99505
15	0.99913	33	0.99473
20	0.99823	34	0.99440
25	0.99707	35	0.99406
26	0.99681	36	0.99371
27	0.99654	37	0.99336
28	0.99626	38	0.99299
29	0.99595	39	0.99262
30	0.99567	40	0.99224

F

BODY COMPOSITION*

This appendix contains the age and gender specific equations to predict body fat percentage based on three girth measurements. There are four charts, one each for young and older men and women. In our experience, it is important to calibrate the tape measure prior to its use. We use a meter stick as the standard and we check the markings on the cloth tape at 10 cm increments. A cloth tape is preferred over a metal one because there is little skin compression when applying a cloth tape to the skin's surface at a relatively constant tension.

To use the charts, measure the three girths at the sites indicated in Chart F-1. The anatomic landmarks and procedures are explained on page 620 and illustrated in Figure I-1. The specific equation to predict percent body fat with its corresponding constant is presented at the bottom of each of the four charts (Charts F-2 to F-5).

CHART F-1.
BODY SITES MEASURED BY THE CIRCUMFERENCE METHOD

Age (Years)	Gender	Site Measured		
		A	B	C
18–26	M	Right upper arm	Abdomen	Right forearm
	F	Abdomen	Right thigh	Right forearm
27–50	M	Buttocks	Abdomen	Right forearm
	F	Abdomen	Right thigh	Right calf

CHART F-2.
CONVERSION CONSTANTS TO PREDICT PERCENT BODY FAT FOR YOUNG MEN

Upper Arm			Abdomen			Forearm		
in	cm	Constant A	in	cm	Constant B	in	cm	Constant C
7.00	17.78	25.91	21.00	53.34	27.56	7.00	17.78	38.01
7.25	18.41	26.83	21.25	53.97	27.88	7.25	18.41	39.37
7.50	19.05	27.76	21.50	54.61	28.21	7.50	19.05	40.72
7.75	19.68	28.68	21.75	55.24	28.54	7.75	19.68	42.08
8.00	20.32	29.61	22.00	55.88	28.87	8.00	20.32	43.44
8.25	20.95	30.53	22.25	56.51	29.20	8.25	20.95	44.80
8.50	21.59	31.46	22.50	57.15	29.52	8.50	21.59	46.15

CHART F-2. *continued*

Upper Arm			Abdomen			Forearm		
in	cm	Constant A	in	cm	Constant B	in	cm	Constant C
8.75	22.22	32.38	22.75	57.78	29.85	8.75	22.22	47.51
9.00	22.86	33.31	23.00	58.42	30.18	9.00	22.86	48.87
9.25	23.49	34.24	23.25	59.05	30.51	9.25	23.49	50.23
9.50	24.13	35.16	23.50	59.69	30.84	9.50	24.13	51.58
9.75	24.76	36.09	23.75	60.32	31.16	9.75	24.76	52.94
10.00	25.40	37.01	24.00	60.96	31.49	10.00	25.40	54.30
10.25	26.03	37.94	24.25	61.59	31.82	10.25	26.03	55.65
10.50	26.67	38.86	24.50	62.23	32.15	10.50	26.67	57.01
10.75	27.30	39.79	24.75	62.86	32.48	10.75	27.30	58.37
11.00	27.94	40.71	25.00	63.50	32.80	11.00	27.94	59.73
11.25	28.57	41.64	25.25	64.13	33.13	11.25	28.57	61.08
11.50	29.21	42.56	25.50	64.77	33.46	11.50	29.21	62.44
11.75	29.84	43.49	25.75	65.40	33.79	11.75	29.84	63.80
12.00	30.48	44.41	26.00	66.04	34.12	12.00	30.48	65.16
12.25	31.11	45.34	26.25	66.67	34.44	12.25	31.11	66.51
12.50	31.75	46.26	26.50	67.31	34.77	12.50	31.75	67.87
12.75	32.38	47.19	26.75	67.94	35.10	12.75	32.38	69.23
13.00	33.02	48.11	27.00	68.58	35.43	13.00	33.02	70.59
13.25	33.65	49.04	27.25	69.21	35.76	13.25	33.65	71.94
13.50	34.29	49.96	27.50	69.85	36.09	13.50	34.29	73.30
13.75	34.92	50.89	27.75	70.48	36.41	13.75	34.92	74.66
14.00	35.56	51.82	28.00	71.12	36.74	14.00	35.56	76.02
14.25	36.19	52.74	28.25	71.75	37.07	14.25	36.19	77.37
14.50	36.83	53.67	28.50	72.39	37.40	14.50	36.83	78.73
14.75	37.46	54.59	28.75	73.02	37.73	14.75	37.46	80.09
15.00	38.10	55.52	29.00	73.66	38.05	15.00	38.10	81.45
15.25	38.73	56.44	29.25	74.29	38.38	15.25	38.73	82.80
15.50	39.37	57.37	29.50	74.93	38.71	15.50	39.37	84.16
15.75	40.00	58.29	29.75	75.56	39.04	15.75	40.00	85.52
16.00	40.64	59.22	30.00	76.20	39.37	16.00	40.64	86.88
16.25	41.27	60.14	30.25	76.83	39.69	16.25	41.27	88.23
16.50	41.91	61.07	30.50	77.47	40.02	16.50	41.91	89.59
16.75	42.54	61.99	30.75	78.10	40.35	16.75	42.54	90.95
17.00	43.18	62.92	31.00	78.74	40.68	17.00	43.18	92.31
17.25	43.81	63.84	31.25	79.37	41.01	17.25	43.81	93.66
17.50	44.45	64.77	31.50	80.01	41.33	17.50	44.45	95.02
17.75	45.08	65.69	31.75	80.64	41.66	17.75	45.08	96.38
18.00	45.72	66.62	32.00	81.28	41.99	18.00	45.72	97.74
18.25	46.35	67.54	32.25	81.91	42.32	18.25	46.35	99.09
18.50	46.99	68.47	32.50	82.55	42.65	18.50	46.99	100.45
18.75	47.62	69.40	32.75	83.18	42.97	18.75	47.62	101.81
19.00	48.26	70.32	33.00	83.82	43.30	19.00	48.26	103.17
19.25	48.89	71.25	33.25	84.45	43.63	19.25	48.89	104.52
19.50	49.53	72.17	33.50	85.09	43.96	19.50	49.53	105.88
19.75	50.16	73.10	33.75	85.72	44.29	19.75	50.16	107.24
20.00	50.80	74.02	34.00	86.36	44.61	20.00	50.80	108.60
20.25	51.43	74.95	34.25	86.99	44.94	20.25	51.43	109.95
20.50	52.07	75.87	34.50	87.63	45.27	20.50	52.07	111.31
20.75	52.70	76.80	34.75	88.26	45.60	20.75	52.70	112.67
21.00	53.34	77.72	35.00	88.90	45.93	21.00	53.34	114.02
21.25	53.97	78.65	35.25	89.53	46.25	21.25	53.97	115.38
21.50	54.61	79.57	35.50	90.17	46.58	21.50	54.61	116.74
21.75	55.24	80.50	35.75	90.80	46.91	21.75	55.24	118.10
22.00	55.88	81.42	36.00	91.44	47.24	22.00	55.88	119.45
			36.25	92.07	47.57			
			36.50	92.71	47.89			
			36.75	93.34	48.22			
			37.00	93.98	48.55			
			37.25	94.61	48.88			
			37.50	95.25	49.21			

CHART F-2. *continued*

Upper Arm			Abdomen			Forearm		
in	cm	Constant A	in	cm	Constant B	in	cm	Constant C
			37.75	95.88	49.54			
			38.00	96.52	49.86			
			38.25	97.15	50.19			
			38.50	97.79	50.52			
			38.75	98.42	50.85			
			39.00	99.06	51.18			
			39.25	99.69	51.50			
			39.50	100.33	51.83			
			39.75	100.96	52.16			
			40.00	101.60	52.49			
			40.25	102.23	52.82			
			40.50	102.87	53.14			
			40.75	103.50	53.47			
			41.00	104.14	53.80			
			41.25	104.77	54.13			
			41.50	105.41	54.46			
			41.75	106.04	54.78			
			42.00	106.68	55.11			

Note: Percent Fat = Constant A + Constant B − Constant C − 10.2

CHART F-3.
CONVERSION CONSTANTS TO PREDICT PERCENT BODY FAT FOR OLDER MEN

Buttocks			Abdomen			Forearm		
in	cm	Constant A	in	cm	Constant B	in	cm	Constant C
28.00	71.12	29.34	25.50	64.77	22.84	7.00	17.78	21.01
28.25	71.75	29.60	25.75	65.40	23.06	7.25	18.41	21.76
28.50	72.39	29.87	26.00	66.04	23.29	7.50	19.05	22.52
28.75	73.02	30.13	26.25	66.67	23.51	7.75	19.68	23.26
29.00	73.66	30.39	26.50	67.31	23.73	8.00	20.32	24.02
29.25	74.29	30.65	26.75	67.94	23.96	8.25	20.95	24.76
29.50	74.93	30.92	27.00	68.58	24.18	8.50	21.59	25.52
29.75	75.56	31.18	27.25	69.21	24.40	8.75	22.22	26.26
30.00	76.20	31.44	27.50	69.85	24.63	9.00	22.86	27.02
30.25	76.83	31.70	27.75	70.48	24.85	9.25	23.49	27.76
30.50	77.47	31.96	28.00	71.12	25.08	9.50	24.13	28.52
30.75	78.10	32.22	28.25	71.75	25.29	9.75	24.76	29.26
31.00	78.74	32.49	28.50	72.39	25.52	10.00	25.40	30.02
31.25	79.37	32.75	28.75	73.02	25.75	10.25	26.03	30.76
31.50	80.01	33.01	29.00	73.66	25.97	10.50	26.67	31.52
31.75	80.64	33.27	29.25	74.29	26.19	10.75	27.30	32.27
32.00	81.28	33.54	29.50	74.93	26.42	11.00	27.94	33.02
32.25	81.91	33.80	29.75	75.56	26.64	11.25	28.57	33.77
32.50	82.55	34.06	30.00	76.20	26.87	11.50	29.21	34.52
32.75	83.18	34.32	30.25	76.83	27.09	11.75	29.84	35.27
33.00	83.82	34.58	30.50	77.47	27.32	12.00	30.48	36.02
33.25	84.45	34.84	30.75	78.10	27.54	12.25	31.11	36.77
33.50	85.09	35.11	31.00	78.74	27.76	12.50	31.75	37.53
33.75	85.72	35.37	31.25	79.37	27.98	12.75	32.38	38.27
34.00	86.36	35.63	31.50	80.01	28.21	13.00	33.02	39.03
34.25	86.99	35.89	31.75	80.64	28.43	13.25	33.65	39.77
34.50	87.63	36.16	32.00	81.28	28.66	13.50	34.29	40.53
34.75	88.26	36.42	32.25	81.91	28.88	13.75	34.92	41.27
35.00	88.90	36.68	32.50	82.55	29.11	14.00	35.56	42.03
35.25	89.53	36.94	32.75	83.18	29.33	14.25	36.19	42.77
35.50	90.17	37.20	33.00	83.82	29.55	14.50	36.83	43.53

CHART F-3. *continued*

Buttocks			Abdomen			Forearm		
in	cm	Constant A	in	cm	Constant B	in	cm	Constant C
35.75	90.80	37.46	33.25	84.45	29.78	14.75	37.46	44.27
36.00	91.44	37.73	33.50	85.09	30.00	15.00	38.10	45.03
36.25	92.07	37.99	33.75	85.72	30.22	15.25	38.73	45.77
36.50	92.71	38.25	34.00	86.36	30.45	15.50	39.37	46.53
36.75	93.34	38.51	34.25	86.99	30.67	15.75	40.00	47.28
37.00	93.98	38.78	34.50	87.63	30.89	16.00	40.64	48.03
37.25	94.61	39.04	34.75	88.26	31.12	16.25	41.27	48.78
37.50	95.25	39.30	35.00	88.90	31.35	16.50	41.91	49.53
37.75	95.88	39.56	35.25	89.53	31.57	16.75	42.54	50.28
38.00	96.52	39.82	35.50	90.17	31.79	17.00	43.18	51.03
38.25	97.15	40.08	35.75	90.80	32.02	17.25	43.81	51.78
38.50	97.79	40.35	36.00	91.44	32.24	17.50	44.45	52.54
38.75	98.42	40.61	36.25	92.07	32.46	17.75	45.08	53.28
39.00	99.06	40.87	36.50	92.71	32.69	18.00	45.72	54.04
39.25	99.69	41.13	36.75	93.34	32.91	18.25	46.35	54.78
39.50	100.33	41.39	37.00	93.98	33.14			
39.75	100.96	41.66	37.25	94.61	33.36			
40.00	101.60	41.92	37.50	95.25	33.58			
40.25	102.23	42.18	37.75	95.88	33.81			
40.50	102.87	42.44	38.00	96.52	34.03			
40.75	103.50	42.70	38.25	97.15	34.26			
41.00	104.14	42.97	38.50	97.79	34.48			
42.25	104.77	43.23	38.75	98.42	34.70			
41.50	105.41	43.49	39.00	99.06	34.93			
41.75	106.04	43.75	39.25	99.69	35.15			
42.00	106.68	44.02	39.50	100.33	35.38			
42.25	107.31	44.28	39.75	100.96	35.59			
42.50	107.95	44.54	40.00	101.60	35.82			
42.75	108.58	44.80	40.25	102.23	36.05			
43.00	109.22	45.06	40.50	102.87	36.27			
43.25	109.85	45.32	40.75	103.50	36.49			
43.50	110.49	45.59	41.00	104.14	36.72			
43.75	111.12	45.85	41.25	104.77	36.94			
44.00	111.76	46.12	41.50	105.41	37.17			
44.25	112.39	46.37	41.75	106.04	37.39			
44.50	113.03	46.64	42.00	106.68	37.62			
44.75	113.66	46.89	42.25	107.31	37.87			
45.00	114.30	47.16	42.50	107.95	38.06			
42.25	114.93	47.42	42.75	108.58	38.28			
45.50	115.57	47.68	43.00	109.22	38.51			
45.75	116.20	47.94	43.25	109.85	38.73			
46.00	116.84	48.21	43.50	110.49	38.96			
46.25	117.47	48.47	43.75	111.12	39.18			
46.50	118.11	48.73	44.00	111.76	39.41			
46.75	118.74	48.99	44.25	112.39	39.63			
47.00	119.38	49.26	44.50	113.03	39.85			
47.25	120.01	49.52	44.75	113.66	40.08			
47.50	120.65	49.78	45.00	114.30	40.30			
47.75	121.28	50.04						
48.00	121.92	50.30						
48.25	122.55	50.56						
48.50	123.19	50.83						
48.75	123.82	51.09						
49.00	124.46	51.35						

Note: **Percent Fat = Constant A + Constant B − Constant C − 15.0**

CHART F-4.
CONVERSION CONSTANTS TO PREDICT PERCENT BODY FAT FOR YOUNG WOMEN

	Abdomen			Thigh			Forearm	
in	cm	Constant A	in	cm	Constant B	in	cm	Constant C
20.00	50.80	26.74	14.00	35.56	29.13	6.00	15.24	25.86
20.25	51.43	27.07	14.25	36.19	29.65	6.25	15.87	26.94
20.50	52.07	27.41	14.50	36.83	30.17	6.50	16.51	28.02
20.75	52.70	27.74	14.75	37.46	30.69	6.75	17.14	29.10
21.00	53.34	28.07	15.00	38.10	31.21	7.00	17.78	30.17
21.25	53.97	28.41	15.25	38.73	31.73	7.25	18.41	31.25
21.50	54.61	28.74	15.50	39.37	32.25	7.50	19.05	32.33
21.75	55.24	29.08	15.75	40.00	32.77	7.75	19.68	33.41
22.00	55.88	29.41	16.00	40.64	33.29	8.00	20.32	34.48
22.25	56.51	29.74	16.25	41.27	33.81	8.25	20.95	35.56
22.50	57.15	30.08	16.50	41.91	34.33	8.50	21.59	36.64
22.75	57.78	30.41	16.75	42.54	34.85	8.75	22.22	37.72
23.00	58.42	30.75	17.00	43.18	35.37	9.00	22.86	38.79
23.25	59.05	31.08	17.25	43.81	35.89	9.25	23.49	39.87
23.50	59.69	31.42	17.50	44.45	36.41	9.50	24.13	40.95
23.75	60.32	31.75	17.75	45.08	36.93	9.75	24.76	42.03
24.00	60.96	32.08	18.00	45.72	37.45	10.00	25.40	43.10
24.25	61.59	32.42	18.25	46.35	37.97	10.25	26.03	44.18
24.50	62.23	32.75	18.50	46.99	38.49	10.50	26.67	45.26
24.75	62.86	33.09	18.75	47.62	39.01	10.75	27.30	46.34
25.00	63.50	33.42	19.00	48.26	39.53	11.00	27.94	47.41
25.25	64.13	33.76	19.25	48.89	40.05	11.25	28.57	48.49
25.50	64.77	34.09	19.50	49.53	40.57	11.50	29.21	49.57
25.75	65.40	34.42	19.75	50.16	41.09	11.75	29.84	50.65
26.00	66.04	34.76	20.00	50.80	41.61	12.00	30.48	51.73
26.25	66.67	35.09	20.25	51.43	42.13	12.25	31.11	52.80
26.50	67.31	35.43	20.50	52.07	42.65	12.50	31.75	53.88
26.75	67.94	35.76	20.75	52.70	43.17	12.75	32.38	54.96
27.00	68.58	36.10	21.00	53.34	43.69	13.00	33.02	56.04
27.25	69.21	36.43	21.25	53.97	44.21	13.25	33.65	57.11
27.50	69.85	36.76	21.50	54.61	44.73	13.50	34.29	58.19
27.75	70.48	37.10	21.75	55.24	45.25	13.75	34.92	59.27
28.00	71.12	37.43	22.00	55.88	45.77	14.00	35.56	60.35
28.25	71.75	37.77	22.25	56.51	46.29	14.25	36.19	61.42
28.50	72.39	38.10	22.50	57.15	46.81	14.50	36.83	62.50
28.75	73.02	38.43	22.75	57.78	47.33	14.75	37.46	63.58
29.00	73.66	38.77	23.00	58.42	47.85	15.00	38.10	64.66
29.25	74.29	39.10	23.25	59.05	48.37	15.25	38.73	65.73
29.50	74.93	39.44	23.50	59.69	48.89	15.50	39.37	66.81
29.75	75.56	39.77	23.75	60.32	49.41	15.75	40.00	67.89
30.00	76.20	40.11	24.00	60.96	49.93	16.00	40.64	68.97
30.25	76.83	40.44	24.25	61.59	50.45	16.25	41.27	70.04
30.50	77.47	40.77	24.50	62.23	50.97	16.50	41.91	71.12
30.75	78.10	41.11	24.75	62.86	51.49	16.75	42.54	72.20
31.00	78.74	41.44	25.00	63.50	52.01	17.00	43.18	73.28
31.25	79.37	41.78	25.25	64.13	52.53	17.25	43.81	74.36
31.50	80.01	42.11	25.50	64.77	53.05	17.50	44.45	75.43
31.75	80.64	42.45	25.75	65.40	53.57	17.75	45.08	76.51
32.00	81.28	42.78	26.00	66.04	54.09	18.00	45.72	77.59
32.25	81.91	43.11	26.25	66.67	54.61	18.25	46.35	78.67
32.50	82.55	43.45	26.50	67.31	55.13	18.50	46.99	79.74
32.75	83.18	43.78	26.75	67.94	55.65	18.75	47.62	80.82
33.00	83.82	44.12	27.00	68.58	56.17	19.00	48.26	81.90
33.25	84.45	44.45	27.25	69.21	56.69	19.25	48.89	82.98
33.50	85.09	44.78	27.50	69.85	57.21	19.50	49.53	84.05
33.75	85.72	45.12	27.75	70.48	57.73	19.75	50.16	85.13
34.00	86.36	45.45	28.00	71.12	58.26	20.00	50.80	86.21
34.25	86.99	45.79	28.25	71.75	58.78			
34.50	87.63	46.12	28.50	72.39	59.30			

CHART F-4. *continued*

Abdomen			Thigh			Forearm		
in	cm	Constant A	in	cm	Constant B	in	cm	Constant C
34.75	88.26	46.46	38.75	73.02	59.82			
35.00	88.90	46.79	29.00	73.66	60.34			
35.25	89.53	47.12	29.25	74.29	60.86			
35.50	90.17	47.46	29.50	74.93	61.38			
35.75	90.80	47.79	29.75	75.56	61.90			
36.00	91.44	48.13	30.00	76.20	62.42			
36.25	92.07	48.46	30.25	76.83	62.94			
36.50	92.71	48.80	30.50	77.47	63.46			
36.75	93.34	49.13	30.75	78.10	63.98			
37.00	93.98	49.46	31.00	78.74	64.50			
37.25	94.61	49.80	31.25	79.37	65.02			
37.50	95.25	50.13	31.50	80.01	65.54			
37.75	95.88	50.47	31.75	80.64	66.06			
38.00	96.52	50.80	32.00	81.28	66.58			
38.25	97.15	51.13	32.25	81.91	67.10			
38.50	97.79	51.47	32.50	82.55	67.62			
38.75	98.42	51.80	32.75	83.18	68.14			
39.00	99.06	52.14	33.00	83.82	68.66			
39.25	99.69	52.47	33.25	84.45	69.18			
39.50	100.33	52.81	33.50	85.09	69.70			
39.75	100.96	53.14	33.75	85.72	70.22			
40.00	101.60	53.47	34.00	86.36	70.74			

Note: Percent Fat = Constant A + Constant B − Constant C − 19.6

CHART F-5.
CONVERSION CONSTANTS TO PREDICT PERCENT BODY FAT FOR OLDER WOMEN

Abdomen			Thigh			Calf		
in	cm	Constant A	in	cm	Constant B	in	cm	Constant C
25.00	63.50	29.69	14.00	35.56	17.31	10.00	25.40	14.46
25.25	64.13	29.98	14.25	36.19	17.62	10.25	26.03	14.82
25.50	64.77	30.28	14.50	36.83	17.93	10.50	26.67	15.18
25.75	65.40	30.58	14.75	37.46	18.24	10.75	27.30	15.54
26.00	66.04	30.87	15.00	38.10	18.55	11.00	27.94	15.91
26.25	66.67	31.17	15.25	38.73	18.86	11.25	28.57	16.27
26.50	67.31	31.47	15.50	39.37	19.17	11.50	29.21	16.63
26.75	67.94	31.76	15.75	40.00	19.47	11.75	29.84	16.99
27.00	68.58	32.06	16.00	40.64	19.78	12.00	30.48	17.35
27.25	69.21	32.36	16.25	41.27	20.09	12.25	31.11	17.71
27.50	69.85	32.65	16.50	41.91	20.40	12.50	31.75	18.08
27.75	70.48	32.95	16.75	42.54	20.71	12.75	32.38	18.44
28.00	71.12	33.25	17.00	43.18	21.02	13.00	33.02	18.80
28.25	71.75	33.55	17.25	43.81	21.33	13.25	33.65	19.16
28.50	72.39	33.84	17.50	44.45	21.64	13.50	34.29	19.52
28.75	73.02	34.14	17.75	45.08	21.95	13.75	34.92	19.88
29.00	73.66	34.44	18.00	45.72	22.26	14.00	35.56	20.24
29.25	74.29	34.73	18.25	46.35	22.57	14.25	36.19	20.61
29.50	74.93	35.03	18.50	46.99	22.87	14.50	36.83	20.97
29.75	75.56	35.33	18.75	47.62	23.18	14.75	37.46	21.33
30.00	76.20	35.62	19.00	48.26	23.49	15.00	38.10	21.69
30.25	76.83	35.92	19.25	48.89	23.80	15.25	38.73	22.05
30.50	77.47	36.22	19.50	49.53	24.11	15.50	39.37	22.41
30.75	78.10	36.51	19.75	50.16	24.42	15.75	40.00	22.77
31.00	78.74	36.81	20.00	50.80	24.73	16.00	40.64	23.14
31.25	79.37	37.11	20.25	51.43	25.04	16.25	41.27	23.50
31.50	80.01	37.40	20.50	52.07	25.35	16.50	41.91	23.86

CHART F-5. *continued*

Abdomen			Thigh			Calf		
in	**cm**	**Constant A**	**in**	**cm**	**Constant B**	**in**	**cm**	**Constant C**
31.75	80.64	37.70	20.75	52.70	25.66	16.75	42.54	24.22
32.00	81.28	38.00	21.00	53.34	25.97	17.00	43.18	24.58
32.25	81.91	38.30	21.25	53.97	26.28	17.25	43.81	24.94
32.50	82.55	38.59	21.50	54.61	26.58	17.50	44.45	25.31
32.75	83.18	38.89	21.75	55.24	26.89	17.75	45.08	25.67
33.00	83.82	39.19	22.00	55.88	27.20	18.00	45.72	26.03
33.25	84.45	39.48	22.25	56.51	27.51	18.25	46.35	26.39
33.50	85.09	39.78	22.50	57.15	27.82	18.50	46.99	26.75
33.75	85.72	40.08	22.75	57.78	28.13	18.75	47.62	27.11
34.00	86.36	40.37	23.00	58.42	28.44	19.00	48.26	27.47
34.25	86.99	40.67	23.25	59.05	28.75	19.25	48.89	27.84
34.50	87.63	40.97	23.50	59.69	29.06	19.50	49.53	28.20
34.75	88.26	41.26	23.75	60.32	29.37	19.75	50.16	28.56
35.00	88.90	41.56	24.00	60.96	29.68	20.00	50.80	28.92
35.25	89.53	41.86	24.25	61.59	29.98	20.25	51.43	29.28
35.50	90.17	42.15	24.50	62.23	30.29	20.50	52.07	29.64
35.75	90.80	42.45	24.75	62.86	30.60	20.75	52.70	30.00
36.00	91.44	42.75	25.00	63.50	30.91	21.00	53.34	30.37
36.25	92.07	43.05	25.25	64.13	31.22	21.25	53.97	30.73
36.50	92.71	43.34	25.50	64.77	31.53	21.50	54.61	31.09
36.75	93.35	43.64	25.75	65.40	31.84	21.75	55.24	31.45
37.00	93.98	43.94	26.00	66.04	32.15	22.00	55.88	31.81
37.25	94.62	44.23	26.25	66.67	32.46	22.25	56.51	32.17
37.50	95.25	44.53	26.50	67.31	32.77	22.50	57.15	32.54
37.75	95.89	44.83	26.75	67.94	33.08	22.75	57.78	32.90
38.00	96.52	45.12	27.00	68.58	33.38	23.00	58.42	33.26
38.25	97.16	45.42	27.25	69.21	33.69	23.25	59.05	33.62
38.50	97.79	45.72	27.50	69.85	34.00	23.50	59.69	33.98
38.75	98.43	46.01	27.75	70.48	34.31	23.75	60.32	34.34
39.00	99.06	46.31	28.00	71.12	34.62	24.00	60.96	34.70
39.25	99.70	46.61	28.25	71.75	34.93	24.25	61.59	35.07
39.50	100.33	46.90	28.50	72.39	35.24	24.50	62.23	35.43
39.75	100.97	47.20	28.75	73.02	35.55	24.75	62.86	35.79
40.00	101.60	47.50	29.00	73.66	35.86	25.00	63.50	36.15
40.25	101.24	47.79	29.25	74.29	36.17			
40.50	102.87	48.09	29.50	74.93	36.48			
40.75	103.51	48.39	29.75	75.56	36.79			
41.00	104.14	48.69	30.00	76.20	37.09			
41.25	104.78	48.98	30.25	76.83	37.40			
41.50	105.41	49.28	30.50	77.47	37.71			
41.75	106.05	49.58	30.75	78.10	38.02			
42.00	106.68	49.87	31.00	78.74	38.33			
42.25	107.32	50.17	31.25	79.37	38.64			
42.50	107.95	50.47	31.50	80.01	38.95			
42.75	108.59	50.76	31.75	80.64	39.26			
43.00	109.22	51.06	32.00	81.28	39.57			
43.25	109.86	51.36	32.25	81.91	39.88			
43.50	110.49	51.65	32.50	82.55	40.19			
43.75	111.13	51.95	32.75	83.18	40.49			
44.00	111.76	52.25	33.00	83.82	40.80			
44.25	112.40	52.54	33.25	84.45	41.11			
44.50	113.03	52.84	33.50	85.09	41.42			
44.75	113.67	53.14	33.75	85.72	41.73			
45.00	114.30	53.44	34.00	86.36	42.04			

***Note:* Percent Fat = Constant A + Constant B − CONSTANT C − 18.4**

Your FITCOMP nutrition program (on the ''a'' diet plan) is designed so you will lose about 1.8 pounds a week and achieve your goal weight of 138.0 pounds in 110 days. Your personalized meal plan provides an optimal blend of carbohydrates (sugars), fats, proteins, vitamins and minerals. Your total food intake each day will equal 1576 Calories. If you faithfully follow your nutrition and exercise program, your weight should decrease according to the pattern shown in the graph below.

Notice that you should weigh 151 pounds on or about Monday, Sep. 16, 1991, and achieve your goal weight of 138.0 pounds during the week of Nov. 3, 1991. GOOD LUCK.

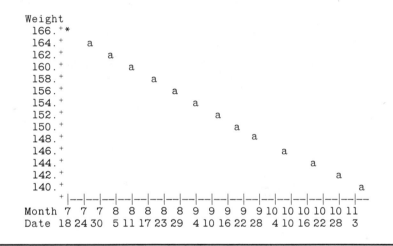

```
      Weight
       166.⁺*
       164.⁺      a
       162.⁺         a
       160.⁺           a
       158.⁺             a
       156.⁺              a
       154.⁺                a
       152.⁺                  a
       150.⁺                    a
       148.⁺                      a
       146.⁺                         a
       144.⁺                           a
       142.⁺                             a
       140.⁺                               a
          ⁺|—|—|—|—|—|—|—|—|—|—|—|—|—|—|—|—|—|—|—
    Month  7  7  7  8  8  8  8  8  9  9  9  9  910 10 10 10 10 11
    Date  18 24 30  5 11 17 23 29  4 10 16 22 28  4 10 16 22 28  3
```

Example of weight loss curve. For this individual who initially weighed 165 lb, the desired weight loss was 27 lb.

Meat (amount for 70 calories)			Alcohol (amount for 80 calories)		
Canadian Bacon	1	ounce	Cognac	1½	ounce
Farmers cheese	1	ounce	Wine red/white	3½	ounce
Ricotta cheese	1	ounce	Sherry	1½	ounce
Mozzarella	1	ounce	Port	1½	ounce
Egg	1		Liquor	1½	ounce
Peanut butter	2	tbsp	Beer	6	ounce
Cottage cheese	¼	cup	Ale	6	ounce
Tuna	¼	cup			
Salmon—canned	¼	cup			
Turkey	1	ounce			
Chicken	1	ounce	Fruit (amount for 40 calories)		
Lamb shoulder	1	ounce			
Lamb roast	1	ounce	Papaya	¾	cup
Lamb chops	1	ounce	Apricot juice	½	cup
Lamb leg	1	ounce	Prune juice	¼	cup
Sirloin	1	ounce	Prunes	2	
Rump steak	1	ounce	Pineapple juice	⅓	cup
Round steak	1	ounce	Orange juice	½	cup
Tenderloin	1	ounce	Banana	½	small
Flank steak	1	ounce	Apple juice	⅓	cup
Chuck steak	1	ounce	Raisins	2	tbsp
Veal—chops	1	ounce	Oranges	1	small
Veal—cutlets	1	ounce	Dates	2	
Pork Roast	1	ounce	Cider—any kind	⅓	cup
Pork Chops	1	ounce	Raspberries	½	cup
Pork—ham	1	ounce	Apple	1	small
Pork—leg	1	ounce	Tangarines	1	medium
Ground round	1	ounce	Pineapple	½	cup
Ground beef	1	ounce	Nectarines	1	small
Scallops	1	ounce	Watermelon	1	cup
Shrimp	1	ounce	Honeydew	⅙	small
Fish—fresh/frzn	1	ounce	Cantaloupe	¼	small
Cornish hen	1	ounce	Apple sauce	½	cup

Treats (amount for 80 calories)		
Popcorn (popped)	3	cup
Oatmeal cookies	1½	2" dia
Choc chip cookie	1½	2" dia
Chocolate fudge	1	sml pc
Candy bar, choc	1	small
Cupcake—icing	1	small

Examples of food items selected from the questionnaire. The vegetable, bread and fat selections are not shown. The caloric values are a close approximation to actual values.

```
****************************************************************
    1576 Calorie Food Plan
    ----------------------------------

    Nutrient Composition
        Carbohydrate...  261 grams or  64.% of your total calories
        Protein.......    71 grams or  17.% of your total calories
        Fat...........    35 grams or  19.% of your total calories
```

	Breakfast	Lunch	Dinner

Day 1	Cooked grits
	  1½ cup
	Milk, skim
	  1 cup
	Cream—light
	  4 tbsp
	Apricot juice
	  1 cup
		Bread—any kind
		  3 slice
		Milk, skim
		  1 cup
		Pork—ham
		  1 ounce
		Cauliflower
		  ½ cup
		Raisins
		  6 tbsp
			Candy bar, choc
			  1 small
			Mashed potato
			  1 cup
			Milk, skim
			  1 cup
			Veal—cutlets
			  2 ounce
			Rhubarb
			  1½ cup
			Radishes
			  no limit
			Diet Margarine
			  4 tsp
			Tangarines
			  2 medium

	Breakfast	Lunch	Dinner

Day 2	Toast
	  3 slice
	Milk, skim
	  1 cup
	Bacon—crisp
	  2 strip
	Prune juice
	  ½ cup
		Bread—any kind
		  3 slice
		Milk, skim
		  1 cup
		Peanut butter
		  2 tbsp
		Celery
		  ½ cup
		Oranges
		  3 small
			Beer
			  6 ounce
			Baked potato
			  1 small
			Milk, skim
			  1 cup
			Pork Roast
			  2 ounce
			Artichokes
			  1½ cup
			Lettuce
			  no limit
			Diet Margarine
			  4 tsp
			Pineapple
			  1 cup

This is an example of the first two days of a 1,576 calorie food plan. The usual procedure is to generate a 14-day plan; because foods are arranged as exchanges within a given food category, each exchange is assigned a specific calorie value. Therefore, one can exchange any one food within a food category with any other food in that category. In addition, any one complete breakfast, lunch, or dinner can be interchanged for any other breakfast, lunch, or dinner. This makes the number of food combinations for a given day equal to 14 factorial.

```
-------------------------------------------------------------------
STEP  Walk for 2½ miles in 47 minutes 24 seconds (18 min 58 sec/mile)
 9    This exercise burns 280 calories.

      Cycle 3½ miles in 19 minutes 26 seconds (10.80 miles/hour)
      Repeat this 2 times. This exercise burns 229 calories.

      Swim 275 yards in 9 minutes 50 seconds (27.96 yards/min)
      Repeat this 5 times. This exercise burns 290 calories.

      The following alternate activities will expend approximately the
      same number of calories as the aerobic activities above expend.
                Racquetball       for      51. minutes
                Circuit Training for      22. minutes
                Squash            for      38. minutes
                Badminton         for      92. minutes
                Basketball        for      34. minutes
                Downhill Skiing   for      40. minutes
                Tennis            for      82. minutes
                Golf              for      47. minutes
                Aerobic Dancing   for      28. minutes
-------------------------------------------------------------------
```

Example for Step 9 of the complementary beginner aerobic exercise plan to accompany the daily meal plans. With this program, individuals proceed to the next week's exercise after they complete their choice of exercise at least 3 times in the same week. Caloric expenditure represents average values for a particular body weight.

This kind of computer-generated output gives the person freedom to exchange activities for any given workout; it offers flexibility and variety in planning workouts to meet individual preferences. The major advantage, however, is the maintenance of caloric equivalency between the different activities that is linked with caloric input from the menus. If inclement weather prohibits jogging or cycling, then swimming or racquetball, for example, can be substituted without altering either the required calorie output (activity) or the required calorie input (food) side of the energy balance equation. In this way, the individual stays in phase with his or her tailor made weight loss curve. The exercise prescription is sensitive to individual differences because it considers age, gender, and current level of physical activity (relative fitness status).

HOW TO ORDER THE COMPUTERIZED MEAL AND EXERCISE PLAN

1. Send $17 (check, money order, or business, school or university purchase order only) to the following address:

Computer Meal and Exercise Plan
P.O. Box 431
Amherst, MA 01004.

2. Make payable to: FITCOMP
3. Outside continental USA add $2 for postage and bank handling (total $19).

Please Note: Questionnaires are processed within 48 hours; you should receive the printout, under normal mail conditions, within 10 days. The $17 cost for the program applies until Dec. 31, 1994. Thereafter, price subject to change and availability. For updated information after this date, write to Computer Plan/Katch, McArdle, Lea & Febiger, 200 Chester Field Parkway, Malvern, Pennsylvania 19355-9725. The computer meal and exercise plan is an exclusive product of Fitcomp ®, Fitness Technologies, Inc., P.O. Box 431, Amherst, MA 01004

More detailed information about the FITCOMP computerized meal and exercise plan can be found in the following article: Katch F.I. and V.L. Katch. Computer technology to evaluate body composition, nutrition, and exercise. *Preventive Medicine. 12:* 619, 1983. See also Katch, F.I., Nutrition for the athlete. *In:* Welsh, R.P. and Shephard, R.J. (eds.). *Current Therapy in Sports Medicine.* B.C. Decker, Toronto, 1985.

Katch, F.I. *FITCOMP* computerized assessment system to evaluate body composition, nutrition, and exercise. In: Katch, F.I. (editor.) *Sport, Health, and Nutrition.* The 1984 Olympic Scientific Congress Proceedings. Volume 2. Human Kinetics Publishers, Champaign, Illinois, 1986.

COMPUTERIZED MEAL PLAN AND EXERCISE QUESTIONNAIRE

Please print with a pen.

1. Name _____
   (First)   (Last)

2. Address _____
   (Street Number and Name)   (Apt. #)

   _____ _____ _____
   City   State   Zip

3. Age _____
   (years)

4. a. _____ Female b. _____ Male

5. Current body weight _____
   (nearest pound)

6. Height (nearest $\frac{1}{4}$ in.)   _____ _____ _____
                                          feet   inches   fraction

7. How much you would like to weigh _____
   (nearest pound)

8. Place an **X** next to the exercises you would like for your program. Select at least one from Group 1. Group 2 are optional.

9. Place an **X** next to the *one* section which best describes your current level of daily physical activity:

   a. _____ **Inactive:** You have a sit-down job and no regular physical activity.

   b. _____ **Relatively Inactive:** Three to four hours of walking or standing per day are usual. You have *no* regular organized physical activity during leisure time.

   c. _____ **Light Physical Activity:** You are sporadically involved in recreational activities such as weekend golf or tennis, occasional jogging, swimming or cycling.

   d. _____ **Moderate Physical Activity:** Usual job activities might include lifting or stair climbing, or you participate regularly in recreational/fitness activities such as jogging, swimming or cycling at least three times per week for 30 to 60 minutes each time.

   e. _____ **Very Vigorous Physical Activity:** You participate in *extensive* physical activity for 60 minutes or more at least four days per week.

Group 1. a. _____ Walking, Jogging, Running; b. _____ Swimming; c. _____ Cycling.

Group 2. d. _____ Racquetball; e. _____ Circuit Weight Training; f. _____ Squash; g. _____ Badminton; h. _____ Basketball; i. _____ Downhill Skiing; j. _____ Tennis; k. _____ Golf; l. _____ Aerobic dancing.

FOOD PREFERENCE LIST

Select the foods you want as part of your daily diet from the Food Groups below. Mark an **X** in the box next to the food items you wish within each

Group. You *must* select at least *one* item from Groups 1 through 18. To ensure menu variety, be sure to select all the foods you would like to eat. If you omit a choice from one of the required groups, the computer will make a selection for you. *Please Note:* The computer *cannot* make vegetarian menus. You may select none, or as many choices as you wish from Groups 19, 20 and 21—these are optional.

Group 1	**A** ☐ milk, skim		**B** ☐ milk, non-fat		**C** ☐ milk, 2%		
Group 2	**A** ☐ yogurt, skim milk	**C** ☐ yogurt, regular milk	**E** ☐ whole milk	**G** ☐ buttermilk
	B ☐ yogurt, 2% milk	**D** ☐ yogurt, fruit	**F** ☐ choc. milk, non-fat	**H** ☐ ice milk
Group 3	**A** ☐ egg	**C** ☐ ricotta cheese	**E** ☐ cheddar cheese	**G** ☐ swiss cheese
	B ☐ mozzarella	**D** ☐ farmers cheese	**F** ☐ american cheese	**H** ☐ canadian bacon
Group 4	**A** ☐ chicken	**D** ☐ corned beef	**G** ☐ crab, canned	**I** ☐ cottage cheese
	B ☐ turkey	**E** ☐ salmon, canned	**H** ☐ oysters	**J** ☐ peanut butter
	C ☐ hot dog	**F** ☐ tuna		
Group 5	**A** ☐ chuck steak	**D** ☐ round steak	**G** ☐ lamb leg	**I** ☐ lamb roast
	B ☐ flank steak	**E** ☐ rump steak	**H** ☐ lamb chops	**J** ☐ lamb shoulder
	C ☐ tenderloin	**F** ☐ sirloin		
Group 6	**A** ☐ cornish hen	**E** ☐ ground beef	**I** ☐ pork chops	**M** ☐ veal cutlets
	B ☐ fish, fresh and frzn.	**F** ☐ ground round	**J** ☐ pork roast	**N** ☐ veal chops
	C ☐ shrimp	**G** ☐ pork leg	**K** ☐ pork shoulder	**O** ☐ veal roast
	D ☐ scallops	**H** ☐ pork/ham	**L** ☐ veal shoulder	
Group 7	**A** ☐ avocado	**C** ☐ almonds	**E** ☐ peanuts, dry roast	**G** ☐ cream cheese
	B ☐ olives	**D** ☐ pecans	**F** ☐ walnuts	**H** ☐ bacon, crisp
Group 8	**A** ☐ diet margarine			
Group 9	**A** ☐ cream, light	**C** ☐ 1000 island dressing	**E** ☐ mayonnaise	**G** ☐ blue cheese dressing
	B ☐ french dressing	**D** ☐ italian dressing	**F** ☐ sour cream	**H** ☐ tartar sauce
Group 10	**A** ☐ raisin bread	**D** ☐ toast	**G** ☐ puffed cereal	**I** ☐ cooked grits
	B ☐ bagel	**E** ☐ bran flakes	**H** ☐ cooked cereal	**J** ☐ donut, plain
	C ☐ english muffin	**F** ☐ cereal, dry		
Group 11	**A** ☐ bread, any kind	**C** ☐ arrowroot crackers	**E** ☐ matzo	**G** ☐ plain muffin
	B ☐ cooked barley	**D** ☐ graham crackers	**F** ☐ soda crackers	**H** ☐ cooked rice
Group 12	**A** ☐ cooked spaghetti	**F** ☐ corn, off cob	**J** ☐ peas	**N** ☐ biscuit
	B ☐ cooked noodles	**G** ☐ corn, on cob	**K** ☐ baked potato	**O** ☐ corn bread
	C ☐ cooked macaroni	**H** ☐ lima beans	**L** ☐ mashed potato	**P** ☐ corn muffins
	D ☐ beans, cooked	**I** ☐ parsnips	**M** ☐ squash	**Q** ☐ yam
	E ☐ lentils, cooked			
Group 13	**A** ☐ apple juice	**D** ☐ grapefruit	**G** ☐ pineapple juice	**J** ☐ apricot juice
	B ☐ banana	**E** ☐ grape juice	**H** ☐ prunes	**K** ☐ papaya
	C ☐ grapefruit juice	**F** ☐ orange juice	**I** ☐ prune juice	
Group 14	**A** ☐ apple	**E** ☐ blueberries	**I** ☐ dates	**M** ☐ pear
	B ☐ apricot	**F** ☐ raspberries	**J** ☐ mango	**N** ☐ plum
	C ☐ apricot, dried	**G** ☐ strawberries	**K** ☐ orange	**O** ☐ raisins
	D ☐ blackberries	**H** ☐ cider, any kind	**L** ☐ peach	
Group 15	**A** ☐ apple sauce	**D** ☐ cantaloupe	**F** ☐ watermelon	**H** ☐ pineapple
	B ☐ cherries	**E** ☐ honeydew	**G** ☐ nectarine	**I** ☐ tangerine
	C ☐ grapes			
Group 16	**A** ☐ cucumber	**C** ☐ tomato	**E** ☐ carrots	**G** ☐ celery
	B ☐ vegetable juice	**D** ☐ tomato juice	**F** ☐ green pepper	**H** ☐ cauliflower
Group 17	**A** ☐ asparagus	**C** ☐ beets	**E** ☐ brussel sprouts	**G** ☐ eggplant
	B ☐ bean sprouts	**D** ☐ broccoli	**F** ☐ cabbage	
Group 18	**A** ☐ collards	**E** ☐ turnip greens	**H** ☐ string beans	**K** ☐ sauerkraut
	B ☐ kale	**F** ☐ mushrooms	**I** ☐ artichokes	**L** ☐ turnips
	C ☐ mustard greens	**G** ☐ okra	**J** ☐ rutabaga	**M** ☐ zucchini
	D ☐ spinach			

Optional Choices

Group 19	**A** ☐ lettuce	**C** ☐ chicory	**E** ☐ escarole	**G** ☐ watercress			
	B ☐ radishes	**D** ☐ endive	**F** ☐ parsley				
Group 20	**A** ☐ ale	**C** ☐ liquor	**E** ☐ sherry	**G** ☐ cognac			
	B ☐ beer	**D** ☐ port	**F** ☐ wine, red/white				
Group 21	**A** ☐ cake, angel food	**D** ☐ cupcake, with icing	**G** ☐ marshmallows, reg.	**J** ☐ sugar cookies			
	B ☐ cake, fruit	**E** ☐ candy bar, choc.	**H** ☐ choc. chip cookies	**K** ☐ pudding			
	C ☐ cake, pound	**F** ☐ chocolate fudge	**I** ☐ oatmeal cookies	**L** ☐ popcorn, popped			

FREQUENTLY CITED JOURNALS IN EXERCISE PHYSIOLOGY

Exercise physiology encompasses numerous scientific areas of inquiry. The following list of journals (with their common abbreviations) has been particularly useful in our own library searches on various topics of interest.

Journal	Abbreviation	Journal	Abbreviation
Acta Medica Scandinavica	Acta Med Scand	Canadian Journal of Applied Sports Sciences	Can J Appl Sport Sci
Acta Physiologica Scandinavica	Acta Physiol Scand	Canadian Medical Association	
American Journal of Clinical Nutrition	Am J Clin Nutr	Journal	Can Med Assoc J
American Heart Journal	Am Heart J	Circulation Research	Circ Res
American Journal of Anatomy	Am J Anat	Circulation: Journal of the	
American Journal of Cardiology	Am J Cardiol	American Heart Association	Circulation
		Clinical Biomechanics	Clin Biomech
American Journal of Epidemiology	Am J Epidemiol	Clinical Chemistry	Clin Chem
American Journal of Human Biology	Am J Hum Biol	Clinical Neurophysiology plus Electroencephalography	CNEEG
		Clinical Science	Clin Sci
American Journal of Obstetrics and Gynecology	Am J Obstet Gynecol	Clinical Sports Medicine	Clin Sports Med
		Diabetes	Diabetes
American Journal of Physical Anthropology	Am J Phys Anthrop	Diabetologia	Diabetologia
American Journal of Physics	Am J Physics	Endocrinology	Endocrinology
		Ergonomics	Ergonomics
American Journal of Physiology	Am J Physiol	European Journal of Applied Physiology	Eur J Appl Physiol
American Journal of Public Health	Am J Public Health	Experientia	Experientia
		Experimental Brain Research	Exp Brain Res
American Journal of Sports Medicine	Am J Sports Med	Federation Proceedings	Fed Proc
		Fertility and Sterility	Fertil Steril
Annals of Human Biology	Ann Hum Biol	Geriatrics	Geriatrics
Annals of Internal Medicine	Ann Intern Med	Growth	Growth
Appetite	Appetite	Human Biology	Hum Biol
Archives of Environmental Health	Arch Environ Health	Human Movement Science	Hum Mov Sci
		International Journal of Obesity	Int J Obes
Atherosclerosis	Atherosclerosis		
Aviation and Environmental Medicine	Aviat Environ Med	International Journal of Sports Medicine	Int J Sports Med
Brain: Journal of Neurology	Brain	International Journal for Vitamin and Nutrition Research	Int J Vitam Nutr Res
British Heart Journal	Br Heart J	Journal of Applied Physiology	J Appl Physiol
British Journal of Nutrition	Br J Nutr	Journal of Applied Sport Science Research	J Appl Sport Sci Res
British Journal of Sports Medicine	Br J Sports Med		
British Medical Journal	Br Med J	Journal of Biological Chemistry	J Biol Chem

Journal	Abbreviation	Journal	Abbreviation
Journal of Biomechanics	J Biomech	Lancet	Lancet
Journal of Bone and Joint Surgery	J Bone Joint Surg	Medicine and Science in Sports and Exercise	Med Sci Sports Exerc
Journal of Clinical Endocrinology and Metabolism	J Clin Endocrinol Metab	Muscle and Nerve	Muscle/Nerve
Journal of Clinical Investigation	J Clin Invest	Nature	Nature
Journal of Gerontology	J Gerontol	Neuroscience Letters	Neurosci Lett
Journal of Human Movement Studies	J Hum Mov Studies	New England Journal of Medicine	N Engl J Med
Journal of Laboratory and Clinical Medicine	J Lab Clin Med	Nutrition Abstracts and Reviews	Nutr Abstr Rev
Journal of Lipid Research	J Lipid Res	Nutriton and Metabolism	Nutr Metab
Journal of Neurophysiology	J Neurophysiol	Nurition Reviews	Nutr Rev
Journal of Nutrition	J Nutr	Pediatric Exercise Science	Ped Exerc Sci
Journal of Parenteral and Enteral Nutrition	JPEN	Pediatrics	Pediatrics
Journal of Pediatrics	J Pediatr	Physical Therapy Reviews	Phys Ther Rev
Journal of Physical and Medical Rehabilitation	J Phys Med Rehabil	Physician and Sportsmedicine	Phys Sportsmed
Journal of Physiology	J Physiol	Physiological Reviews	Physiol Rev
Journal of Sports Medicine and Physical Fitness	Sports Med Phys Fitness	Preventive Medicine	Prev Med
		Proceedings of the Nutrition Society	Proc Nutr Soc
Journal of Sports Psychology	J Sports Psychol	Psychosomatic Medicine	Psychosom Med
Journal of the American Dietetic Association	J Am Diet Assoc	Public Health Reports	Pub Health Rep
		Radiology	Radiology
JAMA (Journal of the American Medical Association)	JAMA	Research Quarterly for Exercise and Sports	Res Q Exerc Sport
Journal of Sports Medicine	J Sports Med	Scandinavian Journal of Sports Science	Scand J Sports Sci
		Science	Science
		Scientific American	Sci Am
		Sports Medicine	Sports Med

APPENDIX

I

THE BODY PROFILE

There are three aspects to the body profile:
Part 1. Anthropometric measurements
Part 2. Computations
Part 3. Graphic analysis

PART

1 Anthropometric Measurements

The muscular and nonmuscular girth sites are measured with a woven-cloth or plastic-coated tape. The tape should be checked for accuracy prior to taking the measurements by calibrating it at each 10-cm distance from 10 to 120 cm using a meter stick as the criterion. Reliability of the well established measurements in the hands of experienced investigators usually exceeds $r = 0.95$ at each site, with no significant differences between test and retest. Two measurements are usually made at each site, and their average is used in the subsequent calculations. One key to successful technique is consistency in locating the anatomic landmarks and having a "feel" or "touch" in making the measurements. This latter concept is easily appreciated by the person who has made 60 to 100 duplicate sets of measurements. In our experience, a tester can become relatively proficient in anthropometric assessment after making duplicate measurements on a minimum of 30 individuals.

There are 12 measurement sites; these include the muscular component (six sites) and the nonmuscular component (six sites). Bilateral paired measurements are made for the extremities, and an average of the paired scores serves as the criterion score for those sites. Also, an average is used for the two abdominal measures (waist and umbilicus). Figure I-1 shows the anatomic locations of the 12 measurement sites, and Table I-1 describes the landmarks for the muscular and nonmuscular girths.

830

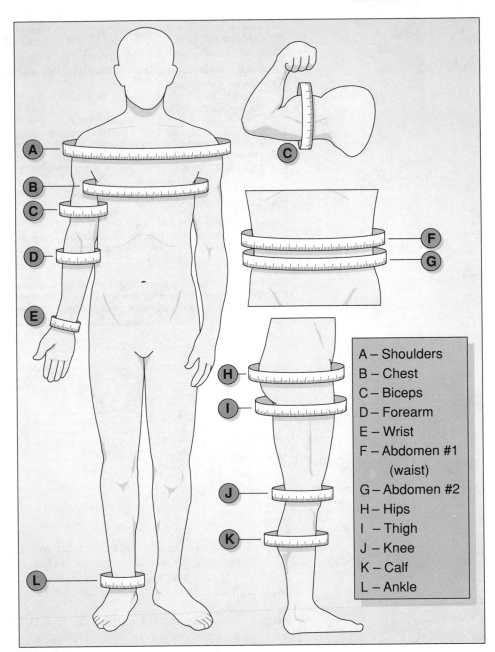

FIG. I-1.
Anatomic locations of the 12 girth sites.

In addition to the anthropometric measurements, body mass is measured in undergarments without shoes on a calibrated balance scale with an accuracy of ±25 grams, and stature is measured without shoes to an accuracy of ±1 millimeter (0.1 cm) while standing erect on a hard surface.

TABLE I-1
ANATOMIC LANDMARKS FOR THE 6 MUSCULAR AND
6 NONMUSCULAR GIRTH SITES

Muscular Component
1. Shoulders: Maximal protrusion of the bideltoid muscles and the prominence of the sternum at the junction of the second rib.
2. Chest: For men, about 2.5 cm above the nipple line; for women, at the axillary level. For both men and women, the tape is placed in position with the arms held horizontally. The arms are then lowered and the measurement is recorded at the mid-tidal level of respiration.
3. Biceps: Maximal circumference with the arm fully flexed and fist clenched.
4. Forearm: Maximal circumference with the arm extended in front of the body and parallel to the floor with the palm supinated.
5. Thigh: Just below the gluteal furrow, about $\frac{2}{3}$ the distance from the mid-knee to the crotch.
6. Calf: Maximal circumference.

Nonmuscular Component
1. Waist: Minimal circumference just below the rib cage.
2. Umbilicus abdomen: Level of the iliac crests at the navel.
3. Hips: Maximal protrusion of the buttocks and, anteriorly, the symphysis pubis. The heels are kept together.
4. Knee: Middle of the patella with the knee relaxed in slight flexion.
5. Wrist: Circumference distal to the styloid processes of the radius and ulna.
6. Ankle: Minimal circumference, slightly above the malleoli.

P   A   R   T

2 | **Computations**

Step 1. Measure stature to the nearest millimeter (0.1 cm). Measure stature twice and use the average value in the subsequent calculations. Express stature in meters in step 6.
Step 2. Measure body mass to an accuracy of ± 25 g, and express mass in kg.
Step 3. Measure the 12 girths to an accuracy of ± 0.1 cm following the procedures described in Table I-1.
Step 4. Divide each muscular and nonmuscular girth by its k constant listed in Table I-2.
Step 5. For each girth, square the result from Step 4.
Step 6. Multiply Step 5 by stature expressed in decimeters (e.g., 173.8 cm becomes 17.38 dm).
Step 7. Compute the average of the 6 muscular girths.
Step 8. Compute the average of the 5 nonmuscular girths (recall that the abdomen is an average of the waist and umbilicus girths).
Step 9. Calculate the percentage deviation for the muscular component girths by using the score from step 6 for each girth site:

$$(\text{step } 6 - \text{step } 8) \div (\text{step } 8 \times 100)$$

TABLE I-2.
k CONSTANTS FOR MEN AND WOMEN

Men	k Constant	Women
	Muscular	
55.4	Shoulders	55.0
45.9	Chest	44.5
15.85	Biceps	14.4
13.45	Forearm	13.0
27.4	Thigh	30.1
17.9	Calf	18.4
	Nonmuscular	
39.2	Abdomen	38.7
46.7	Hips	50.8
18.3	Knee	18.8
8.65	Wrist	8.20
11.25	Ankle	11.1

TABLE I-3
COMPUTATIONS FOR THE BODY PROFILE

Subject: Dr. Albert Behnke, age 86 (measured by FIK 8/17/89)
Step 1.   Stature: 17.38 dm (1.738 m)
Step 2.   Mass: 72.56 kg
Step 3.   Girths (using calibrated cloth tape)

Muscular Component

k	Site	cm	Step 4 cm/k	Steps 5, 6 $(cm/k)^2 \times$ Stature	Step 9 % dev Step 8 − Step 6/Step 8
55.4	Shoulders	101.0	1.823	57.8	−31.0
45.9	Chest	94.9	2.068	74.3	−11.3
15.85	Biceps	32.9	2.076	74.9	−10.6
13.45	Forearm	27.3	2.030	71.6	−14.6
27.40	Thigh	51.4	1.876	61.2	−27.0
17.90	Calf	36.5	2.039	72.3	−13.7
		$\Sigma = 344$		**Step 7** $\bar{X} = 68.7$	

Nonmuscular Component

k	Site	cm	cm/k	$(cm/k)^2 \times$ Stature	Step 10 % dev (Step 6 − Step 7) ÷ Step 7
—	Waist	87.6	—		
—	Umbilicus	92.7	—		
39.2	Abdomen (average)	90.2	2.301	92.0	33.9
46.7	Hips	92.3	1.976	67.9	−1.2
18.3	Knee	42.6	2.328	94.2	37.1
8.65	Wrist	18.4	2.127	78.6	14.4
11.25	Ankle	25.1	2.231	86.5	25.9
		$\Sigma = 268.6$		**Step 8** $\bar{X} = 83.8$	

Step 11. Predicted body mass = $(612.6/100)^2 \times 17.38 \times 0.111 = 72.40$ kg. This predicted value of 72.40 kg for body mass is within 0.2% of the measured body mass of 72.56 kg. The sum of 11 girths (612.6) includes all of the muscular girths and the 5 nonmuscular girths (abdomen average, hips, knees, wrists, ankles).

Step 10. Repeat step 9 for the nonmuscular girths as follows:

$$(\text{Step } 6 - \text{Step } 7) \div (\text{Step } 7 \times 100)$$

Step 11. Determine the relative accuracy of the girth measurements by using the following equation to predict body mass:

Predicted body mass, kg = $(\text{sum } 11 \text{ girths}/100)^2 \times \text{stature, dm} \times 0.11$

If accuracy is high in making the girth measurements, the predicted body mass for adults should be within 1.5 kg of the actual body mass. This is because there is an inherent relationship between body mass and its envelope (girth) measurements. If the predicted body mass deviates by more than 1.5 kg, we suggest that a careful review be made of the accuracy of the girths and stature.

The step-by-step procedure for computing the body profile is shown in Table I-3. We were honored to be able to use the measurements on Dr. Albert Behnke, the creator of the major ideas and concepts that led to the development of the body profile.

P A R T

3 Graphic Analysis

In our experience, a graphic representation of the muscular and nonmuscular percent deviations (steps 9 and 10 in Table I-3) is a convenient way to display the pertinent results. Figure I-2 presents the data from Table I-3, the results for subgroups 9 and 10 in Table 28–8 (Amherst College data of 1882 contrasted with 1886), and growth data for two of one of the author's children, who were both measured at 68 and 101 months of age.

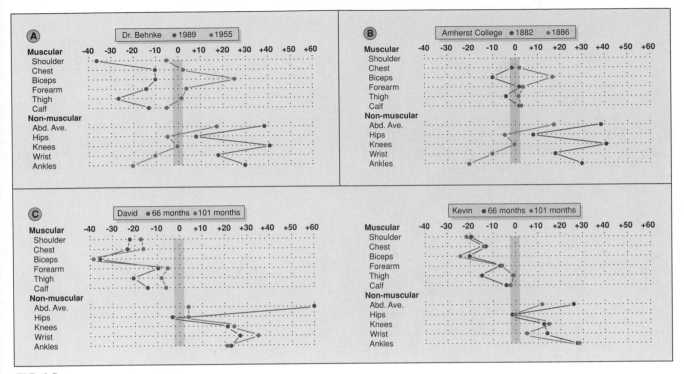

FIG. I-2.
A. Comparison of the body profiles for Dr. Behnke using the 1989 data from Table I-3 and from anthropometric measurements taken in 1955. **B.** Comparison of the body profiles for the Amherst College data from 1882 and 1886 (groups 9 and 10 in Table 28–8). **C.** Body profiles for David and Kevin Katch at ages 66 and 101 months.

INDEX

Note: Page numbers in italics indicate illustrations; page numbers followed by "t" indicate tables; page numbers followed by "n" indicate notes.

A band, 351, *351–352*
Acclimatization
  to altitude, 531–538, 532t
    time required for, 538
  to cold, 572–573
  to heat, 564–565, 565t, *565*
Accommodating resistance exercise, 467
Acetylcholine
  in blood flow regulation, 322
  in heart rate regulation, 316
  in nerve impulse transmission, 372–373, *374*
Acetyl-CoA, 112–113, *113*
  formation of, from fatty acid, 118
Acid(s). *See also* particular acids
  defined, 285
Acid-base balance. *See also* Alkalosis;
    Hydrogen ion concentration
  alterations in
    anaerobic capacity and, 210
  hyperventilation's effect on, 584
  readjustment of, in response to altitude, 535
  regulation of, 285–289
    during exercise, *288*, 288
Acidosis
  bicarbonate buffer's action in, 286
  defined, 285
ACSM (American College of Sports Medicine)
  position statement of, on anabolic steroids,
    499–500
ACTH. *See* Corticotropin
Actin, 350
  in muscle contraction, 355–357, *356*, *358*
  orientation of, in sarcomere, 352, *352–353*
α-Actinin, 350
β-Actinin, 350
Actomyosin, 357
Adenosine, 101
  in myocardial blood flow regulation, 307
Adenosine 3′,5′-cyclic
    monophosphate. *See* Cyclic adenosine
    monophosphate
Adenosine diphosphate (ADP), 102
  energy metabolism regulation by, 120
Adenosine monophosphate (AMP), 102
  cyclic. *See* Cyclic adenosine monophosphate
Adenosine triphosphate (ATP), 54, 101–102,
    103
  changes in resting concentrations of, due to
    training, 429t
  depletion of
    oxygen deficit and, *129*, 129–130
  in muscle function, 357, *358*
  quantity of, in body, 103
  storage of, in muscle, 123

  structure of, 101, *102*
  synthesis of
    from creatine phosphate, 102–103, *103*
    in complete oxidation of glucose, 108, *115*,
      115–116
    in cytoplasmic NADH oxidation, 110
    in fat catabolism, *117*, 117–118
    in glycolysis, 110
    in respiratory chain, *104–105*, 105–106
Adenosine triphosphate-creatine
    phosphate system, 123–125, *124*
  contribution of, to work output
    in short tests, 133, *206*
    with varying exercise duration, 199–200,
      200
  performance evaluation of, 200–205
  physiologic evaluation of, 205
  training of, 440. *See also* Resistance training
    physiologic effects of, *428*, 428t–429t, *428–*
      429
Adenosinetriphosphatase (ATPase), 102
  myofibrillar, 357
ADH. *See* Antidiuretic hormone
Adipocyte(s). *See* Fat cell(s)
ADP. *See* Adenosine diphosphate
Adrenal gland(s), 396–397, *397*
  in blood flow regulation, 322
  secretions of, 390t–391t, 397–398, *397–400*.
    See also particular hormones
  training's effects on, 406t, 408–410
Adrenergic fiber(s)
  in blood flow regulation, 322
Adrenocorticotropic hormone. *See*
    Corticotropin
Aerobic capacity. *See* Maximal oxygen con-
    sumption
Aerobic energy system, 104–107, 112–114
  contribution of, to work output
    in short tests, *206*, 207
    with varying exercise duration, *132*, 132–
      133, 135t, *200*, 200
  evaluation of, 211–227
  in exercise, 127–131
  slow-twitch muscle fibers in, 131
  tests of, 213–216
  training of. *See* Aerobic training
Aerobic metabolism, 106–107. *See also* Aerobic
    energy system
Aerobic power
  maximal. *See* Maximal oxygen consumption
Aerobic training, 441–445
  adaptations with
    cardiovascular, 329, 343
    muscular, 429–430

  blood lactate threshold and, 126
  continuous, 444
  fartlek, 444–445
  goals of, 441, *442*
  interval, 441–444, 443t
  resistance training's interference with, 461,
    478
  specificity of, 425t, 425–426
Aerosinusitis
  in diving, *588*, 590
Aerotitis
  in diving, *588*, 589–590
"Afterglow"
  contribution of, to energy expenditure, 685
Age. *See also* Child(ren); Elderly
  maximal oxygen consumption and, 220–221,
    221
  physiologic function and, 701–708
Air
  alveolar. *See* Alveolar air
  composition of, 254
  partial pressure of gases in, 254–255, 255t
  pressure of, 254
  tracheal, 255
Air embolism
  in diving, 587–588, *588*
Air Force diet, 679t
Air resistance
  effects of, on energy expenditure in running,
    187, 188
  on treadmill vs. track, 189
Airway resistance
  smoking's effects on, 282
Alanine
  catabolism of, 118
  glucose derivation from, *36–37*, 36–37
  structure of, *29*
Alanine-glucose cycle, *36–37*, 36–37
Albumin
  respiratory quotient for, 151
Alcohol
  net energy value of, 89
Alcoholic beverage(s)
  nutritive value of, 779–780
Aldosterone, 391t, *398*, 398–399
  in response to heat stress, 553
  in sodium regulation, 58, 399
  training's effects on, 408
Alkaline reserve
  anaerobic training's effects on, 210
  loss of, at high altitude, 535
Alkalosis
  bicarbonate buffer's action in, 286

Alkalosis (*Continued*)
defined, 285
induced, as ergogenic aid, 210, 506t, 506–507
All-or-none principle, 375
Allosteric modulation, 388
Alpha-actinin, 350
Alpha-ketoglutaric acid
formation of, from glutamine, 118
Altitude, 529–546
high
defined, 529
immediate responses to, 531–534, 532t
longer-term adjustments to, 532t, 534–538
metabolic, physiologic and exercise capacities at, 539–540
stress of, 529–531, *530*
training at, 541t, 541–542
sea level performance and, 540–542
sea level training vs., 542, *542*
wind velocity's effect on energy expenditure at, 188
Alveolar air
composition of, 255t, 255, 259
partial pressure of gases in, 255t, 255–256
Alveolar ventilation, 247t, 247–250
maintenance of, during exercise, 249–250
Alveolus(-i), 235, *236*, 236–237
Amenorrhea. *See* Menstrual cycle
American College of Sports Medicine (ACSM)
position statement of, on anabolic steroids, 499–500
Amino acid(s), 29, *29. See also* Protein(s)
essential, 29–30
nonessential, 29
supplementation with, 32–33
Amino radical, 29, *29*
γ Aminobutyric acid (GABA)
in nerve impulse inhibition, 374
AMP (adenosine monophosphate), 102
cyclic. *See* Cyclic adenosine monophosphate
Amphetamine(s), 502–503, 503t
Anabolic steroid(s), 498–501, 500t
Anabolism, 33
Anaerobic capacity
individual differences in, *210*, 210–211
measurement of, 207
Anaerobic energy system(s). *See* Adenosine triphosphate-creatine phosphate system;
Glycolytic energy system
Anaerobic glycolysis. *See* Glycolysis
Anaerobic power
measurement of, 207
Anaerobic threshold. *See* Onset of blood lactate accumulation
Anaerobic training, 440–441. *See also* Resistance training
interval, 440, 443–444
lactate-producing capacity and, 127
physiologic effects of, *428*, 428t–429t, 428–429
Androgen(s), 400
Androstenedione, 400
Anemia
exercise-induced, 56–57, *57*
iron deficiency. *See* Iron deficiency anemia
pernicious, 46t
Angina pectoris, 307, 713, 725
Angiocardiography
radionuclide
exercise testing with, 728
Angiotensin, 399
Annulospiral nerve fiber, *378, 379*

Anthropometry
body composition assessment by, 616–623, 830–835
markers predictive of obesity in, 646
quantification of physique by comparative data on groups, 644t, 644–646, 646t
in champion athletes, 634–644
Antidiuretic hormone (ADH), 390t, 395
in response to heat stress, 553
training's effects on, 406t, 407–408
Aorta, *293–294, 294*, 295
Aortic body(-ies)
in ventilatory control, 271–272, *272*
Archimedes, 607–608
Arm exercise
blood pressure in, 304t, 304
cardiovascular adjustments to, 339–340, *340*
maximum heart rate for, 434, 437–438
stress testing with, 728
Arm x-ray
body fat assessment using, 624, *625*
Arrhythmia(s)
exercise-induced, 726–727
Arterial pressure. *See also* Blood pressure
mean, 296
Arteriole(s), *295*, 295
Arteriovenous oxygen difference
$(a - \bar{v}O_2)$, 264, 337–339
in exercise, *338*
training's effects on, 339
measurement of, in Fick cardiac output measurement, 327
training's effects on, 431
Artery(-ies), *295*, 295–296
Aspartate
catabolism of, 118
Asthma
exercise and, 245–246
Astrand protocol, *215*
Atherosclerosis, 23, 712–713, *713. See also*
Coronary heart disease
"Athlete's heart," 340–343
Athletic drink(s). *See also* Sugar drink(s)
for mineral replacement, 59
Atom(s), 5
ATP. *See* Adenosine triphosphate
ATPase. *See* Adenosinetriphosphatase
Atrial natriuretic hormone, 387t
Atrioventricular bundle (A-V bundle), *314*, 314
Atrioventricular node (A-V node), *314*, 314
Atrioventricular valve(s), 294
Atrium(-a), *294*, 294
Atwater general factor(s), 89
A-V bundle (atrioventricular bundle), *314*, 314
A-V node (atrioventricular node), *314*, 314
$a - \bar{v}O_2$. *See* Arteriovenous oxygen difference
Axon, *371, 372*

Back
lower
protection of, during resistance training, 63
resistance training for pain syndrome involving, 463
Balke treadmill protocol, *215*, 727
Ballet dancer(s)
body profiles of, 644t, 645
Barbiturate(s)
use of, with amphetamine abuse, 502–503
Baroreceptor(s)
in heart rate regulation, 316
Basal ganglia, 369
Basal metabolic rate (BMR), 158–162. *See also*
Resting metabolic rate

Base(s)
defined, 285
Behnke, Albert, 599
Bends, *591–592*, 591–593
Benzedrine (amphetamine sulfate), 502
Beriberi, 44, 46t
Beta oxidation, *117*, 118
Beta-actinin, 350
Beta-endorphin, 389, 411
Beta-lipotrophin, 411
BIA (bioelectric impedance analysis)
body composition assessment using, 623
Bicarbonate
carbon dioxide transport as, 267–268
Bicarbonate buffer system, 285–286
Bicarbonate loading, 210, 506t, 506–507
Bicuspid valve, *294*, 294
Bicycle test(s)
maximal oxygen consumption measurement with
correlation of, with other aerobic power tests, 214t, 214
of glycolytic power, *206*, 207
of immediate nonaerobic power, 204t
stress testing with, 728
Bicycling
effects of, on body composition, 649
tests using. *See* Bicycle test(s)
Bioelectric impedance analysis (BIA)
body composition assessment using, 623
Bioenergetics. *See* Energy transfer
Biologic value, 30
Biopsy technique
for evaluation of intramuscular nutrients during exercise, 12
Biotin
functions, sources and RDA for, 47t
Blackout
in breathhold diving, 583–584
Blood
major components of, *263*
Blood boosting. *See* Blood doping
Blood doping, 508–510, 510t, *510*
hormonal, 510–511
Blood flow
coronary
regulation of, during exercise, 307
distribution of
at rest, 334t, 334
during exercise, 334–335, 335t
in response to cold, 549
in response to heat, 552–553, 557
regulation of, 319–322
training's effects on, 431
velocity of
vascular cross-sectional area and, 297–298, *298*
Blood lactate threshold. *See* Onset of blood lactate accumulation
Blood pressure, 296
diastolic, 296
exercise's effects on, 301–305, 302t–304t, *304*
high. *See* Hypertension
in body inversion, 305
in exercise stress testing, 727
mean arterial, 296
measurement of, *297*
obesity and, 659
systolic, 296
training's effects on, 431
variations in, in systemic circulation, *298*, 298–299

Blood sugar. *See* Glucose
Blood vessel(s), 295–301
  walls of, *295*
Blood volume
  altitude's effect on, 535–536
  reduction of, with sweat-induced fluid loss,
    63–64
  training's effects on, 430
BMR (basal metabolic rate), 158–162. *See also*
    Resting metabolic rate
Body builder(s)
  body profiles of, 644t, 645
  physique of, 642–644, 643t
Body composition, 597–739. *See also* Body fat;
    Physique
  age-related changes in, 706–707, *707*
  altitude's effects on, 538
  assessment of, 599–633
    by bioelectric impedance analysis, 623
    by fatfold measurement, 616–620
    by girth measurement, 620–623, 813–819
    by hydrostatic weighing, 607–616, 615t,
      707, 812
    common techniques for, 607–624
  during weight loss
    effects of exercise and diet on, 650–651,
      651
  maximal oxygen consumption and, 218t,
    218–219
  of professional athletes, 603, 604t
  of reference man and woman, 600–603, *601*
  resting metabolic rate and, 160
  training's effects on, 433, 647t–648t, 647–
    650, 650t
Body density
  body fat estimation from, 609–610
    limitations of, 608–609, 610–611
  calculation of, 609, 616
    temperature correction for water density
      in, 812
  upper and lower limits of, 608
Body fat
  as protection and insulation for vital organs,
    26
  cold stress and, 572
  distribution of
    for reference woman, 600, *602*
    health risks and, 660
  essential, 600
    calculation of mass of, 611
  estimation of
    by computerized tomography, 624, *626*
    by magnetic resonance imaging, 624, *626*
    by surface area method, 620
    by ultrasound technique, 623–624
    from arm radiograph, 624, *625*
    from body density, 609–610
    from changes in abdominal girth, 622
    from fatfold measurement, 618–620
    from girth measurement, 621–622, 813–
      819
    from hydrostatic weighing, 608–611
  functions of, 25–26
  lower limits for, 603–605
  mass of
    calculation of, 611
  menstrual function and, 605–607
  nitrogen elimination and, 591, *591*
  percent of
    age-related changes in, 668, *707*
    average, 26, 627, 628t
    desirable, 627
    obesity defined by, 659–660
  sex-specific, 600

storage, 600
  calculation of mass of, 611
  energy reserve of, 25–26, 116, 611
Body inversion
  blood pressure and, 305
Body mass. *See also* Weight gain; Weight loss
  changes in, at high altitude, 538
  desirable, 627–628
  effects of, on energy expenditure, 169, *170*,
    178t, 178
  lean. *See* Lean body mass
  minimal, 603–605, 605t
  stability of, 674–675
Body position
  blood pressure and, 305
  cardiac output and, 332t, 332–333
Body profile, 622–623, 830–835
  comparative data for ten groups, 644t, 644–
    646
Body temperature
  during exercise, 557–558, *558*
  in ventilatory control, 275
  oral vs. rectal, 569
  recovery oxygen consumption and, 137
Body volume
  measurement of, 611–616
Body weight. *See* Body mass; Weight gain;
    Weight loss
Bohr effect, 264
Bone
  effects of mechanical stress on, 54
Bone density
  muscular strength and, *483*, 483–484
Bone mass
  age-related changes in, 707–708
"Bonking," 14, 516
Bottle buoyancy
  body volume measurement by, 612, *613*
Boyle's law, 581
Bradycardia
  defined, 316
  training-induced, 329
    mechanisms of, 316
Brain
  blood flow to, 334t–335t, 335
  carbohydrate as fuel for, 12
Breathhold diving, 583–585
  hyperventilation and, 273, 583–584
Breathholding
  duration of, 273, 583
  ventilatory control mechanisms in, 273
Breathing. *See also* Respiration; Ventilation
  energy cost of, *282*, 282–283
    smoking and, 282–283, 283t
Breathing capacity. *See* Pulmonary function
Broad jump
  standing, 201–202
Bronchiole(s), 235
Bronchus(-i), 235
Bruce treadmill protocol, *215*, 727
Buffering, 285–288
  chemical systems for, 286–287
    relative power of, 287t, 287
  defined, 285
  loss of capacity for, at high altitude, 535
  physiologic systems for, 287–288
Buffering solution(s)
  as ergogenic aids, 505–507, 506t
Bundle of His, *314*, 314

C protein, 351
Cable tensiometry, 452–453, *453*
Caffeine, 503–505

Calcium
  functions of, 51t, 52–54
  in muscle function, 353, 354, 357–358, *358*
  osteoporosis and, 53–54
  recommended intake of, 51t, 53
  replacement of, after intense exercise, 59
  sources of, 51t, 53
  supplemental, for osteoporosis prevention, 53
Calisthenics
  effects of, on body composition, 648–649
Calorie, 85
Calorimetry
  direct
    body heat production measured by, 145,
      146
    energy value of food measured by, 85–86,
      86
    indirect vs., 150
  indirect
    body heat production measured by, 99,
      145–150
    computerized systems approach to, *155*,
      156
    direct vs., 150
cAMP. *See* Cyclic adenosine monophosphate
Cancer
  risk of
    delayed menarche and, 607
Capillary(-ies), *295*, 296–298
Carbamino compound(s)
  carbon dioxide transport as, 268
Carbaminohemoglobin, 268
Carbohydrate(s), 6–19
  catabolism of
    fat catabolism and, 12, 120
  content of common foods, *10*
  conversion of, to storage fat, efficiency of,
    675
  digestibility coefficient of, 88t, 88
  energy release from, 107–116
  for pregame meal, 77–78
  functions of, 11–12
  heat of combustion of, 86, 88t
  in typical American diet, 10, 11
  kinds and sources of, 6–10
  nature of, 6
  needs for, in prolonged, severe training, 71–
    73, *72*
  net energy value of, 88t, 89
  protein-sparing effect of, 11–12
    during exercise, 35, 35–36
  recommended intake of, 10–11
    for exercise, 70, *71*
  respiratory quotient for, 150–151
  storage of, 9
    energy reserve of, 116
  supplemental
    during exercise, 15–17, *16*
    prior to exercise, 17–18, 78
Carbohydrate balance
  during exercise, 12–18
Carbohydrate loading, 516–520
  classic procedure for, 517t, 517
  modified procedure for, *519*, 520
  negative aspects of, 519–520
  sample diets for, 518t
Carbon, 6
  as percentage of body mass, 5
Carbon dioxide
  partial pressure of. *See* Carbon dioxide partial
    pressure
  production of
    calculation of, 801

Carbon dioxide (*Continued*)
  solubility of, 256
  transport of, 266–268, *267*
    as bicarbonate, 267–268
    as carbamino compounds, 268
    in solution, 267
Carbon dioxide partial pressure
    (P_{CO_2})
  alveolar, 255
    at high altitude, 535
    oxygen consumption and, during exercise, *274*, 274
  arterial, 259
    in ventilatory control, 272–273
  in ambient air, 254
  in tissues, 257
Carbon dioxide rebreathing method
  for cardiac output measurement, 328
Carbon monoxide poisoning
  in diving, 593–594
Carbonic acid
  formation of, in blood, 267, 268
  in acid-base regulation, 285–286
Carbonic anhydrase, 267
Carboxyl group, *29*, 29
Cardiac hypertrophy
  in disease, 341
  in response to training, 340–343, 341t–342t, 430
    specificity of, 342t, 342–343
Cardiac impulse
  normal conduction of, 314, *314*
Cardiac index
  resting
    age-related decline in, 701
Cardiac output
  altitude's effects on, 533
  at rest, 329
  blood pressure and, *304*, 304
  determinants of, 326
  distribution of. *See* Blood flow, distribution of
  during exercise, 330–333
    altitude's effects on, 539–540
    heat's effects on, 557
  maximum
    age-related changes in, 705
    maximal oxygen consumption and, *336*, 336–337
    training's effects on, 330, 431
  measurement of, 326–328
  of racehorses, 330
  oxygen consumption and, 336–337, *337*
  oxygen transport and, 336–337
  sex differences in, 337
  submaximal
    training's effects on, 337, 431
  ventilatory response to exercise and, 277
Cardiac vein(s), 305, *306*
Cardiovascular disease. *See* Coronary heart disease
Cardiovascular system, 292–312, *293*
  components of, 292–301
  functional capacity of, 326–347
    age-related changes in, 704–706
    general nature of training-induced improvements in, 426
  regulation of, 313–325
  response of, to altitude, 533t, 533, 538
  response of, to cold, 549
  response of, to exercise, 322t, 323
    altitude's effects on, 539–540
    warm-up's effect on, 512–513
  response of, to heat, 552–553, 557

Carotene(s), 45
Carotid artery palpation
  heart rate changes due to, 316–317
Carotid body(-ies)
  in ventilatory control, 271–272, *272*
Catecholamine(s). *See also* Epinephrine; Norepinephrine
  in heart rate regulation, 316
Cellular oxidation, 104–107
Cellulose, 8–9
Central nervous system. *See also* Brain
  carbohydrate as fuel for, 12
Cerebellum, 369
Cheilosis, 46t
Chemical bonding, 5
Chemistry
  clinical
    SI conversion table for common values in, 758t–760t
Chemoreceptor(s)
  in ventilatory control, 271–272, *272*
    during exercise, 274
Child(ren)
  body density of, 610–611
  coronary heart disease risk factors in, 718–720, 720t, *721*
  maximal oxygen consumption in, 220–221, *221*
  resistance training for, 461
Chloride shift, 267–268
Chlorine
  functions of, 57
Cholecystokinin, 387t, 405
Cholesterol, 22–25
  content of common foods, 24t
  endogenous, 23
  exogenous, 23
  fiber and, 8–9
  functions of, 23
  in lipoproteins, 22
  recommended daily intake of, 24–25
  serum levels of
    acceptable, 24
    heart disease and, 23–24, 715t, 715–717
  sources of, 23
  steroid use and, 501
  synthesis of, 23
Cholinergic fiber(s)
  in blood flow regulation, 322
Cholinesterase
  in nerve impulse transmission, 373
Chromium
  functions, sources and RDA for, 51t
  loss of, with strenuous exercise, 59
Cigarette smoking
  coronary heart disease and, 717
  energy cost of breathing with, 282–283, 283t
Circuit resistance training, 484–485
Circulation. *See* Blood flow, distribution of
Circumference measurement. *See* Girth measurement
Citric acid
  in Krebs cycle, 113
Citric acid cycle. *See* Krebs cycle
Climate. *See also* Cold; Heat
  resting metabolic rate and, 163
Clothing
  effects of, on thermoregulation, 553–555
Coefficient of digestibility, 87–89, 88t
Coenzyme(s), 98–99
  vitamins as, 45
Cold
  acclimatization to, 572–573
  assessment of, for safe exercise, 573–574, *574*

  exercise in, 571–575
    body fat and, 572
    respiratory tract during, 574
  metabolic rate in, 163
  thermoregulation in, 549–550
Cold receptor(s)
  cutaneous, 548
Cold treatment(s)
  effectiveness of, for reducing heat stress, 561
College student(s)
  body profiles of
    in nineteenth century, 644t, 646
    in twentieth century, 644t, 645
Combustion
  heat of. *See* Heat of combustion
Computer(s)
  indirect calorimetry using, *155*, 156
  meal and exercise plan using, 78–79, 820–827
  muscular strength measurement using, 455
Computerized tomography (CT)
  body fat assessment using, 624, *626*
Conditioning. *See* Training
Conduction
  heat loss by, 551
Connective tissue damage
  muscle soreness and, 489–490
Conservation of energy, 92
Convection
  heat loss by, 551
Conversion factor(s), 745t–756t
"Cooling-down." *See* Recovery, active
Cooper test, 223
Copper
  functions, sources and RDA for, 51t
Cori cycle, *112*, 112
Coronary artery(-ies), 305, *306*
Coronary blood flow
  regulation of, during exercise, 307
Coronary heart disease, 712–721
  calculating risk of, 720–721, 722t
  cellular changes in, 712–713, *713*
  cholesterol and, 23–24, 715t, 715–717
  dietary fat and, 21
  epidemiology of, *714*, 714
  exercise stress testing in evaluation for, 721–730
  exercise-induced indicators of, 724–727
  high density lipoproteins' protective effect against, 22
  obesity and, 658–659, 717
  protective effects of exercise in, 711, 717–718
    mechanisms of, 718
  risk factors for, 714–718
    in children, 718–720, 720t, *721*
    interaction of, 718, *719*
  therapeutic effects of exercise in, 732
Coronary sinus, 305, *306*
Corticospinal tract, 367–368
Corticosterone, 390t
Corticotropin (ACTH), 390t, 394
  training's effects on, 406, 406t
Corticotropin releasing factor, 400
Cortisol, 390t, 399–400
  training's effects on, 406t, 408
Cotton
  for warm-weather clothing, 554
Coughing
  post-exercise, 246
CP. *See* Creatine phosphate
Creatine
  changes in resting concentrations of, due to training, 429t
Creatine kinase, 103

Creatine phosphate (CP), 54, 102–103, *103. See also* Adenosine triphosphate-creatine phosphate system
changes in resting concentrations of, due to training, 429t
depletion of
oxygen deficit and, *129,* 129–130
storage of, in muscle, 123
CT (computerized tomography)
body fat assessment using, *624, 626*
Cycle ergometry. *See* Bicycle test(s)
Cyclic adenosine monophosphate (cAMP), 386–387
in lipase activation, 117
Cytochrome(s), 55
in electron transport, 104
Cytochrome oxidase (cytochrome A₃)
in respiratory chain, 104

Dead space
anatomic, 247, *249*
tidal volume and, 248
in snorkeling, 583
physiologic, 248, *249*
Deamination, 34
Decompression sickness, *591–592,* 591–593
Dehydration. *See also* Water replacement
at high altitude, 534
consequences of, 559
diuretic-induced, 560
in heat, 558–560
physiologic and performance decrements due to, 559–560
Dehydroepiandrosterone (DHEA), 400
Dehydroepiandrosterone sulfate (DHEAS), 400
Dehydrogenase enzyme(s)
in cellular oxidation, 104
Dendrite(s), *371, 372*
Density, 608n
body. *See* Body density
bone
muscular strength and, *483,* 483–484
water
temperature correction for, 812
Detraining, 427t, 427
Dextroamphetamine sulfate (Dexedrine), 502
Dextrose. *See* Glucose
DHEA (dehydroepiandrosterone), 400
DHEAS (dehydroepiandrosterone sulfate), 400
Diabetes mellitus, 388, 403–404
complications of, 403
recommendations for exercise in, 403–404
training in, 410–411
Diabetic coma, 403
Diaphragm, *237,* 238
Diastole, 296
enhanced filling during stroke volume and, 332–333
Diet(s). *See also* Food
carbohydrate stores and, 10
cholesterol levels and, 23
computerized meal planning with, 77–78, 820–827
efficiency of fat storage and, 675
fats in, 20–21, *25,* 25
for weight control, 676–682, 679t
exercise with, 650–651, *651,* 687–688
long-term maintenance of, 678, *680*
Four-Food-Group Plan for, 73t, 73, 74t
high-carbohydrate
respiratory exchange ratio with gain in body fat on, 154

high-carbohydrate/low-fat, for weight control, 679t
high-protein, 677
obligatory water loss in urine and, 62
rationale for, for weight loss, 162
high-protein/low-carbohydrate, for weight control, 679t
in acclimatization to altitude, 534
ketogenic (low-carbohydrate/high-fat), 676–677, 679t
liquid protein, 677
low-carbohydrate
harmful effects of, 14–15, 34
low-carbohydrate/high-protein, for weight control, 679t
low-fat
harmful effects of, 71
muscle glycogen stores and, 14–15, *15*
one-food-centered, for weight control, 679t
optimal
defined, 68
for exercise, 68–81
protein sources in, *31*
starvation, 677–678
vegetarian. *See* Vegetarian diet(s)
Diet-induced thermogenesis, 162–163
in obesity, 658
Diffusion, 97
facilitated, 401
Digestibility
coefficient of, 87–89, 88t
2,3-Diphosphoglycerate (2,3-DPG), 264–265
changes in red blood cell concentration of, at high altitude, 538
Disaccharide(s), 7
Diuretic(s)
dehydration induced by, 560
for hypertension, 58
Diving, 580–595
breathhold, 583–585
hyperventilation and, 273, 583–584
depth of
inspiratory capacity and, 582–583
limits on, with breathholding, 584–585
limits on, with pure oxygen, 593t
limits on, with snorkeling, 582–583
pressure and, 580, 581t, *582*
zero decompression limits for, *592,* 592
scuba. *See* Scuba diving
skin, 581–583
with pure oxygen, 586–587, 593
depth-time limits for, 593t
Docosahexenoic acid, 20
Double product
as estimate of myocardial work, 307–309
Douglas bag, 148–150, *149*
''Downer(s)''
use of, with amphetamine abuse, 502–503
Down-regulation, 386
2,3-DPG. *See* 2,3-Diphosphoglycerate
Drag
in swimming, 191–192, *192*
Drinking Man's diet, 679t
Drug(s). *See also* particular drugs
as aids to performance, 497–507
Dynamometry, 453, *454*
isokinetic, 455
Dynorphin, 411

Ear(s)
pressure problems with, in diving, 585, *588,* 589–590
Earplug(s)
in diving, 590

ECG. *See* Electrocardiography
Economy of movement, 175–176
Ectomorphy, 634
Edema
due to venous pooling, 300
pulmonary
high-altitude (HAPE), 534
Efficiency, 175
mechanical, in exercise, 176–177
Eicosapentaenoic acid, 21
Elderly, 698–739
aerobic capacity of, *221,* 221, 704–705
applicability of cholesterol and heart disease studies to, 24
body density of, 610–611
exceptional performance capacity in, 705–706, *706*
heat tolerance in, *566,* 566–567
trainability of, 701–702, *702,* 703t, 705, 708–709, *709*
Electrocardiography (ECG), 314–315, *315*
findings on, in coronary heart disease, 725–726, *726*
normal tracing on, *315,* 725–726, *726*
Electrolyte(s)
functions of, 57–58
loss of, during hot-weather exercise, 58–59, 564
replacement of, 59, 563–564
Electromyography (EMG), 471–474, *472–473*
Electron transport, *104,* 104
efficiency of, 106
Ellestad protocol, *215*
Embolism
air, in diving, 587–588, *588*
EMG (electromyography), 471–474, *472–473*
Employment
fitness testing for
specificity in, 471
Endergonic reaction(s), 93–94
example of, *96,* 96
Endocrine gland(s), 390t–392t. *See also* particular glands
defined, 384–385
Endocrine system, 384–417, *385. See also* particular glands
organization of, 384–389, 390t–392t
training's effects on function of, 405–411, 406t
Endomorphy, 634
Endomysium, 348, *349*
Endorphin(s), 390t, 411–412
Endurance
defined, 756
Endurance performance. *See under* Performance
Endurance run(s)
maximal oxygen consumption prediction using, 222–223
Endurance training. *See* Aerobic training
Energy, 83–232. *See also* Energy transfer
carbohydrate as source of, 11, 107–116
conservation of, 92
defined, 92, 756
expenditure of. *See* Energy expenditure
fats as source and reserve of, 25–26, 116–118
forms of, *95,* 95
immediate. *See* Adenosine triphosphate-creatine phosphate system
interconversions of, *95–97,* 95–97
kinetic, 92–93, *93*
long-term. *See* Aerobic energy system
phosphate bond, 101–107

Energy (*Continued*)
  potential, 92–93, *93*
    degradation of, 94, *95*
  processes that conserve, 93–94
  processes that release, 93–94
  protein as source of, 34, 118
    during exercise, 35–36
  release of, from food, 107–121, *108*
    from carbohydrate, 107–116
    from fat, 116–118
    from protein, 118
  short-term. *See* Glycolytic energy system
  storage of, in body fat, 25–26
    calculation of, 116, 611
  systems for delivery and utilization of, 233–417
Energy balance, 675–676
Energy balance equation, 675–676
Energy capacity
  enhancement of, 421–526. *See also* Training
Energy cost of breathing, *282*, 282–283
  smoking and, 282–283, 283t
Energy expenditure
  classification of physical activities by, 165–167, 166t
  components of, 158, *159*
  daily rates of, 167t–168t, 167–168
  during running, 181–190. *See also under* Running
  during swimming, 190–195. *See also under* Swimming
  during walking, *177*, 177–181. *See also under* Walking
  effect of body mass on, 169, *170*
  estimation of, 145–157
    at rest, 161–162
    factors complicating, 175
    methods for, 145–150
    use of heart rate for, 169–172, *171*
  factors affecting, 162–165
  for various activities, 168–169, 804–811
  for various occupations, 168, 168t
  gross, 174
  in physical activity, 165–172
  net, 174
  resting, 158–165
    estimation of, 161–162
  weight gain and, 657–658
Energy metabolism. *See* Energy expenditure;
    Energy transfer
Energy system(s)
  aerobic. *See* Aerobic energy system
  anaerobic. *See* Adenosine triphosphate-creatine phosphate system; Glycolytic energy system
  immediate. *See* Adenosine triphosphate-creatine phosphate system
  long-term. *See* Aerobic energy system
  muscle fiber type and, 131–132
  relative contribution of, to work output
    in short tests, *206*, 207
    with varying exercise duration, *132*, 132–133, 135t, 199–200, *200*
  short-term. *See* Glycolytic energy system
Energy transfer, 92–100. *See also* Energy expenditure
  factors affecting rate of, 97–99
  in biosynthesis, 93
  in exercise, 123–144
  in the body, 101–122
  measurement of, in humans, 99
  oxygen's role in, 106–107
  regulation of, 120

systems for. *See* Energy system(s); particular energy systems
Energy value of food, 85–91
  gross, 85–87
  measurement of, 85–90
  net, 87–89, 88t
    for common foods, 88t
Enkephalin(s), 411
Enterogastrin, 387t
Entropy, 95
Environmental stress. *See also* Altitude; Cold; Diving; Heat
  work performance and, 527–595
Enzyme(s), 98
Epimysium, 348, *349*
Epinephrine, 391t, 397–398
  chemical structure of, *397*
  in blood flow regulation, 322
  in glycogenolysis, 108
  in heart rate regulation, 316
  in lipase activation, 116–117
  in response to cold stress, 549
  training's effects on, 406t, 408–410, *409*
EPSP (excitatory postsynaptic potential), 374
Ergogenic aid(s), 497–526. *See also* particular techniques
Ergometry
  arm. *See* Arm exercise
  bicycle. *See* Bicycle test(s)
ERV. *See* Expiratory reserve volume
Erythrocythemia
  induced. *See* Blood doping
Erythropoietin, 386, 387t
  as ergogenic aid, 510–511
Estradiol, 400, 405
Estrogen, 391t, 400, 405
  osteoporosis and, 53
Eustachian tube(s)
  equalizing air pressure in, in diving, *588*, 589–590
Evaporation
  heat loss by, 551–552, *553*
Excess metabolite theory of muscle soreness, 488–489
Excitation, 373–374
Excitation-contraction coupling, 357–358, *358*
Excitatory postsynaptic potential (EPSP), 374
Exercise. *See also* Physical activity(-ies); Training
  acid-base regulation in, *288*, 288
  altitude and, 529–546. *See also* Altitude
  anaerobic
    respiratory exchange ratio in, 154
  anemia due to, 56–57, *57*
  arm. *See* Arm exercise
  asthma and, 245–246
  blood distribution during, 319, 334–335, 335t
  blood pressure and, 301–305, 302t–304t, *304*
  body composition during weight loss and, 650–651, *651*
  carbohydrate balance in, 12–18
  cardiac output during, 330–333
  cardiovascular system's response to, 322t, 323
    altitude's effects on, 539–540
  "conversational," 435
  coronary heart disease and, 711, 717–718
    mechanisms of protection in, 718
    therapeutic benefits in, 732
  coughing after, 246
  defined, 698, 756
  diet-induced thermogenesis and, 163

duration of
  effects of, on body composition, 647–648, 648t
  for fitness maintenance, 439–440
  for weight loss, 439, 686
  glucose uptake by muscle and, *13*, 13
  injuries and, 438
  relative contribution of aerobic and anaerobic energy and, *132*, 132–133, 135t, 199–200, *200*, *206*, 207
  training effectiveness and, 438
economy of movement in, 175–176
electron transfer and oxidative phosphorylation during, 106
energy transfer in, 123–144
fat balance in, 27, *28*
food intake and, 69t, 74–76, *76*, 77t, 683–684, 684t
for weight control, 682–687
  diet with, 687–688
  dose-response relationship in, 686
  effectiveness of, 685–686, 687t
frequency of
  body composition changes and, 649–650
  for fitness maintenance, 439–440
  for weight loss, 439, 687
  training effectiveness and, 438–439
graded
  effects of, on blood pressure, *304*, 304
hematologic characteristics in response to, 56–57, *57*
hormone secretion and, 389–405, 390t–392t
intense
  carbohydrate balance in, *13*, 13
intensity of
  body temperature and, *558*, 558
  ceiling for training effects of, 435–436
  defined, 756
  for fitness maintenance, 440
  for weight loss, 439
  glucose uptake by muscle and, *13*, 13
  lactic acid accumulation and, 125–127, *126*
  measures of, 434
  training effectiveness and, 434–438
intermittent, 139–141, 140t. *See also* Interval training
iron loss due to, 56–57
isometric
  contraindications for, 240
longevity and, 709–712, *710–712*
mechanical efficiency in, 176–177
menstrual alterations and. *See under* Menstrual cycle
mode of
  training effectiveness and, 439
  variations in maximal oxygen consumption with, 216
moderate
  carbohydrate balance in, 14
muscle metabolism during
  magnetic resonance spectroscopy studies of, 123–125, *124*
non-steady-rate
  optimal recovery from, 138–139
  pulmonary ventilation during, 278–281
optimal nutrition for, 68–81
oral glucose administration before, 17–18, 78
oral glucose administration during, 15–17, *16*
osteoporosis and, 54
oxygen consumption during. *See under* Oxygen consumption
preliminary, as ergogenic aid, 511–513
progressive resistance (PRE), 462
protein balance in, *35–37*, 35–38

pulmonary ventilation during, 278–285
  regulation of, 273–277, *276*
resistance. *See* Resistance exercise
safety of, 700
steady-rate
  effects of, on blood pressure, 303
  optimal recovery from, 138
  ventilation in, 278, *279*
sudden strenuous
  warm-up and, 512–513
sustained
  carbohydrate balance in, 14
  factors determining capacity for, 128
thermoregulation during
  in cold, 571–575
  in heat, 556–571
water requirement in, 63–64
Exercise capacity
  aerobic. *See* Maximal oxygen consumption
  anaerobic
    individual differences in, *210*, 210–211
    measurement of, 207
  generality of, 199, *200*
  individual differences in, 199–232
  specificity of, 199, *200*
"Exercise high"
  endogenous opioids in, 411
Exercise plan(s)
  computerized, with computerized meal plan, 78–79, 820–827
Exercise prescription, 730–732, *731–732*
Exercise program(s), 732–733
  levels of supervision in, 733t, 733
Exercise stress testing, 721–730
  accuracy of, 729
  exercise prescription based on, 730–732, *731*
  false-negative results in, 723, 728, 729
  false-positive results in, 728, 729
  guidelines for, 729–730
  indications for, 724, 725t
  indicators of coronary heart disease on, 724–727
  protocols for, 727–728
  purposes of, 723–724
  safety of, 728–729
Exergonic reaction(s), 93–94
  example of, 96–97, *97*
Exertion
  perceived, in exercise prescription, *732*, 732
Exhalation, *237*, 238–239
Exhaustion
  in graded exercise tests, 213–214
Exocrine gland(s)
  defined, 385
Expiration, *237*, 238–239
Expiratory neuron(s), 270
Expiratory reserve volume (ERV), *240*, 241
  average values for, 242t
Extrapyramidal tract(s), 368–369

Facilitated diffusion, 401
FAD (flavin adenine dinucleotide)
  in cellular oxidation, 104
Fartlek training, 444–445
Fasciculus, 348, *349*
Fasting
  for weight loss, 679t
  protein-sparing modified, for weight loss, 679t
  therapeutic, 677–678
Fat(s), 19–28
  as hunger depressor, 27
  as vitamin carrier, 26–27
  body. *See* Body fat

catabolism of, 116–118
  caffeine's effects on, 504
  carbohydrate catabolism and, 12, 120
  total energy transfer from, *117*, 118
compound, 21–22
derived, 22–24
digestibility coefficient of, 88, 88t
heat of combustion of, 86, 88t
hydrogenated, 20
in diet, 20–21, 22t, *25*, 25
nature of, 19
net energy value of, 88t, 89
neutral, 20–21
recommended intake of, 21
  for exercise, *70*, 70–71
respiratory quotient for, 151
saturated
  cholesterol synthesis and, 23
  content of common foods, 24t
simple, 20–21
transport of, in blood, 21–22
Fat balance
  in exercise, 27, *28*
Fat cell(s), 116–117
  development of, 666–668, *667*
  in normal vs. obese individuals, 662–665, *664*
  size and number of
    age changes in, 664t, 664–665
    effects of weight gain on, 666
    effects of weight loss on, *665*, 665–666
    in development of obesity, 662–665
    modification of, 668–674
    nutrition and, 668–669, *669*
    obesity defined by, 660–662, *661*
    physical activity and, 669–671, *670–671*
    upper limit on size of, 662, 666, *667*
Fat cell hyperplasia, 660
  in development of obesity, 662–665
Fat cell hypertrophy, 660
  in development of obesity, 662–665
Fatfold measurement
  body composition assessment using, 616–620, *617*
  limitations of, 619–620
  usefulness of, 618–620
  changes in, with exercise program, 618, 618t
Fat-free mass
  lean body mass vs., 600–602
Fatigue
  glycogen depletion and, 14, 516
  mediation of, 111
  neuromuscular, 377
  rate of, on Wingate test, 207
  specificity of response to resistance training and, 467, *468*
  ventilatory, 245
Fatty acid(s), 20
  catabolism of, *117*, 118
  essential, 21
  free
    release of, from triglyceride, 116
  monounsaturated, 20
  muscle uptake of, during exercise, 27, *28*
  omega-3, 21
  polyunsaturated, 20
  saturated, *20*, 20
  unsaturated, *20*, 20
Feces
  composition of, 63
  water loss in, 63
Female(s). *See* Sex differences; Women
Ferritin, 55

Fetus
  effects of maternal exercise on, 164–165
$FEV_{1.0}$ (forced expiratory volume), 243
Fiber, 8–9
  content of common foods, 9t
  net energy value of food and, 88
  suggested intake of, 9
Fibril(s), 350, *351*
Fick equation, 326
Fick method
  for cardiac output measurement, 326–328, *327*
Filament(s), 350, *351*
Fish oil
  heart disease risk and, 21
Fitness
  assessment of
    prior to training, 445–446
    specificity in, 471
  defined, 698
  health-related
    defined, 698
  initial level of
    training effectiveness and, 434
  maintenance of
    exercise frequency and level for, 439–440
  Public Health Service objectives for, 700
Flavin adenine dinucleotide (FAD)
  in cellular oxidation, 104
Flower-spray ending(s), *378*, 379
Fluorine
  functions, sources and RDA for, 51t
Fluoxymesterone
  liver disease and, 500t
Folacin
  functions of, 46t
  RDA for, 46t
  sources of, 46t, 48
Follicle-stimulating hormone (FSH), 390t, 394–395
  training's effects on, 406t, 406–407, *407*
Food. *See also* Diet(s)
  energy release from. *See under* Energy
  energy value of, 85–91
    calculation of, 89t, 89–90
    gross, 85–87
    net, 87–90, 88t
    water and, 60
  intake of
    adipose tissue cellularity and, 668–669, *669*
    at pre-event meal, 76–78
    exercise and, 69t, 74–76, *76*, 77t, 683–684, 684t
    obesity and, 683
  nutritive value of, 761–795
  use of tables of, 90
Football player(s)
  physique of, 638–640, 639t
Football uniform(s), 554–555, *555*
Footwear
  effects of, on energy expenditure during walking, 179
Force
  defined, 756
Force platform(s), 455
Forced expiratory volume ($FEV_{1.0}$), 243
Forced vital capacity (FVC), *240*, 241
40-yard dash
  correlation of, with other power tests, 204t, 204
Four-Food-Group Plan, 73t, 73
  daily menus based on, 74t

Frank-Starling mechanism
exercise-induced increases in stroke volume and, 332
FRC. *See* Functional residual capacity
Fructose, 7
as pre-exercise supplement, 18
as substitue for sucrose, 11
structure of, 6
Fructose 1,6-diphosphate, 110
Fructose 6-phosphate
formation of, 110
Fruit sugar. *See* Fructose
FSH. *See* Follicle-stimulating hormone
Functional residual capacity (FRC), *240*, 241
average values for, 242t
FVC (forced vital capacity), *240*, 241

GABA (gamma aminobutyric acid)
in nerve impulse inhibition, 374
Galactose, 7
structure of, 6
Gamma aminobutyric acid (GABA)
in nerve impulse inhibition, 374
Gamma efferent fiber(s), *378*, 379
Gas(es). *See also* particular gases
movement of, in air and fluids, *256*, 256
partial pressure of, 254–256
in ambient air, 254–255, 255t
pressurized
problems in breathing, 587–594, *588*
solubility of, 256
volume of
diving depth and, 581t, 581, *582*
standardization of, 796–799, 797t–799t
Gas exchange, 254–259
in lungs, 257, *258*
in tissues, 257–259, *258*
Gastric emptying
in water replacement, 18, 561–562
Gastrin, 386, 387t, 405
Gender. *See* Sex (gender) differences
Generality concept, 199, *200*
GH. *See* Growth hormone
Girth measurement
body composition assessment using, 620–623, 813–819
body profile construction from, 622–623, 830–835
comparative data for groups, 644t, 644–646, 646t
Gland(s). *See also* particular glands
endocrine, 390t–392t
defined, 384–385
exocrine, defined, 385
Glottis
in Valsalva maneuver, *239*, 239
Glucagon, 391t, 404–405
in lipase activation, 116–117
training's effects on, 406t, 410
Glucocorticoid(s), 399–400
Glucolipid(s), 21
Gluconeogenesis, 7
glycerol in, 118
glycogen depletion and, 10
Glucose, 7
catabolism of
anaerobic. *See* Glycolysis
stages of, 108
total energy transfer from, 108, *115*, 115–116
metabolism of
hormonal influences on, 388
insulin in, 388, 401–403, *402*

muscle uptake of
exercise duration and intensity and, *13*, 13
oral administration of
during exercise, 15–17, *16*
prior to exercise, 17–18, 78
phosphorylation of, 108–109
structure of, *6*, 7
Glucose 6-phosphate
formation of, 108–109
phosphorylation of, 110
Glucose polymer(s)
for fluid replacement, 18, 562
Glutamine
catabolism of, 118
Glycerol, 20
catabolism of, *117*, 117–118
in gluconeogenesis, 118
Glycine
in nerve impulse inhibition, 374
Glycogen, 9–10
body storage of, 9
diet's effects on, 10, 14–15, *15*
training's effects on, 428t–429t, 429
depletion of
as measure of glycolytic capacity, 208–209, *209*
fatigue and, 14, 516
in carbohydrate loading procedure, 517
with prolonged, severe training, 71–72, *72*
formation of, 109
restoration of pre-exercise levels of
in contemporary concepts of oxygen debt, 137
in lactic acid theory of oxygen debt, 135, 136
use of, during exercise, 9–10
water content of, 26
Glycogen supercompensation. *See* Carbohydrate loading
Glycogen synthetase
in glycogen formation, 109
Glycogenolysis, 10, 108
Glycolysis, 108–112, *109*. *See also* Glycolytic energy system
aerobic, 111
glycerol catabolism via, *117*, 117–118
hydrogen release in, 110
lactic acid formation in, 110–112, 125–127, *126*
net energy transfer from, 110
substrate-level phosphorylation in, 110
Glycolytic energy system, 125–127, *126*
contribution of, to work output
in short tests, *206*, 207
with varying exercise duration, *132*, 132–133, 135t, 199–200, *200*
dependence on, in exercise in heat, 557
evaluation of, 205–209
individual differences in, *210*, 210–211
training of, 441. *See also* Resistance training
physiologic effects of, *428*, 428t–429t, 428–429
Goiter, 51t
Golgi tendon organ(s), 380–381, *381*
Gonadotropic hormone(s), 390t, 394–395
training's effects on, 406t, 406–407, *407*
Gram molecular weight
defined, 94
Growth hormone (GH), 389–393, 390t, 501–502
classification of actions of, *393*
exercise and tissue synthesis and, 393–394
in lipase activation, 116–117
training's effects on, 405, 406t

Guar gum
cholesterol and, 8

H zone, 351, *351–352*
Haldane effect, 268
HAPE (high-altitude pulmonary edema), 534
Harbor protocol, *215*, 216
Head
heat loss through, 553
Health
defined, 698
Heart, 292–295, *294*. *See also* Myocardium
"athlete's," 340–343
autonomic innervation of, 315, *315*
blood flow to, during exercise, 334t–335t, 335
blood supply of, 305, *306*
Heart disease. *See* Coronary heart disease
Heart rate
anticipatory, 317–319, 318t, *318*
energy expenditure estimation from, 169–172, *171*
factors influencing, 171–172
in exercise stress testing, 727
maximal oxygen uptake prediction based on, 223–227, *224*, 226t, *227*
maximum
age-related changes in, 705
estimation of, 436, *437*
exercise prescription based on, 434–435, 435t
for arm exercise, 434, 437–438
maximal oxygen consumption and, 434, 435t
pulse rate and, 296
regulation of, 313–319
extrinsic, 313, 315–319
intrinsic, 313–315, *314*
training's effects on, 316
"training sensitive zone" for, 436–437, *437*
training's effects on, *333*, 333, 430, 435, *436*
voluntary control over, 319
Heat
acclimatization to, 564–565, 565t, *565*
assessment of, for safe exercise, 569–570, 570t
dehydration in, 558–560
excessive
complications from, 568–569
exercise in, 556–571
circulatory adjustments in, 557
metabolic rate in, 163
thermoregulation in, *550*, 550–553
tolerance of
factors modifying, 564–568
training's effects on, 433
Heat cramp(s), 568
Heat exhaustion, 569
Heat illness, 568–569
Heat loss
mechanisms of, 550–553
Heat of combustion, 86
for carbohydrate, 86
for common foods, 88t
for fat, 86
for protein, 86–87
in body, 87
Heat production
methods of measurement of, 145–150
Heat stress index, 569–570, 570t
Heat stroke, 569
Height-weight table(s)
limitations of, 599

Helium method
  for residual lung volume measurement, 241
Hematocrit, 263
  altitude's effect on, 535–538, *537*
Hematology
  SI conversion table for common values in,
    758t–760t
Hemoglobin, 260
  altitude's effect on, 535–538, *537*
  carbon dioxide reaction with, 268
  in acid-base regulation, 287
  oxygen transport by, 260–265
  oxygen-carrying capacity of, 261–262
  saturation of
    oxygen partial pressure and, *262*, 262–263
Hemosiderin, 55
Henry's Law, 256
Heredity
  maximal aerobic power and, 216–217
High school student(s)
  body profiles of, 644t, 645–646, 646t
High-altitude pulmonary edema (HAPE), 534
High-energy bond(s), 101
High-energy phosphate(s). *See* Adenosine tri-
    phosphate; Creatine phosphate
Hill, Archibald Vivian, 135
His bundle, *314*, 314
"Hitting the wall," 14, 516
Hormone(s), 386. *See also* particular hormones
  actions of, 388
  factors determining levels of, 388–389
  polypeptide, 386
  produced by nonendocrine organs, 387t
  releasing, 393
  resting and exercise-induced secretion of,
    389–405, 390t–392t
  steroid, 386
  target cell specificity of, 386–388
  training's effects on, 405–411, 406t
Hormone receptor(s), 386
Horse(s)
  cardiac output of, 330
Humidity, 63
  evaporative heat loss and, 552
  sweat rate and, 559, *560*
Hunger
  fat and, 27
Hydrocortisone. *See* Cortisol
Hydrogen
  as percentage of body mass, 5
  oxidation of, 104, *104–105*
  release of, in glycolysis, 110
  temporary storage of, in glycolysis, 111
Hydrogen ion concentration. *See also* Acid-base
    balance
  in ventilatory control, 272–273
  normal, 285
Hydrogenation, 20
Hydrolysis
  of adenosine triphosphate, 102
  of fat, 20
Hydrostatic weighing
  body composition assessment by, 607–616,
    615t
  accuracy of, with osteoporosis, 707
  limitations of, 608–609, 610–611
  temperature correction for water density
    in, 812
Hydroxyl
  defined, 6
Hyperhydration, 561
Hyperlipidemia, 715–716
Hyperlipoproteinemia, 716

Hyperoxia
  as ergogenic aid, 513–516, *515*
Hyperpnea
  during exercise
    regulation of, 273–277
Hypertension, 301–302
  cardiac hypertrophy in, 341
  sodium-induced, 58
  training's effects on, 301–302, 302t
Hyperventilation
  breathhold time and, 273
  hazards of, in breathhold diving, 273, 583–
    584
  in response to altitude, 532
  respiratory exchange ratio in, 154
Hypervitaminosis A, 45, 49
Hypoglycemia
  symptoms of, 12
Hypophysis. *See* Pituitary gland
Hypothalamus, 389
  hormone secretion by, 405
  in thermoregulation, 548–549

I band, 351, *351–352*
Indicator dilution method
  for cardiac output measurement, 328
Individual differences
  in exercise capacity, 199–232
  in response to training, 426–427
Inferior vena cava, *293–294*, 298
Inhibition, 374
  reflex, *381*, 381
Inhibitory postsynaptic potential (IPSP), 374
Inspiration, *237*, 238
Inspiratory capacity
  average values for, 242t
  in diving
    depth and, 582–583
Inspiratory neuron(s), 270
Inspiratory reserve volume (IRV), *240*, 240
  average values for, 242t
Insulin, 391t, 400–404, *402*
  levels of, during exercise and recovery, 402–
    403, *403*
  training's effects on, 406t, 410
  zinc and, 52
Intercostal muscle(s)
  in inspiration, 238
Interneuron(s), 369
Interval training, 140t, 140–141
  aerobic, 441–444, 443t
  anaerobic, 440, 443–444
Intestine
  hormone secretion by, 405
Intrapulmonic pressure, 238
Iodine, 52
  functions, sources and RDA for, 51t
IPSP (inhibitory postsynaptic potential), 374
Iron
  absorption of
    iron source and, 56
  deficiency of, 52. *See also* Iron deficiency
    anemia
  functions of, 51t, 55–57
  loss of, due to exercise, 56–57
  Recommended Dietary Allowances for, 51t,
    52t
  sources of, 51t, 52
  supplemental, 57
    hematologic adjustment to altitude and,
      536–538, *537*
Iron deficiency anemia, 51t, 55–57
  exercise capacity in, 261t, 261–262
  in women, 55–56

oxygen-carrying capacity of hemoglobin in,
    261
IRV. *See* Inspiratory reserve volume
Islet(s) of Langerhans, 400
Isokinetic dynamometry, 455
Isokinetic resistance training, 467–470, *469*
Isometric exercise
  contraindications for, 240
Isometric muscular contraction(s), *460*, 461
  electromyography during, 471
Isometric strength training, 463–464
Isovolumetric contraction period, 294

Jogging, 181–182. *See also* Running
  effects of, on body composition, 647t–648t,
    647–649
Joule, 85
Journal(s)
  frequently cited, 828–829
Jumping power test(s), 201–204, *203*

Katch test, 207
kcal (kilocalorie), 85
Keto acid(s)
  short-chain
    metabolism of, 151
Ketogenic diet(s), 676–677, 679t
α-Ketoglutaric acid
  formation of, from glutamine, 118
Ketone body(-ies), 12
Ketosis, 12, 120
  cortisol secretion and, 399
Kidney(s)
  blood flow to, 334t
    during exercise, 319, 335t, 335
  hormone secretion by, 392t, 399
  in acid-base regulation, 288
Kilocalorie (kilogram calorie; kcal), 85
Kilojoule (kJ), 85
Krebs cycle, 112–114, *113–114*, *119*, 119–120

Lactate. *See* Lactic acid
Lactate threshold. *See* Onset of blood
    lactate accumulation
Lactate-producing capacity
  anaerobic training's effects on, 127
Lactic acid
  accumulation of. *See also* Onset of blood lac-
    tate accumulation
    exercise intensity and, 125–127, *126*
    in exercise in heat, 557
    oxygen deficit and, *129*, 130
  blood levels of, as measure of glycolytic ca-
    pacity, 207–208, *208*
  buffering of, 286
  formation of, in glycolysis, 110–112, 125–
    127, *126*
  metabolism of, during recovery, 136
    oxygen debt and, 135–136
  procedures for facilitating, 138–139, *139*
Lactic acid energy system. *See* Glycolytic en-
    ergy system
Lactic acid theory of oxygen debt, 135
  problems with, 136
Lactic dehydrogenase
  in lactic acid formation, 111
Lactose, 7
Lactovegetarian diet(s). *See* Vegetarian diet(s)
Lean body mass (LBM), 603
  calculation of, 611
  fat-free mass vs., 600–602
Leanness. *See also* Body fat
  minimal standards for, 603–605

Leucine
structure of, *29*
Leucine enkephalin, 411
LH. *See* Luteinizing hormone
Linen
for warm-weather clothing, 554
Linoleic acid, 21
Lipase
hormones in activation of, 116–117
in triglyceride hydrolysis, 116
Lipid(s). *See* Fat(s)
Lipid abnormality(-ies)
coronary heart disease and, 715t, 715–717
Lipolysis, 116
Lipoprotein(s), 21–22, 716t, 716–717
plasma
steroid use and, 501
β-Lipotrophin, 411
Liquid meal(s)
pregame, 78
Liquid protein diet(s), 677
Liver
hormone secretion by, 405
Liver disease
steroid use and, 500t
Longevity
defined, 698
exercise and, 709–712, *710–712*
Low back pain syndrome
resistance training for, 463
Lung(s)
anatomy of, 235–236
function of. *See* Pulmonary function
gas exchange in, 257, *258*
oxygen partial pressure in, 263
surface area of, 235–236, *236*
Lung disease
energy cost of breathing in, 282
physiologic dead space in, 248
pulmonary function in, 242–243
Lung squeeze
in breathhold diving, 584–585
Lung volume(s), 240–243
average values for, 242t, *242*
dynamic, 242–243
static, *240*, 240–242
Luteinizing hormone (LH), 390t, 394–395
training's effects on, 406t, 406–407, *407*

M line, 351, *352, 353*, 353
M protein, 350
Macronutrient(s), 5. *See also* Carbohydrate(s);
Fat(s); Protein(s)
Magnesium
functions of, 51t, 55
replacement of, after intense exercise, 59
sources and RDA for, 51t
Magnetic resonance imaging (MRI)
body fat assessment using, 624, *626*
Magnetic resonance spectroscopy
exercise muscle metabolism studies with,
123–125, *124*
Male(s). *See* Sex differences
Maltodextrin(s). *See also* Glucose
polymer(s)
in fluid replacement, 562
Maltose, 7
Man. *See* Sex differences
Mandibular orthopedic repositioning appliance
(MORA), 507–508
Mask squeeze
in scuba diving, *588, 589*

Mass
body. *See* Body mass
defined, 756
Mass action effect, 98
Master "2-step" bench stepping test, 723
Max V̇O₂. *See* Maximal oxygen consumption
Maximal aerobic power. *See* Maximal
oxygen consumption
Maximal oxygen consumption (max V̇O₂), *130*,
130–131
age-related changes in, 220–221, *221*, 704–
705
altitude's effects on, 539
arteriovenous oxygen difference and, 338
criteria for, 211–213, *213*
factors affecting, 216–221
maximum cardiac output and, *336*, 336–337
maximum heart rate and, 434, 435t
measurement of, 211–216
tests for, 213–216
of athletes vs. sedentary subjects, 211, *212*
on return to sea level after altitude training,
540–541
onset of blood lactate accumulation and,
280–281, *281*
prediction of
heart rate in, 223–227, *224*, 226t, *227*
tests for, 221–227
pulmonary function and, 244
stroke volume and, 331t, *331*
training for improvement of. *See* Aerobic
training
ventilation as limiting factor on, 281
Maximum voluntary ventilation (MVV), 243
Mayo diet, 679t
Meal(s). *See also* Diet(s); Food
based on Four-Food-Group Plan, 74t
computerized plan for, 78–79, 820–827
pre-event, 76–78
Mean arterial pressure, 296
Measurement
terminology of, 756–757
units of, 743–760
Mechanoreceptor(s)
in ventilatory control during exercise, 275
Medulla
in ventilatory control, 270, *271*
Megavitamin(s), 49
Men. *See* Sex differences
Menarche
delayed. *See* Menstrual cycle
Menstrual cycle
changes in body mass with
accuracy of body composition assessment
and, 615
endorphins in regulation of, 411
exercise-induced alterations in, 406–407
body fat and, 605–607
bone loss and, 54
cancer risk and, 607
endorphins and, 411
prolactin and, 394
heat tolerance and, 568
Mesomorphy, 634
MET(s), 166–167
classification of physical activity by, 166t
Metabolic mill, *119*, 119–120
Metabolic rate
basal (BMR), 158–162
exercise's effect on, 547
resting. *See* Resting metabolic rate
shivering's effect on, 547
Metarteriole(s), 296

Methandrostenolone
liver disease and, 500t
Methionine enkephalin, 411
Methyltestosterone
liver disease and, 500t
Metric system, 743–744
Micronutrient(s), 44. *See also* Mineral(s); Vita-
min(s); particular nutrients
Mineral(s), 50–60. *See also* particular minerals
exercise performance and, 58–59
functions of, 51t, 52–58
in catabolism and anabolism, 52, *52*
major, 50
nature of, 50
sources and RDAs for, 51t
supplemental
value of, 50 52
trace, 50
Mineralocorticoid(s), 398–399
Minimal body mass, 603–605, 605t
Minute ventilation (V̇ₑ), 246–247
alveolar ventilation and, 247t, 247–248
Mitochondrion(-a), 104
training-induced changes in, 429
Mitral valve, *294*, 294
Mole
defined, 94, 757
Molecule(s), 5
chemical bonding of, 5
Molybdenum
functions, sources and RDA for, 51t
Monosaccharide(s), 6–7
MORA (mandibular orthopedic repositioning
appliance), 507–508
Motivation
anaerobic capacity and, 211
performance on aerobic power tests and,
213–214
Motoneuron(s), *370*
alpha, 368
anterior, *371*, 371–372
gamma efferent, 372
Motor endplate, *371*, 372–374, *373*
Motor unit(s), *370*, 374–377
characteristics of, 375–376, *376*
training's effects on, 375
firing pattern of, *376*, 376–377
types of, 375–376, *376*
Mountain sickness
acute, 534
MRI. *See* Magnetic resonance imaging
Muscle(s), 348–366
"all-or-none" principle of action of, 375
atrophy of, during inactivity, 34
blood flow to, 334t
during exercise, 334, 335t, 349–350
capillary density in
training's effects on, 339, 350
cellular adaptations in
at high altitude, 538
due to training, *432*, 432–433
chemical composition of, 349
contraction of. *See* Muscular contraction(s)
gross structure of, 348–350, *349*
insertion of, 348
metabolism in, during exercise
aerobic training's effects on, 339
magnetic resonance spectroscopy studies
of, 123–125, *124*
nerve supply to, 370–377
origin of, 348
relaxation of
chemical and mechanical events in, 358

strength of. *See* Muscular strength
ultrastructure of, 350–354, *351–354*
Muscle biopsy
for evaluation of intramuscular nutrients
during exercise, 12
Muscle fiber(s), 348–349, *349*
age-related changes in, 701
fast-twitch (FG), 131, 359
glycogen depletion in, 209
myosin ATPase in, 357
subdivisions of, 360
in muscle spindle, *378*, 379
intrafusal, 379
longitudinal splitting of, with resistance
training, 478
racial influence on composition of, 480
slow-twitch (SO), 131, 359–360
glycogen depletion in, 209
training-induced changes in, 430, 480, *481*
type I. *See* Muscle fiber(s), slow-twitch
type II. *See* Muscle fiber(s), fast-twitch
types of, 131–132, 359–363
differences in, between athletic groups,
360–362, *361*
force-velocity relationship and, 469–470,
470
heredity's influence on distribution of, 217
training's effects on distribution of, 362–
363, 363t
Muscle mass
age-related changes in, 701
factors affecting, *475*, 475
Muscle soreness, 486–490
connective tissue damage in, 489–490
delayed onset (DOMS), *486*, 486
cellular damage in, 486–488, 487t
experimental studies of, 488–490
Muscle spindle(s), 378–380
structural organization of, *378*, 378–379
Muscular contraction(s)
ballistic
electromyography during, 471–473, *472*
bidirectional
neuromuscular facilitation in, *473*, 473–
474
chemical and mechanical events in, 355–359
concentric, *460*, 460–461
dynamic, 460
eccentric, *460*, 461
muscle soreness due to, 486–488, 487t
electromyography during, 471–474, *472–473*
gradation of force of, 375
isometric, *460*, 461
electromyography during, 471
sequence of events in, 358–359
types of, *460*, 460–461
muscle soreness and, 486, 487t, 487–488
Muscular hyperplasia, 478–480
Muscular hypertrophy
with resistance training, 477–478, *479*
sex differences in, 482–483
Muscular strength
absolute
sex differences in, 457–458
aging and, 701–702
body size and composition and
sex differences in, 458–459, *459*
bone density and, *483*, 483–484
factors modifying expression of, *476*, 476–
481
measurement of, 452–456
considerations in, 456
muscle cross section and, 456–457, *457*
muscular factors affecting, 477t, 477–480

neural factors affecting, 476–477
neuromuscular factors affecting, 480–481,
481
psychologic factors affecting, 476–477
sex differences in, 456–459, *457*, *459*
training for, 452–496. *See also* Resistance
training
methods for, 459–474
MVV (maximum voluntary ventilation), 243
Myelin sheath, *371*, *372*
Myocardial blood flow
regulation of, during exercise, 307
Myocardial infarction, 307, 713
Myocardial ischemia, 713, 725
Myocardial perfusion imaging
exercise testing with, 728
Myocardium, 292. *See also* Heart
metabolism in, *308*, 309
oxygen utilization by, 306–307
rate-pressure product as estimate of work of,
307–309
Myofibril(s), 350, *351*
Myofibrillar adenosinetriphosphatase,
357
Myofilament(s), 350, *351*
Myoglobin, 55, 265
dissociation curve for, *262*, 265
muscle content of, 349
training's effects on, 429
Myosin, 350
in muscle contraction, 355–357, *356*, *358*
orientation of, in sarcomere, *352*, *352–353*
Myosin adenosinetriphosphatase, 357

NAD⁺. *See* Nicotinamide adeninedinucleotide
NADH (reduced nicotinamide adeninedinucleotide)
adenosine triphosphate generation from, 110
Narcosis
nitrogen, 590–591
Naughton protocol, *215*
Nerve fiber(s)
servicing muscle spindle, *378*, 379
speed of conduction in
age-related decline in, 701
myelin sheath thickness and, 372
type A α, 372
Nervous system. *See also* Parasympathetic
nervous system; Sympathetic nervous
system
age-related changes in function of, 701, 702–
704, *704*
central. *See also* Brain
carbohydrate as fuel for, 12
in control of movement, 367–383
multiple feedback control in, 369
organization of, 367–370, *368*
Neurohypophysis. *See* Pituitary gland, posterior
Neurolemma, *371*, 372
Neuromuscular junction, *371*, 372–374, *373*
Niacin. *See* Nicotinic acid
Nicotinamide adeninedinucleotide
(NAD⁺), 98
in cellular oxidation, 104
reduced (NADH)
adenosine triphosphate generation from,
110
Nicotinic acid (niacin)
functions, sources and RDA for, 46t
harmful effects of megadoses of, 49
Night blindness, 44, 47t
Nitrogen
as percentage of body mass, 5

partial pressure of, in ambient air, 255
urinary
measurement of, in respiratory quotient
calculation, 151–152
Nitrogen balance, 34
Nitrogen narcosis, 590–591
Nitrogen partial pressure
in ambient air, 255
NMR. *See* Magnetic resonance imaging; Magnetic resonance spectroscopy
Node(s) of Ranvier, *371*, *372*
Nonaerobic
defined, 102n
Norepinephrine, 391t, 397–398
chemical structure of, *397*
in blood flow regulation, 322
in heart rate regulation, 316
in lipase activation, 116–117
in response to cold stress, 549
training's effects on, 406t, 408–410, *409*
Norethandrolone
liver disease and, 500t
Nuclear bag fiber(s), *378*, 379
Nuclear chain fiber(s), *378*, 379
Nuclear magnetic resonance. *See* Magnetic resonance imaging; Magnetic resonance
spectroscopy
Nucleoprotein(s), 33
Nutrient(s). *See also* particular nutrients
basic structure of, 5–6
interconversions among, *119*, 119–120
recommended intake of, for exercise, 69–71,
70
Nutrition, 3–81. *See also* Diet(s); Food

Obesity, 656–674
anthropometric markers predictive of, 646
coronary heart disease and, 658–659, 717
criteria of, 659–662
development of, 656–658
food intake levels and, 683
health risks of, 658–659
heat tolerance and, 568
predisposing factors for, 657
types of, 660, 662, *663*
OBLA. *See* Onset of blood lactate accumulation
Occupation(s)
energy expenditure in, 168t, 168
Occupational testing
specificity in, 471
Oil(s), 20
Oligosaccharide(s), 7
Olympic competitor(s)
quantification of physique in, 634–636, 635t
One-repetition maximum (1-RM), 454
Onset of blood lactate accumulation (OBLA),
126, 279–280
assessment of, 280
endurance performance and, 280–281
specificity of, to exercise task, 280
Opioid peptide(s), 411–412
Organic acid, *29*, 29
Osteomalacia, 47t
Osteoporosis, 51t, 53–54, 707–708
muscular strength and, *483*, 483–484
risk factors for, 54
Ovary(-ies)
hormone secretion by, 391t, 400, 405
Overload principle, 424–425
applied to strength training, 459–460
Overweight, 599. *See also* Obesity
Ovolactovegetarian diet(s). *See* Vegetarian
diet(s)

Oxaloacetic acid
  formation of, from aspartate, 118
  in fatty acid metabolism, 120
  in Krebs cycle, 113
β-Oxidation, *117*, 118
Oxidative metabolism, 106–107. *See also* Aerobic energy system
Oxidative phosphorylation, 105–106
Oxygen
  as percentage of body mass, 5
  fat catabolism and, 118
  partial pressure of. *See* Oxygen partial pressure
  pure
    scuba diving with, 586–587, 593t, *593*
  role of, in energy metabolism, 106–107
  solubility of, 256
  transport of, 260–266
    by hemoglobin, 260–265
      cardiac output and, 336–337
    in solution, 260
  uptake of. *See* Oxygen consumption
Oxygen consumption
  as measure of energy metabolism, 145–146
  at rest, for various tissues, 162t, *162*
  cardiac output and, 336–337, *337*
  during exercise, 127–130, *128*
    carbon dioxide partial pressure and, *274*, 274
    in cold, 549, 572
    oxygen partial pressure and, 273–274, *274*
    with arms vs. legs, 339–340, *340*
  during recovery. *See* Oxygen debt
  maximal. *See* Maximal oxygen consumption
  measurement of
    by closed-circuit spirometry, 146–147, *147*
    by open-circuit spirometry, 147–150, *148–149*, 799–801
  myocardial, 306–307
    rate-pressure product as estimate of, 307–309
  peak (peak & V̇O_2), 211
  steady rate, 127–128
Oxygen cost of breathing, *282*, 282–283
  smoking and, 282–283, 283t
Oxygen debt, 133–141, *134*
  alactacid, 136
  implications of, for exercise and recovery, 137–139
  lactacid, 136
  metabolic dynamics of, 135–137
Oxygen deficit, *128–129*, 128–130
Oxygen dilution method
  for residual lung volume measurement, 241–242
Oxygen extraction. *See* Arteriovenous oxygen difference
Oxygen inhalation
  as ergogenic aid, 513–516, *515*
Oxygen partial pressure (PO_2)
  altitude and, 529, *530*
  alveolar, 255
  arterial
    alveolar vs., 257n
    hemoglobin saturation and, *262*, 262–263
    in ambient air, 254
    in lungs, 263
    in ventilatory control, 271–272
    oxygen consumption and, during exercise, 273–274, *274*
  tissue, 257, 263–264
  tracheal, 255
Oxygen poisoning
  in diving, 593

Oxygen saturation
  altitude and, 529–531, *530*
Oxygen uptake. *See* Oxygen consumption
Oxygen-carrying capacity, 261–262
  increase in, at high altitude, 535–538, *537*
  normal, 336
Oxyhemoglobin dissociation curve, *262*, 262–263
  changes in, at high altitude, 538
Oxymetholone
  liver disease and, 500t
Oxytocin, 390t, 395

P wave, *315*, 315
Pacemaker, 313, *314*
Pacinian corpuscle(s), 381–382
Palmitic acid
  respiratory quotient for, 151
  structure of, 87
Pancreas, 400, *401*
  exocrine function of, 400
  hormone secretion by, 391t, 400–405, *401*
    training's effects on, 406t, 410–411
Pangamic acid, 505
Pantothenic acid
  functions, sources and RDA for, 46t
Parasympathetic nervous system
  distribution of, in heart, *315*
  influence of, on heart rate, 316
Parathormone, 391t, 405
Parathyroid gland
  hormone secretion by, 391t, 405
Partial pressure, 254–256. *See also* particular gases
  of gases in ambient air, 254–255, 255t
PCO_2. *See* Carbon dioxide partial pressure
Peak oxygen consumption (peak V̇O_2), 211
Pectin
  cholesterol and, 8
Pellagra, 46t
Pennington diet, 679t
Pep pill(s), 502–503, 503t
Perceived exertion
  rating of, in exercise prescription, *732*, 732
Performance
  altitude's effects on, 540
  endurance
    factors influencing, 281
    onset of blood lactate accumulation and, 280–281
  environmental stress and, 527–595
  lung function and, 243–245, 244t
  minerals and, 58–59
  special aids to, 497–526
  training's effects on, *433*, 433
  vitamins and, 48
Perimysium, 348, *349*
Periosteum, 348
Pernicious anemia, 46t
Perspiration. *See also* Sweat
  insensible, 63
PFK (phosphofructokinase)
  in glycolysis, 110
pH. *See* Acid-base balance; Hydrogen ion concentration
Phlebitis, 300
Phosphate(s)
  high-energy. *See* Adenosine triphosphate; Creatine phosphate
Phosphate bond energy, 101–107
Phosphate buffer system, 287
Phosphofructokinase (PFK)
  in glycolysis, 110

3-Phosphoglyceraldehyde
  oxidation of, in glycolysis, 111
Phospholipid(s), 21
Phosphoric acid
  in acid-base regulation, 287
Phosphorus
  functions of, 51t, 54
  sources and RDA for, 51t
Phosphorylase
  in glycogenolysis, 108
Phosphorylation, 103
  oxidative, 105–106
  substrate-level, 110
Photosynthesis
  energy conversion in, *96*, 96
Phylloquinone, 47t
Physical activity(-ies). *See also* Exercise
  adipose tissue cellularity and, 669–671, *670–671*
  aging and, 698–739
  assessment of participation in, *699*, 699–700
  classification of
    by energy expenditure, 165–167, 166t
    by predominant energy pathways, 423, *424*
  defined, 698
  energy expenditure in, 162, 165–172, 804–811
    weight loss and, 684–685
  epidemiology of, 698–700, *699*
  heat generation by, in response to cold, 549
  weight gain and, 657–658
Physical fitness. *See* Fitness
Physical working capacity
  maximum, in graded stress testing, 723
Physique, 634–646. *See also* Body composition
  comparative data on, 644t, 644–646
  of football players, 638–640, 639t
  of high school wrestlers, 640–642
  of long-distance runners, 636–638, 637t–638t
  of Olympic competitors, 634–636, 635t
  of triathletes, 638
  of weight lifters and body builders, 642–644, 643t
  racial differences in, among Olympic competitors, 636
  sex differences in
    in body builders, 643t, 643–644
    in Olympic competitors, 635t, 636
Pituitary gland, 389, *392*
  anterior
    secretions of, 389–395, 390t
    training's effect on secretions of, 405–407, 406t
  posterior
    secretions of, 390t, 395
    training's effect on secretions of, 406t, 407–408
Plaque
  in atherosclerosis, 23
Plyometric training, 470–471
Pneumothorax
  in scuba diving, *588*, 588–589
PO_2. *See* Oxygen partial pressure
Poiseuille's law, 320
Polycythemia
  in response to altitude, 536
Polypropylene
  for cold-weather clothing, 553
Polysaccharide(s), 7–10
  animal, 9–10
  plant, 7–9
POMC (proopiomelanocortin), 389

Ponderal equivalent(s)
body profile comparisons using, 645–646, 646t
Ponderal somatogram. *See* Body profile
Pons
in ventilatory control, 271
Postsynaptic membrane, 372, *373*
Potassium
functions of, 57–58
replacement of, 59, 564
Power
defined, 201, 756–757
Power output
formula for, 201
Power test(s), 201–204
interrelationships among, 204t, 204–205
PRE (progressive resistance exercise), 462
Precapillary sphincter, 297
Pregame meal, 76–78
Pregnancy
energy expenditure during, 164–165
Premature ventricular contraction(s) (PVC)
exercise-induced, 726
Pressure-volume relationship(s)
diving depth and, 580–581, 581t, *582*
Presynaptic terminal(s), 372
Pritikin diet, 679t
PRL. *See* Prolactin
Progesterone, 391t, 400, 405
Progressive resistance exercise (PRE), 462
Progressive resistance weight training, 461–463
Prolactin (PRL), 390t, 394
training's effects on, 406t, 406
Proopiomelanocortin (POMC), 389
Proprioceptor(s), 378–382
Prostaglandin(s), 386, 387t
Protein(s), 29–38
buffering function of, 33, 287
calorigenic effect of, 162
carbohydrate vs., for pregame meal, 77–78
carbohydrate's role in preserving stores of, 11–12
during exercise, *35*, 35–36
complete, 30
digestibility coefficient of, 88t, 88–89
energy release from, 34, 118
function of, 33
globular, 33
harmful effects of excessive intake of, 31
heat of combustion for, 86–87, 88t
in body, 87
in diet, 31, *31*
obligatory water loss in urine and, 62
incomplete, 30
kinds of, 29–31
nature of, 29
net energy value of, 88t, 89
predigested
supplementation with, 32–33
recommended intake of, 31–33, 32t
adequacy of RDA, 35–38
for exercise, 69–70, *70*
respiratory quotient for, 151
sources of, 30t, 30
structural, 33
supplemental, 32–33
Protein balance
in exercise and training, *35–37*, 35–38
Provitamin(s), 45
"Psyching," 374, 477
Public Health Service
national objectives of, for physical fitness, 700
Pulmonary dead space. *See* Dead space

Pulmonary disease. *See* Lung disease
Pulmonary edema
high-altitude (HAPE), 534
Pulmonary function
age-related changes in, 701, 704
assessment of, 240–243
exercise performance and, 243–245, 244t
training's effects on, 431–432
Pulmonary ventilation, 270–291
as limit on aerobic power, 281
at onset and end of exercise, 274–275, *275*
defined, 235
during exercise, 278–285
non-steady-rate, 278–281
steady-rate, 278, *279*
energy demands and, 278–281
in acid-base regulation, 287
regulation of, 270–277, *271*
during exercise, 273–277, *276*
humoral factors in, 271–273
neural factors in, 270–271
training's effects on, 283–284, *284*
Pulse rate
heart rate and, 296
Purkinje system, *314*, 314
PVC (premature ventricular contraction)
exercise-induced, 726
Pyramidal tract, 367–368
Pyridoxine. *See* Vitamin B$_6$
Pyruvic acid
breakdown of, in Krebs cycle, 112–114, *113–114*
formation of, 110
from alanine, 118

QNF (quadruple neuromuscular facilitation), 474
QRS complex, *315*, 315
Quadruple neuromuscular facilitation (QNF), 474
Queens College Step Test, 225–227

R (respiratory exchange ratio), 154
Racial differences
in adipose tissue metabolism, 656
in body density, 610
in muscle fiber composition, 480
in physique, among Olympic competitors, 636
Radiation
heat loss by, 550–551
Radionuclide angiocardiography
exercise testing with, 728
Ramp test, 216
Rapture of the deep, 590–591
Rate-pressure product (RPP)
as estimate of myocardial work, 307–309
Rating of perceived exertion (RPE)
in exercise prescription, *732*, 732
Recommended Dietary Allowance(s) (RDA).
See also under particular nutrients
adequacy of, 31–32
Recovery
active ("cooling down"), 137–138, 138–139, *139*, 446
prevention of venous pooling by, 300
blood pressure during, 305
metabolism during
contribution of, to energy expenditure, 685
optimal
from non-steady-rate exercise, 138–139
from steady-rate exercise, 138
oxygen consumption during. *See* Oxygen debt

passive, 138
procedures for speeding, 137–139
Red blood cell reinfusion. *See* Blood doping
Reference man
body composition of, 600–602, *601*
density of, 602
lean body mass of, 603
Reference woman
body composition of, 600–603, *601–602*
density of, 602–603
minimal body mass of, 603
Reflex arc, 369–370, *370*
Reflex inhibition, *381*, 381
Refractory period
of heart, 315
Rehydration. *See* Water replacement
Renin, 392t, 399
Residual lung volume (RV), *240*, 241–242
average values for, 242t
body density and, 611
in body volume measurement, 612, 613, 615
Resistance
to blood flow, 319–320
Resistance exercise. *See also* Resistance training
accommodating, 467
effects of
on blood pressure, 302–303
on varicose veins, 300
progressive (PRE), 462
Resistance training, 452–496
adaptations with, 475–491, 477t
cardiovascular, 343
muscular, *476*, 477–480
neural, *476*, 477
sex differences in, 481–484
aerobic training's interaction with, 461, 478
aging and, 707, *708*
caloric cost for various methods of, 485t, 485–486
categories of equipment for, 455
circuit, 484–485
dynamic vs. static methods for, 464–467
effects of, on body composition, 649, 650t
during weight loss, *651*, 651
for children, 461
for weight gain in athletes, 652
isokinetic, 467–470, *469*
metabolic stress of, 484–486
specificity of, 464–465
exceptions to, 465–467, *466*
variable resistance machine for, 468–469
Respiration. *See also* Breathing; Ventilation
energy conversion in, 96–97, *97*
water loss through, 63
Respiratory center, 270–271, *271*
Respiratory chain, 104, *105*
cytoplasmic NADH oxidation by adenosine triphosphate formation in, 110
Respiratory disease. *See* Lung disease
Respiratory exchange ratio (R), 154
Respiratory function. *See* Pulmonary function
Respiratory quotient (RQ), 150–153
calculation of, 801
for carbohydrate, 150–151
for fat, 151
for mixed diet, 153
for protein, 151
nonprotein, 151–152
thermal equivalent of oxygen for, 152, 153t
Respiratory tract
anatomy of, 235–237, *236*
during exercise in cold, 574

Rest
    energy expenditure at, 158–165
        estimation of, 161–162
    oxygen consumption of various tissues at, 162t, 162
Resting metabolic rate, 158
    age and, *160*
    body size and, 159–161
    lowering of, during weight loss, 680–682, *681*
    sex and, *160*, 160
Reticular formation, 369
Reticulospinal tract, 369
Retinol. *See* Vitamin A
Reversibility principle, 427t, 427
Riboflavin. *See* Vitamin B$_2$
Rickets, 47t, 51t
Rigor mortis, 358
RISKO, 720–721, 722t
1-RM (one-repetition maximum), 454
RPE (rating of perceived exertion)
    in exercise prescription, 732, *732*
RPP (rate-pressure product)
    as estimate of myocardial work, 307–309
RQ. *See* Respiratory quotient
Rubrospinal tract, 369
Runner(s)
    long-distance
        physique of, 636–638, 637t–638t
Running. *See also* Jogging
    economy of movement in
        age and, *186*, 187
        training and, 187–188
    effects of, on body composition, 649
    efficiency of
        walking efficiency vs., 180, *180–181*
    endurance
        maximal oxygen consumption prediction using, 222–223
    energy expenditure in, 181–190
        air resistance and, *187*, 188, 189
        energy cost of swimming vs., 190
        on treadmill vs. track, 188–189, 189t
        speed and, 182–183, 183t
    impact force on leg in, 179
    jogging vs., 181–182
    marathon
        energy expenditure in, 189–190
    speed of
        energy expenditure and, 182–183, 183t
        stride pattern and, 183, *184*
    stride frequency in
        optimal, 186, *186*
    stride length in
        optimal, 184–186, *185*
        speed and, 183, *184*
    with hand or ankle weights, 179
RV. *See* Residual lung volume

S-A node (sinoatrial node), 313, *314*
Salt supplement(s)
    during hot-weather exercise, 59, 564
Sarcolemma, 348, *349*
Sarcomere, 351, *351–352*
    contractile protein orientation within, *352–353*, 352–353
Sarcoplasm, 348, *349*
Sarcoplasmic reticulum, 349, 353–354, *354*
Sargent jump-and-reach test, 201–202
    correlation of, with other power tests, 204t
Scalene muscle(s)
    in inspiration, 238
Schwann cell, 372
Scuba diving, 585–594
    air escape during ascent in, 581, 587, 589

closed-circuit, 586–587
    maximum recommended depth range for, 591
    open-circuit, *585–586*, 585–586
    precautions in filling tanks for, 594
    special problems in, 587–594, *588*
Scurvy, 47t
Secretin, 387t, 405
Selenium
    functions, sources and RDA for, 51t
Semilunar valve(s), *294*, 294
Sensitivity
    defined, 729
Setpoint theory, 678–682
Sex (gender) differences
    in aerobic capacity, 217–220, 218t, *220*
    in body composition, 600–603, *601*
        energy expenditure in swimming and, 194–195
    in cardiac output, 337
    in muscular strength, 456–459, *457, 459*
    in physique
        in body builders, 643t, 643–644
        in Olympic competitors, 635t, 636
    in response to resistance training, 481–484
    in thermoregulation, 567–568
Shivering
    during exercise in cold, 549
Shoe(s)
    effects of, on energy expenditure during walking, 179
Shunting
    of blood during exercise, 334–335
SI units, 745, 757–760, 757t–760t
Sinoatrial node (S-A node), 313, *314*
Sinus(es)
    problems with, in diving, 585, *588*, 590
Siri equation, 609–610
Skeletal muscle. *See* Muscle(s)
Skin
    water loss through, 63
    in thermoregulation, 551–552
Skin friction drag, 191
Skin wetting
    effectiveness of, for reducing heat stress, 561
Sliding-filament theory, *355–356*, 355–357
Small intestine
    hormone secretion by, 405
Smoking. *See* Cigarette smoking
Snorkeling, 581–583
Sodium
    aldosterone in regulation of, 399
    functions of, 51t, 57–58
    hypertension due to, 58
    in typical American diet, 58
    recommended intake of, 51t, 58
    replacement of, in hot-weather exercise, 59, 563–564
    restriction of intake of, for hypertension, 58
    sources of, 51t, 58
Sodium bicarbonate
    in acid-base regulation, 285–286
    pre-event ingestion of performance and, 210, 506t, 506–507
Sodium phosphate
    in acid-base regulation, 287
Solubility
    of gases, 256
Somatoliberin, 405
Somatomedin, 405
Somatostatin
    exercise's effects on, 393
Somatotrophic hormone (somatotropin). *See* Growth hormone

Somatotyping, 634
Spasm hypothesis of muscle soreness, 488, 489
Spatial summation, 374
Specific dynamic action effect. *See* Thermic effect of food
Specific gravity, 608
Specificity principle, 199, *200*, 425–426
Speed
    defined, 756
Sphygmomanometer, *297*
Spirometer
    portable, 147–148, *148*
Spirometry
    closed-circuit, 146–147, *147*
    open-circuit, 147–150, *148–149*
        metabolic computations in, 796–803
Sport diving. *See* Diving
Sports anemia, 56–57, *57*
Sports drink(s). *See also* Sugar drink(s)
    for mineral replacement, 59
Spot reduction, 671–672
Stair-sprinting power test(s), 201, *202*
    correlation of, with other power tests, 204t, 204
Stanazolol
    liver disease and, 500t
Starch, 7–8
Starling's law of the heart, 332
Starvation diet(s), 677–678
Static stretching
    for muscle soreness, 488, 489
Stearin, 19
Step test
    maximal oxygen uptake prediction based on heart rate after, 225–227, *227*
    percentile norms for, 226t
Steroid(s), 498–501, 500t
    endogenous, 386. *See also* particular hormones
"Sticking point," *468*, 468
Stiffness. *See* Muscle soreness
Strength. *See* Muscular strength
Stress testing. *See* Exercise stress testing
Stretch reflex, 379–380, *380*
Stretching
    static, for muscle soreness, 488, 489
Stroke volume, 326
    at rest
        training's effects on, 329, 430–431
    during exercise
        altitude's effects on, 539–540
        training's effects on, *330*, 330–331, 331t, 430–431
    mechanisms regulating, 331–333
Substance
    defined, 5
Substrate
    defined, 98
Substrate utilization
    by heart, *308*, 309
Sucrose, 7
Sugar(s). *See also* Fructose; Glucose
    simple, 7
        effects of excessive intake of, 11
        strategies for reducing intake of, 11
Sugar drink(s)
    use of
        during exercise, 15–17, *16*
        prior to exercise, 117–118
    water uptake and, 18, 562
Summation
    spatial, 374
    temporal, 374
Superior vena cava, *293–294*, 298

Surface area
  estimation of, from stature and body mass,
    161, *161*
Surface area law, 159
Sweat
  age and, 567
  cooling effect of, 551–552, 553
  in heat acclimatization, 564, *565*
  iron loss in, 56
  nitrogen excretion in, during exercise, *35*, 35
  rate of
    humidity and, 559, *560*
    salt content of, 59
    sex and, 567
    water loss in, 63–64, 559
Swim-mill, 190–191, *191*
Swimming
  drag in, 191–192, *192*
  endurance
    energy expenditure in, 195
  energy expenditure in, 190–195
    buoyance and, 194–195
    energy cost of running vs., 190
    methods of measurement of, 190–191, *191*
    skill and stroke and, 192, *193*
    water temperature and, 192, *194*
  hyperventilation to extend breathhold time
    in, 273
  ventilatory equivalent during, 278
Swimming flume, 190–191, *191*
Sympathetic nervous system
  distribution of, in heart, *315*
  in blood flow regulation, 320–322, *321*
  influence of, on heart rate, 316
Sympathoadrenal response to training, *409*,
  409–410
Synaptic cleft, 372, *373*
Synaptic gutter, 372, *373*
Système International d'Unités, 745, 757–760,
  757t–760t
Systole, 296
  increased ejection during
    stroke volume and, 333

T wave, *315*, 315
T_3. *See* Triiodothyronine
T_4. *See* Thyroxine
Tachycardia
  defined, 316
"Tapering-off." *See* Recovery, active
Tarahumara Indians
  diet of, 71
Tear theory of muscle soreness, 488
Temperature
  body. *See* Body temperature
  environmental. *See* Cold; Heat
  water
    energy expenditure in swimming and, 192,
      194
    water density and, 812
Temporal summation, 374
Temporomandibular joint repositioning, 507–
  508
Tendon(s), 348, *349*
Tensiometry
  cable, 452–453, *453*
Terrain
  effects of, on energy expenditure during
    walking, 178, 179t
Testis(-es)
  hormone secretion by, 392t, 400, 405
Testosterone, 392t, 405
  in females, 482
  training's effects on, 406t, 406–407, *407*

Thallium-201 imaging
  exercise testing with, 728
Therapeutic fasting, 677–678
Thermal receptor(s)
  peripheral, 548–549
Thermal stress, 547–579. *See also* Cold; Heat
Thermic effect of food, 162–163
  in obesity, 658
Thermodynamics
  first law of, 92
  second law of, 94, 95
Thermogenesis
  diet-induced, 162–163
  in obesity, 658
Thermoregulation
  clothing's effects on, 553–555
  during exercise
    in cold, 571–575
    in heat, 556–571
  hypothalamus in, 548–549
  in cold, 549–550
    during exercise, 571–575
  in heat, *550*, 550–553
    during exercise, 556–571
  mechanisms of, 547–556, *548*
Thiamine
  functions, sources and RDA for, 46t
Thinness
  underweight and, 605
Thoracic squeeze
  in breathhold diving, 584–585
Thrombus
  coronary, 307, 713. *See also* Coronary heart
    disease
Thyroid gland, 395
  secretions of, 391t, 395–396
    training's effects on, 406t, 408
Thyrotropin (thyroid-stimulating hormone;
    TSH), 390t, 394
  in thyroid hormone regulation, *396*, 396
Thyroxine (T_4), 391t, 395–396
  feedback control of, *396*, 396
  in response to cold, 550
  iodine and, 52
  training's effects on, 406t, 408
Tidal volume (TV), *240*, 240
  average values for, 242t
  changes in, with exercise, *249*, 249–250
Tissue(s)
  gas transfer in, 257–259, *258*
  oxygen partial pressure in, 263–264
TLC. *See* Total lung capacity
Tobacco. *See* Cigarette smoking
Tocopherol. *See* Vitamin E
Torque
  defined, 757
Total lung capacity (TLC), *240*, 241
  average values for, 242t
Trachea, 235
  partial pressure of gases in air in, 255
Training, 421–526. *See also* Exercise
  adaptive changes in active muscle due to,
    432, 432–433
  aerobic. *See* Aerobic training
  altitude, 541t, 541–542
    sea level performance and, 540–542
    sea level training vs., 542, *542*
  anaerobic. *See* Anaerobic training; Resistance
    training
  buffering capacity and, 288
  cardiac hypertrophy in response to, 340–343,
    341t–342t, 430
    specificity of, 342t, 342–343
  continuous, 444

  duration of, *438*, 438
  effects of
    on anaerobic capacity, *210*, 210
    on blood lactate threshold, 126, 127
    on body composition, 433, 647t–648t, 647–
      650, 650t
    on cardiac output, 330, 337, 431
    on endocrine function, 405–411, 406t
    on heart rate, *333*, 333, 430, 435, *436*
    on heart rate regulation, 316
    on maximal oxygen consumption, 217
    on motor unit characteristics, 375
    on muscle blood supply, 339, 350
    on muscle fiber type distribution, 362–363,
      363t
    on myocardial substrate utilization, *308*,
      309
    on stroke volume, 329, *330*, 330–331, 331t,
      430–431
    on ventilation, 245, 283–284, *284*
  endurance. *See* Aerobic training
  factors affecting, 433–440
  fartlek, 444–445
  for muscular strength, 452–496. *See also* Re-
    sistance training
  heat tolerance and, 565–566
  improvements expected from
    initial fitness level and, 434
  initiating program of, 445–446
  interval, 140t, 140–141
    aerobic, 441–444, 443t
    anaerobic, 440, 443–444
  isometric strength, 463–464
  leadership for programs of, 446
  methods of, 440–445
  of elderly, 701–702, *702*, 703t, 705, 708–709,
    709
  over-distance, 444
  oxygen deficit and, *128*, 130
  physiologic consequences of, 428t, 428–433
    practical implications of, 432–433
  plyometric, 470–471
  principles of, 423–427
  resistance. *See* Resistance training
  running economy and, 187–188
  weight, 461–463. *See also* Resistance training
"Training-sensitive zone"
  for exercise heart rate, 436–437, *437*
Transamination, 34
Transport
  active, 97
Treadmill test(s)
  exercise stress testing with
    protocols for, 727
  maximal oxygen consumption measurement
    with, 214
    advantages of, 216
    common protocols for, *215*, 216
    correlation of, with other aerobic power
      tests, 214t, 214
Triad(s), 354
Triathlete(s)
  physique of, 638
Tricarboxylic acid cycle. *See* Krebs
  cycle
Tricuspid valve, *294*, 294
Triglyceride(s), 20
  hydrolysis of, 116
Triiodothyronine (T_3), 391t, 395–396
  training's effects on, 406t, 408
Tropomyosin, 350, 352–353, *353*
Troponin, 350, 352, *353*, 353
TSH. *See* Thyrotropin

T-tubule system, 353–354, *354*
TV. *See* Tidal volume

Ultrasound technique
body fat assessment using, 623–624
Underwater weighing. *See also* Hydrostatic weighing
body volume measurement by, 612–615, *614*
temperature correction for water density in, 812
Underweight, 605
"Upper(s)," 502–503, 503t
Up-regulation, 386
Urea, 34
Urine
water loss in, 62

Vagus nerve
in heart rate regulation, 316
Valsalva maneuver, 239
physiologic consequences of, *239*, 239–240
Variable resistance machine, 468–469
Varicose vein(s), 300
Vasodilatation
in response to exercise, 320
Vasomotor tone, 322
Vasopressin. *See* Antidiuretic hormone
V̇e. *See* Minute ventilation
Vegan diet(s). *See* Vegetarian diet(s)
Vegetarian diet(s)
iron absorption from, 56
protein in, 30–31
vitamin B₁₂ needs in, 48
Vein(s), *295*, 298–301
as blood reservoirs, 299
valves in, *299*, 299
varicose, 300
Velocity
defined, 756
Venous admixture, 257n
Venous pooling, 300
strategies for reducing, 300–301
Venous return, 299
during exercise, 323
stroke volume and, 332
Ventilation, 246–250. *See also* Breathing; Respiration
alveolar, 247t, 247–250
maintenance of, during exercise, 249–250
assessment of. *See* Pulmonary function
maximum voluntary (MVV), 243
mechanics of, *237*, 237–240
minute. *See* Minute ventilation
pulmonary. *See* Pulmonary ventilation
training's effects on, 245, 283–284, *284*
Ventilation-perfusion ratio, 248
Ventilatory equivalent
(V̇e/V̇O₂), 278
Ventilatory system. *See* Respiratory tract
Ventricle(s), *294*, 294–295
Ventricular fibrillation, 726–727
Venule(s), 295
Vertical jump maneuver, 202–204, *203*
Vestibulospinal tract, 368
V̇e/V̇O₂ (ventilatory equivalent), 278
Viscous pressure drag, 191–192
Vital capacity, 241
average values for, 242t
Vitamin(s), 44–50. *See also* particular vitamins
as coenzymes, 99
fat as carrier for, 26–27
fat-soluble, 45
functions, sources and RDAs for, 47t

functions of, 46t–47t, 46–48
kinds of, 45
megadoses of, 49
nature of, 44–45
supplemental
value of, 47–49
water-soluble, 45
functions, sources and RDAs for, 46t–47t
Vitamin A
functions, sources and RDA for, 47t
harmful effects of excessive intake of, 45, 49
Vitamin B₁
functions, sources and RDA for, 46t
Vitamin B₂
functions, sources and RDA for, 46t
harmful effects of megadoses of, 49
Vitamin B₆
functions, sources and RDA for, 46t
harmful effects of megadoses of, 49
supplemental
exercise performance and, 48
Vitamin B₁₂
functions and RDA for, 46t
sources of, 46t, 48
Vitamin B₁₅, 505
Vitamin C
functions and RDA for, 47t
harmful effects of megadoses of, 49
iron absorption and, 56
sources of, 47t, 48
supplemental
exercise performance and, 48
Vitamin D
calcium absorption and, 53
functions, sources and RDA for, 47t
harmful effects of megadoses of, 49
Vitamin D₃
active, 387t
Vitamin E
functions, sources and RDA for, 47t
harmful effects of megadoses of, 49
supplemental, 48
Vitamin K
functions, sources and RDA for, 47t
V̇O₂. *See* Oxygen consumption
Volume
defined, 757

Waist-to-hip ratio, 660
Walking
competitive
energy expenditure during, 179–181, *180*
stride pattern and speed in, *184*, 184
effects of, on body composition, 647–648, 648t, 649
efficiency of
running efficiency vs., 180, *180*–181
energy expenditure during, *177*, 177–181
body mass and, 178t, 178
footwear and, 179
terrain and surface and, 178, 179t
with hand-held or ankle weights, 179
impact force on leg in, 179
tests for maximal oxygen consumption prediction using, 222
Warm-up
as ergogenic aid, 511–513
Water, 60–64
as percentage of body weight, 60
density of
temperature correction for, 812
energy content of food and, 60
extracellular, 60

functions of, in body, 51t, 61
intake of, 61–62, *62*
intracellular, 60
loss of, *62*, 62–63. *See also* Dehydration
in hot-weather exercise, 58–59
in weight loss, *688–689*, 688–689
metabolic, 61–62
requirement for, in exercise, 63–64
sources and RDA for, 51t
Water balance, 61–63, *62*
Water displacement
body volume measurement by, 612
Water replacement, 560–563
assessment of adequacy of, 562–563
recommendations for, 561–562, 563t
sugar drinks in, 18, 562
Water temperature
energy expenditure in swimming and, 192, *194*
Wave drag, 191
WB-GT index (wet bulb-globe temperature index), 570t, *570*
Weighing
underwater. *See also* Hydrostatic weighing
body volume measurement by, 612–615, *614*, 812
Weight
body. *See* Body mass; Weight gain; Weight loss
defined, 756
Weight(s)
energy expenditure during running or walking with, 179
Weight gain, 651–652
effects of, on fat cell size and number, 666
Weight lifter(s). *See also* Body builder(s)
physique of, 642–644
Weight lifting. *See* Weight training
Weight loss, 674–691
body composition during
effects of exercise and diet on, 650–651, *651*
computerized meal and exercise plan for, 78–79, 820–827
diet for, 676–682, 679t
long-term maintenance of, 678, *680*
rationale for high protein content in, 162
diet plus exercise for, 650–651, *651*, 687–688
effects of, on fat cell size and number, *665*, 665–666
exercise for, 682–687
duration, frequency and intensity of, 439, 686
effectiveness of, 685–686, 687t
factors affecting, 688–689, *688–690*
in wrestlers
recommendations for, 641–642
lowering of resting metabolic rate during, 680–682, *681*
methods for, 679t
sites of body fat reduction during, 672–674, *673*
water loss in, *688–689*, 688–689
yo-yo effect in, 682
Weight training, 461–463. *See also* Resistance training
contraindications for, 240
standard
isokinetic training vs., 467–469
"sticking point" in, *468*, 468
Wet bulb-globe temperature index (WB-GT index), 570t, 570
Wilson's disease, 51t
Wind Chill Index, 573–574, *574*

Wingate test, 207
Women. *See also* Sex differences
  applicability of cholesterol and heart disease
    studies to, 24
  steroid use by, 501
Wool
  for cold-weather clothing, 553
Work
  biologic, in humans, 97
  defined, 756
  estimation of, in efficiency calculation, 176

Wrestler(s)
  dehydration in, 559–560
  high school
    physique of, 640–642, 641t
    weight loss recommendations for, 641–642

Xerophthalmia, 47t
X-ray(s)
  arm
    body fat assessment using, 624, *625*

Yo-yo effect
  in weight control, 682

Z line, 351, *351–352*
Zinc
  functions, sources and RDA for, 51t
  insulin synthesis and, 52
  loss of, with strenuous exercise, 59